List of Tables

Basic Engineering Circuit Analysis

4TH EDITION

Basic Engineering Circuit Analysis

J. David Irwin

Professor and Head
Department of Electrical Engineering
Auburn University, Alabama

Macmillan Publishing Company
New York

Maxwell Macmillan Canada
Toronto

Maxwell Macmillan International
New York Oxford Singapore Sydney

Editor: John Griffin
Production Supervisor: Elaine W. Wetterau
Production Manager: Paul Smolenski
Text Designer: Eileen Burke
Cover Designer: Cathleen Norz
Illustrations: York Graphic Services, Inc.

This book was set in Times Roman by York Graphic Services, Inc., and printed and bound by R.R. Donnelley & Sons—Crawfordsville. The cover was printed by R.R. Donnelley & Sons—Crawfordsville.

Printed in the United States of America

Macmillan Publishing Company
866 Third Avenue, New York, New York 10022

Macmillan Publishing Company is part
of the Maxwell Communication Group of Companies.

Maxwell Macmillan Canada, Inc.
1200 Eglinton Avenue East
Suite 200
Don Mills, Ontario M3C 3N1

Library of Congress Cataloging-in-Publication Data

Irwin, J. David
Basic engineering circuit analysis / J. David Irwin.—4th ed.
p. cm.
Includes bibliographical references and index.
ISBN 0-02-359891-3
1. Electric circuit analysis. I. Title.
TK454.I78 1983
621.319′2—dc20 92-4920
CIP

Printing: 1 2 3 4 5 6 7 8 Year: 3 4 5 6 7 8 9 0 1 2

To Edie

Geri Marie
John David, Jr.
Laura Lynne

Preface

Perspective

This book is specifically designed for introductory courses in linear circuit analysis. The book may serve as a primary textbook for electrical engineering majors or as a means of introducing circuit analysis techniques to nonelectrical engineering majors.

In the preparation of this book, enormous effort has been expended in presenting the material in as simple a manner as possible. This method of presentation is designed to aid (1) the teacher in presenting the material in a clear and lucid manner, and (2) the student in grasping the concepts and techniques in order to quickly become proficient in their use. In fact, I have endeavored to present the material in such a way that sufficiently mature students, with the help of the supplements, can learn the material on their own. In addition, a number of practical applications have been included to demonstrate the use of the theory.

Prerequisites

In general, the reader should have a mathematical background through differential and integral calculus and an understanding of the techniques involved in the solution of differential equations with constant coefficients; some basic knowledge of matrices and computer programming would also be helpful.

Philosophy

Pedagogy. The basic philosophy employed in the development of this book can be stated simply as follows: ''To *learn* circuit analysis, one must *do* circuit analysis''; therefore, the book contains 382 examples to illustrate every facet of each solution technique and a total of 1366 problems that enable readers to test their understanding of all concepts and principles. I have tried to keep the mathematics as simple as possible so that the concepts and principles are not obscured in a maze of calculations and numbers. In addition, where appropriate I have worked the same problems in a number of ways to facilitate an understanding and comparison of various techniques, and their relationship to one another. In general, I have tried to present the material in a way that helps both the teacher and the student.

Illustrations. Illustrations are used extensively in order to clarify all the details of a circuit analysis procedure. In addition, a second color highlights variables under investigation in both circuits and graphs.

PSPICE. The computer-aided circuit analysis program PSPICE is employed throughout the book. It is used in Chapters 4, 5, 7, 8, 10, 12, 13, 14, and 18 to determine currents and voltages in a wide variety of circuits, plot time and frequency responses, and compute Fourier series. This is an extremely important feature of the book, as SPICE is extensively used in industry to analyze circuit designs before committing them to production.

This edition of the text employs PSPICE rather than SPICE 2G.5 for the following reasons. PSPICE is available for IBM PC's, clones, MacIntoshes, and several minicomputer and workstation platforms; includes a user-friendly mouse-driven interface; provides excellent graphical output to the screen and printer; supports a variety of laser printers and line printers; and has additional features that allow more flexible and powerful analyses.

The criteria employed by the Accreditation Board for Engineering and Technology (ABET) specifically state ''Students must demonstrate knowledge of the application and use of digital computation techniques to specific engineering problems.'' The employment of PSPICE in this context helps satisfy this standard and, in addition, provides the basis for its use in electronics.

Flexibility. The material is presented in a manner that permits considerable flexibility in its use. For example, one can completely skip selected sections or chapters while proceeding from beginning to end in the development of a coherent presentation.

Summaries. At the end of every chapter the important concepts are summarized in a concise fashion. These summaries serve as a quick reminder for the reader of all the significant principles and techniques contained within the chapter.

Problem Sets. The book contains two types of problem sets. First, there are 289 drill problems. These problems and their answers are placed to immediately reinforce the reader's understanding of the preceding section. Second, there are 1077 problems at the end of the chapters. These problem sets, which are new, are organized to follow the presentation of the material, generally graduated in difficulty, and carefully selected to challenge the reader's understanding of all the material.

Supplements. A for-sale student manual is available that provides hundreds of additional problems and their solutions, that are similar to the ones that appear at the end of the chapters.

For teachers, a manual that contains solutions to all drill problems and end of chapter problems, is available. In addition, a set of transparency masters can also be obtained.

Acknowledgments

I am indebted to a number of colleagues and friends who have helped in the preparation of previous editions of this book. They have been very supportive and I want to reemphasize my appreciation for their help here also. They are Professors E. R. Graf, L. L. Grigsby, C. A. Gross, M. A. Honnell, R. C. Jaeger, J. L. Lowry, M. S. Morse, and C. L. Rogers; students Dr. Travis Blalock, Mr. Kevin Driscoll, Mr. Keith Jones, Mr. George Lindsey, Mr. David Mack, Dr. John Parr, Mr. Monty Rickles, Mr. James Trivitayakhun, Ms. Susan Williamson, and Ms. Jacinda Woodward; and my administrative assistant Ms. Betty Kelley.

I also want to express my appreciation to the following professors who have made numerous suggestions for improving this book: David Anderson, University of Iowa; Richard L. Baker, University of California, Los Angeles; James L. Dodd, Professor Emeritus, Mississippi State University; Earl D. Eyman, University of Iowa; Arvin Grabel, Northeastern University; Paul Gray, University of Wisconsin, Platteville; John Hadjilogiou, Florida Institute of Technology; Ralph Kinney, Louisiana State University; K. S. P. Kumar, University of Minnesota; James Luster, Snow College; Ian McCausland, University of Toronto; Arthur C. Moeller, Marquette University; M. Paul Murray, Mississippi State University; Burks Oakley, II, University of Illinois at Champaign-Urbana; John O'Malley, University of Florida; William R. Parkhurst, The Wichita State University; James Rowland, University of Kansas; Robert N. Sackett, Normandale Community College; Richard Sanford, Clarkson University; Ronald Schultz, Cleveland State University; Janusz Starzyk, Ohio University; and Saad Tabet, Florida State University.

I am very grateful to Ms. Paula Revels and Mr. Les Simonton for their help and support in the preparation of this edition.

Finally, I wish to express my deepest appreciation to my wife, Edie, without whose help and support this book would not have been possible.

J. D. I.

To the Instructor

To obtain free PSPICE software for classroom and student use, please read the following information provided by MicroSim Corporation:

THE DESIGN CENTER™

Free Software

The **Design Center** includes packages containing schematic capture, simulation with our **PSpice** native mixed analog/digital simulator, and graphical waveform analysis of analog and digital circuit designs. Class instructors can receive complimentary evaluation versions for *both* the IBM-PC and Macintosh by submitting a request on company or educational letterhead to:

Product Marketing Department
MicroSim Corporation
20 Fairbanks
Irvine, CA 92718

Duplication of the diskettes for your students is encouraged.

The Standard for Circuit Design

THE MAKERS OF PSPICE

PSpice is a registered trademark of MicroSim Corporation

Contents

Chapter 3
Nodal and Loop Analysis Techniques

Chapter 4
DC PSPICE Analysis

Chapter 5
Additional Analysis Techniques

Basic Engineering Circuit Analysis

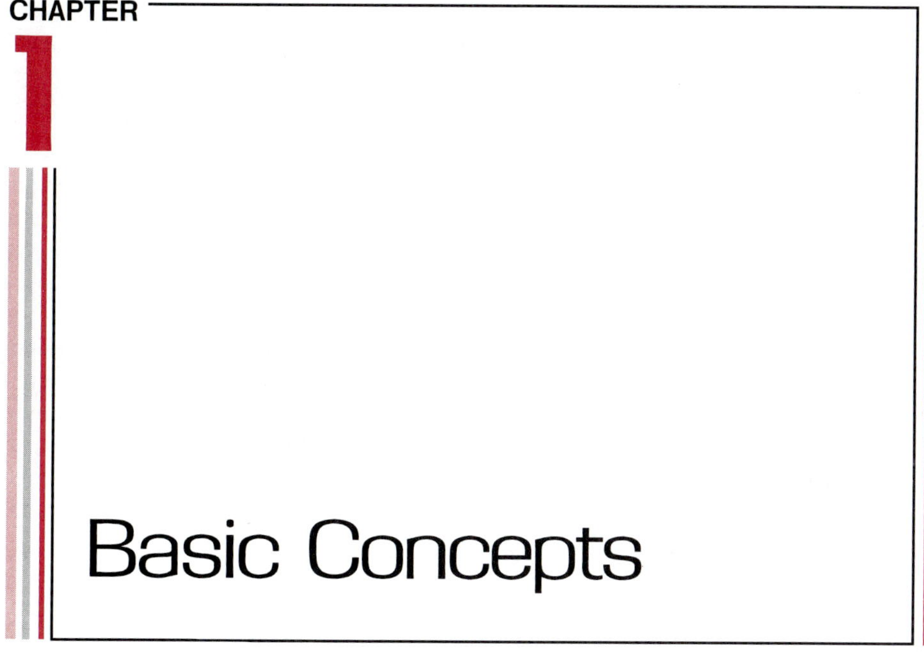

CHAPTER 1 Basic Concepts

It is important to note that the title of this book does not imply that it will be useful only to electrical engineers. Rather, in our time we find that all types of engineers need to be aware of the implications of circuit analysis because they are encountering circuits in many ways: as an integral part of their systems, in their instrumentation, and so on. In addition, many of today's problems are so complicated that they are attacked by a team of specialists with backgrounds in engineering, mathematics, physics, and chemistry. In many instances the electrical components literally permeate an entire system through implications of power, control, instrumentation, monitoring, and the like. Therefore, it is important that as many technical personnel as possible be at least familiar with electric circuit analysis.

1.1 System of Units

The system of units we employ is the international system of units, the Système International des Unités, which is normally referred to as the SI standard system. This system, which is composed of the basic units meter (m), kilogram (kg), second (s), ampere (A), degree kelvin (K), and candela (cd), is defined in all modern physics texts and therefore will not be defined here. However, we will discuss the units in some detail as we encounter them in our subsequent analyses.

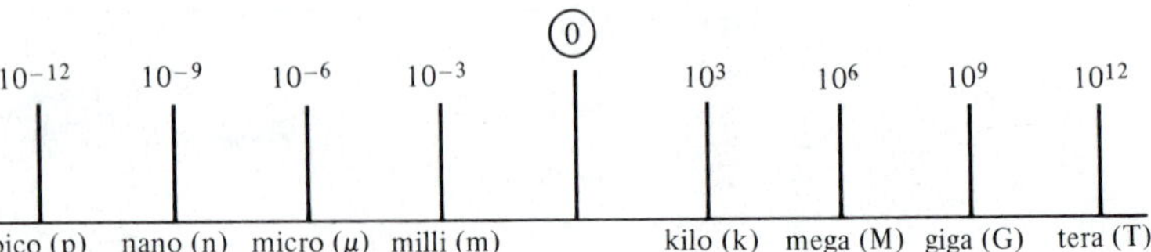

Figure 1.1 Standard SI prefixes.

The standard prefixes that are employed in SI are shown in Fig. 1.1. Note the decimal relationship between these prefixes. These standard prefixes are employed throughout our study of electric circuits.

Only a few years ago a millisecond, 10^{-3} s, was considered to be a short time in the analysis of electric circuits. Advances in technology, however, have led to a state in which we now think of doing such things as performing calculations in nanoseconds or even picoseconds. At the same time, circuits have experienced a phenomenal decrease in size. For example, consider the integrated-circuit chip shown in Fig. 1.2. Such miniaturized circuits are commonplace in calculators, computers, and other electronic equipment.

1.2 Basic Quantities

Before we begin our analysis of electric circuits, we must define the basic terms that we will employ. However, in this chapter and throughout the book our definitions and explanations will be as simple as possible in order to foster an understanding of the use of the material. No attempt will be made to give complete definitions of many of the quantities because such definitions are not only unnecessary at this level but are often confusing. Although most of us have an intuitive concept of what is meant by a circuit, we will simply refer to an *electric circuit* as an interconnection of electrical components.

It is very important at the outset that the reader understand the basic strategy that we will employ in our analysis of electric circuits. This strategy is outlined in Fig. 1.3, and we will be concerned only with the portion of the diagram to the right of the dashed line. In subsequent courses or further study the reader will learn how to model physical devices such as electronic components. Our procedure here will be to employ linear models for our circuit components, then define the variables that are used in the appropriate circuit equations to yield solution values, and finally, interpret the solution values for the variables in order to determine what is actually happening in the physical circuit. The variables may be time varying or constant depending on the nature of the physical parameters they represent.

The most elementary quantity in an analysis of electric circuits is the electric *charge*. We know from basic physics that the nature of charge is based on concepts of atomic theory. We view the atom as a fundamental building block of matter that is composed of a positively charged nucleus surrounded by negatively charged electrons. In the metric system, charge is measured in coulombs (C). The charge on an electron is negative and equal in magnitude to 1.602×10^{-19} C. However, our interest in electric charge is centered around its motion, since charge in motion results in an energy transfer. Of particular interest to us are those situations in which the motion is confined to a definite closed path.

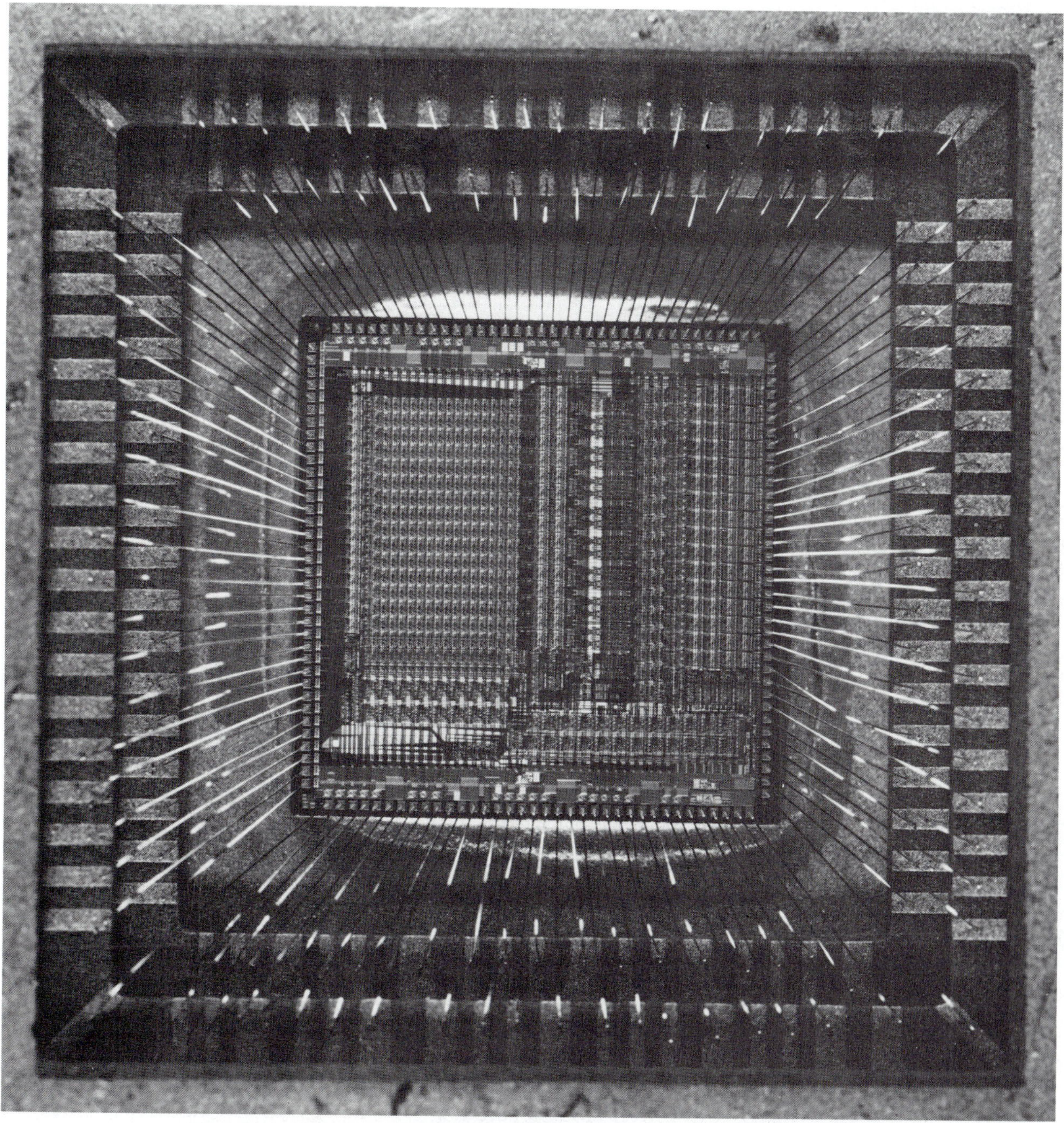

Figure 1.2 Example of a VLSI advanced bipolar chip with 30,000 transistors. This high-speed chip is approximately 0.25 inch on a side and uses a 144-pin-grid-array package. (Courtesy of Honeywell, Inc.)

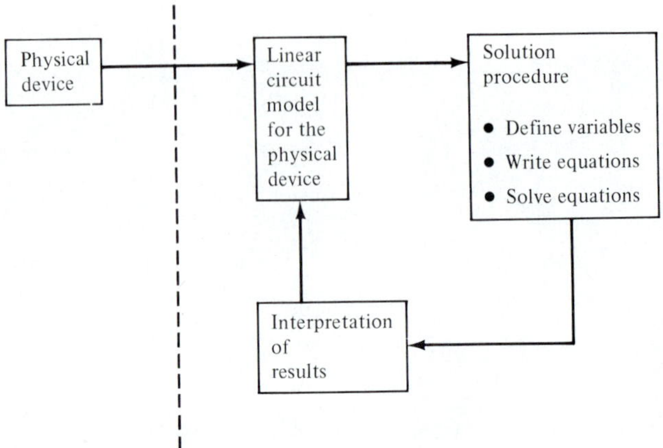

Figure 1.3 Basic strategy employed in circuit analysis.

An electric circuit is essentially a pipeline that facilitates the transfer of charge from one point to another. The time rate of change of charge constitutes an electric *current*. Mathematically, the relationship is expressed as

$$i(t) = \frac{dq(t)}{dt} \qquad \text{or} \qquad q(t) = \int_{-\infty}^{t} i(x)\, dx \tag{1.1}$$

where i and q represent current and charge, respectively (lowercase letters represent time dependency and capital letters are reserved for constant quantities). The basic unit of current is the ampere (A) and 1 ampere is 1 coulomb per second.

Although we know that current flow in metallic conductors results from electron motion, the conventional current flow, which is universally adopted, represents the movement of positive charges. It is important that the reader think of current flow as the movement of positive charge regardless of the physical phenomena that take place. The symbolism that will be used to represent current flow is shown in Fig. 1.4. The variable representing the current in the wire in Fig. 1.4a was defined as I_1 flowing in the wire from left to right in the figure. A set of equations was written for the circuit and a solution value of 2 A was obtained; that is, $I_1 = 2$ A. This means that the physical current in the wire is flowing from left to right, in the direction of our variable, and is 2 A. $I_1 = 2$ A in Fig. 1.4a indicates that at any point in the wire shown, 2 C of charge passes from left to right each second. The same procedure was followed for the wire shown in Fig. 1.4b, and the variable I_2 was defined in the same manner. A set of equations was written and a solution value of -3 A was obtained. An interpretation of this result is that the physical current in

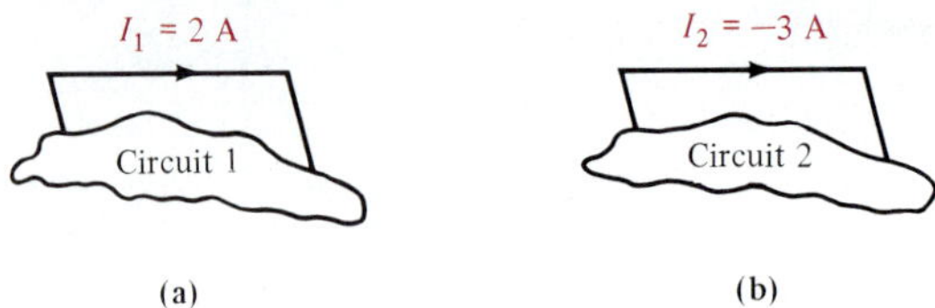

Figure 1.4 Conventional current flow: (a) positive current flow; (b) negative current flow.

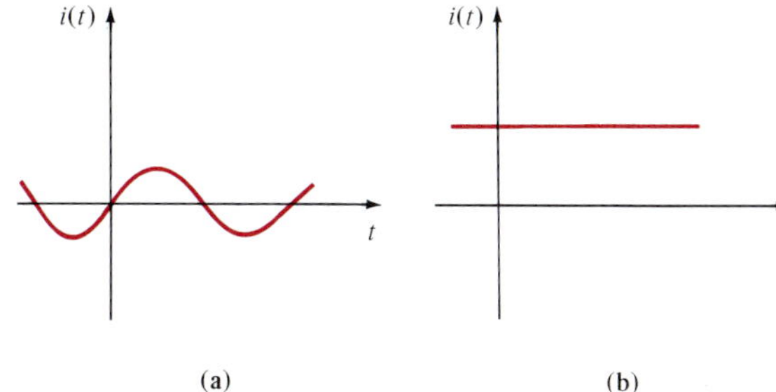

Figure 1.5 Two common types of current: (a) alternating current (ac); (b) direct current (dc).

the wire is from right to left, opposite to our reference direction for I_1; its value is 3 A. $I_2 = -3$ A in Fig. 1.4b indicates that at any point in the wire shown, 3 C of charge passes from right to left each second. Therefore, it is important to specify not only the magnitude of the variable representing the current, but also its direction. After a solution is obtained for the current variable, the actual physical current is known.

There are two types of current that we encounter often in our daily lives, alternating current (ac) and direct current (dc), which are shown as a function of time in Fig. 1.5. *Alternating current* is the common current found in every household, used to run the refrigerator, stove, washing machine, and so on. Batteries, which are used in automobiles or flashlights, are one source of *direct current*. In addition to these two types of currents, which have a wide variety of uses, we can generate many other types of currents. We will examine some of these other types later in the book.

We have indicated that charges in motion yield an energy transfer. Now we define the *voltage* (also called the *electromotive force* or *potential*) between two points in a circuit as the difference in energy level of a unit positive charge located at each of the two points. Work or energy, $w(t)$ or W, is measured in joules (J); 1 joule is 1 newton meter (N · m). Hence voltage [$v(t)$ or V] is measured in volts (V) and 1 volt is 1 joule per coulomb; that is, 1 volt = 1 joule per coulomb = 1 newton meter per coulomb.

If a unit positive charge is moved between two points, the energy required to move it is the difference in energy level between the two points and is the defined voltage. It is extremely important that the variables that are used to represent voltage between two points be defined in such a way that the solution will let us interpret which point is at the higher potential with respect to the other.

In Fig. 1.6a the variable that represents the voltage between points A and B has been

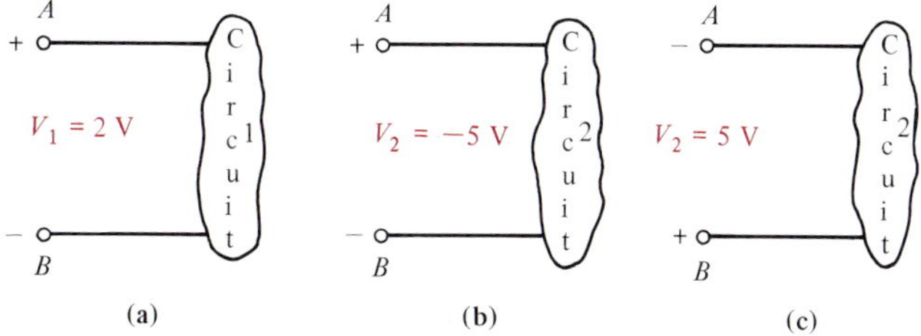

Figure 1.6 Voltage representations.

defined as V_1, and it is assumed that point A is at a higher potential than point B, as indicated by the + and − signs associated with the variable and defined in the figure. The + and − signs define a reference direction for V_1. The equations for the circuit were written and a solution value of 2 V was obtained; that is, $V_1 = 2$ V, as shown on the figure. The physical interpretation of this is that the difference in potential of points A and B is 2 V and point A is at the higher potential. If a unit positive charge is moved from point A through the circuit to point B, it will give up energy to the circuit and have 2 J less energy when it reaches point B. If a unit positive charge is moved from point B to point A, extra energy must be added to the charge by the circuit, and hence the charge will end up with 2 J more energy at point A than it started with at point B.

The same procedure was followed for the circuit in Fig. 1.6b, and the variable V_2 was defined in the same manner. A set of equations was written and a solution value of −5 V was obtained; that is, $V_2 = -5$ V. The physical interpretation of $V_2 = -5$ V is that the potential between points A and B is 5 V and point B is at the higher potential. If the circuit in Fig. 1.6b was reworked with the variable for the voltage defined as shown in Fig. 1.6c, the solution would have been $V_2 = 5$ V. The physical interpretation of this is that the difference in potential of points A and B is 5 V, with point B at the higher potential.

Note that it is important to define a variable with a reference direction so that the answer can be interpreted to give the physical condition in the circuit. We will find that it is not possible in many cases to define the variable so that the answer is positive, and we will also find that it is not necessary to do so. A negative number for a given variable gives exactly the same information as a positive number for a new variable that is the same as the old variable, except that it has an opposite reference direction. Hence, when we define either current or voltage, it is absolutely necessary that we specify both magnitude and direction. Therefore, it is incomplete to say that the voltage between two points is 10 V or the current in a line is 2 A, since only the magnitude and not the direction for the variables has been defined.

At this point we have presented the conventions that we employ in our discussions of current and voltage. Energy is yet another important term of basic significance. Figure 1.7 illustrates the voltage–current relationships for energy transfer. In this figure, the block representing a circuit element has been extracted from a larger circuit for examination. In Fig. 1.7a, energy is being supplied *to* the element by whatever is attached to the terminals. Note that 2 A, that is, 2 C, of charge is moving from point A to point B through the

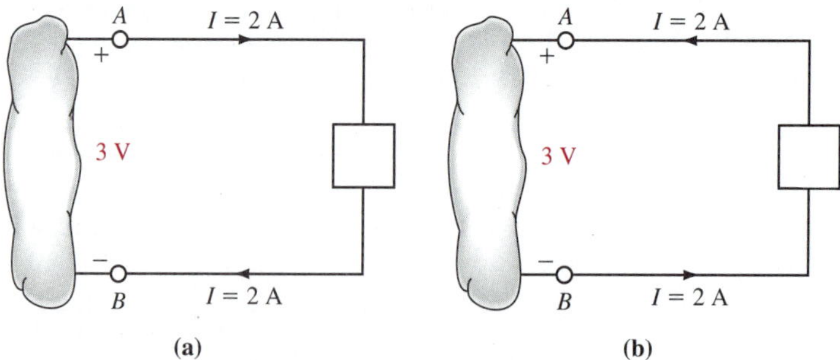

Figure 1.7 Voltage–current relationships for energy absorbed (a) and energy supplied (b).

element each second. Each coulomb loses 3 J of energy as it passes through the element from point A to point B. Therefore, the element is absorbing 6 J of energy per second. Note that when the element is *absorbing* energy, a positive current enters the positive terminal. In Fig. 1.7b energy is being supplied *by* the element to whatever is connected to terminals A-B. In this case, note that when the element is *supplying* energy, a positive current enters the negative terminal and leaves via the positive terminal. In this convention a negative current in one direction is equivalent to a positive current in the opposite direction, and vice versa. Similarly, a negative voltage in one direction is equivalent to a positive voltage in the opposite direction.

EXAMPLE 1.1

Suppose that your car will not start. To determine if the battery is faulty, you turn on the light switch and find that the lights are very dim, indicating a weak battery. You borrow a friend's car and a set of jumper cables. However, how do you connect his car's battery to yours? What do you want his battery to do?

Essentially, his car's battery must supply energy to yours, and therefore it should be connected in the manner shown in Fig. 1.8. Note that the positive current leaves the positive terminal of the good battery (supplying energy) and enters the positive terminal of the weak battery (absorbing energy). Note that the same connections are used when charging a battery.

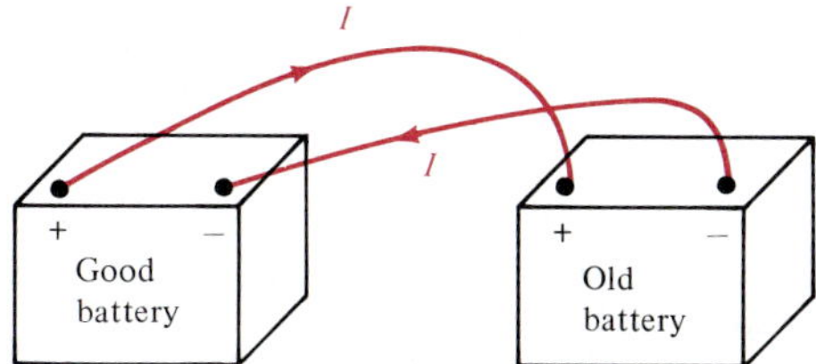

Figure 1.8 Diagram for Example 1.1.

In practical applications there are often considerations other than simply the electrical relations (e.g., safety). Such is the case with jump starting an automobile. Automobile batteries evolve explosive gases that can be ignited accidentally, causing severe physical injury. Be safe—follow the procedure described in your auto owner's manual. ■

We have defined voltage in joules per coulomb as the energy required to move a positive charge of 1 C through an element. If we assume that we are dealing with a differential amount of charge and energy, then

$$v = \frac{dw}{dq} \tag{1.2}$$

Multiplying this quantity by the current in the element yields

$$vi = \frac{dw}{dq}\left(\frac{dq}{dt}\right) = \frac{dw}{dt} = p \tag{1.3}$$

which is the time rate of change of energy or power measured in joules per second, or watts (W). Since, in general, both v and i are functions of time, p is also a time-varying

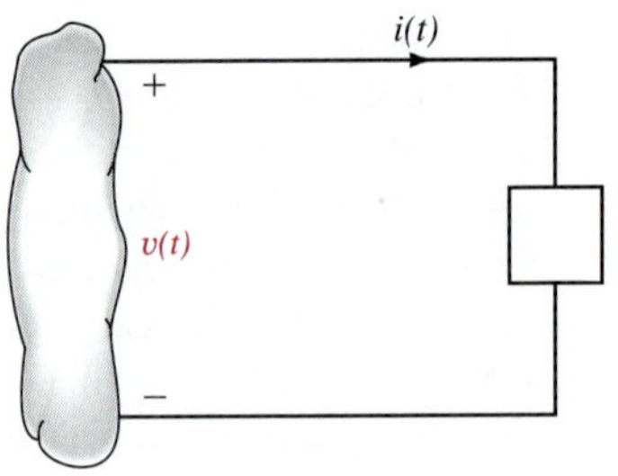

Figure 1.9 Sign convention for power.

quantity. Therefore, the change in energy from time t_1 to time t_2 can be found by integrating Eq. (1.3); that is,

$$w = \int_{t_1}^{t_2} p\,dt = \int_{t_1}^{t_2} vi\,dt \tag{1.4}$$

At this point, let us summarize our sign convention for power. To determine the sign of any of the quantities involved, the variables for the current and voltage should be arranged as shown in Fig. 1.9. The variable for the voltage $v(t)$ is defined as the voltage across the element with the positive reference at the same terminal that the current variable $i(t)$ is entering. This convention is called the *passive sign convention* and will be so noted in the remainder of this book. The product of v and i, with their attendant signs, will determine the magnitude and sign of the power. If the sign of the power is positive, power is being absorbed by the element; if the sign is negative, power is being supplied by the element.

EXAMPLE 1.2

We wish to determine the power absorbed by, or supplied by, the elements in Fig. 1.7.

In Fig. 1.7a, $P = VI = (3\text{ V})(2\text{ A}) = 6\text{ W}$ is absorbed by the element. In Fig. 1.7b, $P = VI = -(3\text{ V})(2\text{ A}) = -6\text{ W}$ is absorbed by the element, or +6 W is supplied by the element. ■

EXAMPLE 1.3

Given the two diagrams shown in Fig. 1.10, determine whether the element is absorbing or supplying power and how much.

In Fig. 1.10a, the power is $P = (2\text{ V})(-4\text{ A}) = -8\text{ W}$. Therefore, the element is supplying power. In Fig. 1.10b, the power is $P = (2\text{ V})(2\text{ A}) = 4\text{ W}$. Therefore, the element is absorbing power. ■

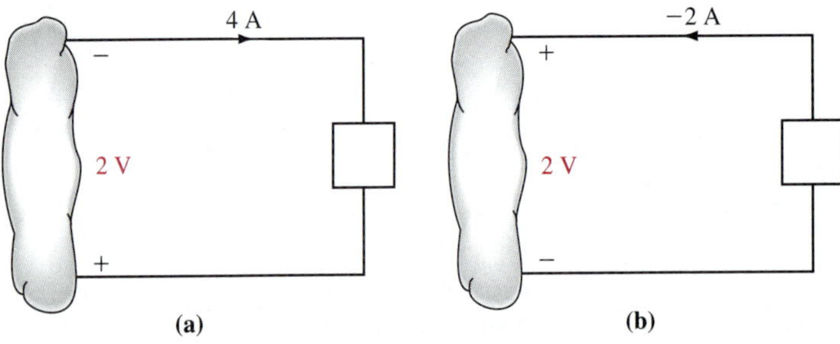

Figure 1.10 Elements for Example 1.3.

DRILL EXERCISE

D1.1. Determine the amount of power absorbed or supplied by the elements in Fig. D1.1.

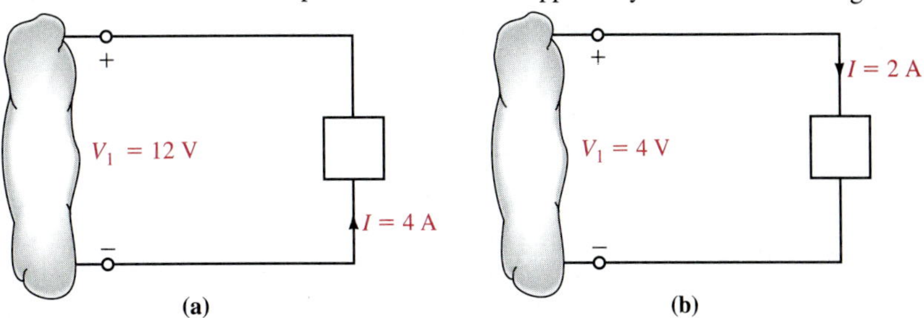

Figure D1.1

Ans: (a) $P = -48$ W; (b) $P = 8$ W.

EXAMPLE 1.4

We wish to determine the unknown voltage or current in Fig. 1.11.

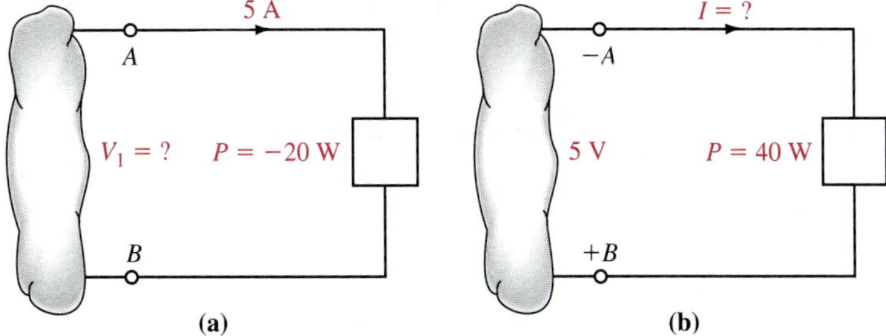

Figure 1.11 Elements for Example 1.4.

In Fig. 1.11a, a power of −20 W indicates that the element is delivering power. Therefore, the current enters the negative terminal (terminal *A*), and from Eq. (1.3) the voltage is 4 V.

In Fig. 1.11b, a power of +40 W indicates that the element is absorbing power and therefore the current should enter the positive terminal *B*. The current thus has a value of −8 A, as shown in the figure. ■

DRILL EXERCISE

D1.2. Determine the unknown variables in Fig. D1.2.

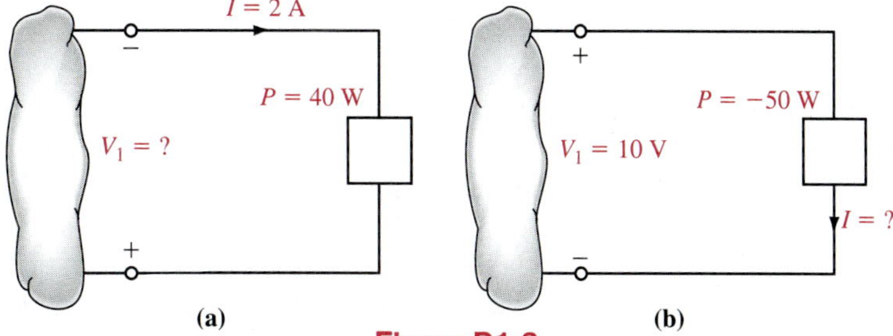

Figure D1.2

Ans: (a) $V_1 = -20$ V; (b) $I = -5$ A.

Finally, it is important to note that these electrical networks satisfy the principle of conservation of energy. For our present purposes this means that the power that is supplied in a network is exactly equal to the power absorbed.

1.3 Circuit Elements

It is important to realize at the outset that in our discussions we will be concerned with the behavior of a circuit element, and we will describe that behavior by a mathematical model. Thus, when we refer to a particular circuit element, we mean the *mathematical model* that describes its behavior.

Thus far we have defined voltage, current, and power. In the remainder of this chapter we will define both independent and dependent current and voltage sources. Although we will assume ideal elements, we will try to indicate the shortcomings of these assumptions as we proceed with the discussion.

In general, the elements we will define are terminal devices that are completely characterized by the current through the element and/or the voltage across it. These elements, which we will employ in constructing electric circuits, will be broadly classified as being either active or passive. The distinction between these two classifications depends essentially upon one thing—whether they supply or absorb energy. As the words themselves imply, an *active* element is capable of generating energy and a *passive* element cannot generate energy.

However, we will show later that some passive elements are capable of storing energy. Typical active elements are batteries, generators, and transistor models. The three common passive elements are resistors, capacitors, and inductors.

In Chapter 2 we will launch an examination of passive elements by discussing the resistor in detail. However, before proceeding with that element, we first present some very important active elements.

1. Independent voltage source.
2. Independent current source.
3. Two dependent voltage sources.
4. Two dependent current sources.

Independent Sources

An *independent voltage source* is a two-terminal element that maintains a specified voltage between its terminals regardless of the current through it. The general symbol for an independent source, a circle, is shown in Fig. 1.12a. As the figure indicates, terminal A is $v(t)$ volts positive with respect to terminal B. The word ''positive'' may be somewhat misleading. What is meant in this case is that $v(t)$ is referenced positive at terminal A and that the physical voltage across the device must be interpreted from the numerical value of $v(t)$. That is, if $v(t)$ at $t = 2$ sec is -10 V, point B is at a higher potential than point A at $t = 2$ sec. The symbol $v(t)$ is normally employed for time-varying voltages. However, if

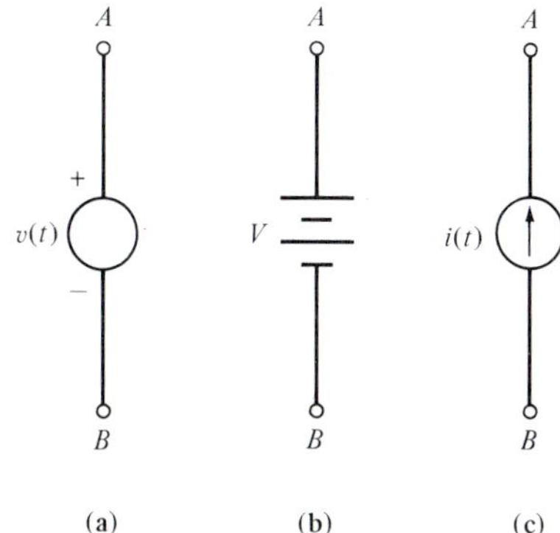

Figure 1.12 Symbols for (a) independent voltage source; (b) constant voltage source; (c) independent current source.

the voltage is time invariant (i.e., constant), the symbol shown in Fig. 1.12b is sometimes used. This symbol, which is used to represent a battery, illustrates that terminal A is V volts positive with respect to terminal B, where the long line on the top and the short line on the bottom indicate the positive and negative terminals, respectively, and thus the polarity of the element.

In contrast to the independent voltage source, the *independent current source* is a two-terminal element that maintains a specified current regardless of the voltage across its terminals. The general symbol for an independent current source is shown in Fig. 1.12c, where $i(t)$ is the specified current and the arrow indicates the positive direction of current flow.

It is important that we pause here to inject a comment concerning a shortcoming of the models. In general, mathematical models approximate actual physical systems only under a certain range of conditions. Rarely does a model accurately represent a physical system under every set of conditions. To illustrate this point, consider the model for the voltage source in Fig. 1.12a. We assume that the voltage source delivers v volts regardless of what is connected to its terminals. Theoretically, we could adjust the external circuit so that an infinite amount of current would flow, and therefore the voltage source would deliver an infinite amount of power. This is, of course, physically impossible. A similar argument could be made for the independent current source. Hence the reader is cautioned to keep in mind that models have limitations and are thus valid representations of physical systems only under certain conditions.

EXAMPLE 1.5

Determine the power absorbed or supplied by the elements in the network in Fig. 1.13.

The current flow is out of the positive terminal of the 24-V source, and therefore this element is supplying (2)(24) = 48 W of power. The current is into the positive terminals of elements 1 and 2, and therefore elements 1 and 2 are absorbing (2)(6) = 12 W and (2)(18) = 36 W, respectively. Note that the power supplied is equal to the power absorbed. ■

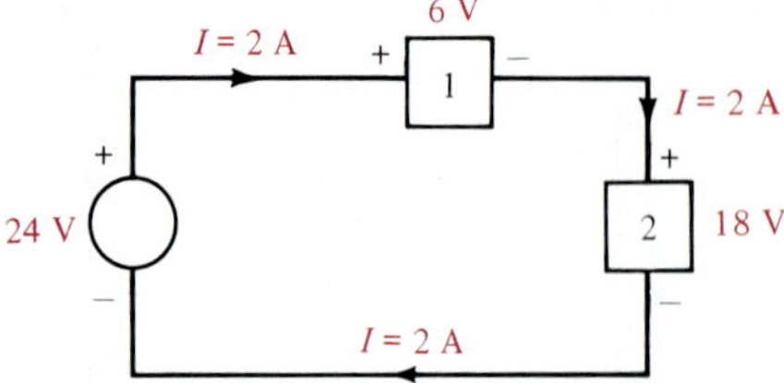

Figure 1.13 Network for Example 1.5.

DRILL EXERCISE

D1.3. Find the power that is absorbed or supplied by the three elements in Fig. D1.3.

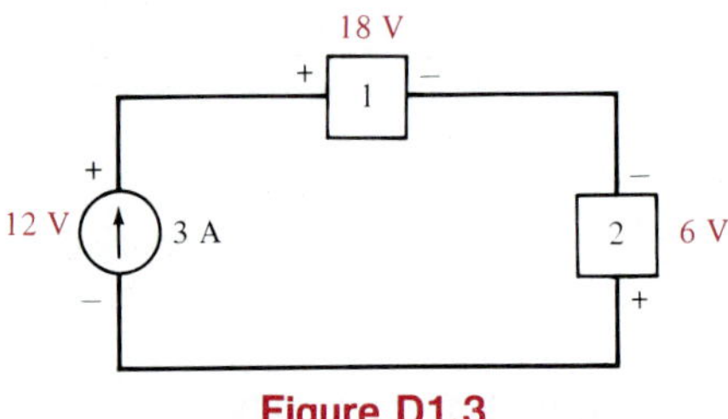

Figure D1.3

Ans: Current source supplies 36 W, element 1 absorbs 54 W, and element 2 supplies 18 W.

Dependent Sources

In contrast to the independent sources, which produce a particular voltage or current completely unaffected by what is happening in the remainder of the circuit, dependent sources generate a voltage or current that is determined by a voltage or current at a specified location in the circuit. These sources are very important because they are an integral part of the mathematical models used to describe the behavior of many electronic circuit elements. Consider, for example, the two commonly used electronic devices shown in Fig. 1.14. The device in Fig. 1.14a is an *N*-channel enhancement-mode metal-

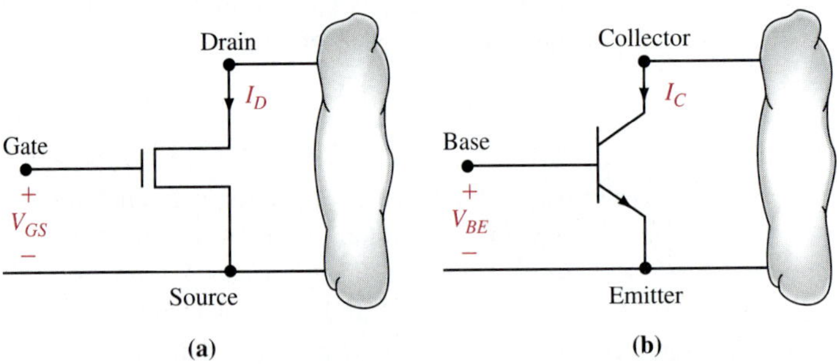

Figure 1.14 Two commonly used electronic devices.

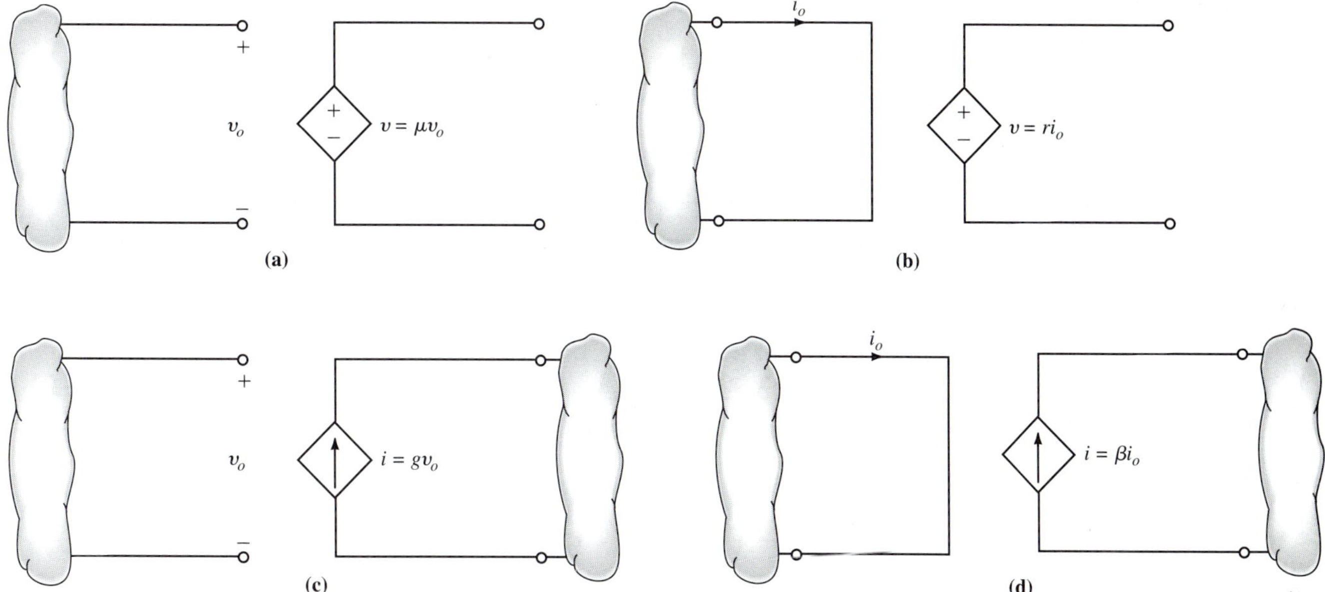

Figure 1.15 Four different types of dependent sources.

oxide-semiconductor field-effect transistor (MOSFET). The device in Fig. 1.14b is an *npn* bipolar transistor. The drain current I_D in the MOSFET is dependent on the gate-to-source voltage V_{GS}. Similarly, the collector current I_C in the bipolar device is controlled by the base-to-emitter voltage V_{BE}.

In contrast to the circle used to represent independent sources, a diamond is used to represent a dependent or controlled source. Figure 1.15 illustrates the four types of dependent sources. The input terminals on the left represent the voltage or current that controls the dependent source, and the output terminals on the right represent the output current or voltage of the controlled source. Note that in Fig. 1.15a and d the quantities μ and β are dimensionless constants because we are transforming voltage to voltage and current to current. This is not the case in Fig. 1.15b and c; hence when we employ these elements a short time later, we must describe the units of the factors r and g.

EXAMPLE 1.6

Given the two networks shown in Fig. 1.16, we wish to determine the outputs.

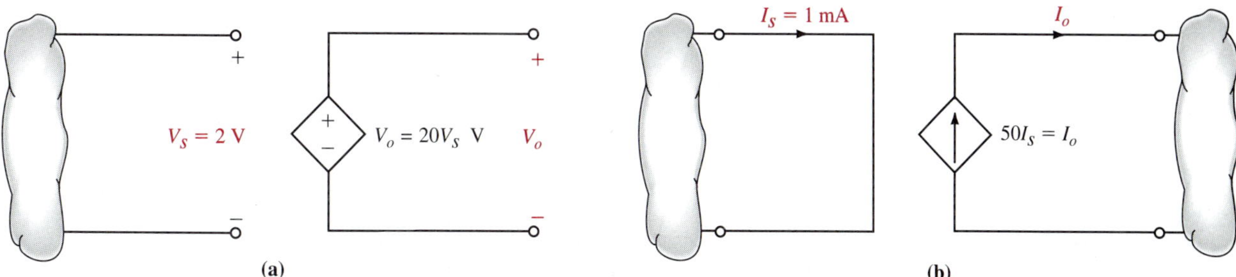

Figure 1.16 Circuits for Example 1.6.

In Fig. 1.16a the output voltage is $V_o = \mu V_S$ or $V_o = 20V_S = (20)(2\text{ V}) = 40\text{ V}$. Note that the output voltage has been amplified from 2 V at the input terminals to 40 V at the output terminals; that is, the circuit is an amplifier with an amplification factor of 20.

In Fig. 1.16b, the output current is $I_o = \beta I_S = (50)(1\text{ mA}) = 50\text{ mA}$; that is, the circuit has a current gain of 50, meaning that the output current is 50 times greater than the input current. ■

DRILL EXERCISE

D1.4. Determine the power supplied by the dependent sources in Fig. D1.4.

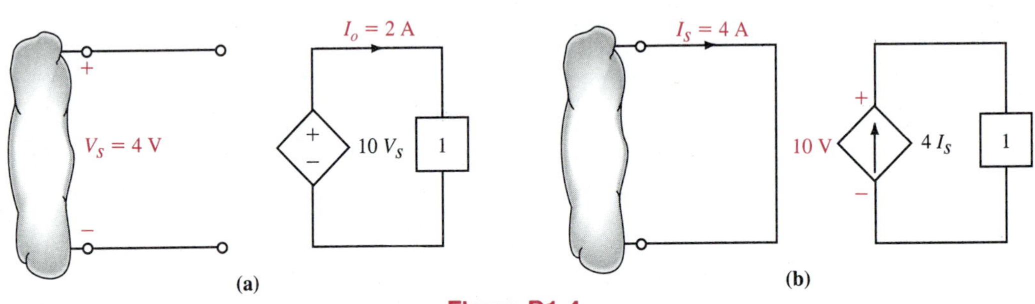

Figure D1.4

Ans: (a) Power supplied = 80 W; (b) power supplied = 160 W.

EXAMPLE 1.7

Let us compute the power that is absorbed or supplied by the elements in the network in Fig. 1.17.

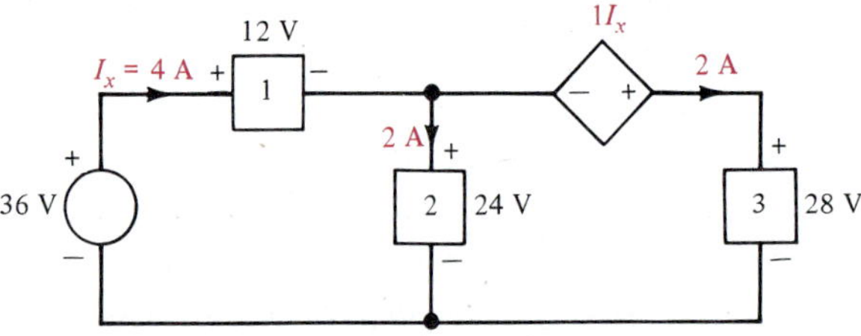

Figure 1.17 Circuit for Example 1.7.

Using the sign convention for power illustrated in Fig. 1.9, we find the following:

$$P_{36\text{V}} = (36)(-4) = -144\text{ W}$$
$$P_1 = (12)(4) = 48\text{ W}$$
$$P_2 = (24)(2) = 48\text{ W}$$
$$P_{DS} = (1I_x)(-2) = (4)(-2) = -8\text{ W}$$
$$P_3 = (28)(2) = 56\text{ W}$$

Note that the 36-V source and the dependent source are supplying power to the network and the remaining elements are absorbing power. Furthermore, energy is conserved, since the power supplied to the network is identically equal to the power absorbed by the network. ■

DRILL EXERCISE

D1.5. Find the power that is absorbed or supplied by the circuit elements in the network in Fig. D1.5.

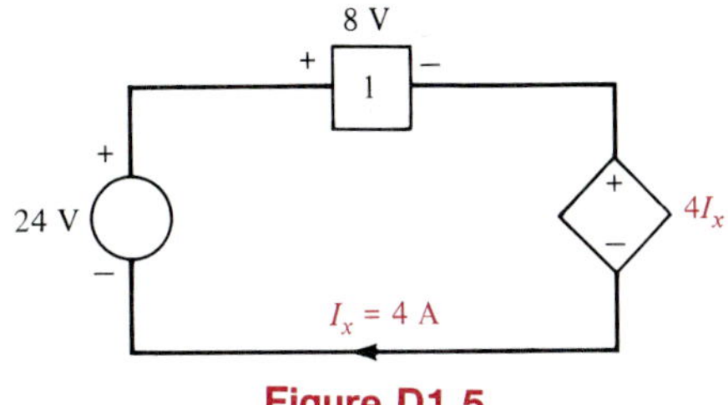

Figure D1.5

Ans: $P_{24V} = 96$ W supplied, $P_1 = 32$ W absorbed, $P_{4Ix} = 64$ W absorbed.

As we conclude this chapter, a number of important comments are in order. The reader must thoroughly understand our approach as outlined in Fig. 1.3 because in the following chapters we will simply define and refer to voltages and currents at specific locations within a circuit. Before any equations are written, readers should think carefully about what kinds of answers they expect. For example, if all the sources are constant, all the voltages and currents in the circuit will be constant, and therefore capital letters should be used for all the variables representing the voltages and currents. It is also useful for readers to think ahead about how variables are related to one another and define the variables so that a minimum number of minus signs are used. Although this is not important in the final solution, as indicated earlier, it does force readers to think about the problem from beginning to end before attacking it so that the solution will not come as a surprise.

1.4 Summary

We have introduced the basic strategy for an analysis of electric circuits. This strategy involves the use of linear models to represent the various circuit elements, the definition of variables used in the circuit equations, and the interpretation of the solution values of the variables to determine the actual values of the quantities which they represent in the physical network.

The system of units that has been adopted is the SI standard. Charge, current, voltage, power, and energy have been defined and the interrelationships among these quantities have been examined. Current and voltage sources, both dependent and independent, have been discussed, and the passive sign convention that will be used throughout the text has been presented.

It should be obvious at this point that there is ambiguity in our notation for variables and units. ''V'' is used both as a symbol for a voltage and as the unit abbreviation for volt. Similarly, ''W'' is used as the symbol for energy and as the abbreviation for watt, the unit of power. Thus we may find ourselves with expressions such as $V = 12$ V or $P = 4$ W which, when taken out of context, may be misleading; however, if we follow the associated discussions, the meaning will be clear.

KEY POINTS

- p $= 10^{-12}$ k $= 10^{3}$
  n $= 10^{-9}$ M $= 10^{6}$
  $\mu = 10^{-6}$ G $= 10^{9}$
  m $= 10^{-3}$ T $= 10^{12}$
- The passive sign convention is defined in Fig. 1.9.
- Power is being absorbed by an element when the sign of the power is positive using the passive sign convention.
- Power is being supplied by an element when the sign of the power is negative using the passive sign convention.
- The electrical networks considered here satisfy the principle of conservation of energy.
- An ideal independent voltage (current) source is a two-terminal element that maintains a specified voltage (current) between its terminals regardless of the current (voltage) through (across) the element.
- Dependent or controlled sources generate a voltage or current that is determined by a voltage or current at a specified location in the circuit.

PROBLEMS

1.1. The current in a conductor is known to be 12 A. How many coulombs of charge pass any given point in a time interval of 1.5 min?

1.2. In a given conductor, a charge of 600 C passes any point in 12-sec intervals. We wish to determine the current in the conductor.

1.3. A 12-V car battery supplies 6 J of energy in a specific time interval. How much charge is moved during this period?

1.4. The charge entering the positive terminal of an element is $q(t) = -30e^{-4t}$ mC. If the voltage across the element is $120e^{-2t}$ V, determine the energy delivered to the element in the time interval $0 < t < 50$ msec.

1.5. The charge entering the positive terminal of an element is given by the expression $q(t) = -12e^{-2t}$ mC. The power delivered to the element is $p(t) = 2.4e^{-3t}$ W. Compute the current in the element, the voltage across the element, and the energy delivered to the element in the time interval $0 < t < 100$ msec.

1.6. The current that enters an element is shown in Fig. P1.6. Find the charge that enters the element in the time interval $0 < t < 4$ sec.

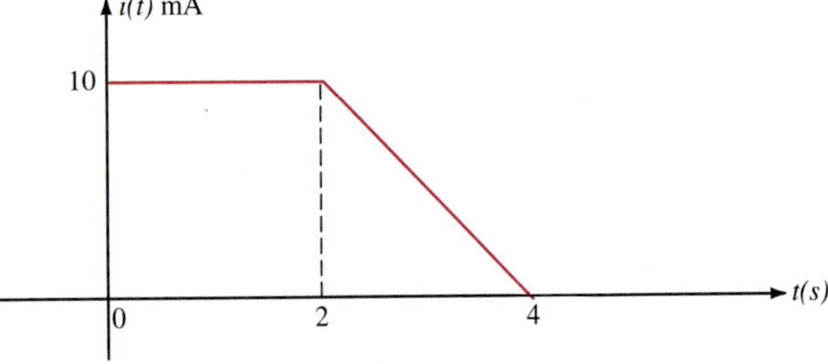

Figure P1.6

1.7. The voltage across an element is $12e^{-2t}$ V. The current entering the positive terminal of the element is $2e^{-2t}$ A. Find the energy absorbed by the element in 1.5 sec.

1.8. Determine how much power is absorbed by the circuit in Fig. P1.8 if

(a) $V_1 = 10$ V and $I = 3$ A.

(b) $V_1 = 4$ V and $I = -4$ A.

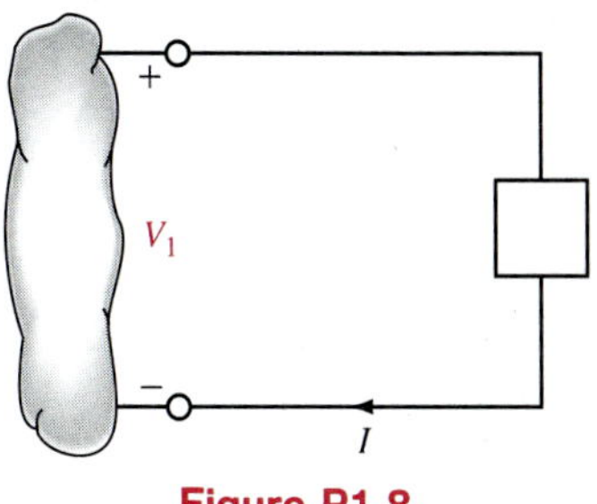

Figure P1.8

1.9. Determine the magnitude and direction of the current in the elements in Fig. P1.9.

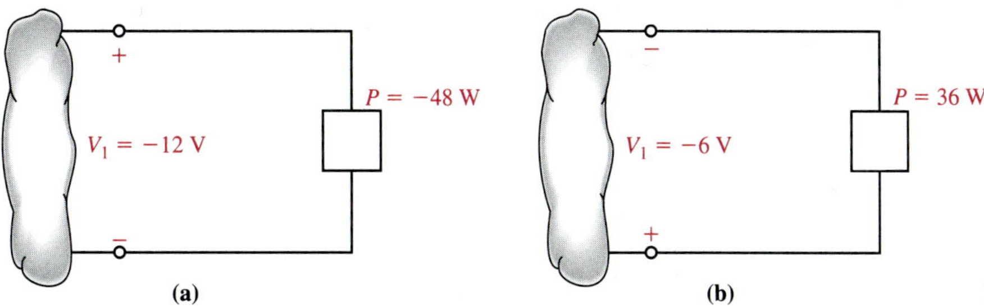

Figure P1.9

1.10. Determine the magnitude and direction of the voltage across the elements in Fig. P1.10.

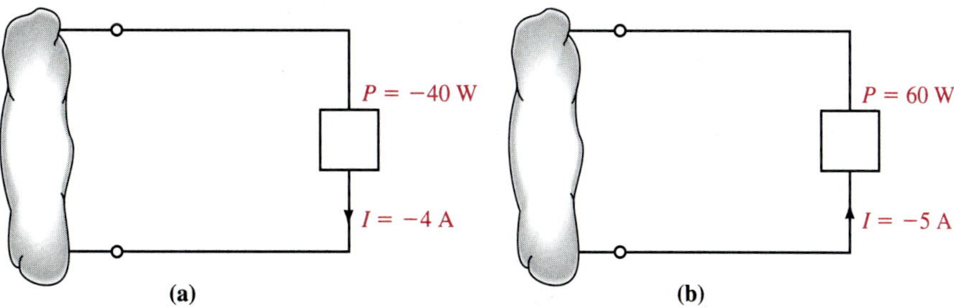

Figure P1.10

1.11. Two elements are connected in series as shown in Fig. P1.11. Element 1 supplies 24 W of power. Is element 2 absorbing or supplying power, and how much?

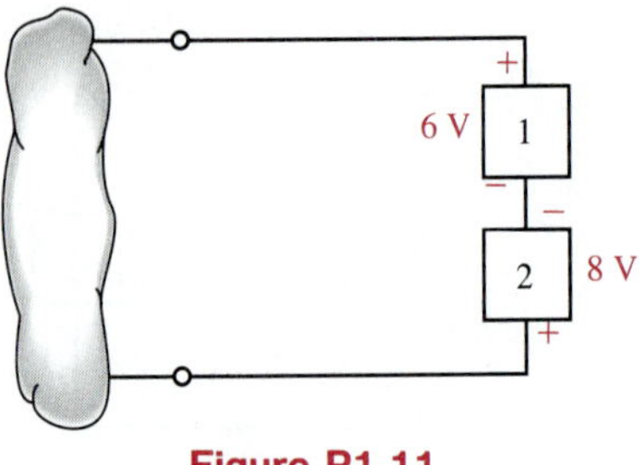

Figure P1.11

1.12. Two elements are connected in series as shown in Fig. P1.12. Element 1 supplies 36 W of power. Is element 2 absorbing or supplying power, and how much?

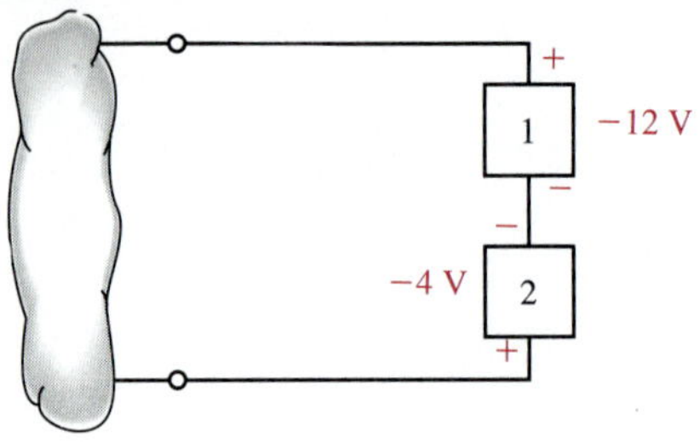

Figure P1.12

1.13. Is the source V_s in the network in Fig. P1.13 absorbing or supplying power, and how much?

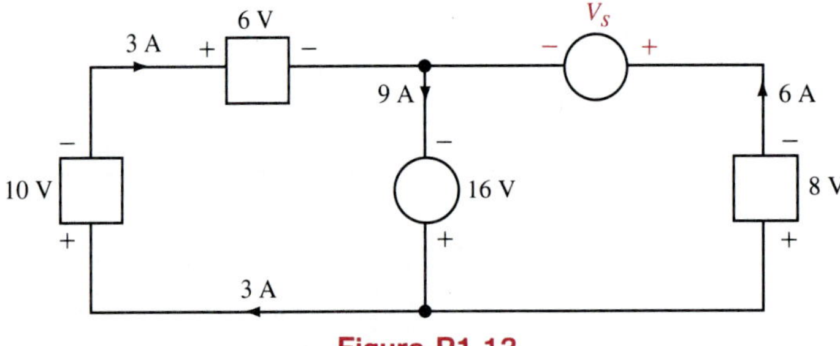

Figure P1.13

1.14. Determine the power that is absorbed or supplied by the elements in Fig. P1.14.

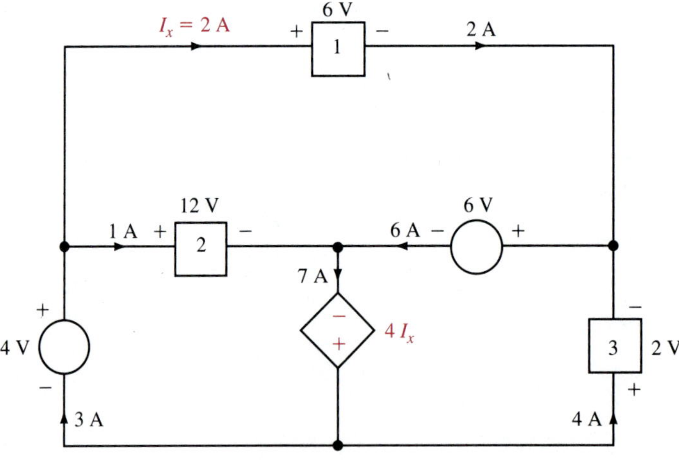

Figure P1.14

1.15. Find V_x in the network in Fig. P1.15.

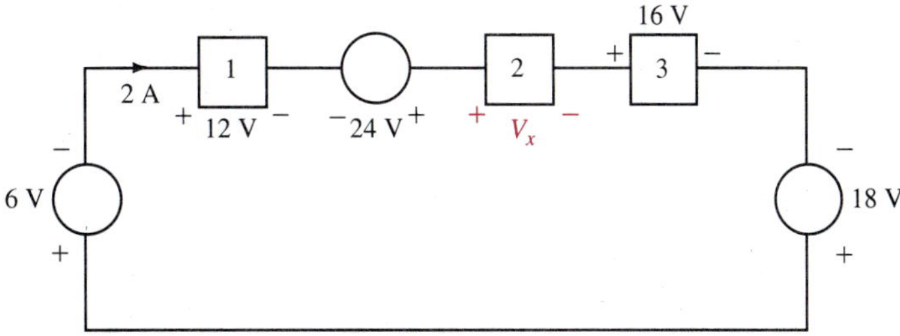

Figure P1.15

1.16. Find I_x in the network in Fig. P1.16.

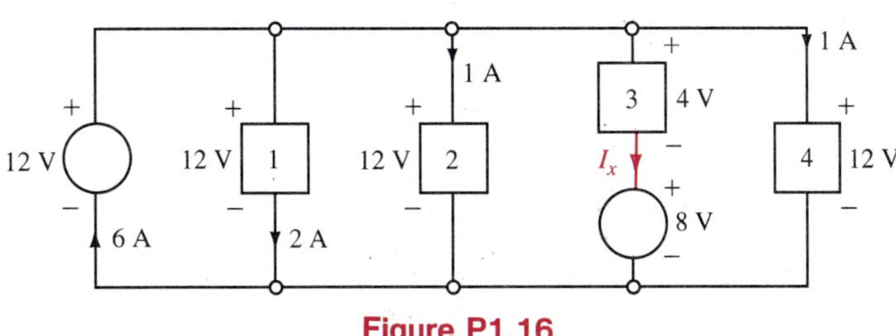

Figure P1.16

1.17. Find V_x in the network in Fig. P1.17.

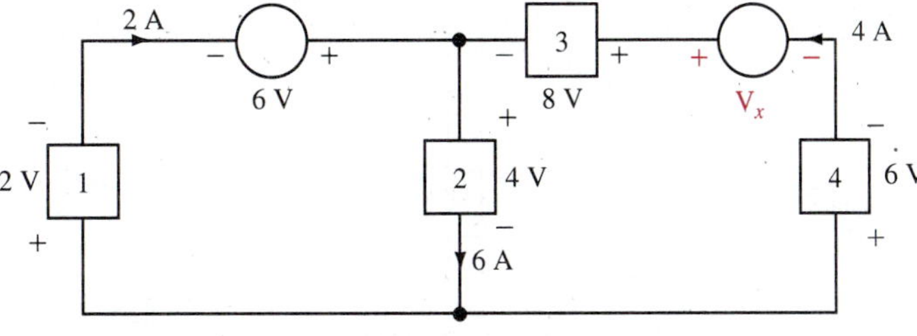

Figure P1.17

1.18. Find V_x in the network in Fig. P1.18.

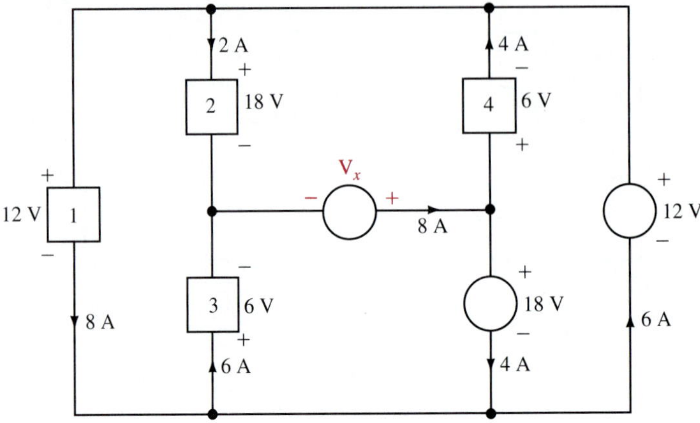

Figure P1.18

1.19. Find I_o in the network in Fig. P1.19.

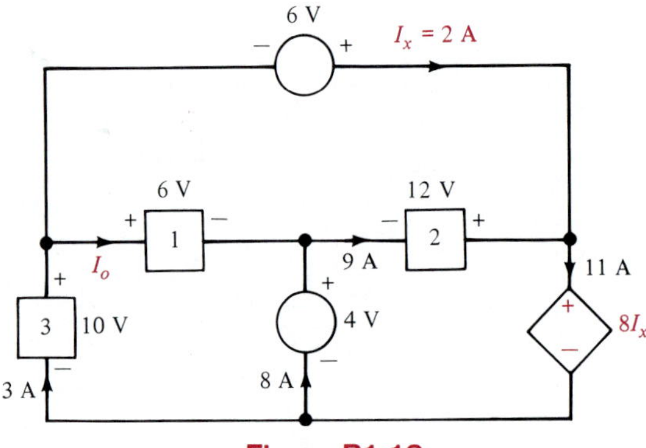

Figure P1.19

1.20. Find I_o in the network in Fig. P1.20.

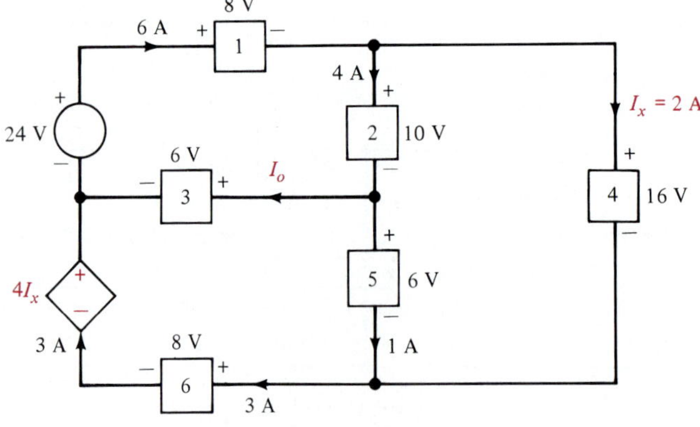

Figure P1.20

CHAPTER

2

Resistive Circuits

In this chapter we introduce some of the basic concepts and laws that are fundamental to circuit analysis. In general, we will restrict our activities to *analysis,* that is, to the determination of a specific voltage, current, or power somewhere in a network. The techniques we introduce have wide application in circuit analysis, even though we discuss them within the framework of simple networks.

2.1 Ohm's Law

Ohm's law is named for the German physicist Georg Simon Ohm, who is credited with establishing the voltage–current relationship for resistance. As a result of his pioneering work, the unit of resistance bears his name.

Ohm's law states that *the voltage across a resistance is directly proportional to the current flowing through it*. The resistance, measured in ohms, is the constant of proportionality between the voltage and current.

A circuit element whose electrical characteristic is primarily resistive is called a resistor and is represented by the symbol shown in Fig. 2.1. A resistor is a physical device that can be purchased in certain standard values in an electronic parts store. These resistors, which find wide use in a variety of electrical applications, are normally carbon-composition or wirewound. In addition, resistors can be fabricated using thick oxide or thin metal

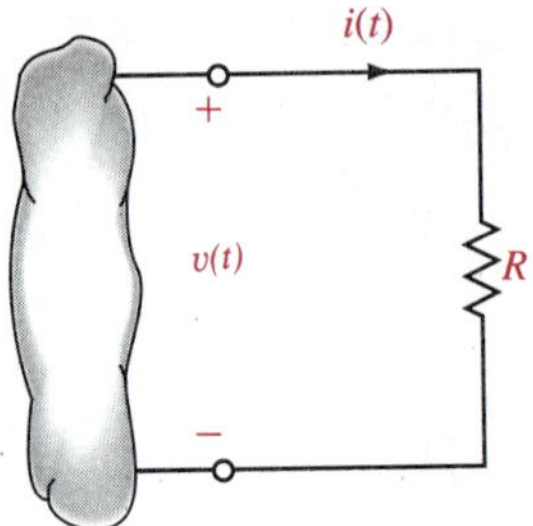

Figure 2.1 Symbol for the resistor.

films for use in hybrid circuits, or they can be diffused in semiconductor integrated circuits.

The mathematical relationship of Ohm's law is illustrated by the equation

$$v(t) = Ri(t) \qquad \text{where } R \geq 0 \tag{2.1}$$

or equivalently, by the voltage–current characteristic shown in Fig. 2.2a. Note carefully the relationship between the polarity of the voltage and the direction of the current. In addition, note that we have tacitly assumed that the resistor has a constant value and therefore that the voltage–current characteristic is linear.

The symbol Ω is used to represent ohms, and therefore

$$1\ \Omega = 1\ \text{V/A}$$

Although in our analysis we will always assume that the resistors are *linear* and are thus described by a straight-line characteristic that passes through the origin, it is important that readers realize that some very useful and practical elements do exist that exhibit a *nonlinear* resistance characteristic; that is, the voltage–current relationship is not a straight line. Diodes, which are used extensively in electric circuits, are examples of nonlinear resistors. A typical diode characteristic is shown in Fig. 2.2b.

Since a resistor is a passive element, the proper current–voltage relationship is illustrated in Fig. 2.1. The power supplied to the terminals is absorbed by the resistor. Note that the charge moves from the higher to the lower potential as it passes through the

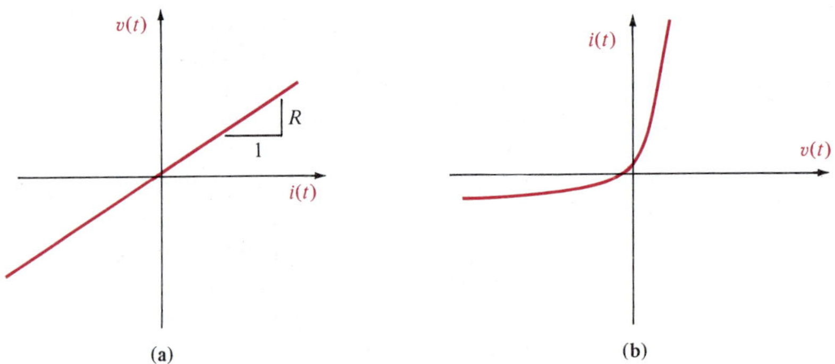

Figure 2.2 Graphical representation of the voltage–current relationship for (a) a linear resistor and (b) a diode.

resistor and the energy absorbed is dissipated by the resistor in the form of heat. As indicated in Chapter 1, the rate of energy dissipation is the instantaneous power, and therefore

$$p(t) = v(t)i(t) \tag{2.2}$$

which, using Eq. (2.1), can be written as

$$p(t) = Ri^2(t) = \frac{v^2(t)}{R} \tag{2.3}$$

This equation illustrates that the power is a nonlinear function of either current or voltage and that it is always a positive quantity.

Conductance, represented by the symbol G, is another quantity with wide application in circuit analysis. By definition, conductance is the inverse of resistance; that is,

$$G = \frac{1}{R} \tag{2.4}$$

The unit of conductance is the siemens, and the relationship between units is

$$1 \text{ S} = 1 \text{ A/V}$$

Using Eq. (2.4), we can write two additional expressions,

$$i(t) = Gv(t) \tag{2.5}$$

and

$$p(t) = \frac{i^2(t)}{G} = Gv^2(t) \tag{2.6}$$

Equation (2.5) is another expression of Ohm's law.

Two specific values of resistance, and therefore conductance, are very important: $R = 0$ and $R = \infty$. If the resistance $R = 0$, we have what is called a *short circuit*. From Ohm's law,

$$\begin{aligned} v(t) &= Ri(t) \\ &= 0 \end{aligned}$$

Therefore, $v(t) = 0$, although the current could theoretically be any value. If the resistance $R = \infty$, we have what is called an *open circuit,* and from Ohm's law,

$$\begin{aligned} i(t) &= \frac{v(t)}{R} \\ &= 0 \end{aligned}$$

Therefore, the current is zero regardless of the value of the voltage across the open terminals.

EXAMPLE 2.1

In the circuit shown in Fig. 2.3a, determine the current and the power absorbed by the resistor.

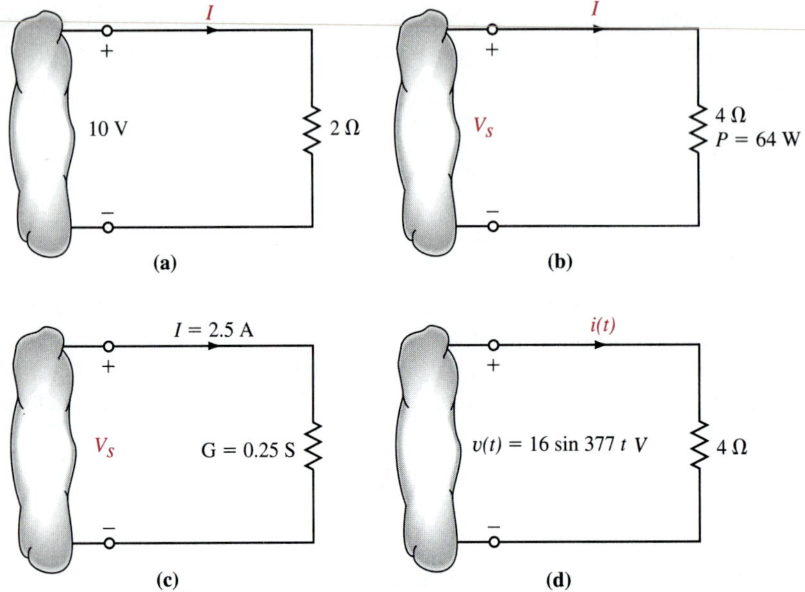

Figure 2.3 Circuits for Examples 2.1 to 2.4.

Using Eq. (2.1), we find the current to be

$$I = \frac{V}{R} = \frac{10}{2} = 5 \text{ A}$$

and from Eq. (2.2) or (2.3), the power absorbed by the resistor is

$$\begin{aligned} P &= VI = (10)(5) = 50 \text{ W} \\ &= RI^2 = (2)(5)^2 = 50 \text{ W} \\ &= \frac{V^2}{R} = \frac{(10)^2}{2} = 50 \text{ W} \end{aligned}$$

■

EXAMPLE 2.2

The power absorbed by the 4-Ω resistor in Fig. 2.3b is 64 W. Determine the voltage and the current.

Using Eq. (2.3), we can immediately determine either of the unknowns.

$$64 \ W = (4)I^2$$

$$I = 4 \text{ A}$$

and

$$64 = \frac{V_S^2}{4}$$

$$V_S = 16 \text{ V}$$

Note, however, that once I is determined, V_S could be derived from Ohm's law. Note carefully that $I = -4$ A and $V_S = -16$ V also satisfy the mathematical equations above. ■

EXAMPLE 2.3

Given the circuit in Fig. 2.3c, we wish to determine the voltage across the terminals and the power absorbed by the resistance.

From Eq. (2.5), the voltage is

$$V_S = \frac{I}{G}$$

$$= \frac{2.5}{0.25} = 10 \text{ V}$$

The power is determined from Eq. (2.6) as

$$P = \frac{I^2}{G} = \frac{(2.5)^2}{0.25} = 25 \text{ W}$$

■

EXAMPLE 2.4

Given the circuit in Fig. 2.3d with a sinusoidal input, determine the resultant current and the power absorbed by the resistor.

From Ohm's law,

$$i(t) = \frac{v(t)}{R}$$

$$= \frac{16 \sin 377t}{4}$$

$$= 4 \sin 377t \text{ A}$$

Therefore, both the voltage and current are sinusoidal. The power is

$$p(t) = v(t)i(t)$$

$$= (16 \sin 377t)(4 \sin 377t) \text{ A}$$

$$= 64 \sin^2 377t \text{ W}$$

Note that although the voltage and current are negative during the intervals when the sine function is negative, the power is always a positive quantity. ■

DRILL EXERCISE

D2.1. Given the network in Fig. D2.1, determine the voltage V_S across the resistor and the power absorbed by the resistor.

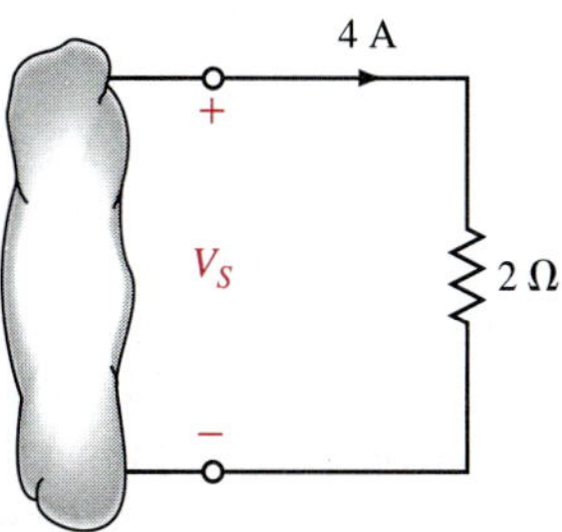

Figure D2.1

Ans: $V_S = 8$ V, $P = 32$ W.

D2.2. Given the network in Fig. D2.2, determine the resistance R and the voltage V_S.

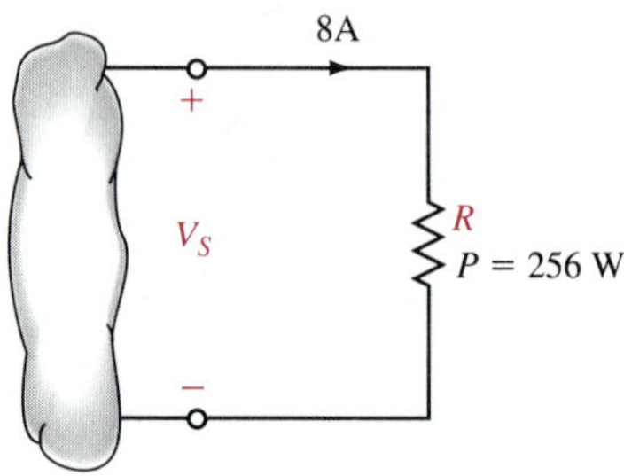

Figure D2.2

Ans: $R = 4\ \Omega$, $V_S = 32$ V.

2.2 Kirchhoff's Laws

The previous circuits that we have considered have all contained a single resistor and were analyzed using Ohm's law. At this point we begin to expand our capabilities to handle more complicated networks which result from an interconnection of two or more of these simple elements. We will assume that the interconnection is performed by electrical conductors (wires) that have zero resistance, that is, perfect conductors. Because the wires have zero resistance, the energy in the circuit is in essence lumped in each element, and we employ the term *lumped-parameter circuit* to describe the network.

To aid us in our discussion, we will define a number of terms that will be employed throughout our analysis. As will be our approach throughout this text, we will use examples to illustrate the concepts and define the appropriate terms. For example, the circuit shown in Fig. 2.4a will be used to describe the terms *node, loop,* and *branch*. A *node* is simply a point of connection of two or more circuit elements. The nodes in the circuit in Fig. 2.4a are exaggerated in Fig. 2.4b for clarity. The reader is cautioned to compare the two figures carefully and note that although one node can be spread out with perfect conductors, it is still only one node. For example, node 5 consists of the entire bottom connector of the circuit. In other words, if we start at some point in the circuit and move along perfect conductors in any direction until we encounter a circuit element, the total path we cover represents a single node. Therefore, we can assume that a node is one end of a circuit element together with all the perfect conductors that are attached to it. Examining the circuit, we note that there are numerous paths through it. A *loop* is simply any *closed path* through the circuit in which no node is encountered more than once. For example, starting from node 1, one loop would contain the elements R_1, R_2, R_3, and $v(t)$; another loop would contain R_2, $i(t)$, R_5, and R_6; and so on. Finally, a *branch* is a portion of a circuit containing only a single element and the nodes at each end of the element. The circuit in Fig. 2.4 contains eight branches.

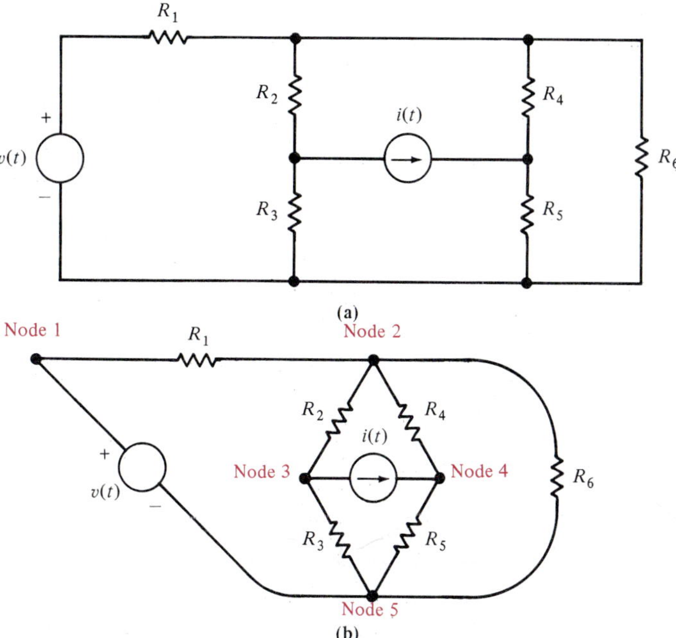

Figure 2.4 Circuits used to illustrate terms: (a) example circuit; (b) circuit in (a) with nodes illustrated.

Given the previous definitions, we are now in a position to consider Kirchhoff's laws, named after the German scientist Gustav Robert Kirchhoff. These two laws are quite simple but extremely important. We will not attempt to prove them because the proofs are beyond our current level of understanding. However, we will demonstrate their usefulness and attempt to make the reader proficient in their use. The first law is *Kirchhoff's current law,* which states that *the algebraic sum of the currents entering any node is zero.* In mathematical form the law appears as

$$\sum_{j=1}^{N} i_j(t) = 0 \tag{2.7}$$

where $i_j(t)$ is the jth current entering the node through branch j and N is the number of branches connected to the node. To understand the use of this law, consider the node shown in Fig. 2.5. Applying Kirchhoff's current law to this node yields

$$i_1(t) + [-i_2(t)] + i_3(t) + i_4(t) + [-i_5(t)] = 0$$

We have assumed that the algebraic signs of the currents entering the node are positive and therefore that the signs of the currents leaving the node are negative.

If we multiply the foregoing equation by -1, we obtain the expression

$$-i_1(t) + i_2(t) - i_3(t) - i_4(t) + i_5(t) = 0$$

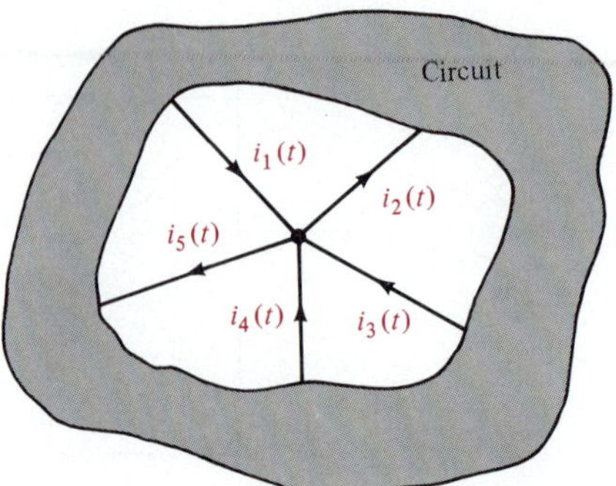

Figure 2.5 Currents at a node.

which simply states that *the algebraic sum of the currents leaving a node is zero*. Alternatively, we can write the equation as

$$i_1(t) + i_3(t) + i_4(t) = i_2(t) + i_5(t)$$

which states that *the sum of the currents entering a node is equal to the sum of the currents leaving the node*. Both of these italicized expressions are alternative forms of Kirchhoff's current law.

Once again it must be emphasized that the latter statement means that the sum of the *variables* that have been defined entering the node is equal to the sum of the *variables* that have been defined leaving the node, not the actual currents. For example, $i_j(t)$ may be defined entering the node, but if its actual value is negative, there will be positive charge leaving the node.

Note carefully that Kirchhoff's current law states that the *algebraic* sum of the currents either entering or leaving a node must be zero. We now begin to see why we stated in Chapter 1 that it is critically important to specify both the magnitude and the direction of a current.

EXAMPLE 2.5

Using Kirchhoff's current law, we want to determine the value of I_1 in Fig. 2.6a.

The algebraic sum of the currents entering the node is

$$I_1 + 5 - 2 - 1 = 0$$

or

$$I_1 = -2 \text{ A}$$

This equation indicates that the magnitude of I_1 is 2 A, but the direction is opposite to that which was defined, and therefore 2 A is leaving the node. ■

EXAMPLE 2.6

In Fig. 2.6b, we wish to determine the currents I_1 and I_2.

The equations for Kirchhoff's current law at nodes 1 and 2, respectively, are

$$2 - 4 - 2 - I_1 = 0$$

$$I_1 - 8 + 3 - I_2 = 0$$

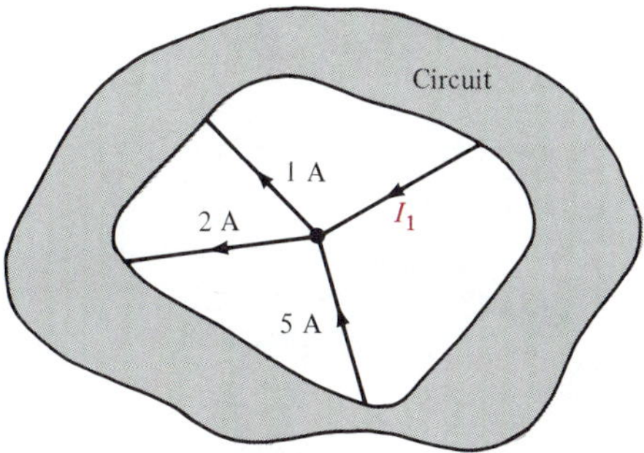

(a) Single-node example.

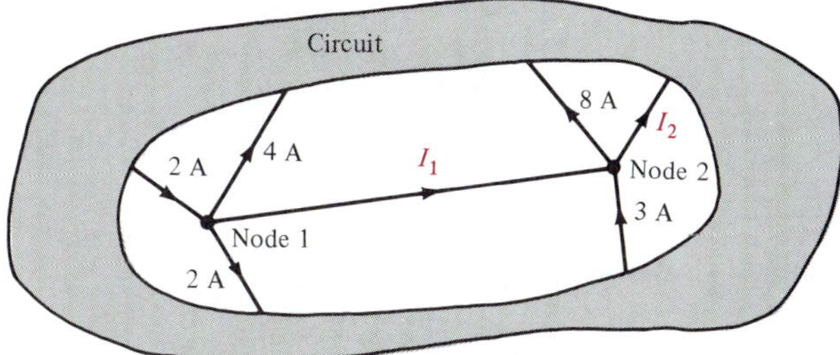

(b) Multiple-node example.

Figure 2.6 Illustration nodes for Kirchhoff's current law: (a) single-node example; (b) multiple-node example.

From the first equation, $I_1 = -4$ A. Since it was assumed that I_1 was leaving node 1, the negative sign illustrates that positive current is actually entering node 1. Using this value of I_1 in the second equation, we find that $I_2 = -9$ A. ■

Finally, it is possible to generalize Kirchhoff's current law to include a closed surface. By a closed surface we mean some set of elements completely contained within the surface that are interconnected. Since the current entering each element within the surface is equal to that leaving the element (i.e., the element stores no net charge), it follows that the current entering an interconnection of elements is equal to that leaving any surface enclosing that interconnection. Therefore, Kirchhoff's current law can also be stated: *The algebraic sum of the currents entering any closed surface is zero.*

EXAMPLE 2.7

To illustrate this generalization of Kirchhoff's current law stated above, we need only consider the multiple-node arrangement discussed in Example 2.6 and illustrated in Fig. 2.6b.

Now we apply Kirchhoff's current law to the surface. Assuming that the currents entering the surface are positive and those leaving the surface are negative, we can write

$$-2 + 4 + 8 + I_2 - 3 + 2 = 0$$

$$I_2 = -9 \text{ A}$$

which is, of course, what we obtained for I_2 in Example 2.6. Note however, that we did not need to solve for I_1 to determine I_2. ■

DRILL EXERCISE

D2.3. Find I_1 in the network in Fig. D2.3.

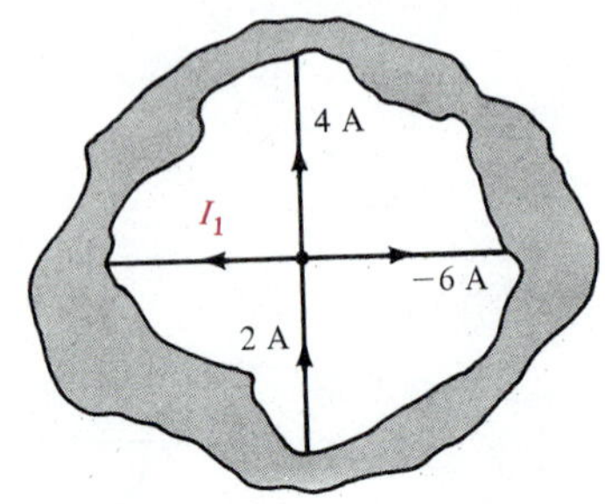

Figure D2.3

Ans: $I_1 = 4$ A.

D2.4. Find the currents I_1, I_2, and I_3 in the network in Fig. D2.4.

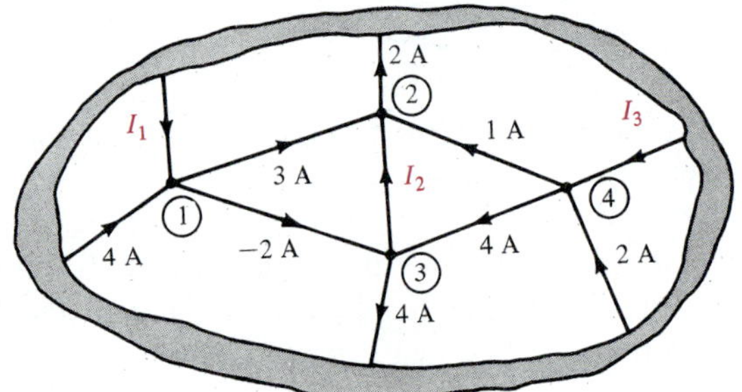

Figure D2.4

Ans: $I_1 = -3$ A, $I_2 = -2$ A, $I_3 = 3$ A.

Kirchhoff's second law, called *Kirchhoff's voltage law,* states that *the algebraic sum of the voltages around any loop is zero*. As was the case with Kirchhoff's current law, we will defer the proof of this law and concentrate on understanding how to apply it. Once again the reader is cautioned to remember that we are dealing only with lumped-parameter circuits. These circuits are conservative, meaning that the work required to move a unit charge around any loop is zero.

Recall that in Kirchhoff's current law, the algebraic sign was required to keep track of whether the currents were entering or leaving a node. In Kirchhoff's voltage law the algebraic sign is used to keep track of the voltage polarity. In other words, as we traverse the circuit, it is necessary to sum to zero the increases and decreases in energy level. Therefore, it is important that we keep track of whether the energy level is increasing or decreasing as we go through each element.

EXAMPLE 2.8

Consider the circuit shown in Fig. 2.7a. In applying Kirchhoff's voltage law we must traverse the circuit and sum to zero the increases and decreases in energy level. Since the network is a single loop, we can traverse the path in either a clockwise or counterclockwise direction. Furthermore, we can consider an increase in potential energy level as positive and a decrease in energy level as negative, or vice versa. Based on these arguments, we have four possible choices for employing Kirchhoff's voltage law in this case. They are outlined in the following table.

Convention	Increase in Energy Level	Decrease in Energy Level	Direction for Traversing the Circuit
1	+	−	CW
2	+	−	CCW
3	−	+	CW
4	−	+	CCW

Starting at point a, the results obtained by using these four conventions are as follows:

Convention 1

$$-V_{R_1} + 5 - V_{R_2} + 15 - V_{R_3} + 30 = 0$$

Convention 2

$$-30 + V_{R_3} - 15 + V_{R_2} - 5 + V_{R_1} = 0$$

Convention 3

$$+ V_{R_1} - 5 + V_{R_2} - 15 + V_{R_3} - 30 = 0$$

Convention 4

$$+ 30 - V_{R_3} + 15 - V_{R_2} + 5 - V_{R_1} = 0$$

Since an equation is not changed by multiplying every term by -1, an examination of the four equations above illustrates that they are all identical. Under these conditions, which convention should we use? Since they are all valid, use the one you like best. We will employ convention 1 for now. However, since the first step in solving an algebraic equation is to place the known quantities on one side and the unknown quantities on the other, we will soon start writing the equations in the form

$$\begin{aligned} + V_{R_1} + V_{R_2} + V_{R_3} &= 5 + 15 + 30 \\ &= 50 \end{aligned}$$

Now suppose that V_{R_1} and V_{R_2} are known to be 18 V and 12 V, respectively. Then any of the equations can be used to find $V_{R_3} = 20$ V. The circuit with all known voltages labeled is shown in Fig. 2.7b.

Finally, we employ the convention V_{ab} to indicate the voltage of point a with respect to point b: that is, the variable for the voltage between point a and point b, with point a

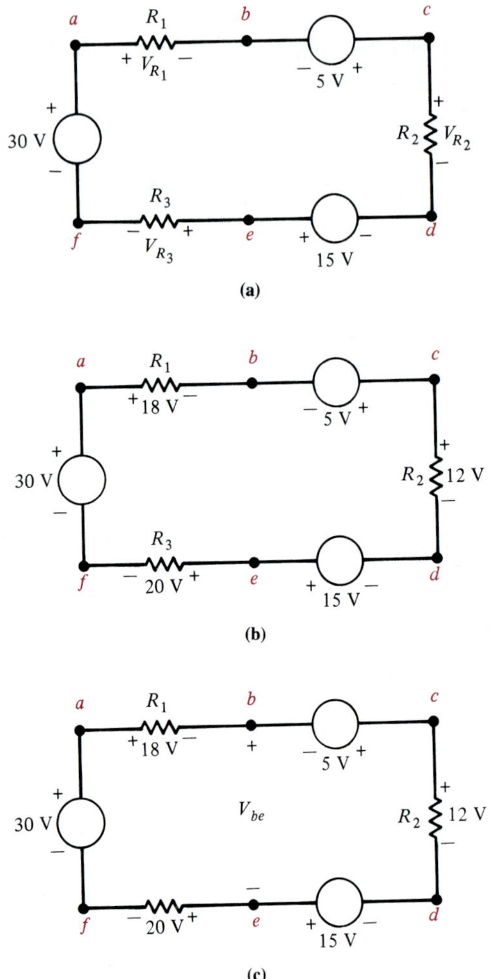

Figure 2.7 Circuits used to illustrate Kirchhoff's voltage law.

considered positive relative to point b. Since the potential is measured between two points, it is convenient to use an arrow between the two points with the head of the arrow located at the positive node. Note that the double-subscript notation, the + and − notation, and the single-headed arrow notation are all the same if the head of the arrow is pointing toward the positive terminal and the first subscript in the double-subscript notation. All of these equivalent forms for labeling voltages are shown in Fig. 2.8. If we employ the results in Fig. 2.8 to the network in Fig. 2.7, we find, for example, that $V_{af} = 30$ V, $V_{fa} = -30$ V, $V_{ab} = 18$ V, $V_{dc} = -12$ V, $V_{ed} = 15$ V, and $V_{ef} = 20$ V. Furthermore, we can apply Kirchhoff's voltage law to the circuit to determine the voltage between any two points. For example, suppose that we want to determine the voltage V_{be} as shown in Fig. 2.7c. Although we could employ any one of the four conventions

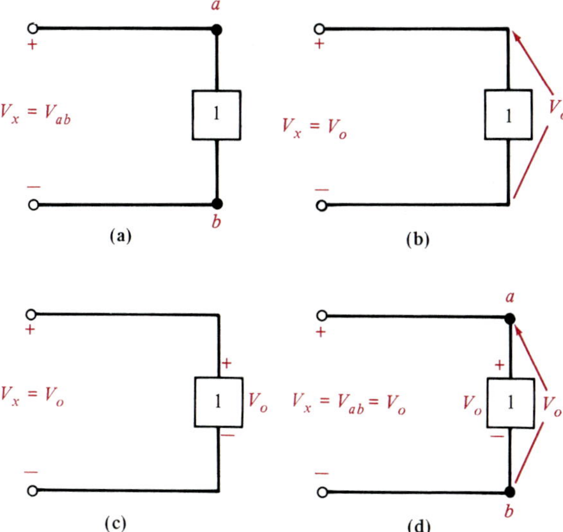

Figure 2.8 Equivalent forms for labeling voltage.

outlined earlier, we will use convention 1 and start at point b. Note that we have a choice of two paths; one is $bcdeb$ and the other is $befab$. For the first path we obtain

$$+5 - 12 + 15 + V_{be} = 0$$

$$V_{be} = -8\text{ V}$$

For the second path the equation is

$$-V_{be} - 20 + 30 - 18 = 0$$

$$V_{be} = -8\text{ V}$$

In a similar manner we can show that $V_{ad} = 25$ V and $V_{fc} = -17$ V. ■

EXAMPLE 2.9

Consider now the network in Fig. 2.9a. In addition to the voltage source, the following voltages are known: $V_{R_1} = 2$ V, $V_{R_2} = 4$ V, $V_{R_4} = 9$ V, and $V_{R_5} = -3$ V. Given this information, let us determine V_{R_3}, V_{ae}, and V_{ec}.

We could, of course, start at any point in the network and employ any one of the conventions outlined earlier. Once again we will employ convention 1 and apply Kirchhoff's voltage law to the loop $abcdefa$.

$$-24 + V_{R_2} - V_{R_3} + V_{R_4} - V_{R_5} + V_{R_1} = 0$$

Substituting the known quantities into this equation yields

$$-24 + 4 - V_{R_3} + 9 - (-3) + 2 = 0$$

or

$$V_{R_3} = -6\text{ V}$$

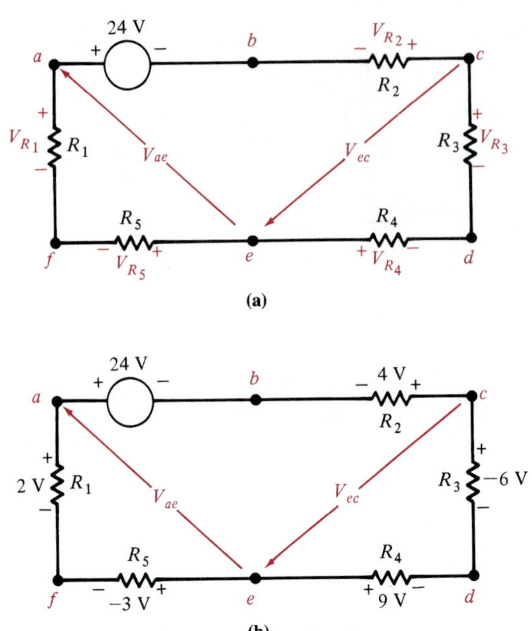

Figure 2.9 Another circuit used to illustrate Kirchhoff's voltage law.

Using the same approach, we can determine V_{ae}. At this point all known voltages are shown in Fig. 2.9b. Recall that we could change the voltages across R_3 and R_5 to positive values if we also reversed the polarity of both V_{R_3} and V_{R_5} (i.e., interchanged the + and − signs at the ends of the resistors). Using the path *abcdea*, we obtain

$$-24 + 4 - (-6) + 9 + V_{ae} = 0$$
$$V_{ae} = 5 \text{ V}$$

Using the path *aefa* yields

$$-V_{ae} - (-3) + 2 = 0$$
$$V_{ae} = 5 \text{ V}$$

Finally, with the information given above, V_{ec} can be determined in a variety of ways. Using path *cdec*, we have

$$-(-6) + 9 - V_{ec} = 0$$
$$V_{ec} = 15 \text{ V}$$

Using path *cefabc* yields

$$V_{ec} - (-3) + 2 - 24 + 4 = 0$$
$$V_{ec} = 15 \text{ V}$$

Using path *ceabc* gives us

$$V_{ec} + V_{ae} - 24 + 4 = 0$$

$$V_{ec} + 5 - 24 + 4 = 0$$

$$V_{ec} = 15 \text{ V}$$

■

In general, the mathematical representation of Kirchhoff's voltage law is

$$\sum_{j=1}^{N} v_j(t) = 0 \tag{2.8}$$

where $v_j(t)$ is the voltage across the jth branch (with the proper reference direction) in a loop containing N voltages. This expression is analogous to Eq. (2.7) for Kirchhoff's current law.

DRILL EXERCISE

D2.5. In the network in Fig. D2.5, V_{R_1} is known to be 4 V. Find V_{R_2} and V_{bd}.

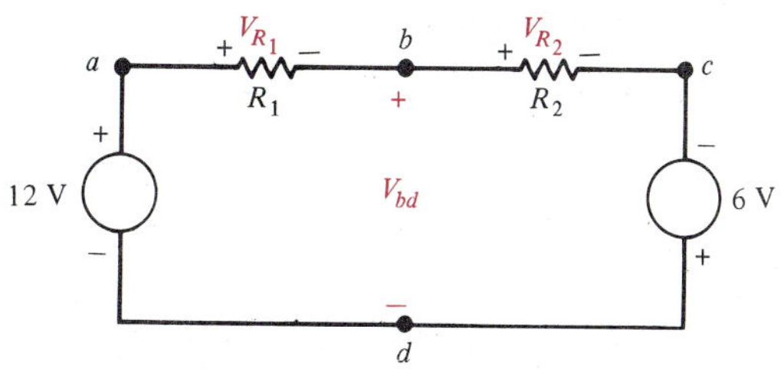

Figure D2.5

Ans: $V_{R_2} = 14$ V, $V_{bd} = 8$ V.

D2.6. In the network in Fig. D2.6, find V_{R_4}, V_{bf}, and V_{ec}.

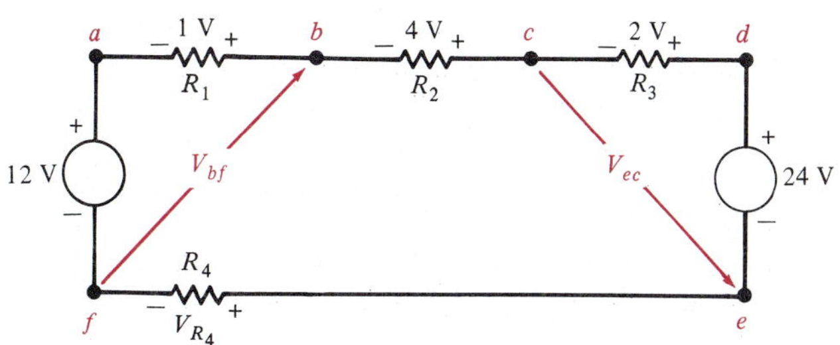

Figure D2.6

Ans: $V_{R_4} = -5$ V, $V_{bf} = 13$ V, $V_{ec} = -22$ V.

Before proceeding with the analysis of simple circuits, it is extremely important that we emphasize a subtle but very critical point. Ohm's law as defined by the equation $V = IR$ refers to the relationship between the voltage and current as defined in Fig. 2.10a.

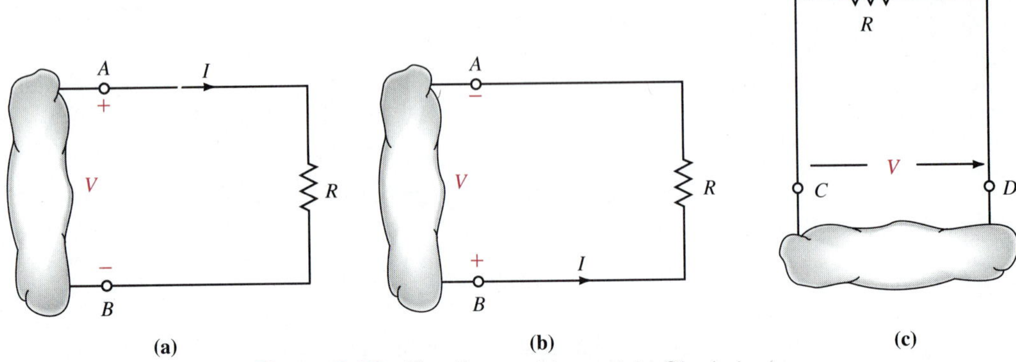

Figure 2.10 Circuits used to explain Ohm's law.

If the direction of either the current or the voltage, but not both, is reversed, the relationship between the current and the voltage would be $V = -IR$. In a similar manner, given the circuit in Fig. 2.10b, $V = IR$ implies that the current I flows from point B through the resistor to point A. Likewise, for the circuit in Fig. 2.10c, $V = IR$ implies that point D is at a higher potential than point C, and therefore the arrow representing the voltage V is from point C to point D.

2.3 Single-Loop Circuits

At this point we can begin to apply the laws we have presented earlier to the analysis of simple circuits. To begin, we examine what is perhaps the simplest circuit—a single closed path, or loop, of elements. The elements of a single loop carry the same current and therefore are said to be in *series*. However, we will apply Kirchhoff's voltage law and Ohm's law to the circuit to determine various quantities in the circuit.

Our approach will be to begin with a simple circuit and then generalize the analysis to more complicated ones. The circuit shown in Fig. 2.11 will serve as a basis for discussion. This circuit consists of an independent voltage source that is in series with two resistors. We have assumed that the current flows in a clockwise direction. If this assumption is correct, the solution of the equations that yields the current will produce a positive

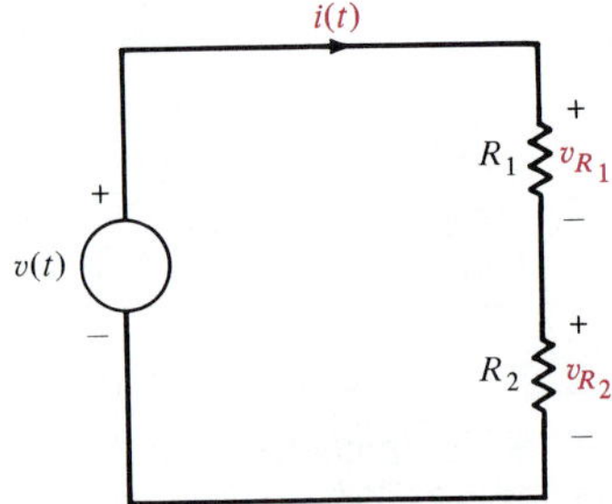

Figure 2.11 Single-loop circuit.

value. If the current is actually flowing in the opposite direction, the value of the current variable will simply be negative, indicating that the current is flowing in a direction opposite to that assumed. We have also made voltage polarity assignments for v_{R_1} and v_{R_2}. These assignments have been made using the convention employed in our discussion of Ohm's law and our choice for the direction of $i(t)$, that is, the convention shown in Fig. 2.10a.

Applying Kirchhoff's voltage law to this circuit yields

$$+v(t) - v_{R_1} - v_{R_2} = 0$$

or

$$v(t) = v_{R_1} + v_{R_2}$$

However, from Ohm's law we know that

$$v_{R_1} = R_1 i(t)$$

$$v_{R_2} = R_2 i(t)$$

Therefore,

$$v(t) = R_1 i(t) + R_2 i(t)$$

Solving the equation for $i(t)$ yields

$$i(t) = \frac{v(t)}{R_1 + R_2} \tag{2.9}$$

Knowing the current, we can now apply Ohm's law to determine the voltage across each resistor:

$$\begin{aligned} v_{R_1} &= R_1 i(t) \\ &= \frac{R_1}{R_1 + R_2} v(t) \end{aligned} \tag{2.10}$$

Similarly,

$$v_{R_2} = \frac{R_2}{R_1 + R_2} v(t) \tag{2.11}$$

Note that the equations satisfy Kirchhoff's voltage law, since

$$+v(t) - \frac{R_1}{R_1 + R_2} v(t) - \frac{R_2}{R_1 + \mathrm{R}_2} v(t) = 0$$

Although simple, Eqs. (2.10) and (2.11) are very important because they describe the operation of what is called a *voltage divider*. In other words, the source $v(t)$ *is divided between the resistors R_1 and R_2 in direct proportion to their resistances.*

EXAMPLE 2.10

Consider the circuit shown in Fig. 2.12. The circuit is identical to Fig. 2.11 except that R_1 is a variable resistor such as the volume control for a radio or television set. Suppose that $V_S = 24$ V, $R_1 = 10\ \Omega$, and $R_2 = 2\ \Omega$.

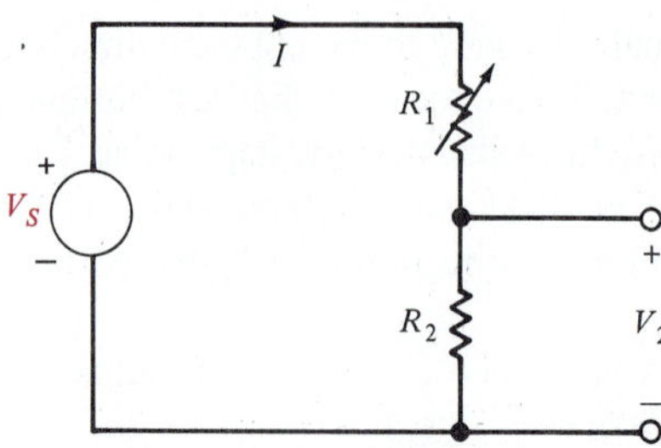

Figure 2.12 Voltage-divider circuit.

Using Eq. (2.11), we find that the voltage V_2 is

$$V_2 = \frac{R_2}{R_1 + R_2} V_S$$

$$= \frac{2}{10 + 2} 24$$

$$= 4 \text{ V}$$

Now suppose that the variable resistor R_1 is changed from 10 to 0.4 Ω. Then

$$V_2 = \frac{2}{0.4 + 2} 24$$

$$= 20 \text{ V}$$

Note that the use of Eq. (2.11) is equivalent to determining the current I and then using Ohm's law to find V_2. Note that the larger voltage is across the larger resistance. This voltage-divider concept and the simple circuit we have employed to describe it are very useful because, as will be shown later, more complicated circuits can be reduced to this form. ■

Let us now consider the power relationship that exists in the circuit of Fig. 2.11. The instantaneous power delivered by the voltage source is

$$p(t) = v(t)i(t)$$

and the instantaneous power absorbed by resistors R_1 and R_2 is

$$p_1(t) = \frac{v_{R_1}^2(t)}{R_1} = \frac{R_1}{(R_1 + R_2)^2} v^2(t)$$

and

$$p_2(t) = \frac{v_{R_2}^2(t)}{R_2} = \frac{R_2}{(R_1 + R_2)^2} v^2(t)$$

respectively. Now note that

$$\begin{aligned} p_1(t) + p_2(t) &= \frac{R_1}{(R_1 + R_2)^2} v^2(t) + \frac{R_2}{(R_1 + R_3)^2} v^2(t) \\ &= \frac{v^2(t)}{R_1 + R_2} \\ &= \frac{v(t)}{R_1 + R_2} v(t) \\ &= v(t)i(t) \\ &= p(t) \end{aligned}$$

This analysis illustrates the conservation of power in the circuit, since the power supplied by the voltage source is completely absorbed by the two resistors.

EXAMPLE 2.11

Determine the instantaneous power absorbed in the resistor R_2 of Example 2.10 when $R_1 = 10\ \Omega$ and when $R_1 = 0.4\ \Omega$.

$$P_2 = \frac{R_2}{(R_1 + R_2)^2} V^2$$

For $R_1 = 10\ \Omega$,

$$\begin{aligned} P_2 &= \frac{2}{(10 + 2)^2} (24)^2 \\ &= 8 \text{ W} \end{aligned}$$

and for $R_1 = 0.4\ \Omega$,

$$\begin{aligned} P_2 &= \frac{2}{(0.4 + 2)^2} (24)^2 \\ &= 200 \text{ W} \end{aligned}$$

■

At this point we wish to extend our analysis to include a multiplicity of voltage sources and resistors. For example, consider the circuit shown in Fig. 2.13a. Here we have assumed that the current flows in a clockwise direction, and we have defined the variable $i(t)$ accordingly. This may or may not be the case, depending on the value of the various voltage sources. Kirchhoff's voltage law for this circuit is

$$-v_{R_1} - v_2(t) + v_3(t) - v_{R_2} - v_4(t) - v_5(t) + v_1(t) = 0$$

or using Ohm's law,

$$(R_1 + R_2)i(t) = v_1(t) - v_2(t) + v_3(t) - v_4(t) - v_5(t)$$

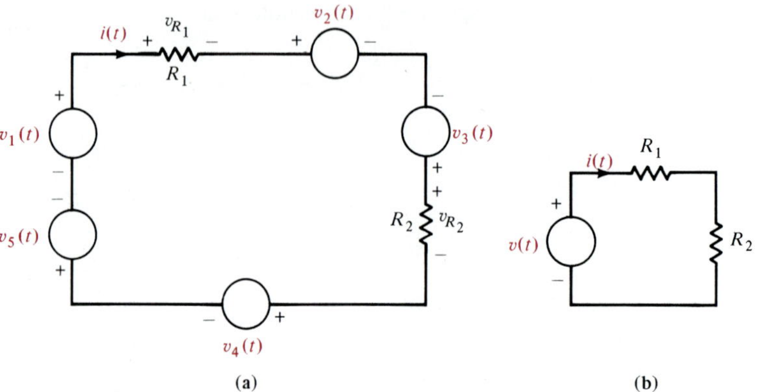

Figure 2.13 Equivalent circuits with multiple sources.

which can be written as

$$(R_1 + R_2)i(t) = v(t)$$

where

$$v(t) = v_1(t) + v_3(t) - [v_2(t) + v_4(t) + v_5(t)]$$

so that under the definitions above, Fig. 2.13a is equivalent to Fig. 2.13b. In other words, the sum of several voltage sources in series can be replaced by one source whose value is the algebraic sum of the individual sources. This analysis can, of course, be generalized to a circuit with N series sources.

Now consider the circuit with N resistors in series as shown in Fig. 2.14a. Applying Kirchhoff's voltage law to this circuit yields

$$\begin{aligned} v(t) &= v_{R_1} + v_{R_2} + \cdots + v_{R_N} \\ &= R_1 i(t) + R_2 i(t) + \cdots + R_N i(t) \end{aligned}$$

and therefore,

$$v(t) = R_S i(t) \tag{2.12}$$

where

$$R_S = R_1 + R_2 + \cdots + R_N \tag{2.13}$$

and hence

$$i(t) = \frac{v(t)}{R_S} \tag{2.14}$$

Note also that for any resistor R_i in the circuit, the voltage across R_i is given by the expression

$$v_{R_i} = \frac{R_i}{R_S} v(t) \tag{2.15}$$

which is the voltage-division property for multiple resistors in series.

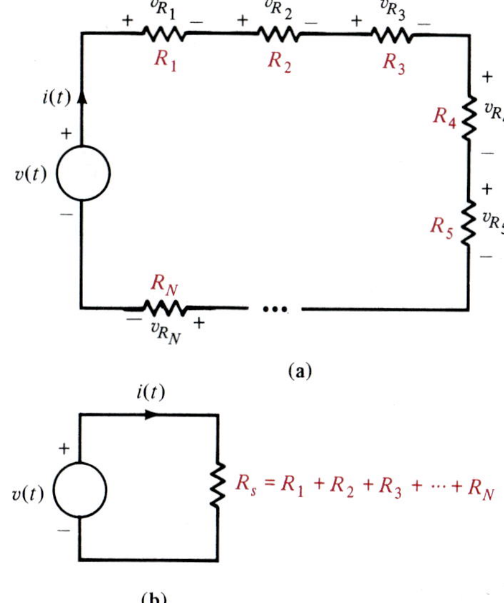

Figure 2.14 Equivalent circuits.

Equation (2.13) illustrates that *the equivalent resistance of N resistors in series is simply the sum of the individual resistances*. Thus, using Eqs. (2.13) and (2.14), we can draw the circuit in Fig. 2.14b as an equivalent circuit for the one in Fig. 2.14a.

The instantaneous power absorbed by the series combination of N resistors is

$$\begin{aligned} p(t) &= \left[\frac{v(t)}{R_S}\right]^2 R_1 + \left[\frac{v(t)}{R_S}\right]^2 R_2 + \cdots + \left[\frac{v(t)}{R_S}\right]^2 R_N \\ &= \frac{R_1}{R_S^2} v^2(t) + \frac{R_2}{R_S^2} v^2(t) + \cdots + \frac{R_N}{R_S^2} v^2(t) \\ &= \frac{v^2(t)}{R_S} \\ &= v(t)i(t) \end{aligned}$$

which is the power supplied by the voltage source.

EXAMPLE 2.12

Given the circuit in Fig. 2.15, we wish to determine the current I, the power absorbed by the resistor R_2, and an equivalent circuit.

If we assume that the current flows in the counterclockwise direction and define the current variable accordingly, Kirchhoff's voltage law yields the equation

$$-36 - 7I - 3I + 12 - 2I = 0$$

or

$$(7 + 3 + 2)I = 12 - 36$$

$$I = -2 \text{ A}$$

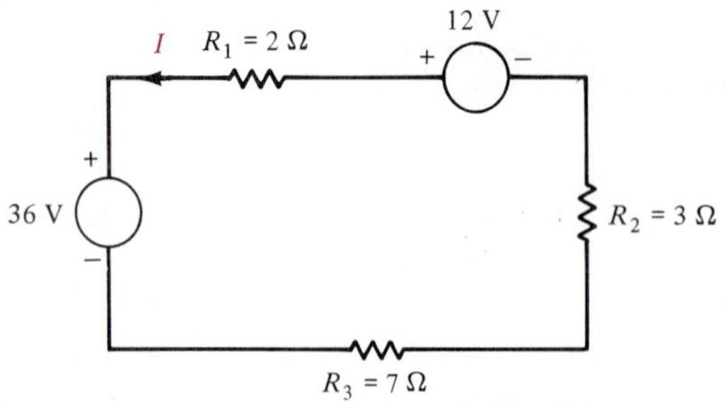

Figure 2.15 Example circuit with multiple sources and resistors.

Therefore, the magnitude of the current is 2 A, but it flows in a clockwise direction. The power absorbed by the resistor R_2 is

$$\begin{aligned} P &= I^2R_2 \\ &= (2)^2(3) \\ &= 12 \text{ W} \end{aligned}$$

The equivalent circuit is as shown in Fig. 2.14b, where

$$v(t) = 24 \text{ V}, \quad R_s = 12\ \Omega, \quad \text{and} \quad i(t) = 2 \text{ A}$$

EXAMPLE 2.13

Given the network in Fig. 2.16, determine the current I and the voltages V_{fb} and V_{be}.

To begin, we have assumed that the current flows in a clockwise direction. If we start at point f and traverse the network in a clockwise direction, Kirchhoff's voltage law states that

$$24 - 1I - 2I - 64 - 3I - 4I = 0^*$$

or

$$1I + 2I + 3I + 4I = -64 + 24$$

We could have written this equation by inspection. The procedure is simply to put the voltage sources on the right side of the equation and assign a positive sign to them if they aid the assumed direction of the current flow and a negative sign if they oppose the assumed direction of the current. The left side of the equation is simply the voltage across the resistors. Solving the equation above for I, we obtain

$$I = -4 \text{ mA}$$

*The equations $v = iR$ and $p = vi$ are valid when v, R, and p are in units of volts (V), amperes (A), ohms (Ω), and watts (W), respectively. However, the units V, mA, kΩ, and mW also constitute a mutually compatible set. Consider $i = 4$ mA flowing through a 6-kΩ resistor:

$$v = iR = (4 \times 10^{-3})(6 \times 10^{+3}) = 24 \text{ V} \qquad (\text{V, A, } \Omega, \text{ W})$$
$$v = iR = (4)(6) = 24 \text{ V} \qquad (\text{V, mA, k}\Omega, \text{ mW})$$

$$p = vi = (24)(4 \times 10^{-3}) = 96 \times 10^{-3} \text{ W} \qquad (\text{V, A, } \Omega, \text{ W})$$
$$p = vi = (24)(4) = 96 \text{ mW} \qquad (\text{V, mA, k}\Omega, \text{ mW})$$

In some situations the (V, mA, kΩ, mW) set results in simpler calculations because of its compatibility with actual element values.

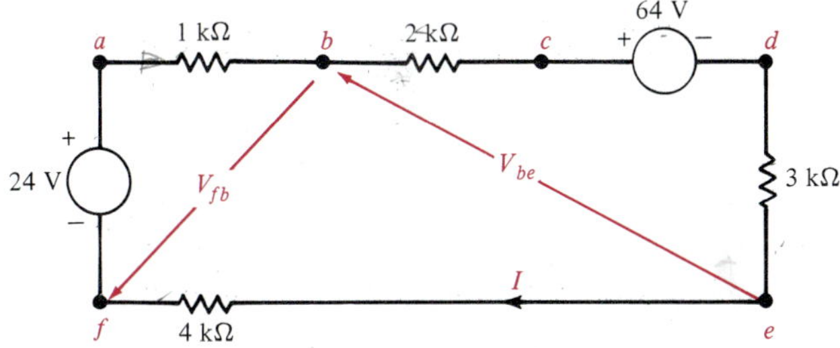

Figure 2.16 Series circuit with multiple components.

Using this value of I, the voltage V_{fb} can be obtained using either path *fabf* or *bcdefb*. In the first case,

$$V_{fb} + 24 - 1I = 0$$

$$V_{fb} = -28 \text{ V}$$

In the second case,

$$-2I - 64 - 3I - 4I - V_{fb} = 0$$

$$V_{fb} = -28 \text{ V}$$

In a similar manner, V_{be} can be obtained using path *bcdeb* or *befab* or the path through the now known voltage V_{fb}. In the first case,

$$-2I - 64 - 3I + V_{be} = 0$$

$$V_{be} = 44 \text{ V}$$

The other paths will yield the same result. ■

EXAMPLE 2.14

The voltage V_A across the 2-Ω resistor in Fig. 2.17 is known to be 8 V. Let us determine the voltages V_1 and V_o.

Upon using Ohm's law, the current in the 2-Ω resistor is found to be

$$V_A = I(2)$$

$$8 = I(2)$$

$$I = 4 \text{ A}$$

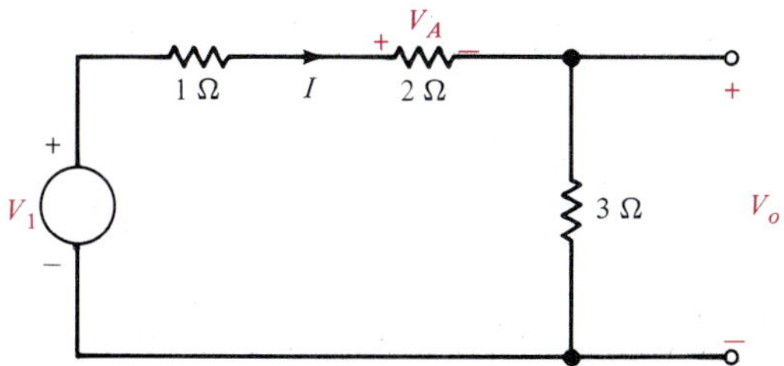

Figure 2.17 Simple series circuit.

Since V_A is positive on the left side of the 2-Ω resistor, the current flows from left to right in this resistor. The current I flows down through the 3-Ω resistor and hence

$$V_o = I(3)$$
$$= 12 \text{ V}$$

Applying Kirchhoff's voltage law around the entire loop yields

$$+V_1 - I(1) - I(2) - I(3) = 0$$
$$V_1 = (6)I$$
$$= 24 \text{ V}$$

or equivalently, $V_1 = I(1) + V_A + V_o$. ■

DRILL EXERCISE

D2.7. Find V_o in the network in Fig. D2.7.

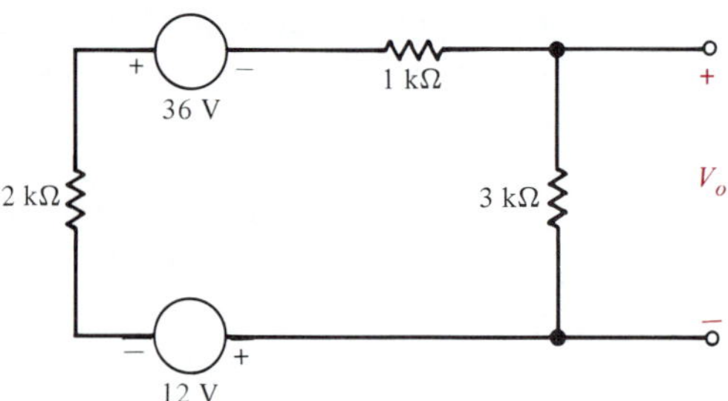

Figure D2.7

Ans: $V_o = -24$ V.

D2.8. Given the network in Fig. D2.8, find the current I, the power absorbed by the 5-Ω resistor, V_{bd}, and V_{be}.

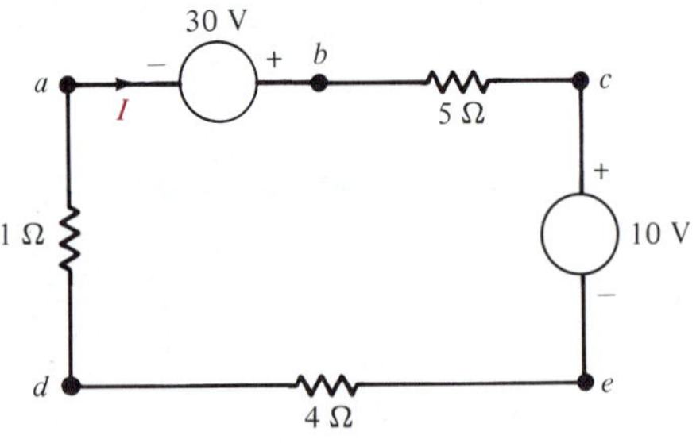

Figure D2.8

Ans: $I = 2$ A, $P_{5\Omega} = 20$ W, $V_{bd} = 28$ V, $V_{be} = 20$ V.

D2.9. Find V_o in the network in Fig. D2.9.

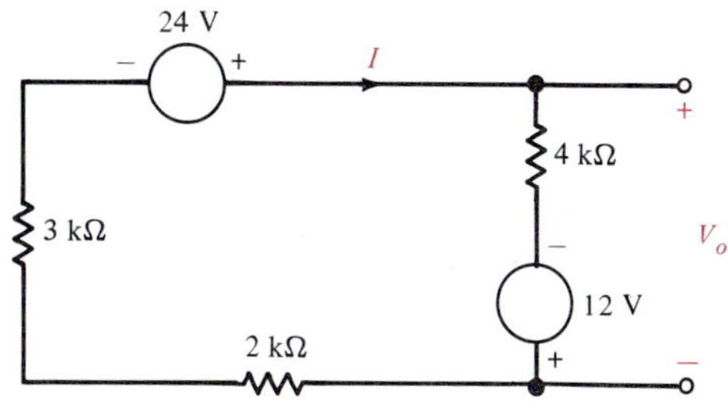

Figure D2.9

Ans: $V_o = 4$ V.

D2.10. V_o is known to be 4 V in the network in Fig. D2.10. Find V_2.

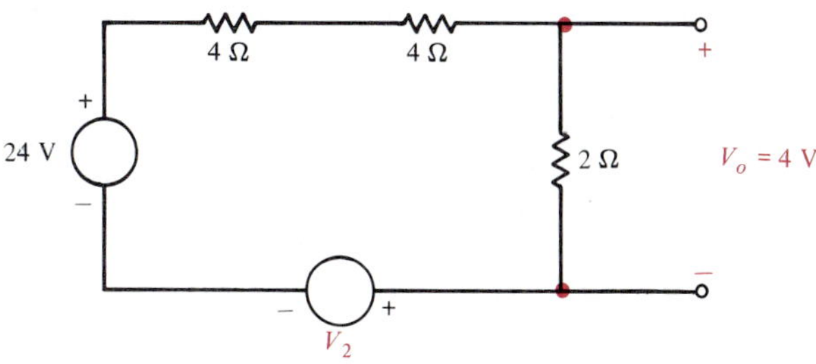

Figure D2.10

Ans: $V_2 = 4$ V.

2.4 Single-Node-Pair Circuits

An important circuit is the single-node-pair circuit. In this case the elements have the same voltage across them, and therefore are in *parallel*. We will, however, apply Kirchhoff's current law and Ohm's law to determine various unknown quantities in the circuit.

Following our approach with the single-loop circuit, we will begin with the simplest case and then generalize our analysis. Consider the circuit shown in Fig. 2.18. Here we have an independent current source in parallel with two resistors.

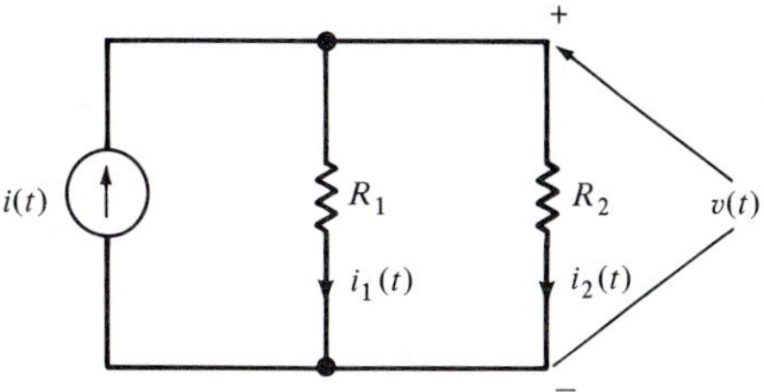

Figure 2.18 Simple parallel circuit.

Since all of the circuit elements are in parallel, the voltage $v(t)$ appears across each of them. The currents $i_1(t)$ and $i_2(t)$ are arbitrarily assigned as shown in the figure. Now applying Kirchhoff's current law to the upper node, we obtain

$$i(t) = i_1(t) + i_2(t)$$

And, employing Ohm's law, we have

$$\begin{aligned} i(t) &= \frac{v(t)}{R_1} + \frac{v(t)}{R_2} \\ &= \left(\frac{1}{R_1} + \frac{1}{R_2}\right)v(t) \\ &= \frac{v(t)}{R_p} \end{aligned}$$

where

$$\frac{1}{R_p} = \frac{1}{R_1} + \frac{1}{R_2} \tag{2.16}$$

$$R_p = \frac{R_1 R_2}{R_1 + R_2} \tag{2.17}$$

Therefore, the equivalent resistance of two resistors connected in parallel is equal to the product of their resistances divided by their sum. Note also that this equivalent resistance R_p is always less than either R_1 or R_2. Hence by connecting resistors in parallel we reduce the overall resistance. In the special case when $R_1 = R_2$, the equivalent resistance is equal to half of the value of the individual resistors.

The manner in which the current $i(t)$ from the source divides between the two branches is called *current division* and can be found from the expressions above. For example,

$$\begin{aligned} v(t) &= R_p i(t) \\ &= \frac{R_1 R_2}{R_1 + R_2} i(t) \end{aligned} \tag{2.18}$$

and

$$\begin{aligned} i_1(t) &= \frac{v(t)}{R_1} \\ &= \frac{R_2}{R_1 + R_2} i(t) \end{aligned} \tag{2.19}$$

and

$$\begin{aligned} i_2(t) &= \frac{v(t)}{R_2} \\ &= \frac{R_1}{R_1 + R_2} i(t) \end{aligned} \tag{2.20}$$

Equations (2.19) and (2.20) are mathematical statements of the current-division rule.

Therefore, the current divides in inverse proportion to the resistances. In other words, to determine the current in the branch containing R_1 we multiply the incoming current $i(t)$ by the opposite resistance R_2 and divide that product by the sum of the two resistors. Note that this description applies to the special case of two resistors in parallel.

If we employ conductance $G_i = 1/R_i$ instead of resistance in the analysis, we can show that

$$i(t) = (G_1 + G_2)v(t)$$

$$G_p = G_1 + G_2 \tag{2.21}$$

and hence that

$$i_1(t) = \frac{G_1}{G_1 + G_2} i(t) \tag{2.22}$$

$$i_2(t) = \frac{G_2}{G_1 + G_2} i(t) \tag{2.23}$$

The current divides as the ratio of the path conductance to the total conductance. This formulation is more general and we will later extend it to handle any number of resistors in parallel.

EXAMPLE 2.15

Consider the circuit shown in Fig. 2.19a. The equivalent circuit is shown in Fig. 2.19b. Given the information on the circuit, we wish to determine the currents and equivalent resistance.

The equivalent resistance for the circuit is

$$R_p = \frac{R_1 R_2}{R_1 + R_2}$$

$$= \frac{(3)(6)}{3 + 6}$$

$$= 2\ \Omega$$

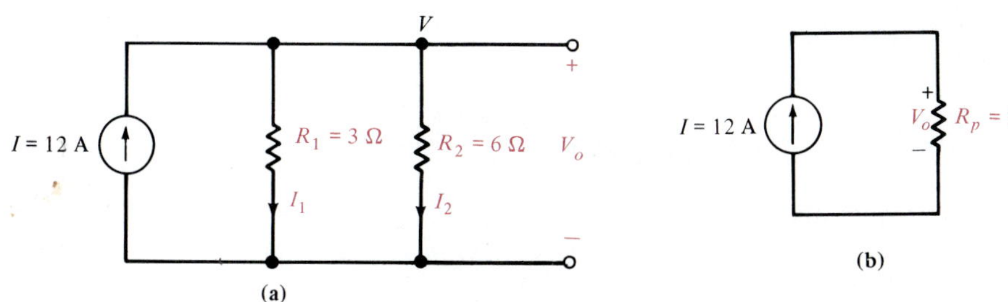

Figure 2.19 Example of a parallel circuit.

Now V_o can be calculated as

$$\begin{aligned} V_o &= R_p I \\ &= (2)(12) \\ &= 24 \text{ V} \end{aligned}$$

Once the voltage V_o is known, Ohm's law can be used to calculate the currents I_1 and I_2.

$$\begin{aligned} I_1 &= \frac{V_o}{R_1} \\ &= \frac{24}{3} \\ &= 8 \text{ A} \end{aligned}$$

and

$$\begin{aligned} I_2 &= \frac{V_o}{R_2} \\ &= \frac{24}{6} \\ &= 4 \text{ A} \end{aligned}$$

Note that these currents satisfy Kirchhoff's current law at both the upper and lower nodes.

$$\begin{aligned} I &= I_1 + I_2 \\ 12 \text{ A} &= 8 \text{ A} + 4 \text{ A} \end{aligned}$$

We can also apply current division to determine I_1 and I_2. For example,

$$\begin{aligned} I_1 &= \frac{R_2}{R_1 + R_2} I = \frac{6}{3 + 6}(12) \\ &= 8 \text{ A} \end{aligned}$$

and

$$I_2 = \frac{3}{3 + 6}(12) = 4 \text{ A}$$

Note that the larger current flows through the smaller resistor, and vice versa. In addition, one should note that if R_1 and R_2 are equal, the current will divide equally between them. ■

The power delivered by the current source in Fig. 2.18 is

$$p(t) = v(t)i(t)$$

The power absorbed by the two resistors is

$$p_1(t) + p_2(t) = i_1^2(t)R_1 + i_2^2(t)R_2$$
$$= \left[\frac{R_2 i(t)}{R_1 + R_2}\right]^2 R_1 + \left[\frac{R_1 i(t)}{R_1 + R_2}\right]^2 R_2$$
$$= \frac{R_2^2 R_1 + R_1^2 R_2}{(R_1 + R_2)^2} i^2(t)$$
$$= \left[\frac{R_1 R_2}{R_1 + R_2} i(t)\right] i(t)$$
$$= v(t) i(t)$$

which is, of course, the power delivered by the current source.

EXAMPLE 2.16

Determine the instantaneous power supplied by the source and absorbed by each resistor in the circuit analyzed in Example 2.15.

The power supplied is

$$P = V_o I$$
$$= (24)(12)$$
$$= 288 \text{ W}$$

The power absorbed in R_1 and R_2 is, respectively,

$$P_1 = I_1^2 R_1$$
$$= (8)^2(3)$$
$$= 192 \text{ W}$$

and

$$P_2 = I_2^2 R_2$$
$$= (4)^2(6)$$
$$= 96 \text{ W}$$

and

$$P = P_1 + P_2$$
$$= 192 + 96$$
$$= 288 \text{ W}$$

■

Let us now extend our analysis to include a multiplicity of current sources and resistors in parallel. For example, consider the circuit shown in Fig. 2.20a. We have assumed

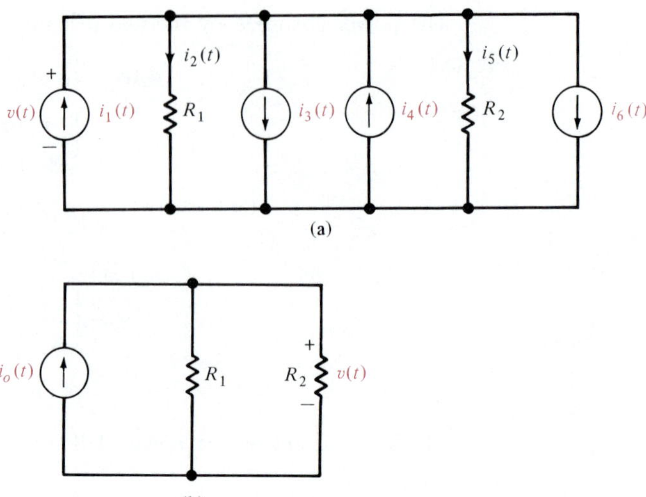

Figure 2.20 Equivalent circuits.

that the upper node is $v(t)$ volts positive with respect to the lower node. Applying Kirchhoff's current law to the upper node yields

$$i_1(t) - i_2(t) - i_3(t) + i_4(t) - i_5(t) - i_6(t) = 0$$

or

$$i_1(t) - i_3(t) + i_4(t) - i_6(t) = i_2(t) + i_5(t)$$

which is equivalent to

$$i_o(t) = \left(\frac{1}{R_1} + \frac{1}{R_2}\right)v(t)$$

$$= \frac{R_1 + R_2}{R_1 R_2}\, v(t)$$

where

$$i_o(t) = i_1(t) - i_3(t) + i_4(t) - i_6(t)$$

so that under the definitions above, the circuit in Fig. 2.20b is equivalent to that in Fig. 2.20a. We could, of course, generalize this analysis to a circuit with N current sources.

Now consider the circuit with N resistors in parallel as shown in Fig. 2.21a. Applying Kirchhoff's current law to the upper node yields

$$i_o(t) = i_1(t) + i_2(t) + \cdots i_N(t)$$

$$= \left(\frac{1}{R_1} + \frac{1}{R_2} + \cdots + \frac{1}{R_N}\right)v(t) \tag{2.24}$$

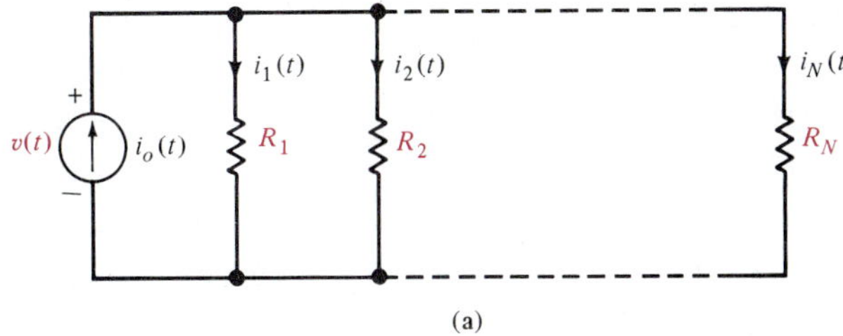

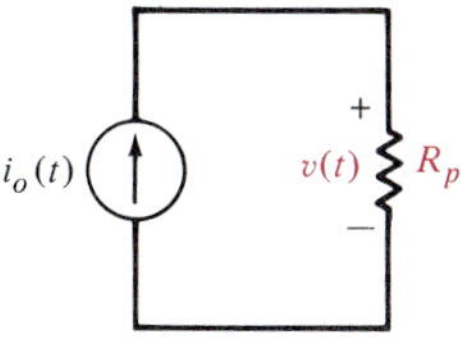

Figure 2.21 Equivalent circuits.

or

$$i_o(t) = (G_1 + G_2 + \cdots + G_N)v(t) \tag{2.25}$$

These two equations can be rewritten as

$$i_o(t) = \frac{v(t)}{R_p} \tag{2.26}$$

$$= G_p v(t) \tag{2.27}$$

where

$$\frac{1}{R_p} = \sum_{i=1}^{N} \frac{1}{R_i} \tag{2.28}$$

and

$$G_p = \sum_{i=1}^{N} G_i \tag{2.29}$$

so that as far as the source is concerned, Fig. 2.21a can be reduced to an equivalent circuit as shown in Fig. 2.21b.

The current division for any branch can be calculated using Ohm's law and the equations above. For example, for the jth branch in the network of Fig. 2.21a,

$$i_j(t) = \frac{v(t)}{R_j}$$

Using Eq. (2.26), we obtain

$$i_j(t) = \frac{R_p}{R_j} i_o(t) \tag{2.30}$$

In a similar manner we find that

$$i_j(t) = \frac{G_j}{G_p} i_o(t) \tag{2.31}$$

Equations (2.30) and (2.31) define the current-division rule for the general case.

EXAMPLE 2.17

Given the circuit in Fig. 2.22, we wish to determine the voltage V_o, the currents in each resistor, and an equivalent circuit.

Employing Kirchhoff's current law, we obtain

$$\left(\tfrac{1}{8} + \tfrac{1}{24} + \tfrac{1}{3}\right)V_o = 12 - 6 + 18$$

$$\tfrac{1}{2}V_o = 24$$

$$V_o = 48 \text{ V}$$

Then

$$I_1 = \tfrac{48}{8}$$

$$= 6 \text{ A}$$

$$I_2 = \tfrac{48}{24}$$

$$= 2 \text{ A}$$

and

$$I_3 = \tfrac{48}{3}$$

$$= 16 \text{ A}$$

Now applying Kirchhoff's current law at the upper node yields

$$-6 + 12 - 6 - 2 - 16 + 18 = 0$$

Since the equivalent source is a current of 24 A entering the upper node and the equivalent resistance is $R_p = 2\ \Omega$, the equivalent circuit consists of a 24-A current source in parallel with a 2-Ω resistor. ■

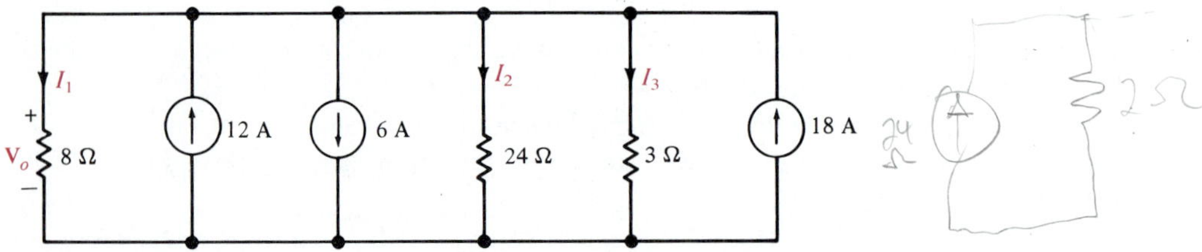

Figure 2.22 Single-node example circuit with multiple sources and resistors.

EXAMPLE 2.18

In the circuit in Fig. 2.23, the power absorbed by the 6-Ω resistor is 24 W. We wish to determine the value of the current source I_o.

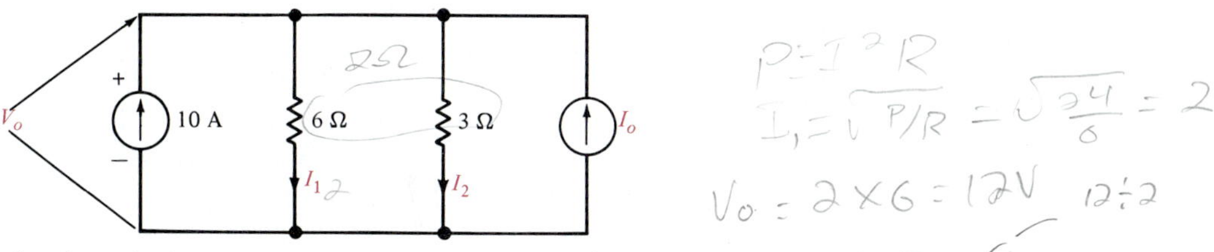

Figure 2.23 Another single-node circuit with multiple sources and resistors.

Since

$$P = I^2R$$

then

$$24 = I_1^2(6)$$

and hence

$$I_1 = \pm 2 \text{ A}$$

Therefore, the voltage V_o is

$$V_o = I_1(6) = \pm 12 \text{ V}$$

The current I_2 can now be found using Ohm's law as

$$I_2 = \frac{V_o}{3}$$

$$= \pm\frac{12}{3}$$

$$= \pm 4 \text{ A}$$

Now applying Kirchhoff's current law at the upper node yields

$$10 - 2 - 4 + I_o = 0 \qquad \text{or} \qquad 10 + 2 + 4 + I_o = 0$$

$$I_o = -4 \text{ A} \qquad\qquad I_o = -16 \text{ A}$$

DRILL EXERCISE

D2.11. Given the network in Fig. D2.11, determine the power absorbed by the 6-Ω resistor.

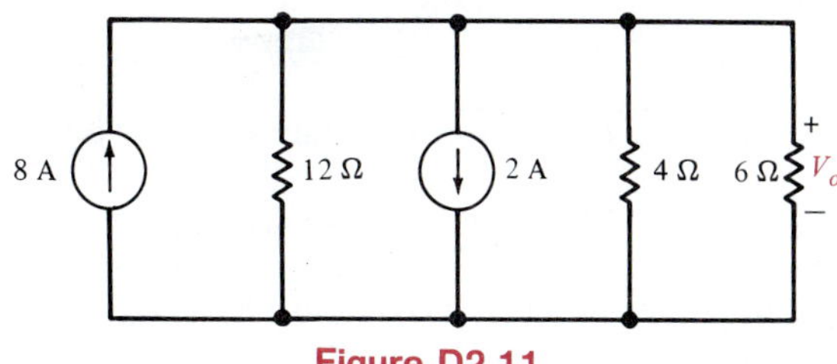

Figure D2.11

Ans: $P_{6\Omega} = 24$ W.

D2.12. In the network in Fig. D2.12, $V_o = 12$ V. Find I_o.

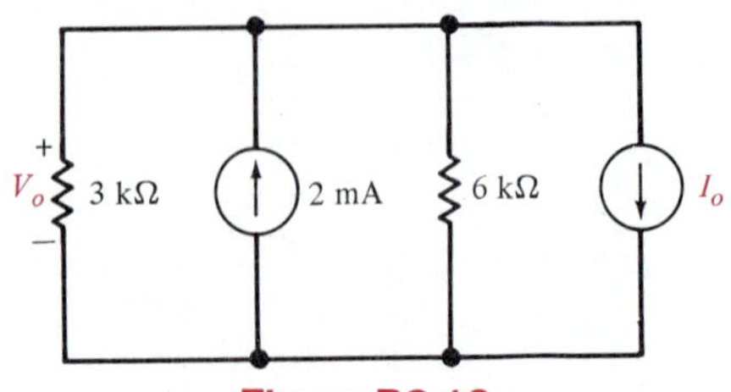

Figure D2.12

Ans: $I_o = -4$ mA.

D2.13. In the network in Fig. D2.13, $I_2 = 4$ mA. Find I_o.

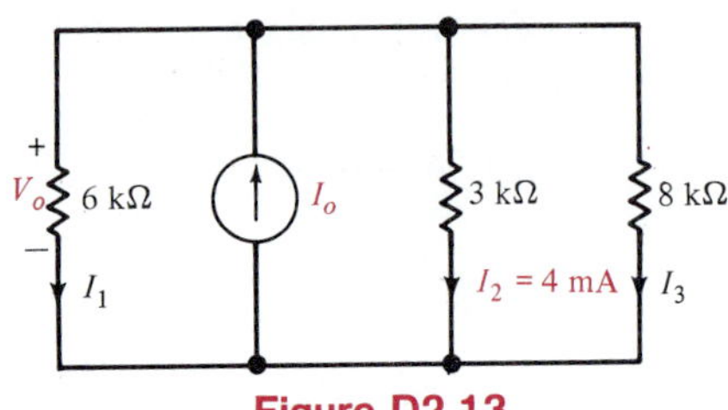

Figure D2.13

Ans: $I_o = 7.5$ mA.

2.5 Series and Parallel Resistor Combinations

We have shown in our earlier developments that the equivalent resistance of N resistors in series is

$$R_s = R_1 + R_2 + \cdots + R_N$$

and the equivalent resistance of N resistors in parallel is found from

$$\frac{1}{R_p} = \frac{1}{R_1} + \frac{1}{R_2} + \cdots + \frac{1}{R_N}$$

Let us now examine some combinations of these two cases.

EXAMPLE 2.19

Let us determine the resistance at terminals A-B of the network shown in Fig. 2.24a.

To determine the equivalent resistance at A-B, we begin at the opposite end of the network and combine resistors as we progress toward terminals A-B. The 1-, 2-, and 3-Ω

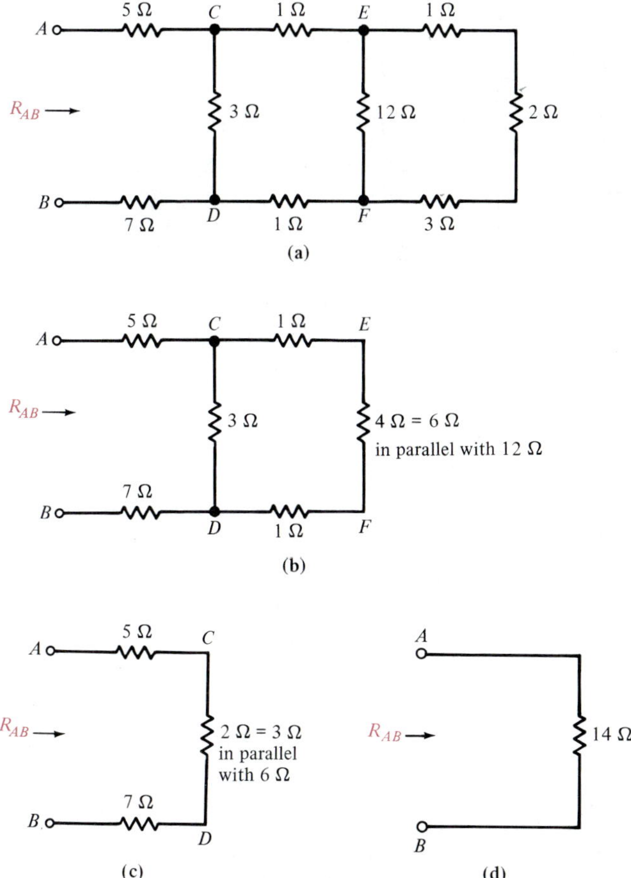

Figure 2.24 Simplification of a resistance network.

resistors connected in series between terminals E and F are equivalent to one 6-Ω resistor, which in turn is in parallel with the 12-Ω resistor. This parallel combination is equivalent to one 4-Ω resistor connected between E and F as shown in Fig. 2.24b. The two 1-Ω resistors and the 4-Ω resistor are in series, and this combination is in parallel with the 3-Ω resistor. Combining these resistors reduces the network to that shown in Fig. 2.24c. Therefore, the resistance at terminals A-B is 14 Ω as shown in Fig. 2.24d. ■

EXAMPLE 2.20

We wish to determine the resistance at terminals A-B in the network in Fig. 2.25a. Once again starting at the opposite end of the network from the terminals and combining resistors as shown in the sequence of circuits in Fig. 2.25, we find that the equivalent resistance at the terminals is 5 kΩ. ■

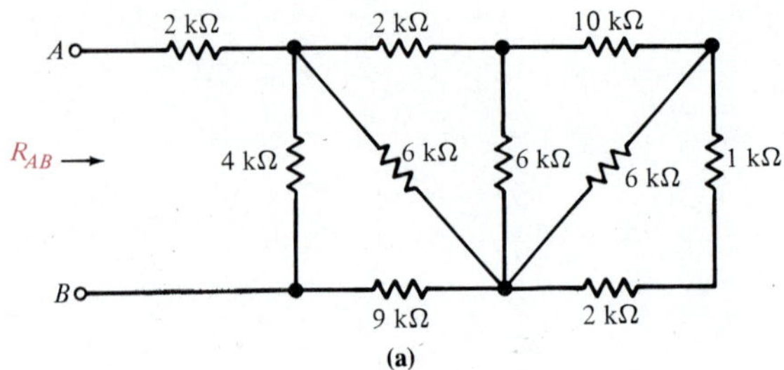

(a)

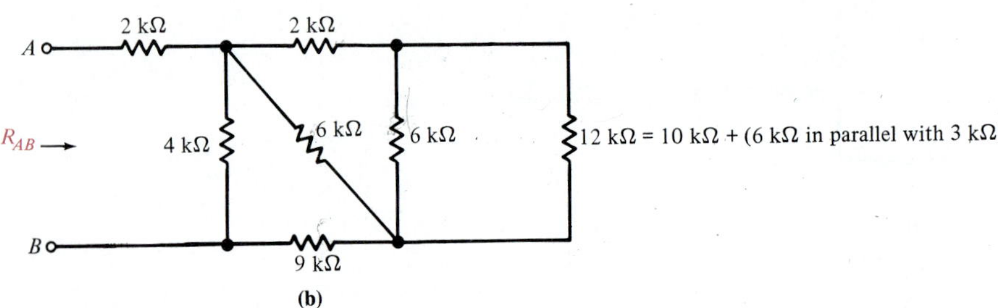

(b)

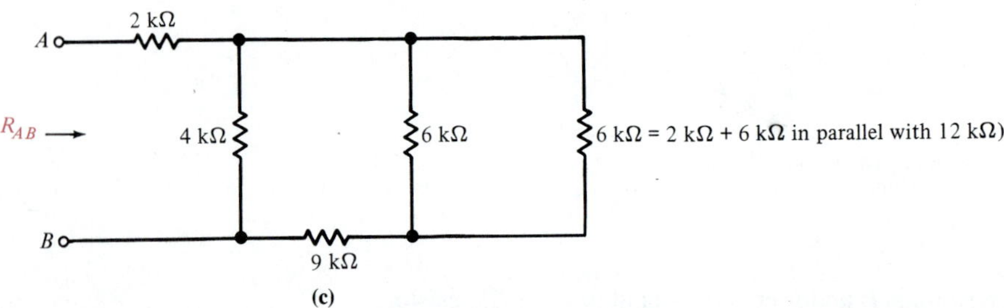

(c)

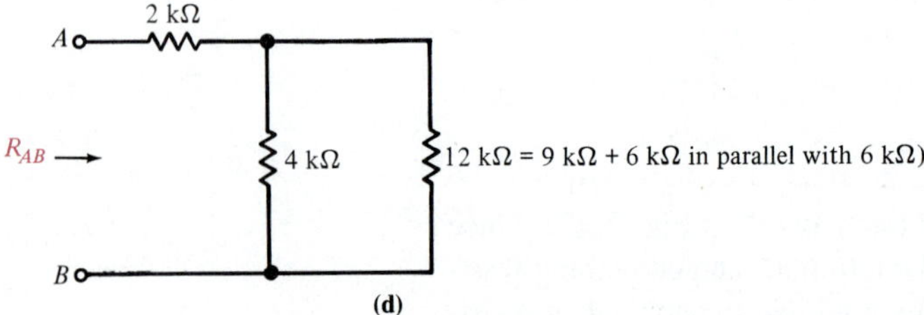

(d)

Figure 2.25 Simplification of a resistance network.

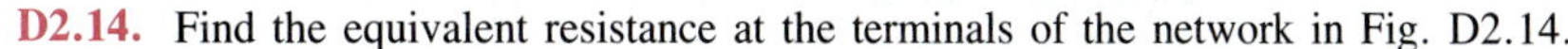

DRILL EXERCISE

D2.14. Find the equivalent resistance at the terminals of the network in Fig. D2.14.

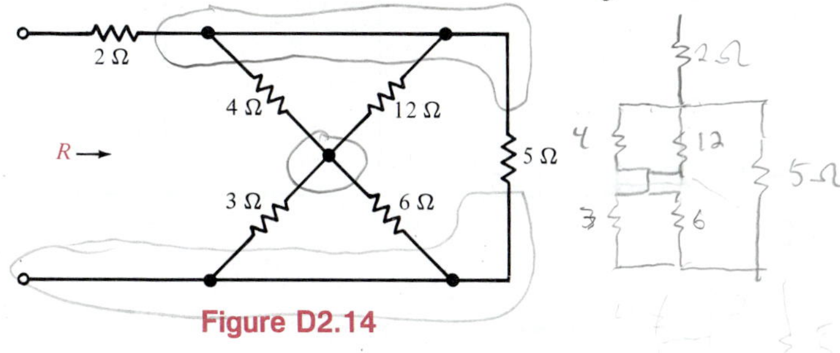

Figure D2.14

Ans: $R = 4.5\ \Omega$.

D2.15. Compute the resistance at the terminals of the network in Fig. D2.15.

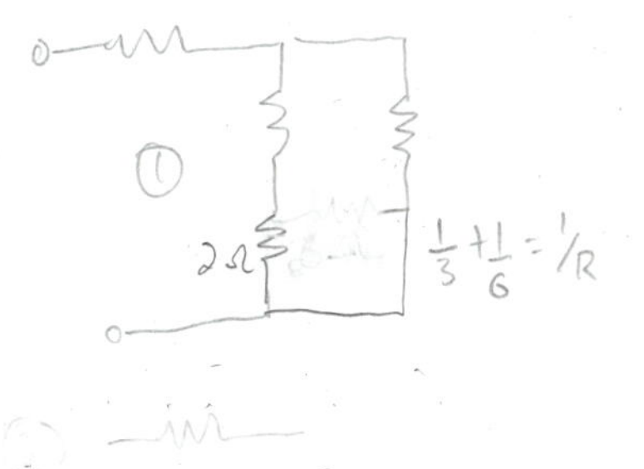

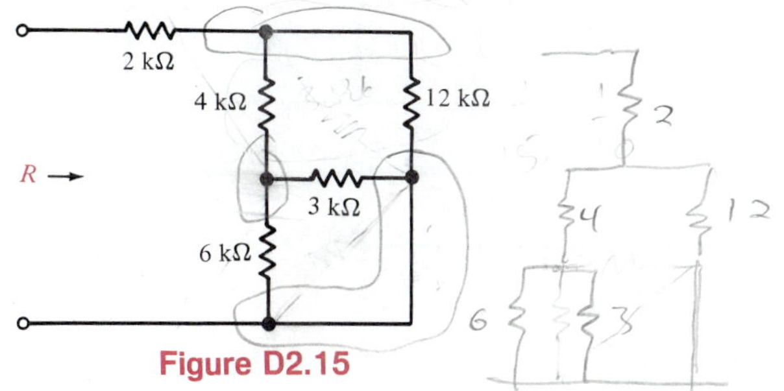

Figure D2.15

Ans: $R = 6\ \text{k}\Omega$.

2.6

Circuits with Series-Parallel Combinations of Resistors

At this point we have learned many techniques that are fundamental to circuit analysis. Now we wish to apply them and show how they can be used in concert to analyze circuits. We will illustrate their application through a number of examples that will be treated in some detail.

EXAMPLE 2.21

We wish to determine all the currents and voltages in the ladder network shown in Fig. 2.26a.

To begin our analysis of the network, we start at the right end of the circuit and combine the resistors to determine the total resistance seen by the 64-V source. This will allow us to calculate the current I_1. Then employing Kirchhoff's current and voltage laws, Ohm's law, and/or current division, we will be able to calculate all currents and voltages in the circuit.

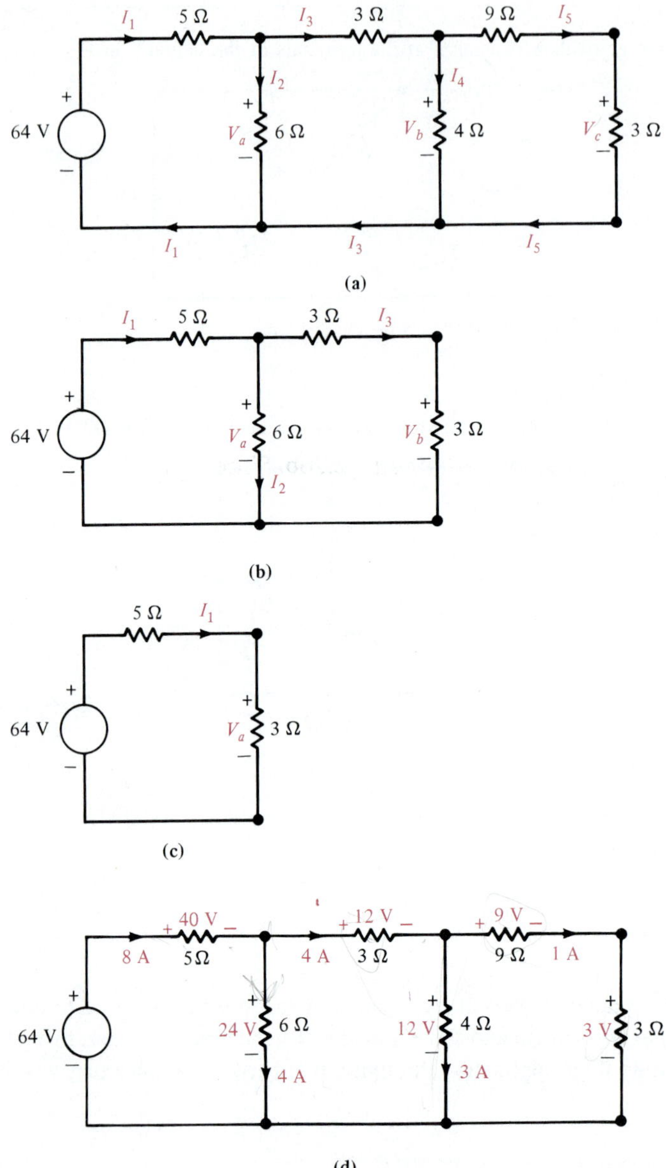

Figure 2.26 Analysis of a ladder network.

At the right end of the circuit the 9- and 3-Ω resistors are in series and thus can be combined into one equivalent 12-Ω resistor. This resistor is in parallel with the 4-Ω resistor, and their combination yields an equivalent 3-Ω resistor, shown at the right edge of the circuit in Fig. 2.26b. In Fig. 2.26b the two 3-Ω resistors are in series and their combination is in parallel with the 6-Ω resistor. Combining all three resistances yields the circuit shown in Fig. 2.26c.

Applying Kirchhoff's voltage law to the circuit in Fig. 2.26c yields

$$I_1(5 + 3) = 64$$

$$I_1 = 8 \text{ A}$$

V_a can be calculated from Ohm's law as

$$V_a = I_1(3)$$

$$= 24 \text{ V}$$

or using Kirchhoff's voltage law,

$$V_a = 64 - 5I_1$$

$$= 64 - 40$$

$$= 24 \text{ V}$$

Knowing I_1 and V_a, we can now determine all currents and voltages in Fig. 2.26b. Since $V_a = 24$ V, the current I_2 can be found using Ohm's law as

$$I_2 = \tfrac{24}{6}$$

$$= 4 \text{ A}$$

Then using Kirchhoff's current law, we have

$$I_1 = I_2 + I_3$$

$$8 = 4 + I_3$$

$$I_3 = 4 \text{ A}$$

Note that I_3 could also be calculated using Ohm's law:

$$V_a = (3 + 3)I_3$$

$$I_3 = \tfrac{24}{6}$$

$$= 4 \text{ A}$$

Applying Kirchhoff's voltage law to the right-hand loop in Fig. 2.26b yields

$$V_a - V_b = 3I_3$$

$$24 - V_b = 12$$

$$V_b = 12 \text{ V}$$

or since V_b is equal to the voltage drop across the 3-Ω resistor, we could use Ohm's law as

$$V_b = 3I_3$$

$$= 12 \text{ V}$$

We are now in a position to calculate the final unknown currents and voltages in Fig. 2.26a. Knowing V_b, we can calculate I_4 using Ohm's law as

$$V_b = 4I_4$$

$$I_4 = \tfrac{12}{4}$$

$$= 3 \text{ A}$$

Then from Kirchhoff's current law, we have

$$I_3 = I_4 + I_5$$

$$4 = 3 + I_5$$

$$I_5 = 1 \text{ A}$$

We could also have calculated I_5 using the current-division rule. For example,

$$I_5 = \frac{4}{4 + (9 + 3)} I_3$$

$$= 1 \text{ A}$$

Finally, V_c can be computed as

$$V_c = I_5(3)$$

$$= 3 \text{ V}$$

Note that Kirchhoff's current law is satisfied at every node and Kirchhoff's voltage law is satisfied around every loop as shown in Fig. 2.26d. ■

EXAMPLE 2.22

Given the circuit in Fig. 2.27 with $V_o = 72$ V, determine all the currents and voltages.

The circuit can be simplified as shown in the progression from Fig. 2.27a to Fig. 2.27b to Fig. 2.27c. Then I_1 can be computed from Ohm's law as

$$I_1 = \frac{V_o}{6 + 2 + 4}$$

$$= 6 \text{ mA}$$

Now using Kirchhoff's voltage law in Fig. 2.27c yields

$$72 = (6)I_1 + V_a + V_b + (4)I_1$$

$$72 - (6)I_1 - (4)I_1 = V_a + V_b$$

$$V_a + V_b = 12 \text{ V}$$

This value could also be calculated from Ohm's law by multiplying the current of 6 mA by the 2-kΩ resistance. From Fig. 2.27b we can obtain I_2 as

$$I_2 = \frac{V_a + V_b}{2 + 2}$$

$$= 3 \text{ mA}$$

Then using Ohm's law, we obtain

$$V_a = I_2(2)$$

$$= 6 \text{ V}$$

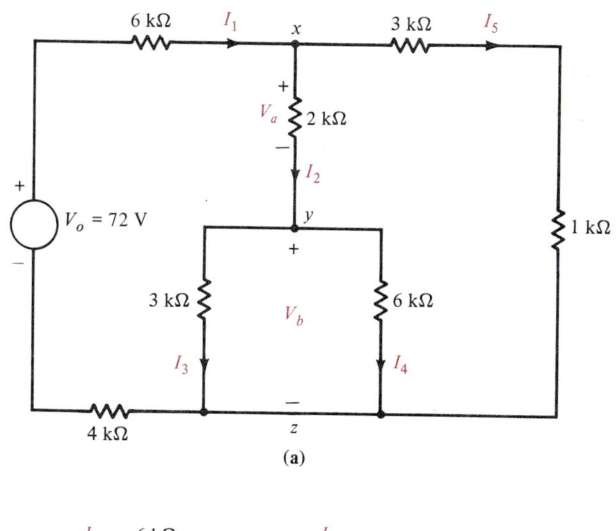

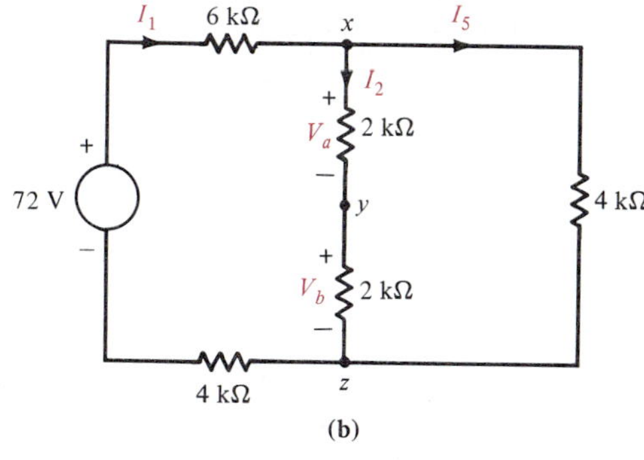

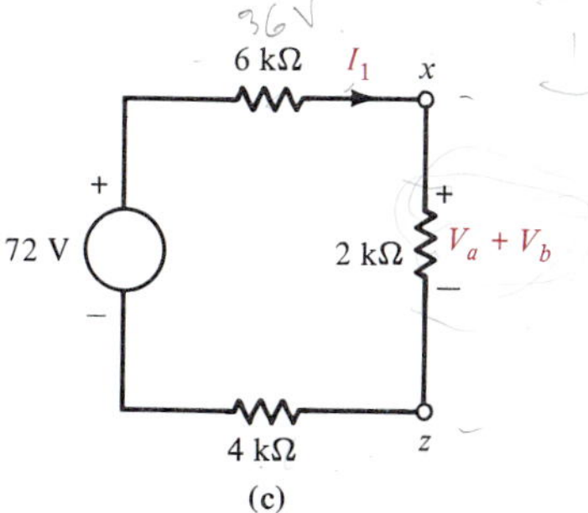

Figure 2.27 Example circuit for analysis.

and

$$V_b = I_2(2)$$
$$= 6 \text{ V}$$

Therefore, point y is 6 V positive with respect to point z, and point x is 6 V positive with respect to point y, or 12 V positive with respect to point z. From Kirchhoff's current law,

$$I_1 = I_2 + I_5$$

$$6 = 3 + I_5$$

$$I_5 = 3 \text{ mA}$$

Since V_b is known, currents I_3 and I_4 can be obtained from Ohm's law as

$$I_3 = \frac{V_b}{3}$$

$$= 2 \text{ mA}$$

and

$$I_4 = \frac{V_b}{6}$$

$$= 1 \text{ mA}$$

Either current could also be calculated using current division on the current I_2. For example, I_3 can be obtained as

$$I_3 = \frac{6}{3 + 6} I_2$$

$$= 2 \text{ mA}$$ ■

EXAMPLE 2.23

Suppose that we are given the circuit in Fig. 2.27a and $I_4 = \frac{1}{2}$ mA, and we want to find the source voltage V_o.

If $I_4 = \frac{1}{2}$ mA, then from Ohm's law, $V_b = 3$ V. V_b can now be used to calculate $I_3 = 1$ mA. Kirchhoff's current law applied at node y yields

$$I_2 = I_3 + I_4$$

$$= 1.5 \text{ mA}$$

Then from Ohm's law, we have

$$V_a = (1.5)(2)$$

$$= 3 \text{ V}$$

Since $V_a + V$ is now known, I_5 can be obtained:

$$I_5 = \frac{V_a + V_b}{3 + 1}$$

$$= 1.5 \text{ mA}$$

Applying Kirchhoff's current law at node x yields

$$I_1 = I_2 + I_5$$

$$= 3 \text{ mA}$$

Now using Fig. 2.27c, we can employ Kirchhoff's voltage law since we now know I_1.

$$\begin{aligned} V_o &= (6)I_1 + (2)I_1 + (4)I_1 \\ &= (12)I_1 \\ &= 36 \text{ V} \end{aligned}$$

EXAMPLE 2.24

Consider the network in Fig. 2.28a. Given that $V_{DE} = V_o = 4$ V, find the value of the voltage source V_S and the voltage across the current source V_{AD}.

By using Kirchhoff's laws and Ohm's law, we can calculate the desired quantities. Since $V_{DE} = 4$ V, using Ohm's law we obtain $I_8 = 2$ A. Applying Kirchhoff's current law

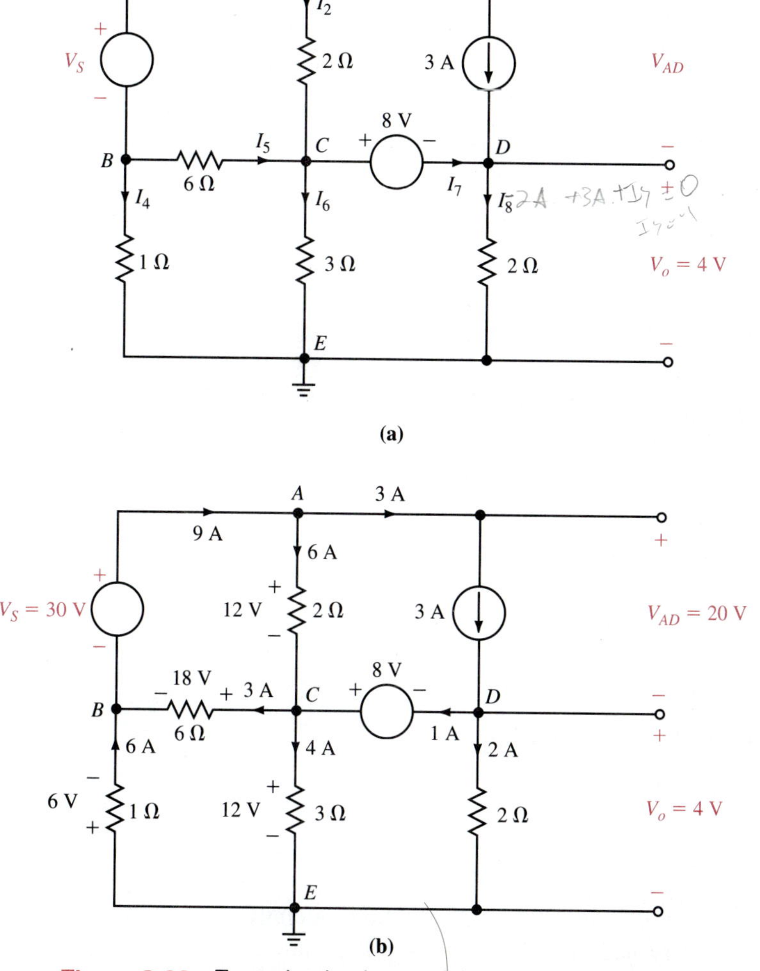

Figure 2.28 Example circuit containing a current source.

at node D yields

$$I_3 + I_7 = I_8$$

Solving for I_7 we obtain

$$I_7 = -1 \text{ A}$$

Then since

$$\begin{aligned} V_{CE} &= V_{CD} + V_{DE} \\ &= 8 + 4 \\ &= 12 \text{ V} \end{aligned}$$

I_6 can be obtained from Ohm's law as 4 A. Kirchhoff's current law at node E yields

$$I_4 + I_6 + I_8 = 0$$

and hence $I_4 = -6$ A. Then since

$$\begin{aligned} V_{CB} &= V_{CE} + V_{EB} \\ &= 12 + (6)(1) \\ &= 18 \text{ V} \end{aligned}$$

Ohm's law yields $I_5 = -3$ A. At node C,

$$I_2 + I_5 = I_6 + I_7$$

Solving for the only unknown I_2 yields $I_2 = 6$ A. Then

$$\begin{aligned} V_{AC} &= (6)(2) \\ &= 12 \text{ V} \end{aligned}$$

The only remaining unknown current is I_1. At node A

$$\begin{aligned} I_1 &= I_2 + I_3 \\ &= 9 \text{ A} \end{aligned}$$

Now Kirchhoff's voltage law around the upper left-hand loop yields

$$V_S - V_{AC} - V_{CB} = 0$$

or

$$V_S = 30 \text{ V}$$

Kirchhoff's voltage law around the upper right-hand loop yields

$$V_{AC} - V_{AD} + 8 = 0$$

or

$$V_{AD} = 20 \text{ V}$$

The circuit with all voltages and currents labeled is shown in Fig. 2.28b. Note carefully that Kirchhoff's current law is satisfied at every node and Kirchhoff's voltage law is satisfied around every loop. ■

DRILL EXERCISE

D2.16. Find all the currents in the network in Fig. D2.16.

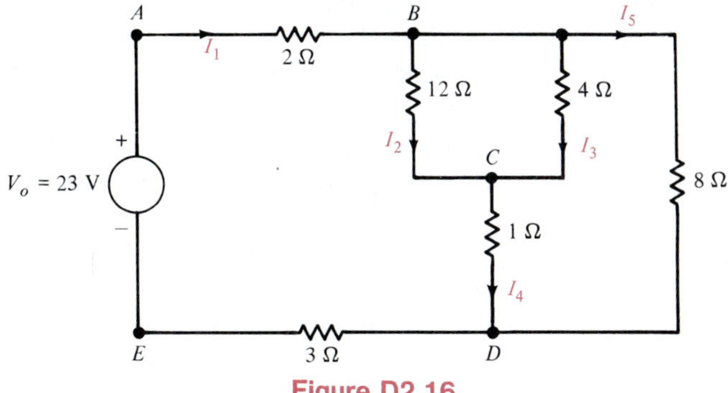

Figure D2.16

Ans: $I_1 = 3$ A, $I_2 = \frac{1}{2}$ A, $I_3 = \frac{3}{2}$ A, $I_4 = 2$ A, $I_5 = 1$ A.

D2.17. Find V_o in the network in Fig. D2.16 if it is known that $I_3 = 6$ A.

Ans: $V_o = 92$ V.

2.7 Wye ⇌ Delta Transformations

In order to provide motivation for this topic, consider the circuit in Fig. 2.29. Note that this network has essentially the same number of elements as contained in our recent examples. However, when we attempt to reduce the circuit to an equivalent network containing the source V_1 and an equivalent resistor R, we find that nowhere is a resistor in series or parallel with another. Therefore, we cannot attack the problem directly using the techniques that we have learned thus far. We can, however, replace one portion of the network with an equivalent circuit, and this conversion will permit us, with ease, to

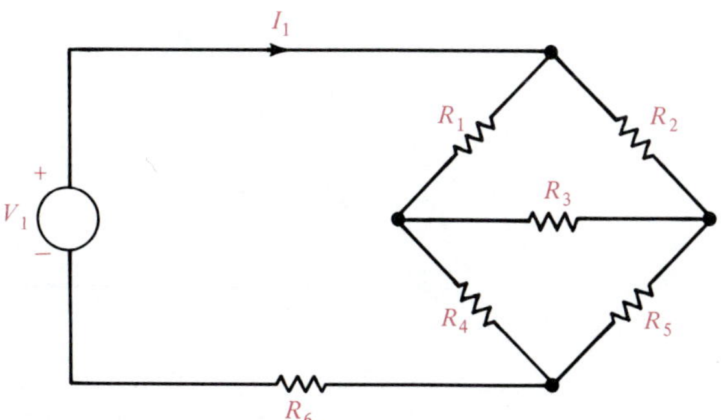

Figure 2.29 Network used to illustrate the need for the wye ⇌ delta transformation.

reduce the combination of resistors to a single equivalent resistance. This conversion is called the wye-to-delta or delta-to-wye transformation.

Consider the networks shown in Fig. 2.30. Note that the resistors in Fig. 2.30a form a Δ (delta) and the resistors in Fig. 2.30b form a Y (wye). If both of these configurations are connected at only three terminals a, b, and c, it would be very advantageous if an equivalence could be established between them. It is, in fact, possible to relate the resistances of one network to those of the other such that their terminal characteristics are the same. This relationship between the two network configurations is called the Y–Δ transformation.

The transformation that relates the resistances R_1, R_2, and R_3 to the resistances R_a, R_b, and R_c is derived as follows. For the two networks to be equivalent at each corresponding pair of terminals, it is necessary that the resistance at the corresponding terminals be equal (e.g., the resistance at terminals a and b with c open circuited must be the same for both networks). Therefore, if we equate the resistances for each corresponding set of terminals, we obtain the following equations:

$$\begin{aligned} R_{ab} &= R_a + R_b = \frac{R_2(R_1 + R_3)}{R_2 + R_1 + R_3} \\ R_{bc} &= R_b + R_c = \frac{R_3(R_1 + R_2)}{R_3 + R_1 + R_2} \\ R_{ca} &= R_c + R_a = \frac{R_1(R_2 + R_3)}{R_1 + R_2 + R_3} \end{aligned} \tag{2.32}$$

Solving this set of equations for R_a, R_b, and R_c yields

$$\begin{aligned} R_a &= \frac{R_1R_2}{R_1 + R_2 + R_3} \\ R_b &= \frac{R_2R_3}{R_1 + R_2 + R_3} \\ R_c &= \frac{R_1R_3}{R_1 + R_2 + R_3} \end{aligned} \tag{2.33}$$

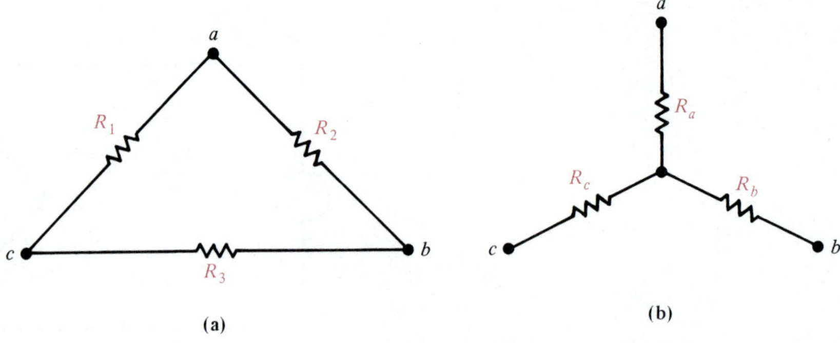

Figure 2.30 Delta and wye resistance networks.

Similarly, if we solve Eq. (2.32) for R_1, R_2, and R_3, we obtain

$$R_1 = \frac{R_aR_b + R_bR_c + R_aR_c}{R_b}$$

$$R_2 = \frac{R_aR_b + R_bR_c + R_aR_c}{R_c} \tag{2.34}$$

$$R_3 = \frac{R_aR_b + R_bR_c + R_aR_c}{R_a}$$

Equations (2.33) and (2.34) are general relationships and apply to any set of resistances connected in a Y or Δ. For the balanced case where $R_a = R_b = R_c$ and $R_1 = R_2 = R_3$, the equations above reduce to

$$R_Y = \tfrac{1}{3}R_\Delta \tag{2.35}$$

and

$$R_\Delta = 3R_Y \tag{2.36}$$

It is important to note that it is not necessary to memorize the formulas in Eqs. (2.33) and (2.34). Close inspection of these equations and Fig. 2.30 illustrates a definite pattern to the relationships between the two configurations. For example, the resistance connected to point a in the wye (i.e., R_a) is equal to the product of the two resistors in the Δ that are connected to point a divided by the sum of all the resistance in the delta. R_b and R_c are determined in a similar manner. Similarly, there are geometrical patterns associated with the equations for calculating the resistors in the delta as a function of those in the wye.

Let us now examine the use of the delta ⇌ wye transformation in the solution of a network problem.

EXAMPLE 2.25

Given the network in Fig. 2.31a, let us determine the current I_o.

Note that the right side of the network is in essence a back-to-back connection of two deltas. Since we want to determine the current in the branch that connects point b to point d in the lower delta, we will apply the transformation to the upper delta connected between points a, b, and c. The resulting network is shown in Fig. 2.31b. Since the 2-kΩ and 4-kΩ resistors are in series and their combination is in parallel with the series combination of the two 6-kΩ resistors yielding a 4-kΩ resistor, the source current I_1 is

$$I_1 = \frac{36}{3 + 4 + 5}$$

$$= 3 \text{ mA}$$

Using current division gives us

$$I_o = \frac{(3)(6 + 6)}{2 + 4 + 6 + 6}$$

$$= 2 \text{ mA}$$

■

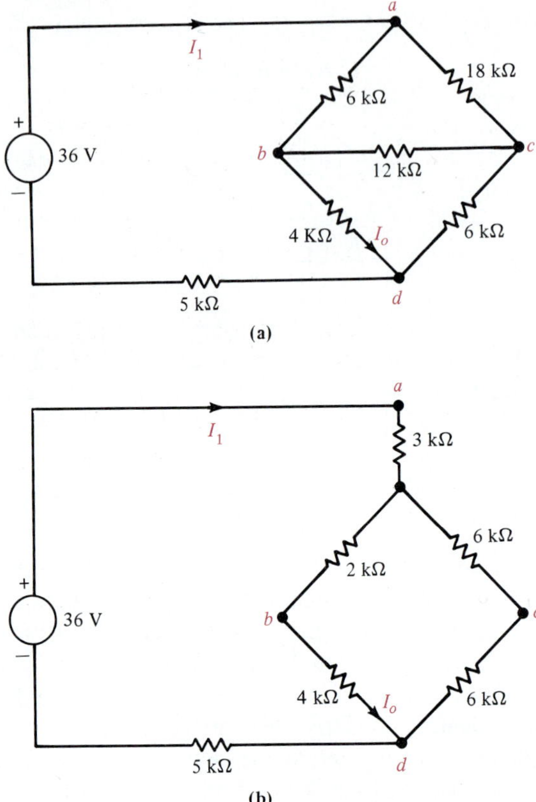

Figure 2.31 Example networks employing a delta-to-wye transformation.

DRILL EXERCISE

D2.18. Find V_o in the network in Fig. D2.18.

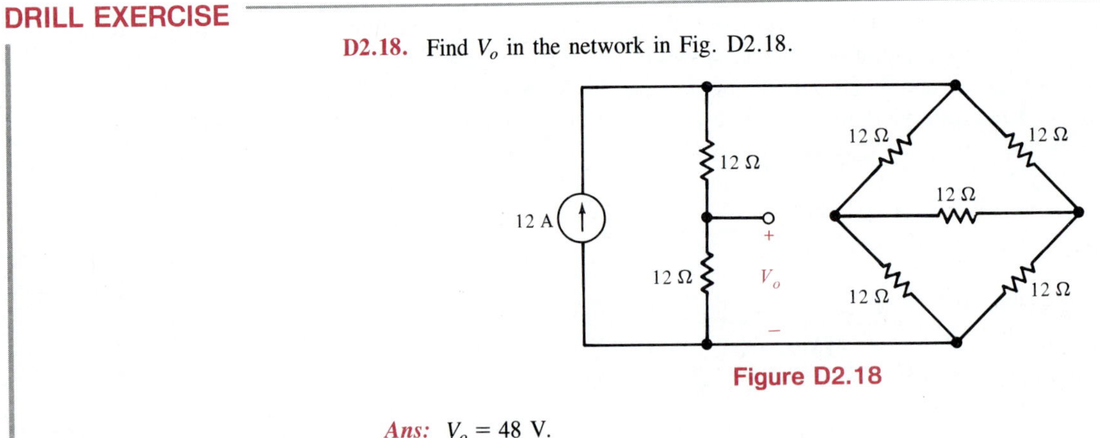

Figure D2.18

Ans: $V_o = 48$ V.

2.8

Circuits with Dependent Sources

In Chapter 1 we outlined the different kinds of dependent sour sources are extremely important because they are used to model phy *npn* and *pnp* bipolar junction transistors (BJTs) and field-effect transis either metal-oxide-semiconductor field-effect transistors (MOSFETs, -gate field-effect transistors (IGFETs). These basic structures are in turn use make analog and digital devices. A typical analog device is an operational amplifier (op-amp). Typical digital devices are random access memories (RAMs), read-only memories (ROMs), and microprocessors. We will now show how to solve simple one-loop and one-node circuits that contain these dependent sources. Although the following examples are fairly simple, they will serve to illustrate the basic concepts.

The following example employs a current-controlled voltage source.

EXAMPLE 2.26

Consider the circuit shown in Fig. 2.32. To determine the voltage V_o across the 5-Ω resistor, we employ Kirchhoff's voltage law:

$$V_S - 5I_1 + 2I_1 - 3I_1 = 0$$

$$24 = I_1(5 + 3 - 2)$$

$$I_1 = 4 \text{ A}$$

Therefore,

$$V_o = (5)(4)$$

$$= 20 \text{ V}$$

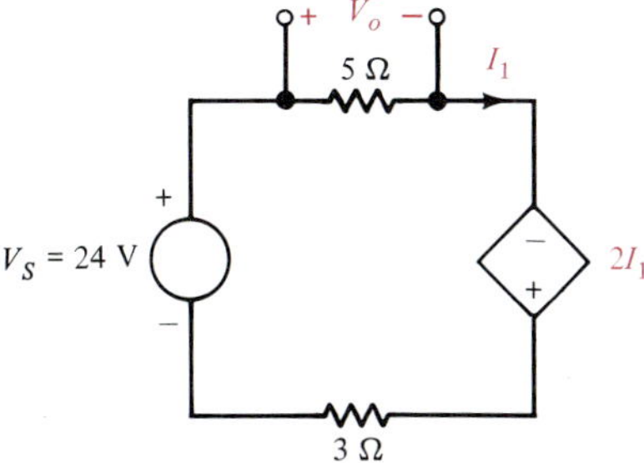

Figure 2.32 Example circuit containing a current-controlled voltage source.

EXAMPLE 2.27

Given the circuit in Fig. 2.33 containing a current-controlled current source, we wish to determine the voltage V_o.

Employing Kirchhoff's current law at the top node, we have

$$\frac{V_S}{3 + 3} + \frac{V_S}{3} = 4I_o - 10$$

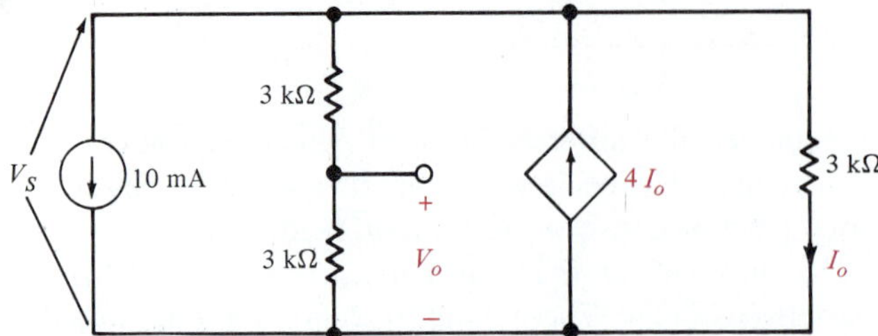

Figure 2.33 Example circuit containing a current-controlled current source.

But

$$I_o = \frac{V_S}{3}$$

Therefore,

$$\frac{V_S}{6} + \frac{V_S}{3} - \frac{4V_S}{3} = -10$$

Solving the equation for V_S, we obtain

$$V_S = 12 \text{ V}$$

The current in the two series-connected 3-kΩ resistors is 12/(3 + 3) = 2 mA. Hence $V_o = (3)(2) = 6$ V. This value could also be calculated readily by noting that the relationship between V_S and V_o is a simple voltage divider through the two equal 3-kΩ resistors.

EXAMPLE 2.28

We wish to solve the circuit in Fig. 2.34 containing a voltage-controlled voltage source to determine the voltage V_o.

Applying Kirchhoff's voltage law to this single-loop circuit yields

$$30 - 2V_o = (1 + 2 + 3)I_1$$

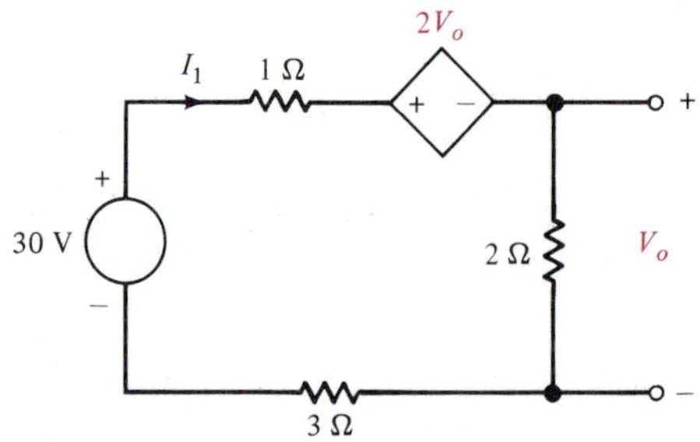

Figure 2.34 Example circuit containing a voltage-controlled voltage source.

where

$$V_o = 2I_1$$

Therefore, the equation above becomes

$$30 - 4I_1 = 6I_1$$

or

$$I_1 = 3 \text{ A}$$

Then

$$V_o = 2I_1 = 6 \text{ V}$$

■

EXAMPLE 2.29

Given the network in Fig. 2.35 containing a voltage-controlled current source, we want to find the voltage V_o.

Applying Kirchhoff's current law yields

$$\left(\frac{V_x}{4}\right) + I_1 - 21 + I_2 = 0$$

or

$$\left(\frac{V_x}{4}\right) + \frac{V_o}{6} + \frac{V_o}{3} = 21$$

But

$$V_x = 2(I_1) = 2\left(\frac{V_o}{6}\right) = \frac{V_o}{3}$$

Therefore,

$$\frac{V_o}{12} + \frac{V_o}{6} + \frac{V_o}{3} = 21$$

Solving for V_o, we obtain

$$V_o = 36 \text{ V}$$

■

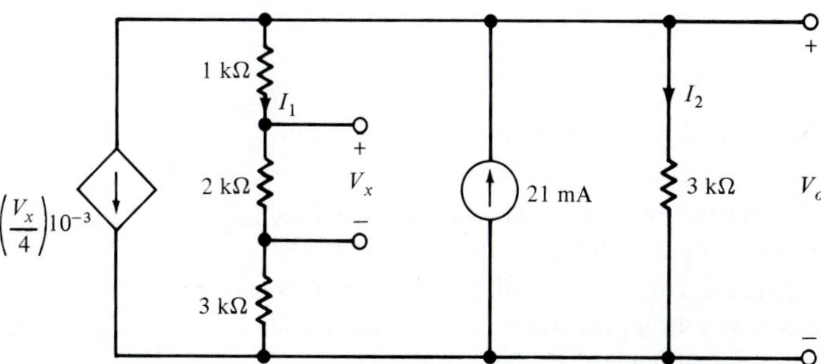

Figure 2.35 Example circuit containing a voltage-controlled current source.

At this point it is perhaps helpful to point out that when analyzing circuits with dependent sources, we first treat the dependent source as though it were an independent source when we write a Kirchhoff's current or voltage law equation. Once the equation is written, we then write the controlling equation that specifies the relationship of the dependent source to the unknown parameter. For instance, the first equation in Example 2.28 treats the dependent source like an independent source. The second equation in the example specifies the relationship of the dependent source to the voltage, which is the unknown in the first equation.

DRILL EXERCISE

D2.19. Find V_o in the network in Fig. D2.19.

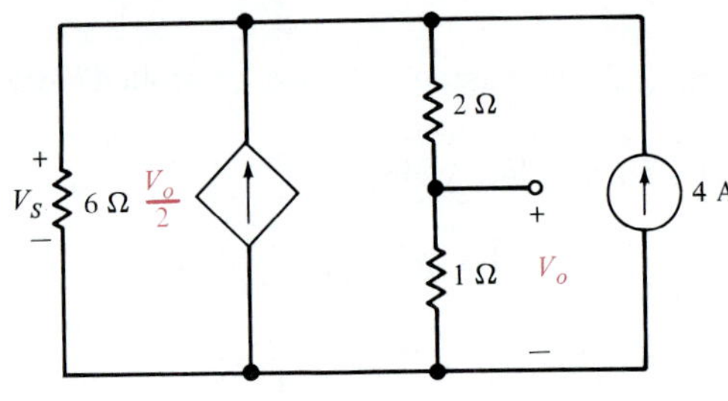

Figure D2.19

Ans: $V_o = 4$ V.

D2.20. Find V_o in the network in Fig. D2.20.

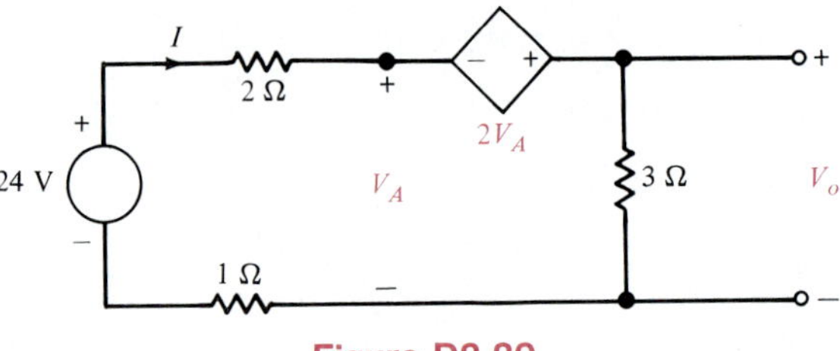

Figure D2.20

Ans: $V_o = 18$ V.

EXAMPLE 2.30

The midband small-signal equivalent circuit for a field-effect transistor (FET) common-source amplifier or bipolar junction transistor (BJT) common-emitter amplifier can be modeled by the circuit shown in Fig. 2.36a. We wish to determine an expression for the gain of the amplifier, which is the ratio of the output voltage to the input voltage.

Note that although this circuit, which contains a voltage-controlled current source, appears to be somewhat complicated, we are actually in a position now to solve it with techniques we have studied up to this point. The loop on the left, or input to the amplifier,

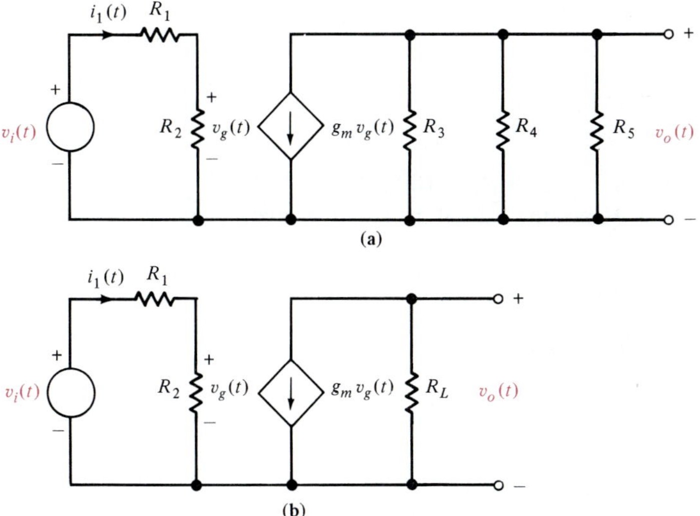

Figure 2.36 Example circuit containing a voltage-controlled current source.

is essentially detached from the output portion of the amplifier on the right. The voltage across R_2 is $v_g(t)$, which controls the dependent current source.

To simplify the analysis, let us replace the resistors R_3, R_4, and R_5 with R_L such that

$$\frac{1}{R_L} = \frac{1}{R_3} + \frac{1}{R_4} + \frac{1}{R_5}$$

Then the circuit reduces to that shown in Fig. 2.36b. Applying Kirchhoff's voltage law to the input portion of the amplifier yields

$$v_i(t) = i_1(t)(R_1 + R_2)$$

and

$$v_g(t) = i_1(t)R_2$$

Solving these equations for $v_g(t)$ yields

$$v_g(t) = \frac{R_2}{R_1 + R_2} v_i(t)$$

From the output circuit, note that the voltage $v_o(t)$ is given by the expression

$$v_o(t) = -g_m v_g(t) R_L$$

Combining this equation with the one above yields

$$v_o(t) = \frac{-g_m R_L R_2}{R_1 + R_2} v_i(t)$$

Therefore, the amplifier gain, which is the ratio of the output voltage to the input voltage, is given by

$$\frac{v_o(t)}{v_i(t)} = -\frac{g_m R_L R_2}{R_1 + R_2}$$

Reasonable values for the circuit parameters in Fig. 2.36a are $R_1 = 100\ \Omega$, $R_2 = 1\ \text{k}\Omega$, $g_m = 0.04$ S, $R_3 = 50\ \text{k}\Omega$, and $R_4 = R_5 = 10\ \text{k}\Omega$. Hence the gain of the amplifier under these conditions is

$$\frac{v_o(t)}{v_i(t)} = \frac{-(0.04)(4.545)(10^3)(1)(10^3)}{(1.1)(10^3)}$$

$$= -165.29$$

Thus the magnitude of the gain is 165.29. ■

2.9 Summary

This chapter has dealt primarily with the fundamental laws that we employ in our analysis of circuits: Ohm's law and Kirchhoff's laws. The resistive circuit element has been introduced and we have shown how to compute an equivalent resistance when these elements are placed in series or parallel. Both voltage division and current division have been presented. We have also illustrated the use of the wye-to-delta transformation in network simplification. We have shown that by employing the fundamental laws, we can analyze some fairly complicated circuits containing both independent and dependent sources.

KEY POINTS

- For a short circuit, the resistance is zero, the voltage across the short is zero, and the current in the short is determined by the rest of the circuit.
- For an open circuit, the resistance is infinite, the current is zero, and the voltage across the open terminals is determined by the rest of the circuit.
- A node is a point of interconnection of two or more circuit elements.
- A loop is any closed path through the circuit in which no node is encountered more than once.
- A branch is a portion of a circuit containing only a single element and the nodes at each end of the element.
- Kirchhoff's current law states that the algebraic sum of the currents entering a node is zero.
- Kirchhoff's voltage law states that the algebraic sum of the voltages around any closed path is zero.

- Current division shows that the current divides among parallel resistors in inverse proportion to the resistance paths.
- Voltage division shows that the voltage is divided among series resistors in direct proportion to their resistances.
- The wye-to-delta transformation allows us to make an equivalent exchange between resistors connected in a delta configuration and resistors connected in a wye configuration.

PROBLEMS

2.1. Find I_o, I_1, and I_2 in the circuit in Fig. P2.1.

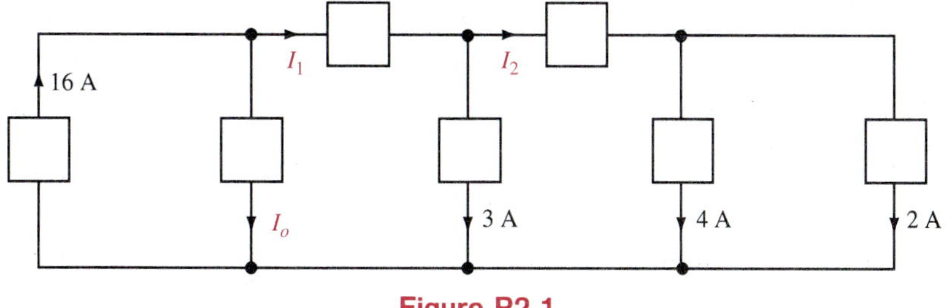

Figure P2.1

2.2. Find I_1, I_2, I_3, and I_4 in the network in Fig. P2.2.

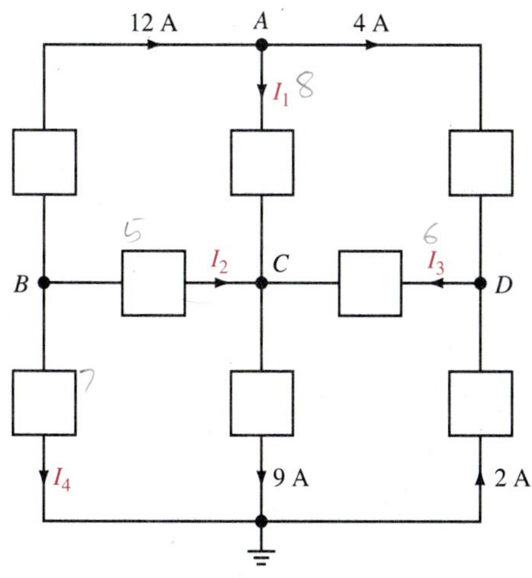

Figure P2.2

2.3. Since Kirchhoff's current law must be satisfied at every partition that cuts the circuit into two pieces, use the partitions to find the unknown currents in the circuit in Fig. P2.3.

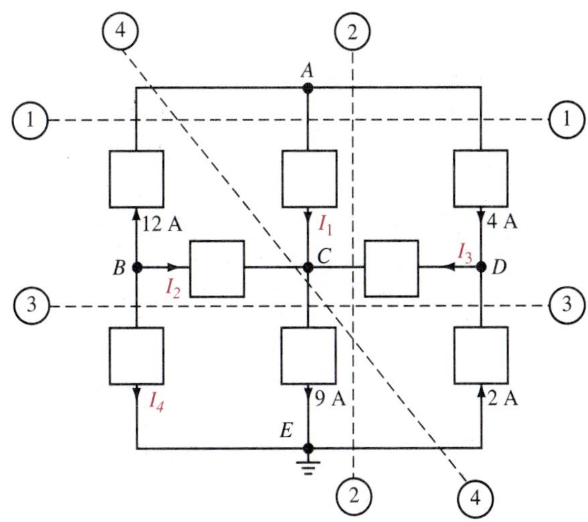

Figure P2.3

2.4. Find the power dissipated in the 2-Ω resistor in the network shown in Fig. P2.4.

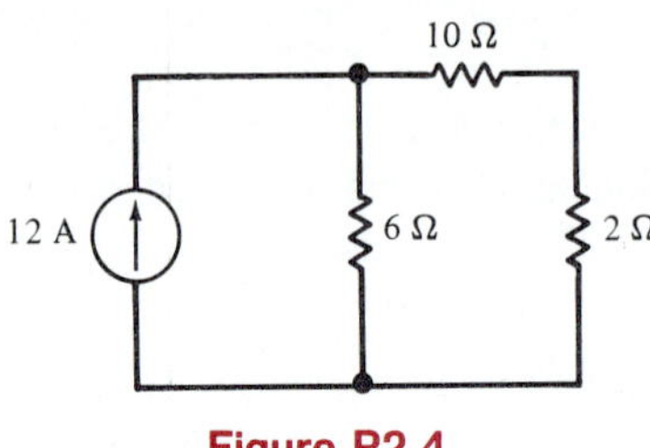

Figure P2.4

2.5. Find the power absorbed in the 1-Ω resistor in the circuit shown in Fig. P2.5.

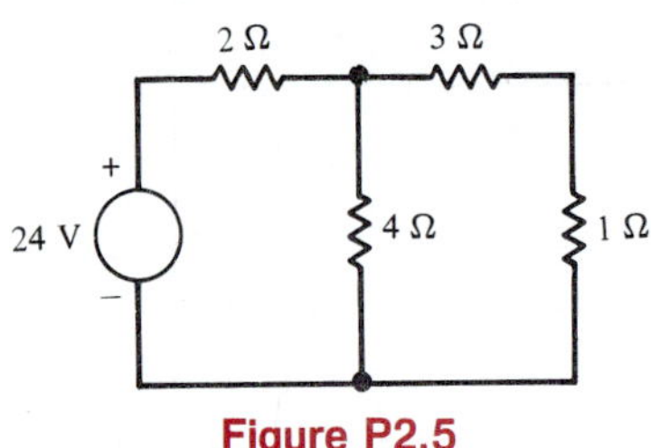

Figure P2.5

2.6. Find I_1 in the circuit shown in Fig. P2.6.

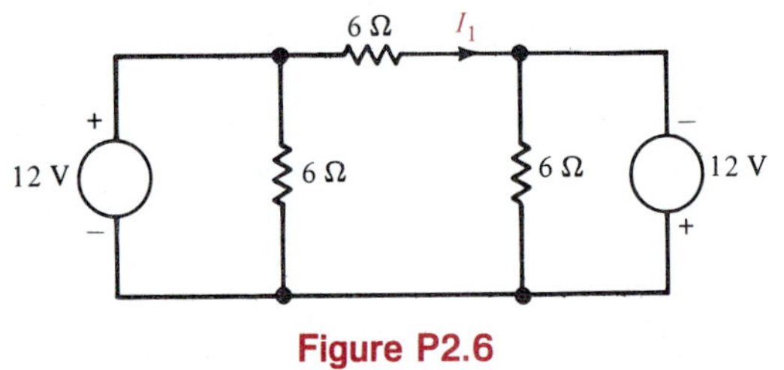

Figure P2.6

2.7. Given the network in Fig. P2.7, find V_o and I_o.

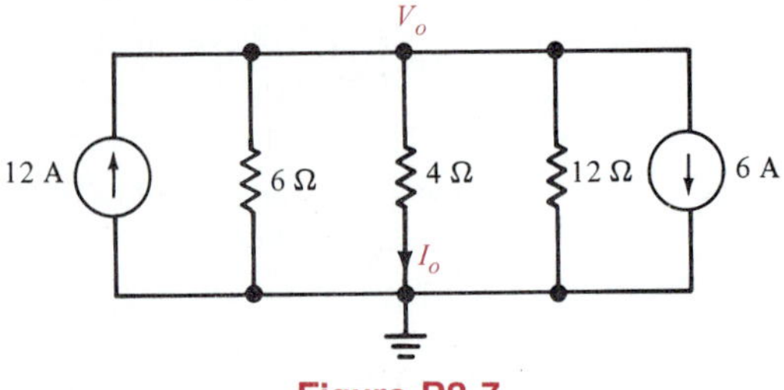

Figure P2.7

2.8. Determine the equivalent conductance of the circuit in Fig. P2.8.

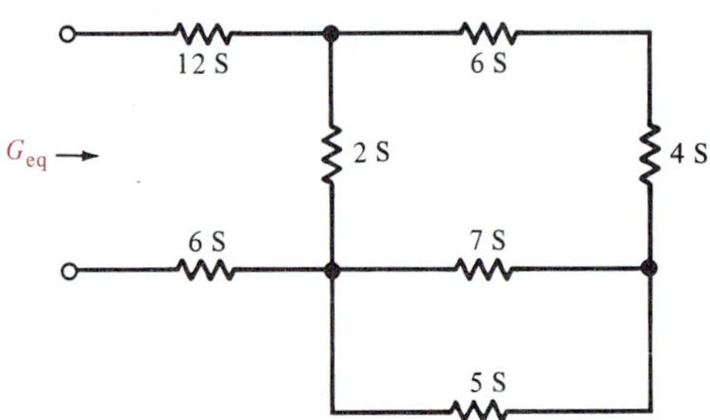

Figure P2.8

2.9. Find the equivalent resistance of the network in Fig. P2.9.

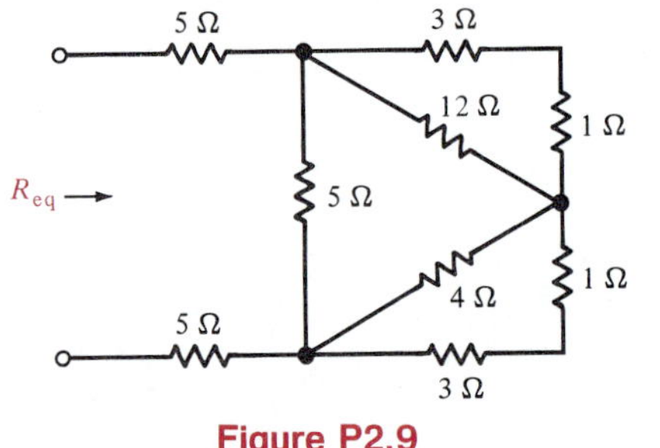

Figure P2.9

2.10. Find I_o in the circuit in Fig. P2.10.

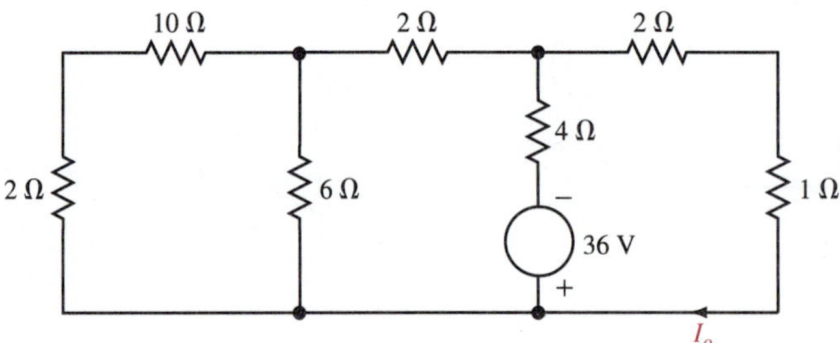

Figure P2.10

2.11. Find V_o in the network in Fig. P2.11.

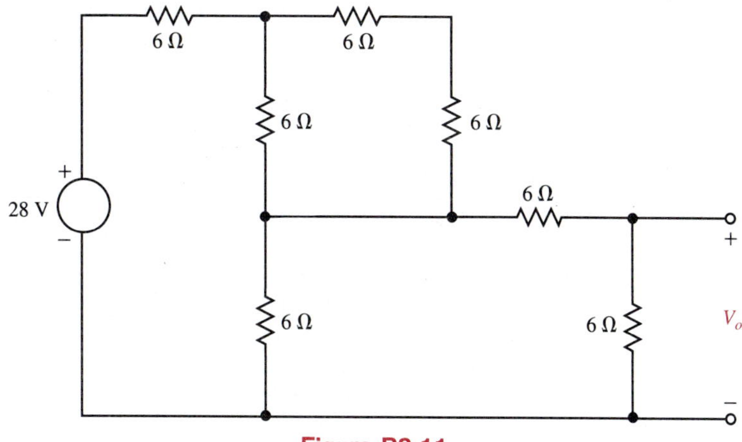

Figure P2.11

2.12. Find V_a, V_b, and V_c in the network in Fig. P2.12.

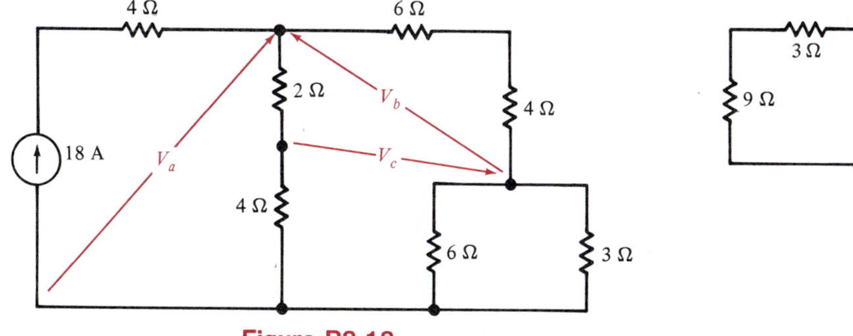

Figure P2.12

2.13. Find V_o in the network in Fig. P2.13.

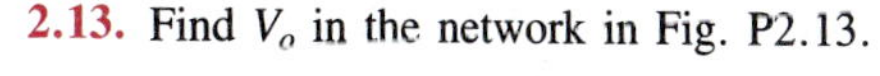

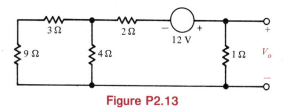

Figure P2.13

2.14. Find I_1 and V_o in the circuit in Fig. P2.14.

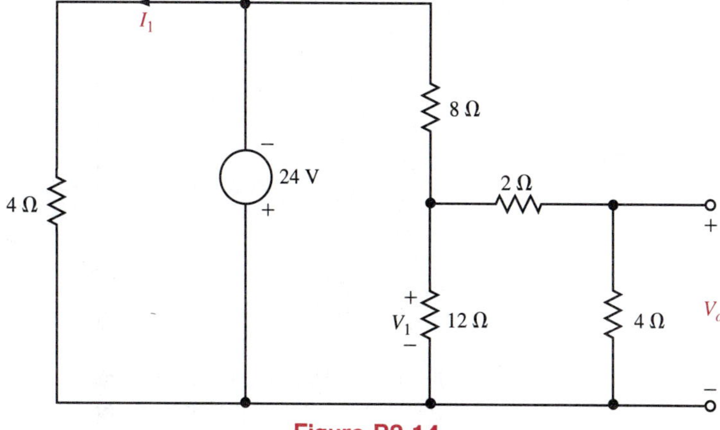

Figure P2.14

2.15. Find I_1 and V_o in the circuit in Fig. P2.15.

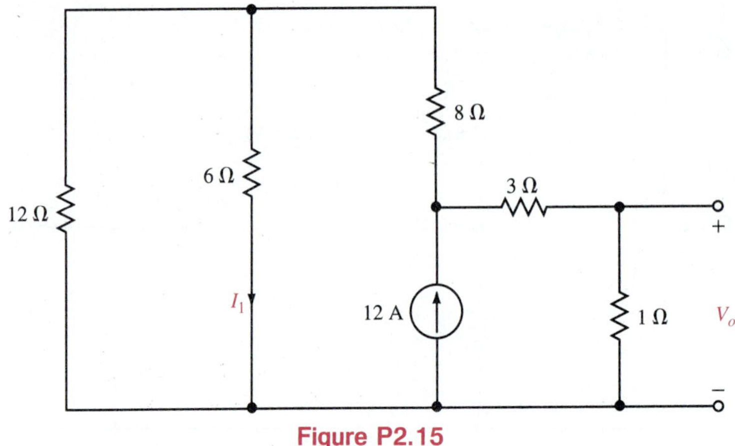

Figure P2.15

2.16. Find I_o in the network in Fig. P2.16.

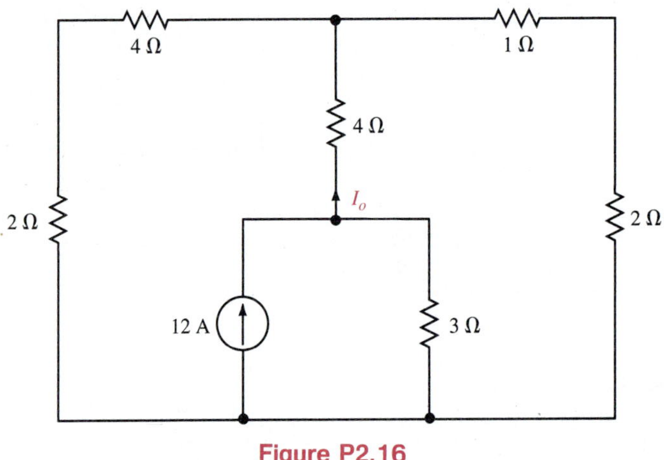

Figure P2.16

2.17. Find V_o in the network in Fig. P2.17.

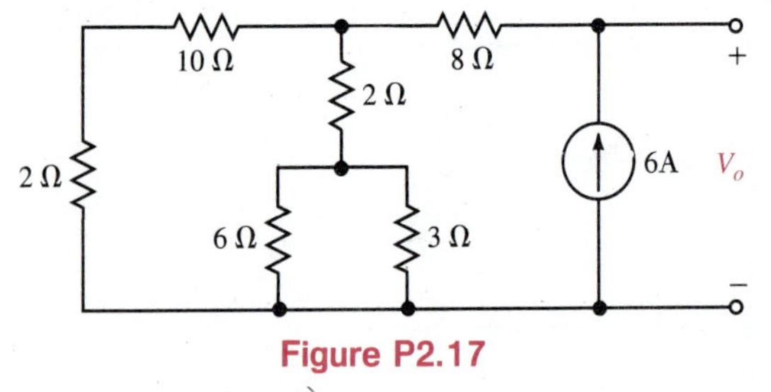

Figure P2.17

2.18. Find I_o in the circuit in Fig. P2.18.

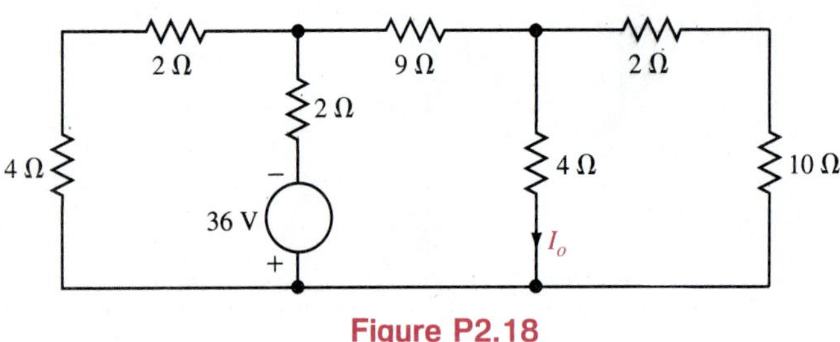

Figure P2.18

2.19. Find I_o in the network in Fig. P2.19.

2.20. Find V_o in the network in Fig. P2.20.

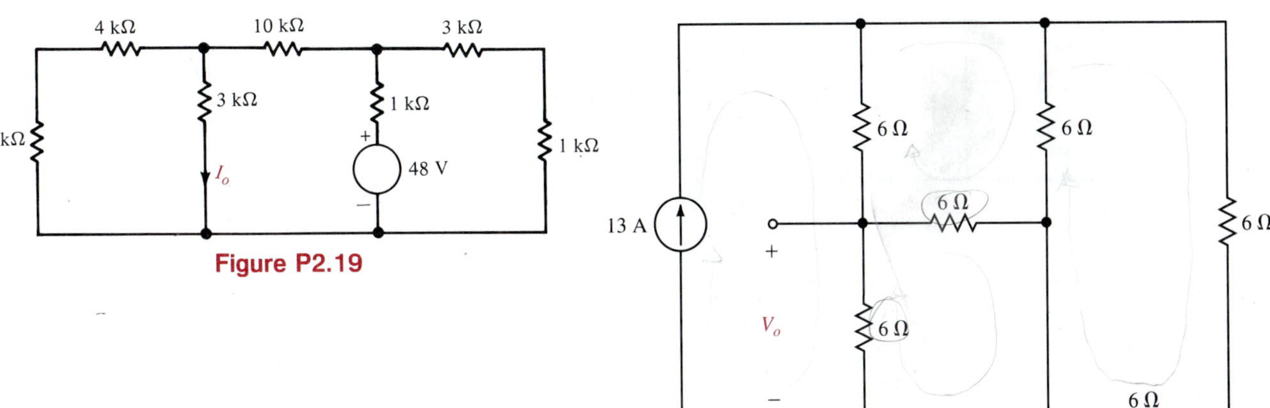

Figure P2.19

Figure P2.20

2.21. Find V_o in the network in Fig. P2.21.

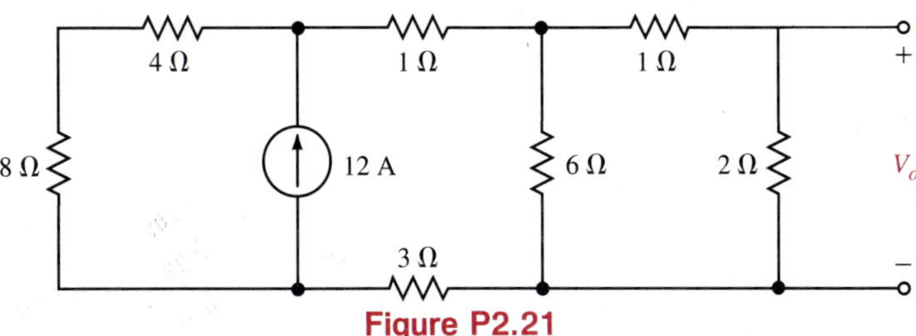

Figure P2.21

2.22. Find I_o in the circuit in Fig. P2.22.

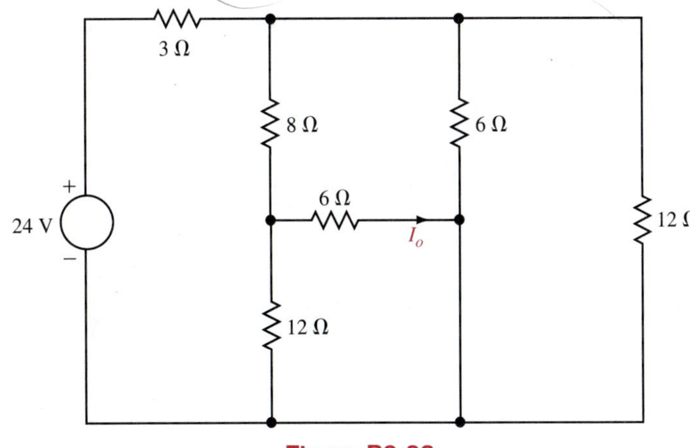

Figure P2.22

2.23. Find I_o in the network in Fig. P2.23.

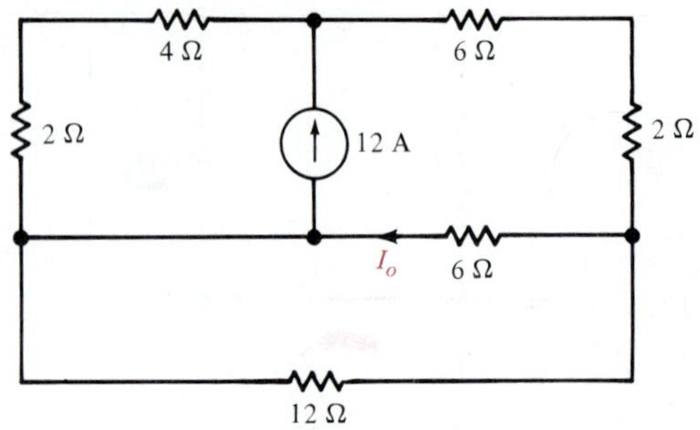

Figure P2.23

2.24. The network in Fig. P2.24 is used in digital-to-analog converters. Find expressions for I_1, I_3, and I_5 in terms of I_o.

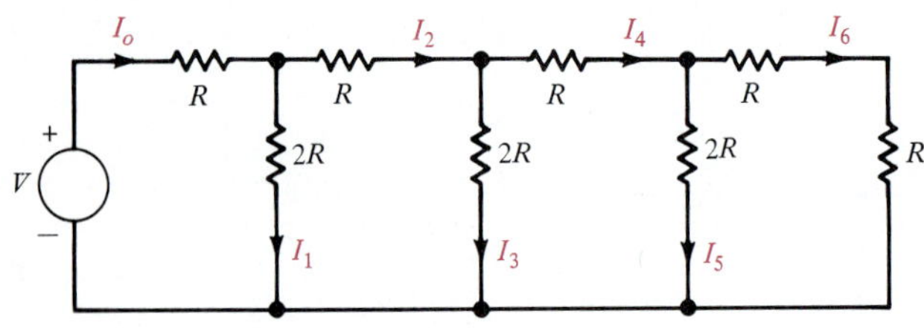

Figure P2.24

2.25. Find V_o in the circuit in Fig. P2.25.

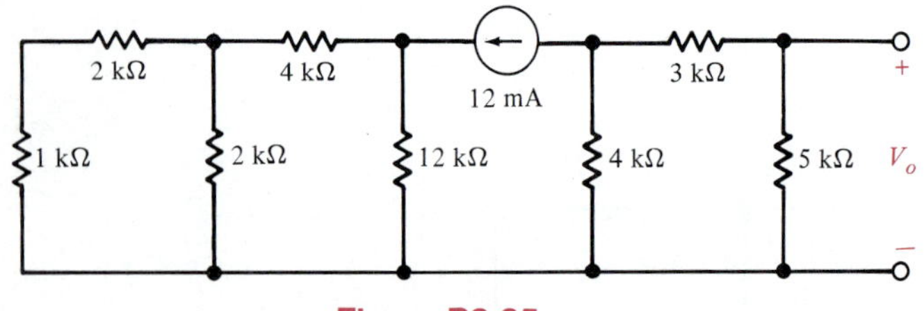

Figure P2.25

2.26. Find I_o in the circuit in Fig. P2.26.

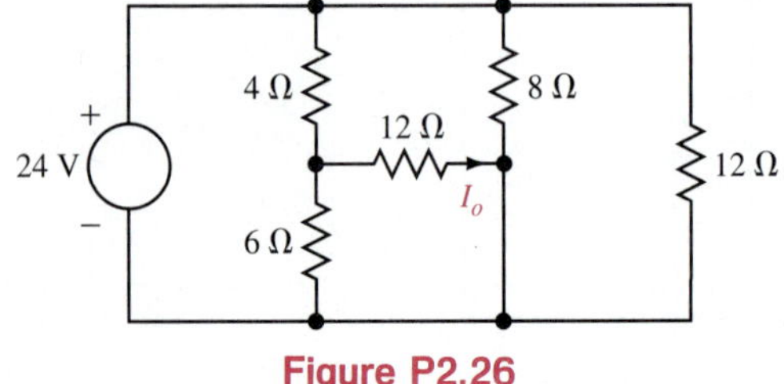

Figure P2.26

2.27. Find I_o in the circuit in Fig. P2.27.

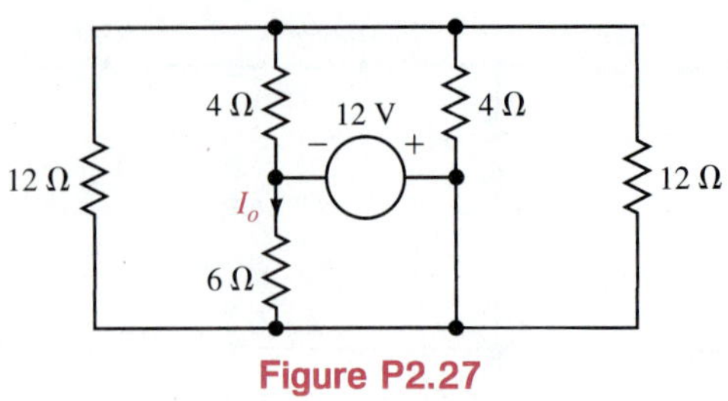

Figure P2.27

2.28. Find I_o in the circuit in Fig. P2.28.

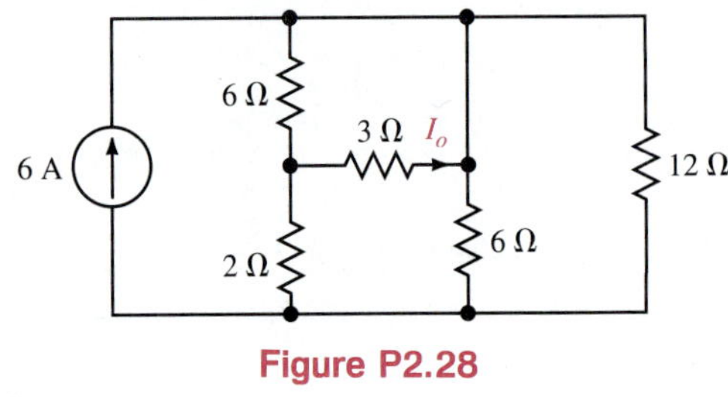

Figure P2.28

2.29. Find I_o in the circuit in Fig. P2.29.

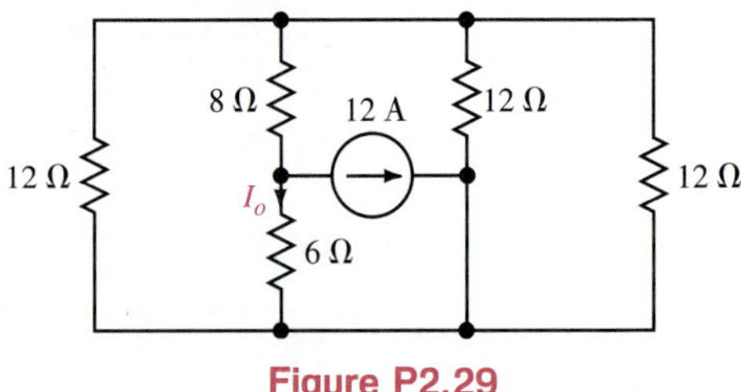

Figure P2.29

2.30. Find I_o in the circuit in Fig. P2.30.

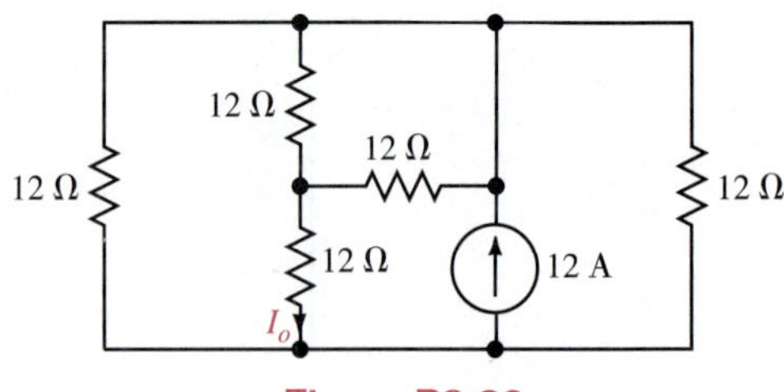

Figure P2.30

2.31. Find I_o in the circuit in Fig. P2.31.

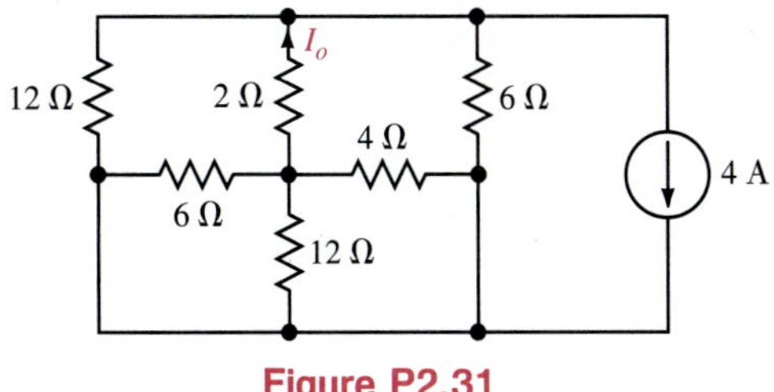

Figure P2.31

2.32. Given that $V_o = 12$ V in the network in Fig. P2.32, find V_S.

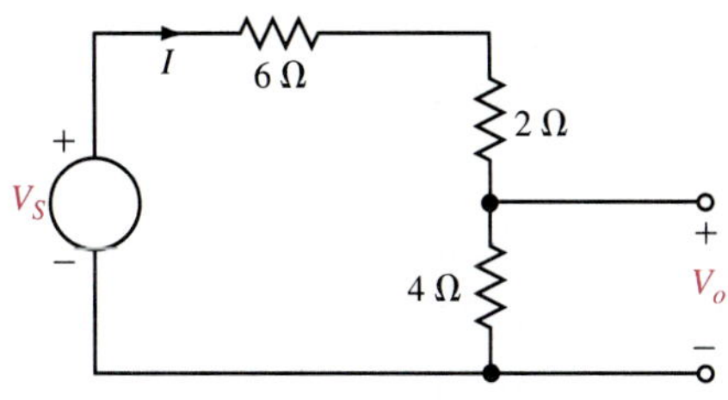

Figure P2.32

2.33. Find V_o in the network in Fig. P2.33.

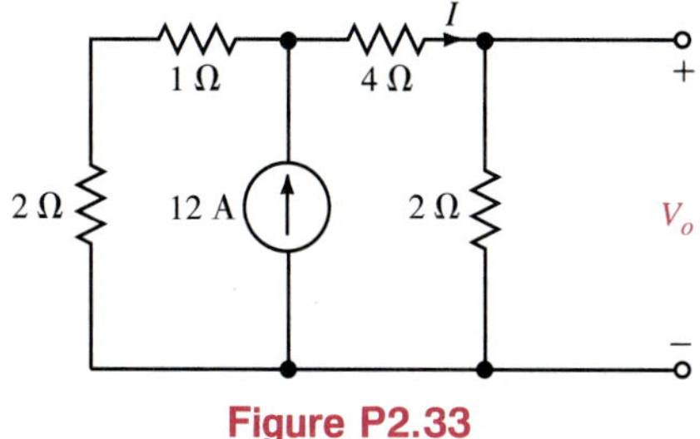

Figure P2.33

2.34. In the network in Fig. P2.34, if $I_1 = 6$ A, find V_o.

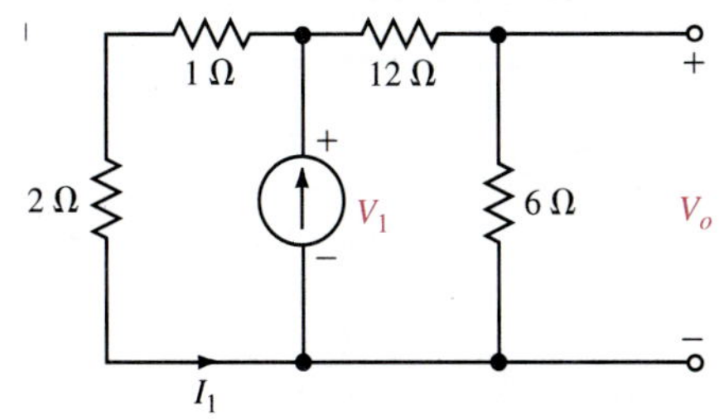

Figure P2.34

2.35. In the circuit in Fig. P2.35, if $P_{6\Omega} = 24$ W, find P_{V_S}

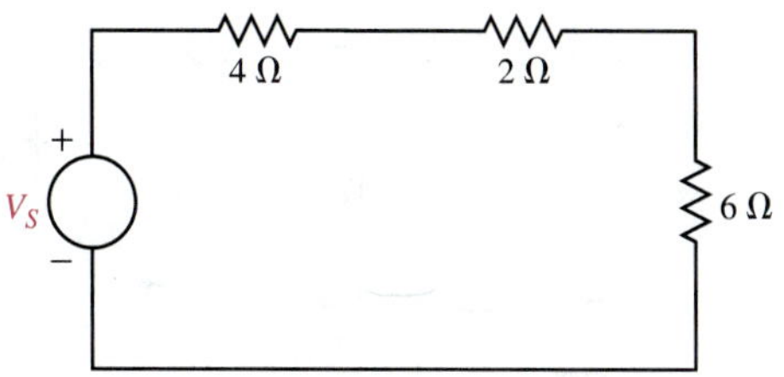

Figure P2.35

2.36. In the network in Fig. P2.36, if $V_o = 12$ V, find P_I.

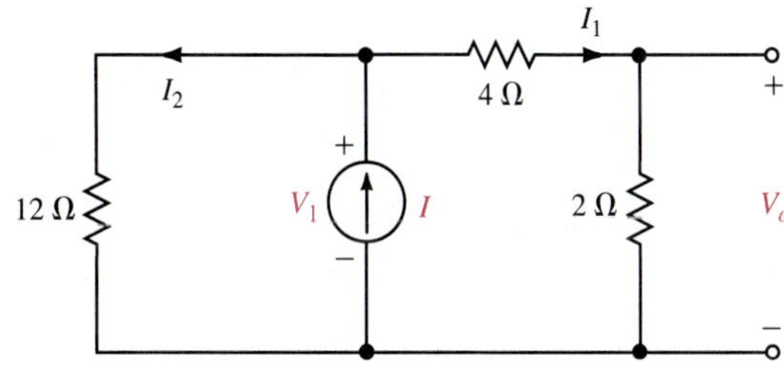

Figure P2.36

2.37. The power absorbed by the 4-Ω resistor in the network in Fig. P2.37 is 64 W. Find V_S.

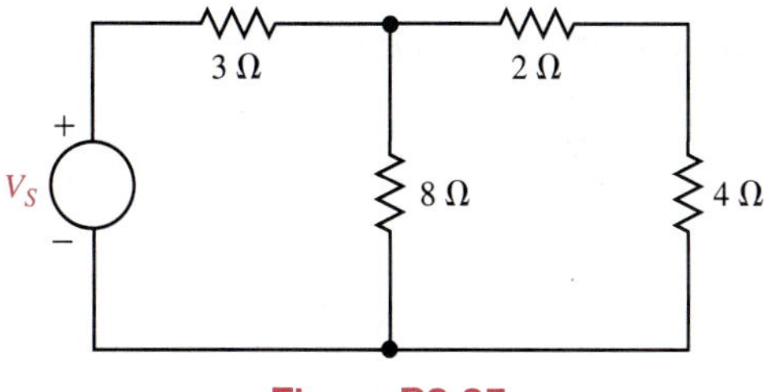

Figure P2.37

2.38. In the network in Fig. P2.38, $I = 4$ A. Find V_S.

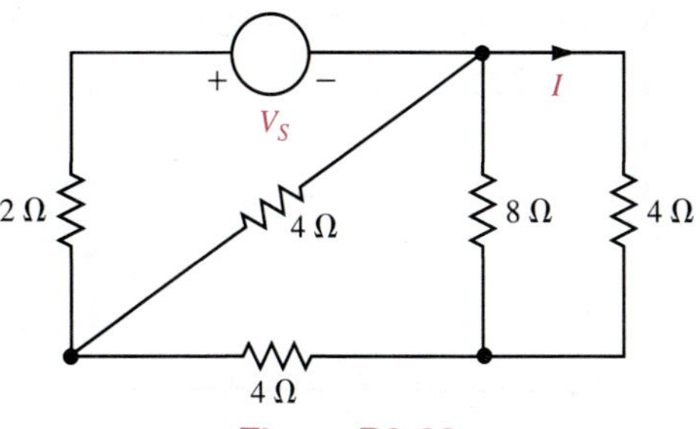

Figure P2.38

2.39. In the network in Fig. P2.39, if $V_o = 12$ V, find V_S.

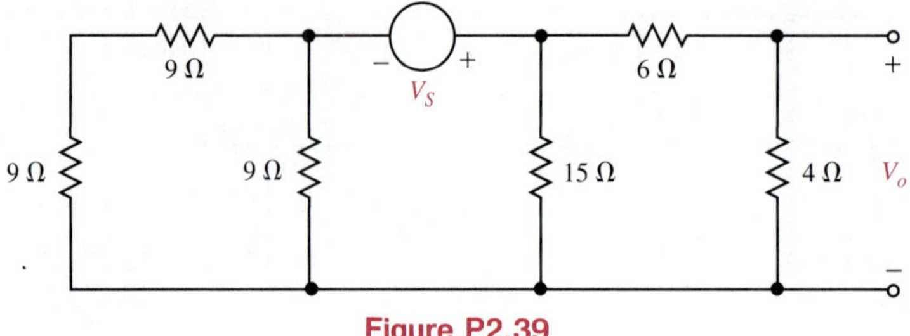

Figure P2.39

2.40. In the network in Fig. P2.40, if $I_o = 4$ A, find I_S.

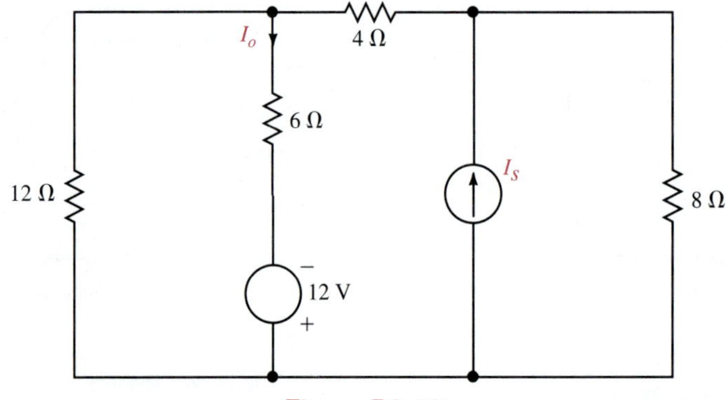

Figure P2.40

2.41. In the network in Fig. P2.41, if $V_o = 4$ V, find R.

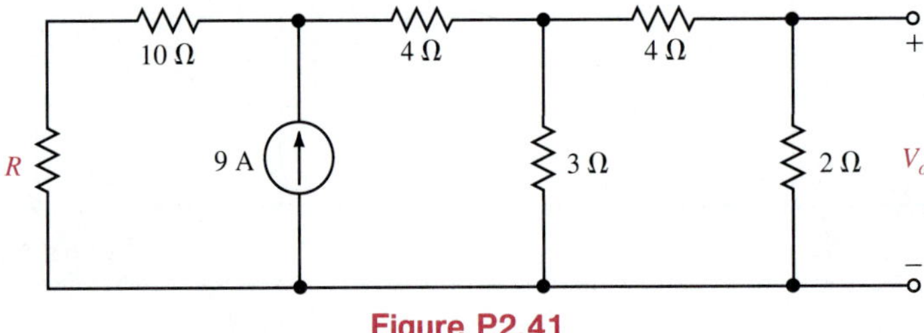

Figure P2.41

2.42. In the network in Fig. P2.42, if $V_o = 12$ V, find R.

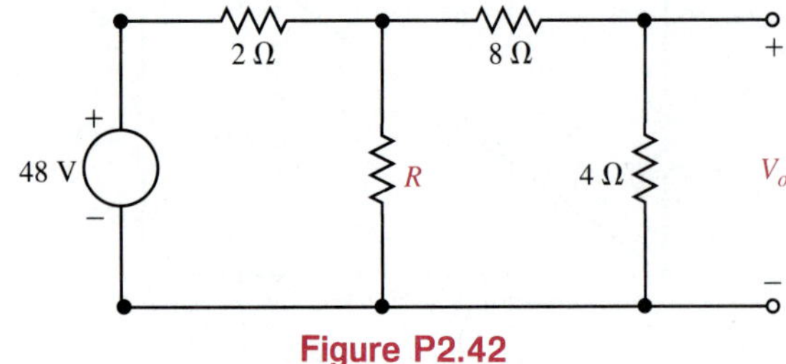

Figure P2.42

2.43. In the network in Fig. P2.43, if $V_1 = 16$ V, find R.

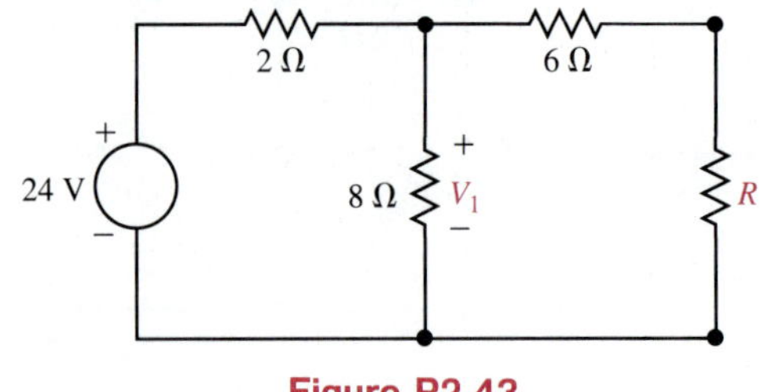

Figure P2.43

2.44. Given the circuit in Fig. P2.44, if $V_o = 12$ V, find I_S.

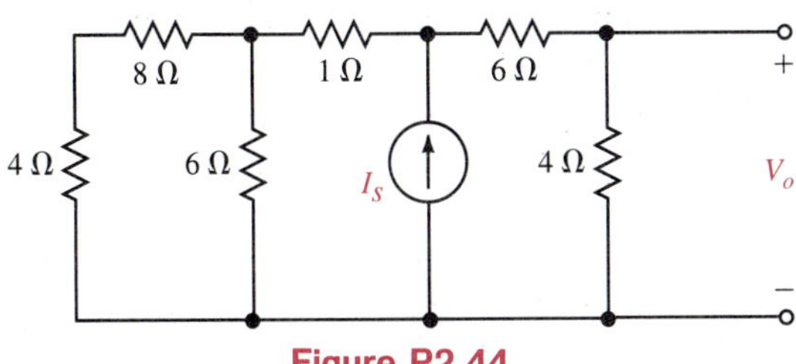

Figure P2.44

2.45. In the network in Fig. P2.45, if $I_o = 2$ A, find V_S.

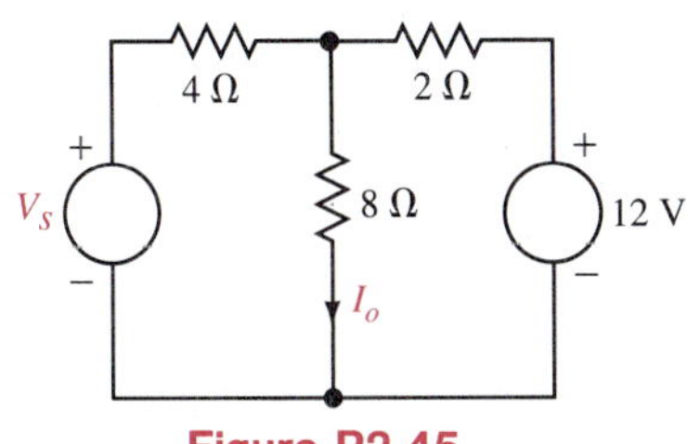

Figure P2.45

2.46. In the network in Fig. P2.46, if $V_o = 12$ V, find I_o.

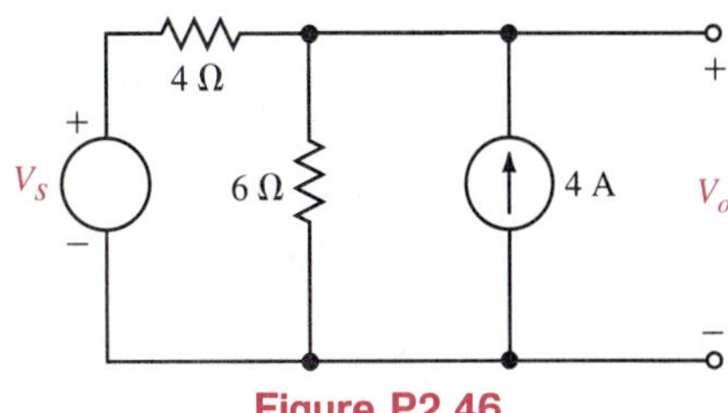

Figure P2.46

2.47. In the network in Fig. P2.47, if $V_o = 12$ V, find V_S.

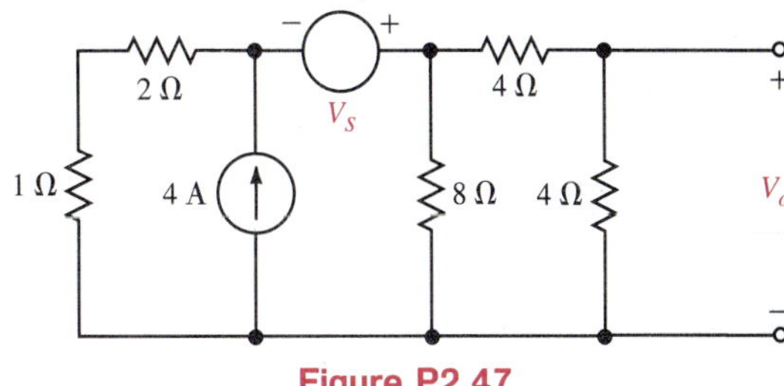

Figure P2.47

2.48. In the network in Fig. P2.48, if $I_o = 2$ A, find I_S.

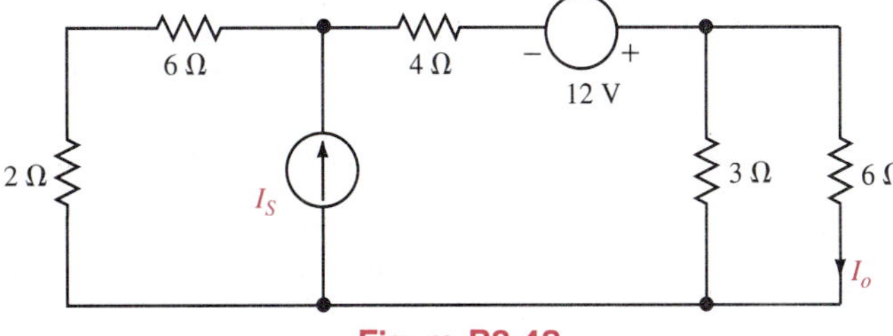

Figure P2.48

2.49. In the network in Fig. P2.49, if $V_o = 12$ V, find V_S.

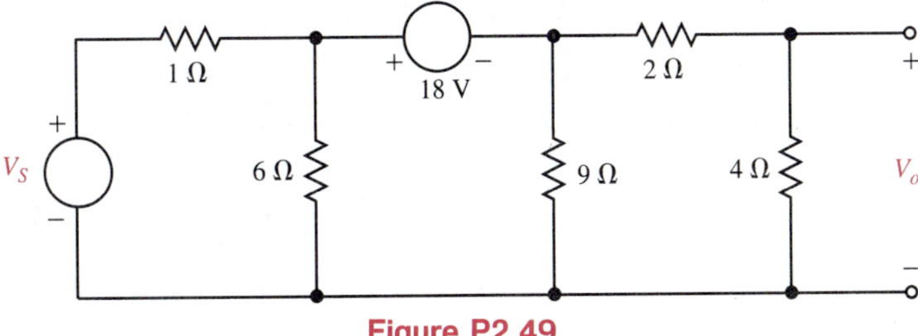

Figure P2.49

2.50. Find the power supplied or absorbed by the source V_S in the network in Fig. P2.50.

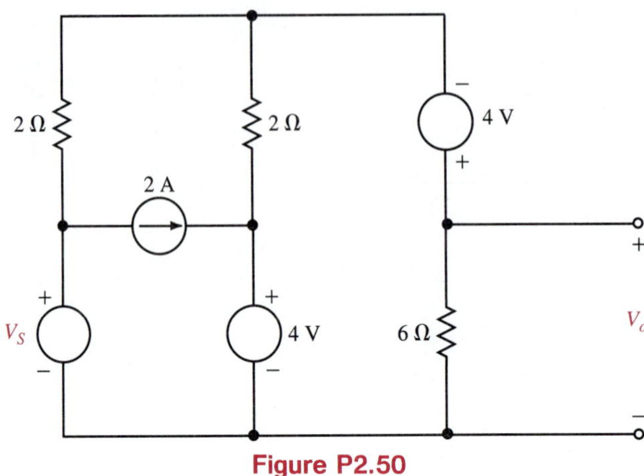

Figure P2.50

2.51. In the network in Fig. P2.51, if $V_o = 4$ V, find R.

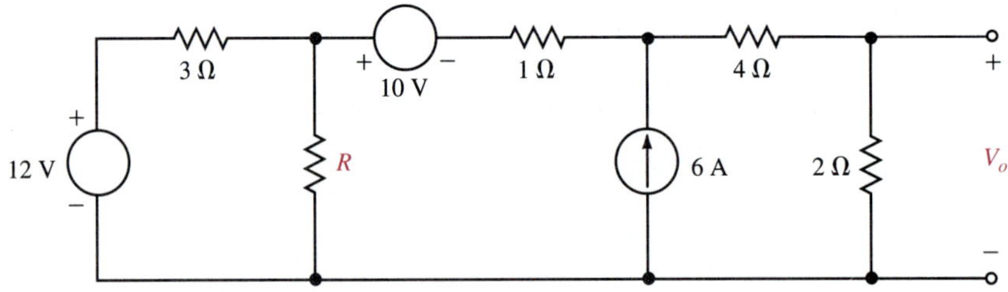

Figure P2.51

2.52. Find V_S in the network in Fig. P2.52.

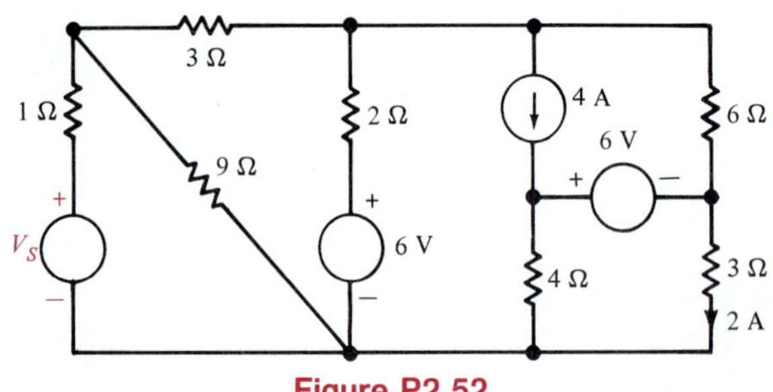

Figure P2.52

2.53. In the network in Fig. P2.53, the 4-A source supplies 24 W of power to the network. Find V_S.

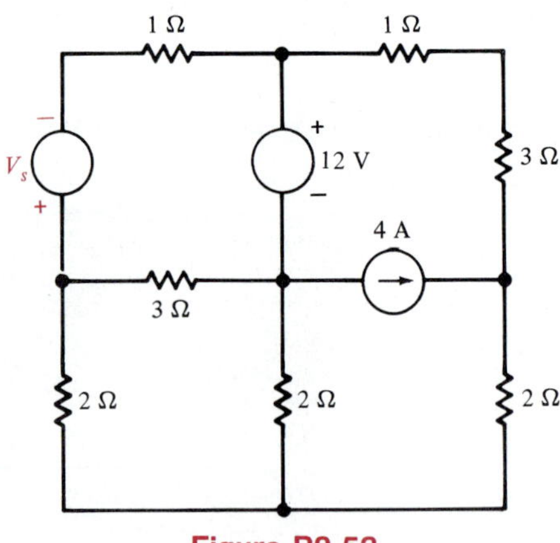

Figure P2.53

2.54. If $I_o = 2$ mA in the network in Fig. P2.54, find I_S.

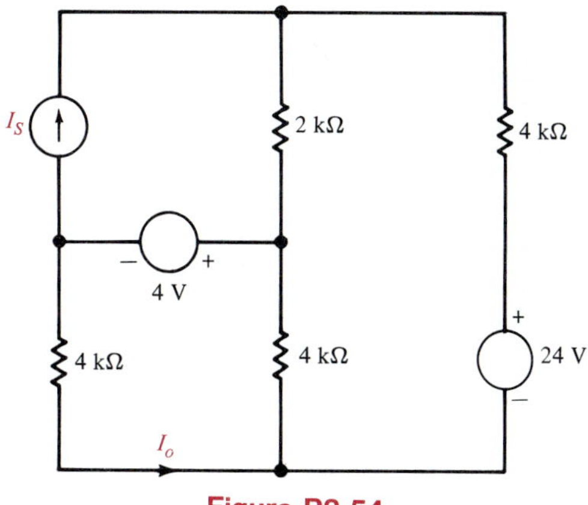

Figure P2.54

2.55. Given $I_o = 2$ mA in the network in Fig. P2.55, find V_o.

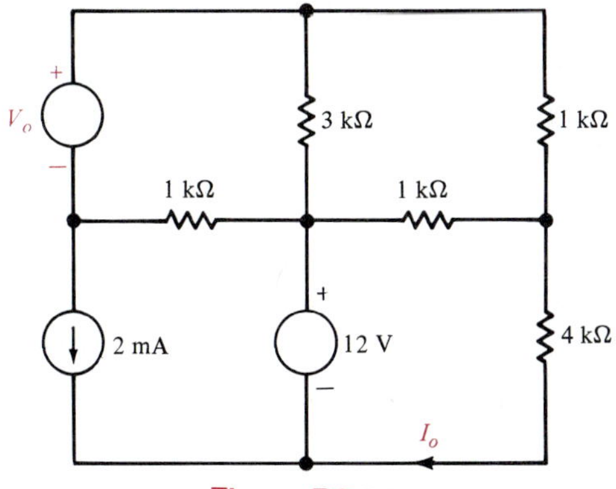

Figure P2.55

2.56. In the network in Fig. P2.56, if the power generated by the 4-A source is 48 W, find V_o.

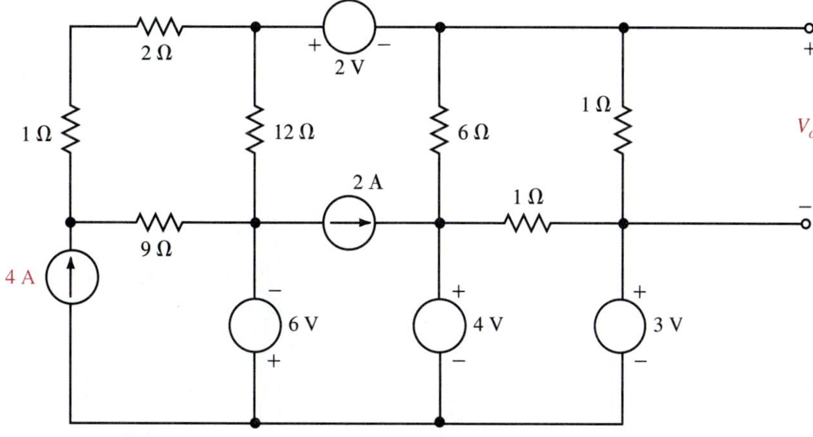

Figure P2.56

2.57. Find I_o in the network in Fig. P2.57.

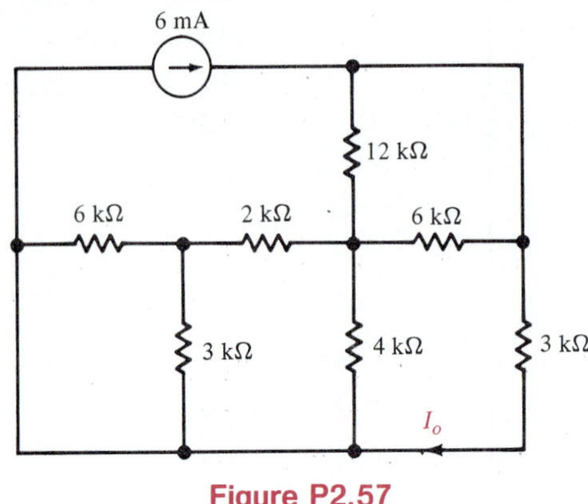

Figure P2.57

2.58. Find the power absorbed by the network in Fig. P2.58.

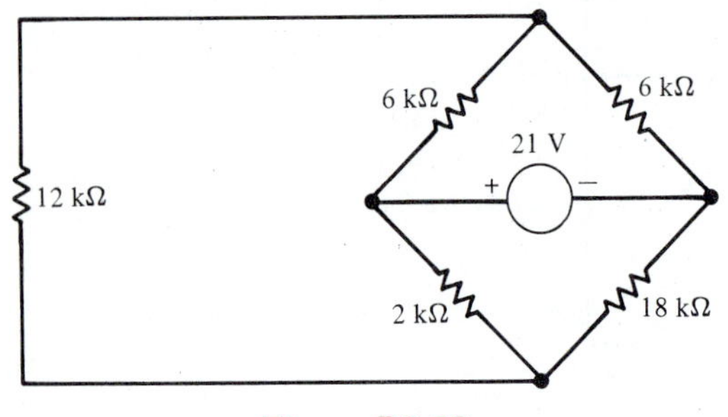

Figure P2.58

2.59. Find I_o in the circuit in Fig. P2.59.

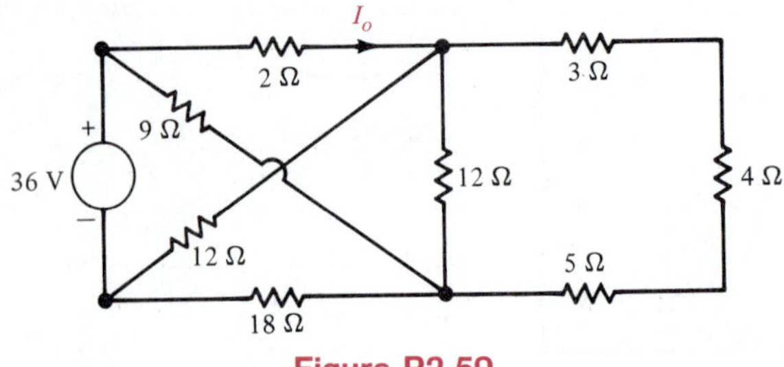

Figure P2.59

2.60. Find I_o in the circuit in Fig. P2.60.

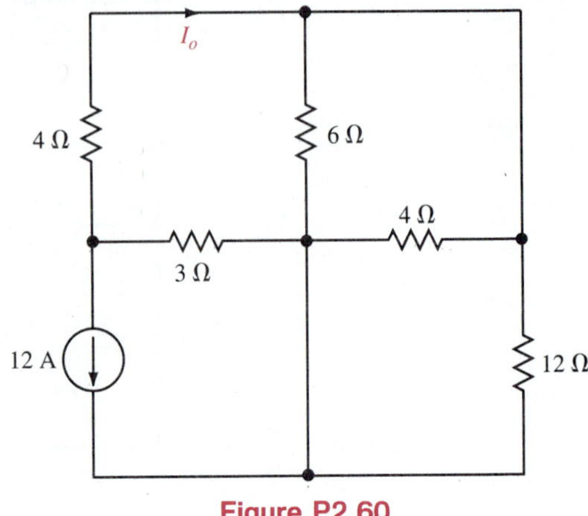

Figure P2.60

2.61. Find I_o in the circuit in Fig. P2.61.

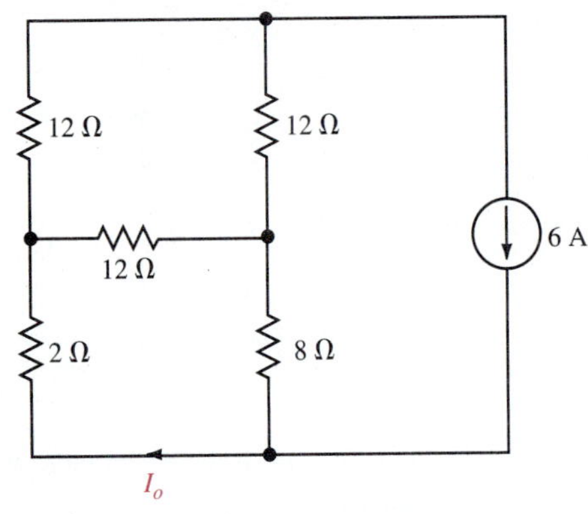

Figure P2.61

2.62. Find I_o in the circuit in Fig. P2.62.

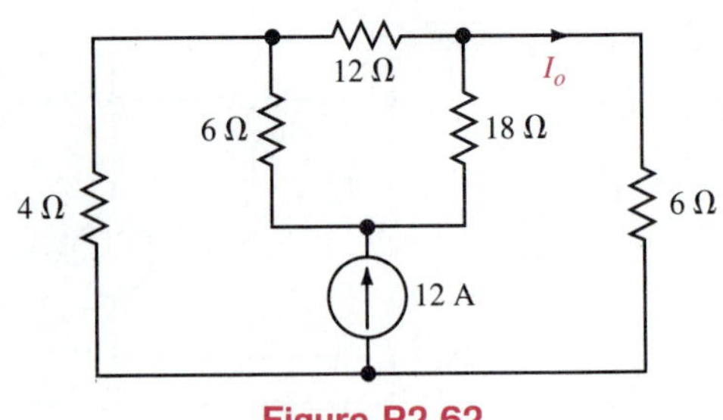

Figure P2.62

2.63. Find I_o in the circuit in Fig. P2.63.

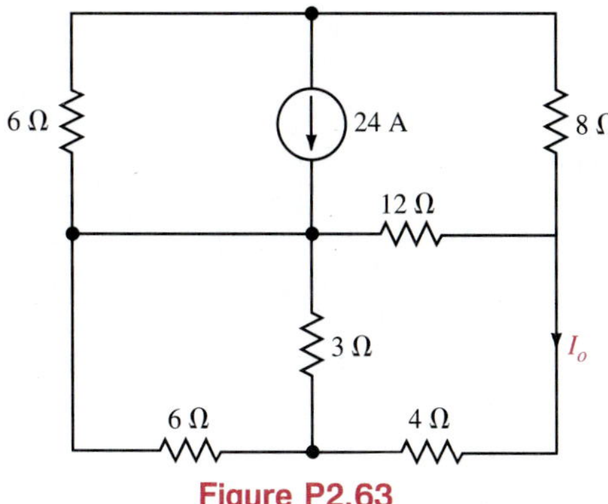

Figure P2.63

2.64. Find I_o in the circuit in Fig. P2.64.

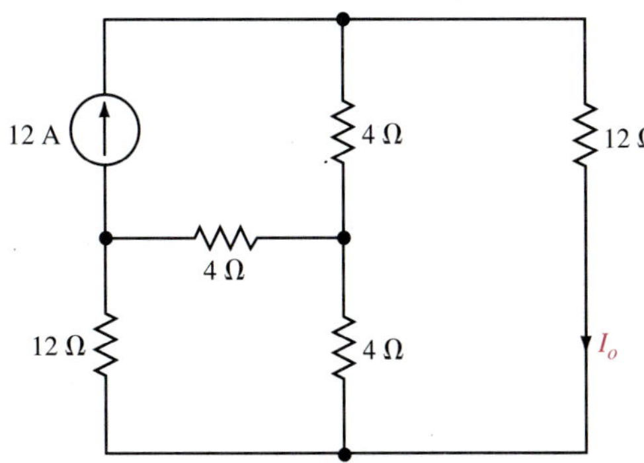

Figure P2.64

2.65. Find I_o in the circuit in Fig. P2.65.

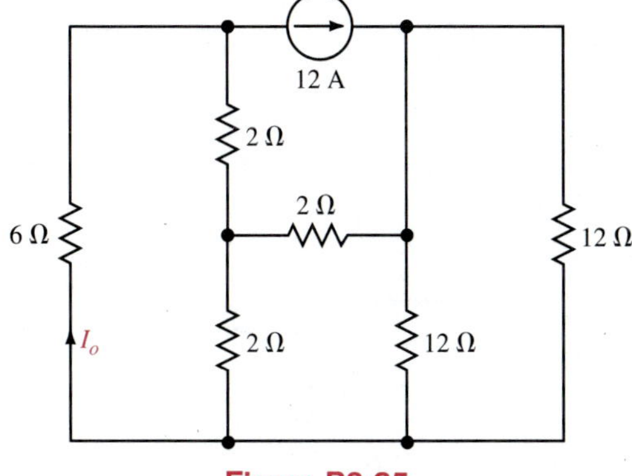

Figure P2.65

2.66. Find I_o in the circuit in Fig. P2.66.

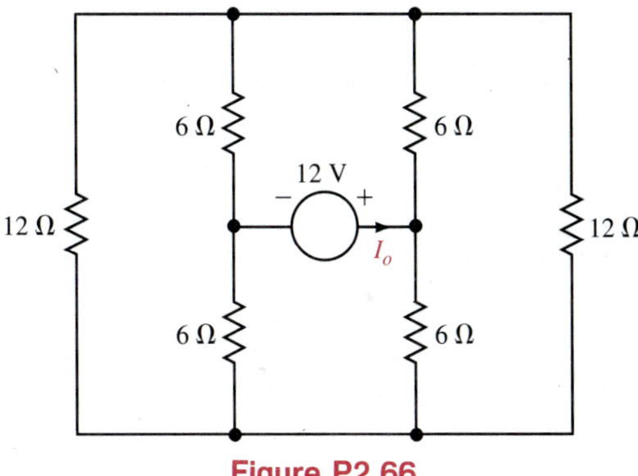

Figure P2.66

2.67. What must be the value of R_o in the network in Fig. P2.67 if the current I_o is 4 A?

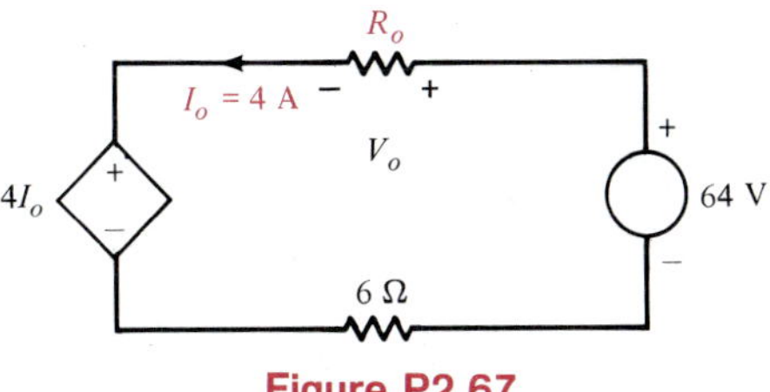

Figure P2.67

2.68. Find V_o in the network in Fig. P2.68.

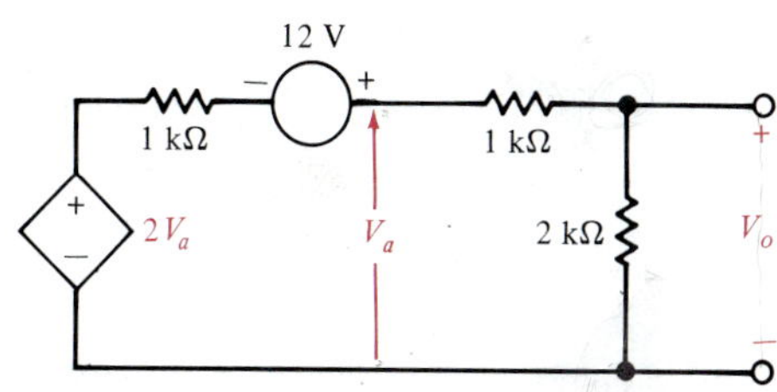

Figure P2.68

2.69. Find the voltage V_o in the circuit shown in Fig. P2.69.

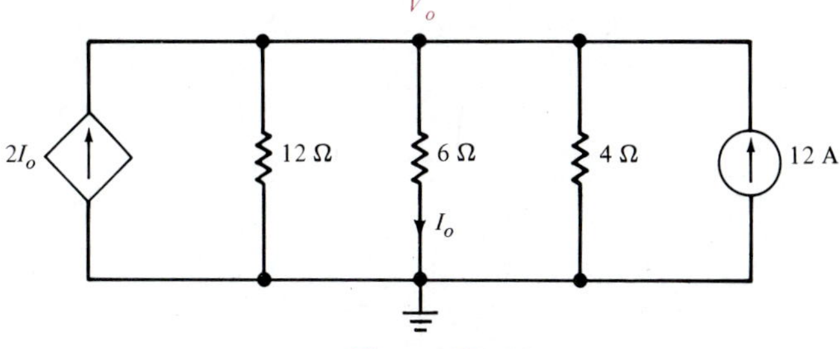

Figure P2.69

2.70. Find the power absorbed by the 12-Ω resistor in the network in Fig. P2.70.

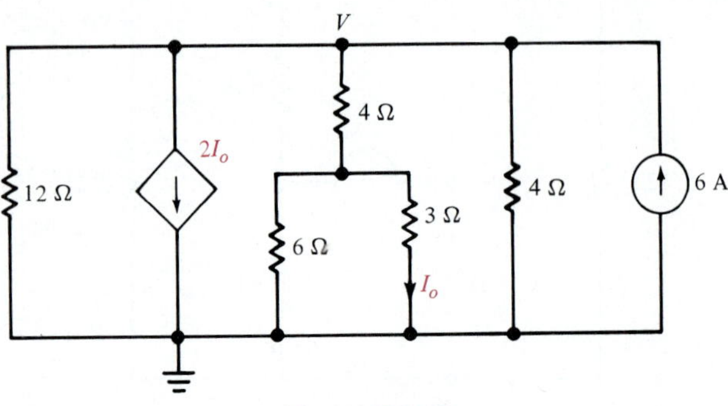

Figure P2.70

2.71. Find the voltage V_o in the network in Fig. P2.71.

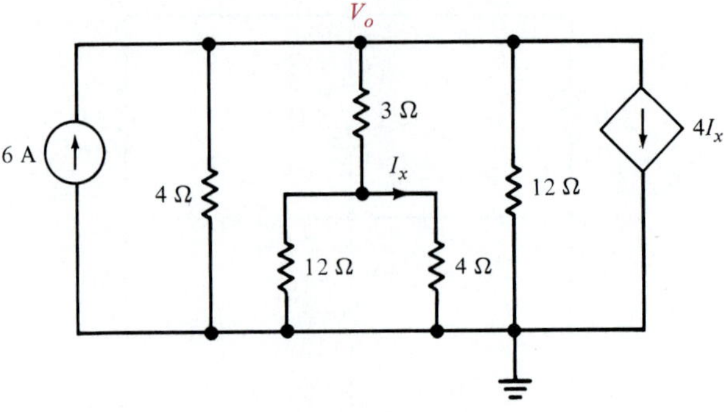

Figure P2.71

CHAPTER

3

Nodal and Loop Analysis Techniques

In Chapter 2 we analyzed the simplest possible circuits, those containing only a single-node pair or a single loop. We found that these circuits can be completely analyzed via a single algebraic equation. In the case of the single-node-pair circuit (i.e., one containing two nodes, one of which is a reference node), once the node voltage is known, we can calculate all the currents. In a single-loop circuit, once the loop current is known, we can calculate all the voltages.

In this chapter we extend our capabilities in a systematic manner so that we can calculate all currents and voltages in circuits that contain multiple nodes and loops. Our analyses are based primarily on two laws with which we are already familiar: Kirchhoff's current law (KCL) and Kirchhoff's voltage law (KVL). In a nodal analysis we employ KCL to determine the node voltages, and in a loop analysis we use KVL to determine the loop currents. We briefly introduce network topology and demonstrate its use in writing KCL and KVL equations. Finally, we discuss a very important circuit known as the operational amplifier.

3.1 Nodal Analysis

In a nodal analysis the variables in the circuit are selected to be the node voltages. The node voltages are defined with respect to a common point in the circuit. One node is selected as the reference node, and all other node voltages are defined with respect to that node. Quite often this node is the one to which the largest number of branches are connected. It is commonly called *ground* because it is said to be at ground-zero potential, and it sometimes represents the chassis or ground line in a practical circuit.

We will select our variables as being positive with respect to the reference node. If one or more of the node voltages is actually negative with respect to the reference node, the analysis will indicate it.

Suppose for a moment that we know all the node voltages in a given network. Each resistive element in the circuit is connected between two nodes, one of which may be the reference node. Nevertheless, the voltage across each resistive element is known. Hence, with reference to Fig. 3.1, we can use Ohm's law, that is,

$$i = \frac{V_m - V_n}{R} \tag{3.1}$$

to calculate the current through any resistive element. In this manner we can determine every voltage and every current in the circuit.

In a double-node circuit (i.e., one containing two nodes, one of which is the reference node), a single equation is required to solve for the unknown node voltage. In the case of an N-node circuit, $N - 1$ linearly independent simultaneous equations are required to determine the $N - 1$ unknown node voltages. These equations are written by employing KCL at $N - 1$ of the N nodes.

Consider, for example, the circuit shown in Fig. 3.2a. This circuit has three nodes. The node at the bottom is selected as the reference node and is so labeled using the ground symbol, ⏚. This reference node is assumed to be at zero potential and the node voltages v_1 and v_2 are defined with respect to this node; that is, if $v_1 = 4$ V and $v_2 = 16$ V, the voltage across R_1 and R_2 is 4 V, the voltage across R_4 and the current source is 16 V, and the voltage across R_3 is $16 - 4 = 12$ V. The right terminal of R_3 is positive with respect to the left terminal, and therefore the current will flow from right to left through R_3 as illustrated in Fig. 3.2a. Thus, in applying Ohm's law to find the current in a resistor, we take the

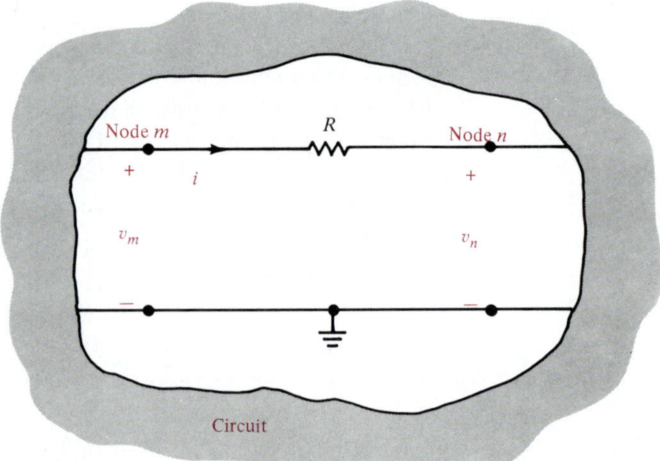

Figure 3.1 Circuit used to illustrate Ohm's law in a multiple-node network.

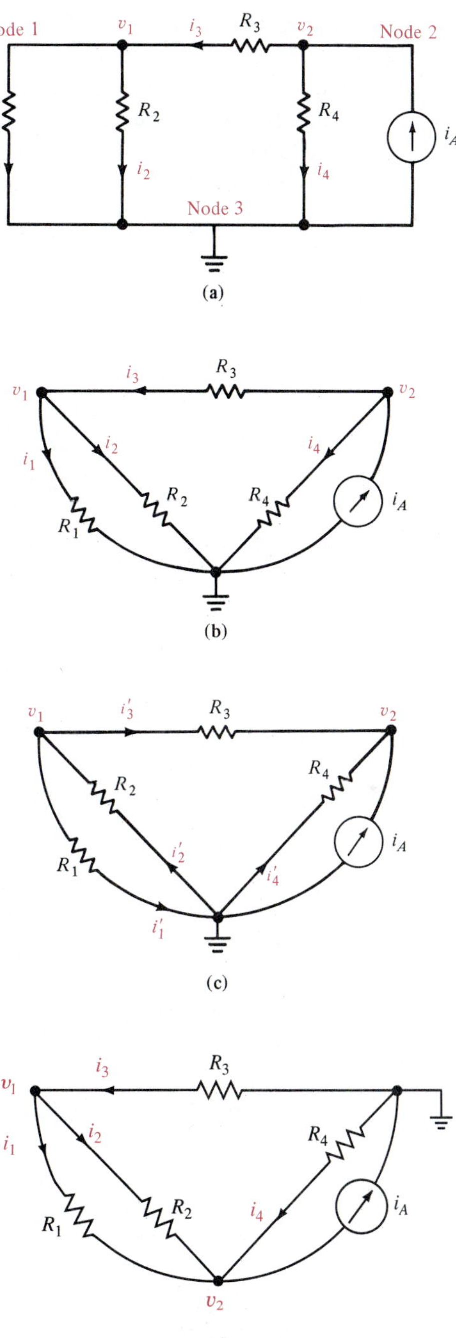

Figure 3.2 Three-node circuit.

difference in potential between the two terminals of the resistor and divide it by the value of the resistor. As shown in Fig. 3.2a, the current will flow from the terminal of higher potential to the terminal of lower potential.

The circuit is redrawn in Fig. 3.2b to indicate the nodes clearly. The branch currents are assumed to flow in the directions indicated in the figures. If one or more of the branch currents are actually flowing in a direction opposite to that assumed, the analysis will simply produce a branch current that is negative.

Applying KCL at node 1 yields

$$i_1 + i_2 - i_3 = 0$$

Using Ohm's law and noting that the reference node is at zero potential, we obtain

$$\frac{v_1 - 0}{R_1} + \frac{v_1 - 0}{R_2} - \frac{v_2 - v_1}{R_3} = 0$$

or

$$v_1\left(\frac{1}{R_1} + \frac{1}{R_2} + \frac{1}{R_3}\right) - v_2\frac{1}{R_3} = 0$$

which can, of course, be written

$$v_1(G_1 + G_2 + G_3) - v_2(G_3) = 0$$

At node 2, KCL for the currents leaving the node is

$$+i_3 + i_4 - i_A = 0$$

Note that since i_A enters the node, $-i_A$ is leaving the node. This equation can be written

$$+\frac{v_2 - v_1}{R_3} + \frac{v_2 - 0}{R_4} - i_A = 0$$

or

$$-v_1\frac{1}{R_3} + v_2\left(\frac{1}{R_3} + \frac{1}{R_4}\right) = i_A$$

which can also be written

$$-v_1(G_3) + v_2(G_3 + G_4) = i_A$$

Therefore, the two equations, which when solved yield the node voltages, are

$$\begin{aligned} v_1\left(\frac{1}{R_1} + \frac{1}{R_2} + \frac{1}{R_3}\right) - v_2\frac{1}{R_3} &= 0 \\ -v_1\frac{1}{R_3} + v_2\left(\frac{1}{R_3} + \frac{1}{R_4}\right) &= i_A \end{aligned} \tag{3.2}$$

Note that the analysis has produced two simultaneous equations in the unknowns v_1 and v_2. They can be solved using any convenient technique. In Appendix A two methods for solving simultaneous equations are presented. The first is called Gaussian elimination, and the second involves the use of matrices. As shown in the appendix, Eq. (3.2) may be rewritten in matrix form as

$$\boldsymbol{A}\boldsymbol{v} = \boldsymbol{i} \tag{3.3}$$

where

$$\boldsymbol{A} = \begin{bmatrix} \dfrac{1}{R_1} + \dfrac{1}{R_2} + \dfrac{1}{R_3} & -\dfrac{1}{R_3} \\ -\dfrac{1}{R_3} & \dfrac{1}{R_3} + \dfrac{1}{R_4} \end{bmatrix}, \quad \boldsymbol{v} = \begin{bmatrix} v_1 \\ v_2 \end{bmatrix}, \quad \text{and} \quad \boldsymbol{i} = \begin{bmatrix} 0 \\ i_A \end{bmatrix}$$

Thus

$$\begin{bmatrix} \dfrac{1}{R_1} + \dfrac{1}{R_2} + \dfrac{1}{R_3} & -\dfrac{1}{R_3} \\ -\dfrac{1}{R_3} & \dfrac{1}{R_3} + \dfrac{1}{R_4} \end{bmatrix} \begin{bmatrix} v_1 \\ v_2 \end{bmatrix} = \begin{bmatrix} 0 \\ i_A \end{bmatrix} \tag{3.4}$$

In general, the solution of Eq. (3.3) is

$$\boldsymbol{v} = \boldsymbol{A}^{-1}\boldsymbol{i} \tag{3.5}$$

where $\boldsymbol{A}^{-1}$ is the inverse of matrix $\boldsymbol{A}$. As a rule, matrix methods used in conjunction with a digital computer are extremely useful in circuit analysis.

The KCL equations at nodes 1 and 2 produced the two linearly independent simultaneous equations

$$i_1 + i_2 - i_3 = 0$$

$$i_3 + i_4 - i_A = 0$$

The KCL equation for the ground (third) node is

$$i_1 + i_2 + i_4 - i_A = 0$$

Note that if we add the first two equations, we obtain the third. Furthermore, any two of the equations can be used to derive the remaining equation. Therefore, in this $N = 3$ node circuit, only $N - 1 = 2$ of the equations are linearly independent and required to determine the $N - 1 = 2$ unknown node voltages.

Note that a nodal analysis employs KCL in conjunction with Ohm's law. Once the direction of the branch currents has been *assumed,* then Ohm's law, as illustrated by Fig. 3.1 and expressed by Eq. (3.1), is used to express the branch currents in terms of the unknown node voltages. We can assume the currents to be in any direction. However, once we assume a particular direction, we must be very careful to write the currents correctly in terms of the node voltages using Ohm's law. For example, suppose we had assumed the direction of the branch currents as shown in Fig. 3.2c. Then noting that when we employ Ohm's law, as illustrated in Fig. 3.1 and expressed in Eq. (3.1), the current in a resistor is equal to the difference in potential between the two terminals of the resistor divided by the value of the resistor, and that the current flows from the terminal of higher potential toward the terminal of lower potential. The branch currents for the circuit in Fig. 3.2c are

$$i_1' = \frac{v_1 - 0}{R_1} \qquad i_3' = \frac{v_1 - v_2}{R_3}$$

$$i_2' = \frac{0 - v_1}{R_2} \qquad i_4' = \frac{0 - v_2}{R_4}$$

KCL applied at the two nodes of the circuit in Fig. 3.2c yields

$$i_1' - i_2' + i_3' = 0$$

and

$$i_3' + i_4' + i_A = 0$$

When we combine these equations for the circuit in Fig. 3.2c, we obtain the same equations for the node voltage that we derived for the network in Fig. 3.2a and b.

EXAMPLE 3.1

Suppose that the network shown in Fig. 3.2 has the following parameters: $R_1 = 0.5\ \Omega$, $R_2 = 1\ \Omega$, $R_3 = 1\ \Omega$, $R_4 = 0.5\ \Omega$, and $i_A = I_A = 11$ A. Then Eq. (3.2) becomes

$$4V_1 - V_2 = 0$$

$$-V_1 + 3V_2 = 11$$

where as stated in Chapter 1, we are employing capital letters because the voltages are constant.

Using Gaussian elimination, we solve the first equation for V_1 in terms of V_2; that is,

$$V_1 = \frac{V_2}{4}$$

This value is then substituted into the second equation to yield

$$-\frac{V_2}{4} + 3V_2 = 11$$

or

$$V_2 = 4 \text{ V}$$

This value for V_2 is now substituted back into the equation for V_1 in terms of V_2, which yields

$$V_1 = \frac{V_2}{4} = 1 \text{ V}$$

The circuit equations can also be solved using matrix analysis. In matrix form the equations are

$$\begin{bmatrix} 4 & -1 \\ -1 & 3 \end{bmatrix} \begin{bmatrix} V_1 \\ V_2 \end{bmatrix} = \begin{bmatrix} 0 \\ 11 \end{bmatrix}$$

and therefore,

$$\begin{bmatrix} V_1 \\ V_2 \end{bmatrix} = \begin{bmatrix} 4 & -1 \\ -1 & 3 \end{bmatrix}^{-1} \begin{bmatrix} 0 \\ 11 \end{bmatrix}$$

To calculate the inverse of $\boldsymbol{A}$, we need the adjoint and the determinant. The adjoint is

$$\text{adj } \boldsymbol{A} = \begin{bmatrix} 3 & 1 \\ 1 & 4 \end{bmatrix}$$

and the determinant is

$$|\boldsymbol{A}| = (3)(4) - (-1)(-1) = 11$$

Therefore,

$$\begin{bmatrix} V_1 \\ V_2 \end{bmatrix} = \frac{1}{11} \begin{bmatrix} 3 & 1 \\ 1 & 4 \end{bmatrix} \begin{bmatrix} 0 \\ 11 \end{bmatrix} = \begin{bmatrix} 1 \\ 4 \end{bmatrix}$$

Thus $V_1 = 1$ V and $V_2 = 4$ V.

Knowing the node voltages, we can determine all the currents from Ohm's law:

$$I_1 = \frac{V_1}{R_1} = \frac{1}{0.5} = 2 \text{ A}$$

$$I_2 = \frac{V_1}{\mathrm{R}_2} = \frac{1}{1} = 1 \text{ A}$$

and

$$\begin{aligned} I_3 &= \frac{V_2 - V_1}{R_3} \\ &= \frac{4 - 1}{1} \\ &= 3 \text{ A} \end{aligned}$$

Note that the positive value of I_3 simply means that the current is actually flowing in the circuit from right to left as indicated by the values of I. The sum of the currents leaving node 1 is

$$I_1 + I_2 - I_3 = 2 + 1 - 3 = 0$$

The current

$$I_4 = \frac{V_2}{R_4} = \frac{4}{0.5} = 8 \text{ A}$$

and the sum of the currents leaving node 2 is

$$+I_3 + I_4 - I_A = 3 + 8 - 11 = 0$$

Since KCL is satisfied at each node, the answers check.

Before leaving this example, let us examine the case in which the node v_2 and the ground node (⏚) in the circuit in Fig. 3.2b are interchanged as shown in Fig. 3.2d. The KCL equations for this case are

$$i_1 + i_2 - i_3 = 0$$

$$i_1 + i_2 + i_4 - i_A = 0$$

or

$$\frac{v_1 - v_2}{0.5} + \frac{v_1 - v_2}{1} - \left(\frac{0 - v_1}{1}\right) = 0$$

$$\frac{v_1 - v_2}{0.5} + \frac{v_1 - v_2}{1} + \left(\frac{0 - v_2}{0.5}\right) - 11 = 0$$

Solving these equations yields $v_1 = -3$ V and $v_2 = -4$ V. Therefore all the currents in the network are identical to those calculated using the circuit in Fig. 3.2b, and the node voltages bear the same relationship to one another. ■

Let us now examine the circuit in Fig. 3.3a as redrawn in Fig. 3.3b. The current directions are assumed as shown in the figures.

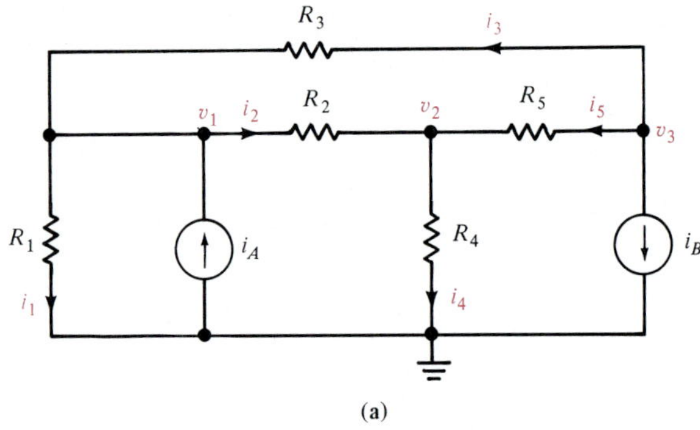

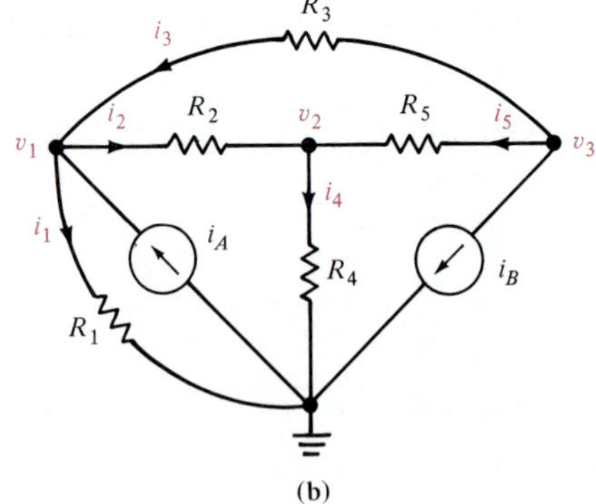

Figure 3.3 Four-node circuit.

At node 1 KCL yields

$$i_1 - i_A + i_2 - i_3 = 0$$

or

$$\frac{v_1}{R_1} - i_A + \frac{v_1 - v_2}{R_2} - \frac{v_3 - v_1}{R_3} = 0$$

$$v_1\left(\frac{1}{R_1} + \frac{1}{R_2} + \frac{1}{R_3}\right) - v_2\frac{1}{R_2} - v_3\frac{1}{R_3} = i_A$$

At node 2 KCL yields

$$-i_2 + i_4 - i_5 = 0$$

or

$$-\frac{v_1 - v_2}{R_2} + \frac{v_2}{R_4} - \frac{v_3 - v_2}{R_5} = 0$$

$$-v_1\frac{1}{R_2} + v_2\left(\frac{1}{R_2} + \frac{1}{R_4} + \frac{1}{R_5}\right) - v_3\frac{1}{R_5} = 0$$

At node 3 the equation is

$$i_3 + i_5 + i_B = 0$$

or

$$\frac{v_3 - v_1}{R_3} + \frac{v_3 - v_2}{R_5} + i_B = 0$$

$$-v_1\frac{1}{R_3} - v_2\frac{1}{R_5} + v_3\left(\frac{1}{R_3} + \frac{1}{R_5}\right) = -i_B$$

Grouping the node equations together, we obtain

$$\begin{aligned}
v_1\left(\frac{1}{R_1} + \frac{1}{R_2} + \frac{1}{R_3}\right) - v_2\frac{1}{R_2} - v_3\frac{1}{R_3} &= i_A \\
-v_1\frac{1}{R_2} + v_2\left(\frac{1}{R_2} + \frac{1}{R_4} + \frac{1}{R_5}\right) - v_3\frac{1}{R_5} &= 0 \\
-v_1\frac{1}{R_3} - v_2\frac{1}{R_5} + v_3\left(\frac{1}{R_3} + \frac{1}{R_5}\right) &= -i_B
\end{aligned} \tag{3.6}$$

Note that our analysis has produced three simultaneous equations in the three unknown node voltages v_1, v_2, and v_3. The equations can also be written in matrix form as

$$\begin{bmatrix} \frac{1}{R_1} + \frac{1}{R_2} + \frac{1}{R_3} & -\frac{1}{R_2} & -\frac{1}{R_3} \\ -\frac{1}{R_2} & \frac{1}{R_2} + \frac{1}{R_4} + \frac{1}{R_5} & -\frac{1}{R_5} \\ -\frac{1}{R_3} & -\frac{1}{R_5} & \frac{1}{R_3} + \frac{1}{R_5} \end{bmatrix} \begin{bmatrix} v_1 \\ v_2 \\ v_3 \end{bmatrix} = \begin{bmatrix} i_A \\ 0 \\ -i_B \end{bmatrix} \tag{3.7}$$

EXAMPLE 3.2

Given that the network shown in Fig. 3.3 has the following parameters, determine the node voltages.

$$R_1 = 0.5\ \Omega \qquad R_3 = 1\ \Omega \qquad R_5 = 0.5\ \Omega \qquad i_B = I_B = 1\ \text{A}$$

$$R_2 = 0.25\ \Omega \qquad R_4 = 1\ \Omega \qquad i_A = I_A = 4\ \text{A}$$

The data indicate that

$$G_1 = 2 \qquad G_3 = 1 \qquad G_5 = 2 \qquad G_2 = 4 \qquad G_4 = 1$$

The node equations are

$$\begin{aligned} 7V_1 - 4V_2 - V_3 &= 4 \\ -4V_1 + 7V_2 - 2V_3 &= 0 \\ -V_1 - 2V_2 + 3V_3 &= -1 \end{aligned}$$

The matrix equation for the circuit is

$$\begin{bmatrix} 7 & -4 & -1 \\ -4 & 7 & -2 \\ -1 & -2 & 3 \end{bmatrix} \begin{bmatrix} V_1 \\ V_2 \\ V_3 \end{bmatrix} = \begin{bmatrix} 4 \\ 0 \\ -1 \end{bmatrix}$$

and therefore,

$$\begin{bmatrix} V_1 \\ V_2 \\ V_3 \end{bmatrix} = \begin{bmatrix} 7 & -4 & -1 \\ -4 & 7 & -2 \\ -1 & -2 & 3 \end{bmatrix}^{-1} \begin{bmatrix} 4 \\ 0 \\ -1 \end{bmatrix}$$

The cofactors that produce the adjoint of the matrix are

$$A_{11} = \begin{vmatrix} 7 & -2 \\ -2 & 3 \end{vmatrix} = 17, \qquad A_{12} = (-1)\begin{vmatrix} -4 & -2 \\ -1 & 3 \end{vmatrix} = 14,$$

$$A_{13} = \begin{vmatrix} -4 & 7 \\ -1 & -2 \end{vmatrix} = 15$$

The remaining cofactors are calculated as

$$A_{21} = 14 \qquad A_{22} = 20 \qquad A_{23} = 18$$

$$A_{31} = 15 \qquad A_{32} = 18 \qquad A_{33} = 33$$

The determinant of the matrix can be calculated as

$$\Delta = |A| = 48$$

Therefore,

$$\begin{bmatrix} V_1 \\ V_2 \\ V_3 \end{bmatrix} = \frac{1}{48} \begin{bmatrix} 17 & 14 & 15 \\ 14 & 20 & 18 \\ 15 & 18 & 33 \end{bmatrix} \begin{bmatrix} 4 \\ 0 \\ -1 \end{bmatrix}$$

Hence

$$V_1 = \frac{1}{48}[(4)(17) - (1)(15)] = +1.104 \text{ V}$$

$$V_2 = \frac{1}{48}[(4)(14) - (1)(18)] = +0.792 \text{ V}$$

$$V_3 = \frac{1}{48}[(4)(15) - (1)(33)] = +0.563 \text{ V}$$

We can now calculate all the currents.

$$I_1 = \frac{V_1}{R_1} = \frac{1.10}{0.5} = 2.21 \text{ A}$$

$$I_2 = \frac{V_1 - V_2}{R_2} = \frac{1.10 - 0.79}{0.25} = 1.25 \text{ A}$$

$$I_3 = \frac{V_3 - V_1}{R_3} = \frac{0.56 - 1.10}{1} = -0.54 \text{ A}$$

$$I_4 = \frac{V_2}{R_4} = \frac{0.79}{1} = 0.79 \text{ A}$$

$$I_5 = \frac{V_3 - V_2}{R_2} = \frac{0.56 - 0.79}{0.5} = -0.46 \text{ A}$$

We can now check our results using KCL and KVL. At node 1 the currents entering the node are

$$-I_1 + I_A - I_2 + I_3 = -2.21 + 4 - 1.25 - 0.54 = 0$$

At node 2 the currents entering the node are

$$I_2 - I_4 + I_5 = 1.25 - 0.79 - 0.46 = 0$$

At node 3 the currents entering the node are

$$-I_3 - I_5 - I_B = 0.54 + 0.46 - 1 = 0$$

■

At this point it is important that we note the symmetrical form of the equations that describe the two previous networks. Equations (3.2) and (3.4) and Eqs. (3.6) and (3.7) exhibit the same type of symmetrical form. The *A* matrix for each network (3.4) and (3.7) is a symmetrical matrix. This symmetry is not accidental. The node equations for networks containing only resistors and independent current sources can always be written in this symmetrical form. We can take advantage of this fact and learn to write the equations by inspection. Note in the first equation of (3.2) that the coefficient of v_1 is the sum of all the conductances connected to node 1 and the coefficient of v_2 is the negative of the conductances connected between node 1 and node 2. The right-hand side of the equation is the sum of the currents entering node 1 through current sources. This equation is KCL at node 1. In the second equation in (3.2), the coefficient of v_2 is the sum of all the conductances connected to node 2, the coefficient of v_1 is the negative of the conductance connected between node 2 and node 1, and the right-hand side of the equation is the sum of the currents entering node 2 through current sources. This equation is KCL at node 2. Similarly, in the first equation in (3.6) the coefficient of v_1 is the sum of the conductances connected to node 1, the coefficient of v_2 is the negative of the conductance connected between node 1 and node 2, the coefficient of v_3 is the negative of the conductance connected between node 1 and node 3, and the right-hand side of the equation is the sum of the currents entering node 1 through current sources. The other two equations in (3.6) are obtained in a similar manner. In general, if KCL is applied to node j with node voltage v_j, the coefficient of v_j is the sum of all the conductances connected to node j and the coefficients of the other node voltages (e.g., v_{j-1}, v_{j+1}) are the negative of the sum of the conductances connected directly between these nodes and node j. The right-hand side of the equation is equal to the sum of the currents entering the node via current sources. Therefore, the left-hand side of the equation represents the sum of the currents leaving node j and the right-hand side of the equation represents the currents entering node j.

EXAMPLE 3.3

Let us apply what we have just learned and write down the equations for the network in Fig. 3.4 by inspection.

The equations are

$$v_1\left(\frac{1}{R_1}+\frac{1}{R_2}\right) - v_2\frac{1}{R_1} = i_A$$

$$-v_1\frac{1}{R_1} + v_2\left(\frac{1}{R_1}+\frac{1}{R_4}\right) - v_4\frac{1}{R_4} = -i_B$$

$$v_3\left(\frac{1}{R_3}+\frac{1}{R_6}\right) - v_5\frac{1}{R_6} = -i_A$$

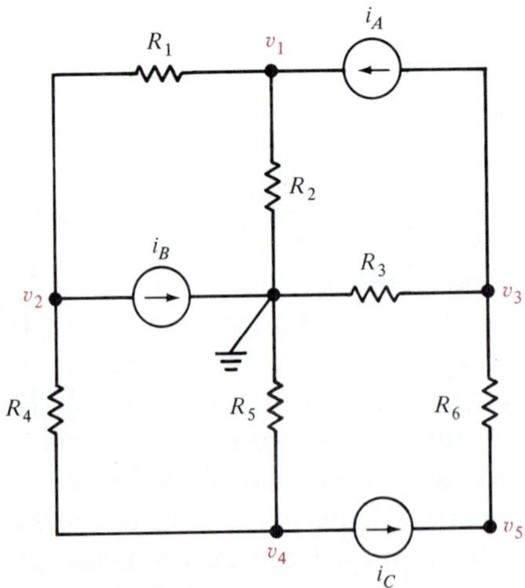

Figure 3.4 Example six-node circuit.

$$-v_2 \frac{1}{R_4} + v_4\left(\frac{1}{R_4} + \frac{1}{R_5}\right) = -i_C$$

$$-v_3 \frac{1}{R_6} + v_5 \frac{1}{R_6} = i_C$$

or in matrix form the equations become

$$\begin{bmatrix} \frac{1}{R_1} + \frac{1}{R_2} & -\frac{1}{R_1} & 0 & 0 & 0 \\ -\frac{1}{R_1} & \frac{1}{R_1} + \frac{1}{R_4} & 0 & -\frac{1}{R_4} & 0 \\ 0 & 0 & \frac{1}{R_3} + \frac{1}{R_6} & 0 & -\frac{1}{R_6} \\ 0 & -\frac{1}{R_4} & 0 & \frac{1}{R_4} + \frac{1}{R_5} & 0 \\ 0 & 0 & -\frac{1}{R_6} & 0 & \frac{1}{R_6} \end{bmatrix} \begin{bmatrix} v_1 \\ v_2 \\ v_3 \\ v_4 \\ v_5 \end{bmatrix} = \begin{bmatrix} i_A \\ -i_B \\ -i_A \\ -i_C \\ i_C \end{bmatrix}$$

Once again, note the symmetry of the $\mathbf{A}$ matrix. ■

DRILL EXERCISE

D3.1. Use nodal analysis to find the node voltages in the network in Fig. D3.1.

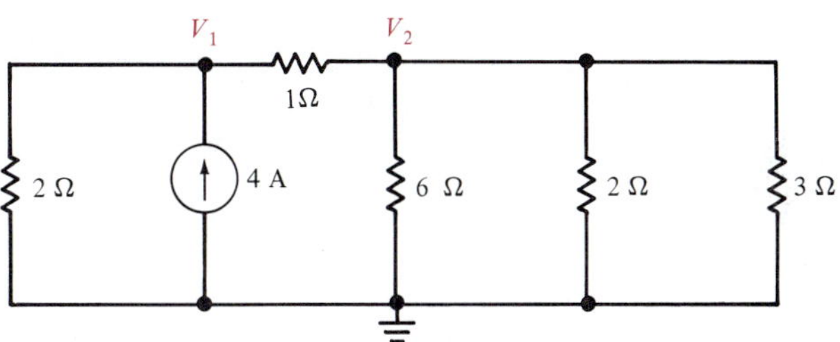

Figure D3.1

Ans: $V_1 = 4$ V, $V_2 = 2$ V.

D3.2. Use nodal analysis to find all the branch currents in the circuit in Fig. D3.2.

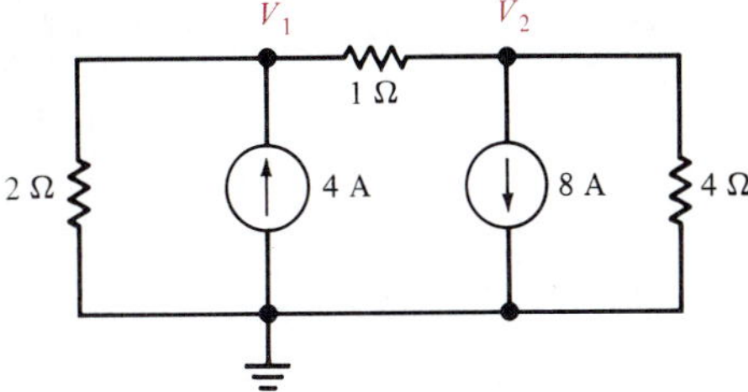

Figure D3.2

Ans: $I_{2\Omega} = \frac{12}{7}$ A ↑, $I_{1\Omega} = \frac{40}{7}$ A →, $I_{4\Omega} = \frac{16}{7}$ A ↑.

D3.3. Find all the branch currents in the network in Fig. D3.3 using nodal analysis.

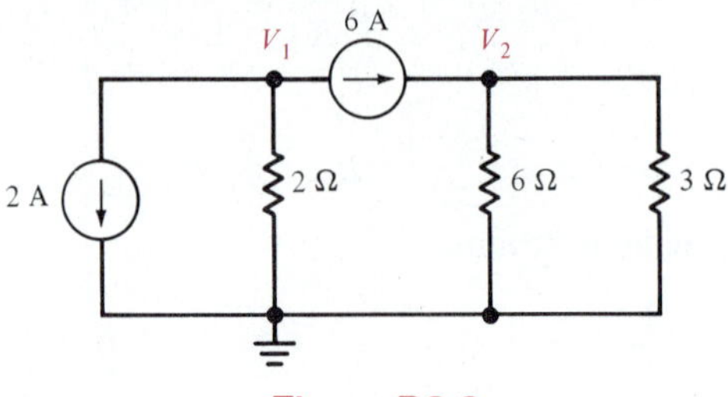

Figure D3.3

Ans: $I_{2\Omega} = 8$ A ↑, $I_{6\Omega} = 2$ A ↓, $I_{3\Omega} = 4$ A ↓.

Node Equations for Circuits Containing Independent Voltage Sources

To introduce this topic, let us consider the circuit shown in Fig. 3.5a and assume that the resistances and source voltages are known quantities. It might appear at first glance that

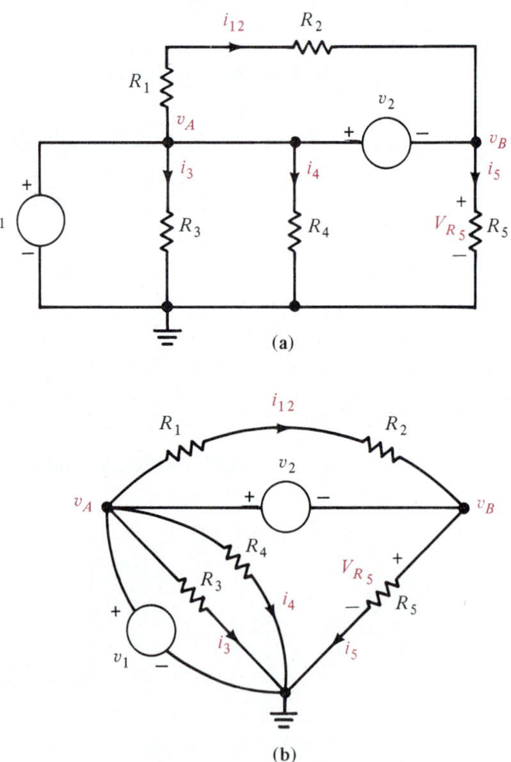

Figure 3.5 Network containing voltage sources between nodes.

there are two unknown node voltages, v_A and v_B. However, let us redraw the circuit as shown in Fig. 3.5b. A close examination of this network shows that the independent voltage sources exist between the nodes, and thus the node voltages are completely defined. For example, $v_A = v_1$ because the source v_1 is connected directly between terminal A and ground. In addition, note that $v_{R_5} = v_B$. KVL around the loop containing v_1, v_2, and v_{R_5} yields

$$+ v_1 - v_2 - v_B = 0$$

Therefore,

$$v_B = v_1 - v_2$$

and since v_1 and v_2 are known quantities, all node voltages are known. The currents can be immediately calculated as

$$i_{12} = \frac{v_2}{R_1 + R_2}$$

$$i_3 = \frac{v_1}{R_3}$$

$$i_4 = \frac{v_1}{R_4}$$

$$i_5 = \frac{v_B}{R_5} = \frac{v_1 - v_2}{R_5}$$

Although this is a very simple example, it is clear that the presence of the voltage sources has simplified the analysis. As a general rule, any time that voltage sources exist between nodes, the node voltage equations that describe the network will be simpler.

Consider the network in Fig. 3.6a. Suppose that we wish to determine the branch current I_2 using node equations. Since $I_2 = V_2/R_2$, we will determine the node voltage V_2. Recall from our previous development that when applying KCL to circuits containing only current sources and resistors, the branch currents were either known source values or could be expressed as the branch voltage divided by the branch resistance. However, the branch current I_S through the voltage source V_S cannot be expressed directly in either of these two forms. Therefore, the KCL equations for the network are

$$I_A = I_1 + I_S$$

and

$$I_S = I_2 + I_B$$

which can be written as

$$I_A = \frac{V_1}{R_1} + I_S \tag{3.8}$$

$$I_S = \frac{V_2}{R_2} + I_B \tag{3.9}$$

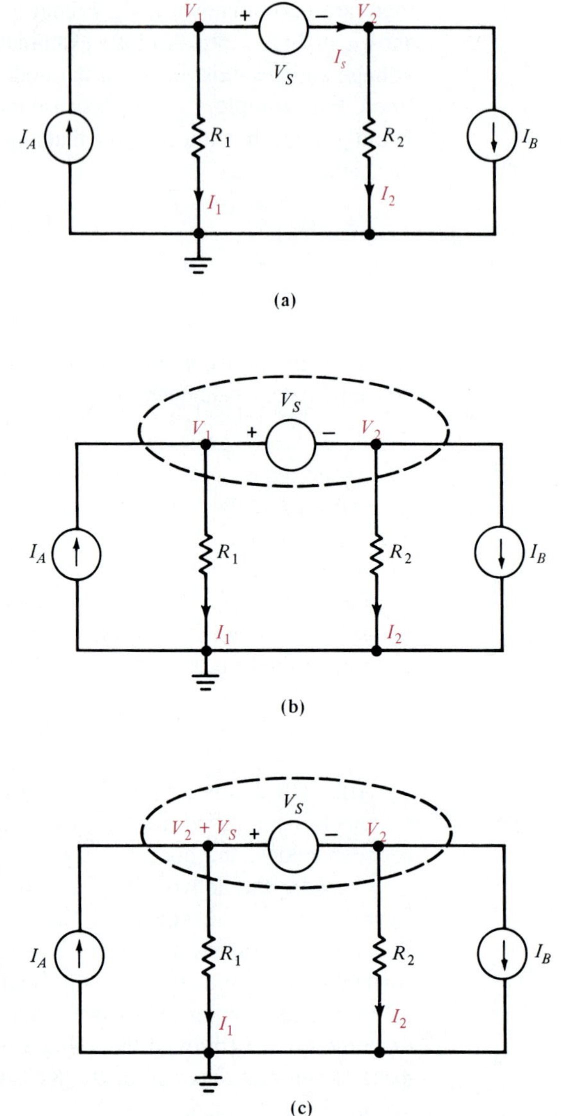

Figure 3.6 Circuits used to illustrate nodal analysis for a supernode.

We now have two linearly independent equations with three unknowns, V_1, V_2, and I_S. We should note, however, that if V_1 or V_2 is known, then the other node voltage, V_2 or V_1, is also known, since the difference in potential between the two nodes is *constrained* by the voltage source V_S. This constraint yields one additional linearly independent equation,

$$V_1 - V_2 = V_S \tag{3.10}$$

Equations (3.8), (3.9), and (3.10) form a set of three linearly independent equations in the three unknowns V_1, V_2, and I_S. However, the network in Fig. 3.6a has three nodes and we stated earlier that given an N-node circuit, $N - 1$ linearly independent equations are required to determine the $N - 1$ node voltages. Eliminating I_S in Eqs. (3.8) and (3.9) yields the equation

$$I_A = \frac{V_1}{R_1} + \frac{V_2}{R_2} + I_B \tag{3.11}$$

Equations (3.10) and (3.11) can now be written as

$$\begin{aligned} V_1 - V_2 &= V_S \\ \frac{V_1}{R_1} + \frac{V_2}{R_2} &= I_A - I_B \end{aligned} \tag{3.12}$$

which are two linearly independent equations in the two unknown node voltages V_1 and V_2. If we solve these equations for V_2, we can determine $I_2 = V_2/R_2$. Note that this procedure required three equations and three unknowns. One of the variables was eliminated immediately, yielding two equations involving the two unknown node voltages. We will now illustrate that these two equations can be written directly. To do so, let us now examine the network in Fig. 3.6b. Note that the voltage source V_S is completely enclosed within the dashed surface. In addition, we know that within that surface if one node voltage can be determined, the other is known immediately; that is, a *constraint* equation for this dashed portion of the network is

$$V_1 - V_2 = V_S \tag{3.13}$$

This equation is one of the two linearly independent equations needed to determine the node voltages. The other equation is obtained by applying KCL to the dashed surface, which is commonly called a *supernode*. We demonstrated in Chapter 2 that KCL must hold for such a surface, and this technique eliminates the problem of dealing with a current through a voltage source. KCL for the supernode is

$$\frac{V_1}{R_1} + \frac{V_2}{R_2} = I_A - I_B \tag{3.14}$$

Note that Eqs. (3.13) and (3.14) are identical to Eq. (3.12). Note that if we avoid considering currents in the voltage sources, fewer equations and fewer unknowns are required for the analysis.

An additional simplification can be made, however, based on the constraint that exists between the node voltages because of the presence of the voltage source. Once again consider the network as it is labeled in Fig. 3.6c. Note that the constraint equation has already been used in that the node voltage V_1 is labeled as $V_2 + V_S$. If we now apply KCL to the supernode, we obtain directly one equation in the unknown V_2; that is,

$$-I_A + \frac{V_2 + V_S}{R_1} + \frac{V_2}{R_2} + I_B = 0 \tag{3.15}$$

This equation, which is equivalent to Eqs. (3.13) and (3.14), will yield V_2 and $I_2 = V_2/R_2$.

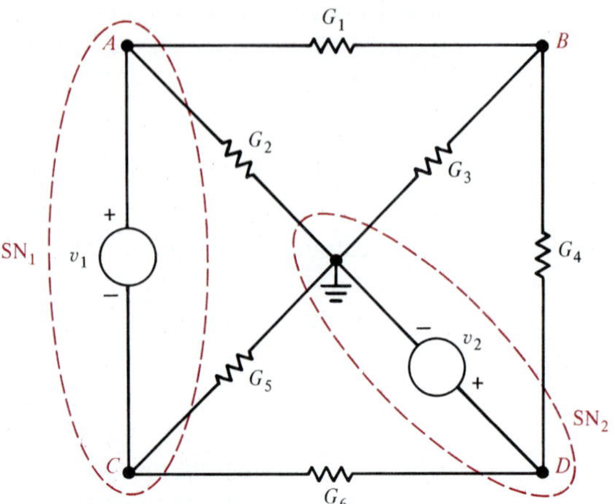

Figure 3.7 Circuit containing voltage sources, illustrating supernodes.

Let us now consider a more complicated circuit such as the one shown in Fig. 3.7. From our previous discussion we recognize immediately that the voltage v_D is known, and that the voltage between nodes A and C is known. These two conditions are thus *constraints* on the circuit and represent two of the basic equations necessary to solve for all the node voltages. Therefore, two of the independent equations required to solve for the four node voltages are

$$\begin{aligned} v_D &= v_2 \\ v_A - v_C &= v_1 \end{aligned} \tag{3.16}$$

The supernodes are labeled SN_1 and SN_2.

If we now examine the circuit in Fig. 3.7, it appears that we have three nodes, two supernodes and node B. Since the node voltage v_D is known, KCL applied at nodes B and SN_1 yields the two remaining required equations. For these nodes the equations are

$$(V_B - V_A)G_1 + V_B G_3 + (V_B - V_D)G_4 = 0$$

$$(V_A - V_B)G_1 + V_A G_2 + V_C G_5 + (V_C - V_D)G_6 = 0$$

which when simplified become

$$\begin{aligned} -V_A G_1 + V_B(G_1 + G_3 + G_4) - V_D G_4 &= 0 \\ V_A(G_1 + G_2) - V_B(G_1) + V_C(G_5 + G_6) - V_D G_6 &= 0 \end{aligned} \tag{3.17}$$

The four equations in (3.16) and (3.17) can be used to determine all the unknown voltages.

EXAMPLE 3.4

Let us determine the current I_o in the network in Fig. 3.8a.

Examining the network, we note that node voltages V_2 and V_4 are known and the node voltages V_1 and V_3 are constrained by the equation

$$V_1 - V_3 = 12$$

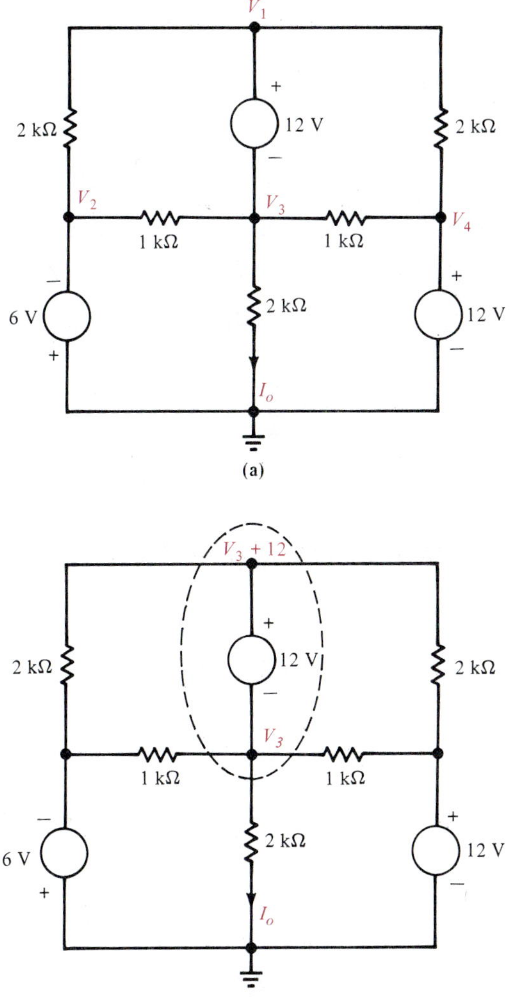

Figure 3.8 Example circuit with supernodes.

The network is redrawn in Fig. 3.8b. Although we could indicate a supernode for both node voltages V_2 and V_4, they have been omitted so that the network does not appear unduly complicated.

Since we want to find the current I_o, V_1 in the supernode containing V_1 and V_3 is written as $V_3 + 12$. The KCL equation at the supernode is then

$$\frac{V_3 + 12 - (-6)}{2} + \frac{V_3 + 12 - 12}{2} + \frac{V_3 - (-6)}{1} + \frac{V_3 - 12}{1} + \frac{V_3}{2} = 0$$

Solving this equation for V_3 yields

$$V_3 = -\tfrac{6}{7} \text{ V}$$

I_o can then be computed immediately as

$$I_o = \frac{-\frac{6}{7}}{2} = -\frac{3}{7} \text{ mA}$$

DRILL EXERCISE

D3.4. Use nodal analysis to find the currents in the two resistors in the network in Fig. D3.4.

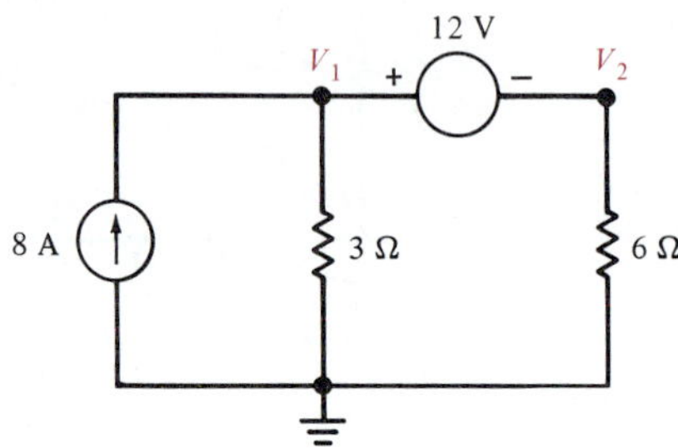

Figure D3.4

Ans: $I_{3\Omega} = \frac{20}{3}$ A ↓, $I_{6\Omega} = \frac{4}{3}$ A ↓.

D3.5. Use nodal analysis to find I_o in the network in Fig. D3.5.

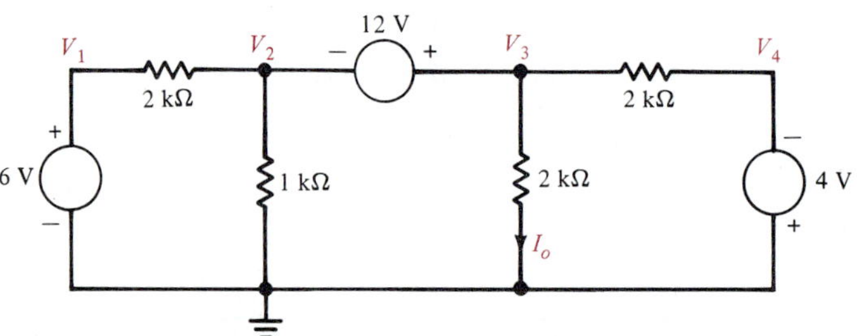

Figure D3.5

Ans: $I_o = 3.8$ mA.

As a final point in this discussion, consider the dashed portion of the network shown in Fig. 3.9. Note that R_5 and node A form a supernode. This supernode exists because the current in R_5 is known and thus the voltage across R_5 is also known.

Node Equations for Circuits Containing Dependent Voltage Sources

Networks containing dependent (controlled) sources are treated in the same manner as described above. There is, however, one important difference in the form of the resulting

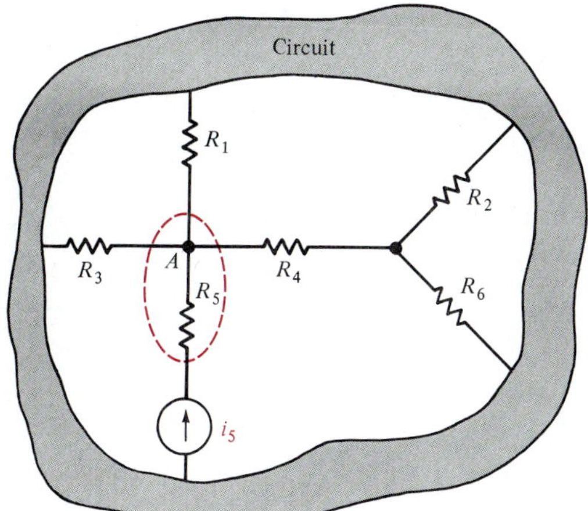

Figure 3.9 Network containing a supernode.

equations: The presence of a dependent source may destroy the symmetrical form of the nodal equations that define the circuit.

Consider the network shown in Fig. 3.10. Note that $v_4 = v_B$ and $v_A = v_2 - v_3$. Therefore, the equations that describe the network are

$$v_4 = v_B$$

$$v_1(G_1 + G_3) - v_2(G_1) - v_4(G_3) = i_A$$

$$-v_1(G_1) + v_2(G_1 + G_2) - v_3(G_2) = - i_A - 3(v_2 - v_3)$$

$$-v_2(G_2) + v_3(G_2 + G_4) = i_B$$

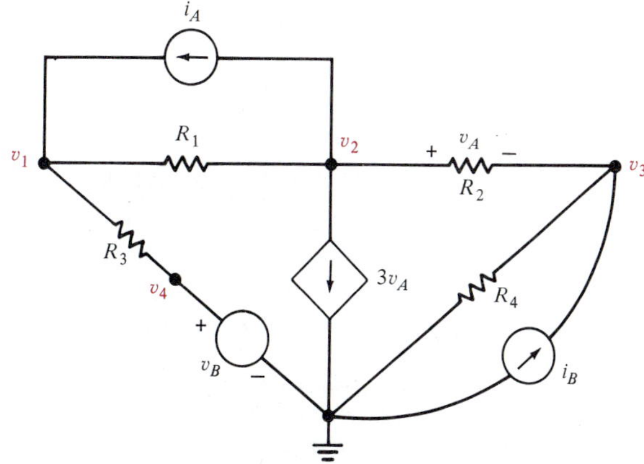

Figure 3.10 Network containing a controlled current source.

EXAMPLE 3.5

Given the following parameters for the network in Fig. 3.10, determine the node voltages.

$$R_1 = 1\ \Omega \qquad R_3 = 2\ \Omega \qquad i_A = I_A = 4\text{ A} \qquad v_B = V_B = 10\text{ V}$$

$$R_2 = 0.5\ \Omega \qquad R_4 = 1\ \Omega \qquad i_B = I_B = 5\text{ A}$$

Substituting these values into the equations above yields

$$\begin{aligned} 1.5V_1 - V_2 \qquad\quad &= 9 \\ -V_1 + 6V_2 - 5V_3 &= -4 \\ -2V_2 + 3V_3 &= 5 \end{aligned}$$

or in matrix form,

$$\begin{bmatrix} 1.5 & -1 & 0 \\ -1 & 6 & -5 \\ 0 & -2 & 3 \end{bmatrix} \begin{bmatrix} V_1 \\ V_2 \\ V_3 \end{bmatrix} = \begin{bmatrix} 9 \\ -4 \\ 5 \end{bmatrix}$$

The cofactors of the coefficient matrix (i.e., the $\mathbf{A}$ matrix) are

$$A_{11} = 8 \qquad A_{12} = 3 \qquad A_{13} = 2$$

$$A_{21} = 3 \qquad A_{22} = 4.5 \qquad A_{23} = 3$$

$$A_{31} = 5 \qquad A_{32} = 7.5 \qquad A_{33} = 8$$

The determinant is $|\mathbf{A}| = 9$, and therefore the solution of the equation is

$$\begin{bmatrix} V_1 \\ V_2 \\ V_3 \end{bmatrix} = \frac{1}{9} \begin{bmatrix} 8 & 3 & 5 \\ 3 & 4.5 & 7.5 \\ 2 & 3 & 8 \end{bmatrix} \begin{bmatrix} 9 \\ -4 \\ 9 \end{bmatrix}$$

Therefore,

$$\begin{bmatrix} V_1 \\ V_2 \\ V_3 \end{bmatrix} = \begin{bmatrix} 9.4444 \\ 5.1666 \\ 5.1111 \end{bmatrix}$$

As a quick check, KCL at node 3 yields

$$\frac{V_3 - V_2}{0.5} + \frac{V_3}{1} - 5 = \frac{5.1111 - 5.1666}{0.5} + \frac{5.1111}{1} - 5 = 0$$

■

Next, let us consider the circuit shown in Fig. 3.11, which contains a dependent voltage source. The nodal equations for this circuit are

$$\frac{v_1 - v_2}{R_2} + \frac{v_1 - v_3}{R_1} = i_A$$

$$v_2 = 10 i_4 = \frac{10 v_3}{R_4}$$

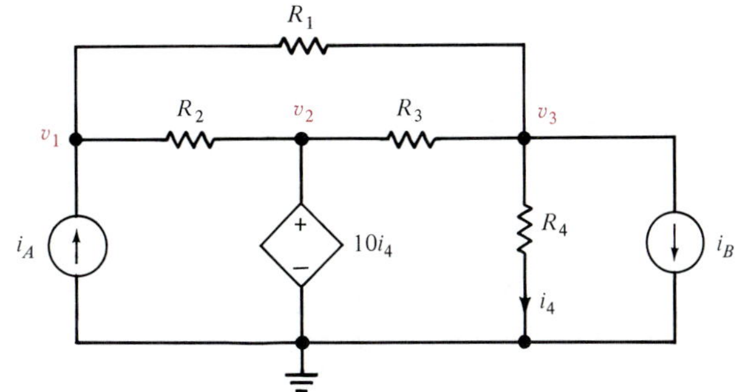

Figure 3.11 Circuit containing a controlled voltage source.

$$\frac{v_3 - v_1}{R_1} + \frac{v_3 - v_2}{R_3} + \frac{v_3}{R_4} = -i_B$$

Grouping the various terms in the equations and substituting for v_2 yields

$$v_1\left(\frac{1}{R_1} + \frac{1}{R_2}\right) - v_3\left(\frac{1}{R_1} + \frac{10}{R_2R_4}\right) = i_A$$

$$-v_1\frac{1}{R_1} + v_3\left(\frac{1}{R_1} + \frac{1}{R_3} + \frac{1}{R_4} - \frac{10}{R_3R_4}\right) = -i_B$$

Once again a simple substitution changes the problem from a three-variable problem to a two-variable problem, since once v_3 is known, v_2 can be calculated directly. A dependent voltage source will reduce the order of the equations in the same manner as an independent voltage source.

EXAMPLE 3.6

Suppose that the parameters in the circuit shown in Fig. 3.11 are

$$R_1 = 0.5\ \Omega \qquad R_3 = 2\ \Omega \qquad i_A = I_A = 10\ \text{A}$$

$$R_2 = 1\ \Omega \qquad R_4 = 10\ \Omega \qquad i_B = I_B = 2\ \text{A}$$

Substituting these values into the equations developed above yields

$$3V_1 - 3V_3 = 10$$

$$-2V_1 + 2.1V_3 = -2$$

or

$$\begin{bmatrix} 3 & -3 \\ -2 & 2.1 \end{bmatrix}\begin{bmatrix} V_1 \\ V_3 \end{bmatrix} = \begin{bmatrix} 10 \\ -2 \end{bmatrix}$$

and therefore,

$$\begin{bmatrix} V_1 \\ V_3 \end{bmatrix} = \begin{bmatrix} 3 & -3 \\ -2 & 2.1 \end{bmatrix}^{-1} \begin{bmatrix} 10 \\ -2 \end{bmatrix} = \frac{1}{0.3} \begin{bmatrix} 2.1 & 3 \\ 2 & 3 \end{bmatrix} \begin{bmatrix} 10 \\ -2 \end{bmatrix}$$

$$= \begin{bmatrix} 50.0000 \\ 46.6666 \end{bmatrix}$$

Since

$$V_2 = \frac{10}{R_4} V_3$$

then

$$V_2 = V_3$$

This means, of course, that no current flows in R_3. Let us check this answer by applying KCL at node 3.

$$\frac{V_3 - V_2}{R_3} + \frac{V_3 - V_1}{R_1} + \frac{V_3}{R_4} + I_B = 0 + \frac{46.6666 - 50.0000}{0.5} + \frac{46.6666}{10} + 2$$

$$= 0$$

■

DRILL EXERCISE

D3.6. Find the node voltages for the network in Fig. D3.6.

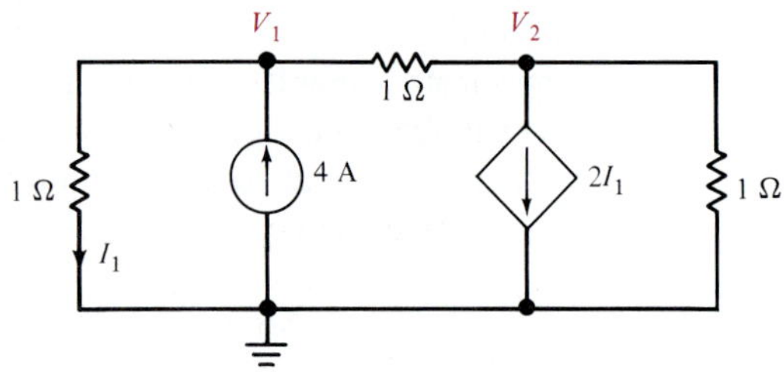

Figure D3.6

Ans: $V_1 = \frac{8}{5}$ V, $V_2 = -\frac{4}{5}$ V.

D3.7. Compute the node voltages for the network in Fig. D3.7.

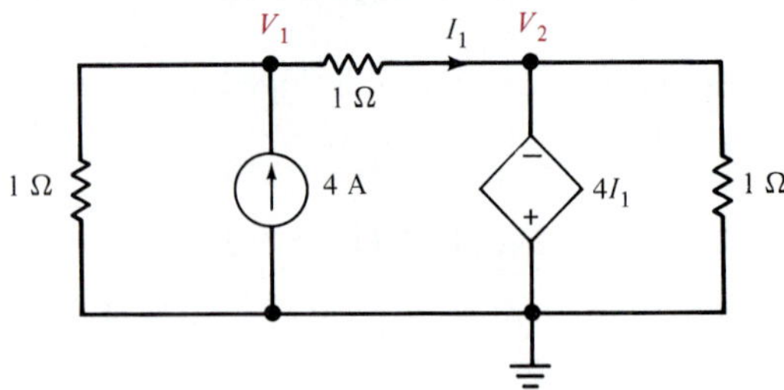

Figure D3.7

Ans: $V_1 = 6$ V, $V_2 = 8$ V.

D3.8. Use nodal analysis to find the voltage across the dependent current source for the network in Fig. D3.8.

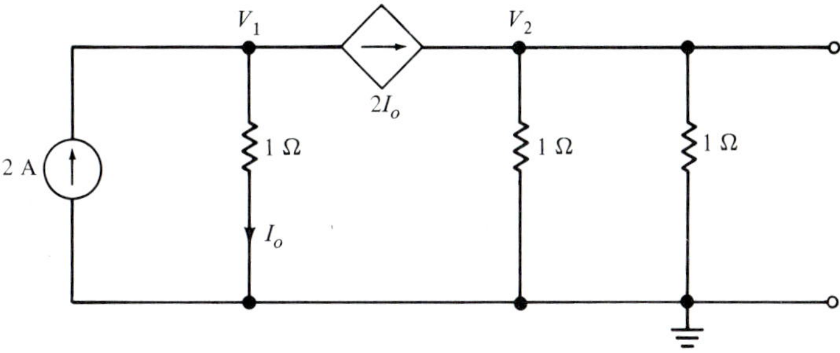

Figure D3.8

Ans: $V_1 - V_2 = 0$ V.

3.2 Loop Analysis

In a nodal analysis the unknown parameters are the node voltages, and KCL is employed to determine them. In contrast to this approach, a loop analysis uses KVL to determine currents in the circuit. Once the currents are known, Ohm's law can be used to calculate the voltages. Recall that, in Chapter 2, we found that a single equation was sufficient to determine the current in a circuit containing a single loop. If the circuit contains N independent loops, we will show that N independent simultaneous equations will be required to describe the network. For now we will assume that the circuits are *planar,* which simply means that we can draw the circuit on a sheet of paper in a way such that no conductor crosses another conductor.

To begin our analysis, consider the circuit shown in Fig. 3.12. Let us also identify two loops, *A-B-E-F-A* and *B-C-D-E-B*. We now define a new set of current variables, called *loop currents,* which can be used to find the physical currents in the circuit. Let us assume that current i_1 flows in the first loop and that current i_2 flows in the second loop. Then the branch current flowing from B to E through R_3 is $i_1 - i_2$. The directions of the currents

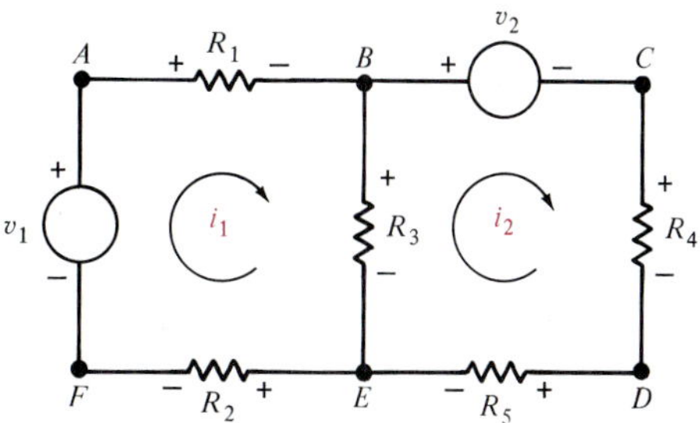

Figure 3.12 Two-loop circuit.

have been assumed. As was the case in the nodal analysis, if the actual currents are not in the direction indicated, the values calculated will be negative.

Applying KVL to the first loop yields

$$-i_1R_1 - (i_1 - i_2)R_3 - i_1R_2 + v_1 = 0$$

or

$$i_1(R_1 + R_2 + R_3) - i_2R_3 = v_1$$

KVL applied to loop 2 yields

$$-v_2 - i_2R_4 - i_2R_5 + (i_1 - i_2)R_3 = 0$$

or

$$-i_1R_3 + i_2(R_3 + R_4 + R_5) = -v_2$$

Therefore, the two simultaneous equations required to solve this two-loop circuit are

$$\begin{aligned} i_1(R_1 + R_2 + R_3) - i_2(R_3) &= v_1 \\ -i_1(R_3) + i_2(R_3 + R_4 + R_5) &= -v_2 \end{aligned} \tag{3.18}$$

or in matrix form,

$$\begin{bmatrix} R_1 + R_2 + R_3 & -R_3 \\ -R_3 & R_3 + R_4 + R_5 \end{bmatrix} \begin{bmatrix} i_1 \\ i_2 \end{bmatrix} = \begin{bmatrix} v_1 \\ -v_2 \end{bmatrix}$$

At this point it is important to define what is called a *mesh*. A *mesh* is nothing more than a special kind of loop in that it does not contain any loops within it. Therefore, as we traverse the path of a mesh, we do not encircle any circuit elements. For example, the network in Fig. 3.12 contains two meshes defined by the paths *A-B-E-F-A* and *B-C-D-E-B*. The path *A-B-C-D-E-F-A* is a loop but it is not a mesh. Since the majority of our analysis in this section will involve writing KVL equations for meshes, we will refer to the currents as mesh currents and the analysis as a *mesh analysis*.

EXAMPLE 3.7

If the circuit in Fig. 3.12 has the following parameters, determine the mesh currents.

$R_1 = 2\ \Omega$ $\quad R_3 = 2\ \Omega$ $\quad R_5 = 2\ \Omega$ $\quad v_2 = V_2 = 36$ V

$R_2 = 4\ \Omega$ $\quad R_4 = 5\ \Omega$ $\quad v_1 = V_1 = 24$ V

The equations for the circuit are

$$\begin{aligned} 8I_1 - 2I_2 &= 24 \\ -2I_1 + 9I_2 &= -36 \end{aligned}$$

or

$$\begin{bmatrix} 8 & -2 \\ -2 & 9 \end{bmatrix} \begin{bmatrix} I_1 \\ I_2 \end{bmatrix} = \begin{bmatrix} 24 \\ -36 \end{bmatrix}$$

Solving for the currents yields

$$\begin{bmatrix} I_1 \\ I_2 \end{bmatrix} = \frac{1}{68}\begin{bmatrix} 9 & 2 \\ 2 & 8 \end{bmatrix}\begin{bmatrix} 24 \\ -36 \end{bmatrix} = \begin{bmatrix} \dfrac{144}{68} \\ -\dfrac{240}{68} \end{bmatrix} = \begin{bmatrix} 2.1176 \\ -3.5294 \end{bmatrix}$$

Note that since I_2 is negative, it is actually flowing in the direction opposite to that which was assumed.

We can check the validity of our answers via KVL. In mesh 1, KVL requires that

$$-I_1R_1 - (I_1 - I_2)R_3 - I_1R_2 + V_1$$
$$= (2.1176)(2) - (2.1176 + 3.5294)(2) - (2.1176)(4) + 24 = 0$$

In mesh 2, KVL requires that

$$-V_2 - I_2R_4 - I_2R_5 + (I_1 - I_2)R_3$$
$$= -36 + (3.5294)(5) + (3.5294)(2) + (2.1176 + 3.5294)(2) = 0$$

The reader is encouraged to check KVL around the entire outside loop *A-B-C-D-E-F-A*.

Once the mesh currents are known, the voltages at any point in the circuit can easily be determined. For example, the voltage at node A with respect to node $E = V_1 - I_1R_2 = I_1R_1 + (I_1 - I_2)R_3$. The voltage at node C with respect to node $E = -V_2 + (I_1 - I_2)R_3 = I_2R_4 + I_2R_5$. The voltage at any point in the circuit with respect to any other point can be obtained in a similar manner. ■

Let us now consider the circuit shown in Fig. 3.13. KVL applied to the closed path *A-B-E-D-C-A* yields

$$+v_1 - i_1R_2 + v_4 - R_4(i_1 - i_3) - R_3(i_1 - i_2) - i_1R_1 = 0$$

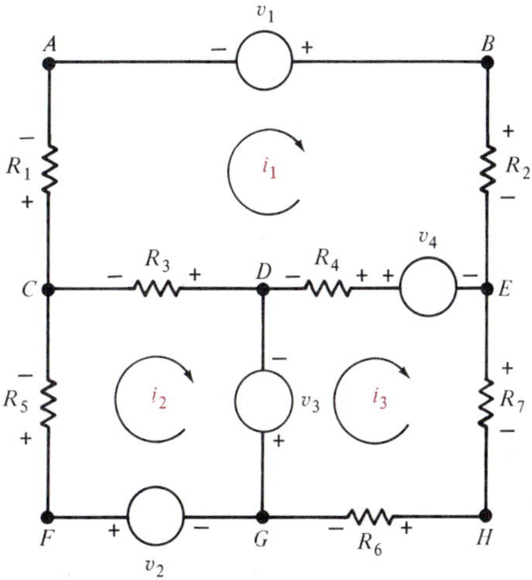

Figure 3.13 Three-loop circuit.

or

$$i_1(R_1 + R_2 + R_3 + R_4) - i_2(R_3) - i_3(R_4) = v_1 + v_4$$

KVL for mesh 2 around the path *C-D-G-F-C* is

$$+R_3(i_1 - i_2) + v_3 + v_2 - i_2R_5 = 0$$

or

$$-i_1(R_3) + i_2(R_3 + R_5) = v_2 + v_3$$

KVL for mesh 3 is

$$+R_4(i_1 - i_3) - v_4 - R_7i_3 - R_6i_3 - v_3 = 0$$

or

$$-i_1(R_4) + i_3(R_4 + R_6 + R_7) = -v_3 - v_4$$

Therefore, the three equations necessary to solve for the three unknown mesh currents are

$$\begin{aligned} i_1(R_1 + R_2 + R_3 + R_4) - i_2(R_3) \quad - i_3(R_4) &= v_1 + v_4 \\ -i_1(R_3) \quad + i_2(R_3 + R_5) &= v_2 + v_3 \\ -i_1(R_4) \quad + i_3(R_4 + R_6 + R_7) &= -v_3 - v_4 \end{aligned} \tag{3.19}$$

or in matrix form,

$$\begin{bmatrix} R_1 + R_2 + R_3 + R_4 & -R_3 & -R_4 \\ -R_3 & R_3 + R_5 & 0 \\ -R_4 & 0 & R_4 + R_6 + R_7 \end{bmatrix} \begin{bmatrix} i_1 \\ i_2 \\ i_3 \end{bmatrix} = \begin{bmatrix} v_1 + v_4 \\ v_2 + v_3 \\ -v_3 - v_4 \end{bmatrix}$$

EXAMPLE 3.8

Given the following parameters for the network shown in Fig. 3.13, determine the mesh currents.

$$\begin{array}{llll} R_1 = 2\ \Omega & R_3 = 3\ \Omega & R_5 = 4\ \Omega & R_7 = 1\ \Omega \\ R_2 = 1\ \Omega & R_4 = 1\ \Omega & R_6 = 2\ \Omega & v_1 = V_1 = 12\text{ V} \\ & & v_2 = V_2 = 6\text{ V} & v_4 = V_4 = 24\text{ V} \\ & & v_3 = V_3 = 18\text{ V} & \end{array}$$

The matrix equation above becomes

$$\begin{bmatrix} 7 & -3 & -1 \\ -3 & 7 & 0 \\ -1 & 0 & 4 \end{bmatrix} \begin{bmatrix} I_1 \\ I_2 \\ I_3 \end{bmatrix} = \begin{bmatrix} 36 \\ 24 \\ -42 \end{bmatrix}$$

The cofactors of the matrix are $A_{11} = 28, A_{12} = 12, A_{13} = 7, A_{21} = 12, A_{22} = 27, A_{23} = 3$, $A_{31} = 7$, $A_{32} = 3$, and $A_{33} = 40$. The determinant of $\mathbf{A}$, $|\mathbf{A}| = 153$. Therefore,

$$\begin{bmatrix} I_1 \\ I_2 \\ I_3 \end{bmatrix} = \frac{1}{153} \begin{bmatrix} 28 & 12 & 7 \\ 12 & 27 & 3 \\ 7 & 3 & 40 \end{bmatrix} \begin{bmatrix} 36 \\ 24 \\ -42 \end{bmatrix} = \begin{bmatrix} 6.5490 \\ 6.2353 \\ -8.8627 \end{bmatrix}$$

Using these data and summing the voltages around mesh 1 yields

$$\begin{aligned} &+V_1 - I_1R_2 + V_4 - R_4(I_1 - I_3) - R_3(I_1 - I_2) - I_1R_1 \\ &\quad = 12 - (6.5490)(1) + 24 - 3(6.5490 - 6.2353) - 1(6.5490 + 8.8627) \\ &\qquad\qquad - 2(6.5490) = 0 \end{aligned}$$

As we saw in Example 3.7, once the mesh currents are known, we can determine all the branch currents and the voltage at any point with respect to any other point. ■

Once again we are compelled to note the symmetrical form of the mesh equations that describe the previous networks. Note also that the $\mathbf{A}$ matrix for each circuit is symmetrical. Since this symmetry is generally exhibited by networks containing resistors and independent voltage sources, we can learn to write the mesh equations by inspection. In the first equation of (3.18) the coefficient of i_1 is the sum of the resistances through which mesh current 1 flows, and the coefficient of i_2 is the negative of the sum of the resistances common to mesh current 1 and mesh current 2. The right-hand side of the equation is the sum of the voltage sources in mesh 1. The sign of the voltage source is positive if it aids the assumed direction of the current flow and negative if it opposes the assumed flow. The first equation is KVL for mesh 1. In the second equation in (3.18), the coefficient of i_2 is the sum of all the resistances in mesh 2, the coefficient of i_1 is the negative of the sum of the resistances common to mesh 1 and mesh 2, and the right-hand side of the equation is the sum of the voltage sources in mesh 2. The equations in (3.19) are obtained in a similar manner. In general, if we assume all of the mesh currents to be in the same direction, then if KVL is applied to mesh j with mesh current i_j, the coefficient of i_j is the sum of the resistances in mesh j and the coefficients of the other mesh currents (e.g., i_{j-1}, i_{j+1}) are the negatives of the resistances common to these meshes and mesh j. The right-hand side of the equation is equal to the sum of the voltage sources in mesh j. These voltage sources have a positive sign if they aid the current flow i_j and a negative sign if they oppose it.

EXAMPLE 3.9

Let us apply this inspection technique, which we have just presented, to the circuit shown in Fig. 3.14.

By inspection, KVL yields

$$\begin{aligned} (R_1 + R_2)i_1 - R_2i_2 &= v_1 + v_3 \\ -R_2i_1 + (R_2 + R_3 + R_4 + R_5)i_2 - R_5i_4 &= -v_2 \\ (R_6 + R_7 + R_8)i_3 - R_8i_4 &= -v_3 - v_4 \\ -R_5i_2 - R_8i_3 + (R_5 + R_8 + R_9)i_4 &= v_4 - v_5 \end{aligned}$$

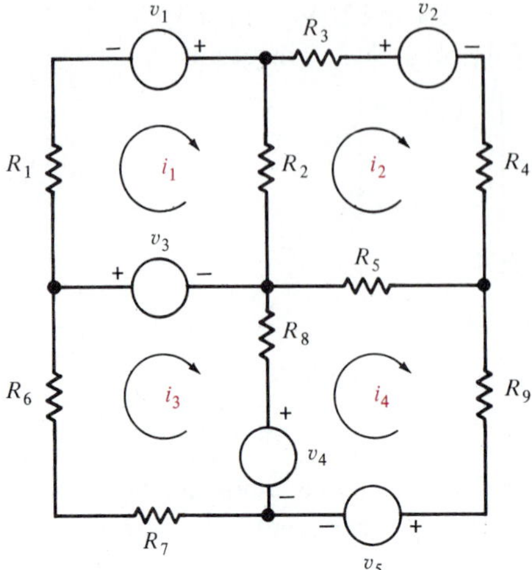

Figure 3.14 Example four-loop circuit.

or in matrix form,

$$\begin{bmatrix} R_1 + R_2 & -R_2 & 0 & 0 \\ -R_2 & R_2 + R_3 + R_4 + R_5 & 0 & -R_5 \\ 0 & 0 & R_6 + R_7 + R_8 & -R_8 \\ 0 & -R_5 & -R_8 & R_5 + R_8 + R_9 \end{bmatrix} \begin{bmatrix} i_1 \\ i_2 \\ i_3 \\ i_4 \end{bmatrix}$$

$$= \begin{bmatrix} v_1 + v_3 \\ -v_2 \\ -v_3 - v_4 \\ v_4 - v_5 \end{bmatrix}$$

■

DRILL EXERCISE

D3.9. Find V_o in the network in Fig. D3.9 using mesh analysis.

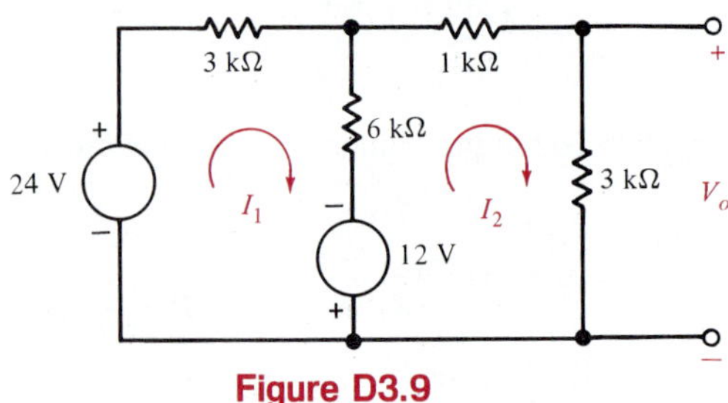

Figure D3.9

Ans: $V_o = 6$ V.

D3.10. Use mesh analysis to compute V_o in the network in Fig. D3.10.

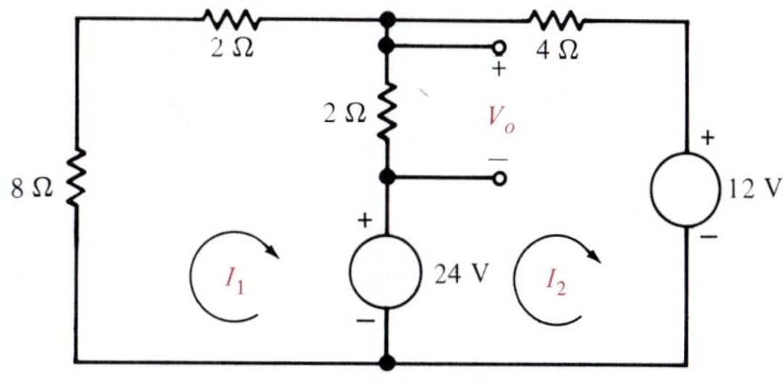

Figure D3.10

Ans: $V_o = -6.35$ V.

Mesh Equations for Circuits Containing Independent Current Sources

Consider the circuit shown in Fig. 3.15. It appears that there are two unknown currents i_1 and i_2. However, since i_1 is assumed to go through the branch containing i_A, i_1 must be equal to i_A. Therefore, since $i_1 = i_A$, the only unknown current is i_2. Hence KVL for the second mesh is

$$-i_1R_2 + i_2(R_2 + R_3 + R_4) = v_1 + v_2$$

or since $i_1 = i_A$,

$$i_2(R_2 + R_3 + R_4) = v_1 + v_2 + i_AR_2$$

Thus i_2 can be computed immediately. Once the mesh currents are known, all voltages in the network can be calculated. For example,

$$v_C = -v_2 + i_2R_4$$

$$v_B = v_C + i_2R_3 = -v_2 + i_2R_4 + i_2R_3$$

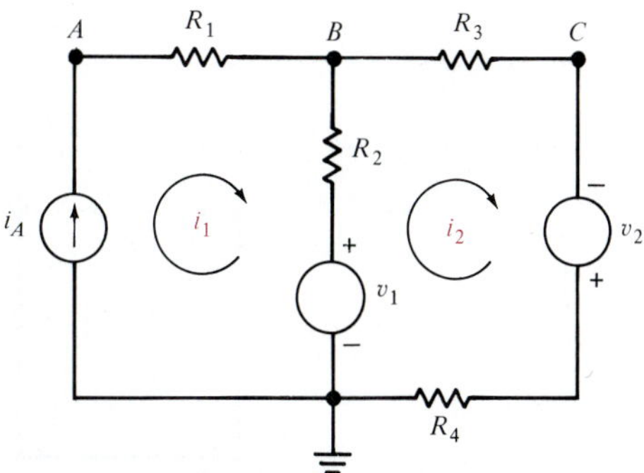

Figure 3.15 Circuit containing an independent current source.

or

$$v_B = v_1 + (i_1 - i_2)R_2$$

and

$$v_A = v_B + i_A R_1$$

Clearly, the presence of a current source has simplified the mesh equations. As a general rule, KVL equations are simpler whenever current sources are present.

Consider now the more complicated network in Fig. 3.16a. Suppose that we wish to calculate V_o using mesh equations. In the previous development in which the networks consisted only of voltage sources and resistors, the branch voltages were either known source values or they could be expressed as the product of the branch current and the branch resistance. However, the branch voltage that we have labeled V_A cannot be expressed directly in either of these two forms. The mesh equations for the network are

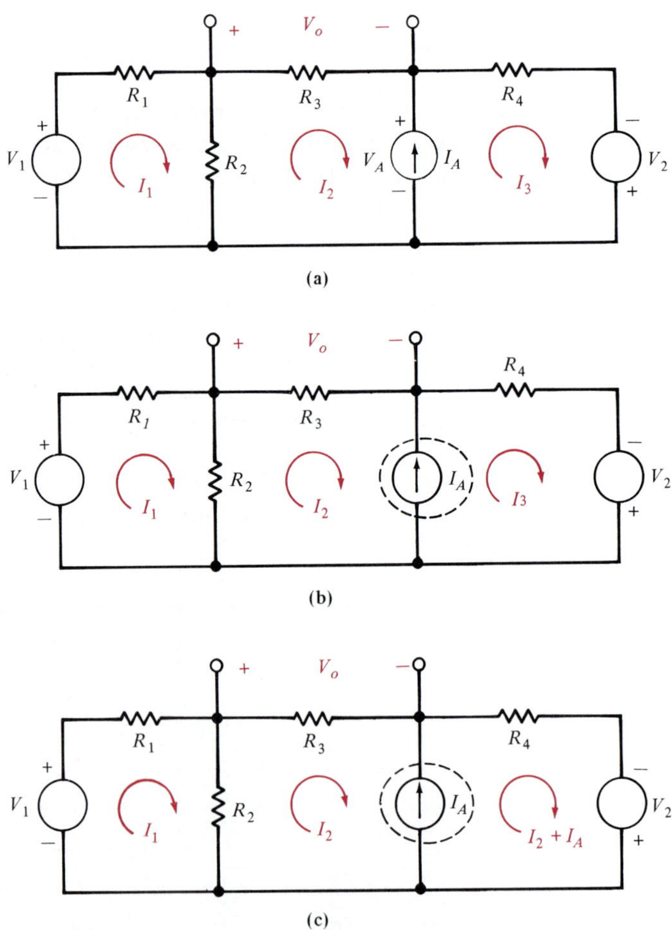

Figure 3.16 Example network containing a current source.

$$V_1 - I_1R_1 - (I_1 - I_2)R_2 = 0 \tag{3.20}$$

$$-(I_2 - I_1)R_2 - R_3I_2 - V_A = 0 \tag{3.21}$$

$$V_A - I_3R_4 + V_2 = 0 \tag{3.22}$$

Hence we have three linearly independent KVL equations in the four unknowns I_1, I_2, I_3, and V_A. Note however, since the two mesh currents I_2 and I_3 go through the current source I_A, the mesh currents are *constrained* by the requirement

$$I_3 - I_2 = I_A \tag{3.23}$$

Equations (3.20) to (3.23) form a set of four linearly independent equations in the four unknowns. Eliminating V_A in Eqs. (3.21) and (3.22) yields the equation

$$-(I_2 - I_1)R_2 - R_3I_2 - I_3R_4 + V_2 = 0 \tag{3.24}$$

Equations (3.20), (3.23), and (3.24) can now be written as

$$\begin{aligned} I_1(R_1 + R_2) - I_2R_2 &= V_1 \\ -I_2 + I_3 &= I_A \\ -I_1R_2 + I_2(R_2 + R_3) + I_3R_4 &= V_2 \end{aligned} \tag{3.25}$$

which are three linearly independent equations in the three unknown mesh currents I_1, I_2, and I_3. If we solve these equations for I_2, we can determine $V_o = R_3I_2$. Note that this procedure required four equations and four unknowns. One of the variables was immediately eliminated, yielding three equations involving the three unknown mesh currents I_1, I_2, and I_3. These equations can be written directly, however, if we are careful in choosing the path around which we apply KVL. To illustrate this fact, let us rework the problem using a loop in place of two meshes. With reference to Fig. 3.16b, we have enclosed the current source within a dashed surface. Within the surface, the currents I_2 and I_3 are *constrained* by the equation

$$I_3 - I_2 = I_A \tag{3.26}$$

This is one of the three linearly independent equations required to solve for the currents. The KVL equation for the first mesh is another equation,

$$V_1 - I_1R_1 - (I_1 - I_2)R_2 = 0 \tag{3.27}$$

The third and final equation is obtained by applying KVL to another loop which does not contain the current source I_A. Such a loop is that containing the elements R_2, R_3, R_4, and V_2. The equation for this loop is

$$-(I_2 - I_1)R_2 - I_2R_3 - I_3R_4 + V_2 = 0 \tag{3.28}$$

Equations (3.26), (3.27), and (3.28) can be written as

$$\begin{aligned} I_1(R_1 + R_2) - I_2R_2 &= V_1 \\ -I_2 + I_3 &= I_A \\ -I_1R_2 + I_2(R_2 + R_3) + I_3R_4 &= V_2 \end{aligned} \tag{3.29}$$

These equations are identical to those in Eq. (3.25).

An additional simplification can be made, however, based on the constraint that is placed upon the mesh currents I_2 and I_3 by the presence of the current source. Once again, consider the network as it is labeled in Fig. 3.16c. Note that the constraint equation has already been employed in that the mesh current for I_3 is labeled as $I_2 + I_A$. If we now write KVL equations for the first mesh and the loop containing R_2, R_3, R_4, and V_2, we obtain directly the two linearly independent equations

$$\begin{aligned} V_1 - I_1R_1 - (I_1 - I_2)R_2 &= 0 \\ -(I_2 - I_1)R_2 - I_2R_3 - (I_2 + I_A)R_4 + V_2 &= 0 \end{aligned} \tag{3.30}$$

which are, of course, equivalent to those in Eq. (3.29). Finally, V_o can be completed using the relationship $V_o = R_3I_2$.

EXAMPLE 3.10

Let us determine all the branch currents in the network in Fig. 3.17a.

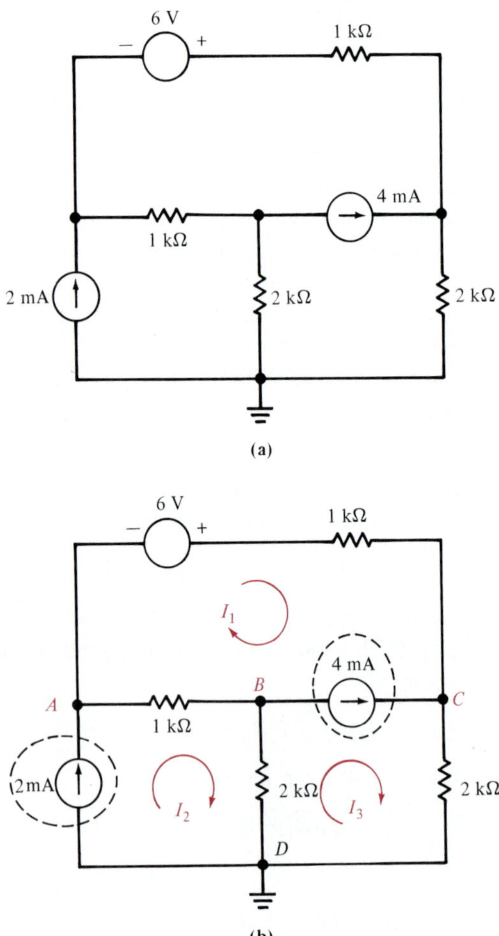

Figure 3.17 Network containing multiple independent current sources.

The network is redrawn in Fig. 3.17b with the mesh currents labeled. Within the dashed surfaces the mesh currents are constrained by the equations

$$I_2 = 2 \text{ mA}$$

$$I_3 - I_1 = 4 \text{ mA}$$

Since the network contains three meshes, three linearly independent equations are required to find the mesh currents. The equations above are two of these equations. The final equation is a KVL equation around a loop that does not traverse a current source. Such a loop is the path *A-C-D-B-A*. The KVL equation for this loop is

$$+6 - (1)I_1 - (2)I_3 - (2)(I_3 - I_2) - (1)(I_1 - I_2) = 0$$

Solving these equations for I_1 yields

$$I_1 = -\tfrac{2}{3} \text{ mA}$$

Then

$$I_3 = I_1 + 4 \text{ mA} = \tfrac{10}{3} \text{ mA}$$

Hence, in addition to the branch currents that are equal to mesh currents, $I_{AB} = \tfrac{8}{3}$ mA and $I_{DB} = \tfrac{4}{3}$ mA. ■

EXAMPLE 3.11

We wish to determine the current I_o in the network in Fig. 3.18a.

The circuit is redrawn in Fig. 3.18b with the mesh currents appropriately labeled. Note that I_2 is constrained to be -2 and I_4 is constrained to be $4 + I_3$. Because of the constraints, only two KVL equations are required to determine I_1 and I_3, and therefore, $I_o = I_3 - I_1$. The two equations can be derived from the mesh *A-B-D-C-A* and the loop *C-D-E-F-C*. The equations are

$$+12 - 1I_1 - 2(I_1 + 2) - 1(I_1 - I_3) = 0$$

$$-2I_3 - 1(I_3 - I_1) - 1(4 + I_3 + 2) - 2(4 + I_3) = 0$$

These equations can be placed in matrix form as

$$\begin{bmatrix} 4 & -1 \\ -1 & 6 \end{bmatrix} \begin{bmatrix} I_1 \\ I_3 \end{bmatrix} = \begin{bmatrix} 8 \\ -14 \end{bmatrix}$$

Solving for the currents, we obtain

$$\begin{bmatrix} I_1 \\ I_3 \end{bmatrix} = \frac{1}{23} \begin{bmatrix} 6 & 1 \\ 1 & 4 \end{bmatrix} \begin{bmatrix} 8 \\ -14 \end{bmatrix} = \begin{bmatrix} \dfrac{34}{23} \\ \dfrac{-48}{23} \end{bmatrix}$$

Therefore,

$$I_o = I_3 - I_1 = \frac{-82}{23} \text{ A}$$

■

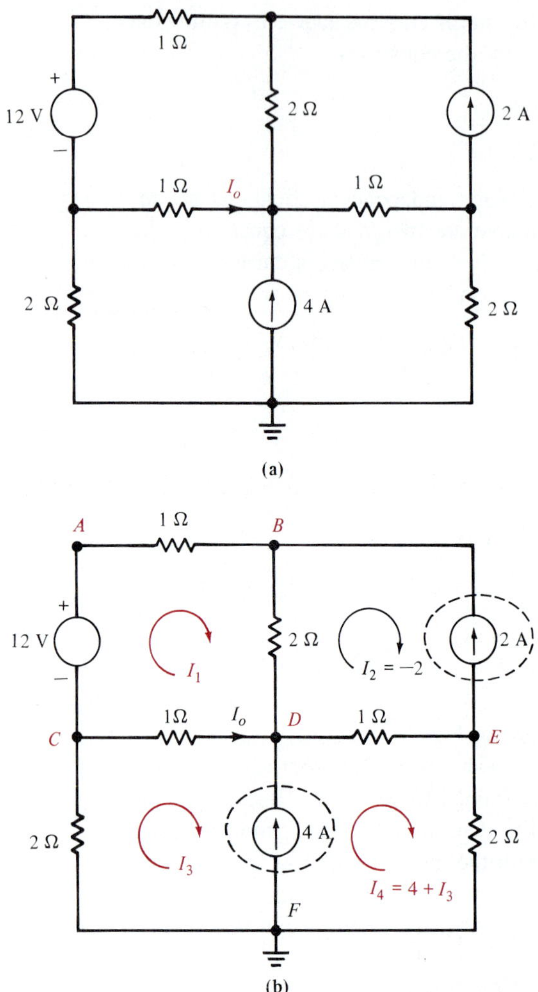

Figure 3.18 Four-mesh network with two independent current sources.

DRILL EXERCISE

D3.11. Use mesh analysis to find V_o in the network in Fig. D3.11.

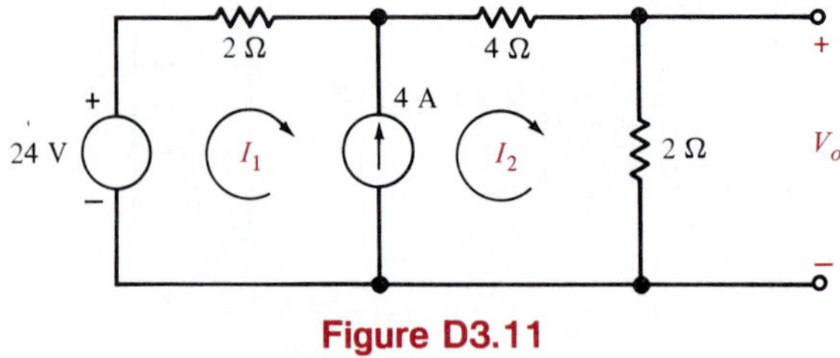

Figure D3.11

Ans: $V_o = 8$ V.

D3.12. Find V_o in the network in Fig. D3.12 using mesh analysis.

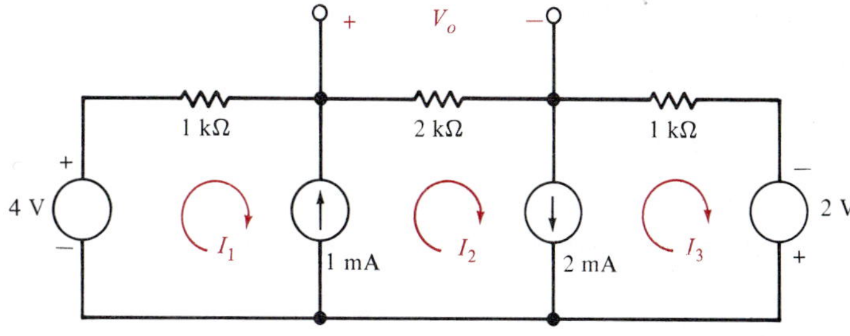

Figure D3.12

Ans: $V_o = 4.5$ V.

Mesh Equations for Circuits Containing Dependent Sources

Although we deal with circuits containing dependent sources in the same manner as the circuits described above, we must recall that the dependent source may destroy the symmetry of the resulting equations. The following simple example will illustrate the application of mesh equations in circuits with dependent sources.

EXAMPLE 3.12

We want to find the output voltage in the circuit shown in Fig. 3.19a.

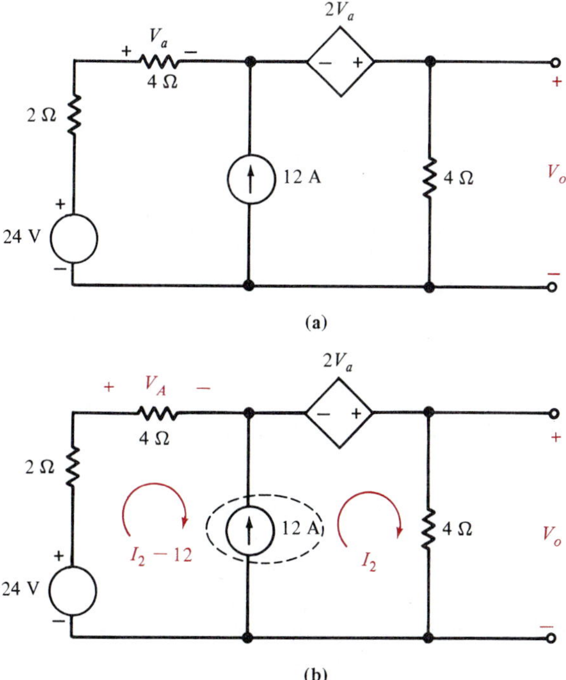

Figure 3.19 Circuit containing a dependent source.

The network is redrawn in Fig. 3.19b to illustrate the constraint on the mesh currents. The KVL equation for the outer loop is

$$+24 - 2(I_2 - 12) - V_a + 2V_a - 4I_2 = 0$$

where

$$V_a = 4(I_2 - 12)$$

Solving these equations, we find that $I_2 = 0$ and therefore $V_o = 0$. ■

EXAMPLE 3.13

Let us determine the output voltage in the network in Fig. 3.20a.

The constraint on the mesh currents is shown in Fig. 3.20b. The KVL equations for the network are

$$+12 - 2I_1 - 1(I_1 + 3I_o) - 2(I_1 + 3I_o - I_o) - 6 = 0$$
$$+6 - 2(I_o - I_1 - 3I_o) - 1I_o - 1I_o = 0$$

The equations in matrix form are

$$\begin{bmatrix} 5 & 7 \\ -2 & -2 \end{bmatrix} \begin{bmatrix} I_1 \\ I_o \end{bmatrix} = \begin{bmatrix} 6 \\ 6 \end{bmatrix}$$

Solving for the currents, we obtain

$$\begin{bmatrix} I_1 \\ I_o \end{bmatrix} = \frac{1}{4} \begin{bmatrix} -2 & -7 \\ 2 & 5 \end{bmatrix} \begin{bmatrix} 6 \\ 6 \end{bmatrix} = \begin{bmatrix} -\frac{27}{2} \\ \frac{21}{2} \end{bmatrix}$$

Therefore, $I_o = \frac{21}{2}$ A and $V_o = \frac{21}{2}$ V. ■

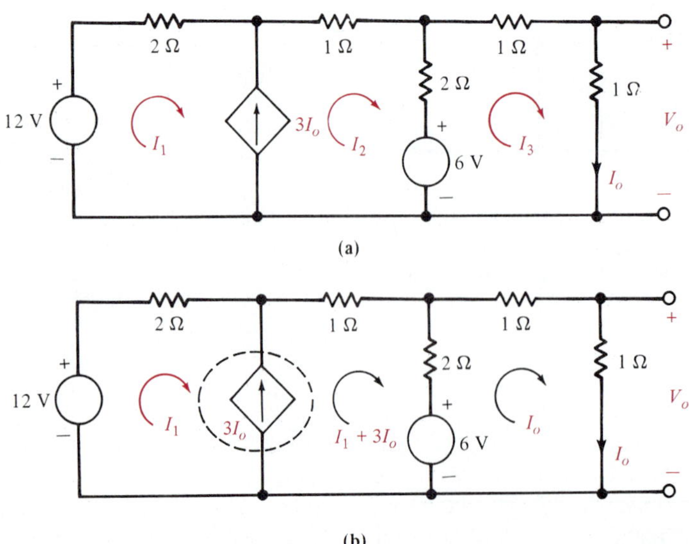

Figure 3.20 Network containing a dependent current source.

DRILL EXERCISE

D3.13. Use mesh analysis to find V_o in the network in Fig. D3.13.

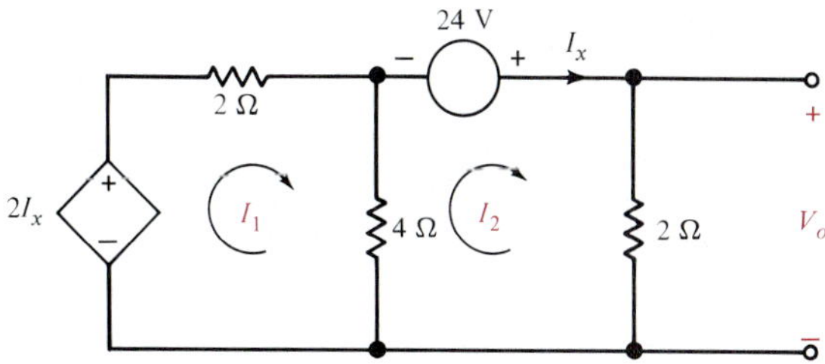

Figure D3.13

Ans: $V_o = 24$ V.

D3.14. Find V_o in the network in Fig. D3.14 using mesh analysis.

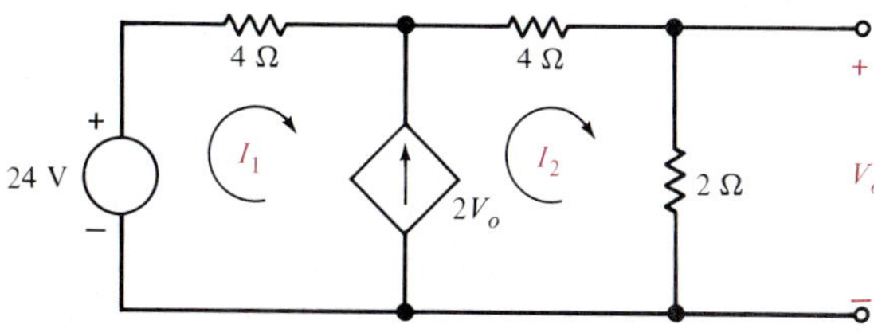

Figure D3.14

Ans: $V_o = -8$ V.

3.3 Circuit Equations via Network Topology

We now generalize the methods of nodal and loop analyses that we have presented. The vehicle that we will employ to accomplish this is *topology*. For our purposes topology refers to the properties that relate to the geometry of the circuit. These properties remain unchanged even if the circuit is bent into any other shape provided that no parts are cut and no new connections are made.

Basic Definitions

To provide motivation for the concepts to be discussed next, consider the network shown in Fig. 3.21. Suppose that we are required to find all the currents and voltages in this circuit. This circuit seems to be much more complicated than those we considered earlier. It is more complicated because it is what is called a *nonplanar* circuit. A *planar* circuit is one that can be drawn on a plane surface with no crossovers; that is, no branch passes over any other branch. And, of course, a nonplanar circuit is one that is not planar.

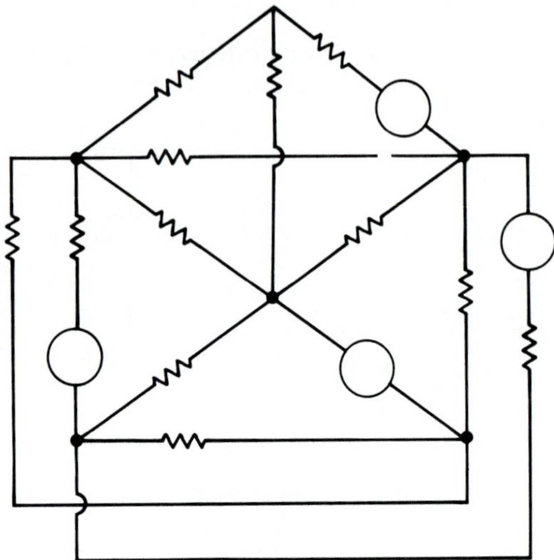

Figure 3.21 Example of a nonplanar circuit.

Let us now present some topological concepts to aid us in our discussion. Since the geometrical properties of a circuit are independent of the circuit elements that are contained in each branch, we will simply represent each network branch by a line segment. Such a drawing of the circuit is called a *graph*.

EXAMPLE 3.14

The graph associated with the circuit shown in Fig. 3.22a is given in Fig. 3.22b

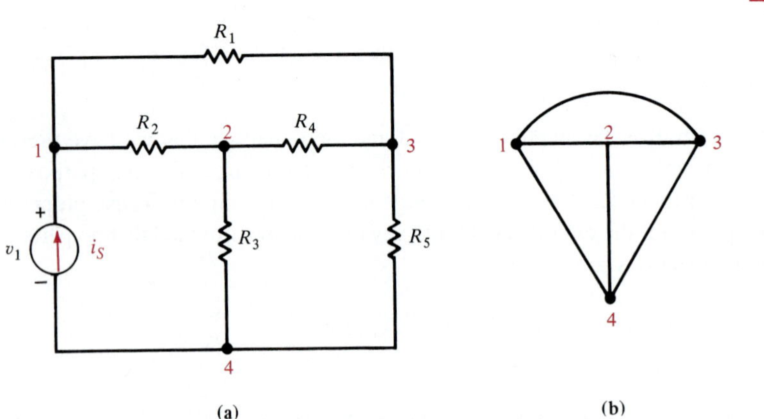

Figure 3.22 Circuit and its associated graph.

As can be seen from the example, the graph consists of a number of interconnected nodes. If a path exists from any node to every other node, the graph is said to be *connected*.

A *tree* of the graph is defined as a set of branches that connects every node to every other node via some path without forming a closed path. In general, there exists a number of different trees for any given circuit.

EXAMPLE 3.15

Three possible trees for the graph shown in Fig. 3.22b are given in Fig. 3.23. ■

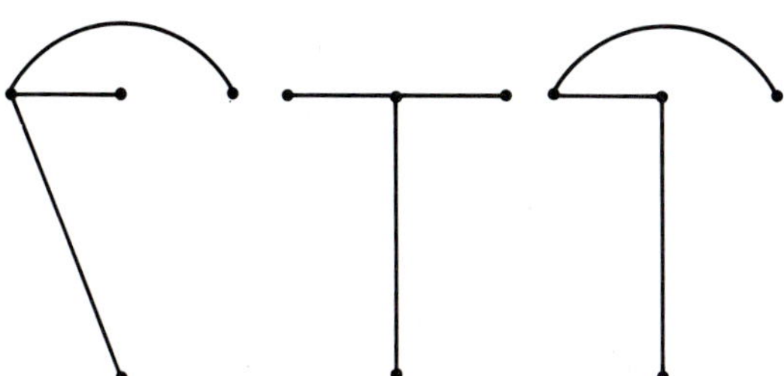

Figure 3.23 Trees for the graph in Fig. 3.22b.

If a graph for a network is known and a particular tree is specified, those branches of the graph that are not part of the tree form what is called the *cotree*. The cotree consists of what we call *links*.

EXAMPLE 3.16

The links that belong to the cotrees that correspond to the trees shown in Fig. 3.23 are shown dashed in Fig. 3.24. ■

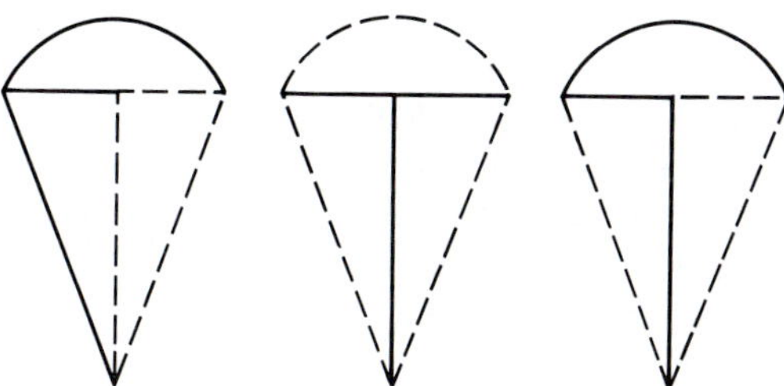

Figure 3.24 Cotrees consisting of links for the trees shown in Fig. 3.23.

Note that it is impossible to specify categorically a particular branch of a graph as a link, since it may be a link for one choice of tree and not for another.

Finally, we wish to define what is called a *cut set*. A cut set is simply a minimum set of branches which, when cut, will divide the graph into two separate parts. Hence it is impossible to go from a node in one part of the graph to a node in another part without passing through a branch of the cut set.

EXAMPLE 3.17

Two cut sets for the graph shown in Fig. 3.22b are shown in Fig. 3.25. Note that in one case node 4 is separated from nodes 1, 2, and 3; in the other case nodes 1 and 2 are separated from nodes 3 and 4. ■

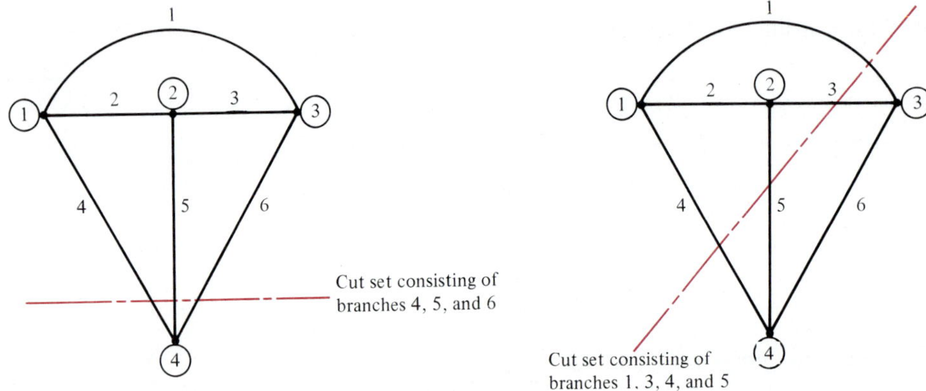

Figure 3.25 Two cut sets for the graph in Fig. 3.22b.

Suppose that we have a graph that represents a circuit, and let us define the following parameters:

$$L = \text{number of links}$$

$$B = \text{number of branches in the graph}$$

$$N = \text{number of nodes in the graph}$$

Now suppose that we remove all the branches from the graph so that only nodes remain. In order to construct a tree, we begin by placing one branch between two nodes. We continue adding branches to the tree without forming any loops. Each time we add an additional branch we add one more node. Note that the addition of every successive branch connects one node, except for the first branch, which connected two nodes. Hence there is one more node than there are branches on the tree. Therefore, in general, a tree consists of $N - 1$ branches. Since the entire graph contains B branches, the number of links is given by the expression

$$L = B - (N - 1) = B - N + 1 \tag{3.31}$$

EXAMPLE 3.18

A specific tree for the network shown in Fig. 3.21 is illustrated in Fig. 3.26. Note that the graph of the network satisfies the relationship in Eq. (3.31).

$$L = B - N + 1$$

$$8 = 13 - 6 + 1$$

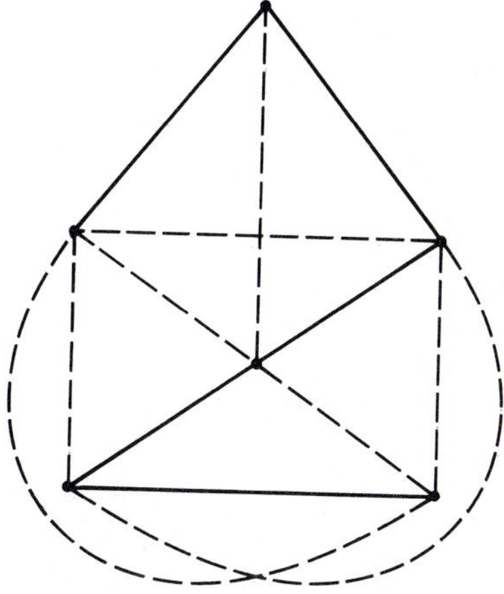

Figure 3.26 Tree and its corresponding links for the network in Fig. 3.21.

DRILL EXERCISE

D3.15. Given the graph for a network in Fig. D3.15a, state whether the tree–cotree combinations in Fig. D3.15b to f form a valid set of tree branches and links.

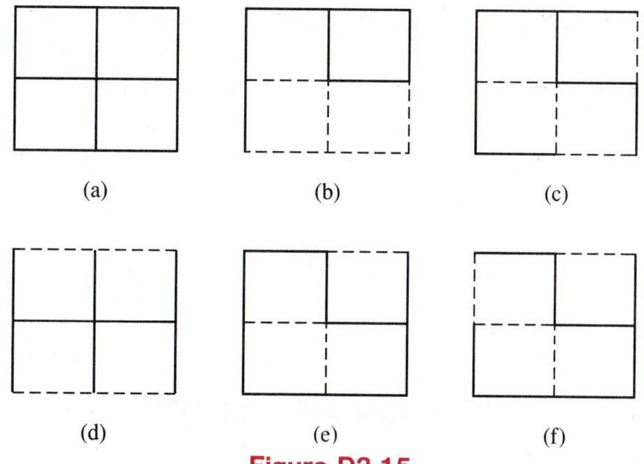

Figure D3.15

Ans: (b) No, (c) yes, (d) yes, (e) no, (f) yes.

D3.16. Compute the proper number of links for the graphs in Fig. D3.15a.

Ans: $L = 4$.

General Nodal Analysis Equations

We have shown that if we had a *linearly independent* set of $N - 1$ nodal equations for an N-node network, we could determine the necessary node voltages. With this in mind, let

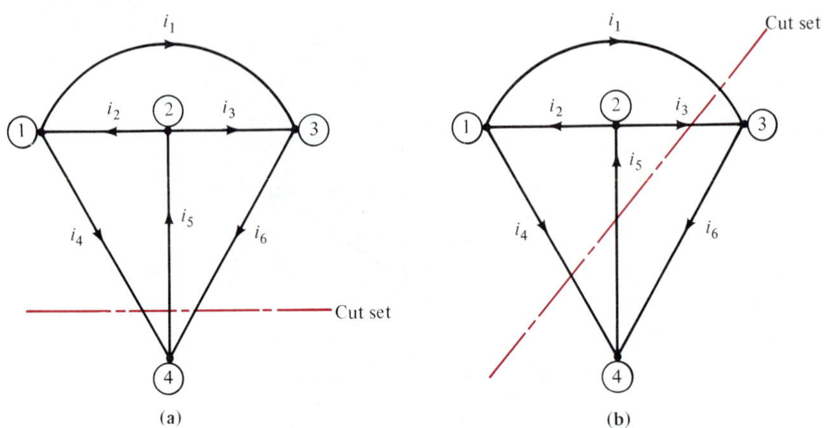

Figure 3.27 Example graph for illustrating cut-set equations.

us redraw the graphs in Fig. 3.25 with assumed directions of currents as shown in Fig. 3.27. KCL for either part of the circuit divided by a cut set is called a *cut-set equation*. Note that each cut-set equation is nothing more than a sum of KCL equations written for all nodes in either part of the circuit. For example, the cut-set equation for Fig. 3.27a is

$$i_4 - i_5 + i_6 = 0$$

which is KCL at node 4, where $i_k(t)$ represents the kth branch. Similarly, the cut-set equation for Fig. 3.27b is

$$i_4 - i_5 + i_3 + i_1 = 0$$

which is the sum of the KCL equations at nodes 3 and 4, or equivalently at nodes 1 and 2:

For node 3: $\quad i_1 + i_3 - i_6 = 0$

For node 4: $\quad i_4 - i_5 + i_6 = 0$

Adding these two equations yields

$$i_4 - i_5 + i_3 + i_1 = 0$$

Because of this relationship between KCL equations and the cut-set equations, we can determine the independent KCL equations for a network by determining the independent cut-set equations. Furthermore, we can prove [1] that the number of independent cut-set equations (and therefore KCL equations) is exactly $N - 1$.

The cut-set equations are derived from a tree that, as we have shown, contains $N - 1$ branches. If any branch of the tree is cut, the tree will be divided into two parts. Each tree branch that is cut, together with the links that connect the two separate parts of the graph, form a cut set. For example, the cut sets for a specific tree for the graph shown in Fig. 3.22 are shown in Fig. 3.28. Note that cut set 1 includes tree branch 5 and links 4 and 6. Cut set 2 includes tree branch 2 and links 4, 3, and 6. Cut set 3 includes tree branch 1 and links 3 and 6. Cut sets containing only a single tree branch are called *fundamental cut sets*, and there are $N - 1$ fundamental sets because there are $N - 1$ tree branches.

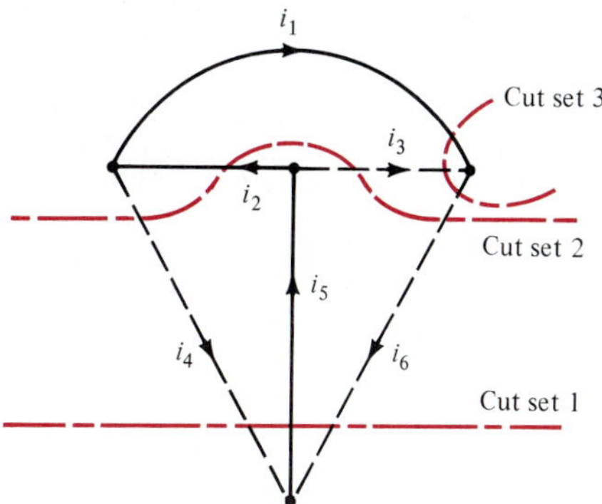

Figure 3.28 Illustration of fundamental cut sets.

Let us now demonstrate the use of topology in determining the nodal equations for a network.

EXAMPLE 3.19

The circuit in Fig. 3.22a is redrawn in Fig. 3.29a with assumed directions for currents. A corresponding tree with fundamental cut sets is shown in Fig. 3.29b. The cut-set equations for cut sets 1, 2, and 3 are

$$i_1 + i_2 - i_S = 0 \tag{3.32}$$

$$i_1 + i_4 - i_5 = 0 \tag{3.33}$$

$$i_S - i_3 - i_5 = 0 \tag{3.34}$$

Subtracting the second equation from the first and adding the third equation yields

$$i_2 - i_4 - i_3 = 0 \tag{3.35}$$

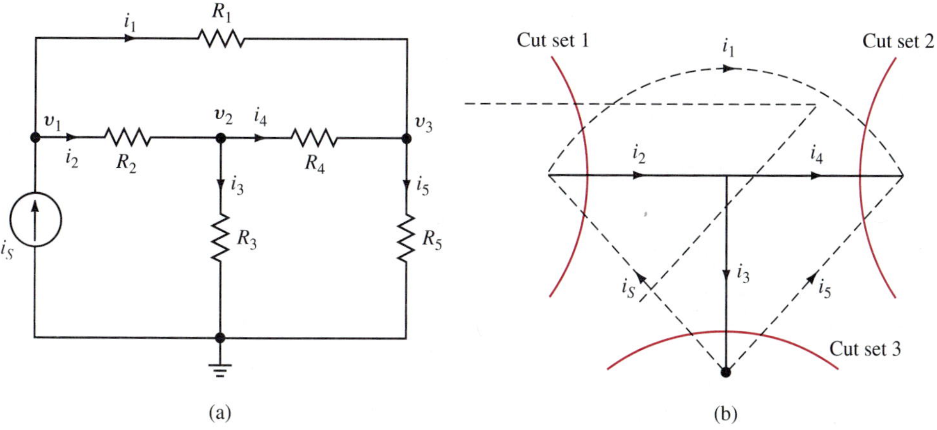

Figure 3.29 Network and graph used in Example 3.19.

Writing the branch currents in terms of the node voltages in equations (3.32), (3.35), and (3.34) yields

$$v_1\left(\frac{1}{R_1}+\frac{1}{R_2}\right)-v_2\left(\frac{1}{R_2}\right)-v_3\left(\frac{1}{R_1}\right)=i_S$$

$$-v_1\left(\frac{1}{R_2}\right)+v_2\left(\frac{1}{R_2}+\frac{1}{R_3}+\frac{1}{R_4}\right)-v_3\left(\frac{1}{R_4}\right)=0$$

$$-v_1\left(\frac{1}{R_1}\right)-v_2\left(\frac{1}{R_4}\right)+v_3\left(\frac{1}{R_1}+\frac{1}{R_4}+\frac{1}{R_5}\right)=0$$

which we recognize as the nodal equations for the network. ■

EXAMPLE 3.20

We wish to derive the nodal equations for the network in Fig. 3.3. A graph for this network is shown in Fig. 3.30.

The cut-set equations for this graph are

$$i_1 - i_A + i_2 - i_3 = 0$$

$$i_3 - i_2 + i_4 + i_B = 0$$

$$i_3 + i_5 + i_B = 0$$

Subtracting the third equation from the second equation and using the resultant equation with the first and third equations yields

$$i_1 + i_2 - i_3 = i_A$$

$$-i_2 + i_4 - i_5 = 0$$

$$i_3 + i_5 = -i_B$$

which are identical to the equations derived earlier.

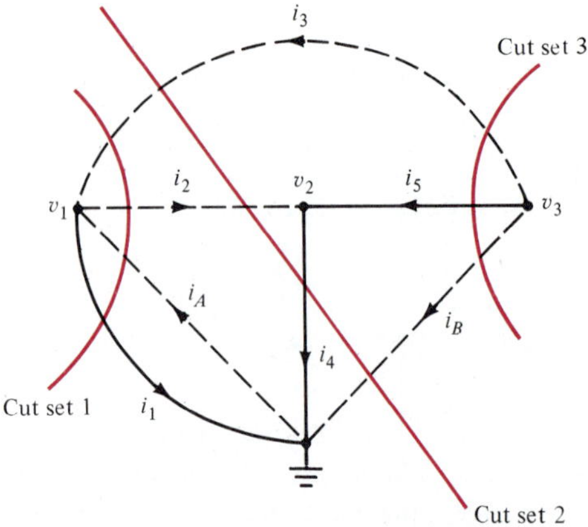

Figure 3.30 Graph and cut sets for the network shown in Fig. 3.3.

Finally, if there are voltage sources present in the network, then, in general, the number of independent KCL equations is $N - 1$—the number of voltage sources.

When using node equations to solve circuits containing voltage sources, the following rules govern the formulation of the equations required.

(1) number of KCL equations = number of topological nodes − 1 (i.e., nodes remaining with current and voltage sources replaced by open and short circuits, respectively, − 1)

(2) number of constraint equations = number of voltage sources

(3) number of unknowns = number of physical nodes − 1

= sum of the numbers of equations (1) and (2) ■

DRILL EXERCISE

D3.17. Given the network in Fig. D3.17a, use the specified graph for the network in Fig. D3.17b to write a proper set of nodal equations for the network.

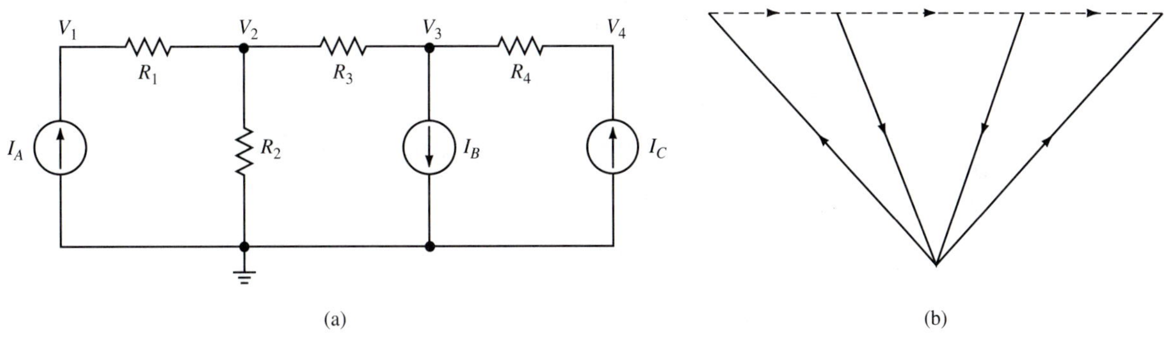

Figure D3.17

Ans:

$$I_A = \frac{V_1 - V_2}{R_1}, \quad \frac{V_1 - V_2}{R_1} = \frac{V_2}{R_2} + \frac{V_2 - V_3}{R_3},$$

$$\frac{V_2 - V_3}{R_3} = \frac{V_3 - V_4}{R_4} + I_B, \quad \frac{V_3 - V_4}{R_4} = -I_C.$$

General Loop Analysis Equations

Consider once again the graph shown in Fig. 3.28. Imagine that we begin only with the tree branches 1, 2, and 5. If we now add one link at a time to the tree, we create a new loop with each link. For example, adding link 3 creates the loop consisting of that link and tree branches 1 and 2. The other links added one at a time create similar loops. Since we have shown that the number of links is equal to $B - N + 1$, we will construct this same

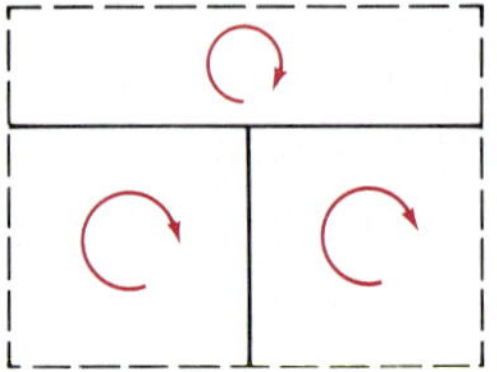

Figure 3.31 Graph illustrating windowpanes for circuit in Fig. 3.29.

number of loops, each one of which contains only one link. The loops constructed in this manner are called the *fundamental loops*. If KVL equations are written for these fundamental loops, these equations will be independent, since each one contains a link voltage that is not present in any other equation. Hence there are at least $B - N + 1$ independent KVL equations, and in fact, we can prove [1] that there are exactly $B - N + 1$ independent KVL equations.

If there are current sources present in the network, then in general, the number of independent KVL equations is $B - N + 1 -$ the number of current sources.

When using loop equations to solve circuits containing current sources, the following rules govern the formulation of the equations required:

(1) number of KVL equations = number of topological loops (i.e., loops remaining with current sources and voltage sources turned off)

(2) number of constraint equations = number of current sources

(3) number of unknowns = number of physical loops

= the sum of the numbers of equations (1) and (2)

Thus far our analysis has been very general. In the special case of planar networks, "windowpanes" can be employed to write the necessary KVL equations. Figure 3.31, which is a redrawn version of Fig. 3.29, illustrates this approach. If all the mesh currents are assumed to be in the same direction and no controlled sources are present, the KVL equations for the windowpanes can be written by inspection, as demonstrated earlier.

EXAMPLE 3.21

Using the topology methods we have just described, we wish to derive the loop equations for the network in Fig. 3.12. A graph of this network is shown in Fig. 3.32a. Branch voltages have been selected and the links together with the appropriate tree branches form the fundamental loops.

The fundamental loop current variables are i_1 and i_2 and the fundamental loop equations are

$$v_1 - v_{R_1} - v_{R_3} - v_{R_2} = 0$$

$$v_{R_3} - v_2 - v_{R_4} - v_{R_5} = 0$$

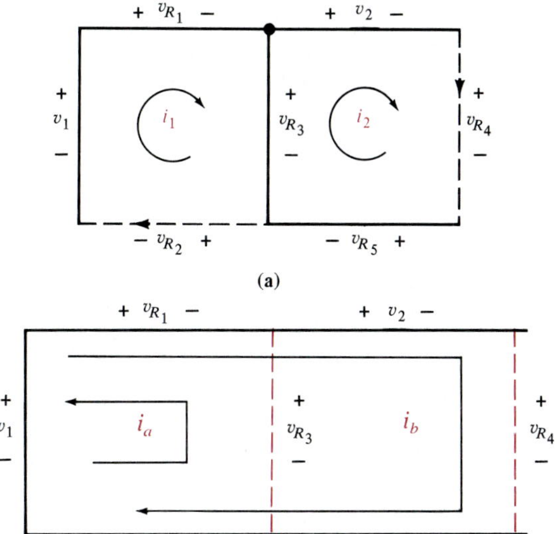

Figure 3.32 Graph for the network in Fig. 3.12 illustrating fundamental loops.

Using the relationship $v = iR$, we obtain

$$i_1R_1 + (i_1 - i_2)R_3 + i_1R_2 = v_1$$
$$-(i_1 - i_2)R_3 + i_2R_4 + i_2R_5 = -v_2$$

or

$$i_1(R_1 + R_2 + R_3) - i_2R_3 = v_1$$
$$-i_1R_3 + i_2(R_3 + R_4 + R_5) = -v_2$$

which are the same as the equations in (3.18).

Suppose, however, that we had selected tree branches and corresponding links for the network as shown in Fig. 3.32b. The fundamental loop currents are i_a and i_b, and the fundamental loop equations in this case are

$$v_1 - v_{R_1} - v_{R_3} - v_{R_2} = 0$$
$$v_1 - v_{R_1} - v_2 - v_{R_4} - v_{R_5} - v_{R_2} = 0$$

Since $v_{R_1} = (i_a + i_b)R_1$, and so on, the equations above can be written in terms of the loop currents as

$$(R_1 + R_2 + R_3)i_a + (R_1 + R_2)i_b = v_1$$
$$(R_1 + R_2)i_a + (R_1 + R_2 + R_4 + R_5)i_b = v_1 - v_2$$

These two linearly independent equations will also yield all the currents and voltages in the network. ■

DRILL EXERCISE

D3.18. Given the network in Fig. D3.18a and the graph for the network in Fig. D3.18b, write the proper set of loop equations that will yield all the currents in the network.

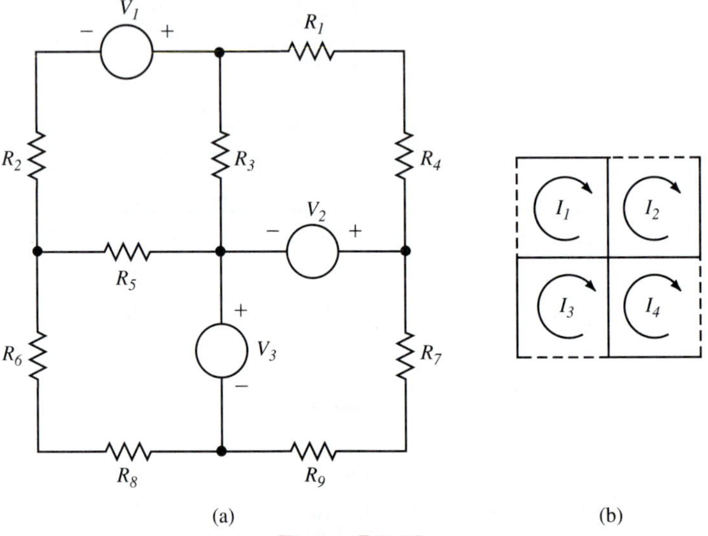

Figure D3.18

Ans: $I_1(R_2 + R_3 + R_5) - I_2(R_3) - I_3(R_5) - I_4(0) = V_1,$

$-I_1(R_3) + I_2(R_1 + R_3 + R_4) - I_3(0) - I_4(0) = -V_2,$

$-I_1(R_5) - I_2(0) + I_3(R_5 + R_6 + R_8) - I_4(0) = -V_3,$

$-I_1(0) - I_2(0) - I_3(0) + I_4(R_7 + R_9) = V_2 + V_3.$

3.4 Circuits with Operational Amplifiers

The operational amplifier, or op-amp as it is commonly known, is an extremely important circuit. It is a versatile interconnection of devices that vastly expands our capabilities in circuit design. It is employed in everything from engine control systems to microwave ovens.

In contrast to the circuit elements we have introduced thus far, the op-amp is modeled as a multiterminal device. It was first introduced in the 1940s for use in analog computers, and now finds wide application in circuit design as a result of the advances made in integrated-circuit technology. Although we will model the op-amp as a fairly simple device, its actual construction involves the use of numerous components, including resistors, capacitors, and transistors.

The circuit symbol for the op-amp is shown in Fig. 3.33a. Only the five principal terminals of the device are shown. The device sometimes has other terminals for fine adjustment of the device's characteristics; however, a discussion of them at this point

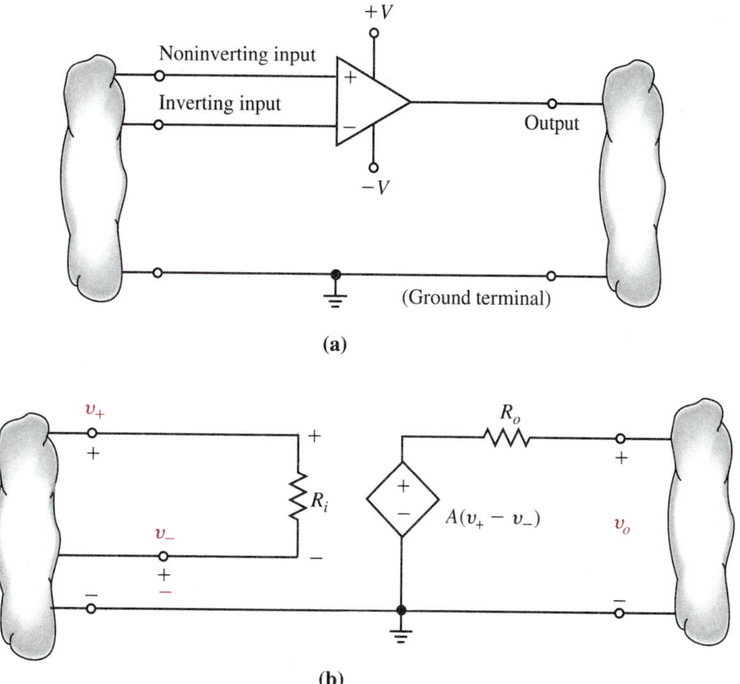

Figure 3.33 Operational amplifier representations.

would only serve to complicate the presentation unnecessarily. The two voltage terminals are for dc power supply (typically, 10 to 15 V); and, in addition, they provide a return path to ground for ac current. The + and − signs for the two input terminals indicate the polarity of the output as compared with the input; that is, a positive input on the *noninverting* input terminal produces a positive output, whereas a positive input on the *inverting* terminal produces a negative output.

Our interest in the op-amp is confined to its input–output characteristics, and therefore our model ignores the supply voltages. Functionally, the op-amp operates like the equivalent circuit shown in Fig. 3.33b. This equivalent circuit implies a unilateral device; that is, the output is determined by the difference in input voltages, but the input is unaffected by voltages applied at the output. The resistance R_i is very large, the gain A is very high, and the output resistance R_o is very low; typical values for these parameters are 10^5 to 10^{12} Ω, 10^5 to 10^7 V/V, and 1 to 50 Ω, respectively. The values of these parameters in this model suggest an even simpler model. R_i is very large and therefore can be assumed to be an open circuit. R_o is very low with respect to the recommended output load connection, and therefore can be assumed to be zero. Finally, the gain A is an extremely large amplification factor and would appear to represent essentially an infinite (∞) gain. At this point in our education, however, the term "infinite gain" causes some concern. Does this mean that if the input is 1 μV, the output is unlimited? Our common sense leads us to question such a condition. If there is no feedback signal path from the output to the inverting input, the output would be limited only by the value of the dc power supply voltages. Obviously, this is not a very useful situation. For this reason, in typical applications, the op-amp is

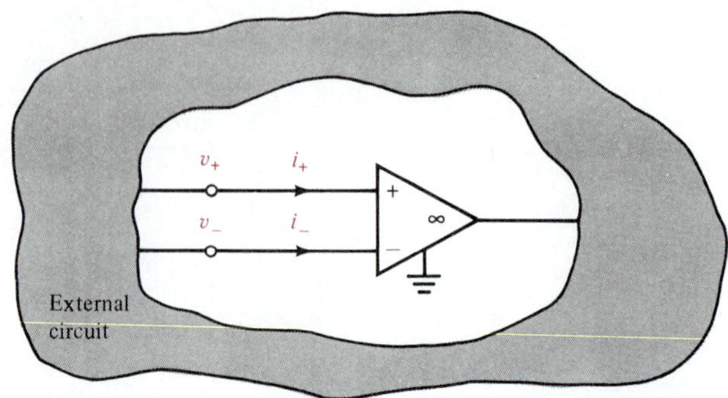

Figure 3.34 Ideal model for an operational amplifier. Model parameters: $i_+ = i_- = 0$, $v_+ = v_-$.

connected to external components with a signal path from the output to the inverting input. With the op-amp connected into a circuit in this manner, we will see that the high gain serves to maintain the input–output voltage relationship despite changes in load characteristics. As long as the amplifier gain A is very large, the circuit functionality will be set by the externally connected components.

Under the assumptions stated above, the ideal model for the op-amp is reduced to that shown in Fig. 3.34. The important characteristics of the model are: (1) Since R_i is extremely large, the input currents to the op-amp are approximately zero (i.e., $i_+ \approx i_- \approx 0$); and (2) if the output voltage is to remain bounded, then as the gain becomes very large and approaches infinity, the voltage across the input terminals must simultaneously become infinitesimally small so that as $A \rightarrow \infty$, $v_+ - v_- \rightarrow 0$ (i.e., $v_+ - v_- = 0$ or $v_+ = v_-$. The difference between these input voltages is often called the *error signal* for the op-amp (i.e., $v_+ - v_- = v_e$).

The ground terminal ($\stackrel{\perp}{=}$) shown on the op-amp is necessary for signal current return, and it guarantees that Kirchhoff's current law is satisfied at both the op-amp and the ground node in the circuit.

In summary, then, our ideal model for the op-amp is simply stated by the following conditions:

$$\begin{aligned} i_+ = i_- &= 0 \\ v_+ &= v_- \end{aligned} \tag{3.36}$$

These simple conditions are extremely important because they form the basis of our analysis of op-amp circuits.

The following example will serve to illustrate the validity of our assumptions when analyzing op-amp circuits.

EXAMPLE 3.22

Consider the op-amp circuit shown in Fig. 3.35a. Our model for the op-amp is shown generically in Fig. 3.35b and specifically in terms of the parameters R_i, A, and R_o in Fig. 3.35c. If the model is inserted in the network in Fig. 3.35a, we obtain the circuit shown in Fig. 3.35d, which can be redrawn as shown in Fig. 3.35e.

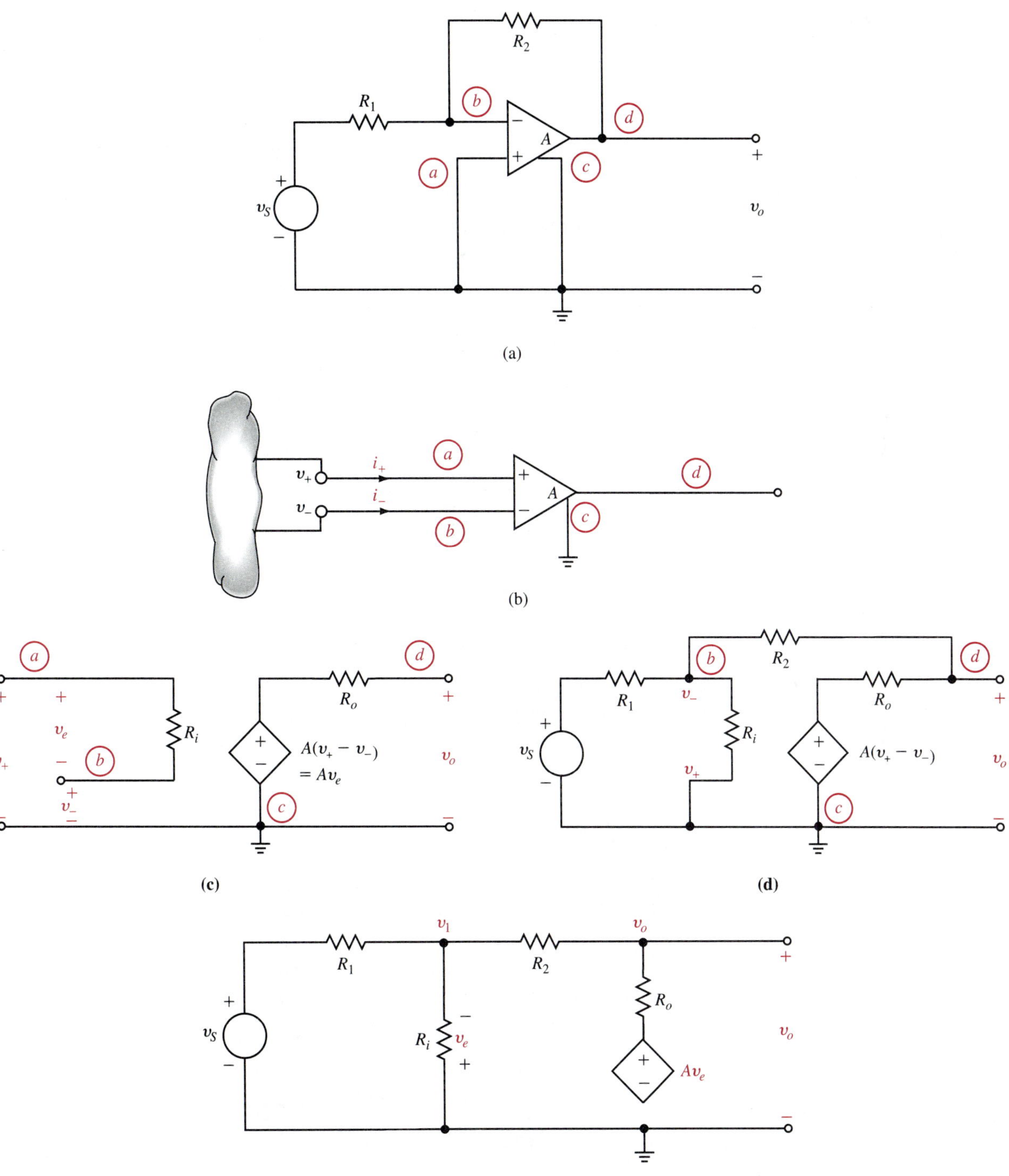

Figure 3.35 Op-amp circuit.

The node equations for the network are

$$\frac{v_1 - v_S}{R_1} + \frac{v_1}{R_i} + \frac{v_1 - v_o}{R_2} = 0$$

$$\frac{v_o - v_1}{R_2} + \frac{v_o - Av_e}{R_o} = 0$$

where $v_e = -v_1$. The equations can be written in matrix form as

$$\begin{bmatrix} \frac{1}{R_1} + \frac{1}{R_i} + \frac{1}{R_2} & -\left(\frac{1}{R_2}\right) \\ -\left(\frac{1}{R_2} - \frac{A}{R_o}\right) & \frac{1}{R_2} + \frac{1}{R_o} \end{bmatrix} \begin{bmatrix} v_1 \\ v_0 \end{bmatrix} = \begin{bmatrix} \frac{v_S}{R_1} \\ 0 \end{bmatrix}$$

Solving for the node voltages, we obtain

$$\begin{bmatrix} v_1 \\ v_o \end{bmatrix} = \frac{1}{\Delta} \begin{bmatrix} \frac{1}{R_2} + \frac{1}{R_o} & \frac{1}{R_2} \\ \frac{1}{R_2} - \frac{A}{R_o} & \frac{1}{R_1} + \frac{1}{R_i} + \frac{1}{R_o} \end{bmatrix} \begin{bmatrix} \frac{v_S}{R_1} \\ 0 \end{bmatrix}$$

where

$$\Delta = \left(\frac{1}{R_1} + \frac{1}{R_i} + \frac{1}{R_2}\right)\left(\frac{1}{R_2} + \frac{1}{R_o}\right) - \left(\frac{1}{R_2}\right)\left(\frac{1}{R_2} - \frac{A}{R_o}\right)$$

Hence

$$v_o = \frac{\left(\frac{1}{R_2} - \frac{A}{R_o}\right)\left(\frac{v_S}{R_1}\right)}{\left(\frac{1}{R_1} + \frac{1}{R_i} + \frac{1}{R_2}\right)\left(\frac{1}{R_2} + \frac{1}{R_o}\right) - \left(\frac{1}{R_2}\right)\left(\frac{1}{R_2} - \frac{A}{R_o}\right)}$$

which can be written as

$$\frac{v_o}{v_S} = \frac{-(R_2/R_1)}{1 - \left[\left(\frac{1}{R_1} + \frac{1}{R_i} + \frac{1}{R_2}\right)\left(\frac{1}{R_2} + \frac{1}{R_o}\right) \Big/ \left(\frac{1}{R_2}\right)\left(\frac{1}{R_2} - \frac{A}{R_o}\right)\right]}$$

If we now employ typical values for the circuit parameters (e.g., $A = 10^5$, $R_i = 10^8\ \Omega$, $R_o = 10\ \Omega$, $R_1 = 1\ \text{k}\Omega$, and $R_2 = 5\ \text{k}\Omega$), the voltage gain of the network is

$$\frac{v_o}{v_S} = -4.9996994 \approx -5.000$$

However, the ideal op-amp has infinite gain, and therefore if we take the limit of the gain equation as $A \to \infty$, we obtain

$$\lim_{A\to\infty}\left(\frac{v_o}{v_S}\right) = -\frac{R_2}{R_1} = -5.000$$

Note that the ideal op-amp yielded a result accurate to within four significant digits of that obtained from an exact solution of a typical op-amp model. These results are easily repeated for the vast array of useful op-amp circuits.

We now analyze the network in Fig. 3.35a using the ideal op-amp model. In this model

$$i_+ = i_- = 0$$

$$v_+ = v_-$$

As shown in Fig. 3.35a, $v_+ = 0$ and therefore $v_- = 0$. If we now write a node equation at the negative terminal of the op-amp, we obtain

$$\frac{v_S - 0}{R_1} + \frac{v_o - 0}{R_2} = 0$$

or

$$\frac{v_o}{v_S} = -\frac{R_2}{R_1}$$

and we have immediately obtained the result derived above. Since it is much easier to employ the ideal op-amp model rather than the nonideal model, unless otherwise stated we will use the ideal op-amp assumptions to analyze circuits that contain operational amplifiers. ■

EXAMPLE 3.23

Consider the op-amp circuit shown in Fig. 3.36. The node equation at the inverting terminal is

$$\frac{v_1 - v_-}{R_1} + \frac{v_o - v_-}{R_2} = i_-$$

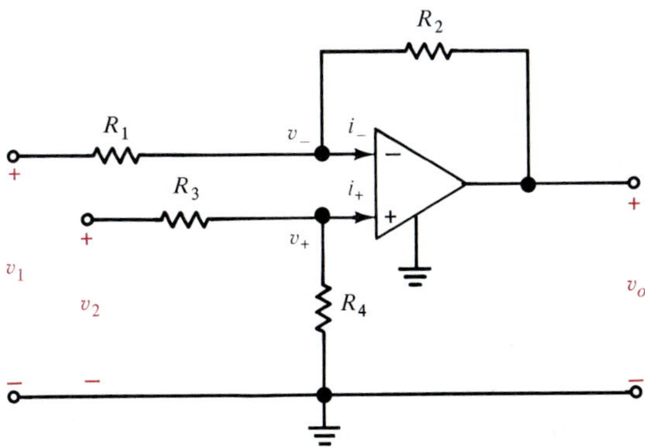

Figure 3.36 Differential amplifier operational amplifier circuit.

At the noninverting terminal KCL yields

$$\frac{v_2 - v_+}{R_3} = \frac{v_+}{R_4} + i_+$$

However, $i_+ = i_- = 0$ and $v_+ = v_-$. Substituting these values into the two equations above yields

$$\frac{v_1 - v_-}{R_1} + \frac{v_o - v_-}{R_2} = 0$$

and

$$\frac{v_2 - v_-}{R_3} = \frac{v_-}{R_4}$$

Solving these two equations for v_o results in the expression

$$v_o = \frac{R_2}{R_1}\left(1 + \frac{R_1}{R_2}\right)\frac{R_4}{R_3 + R_4}v_2 - \frac{R_2}{R_1}v_1$$

Note that if $R_4 = R_2$ and $R_3 = R_1$, the expression reduces to

$$v_o = \frac{R_2}{R_1}(v_2 - v_1)$$

Therefore, this op-amp circuit can be employed to subtract two input voltages. ■

EXAMPLE 3.24

The circuit shown in Fig. 3.37a is a precision differential voltage-gain device. It is used to provide a single-ended input for an analog-to-digital converter. We wish to derive an expression for the output of the circuit in terms of the two inputs.

To accomplish this, we draw the equivalent circuit shown in Fig. 3.37b. Recall that the voltage across the input terminals of the op-amp is approximately zero and the currents into the op-amp input terminals are approximately zero. Note that we can write node equations for node voltages v_1 and v_2 in terms of v_o and v_a. Since we are interested in an expression for v_o in terms of the voltages v_1 and v_2, we simply eliminate the v_a terms from the two node equations. The node equations are

$$\frac{v_1 - v_o}{R_2} + \frac{v_1 - v_a}{R_1} + \frac{v_1 - v_2}{R_G} = 0$$

$$\frac{v_2 - v_a}{R_1} + \frac{v_2 - v_1}{R_G} + \frac{v_2}{R_2} = 0$$

Adding the two equations will eliminate v_a, and hence if v_o is written in terms of v_1 and v_2, we obtain

$$v_o = (v_1 - v_2)\left(1 + \frac{R_2}{R_1} + \frac{2R_2}{R_G}\right)$$

■

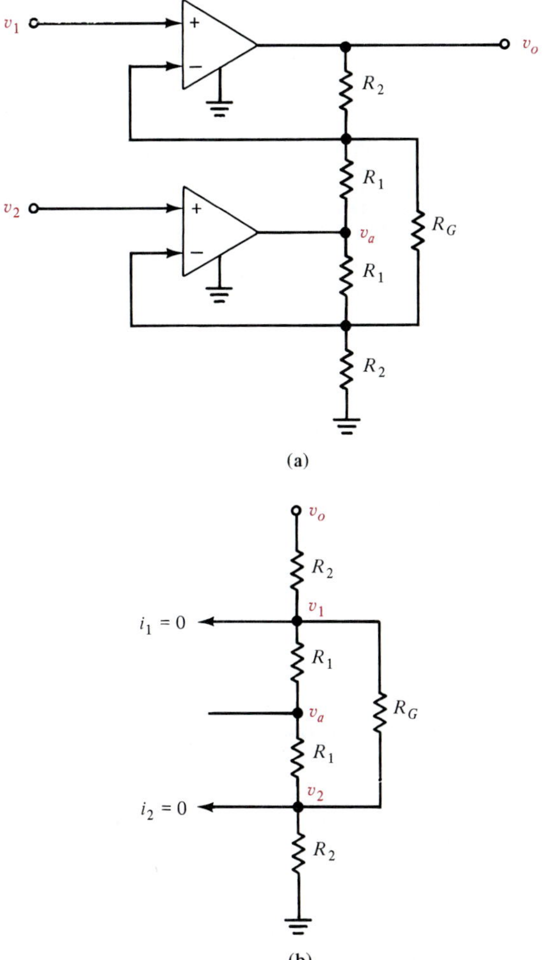

Figure 3.37 Instrumentation amplifier circuit.

EXAMPLE 3.25

We wish to calculate the input resistance of the op-amp circuit shown in Fig. 3.38 without any assumptions about the sizes of the parameters A, R_i, and R_o.

The input resistance is the resistance seen by the input voltage source v_1. Using the op-amp model shown in Fig. 3.33b, we obtain the circuit shown in Fig. 3.38b. The mesh equations for this circuit are

$$v_1 = i_1(R_i + R_1) - i_2 R_1$$

$$-Av_i = -i_1 R_1 + i_2(R_1 + R_2 + R_o)$$

$$v_i = R_i i_1$$

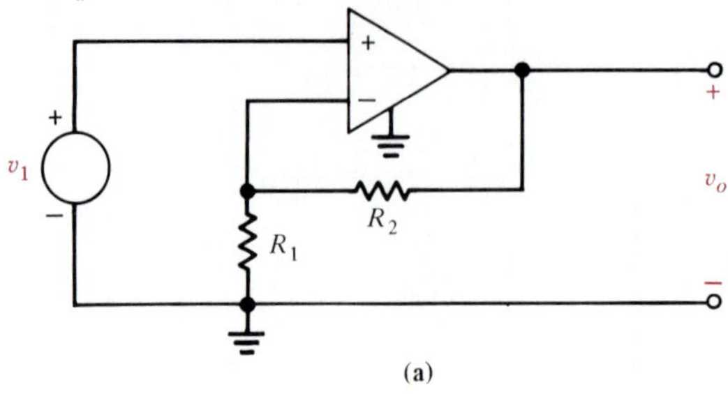

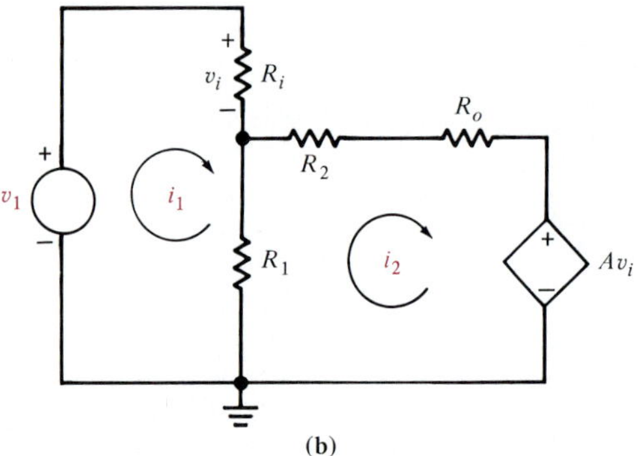

Figure 3.38 Circuit and model for a noninverting operational amplifier.

Solving these equations for i_1 yields

$$i_1 = \frac{v_1(R_1 + R_2 + R_o)}{(R_i + R_1)(R_1 + R_2 + R_o) + R_1(AR_i - R_1)}$$

The input resistance is

$$R_{\text{in}} = \frac{v_1}{i_1} = \frac{(R_i + R_1)(R_1 + R_2 + R_o) + R_1(AR_i - R_1)}{R_1 + R_2 + R_o}$$

$$= R_i + \frac{R_1(R_2 + R_o + AR_i)}{R_1 + R_2 + R_o}$$

Since A and R_i are normally very large and R_o is very small, the expression for R_{in} can be approximated by the equation

$$R_{\text{in}} \simeq \frac{AR_i}{1 + R_2/R_1}$$ ■

DRILL EXERCISE

D3.19. Find I_o in the network in Fig. D3.19.

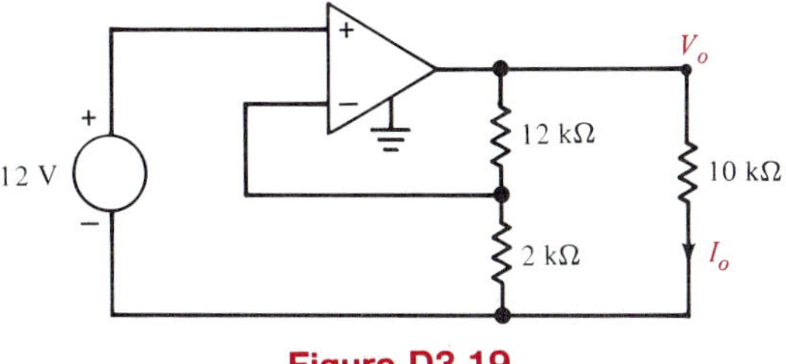

Figure D3.19

Ans: $I_o = 8.4$ mA.

D3.20. Determine the gain of the op-amp circuit in Fig. D3.20.

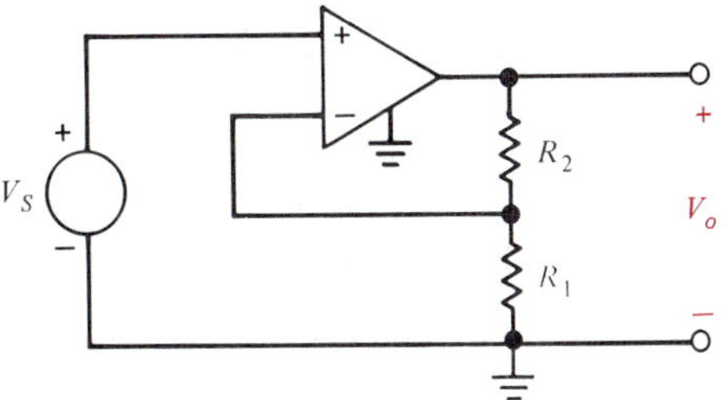

Figure D3.20

Ans: $\dfrac{V_o}{V_S} = 1 + \dfrac{R_2}{R_1}$.

D3.21. Determine both the gain and the output voltage of the op-amp configuration shown in Fig. D3.21.

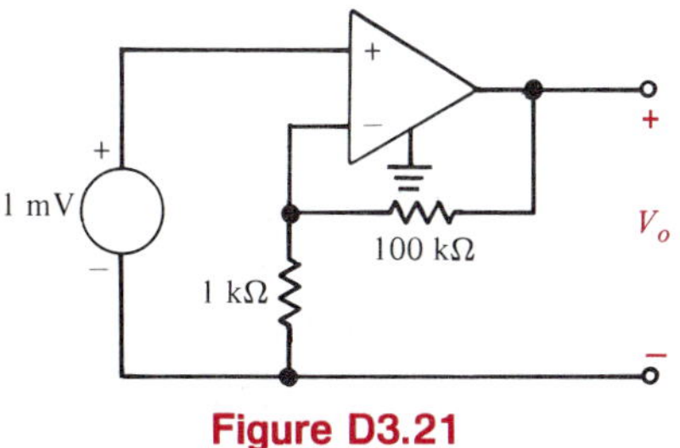

Figure D3.21

Ans: $V_o = 0.101$ V, gain = 101.

An important special case of the noninverting amplifier is the configuration shown in Fig. 3.39. Note carefully that this circuit is equivalent to that shown in Fig. D3.20 with $R_1 = \infty$ (i.e., open circuited) and $R_2 = 0$ (i.e., short circuited). Under these conditions we note that the gain of the circuit is $1 + R_2/R_1 = 1$ (i.e., $v_o = v_S$). Since v_o follows v_S, the circuit is called a *voltage follower*.

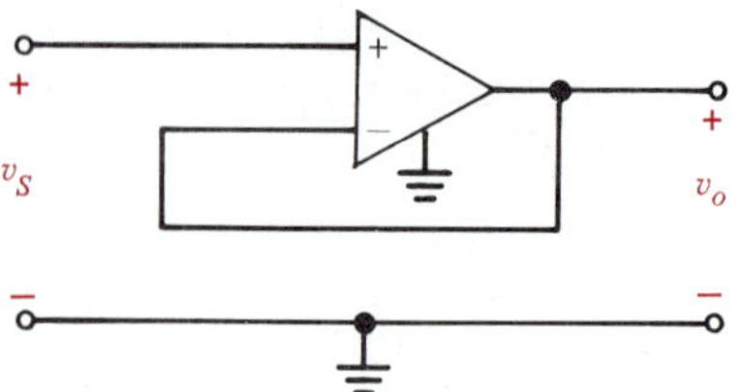

Figure 3.39 Voltage-follower operational amplifier configuration.

An obvious question at this point is: If $v_o = v_S$, why not just connect v_S to v_o via two parallel connecting wires; why do we need to place an op-amp between them? The answer to this question is fundamental and provides us with some insight that will aid us in circuit analysis and design.

Consider the circuit shown in Fig. 3.40a. In this case v_o is not equal to v_S because of the voltage drop across R_S:

$$v_o = v_S - iR_S$$

However, in Fig. 3.40b, the input current to the op-amp is zero and therefore v_S appears at the op-amp input. Since the gain of the op-amp configuration is 1, $v_o = v_S$. In Fig. 3.40a, the resistive network's interaction with the source caused the voltage $_o$ to be less than v_S. In other words, the resistive network loads the source voltage. However, in Fig. 3.40b the op-amp isolates the source from the resistive network, and therefore the voltage follower is referred to as a *buffer amplifier* because it can be used to isolate one circuit from

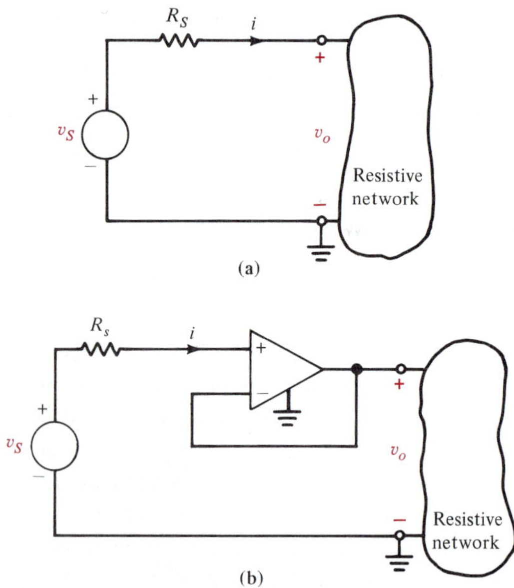

Figure 3.40 Illustration of the isolation capability of a voltage follower.

another. The energy supplied to the resistive network in the first case must come from the source v_S, whereas in the second case it comes from the power supplies that supply the amplifier, and little or no energy is drawn from v_S.

3.5 Summary

Two very important and extremely useful techniques for circuit analysis have been introduced. Both the node voltage and loop current methods produce a set of simultaneous equations which when solved enable us to determine any voltage or current in the network. Procedures for simplifying the analysis were presented and techniques for solving the equations were outlined. In addition, we have introduced both topology as well as the operational amplifier and its equivalent circuit, which we will use in a variety of circuits throughout the book.

KEY POINTS

- When solving an N-node circuit using nodal analysis:
  1. Label all nodes, selecting one as the reference node.
  2. If the circuit contains only independent current sources, write $N - 1$ linearly independent simultaneous equations using KCL to determine the $N - 1$ unknown node voltages.
  3. If the circuit contains an independent voltage source, a supernode enclosing this source is created and a constraint equation for this source is written in addition to the KCL equations for the supernode and the remaining nodes.
  4. If the circuit contains a dependent source, treat the dependent source as though it were an independent source when writing the $N - 1$ independent node equations, and then write one additional equation for the controlling parameter of the dependent source.
  5. Solve the equations using any convenient method.
- When solving an N-loop circuit using loop analysis:
  1. Label N loop currents for N distinct closed paths through the network.
  2. If the circuit contains only independent voltage sources, write N linearly independent simultaneous equations using KVL to determine the N unknown loop currents.
  3. If the circuit contains an independent current source, a constraint equation is written for the current that flows through it; the current source is then replaced by an open circuit and another equation is written for the newly created loop.
  4. If the circuit contains a dependent source, treat the dependent source as though it were an independent source when writing the N-loop equations, and then write one additional equation for the controlling parameter of the dependent source.
  5. Solve the equations using any convenient method, such as Cramer's rule, matrices, or substitution.
- For an ideal op-amp, $i_+ = i_- = 0$ and $v_+ = v_-$.

PROBLEMS

3.1. Find V_o in the network in Fig. P3.1.

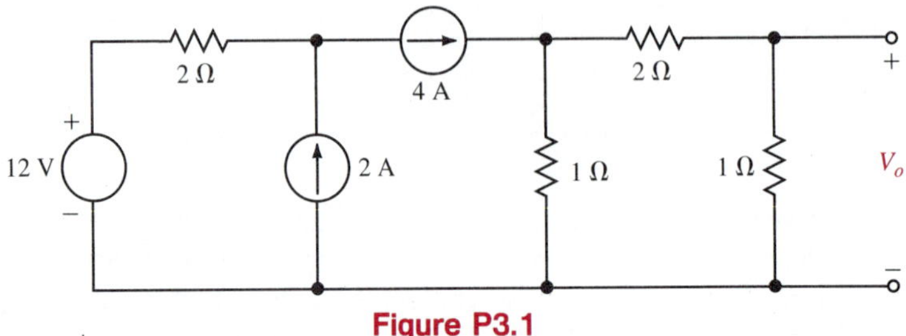

Figure P3.1

3.2. Find V_o in the network in Fig. P3.2.

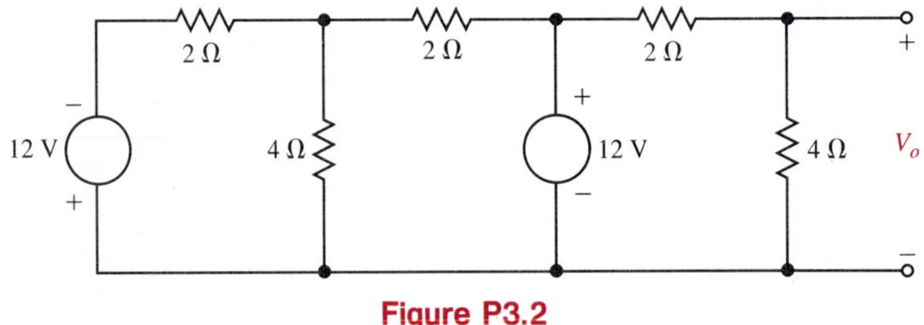

Figure P3.2

3.3. Use nodal analysis to find I_o in the network in Fig. P3.3.

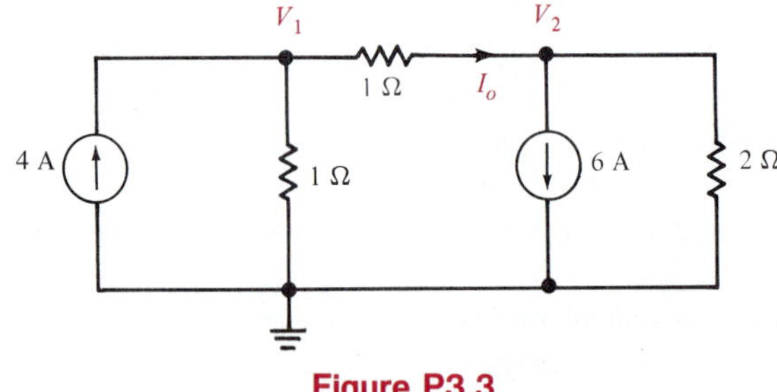

Figure P3.3

3.4. Find I_o in the network in Fig. P3.4 using nodal analysis.

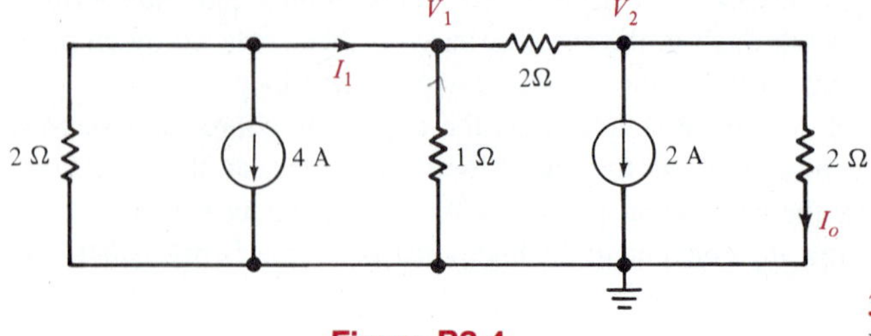

Figure P3.4

3.5. Find I_1 in the network in Fig. P3.4 using nodal analysis.

3.6. Use nodal analysis to find I_o in the network in Fig. P3.6.

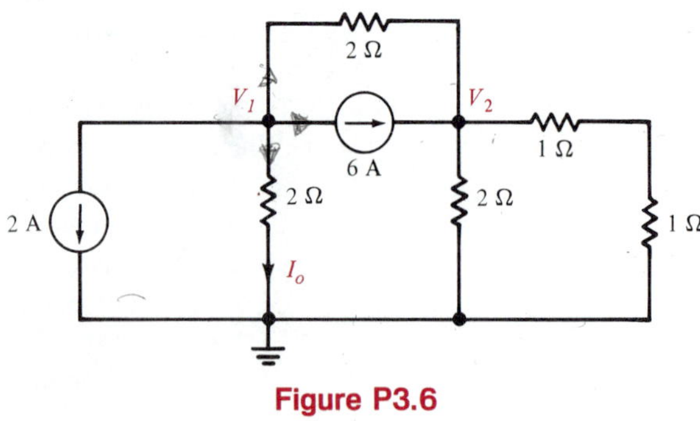

Figure P3.6

3.7. Use node equations to determine V_o in the circuit shown in Fig. P3.7.

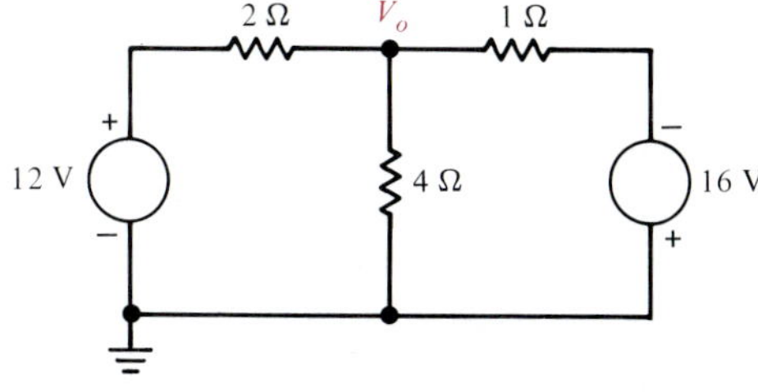

Figure P3.7

3.8. Use node equations to find V_2 in the circuit shown in Fig. P3.8.

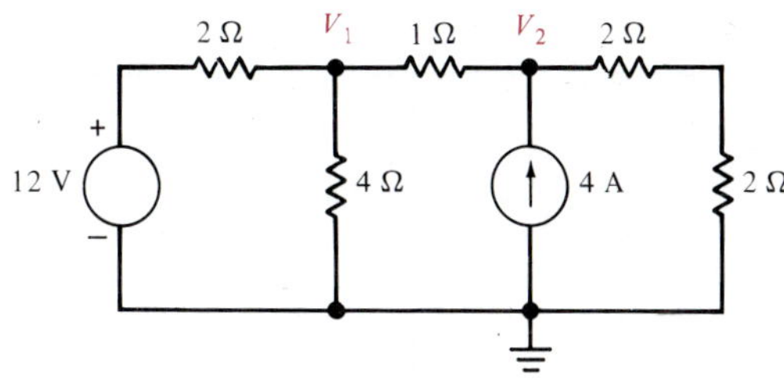

Figure P3.8

3.9. Find I_o and I_x in the network in Fig. P3.9 using nodal analysis.

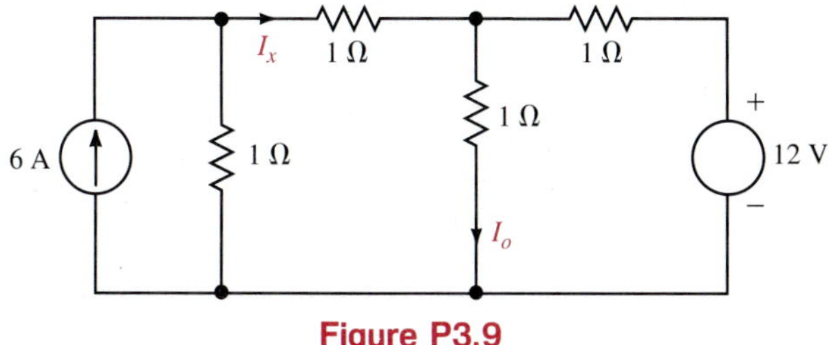

Figure P3.9

3.10. Use nodal analysis to find I_o in the network in Fig. P3.10.

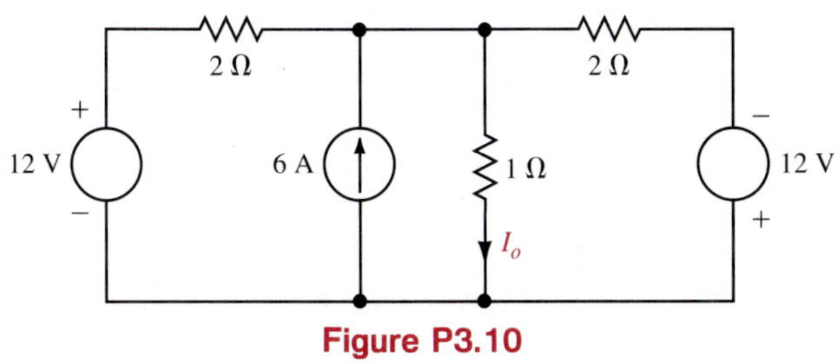

Figure P3.10

3.11. Use nodal analysis to find V_o in the network in Fig. P3.11.

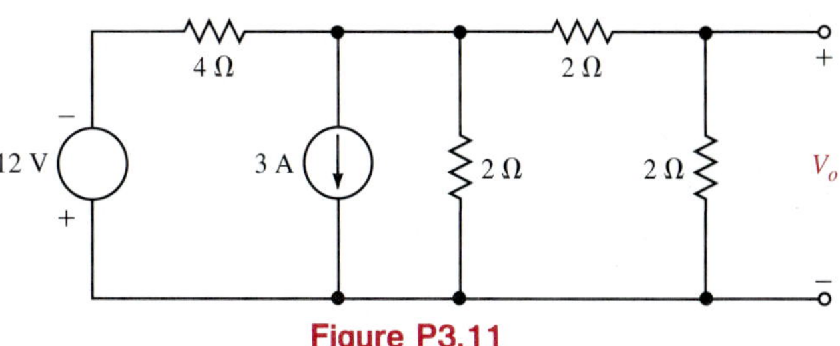

Figure P3.11

3.12. Use node equations to find V_2 in the circuit shown in Fig. P3.12.

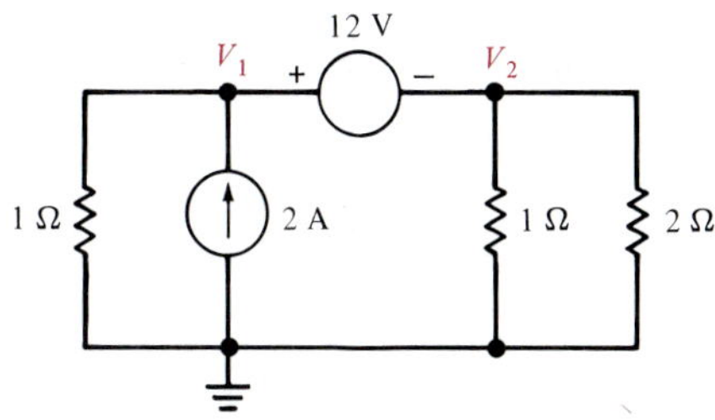

Figure P3.12

3.13. Use nodal analysis to find I_o in the network in Fig. P3.13.

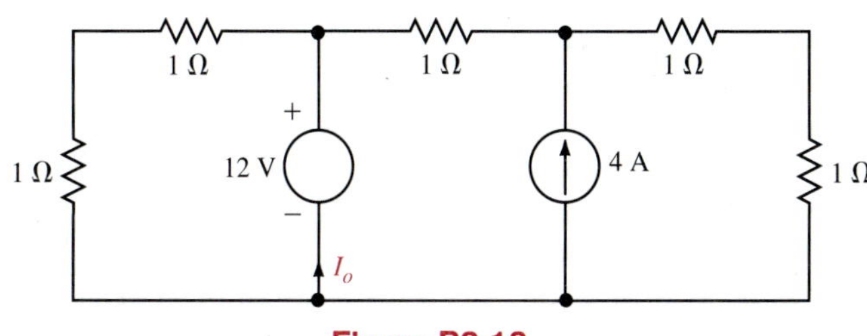

Figure P3.13

3.14. Use nodal analysis to find I_o in the network in Fig. P3.14.

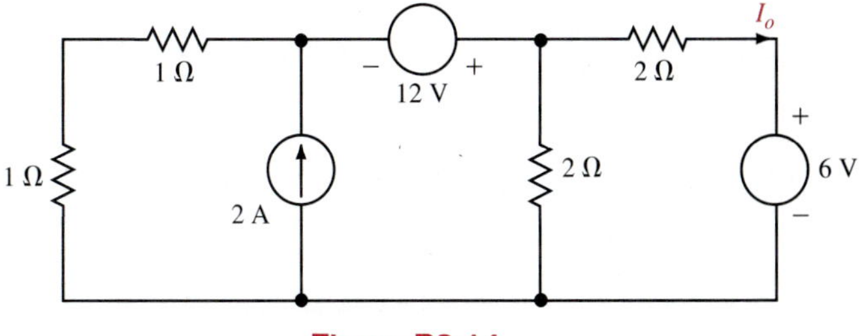

Figure P3.14

3.15. Use nodal analysis to find I_o in the network in Fig. P3.15.

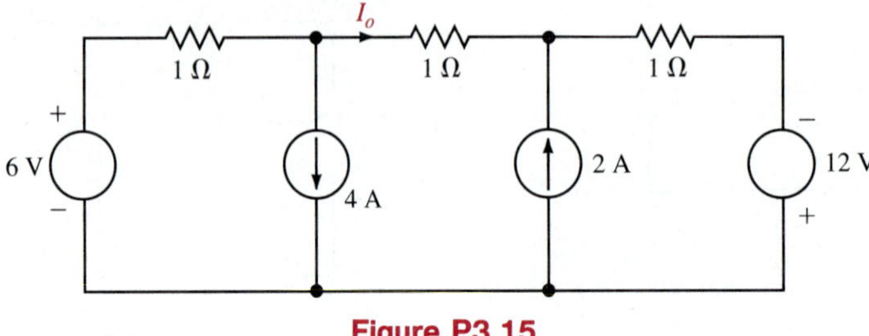

Figure P3.15

3.16. Use nodal analysis to find I_o in the network in Fig. P3.16.

3.17. Use node equations to find I_o in the circuit in Fig. P3.17.

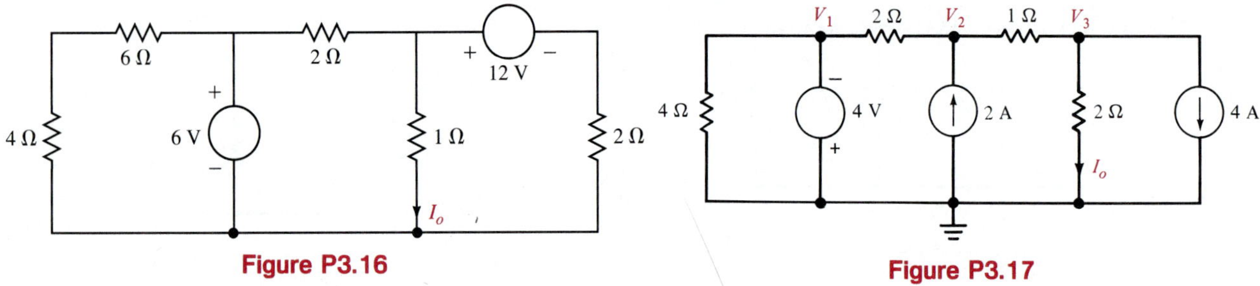

Figure P3.16

Figure P3.17

3.18. Use nodal analysis to find I_o in the network in Fig. P3.18.

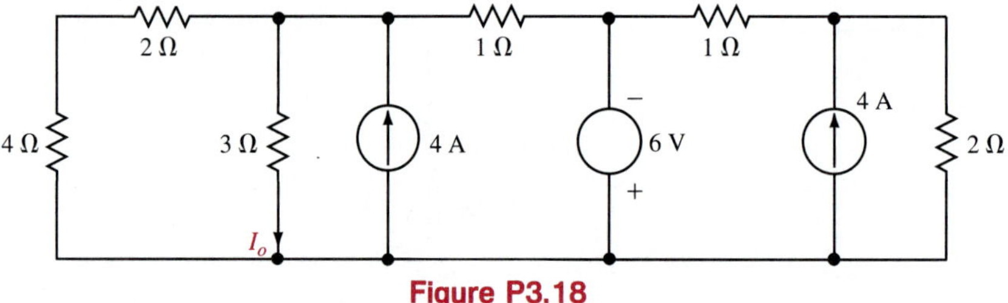

Figure P3.18

3.19. Use nodal analysis to find V_o in the network in Fig. P3.19.

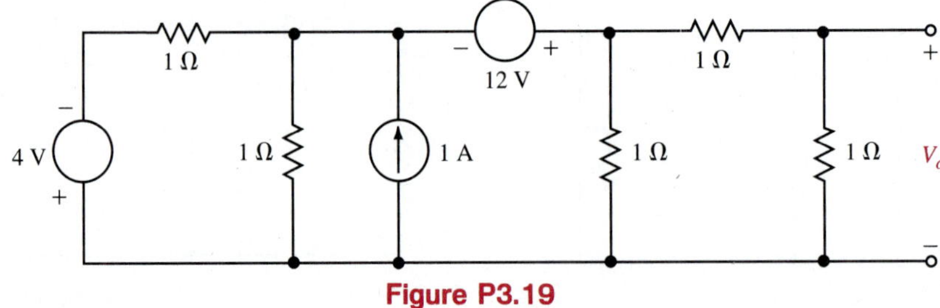

Figure P3.19

3.20. Find all node voltages in the circuit shown in Fig. P3.20 using node equations.

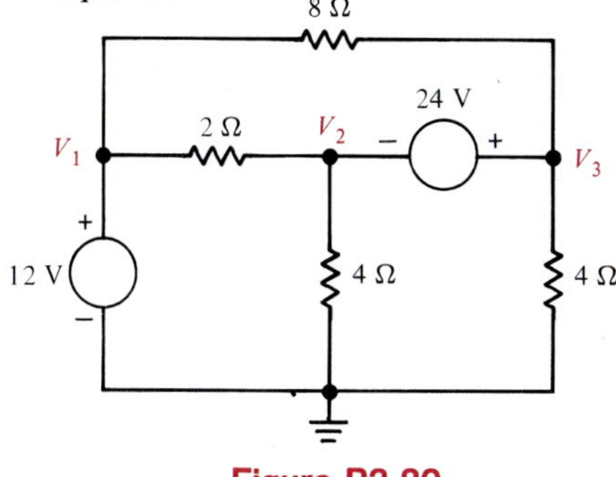

Figure P3.20

3.21. Find I_o in the network in Fig. P3.21 using nodal analysis.

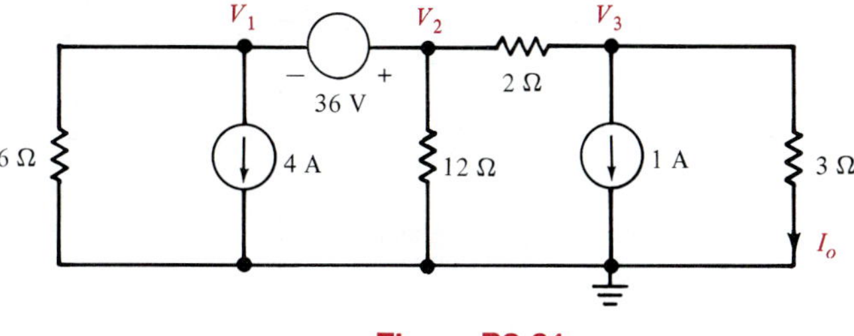

Figure P3.21

3.22. Use nodal analysis to find I_o in the network in Fig. P3.22.

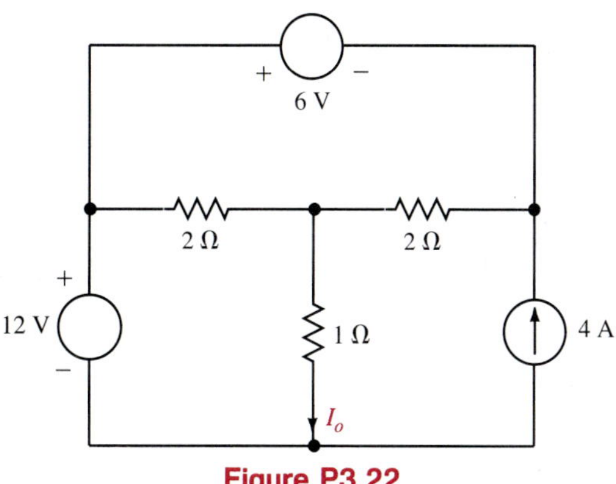

Figure P3.22

3.23. Use nodal analysis to find V_o in the network in Fig. P3.23.

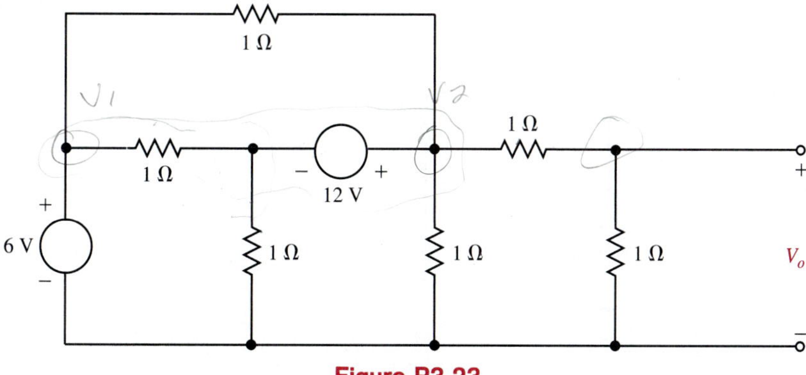

Figure P3.23

3.24. Use nodal analysis to find V_o in the circuit in Fig. P3.24.

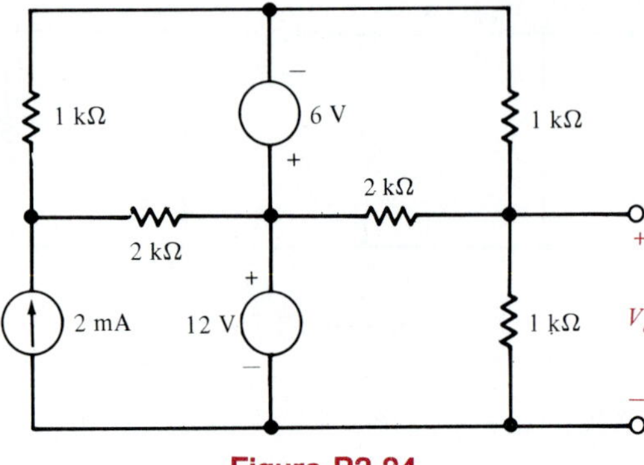

Figure P3.24

3.25. Use nodal analysis to find I_o in the network in Fig. P3.25

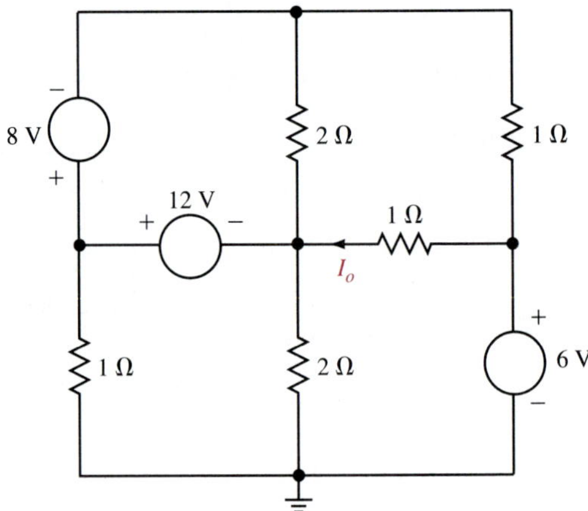

Figure P3.25

3.26. Use nodal analysis to find I_o in the network in Fig. P3.26.

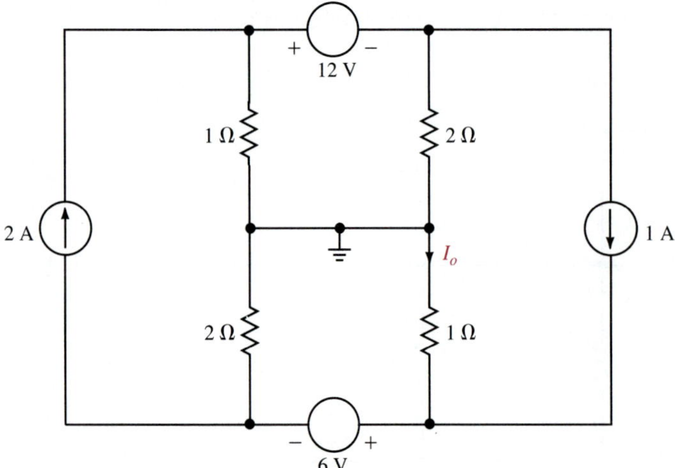

Figure P3.26

3.27. Use nodal analysis to find I_o in the network in Fig. P3.27.

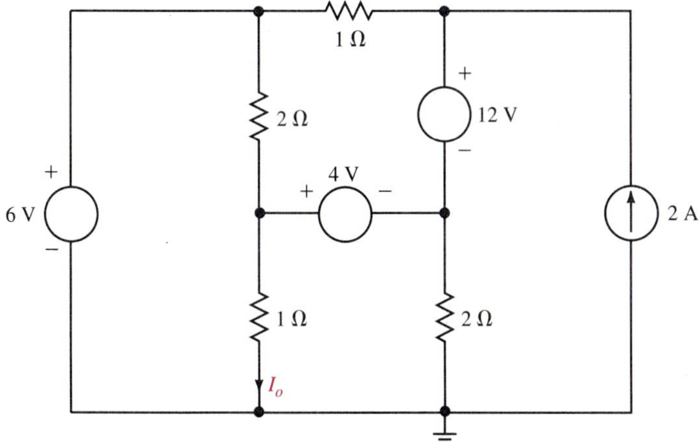

Figure P3.27

3.28. Use nodal analysis to find I_o in the network in Fig. P3.28.

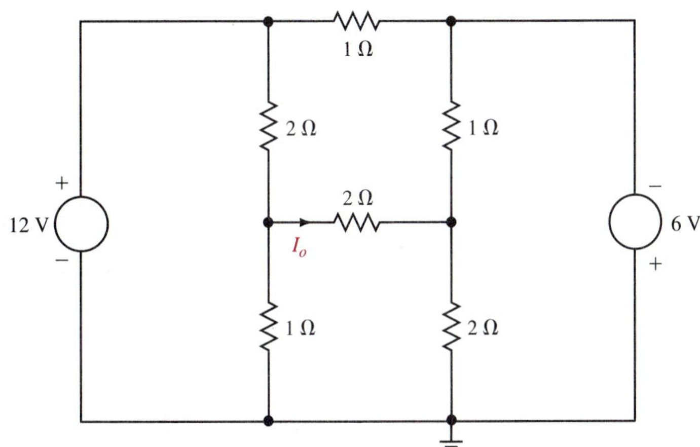

Figure P3.28

3.29. Find V_o in the network in Fig. P3.29 using node equations.

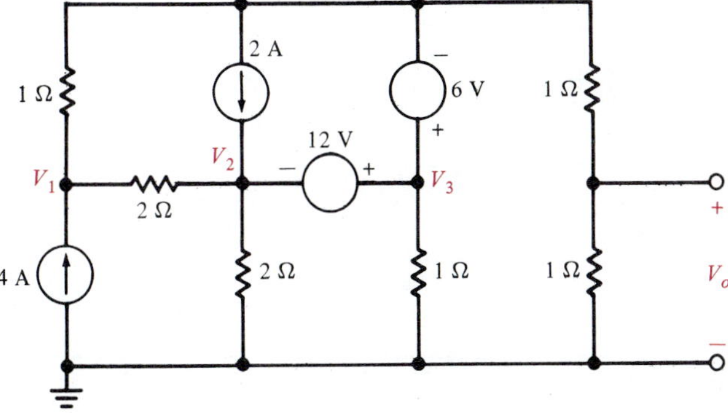

Figure P3.29

3.30. Find I_o in the network in Fig. P3.30 using node equations.

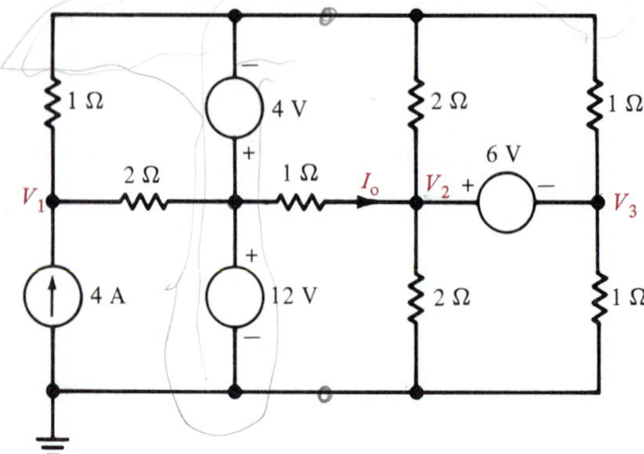

Figure P3.30

3.31. Use nodal analysis to find V_o in the network in Fig. P3.31.

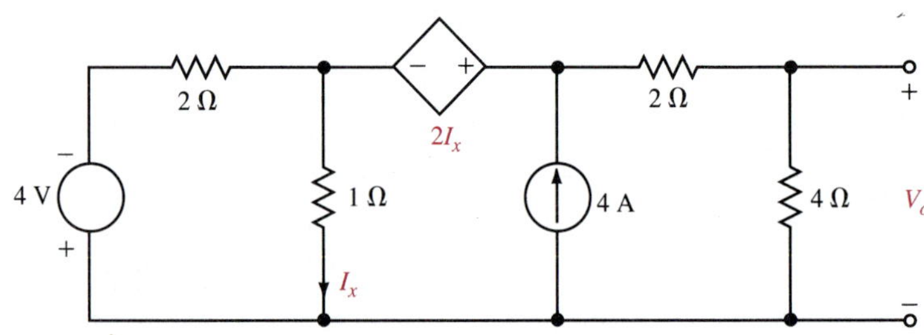

Figure P3.31

3.32. Use nodal analysis to find V_o in the network in Fig. P3.32.

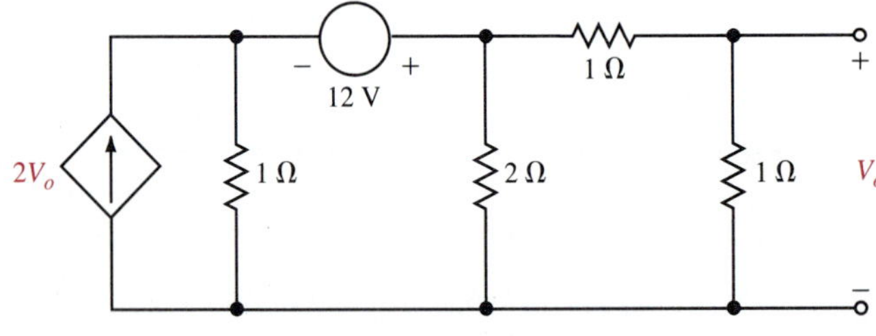

Figure P3.32

3.33. Find V_o in the network in Fig. P3.33 using node equations.

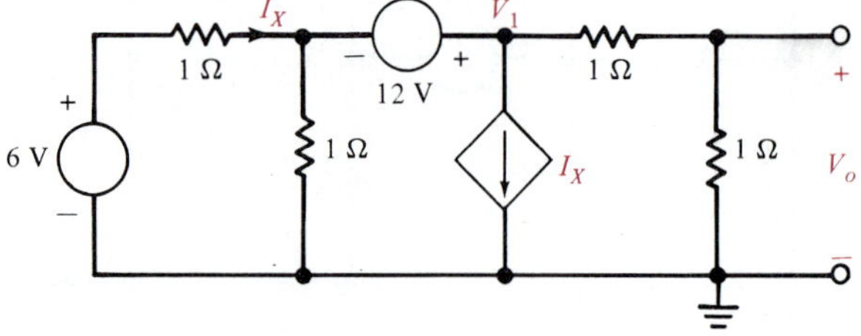

Figure P3.33

3.34. Calculate the voltage V_2 in Fig. P3.34 using node equations.

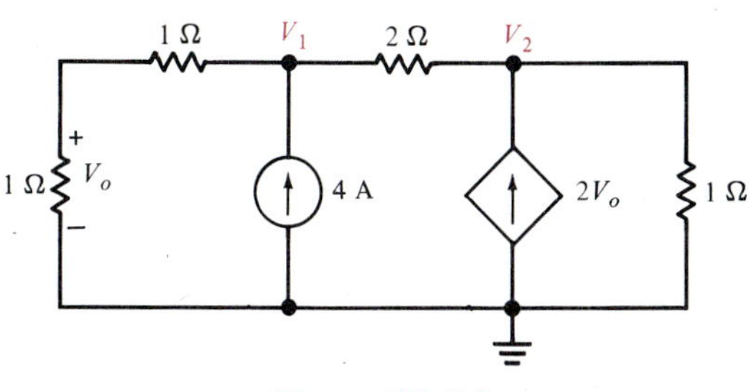

Figure P3.34

3.35. Find V_o in Fig. P3.35, using node equations.

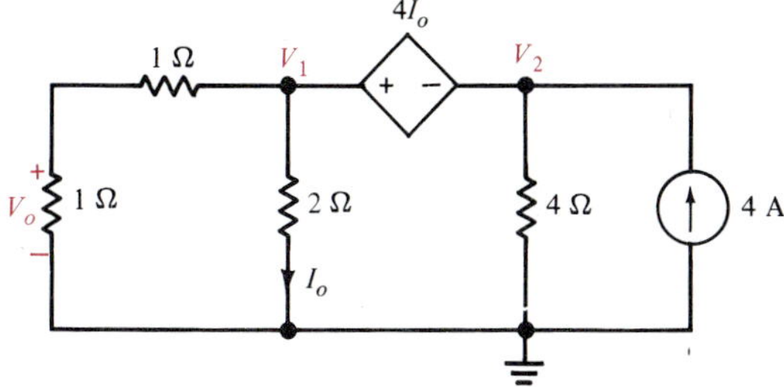

Figure P3.35

3.36. Find V_o in the network in Fig. P3.36 using nodal analysis.

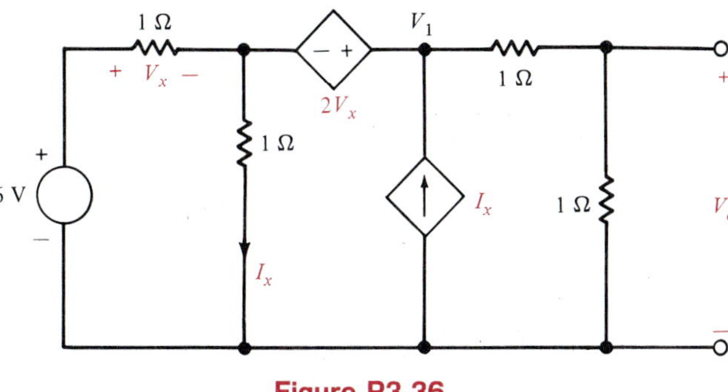

Figure P3.36

3.37. Determine I_o in the circuit shown in Fig. P3.37.

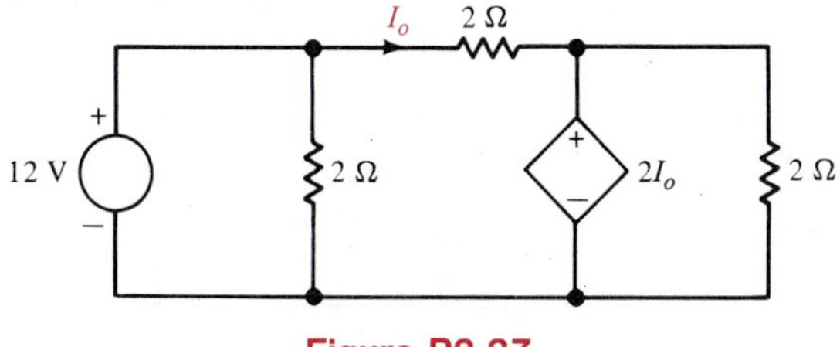

Figure P3.37

3.38. Use nodal analysis to find I_o in the network in Fig. P3.38.

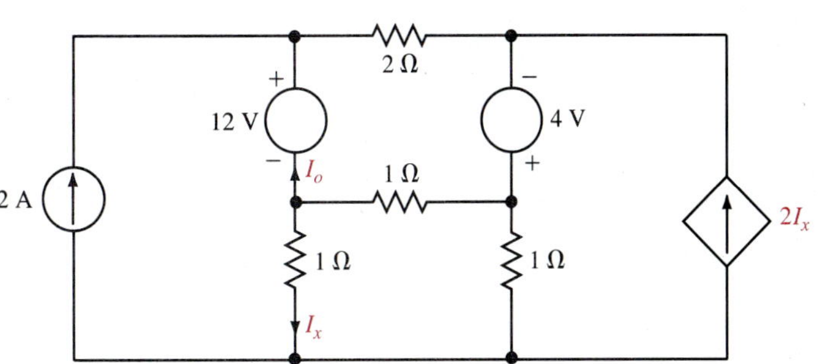

Figure P3.38

3.39. Use nodal analysis to find I_o in the network in Fig. P3.39.

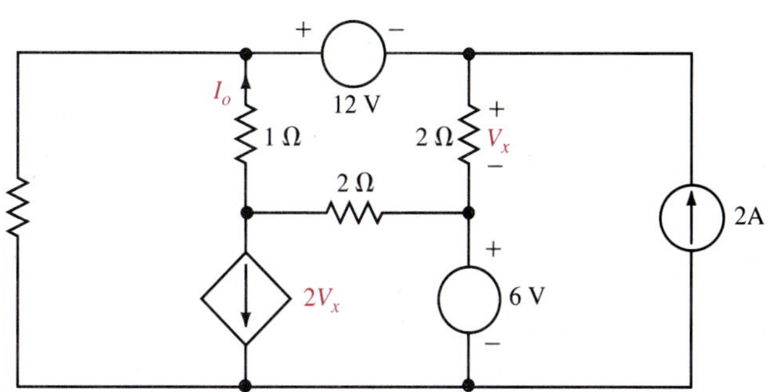

Figure P3.39

3.40. Use mesh equations to calculate the power absorbed by the 5-Ω resistor in Fig. P3.40.

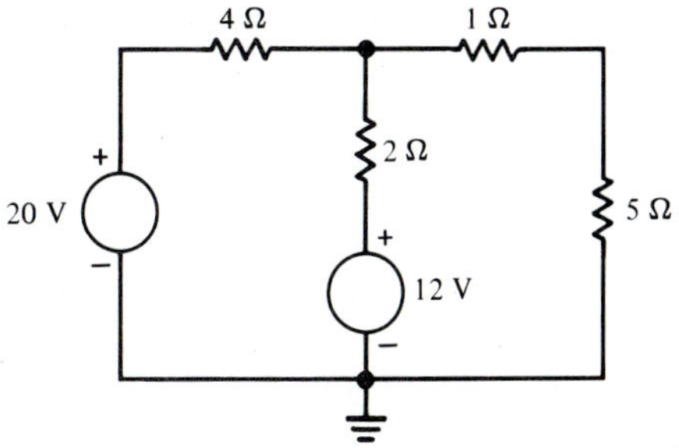

Figure P3.40

3.41. Use mesh equations to find V_o in the circuit in Fig. P3.41.

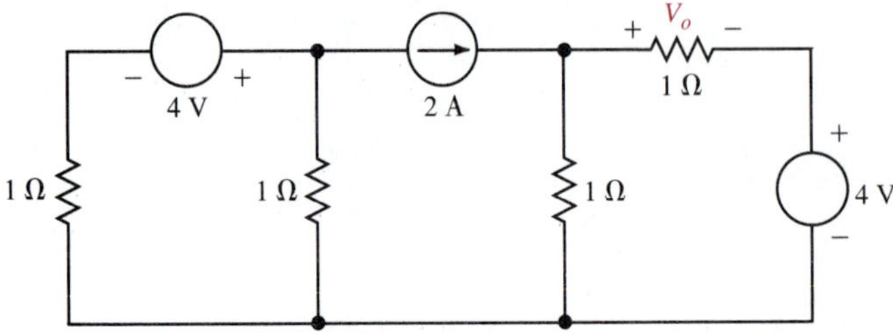

Figure P3.41

3.42. Use mesh equations to find I_o in the network in Fig. P3.42.

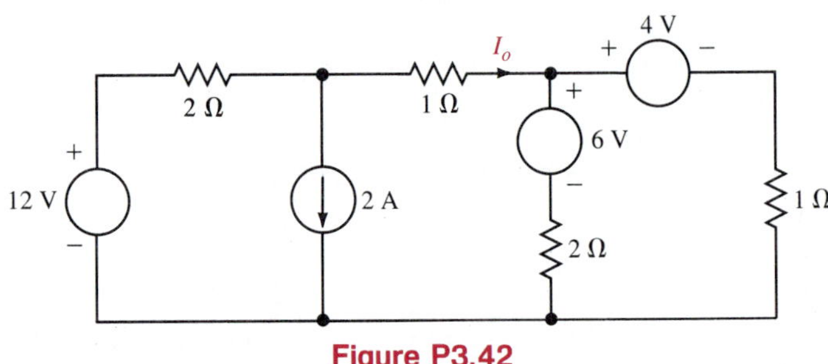

Figure P3.42

3.43. Use loop analysis to find I_o in the network in Fig. P3.43.

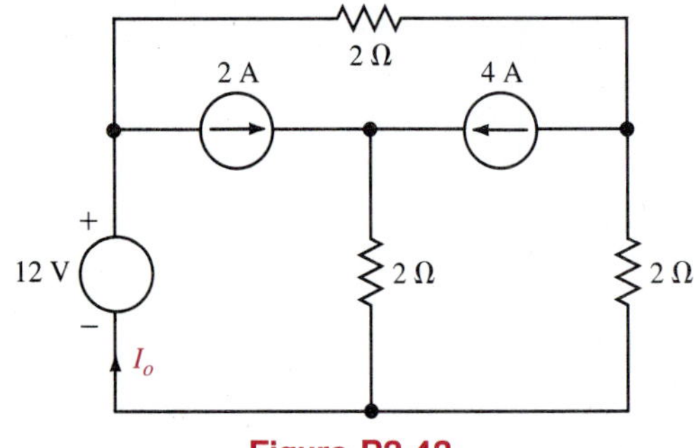

Figure P3.43

3.44. Use mesh equations to find I_o in the network in Fig. P3.44.

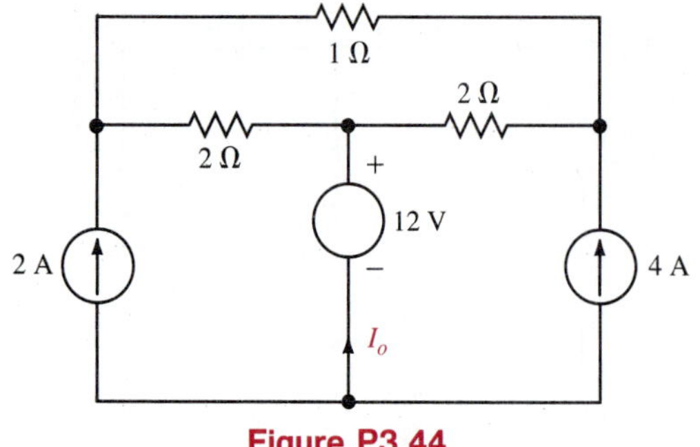

Figure P3.44

3.45. Use mesh equations to find I_o in the network in Fig. P3.45.

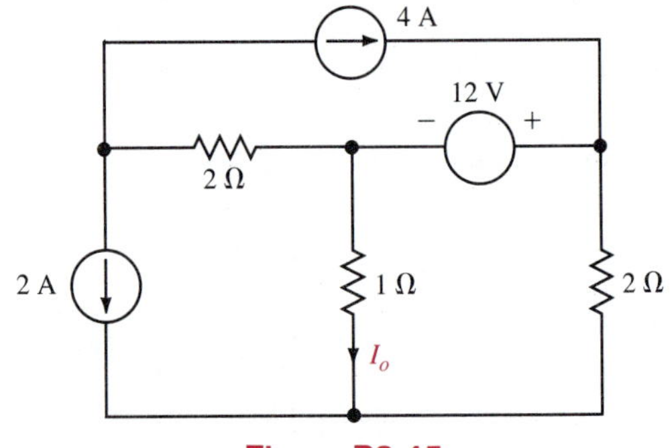

Figure P3.45

3.46. Use loop analysis to find V_o in the circuit in Fig. 3.46.

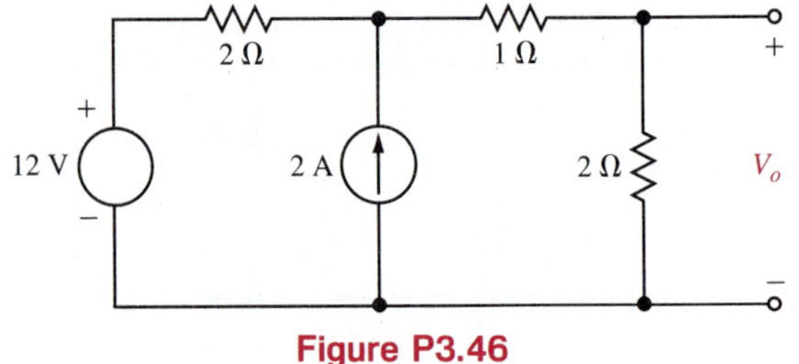

Figure P3.46

3.47. Use mesh equations to find V_o in the circuit in Fig. P3.47.

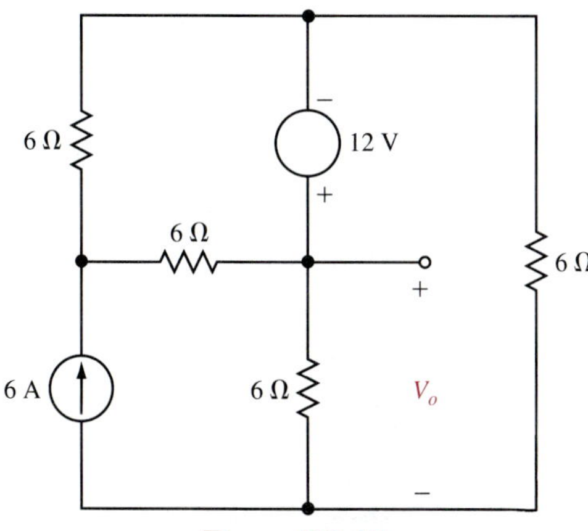

Figure P3.47

3.48. Use mesh equations to find V_o in the network shown in Fig. P3.48.

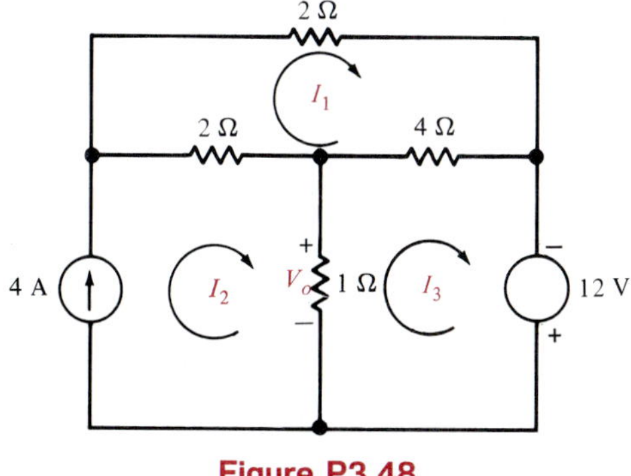

Figure P3.48

3.49. Use mesh equations to find V_o in the circuit in Fig. P3.49.

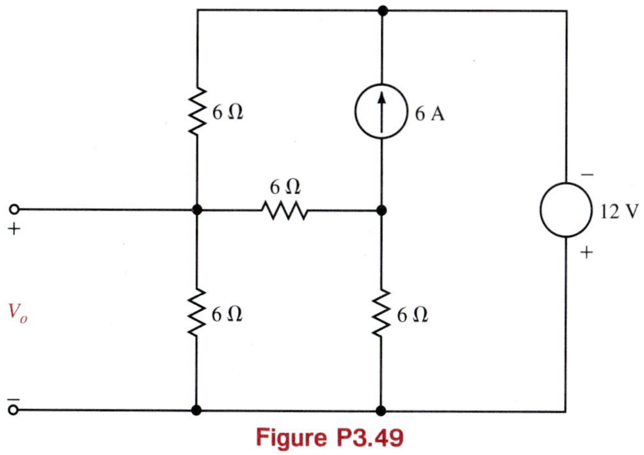

Figure P3.49

3.50. Use loop equations to find the power dissipated in the 2-Ω resistor in the circuit in Fig. P.3.50.

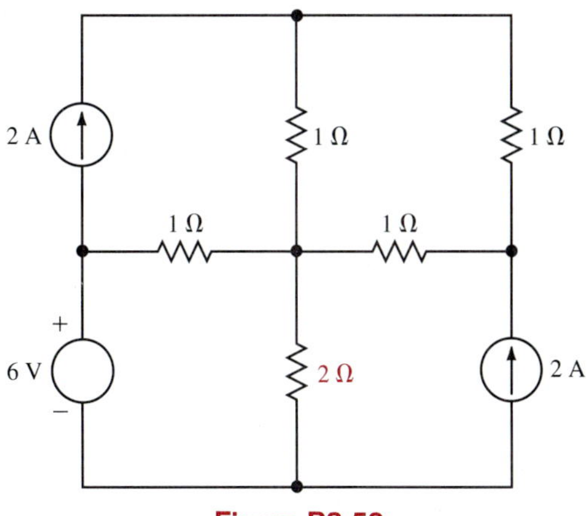

Figure P3.50

3.51. Use loop equations to find I_o in the circuit in Fig. P3.51.

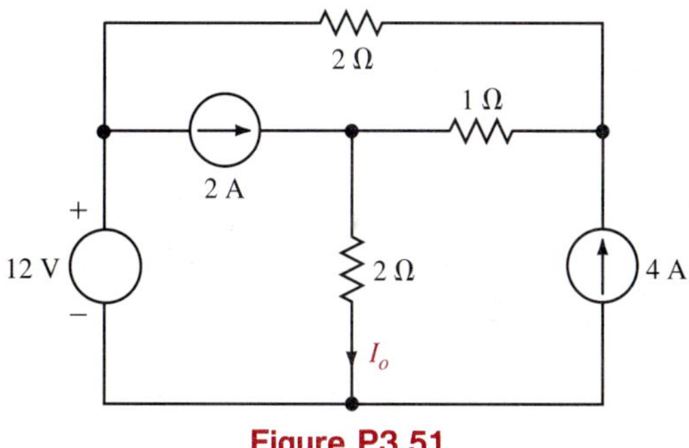

Figure P3.51

3.52. Use loop equations to find I_o in the network in Fig. P3.52.

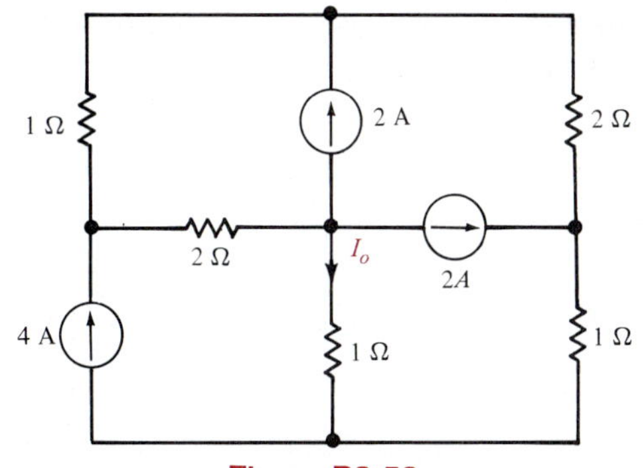

Figure P3.52

3.53. Use loop equations to find I_o in the network in Fig. P3.53.

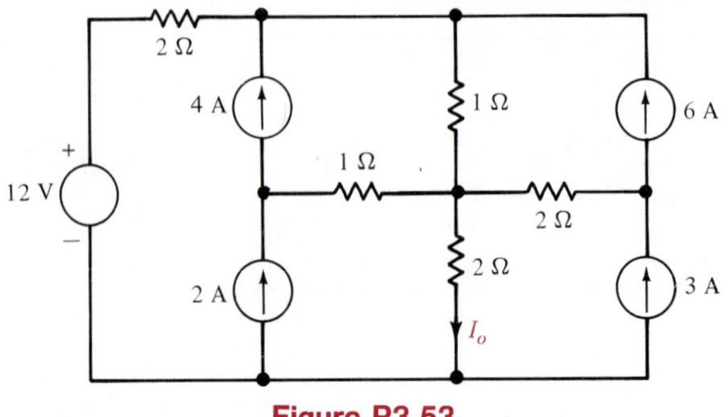

Figure P3.53

3.54. Use loop analysis to find V_o in the network in Fig. P3.54.

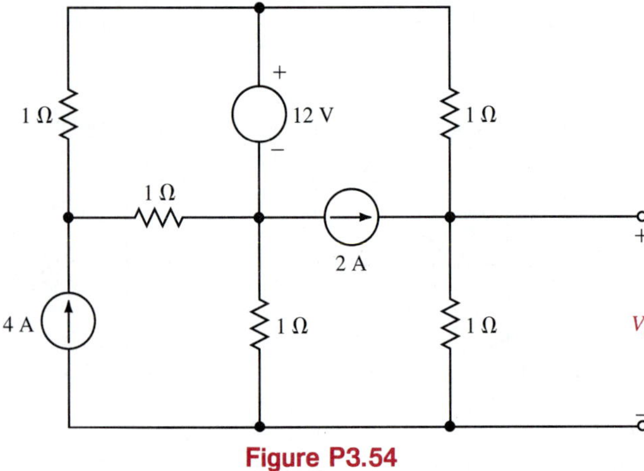

Figure P3.54

3.55. Use mesh equations to find V_o in the circuit in Fig. P3.55.

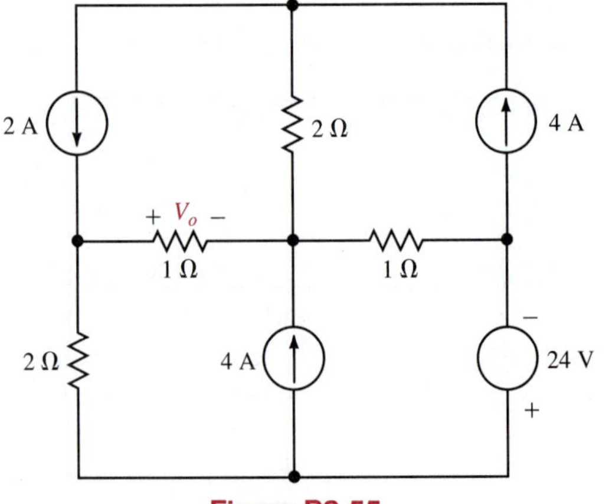

Figure P3.55

3.56. Use loop analysis to find V_o in the circuit in Fig. P3.56.

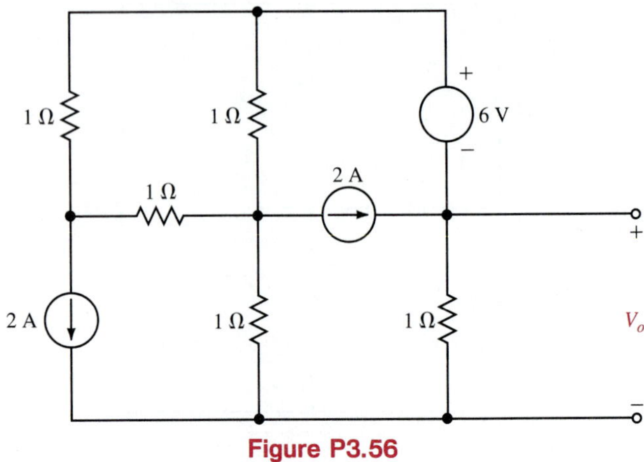

Figure P3.56

3.57. Use loop analysis to find I_o in the circuit in Fig. P3.57.

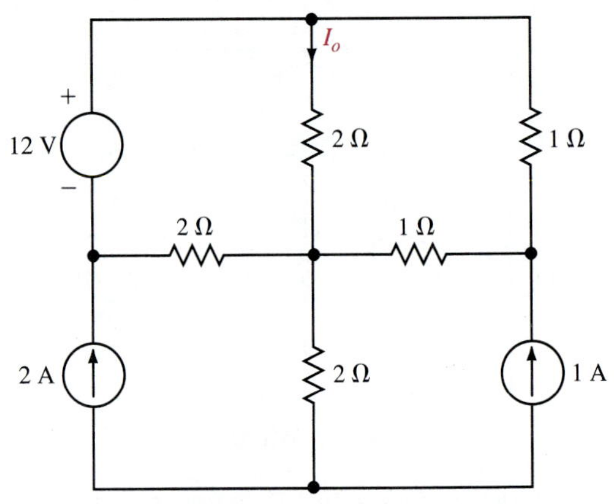

Figure P3.57

Find V_o in the circuit in Fig. P3.58 using loop analysis.

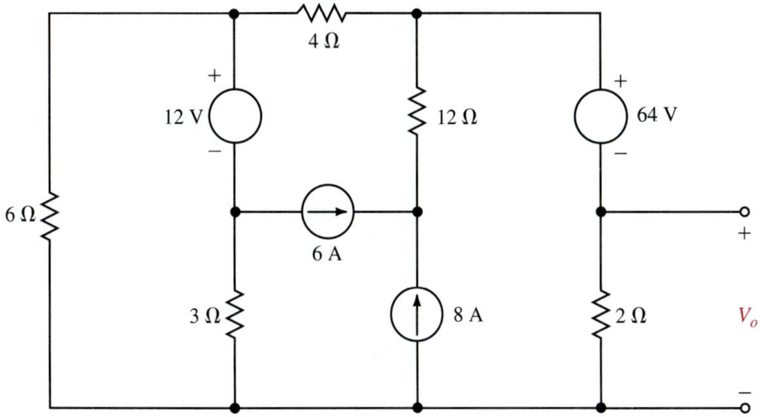

Figure P3.58

3.59. Use loop analysis to find V_o in the circuit in Fig. P3.59.

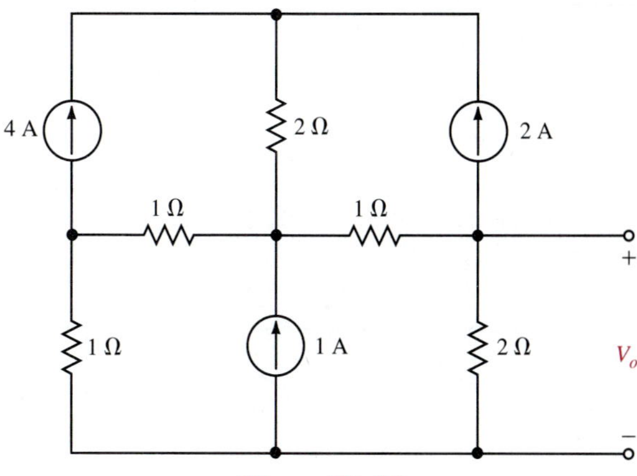

Figure P3.59

3.60. Use loop analysis to find I_o in the network in Fig. P3.60.

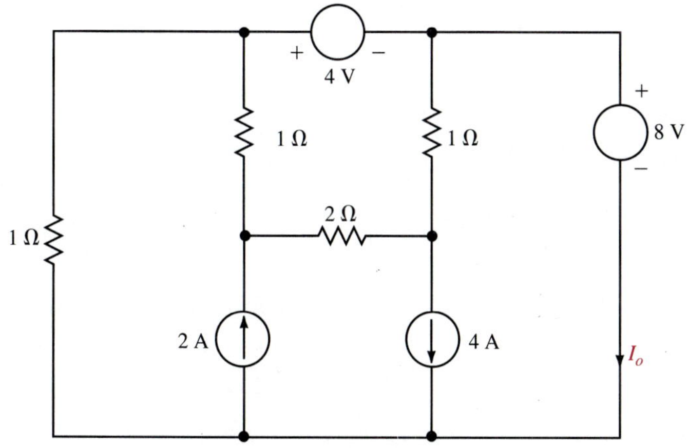

Figure P3.60

3.61. Use mesh equations to find V_o in the network in Fig. P3.61.

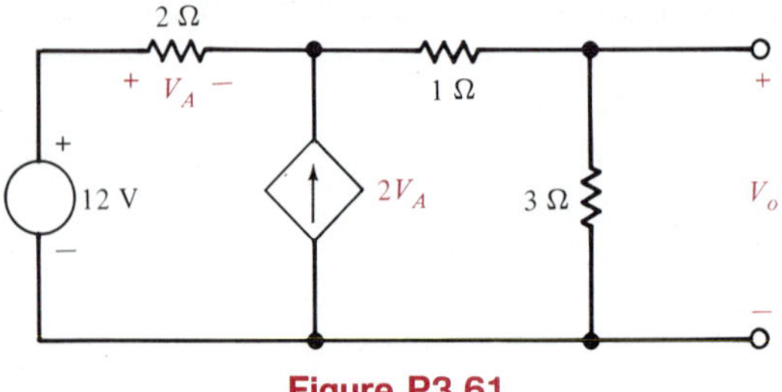

Figure P3.61

3.62. Find V_o in the circuit in Fig. P3.62.

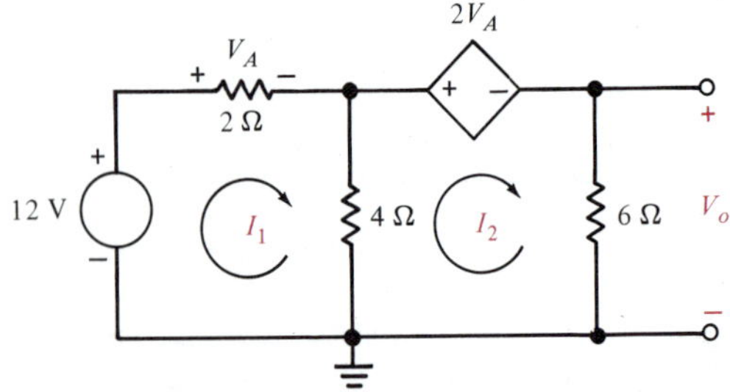

Figure P3.62

Determine the voltage across the 3-Ω resistor in the network in Fig. P3.63 using mesh analysis.

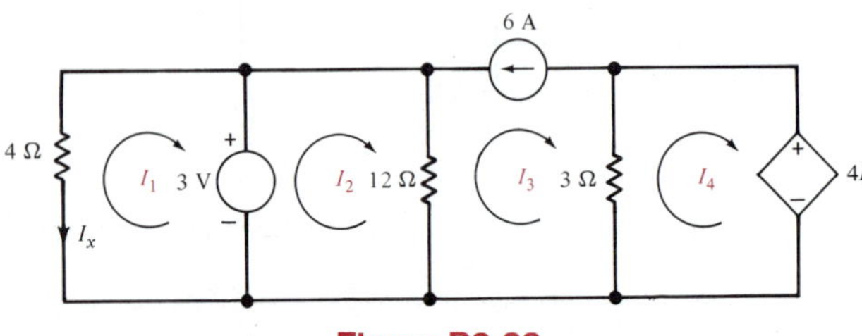

Figure P3.63

3.64. Find V_o in the network in Fig. P3.64.

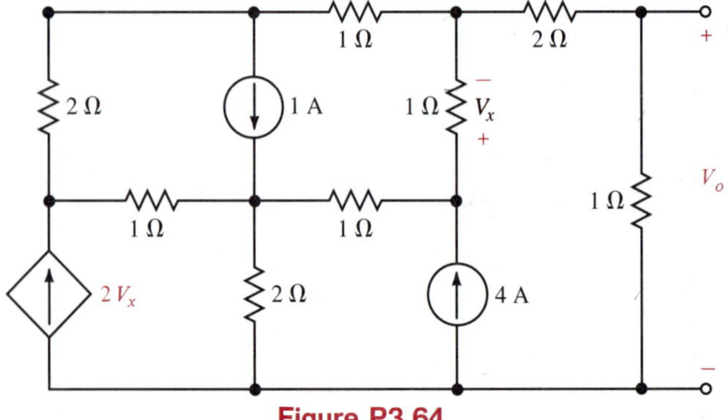

Figure P3.64

3.65. Use loop analysis to find V_o in the network in Fig. P3.65.

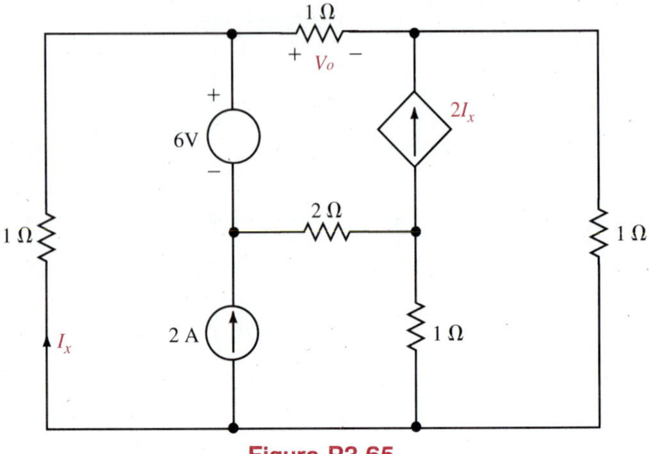

Figure P3.65

3.66. Draw a network defined by the following mesh equations.

$$(12\text{ k}\Omega)I_1 - (4\text{ k}\Omega)I_3 = -28\text{ V}$$

$$(3\text{ k}\Omega)I_2 - (2\text{ k}\Omega)I_3 = 16\text{ V}$$

$$I_3 = -\frac{4\text{ V}}{1\text{ k}\Omega}$$

3.67. Draw a circuit defined by the following mesh equations.

$$I_1 = -\frac{6\text{ V}}{1\text{ k}\Omega}$$

$$-(4\text{ k}\Omega)I_1 + (12\text{ k}\Omega)I_2 - (2\text{ k}\Omega)I_4 = 0$$

$$(4\text{ k}\Omega)I_3 - (1\text{ k}\Omega)I_4 = -12\text{ V}$$

$$I_4 = -\frac{4\text{ V}}{1\text{ k}\Omega}$$

3.68. Draw a network defined by the following node equations.

$$V_1\left(\frac{1}{8\text{ k}\Omega} + \frac{1}{1\text{ k}\Omega}\right) - V_2\left(\frac{1}{1\text{ k}\Omega}\right) = -\frac{2\text{ V}}{1\text{ k}\Omega} - \frac{1\text{ V}}{1\text{ k}\Omega}$$

$$-V_1\left(\frac{1}{1\text{ k}\Omega}\right) + V_2\left(\frac{1}{1\text{ k}\Omega} + \frac{1}{2\text{ k}\Omega}\right) = \frac{2\text{ V}}{1\text{ k}\Omega} + \frac{4\text{ V}}{1\text{ k}\Omega}$$

3.69. Draw a network defined by the following node equations.

$$V_1\left(\frac{1}{1\text{ k}\Omega} + \frac{1}{2\text{ k}\Omega} + \frac{1}{2\text{ k}\Omega}\right) - V_2\left(\frac{1}{2\text{ k}\Omega}\right) = \frac{4\text{ V}}{1\text{ k}\Omega} + \frac{2\text{ V}}{1\text{ k}\Omega}$$

$$-V_1\left(\frac{1}{2\text{ k}\Omega}\right) + V_2\left(\frac{1}{2\text{ k}\Omega} + \frac{1}{4\text{ k}\Omega}\right) = -\frac{1\text{ V}}{1\text{ k}\Omega}$$

3.70. Draw a network defined by the following node equations.

$$V_1\left(\frac{1}{1\text{ k}\Omega}+\frac{1}{2\text{ k}\Omega}+\frac{1}{4\text{ k}\Omega}\right)-V_2\left(\frac{1}{2\text{ k}\Omega}\right)-V_3\left(\frac{1}{1\text{ k}\Omega}\right)=0$$

$$V_2 = 6\text{ V}$$

$$-V_1\left(\frac{1}{1\text{ k}\Omega}\right)-V_2\left(\frac{1}{1\text{ k}\Omega}\right)+V_3\left(\frac{1}{1\text{ k}\Omega}+\frac{1}{1\text{ k}\Omega}+\frac{1}{2\text{ k}\Omega}\right)=0$$

3.71. For the graphs in Fig. P3.71, determine the number of tree branches and the number of links.

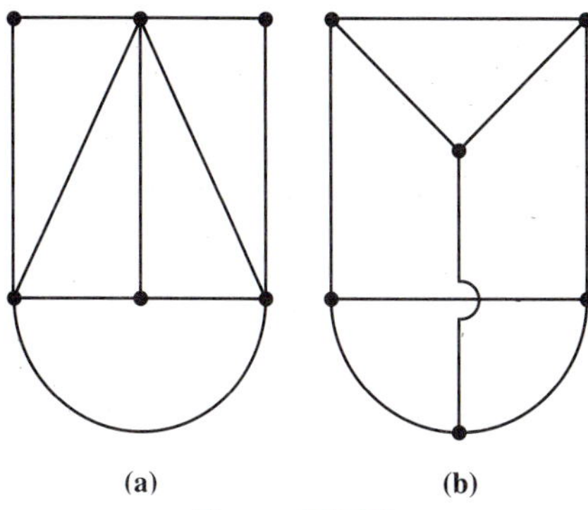

Figure P3.71

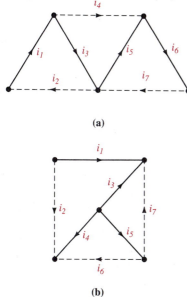

3.72. For the graphs shown in Fig. P3.72, with a specified tree, find an independent set of KCL equations using cut sets.

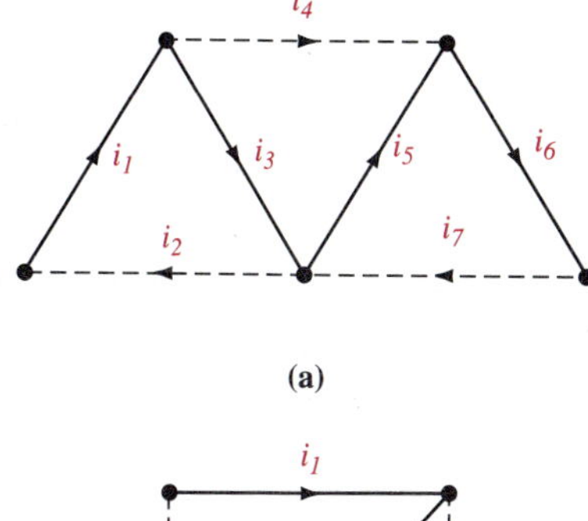

(b)
Figure P3.72

3.73. For the graph shown in Fig. P3.73 with a specified tree, find an independent set of KCL equations using cut sets.

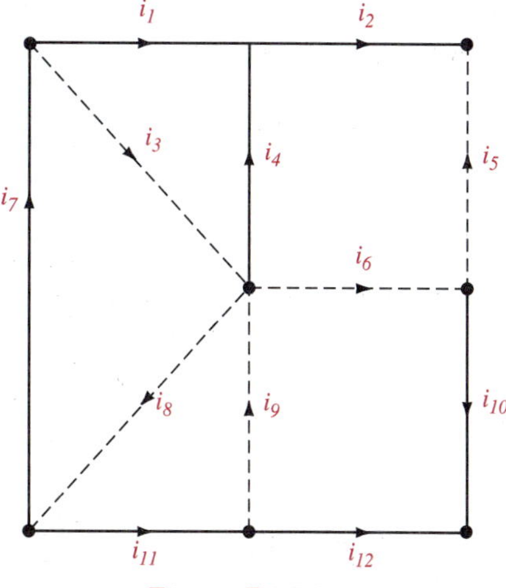

Figure P3.73

3.74. For the graphs shown in Fig. P3.74 with a specified tree, find an independent set of KVL equations using fundamental loops.

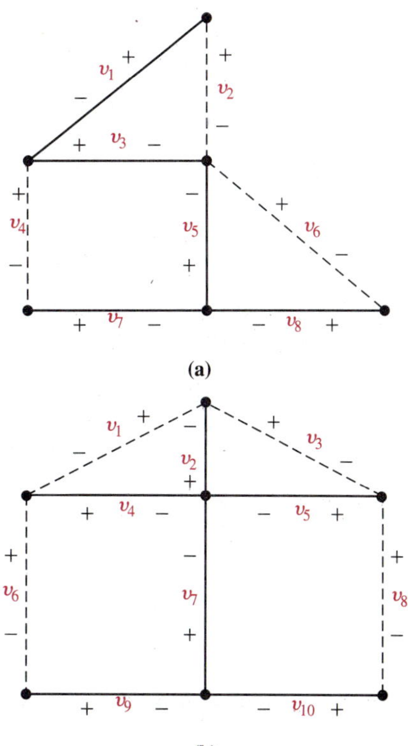

Figure P3.74

3.75. For the graphs shown in Fig. P3.75 with a specified tree, find an independent set of KVL equations using fundamental loops.

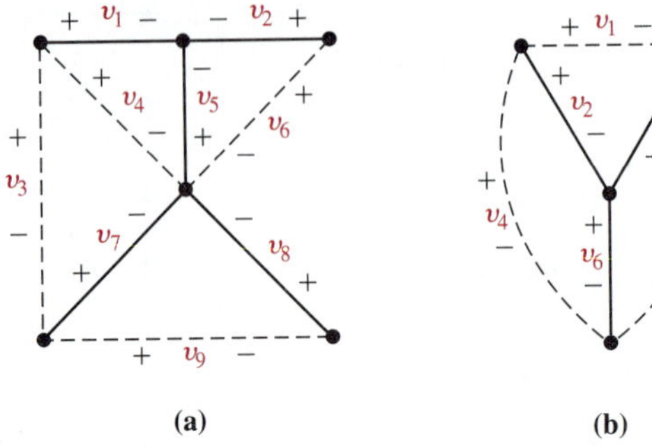

Figure P3.75

3.76. The tree for the network in Fig. P3.76a is given in Fig. P3.76b. Write the KCL equations in terms of the unknown node voltages.

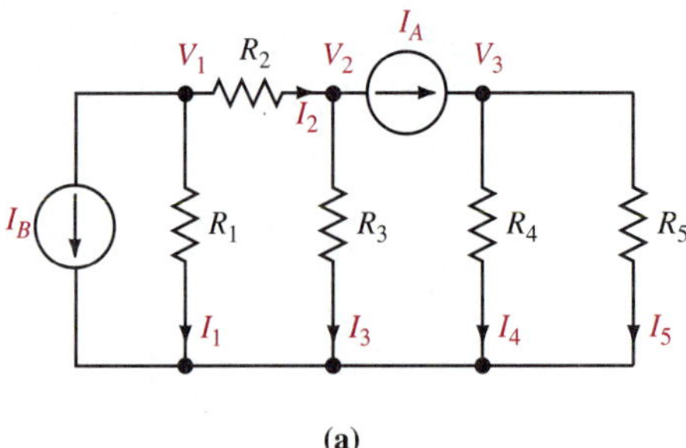

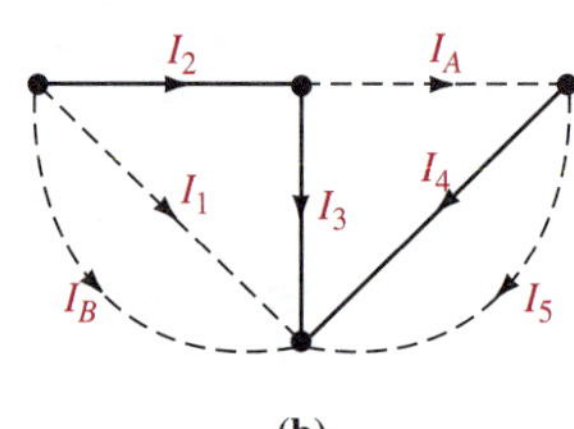

Figure P3.76

3.77. Given the tree in Fig. P3.77b for the network in Fig. P3.77a, write the loop equations necessary to solve for all the unknown voltages.

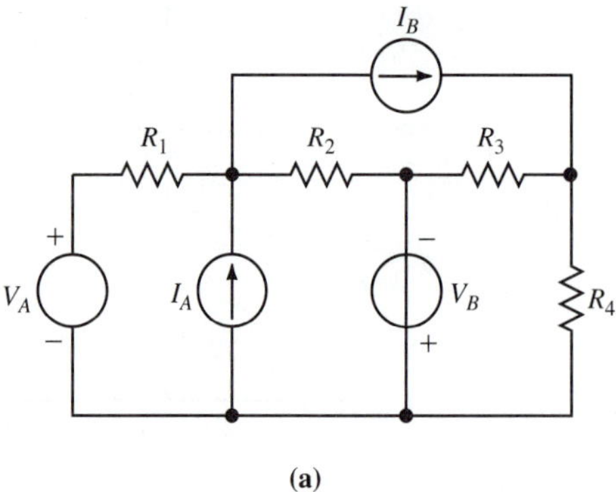

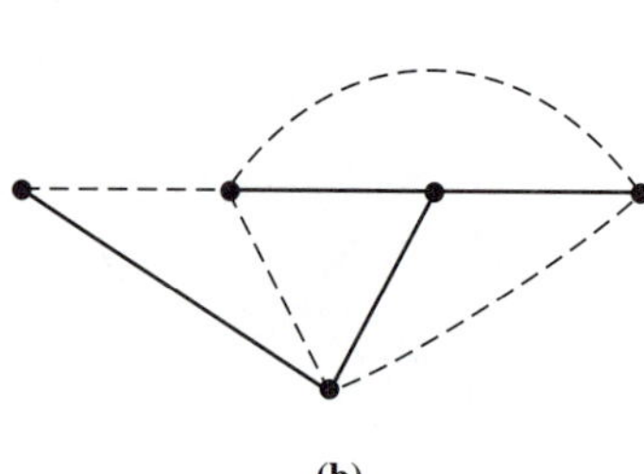

Figure P3.77

3.78. Given a box of 10-kΩ resistors and an op-amp, design a circuit that will have an output voltage of

$$v_o = -2v_1 - 4v_2$$

3.79. Find the expression for the output voltage in the op-amp circuit shown in Fig. P3.79.

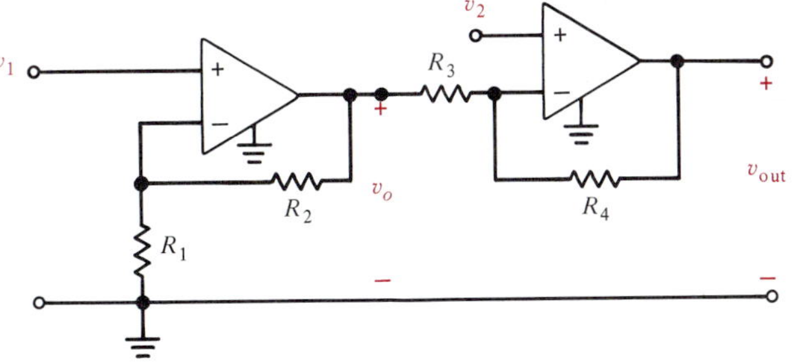

Figure P3.79

3.80. The network in Fig. P3.80 is a current-to-voltage converter or transconductance amplifier. Find v_o/i_s for this network.

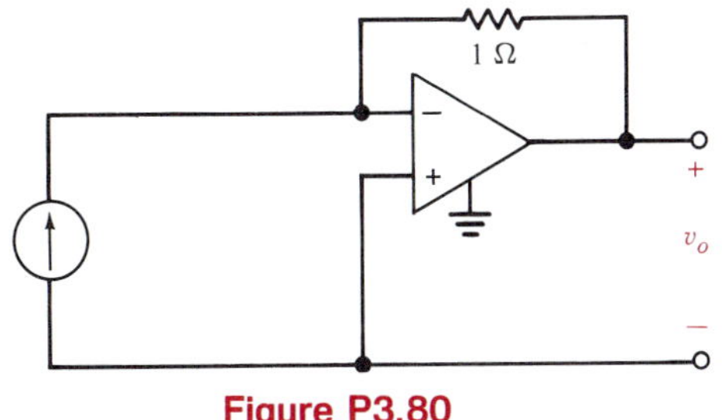

Figure P3.80

3.81. Using the equivalent circuit in Fig. 3.33, find the input resistance $R_i = v_1/i_1$ for the op-amp circuit shown in Fig. P3.81. Show that, for R_o small and A large, R_i reduces to R_1.

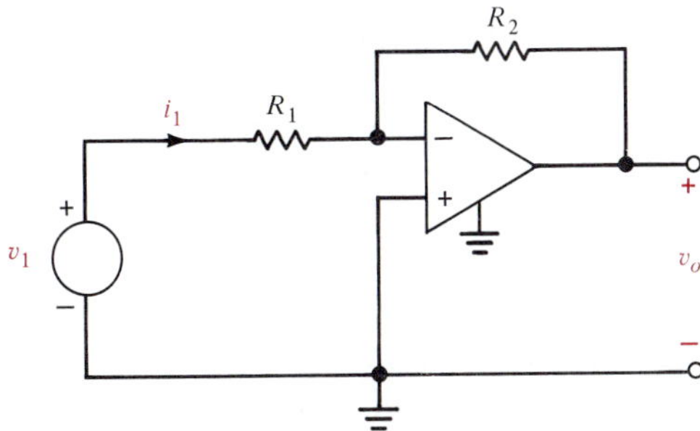

Figure P3.81

3.82. Using node equations, determine the gain v_o/v_1 of the circuit shown in Fig. P3.82.

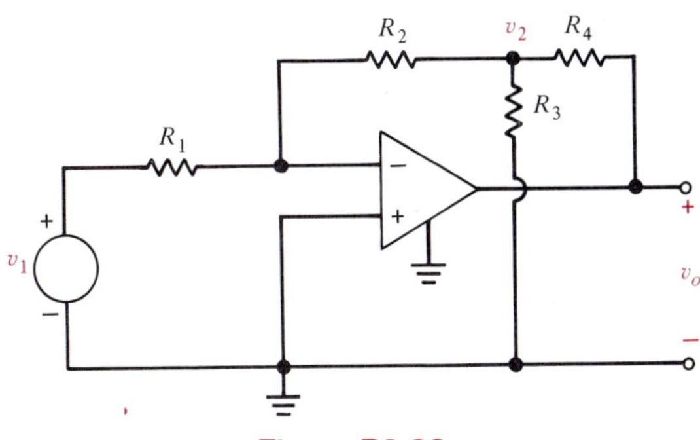

Figure P3.82

3.83. Find v_o in the network in Fig. P3.83 and explain what effect R_1 has on the output.

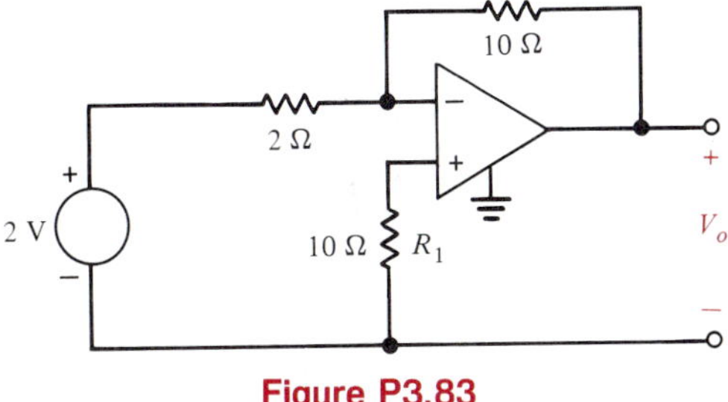

Figure P3.83

3.84. Find v_o in the network in Fig. P3.84.

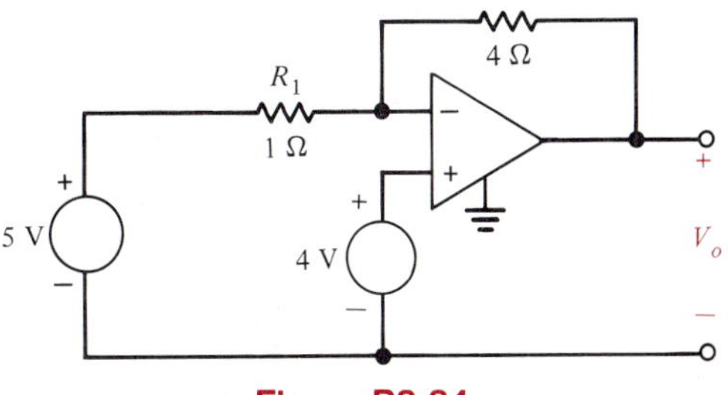

Figure P3.84

CHAPTER

4

DC PSPICE Analysis

4.1 Introduction

SPICE was developed in the Department of Electrical Engineering and Computer Science at the University of California at Berkeley. This program applies the computational power of the digital computer to the analysis of complicated circuits.

Although we demonstrate its use here with simple examples that the reader can easily check using techniques presented in this book, the real power of the program lies in the ease with which it can be applied to very complicated circuits that would appear intractable if they had to be solved by hand calculations.

Over the years SPICE has become the industry standard, and corporations routinely perform a SPICE analysis on their circuit designs before placing them in the production stream.

Unfortunately, SPICE is no panacea or substitute for a thorough understanding of circuit analysis, and therefore we cannot just turn the problem over to the computer. It is perhaps this ''let the computer figure it out'' philosophy that resulted in the acronym GIGO, which stands for ''garbage in, garbage out.''

SPICE is capable of performing dc, ac, or transient analysis. In the dc analysis case, a set of simultaneous linear or nonlinear equations with real coefficients is solved; in the ac analysis case, a set of simultaneous linear complex equations is solved; and in the transient analysis case, the solution is obtained by solving a set of simultaneous nonlinear integrodifferential equations.

PSPICE is a version of SPICE that runs on a PC or PC-compatible machine. The new version of PSPICE, number 4.05, is the one employed in this text, and the procedure for installing it on a machine is described in detail in Appendix B.

4.2 Elements of the Program

In general, the use of PSPICE in the analysis of an electric circuit requires the following steps.

1. Select the type of analysis to be performed: dc, ac, or transient analysis.
2. Draw the circuit in PSPICE format using only the building blocks that are allowed in the type of analysis desired.
3. Write the PSPICE statements describing the circuit.
4. Write the PSPICE statements that tell the program how to analyze the circuit and describe the type of solution desired.

The PSPICE statements may be subdivided into the following five categories:

1. Title and comment statements.
2. Data statements.
3. Solution control statements.
4. Output specification statements.
5. End statement.

Title and Comment Statements

The *title statement* is the first statement in any PSPICE program and typically contains a title descriptive enough to identify the particular problem under investigation. The computer will print this description as a heading for the output. The title statement *must* be present, since PSPICE always uses the first line as a title.

The *comment statement* is basically an aid to the programmer and appears as a part of the output listing. It is characterized by an asterisk as the first character, with the remaining characters containing information describing what the program is doing at that point. These statements are extremely useful when referring back to a program that has not been examined in some time.

Data Statements

Describing a circuit to someone verbally, without the use of a circuit diagram, would appear to be a difficult task at best and an insurmountable one if the circuit is very complicated. Hence, in describing the circuit to the computer, a systematic approach must be employed that clearly itemizes the exact manner in which the circuit elements are interconnected.

The fundamental building block in a PSPICE analysis is a circuit branch, and the standard branch in PSPICE is shown in Fig. 4.1. Note that a double subscript is employed to describe the branch voltages. The first subscript is assumed to be the positive node for the

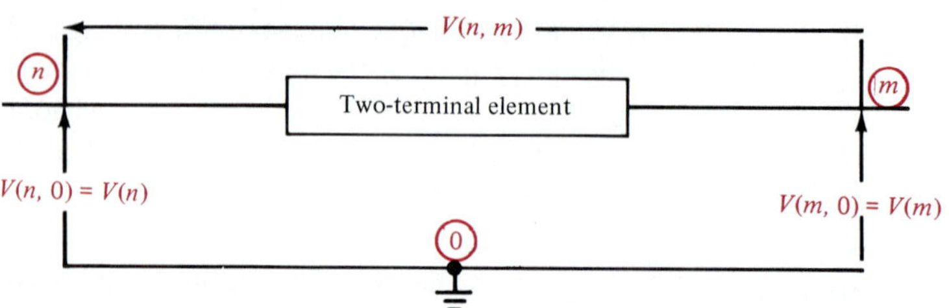

Figure 4.1 Standard PSPICE circuit branch.

branch voltage. The node voltages are denoted by $V(n, 0)$ or, for simplicity, $V(n)$. The branch current is referenced so that a positive current in a resistor produces a positive branch voltage, and therefore the reference direction for the current is from the first listed subscript on the branch voltage to the second.

Each branch may contain any two-terminal element, such as a resistor, current source, or voltage source. The values of the current or voltage sources may be zero, but the resistor value must be nonzero.

The branches are connected together at nodes which are identified to PSPICE with numbers. The numbers of the nodes must be nonnegative but need not be sequential. The reference ground node must be numbered zero. The circuit cannot contain a loop of voltage sources or a cut set (see Chapter 3) of current sources. Each node in the circuit must have a dc path to ground, and every node must have at least two connecting branches.

One of the main problems encountered in describing a circuit and the desired solutions is keeping track of the reference directions of all voltages and currents. All reference directions in PSPICE are fixed by the programmer and specified by the order in which the node connections are listed in the data statements.

A free format is used on all data statements. The fields must be separated by one or more blanks, an equal sign, or an opening or closing parenthesis. A statement may be continued by entering a plus sign as the first character of the continuation statement, and PSPICE will continue reading beginning with the second character.

Each data segment is composed of three fields:

1. The element name.
2. The circuit nodes to which the element is connected.
3. The value of the parameters that determine the electrical characteristic of the element.

The element name field must begin with a letter of the alphabet and cannot contain any delimiters. The first letter of the name specifies the element type:

R	resistor
V	independent voltage source
I	independent current source
G	voltage-controlled current source
E	voltage-controlled voltage source
F	current-controlled current source
H	current-controlled voltage source

A name can contain from one to eight characters. For example, a resistor name must begin with the letter R and can be followed by one to seven additional characters. Hence R, R1348, RINPUT, ROUT3, and RAIB66DG are all valid resistor names. In the discussion that follows, XXXXXXX, YYYYYYY, and ZZZZZZZ will denote arbitrary alphanumeric strings.

The number field may be an integer field such as 12 or −44, a floating-point field such as 3.142 or 1.4146, an integer or floating-point followed by an exponent such as 1E-14 or 2.65E3, or an integer or floating-point followed by one of the following scale factors:

T = 1E12	G = 1E9	MEG = 1E6
K = 1E3	M = 1E-3	U = 1E-6
N = 1E-9	P = 1E-12	F = 1E-15

Letters immediately following a number that are not scale factors are ignored, and letters immediately following a scale factor are ignored. Hence 10, 10V, 10VOLTS, and 10HZ all represent the same scale factor. Note also that 1000, 1000.0, 1000HZ, 1E3, 1.OE3, 1KHZ, and 1K all represent the same number. Note, however, that

1. Commas are not allowed.
2. Missing signs are assumed + by the computer.
3. A missing decimal point is placed at the end of the series of numbers.

The following is a detailed description of the format of the data statements that are used to describe each type of branch in a network.

Branch Statements for Resistive Elements. The general form for resistive elements is

```
RXXXXXXX N1 N2 VALUE
```

Here XXXXXXX denotes an arbitrary alphanumeric string that uniquely identifies the particular element. N1 and N2 are the circuit nodes to which the element is connected. VALUE is the value of the resistance in ohms. The value of a resistor can be positive or negative but cannot be zero.

EXAMPLE 4.1

The branch statements for the two resistors shown in Fig. 4.2a are

```
R1  4  6   2
R2  6  3  10
```

■

Branch Statements for Independent Sources. Branch statements for independent sources have the form

```
BXXXXXXX N+ N- DC (DC VALUE)
```

B is the letter V for voltage sources or I for current sources. N + and N − are the positive and negative nodes, respectively. Positive current, for either voltage or current sources, is assumed to flow from the positive node to the negative node through the source. DC VALUE is the dc value of the source. If the dc value is zero, this value may be omitted.

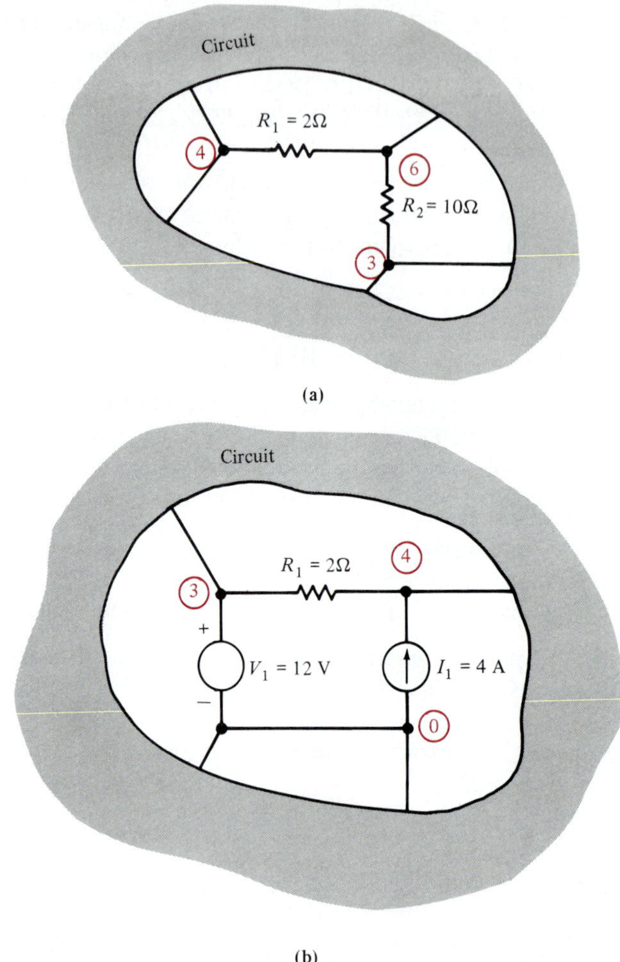

Figure 4.2 Circuits used to explain branch statements.

EXAMPLE 4.2

The branch statements for the elements shown in Fig. 4.2b are

```
R1  4  3  2
V1  3  0  DC  12
I1  0  4  DC   4
```

■

Branch Statements for Linear Dependent Sources. A linear dependent source in PSPICE is defined as a voltage source or current source whose voltage or current is a constant times either a specified branch voltage or the current flowing through a voltage source. The latter point introduces an interesting idiosyncracy of PSPICE. In order to employ the current in a particular branch, we must introduce a dummy voltage source in series with this particular branch so that the program can calculate the current in this voltage source.

Branch statements for linear voltage-controlled sources have the general form

```
BXXXXXXX N+ N- NC+ NC- VALUE
```

B is either the letter G or E, where $I_S = GV$ for voltage-controlled current sources and $V_S = EV$ for voltage-controlled voltage sources. N + and N − are the positive and negative nodes, respectively. Current flow is from the positive node, through the source, toward the negative node. NC + and NC − are the positive and negative controlling nodes, respectively. VALUE is either the value of the voltage gain E or the transconductance G.

EXAMPLE 4.3

The branch statement for the dependent source in Fig. 4.3a is

```
G1 8 11 3 5 4
```

■

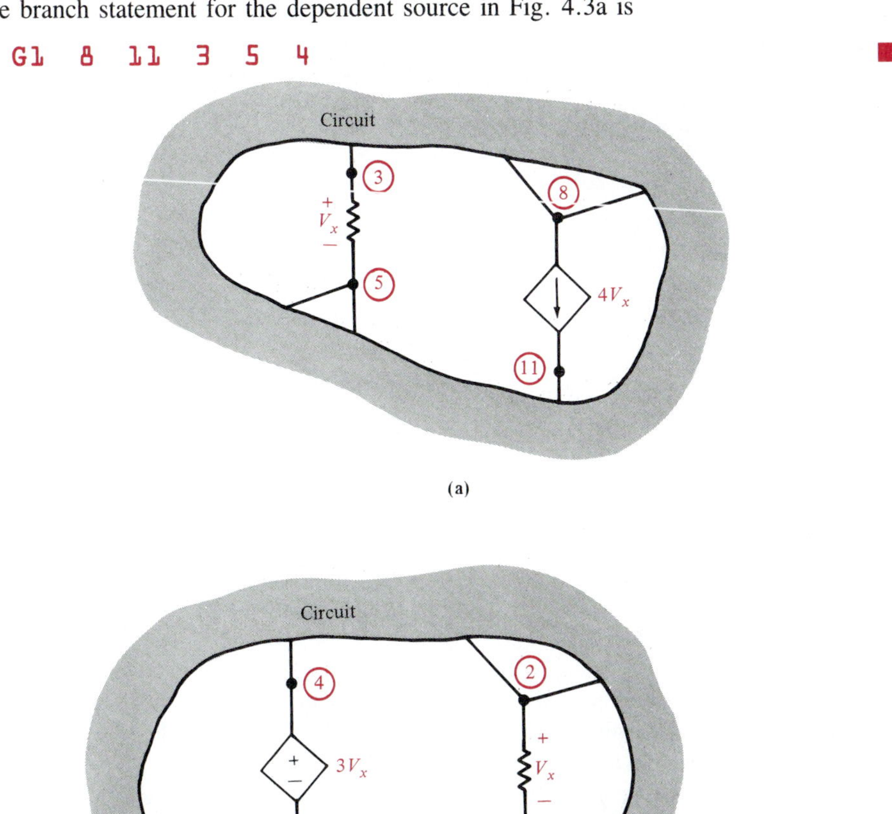

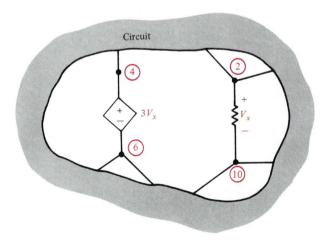

Figure 4.3 Circuits used to explain branch statements for dependent sources.

EXAMPLE 4.4

The branch statement for the dependent source in Fig. 4.3b is

```
E1  4  6  2  10  3
```

■

Branch statements for linear current-controlled sources have the form

```
BXXXXXXX N+ N- VNAME VALUE
```

B is either the letter F or H, where I_S = FI for current-controlled current sources and V_S = HI for current-controlled voltage sources. N + and N − are the positive and negative nodes, respectively. Current flow is from the positive node, through the source, to the negative node. VNAME is the name of the voltage source through which the controlling current flows. Positive controlling current flows into the positive node, through the source VNAME, and out of the negative node. VALUE is either the value of current gain F or the transresistance H.

EXAMPLE 4.5

The branch statement for the dependent source in Fig. 4.4a is

```
F1  5  12  VX  7
```

■

EXAMPLE 4.6

The branch statement for the dependent source in Fig. 4.4b is

```
H1  15  1  VX  20
```

■

In Examples 4.5 and 4.6 note that V_x is the dummy voltage source inserted into the network in order to obtain the control current I_x.

Solution Control Statements

Three of the dc analysis options are specified by the statements .OP, .DC, and .TF. The .OP statement will cause all the dc node voltages and currents in voltage sources to be listed as part of the output.

The dc analysis may be performed for a range of source voltages or currents using the .DC statement. The general form of this statement is

```
.DC SCRNAM VSTART VSTOP VINCR
```

SCRNAM is the name of an independent voltage or current source to be varied. VSTART, VSTOP, and VINCR are the starting, final, and incremental values, respectively. A second source may optionally be specified with associated sweep parameters. The form for this statement is

```
.DC SCRNAM VSTART VSTOP VINCR SCR2 START2 STOP2
INCR2
```

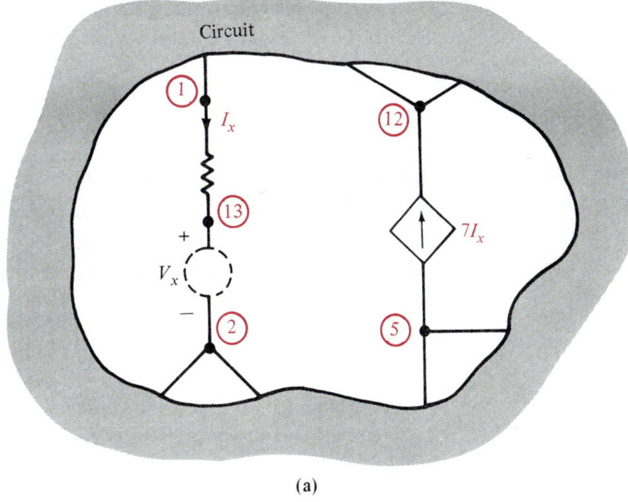

(a)

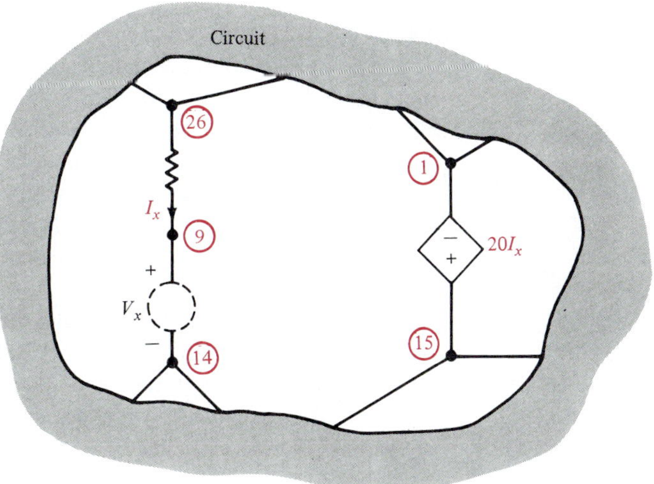

(b)

Figure 4.4 Circuits used to explain branch statements for dependent sources.

In this case, the first source will be swept over its range for each value of the second source. Some typical examples of the .DC statement are

```
.DC V1 1.25 10.50 0.25
.DC V0 0 50 1 I1 1M 10M 1M
```

The dc analysis option may also be used to evaluate dc small-signal transfer functions, that is, the ratio of an output variable to an input variable. Input resistance and output resistance will also be computed automatically as part of the solution:

```
.TF OUTVAR INSRC
```

OUTVAR is the small-signal output variable. INSRC is the small-signal input source. Two typical example statements are

```
.TF V(4, 1) VIN
.TF I(VOUT) VIN
```

Recall that the voltage between any two nodes n and m is referred to as $V(n, m)$. If the second node is zero, it may be omitted. For the equations above, PSPICE would compute the dc small-signal value of the ratio of $V(4, 1)$ to VIN, the input resistance at VIN, and the output resistance at VOUT. If no analysis is specified, PSPICE will perform an .OP analysis by default.

Output Specification Statements

Printed output is requested using the .PRINT statement with the format

```
.PRINT DC OV1 OV2 . . . OV8
```

OV1, OV2, . . . , OV8 are the current or voltage output variables desired with a maximum of eight output variables per statement. However, in a dc analysis, a print statement is no longer needed with PSPICE. All node voltages and currents in voltage sources are automatically printed. For currents the reference direction is from the positive node, through the source, to the negative node, as shown in Fig. 4.5.

Probe is an additional feature of PSPICE that allows the users to plot PSPICE results on the screen of a PC and to an attached printer. By including the **.PROBE** statement in a PSPICE file, PSPICE will automatically enter the Probe program upon completion of the PSPICE analysis.

The initial screen in Probe displays a blank graph above a menu line containing various options. The default option is "Add_trace". Press the Enter key. The blank graph remains on the screen, but the menu line is replaced by a request to enter the variables to be plotted. The user may now enter a variable or a list of variables separated by spaces. For example, "V(1) V(2) V(2,3) I(VS)" will plot the node voltages at nodes 1 and 2, the

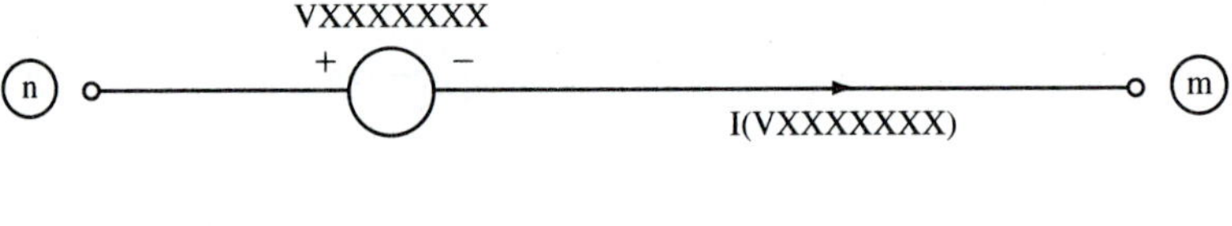

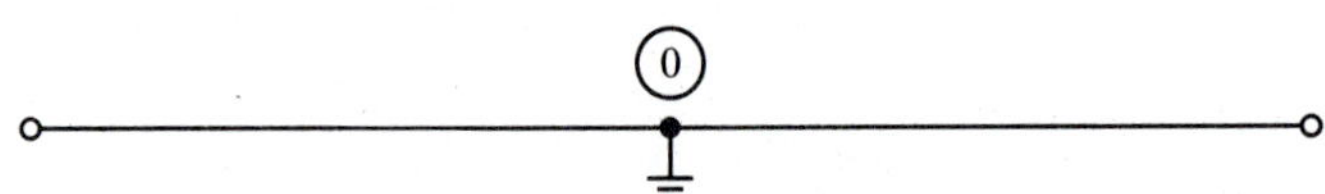

Figure 4.5 Current referenced to an independent voltage source.

node voltage between node 2 and 3, and the branch current in the branch containing VS. Also, ''V(1) * I(VS)'' will plot the product of the voltage at node 1 and the current through the branch containing VS. Press the enter key after typing the variables to be plotted. The ''Hard_copy'' option will plot the graph on the attached printer. The ''Exit'' option must be chosen to leave the Probe program. The arrow keys are used to select the desired option from the menu line.

End Statement

The programmer must tell PSPICE when the end of the input data (the program) has been reached. This is done by placing a statement at the end of the program with the format .END.

The following examples will serve to illustrate the techniques we have described.

4.3 Applications

Circuits Containing Only Independent Sources

EXAMPLE 4.7

Let us determine the node voltages in the circuit shown in Fig. 4.6. In this first PSPICE example we will describe in detail the steps employed in using PSPICE to obtain a solution.

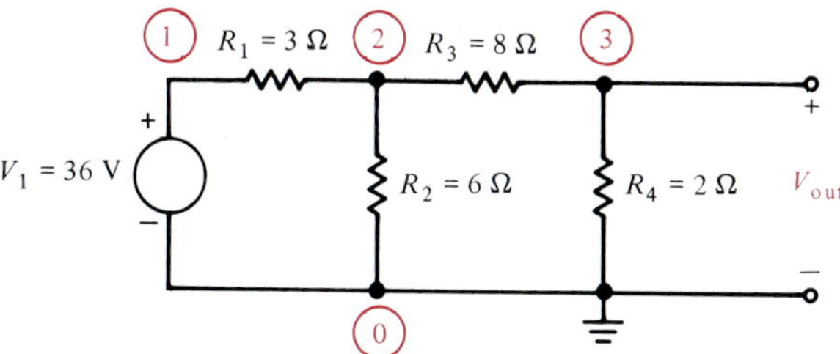

Figure 4.6 Example 4.7 dc circuit.

To use PSPICE, type 'PS' at the 'C:\ > ' prompt. The screen displays the 'PSPICE Control Shell' with the menu line at the top and the information line at the bottom. The ''Files'' option is highlighted. Press the Enter key. This activates the ''Files'' submenu. The ''Current File'' option is highlighted. Press the Enter key. Type in the name of your PSPICE file. If this will be a new circuit, select a name for new PSPICE file. The ''PSPICE Control Shell'' is once again displayed. Press the Enter key to display the ''Files'' submenu. The ''Edit'' option is highlighted. Press the Enter key. This editor is now activated.

The PSPICE program is

```
FIRST PSPICE EXAMPLE
V1   1   0   DC   36
R1   1   2    3
R2   2   0    6
R3   2   3    8
R4   3   0    2
* NODE VOLTAGES V1, V2, AND V3 AS VOUT
.END
```

Upon completion of the program, press the Esc (escape) key. Then press the "S" key to save your file. The "PSPICE Control Shell" is again displayed. Press the right arrow key to highlight the "Analysis" option. Press the Enter key to display the "Analysis" submenu. The "Run PSPICE" option is highlighted. Press the Enter key to run PSPICE on your newly created PSPICE circuit. Upon completion of the PSPICE analysis, the "PSPICE Control Shell" is displayed. Press the Enter key to display the "Files" submenu. Use the down arrow key to highlight the "Browse Output" option. Press the Enter key. The first 20 lines of your output file are displayed. Use the arrow keys, the Page Up, and the Page Down keys to display other portions of your output. Press the Esc key to return to the "Files" submenu. To exit from PSPICE, use the arrow keys to highlight the "Quit" menu option. Select the "Exit to DOS" option from the "Quit" submenu. To save your newly created file, press "S".

The output listing from the computer includes the following:

```
NODE   VOLTAGE     NODE   VOLTAGE     NODE   VOLTAGE
(1)    36.0000     (2)    20.0000     (3)    4.0000
```

Output voltage is always listed in volts and all node voltages are printed automatically. ■

EXAMPLE 4.8

Let us determine the output voltage V_o in the network in Fig. 4.7a. The circuit is redrawn in Fig. 4.7b and labeled to facilitate the PSPICE analysis. Note that in this case V(4) is V_o.

The PSPICE program for this circuit is as follows:

```
SECOND PSPICE EXAMPLE
V1    1   0   DC   12
I1    2   1   DC    2
I2    0   3   DC    4
R1    1   3    2
R2    2   0    2
R3    2   3    1
R4    3   4    1
R5    4   0    2
* Vo IS THE SAME AS NODE VOLTAGE V(4)
.END
```

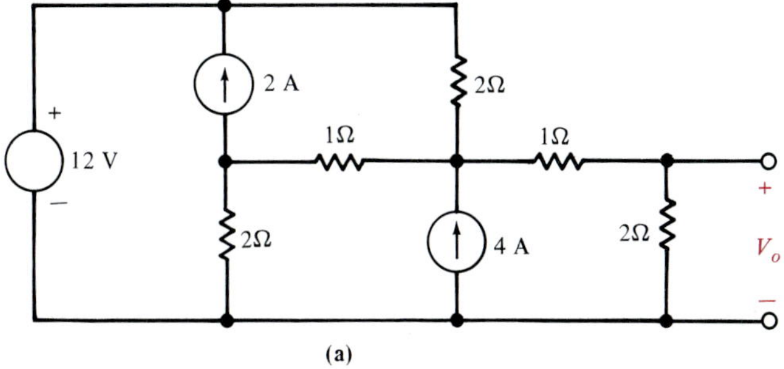

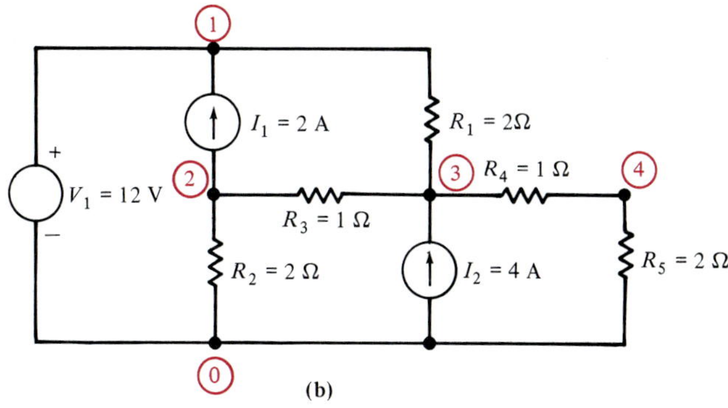

Figure 4.7 Circuit used in Example 4.8.

The output listing from the computer includes the following:

```
NODE   VOLTAGE     NODE   VOLTAGE     NODE   VOLTAGE     NODE   VOLTAGE
(1)    12.0000     (2)    3.6190      (3)    7.4286      (4)    4.9524

VOLTAGE SOURCE CURRENTS
NAME        CURRENT
V1          -2.857D-01
```

■

Note that in addition to the node voltages, the output lists the currents in all voltage sources. Remember that the current reference direction is from the positive node through the voltage source to the negative node.

EXAMPLE 4.9

We wish to find the current I_o in the network in Fig. 4.8a. The network is labeled for a PSPICE analysis in Fig. 4.8b and the dummy voltage source inserted in order to obtain I_o.

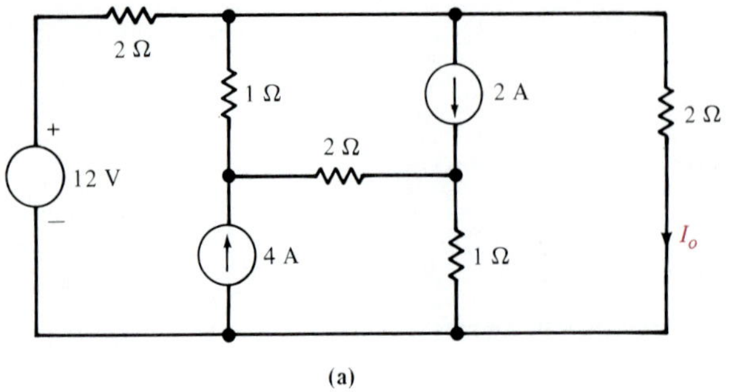

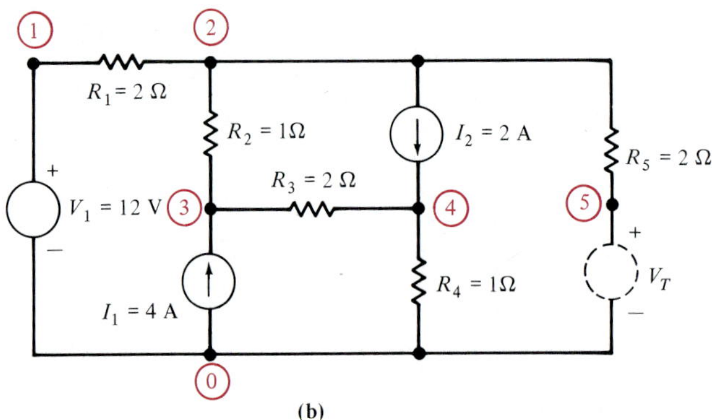

Figure 4.8 Circuit used in Example 4.9.

The PSPICE program for the circuit is

```
THIRD PSPICE EXAMPLE
V1    1   0   DC 12
VT    5   0   DC  0 ;DUMMY SOURCE USED TO CALCULATE Io.
I1    0   3   DC  4
I2    2   4   DC  2
R1    1   2    2
R2    2   3    1
R3    3   4    2
R4    4   0    1
R5    2   5    2
* Io IS THE SAME AS I(VT)
.END
```

The output listing from the computer includes the following:

```
NODE  VOLTAGE    NODE  VOLTAGE    NODE  VOLTAGE    NODE  VOLTAGE
(1)   12.0000    (2)   6.0000     (3)   8.0000     (4)   4.0000
(5)   0.0000
```

```
VOLTAGE SOURCE CURRENTS
NAME          CURRENT
V1           -3.000E+00
VT            3.000E+00
```

■

DRILL EXERCISE

D4.1. Given the circuit in Fig. D4.1, use PSPICE to determine all node voltages and the current in the 1-Ω resistor.

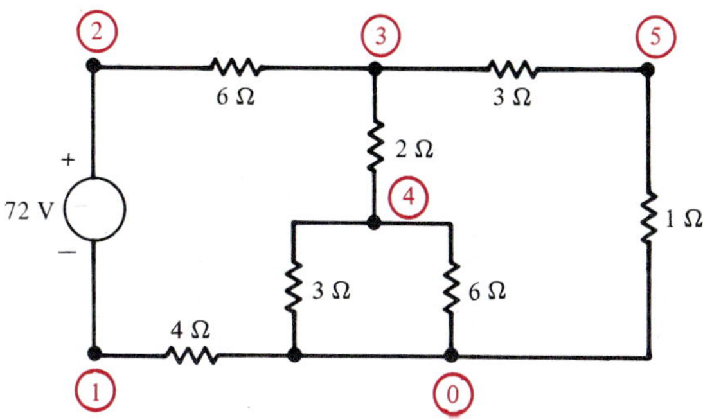

Figure D4.1

Ans: $V_1 = -24$ V, $V_2 = 48$ V, $V_3 = 12$ V, $V_4 = 6$ V, $V_5 = 3$ V, $I_{1\Omega} = 3$ A.

Circuits Containing Dependent Sources

The following examples will indicate the use of PSPICE for problems containing dependent sources.

EXAMPLE 4.10

Let us compute the voltage V_o in the network in Fig. 4.9a. The network is redrawn for a PSPICE analysis in Fig. 4.9b. In this case V(3) = V_o.

The PSPICE program is

```
FOURTH PSPICE EXAMPLE
V1  4  0  DC  6
G1  0  1   1  2 2
R1  1  4   1
R2  1  2  .5
R3  2  0   1
R4  2  3   1
R5  3  0   1
* Vo IS THE SAME AS NODE VOLTAGE V(3)
.END
```

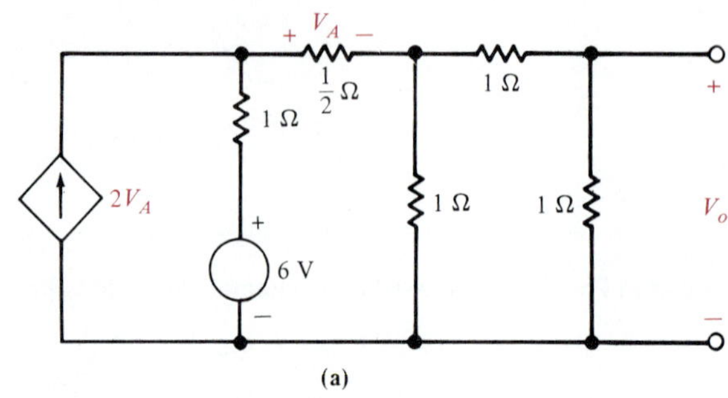

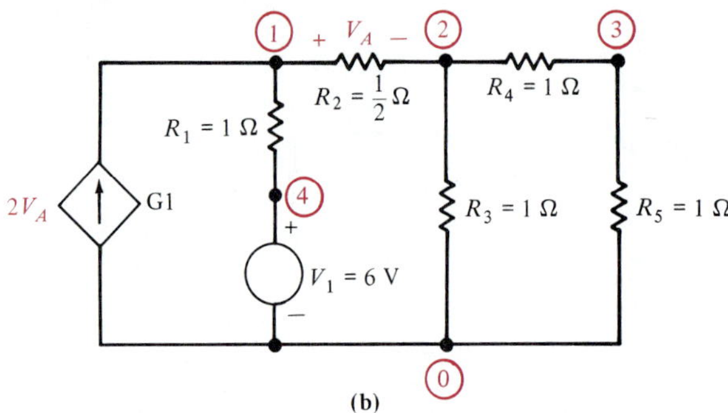

Figure 4.9 Circuit used in Example 4.10.

The computer output listing includes the following:

```
V1              V(3)
6.000E+00    1.714E+00
```

■

EXAMPLE 4.11

Let us determine the voltage V_o in the network in Fig. 4.10a. The network, as drawn in Fig. 4.10b, illustrates that V_o = V(4).

The PSPICE program is

```
FIFTH PSPICE EXAMPLE
V1   1   0   DC   6
VX   5   0   DC   0 ;DUMMY SOURCE USED TO CALCULATE Ix FOR F1
E1   3   2    1   2   2
F1   0   3   VX   1
R1   1   2    1
R2   2   5    1
R3   3   4    1
R4   4   0    1
* Vo IS THE SAME AS NODE VOLTAGE V(4)
.END
```

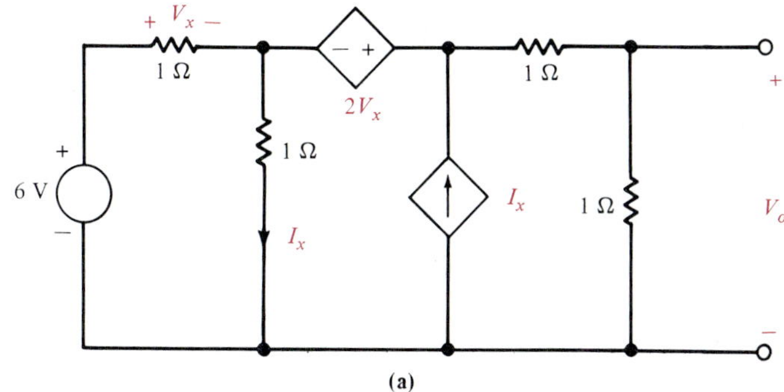

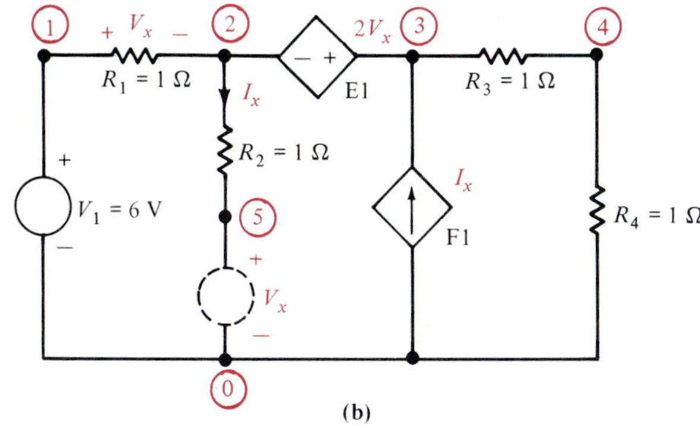

Figure 4.10 Circuit used in Example 4.11.

The computer output listing includes the following:

```
V1            V(4)
6.000E+00    6.000E+00
```

■

EXAMPLE 4.12

We wish to determine the voltage V_o in the network in Fig. 4.11a. The network labeled for PSPICE is shown in Fig. 4.11b.

The PSPICE program is

```
SIXTH PSPICE EXAMPLE
V1  2  3  DC  12
VX  5  0  DC   0 ;DUMMY SOURCE USED TO CALCULATE Ix FOR H1
H1  0  1  VX   2
R1  1  2   1
R2  2  0   2
R3  3  5   2
R4  3  4   1
R5  4  0   1
* Vo IS THE SAME AS NODE VOLTAGE V(4)
.END
```

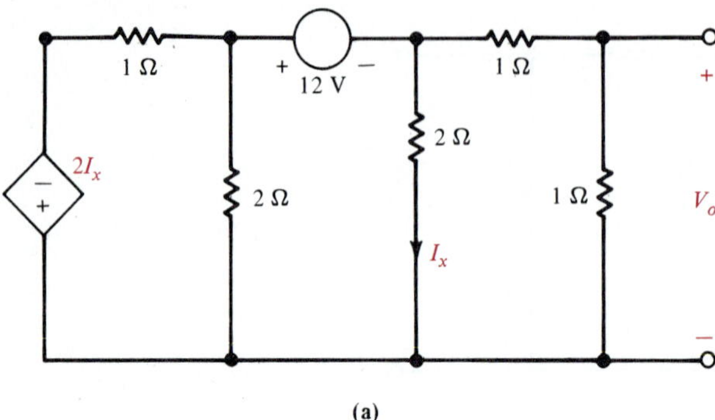

(a)

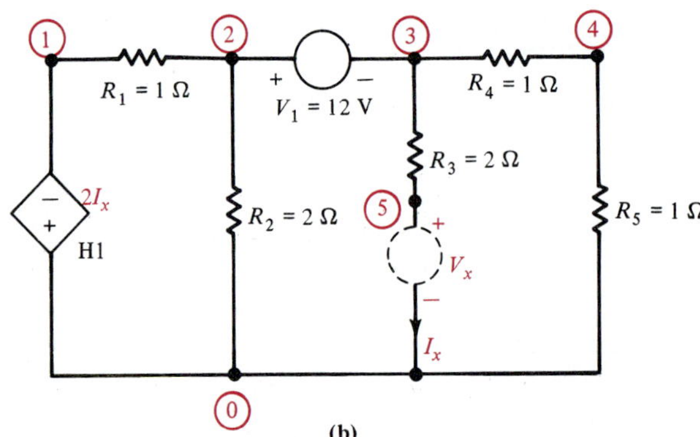

(b)

Figure 4.11 Circuit used in Example 4.12.

The computer output listing includes

```
NODE     VOLTAGE
(4)      -2.5714
```

EXAMPLE 4.13

Let us determine the voltage V_o in the network in Fig. 4.12a. The network is redrawn for the PSPICE analysis in Fig. 4.12b.

The PSPICE program is

```
SEVENTH PSPICE EXAMPLE
V1   1   0   DC    6
V2   4   3   DC   12
VX   2   3   DC    0 ;DUMMY SOURCE USED TO CALCULATE Ix FOR F1
F1   4   0   VX    1
R1   1   2    1
R2   3   0    1
R3   4   5    1
R4   5   0    1
* Vo IS THE SAME AS NODE VOLTAGE V(5)
.END
```

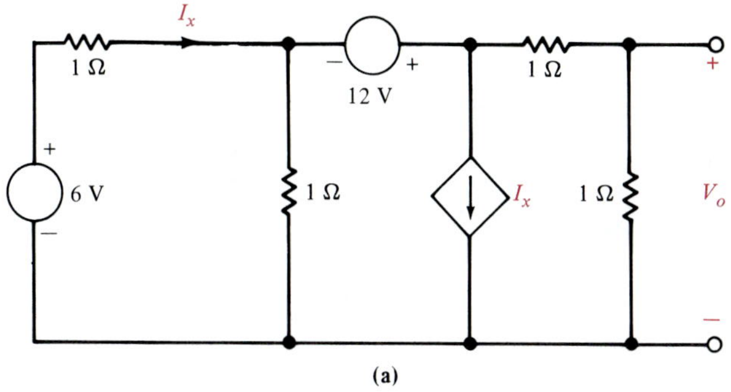

(a)

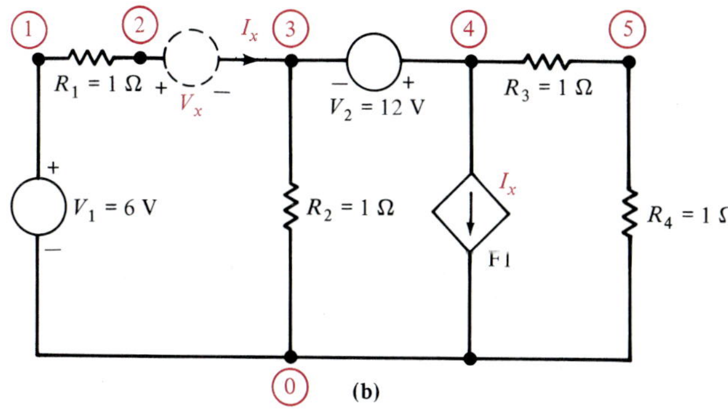

(b)

Figure 4.12 Circuit used in Example 4.13.

The computer output listing includes

```
V1              V(5)
6.000E+00       4.000E+00
```

DRILL EXERCISE

D4.2. Compute the node voltages in the circuit in Fig. D4.2.

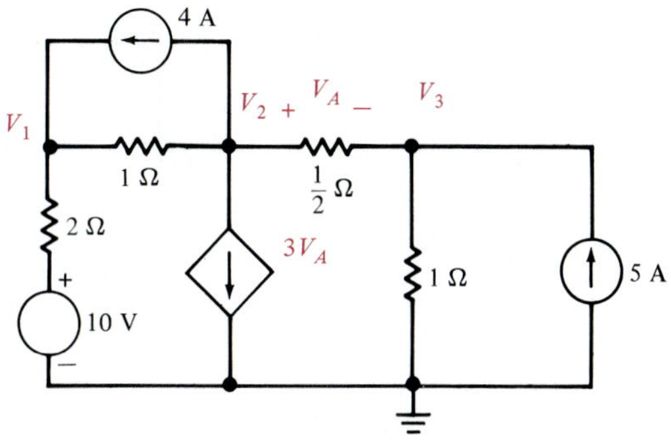

Figure D4.2

Ans: $V_1 = 9.44$ V, $V_2 = 5.17$ V, $V_3 = 5.11$ V.

D4.3. Compute the node voltages in the network in Fig. D4.3.

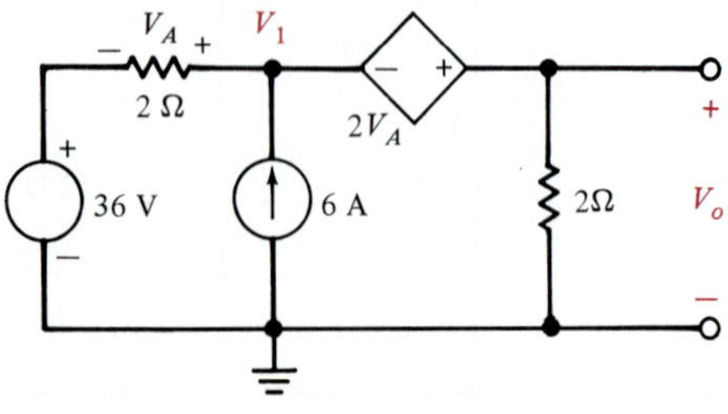

Figure D4.3

Ans: $V_1 = 30$ V, $V_o = 18$ V.

D4.4. Compute the node voltages in the network in Fig. D4.4.

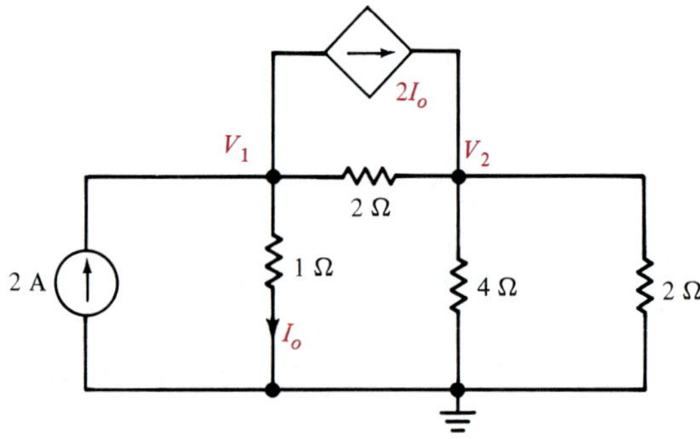

Figure D4.4

Ans: $V_1 = 0.8$ V, $V_2 = 1.6$ V.

D4.5. Given the network in Fig. D4.5, compute the node voltages and the current I_x.

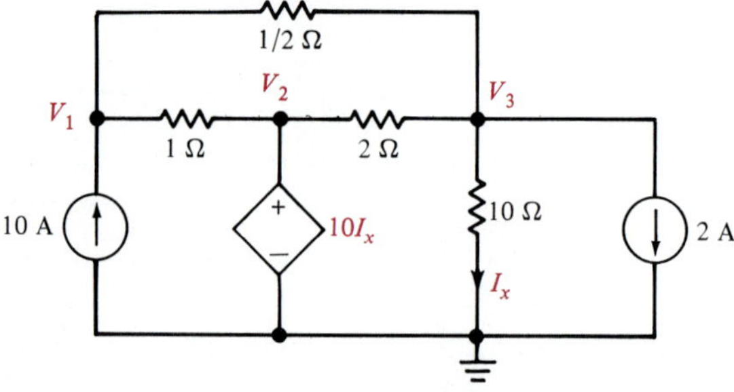

Figure D4.5

Ans: $V_1 = 50$ V, $V_2 = 46.67$ V, $V_3 = 46.67$ V, $I_x = 4.67$ A.

Circuits Containing Operational Amplifiers

We will now apply PSPICE to networks containing op-amps. The model we will employ is shown in Fig. 3.33b. The typical parameter values we will use are R_i = 1 MEGΩ, $A = 10^5$, and $R_o = 50\ \Omega$.

EXAMPLE 4.14

Let us determine the output voltage of the op-amp circuit shown in Fig. 4.13a. The network is redrawn in Fig. 4.13b. The PSPICE program and output listing are

```
BASIC INVERTING CONFIGURATION OF AN OP-AMP
V1    1    0    DC    10M
E1    4    0    0     2   1E5
R1    1    2    10K
R2    2    0    1MEG
R3    2    3    50K
R4    3    4    50
* Vo IS THE SAME AS NODE VOLTAGE V(3)
.END
```

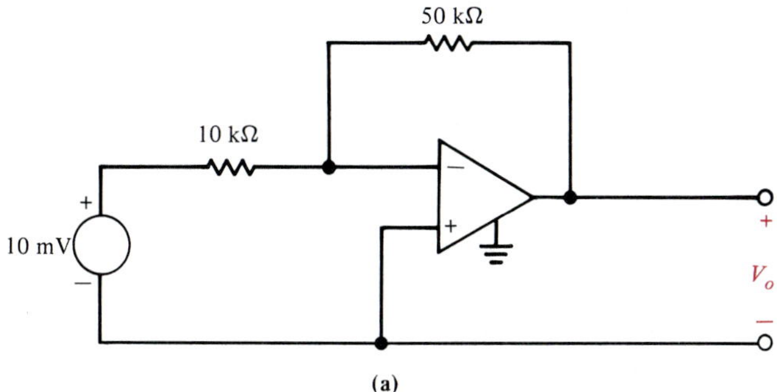

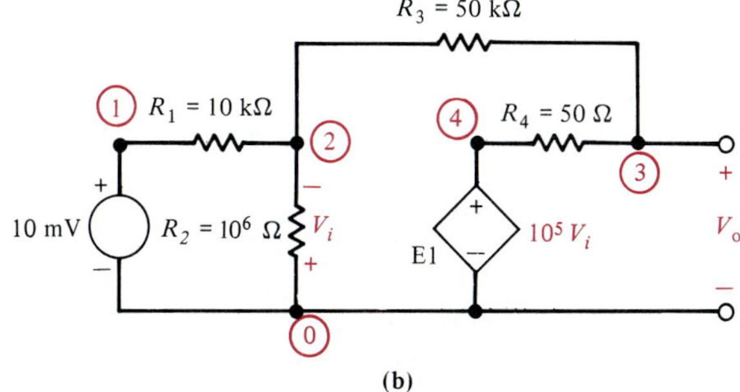

Figure 4.13 Circuit used in Examples 4.14 and 4.15.

The program output includes

```
V1              V(3)
1.000E-02       -5.000E-02
```

■

EXAMPLE 4.15

Given the op-amp network in Example 4.14, let us change the source voltage to 1 V and compute the output voltage, the gain of the circuit, and the input resistance and output resistance of the network.

The PSPICE program and output listing are as follows:

```
USE OF .TF ON INVERTING OP-AMP
V1    1    0    DC    1
E1    4    0    0    2    1E5
R1    1    2    10K
R2    2    0    1MEG
R3    2    3    50K
R4    3    4    50
* .TF GIVES THE GAIN, INPUT RESISTANCE, AND
*OUTPUT RESISTANCE
.TF V(3) V1
* Vo IS THE SAME AS NODE VOLTAGE V(3)
.END
```

The program output includes

```
V1              V(3)
1.000E+00      -5.000E+00
V(3)/V1 = -5.000E+00
INPUT RESISTANCE AT V1 = 1.000E+04
OUTPUT RESISTANCE AT V(3) = 3.025E-03
```

DRILL EXERCISE

D4.6. Compute the output voltage, gain, and input and output resistances for the network in Fig. D4.6.

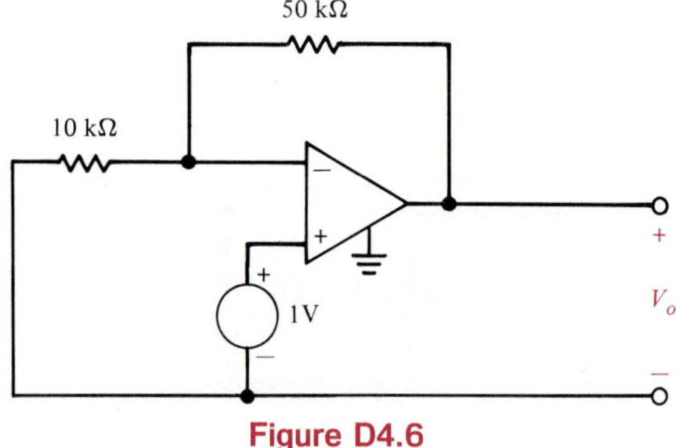

Figure D4.6

Ans: $V_o = 6$ V, gain = 6, $R_i = 1.67 \times 10^{11}$ Ω, $R_o = 3 \times 10^{-4}$ Ω.

4.4 Summary

In this chapter we have introduced the computer-aided analysis program PSPICE. We have presented and explained the various computer statements used to input data, describe the type of solution desired, and specify the output. Finally, we have applied PSPICE to circuits that contain independent sources, dependent sources, and op-amps.

KEY POINTS

- The PSPICE program consists of the following five categories of statements:
  1. Title and comment statements.
  2. Data statements.
  3. Solution control statements.
  4. Output specification statements.
  5. End statement.
- Helpful hints when programming in PSPICE:
  1. The very first available line of a program is taken as the title. If you skip a line before typing the title, the title will be taken to be a data statement and the program will not run.
  2. The number ''zero'' and the letter ''oh'' are different characters. If you inadvertently type ''oh'' where ''zero'' is intended, your programs may not run and you will experience great difficulty in locating the error.
  3. The lack of understanding of current and voltage reference direction conventions causes more errors in programming and result interpretation than does any other single factor. Be forewarned!

PROBLEMS

4.1. Find I_1 in the circuit shown in Fig. P4.1.

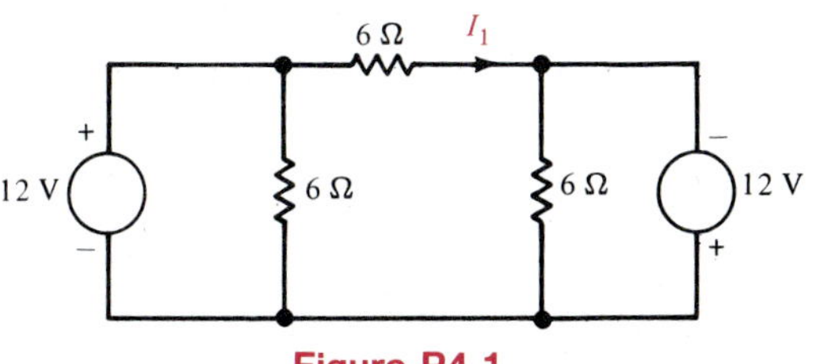

Figure P4.1

4.2. Find V_o in the network in Fig. P4.2.

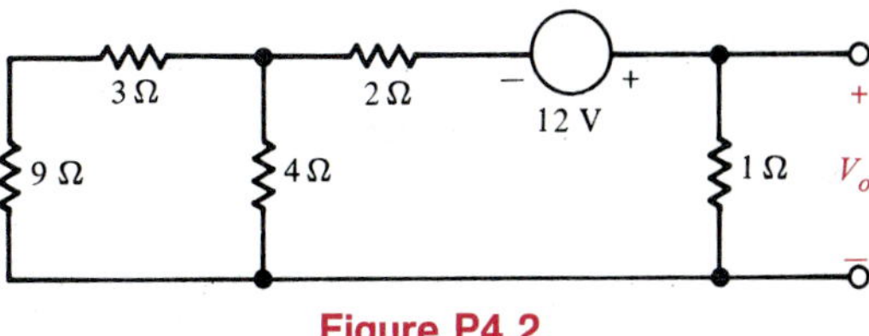

Figure P4.2

4.3. Find I_o in the circuit in Fig. P4.3.

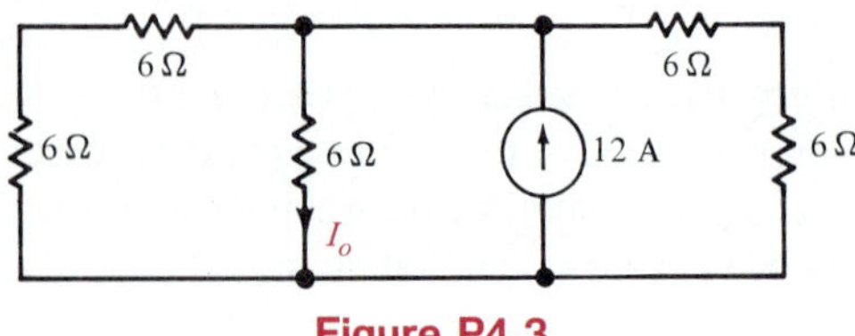

Figure P4.3

4.4. Determine I_o in the network in Fig. P4.4.

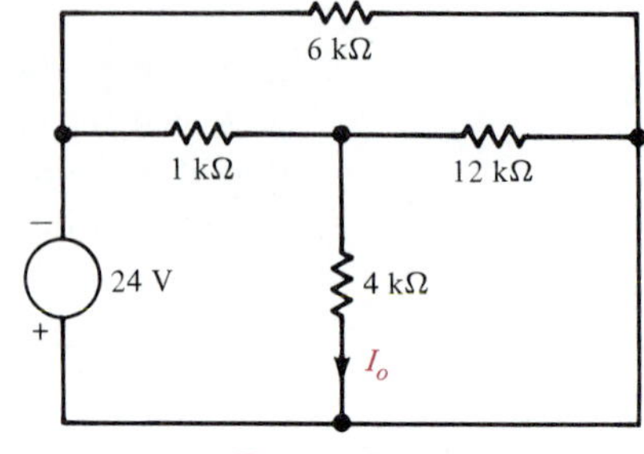

Figure P4.4

4.5. Find V_o in the network in Fig. P4.5.

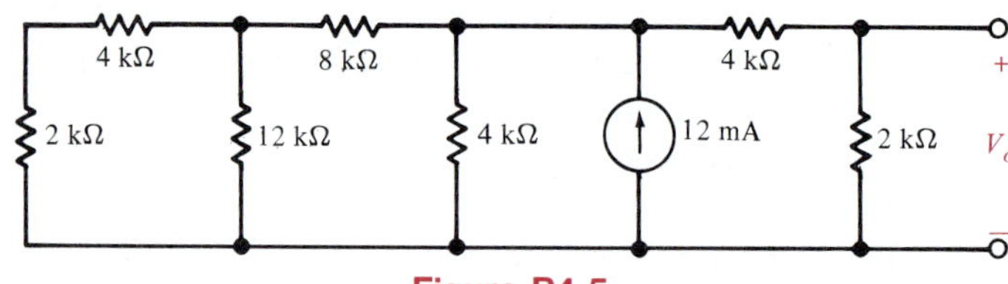

Figure P4.5

4.6. Determine I_o in the circuit in Fig. P4.6.

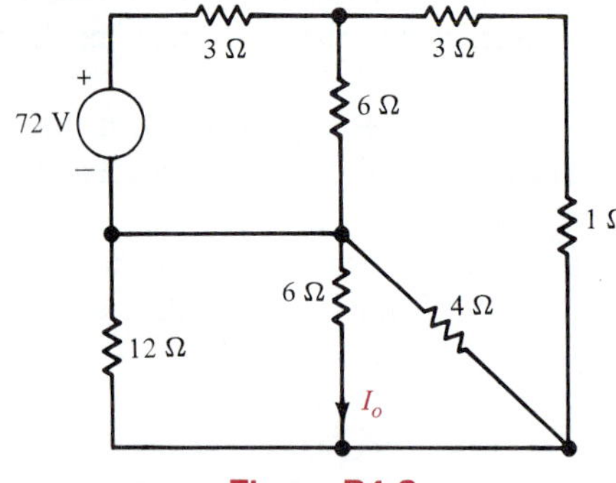

Figure P4.6

4.7. Find I_o in the network in Fig. P4.7.

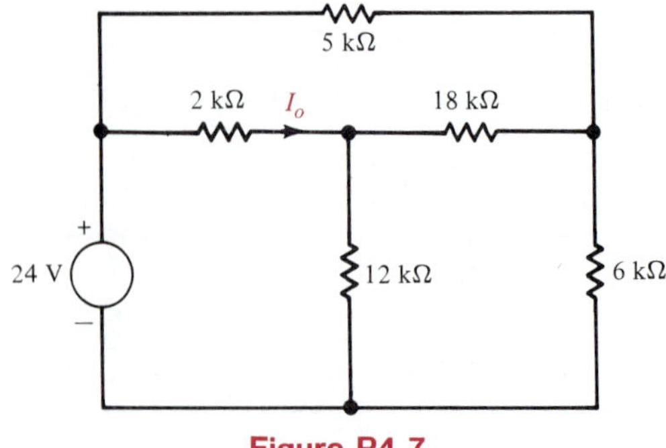

Figure P4.7

4.8. Find I_o in the network in Fig. P4.8.

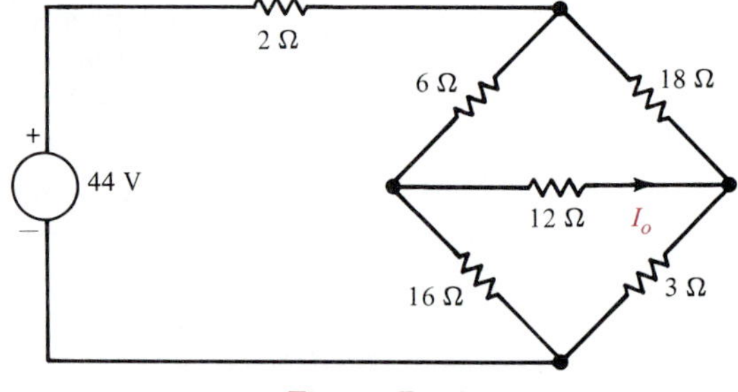

Figure P4.8

4.9. Find I_o in the network in Fig. P4.9.

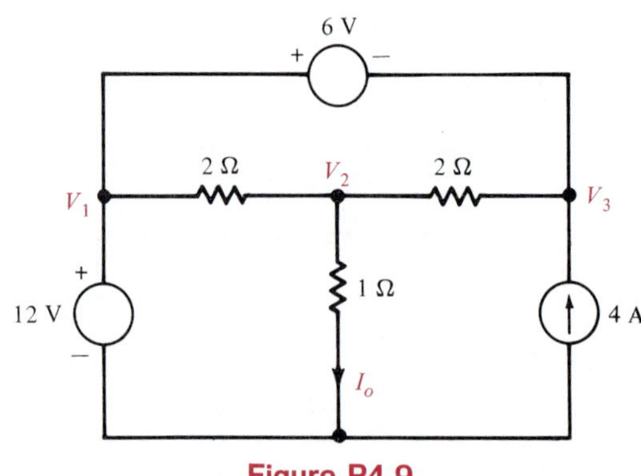

Figure P4.9

4.10. Find V_o in the circuit in Fig. P4.10.

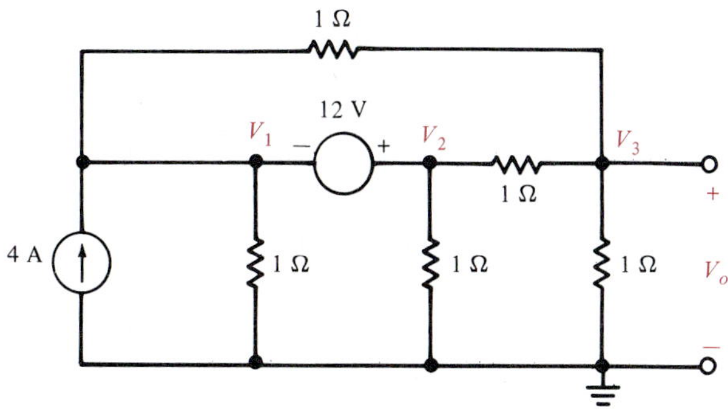

Figure P4.10

4.11. Determine I_o in the network in Fig. P4.11.

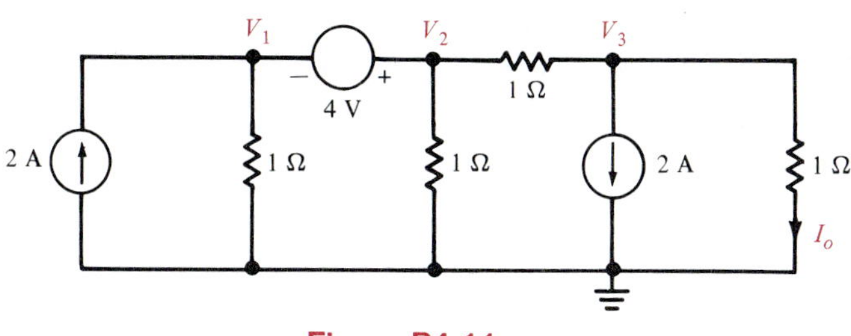

Figure P4.11

4.12. Find V_o in the circuit in Fig. P4.12.

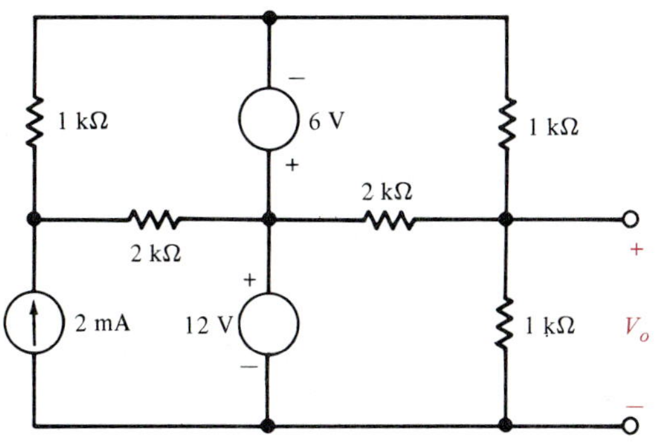

Figure P4.12

4.13. Compute V_o in the network in Fig. P4.13.

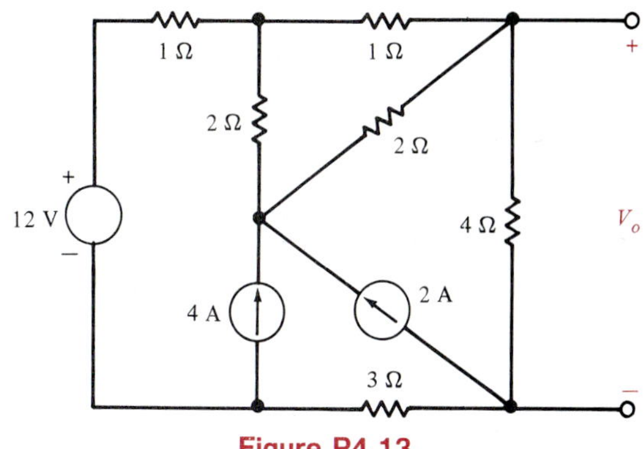

Figure P4.13

4.14. Find I_o in the circuit in Fig. P4.14.

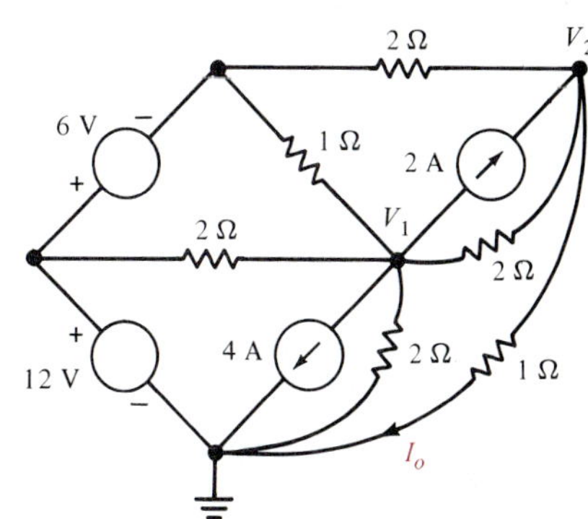

Figure P4.14

4.15. Determine V_o in the network in Fig. P4.15.

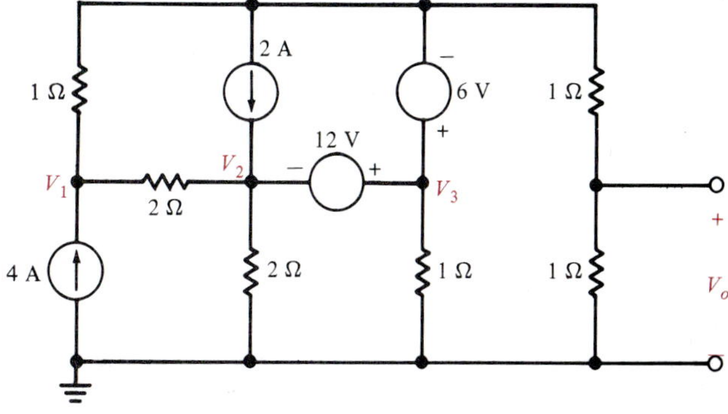

Figure P4.15

4.16. Find the current in each branch of the network in Fig. P4.16.

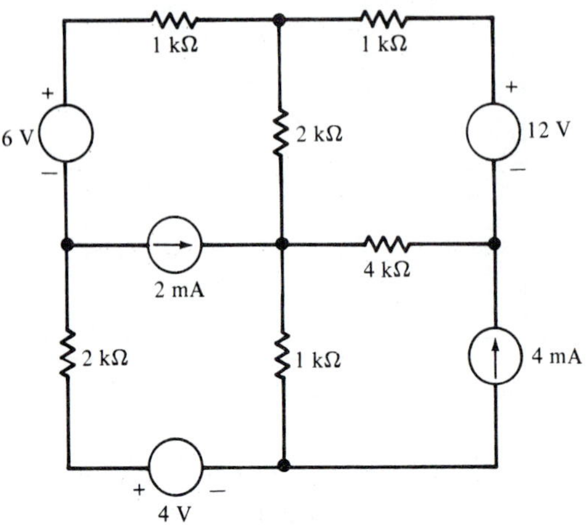

Figure P4.16

4.17. Compute V_o in the network in Fig. P4.17.

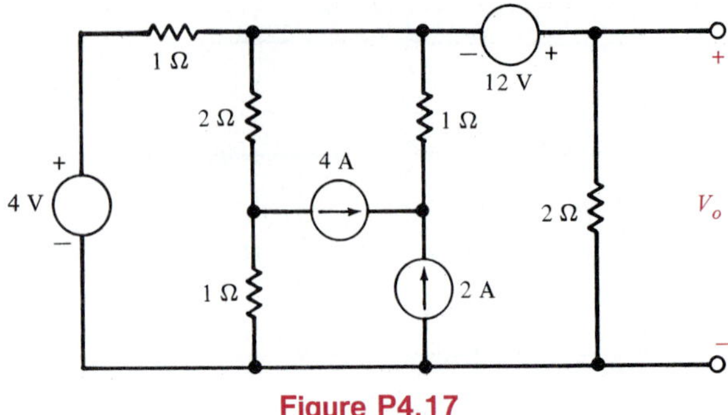

Figure P4.17

4.18. Compute V_o in the network in Fig. P4.18.

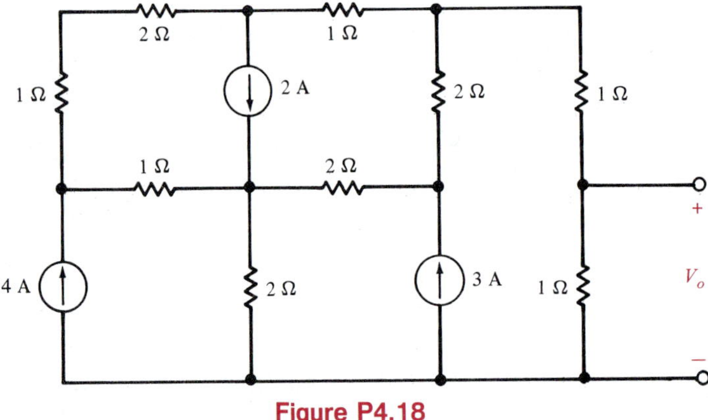

Figure P4.18

4.19. Determine V_o in the circuit in Fig. P4.19.

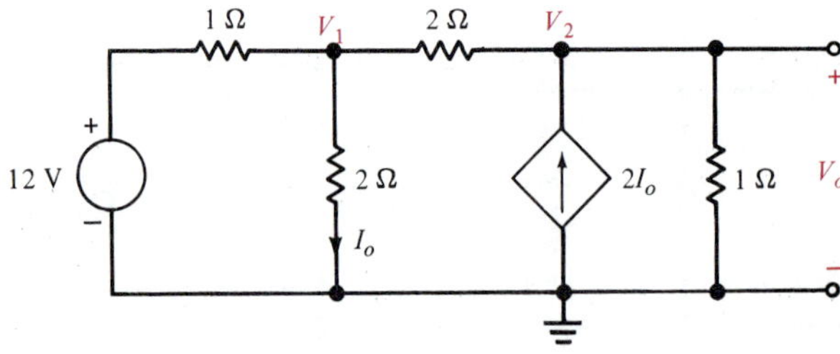

Figure P4.19

4.20. Find V_o in the network in Fig. P4.20.

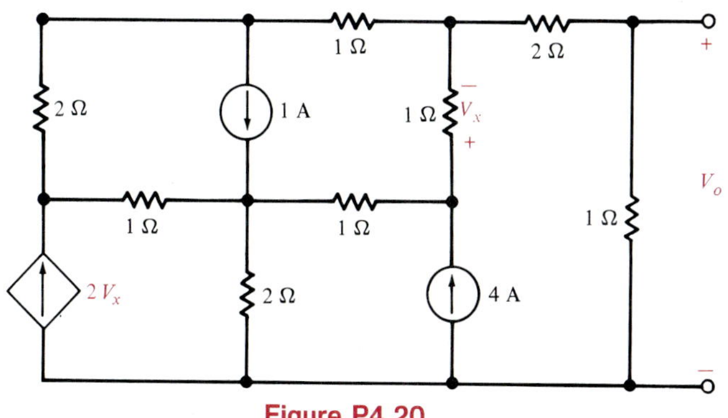

Figure P4.20

4.21. Compute all the node voltages, the current I_x, and the current in the 10-V source in the network in Fig. P4.21.

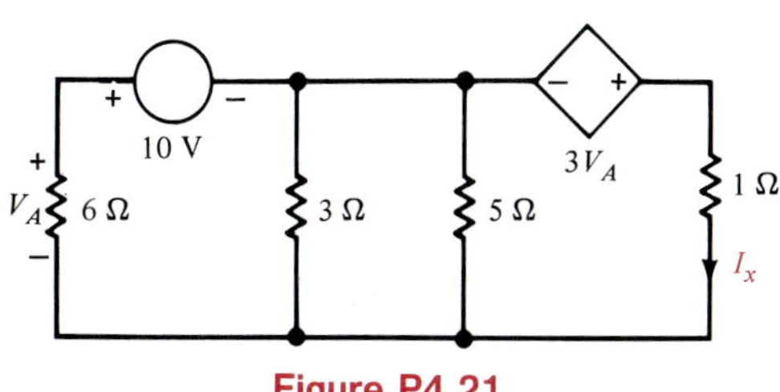

Figure P4.21

4.22. Compute all the node voltages, the currents I_x and I_a, and the current in the 10-V source in the network in Fig. P4.22.

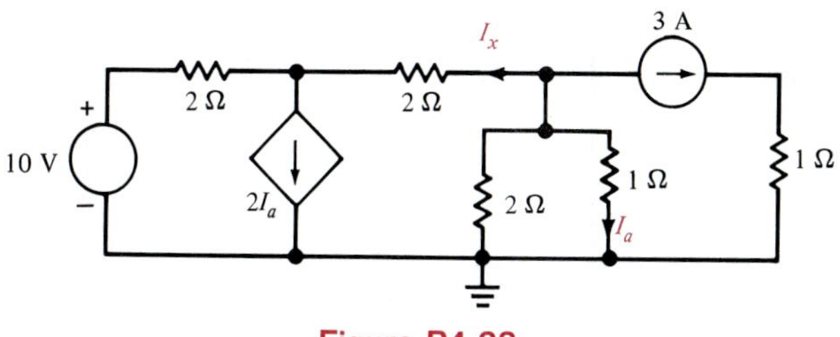

Figure P4.22

4.23. Compute the voltage V_o in the network in Fig. P4.23.

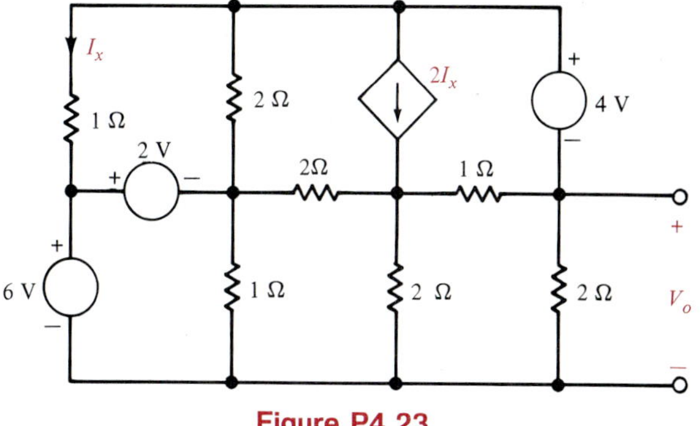

Figure P4.23

4.24. Compute V_o in the network in Fig. P4.24.

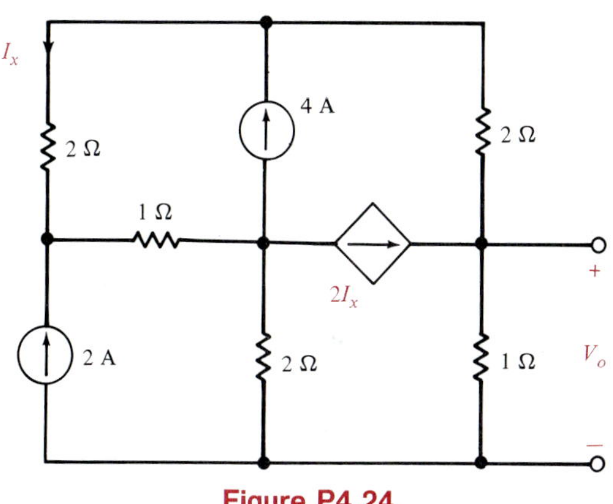

Figure P4.24

4.25. Compute V_o in the network in Fig. P4.25.

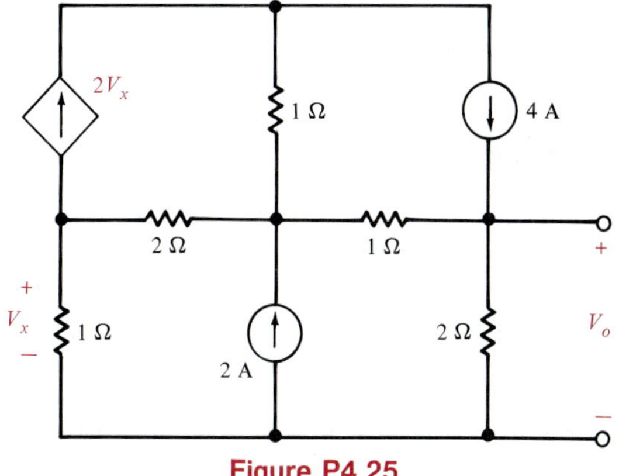

Figure P4.25

4.26. Compute V_o in the network in Fig. P4.26.

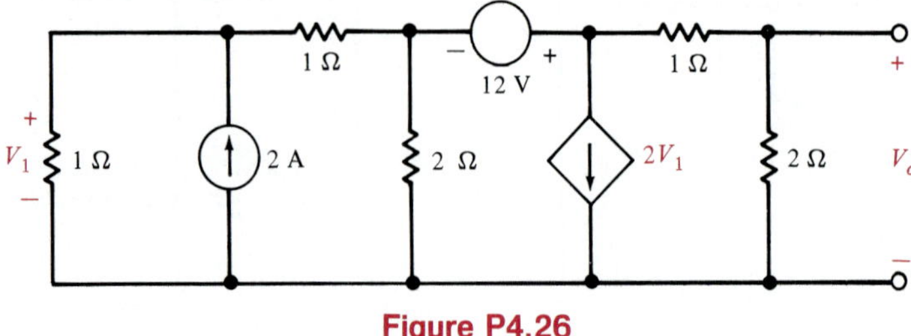

Figure P4.26

4.27. Compute V_o in the network in Fig. P4.27.

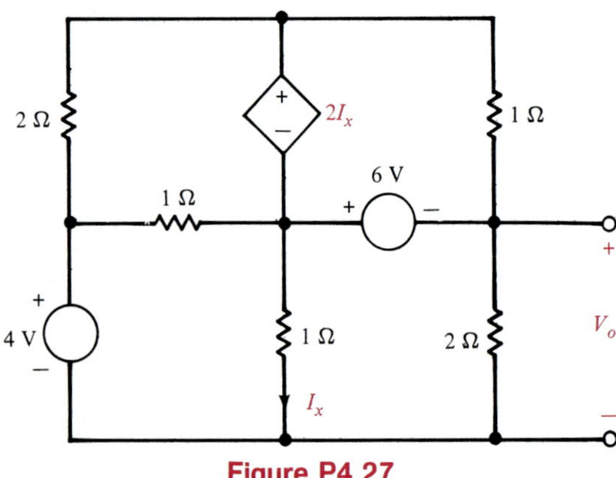

Figure P4.27

4.28. Compute V_o in the network in Fig. P4.28.

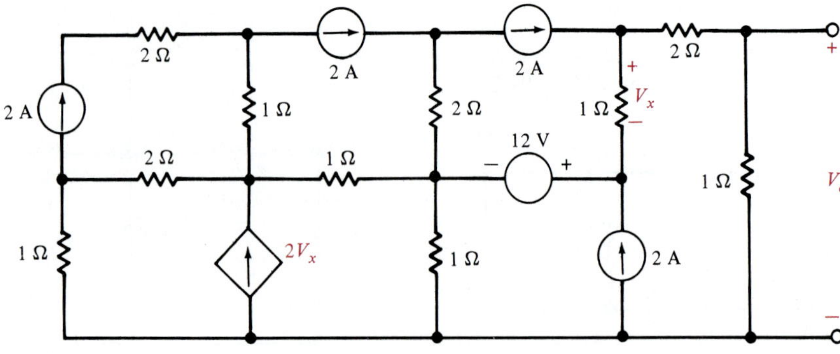

Figure P4.28

4.29. Compute I_1, I_2, and V_o in the network shown in Fig. P4.29.

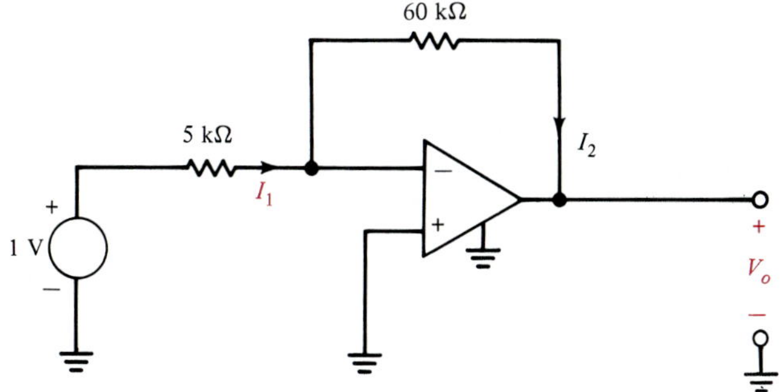

Figure P4.29

4.30. Given the network in Fig. P4.30, (a) compute the output voltage V_o and (b) determine what single resistor R_2 could be used to replace the T network in the feedback loop.

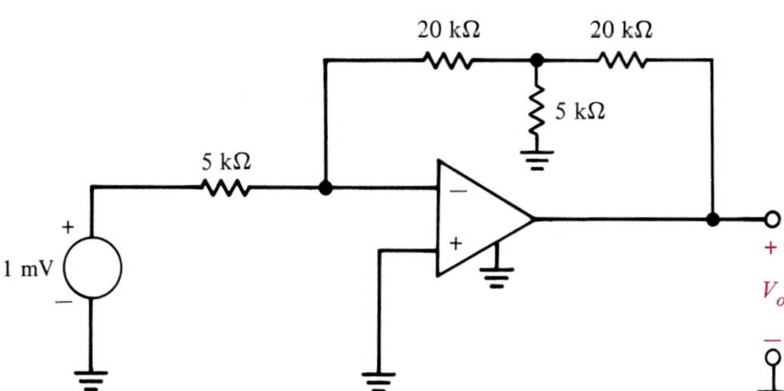

Figure P4.30

4.31. Given the network in Fig. P4.31, compute V_o and calculate the gain of the circuit.

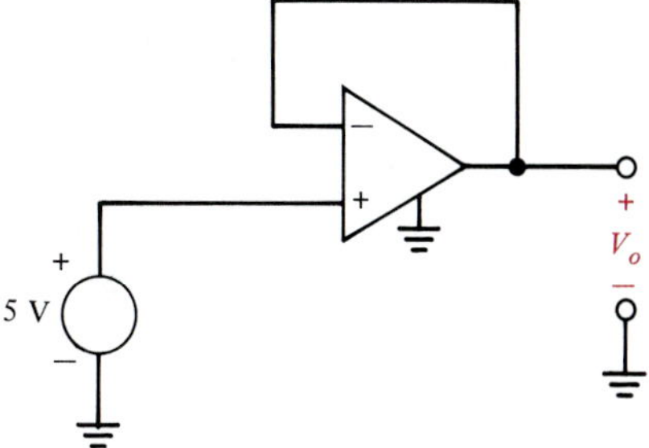

Figure P4.31

4.32. The network in Fig. P4.32 is a difference amplifier. Compute the voltage V_o.

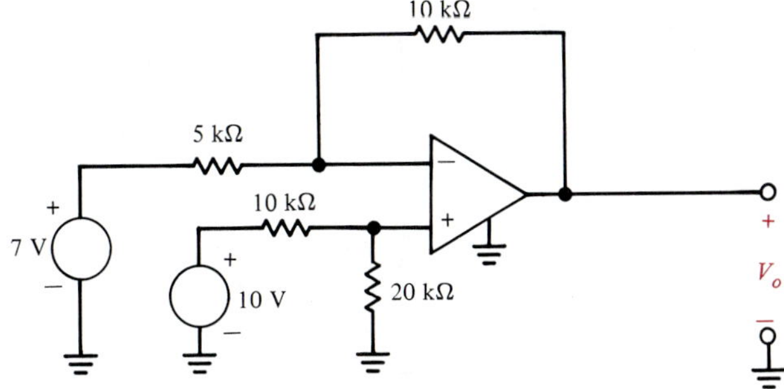

Figure P4.32

4.33. Given the network in Fig. P4.33, compute V_1, V_o, the gain, and the input and output resistances.

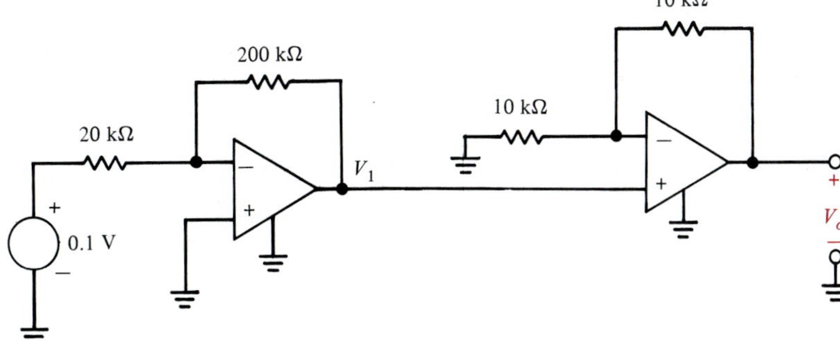

Figure P4.33

4.34. Given the network in Fig. P4.34, compute V_1, V_2, V_o, the gain of the circuit, and the input and output resistance.

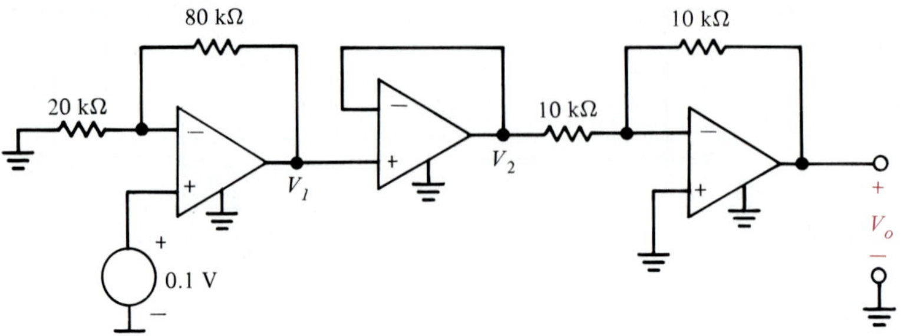

Figure P4.34

4.35. Compute V_o in the network in Fig. P4.35.

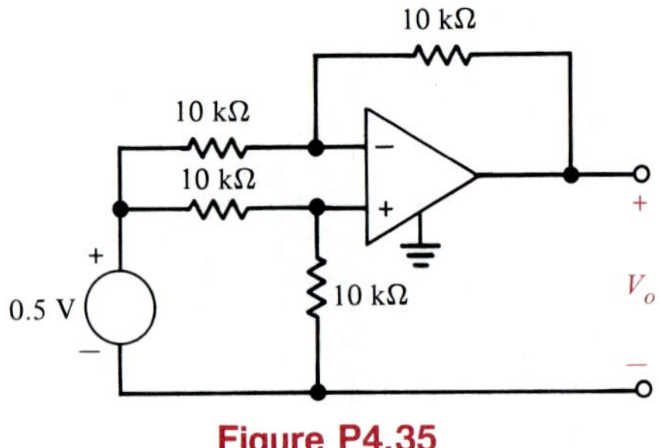

Figure P4.35

4.36. Compute the voltages V_2 and V_o in the network in Fig. P4.36.

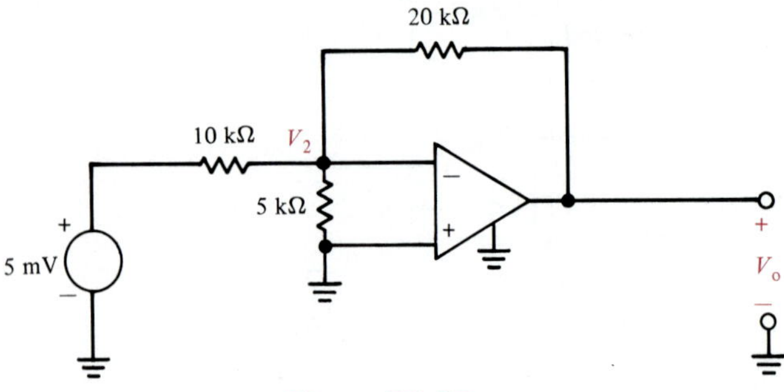

Figure P4.36

4.37. Compute the output voltage V_o in the network in Fig. P4.37 and use the result to calculate the gain of the network.

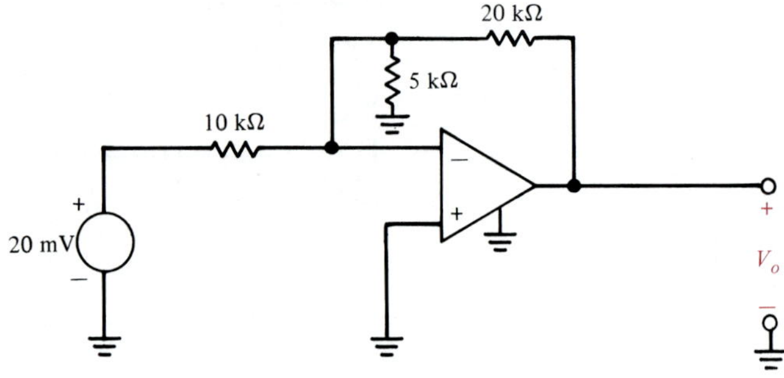

Figure P4.37

4.38. Compute V_o in the network in Fig. P4.38.

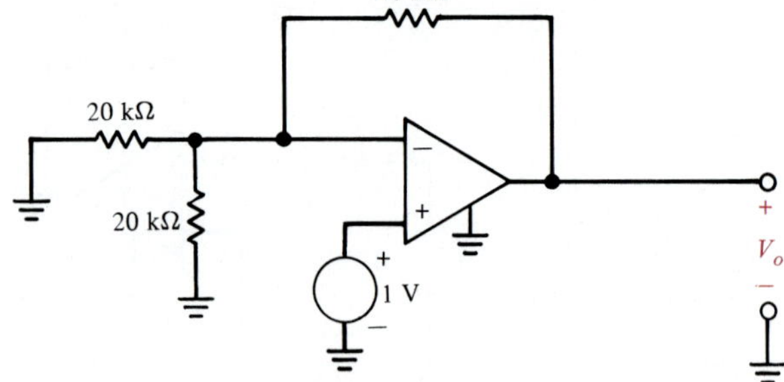

Figure P4.38

4.39. Compute V_o in the network in Fig. P4.39.

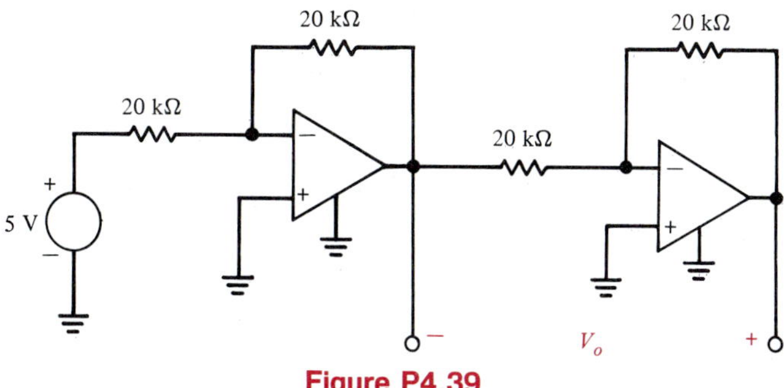

Figure P4.39

4.40. Given the following PSPICE program listing for a network, draw the network.

```
VS1 3 2 DC 8
VS2 5 4 DC 4
IS 0 1 DC 2
R1 1 0 2
R2 1 2 1
R3 1 4 5
R4 2 0 4
R5 3 5 3
R6 5 0 1
.END
```

4.41. Given the following PSPICE program listing for a network, draw the network.

```
VS1 1 3 DC 4
VS2 2 0 DC 12
IS 2 4 DC 6
VDUM 4 5 DC 0
R1 1 0 2
R2 1 2 3
R3 3 2 4
R4 3 4 2
R5 2 4 1
R6 5 0 3
R7 2 0 6
.END
```

4.42. Given the following PSPICE program listing for a network, draw the network.

```
VS 3 2 DC 15
VDUM 4 5 DC 0
IS 0 2 DC 5
HS 1 0 VDUM 3
R1 1 0 2
R2 1 2 5
R3 2 0 4
R4 3 0 3
R5 3 4 4
R6 5 0 1
.END
```

4.43. Given the following PSPICE program listing for a network, draw the network.

```
VS 2 1 DC 10
IS 3 0 DC 3
GS 1 4 4 3 3
R1 1 0 2
R2 2 0 1
R3 2 3 4
R4 3 4 1
R5 4 0 3
.END
```

4.44. Given the following PSPICE program listing for a network, draw the network.

```
VS1 5 2 DC 26
VS2 0 4 DC 6
VDUM 2 3 DC 0
FS 5 6 VDUM 4
R1 1 0 1
R2 1 2 2
R3 2 0 4
R4 3 4 2
R5 6 4 1
R6 4 0 1
.END
```

4.45. Given the following PSPICE program listing for a network, draw the network.

```
VS1 1 0 DC 14
VS2 0 2 DC 7
IS 2 0 DC 6
GS 1 3 5 0 5
ES 2 1 5 0 2
R1 3 4 3
R2 4 1 1
R3 4 5 2
R4 1 0 4
R5 5 2 2
R6 5 0 1
.END
```

CHAPTER

5

Additional Analysis Techniques

In this chapter we introduce a number of theorems and concepts that will provide both insight and understanding as we strive to improve our capabilities in circuit analysis. Although each theorem is itself a new tool, which may provide an alternative method for analysis, it will not always lead to a simpler or shorter solution. The theorems will, however, allow us to view the analysis procedure from a perspective that we are unable to visualize at the present time.

As a general rule, the network theorems are useful in simplifying a network to obtain fewer variables and fewer equations. For example, when we combine two resistors in series, we reduce the number of node equations by one, and when we combine two resistors in parallel, we reduce the number of loop equations by one.

The concept of maximum power transfer is presented, and the manner in which to adjust a load in order to effect a transfer of maximum power is discussed.

Sensitivity analysis is also introduced. A sensitivity analysis tells us how the variation of one network component will affect a particular network variable.

5.1 Network Theorems

Linearity

All the circuits we have analyzed thus far and all the circuits we will treat in this book are linear circuits. We have shown earlier that the resistor is a linear element because its current–voltage relationship is a linear characteristic; that is,

$$v(t) = Ri(t) \tag{5.1}$$

Linearity requires both additivity and homogeneity (scaling). In the case of a resistive element, if $i_1(t)$ is applied, the voltage across the resistor is

$$v_1(t) = Ri_1(t) \tag{5.2}$$

Similarly, if $i_2(t)$ is applied, then

$$v_2(t) = Ri_2(t) \tag{5.3}$$

However, if $i_1(t) + i_2(t)$ is applied, the voltage across the resistor is

$$v(t) = R[i_1(t) + i_2(t)] = Ri_1(t) + Ri_2(t) = v_1(t) + v_2(t) \tag{5.4}$$

This demonstrates the additive property. In addition, if the current is scaled by a constant K_1, the voltage is also scaled by the constant K_1, since

$$RK_1i(t) = K_1Ri(t) = K_1v(t) \tag{5.5}$$

This demonstrates homogeneity.

Dependent sources are linear if their output current or voltage is proportional to the sum of the first power of current or voltage variables present in the circuit. In other words, $v(t)$ or $i(t)$ may be of the form $a_1i_1(t)$ or $a_2v_2(t)$ or $a_1i_1(t) + a_2v_2(t)$. However, if either $v(t)$ or $i(t)$ is given by an expression of the form $a_1i_1^2(t)$ or $a_2i_1(t)v_2(t)$, the dependent source is nonlinear.

We have shown in the preceding chapters that a circuit that contains only independent sources, linear dependent sources, and resistors is described by equations of the form

$$a_1v_1(t) + a_2v_2(t) + \cdots + a_nv_n(t) = i(t) \tag{5.6}$$

or

$$b_1i_1(t) + b_2i_2(t) + \cdots + b_ni_n(t) = v(t)$$

Note that if the independent sources are multiplied by a constant, the node voltages or loop currents are also multiplied by the same constant. Thus we define a linear circuit as one composed of only independent sources, linear dependent sources, and linear elements. Capacitors and inductors, which we will examine in Chapter 6, are also circuit elements that have a linear input–output relationship provided that their initial stored energy is zero.

EXAMPLE 5.1

We wish to illustrate the property of linearity for the circuit given in Fig. 5.1.

Suppose that we wish to calculate I_1 for the voltage sources given and also for the case in which they are doubled in value. The two loop equations are

$$6I_1 - 3I_2 = 30$$

$$-3I_1 + 9I_2 = -15$$

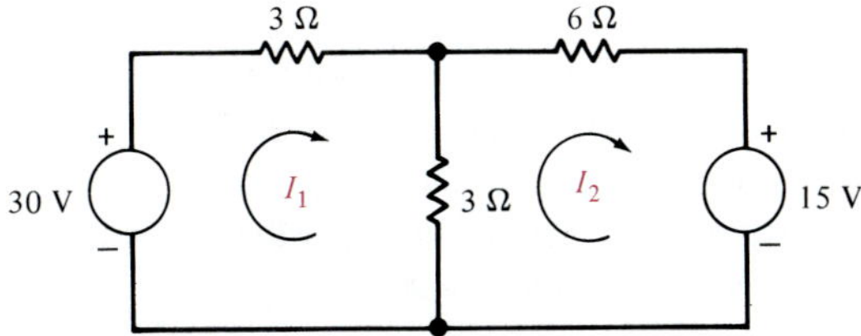

Figure 5.1 Circuit for Example 5.1.

Using a matrix approach, we have

$$\begin{bmatrix} I_1 \\ I_2 \end{bmatrix} = \begin{bmatrix} 6 & -3 \\ -3 & 9 \end{bmatrix}^{-1} \begin{bmatrix} 30 \\ -15 \end{bmatrix}$$

If the value of each voltage source is doubled, then

$$\begin{bmatrix} I_1 \\ I_2 \end{bmatrix} = \begin{bmatrix} 6 & -3 \\ -3 & 9 \end{bmatrix}^{-1} \begin{bmatrix} 60 \\ -30 \end{bmatrix} = 2 \begin{bmatrix} 6 & -3 \\ -3 & 9 \end{bmatrix}^{-1} \begin{bmatrix} 30 \\ -15 \end{bmatrix}$$

and therefore, when the source values are doubled, the value of both I_1 and I_2 are doubled also. ■

EXAMPLE 5.2

Calculate the value of the current I in Fig. 5.2 if $V_S = 6$ V and $V_S = 30$ V.

Applying KVL yields

$$V_S + 2V_x = 6I$$

where

$$V_x = 2I$$

Therefore,

$$\frac{V_S}{2} = I$$

If $V_S = 6$ V, then $I = 3$ A, and if $V_S = 30$ V $= (5)(6)$ V, then $I = (5)(3)$ A $= 15$ A. ■

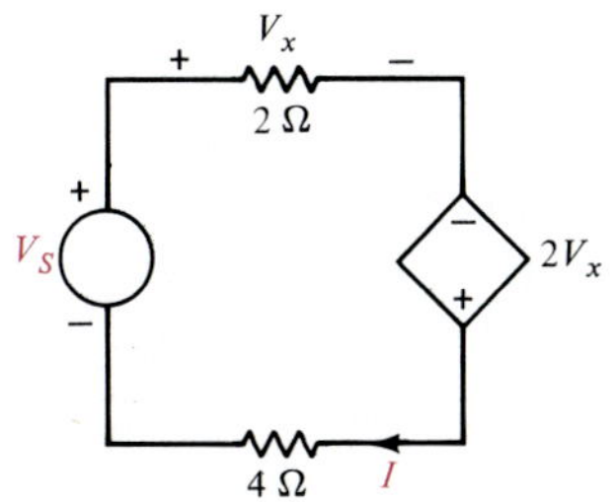

Figure 5.2 Circuit for Example 5.2.

EXAMPLE 5.3

For the circuit shown in Fig. 5.3, we wish to determine the output voltage V_{out}. However, rather than approach the problem in a straightforward manner and calculate I_o, then I_1, then I_2, and so on, we will use linearity and simply assume that the output voltage is $V_{out} = 1$ V. This assumption will yield a value for the source voltage. We will then use the actual value of the source voltage and linearity to compute the actual value of V_{out}.

If we assume that $V_{out} = V_3 = 1$ V, then

$$I_4 = \frac{V_3}{2} = 0.5 \text{ A}$$

V_2 can then be calculated as

$$V_2 = 4I_4 + V_3$$
$$= 3 \text{ V}$$

Hence

$$I_3 = \frac{V_2}{3} = 1 \text{ A}$$

Using KCL gives us

$$I_2 = I_3 + I_4$$
$$= 1.5 \text{ A}$$

Then

$$V_1 = 2I_2 + V_2$$
$$= 6 \text{ V}$$

I_1 is then computed as

$$I_1 = \frac{V_1}{4}$$
$$= 1.5 \text{ A}$$

Applying KCL again, we have

$$I_o = I_1 + I_2$$
$$= 3 \text{ A}$$

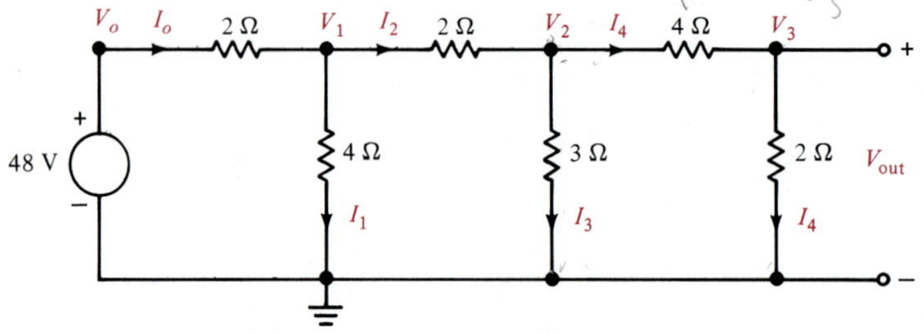

Figure 5.3 Circuit for Example 5.3.

and finally,

$$V_o = 2I_o + V_1$$
$$= 12 \text{ V}$$

Therefore, assuming that $V_{out} = 1$ V yields a source voltage of 12 V. However, the actual source voltage is 48 V and hence the actual output voltage is 1 V $(\frac{48}{12}) = 4$ V. ■

DRILL EXERCISE

D5.1. Use linearity and the assumption that $V_o = 1$ V to compute the correct voltage V_o in the network in Fig. D5.1 if $I_o = 4$ A.

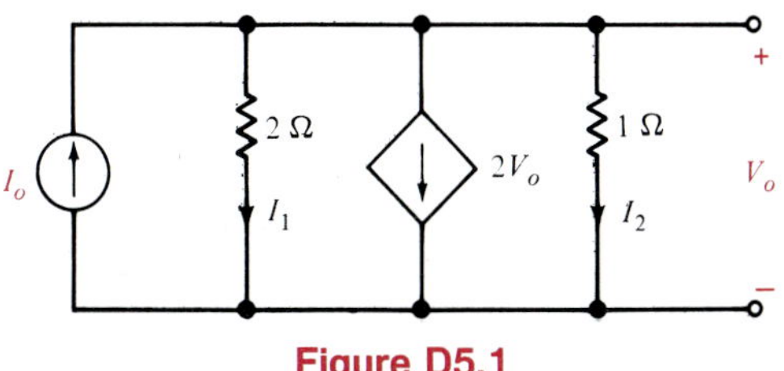

Figure D5.1

Ans: $V_o = \frac{8}{7}$ V.

D5.2. Use linearity and the assumption that $I_o = 1$ A to compute the correct current I_o in the network in Fig. D5.2 if $I = 12$ A.

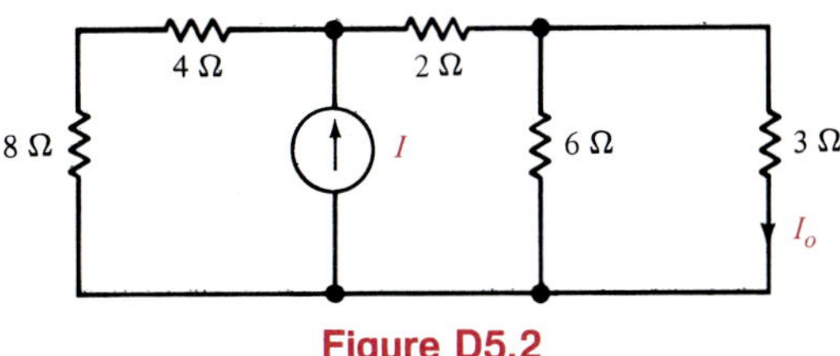

Figure D5.2

Ans: $I_o = 6$ A.

D5.3. Use linearity and the assumption that $I_o = 1$ A to determine the actual value of I_o in the network in Fig. D5.3 if $V = 24$ V.

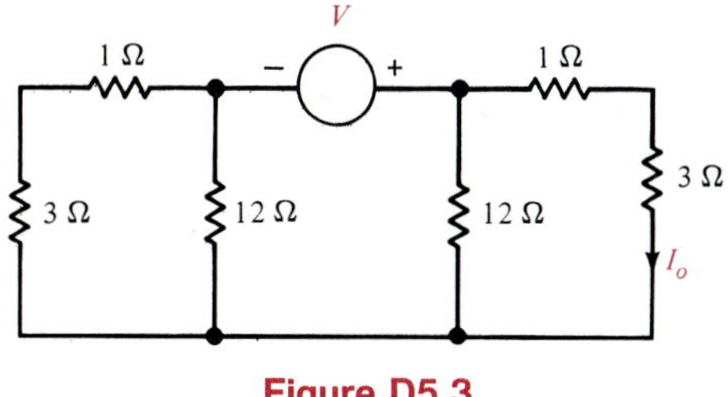

Figure D5.3

Ans: $I_o = 3$ A.

Superposition

In order to provide motivation for this subject, it is instructive to reexamine a problem that we have just considered.

EXAMPLE 5.4

The circuit given in Fig. 5.1 is redrawn in Fig. 5.4, where the values of the voltage sources are unspecified. The value of the current $i_1(t)$ can be computed using KVL as shown in Example 5.1 as

$$\begin{bmatrix} i_1(t) \\ i_2(t) \end{bmatrix} = \frac{1}{45}\begin{bmatrix} 9 & 3 \\ 3 & 6 \end{bmatrix}\begin{bmatrix} v_1(t) \\ -v_2(t) \end{bmatrix}$$

$$i_1(t) = \frac{1}{45}[9v_1(t) - 3v_2(t)]$$

$$= \frac{v_1(t)}{5} - \frac{v_2(t)}{15}$$

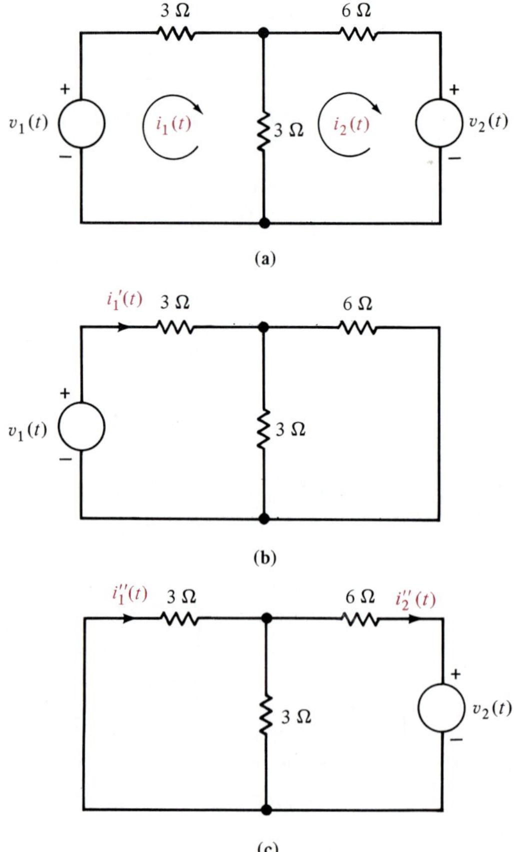

Figure 5.4 Illustration of superposition via an example.

In other words, the current $i_1(t)$ has a component due to $v_1(t)$ and a component due to $v_2(t)$. In view of the fact that $i_1(t)$ has two components, one due to each independent source, it would be interesting to examine what each source acting alone would contribute to $i_1(t)$. For $v_1(t)$ to act alone, $v_2(t)$ must be zero. As we pointed out in Chapter 2, $v_2(t) = 0$ means that the source $v_2(t)$ is replaced with a short circuit. Therefore, to determine the value of $i_1(t)$ due to $v_1(t)$ only, we employ the circuit in Fig. 5.4b and refer to this value of $i_1(t)$ as $i_1'(t)$.

$$i_1'(t) = \frac{v_1(t)}{3 + \dfrac{(3)(6)}{3 + 6}}$$

$$= \frac{v_1(t)}{5}$$

Let us now determine the value of $i_1(t)$ due to $v_2(t)$ acting alone and refer to this value as $i_1''(t)$. We employ the circuit shown in Fig. 5.4c and compute this value as

$$i_2''(t) = -\frac{v_2(t)}{6 + \dfrac{(3)(3)}{3 + 3}} = -\frac{2v_2(t)}{15}$$

Then, using current division, we obtain

$$i_1''(t) = -\frac{2v_2(t)}{15}\left(\frac{3}{3 + 3}\right)$$

$$= -\frac{v_2(t)}{15}$$

Now if we add the values $i_1'(t)$ and $i_1''(t)$, we obtain the value computed directly:

$$i_1(t) = i_1'(t) + i_1''(t)$$

$$= \frac{v_1(t)}{5} - \frac{v_2(t)}{15}$$

Note that we have *superposed* the value of $i_1'(t)$ on $i_1''(t)$, or vice versa, to determine the total value of the unknown current. ■

What we have demonstrated in Example 5.4 is true in general for linear circuits and is a direct result of the property of linearity. The *principle of superposition,* which provides us with this ability to reduce a complicated problem to several easier problems—each containing only a single independent source—states that

> In any linear circuit containing multiple independent sources, the current or voltage at any point in the network may be calculated as the algebraic sum of the individual contributions of each source acting alone.

When determining the contribution due to an independent source, any remaining voltage sources are made zero by replacing them with short circuits, and any remaining current sources are made zero by replacing them with open circuits; however, dependent sources are not made zero and remain in the circuit. Thus superposition will not always lead to a simpler solution. The following examples will illustrate the technique.

EXAMPLE 5.5

We wish to determine the voltage V_o shown in Fig. 5.5a.

We will solve the problem using superposition. If we let V_o' be the component of V_o due to the 64-V, source, then when the 12-V and 2-A sources are made zero, the network reduces to that shown in Fig. 5.5b. Combining resistors, the network is further reduced to that shown in Fig. 5.5c. Therefore,

$$V' = \frac{64}{4 + \frac{12}{5}}\left(\frac{12}{5}\right)$$

$$= 24 \text{ V}$$

From Fig. 5.5b we now see that

$$V_o' = \frac{24}{2 + 3 + 1}(1)$$

$$= 4 \text{ V}$$

The contribution of the 12-V source to V_o, which we will call V_o'', is computed from Fig. 5.5d. Combining the resistors in this network reduces the network to that shown in Fig. 5.5e. V_1'' is clearly -6 V. Using this value of V_1'' in Fig. 5.5d, we can immediately obtain $V_o'' = -1$ V.

V_o''', the component of V_o due to the 2-A current source, is calculated from Fig. 5.5f. Once again, combining resistors, the network is reduced to that shown in Fig. 5.5g. Using current division, we find that $V_o''' = 1$ V.

Now applying superposition, we obtain

$$V_o = V_o' + V_o'' + V_o'''$$

$$= 4 \text{ V}$$

It is interesting to note that solving the problem by superposition actually permits us to see the contribution that each source makes to the output voltage. ■

EXAMPLE 5.6

Let us determine the voltage V_o in the network in Fig. 5.6a.

Employing the principle of superposition, we first solve the circuit in Fig. 5.6b. The equations for this circuit are

$$2V_x' = 4(I_1' - I_2') + 2I_1'$$

$$V_x' = -4(I_1' - I_2')$$

$$I_2' = -3 \text{ A}$$

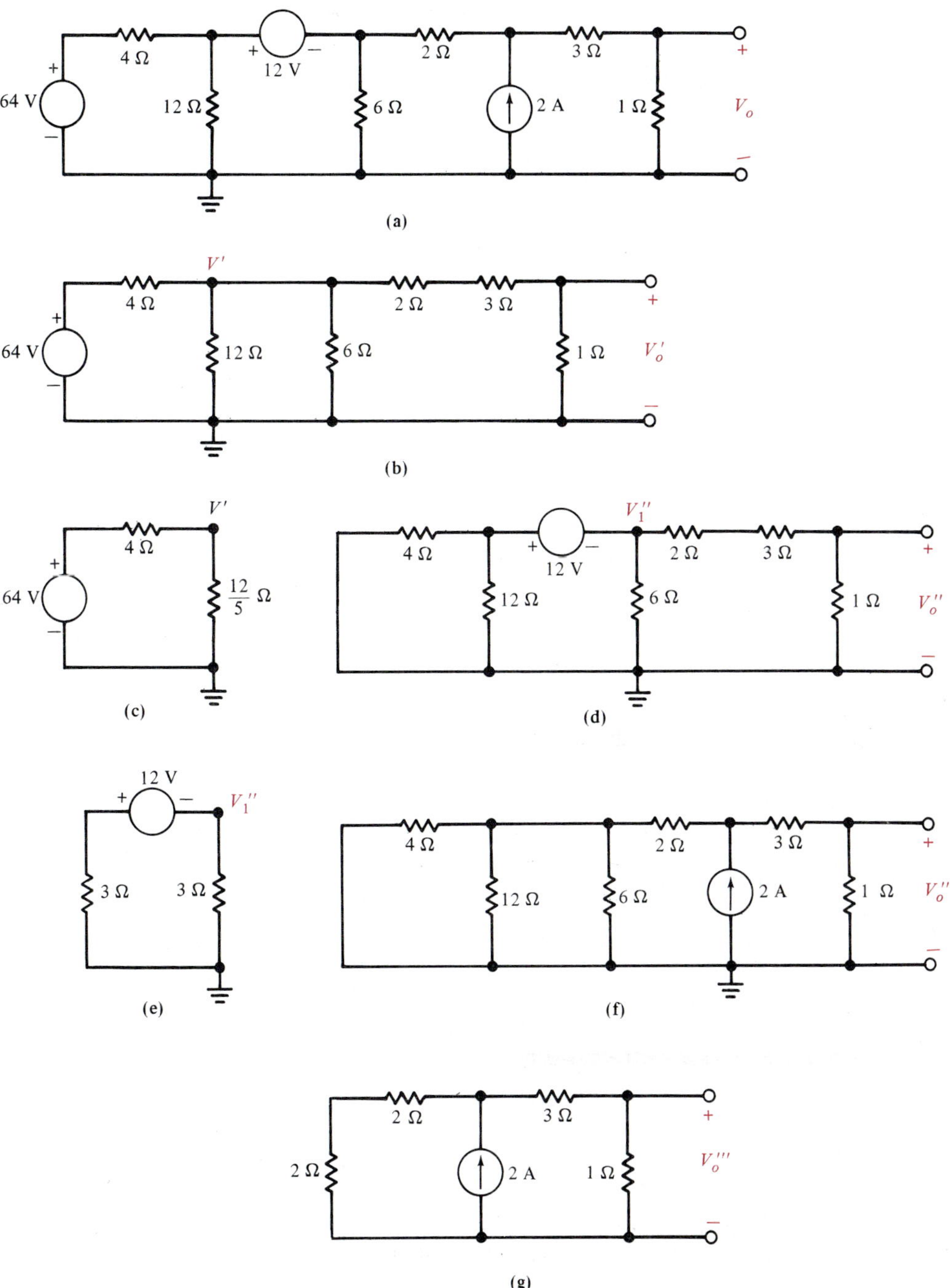

Figure 5.5 Circuits employed in Example 5.5.

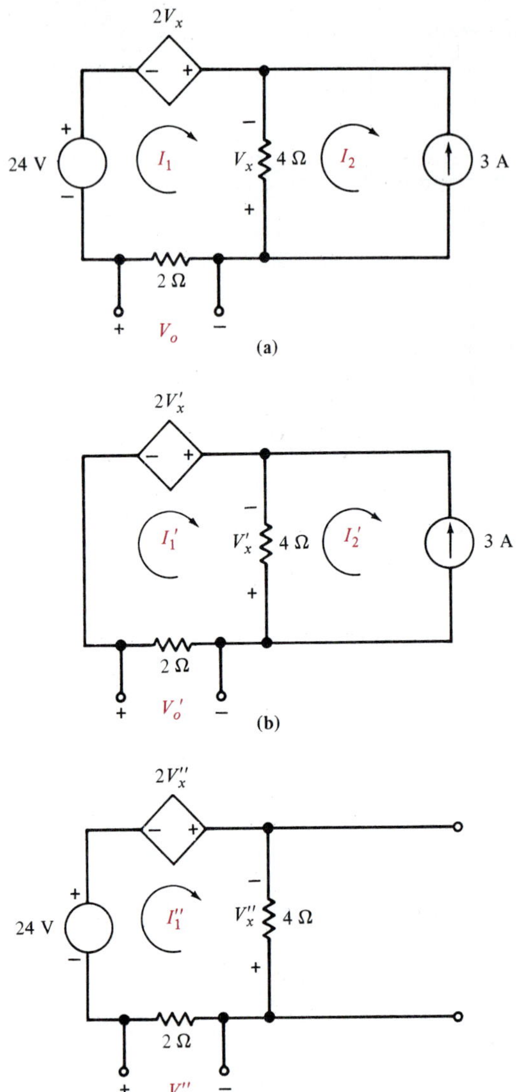

Figure 5.6 Illustration of superposition for a circuit containing a dependent source.

These equations yield $I_1' = -36/14$ A, and therefore,

$$V_o' = \frac{72}{14} \text{ V}$$

For the circuit in Fig. 5.6c the equations are

$$24 + 2V_x'' = 6I_1''$$

$$V_x'' = -4I_1''$$

From these equations we obtain $I_1'' = 24/14$ A, and hence

$$V_o'' = -\frac{48}{14}\text{ V}$$

Therefore,

$$V_o = V_o' + V_o''$$
$$= \frac{24}{14}\text{ V}$$

Note that although we have obtained the correct answer, the problem could have been solved more easily by mesh equations. ■

EXAMPLE 5.7

Consider the differential amplifier op-amp circuit shown in Fig. 5.7. Employing superposition, we first set $v_2 = 0$. We note that since $i_+ = 0$, there can be no current in R_1 or R_2 and therefore $v_+ = 0$. The circuit is therefore an inverting amplifier and hence

$$v_o' = -\frac{R_2}{R_1} v_1$$

Then setting, $v_1 = 0$, the circuit equations are

$$\frac{v_2 - v_+}{R_1} = \frac{v_+}{R_2} + i_+$$

$$\frac{v_o'' - v_-}{R_2} = \frac{v_-}{R_1} + i_-$$

However, $i_- = i_+ = 0$ and $v_+ = v_-$. Substituting these values into the equations yields

$$v_o'' = \frac{R_2}{R_1} v_2$$

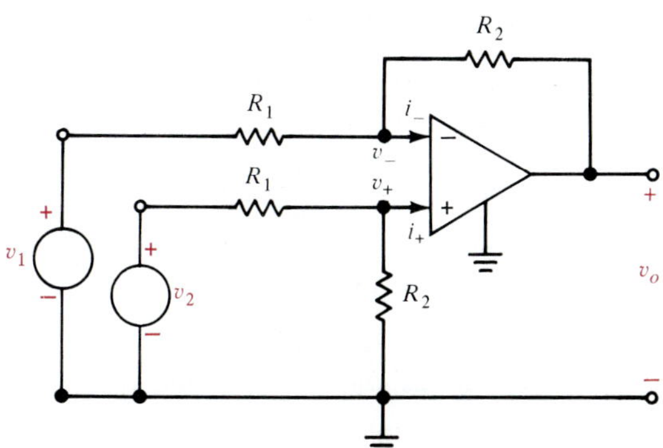

Figure 5.7 Operational amplifier circuit for a differential amplifier.

Therefore,

$$v_o = v_o' + v_o''$$

$$= \frac{R_2}{R_1}(v_2 - v_1)$$

which is the same value as that obtained in Chapter 3. ■

Superposition can be applied to a circuit with any number of dependent and independent sources. In fact, superposition can be applied to such a network in a variety of ways. For example, a circuit with three independent sources can be solved using each source acting alone, as we have demonstrated above, or we could use two at a time and sum the result with that obtained from the third acting alone. In addition, the independent sources do not have to assume their actual value or zero. However, it is mandatory that the sum of the different values chosen add to the total value of the source.

Superposition is a fundamental property of linear equations, and therefore can be applied to any effect that is linearly related to its cause. In this regard it is important to point out that although superposition applies to the current and voltage in a linear circuit, it cannot be used to determine power, because power is a nonlinear function. As a quick illustration, note that the power in the 2-Ω resistor in Example 5.6 is $P = I^2R = (-6/7)^2(2) = 1.47$ W. However, superposition would have yielded $P = (-18/7)^2(2) + (12/7)^2(2) = 19.1$ W, which is incorrect.

DRILL EXERCISE

D5.4. Compute V_o in the network in Fig. D5.4 using superposition.

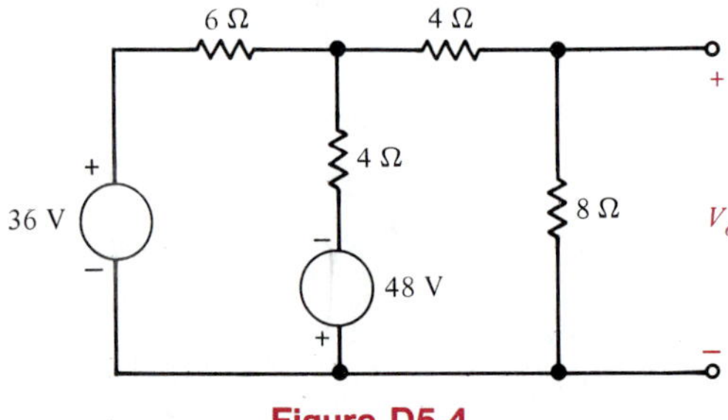

Figure D5.4

Ans: $V_o = -8$ V.

D5.5. Compute V_o in the network in Fig. D5.5 using superposition.

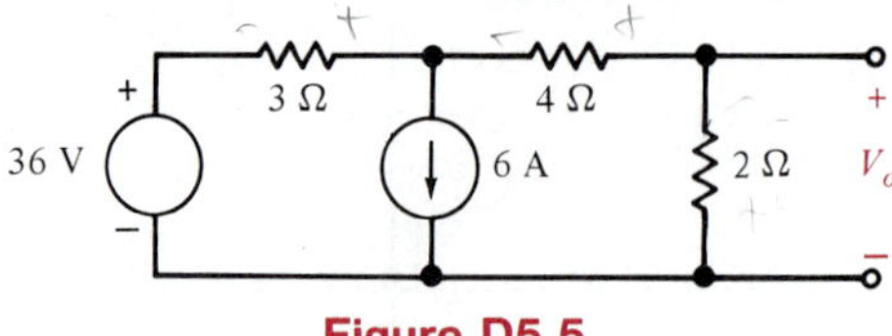

Figure D5.5

Ans: $V_o = 4$ V.

D5.6. Find V_o in the network in Fig. D5.6 using superposition.

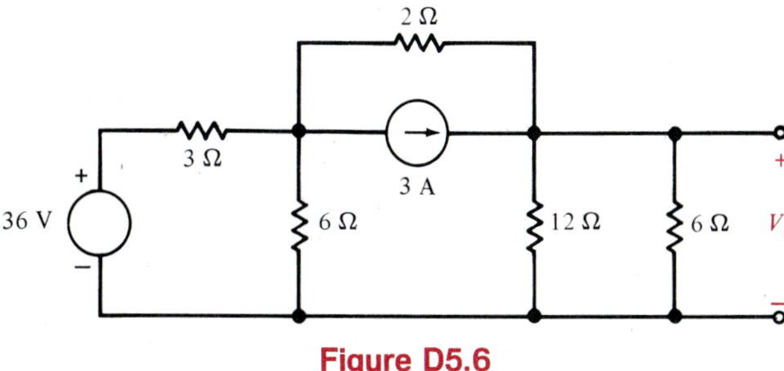

Figure D5.6

Ans: $V_o = 15$ V.

Source Transformation

Before we begin discussing source transformations, it is necessary that we point out that real sources differ from the ideal models we have presented thus far. In general, a practical voltage source does not produce a constant voltage regardless of the load resistance or the current it delivers, and a practical current source does not deliver a constant current regardless of the load resistance or the voltage across its terminals. Practical sources contain internal resistance, and therefore the models shown in Fig. 5.8a and b more

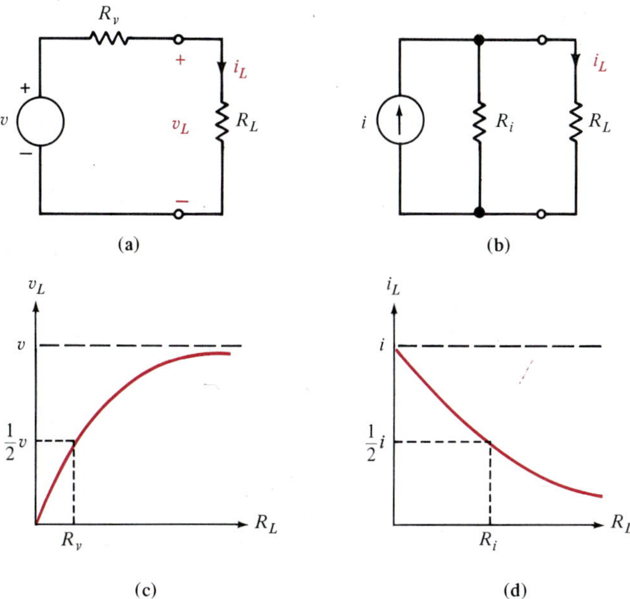

Figure 5.8 Effect of internal resistance on actual voltage and current sources.

closely represent actual sources. Note that the power delivered by the practical voltage source is given by the expression

$$P_L = i_L^2 R_L$$
$$= \left(\frac{v}{R_v + R_L}\right)^2 R_L$$

which can be written as

$$P_L = \frac{v^2}{R_L}\left(\frac{1}{1 + R_v/R_L}\right)^2$$

and therefore if $R_L >> R_v$, then

$$P_L = \frac{v^2}{R_L}$$

which is the power delivered by the ideal voltage source. Similarly, the power delivered by the practical current source is computed using current division as

$$P_L = i_L^2 R_L$$
$$= \left(\frac{iR_i}{R_i + R_L}\right)^2 R_L$$

which can be written as

$$P_L = i^2 R_L \left(\frac{1}{1 + R_L/R_i}\right)^2$$

and hence if $R_i >> R_L$, then

$$P_L = i^2 R_L$$

which is the power delivered by the ideal current source.

The graphs shown in Fig. 5.8c and d illustrate the effect of the internal resistance on the voltage source and current source, respectively. The graphs indicate that the output voltage approaches the ideal source voltage only for large values of load resistance and, therefore, small values of current. Similarly, the load current is approximately equal to the ideal source current only for values of the load resistance R_L that are small in comparison to the internal resistance R_i.

With this material on practical sources as background information, we now ask if it is possible to exchange one source model for another; that is, exchange a voltage source model for a current source model, or vice versa. We could exchange one source for another provided that they are equivalent; that is, each source produces exactly the same voltage and current for any load that is connected across its terminals.

Let us examine the two circuits shown in Fig. 5.9. In order to determine the conditions required for the two sources to be equivalent, let us examine the terminal conditions of each. For the network in Fig. 5.9a,

$$i = i_L + \frac{v_L}{R_i}$$

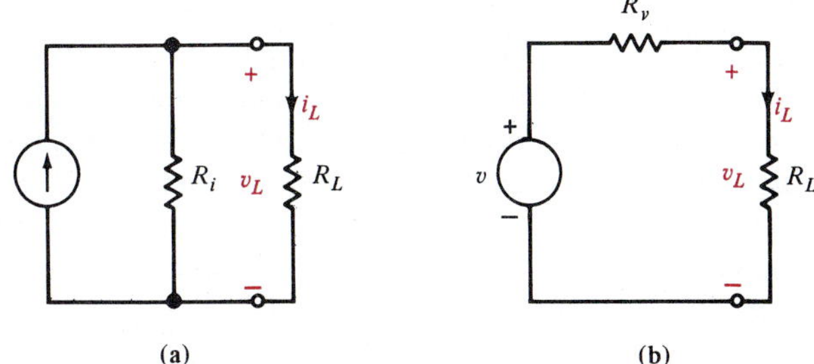

Figure 5.9 Circuits used to determine conditions for a source exchange.

or

$$iR_i = R_i i_L + v_L$$

For the network in Fig. 5.9b,

$$v = i_L R_v + v_L$$

For the two networks to be equivalent, their terminal characteristics must be identical; that is,

$$v = iR_i \qquad \text{and} \qquad R_i = R_v \tag{5.7}$$

The relationships specified in Eq. (5.7) and Fig. 5.9 are extremely important and the reader should not fail to grasp their significance. What these relationships tell us is that if we have embedded within a network a current source i in parallel with a resistor R, we can replace this combination with a voltage source of value $v = iR$ in series with the resistor R. The reverse is also true; that is, a voltage source v in series with a resistor R can be replaced with a current source of value $i = v/R$ in parallel with the resistor R. Parameters within the circuit (e.g., an output voltage) are unchanged under these transformations.

The following examples will demonstrate the utility of a source exchange.

EXAMPLE 5.8

Given the nonideal voltage source in Fig. 5.10a, determine the equivalent nonideal current source.

The current source can be obtained using the expressions in Eq. (5.7) and is shown in Fig. 5.10b. The reader is cautioned to keep the polarity of the voltage source and the direction of the current source in agreement, as shown in Fig. 5.10. ■

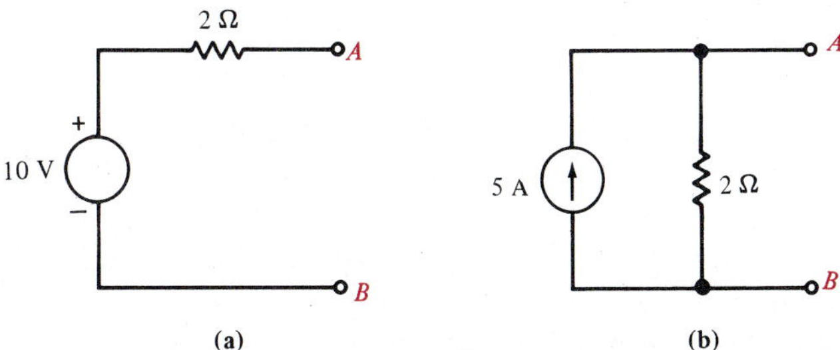

Figure 5.10 Two equivalent sources.

EXAMPLE 5.9

Consider the network in Fig. 5.11a. We wish to compute the current I_o. Note that since node voltages V_1 and V_3 are known, one KCL equation at the center node will yield V_2 and therefore I_o. However, two mesh equations are required to find I_o and two separate circuits must be analyzed to determine I_o using superposition.

To solve the problem using source transformation, we first transform the 60-V source and 6-Ω resistor into a 10-A current source in parallel with the 6-Ω resistor as shown in Fig. 5.11b. Next, the 3-Ω resistor in series with the 15-V source are transformed into a 5-A current source in parallel with the 3-Ω resistor, as shown in Fig. 5.11c. Combining the resistors and current sources yields the network in Fig. 5.11d. Now employing current division, we obtain $I_o = 2$ A. ■

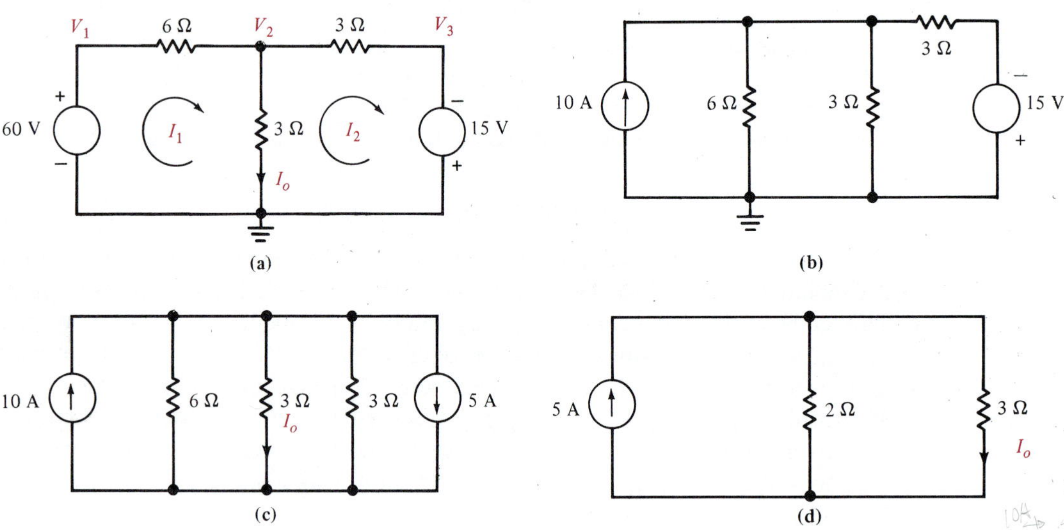

Figure 5.11 Circuits employed in Example 5.9.

EXAMPLE 5.10

Let us demonstrate how to find V_o in the network in Fig. 5.5a using repeated application of source transformation.

If we begin at the left end of the network in Fig. 5.5a, the series combination of the 64-V source and the 4-Ω resistor is converted to a 16-A current source in parallel with a 4-Ω resistor. If we also combine the 4- and 12-Ω resistors that are now in parallel, we obtain the circuit in Fig. 5.12a. To continue the circuit reduction, the parallel combination of the 16-A current source and the 3-Ω resistor are converted to a 48-V source in series with a 3-Ω resistor, and these elements are then combined with the 12-V source to yield the network in Fig. 5.12b. The 36-V source in series with the 3-Ω resistor is converted to a 12-A current source and parallel 3-Ω resistor, which when combined with the 6-Ω resistor yields the network in Fig. 5.12c. When the 12-A current source is transformed to

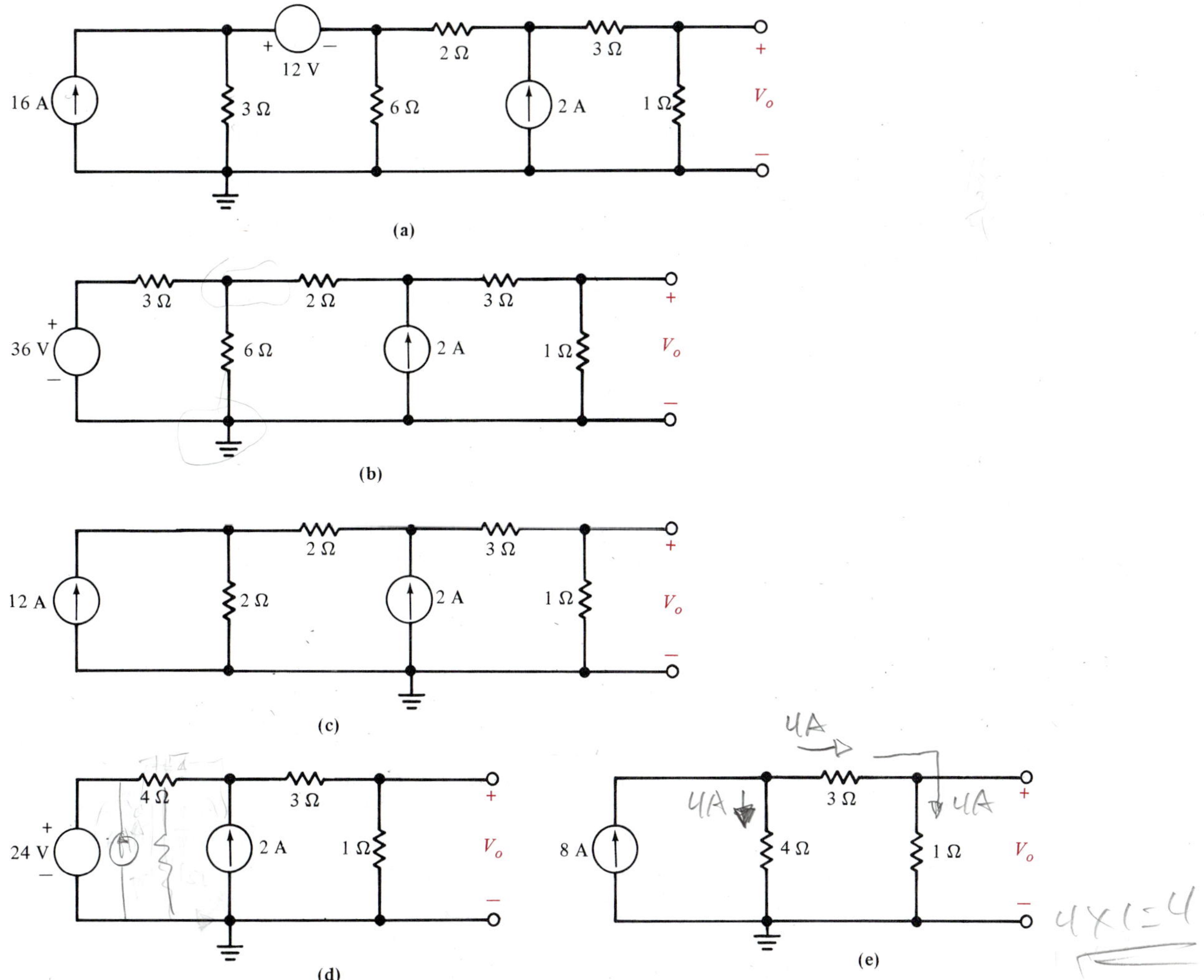

Figure 5.12 Application of source transformation to the network in Fig. 5.5a.

a voltage source and the two series resistors are added, we obtain the network in Fig. 5.12d. Finally, transforming the 24-V source and 4-Ω resistor into a current source and parallel resistor, and then combining the parallel current sources yields the circuit shown in Fig. 5.12e. Employing current division on the latter network, we find that $V_o = 4$ V. ■

Note that this systematic, sometimes tedious, transformation allows us to reduce the network methodically to a simpler equivalent form with respect to some other circuit element.

DRILL EXERCISE

D5.7. Find V_o in the circuit in Fig. D5.4 using source transformation.

Ans: $V_o = -8$ V.

D5.8. Find V_o in the network in Fig. D5.5 using source transformation.

Ans: $V_o = 4$ V.

D5.9. Find V_o in the circuit in Fig. D5.6 using source transformation.

Ans: $V_o = 15$ V.

Thévenin's and Norton's Theorems

Thus far we have presented a number of techniques for circuit analysis. At this point we will add two theorems to our collection of tools that will prove to be extremely useful. The theorems are named after their authors, M. L. Thévenin, a French engineer, and E. L. Norton, a scientist formerly with Bell Telephone Laboratories.

Suppose that we are given a circuit and that we wish to find the current, voltage, or power that is delivered to some resistor of the network which we will call the load. *Thévenin's theorem* tells us that we can replace the entire network, exclusive of the load, by an equivalent circuit that contains only an independent voltage source in series with a resistor in such a way that the current–voltage relationship at the load is unchanged. *Norton's theorem* is identical to the statement above except that the equivalent circuit is an independent current source in parallel with a resistor.

Note that this is a very important result. It tells us that if we examine any network from a pair of terminals, we know that with respect to those terminals, the entire network is equivalent to a simple circuit consisting of an independent voltage source in series with a resistor or an independent current source in parallel with a resistor.

In developing the theorems, we will assume that the circuit shown in Fig. 5.13a can be split into two parts, as shown in Fig. 5.13b. In general, circuit B is the load and may be linear or nonlinear. Circuit A is the balance of the original network exclusive of the load and must be linear. As such, circuit A may contain independent sources, dependent sources and resistors, or any other linear element. We require, however, that a dependent source and its control variable appear in the same circuit.

Circuit A delivers a current i to circuit B and produces a voltage v_o across the input terminals of circuit B. From the standpoint of the terminal relations of circuit A, we can replace circuit B by a voltage source of v_o volts (with the proper polarity), as shown in Fig. 5.13c. Since the terminal voltage is unchanged and circuit A is unchanged, the terminal current i is unchanged.

Now applying the principle of superposition to the network shown in Fig. 5.13c, the total current i shown in the figure is the sum of the currents caused by all the sources in circuit A and the source v_o which we have just added. Therefore, via superposition the current i can be written

$$i = i_o + i_{sc} \tag{5.8}$$

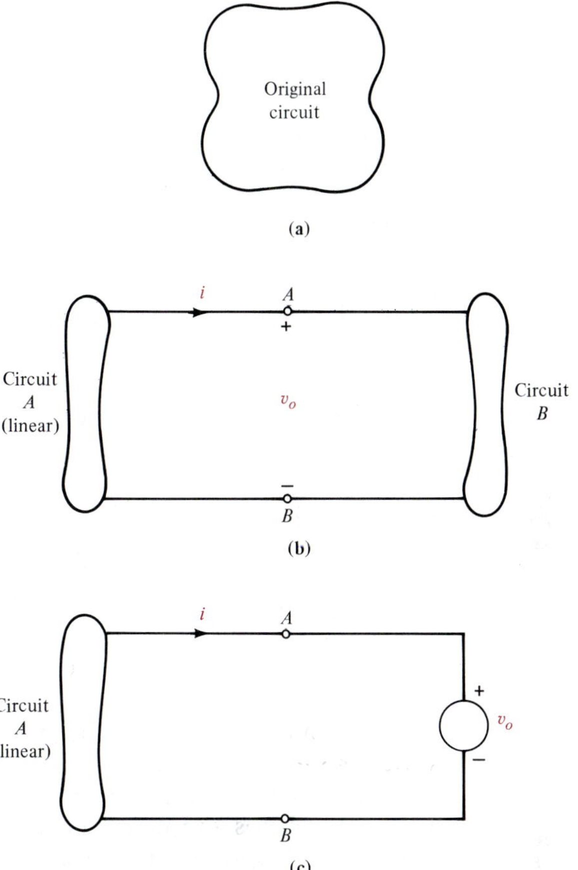

Figure 5.13 Concepts used to develop Thévenin's theorem.

where i_o is the current due to v_o with all independent sources in circuit A made zero (i.e., voltage sources replaced by short circuits and current sources replaced by open circuits), and i_{sc} is the short-circuit current due to all sources in circuit A with v_o replaced by a short circuit.

The terms i_o and v_o are related by the equation

$$i_o = \frac{-v_o}{R_{Th}} \tag{5.9}$$

where R_{Th} is the equivalent resistance looking back into circuit A from terminals A-B with all independent sources in circuit A made zero.

Substituting Eq. (5.9) into Eq. (5.8) yields

$$i = -\frac{v_o}{R_{Th}} + i_{sc} \tag{5.10}$$

This is a general relationship and, therefore, must hold for any specific condition at terminals *A-B*. As a specific case, suppose that the terminals are open-circuited. For this condition $i = 0$ and v_o is equal to the open-circuit voltage v_{oc}. Thus Eq. (5.10) becomes

$$i = 0 = \frac{-v_{oc}}{R_{Th}} + i_{sc} \tag{5.11}$$

Hence

$$v_{oc} = R_{Th} i_{sc} \tag{5.12}$$

This equation states that the open-circuit voltage is equal to the short-circuit current times the equivalent resistance looking back into circuit *A* with all independent sources made zero. We refer to R_{Th} as the Thévenin equivalent resistance.

Substituting Eq. (5.12) into Eq. (5.10) yields

$$i = \frac{-v_o}{R_{Th}} + \frac{v_{oc}}{R_{Th}}$$

or

$$v_o = v_{oc} - R_{Th} i \tag{5.13}$$

Let us now examine the circuits that are described by these equations. The circuit represented by Eq. (5.13) is shown in Fig. 5.14a. The fact that this circuit is equivalent at terminals *A-B* to circuit *A* in Fig. 5.13 is a statement of *Thévenin's theorem*. The circuit represented by Eq. (5.10) is shown in Fig. 5.14b. The fact that this circuit is equivalent at terminals *A-B* to circuit *A* in Fig. 5.13 is a statement of *Norton's theorem*.

Note carefully that the circuits in Fig. 5.14 together with the relationship in Eq. (5.12) represent a source transformation.

The manner in which these theorems are applied depends on the structure of the original network under investigation. For example, if only independent sources are present, we can calculate the open-circuit voltage or short-circuit current and the Thévenin equivalent resistance. However, if dependent sources are also present, the Thévenin equivalent will be determined by calculating v_{oc} and i_{sc}, since this is normally the best

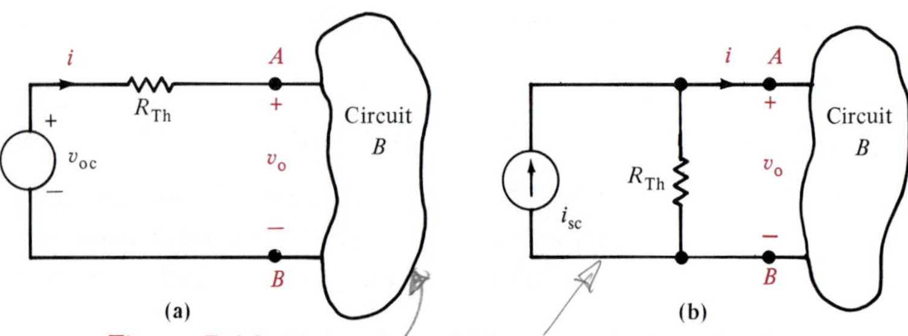

Figure 5.14 Thévenin and Norton equivalent circuits.

approach for determining R_{Th} in a network containing dependent sources. Finally, if circuit A contains no *independent* sources, then both v_{oc} and i_{sc} will necessarily be zero. (Why?) Thus we cannot determine R_{Th} by $v_{\text{oc}}/i_{\text{sc}}$, since the ratio is indeterminate. We must look for another approach. Notice that if $v_{\text{oc}} = 0$, then the equivalent circuit is merely the unknown resistance R_{Th}. If we apply an external source to circuit A—a test source v_t—and determine the current, i_t, which flows into circuit A from v_t, then R_{Th} can be determined from $R_{\text{Th}} = v_t/i_t$. Although the numerical value of v_t need not be specified, we could let $v_t = 1$ V and then $R_{\text{Th}} = 1/i_t$. Alternatively, we could use a current source as a test source and let $i_t = 1$ A, then $v_t = (1)R_{\text{Th}}$.

A variety of examples are now presented to demonstrate the utility of these theorems.

EXAMPLE 5.11

Consider the network in Fig. 5.15a. We wish to find V_o using both Thévenin's and Norton's theorems.

If we break the network at points A-B, then the open-circuit voltage is shown in Fig. 5.15b. Using voltage division yields

$$V_{\text{oc}} = 60\left(\frac{4}{4 + 8 + 4}\right) = 15 \text{ V}$$

The Thévenin equivalent resistance, obtained by looking into the open-circuit terminals A-B and replacing the voltage source with a short circuit, as shown in Fig. 5.15c is 3 Ω. If the Thévenin equivalent circuit consisting of V_{oc} in series with R_{Th} is now attached to the remainder of the original circuit at terminals A-B, the network is reduced to that shown in Fig. 5.15d. From this latter network we find that

$$V_o = 15\left(\frac{7}{3 + 5 + 7}\right) = 7 \text{ V}$$

We could also determine V_o using Norton's theorem. Once again the network is broken at terminals A-B. The short-circuit current is shown in Fig. 5.15e. Since no current will flow in the 4-Ω resistor in parallel with the short circuit,

$$I_{\text{sc}} = \frac{60}{4 + 8} = 5 \text{ A}$$

The Thévenin equivalent resistance was computed in Fig. 5.15c to be $R_{\text{Th}} = 3$ Ω. If the Norton equivalent circuit consisting of the short-circuit current source in parallel with the Thévenin equivalent resistance is now attached to the remainder of the original network at terminals A-B, the resultant network is shown in Fig. 5.15f. Using current division, we find that

$$V_o = 5\left(\frac{3}{3 + 5 + 7}\right)(7)$$

$$= 7 \text{ V}$$

■

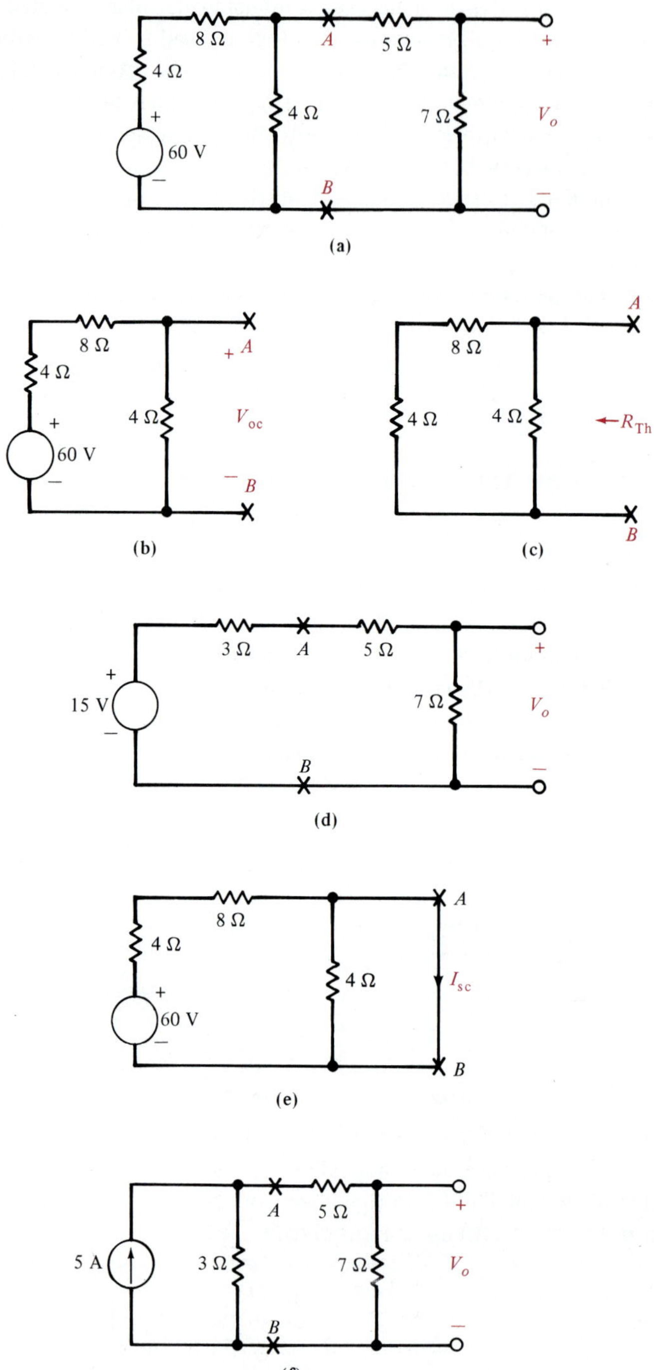

Figure 5.15 Circuits employed in Example 5.11.

Let us examine for a moment some of the salient features of Example 5.11. Note that in applying Thévenin's theorem, there is no point in breaking the network to the left of the center 4-Ω resistor, since the 60-V source and the 4- and 8-Ω resistors are already a Thévenin equivalent. In addition, we could have broken the network to the right of the 5-Ω resistor. Breaking the network at this point would not change the open-circuit voltage since there is no current in the 5-Ω resistor. However, the R_{Th} would increase by 5 Ω, to a value of 8 Ω. When the new Thévenin equivalent circuit is attached to the 7-Ω load, the resulting network would again be that shown in Fig. 5.15d. Similar arguments could be made for the analysis using Norton's theorem.

Finally, it is extremely important to note that once the network has been simplified using a Thévenin or Norton equivalent, we simply have a new network with which we can again apply the theorems. The following example illustrates this approach.

EXAMPLE 5.12

Let us use Thévenin's and Norton's theorems to find V_o in the network in Fig. 5.5a.

Applying Thévenin's theorem, we first break the network at the points *A-B* as shown in Fig. 5.16a. The voltage across the 12-Ω resistor is 48 V, and therefore $V_{oc_1} = -12 + 48 = 36$ V. R_{Th_1} computed from Fig. 5.16b is 3 Ω. If we attach this Thévenin equivalent (V_{oc_1} and R_{Th_1}) to the remainder of the network, we obtain the network in Fig. 5.16c. If we now break the network at the terminals *C-D*, the open-circuit voltage V_{oc_2} as shown in Fig. 5.16d is

$$V_{oc_2} = 36\left(\frac{6}{3+6}\right) = 24 \text{ V}$$

and R_{Th_2} is calculated from Fig. 5.16e as $R_{Th_2} = 4$ Ω. Connecting this Thévenin equivalent (V_{oc_2} and R_{Th_2}) to the remainder of the network yields the network in Fig. 5.16f. Once again, breaking the network at the terminals *E-F*, we obtain the network in Fig. 5.16g. From this network we calculate $V_{oc_3} = (2)(4) + 24 = 32$ V. R_{Th_3}, as shown in Fig. 5.16h, is 7 Ω. Finally, attaching the Thévenin equivalent (V_{oc_3} and R_{Th_3}) to the remainder of the network, which at this point is only the 1-Ω load, yields the network in Fig. 5.16i. V_o can now easily be computed to be $V_o = 4$ V.

In applying Norton's theorem, we again break the network at points *A-B*, as shown in Fig. 5.17a. In this network note that the voltage across the 4-Ω resistor is $64 - 12 = 52$ V, and therefore the current in this resistor flowing from left to right is 13 A. Since the voltage across the 12-Ω resistor is 12 V, the current flowing down through the resistor is 1 A. Using KCL, $I_{sc_1} = 13 - 1 = 12$ A. R_{Th_1} was computed in Fig. 5.16b as 3 Ω. The Norton equivalent (I_{sc_1} and R_{Th_1}) is attached to the remainder of the network in Fig. 5.17b. Breaking the network at points *C-D* yields the circuit in Fig. 5.17c. Applying current division yields $I_{sc_2} = 6$ A. R_{Th_2} computed from Fig. 5.17d is 4 Ω. Attaching this Norton equivalent (I_{sc_2} and R_{Th_2}) to the remainder of the network, we obtain the circuit in Fig. 5.17e. Adding the current sources and applying current divison, we find that the 8 A of current will split equally between the two 4-Ω paths and therefore $V_o = 4$ V. ■

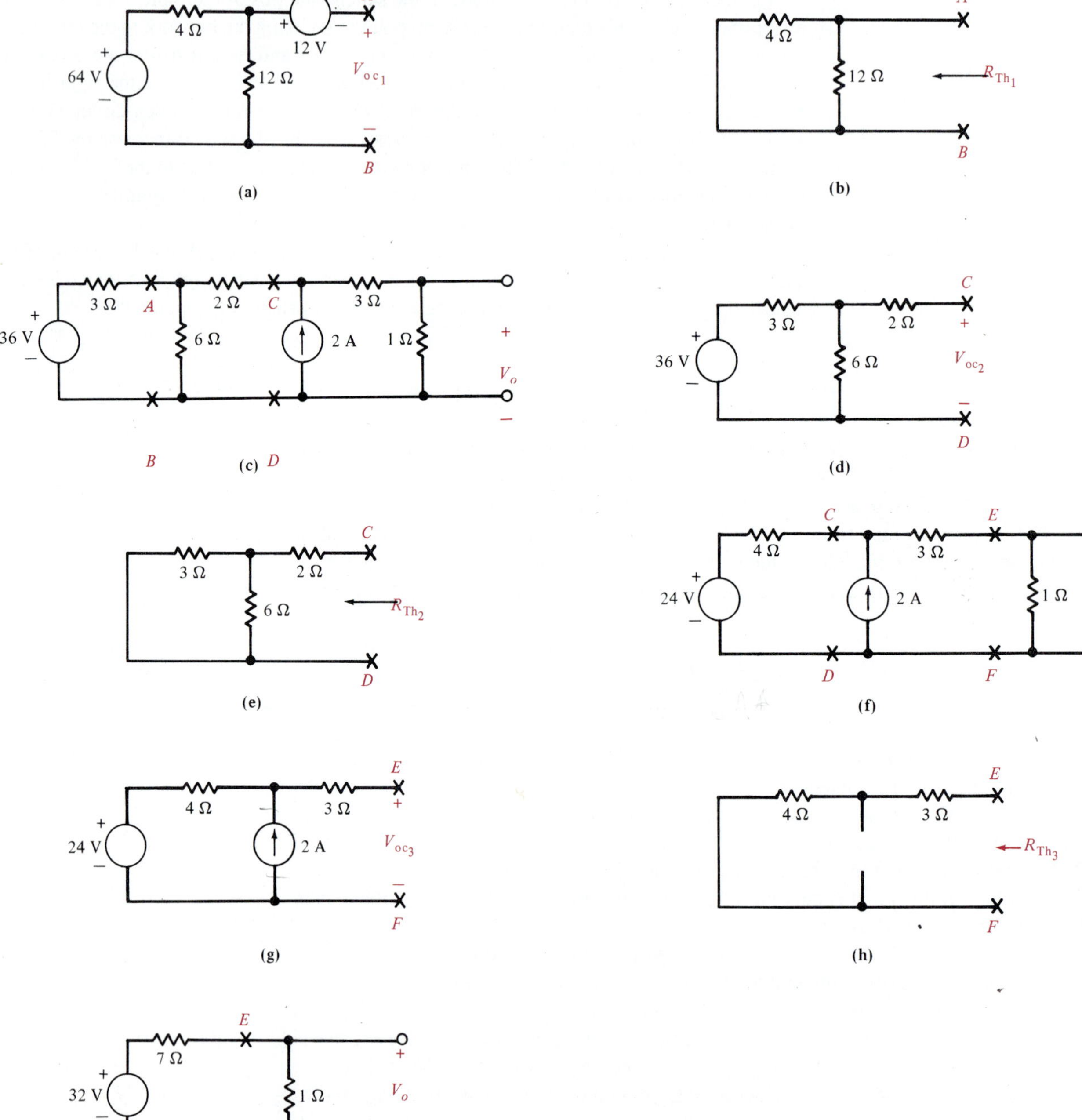

Figure 5.16 Circuits used in Example 5.12.

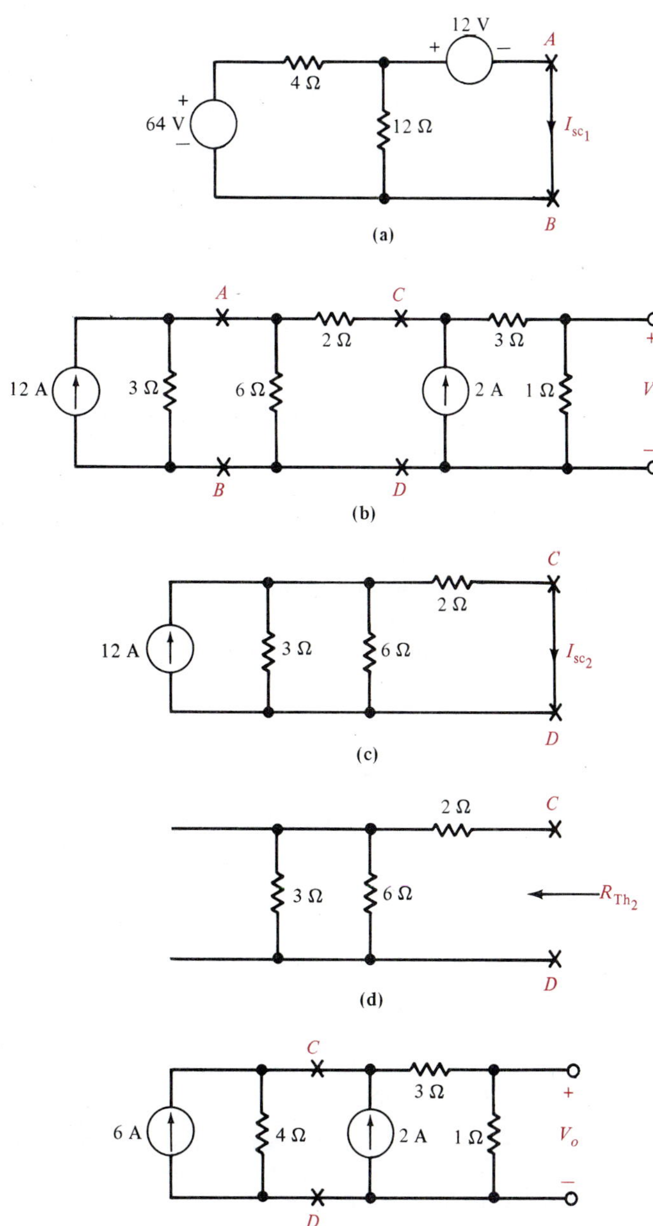

Figure 5.17 Networks employed in Example 5.12.

DRILL EXERCISE

D5.10. Use Thévenin's theorem to find V_o in the circuit in Fig. D5.4.

Ans: $V_o = -8$ V.

D5.11. Use Thévenin's theorem to find V_o in the circuit in Fig. D5.5.

Ans: $V_o = 4$ V.

D5.12. Use Thévenin's theorem to find V_o in the network in Fig. D5.6.

Ans: $V_o = 15$ V.

D5.13. Use Norton's theorem to find V_o in the circuit in Fig. D5.4.

Ans: $V_o = -8$ V.

D5.14. Use Norton's theorem to find V_o in the network in Fig. D5.5.

Ans: $V_o = 4$ V.

In the previous examples we have illustrated the manner in which Thévenin's and Norton's theorems can be applied to networks containing only independent sources. Let us now apply the theorems to networks that contain only dependent sources. As stated earlier, the Thévenin or Norton equivalent of such a network is R_{Th}, and the following examples will serve to illustrate how to determine this Thévenin equivalent resistance.

EXAMPLE 5.13

We wish to determine the Thévenin equivalent of the network in Fig. 5.18a at the terminals A-B. Our approach to this problem will be to apply a 1-V source at the terminals as shown in Fig. 5.18b, then compute the current I_o and $R_{\text{Th}} = 1/I_o$.

The equations for the network in Fig. 5.18b are as follows. KVL around the outer loop specifies that

$$V_1 + V_x = 1$$

The KCL equation at the node labeled V_1 is

$$\frac{V_1}{1} + \frac{V_1 - 2V_x}{2} + \frac{V_1 - 1}{1} = 0$$

Once again, as noted in Chapter 3, we have an equation with the voltage in volts and the resistance in kilohms; therefore, the current will be in milliamperes.

Solving the equations for V_x yields $V_x = \frac{3}{7}$ V. Knowing V_x, we can compute the currents I_1, I_2, and I_3. Their values are

$$I_1 = \frac{V_x}{1} = \frac{3}{7} \text{ mA}$$

$$I_2 = \frac{1 - 2V_x}{1} = \frac{1}{7} \text{ mA}$$

$$I_3 = \frac{1}{2} = \frac{1}{2} \text{ mA}$$

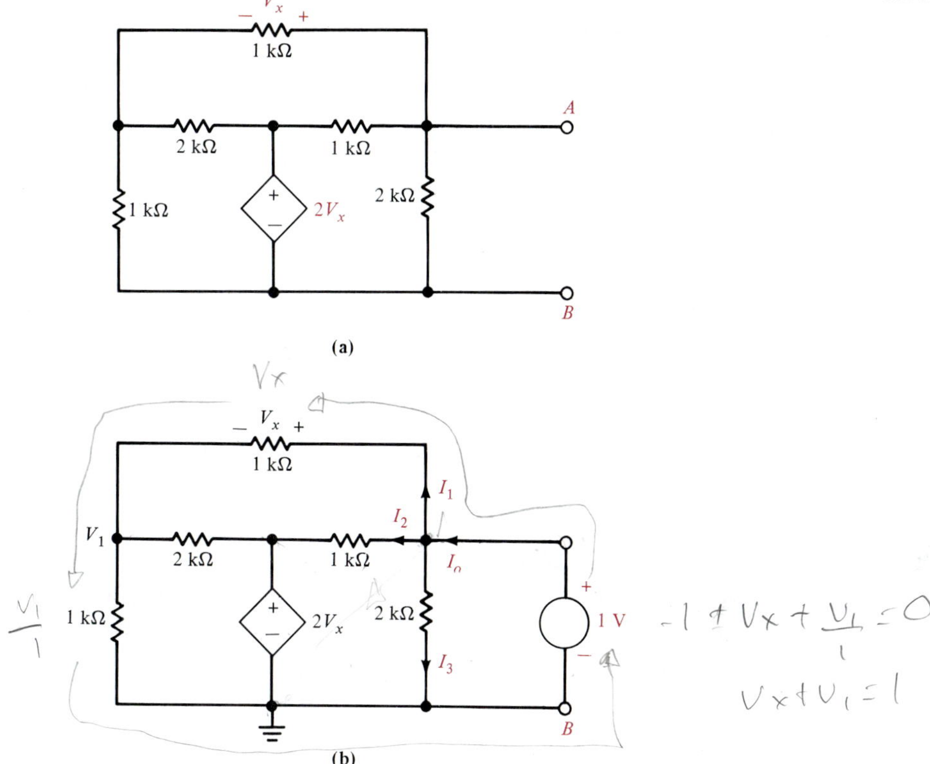

Figure 5.18 Networks employed in Example 5.13.

Therefore,

$$I_o = I_1 + I_2 + I_3$$
$$= \frac{15}{14} \text{ mA}$$

and

$$R_{\text{Th}} = \frac{1}{I_o}$$
$$= \frac{14}{15} \text{ k}\Omega$$

EXAMPLE 5.14

Let us determine R_{Th} at the terminals A-B for the network in Fig. 5.19a. Our approach to this problem will be to apply a 1-mA current source at the terminals A-B, compute the terminal voltage V_2 as shown in Fig. 5.19b, and then $R_{\text{Th}} = V_2/0.001$.

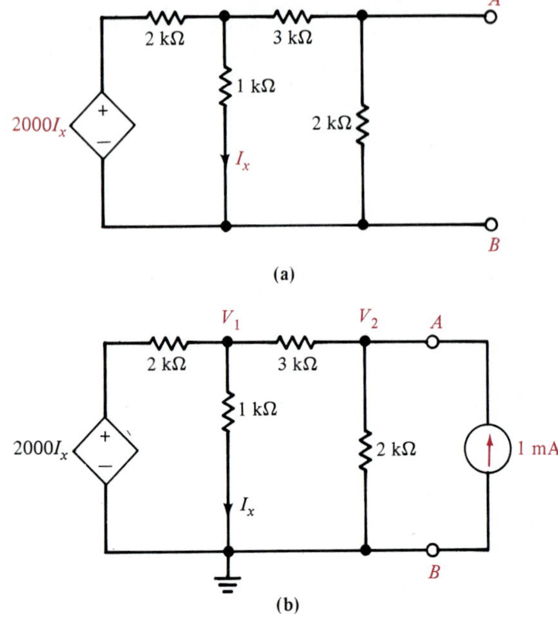

Figure 5.19 Networks used in Example 5.14.

The node equations for the network are

$$\frac{V_1 - 2000I_x}{2000} + \frac{V_1}{1000} + \frac{V_1 - V_2}{3000} = 0$$

$$\frac{V_2 - V_1}{3000} + \frac{V_2}{2000} = \frac{1}{1000}$$

and

$$I_x = \frac{V_1}{1000}$$

These equations can be placed in the following form:

$$\begin{bmatrix} \frac{5}{6000} & -\frac{1}{3000} \\ -\frac{1}{3000} & \frac{5}{6000} \end{bmatrix} \begin{bmatrix} V_1 \\ V_2 \end{bmatrix} = \begin{bmatrix} 0 \\ \frac{1}{1000} \end{bmatrix}$$

Therefore,

$$\begin{bmatrix} V_1 \\ V_2 \end{bmatrix} = \frac{12(1000)^2}{7} \begin{bmatrix} \frac{5}{6000} & \frac{1}{3000} \\ \frac{1}{3000} & \frac{5}{6000} \end{bmatrix} \begin{bmatrix} 0 \\ \frac{1}{1000} \end{bmatrix}$$

and

$$V_2 = \frac{10}{7}\text{ V}$$

Hence

$$R_{\text{Th}} = \frac{V_2}{0.001}$$

$$= \frac{10}{7}\text{ k}\Omega$$

EXAMPLE 5.15

Using the operational amplifier model shown in Fig. 3.33b, find the output resistance of the circuit shown in Fig. 5.20a. The output resistance is the Thévenin equivalent resist-

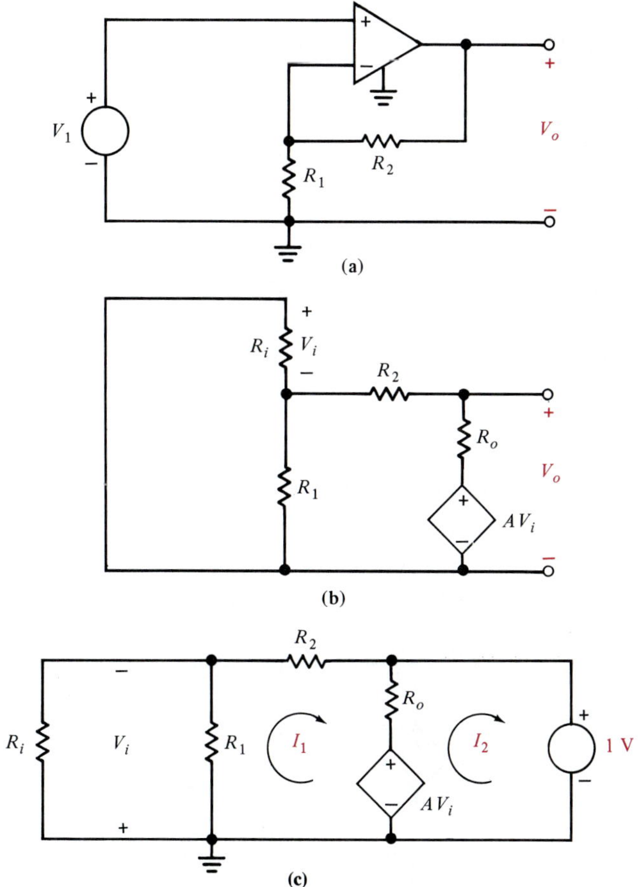

Figure 5.20 Circuits for determining the output resistance of an operational amplifier circuit.

ance at the output terminals. Using the model, the equivalent circuit is shown in Fig. 5.20b.

The Thévenin equivalent resistance at the output terminals can be obtained by assuming a voltage source of 1 V connected to the terminals and determining the resultant current it produces. The circuit in Fig. 5.20b has been redrawn in Fig. 5.20c with this connection.

The circuit equations for this configuration are

$$-AV_i = I_1(R_{i1} + R_2 + R_o) - I_2 R_o$$
$$AV_i - 1 = -I_1 R_o + I_2 R_o$$
$$V_i = R_{i1} I_1$$

where R_{i1} is the parallel combination of R_i and R_1. Solving these equations for I_2 yields

$$I_2 = \frac{-(R_{i1} + AR_{i1} + R_2 + R_o)}{R_o(R_{i1} + R_2)}$$

Then

$$R_{out} = \frac{1}{-I_2} = \frac{R_o(R_{i1} + R_2)}{R_{i1} + AR_{i1} + R_2 + R_o}$$

Since R_i is normally much greater than R_1, $R_{i1} \approx R_1$ and thus

$$R_{out} = \frac{R_o(R_1 + R_2)}{R_1 + AR_1 + R_2 + R_o}$$

Since A is very large, AR_1 is normally much greater than $R_1 + R_2 + R_o$. Therefore, R_{out} can be approximated by the expression

$$R_{out} = \frac{R_o(R_1 + R_2)}{AR_1}$$
$$= \frac{R_o(1 + R_2/R_1)}{A}$$

Note that this equation indicates that the output resistance of this configuration is much lower than the basic op-amp whose output resistance is R_o. ■

DRILL EXERCISE

D5.15. For the circuit in Fig. D5.15, find the Thévenin equivalent at the terminals *A-B*.

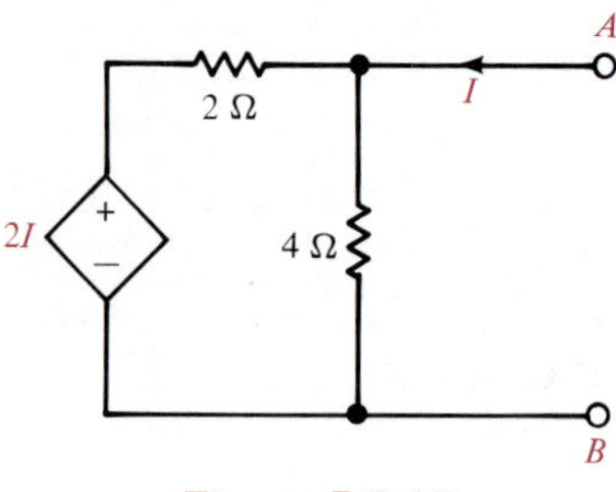

Figure D5.15

Ans: $R_{Th} = \frac{8}{3}\ \Omega$.

D5.16. For the network in Fig. D5.16, find the Thévenin equivalent at the terminals *A-B*.

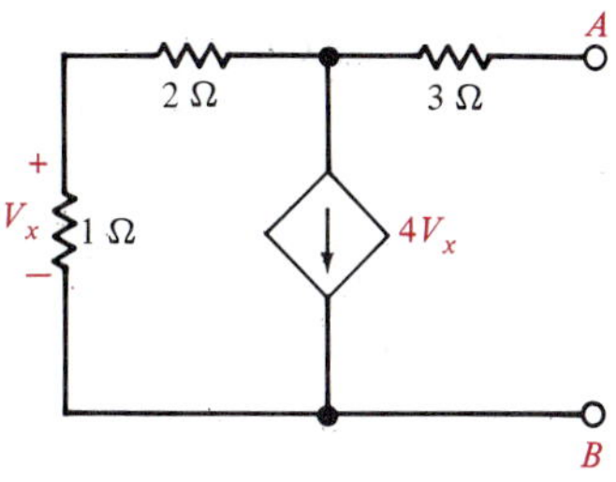

Figure D5.16

Ans: $R_{Th} = \frac{18}{5}\ \Omega$.

We have now examined the methods with which to apply Thévenin's and Norton's theorems to networks containing only independent sources or dependent sources. We now illustrate the application of the theorems to circuits that contain both elements.

EXAMPLE 5.16

Let us compute the voltage V_o in the network in Fig. 5.21a using Thévenin's theorem.

To begin, we break the network at points *A-B*. Recall that when the original network is split into circuit *A* and circuit *B*, the dependent source and its controlling variable must be contained in the same network. The open-circuit voltage is calculated from the network in Fig. 5.21b. Note that the dependent source is now labeled as $2I'_x$ since the network in Fig. 5.21b is different from that in Fig. 5.21a. The KCL equation for the supernode around the 12-V source is

$$\frac{V_{oc} + 12 - (-2I'_x)}{1} + \frac{V_{oc} + 12}{2} + \frac{V_{oc}}{2} = 0$$

where

$$I'_x = \frac{V_{oc}}{2}$$

Solving for V_{oc}, we obtain

$$V_{oc} = -6\text{ V}$$

At this point we wish to determine the Thévenin equivalent resistance. However, because of the presence of the dependent source, we cannot simply zero the independent sources and look back into the network at terminals *A-B* to compute R_{Th}. We will now illustrate two methods for determining this resistance. In the first method we will find the short-circuit current and use Eq. (5.12) to find R_{Th}, and in the second method we will zero the independent sources and apply an external source to determine R_{Th} directly. I_{sc} can be calculated from the network in Fig. 5.21c. Since the terminals *A-B* are shorted, I''_x will be zero and the network reduces to that shown in Fig. 5.21d. Therefore,

$$I_{sc} = \frac{-12}{\frac{2}{3}}$$

$$= -18\text{ A}$$

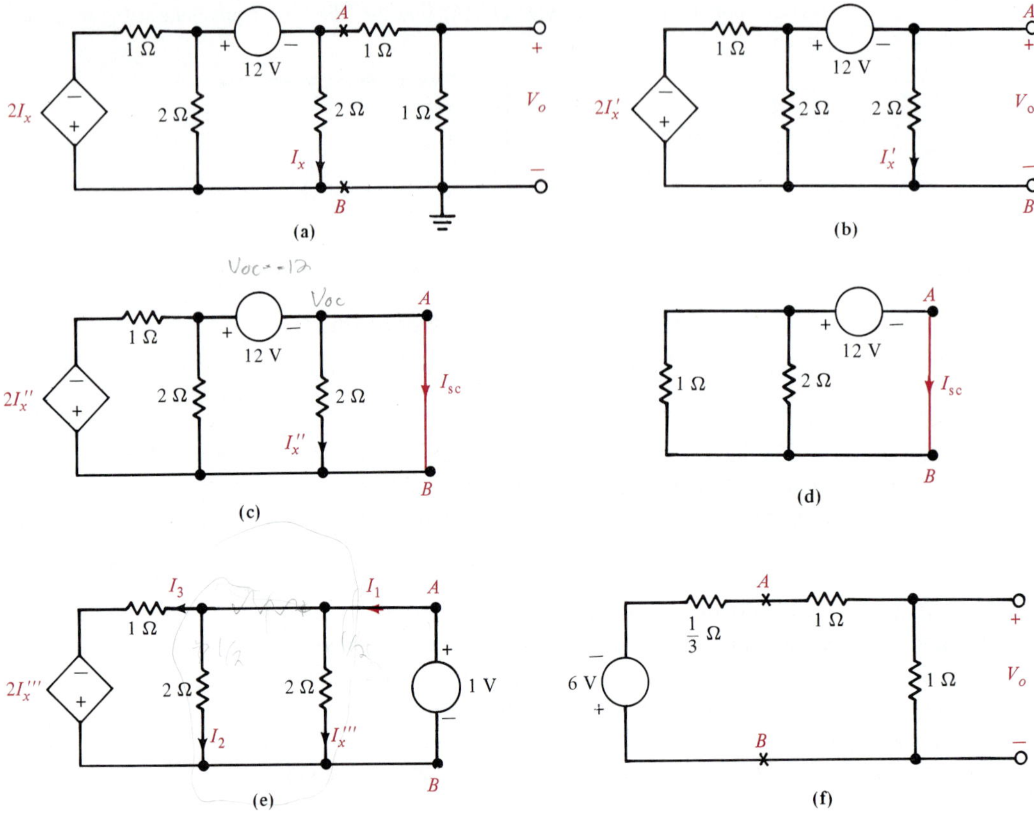

Figure 5.21 Circuits used in Example 5.16.

Then using Eq. (5.12) gives us

$$R_{\text{Th}} = \frac{V_{\text{oc}}}{I_{\text{sc}}}$$

$$= \tfrac{1}{3}\,\Omega$$

R_{Th} can also be calculated by employing an external source as shown in Fig. 5.21e. The branch currents in this network can be computed immediately as

$$I_x''' = \tfrac{1}{2}\text{ A}$$

$$I_2 = \tfrac{1}{2}\text{ A}$$

$$I_3 = \frac{1 - (-2I_x''')}{1}$$

$$= 2\text{ A}$$

Therefore,

$$I_1 = I_x''' + I_2 + I_3$$

$$= 3A$$

and hence

$$R_{\text{Th}} = \tfrac{1}{3}\,\Omega$$

Finally, connecting the Thévenin equivalent circuit at terminals A-B back to the remainder of the original network and noting carefully the polarity of the voltage source, we obtain the network in Fig. 5.21f. From this network we obtain

$$V_o = \frac{-6}{2 + \frac{1}{3}}(1)$$
$$= -\tfrac{18}{7}\text{ V}$$

Note that this answer is the same as that obtained in Example 4.12. ■

EXAMPLE 5.17

We wish to determine V_o in the network in Fig. 5.22a.

If we break the network at the points A-B, the open-circuit voltage can be calculated from the circuit in Fig. 5.22b. The mesh current I_o can be determined from the KVL equation

$$(I_o - 2V_A')(1) + \tfrac{1}{2}I_o + 1I_o = 6$$

where

$$V_A' = \frac{I_o}{2}$$

and therefore

$$V_{\text{oc}} = 4\text{ V}$$

R_{Th} can be determined from Fig. 5.22c. The KCL equation for the node labeled V_1 is

$$\frac{V_1}{1} + \frac{V_1 - 1}{\frac{1}{2}} = 2V_A''$$

where

$$V_A'' = V_1 - 1$$

Solving the equations for V_A yields $V_A = -1$ V. Therefore, $I_2 = 2$ A, and since $I_3 = 1$ A,

$$I_o = I_2 + I_3$$
$$= 3\text{ A}$$

Hence

$$R_{\text{Th}} = \tfrac{1}{3}\,\Omega$$

Connecting the Thévenin equivalent circuit back to the remainder of the original network as shown in Fig. 5.22d, we find that $V_o = \frac{12}{7}$ V. Note that this is the same as that obtained in Example 4.10. ■

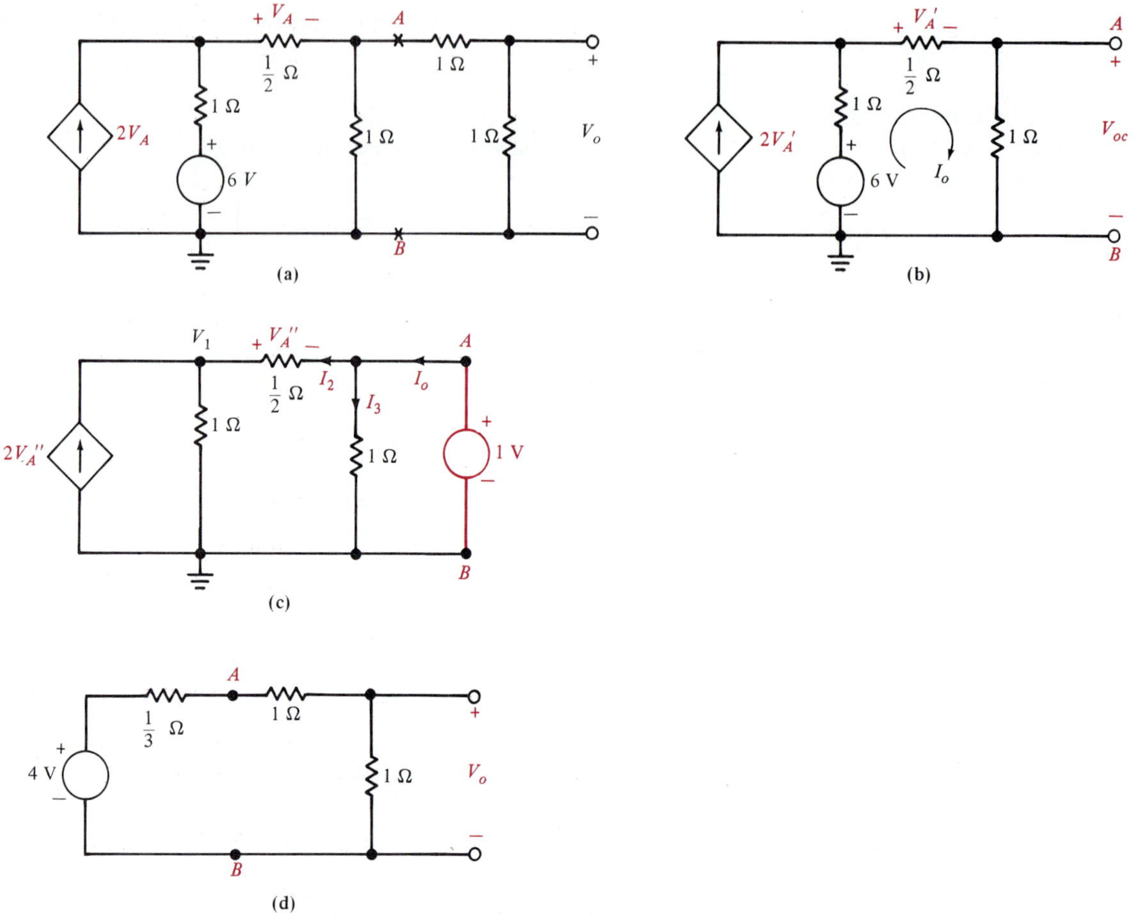

Figure 5.22 Circuits used in Example 5.17.

DRILL EXERCISE

D5.17. Find the Thévenin equivalent circuit at the terminals *A-B* for the network in Fig. D5.17.

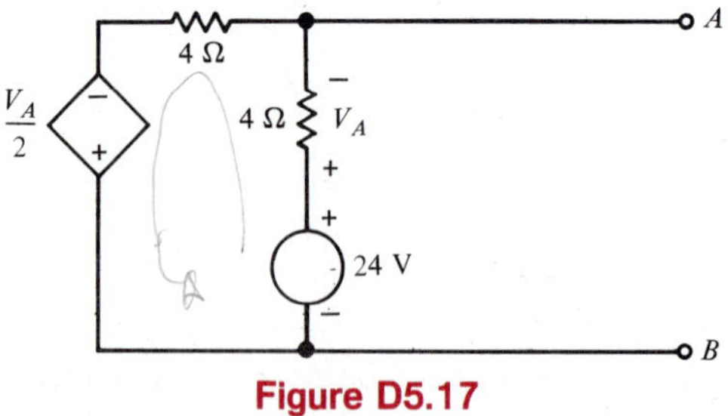

Figure D5.17

Ans: $V_{oc} = 8$ V, $R_{Th} = \frac{8}{3}\ \Omega$.

D5.18. Find the Thévenin equivalent circuit at the terminals *A-B* for the network in Fig. D5.18.

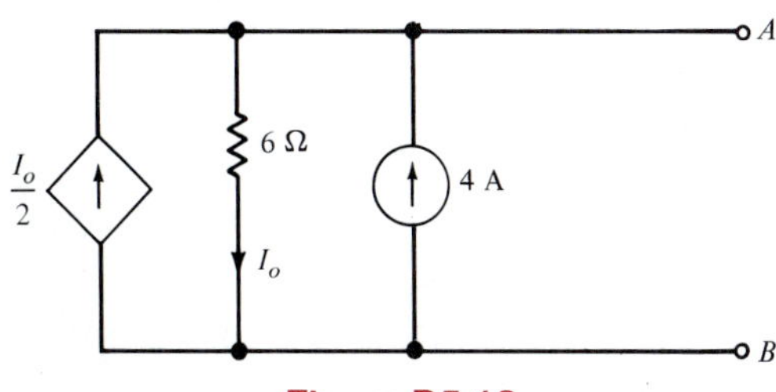

Figure D5.18

Ans: $V_{oc} = 48$ V, $R_{Th} = 12\ \Omega$.

At this point, it is important once again for the reader to remember that all types of electronic devices are modeled with dependent sources, and therefore the analyses here are similar to those involved in both the analysis and design of electronic circuits. In addition, it is interesting to note that as we systematically reduce a network using source transformation, we are actually flipping back and forth between Thévenin and Norton equivalents.

5.2 Maximum Power Transfer

In circuit analysis we are sometimes interested in determining the maximum power that can be delivered to a load. By employing Thévenin's theorem, we can determine the maximum power that a circuit can supply and the manner in which to adjust the load to effect maximum power transfer.

Suppose that we are given the circuit shown in Fig. 5.23. The power that is delivered to the load is given by the expression

$$P_{\text{load}} = i^2 R_L$$

$$= \left(\frac{v}{R + R_L}\right)^2 R_L$$

We want to determine the value of R_L that maximizes this quantity. Hence we differentiate this expression with respect to R_L and equate the derivative to zero.

$$\frac{dP_{\text{load}}}{dR_L} = \frac{(R + R_L)^2 v^2 - 2v^2 R_L (R + R_L)}{(R + R_L)^4}$$

which yields

$$R_L = R$$

In other words, maximum power transfer takes place when the load resistance $R_L = R$. Although this is a very important result, we have derived it using the simple network in Fig. 5.23. However, we should recall that v and R in Fig. 5.23 could represent the Thévenin equivalent circuit for any linear network.

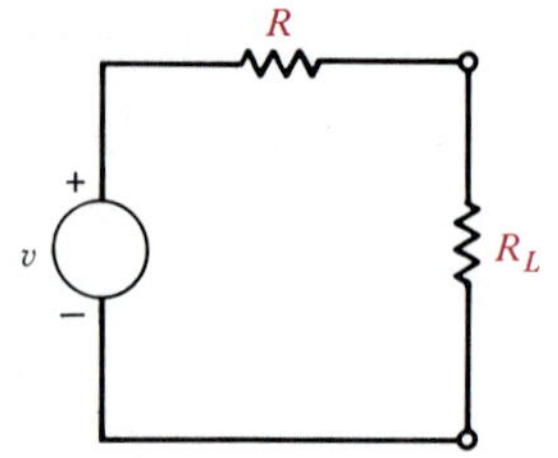

Figure 5.23 Equivalent circuit for examining maximum power transfer.

EXAMPLE 5.18

Let us determine the value of R_L for maximum power transfer in the network in Fig. 5.24a. After finding this value of R_L, we will find the maximum power that can be transferred to this load.

If we break the network at the load, we obtain the circuit shown in Fig. 5.24b. At this point it is important to note that we simply have another network, and therefore all the techniques that we have learned can be applied to solve it. Note that we can form a supernode around the dependent source. The voltage with respect to the reference node at the positive terminal of the dependent source is V_{oc}, and therefore the voltage with respect to the reference node at the negative terminal of the dependent source is $V_{oc} - 2I$. KCL applied to the supernode is

$$\frac{V_{oc} - 2I'}{4} - 12 + \frac{V_{oc}}{2} = 0$$

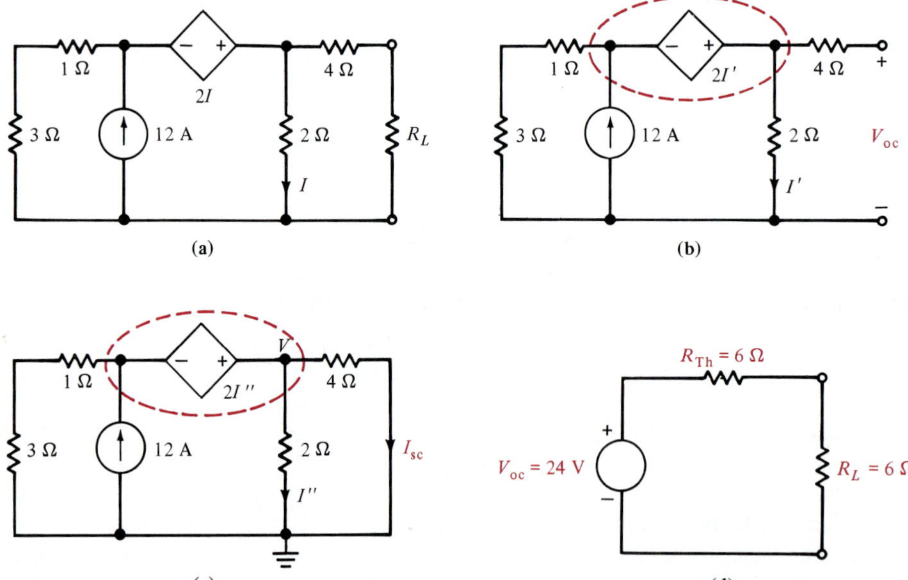

Figure 5.24 Example circuit for determining maximum power transfer.

where

$$I' = \frac{V_{oc}}{2}$$

Solving these equations yields

$$V_{oc} = 24 \text{ V}$$

I_{sc} can be derived in a similar manner using the network in Fig. 5.24c. KCL for the supernode is

$$\frac{V - 2I''}{4} - 12 + \frac{V}{2} + \frac{V}{4} = 0$$

where

$$I'' = \frac{V}{2}$$

Therefore,

$$V = 16 \text{ V}$$

and hence

$$I_{sc} = \frac{V}{4} = 4 \text{ A}$$

The Thévenin equivalent resistance is then

$$R_{Th} = \frac{V_{oc}}{I_{sc}} = \frac{24}{4} = 6 \ \Omega$$

The Thévenin equivalent network is shown in Fig. 5.24d, and hence the maximum power transferred to the load is

$$P_L = \left(\frac{24}{6 + 6}\right)^2 (6) = 24 \text{ W}$$

■

DRILL EXERCISE

D5.19. In the network in Fig. D5.19, find R_L for maximum power transfer and the maximum power absorbed by the load.

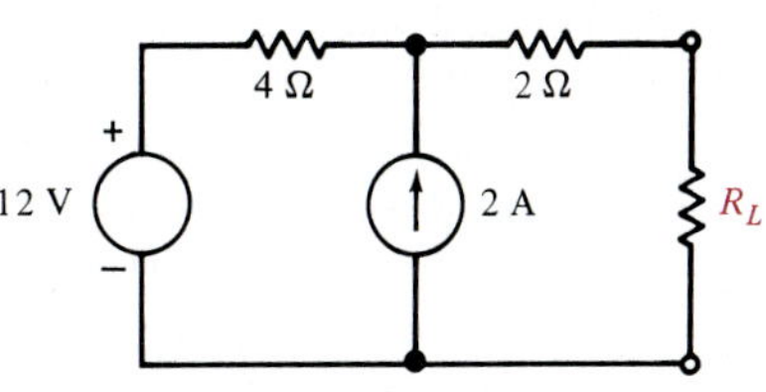

Figure D5.19

Ans: $R_L = 6\ \Omega$, $P_L = \frac{50}{3}$ W.

D5.20. In the network in Fig. D5.20, find R_L for maximum power transfer and the maximum power absorbed by the load.

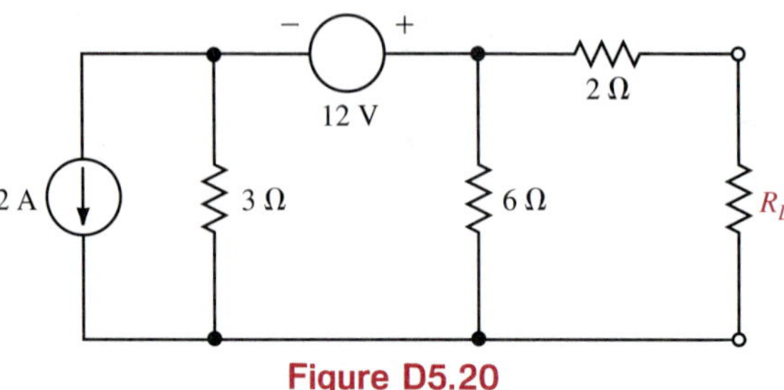

Figure D5.20

Ans: $R_L = 4\ \Omega$, $P_L = 1$ W.

As we proceed throughout the text, we will continually introduce appropriate new PSPICE statements that will expand our capabilities. The first statement of this type is the .MODEL statement.

The .MODEL statement defines a set of device parameters which can be referenced by passive or active elements in the circuit. The general format of the statement is as follows:

```
.MODEL <name> <type> (<parameter name>=<value>)
```

<name> is the name of the model that devices use to reference the model. <type> is the device type. Some valid device types are RES, CAP, IND, D, and NPN for resistor, capacitor, inductor, diode, and NPN bipolar transistor, respectively. The <parameter name> and <value> are optional. Their meaning depends upon the type of device specified.

The .MODEL statement is used in conjunction with the .DC statement, and a device declaration allows the circuit to be analyzed at specified values of the device within a given range. For example, the following PSPICE statements

```
RL 3 0 RMOD 5
.MODEL RMOD RES(R=1)
.DC RES RMOD(R) 4 10 2
```

perform a set of dc analyses on the circuit with a different value of R_L for each analysis. The first statement declares resistor R_L to be connected between nodes 3 and 0, have a value of 5 ohms, and use model RMOD to describe its characteristics further. The second statement declares that RMOD is a model of type RES, or resistor, and has a scaling factor, R, of 1. The scaling factor is used to scale the declared value of any resistor using the model, RMOD. The third statement uses ‘RES RMOD(R)’ in place of the source variable name. This last statement causes a dc sweep analysis to be performed on the circuit varying all resistors employing RMOD. The variable to be swept through the specified range is given in parentheses. In this case, the scaling variable, R, is to be swept from 4 to 10 using an increment of 2. The first dc analysis is performed with R_L equal 20 ohms; the second analysis is performed with R_L equal 30 ohms; and so on. The final analysis is performed with R_L equal 50 ohms.

See Reference[2] for more information on the specifics of each of these statements.

We will now use the .MODEL and .PROBE statements to generate a plot in the context of a maximum power transfer problem.

EXAMPLE 5.19

We will apply PSPICE to the network in Fig. 5.25 to find R_L for maximum power transfer and the maximum power absorbed by the load. The PSPICE program for this problem is

```
MAXIMUM POWER TRANSFER EXERCISE 2
V1 1 0 DC 12
I1 2 0 DC 2
R1 1 2 2
R2 2 0 2
R3 2 3 2
RL 3 0 RMOD 1
.MODEL RMOD RES(R=1)
.DC RES RMOD(R) 1 10 1
.PROBE
.END
```

To determine the value of R_L at which maximum power transfer is achieved, the value of R_L will be stepped from 1 ohm to 10 ohms in increments of 1 ohm. This is accomplished with the following statements.

```
RL 3 0 RMOD 1
.MODEL RMOD RES(R=1)
.DC RES RMOD(R) 1 10 1
```

The resistor, R_L, is declared to use the model, RMOD, and to have a value of 1 ohm. The .MODEL statement declares the model, RMOD, to be of type RES with a scaling factor of 1 ($R = 1$). The .DC statement performs a series of dc analyses on the circuit, stepping the value of the scaling factor, R, from 1 to 10 in increments of 1 for each analysis. Since the effective resistance of R_L is the value of R_L (1 ohm) multiplied by the scaling factor, R, the effective resistance of R_L changes from 1 ohm to 10 ohms in increments of 1 ohm for each analysis.

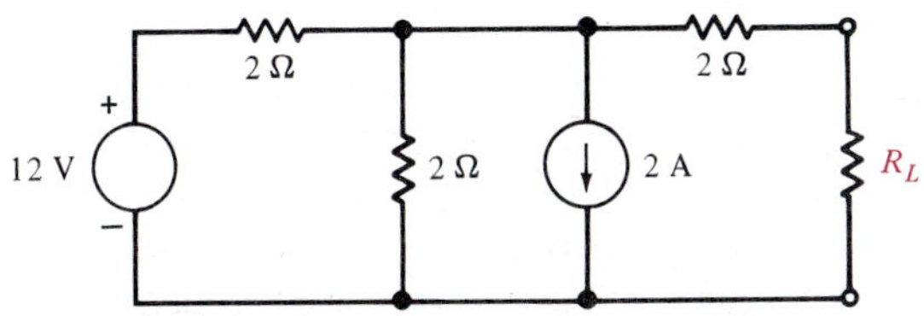

Figure 5.25 PSPICE example circuit.

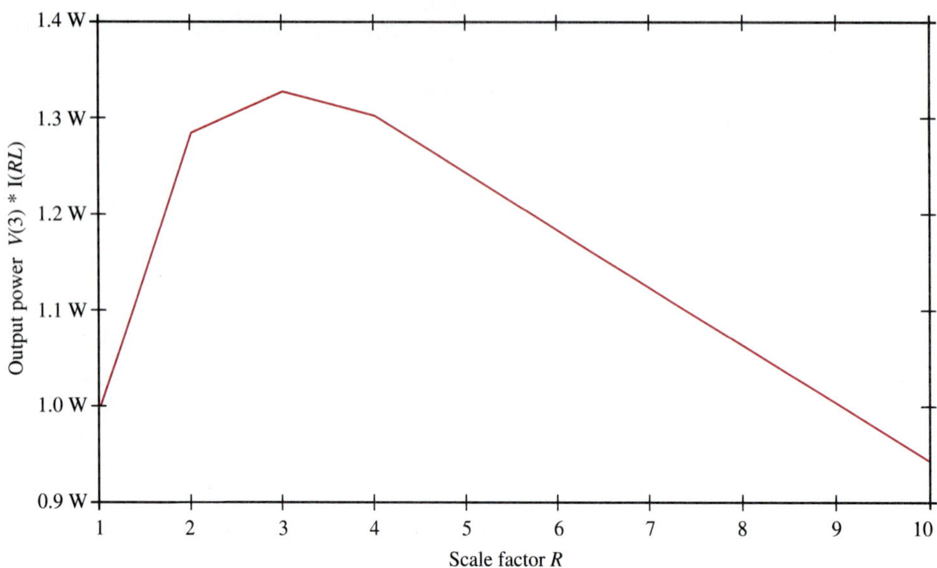

Figure 5.26 Power transfer curve for the network in Example 5.19.

The results are obtained by plotting the value of the power absorbed by R_L as a function of the effective value of R_L. In PSPICE, the Probe program is entered upon completion of all dc analyses. The screen displays a blank graph above a menu line. The option, 'Add_trace' is highlighted. Press enter. Type 'I(RL)*V(3)' and press the Enter key in response to the 'Enter variables or expressions' prompt. The plot of the power absorbed by R_L, current through R_L multiplied by the voltage across R_L, appears on the screen. The maximum power point may then be obtained from the curve. It occurs at R equal 3. Since the initial value of R_L is 1 ohm, the value of R_L for maximum power transfer is 1 ohm multiplied by a scaling factor of 3. Therefore, maximum power transfer occurs at R_L equal 3 ohms as shown in Fig. 5.26. A hard copy may be obtained by using the arrow keys to highlight 'Hard_copy' from the menu. Press the Enter key once the hard-copy option is highlighted. The user may then exit from the Probe program by selecting the 'Exit' option from the menu line in the same manner. ■

5.3 Sensitivity Analysis

In the course of the analysis of a circuit, it is often useful to know how the variation of a particular circuit component will affect a particular circuit voltage. This relationship can be expressed as a "sensitivity" of one parameter to another. Mathematically, this relationship can be expressed as a derivative of the voltage function with respect to the parameter of interest. For example, consider the simple voltage-divider network shown in

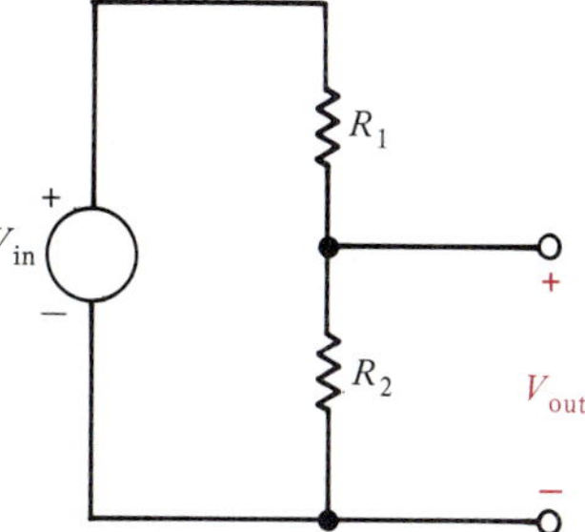

Figure 5.27 Simple voltage-divider network.

Fig. 5.27. The output voltage can be expressed as

$$V_{out} = \frac{(V_{in})(R_2)}{R_1 + R_2}$$

The sensitivity of the voltage V_{out} to resistor R_2 is expressed as the following derivative function:

$$\frac{dV_{out}}{dR_2} = \frac{d}{dR_2}\left[\frac{(V_{in})(R_2)}{R_2 + R_1}\right] = \frac{(V_{in})(R_1)}{(R_2 + R_1)^2}$$

For $V_{in} = 1$ V, $R_1 = 3\ \Omega$, and $R_2 = 1\ \Omega$, this corresponds to a value of $\frac{3}{16}$ V/Ω. An important point to keep in mind is that the derivative is a very localized function. In other words, the sensitivity calculation is true only for small variations of the component values of interest. For instance, if R_2 is increased by 2 Ω, to 3 Ω, the actual value of V_{out} will be $\frac{3}{6}$ or 0.5 V. The sensitivity calculation would predict an increase in V_{out} of 2(3/16) V or 0.38 V. This, added to the starting value of 0.25 V, yields a final value of 0.68 V. Obviously, this is not in very close agreement with the actual value of 0.5 V. If, however, R_1 is increased by only 0.1 Ω, to 1.1 Ω, the sensitivity calculation yields a V_{out} value of 0.269 V, while the actual value would be 0.268 V, substantially closer agreement.

Even with the aforementioned limitation, sensitivity calculations often yield results important in circuit design. In a complex circuit, it is often difficult to predict how the variation of a particular component will effect some other circuit parameter. By calculating sensitivities for the various components of interest, one can extract a quantitative measure (a number) of the ability to manipulate a given parameter. By comparing the relative magnitudes of the calculated sensitivities, one can devise a strategy for the efficient manipulation of the various circuit parameters.

Unfortunately, the calculation of sensitivities can be very laborious even for fairly simple circuits. However, the PSPICE computer package introduced earlier can generate these sensitivities for a complete circuit with minimal effort on the engineer's part.

Sensitivity can be computed using the following control statement:

```
.SENS OV1 OV2 . . . OVN
```

Through the use of this statement SPICE will determine the dc small-signal sensitivities of each output variable, OV1 ··· OVN, with respect to every circuit parameter.

The following is a PSPICE analysis of the voltage divider discussed above.

```
SENSITIVITY ANALYSIS OF SIMPLE VOLTAGE DIVIDER
VIN 1 0 DC 1
R1  1 2 3
R2  2 0 1
*PRINT DC SENSITIVITY OF V(2) TO ALL COMPONENTS
.SENS V(2)
.END
```

```
DC SENSITIVITIES OF OUTPUT V(2)

ELEMENT    ELEMENT       ELEMENT         NORMALIZED
 NAME       VALUE      SENSITIVITY      SENSITIVITY
               (VOLTS/UNIT)    (VOLTS/PERCENT)

  R1      3.000E+00    -6.250E-02       -1.875E-03
  R2      1.000E+00     1.875E-01        1.875E-03
  VIN     1.000E+00     2.500E-01        2.500E-03
```

The element sensitivity column is the sensitivity as calculated above. The normalized sensitivity is the element sensitivity multiplied by the element value and divided by 100 (to form a percentage). Note that all element sensitivities are calculated. For complex circuits, the amount of output can be quite voluminous.

EXAMPLE 5.20

Use PSPICE to calculate the output voltage sensitivities for each of the circuit elements for the circuit in Fig. 5.28a. Use $R_1 = 1\ \text{k}\Omega$, $R_2 = 500\ \Omega$, $R_3 = 500\ \Omega$, $R_4 = 10\ \text{k}\Omega$, and $V_{in} = 1$ V.

The equivalent circuit labeled for PSPICE is shown in Fig. 5.28b. Using this equivalent circuit, PSPICE analysis is applied and the following results are obtained.

```
OP AMP SENSITIVITY CALCULATION
V1    1 0 1
EOUT 4 0 2 0 -10000
R1    1 2 1K
R2    2 3 500
R3    3 4 500
R4    3 0 10K
.SENS V(4)
.END
```

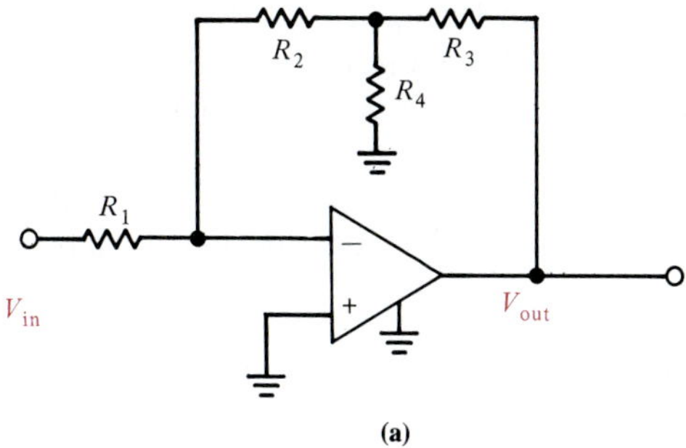

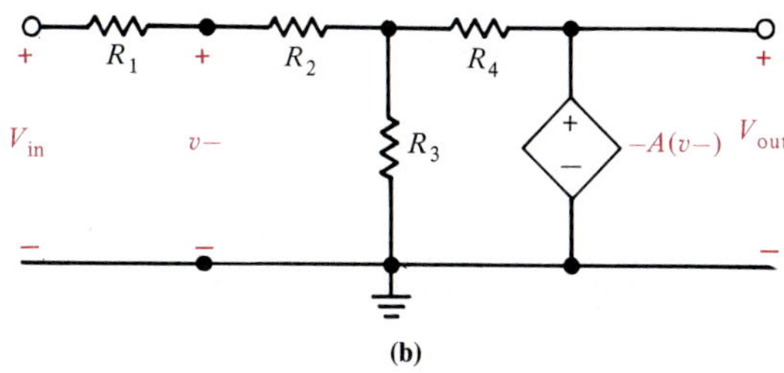

Figure 5.28 Circuits used in Example 5.19.

DC SENSITIVITIES OF OUTPUT V(4)

ELEMENT NAME	ELEMENT VALUE	ELEMENT SENSITIVITY (VOLTS/UNIT)	NORMALIZED SENSITIVITY (VOLTS/UNIT)
R1	1.000E+03	1.025E-03	1.025E-02
R2	5.000E+02	-1.050E-03	-5.248E-03
R3	5.000E+02	-1.050E-03	-5.248E-03
R4	1.000E+04	2.499E-06	2.499E-04
V1	1.000E+00	-1.025E + 00	-1.025E-02

If very fine adjustment of the output voltage is desired, note that R_4 is the best resistor to adjust to set the output voltage. This is true since the output voltage is the least sensitive to this component. ■

5.4 Summary

In this chapter we have presented a number of very powerful tools that have wide applications in circuit analysis. Specifically, we have introduced the principle of superposition, which is based upon linearity and allows us to treat each source independently and then algebraically add the response due to each to determine the total response. We have also shown that source transformation and Thévenin's and Norton's theorems quite often allow us to simplify circuit analysis by solving a series of simple problems rather than a more complicated one. The concept of maximum power transfer from a network to a resistive load was presented.

Finally, we have introduced the concept of sensitivity, which provides us with important information concerning both the design and operation of a network.

KEY POINTS

- In a linear network containing multiple independent sources, the principle of superposition allows us to compute any current or voltage in the network as the algebraic sum of the individual contributions of each source acting alone.
- Superposition is a linear property and does not apply to nonlinear functions such as power.
- Source transformation permits us to replace a voltage source V in series with a resistance R by a current source $I = V/R$ in parallel with the resistance R. The reverse is also true.
- Using Thévenin's theorem, we can replace some portion of a network at a pair of terminals with a voltage source V_{oc} in series with a resistor R_{Th}. V_{oc} is the open-circuit voltage at the terminals, and R_{Th} is the Thévenin equivalent resistance obtained by looking into the terminals with all independent sources made zero.
- Using Norton's theorem, we can replace some portion of a network at a pair of terminals with a current source I_{sc} in parallel with a resistor R_{Th}. I_{sc} is the short-circuit current at the terminals and R_{Th} is the Thévenin equivalent resistance.
- Maximum power transfer can be achieved by selecting the load R_L to be equal to R_{Th} looking into the load terminals.
- The sensitivity of X with respect to Y tells us how variations in Y affect X.

PROBLEMS

5.1. Determine I_o in the network in Fig. P5.1. Use linearity and assume that $I_o = 1$ A.

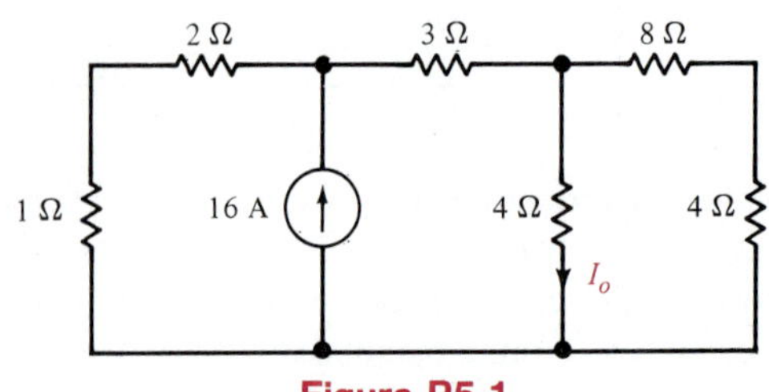

Figure P5.1

5.2. Determine V_o in the network in Fig. P5.2. Use linearity and assume that $V_o = 1$ V.

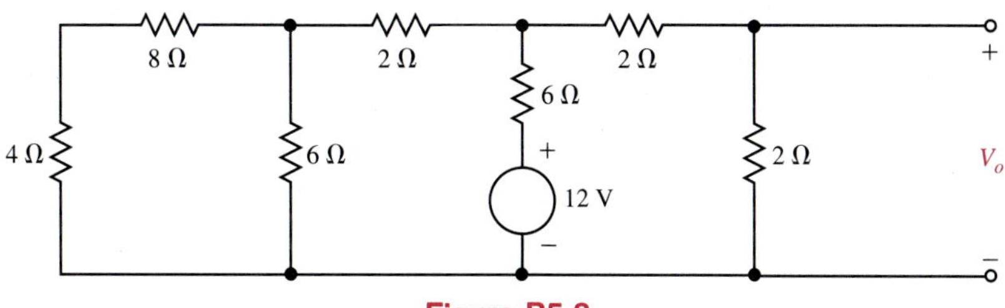

Figure P5.2

5.3. Find I_o in the circuit in Fig. P5.3. Use linearity and assume that $I_o = 1$ A.

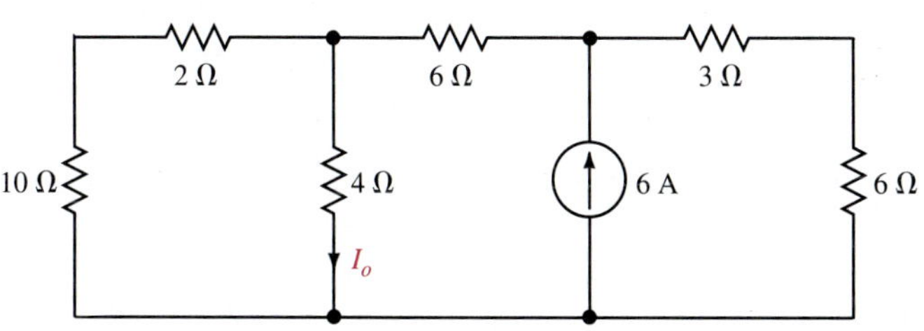

Figure P5.3

5.4. Assume that $I_o = 1$ A and use linearity to find the actual value of I_o in Fig. P5.4.

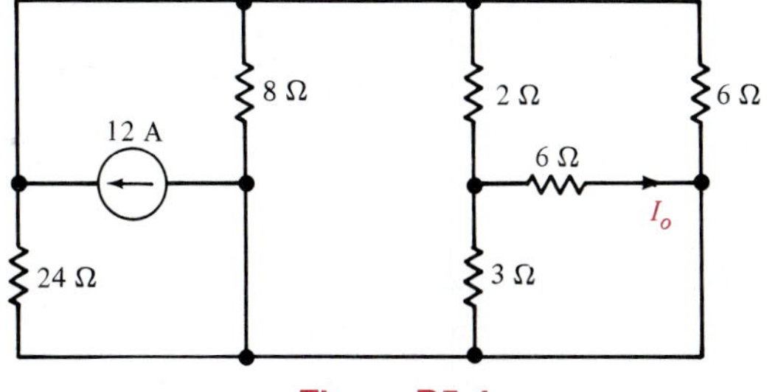

Figure P5.4

5.5. Use superposition to find V_o in Fig. P5.5.

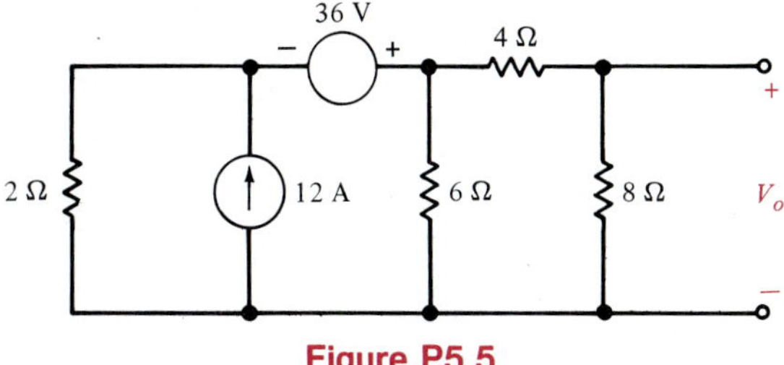

Figure P5.5

5.6. Use superposition to find I_o in the network in Fig. P5.6.

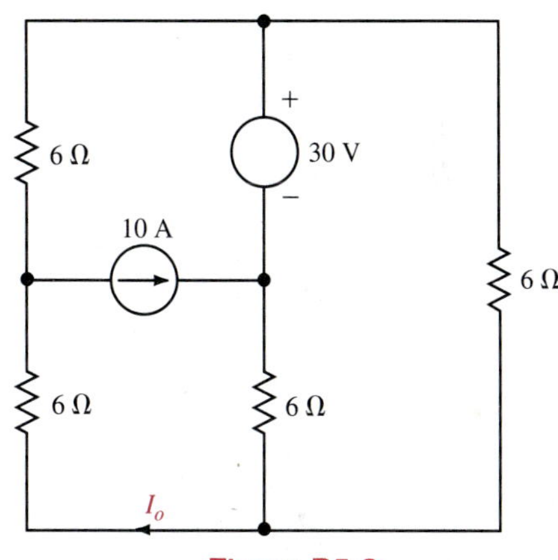

Figure P5.6

5.7. Use superposition to find V_o in the network in Fig. P5.7.

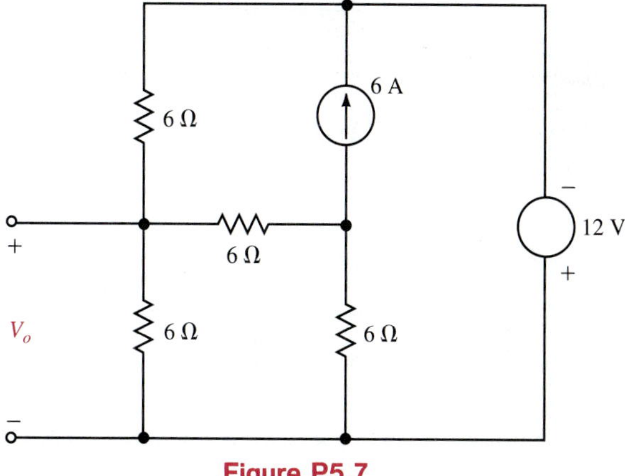

Figure P5.7

5.8. Use superposition to find I_o in the network in Fig. P5.8.

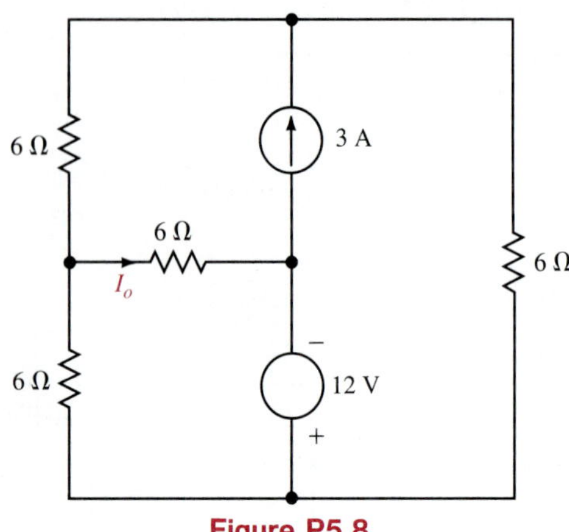

Figure P5.8

5.9. Use superposition to find I_o in the network in Fig. P5.9.

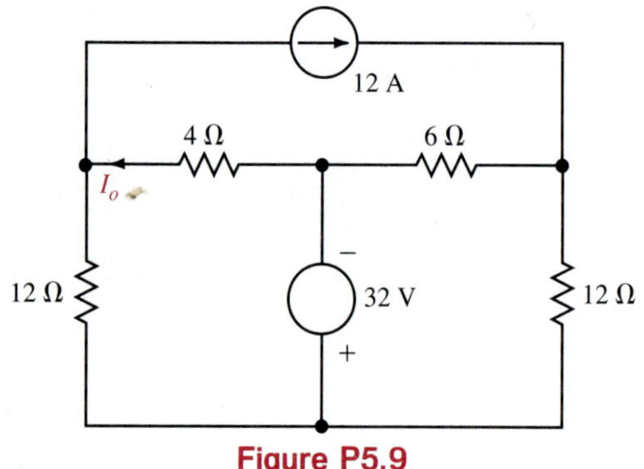

Figure P5.9

5.10. Use superposition to find V_o in the network in Fig. P5.10.

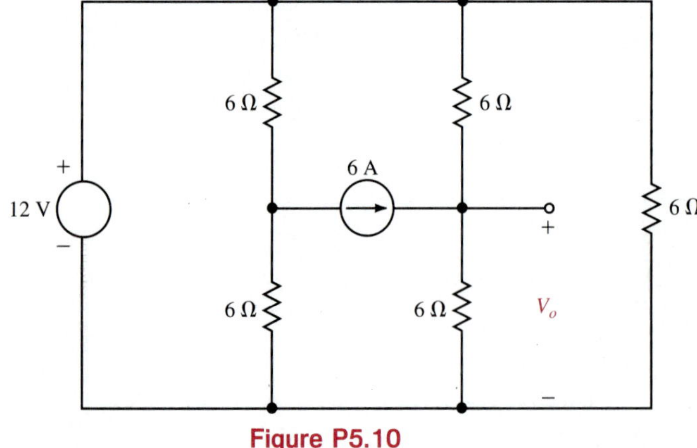

Figure P5.10

5.11. Use superposition to find I_o in the network in Fig. P5.11.

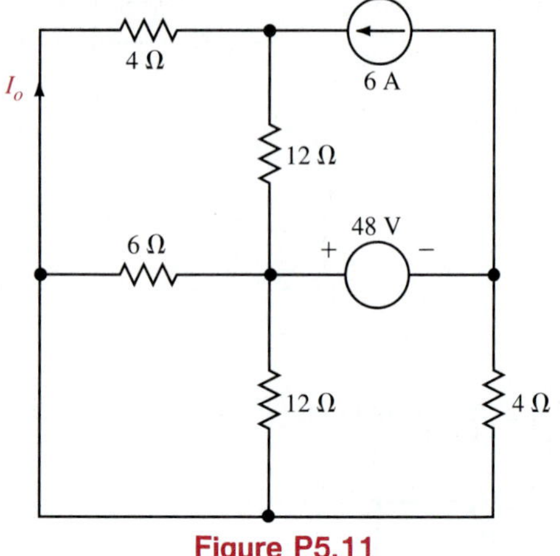

Figure P5.11

5.12. Find V_o and I_o in the network in Fig. P5.12 using superposition.

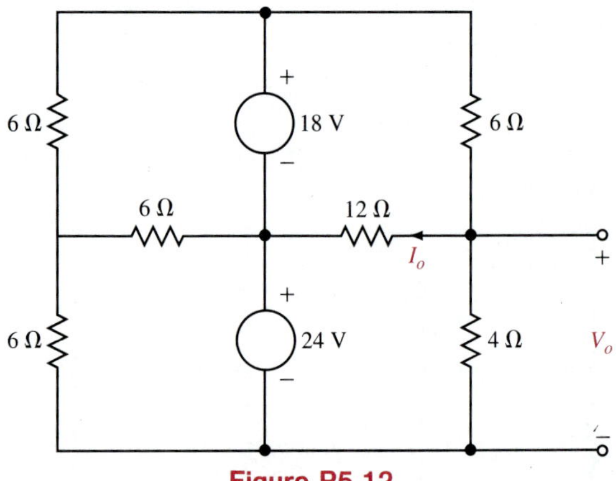

Figure P5.12

5.13. Use superposition to find V_o in the network in Fig. P5.13.

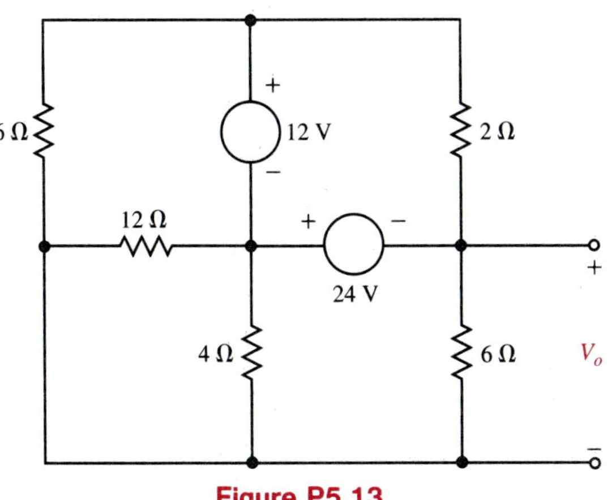

Figure P5.13

5.14. Use superposition to find V_o in the network in Fig. P5.14.

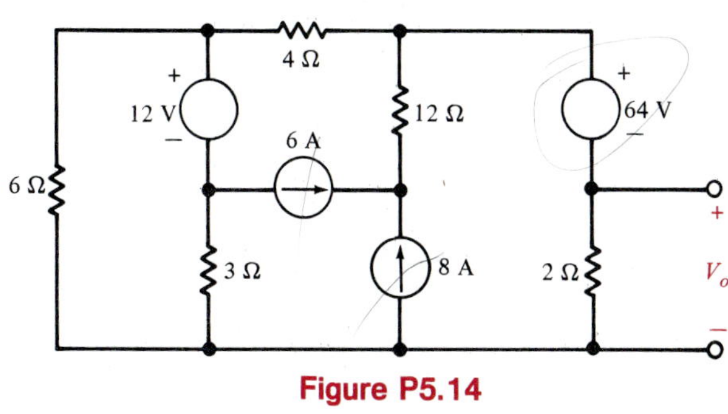

Figure P5.14

5.15. Using superposition, find an expression for the output voltage v_o in the circuit shown in Fig. P5.15.

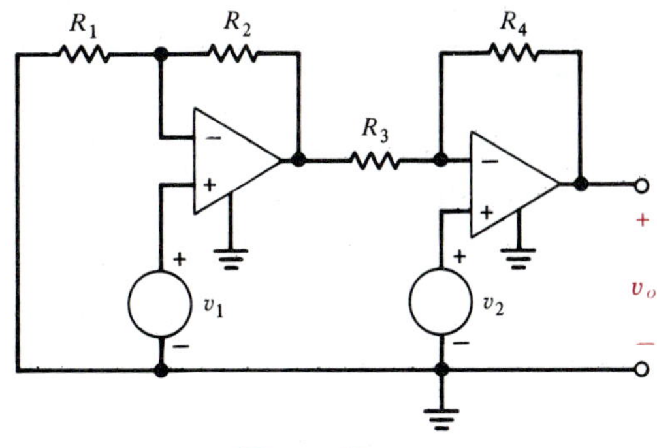

Figure P5.15

5.16. Use source transformation to find I_o in the network in Fig. P5.16.

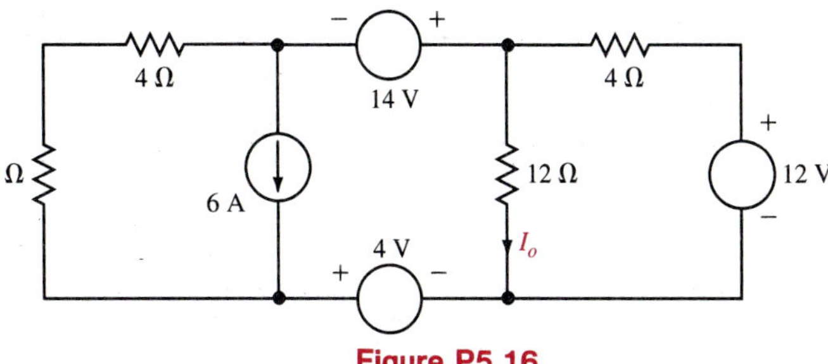

Figure P5.16

5.17. Use source transformation to find I_o in the network in Fig. P5.17.

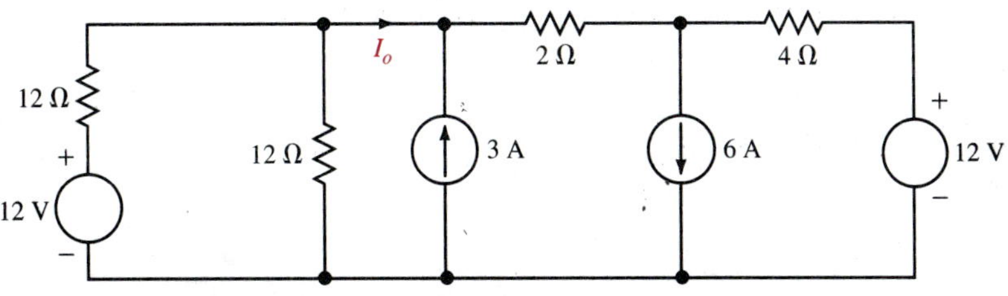

Figure P5.17

5.18. Use source transformation to find I_o in the network in Fig. P5.18.

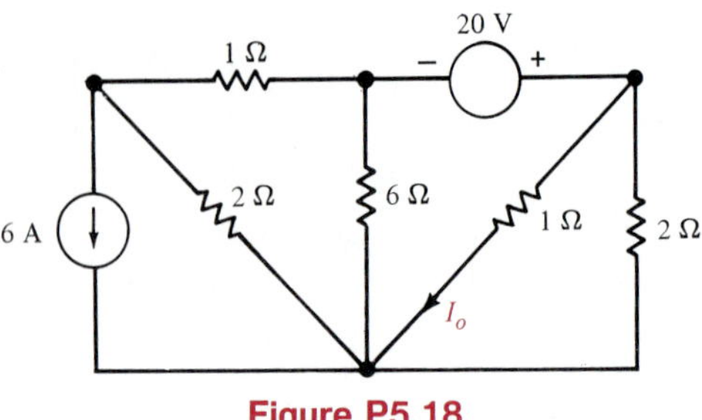

Figure P5.18

5.19. Use source transformation to find I_o in the circuit in Fig. P5.19.

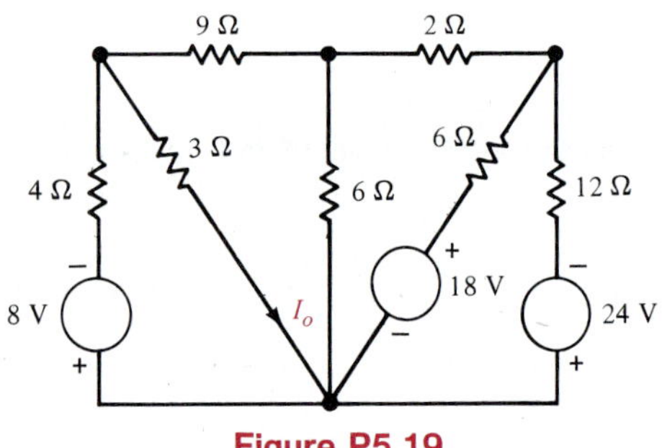

Figure P5.19

5.20. Use source transformation to find V_o in the network in Fig. P5.20.

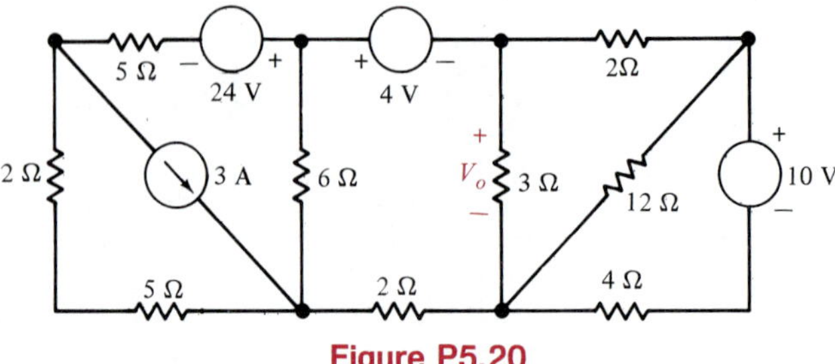

Figure P5.20

5.21. Use source transformation to find I_o in the network in Fig. P5.21.

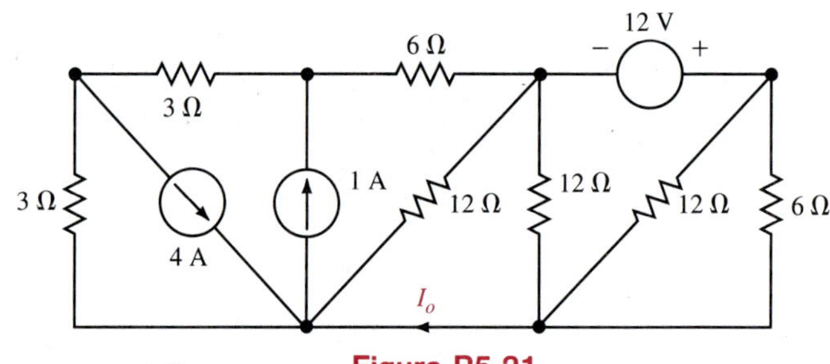

Figure P5.21

5.22. Use source transformation to find I_o in the network in Fig. P5.22.

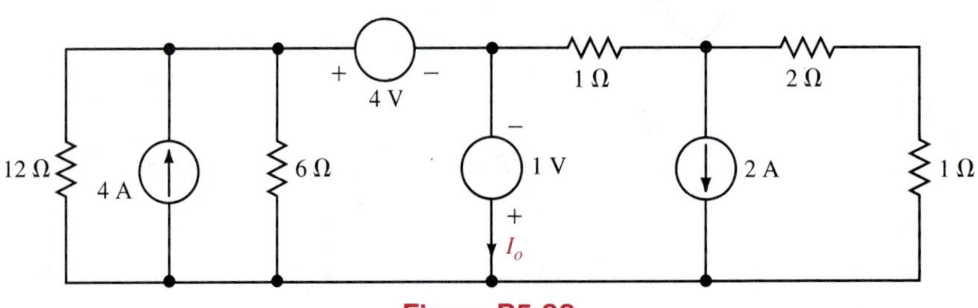

Figure P5.22

5.23. Use source transformation to find I_o in the network in Fig. P5.23.

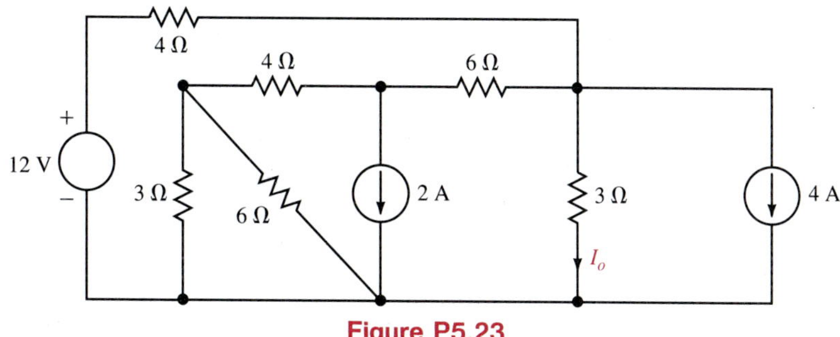

Figure P5.23

5.24. Use Thévenin's theorem to find V_o in the network in Fig. P5.24.

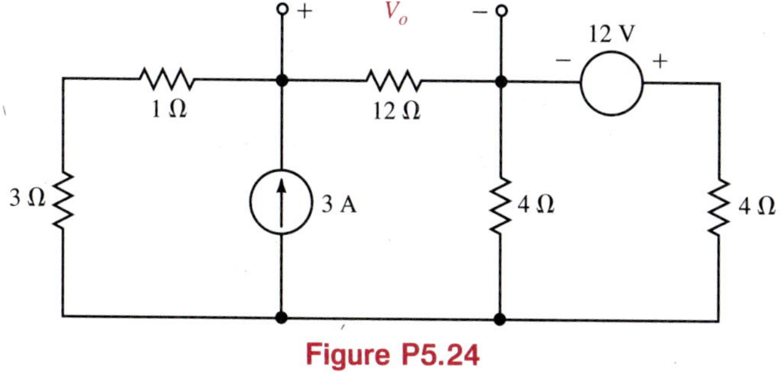

Figure P5.24

5.25. Use Thévenin's theorem to find V_o in the network in Fig. P5.25.

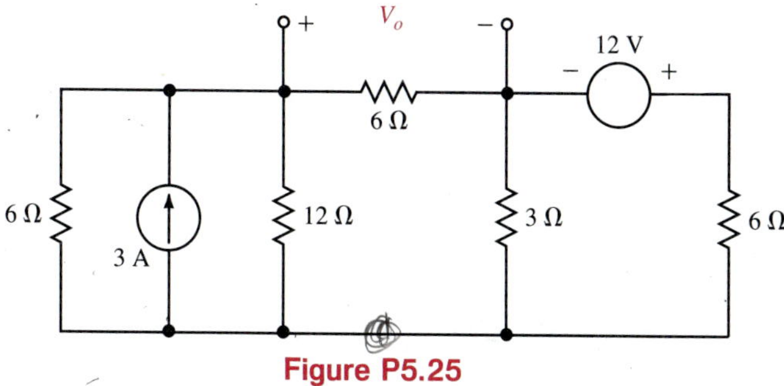

Figure P5.25

5.26. Find V_o in the network in Fig. P5.26 using Thévenin's theorem.

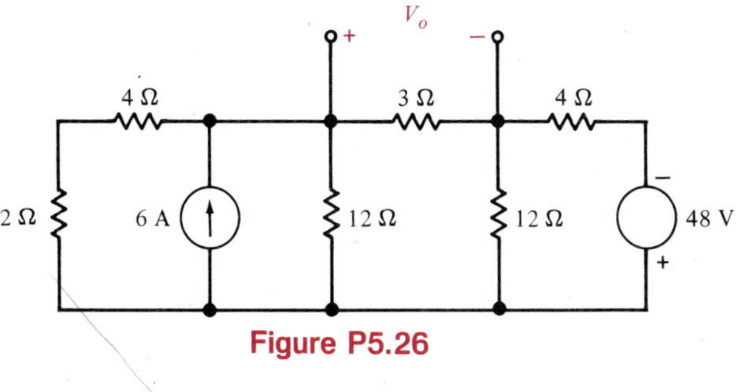

Figure P5.26

5.27. Use Thévenin's theorem to find V_o in the network in Fig. P5.27.

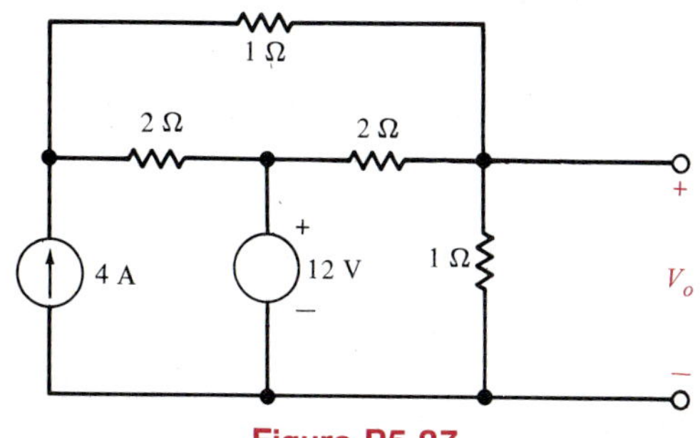

Figure P5.27

5.28. Use Thévenin's theorem to find I_o in the circuit in Fig. P5.28.

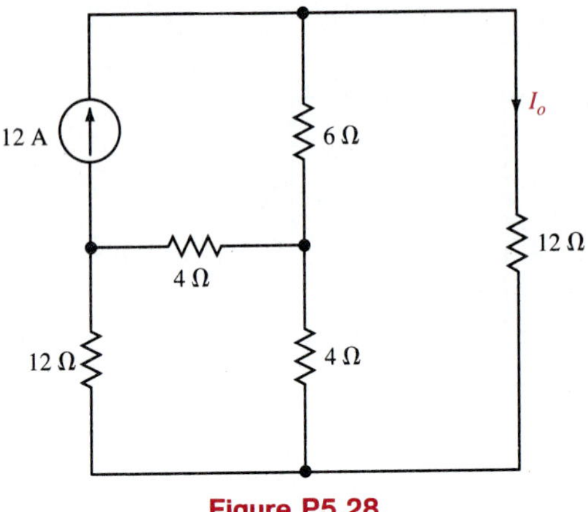

Figure P5.28

5.29. Use Thévenin's theorem to find I_o in the circuit in Fig. P5.29.

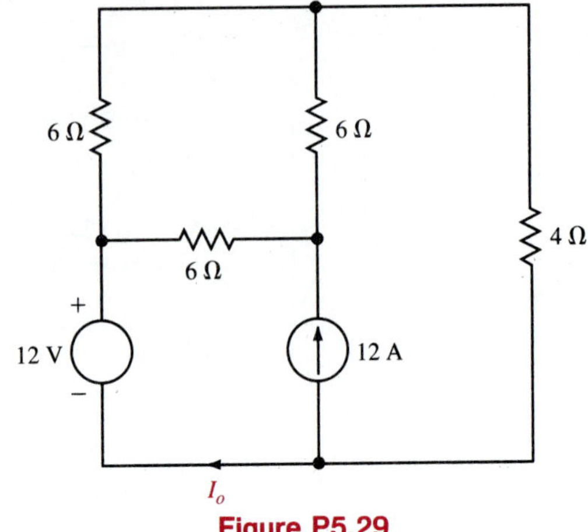

Figure P5.29

5.30. Solve Problem 5.8 using Thévenin's theorem.

5.31. Solve Problem 5.6 using Thévenin's theorem.

5.32. Use Thévenin's theorem to find I_o in the circuit in Fig. P5.32.

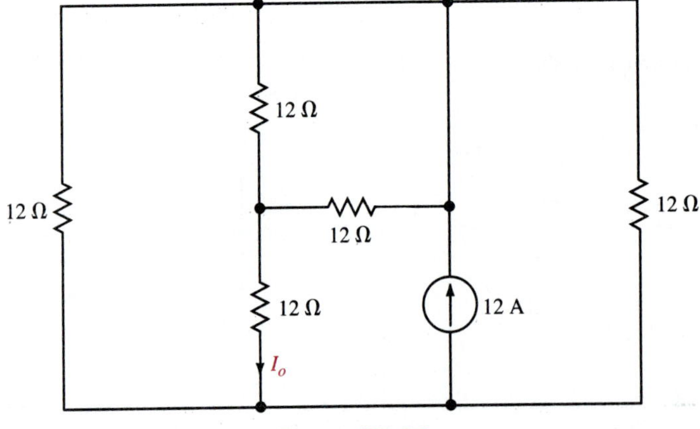

Figure P5.32

5.33. Use Thévenin's theorem to find I_o in the circuit in Fig. P5.33.

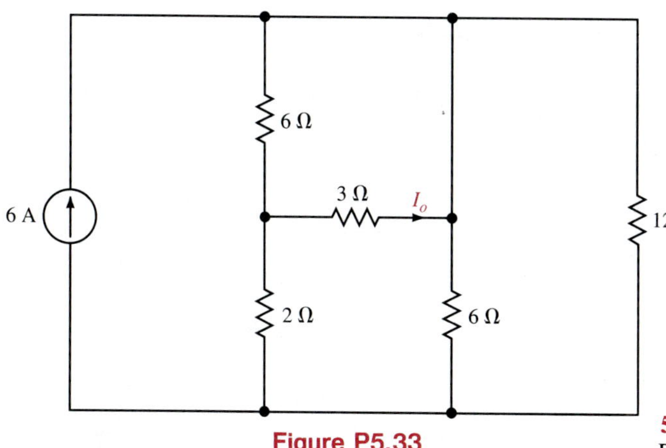

Figure P5.33

5.34. Use Thévenin's theorem to find I_o in the circuit in Fig. P5.34.

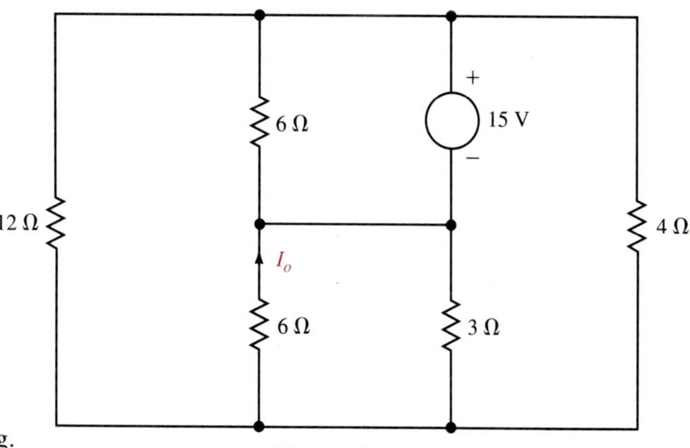

Figure P5.34

5.35. Use Thévenin's theorem to find I_o in the circuit in Fig. P5.35.

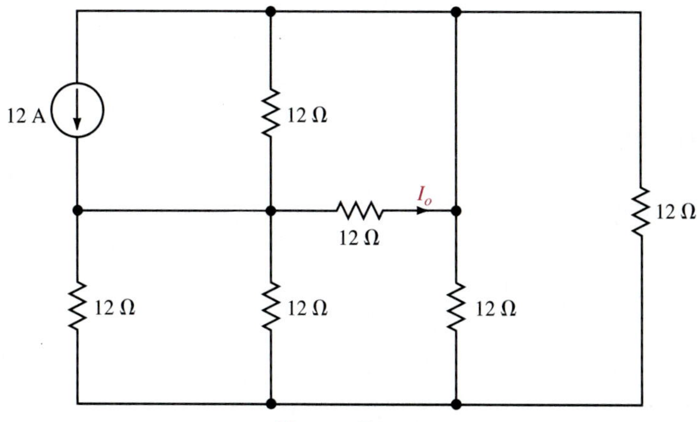

Figure P5.35

5.36. Use a repeated application of Thévenin's theorem to find V_o in the network in Fig. P5.36.

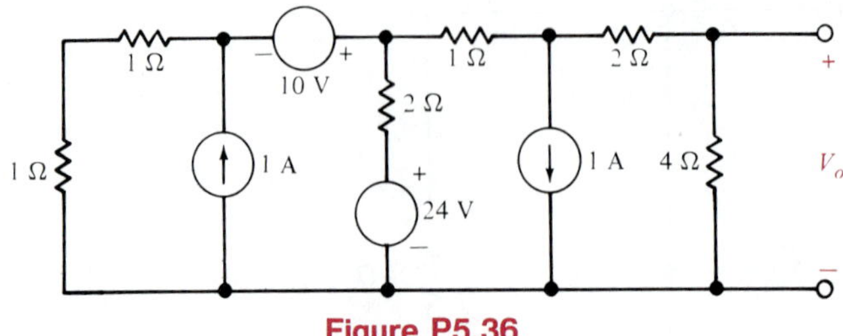

Figure P5.36

5.37. Use a repeated application of Thévenin's theorem to find V_o in the network in Fig. P5.37.

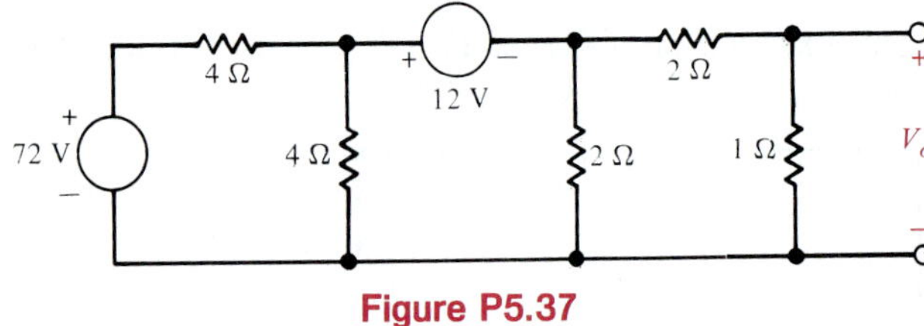

Figure P5.37

5.38. Apply Thévenin's theorem to find V_o in the circuit in Fig. P5.38.

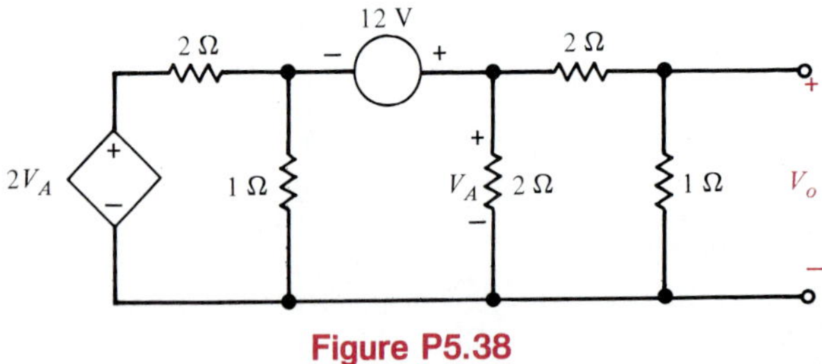

Figure P5.38

5.39. Use Thévenin's theorem to compute the current in R_L in Fig. P5.39.

Figure P5.39

5.40. Use Thévenin's theorem to find V_o in Fig. P5.40.

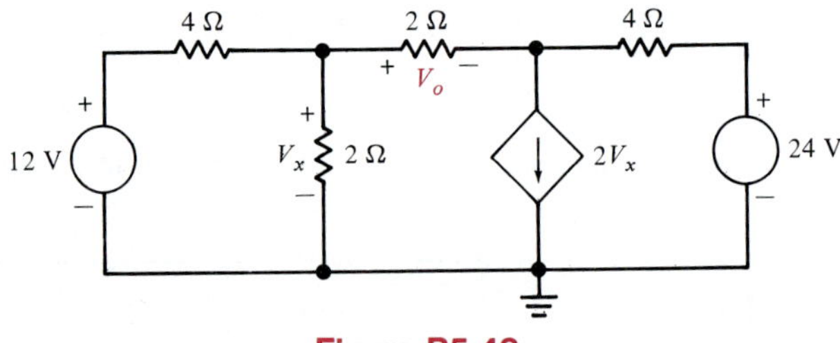

Figure P5.40

5.41. Use Thévenin's theorem to find V_o in the network in Fig. P5.41.

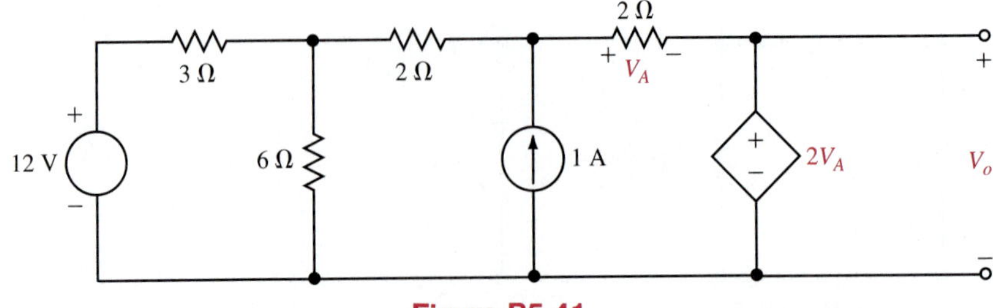

Figure P5.41

5.42. Use Thévenin's theorem to find V_o in the network in Fig. P5.42.

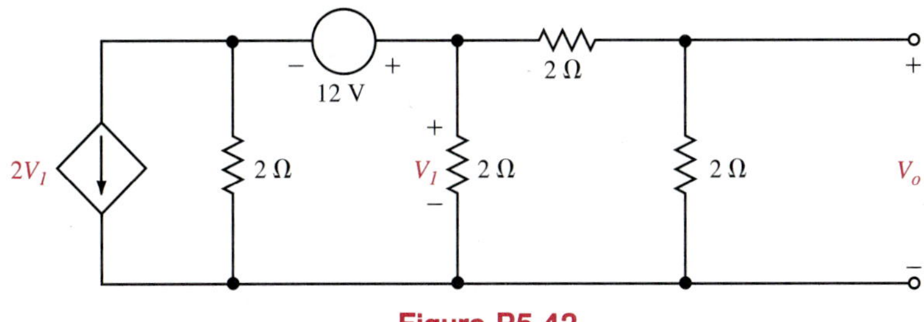

Figure P5.42

5.43. Use Thévenin's theorem to find V_o in the network in Fig. P5.43.

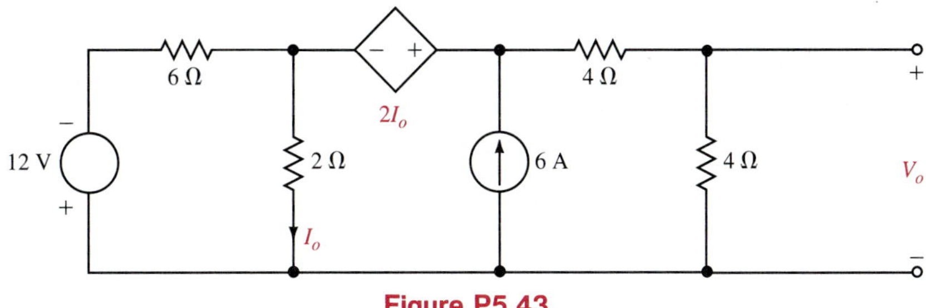

Figure P5.43

5.44. Use Thévenin's theorem to find V_o in the network in Fig. P5.44.

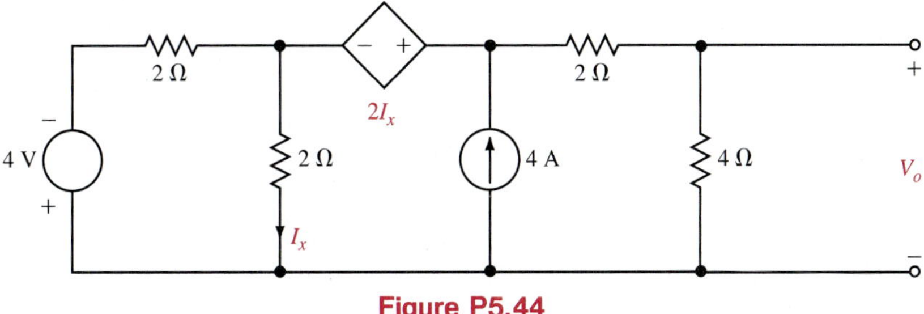

Figure P5.44

5.45. Use Thévenin's theorem to find V_o in the network in Fig. P5.45.

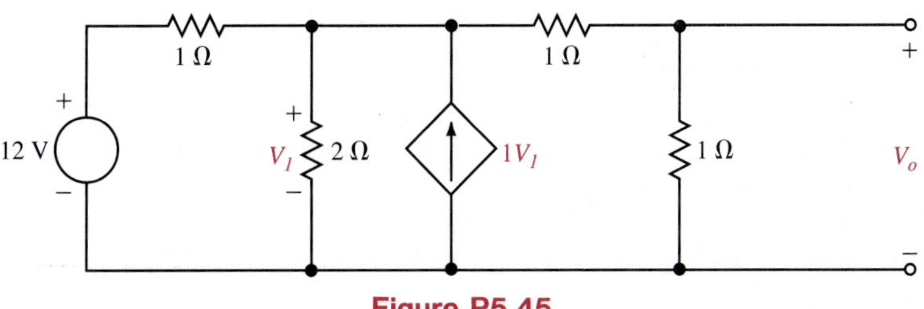

Figure P5.45

5.46. Use Thévenin's theorem to find V_o in the network in Fig. P5.46.

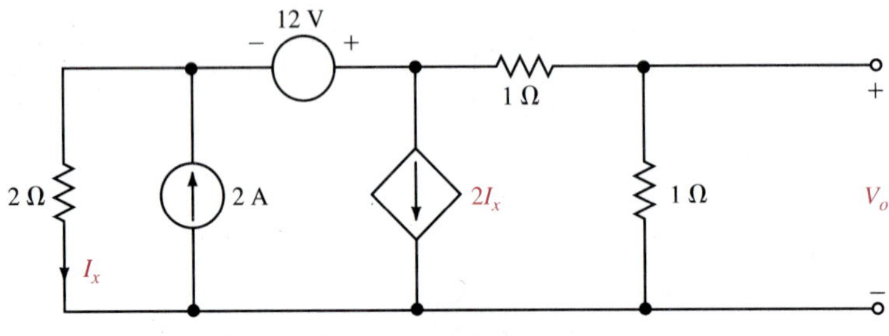

Figure P5.46

5.47. Use Norton's theorem to find I_o in the network in Fig. P5.47.

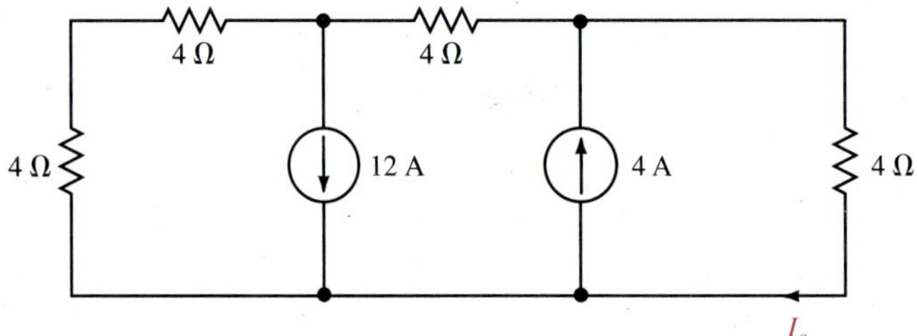

Figure P5.47

5.48. Use Norton's theorem to find I in the network in Fig. P5.48.

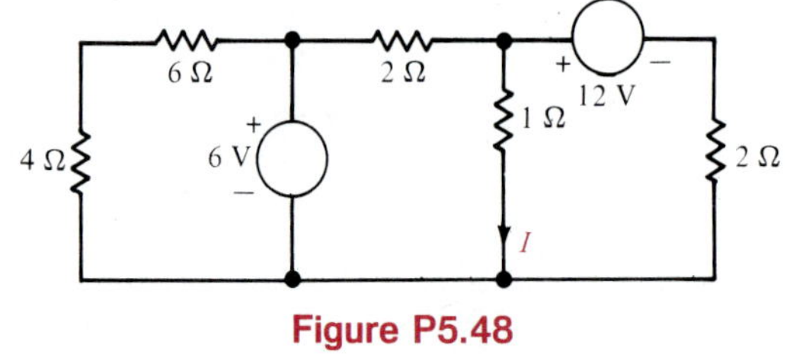

Figure P5.48

5.49. Apply Norton's theorem to find I_o in the network in Fig. P5.49.

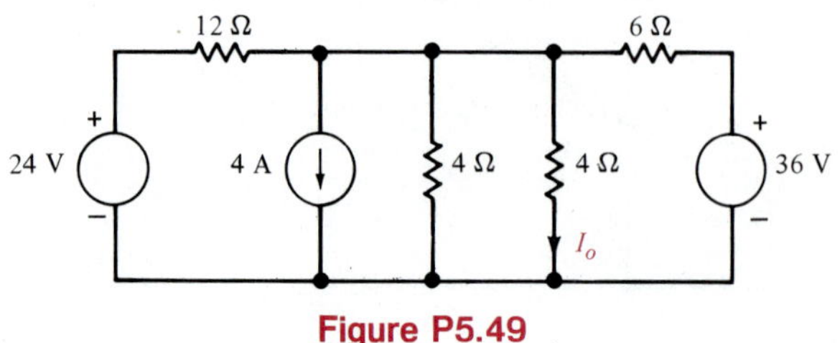

Figure P5.49

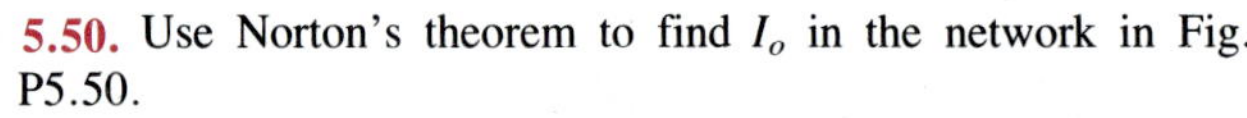

5.50. Use Norton's theorem to find I_o in the network in Fig. P5.50.

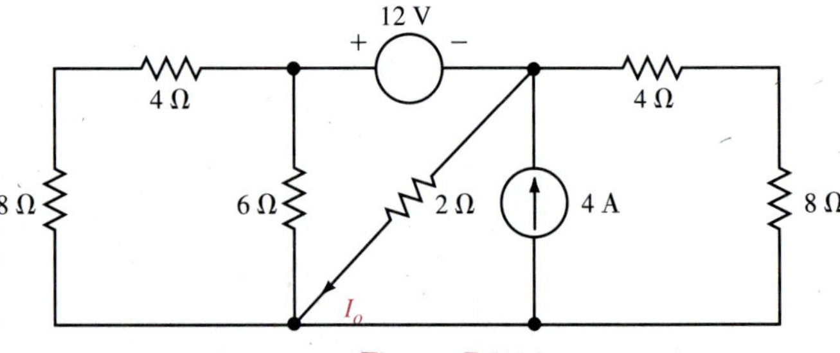

Figure P5.50

5.51. Use Norton's theorem to find I_o in the network in Fig. P5.51.

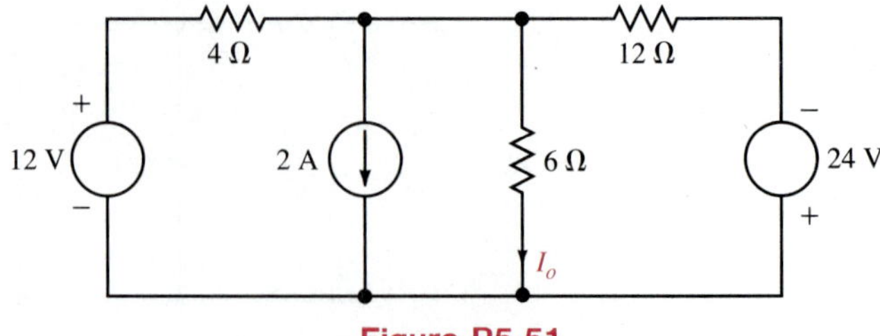

Figure P5.51

5.52. Use Norton's theorem to find I_o in the network in Fig. P5.52.

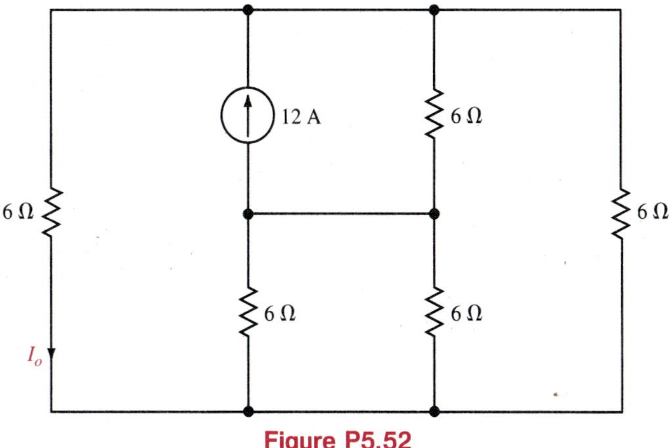

Figure P5.52

5.53. Use Norton's theorem to find I_o in the network in Fig. P5.53.

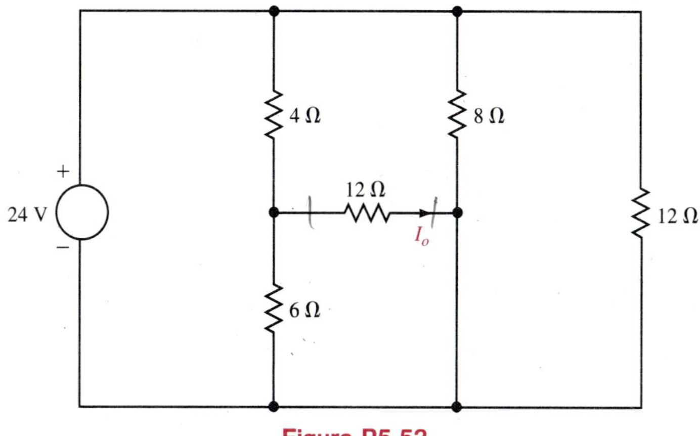

Figure P5.53

5.54. Solve Problem 5.34 using Norton's theorem.

5.55. Find the Thévenin equivalent of the network in Fig. P5.55 at the terminals *A-B*.

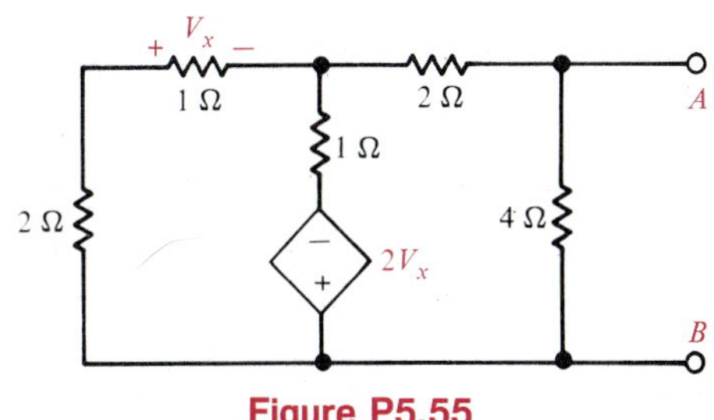

Figure P5.55

5.56. Find the Thévenin equivalent of the network in Fig. P5.56 at the terminals *A-B* using a 1-mA current source.

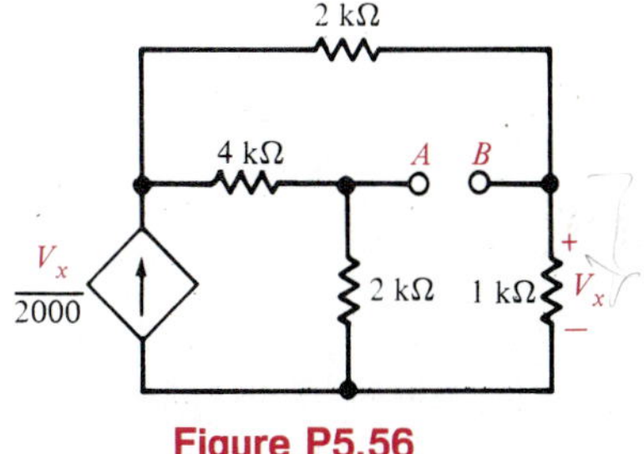

Figure P5.56

5.57. Repeat Problem 5.56 using a 1-V voltage source.

5.58. Determine the value of R_L in Fig. P5.58 for maximum power transfer.

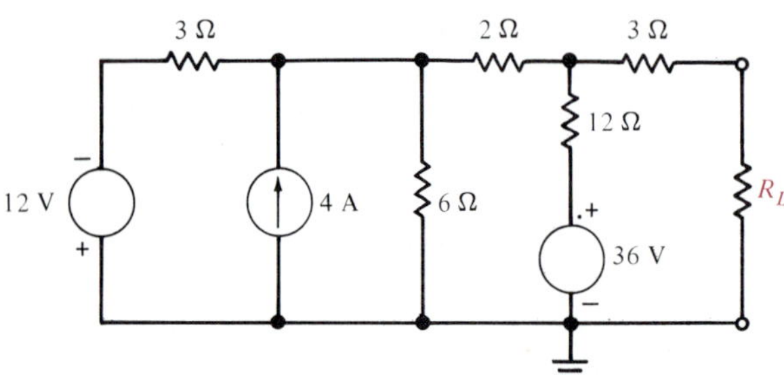

Figure P5.58

5.59. In the circuit in Fig. P5.59, find R_L for maximum power transfer and the maximum power transferred to R_L.

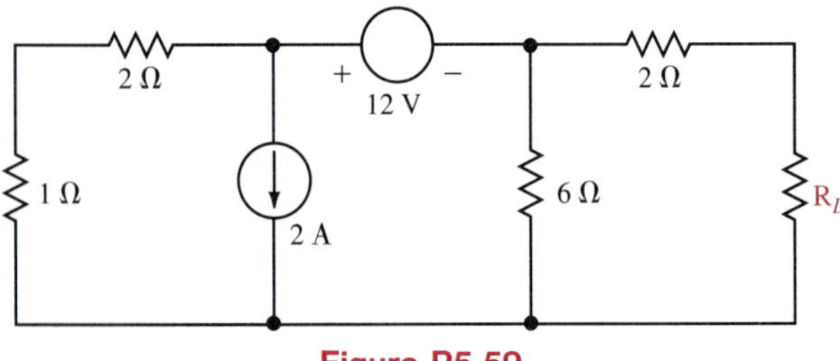

Figure P5.59

5.60. Determine the value of R_i in Fig. P5.60 so that maximum power is transferred to the 6-Ω load.

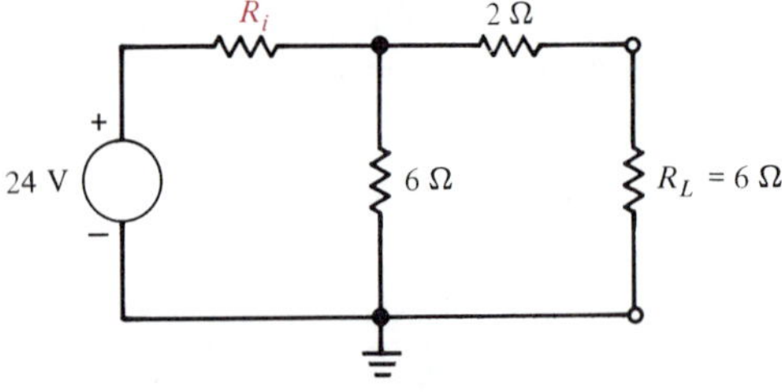

Figure P5.60

5.61. Determine the value of R_L in Problem 5.39 for maximum power transfer.

5.62. Given the linear circuit in Fig. P5.62, it is known that when a 2-Ω load is connected to the terminals *A-B* the load current is 10 A. If a 10-Ω load is connected to the terminals, the load current is 6 A. Find the current in a 20-Ω load.

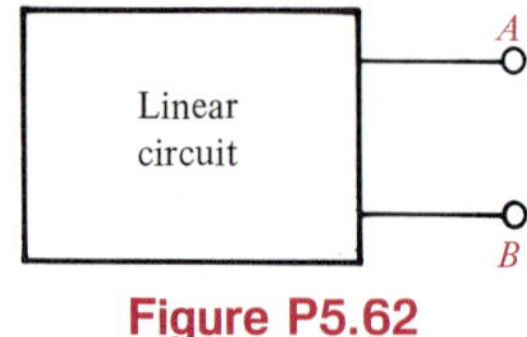

Figure P5.62

5.63. If an 8-Ω load is connected to the terminals of the network in Fig. P5.63, $V_{AB} = 16$ V. If a 2-Ω load is connected to the terminals, $V_{AB} = 8$ V. Find V_{AB} if a 20-Ω load is connected to the terminals.

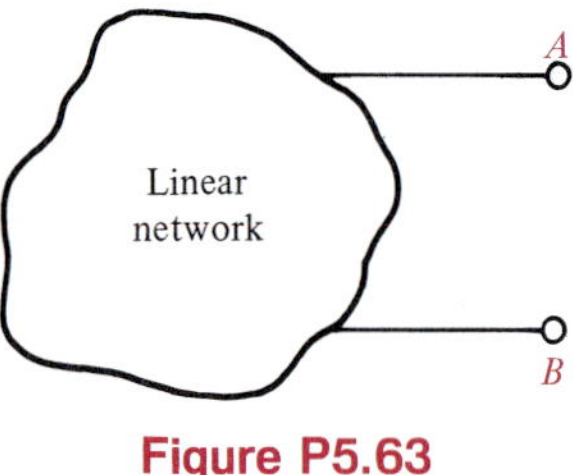

Figure P5.63

5.64. Given the network in Fig. P5.64, use the model statement and PSPICE to vary the resistor R_L from 4 Ω to 6 Ω in steps of 0.1 Ω, and plot the power absorbed by R_L as a function of the model parameter R. What is the value of R_L for maximum power transfer?

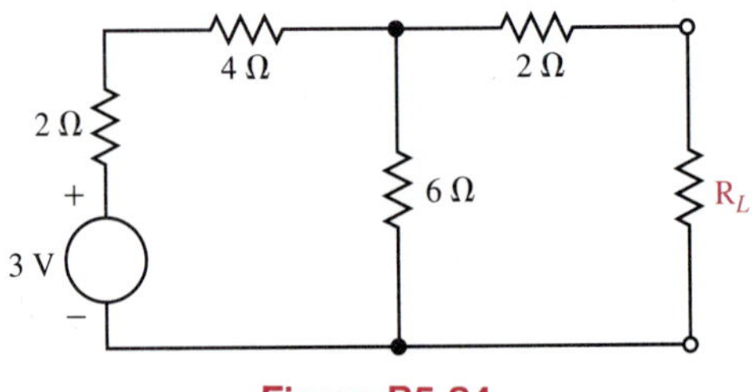

Figure P5.64

5.65. For the network shown in Fig. P5.65, calculate the sensitivity of the output voltage to all the resistors in the network and verify the calculations with PSPICE.

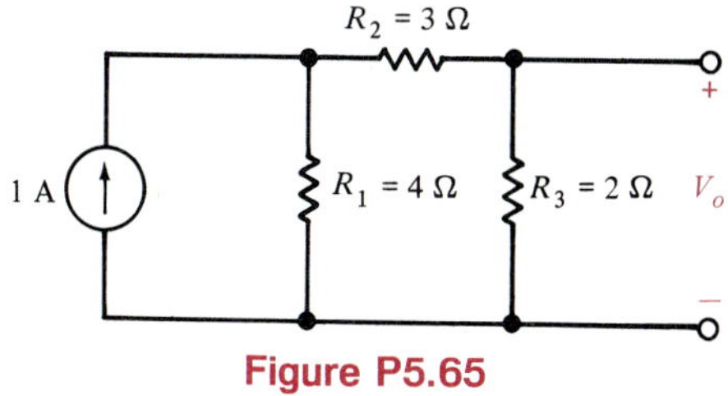

Figure P5.65

5.66. Use PSPICE to compute the sensitivities of V_o with respect to all the circuit parameters for the network in Fig. P5.66.

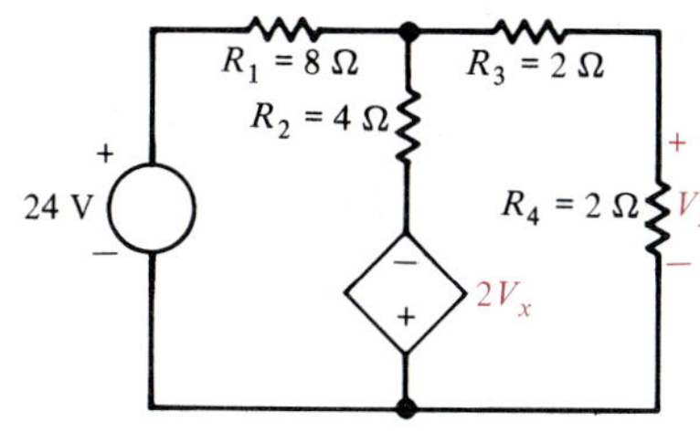

Figure P5.66

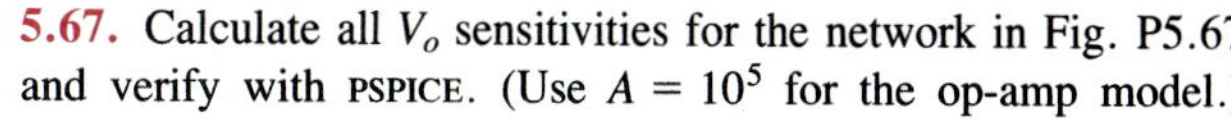

5.67. Calculate all V_o sensitivities for the network in Fig. P5.67 and verify with PSPICE. (Use $A = 10^5$ for the op-amp model.)

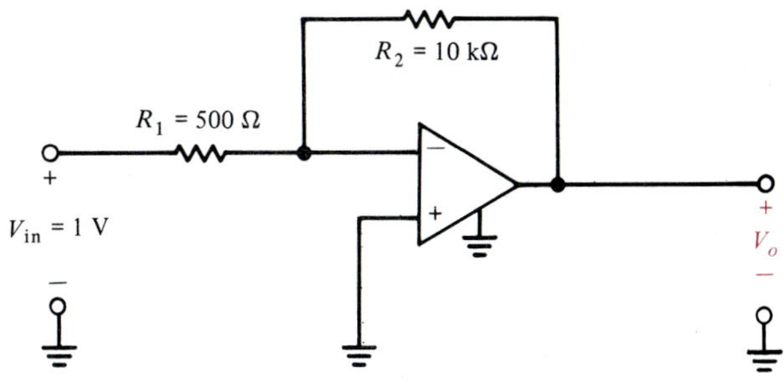

Figure P5.67

5.68. Use PSPICE to compute all the V_o sensitivities in the network in Fig. P5.68.

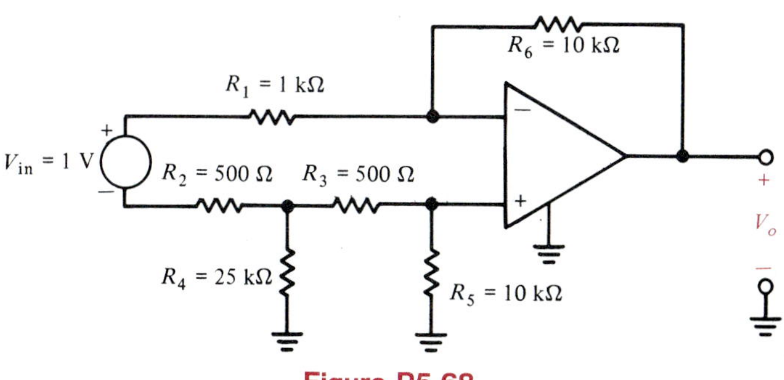

Figure P5.68

CHAPTER

6

Capacitance and Inductance

Our analysis of electric circuits thus far has been confined to circuits that contain only independent sources, dependent sources, and resistors. At this point we introduce two additional circuit elements: the capacitor and the inductor. They are linear elements; however, in contrast to the resistor, their terminal characteristics are described by linear differential equations. Another distinctive feature of these new elements is their ability to absorb energy from the circuit, store it temporarily, and later, return it. Elements that possess this energy storage capability are referred to simply as *storage elements*.

6.1 Capacitors

A *capacitor* is a circuit element that consists of two conducting surfaces separated by a nonconducting, or *dielectric,* material. A simplified capacitor and its electrical symbol are shown in Fig. 6.1.

There are many different kinds of capacitors and they are categorized by the type of dielectric material that is used between the conducting plates. Although any good insulator can serve as a dielectric, each type has characteristics that make it more suitable for particular applications.

For general applications in electronic circuits (e.g., coupling between stages of amplification) the dielectric material may be air, vacuum, paper impregnated with oil or wax, Mylar, polystyrene, mica, glass, or ceramics.

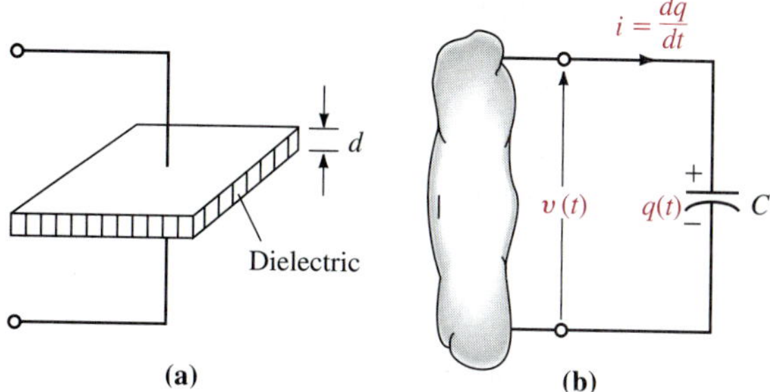

Figure 6.1 Capacitor and its electrical symbol.

Ceramic dielectric capacitors constructed of barium titanates have a large capacitance-to-volume ratio because of their high dielectric constant. Mica, glass, and ceramic dielectric capacitors will operate satisfactorily at high frequencies.

Aluminum electrolytic capacitors, which consist of a pair of aluminum plates separated by a moistened borax paste electrolyte, can provide high values of capacitance in small volumes. They are typically used for filtering, bypassing, and coupling, and in power supplies and motor-starting applications. Tantalum electrolytic capacitors have lower losses and more stable characteristics than those of aluminum electrolytic capacitors.

In addition to these capacitors, which we deliberately insert in a network for specific applications, stray capacitance is present any time there is a difference in potential between two conducting materials separated by a dielectric. Because this stray capacitance can cause unwanted coupling between circuits, extreme care must be exercised in the layout of electronic systems on printed circuit boards.

If a source is connected to the capacitor, positive charges will be transferred to one plate and negative charges to the other. The charge on the capacitor is proportional to the voltage across it such that

$$q = Cv \tag{6.1}$$

where C is the proportionality factor known as the capacitance of the element in coulombs per volt or farads (F). The unit "farad" is named after Michael Faraday, a famous English physicist.

The charge differential between the plates creates an electric field that stores energy. Because of the presence of the dielectric, the conduction current that flows in the wires that connect the capacitor to the remainder of the circuit cannot flow internally between the plates. However, via electromagnetic field theory it can be shown that this conduction current is equal to the displacement current that flows between the plates of the capacitor and is present any time that an electric field or voltage varies with time.

Our primary interest is in the current–voltage terminal characteristics of the capacitor. Since the current is

$$i = \frac{dq}{dt}$$

then for a capacitor

$$i = \frac{d}{dt} Cv$$

which for constant capacitance is

$$i = C \frac{dv}{dt} \tag{6.2}$$

Equation (6.2) can be rewritten as

$$dv = \frac{1}{C} i \, dt$$

Now integrating this expression from $t = -\infty$ to some time t and assuming $v(-\infty) = 0$ yields

$$v(t) = \frac{1}{C} \int_{-\infty}^{t} i(x) \, dx \tag{6.3}$$

where $v(t)$ indicates the time dependence of the voltage. Equation (6.3) can be expressed as two integrals, so that

$$\begin{aligned} v(t) &= \frac{1}{C} \int_{-\infty}^{t_0} i(x) \, dx + \frac{1}{C} \int_{t_0}^{t} i(x) \, dx \\ &= v(t_0) + \frac{1}{C} \int_{t_0}^{t} i(x) \, dx \end{aligned} \tag{6.4}$$

where $v(t_0)$ is the voltage due to the charge that accumulates on the capacitor from time $t = -\infty$ to time $t = t_0$.

The energy stored in the capacitor can be derived from the power that is delivered to the element. This power is given by the expression

$$p(t) = v(t)i(t) = Cv(t) \frac{dv(t)}{dt} \tag{6.5}$$

and hence the energy stored in the electric field is

$$\begin{aligned} w_c(t) &= \int_{-\infty}^{t} Cv(x) \frac{dv(x)}{dx} \, dx = C \int_{-\infty}^{t} v(x) \frac{dv(x)}{dx} \, dx \\ &= C \int_{v(-\infty)}^{v(t)} v(x) \, dv(x) = \tfrac{1}{2} C v^2(x) \Big|_{v(-\infty)}^{v(t)} \\ &= \tfrac{1}{2} C v^2(t) \qquad \text{joules} \end{aligned} \tag{6.6}$$

since $v(t = -\infty) = 0$. The expression for the energy can also be written using Eq. (6.1) as

$$w_c(t) = \frac{1}{2} \frac{q^2(t)}{C} \tag{6.7}$$

Equations (6.6) and (6.7) represent the energy stored by the capacitor, which, in turn, is equal to the work done by the source to charge the capacitor.

The polarity of the voltage across a capacitor being charged is shown in Fig. 6.1b. In the ideal case the capacitor will hold the charge for an indefinite period of time if the source is removed. If at some later time an energy-absorbing device (e.g., a flash bulb) is connected across the capacitor, a discharge current will flow from the capacitor, and therefore, the capacitor will supply the energy stored to the device.

Although we will model the capacitor as an ideal device, in practice there is normally a very large leakage resistance in parallel which provides a conduction path between the plates. It is through this parallel resistance that the realistic capacitor slowly discharges itself.

Capacitors may be fixed or variable and typically range from thousands of microfarads (μF) to a few picofarads (pF). We employ them in numerous situations to obtain specific circuit performance characteristics. Unfortunately, they can also have a detrimental effect on circuit behavior when they are inherently present in the form of stray or parasitic capacitance, which can occur between any two conducting surfaces in a circuit.

EXAMPLE 6.1

If the charge accumulated on two parallel conductors charged to 12 V is 600 pC, what is the capacitance of the parallel conductors?

Using Eq. (6.1), we find that

$$C = \frac{Q}{V} = \frac{(600)(10^{-12})}{12} = 50 \text{ pF}$$

■

EXAMPLE 6.2

Given a capacitor that is initially uncharged, determine the voltage across the capacitor as a function of time if it is subjected to the current pulse shown in Fig. 6.2a.

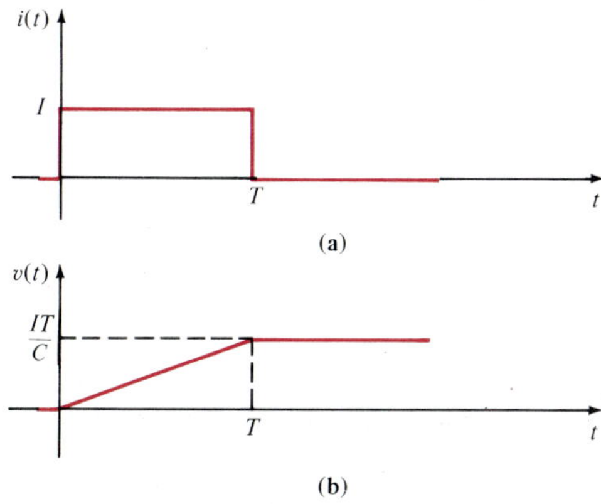

Figure 6.2 Current and voltage waveforms for an initially uncharged capacitor.

Using Eq. (6.4) and the fact that $v(t_0) = v(0) = 0$, we have

$$v(t) = 0 + \frac{1}{C}\int_0^t I\,dx \qquad 0 \le t \le T$$

$$= \frac{It}{C} \qquad 0 \le t \le T$$

For $t > T$,

$$v(t) = v(T) + \frac{1}{C}\int_T^t (0)\,dt = \frac{IT}{C}$$

When current no longer flows, the voltage does not change. Therefore, the voltage waveform is shown in Fig. 6.2b. ■

EXAMPLE 6.3

The voltage across a 5-μF capacitor has the waveform shown in Fig. 6.3a. Determine the current waveform.

Note that

$$v(t) = \frac{24}{6 \times 10^{-3}}t \qquad 0 \le t \le 6 \text{ ms}$$

$$= \frac{-24}{2 \times 10^{-3}}t + 96 \qquad 6 \le t \le 8 \text{ ms}$$

$$= 0 \qquad 8 \text{ ms} \le t$$

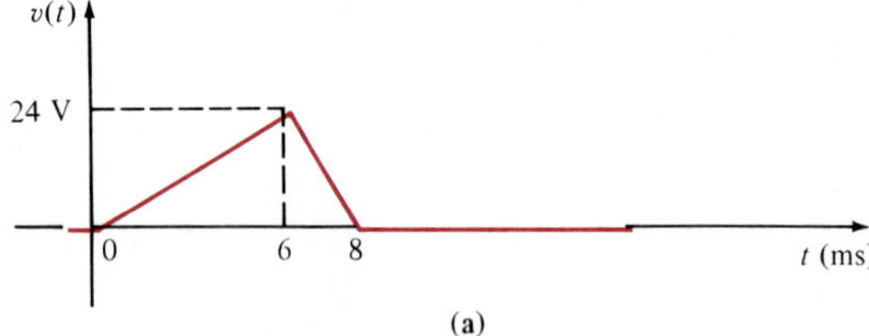

(a)

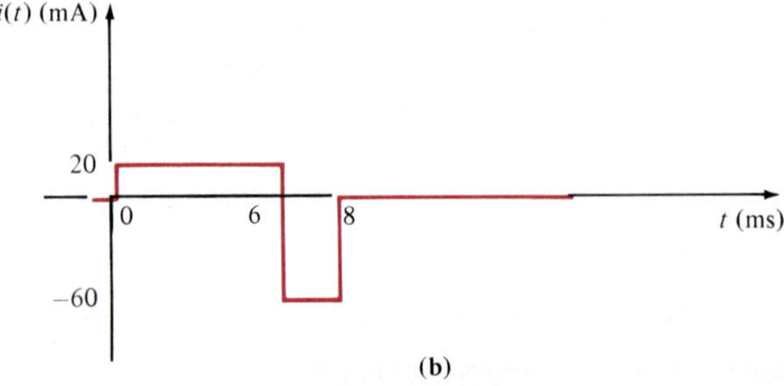

(b)

Figure 6.3 Voltage and current waveforms for a 5-μF capacitor.

Using Eq. (6.2), we find that

$$i(t) = C\frac{dv(t)}{dt}$$

$$= 5 \times 10^{-6}(4 \times 10^{3}) \qquad 0 \le t \le 6 \text{ ms}$$

$$= 20 \text{ mA} \qquad 0 \le t \le 6 \text{ ms}$$

$$i(t) = 5 \times 10^{-6}(-12 \times 10^{3}) \qquad 6 \le t \le 8 \text{ ms}$$

$$= -60 \text{ mA} \qquad 6 \le t \le 8 \text{ ms}$$

and

$$i(t) = 0 \qquad 8 \text{ ms} \le t$$

Therefore, the current waveform is as shown in Fig. 6.3b and $i(t) = 0$ for $t > 8$ ms. ■

EXAMPLE 6.4

Determine the energy stored in the electric field of the capacitor in Example 6.3 at $t = 6$ ms.

Using Eq. (6.6), we have

$$w(t) = \tfrac{1}{2}Cv^2(t)$$

At $t = 6$ ms,

$$w(6 \text{ ms}) = \tfrac{1}{2}(5 \times 10^{-6})(24)^2$$

$$= 1440 \ \mu\text{J}$$

■

The equations and examples above illustrate a number of salient features of the capacitor. An ideal capacitor only stores energy; it does not dissipate energy as a resistor does. If the voltage across a capacitor is constant (i.e., nontime varying), the current through it is zero, and therefore to direct current, a capacitor looks like an open circuit. Although the capacitor current is zero, the capacitor can still store a finite amount of energy. An instantaneous jump in the voltage across a capacitor is not physically realizable, because it requires the movement of a finite amount of charge in zero time, which is an infinite current.

DRILL EXERCISE

D6.1. A 10-μF capacitor has an accumulated charge of 500 nC. Determine the voltage across the capacitor.

Ans: 0.05 V.

D6.2. The voltage across a 4-μF capacitor is shown in Fig. D6.2. Determine the waveform for the capacitor current.

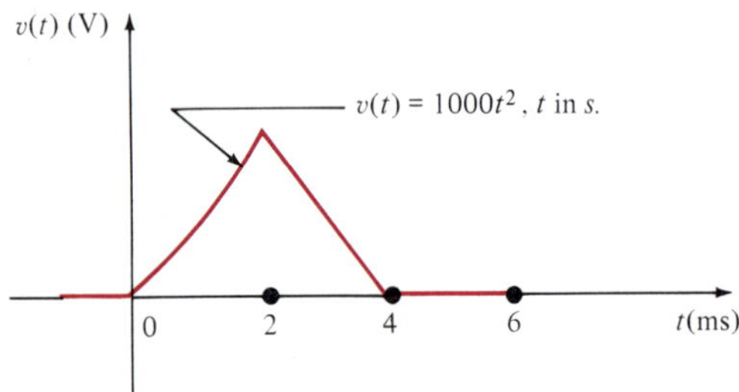

Figure D6.2

Ans:

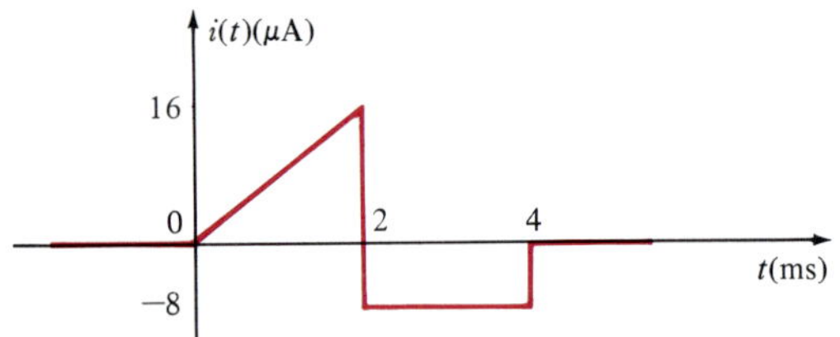

D6.3. Compute the energy stored in the electric field of the capacitor in Drill Exercise D6.2 at $t = 2$ ms.

Ans: $W = 32$ pJ.

6.2 Inductors

An *inductor* is a circuit element that consists of a conducting wire usually in the form of a coil. Two typical inductors and their electrical symbols are shown in Fig. 6.4. Inductors are typically categorized by the type of core on which they are wound. For example, the core material may be air or any nonmagnetic material, iron, or ferrite. Inductors made with air or nonmagnetic materials are widely used in radio, television, and filter circuits. Iron-core inductors are used in electrical power supplies and filters. Ferrite-core inductors are widely used in high-frequency applications. Note that in contrast to the magnetic core that confines the flux, as shown in Fig. 6.4b, the flux lines for nonmagnetic inductors extend beyond the inductor itself, as illustrated in Fig. 6.4a. Like stray capacitance, stray inductance can result from any element carrying current surrounded by flux linkages.

From a historical standpoint, developments that lead to the mathematical model we employ to represent the inductor are as follows. It was first shown that a current-carrying conductor would produce a magnetic field. It was later found that the magnetic field and the current that produced it were linearly related. Finally, it was shown that a changing

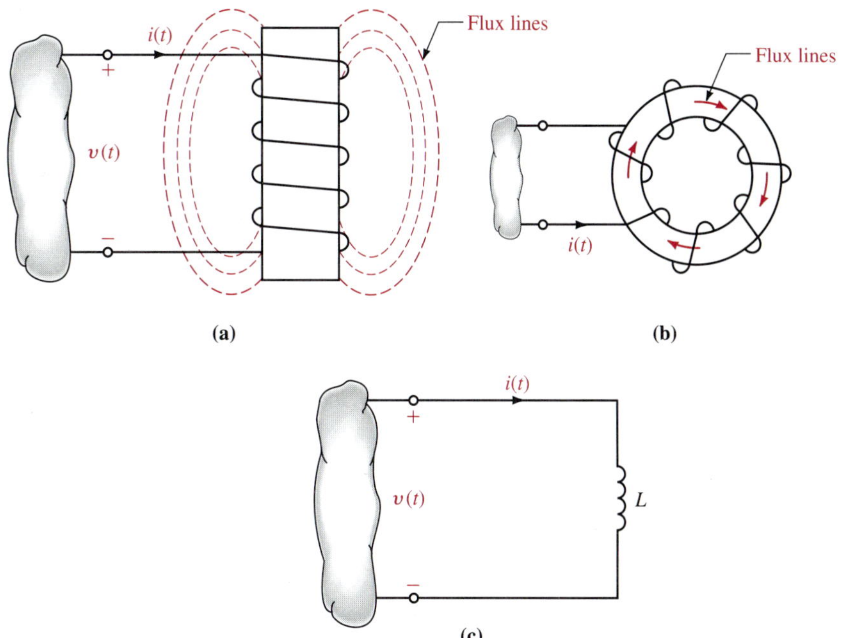

Figure 6.4 Two inductors and their electrical symbols.

magnetic field produced a voltage that was proportional to the time rate of change of the current that produced the magnetic field; that is,

$$v(t) = L\frac{di(t)}{dt} \tag{6.8}$$

The constant of proportionality L is called the inductance, and is measured in the unit ''henry,'' named after the American inventor Joseph Henry, who discovered the relationship. As seen in Eq. (6.8), 1 henry (H) is dimensionally equal to 1 volt-second per ampere.

Following the development of the mathematical equations for the capacitor, we find that the expression for the current in an inductor is

$$i(t) = \frac{1}{L}\int_{-\infty}^{t} v(x)\,dx \tag{6.9}$$

which can also be written as

$$i(t) = i(t_0) + \frac{1}{L}\int_{t_0}^{t} v(x)\,dx \tag{6.10}$$

The power delivered to the inductor can be used to derive the energy stored in the element. This power is equal to

$$\begin{aligned} p(t) &= v(t)i(t) \\ &= \left[L\frac{di(t)}{dt}\right]i(t) \end{aligned} \tag{6.11}$$

Therefore, the energy stored in the magnetic field is

$$w_L(t) = \int_{-\infty}^{t} \left[L \frac{di(x)}{dx}\right] i(x)\,dx$$

Following the development of Eq. (6.6), we obtain

$$w_L(t) = \tfrac{1}{2}Li^2(t) \qquad \text{joules} \tag{6.12}$$

The inductor, like the resistor and capacitor, is a passive element. The polarity of the voltage across the inductor is shown in Fig. 6.4.

Like the capacitor, the inductor is an energy storage device. However, a practical inductor cannot store energy as well as can a practical capacitor. The inductor always has some winding resistance in the coil, which quickly dissipates energy.

Practical inductors typically range from a few microhenrys to tens of henrys. From a circuit design standpoint it is important to note that inductors cannot be easily fabricated on an integrated-circuit chip, and therefore chip designs typically employ only active electronic devices, resistors and capacitors, which can be easily fabricated in microcircuit form.

EXAMPLE 6.5

The current in a 10-mH inductor has the waveform shown in Fig. 6.5a. Determine the voltage waveform.

Using Eq. (6.8) and noting that

$$i(t) = \frac{20 \times 10^{-3}t}{2 \times 10^{-3}} \qquad 0 \le t \le 2 \text{ ms}$$

$$i(t) = \frac{-20 \times 10^{-3}t}{2 \times 10^{-3}} + 40 \times 10^{-3} \qquad 2 \le t \le 4 \text{ ms}$$

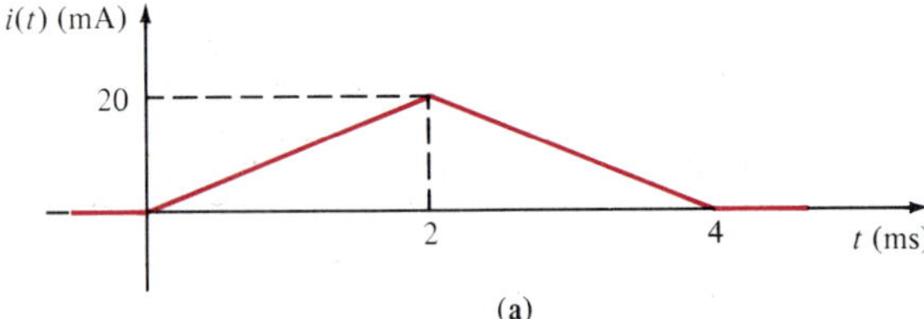

(a)

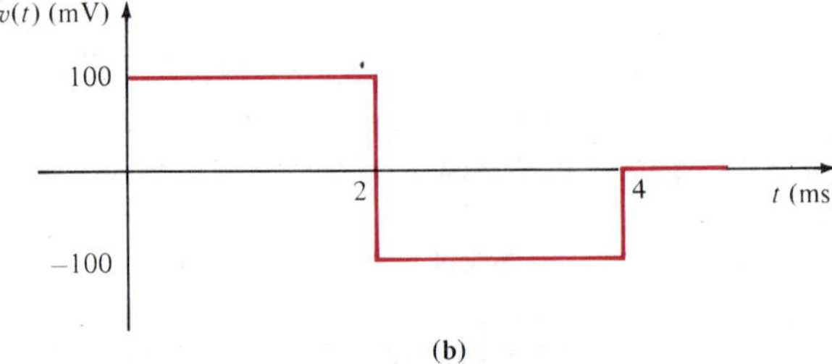

(b)

Figure 6.5 Current and voltage waveforms for a 10-mH inductor.

and

$$i(t) = 0 \qquad 4 \text{ ms} \le t$$

Using Eq. (6.8), we find that

$$v(t) = (10 \times 10^{-3}) \frac{20 \times 10^{-3}}{2 \times 10^{-3}} \qquad 0 \le t \le 2 \text{ ms}$$
$$= 100 \text{ mV}$$

and

$$v(t) = (10 \times 10^{-3}) \frac{-20 \times 10^{-3}}{2 \times 10^{-3}} \qquad 2 \le t \le 4 \text{ ms}$$
$$= -100 \text{ mV}$$

and $v(t) = 0$ for $t > 4$ ms. Therefore, the voltage waveform is shown in Fig. 6.5b. ■

EXAMPLE 6.6

The current in a 2-mH inductor is

$$i(t) = 2 \sin 377t \text{ A}$$

Determine the voltage across the inductor and the energy stored in the inductor.

From Eq. (6.8), we have

$$v(t) = L\frac{di(t)}{dt}$$
$$= (2 \times 10^{-3}) \frac{d}{dt}(2 \sin 377t)$$
$$= 1.508 \cos 377t \text{ V}$$

and from Eq. (6.12),

$$w_L(t) = \tfrac{1}{2}Li^2(t)$$
$$= \tfrac{1}{2}(2 \times 10^{-3})(2 \sin 377t)^2$$
$$= 0.004 \sin^2 377t \text{ J}$$

■

The previous material illustrates a number of important features of the inductor. An ideal inductor only stores energy; it does not dissipate any energy. As has been indicated, this is not the case for physically realizable inductors. A physical coil that has significant winding resistance can be modeled as a resistor in series with an ideal inductor. Generally, when we speak of an inductor we are referring to that ideal, lossless element.

From Eq. (6.8) we note that if the current is constant, the voltage across an inductor is zero. Hence to direct current, the inductor looks like a short circuit. The equation also indicates that an instantaneous change in current would require an infinite voltage. Therefore, it is not possible to change instantaneously the current in an inductor.

DRILL EXERCISE

D6.4. The current in a 5-mH inductor has the waveform shown in Fig. D6.4. Compute the waveform for the inductor voltage.

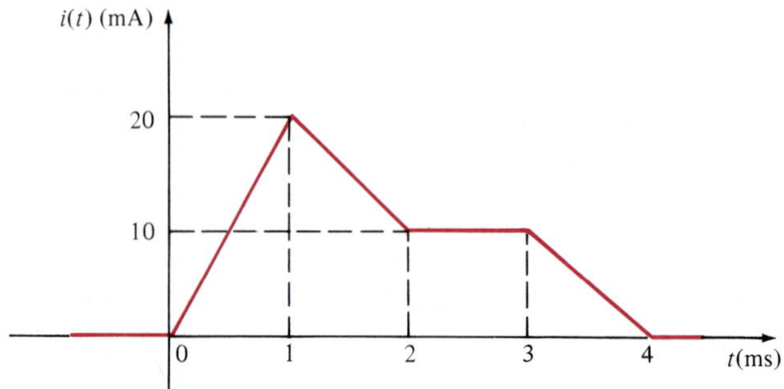

Figure D6.4

Ans:

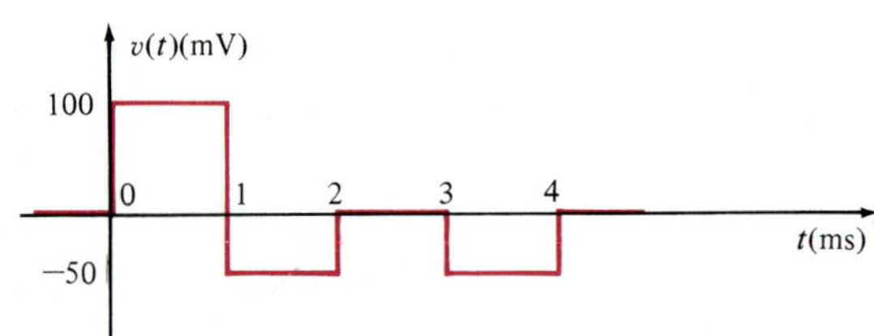

D6.5. Compute the energy stored in the magnetic field of the inductor in Drill Exercise D6.4 at $t = 1.5$ ms.

Ans: $W = 562.5$ nJ.

6.3 Capacitor and Inductor Combinations

Series Capacitors

If a number of capacitors are connected in series, their equivalent capacitance can be calculated using KVL. Consider the circuit shown in Fig. 6.6a. For this circuit

$$v(t) = v_1(t) + v_2(t) + v_3(t) + \cdots + v_N(t) \tag{6.13}$$

but

$$v_i(t) = \frac{1}{C_i}\int_{t_0}^{t} i(t)\,dt + v_i(t_0) \tag{6.14}$$

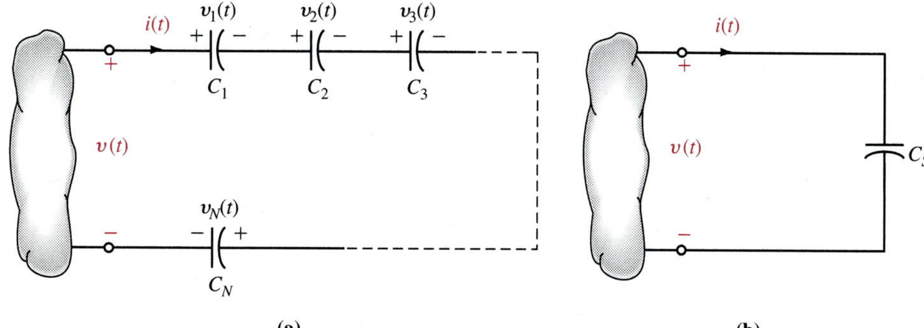

Figure 6.6 Equivalent circuit for N series-connected capacitors.

Therefore, Eq. (6.13) can be written as follows using Eq. (6.14):

$$v(t) = \left(\sum_{i=1}^{N} \frac{1}{C_i}\right) \int_{t_0}^{t} i(t)\, dt + \sum_{i=1}^{N} v_i(t_0) \tag{6.15}$$

$$= \frac{1}{C_s} \int_{t_0}^{t} i(t)\, dt + v(t_0) \tag{6.16}$$

where

$$v(t_0) = \sum_{i=1}^{N} v_i(t_0)$$

and

$$\frac{1}{C_s} = \sum_{i=1}^{N} \frac{1}{C_i} = \frac{1}{C_1} + \frac{1}{C_2} + \cdots + \frac{1}{C_N} \tag{6.17}$$

Thus the circuit in Fig. 6.6b is equivalent to that in Fig. 6.6a under the conditions stated above.

It is also important to note that since the same current flows in each of the series capacitors, each capacitor gains the same charge in the same time period. The voltage across each capacitor will depend on this charge and the capacitance of the element.

EXAMPLE 6.7

Determine the equivalent capacitance and the initial voltage for the circuit shown in Fig. 6.7. Note that these capacitors must have been charged before they were connected in series or else the charge of each would be equal and the voltages would be in the same direction.

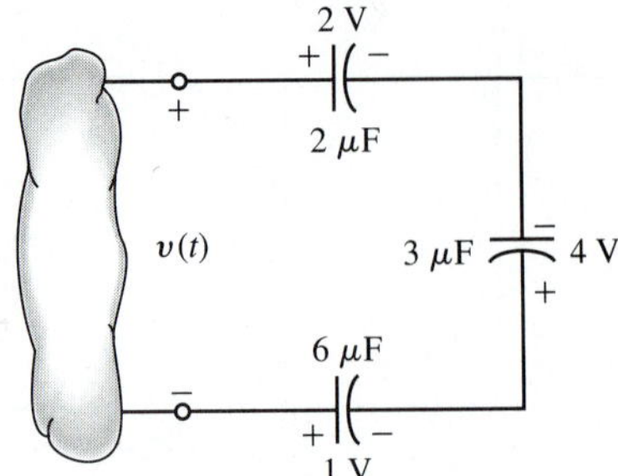

Figure 6.7 Circuit containing multiple capacitors with initial voltages.

The equivalent capacitance is

$$\frac{1}{C_s} = \frac{1}{2} + \frac{1}{3} + \frac{1}{6}$$

where all capacitance values are in microfarads.

Therefore, $C_s = 1\ \mu\text{F}$ and, as seen from the figure, $v(t_0) = -3$ V. Note that the total energy stored in the circuit is

$$\begin{aligned} w(t_0) &= \tfrac{1}{2}[2 \times 10^{-6}(2)^2 + 3 \times 10^{-6}(-4)^2 + 6 \times 10^{-6}(-1)^2] \\ &= 31\ \mu\text{J} \end{aligned}$$

However, the energy recoverable at the terminals is

$$\begin{aligned} w_C(t_0) &= \tfrac{1}{2}C_s v^2(t) \\ &= \tfrac{1}{2}[1 \times 10^{-6}(-3)^2] \\ &= 4.5\ \mu\text{J} \end{aligned}$$

■

EXAMPLE 6.8

Two previously uncharged capacitors are connected in series and then charged with a 12-V source. One capacitor is 30 μF and the other is unknown. If the voltage across the 30-μF capacitor is 8 V, find the capacitance of the unknown capacitor.

The charge on the 30-μF capacitor is

$$Q = CV = (30\ \mu\text{F})(8\ \text{V}) = 240\ \mu\text{C}$$

Since the same current flows in each of the series capacitors, each capacitor gains the same charge in the same time period.

$$C = \frac{Q}{V} = \frac{240\ \mu\text{C}}{4\ \text{V}} = 60\ \mu\text{F}$$

■

Parallel Capacitors

To determine the equivalent capacitance of N capacitors connected in parallel, we employ KCL. As can be seen from Fig. 6.8a,

$$\begin{aligned} i(t) &= i_1(t) + i_2(t) + i_3(t) + \cdots + i_N(t) \qquad (6.18) \\ &= C_1 \frac{dv(t)}{dt} + C_2 \frac{dv(t)}{dt} + C_3 \frac{dv(t)}{dt} + \cdots + C_N \frac{dv(t)}{dt} \\ &= \left(\sum_{i=1}^{N} C_i \right) \frac{dv}{dt} \\ &= C_p \frac{dv(t)}{dt} \qquad (6.19) \end{aligned}$$

where

$$C_p = C_1 + C_2 + C_3 + \cdots + C_N \qquad (6.20)$$

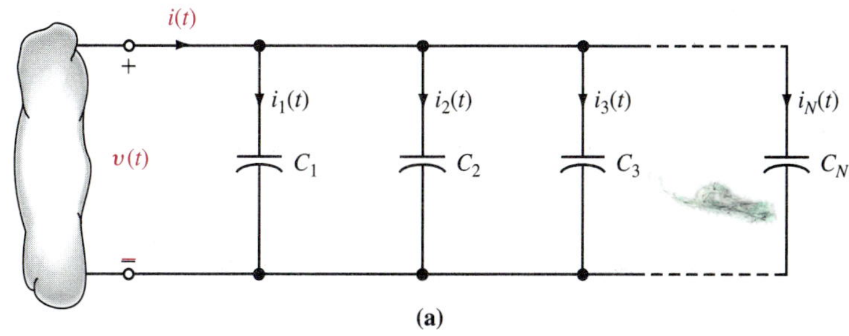

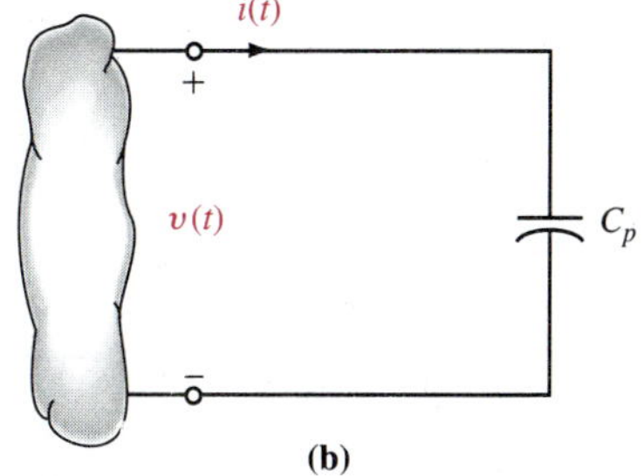

Figure 6.8 Equivalent circuits for N capacitors connected in parallel.

EXAMPLE 6.9

Determine the equivalent capacitance at terminals A-B of the circuit shown in Fig. 6.9.

$$\begin{aligned} C_p &= 4\ \mu\text{F} + 6\ \mu\text{F} + 2\ \mu\text{F} + 3\ \mu\text{F} \\ &= 15\ \mu\text{F} \end{aligned}$$

■

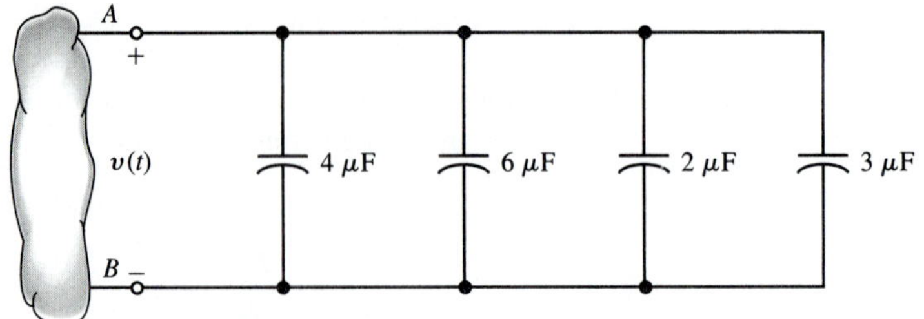

Figure 6.9 Circuit containing multiple capacitors in parallel.

DRILL EXERCISE

D6.6. Two capacitors are connected as shown in Fig. D6.6. Determine the value of C_1.

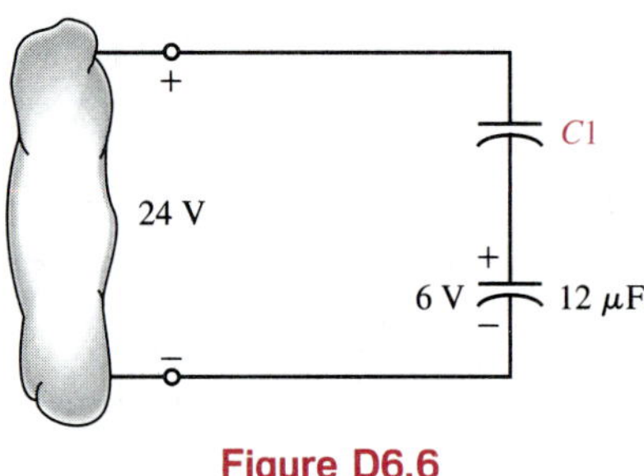

Figure D6.6

Ans: $C_1 = 4\ \mu\text{F}$.

D6.7. Compute the equivalent capacitance of the network in Fig. D6.7.

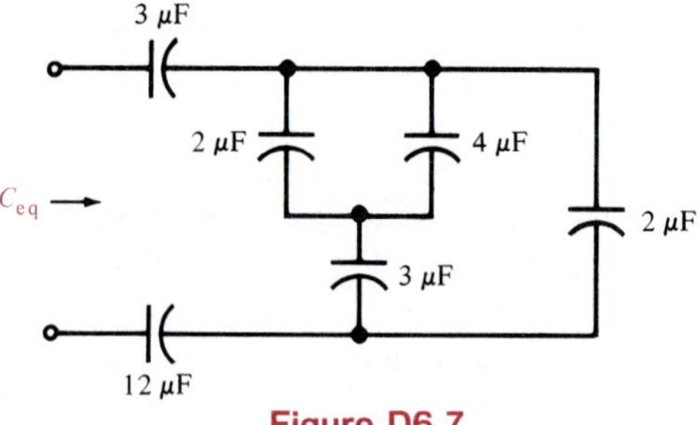

Figure D6.7

Ans: $C_{eq} = 1.5\ \mu\text{F}$.

D6.8. Compute the equivalent capacitance of the network in Fig. D6.8 if all the capacitors are 4 μF.

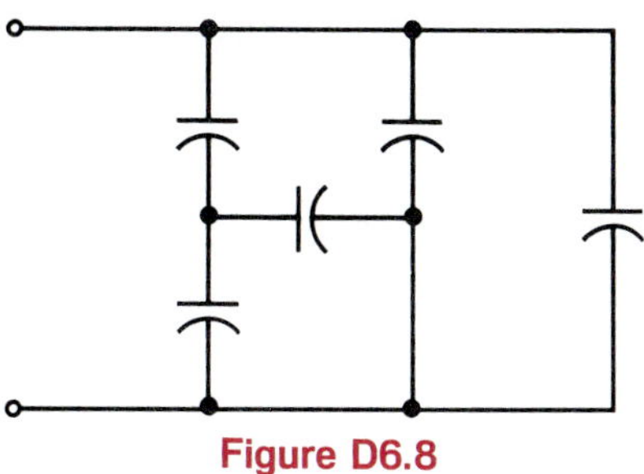

Figure D6.8

Ans: $C_{eq} = \frac{32}{3}$ μF.

Series Inductors

If N inductors are connected in series, the equivalent inductance of the combination can be determined as follows. Referring to Fig. 6.10a and using KVL, we see that

$$v(t) = v_1(t) + v_2(t) + v_3(t) + \cdots + v_N(t) \tag{6.21}$$

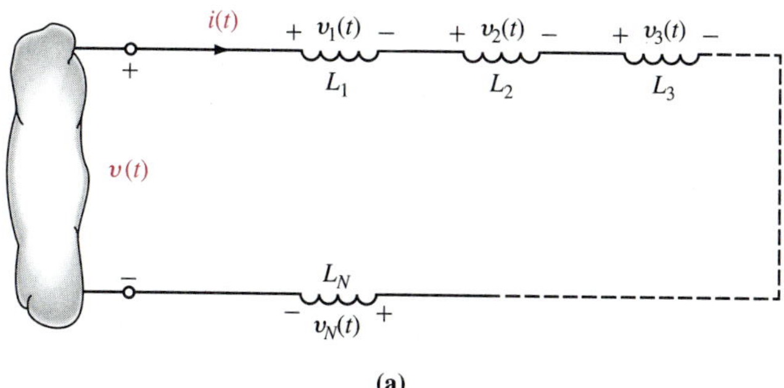

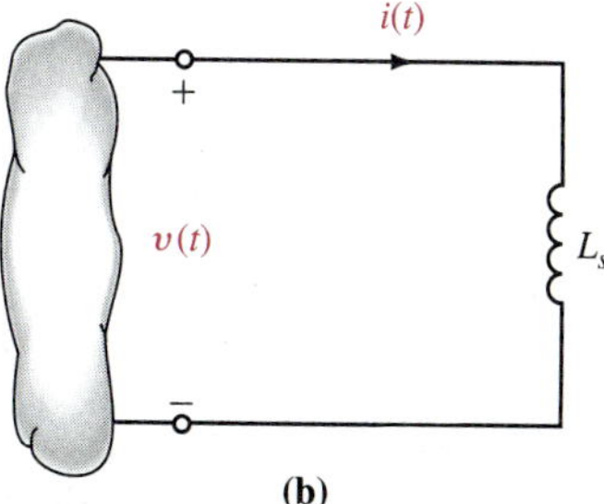

Figure 6.10 Equivalent circuit for N series-connected inductors.

and therefore,

$$v(t) = L_1 \frac{di(t)}{dt} + L_2 \frac{di(t)}{dt} + L_3 \frac{di(t)}{dt} + \cdots + L_N \frac{di(t)}{dt} \tag{6.22}$$

$$= \left(\sum_{i=1}^{N} L_i \right) \frac{di(t)}{dt}$$

$$= L_s \frac{di(t)}{dt} \tag{6.23}$$

where

$$L_s = \sum_{i=1}^{N} L_i \tag{6.24}$$

EXAMPLE 6.10

The equivalent inductance of the circuit shown in Fig. 6.11 is

$$L_s = 1 \text{ H} + 2 \text{ H} + 4 \text{ H}$$

$$= 7 \text{ H}$$

■

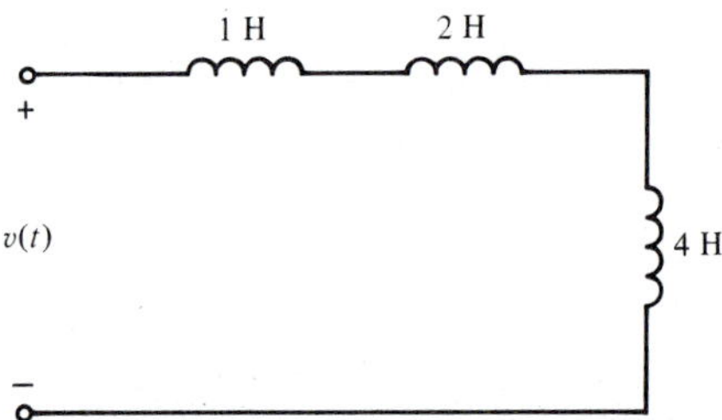

Figure 6.11 Circuit containing multiple inductors.

Parallel Inductors

Consider the circuit shown in Fig. 6.12a, which contains N parallel inductors. Using KCL, we can write

$$i(t) = i_1(t) + i_2(t) + i_3(t) + \cdots + i_N(t) \tag{6.25}$$

However,

$$i_j(t) = \frac{1}{L_j} \int_{t_0}^{t} v(x)\, dx + i_j(t_0) \tag{6.26}$$

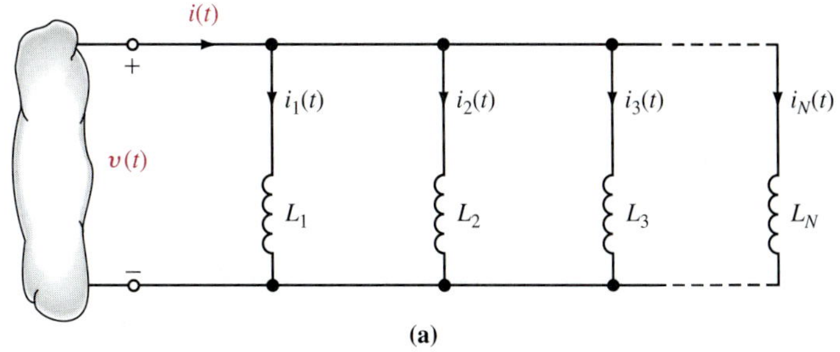

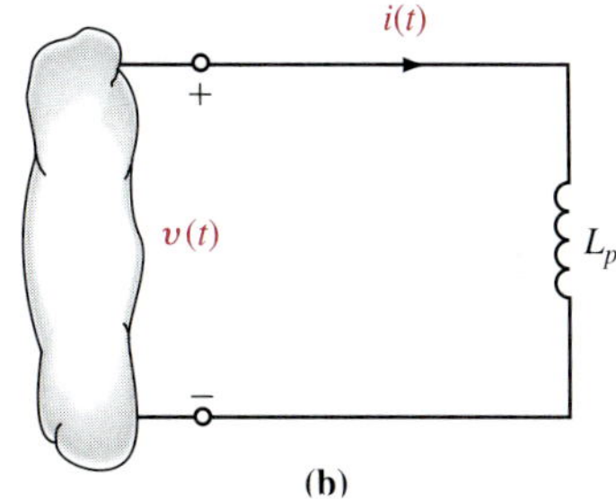

Figure 6.12 Equivalent circuits for *N* inductors connected in parallel.

Substituting this expression into Eq. (6.25) yields

$$i(t) = \left(\sum_{j=1}^{N} \frac{1}{L_j}\right) \int_{t_0}^{t} v(x)\,dx + \sum_{j=1}^{N} i_j(t_0) \tag{6.27}$$

$$= \frac{1}{L_p} \int_{t_0}^{t} v(x)\,dx + i(t_0) \tag{6.28}$$

where

$$\frac{1}{L_p} = \frac{1}{L_1} + \frac{1}{L_2} + \frac{1}{L_3} + \cdots + \frac{1}{L_N} \tag{6.29}$$

and $i(t_0)$ is equal to the current in L_p at $t = t_0$. Thus the circuit in Fig. 6.12b is equivalent to that in Fig. 6.12a under the conditions stated above.

EXAMPLE 6.11

Determine the equivalent inductance and the initial current for the circuit shown in Fig. 6.13.

The equivalent inductance is

$$\frac{1}{L_p} = \frac{1}{12} + \frac{1}{6} + \frac{1}{4}$$

where all inductance values are in millihenrys.

$$L_p = 2 \text{ mH}$$

and the initial current is $i(t_0) = -1$ A. ■

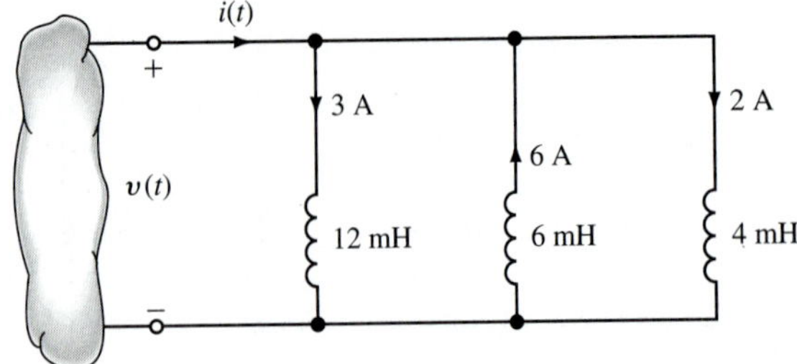

Figure 6.13 Circuit containing multiple inductors with initial currents.

The previous material indicates that capacitors combine like conductances, whereas inductances combine like resistances.

DRILL EXERCISE

D6.9. Compute the equivalent inductance of the network in Fig. D6.9 if all inductors are 4 mH.

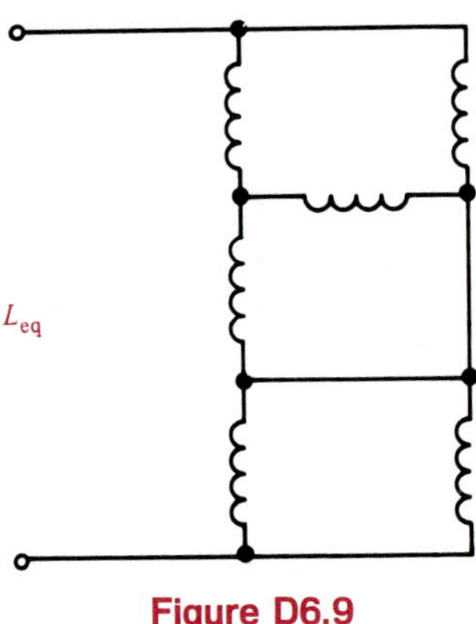

Figure D6.9

Ans: 4.4 mH.

D6.10. Determine the equivalent inductance of the network in Fig. D6.10 if all inductors are 6 mH.

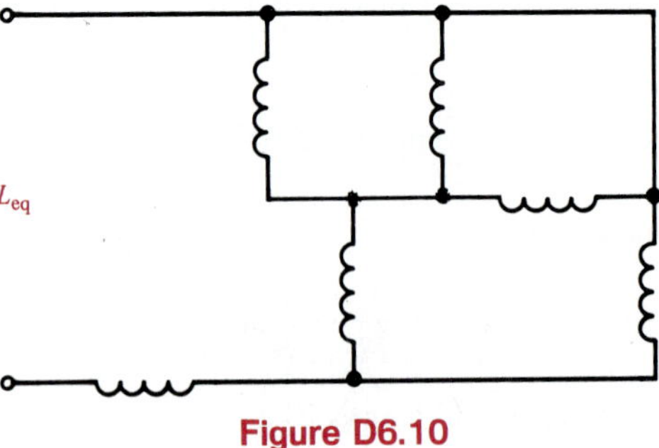

Figure D6.10

Ans: 9.429 mH.

6.4 Summary

Two additional ideal circuit elements were introduced in this chapter: the capacitor and the inductor. We showed that in sharp contrast to the resistor, both of these new elements are capable of storing energy, and the energy relationships for these elements were determined. We illustrated how to compute equivalent capacitance when capacitors are interconnected in series or parallel and how to determine equivalent inductance when these elements are interconnected.

KEY POINTS

- Capacitors and inductors are capable of storing energy.
- A capacitor looks like an open circuit to dc.
- An inductor looks like a short circuit to dc.
- When determining the equivalent capacitance of a number of interconnected capacitors, capacitors in series combine like resistors in parallel and capacitors in parallel combine like resistors in series.
- When determining the equivalent inductance of a number of interconnected inductors, inductors combine like resistors whether they are connected in series or parallel.

PROBLEMS

6.1. An uncharged 100-μF capacitor is charged by a constant current of 1 mA. Find the voltage across the capacitor after 4 sec.

6.2. A 25-μF capacitor initially charged to -10 V is charged by a constant current of 2.5 μA. Find the voltage across the capacitor after $2\frac{1}{2}$ min.

6.3. The voltage across a 100-μF capacitor is given by the expression $v(t) = 120 \sin 377t$ V. Find (a) the current in the capacitor and (b) the expression for the energy stored in the element.

6.4. The energy that is stored in a 25-μF capacitor is $w(t) = 12 \sin^2 377t$ J. Find the current in the capacitor.

6.5. Two capacitors are connected in series as shown in Fig. P6.5.

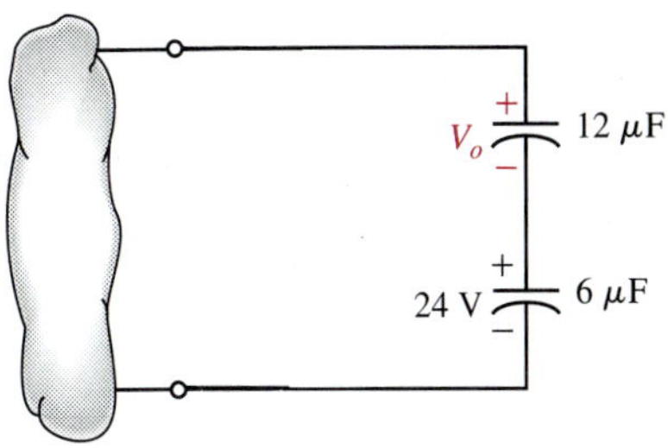

Figure P6.5

If the capacitors are then charged and the voltage across the 6-μF capacitor is 24 V, find the voltage across the 12-μF capacitor.

6.6. The voltage across a 0.1-F capacitor is given by the waveform in Fig. P6.6. Find the waveform for the current in the capacitor.

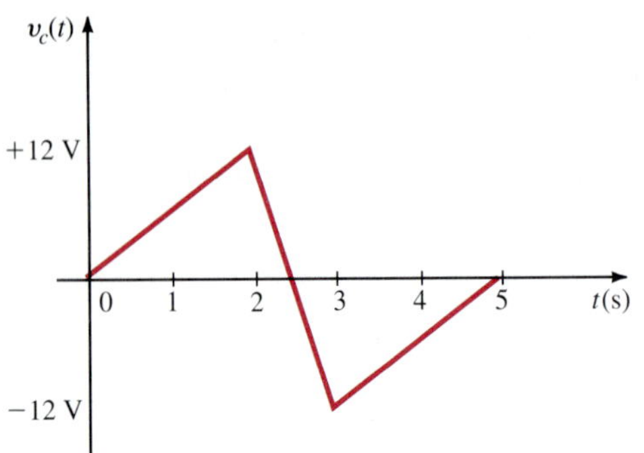

Figure P6.6

6.7. Draw the waveform for the current in a 3-μF capacitor when the voltage across the capacitor is given in Fig. P6.7.

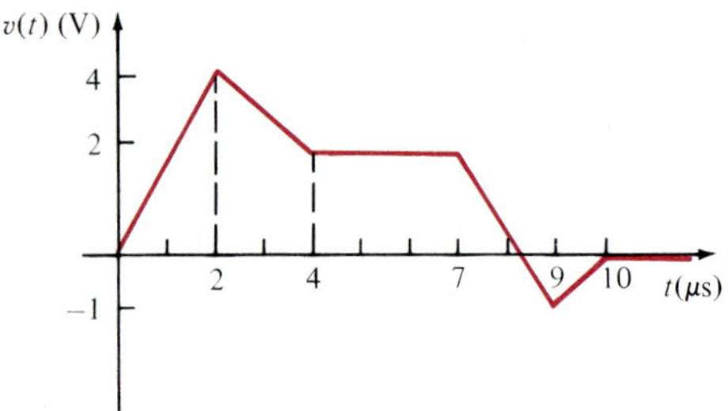

Figure P6.7

6.8. The voltage across a 50-μF capacitor is shown in Fig. P6.8. Determine the current waveform.

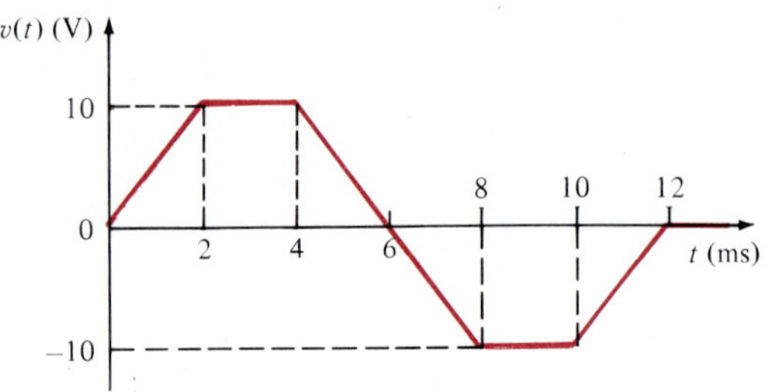

Figure P6.8

6.9. Draw the waveform for the current in a 12-μF capacitor when the capacitor voltage is as described in Fig. P6.9.

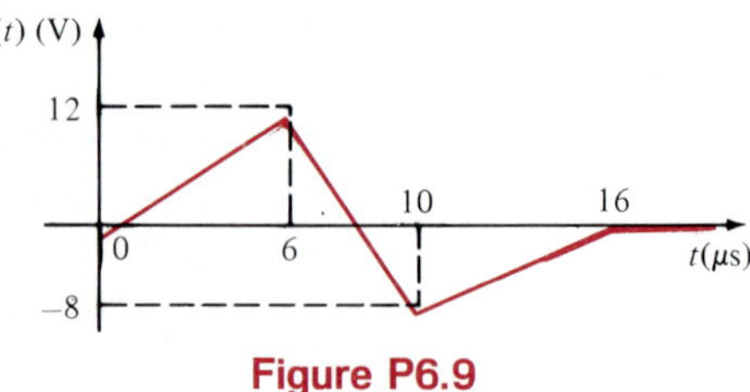

Figure P6.9

6.10. The voltage across a 6-μF capacitor is given by the waveform in Fig. P6.10. Plot the waveform for the capacitor current.

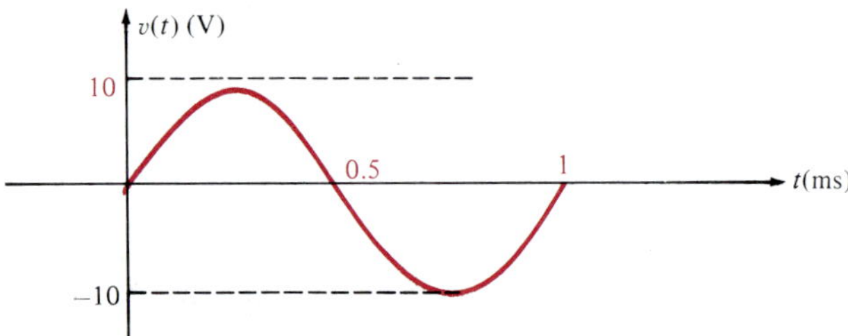

Figure P6.10

6.11. The current flowing through a 100-mH inductor is $i(t) = 2 \sin 377t$ A. Find (a) the voltage across the inductor and (b) the expression for the energy stored in the element.

6.12. The energy stored in a 50-mH inductor is given by the expression $w(t) = 1.2 \sin^2 377t$ J. Determine the voltage across the inductor.

6.13. The current in a 25-mH inductor is given by the expressions

$$i(t) = 0 \qquad t < 0$$

$$i(t) = 10(1 - e^{-t}) \text{ mA} \qquad t > 0$$

Find (a) the voltage across the inductor and (b) the expression for the energy stored in it.

6.14. Given the data in the previous problem, find the voltage across the inductor and the energy stored in it after 1 sec.

6.15. The current

$$i(t) = 0 \qquad t < 0$$

$$i(t) = 100e^{-t/10} \text{ A} \qquad t > 0$$

flows through a 150-mH inductor. Find both the voltage across the inductor and the energy stored in it after 5 sec.

6.16. The current in a 50-mH inductor is specified as follows.

$$i(t) = 0 \qquad t < 0$$

$$i(t) = 2te^{-4t} \text{ A} \qquad t > 0$$

Find (a) the voltage across the inductor, (b) the time at which the current is a maximum, and (c) the time at which the voltage is a minimum.

6.17. The current in a 10-mH inductor is shown in Fig. P6.17. Find the voltage across the inductor.

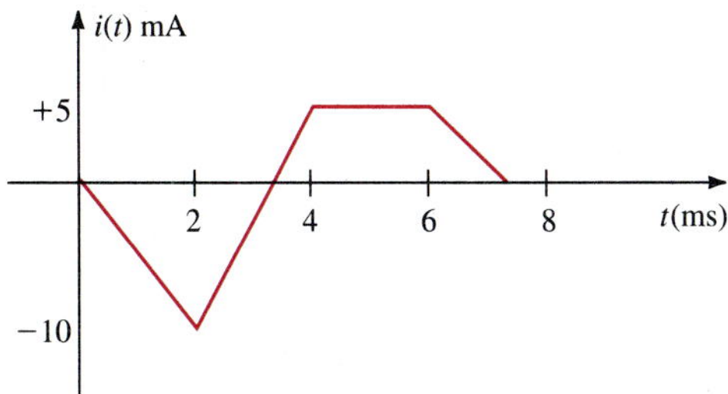

Figure P6.17

6.18. The current in a 50-mH inductor is given in Fig. P6.18. Sketch the inductor voltage.

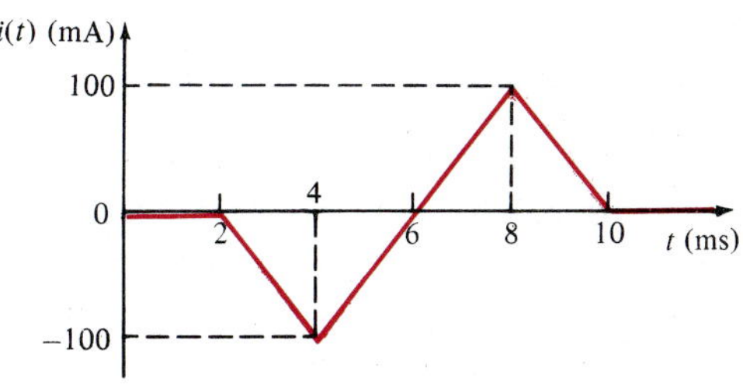

Figure P6.18

6.19. The current in a 16-mH inductor is given by the waveform in Fig. P6.19. Find the waveform for the voltage across the inductor.

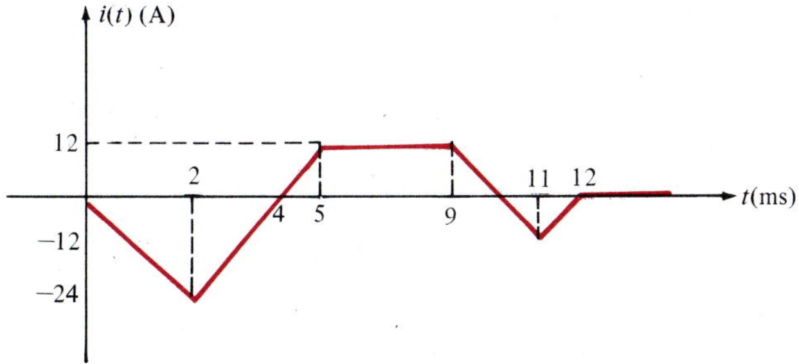

Figure P6.19

6.20. Draw the waveform for the voltage across a 10-mH inductor when the inductor current is given by the waveform shown in Fig. P6.20.

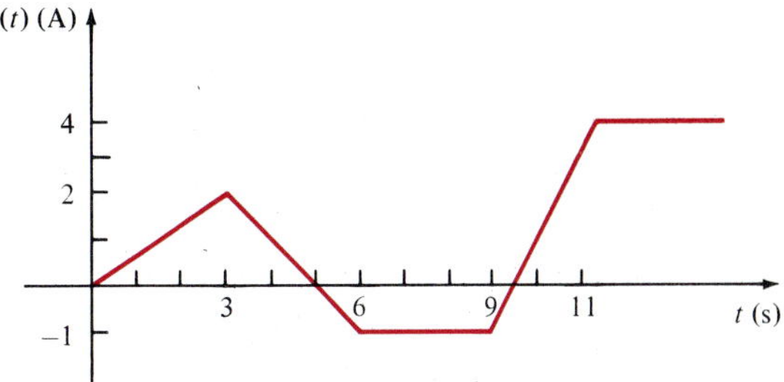

Figure P6.20

6.21. The current in a 4-mH inductor is given by the waveform in Fig. P6.21. Plot the voltage across the inductor.

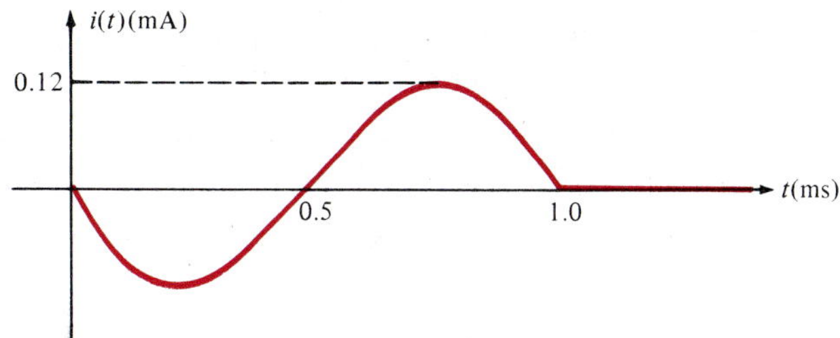

Figure P6.21

6.22. Find the voltage across each of the capacitors if the open-circuit voltage in Fig. P6.22 is 30 V.

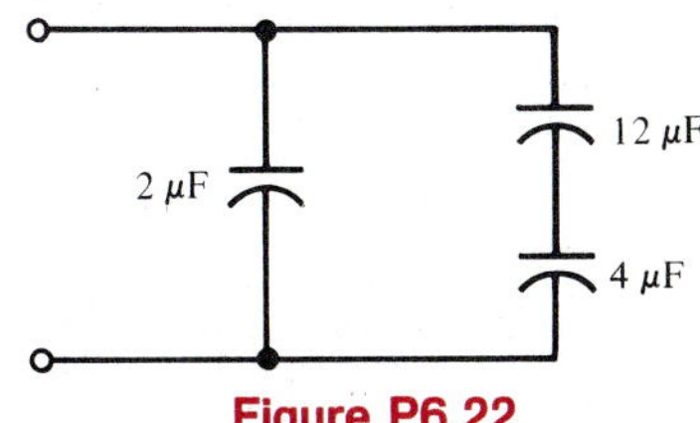

Figure P6.22

6.23. If the open-circuit voltage in the network in Fig. P6.23 is 28 V, find the voltage across each capacitor.

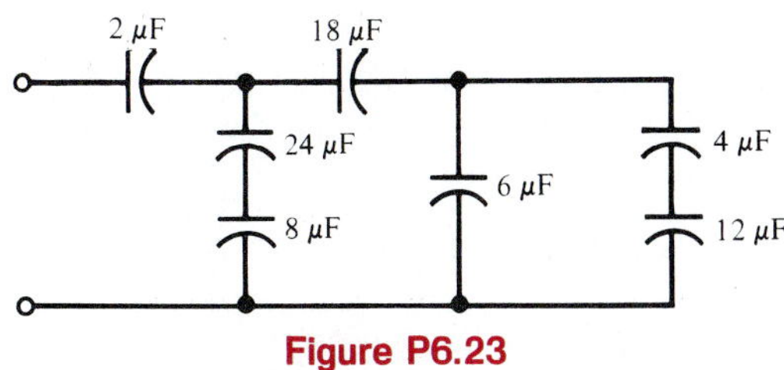

Figure P6.23

6.24. Given four 2-μF capacitors, find the maximum value and minimum value that can be obtained by interconnecting the capacitors in series/parallel combinations.

6.25. Select the value of C to produce the desired total capacitance of $C_T = 2\ \mu$F in the circuit in Fig. P6.25.

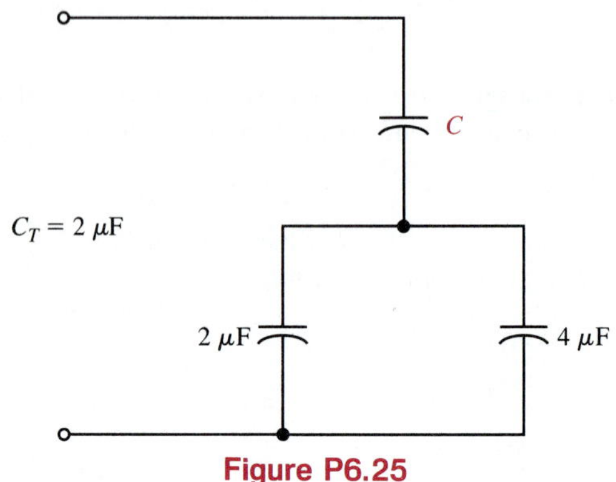

Figure P6.25

6.26. Select the value of C to produce the desired total capacitance of $C_T = 1\ \mu$F in the circuit in Fig. P6.26.

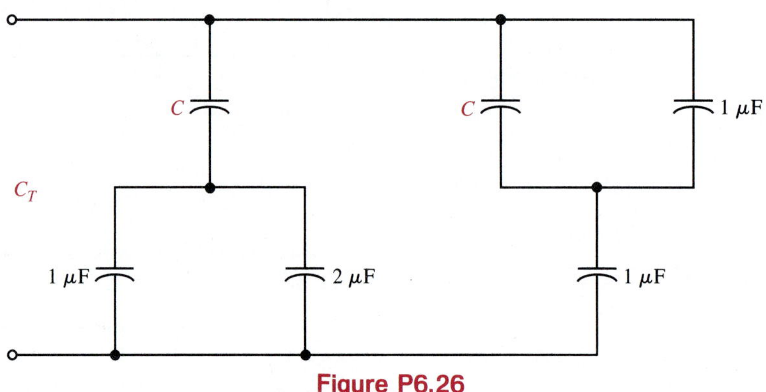

Figure P6.26

6.27. Determine the total capacitance of the network in Fig. P6.27.

4 μF
1 μF
2 μF
C_T
4 μF
2 μF
12 μF
4 μF

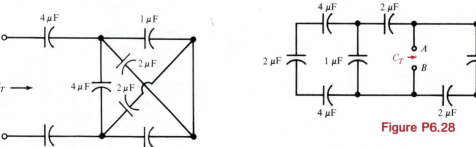

Figure P6.27

6.28. Find the total capacitance of the network in Fig. P6.28.

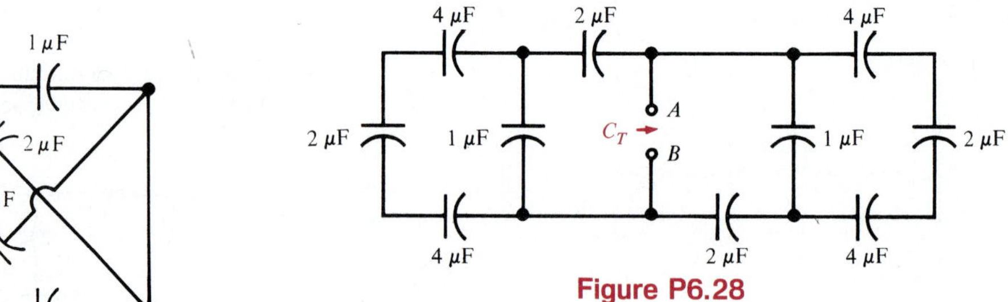

Figure P6.28

6.29. Find the total capacitance C_T of the network in Fig. P6.29.

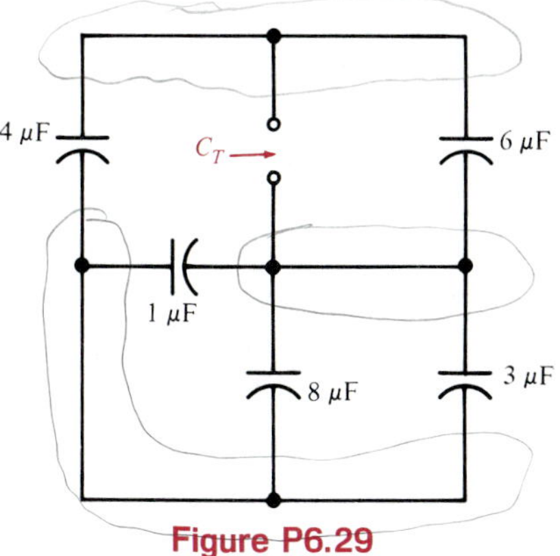

Figure P6.29

6.30. Find the total capacitance C_T of the network in Fig. P6.30.

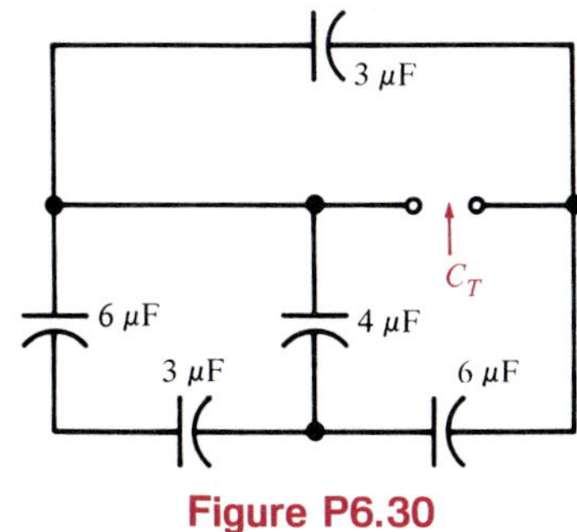

Figure P6.30

6.31. Find the equivalent capacitance at terminals *A-B* in Fig. P6.31.

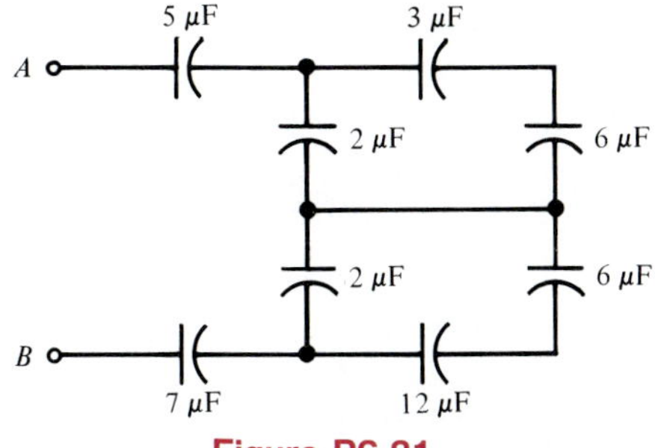

Figure P6.31

6.32. Find C_T in the network in Fig. P6.32 if (a) the switch is open and (b) the switch is closed.

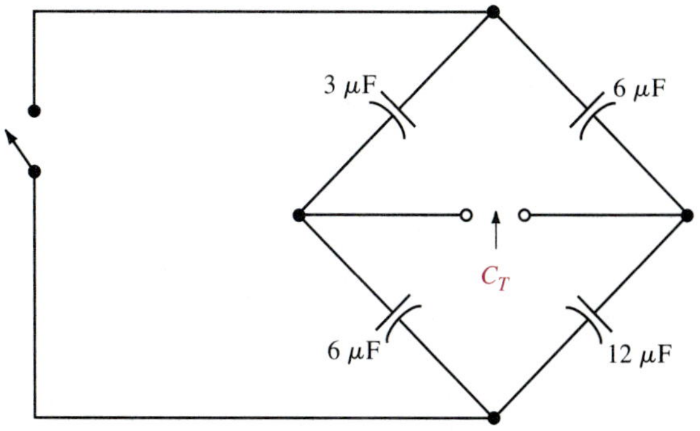

Figure P6.32

6.33. Find the capacitance between points 1 and 2 in the network in Fig. P6.33.

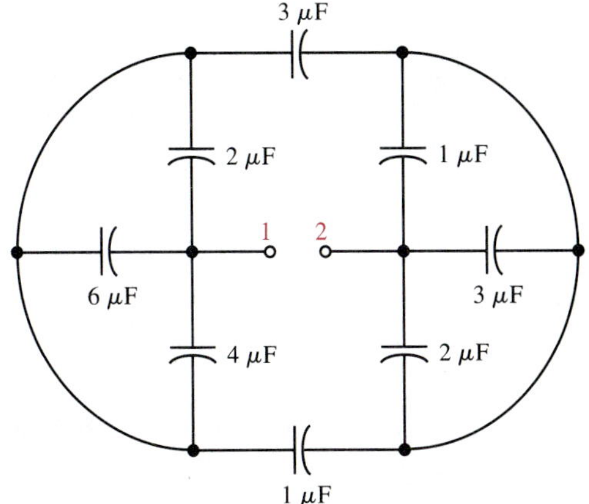

Figure P6.33

6.34. In the network in Fig. P6.34, find the capacitance C_T if (a) the switch is open and (b) if the switch is closed.

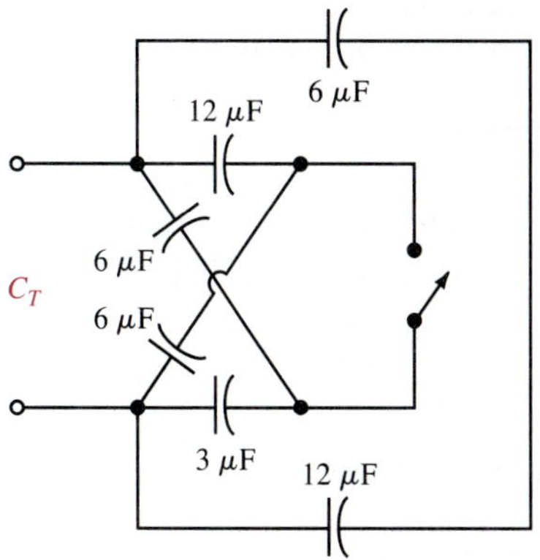

Figure P6.34

6.35. If all the capacitors in Fig. P6.35 are 6 μF, find C_{eq}.

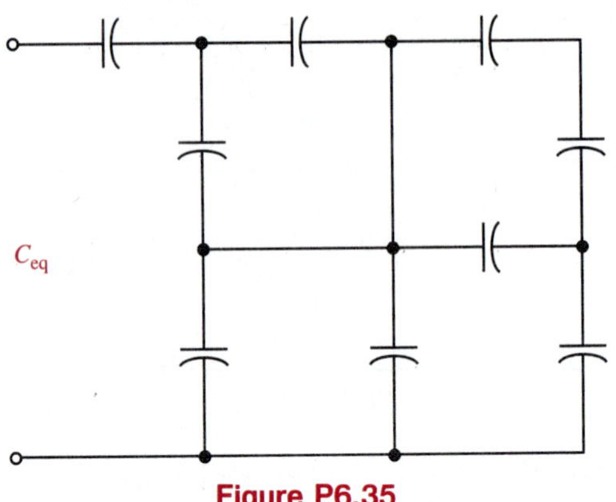

Figure P6.35

6.36. Two capacitors C_1 and C_2 are connected in series. If the voltage across the combination is $v_o(t)$, show that the voltage division between the capacitors is

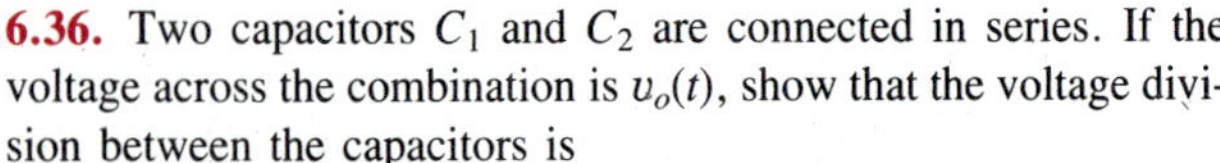

$$v_1 = \frac{C_2}{C_1 + C_2} v_o \quad \text{and} \quad v_2 = \frac{C_1}{C_1 + C_2} v_o$$

where v_1 is the voltage across C_1 and v_2 is the voltage across C_2.

6.37. If the voltage across the network in Fig. P6.37 is 20 V, determine the voltage across each inductor.

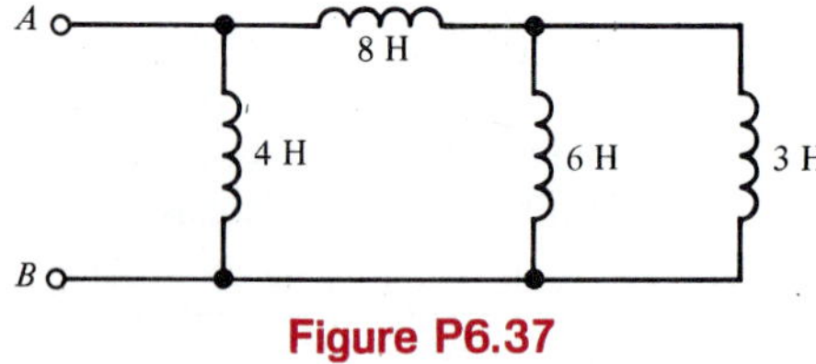

Figure P6.37

6.38. If the current flowing into the circuit in Fig. P6.37 is 7 A, determine the current passing through each inductor.

6.39. Given the network shown in Fig. P6.39, find

(a) The equivalent inductance at terminals *A-B* with terminals *C-D* short-circuited.

(b) The equivalent inductance at terminals *C-D* with terminals *A-B* open-circuited.

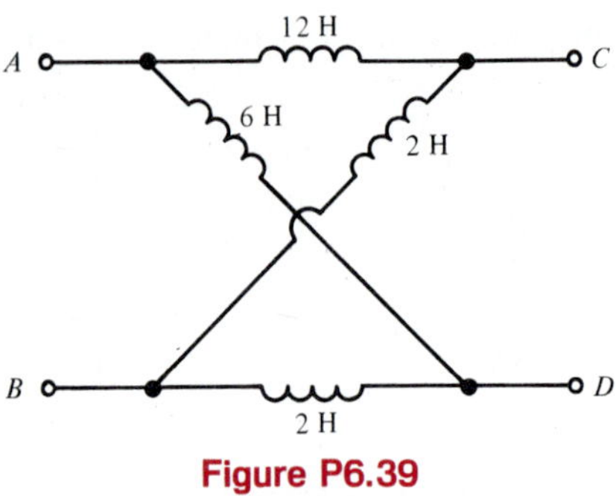

Figure P6.39

6.40. Find the total inductance at the terminals of the network in Fig. P6.40.

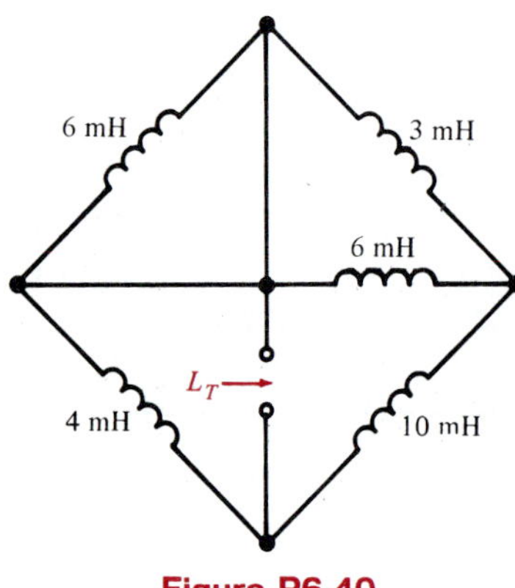

Figure P6.40

6.41. Find the inductance L_T for the network in Fig. P6.41.

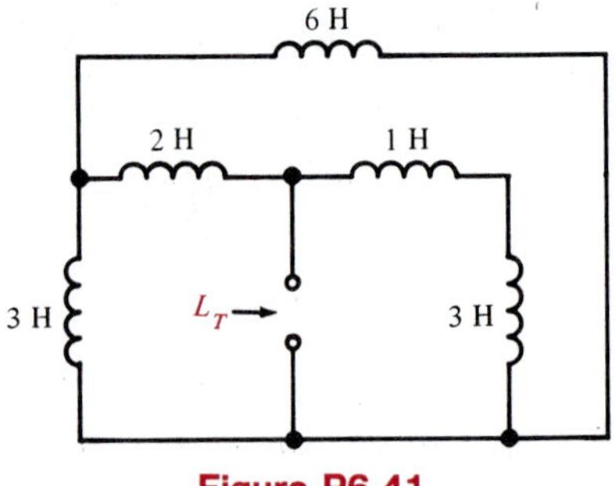

Figure P6.41

6.42. Find L_T in the network in Fig. P6.42 (a) with the switch open and (b) with the switch closed. All inductors are 12 mH.

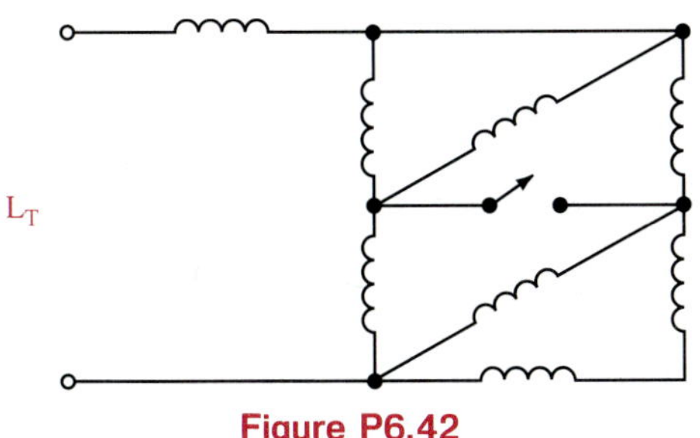

Figure P6.42

6.43. Find L_T in the network in Fig. P6.43. All inductors are 12 mH.

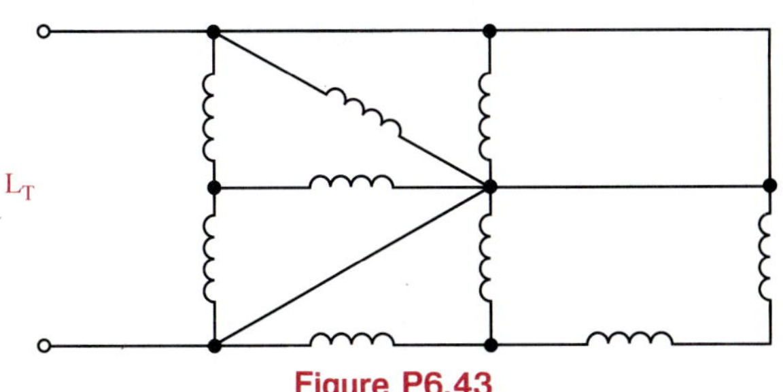

Figure P6.43

6.44. Find the value of L in the network in Fig. P6.44 so that the total inductance L_T will be 2 mH.

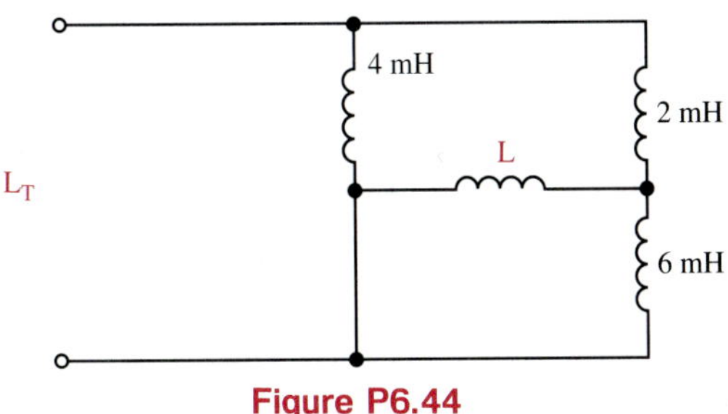

Figure P6.44

6.45. Find the value of L in the network in Fig. P6.45 so that the value of L_T will be 2 mH.

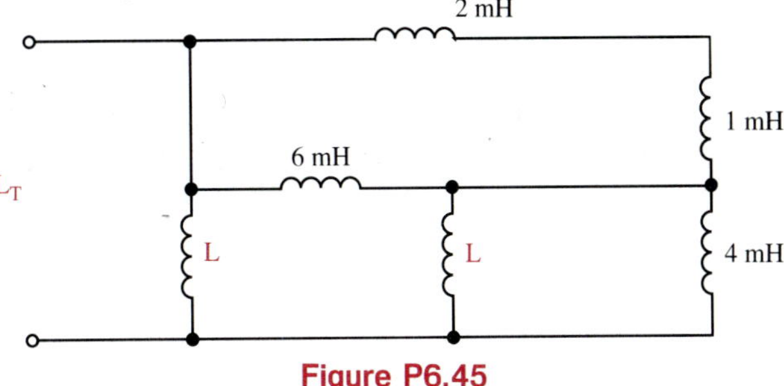

Figure P6.45

6.46. Two inductors L_1 and L_2 are connected in parallel. The current entering the combination is $i_o(t)$. Derive the expressions for the current division between the inductors.

CHAPTER

7

RC and *RL* Circuits

In this chapter we perform what is normally referred to as a transient analysis of a network that contains a single storage element. We examine the behavior of a circuit as a function of time after a sudden change in the network occurs due to switches opening or closing. The time constant of a network is introduced, and a step-by-step procedure for obtaining a transient solution of a network is presented. Finally, we present the techniques required to perform a PSPICE transient analysis.

7.1 Development of the Fundamental Procedures

In our study of *RC* and *RL* circuits we will show that the solution of these circuits, that is, finding a voltage or current, requires us to solve a first-order differential equation of the form

$$\frac{dx(t)}{dt} + ax(t) = f(t) \tag{7.1}$$

Although there are a number of techniques for solving an equation of this type—for example, separation of variables, integrating factor method, or simply guessing a solution—we will employ the procedure outlined below to obtain a general solution that is applicable to the *RC* and *RL* circuit problems we analyze in this chapter.

A fundamental theorem of differential equations states that if $x(t) = x_p(t)$ is any solution to Eq. (7.1), and $x(t) = x_c(t)$ is any solution to the homogeneous equation

$$\frac{dx(t)}{dt} + ax(t) = 0 \tag{7.2}$$

then

$$x(t) = x_p(t) + x_c(t) \tag{7.3}$$

is a solution to the original Eq. (7.1). The term $x_p(t)$ is called the *particular integral solution*, or forced response, and $x_c(t)$ is called the *complementary solution*, or natural response.

At the present time we confine ourselves to the situation in which $f(t) = A$ (i.e., some constant). The general solution of the differential equation then consists of two parts that are obtained by solving the two equations

$$\frac{dx_p(t)}{dt} + ax_p(t) = A \tag{7.4}$$

$$\frac{dx_c(t)}{dt} + ax_c(t) = 0 \tag{7.5}$$

Since the right-hand side of Eq. (7.4) is a constant, it is reasonable to assume that the solution $x_p(t)$ must also be a constant. Therefore, we assume that

$$x_p(t) = K_1 \tag{7.6}$$

Substituting this constant into Eq. (7.4) yields

$$K_1 = \frac{A}{a} \tag{7.7}$$

Examining Eq. (7.5), we note that

$$\frac{dx_c(t)/dt}{x_c(t)} = -a \tag{7.8}$$

This equation is equivalent to

$$\frac{d}{dt}[\ln x_c(t)] = -a$$

Hence

$$\ln x_c(t) = -at + c$$

and therefore,

$$x_c(t) = K_2 e^{-at} \tag{7.9}$$

Therefore, a solution of Eq. (7.1) is

$$x(t) = x_p(t) + x_c(t)$$

$$= \frac{A}{a} + K_2 e^{-at} \tag{7.10}$$

The constant K_2 can be found if the value of the independent variable $x(t)$ is known at one instant of time.

Once the solution in Eq. (7.10) is obtained, certain elements of the equation are given names that are commonly employed in electrical engineering. For example, the term A/a is referred to as the *steady-state solution*: the value of the variable $x(t)$ as $t \to \infty$ when the second term becomes negligible. The constant $1/a$ is called the *time constant* of the circuit. Note that the second term in Eq. (7.10) is a decaying exponential that has a value, if $a > 0$, of K_2 for $t = 0$ and a value of 0 for $t = \infty$. The rate at which this exponential decays is determined by the time constant $T_c = 1/a$. A graphical picture of this effect is shown in Fig. 7.1a. As can be seen from the figure, the value of $x_c(t)$ has fallen from K_2 to a value of $0.368K_2$ in one time constant, a drop of 63.2%. In two time constants the value of $x_c(t)$ has fallen to $0.135K_2$, a drop of 63.2% from the value at time $t = T_c$. This means that the gap between a point on the curve and the final value of the curve is closed by 63.2% each time constant. Finally, after five time constants, $x_c(t) = 0.0067K_2$, which is less than 1%.

An interesting property of the exponential function shown in Fig. 7.1a is that the initial slope of the curve intersects the time axis at a value of $t = T_c$. In fact, we can take any point on the curve, not just the initial value, and find the time constant by finding the

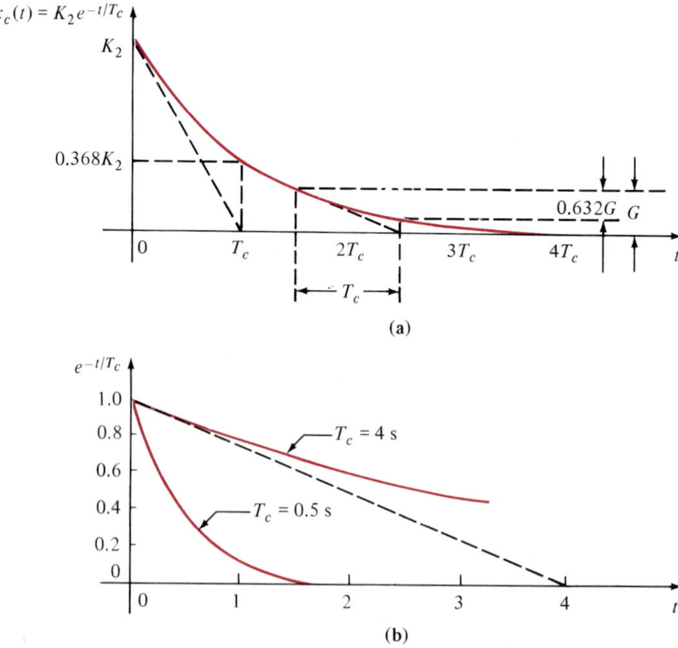

Figure 7.1 Time-constant illustrations.

time required to close the gap by 63.2%. Finally, the difference between a small time constant (i.e., fast response) and a large time constant (i.e., slow response) is shown in Fig. 7.1b. These curves indicate that if the circuit has a small time constant, it settles down quickly to a steady-state value. Conversely, if the time constant is large, more time is required for the circuit to settle down or reach steady state. In any case, note that the circuit response essentially reaches steady state within five time constants (i.e., $5T_c$).

Note that the previous discussion has been very general, in that no particular form of the circuit has been assumed—only that it results in a first-order differential equation. Next, we will study two specific circuits and then outline a method for handling these circuits in general. We will then look at more specific cases using the general method.

Consider the circuit shown in Fig. 7.2a. At time $t = 0$ the switch closes. The equation that describes the circuit for time $t > 0$ is

$$\frac{1}{C}\int i(x)\,dx + Ri(t) = V_S$$

Taking the derivative of this equation with respect to t yields

$$\frac{i(t)}{C} + R\frac{di(t)}{dt} = 0$$

or

$$\frac{di(t)}{dt} + \frac{1}{RC}i(t) = 0$$

Following our previous development, we assume that the solution of this first-order differential equation is of the form

$$i(t) = K_2 e^{-t/T_c}$$

Substituting this solution into the differential equation yields

$$-\frac{K_2 e^{-t/T_c}}{T_c} + \frac{1}{RC}K_2 e^{-t/T_c} = 0$$

$$\left(-\frac{1}{T_c} + \frac{1}{RC}\right)K_2 e^{-t/T_c} = 0$$

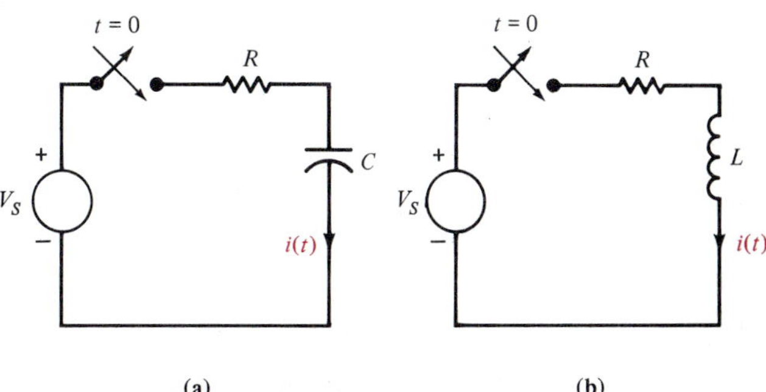

Figure 7.2 *RC* and *RL* circuits.

Our assumed solution is valid if $K_2e^{-t/T_c} = 0$ or $T_c = RC$. The first case implies that $i(t) = 0$ for all t and therefore is disregarded. Therefore,

$$T_c = RC$$

and hence the solution is

$$i(t) = K_2e^{-t/RC}$$

The constant K_2 is chosen so that the complete solution satisfies the particular conditions of the circuit.

The circuit in Fig. 7.2b can be examined in a manner similar to that employed for the circuit in Fig. 7.2a. The equation that describes the circuit for $t > 0$ is

$$L\frac{di(t)}{dt} + Ri(t) = V_S$$

or

$$\frac{di(t)}{dt} + \frac{R}{L}i(t) = \frac{V_S}{L}$$

From our earlier discussions we assume a solution of the form

$$i(t) = K_1 + K_2e^{-t/T_c}$$

Substituting this expression into the differential equation, we obtain

$$-\frac{1}{T_c}K_2e^{-t/T_c} + \frac{R}{L}(K_1 + K_2e^{-t/T_c}) = \frac{V_S}{L}$$

Equating the constant and exponential terms yields

$$\frac{R}{L}K_1 = \frac{V_S}{L}$$

$$\left(-\frac{1}{T_c} + \frac{R}{L}\right)K_2e^{-t/T_c} = 0$$

Our previous analysis indicates that

$$K_1 = \frac{V_S}{R}$$

and

$$T_c = \frac{L}{R}$$

Therefore, the solution is

$$i(t) = \frac{V_S}{R} + K_2e^{-(R/L)t}$$

where once again the constant K_2 is chosen so that the complete solution will satisfy the particular conditions of the circuit.

The importance of the two circuits in Fig. 7.2 stems from the fact that we will find that by employing Thévenin's theorem, we can reduce complicated circuits to these forms, which yield the circuit time constant upon which the circuit response is based.

In general, when the input to an *RC* or *RL* circuit, which contains only a single storage element, is a dc voltage or current, the solution of the differential equation that describes an unknown current or voltage *anywhere in the network* can be written as

$$x(t) = K_1 + K_2 e^{-t/T_c} \tag{7.11}$$

Note that this was the case in the development above and the value of K_1, the steady-state solution, was obtained directly from the differential equation. In our analysis of electrical circuits it is more convenient to determine the constants from an analysis of a modified circuit, as will be shown later.

From Eq. (7.11) we note that as $t \to \infty$, $e^{-at} \to 0$ and $x(t) = K_1$. Therefore, if the circuit is solved for the variable $x(t)$ in steady state (i.e., $t \to \infty$) with the capacitor replaced by an open circuit [v is constant and therefore $i = C(dv/dt) = 0$] or the inductor replaced by a short circuit [i is constant and therefore $v = L(di/dt) = 0$], then the variable $x(t) = K_1$. Note that since the capacitor or inductor has been removed, the circuit is a dc circuit with constant sources and resistors, and therefore only dc analysis is required in the steady-state solution.

The constant K_2 in Eq. (7.11) can also be obtained via the solution of a dc circuit in which a capacitor is replaced by a voltage source or an inductor is replaced by a current source. The value of the voltage source for the capacitor or the current source for the inductor is a known value at one instant of time. In general, we will use the initial condition value since it is generally the one known, but the value at any instant could be used. This value can be obtained in numerous ways and is often specified as input data in a statement of the problem. However, a more likely situation is one in which a switch is thrown in the circuit and the initial value of the capacitor voltage or inductor current is determined from the previous circuit (i.e., the circuit before the switch is thrown). It is normally assumed that the previous circuit has reached steady state, and therefore the voltage across the capacitor or the current through the inductor can be found in exactly the same manner as was used to find K_1.

Finally, the value of the time constant can be found by determining the Thévenin equivalent resistance at the terminals of the storage element. Then $T_c = R_{\text{Th}}C$ for an *RC* circuit, and $T_c = L/R_{\text{Th}}$ for an *RL* circuit.

Let us now reiterate this procedure in a step-by-step fashion.

Step 1. We assume a solution for the variable $x(t)$ of the form $x(t) = K_1 + K_2 e^{-t/T_c}$.

Step 2. Assuming that the original circuit has reached steady state before a switch was thrown (thereby producing a new circuit), draw this previous circuit with the capacitor replaced by an open circuit or the inductor replaced by a short circuit. Solve for the voltage across the capacitor, $v_C(0-)$, or the current through the inductor, $i_L(0-)$, prior to switch action.

Step 3. Assuming that the energy in the storage element cannot change in zero time, draw the circuit, valid only at $t = 0+$. The switches are in their new positions and the capacitor is replaced by a voltage source with a value of $v_C(0+) = v_C(0-)$ or the inductor is replaced by a current source with value $i_L(0+) = i_L(0-)$. Solve for the initial value of the variable $x(0+)$.

Step 4. Assuming that steady state has been reached after the switches are thrown, draw the equivalent circuit, valid for $t > 5T_c$, by replacing the capacitor by an open circuit or the inductor by a short circuit. Solve for the steady-state value of the variable

$$x(t)\Big|_{t>5T_c} \doteq x(\infty)$$

Step 5. Since the time constant for all voltages and currents in the circuit will be the same, it can be obtained by reducing the entire circuit to a simple series circuit containing a voltage source, resistor, and a storage element (i.e., capacitor or inductor) by forming a simple Thévenin equivalent circuit at the terminals of the storage element. This Thévenin equivalent circuit is obtained by looking into the circuit from the terminals of the storage element. The time constant for a circuit containing a capacitor is $T_c = R_{Th}C$, and for a circuit containing an inductor it is $T_c = L/R_{Th}$.

Step 6. Using the results of steps 3, 4, and 5, we can evaluate the constants in step 1 as

$$x(0+) = K_1 + K_2$$

$$x(\infty) = K_1$$

and therefore, $K_1 = x(\infty)$, $K_2 = x(0+) - x(\infty)$, and hence the solution is

$$x(t) = x(\infty) + [x(0+) - x(\infty)]e^{-t/T_c}$$

Keep in mind that this solution form applies only to a first-order circuit having constant, dc sources. If the sources are not dc, the forced response will be different. Generally, the forced response is of the same form as the forcing functions (sources) and their derivatives.

7.2 Source-Free Circuits

We begin our analysis of source-free networks by examining the circuit, which consists of an initially charged capacitor in series with a resistor as shown in Fig. 7.3a. It is assumed that the capacitor is charged to a voltage $v(t) = V_o$ at $t = 0$. Since no independent sources are present in the network, the circuit response is dependent only on the passive circuit elements and the initial voltage. The currents and voltages in the circuit for $t > 0$ can be obtained via KCL. Employing KCL at the top node yields

$$C\frac{dv(t)}{dt} + \frac{v(t)}{R} = 0$$

or

$$\frac{dv(t)}{dt} + \frac{1}{RC}v(t) = 0$$

Let us now apply the step-by-step procedure above to the analysis of this circuit.

Step 1. $v(t)$ is of the form

$$v(t) = K_1 + K_2e^{-t/T_c}$$

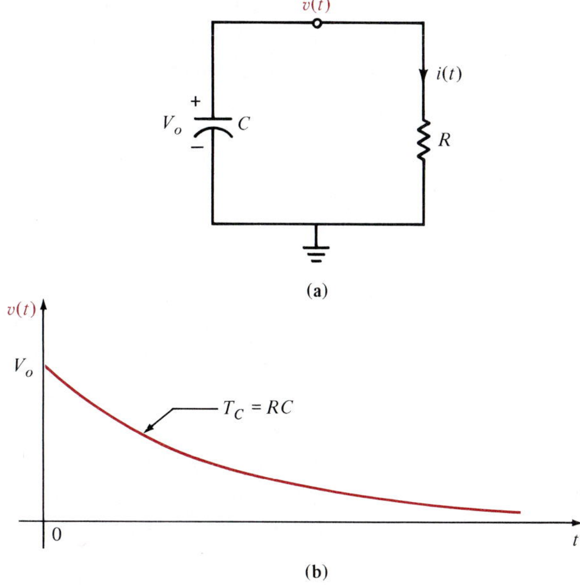

Figure 7.3 Source-free *RC* circuit and its response to an initial condition.

Step 2. The initial voltage across the capacitor is given as $v_C(0-) = V_o$.

Step 3. $v_C(0-) = v_C(0+) = V_o$. Since our unknown is $v(t)$, which is equal to the voltage across the capacitor $v_C(t)$, $v(0+) = V_o$.

Step 4. If the capacitor is replaced by an open circuit, the voltage $v(t)$ across the resistor will be zero. Hence for $t > 5T_c$, $v(\infty) = 0$.

Step 5. The Thévenin equivalent resistance looking into the circuit from the capacitor terminals is obviously $R_{\text{Th}} = R$. Therefore, the circuit time constant is $T_c = RC$.

Step 6. Using the results above, we find that

$$v(\infty) = 0 = K_1$$

$$v(0+) = V_o = K_1 + K_2$$

and therefore,

$$v(t) = V_o e^{-t/RC}$$

Note that the current in the resistor has the same form as the voltage and is given by the expression

$$i(t) = \frac{v(t)}{R} = \frac{V_o}{R} e^{-t/RC}$$

At time $t = 0$ the energy stored in the capacitor is

$$w_C(0+) = \frac{1}{2} C V_o^2 \qquad \text{joules}$$

As time increases, the voltage decreases as shown in Fig. 7.3b, and hence the energy stored in the capacitor also decreases. As this energy decreases, it is absorbed by the resistor. The total energy absorbed by the resistor is

$$\begin{aligned} w_R(\infty) &= \int_0^\infty p_R(x)\, dx \\ &= \int_0^\infty \left(\frac{V_o}{R} e^{-x/RC}\right)^2 R\, dx \\ &= \frac{V_o^2}{R} \int_0^\infty e^{-2x/RC}\, dx \\ &= \frac{1}{2} C V_o^2 \qquad \text{joules} \end{aligned}$$

This is, of course, equal to the initial energy stored in the capacitor.

It is interesting to note that if KVL had been employed to determine the describing equation for the circuit, the result would have been

$$\frac{1}{C} \int_0^t i(x)\, dx - V_o + i(t)R = 0$$

Differentiating the equation yields

$$\frac{i(t)}{C} + R\frac{di(t)}{dt} = 0$$

since

$$i(t) = \frac{v(t)}{R}$$

Then

$$\frac{v(t)}{RC} + \frac{dv(t)}{dt} = 0$$

which is, of course, the original equation in our analysis.

EXAMPLE 7.1

Consider the circuit shown in Fig. 7.4a. Assuming that the switch has been in position 1 for a long time, at time $t = 0$ the switch is moved to position 2. We wish to calculate the current $i(t)$ for $t > 0$.

Step 1. We assume the current to be of the form

$$i(t) = K_1 + K_2 e^{-t/T_c}$$

Step 2. In steady state prior to switch action, the initial capacitor voltage is found from Fig. 7.4b as

$$v_C(0-) = \frac{(12)(3)}{6+3} = 4 \text{ V}$$

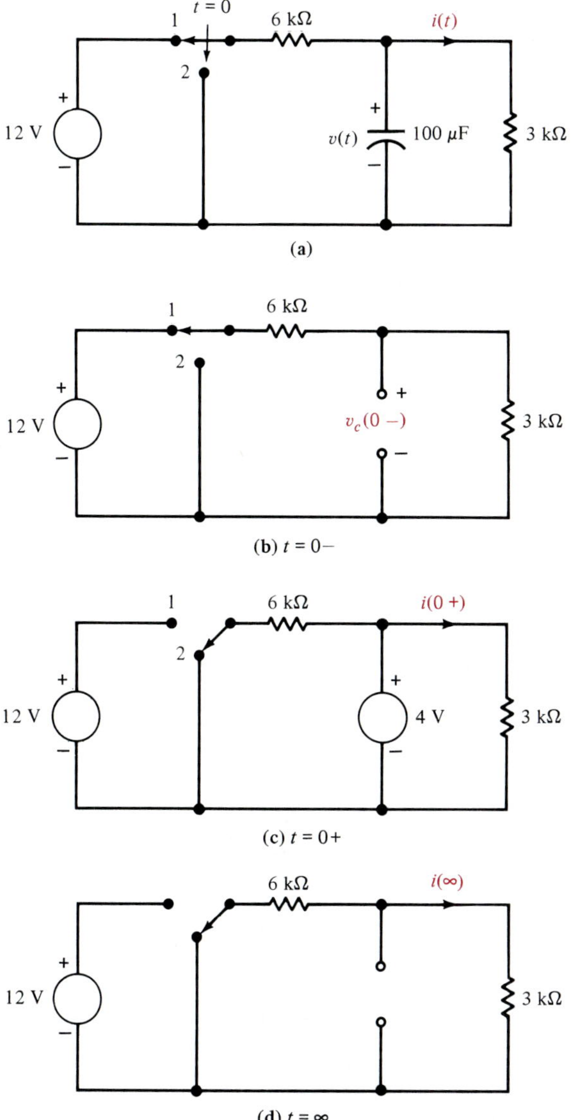

Figure 7.4 Analysis of an *RC* transient circuit.

Step 3. The new circuit, valid only for $t = 0+$, is shown in Fig. 7.4c. The value of the voltage source that replaces the capacitor is $v_C(0-) = v_C(0+) = 4$ V. Therefore, once again, since the voltage is in volts and the resistance is in kilohms, the current will be in milliamperes.

$$i(0+) = \frac{4}{3} \text{ mA}$$

Step 4. The equivalent circuit, valid for $t > 5T_c$, is shown in Fig. 7.4d. Since there are no sources present, the value of $i(\infty) = 0$.

Step 5. The Thévenin equivalent resistance, obtained by looking into the circuit from the open-circuit terminals of the capacitor in Fig. 7.4d, is

$$R_{\text{Th}} = \frac{(6)(3)}{6 + 3} = 2\text{ k}\Omega$$

Therefore, the circuit time constant is

$$\begin{aligned} T_c &= R_{\text{Th}}C \\ &= (2)(10^3)(100)(10^{-6}) \\ &= 0.2\text{ sec} \end{aligned}$$

Step 6. From the previous analysis

$$K_1 = i(\infty) = 0$$

$$K_2 = i(0+) - i(\infty) = \tfrac{4}{3}\text{ mA}$$

and hence

$$\begin{aligned} i(t) &= \tfrac{4}{3}e^{-t/0.2} \\ &= 1.33e^{-5t}\text{ mA} \end{aligned}$$

DRILL EXERCISE

D7.1. Find $v_C(t)$ for $t > 0$ in the circuit shown in Fig. D7.1.

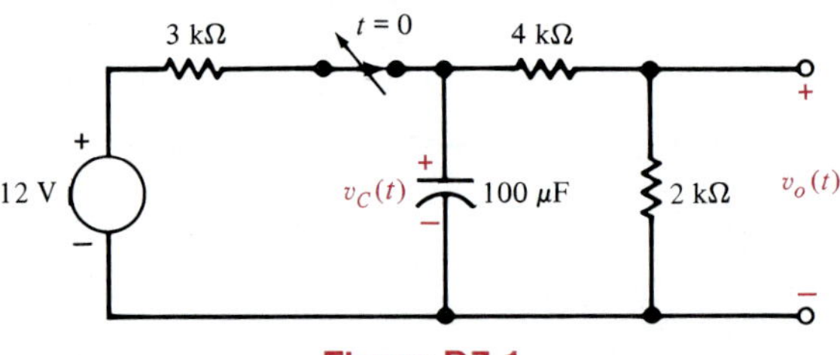

Figure D7.1

Ans: $v_C(t) = 8e^{-t/0.6}$ V.

D7.2. Find $v_o(t)$ for $t > 0$ in Fig. D7.1.

Ans: $v_o(t) = \tfrac{8}{3}e^{-t/0.6}$ V.

Let us next consider a source-free *RL* circuit as shown in Fig. 7.5a. It is assumed that the inductor has an initial current of $i(t) = I_o$ at $t = 0$. Applying KVL around the loop yields

$$L\frac{di(t)}{dt} + Ri(t) = 0$$

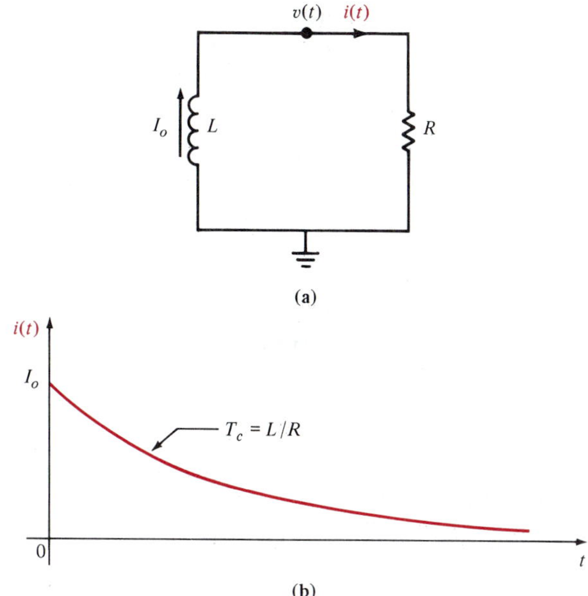

Figure 7.5 Source-free *RL* circuit and its response to an initial condition.

or

$$\frac{di(t)}{dt} + \frac{R}{L}i(t) = 0$$

Once again we employ the step-by-step procedure to analyze this circuit:

Step 1. $i(t)$ is of the form $K_1 + K_2 e^{-t/T_c}$.

Step 2. The initial current through the inductor is $i_L(0-) = I_o$.

Step 3. $i(0+) = i_L(0+) = i_L(0-) = I_o$.

Step 4. If the inductor is replaced by a short circuit, $v(t)$ for $t > 5T_c$ will be zero and hence $i(\infty) = 0$.

Step 5. The Thévenin equivalent resistance looking into the circuit from the inductor terminals is $R_{\text{Th}} = R$; therefore, the circuit time constant is $T_c = L/R$.

Step 6. The results above show that

$$i(\infty) = K_1 = 0$$

Thus

$$i(0+) = K_1 + K_2 = I_o$$

and hence

$$i(t) = I_o e^{-Rt/L}$$

At time $t = 0$ the energy stored in the inductor is

$$w_L(0+) = w_L(0-) = \tfrac{1}{2}LI_o^2 \qquad \text{joules}$$

As time increases the current decreases, as shown in Fig. 7.5b, and hence the energy stored in the inductor will also decrease. This energy is absorbed by the resistor. The energy that the resistor absorbs is given by the expression

$$\begin{aligned} w_R(\infty) &= \int_0^\infty p_R(t)\,dt \\ &= RI_o^2 \int_0^\infty e^{-2Rt/L}\,dt \\ &= \frac{1}{2} LI_o^2 \qquad \text{joules} \end{aligned}$$

This value coincides with that for the energy initially stored in the inductor.

EXAMPLE 7.2

Let us examine the circuit shown in Fig. 7.6a. At $t = 0$ the switch is opened and we wish to calculate the current $i(t)$ for $t > 0$.

Step 1. The current is assumed to be of the form

$$i(t) = K_1 + K_2 e^{-t/T_c}$$

Step 2. In the steady-state mode prior to the time the switch opens, the initial inductor current can be found from Fig. 7.6b as

$$\begin{aligned} i_L(0-) &= \frac{36}{2 + \dfrac{(6)(12)}{6+12}} \left(\frac{12}{18}\right) \\ &= 4\text{ A} \end{aligned}$$

Step 3. The new circuit, valid only for $t = 0+$, is shown in Fig. 7.6c. The value of the current source that replaces the inductor is $i_L(0-) = i_L(0+) = 4$ A. Therefore,

$$i(0+) = -4\text{ A}$$

Step 4. The equivalent circuit, valid only for $t > 5T_c$, is given in Fig. 7.6d. Since there are no sources present, the value of $i(\infty) = 0$.

Step 5. The Thévenin equivalent resistance found by looking into the circuit from the inductor terminals is

$$\begin{aligned} R_{\text{Th}} &= 6 + 4 + 8 \\ &= 18\ \Omega \end{aligned}$$

Therefore, the circuit time constant is

$$\begin{aligned} T_c &= \frac{L}{R_{\text{Th}}} \\ &= \frac{2}{18} \\ &= \frac{1}{9}\text{ sec} \end{aligned}$$

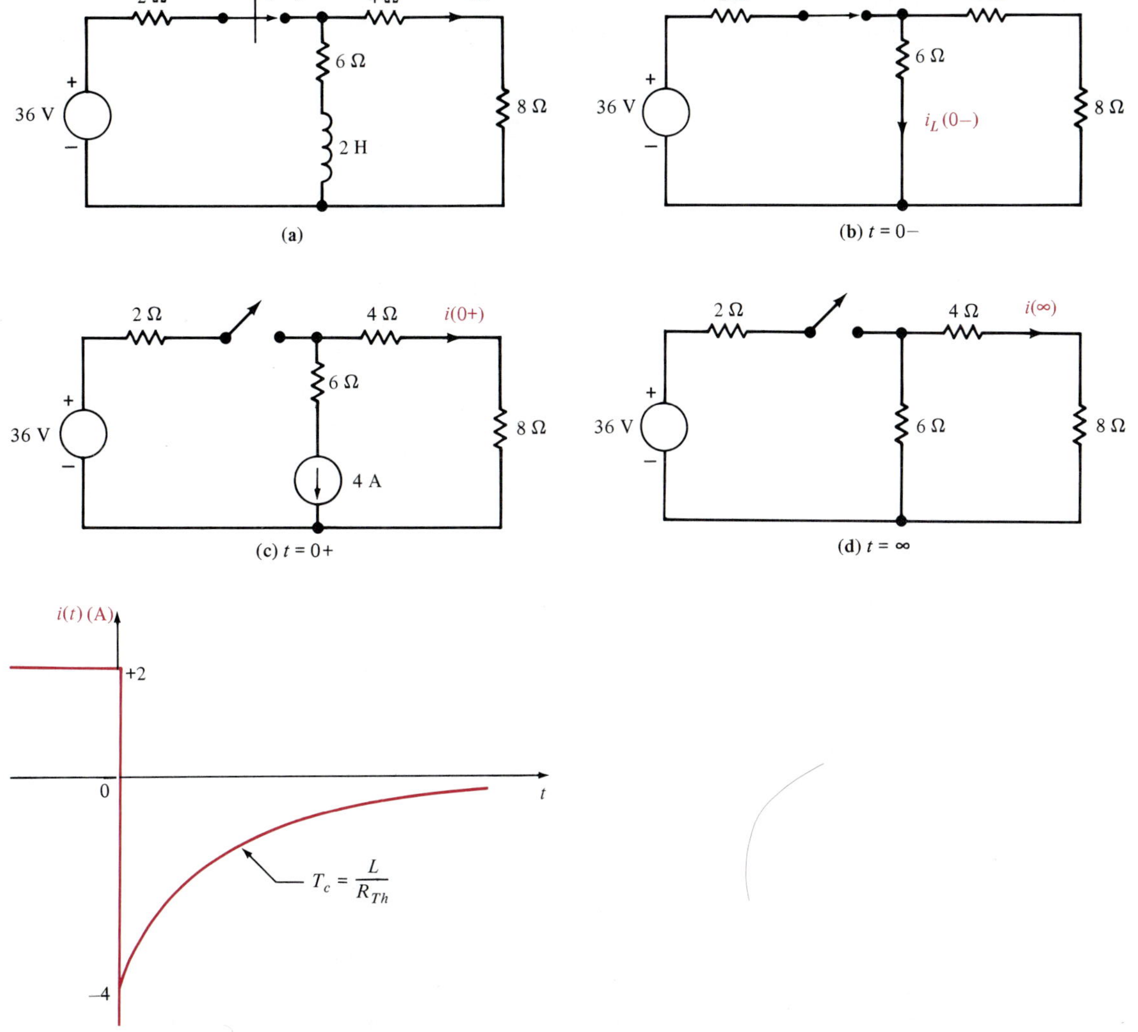

Figure 7.6 Analysis of an *RL* circuit.

Step 6. From the foregoing analysis,

$$K_1 = i(\infty) = 0$$

Then

$$K_2 = i(0+) - i(\infty) = -4$$

and hence

$$i(t) = -4e^{-9t} \text{ A}$$

Note that although the current in the inductor is continuous at $t = 0$, the current $i(t)$ in the 4- and 8-Ω resistors has jumped abruptly at $t = 0$, as shown in Fig. 7.6e. ■

DRILL EXERCISE

D7.3. In the circuit shown in Fig. D7.3, the switch opens at $t = 0$. Find $i_1(t)$ for $t > 0$.

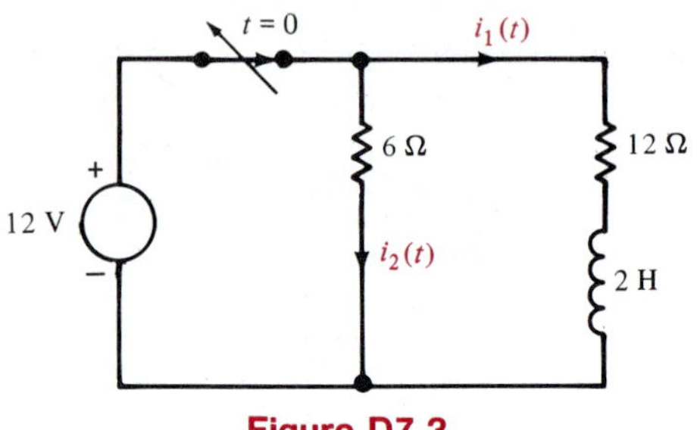

Figure D7.3

Ans: $i_1(t) = 1e^{-9t}$ A.

D7.4. Determine the expression for $i_2(t)$, $t > 0$, in Fig. D7.3.

Ans: $i_2(t) = -1e^{-9t}$ A.

D7.5. In Fig. D7.3, if the switch was open for $t < 0$ and then closes at $t = 0$, find $i_1(t)$ for $t > 0$.

Ans: $i_1(t) = 1 - e^{-6t}$ A.

7.3 Circuits with Constant and Nonconstant Forcing Functions

In Section 7.2 we were concerned with the natural response of circuits to certain initial conditions. In every case the variable $x(t) \rightarrow 0$ as $t \rightarrow \infty$. In this section we examine circuits that contain constant forcing functions in addition to stored initial conditions. To solve these circuits, we will simply employ the same general procedure outlined above. In this case, however, we find that $K_1 = x(\infty) \neq 0$; that is, the steady-state value is some constant other than zero.

EXAMPLE 7.3

Consider the circuit shown in Fig. 7.7a. The circuit is in steady state prior to time $t = 0$, when the switch is closed. Let us calculate the current $i(t)$ for $t > 0$.

Step 1. $i(t)$ is of the form $K_1 + K_2e^{-t/T_c}$.

Step 2. The initial voltage across the capacitor is calculated from Fig. 7.7b as

$$v_C(0-) = 36 - (2)(2)$$
$$= 32 \text{ V}$$

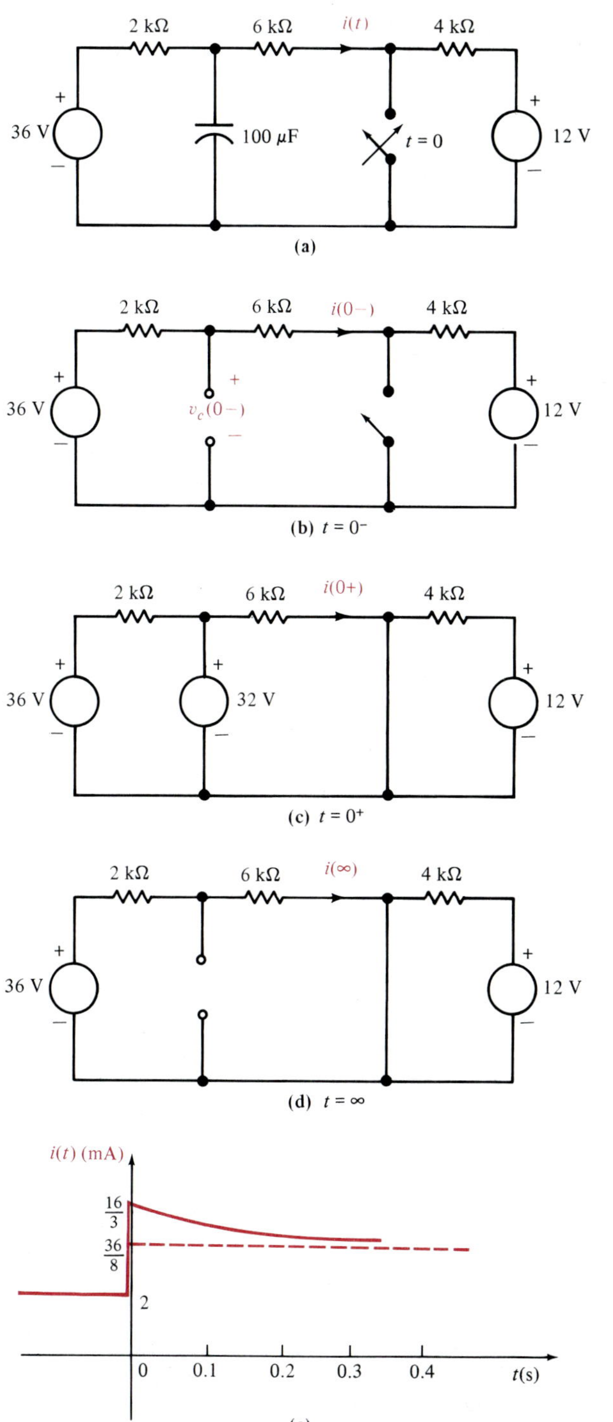

Figure 7.7 Analysis of an *RC* transient circuit with a constant forcing function.

Step 3. The new circuit, valid only for $t = 0+$, is shown in Fig. 7.7c.

The value of the voltage source that replaces the capacitor is $v_C(0-) = v_C(0+) =$ 32 V. Hence

$$i(0+) = \frac{32}{6} = \frac{16}{3} \text{ mA}$$

Step 4. The equivalent circuit, valid for $t > 5T_c$, is shown in Fig. 7.7d. The current $i(\infty)$ caused by the 36-V source is

$$i(\infty) = \frac{36}{2+6} = \frac{36}{8} \text{ mA}$$

Step 5. The Thévenin equivalent resistance, obtained by looking into the open-circuit terminals of the capacitor in Fig. 7.7d, is

$$R_{\text{Th}} = \frac{(2)(6)}{2+6} = \frac{3}{2} \text{ k}\Omega$$

Therefore, the circuit time constant is

$$\begin{aligned} T_c &= R_{\text{Th}}C \\ &= \left(\frac{3}{2}\right)(10^3)(100)(10^{-6}) \\ &= 0.15 \text{ sec} \end{aligned}$$

Step 6

$$\begin{aligned} K_1 &= i(\infty) = \frac{36}{8} \text{ mA} \\ K_2 &= i(0+) - i(\infty) = i(0+) - K_1 \\ &= \frac{16}{3} - \frac{36}{8} \\ &= \frac{5}{6} \text{ mA} \end{aligned}$$

Therefore,

$$i(t) = \frac{36}{8} + \frac{5}{6}e^{-t/0.15} \text{ mA}$$

Once again we see that although the voltage across the capacitor is continuous at $t = 0$, the current $i(t)$ in the 6-kΩ resistor jumps at $t = 0$ from 2 mA to $5\frac{1}{3}$ mA, and finally decays to $4\frac{1}{2}$ mA. ■

EXAMPLE 7.4

The circuit shown in Fig. 7.8a is assumed to have been in a steady-state condition prior to switch closure at $t = 0$. We wish to calculate the voltage $v(t)$ for $t > 0$.

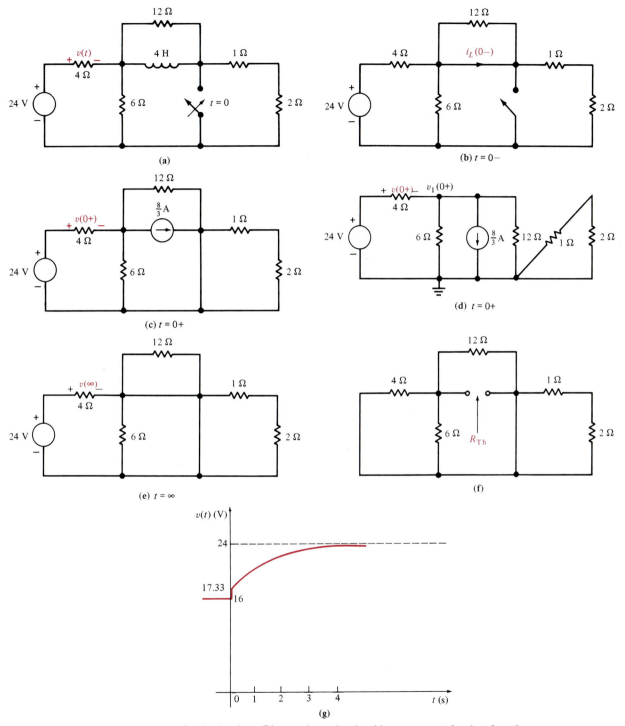

Figure 7.8 Analysis of an *RL* transient circuit with a constant forcing function.

Step 1. $v(t)$ is of the form $K_1 + K_2 e^{-t/T_c}$.
Step 2. In Fig. 7.8b we see that

$$i_L(0-) = \frac{24}{4 + \dfrac{(6)(3)}{6+3}}\left(\frac{6}{6+3}\right)$$

$$= \frac{8}{3}\text{ A}$$

Step 3. The new circuit, valid only for $t = 0+$, is shown in Fig. 7.8c, which is equivalent to the circuit shown in Fig. 7.8d. The value of the current source that replaces the inductor is $i_L(0-) = i_L(0+) = \frac{8}{3}$ A. The node voltage $v_1(0+)$ can be determined from the circuit in Fig. 7.8d using a single-node equation, and $v(0+)$ is equal to the difference between the source voltage and $v_1(0+)$. The equation for $v_1(0+)$ is

$$\frac{v_1(0+) - 24}{4} + \frac{v_1(0+)}{6} + \frac{8}{3} + \frac{v_1(0+)}{12} = 0$$

or

$$v_1(0+) = \frac{20}{3}\text{ V}$$

Then

$$v(0+) = 24 - v_1(0+)$$

$$= \frac{52}{3}\text{ V}$$

Step 4. The equivalent circuit for the steady-state condition after switch closure is given in Fig. 7.8e. Note that the 6-, 12-, 1-, and 2-Ω resistors are shorted, and therefore $v(\infty) = 24$ V.

Step 5. The Thévenin equivalent resistance is found by looking into the circuit from the inductor terminals. This circuit is shown in Fig. 7.8f. Note carefully that R_{Th} is equal to the 4-, 6-, and 12-Ω resistors in parallel. Therefore, $R_{Th} = 2\ \Omega$, and the circuit time constant is

$$T_c = \frac{L}{R_{Th}} = \frac{4}{2} = 2\text{ sec}$$

Step 6. From the previous analysis we find that

$$K_1 = v(\infty) = 24$$

$$K_2 = v(0+) - v(\infty) = -\frac{20}{3}$$

and hence that

$$v(t) = 24 - \frac{20}{3}e^{-t/2}\text{ V}$$

From Fig. 7.8b we see that the value of $v(t)$ before switch closure is 16 V. Therefore, the circuit response $v(t)$, as a function of time, is shown in Fig. 7.8g. ■

DRILL EXERCISE

D7.6. Consider the network in Fig. D7.6. The switch opens at $t = 0$. Find $v_o(t)$ for $t > 0$.

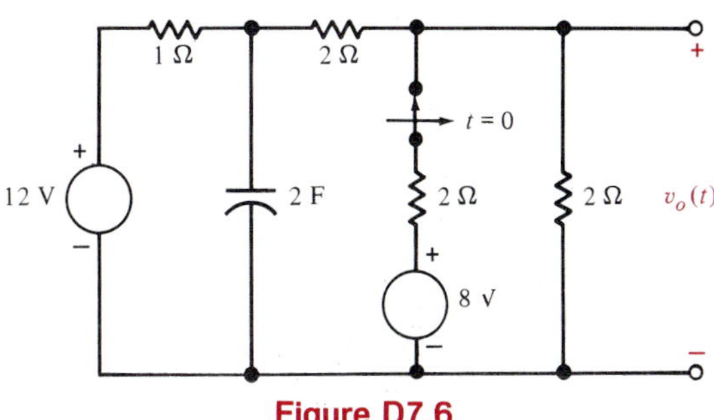

Figure D7.6

Ans: $v_o(t) = \frac{24}{5} + \frac{1}{5}e^{-(5/8)t}$ V.

D7.7. Consider the network shown in Fig. D7.7. If the switch opens at $t = 0$, find the output voltage $v_o(t)$ for $t > 0$.

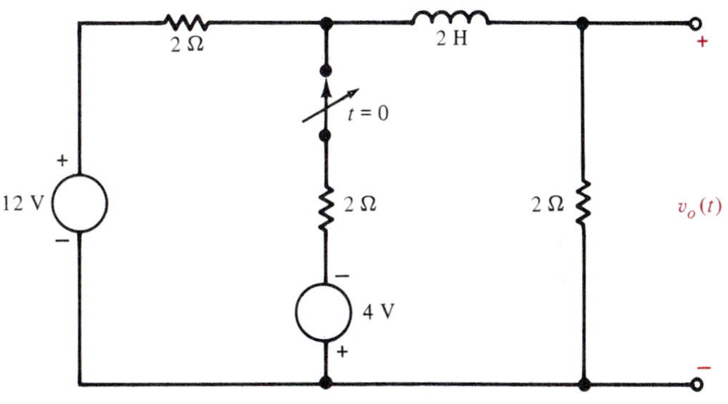

Figure D7.7

Ans: $v_o(t) = 6 - \frac{10}{3}e^{-2t}$ V.

EXAMPLE 7.5

The circuit shown in Fig. 7.9a has reached steady state with the switch in position 1. At time $t = 0$ the switch moves from position 1 to position 2. We want to calculate $v_o(t)$ for $t > 0$.

Step 1. $v_o(t)$ is of the form $K_1 + K_2e^{-t/T_c}$.

Step 2. Using the circuit in Fig. 7.9b, we can calculate $i_L(0-)$.

$$i_A = \frac{12}{4} = 3 \text{ A}$$

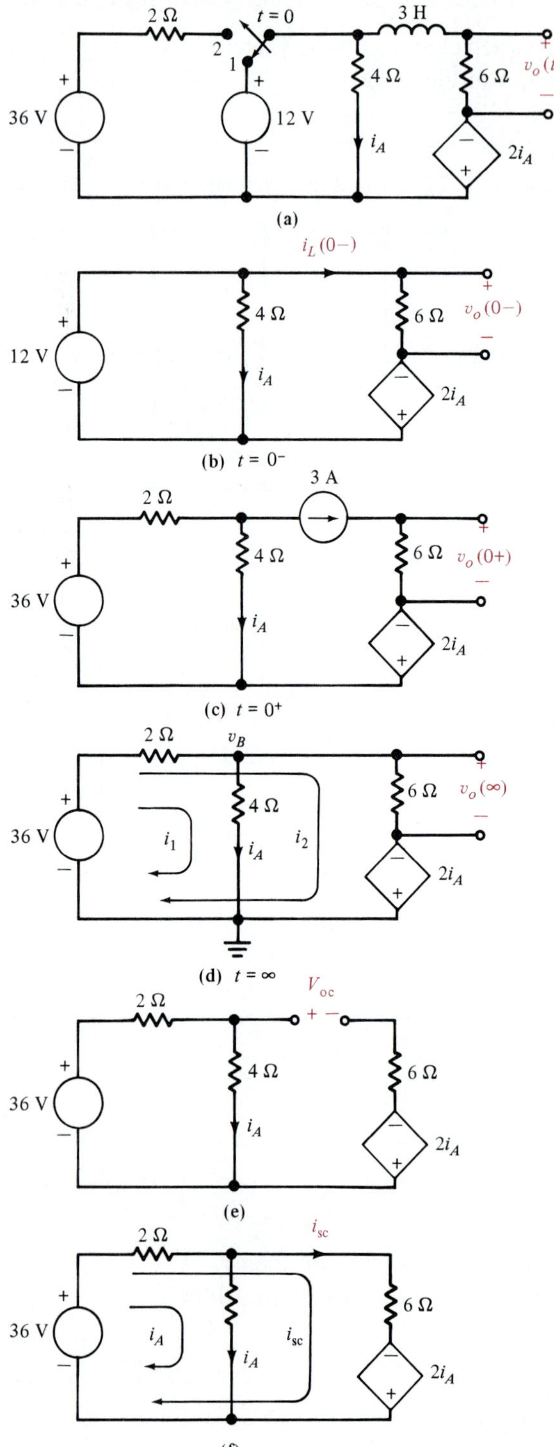

Figure 7.9 Analysis of an *RL* transient circuit containing a dependent source.

Then

$$i_L(0-) = \frac{12 + 2i_A}{6} = \frac{18}{6} = 3 \text{ A}$$

Step 3. The new circuit, valid only for $t = 0+$, is shown in Fig. 7.9c. The value of the current source that replaces the inductor is $i_L(0-) = i_L(0+) = 3$ A. Because of the current source

$$v_o(0+) = (3)(6) = 18 \text{ V}$$

Step 4. The equivalent circuit for the steady-state condition after switch closure is given in Fig. 7.9d. Using the voltages and currents defined in the figure, we can compute $v_o(\infty)$ in a variety of ways. For example, using node equations we can find $v_o(\infty)$ from

$$\frac{v_B - 36}{2} + \frac{v_B}{4} + \frac{v_B + 2i'_A}{6} = 0$$

$$i'_A = \frac{v_B}{4}$$

$$v_o(\infty) = v_B + 2i'_A$$

or using loop equations,

$$36 = 2(i_1 + i_2) + 4i_1$$

$$36 = 2(i_1 + i_2) + 6i_2 - 2i_1$$

$$v_o(\infty) = 6i_2$$

Using either approach, we find that $v_o(\infty) = 27$ V.

Step 5. The Thévenin equivalent resistance can be obtained via v_{oc} and i_{sc}, because of the presence of the dependent source. From Fig. 7.9e we note that

$$i''_A = \frac{36}{2 + 4} = 6 \text{ A}$$

Therefore,

$$v_{oc} = (4)(6) + 2(6)$$

$$= 36 \text{ V}$$

From Fig. 7.9f we can write the following loop equations:

$$36 = 2(i'''_A + i_{sc}) + 4i'''_A$$

$$36 = 2(i'''_A + i_{sc}) + 6i_{sc} - 2i'''_A$$

Solving these equations for i_{sc} yields

$$i_{sc} = \frac{36}{8} \text{ A}$$

Therefore,

$$R_{Th} = \frac{v_{oc}}{i_{sc}} = \frac{36}{36/8} = 8\ \Omega$$

Hence the circuit time constant is

$$T_c = \frac{L}{R_{Th}} = \frac{3}{8}\ \text{sec}$$

Step 6. Using the information computed above, we can derive the final equation for $v_o(t)$:

$$K_1 = v_o(\infty) = 27$$

$$K_2 = v_o(0+) - v_o(\infty) = 18 - 27 = -9$$

Therefore,

$$v_o(t) = 27 - 9e^{-t/(3/8)}\ \text{V}$$

■

DRILL EXERCISE

D7.8. If the switch in the network in Fig. D7.8 closes at $t = 0$, find $v_o(t)$ for $t > 0$.

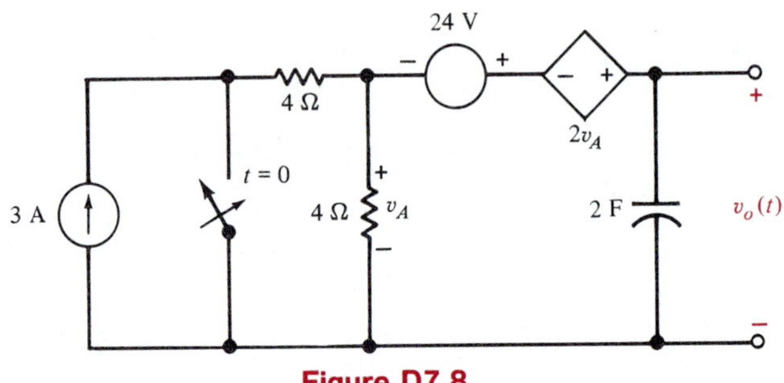

Figure D7.8

Ans: $v_o(t) = 24 + 36e^{-(t/12)}$ V.

At this point it is appropriate to state that not all switch action will always occur at time $t = 0$. It may occur at any time t_0. In this case the results of the step-by-step analysis yield the following equations:

$$x(t_0) = K_1 + K_2$$

$$x(\infty) = K_1$$

and

$$x(t) = x(\infty) + [x(t_0) - x(\infty)]e^{-(t-t_0)/T_c} \qquad t > t_0$$

The function is essentially time shifted by t_0 seconds.

Finally, it should be noted that if more than one independent source is present in the network, we can simply employ superposition to obtain the total response.

Thus far we have considered only sources that were constant. Although this represents a very important type of problem, it is, by no means, the only type nor is it necessarily the most important.

Returning to Eq. (7.1), we ask: "What is the solution if $f(t)$ is not a constant and the network equation is actually as follows?"

$$\frac{dx(t)}{dt} + ax(t) = f(t) \tag{7.12}$$

Recall that $x(t)$ consists of two parts: the natural response (complementary solution) and the forced response (particular solution). That is,

$$x(t) = x_c(t) + x_p(t) \tag{7.13}$$

where $x_c(t)$ and $x_p(t)$ must satisfy the equations

$$\frac{dx_c(t)}{dt} + ax_c(t) = 0 \tag{7.14}$$

and

$$\frac{dx_p(t)}{dt} + ax_p(t) = f(t) \tag{7.15}$$

Thus $x_c(t)$ must still be of the form

$$x_c(t) = k_2 e^{-at} \tag{7.16}$$

However, since $f(t)$ is no longer a constant, $x_p(t)$ will, in general, not be one either. Although there are many different mathematical methods for determining the appropriate form for $x_p(t)$, we will employ a deductive approach here. A method that is systematic and often more efficient will be introduced in Chapter 17.

A careful inspection of Eq. (7.15) suggests that $x_p(t)$ must consist of functional forms such as $f(t)$ and its first derivative.

EXAMPLE 7.6

Consider the circuit in Fig. 7.10. The capacitor has an initial charge and an exponentially decaying source is applied at $t = 0$. We wish to determine both $v_o(t)$ and $i(t)$ for $t > 0$.

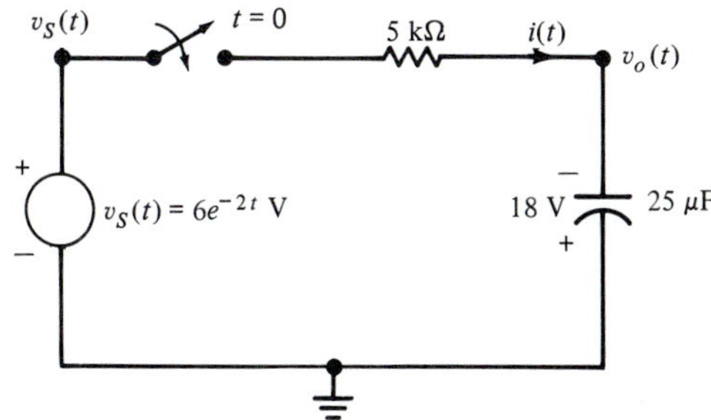

Figure 7.10 *RC* circuit with a nonconstant forcing function.

A nodal analysis yields the equation

$$C\frac{dv_o(t)}{dt} + \frac{v_o(t) - v_S(t)}{R} = 0 \qquad t > 0$$

or

$$\frac{dv_o(t)}{dt} + 8v_o(t) = 48e^{-2t}$$

Now the natural response (complementary solution) must, as before, satisfy the homogeneous equation

$$\frac{dv_{oc}(t)}{dt} + 8v_{oc}(t) = 0$$

which leads to

$$v_{oc}(t) = K_2e^{-8t}$$

The forced response (particular solution) must satisfy the equation

$$\frac{dv_{op}(t)}{dt} + 8v_{op}(t) = 48e^{-2t}$$

We deduce that $v_{op}(t)$ must be of the form of the forcing function, e^{-2t}, and its derivative, which in this case is also e^{-2t}. Thus we assume that

$$v_{op}(t) = K_1e^{-2t}$$

and by substituting this assumed solution back into the differential equation we can determine the required value of K_1.

$$\frac{d}{dt}(K_1e^{-2t}) + 8(K_1e^{-2t}) = 48e^{-2t}$$

$$(-2 + 8)K_1e^{-2t} = 48e^{-2t}$$

$$K_1 = 8$$

Therefore, the total solution is

$$v_o(t) = 8e^{-2t} + K_2e^{-8t}$$

However, since $v_o(0) = -18$, K_2 can be obtained from the expression for $v_o(t)$ as

$$v_o(0) = -18 = 8e^{-2(0)} + K_2e^{-8(0)}$$

or

$$K_2 = -26$$

and then

$$v_o(t) = 8e^{-2t} - 26e^{-8t}\text{ V} \qquad t > 0$$

To determine $i(t)$, we note that

$$i(t) = C\frac{dv_o(t)}{dt}$$

$$= 25 \times 10^{-6}(-16e^{-2t} + 208e^{-8t})$$

$$= -0.4e^{-2t} + 5.2e^{-8t} \text{ mA} \qquad t > 0$$

■

7.4 Pulse Response

Thus far we have examined networks in which a voltage or current source is suddenly applied. As a result of this sudden application of a source, voltages or currents in the circuit are forced to change abruptly. A forcing function whose value changes in a discontinuous manner or has a discontinuous derivative is called a *singular function.* Two such singular functions that are very important in circuit analysis are the unit impulse function and the unit step function. We will defer a discussion of the former until a later chapter and concentrate on the latter.

The *unit step function* is defined by the following mathematical relationship:

$$u(t) = \begin{cases} 0 & t < 0 \\ 1 & t > 0 \end{cases}$$

In other words, this function, which is dimensionless, is equal to zero for negative values of the argument and equal to 1 for positive values of the argument. It is undefined for a zero argument where the function is discontinuous. A graph of the unit step is shown in Fig. 7.11a. The unit step is dimensionless, and therefore a voltage step of V_o volts or a current step of I_o amperes is written as $V_o u(t)$ and $I_o u(t)$, respectively. Equivalent circuits for a voltage step are shown in Fig. 7.11b and c. Equivalent circuits for a current step are shown in Fig. 7.11d and e. If we use the definition of the unit step, it is easy to generalize this function by replacing the argument t by $t - t_0$. In this case

$$u(t - t_0) = \begin{cases} 0 & t < t_0 \\ 1 & t > t_0 \end{cases}$$

A graph of this function is shown in Fig. 7.11f. Note that $u(t - t_0)$ is equivalent to delaying $u(t)$ by t_0 seconds, so that the abrupt change occurs at time $t = t_0$.

Step functions can be used to construct one or more pulses. For example, the voltage pulse shown in Fig. 7.12a can be formulated by initiating a unit step at $t = 0$ and subtracting one that starts at $t = T$ as shown in Fig. 7.12b. The equation for the pulse is

$$v(t) = A[u(t) - u(t - T)]$$

If the pulse is to start at $t = t_0$ and have width T, the equation would be

$$v(t) = A\{u(t - t_0) - u[t - (t_0 + T)]\}$$

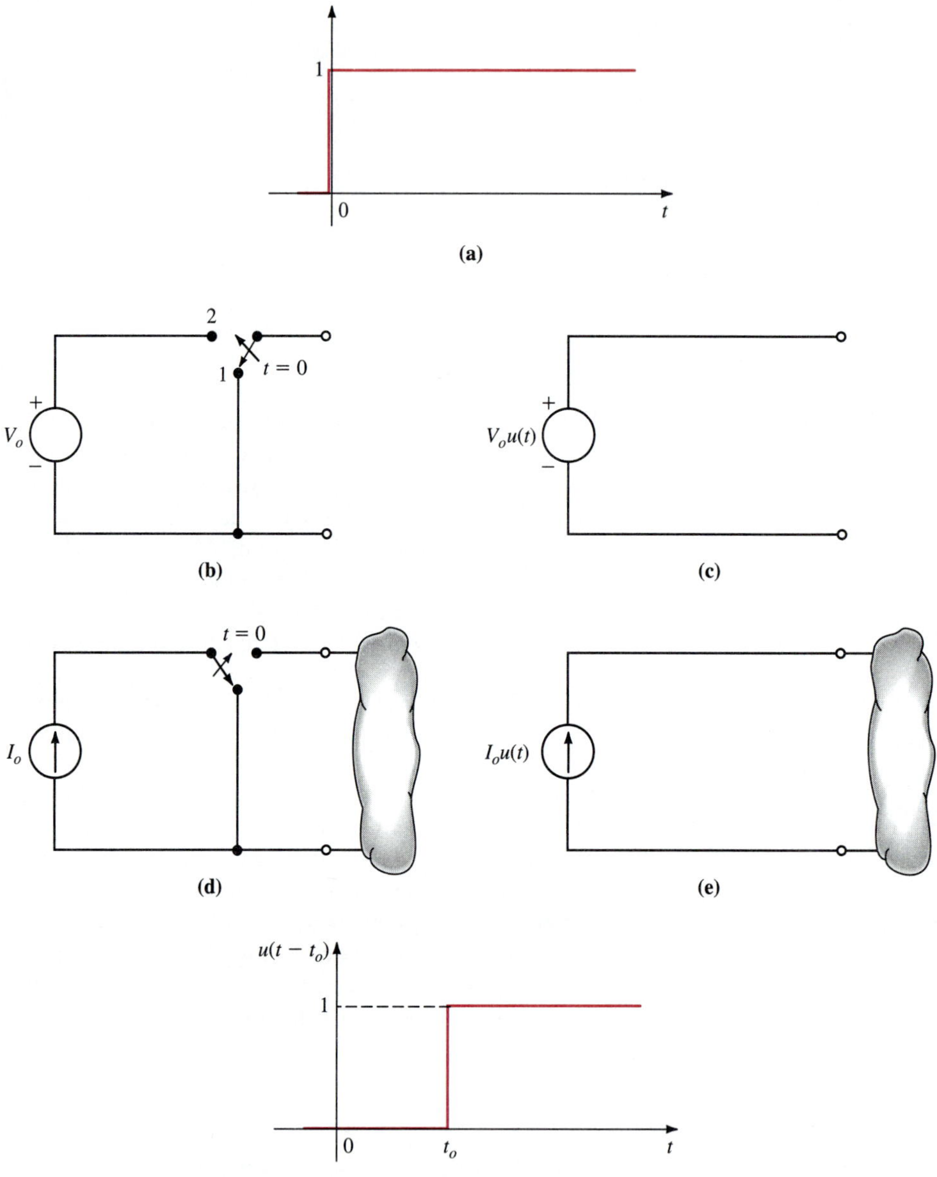

Figure 7.11 Graphs and models of the unit step function.

Using this approach, we can write the equation for a pulse starting at any time and ending at any time. Similarly, using this approach, we could write the equation for a series of pulses, called a *pulse train*, by simply forming a summation of pulses constructed in the manner illustrated above.

The following example will serve to illustrate many of the concepts we have just presented.

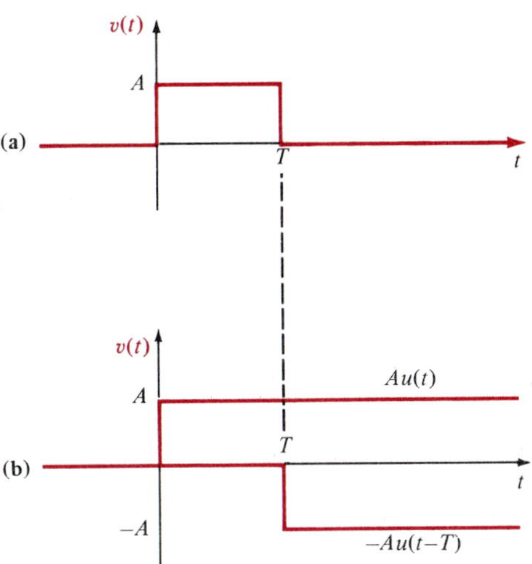

Figure 7.12 Construction of a pulse via two step functions.

EXAMPLE 7.7

Consider the circuit shown in Fig. 7.13a. The input function is the voltage pulse shown in Fig. 7.13b. Since the source is zero for all negative time, the initial conditions for the network are zero [i.e., $v_C(0-) = 0$]. The response $v_o(t)$ for $0 < t < 0.3$ is due to the application of the constant source at $t = 0$ and is not influenced by any source changes that will occur later. At $t = 0.3$ sec the forcing function becomes zero and therefore $v_o(t)$ for $t > 0.3$ sec is the source-free or natural response of the network.

Since the output voltage $v_o(t)$ is a voltage division of the capacitor voltage, and the initial voltage across the capacitor is zero, we know that $v_o(0+) = 0$.

If no changes were made in the source after $t = 0$ the steady-state value of $v_o(t)$ [i.e., $v_o(\infty)$] due to the application of the unit step at $t = 0$, would be

$$v_o(\infty) = \frac{9}{6 + 4 + 8}(8)$$

$$= 4 \text{ V}$$

The Thévenin equivalent resistance is

$$R_{\text{Th}} = \frac{(6)(12)}{6 + 12}$$

$$= 4 \text{ k}\Omega$$

Therefore, the circuit time constant T_c is

$$T_c = R_{\text{Th}}C$$

$$= (4)(10^3)(100)(10^{-6})$$

$$= 0.4 \text{ sec}$$

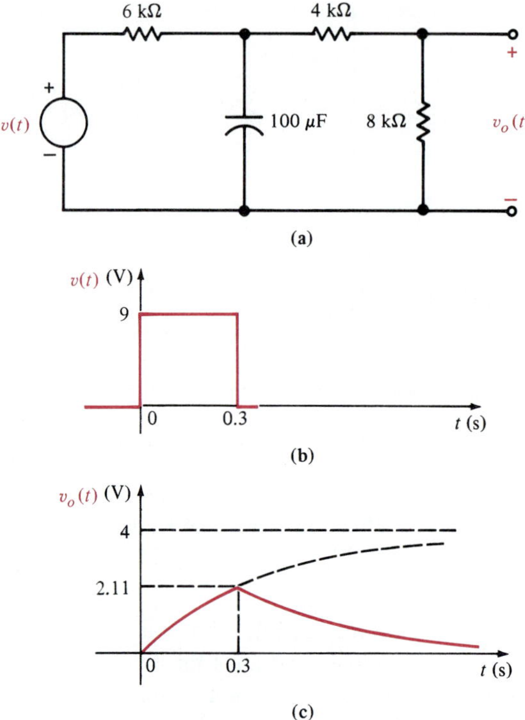

Figure 7.13 Pulse response of a network.

Therefore, the response $v_o(t)$ for the period $0 < t < 0.3$ sec is

$$v_o(t) = 4 - 4e^{-t/0.4} \text{ V} \qquad 0 < t < 0.3 \text{ sec}$$

The capacitor voltage can be calculated by realizing that using voltage division, $v_o(t) = \frac{2}{3}v_C(t)$. Therefore,

$$v_C(t) = \tfrac{3}{2}(4 - 4e^{-t/0.4}) \text{ V}$$

Since the capacitor voltage is continuous,

$$v_C(0.3-) = v_C(0.3+)$$

and therefore,

$$\begin{aligned} v_o(0.3+) &= \tfrac{2}{3}v_C(0.3-) \\ &= 4(1 - e^{-0.3/0.4}) \\ &= 2.11 \text{ V} \end{aligned}$$

Since the source is zero for $t > 0.3$ sec the final value for $v_o(t)$ as $t \to \infty$ is zero. Therefore, the expression for $v_o(t)$ for $t > 0.3$ s is

$$v_o(t) = 2.11e^{-(t-0.3)/0.4} \text{ V} \qquad t > 0.3 \text{ sec}$$

The term $e^{-(t-0.3)/0.4}$ indicates that the exponential decay starts at $t = 0.3$ sec. The complete solution can be written by means of superposition as

$$v_o(t) = 4(1 - e^{-t/0.4})u(t) - 4(1 - e^{-(t-0.3)/0.4})u(t - 0.3) \text{ V}$$

or equivalently, the complete solution is

$$v_o(t) = \begin{cases} 0 & t < 0 \\ 4(1 - e^{-t/0.4}) \text{ V} & 0 < t < 0.3 \text{ sec} \\ 2.11e^{-(t-0.3)/0.4} \text{ V} & 0.3 \text{ sec} < t \end{cases}$$

which in mathematical form is

$$v_o(t) = 4(1 - e^{-t/0.4})[u(t) - u(t - 0.3)] + 2.11e^{-(t-0.3)/0.4}u(t - 0.3) \text{ V}$$

Note that the term $[u(t) - u(t - 0.3)]$ acts like a gating function that captures only the part of the step response that exists in the time interval $0 < t < 0.3$ sec. The output as a function of time is shown in Fig. 7.13c. ■

DRILL EXERCISE

D7.9. The voltage source in the network in Fig. D7.9a is shown in Fig. D7.9b. The initial current in the inductor must be zero. (Why?) Determine the output voltage $v_o(t)$ for $t > 0$.

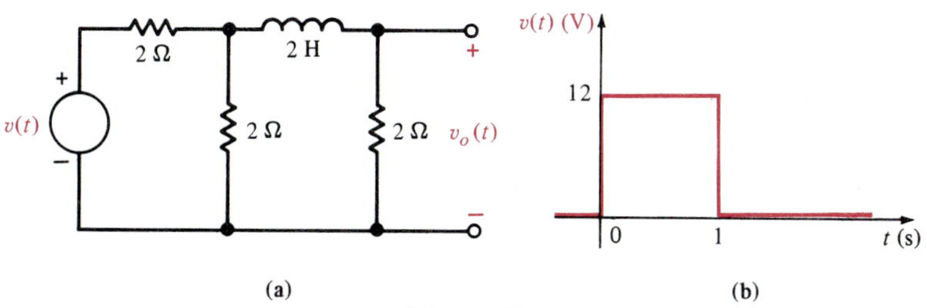

Figure D7.9

Ans: $v_o(t) = 0$ for $t < 0$, $4(1 - e^{-(3/2)t})$ V for $0 \le t \le 1$, and $3.11e^{-(3/2)(t-1)}$ V for $1 < t$.

D7.10. The current source in the network in Fig. D7.10a is defined in Fig. D7.10b. The initial voltage across the capacitor must be zero. (Why?) Determine the current $i_o(t)$ for $t > 0$.

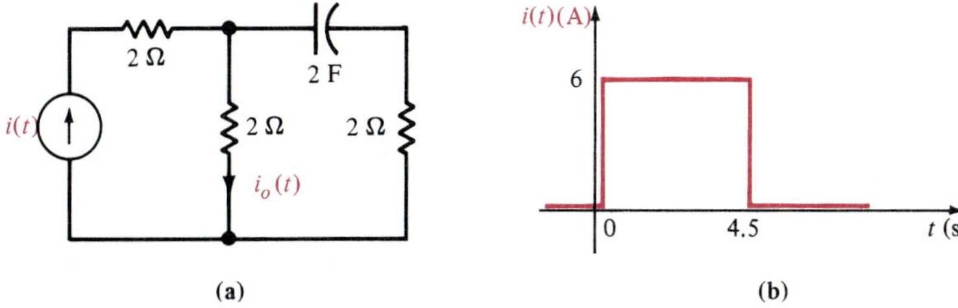

Figure D7.10

Ans: $i_o(t) = 0$ for $t < 0$, $6 - 3e^{-t/8}$ A for $0 \le t \le 4.5$, and $1.29e^{-(1/8)(t-4.5)}$ A for $t > 4.5$.

We have examined the response of a network to only a small number of the many possible input sources. Another very important input forcing function—the sinusoidal source—will be examined in detail in Chapter 9. In Chapter 17 we will demonstrate that a variety of input sources can be simultaneously handled with ease using transform methods.

7.5 *RC* Operational Amplifier Circuits

Two very important *RC* op-amp circuits are the differentiator and the integrator. These circuits are derived from the circuit for an inverting op-amp by replacing the resistors R_1 and R_2, respectively, by a capacitor. Consider, for example, the circuit shown in Fig. 7.14a. The circuit equations are

$$C_1 \frac{d}{dt}(v_1 - v_-) + \frac{v_o - v_-}{R_2} = i_-$$

However, $v_- = 0$ and $i_- = 0$. Therefore,

$$v_o(t) = -R_2 C_1 \frac{dv_1(t)}{dt} \tag{7.17}$$

and thus the output of the op-amp circuit is proportional to the derivative of the input.

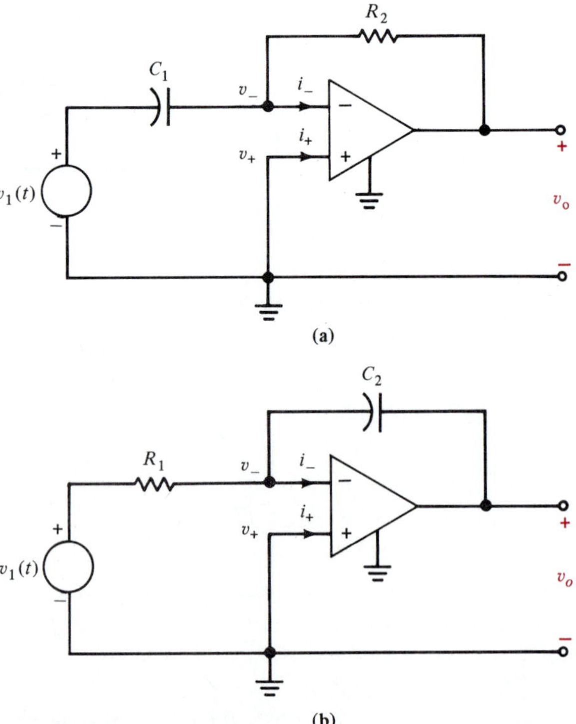

Figure 7.14 Differentiator and integrator operational amplifier circuits.

The circuit equations for the op-amp configuration in Fig. 7.14b are

$$\frac{v_1 - v_-}{R_1} + C_2 \frac{d}{dt}(v_o - v_-) = i_-$$

but since $v_- = 0$ and $i_- = 0$, the equation reduces to

$$\frac{v_1}{R_1} = -C_2 \frac{dv_o}{dt}$$

or

$$v_o(t) = \frac{-1}{R_1 C_2} \int_{-\infty}^{t} v_1(x)\, dx$$

$$= \frac{-1}{R_1 C_2} \int_{0}^{t} v_1(x)\, dx + v_o(0) \qquad (7.18)$$

If the capacitor is initially discharged, then $v_o(0) = 0$ and hence

$$v_o(t) = \frac{-1}{R_1 C_2} \int_{0}^{t} v_1(x)\, dx \qquad (7.19)$$

and thus the output voltage of the op-amp circuit is proportional to the integral of the input voltage.

EXAMPLE 7.8

The waveform in Fig. 7.15a is applied at the input of the differentiator circuit shown in Fig. 7.14a. If $R_2 = 1\ \text{k}\Omega$ and $C_1 = 2\ \mu\text{F}$, determine the waveform at the output of the op-amp.

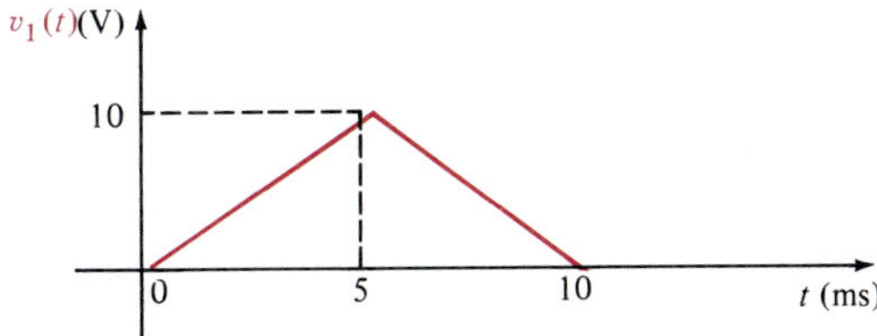

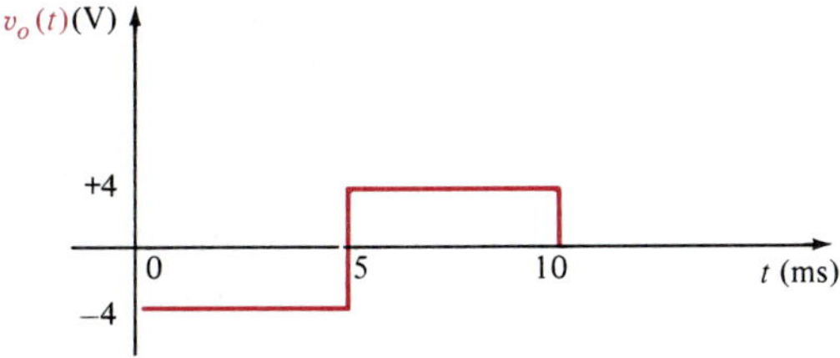

Figure 7.15 Input and output waveforms for a differentiator circuit.

Using Eq. (7.17), we find that the op-amp output is

$$v_o(t) = -R_2C_1 \frac{dv_1(t)}{dt}$$

$$= -(2)10^{-3} \frac{dv_1(t)}{dt}$$

$dv_1(t)/dt = (2)10^3$ for $0 \leq t < 5$ ms, and therefore,

$$v_o(t) = -4 \text{ V} \qquad 0 \leq t < 5 \text{ ms}$$

$dv_1(t)/dt = -(2)10^3$ for $5 \leq t < 10$ ms, and therefore,

$$v_o(t) = 4 \text{ V} \qquad 5 \leq t < 10 \text{ ms}$$

Hence the output waveform of the differentiator is shown in Fig. 7.15b. ■

EXAMPLE 7.9

If the integrator shown in Fig. 7.14b has the parameters $R_1 = 5$ kΩ and $C_2 = 0.2$ μF, determine the waveform at the op-amp output if the input waveform is given as in Fig. 7.16a and the capacitor is initially discharged.

The integrator output is given by the expression

$$v_o(t) = \frac{-1}{R_1C_2} \int_0^t v_1(x)\, dx$$

which with the given circuit parameters is

$$v_o(t) = -10^3 \int_0^t v_1(x)\, dx$$

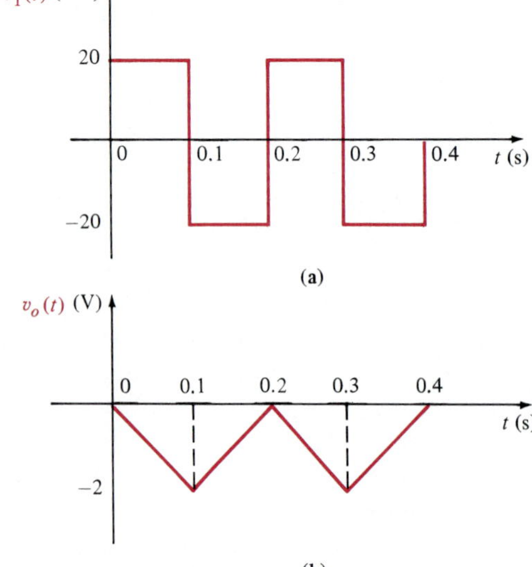

Figure 7.16 Input and output waveforms for an integrator circuit.

In the interval $0 \le t < 0.1$ sec, $v_1(t) = 20$ mV. Hence

$$v_o(t) = -10^3(20)10^{-3}t \qquad 0 \le t < 0.1 \text{ sec}$$
$$= -20t$$

At $t = 0.1$ sec, $v_o(t) = -2$ V. In the interval from 0.1 to 0.2 s, the integrator produces a positive slope output of $20t$ from $v_o(0.1) = -2$ V to $v_o(0.2) = 0$ V. This waveform from $t = 0$ to $t = 0.2$ sec is repeated in the interval $t = 0.2$ to $t = 0.4$ sec, and therefore, the output waveform is shown in Fig. 7.16b. ■

DRILL EXERCISE

D7.11. The waveform in Fig. D7.11 is applied to the input terminals of the op-amp differentiator circuit. Determine the differentiator output waveform if the op-amp circuit parameters are $C_1 = 2$ F and $R_2 = 2\ \Omega$.

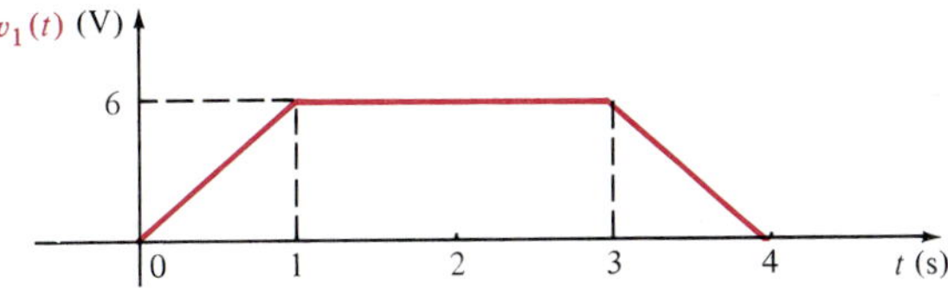

Figure D7.11

Ans:

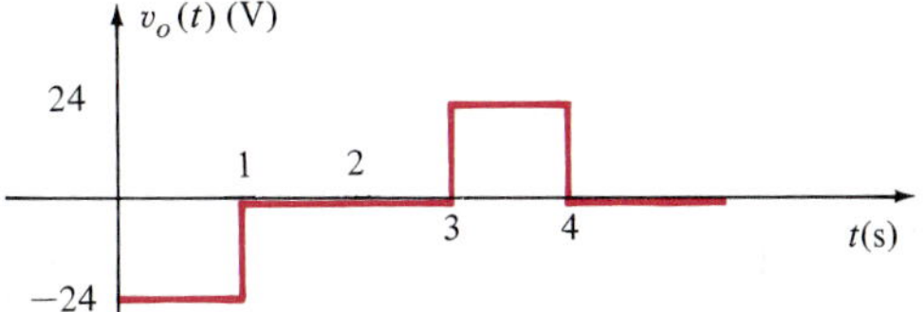

7.6 Transient Circuit Analysis Using PSPICE

As we begin our discussion of this topic, it is important to point out that all the material in Chapter 4 applies here also. The circuit is described in exactly the same manner and, in fact, a dc analysis is performed automatically prior to transient analysis to determine the circuit's initial conditions. The independent variable in transient analysis is always time, and the circuit variables are tabulated and/or plotted as functions of time. It is also important to note that the circuit under analysis cannot contain a loop of voltage sources and/or inductors and cannot contain a cut set (see Chapter 3) of current sources and/or capacitors.

We will now introduce some new data, solution control, and output specification statements that will be useful in transient analysis. The examples will indicate the ease with which we can perform analyses that would be difficult to treat mathematically.

Branch Statements for Inductors and Capacitors

The general statement form for these elements is

```
BXXXXXXX N1 N2 VALUE
```

B is the letter L or C, depending on the type of element. XXXXXXX denotes an arbitrary alphanumeric string that uniquely identifies the particular element. N1 and N2 are the circuit nodes to which the element is connected. VALUE is the value of the element. Capacitance is in farads, and inductance is in henrys.

Branch Statements for Time-Varying Sources

Any independent source can be assigned a time-dependent value for transient analysis. If a source is assigned a time-dependent value, the time-zero value will be used for any dc analysis that is requested. The time-varying sources may be of four different types: pulse, exponential, sinusoidal, or piecewise linear. The type of source is specified on the data statement that defines the source.

The data statement for *pulse sources* is of the form

```
VXXXXXXX N+ N- PULSE (V1 V2 TD TR TF PW PER)
```

V1 is the initial value of the source in volts or amperes, and V2 is the value of the source during the pulse. TD is the delay time between time zero and the start of the pulse in seconds. TR, TF, and PW are the rise time, fall time, and width of the pulse in seconds. PER is the period in seconds for a periodic pulse train. Intermediate points on the pulse are determined by linear interpolation. Therefore, Table 7.1 describes the single pulse shown in Fig. 7.17.

Table 7.1

Time	Value
0	V1
TD	V1
TD + TR	V2
TD + TR + PW	V2
TD + TR + PW + TF	V1

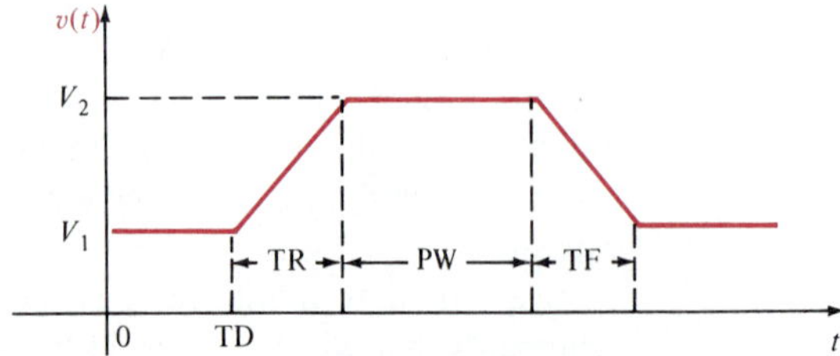

Figure 7.17 General form of a pulse.

The default values and units for the parameters of the pulse source are summarized in Table 7.2.

Table 7.2

Parameter	Default Value	Units
V1—initial voltage		volts or amperes
V2—pulsed value		volts or amperes
TD—delay time	0.0	seconds
TR—rise time	TSTEP	seconds
TF—fall time	TSTEP	seconds
PW—pulse width	TSTOP	seconds
PER—period	TSTOP	seconds

Exponential sources are described using the format

```
VXXXXXXX N+ N- EXP(V1 V2 TD1 TAU1 TD2 TAU2)
```

V1 is the initial value in volts or amperes. V2 is the peak value in volts or amperes. TD1 and TAU1 are the rise-delay-time and rise-time constants, respectively. TD2 and TAU2 are the fall-delay-time and fall-time constants, respectively. All times are expressed in seconds. The waveform shown in Fig. 7.18 is described by the expressions

```
Time 0 to TD1:      V1
Time TD1 to TD2:    V1 + (V2 - V1) *
                        (1 - EXP( - (TIME - TD1)/TAU1))
Time TD2 to TSTOP:  V1 + (V2 - V1) *
                        (1 - EXP( - (TIME - TD1)/TAU1))
                        + (V1 - V2) *
                        (1 - EXP( - (TIME - TD2)/TAU2))
```

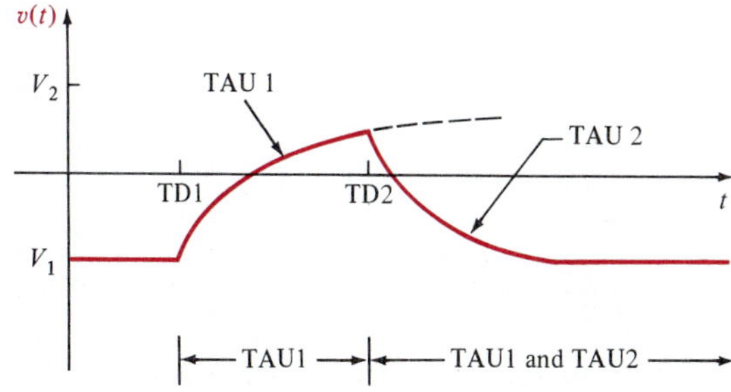

Figure 7.18 General form of an exponential source.

The default values and units for the parameters of the exponential source are summarized in Table 7.3.

Table 7.3

Parameter	Default Value	Units
V1—initial voltage		volts or amperes
V2—pulsed value		volts or amperes
TD1—rise delay time	0.0	seconds
TAU1—rise time constant	TSTEP	seconds
TD2—rise delay time	TD1 + TSTEP	seconds
TAU2—fall time constant	TSTEP	seconds

The data statement for *sinusoidal sources* of the form shown in Fig. 7.19 is

```
VXXXXXXX N+ N- SIN(VO VA FREQ TD THETA)
```

The shape of this waveform is described by the equation

Time 0 to TD: `VO`
Time TD to TSTOP: `VO + VA*EXP(-(TIME-TD)*THETA)* SINE(2PI*FREQ*(TIME + TD))`

VO is the offset in volts or amperes. VA is the amplitude in volts or amperes. FREQ is the frequency in hertz. TD is the delay in seconds, and THETA is the damping factor in $(\text{seconds})^{-1}$.

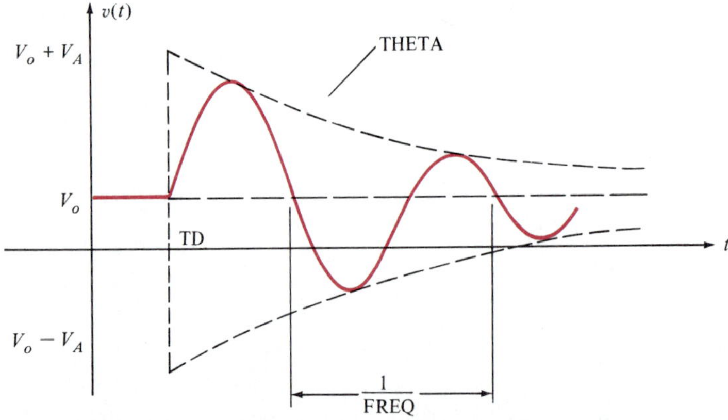

Figure 7.19 General form of a sinusoidal source.

The default values and units for parameters of the sinusoidal source are summarized in Table 7.4.

Table 7.4

Parameter	Default Value	Units
VO—offset		volts or amperes
VA—amplitude		volts or amperes
FREQ—frequency	1/TSTOP	hertz
TD—delay	0.0	seconds
THETA—damping factor	0.0	1 per second

Piecewise linear sources are described using the format

```
VXXXXXXX N+ N- PWL (T1 V1 T2 V2 T3 V3 . . .)
```

As shown in Fig. 7.20, each pair of values (TI, VI) specifies that the value of the source is VI volts or amperes at time TI. The value of the source at intermediate values of time is determined by linear interpolation. Values of TI must always be increasing in time.

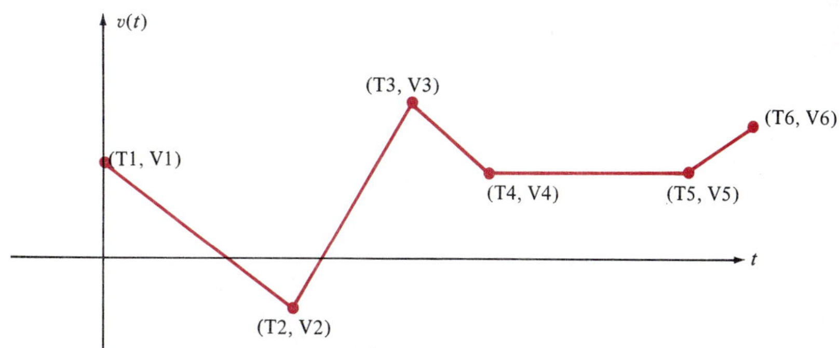

Figure 7.20 General form of a piecewise linear source.

Solution Control Statements

Transient analysis in PSPICE is invoked using a .TRAN statement of the form

```
.TRAN TSTEP TSTOP TSTART TMAX
```

TSTEP is the printing or plotting increment for line printer output. TSTART and TSTOP are the start and stop times for the analysis. If TSTART is omitted, it is assumed to be zero. The transient analysis always begins at time zero. In the interval [0, TSTART], the circuit is analyzed, but no outputs are stored. In the interval [TSTART, TSTOP], analysis continues and the outputs are stored. TMAX is the maximum time step that PSPICE will use and defaults to a value of (TSTOP-TSTART)/50 or TSTEP, whichever is smaller.

If initial conditions for the capacitive and inductive elements are known and are to be employed in the PSPICE analysis, the .TRAN statement includes the keyword UIC (use initial conditions) as shown below.

```
.TRAN TSTEP TSTOP TSTART TMAX UIC
```

If this keyword is specified, PSPICE will use the initial values that can be added to the data statements in the form

```
CXXXXXXX N+ N- VALUE IC= INCOND
LYYYYYYY N+ N- VALUE IC= INCOND
```

For the capacitor, the initial condition is the initial (time-zero) value of the capacitor voltage in volts. For the inductor, the initial condition is the initial (time-zero) value of inductor current in amperes that flows from N + , through the inductor, to N − . When employing the UIC feature, PSPICE uses the initial values on the data statements as the initial transient condition and proceeds with the analysis.

If UIC is not included in the .TRAN statement, PSPICE will perform a $t = 0$ dc analysis to determine the initial conditions.

Output Specification Statements

The .PRINT and .PROBE statements employed in the dc analysis are useful here also. In a transient analysis they take the form

```
.PRINT TRAN OV1 . . . OV8
.PROBE
```

EXAMPLE 7.10

We wish to plot $v_o(t)$ and $i_o(t)$ for $t > 0$ in the network in Fig. 7.21a.

The initial conditions are determined from the circuit in Fig. 7.21b. The PSPICE program used to compute the initial capacitor voltage is

```
CIRCUIT USED TO DETERMINE INITIAL CONDITIONS
*DC VOLTAGE SOURCE
V1 1 3 DC 12
*DC CURRENT SOURCE
I1 0 2 DC 2
*RESISTOR VALUES
R1 1 2 2
R2 2 3 1
R3 3 0 1
*CAPACITOR VALUE
C1 3 0 1
.END
```

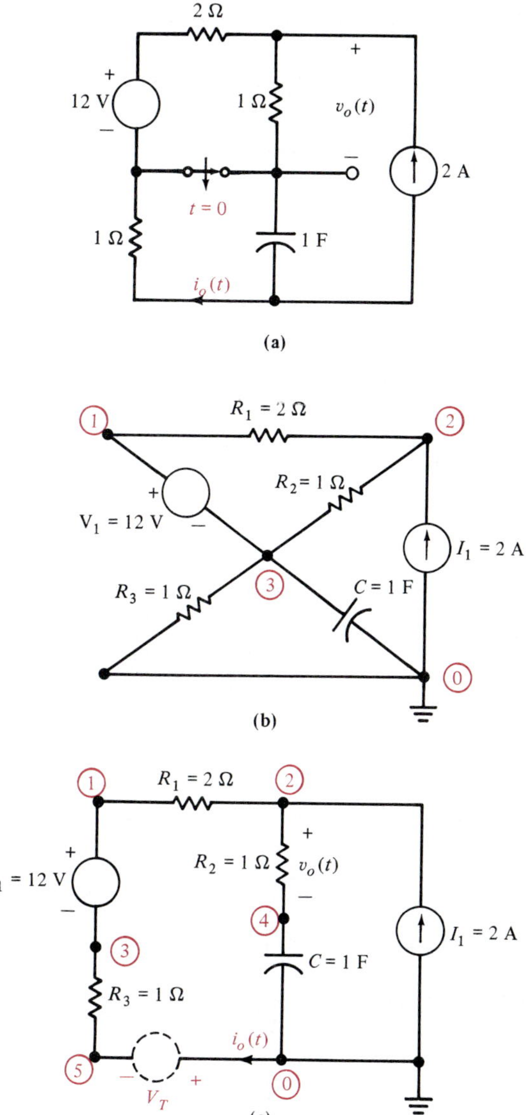

Figure 7.21 Networks used in Example 7.10.

If we select ‘Browse Output’ from the ‘Files’ submenu, we can review the program results, which illustrate that the initial capacitor voltage is 2 V.

Now we know that the initial capacitor voltage is 2 V. The network for $t > 0$ is shown in Fig. 7.21c. The PSPICE program that will plot the output over a 20-sec interval using a step size of 200 ms is listed below.

```
CIRCUIT USED TO COMPUTE OUTPUT RESPONSE
*CIRCUIT CONDITIONS AT TIME T=0+
*DC VOLTAGE SOURCE
V1 1 3 DC 12
*TEST SOURCE
VT 0 5 DC 0
*DC CURRENT SOURCE
I1 0 2 DC 2
*RESISTOR VALUES
R1 1 2 2
R2 2 4 1
R3 3 5 1
*CAPACITOR VALUE
C1 4 0 1 IC=2.0
.TRAN 200MS 20S UIC
.PROBE
.END
```

Upon completion of the transient analysis, the screen will display a blank graph and a menu line below the graph. Press Enter. This selects the 'Add_trace' option from the menu. Type 'I(VT) V(2,4)' and press the Enter key. A plot of the selected current and voltage appears on the screen. A hard copy may be obtained by using the arrow keys to highlight 'Hard_copy' from the menu. Press the Enter key once the hard-copy option is highlighted. The user may then exit from the probe program by selecting the 'Exit' option from the menu line in the same manner.

The plots for $v_o(t)$ and $i_o(t)$ are shown in Fig. 7.22a and b, respectively. ■

EXAMPLE 7.11

Given the network in Fig. 7.23a, plot $v_o(t)$ and $i_o(t)$ over the interval $0 \leq t \leq 10$ sec using 100-ms steps.

The network in Fig. 7.23b is used to determine the initial inductor current. The PSPICE program that will yield the initial condition is

```
COMPUTE INITIAL CONDITION
*DC VOLTAGE SOURCE
V1 1 0 DC 12
*TEST SOURCE TO DETERMINE CURRENT THROUGH L
VT 3 0 DC 0
*RESISTOR VALUES
R1 1 2 2
R2 2 0 2
R3 2 0 2
*INDUCTOR VALUE
L1 2 3 2
.END
```

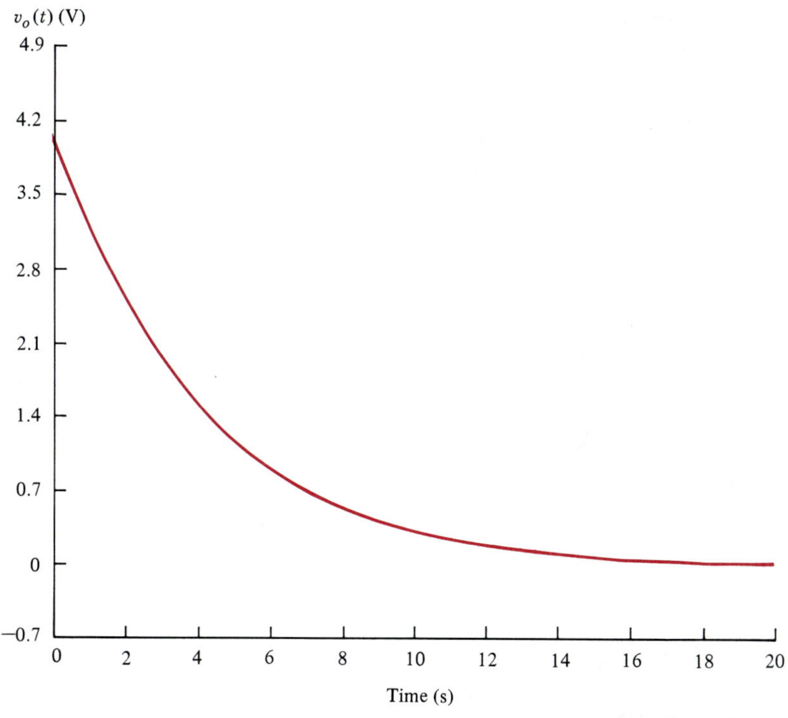

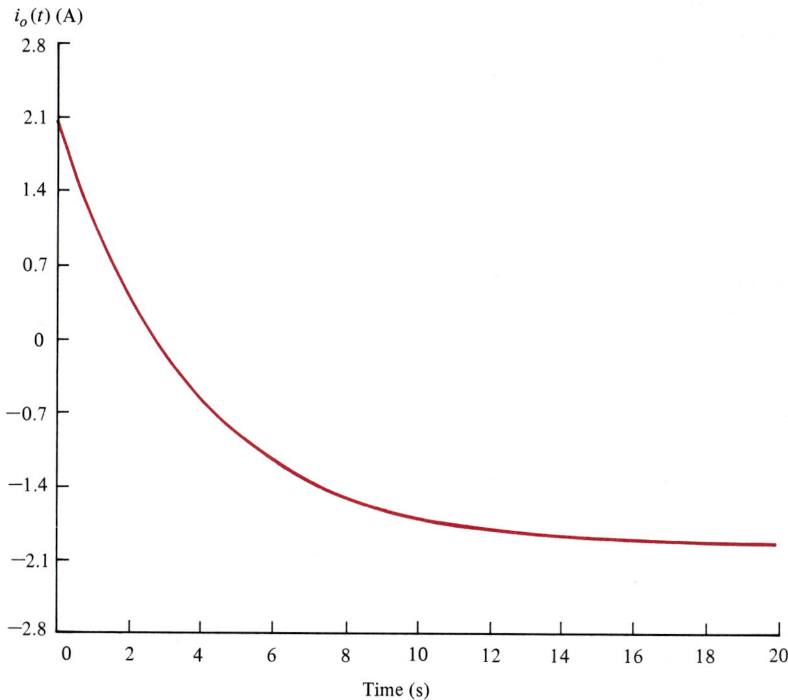

Figure 7.22 Output plots for Example 7.10.

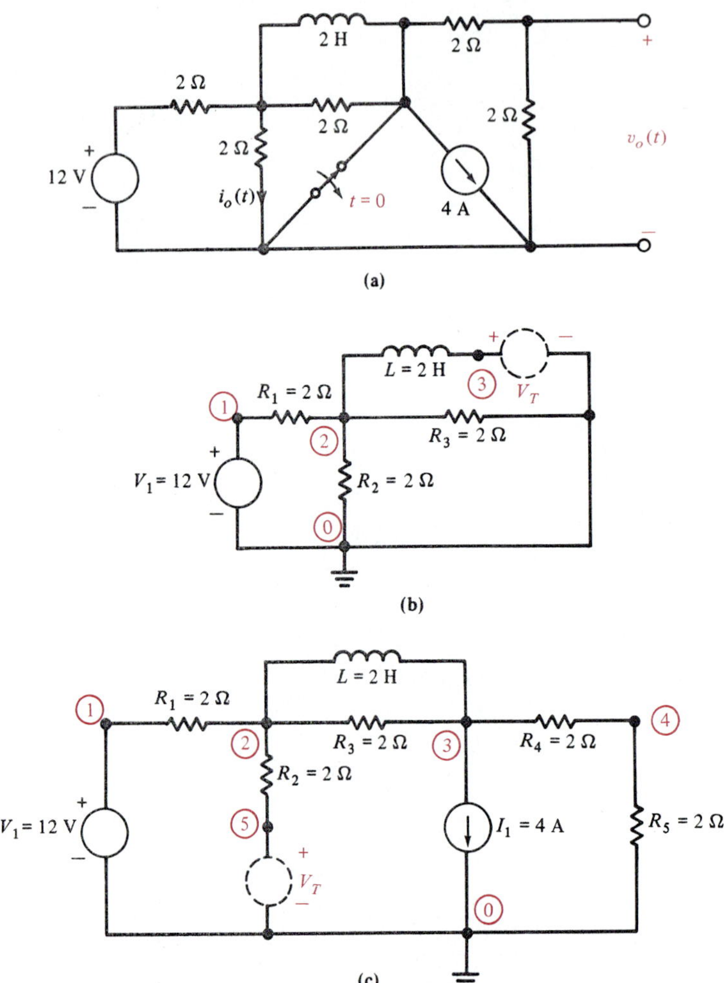

Figure 7.23 Networks used in Example 7.11.

The program output includes the following:

```
VOLTAGE SOURCE CURRENTS
NAME         CURRENT
V1          -6.000D+00
VT           6.000D+00
```

The network for $t > 0$ is shown in Fig. 7.23c. The PSPICE program that will plot the outputs given an initial inductor current of 6 A is listed below.

```
OUTPUT RESPONSE PROGRAM
*CIRCUIT CONDITIONS AT TIME T=0+
*DC VOLTAGE SOURCE
V1 1 0 DC 12
```

```
*DC CURRENT SOURCE
I1 3 0 DC 4
*TEST SOURCE TO DETERMINE SPECIFIED VOLTAGE
VT 5 0 DC 0
*RESISTOR VALUES
R1 1 2 2
R2 2 5 2
R3 2 3 2
R4 3 4 2
R5 4 0 2
*INDUCTOR VALUE
L1 2 3 2 IC=6.0
.TRAN 100MS 10S UIC
.PROBE
.END
```

The plots for $v_o(t)$ and $i_o(t)$ are shown in Fig. 7.24a and b, respectively. ■

DRILL EXERCISE

D7.12. Using PSPICE, plot $v_o(t)$ for $t > 0$ in the network in Fig. D7.6.

Ans:

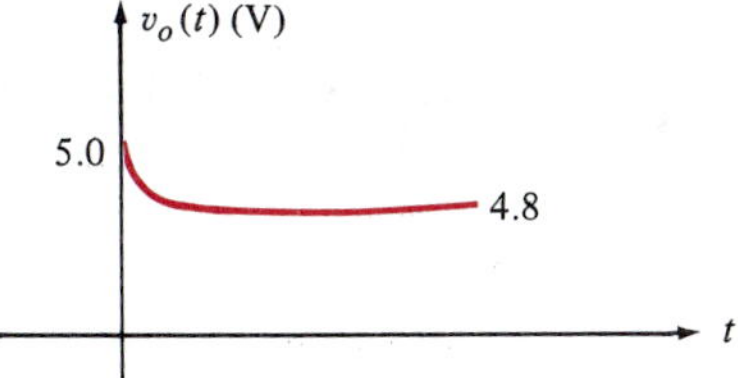

D7.13. Given the network in Fig. D7.7, plot $v_o(t)$ for $t > 0$ using PSPICE.

Ans:

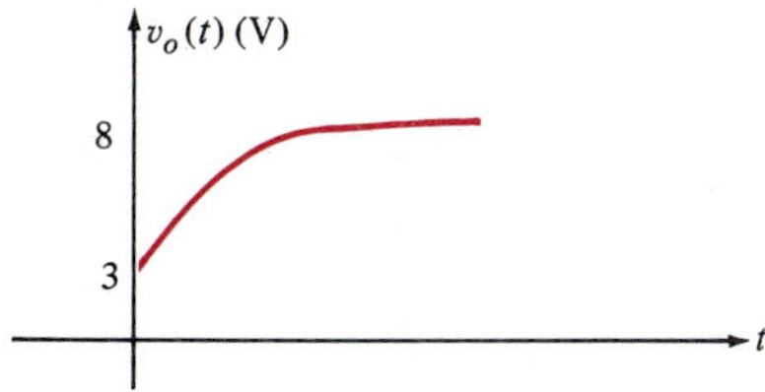

EXAMPLE 7.12

Consider the network in Fig. 7.25a. The input signal for the network is shown in Fig. 7.25b and is given by the equation

$$v_S(t) = 10(1 - e^{-t/0.5}) \text{ V} \qquad 0 \leq t \leq 2 \text{ sec}$$

$$v_S(t) = 10(1 - e^{-t/0.5}) + 10e^{-(t-2)/2} \text{ V} \qquad 2 \text{ sec} \leq t$$

We wish to plot the output signal $v_o(t)$ over a 10-sec interval using a 50-ms step size.

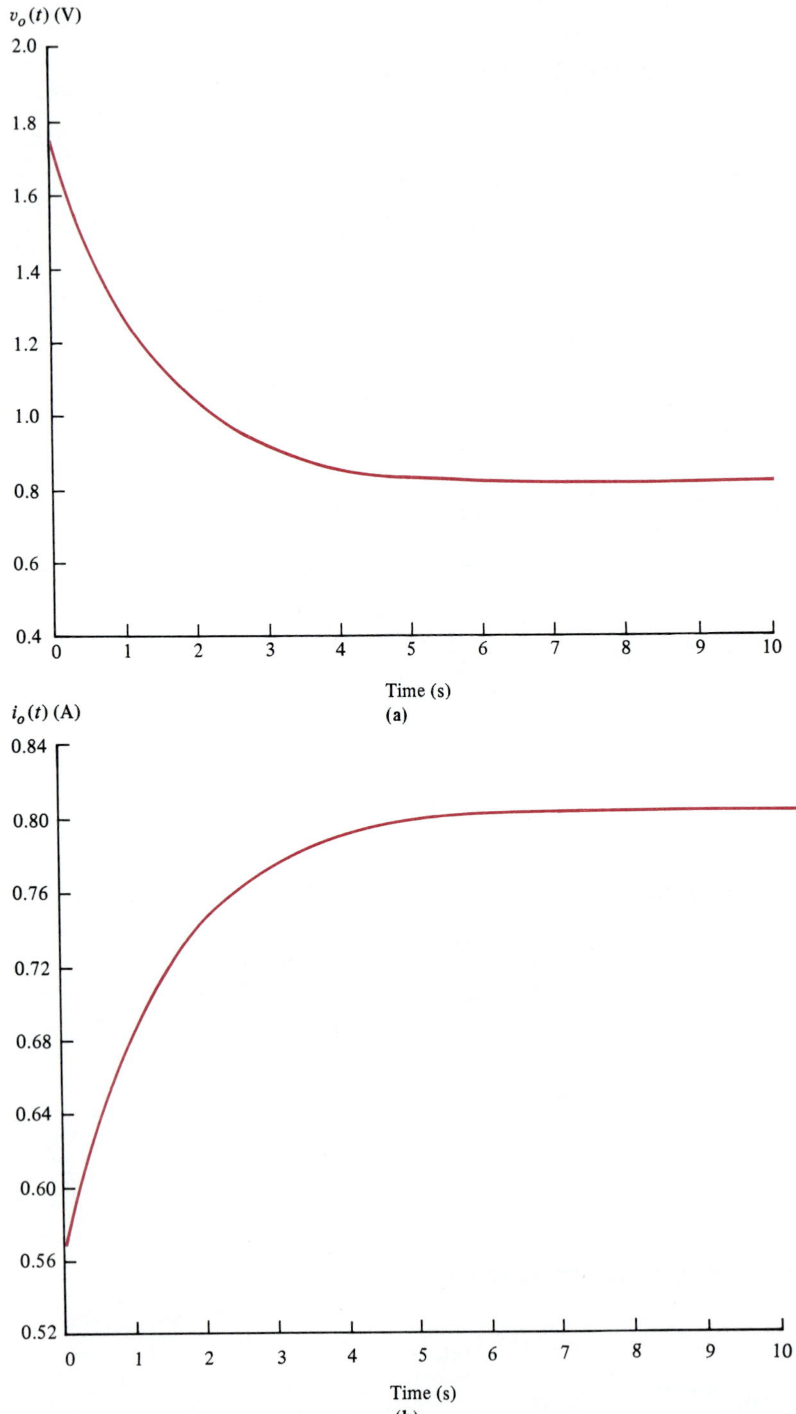

Figure 7.24 Output plots for Example 7.11.

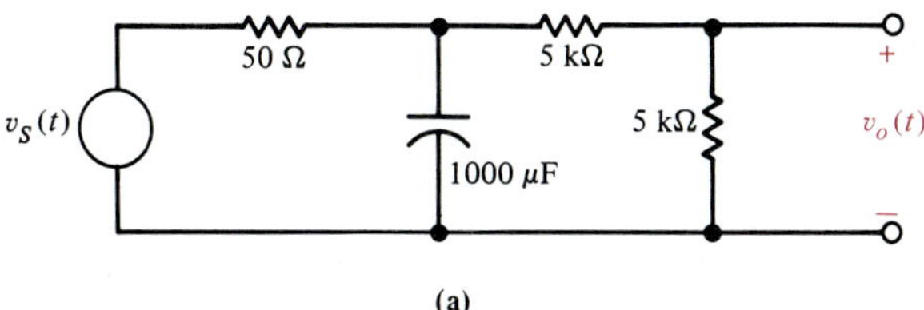

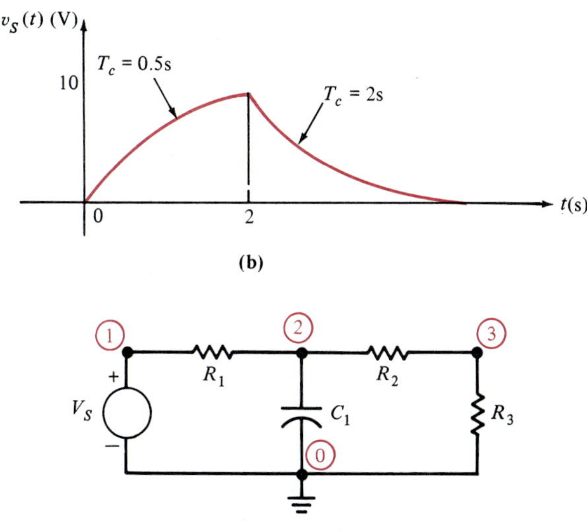

Figure 7.25 Figures used in Example 7.12.

The network is drawn for a PSPICE analysis in Fig. 7.25c. The program for this network is

```
EXAMPLE 7.12
VS 1 0 EXP(0 10 0 .5 2 2)
R1 1 2 50
R2 2 3 5K
R3 3 0 5K
C1 2 0 1000U
* ANALYSIS
.TF V(3) VS
.TRAN .05 10
.PROBE
.END
```

A plot of both the input and output voltages is shown in Fig. 7.26. ■

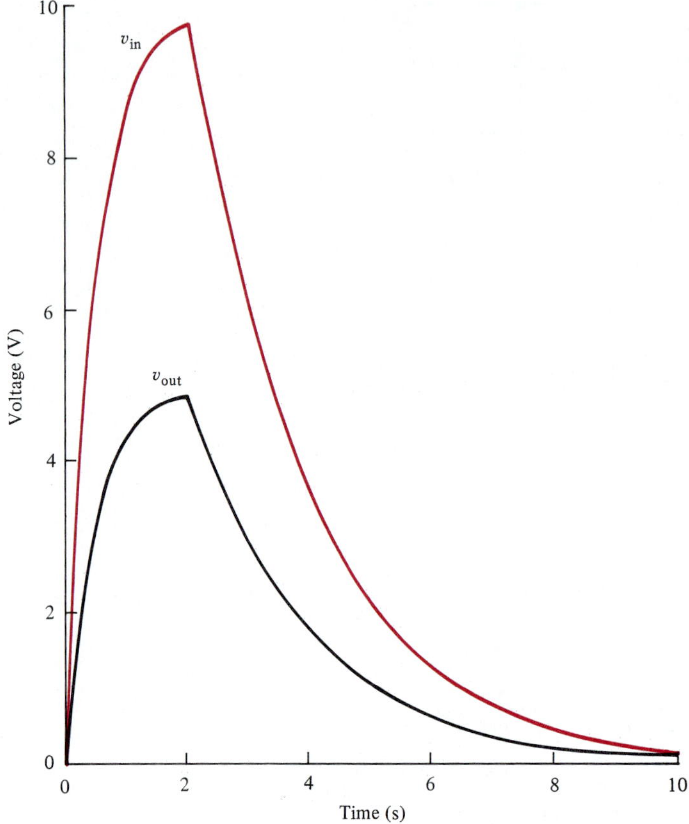

Figure 7.26 Input and output plots for Example 7.12.

EXAMPLE 7.13

Given the network in Fig. 7.25a, let us plot the output response over a 2-sec interval with a step size of 10 ms if the input is given by the expression

$$v_S(t) = 10e^{-2t} \sin [2\pi(4)t] \text{ V}$$

The program for the network in Fig. 7.25c with the new input signal is

```
EXAMPLE 7.13
VS 1 0 SIN(0 10 4 0 2 0)
R1 1 2 50
R2 2 3 5K
R3 3 0 5K
C1 2 0 1000U
```

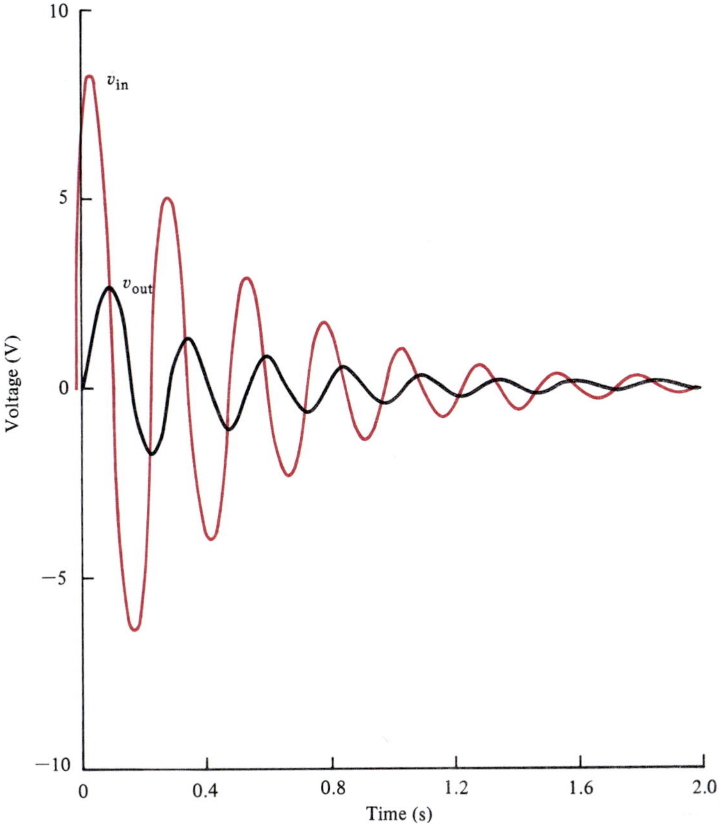

Figure 7.27 Input and output plots for Example 7.13.

```
* ANALYSIS
.TF V(3) VS
.TRAN .01 2
.PROBE
.END
```

Plots of both the input and output are shown in Fig. 7.27. ■

EXAMPLE 7.14

For the network in Fig. 7.25a, let us plot the output response over a 2-sec interval with a step size of 10 ms if the input to the network is as shown in Fig. 7.28.

The program for the network in Fig. 7.25c with the input signal shown in Fig. 7.28 is

```
EXAMPLE 7.14
VS 1 0 PULSE(0 10 0 1 0 .1US 1)
R1 1 2 50
R2 2 3 5K
R3 3 0 5K
C1 2 0 1000U
```

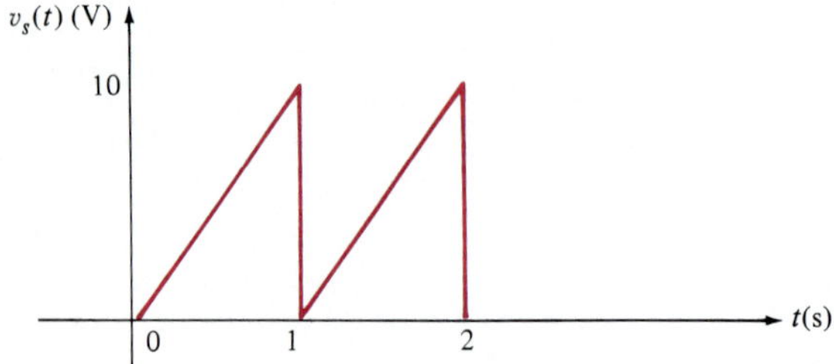

Figure 7.28 Sawtooth input waveform.

```
* ANALYSIS
.TF V(3) VS
.TRAN .01 2
.PROBE
.END
```

Plots of both the input and output are shown in Fig. 7.29. ■

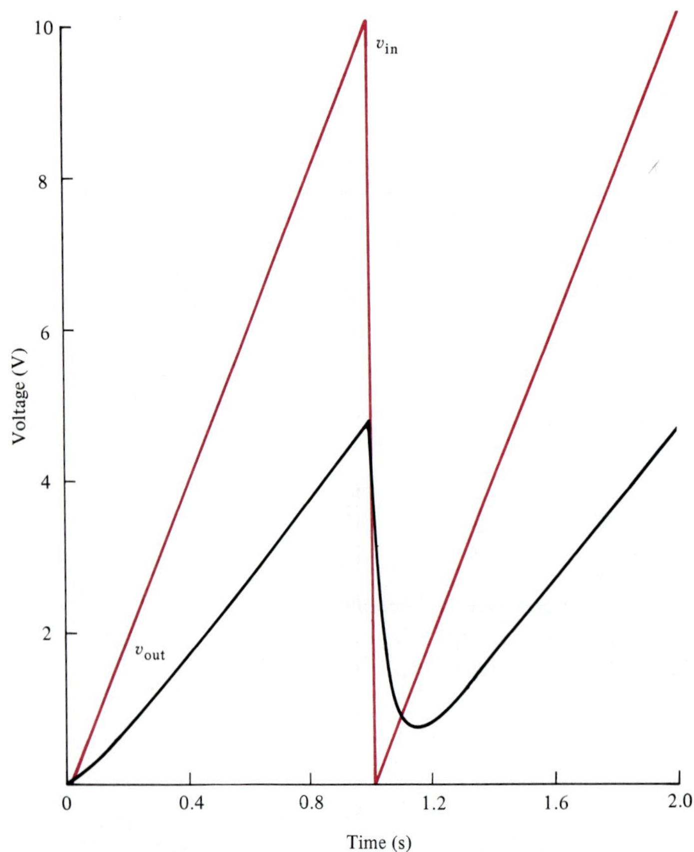

Figure 7.29 Input and output plots for Example 7.14.

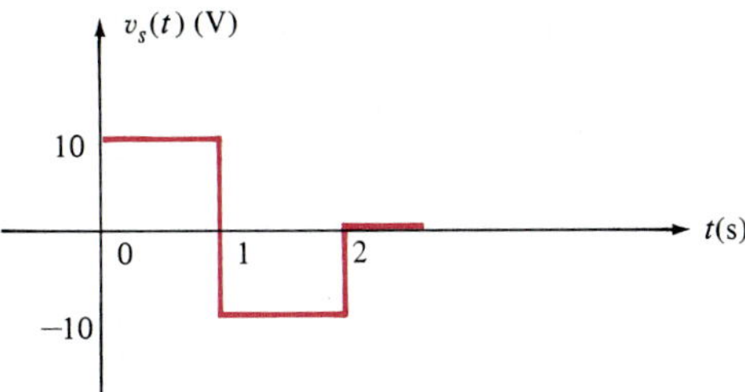

Figure 7.30 Pulse input waveform.

EXAMPLE 7.15

For the network in Fig. 7.25a, let us plot the output response over a 2-sec interval with a step size of 10 ms if the input to the network is that shown in Fig. 7.30.

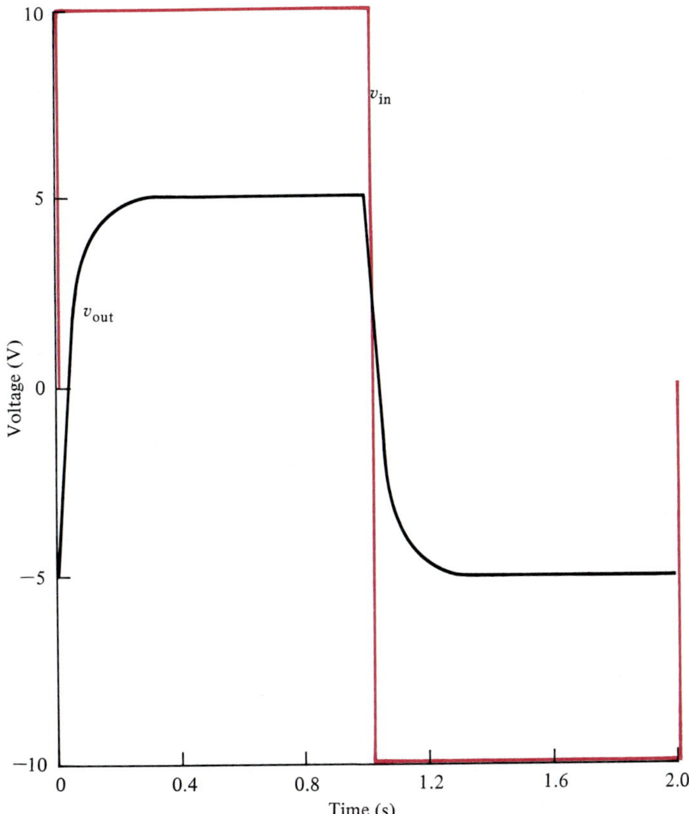

Figure 7.31 Input and output plots for Example 7.15.

The program for the network in Fig. 7.25c with the input waveform shown in Fig. 7.30 is

```
EXAMPLE 7.15
VS 1 0 PULSE (-10 10 0 0 0 1 2)
R1 1 2 50
R2 2 3 5K
R3 3 0 5K
C1 2 0 1000U
* ANALYSIS
.TF V(3) VS
.TRAN .01 2
.PROBE
.END
```

Plots of both the input and output signals are shown in Fig. 7.31. ■

DRILL EXERCISE

D7.14. Given the network in Fig. D7.14a, plot the output voltage over a 10-sec interval using a 50-ms step size if the input voltage is shown in Fig. D7.14b.

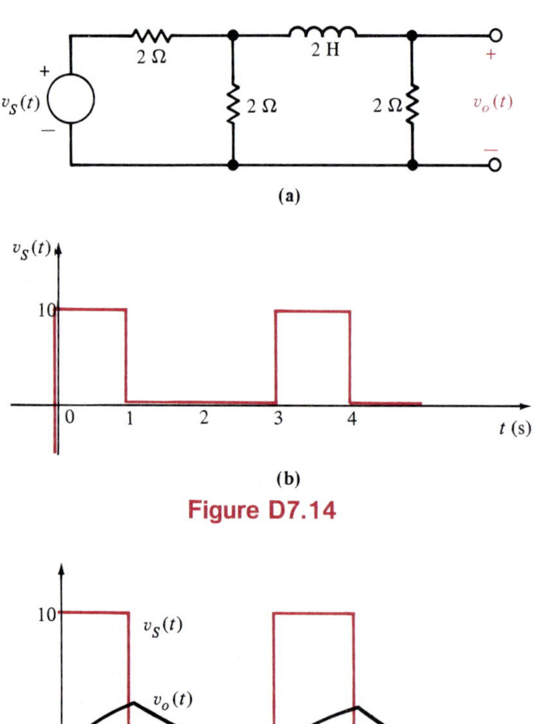

Figure D7.14

Ans:

10 $v_S(t)$ $v_o(t)$ 1 2 3 4 t (s)

D7.15. For the network in Fig. D7.14a, plot the output voltage over a 10-sec interval using a 50-ms step size if the input voltage is

$$\begin{aligned} v_S(t) &= -5 \text{ V} && 0 \le t \le 1 \text{ sec} \\ &= -5 + 15(1 - e^{-(t-1)/2}) \text{ V} && 1 \le t \le 3 \text{ sec} \\ &= -5 + 15(1 - e^{-(t-1)/2}) - 15e^{-(t-3)/1} \text{ V} && 3 \text{ sec} \le t \end{aligned}$$

Ans:

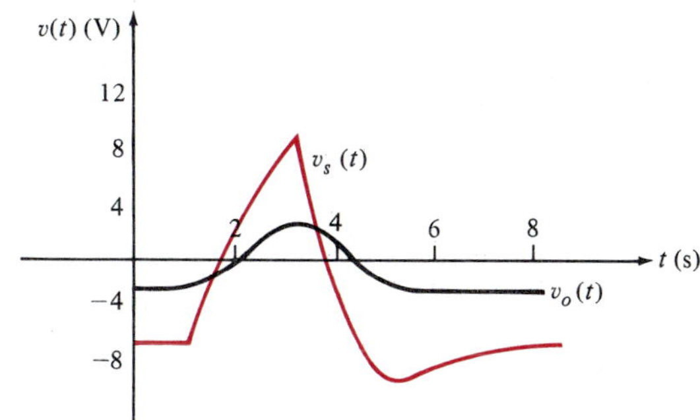

D7.16. For the network in Fig. D7.14a, plot the output voltage over a 4-sec interval using a 20-ms step size if the input voltage is

$$v_S(t) = 10e^{-t} \sin 2\pi t \text{ V}$$

Ans:

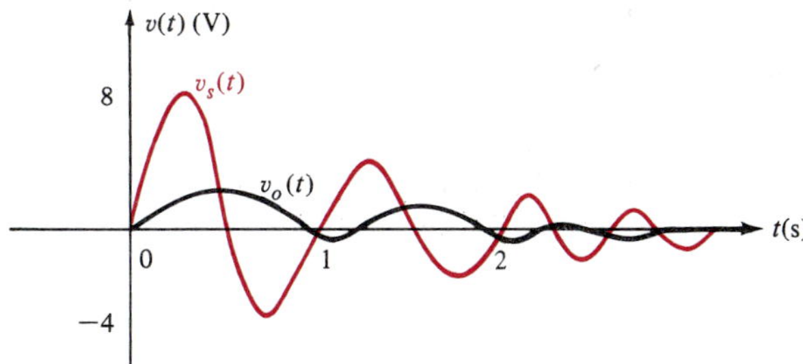

D7.17. For the network in Fig. D7.14a, plot the output voltage over a 10-sec interval using a 50-ms step size if the input is as shown in Fig. D7.17.

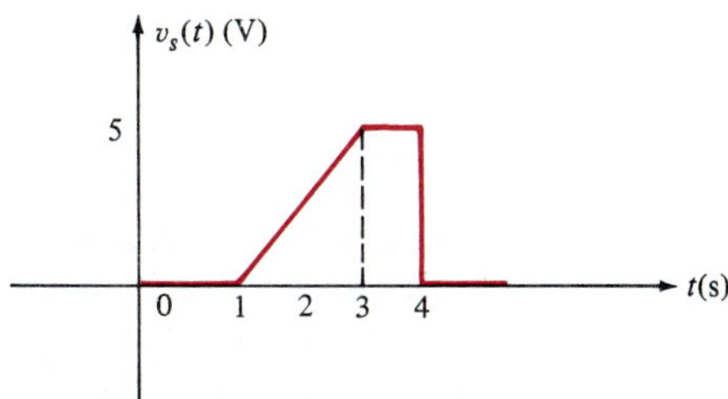

Figure D7.17

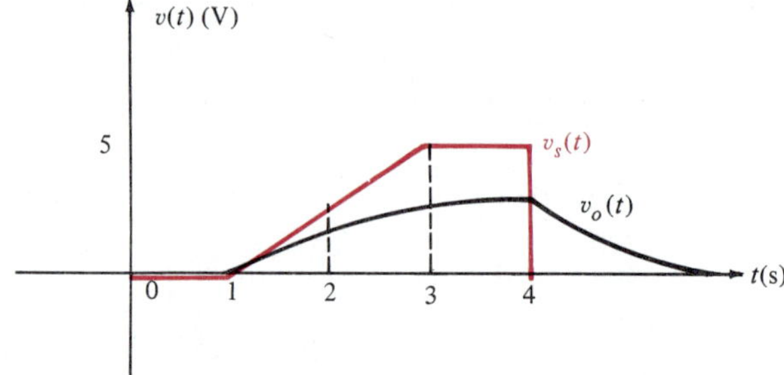

7.7 Summary

We have shown that the solution of *RC* and *RL* circuits which contain only a single energy storage element (i.e., *C* or *L*) involve the solution of a first-order differential equation. A general solution method was developed and a step-by-step procedure outlined for determining a solution. Both the natural and forced response of these networks were examined. It was found that the circuit response in the form of a decaying exponential was controlled by the circuit time constant, and that the response essentially decays to its final value in five time constants. The time constants for *RC* and *RL* circuits are $R_{Th}C$ and L/R_{Th}, respectively, where R_{Th} is the Thévenin equivalent resistance seen from the terminals of the storage element.

Finally, the application of PSPICE to *RC* and *RL* problems was presented.

KEY POINTS

- The voltage or current anywhere in an *RC* or *RL* circuit is obtained by solving a first-order differential equation.
- A solution to the equation

$$\frac{dx(t)}{dt} + ax(t) = A \text{ is } x(t) = \frac{A}{a} + k_2e^{-at}$$

  where A/a is referred to as the steady-state solution and $1/a$ is called the time constant.
- The function e^{-at} decays to a value that is less than 1% of its initial value after a period of $5/a$ seconds.
- If a network has a small time constant, its response to some input will quickly settle to its steady-state value; however, if the network time constant is large, a long time is required for the network to reach steady state.
- An *RC* circuit has a time constant of $R_{Th}C$ seconds and an *RL* circuit has a time constant of L/R_{Th} seconds.

PROBLEMS

7.1. In the network in Fig. P7.1, find $v_c(t)$ for $t > 0$.

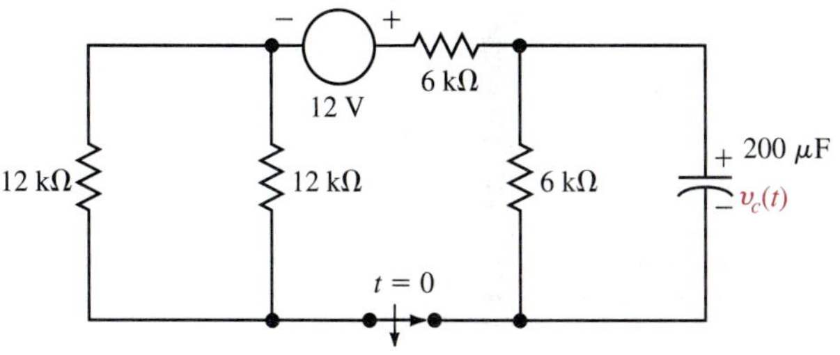

Figure P7.1

7.2. In the network in Fig. P7.2, find $v_c(t)$ for $t > 0$.

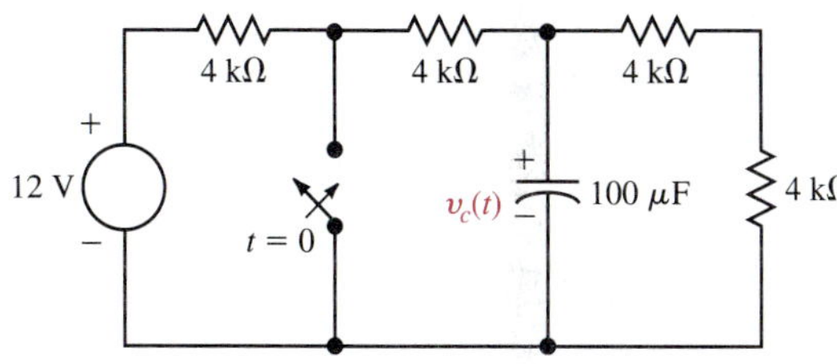

Figure P7.2

7.3. In the circuit in Fig. P7.3, find $i_L(t)$ for $t > 0$.

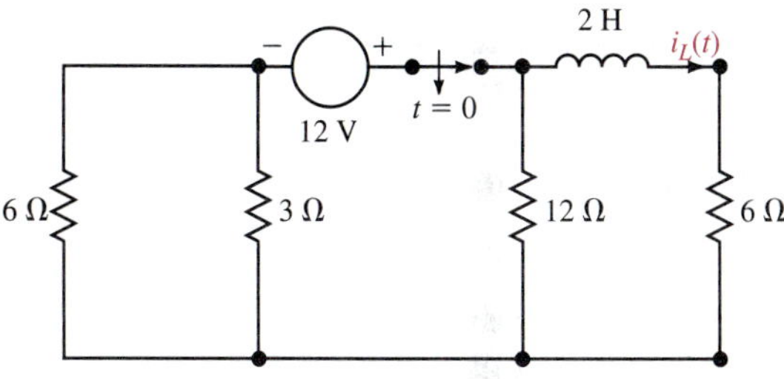

Figure P7.3

7.4. In the circuit in Fig. P7.4, find $v_o(t)$ for $t > 0$.

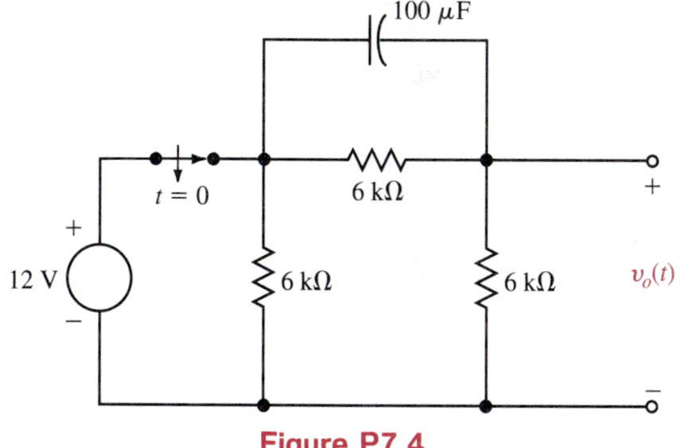

Figure P7.4

7.5. In the network in Fig. P7.5, find $i_o(t)$ for $t > 0$.

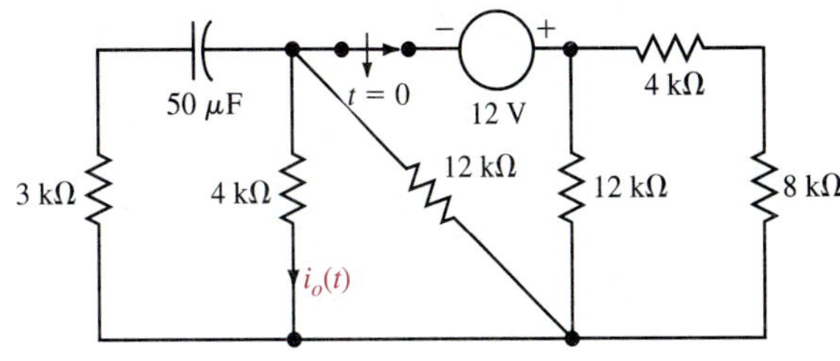

Figure P7.5

7.6. In the network in Fig. P7.6, find $i_o(t)$ for $t > 0$.

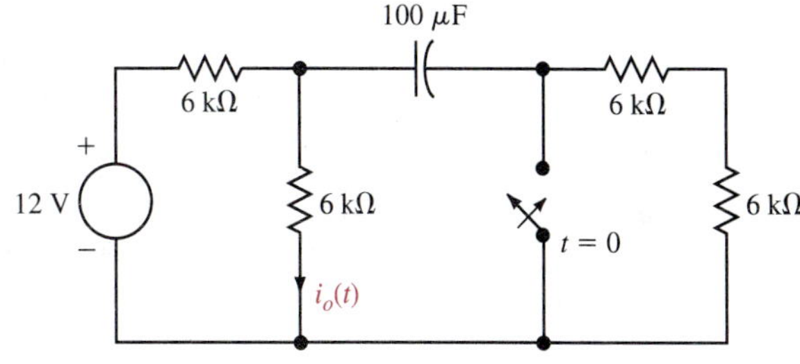

Figure P7.6

7.7. In the network in Fig. P7.7, find $i_o(t)$ for $t > 0$.

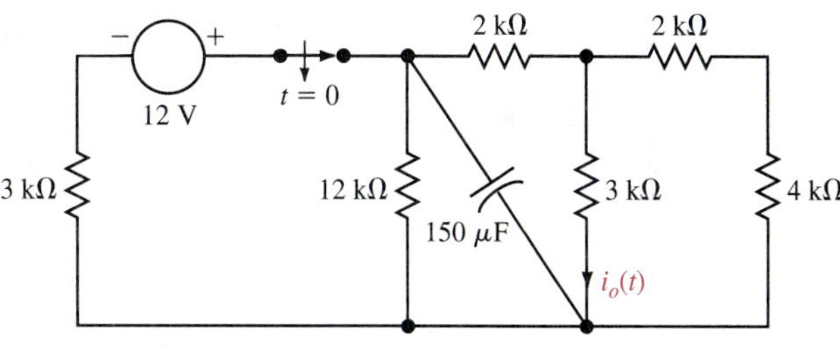

Figure P7.7

7.8. In the circuit in Fig. P7.8, find $v_o(t)$ for $t > 0$.

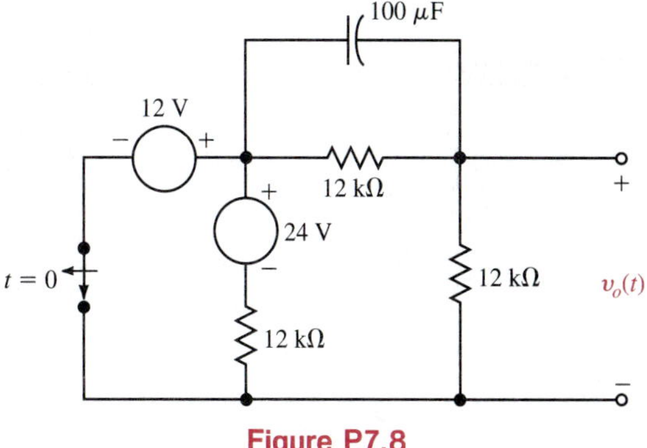

Figure P7.8

7.9. In the network in Fig. P7.9, find $i_o(t)$ for $t > 0$.

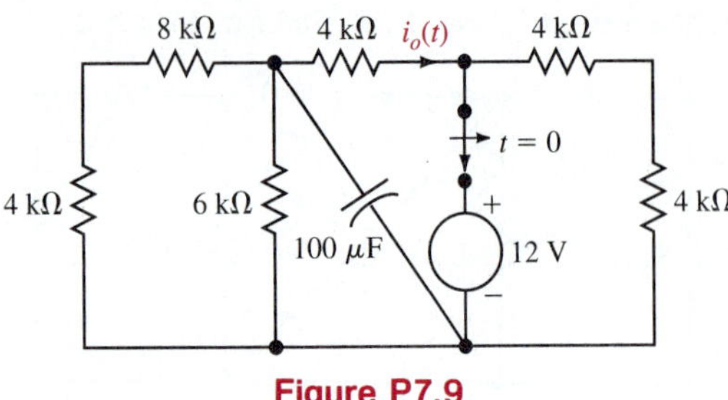

Figure P7.9

7.10. In the circuit in Fig. P7.10, find $v_o(t)$ for $t > 0$.

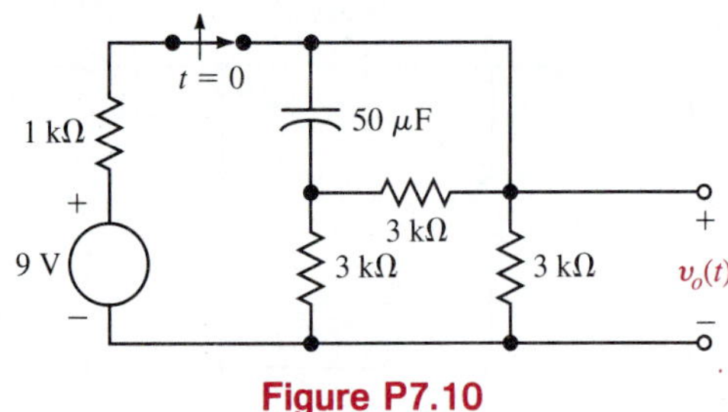

Figure P7.10

7.11. In the network in Fig. P7.11, find $i_o(t)$ for $t > 0$.

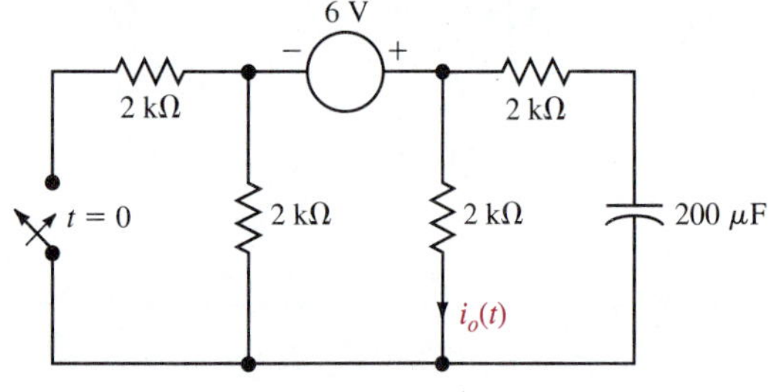

Figure P7.11

7.12. In the network in Fig. P7.12, find $i_o(t)$ for $t > 0$.

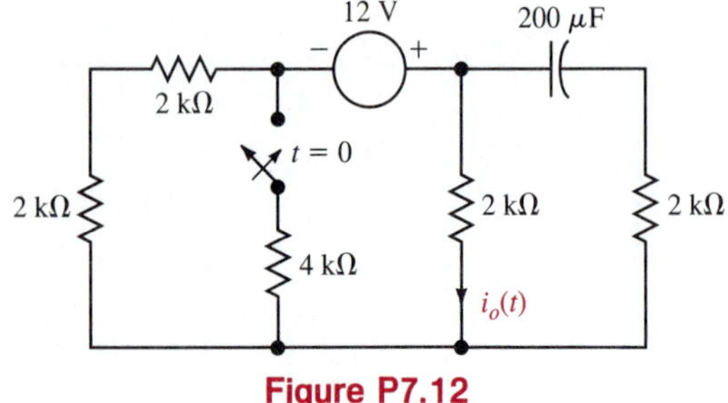

Figure P7.12

7.13. In the circuit in Fig. P7.13, find $v_o(t)$ for $t > 0$.

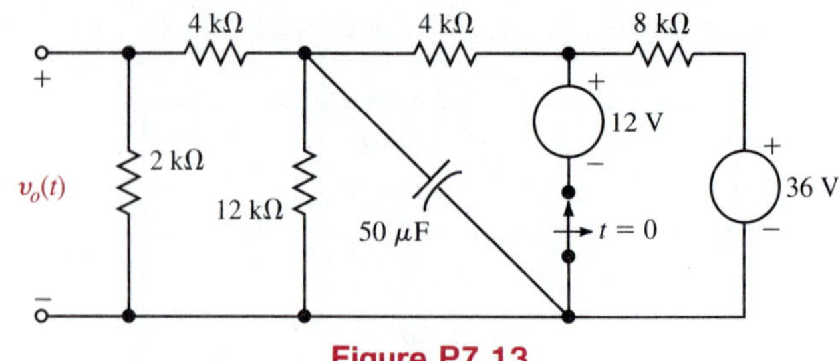

Figure P7.13

7.14. In the circuit in Fig. P7.14, find $v_o(t)$ for $t > 0$.

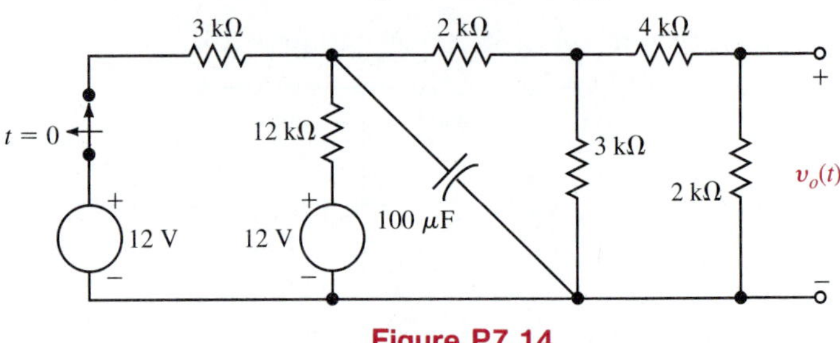

Figure P7.14

7.15. In the network in Fig. P7.15, find $i_o(t)$ for $t > 0$.

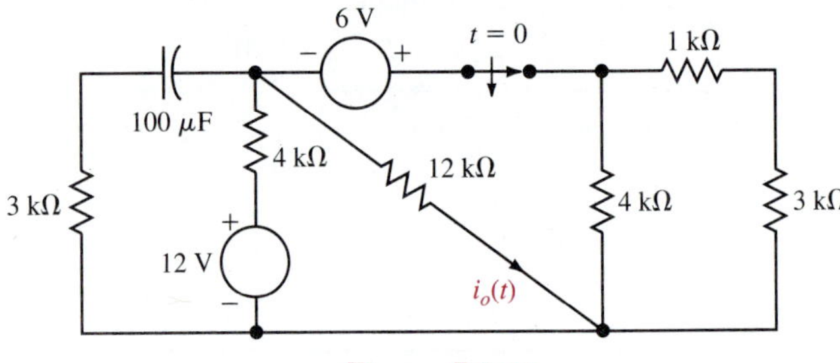

Figure P7.15

7.16. In the circuit in Fig. P7.16, find $v_o(t)$ for $t > 0$.

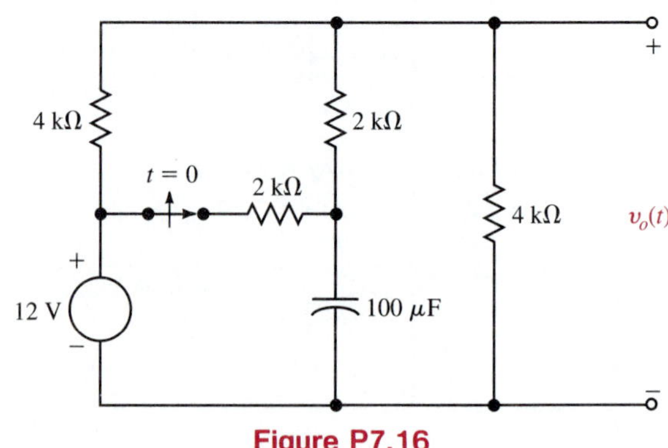

Figure P7.16

7.17. In the circuit in Fig. P7.17, find $v_o(t)$ for $t > 0$.

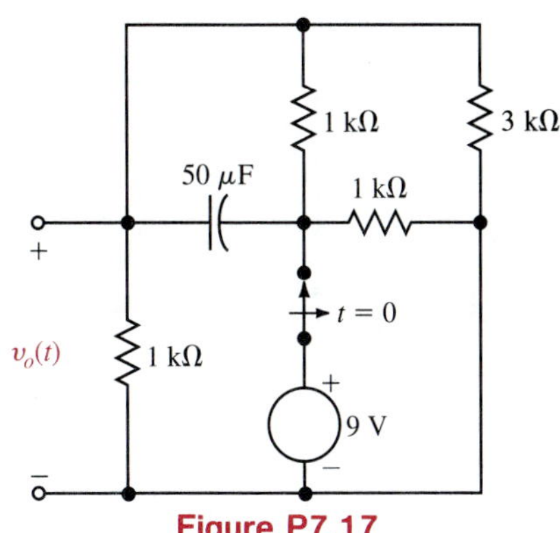

Figure P7.17

7.18. In the network in Fig. P7.18, find $i_o(t)$ for $t > 0$.

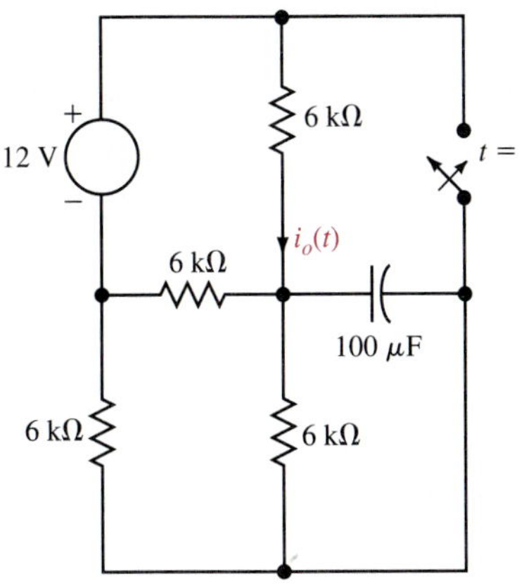

Figure P7.18

7.19. In the circuit in Fig. P7.19, find $v_o(t)$ for $t > 0$.

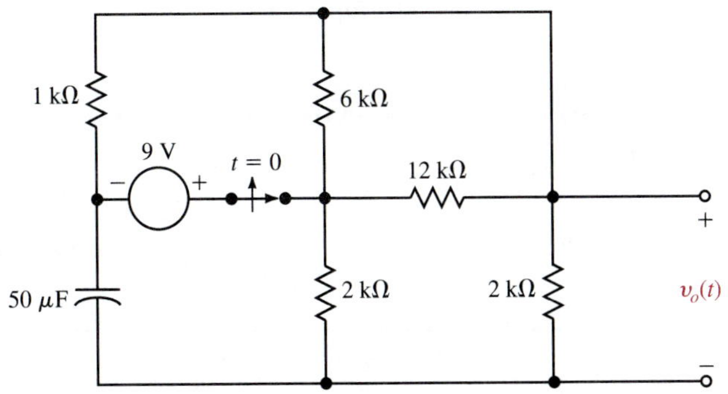

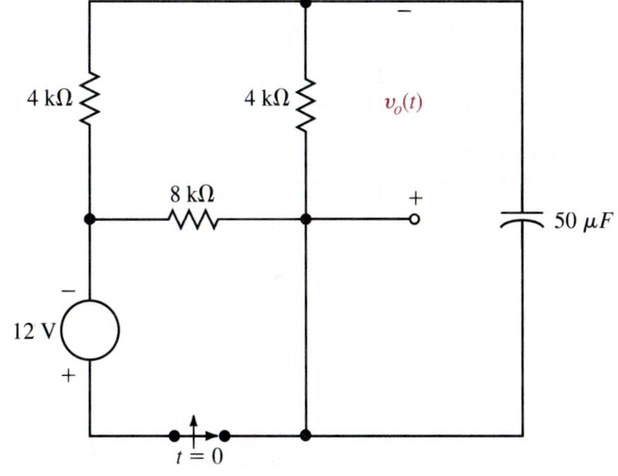

Figure P7.19

7.20. In the circuit in Fig. P7.20, find $v_o(t)$ for $t > 0$.

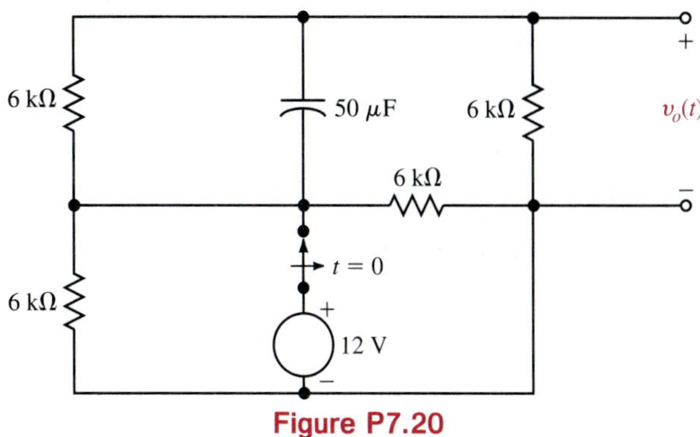

Figure P7.20

7.21. In the network in Fig. P7.21, find $i_o(t)$ for $t > 0$.

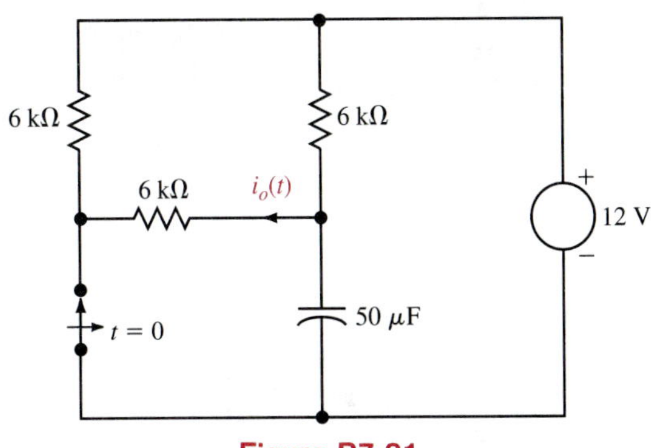

Figure P7.21

7.22. In the circuit in Fig. P7.22, find $v_o(t)$ for $t > 0$.

Figure P7.22

7.23. In the network in Fig. P7.23, find $i_o(t)$ for $t > 0$.

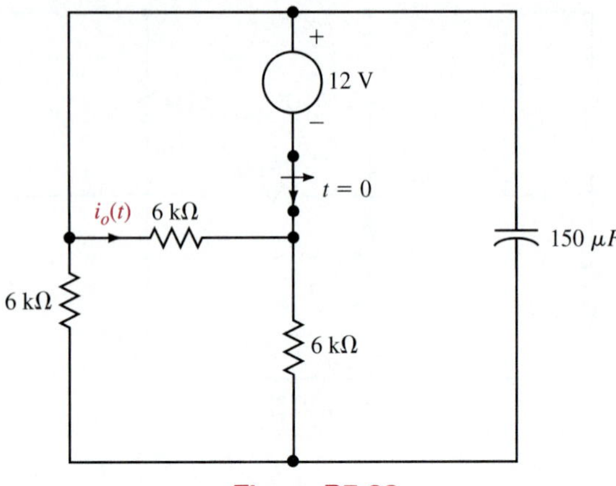

Figure P7.23

7.24. In the network in Fig. P7.24, find $i_o(t)$ for $t > 0$.

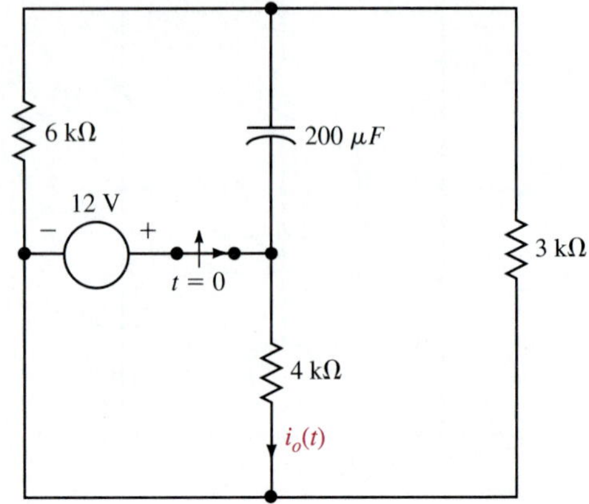

Figure P7.24

7.25. In the circuit in Fig. P7.25, find $v_o(t)$ for $t > 0$.

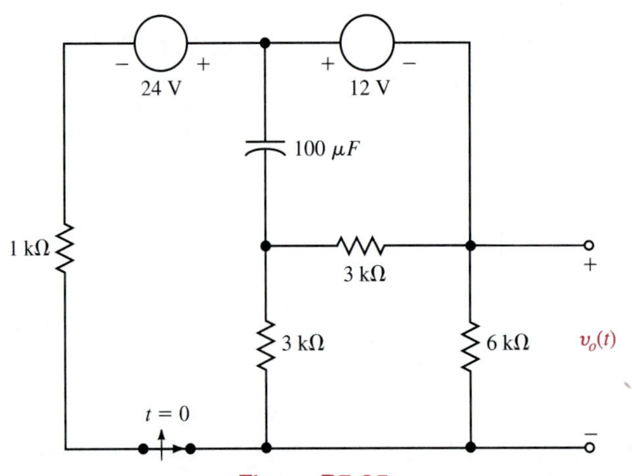

Figure P7.25

7.26. Find $v_o(t)$ for $t > 0$ in the circuit in Fig. P7.26.

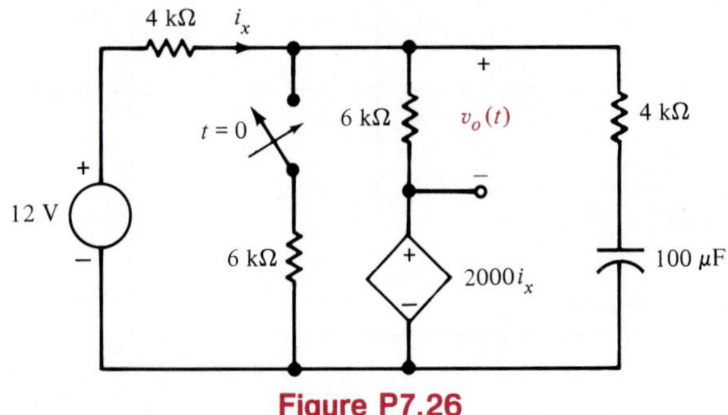

Figure P7.26

7.27. In Fig. P7.27, find $v_o(t)$ for $t > 0$.

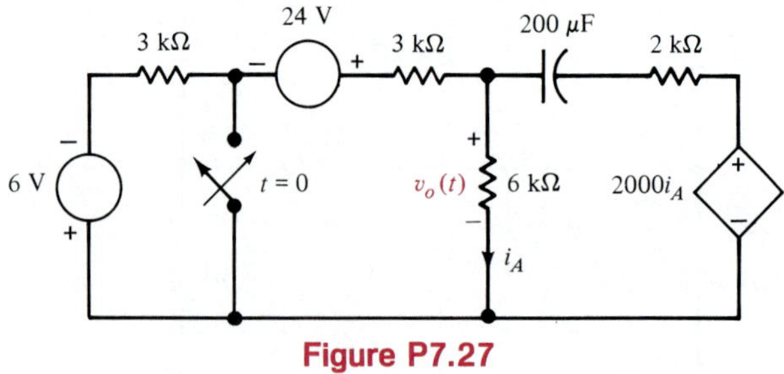

Figure P7.27

7.28. Find $i_o(t)$ for $t > 0$ in the network in Fig. P7.28.

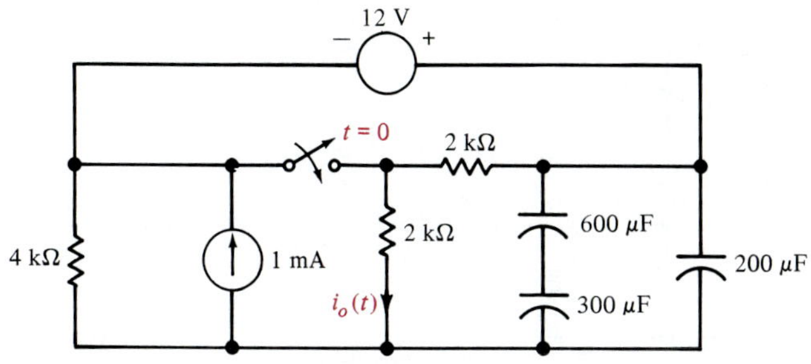

Figure P7.28

7.29. Find $i_o(t)$ for $t > 0$ in the network in Fig. P7.29.

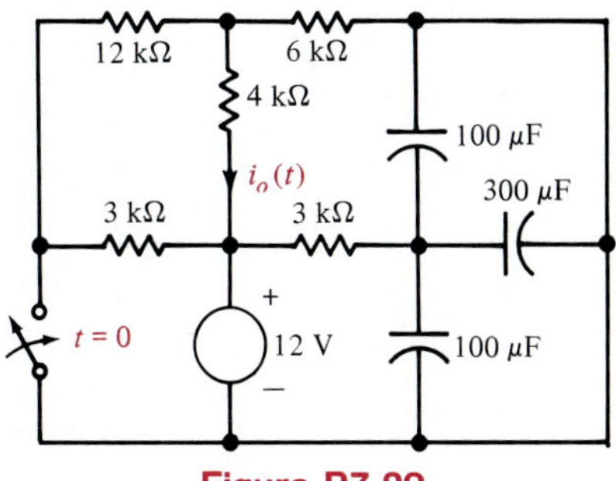

Figure P7.29

7.30. Find $i_o(t)$ for $t > 0$ in the network in Fig. P7.30.

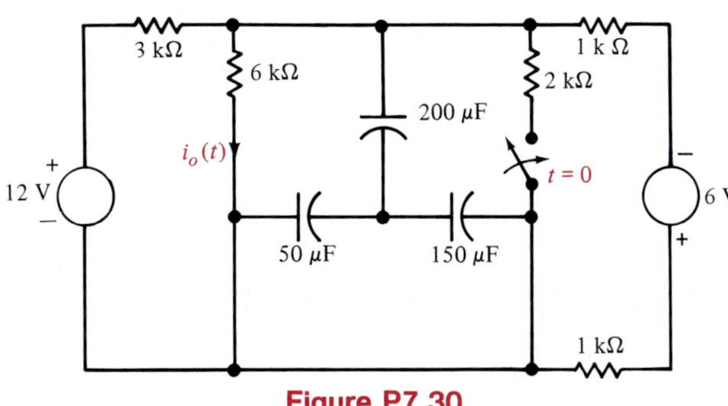

Figure P7.30

7.31. In the circuit in Fig. P7.31, find $v_o(t)$ for $t > 0$.

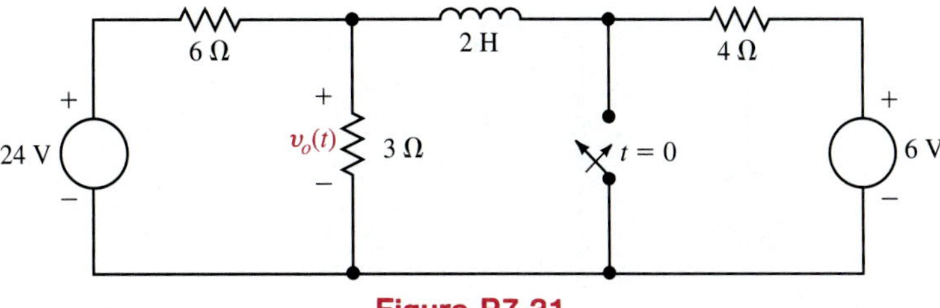

Figure P7.31

7.32. In the circuit in Fig. P7.32, find $v_o(t)$ for $t > 0$.

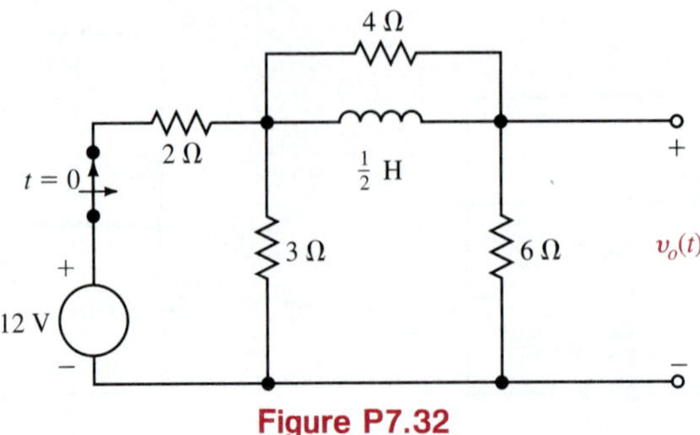

Figure P7.32

7.33. In the network in Fig. P7.33, find $i_o(t)$ for $t > 0$.

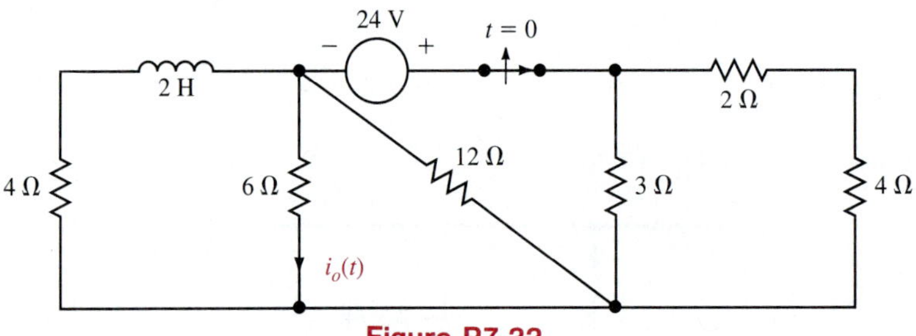

Figure P7.33

7.34. In the circuit in Fig. P7.34, find $v_o(t)$ for $t > 0$.

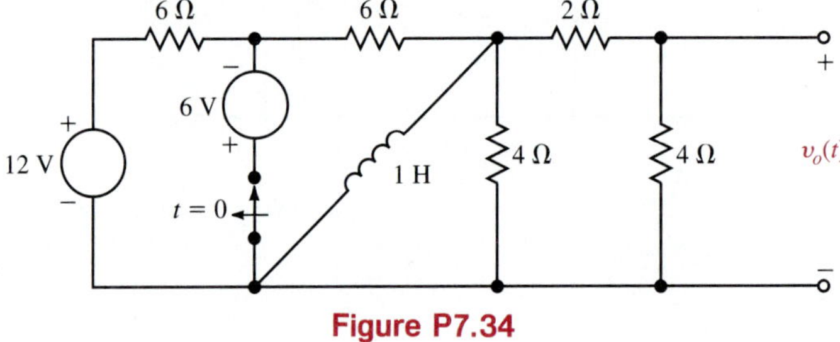

Figure P7.34

7.35. In the network in Fig. P7.35, find $i_o(t)$ for $t > 0$.

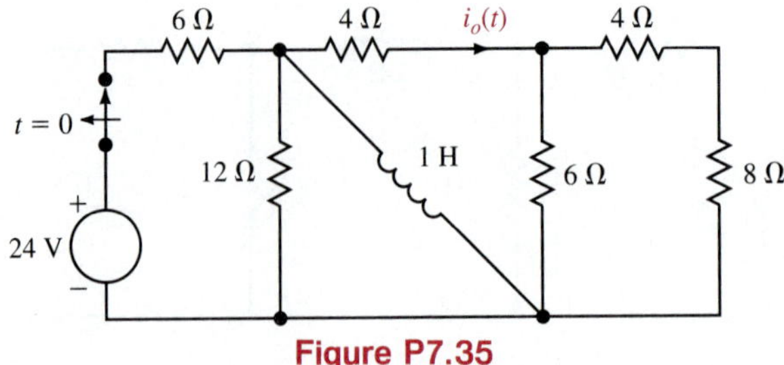

Figure P7.35

7.36. In the circuit in Fig. P7.36, find $v_o(t)$ for $t > 0$.

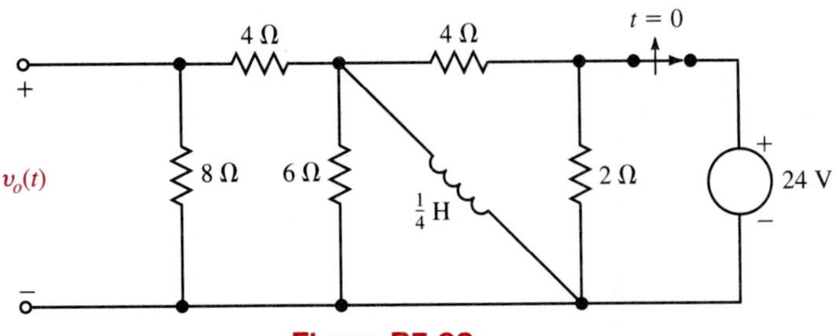

Figure P7.36

7.37. In the network in Fig. P7.37, find $i_o(t)$ for $t > 0$.

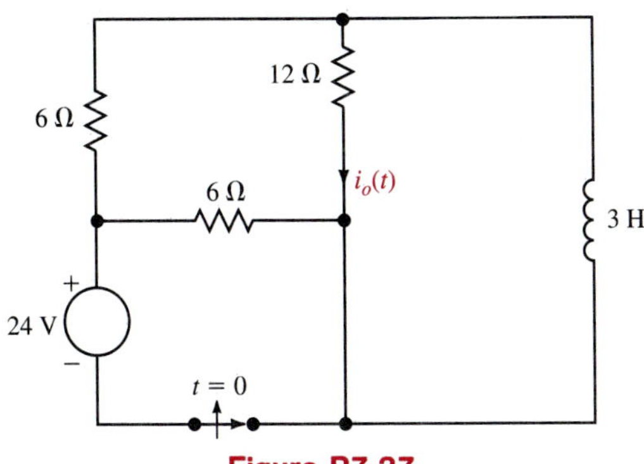

Figure P7.37

7.38. In the circuit in Fig. P7.38, find $v_o(t)$ for $t > 0$.

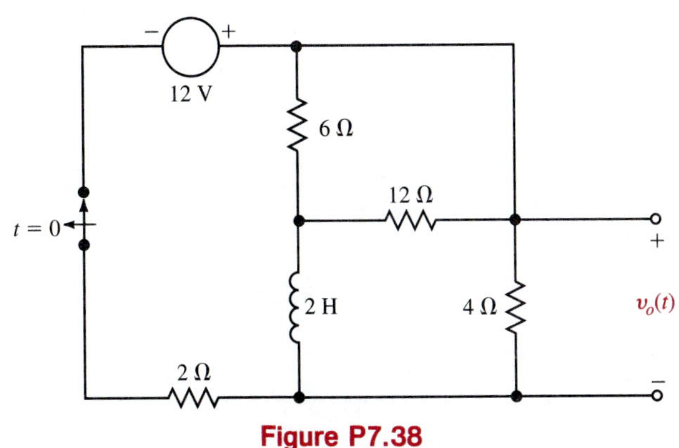

Figure P7.38

7.39. In the circuit in Fig. P7.39, find $v_o(t)$ for $t > 0$.

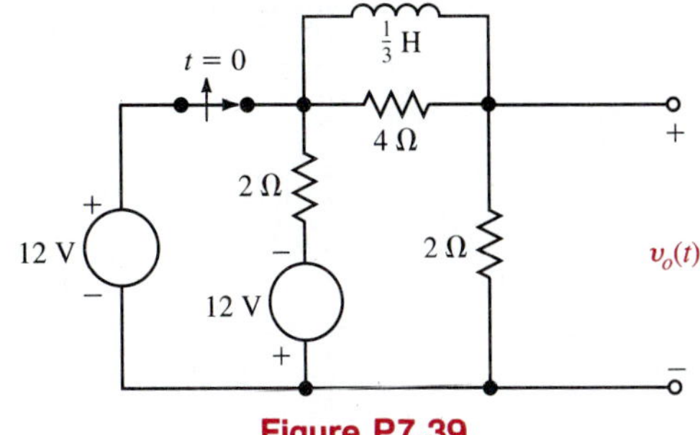

Figure P7.39

7.40. In the network in Fig. P7.40, find $i_o(t)$ for $t > 0$.

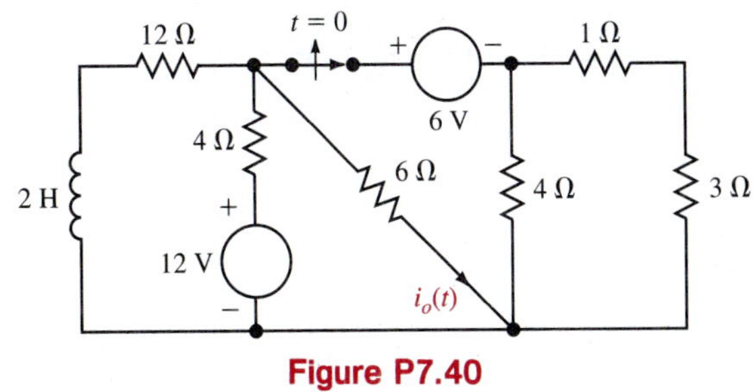

Figure P7.40

7.41. In the network in Fig. P7.41, find $i_o(t)$ for $t > 0$.

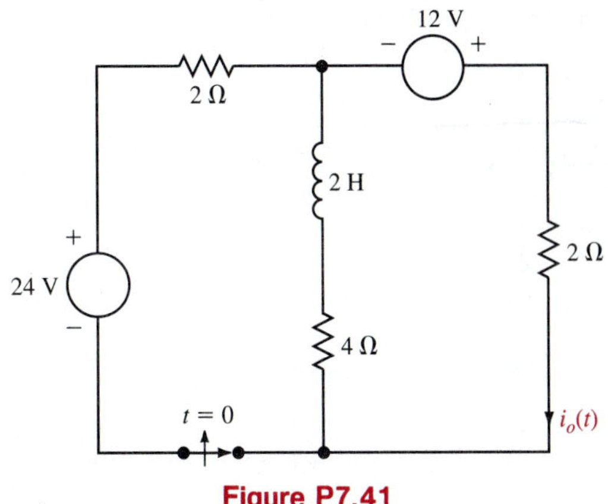

Figure P7.41

7.42. In the network in Fig. P7.42, find $i_o(t)$ for $t > 0$.

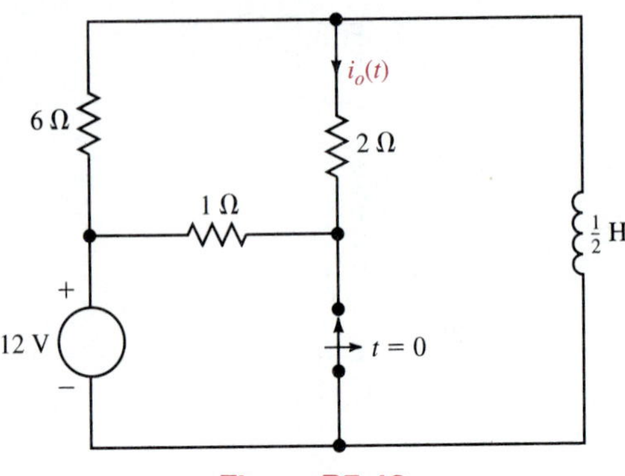

Figure P7.42

7.43. In the network in Fig. P7.43, find $i_o(t)$ for $t > 0$.

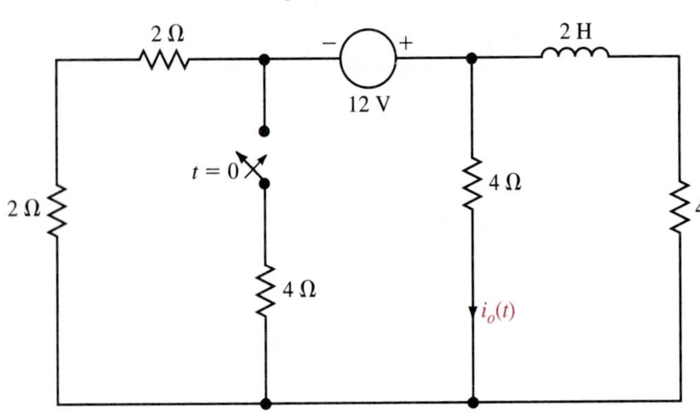

Figure P7.43

7.44. In the network in Fig. P7.44, find $i_o(t)$ for $t > 0$.

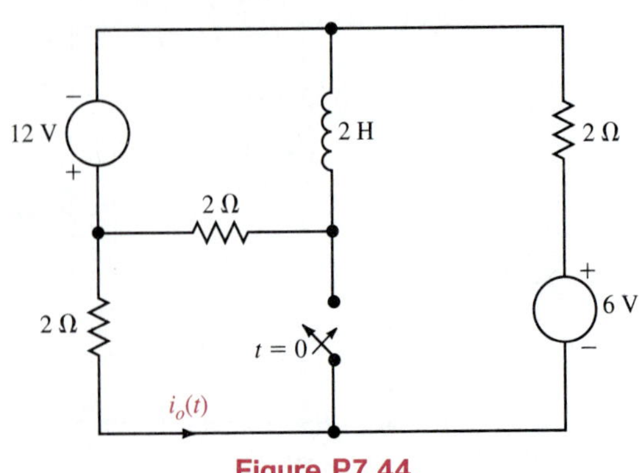

Figure P7.44

7.45. In the circuit in Fig. P7.45, find $v_o(t)$ for $t > 0$.

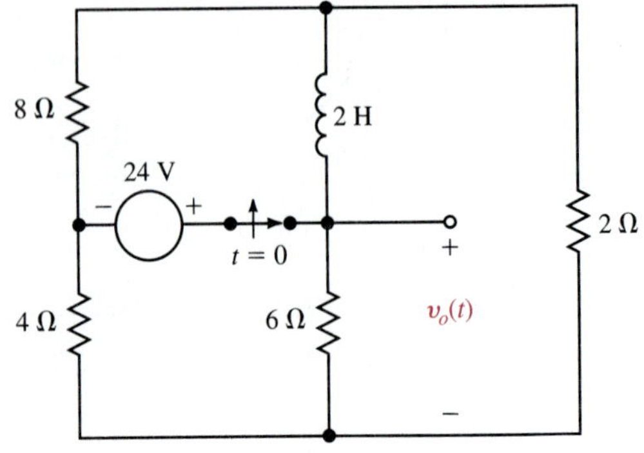

Figure P7.45

7.46. In the circuit in Fig. P7.46, find $v_o(t)$ for $t > 0$.

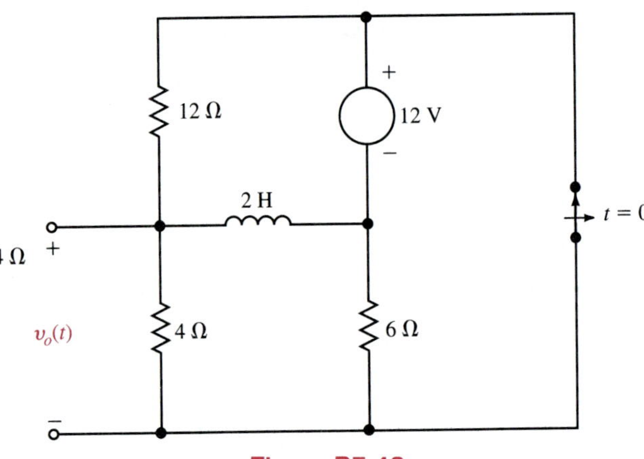

Figure P7.46

7.47. In the network in Fig. P7.47, find $i_o(t)$ for $t > 0$.

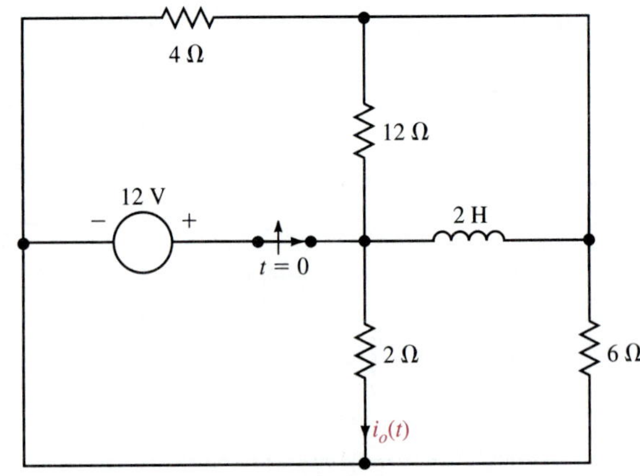

Figure P7.47

7.48. In the circuit in Fig. P7.48, find $v_o(t)$ for $t > 0$.

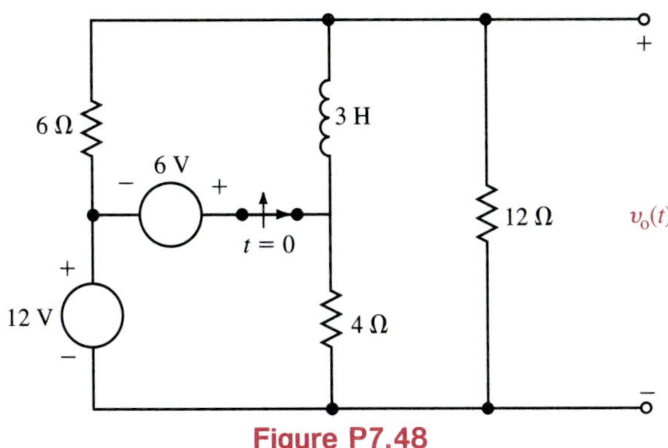

Figure P7.48

7.49. In the circuit in Fig. P7.49, find $v_o(t)$ for $t > 0$.

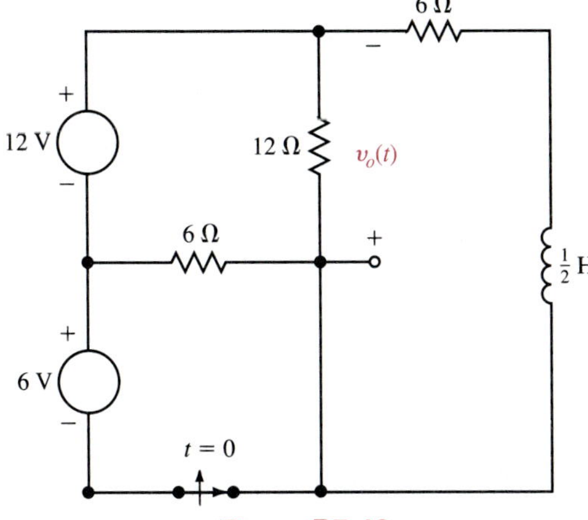

Figure P7.49

7.50. In the circuit in Fig. P7.50, find $v_o(t)$ for $t > 0$.

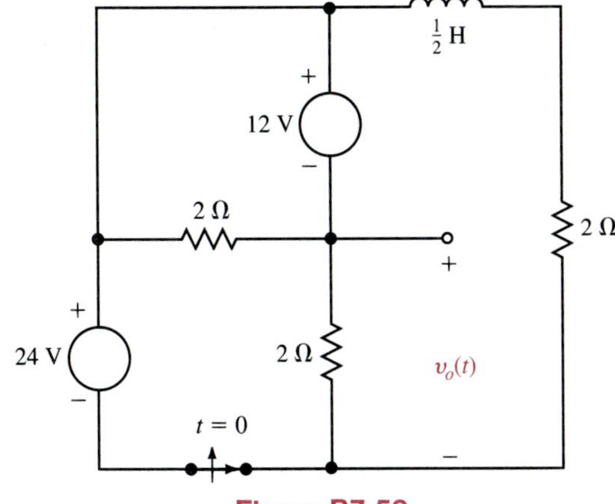

Figure P7.50

7.51. In the circuit in Fig. P7.51, find $v_o(t)$ for $t > 0$.

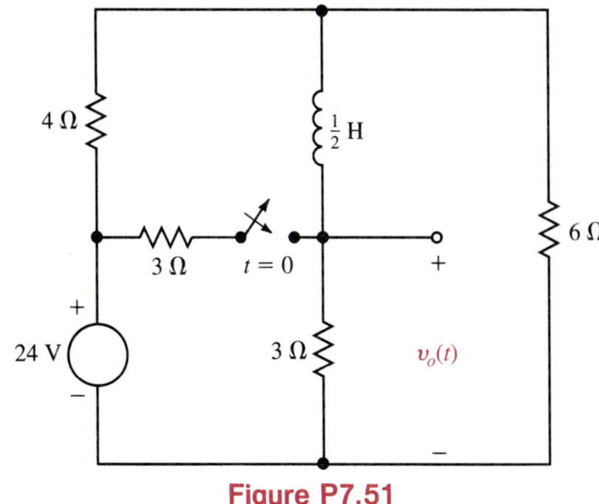

Figure P7.51

7.52. Determine the expression for $v_o(t)$, $t > 0$, in the network in Fig. P7.52.

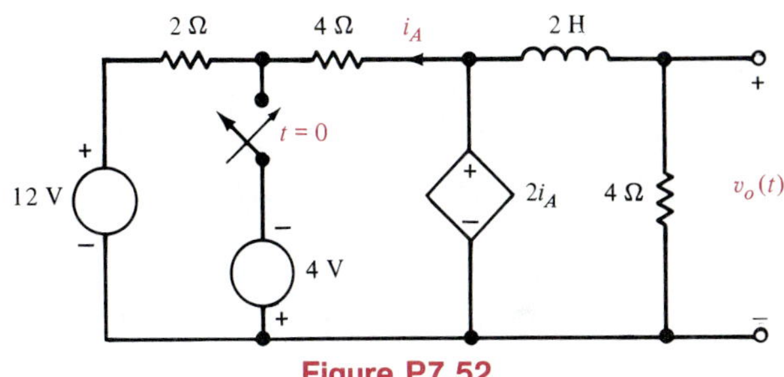

Figure P7.52

7.53. Find $i_o(t)$ for $t > 0$ in the network in Fig. P7.53.

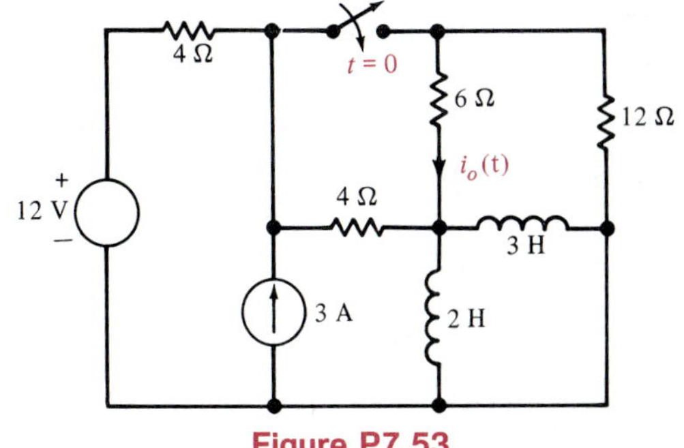

Figure P7.53

7.54. Find $v_o(t)$ for $t > 0$ in the network in Fig. P7.54. The initial capacitor voltage is shown in the network.

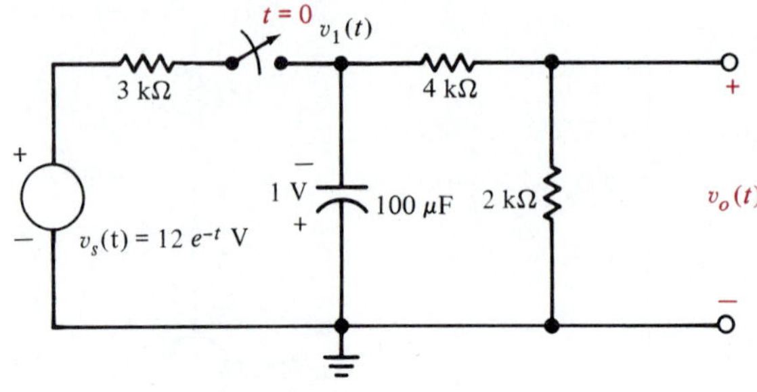

Figure P7.54

7.55. Find $v_o(t)$ for $t > 0$ in the network in Fig. P7.55.

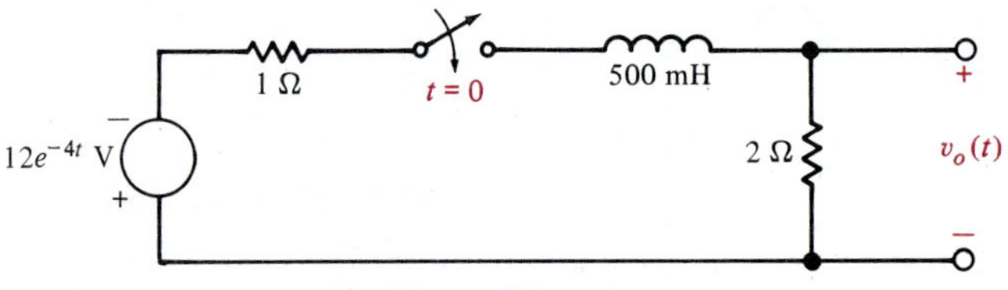

Figure P7.55

7.56. Find $v_o(t)$ for $t > 0$ in the network in Fig. P7.56.

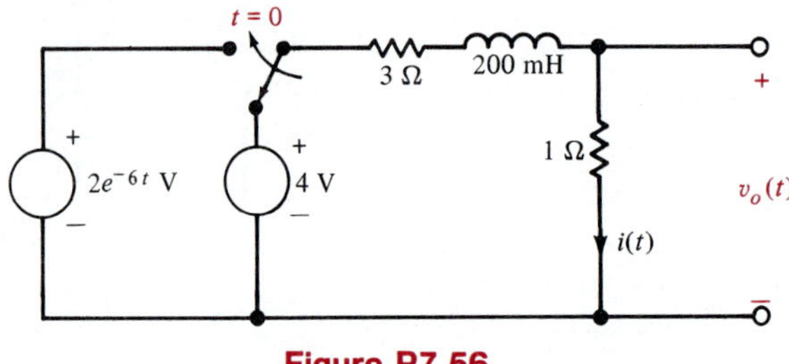

Figure P7.56

7.57. Determine the equation for the voltage $v_o(t)$, $t > 0$, in Fig. P7.57a when subjected to the input pulse shown in Fig. P7.57b.

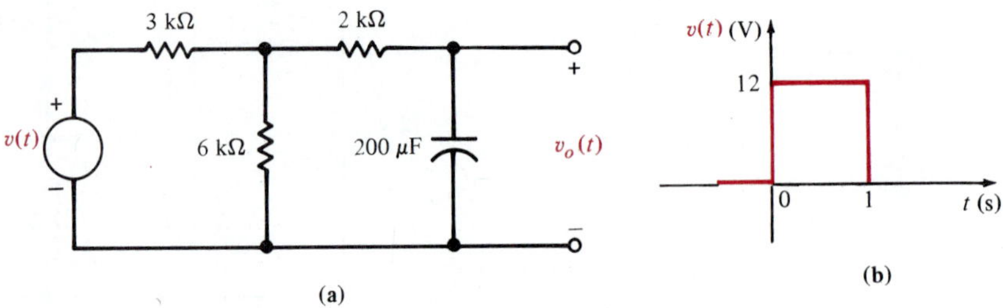

Figure P7.57

7.58. Determine the relationship between the input and output voltages for the circuits shown in Fig. P7.58.

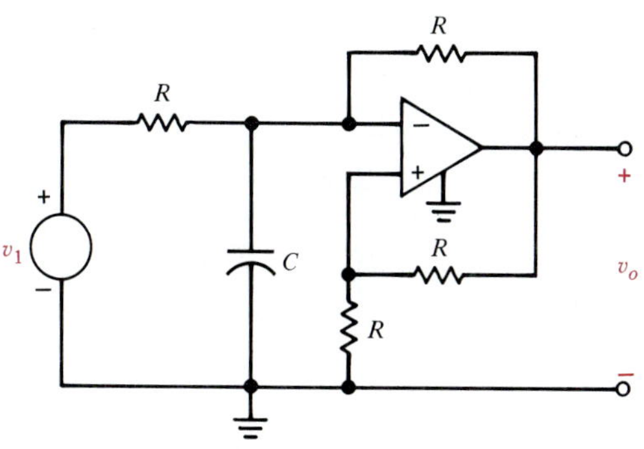

Figure P7.58

7.59. Given the network in Fig. P7.26, plot $v_o(t)$ for $0 \leq t \leq 3$ sec using a 30-ms step size using PSPICE.

7.60. For the network in Fig. P7.27, plot $v_o(t)$ for $0 \leq t \leq 4$ sec using a step size of 40 ms using PSPICE.

7.61. Given the network in Fig. P7.61a, plot $i_o(t)$ over a 4-sec interval using a 20-ms step size if the input source $i_S(t)$ is as shown in Fig. 7.61b.

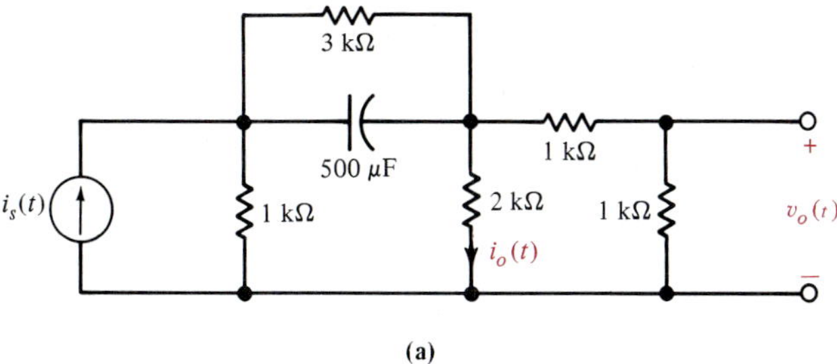

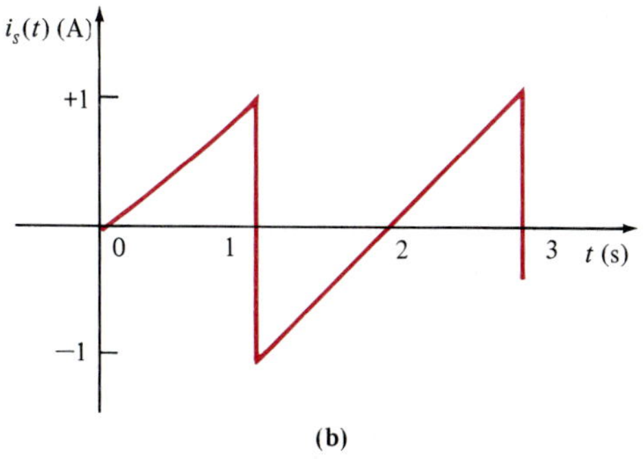

Figure P7.61

7.62. For the network in Fig. P7.61a, plot the current $i_o(t)$ over a 10-sec interval using a 50-ms step size if the input $i_S(t)$ is given by the expression

$$\begin{aligned} i_s(t) &= 10(1 - e^{-t/4}) \text{ A} && 0 \leq t \leq 2 \text{ sec} \\ &= 10(1 - e^{-t/4}) - 10e^{-(t-2)} \text{ A} && 2 \text{ sec} \leq t \end{aligned}$$

7.63. For the network in Fig. P7.61a, plot the voltage $v_o(t)$ over a 2-sec interval using a 10-ms step size if the input $i_S(t)$ is

$$i_S(t) = 10e^{-2t} \sin 4\pi t \text{ A}$$

7.64. For the network in Fig. P7.61a, plot the current $i_o(t)$ over a 4-sec interval using a 20-ms step size if the input $i_S(t)$ is as shown in Fig. P7.64.

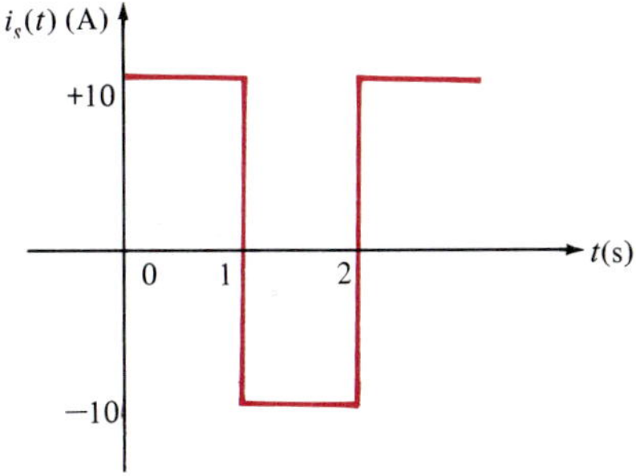

Figure P7.64

7.65. Find the output voltage $v_o(t)$ in the network in Fig. P7.65 if the input voltage is

$$v_i(t) = 5[u(t) - u(t - 0.05)] \text{ V}$$

1 μF

+

$v_i(t)$

−

100 kΩ

+

$v_o(t)$

−

Figure P7.65

7.66. Given the network in Fig. P7.66, plot $v(t)$ over a 10-sec interval using a 50-ms step size if the input source $v_S(t)$ is given by

$$v_S(t) = 15(1 - e^{-t/2})\text{ V} \qquad 0 \le t \le 2 \text{ sec}$$

$$= 15(1 - e^{-t/2}) - 15e^{-(t-2)}\text{ V} \qquad 2 \text{ sec} \le t$$

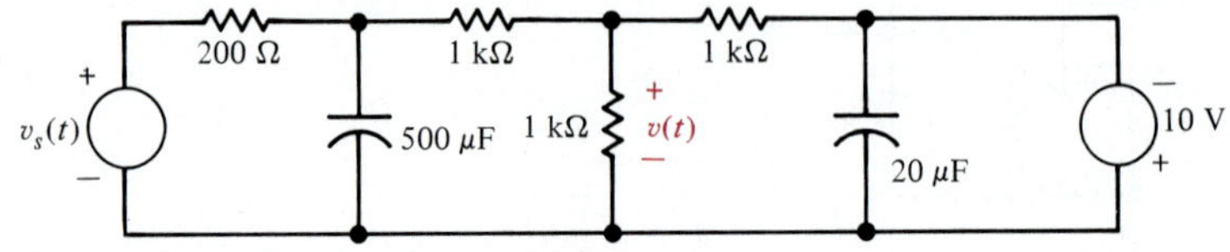

Figure P7.66

7.67. For the network in Fig. P7.66, plot the output $v(t)$ over a 2-sec interval using a 10-ms step size if the input is $v_S(t) = 10e^{-2t}$ sin $4\pi t$ V.

7.68. For the network in Fig. P7.66, plot the voltage $v(t)$ over a 2-sec interval using a 10-ms step size if the input voltage $v_S(t)$ is as shown in Fig. P7.68.

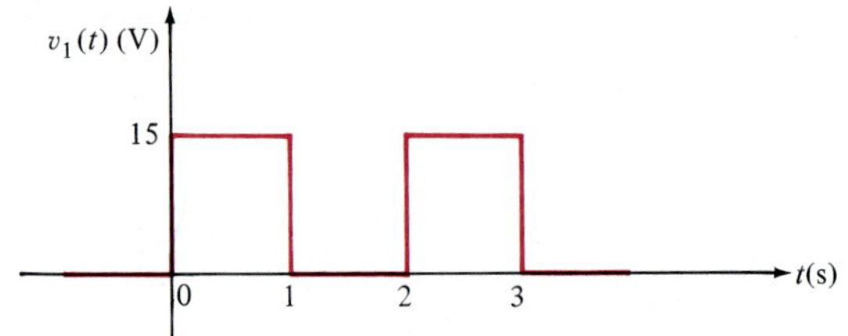

Figure P7.68

7.69. For the network in Fig. P7.66, plot the voltage $v(t)$ over a 6-sec interval using a 30-ms step size if the input voltage $v_S(t)$ is as shown in Fig. P7.69.

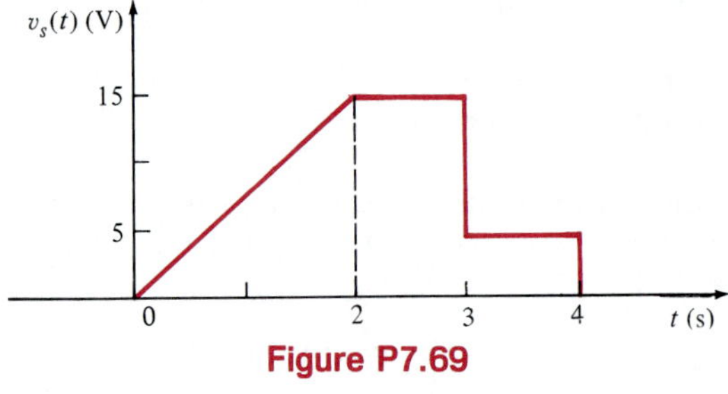

Figure P7.69

7.70. Use PSPICE to determine the pulse response of the network in Fig. P7.70a if the input is as shown in Fig. P7.70b. Model the op-amp with $A = 10^5$, $R_i = \infty$, and $R_o = 0$.

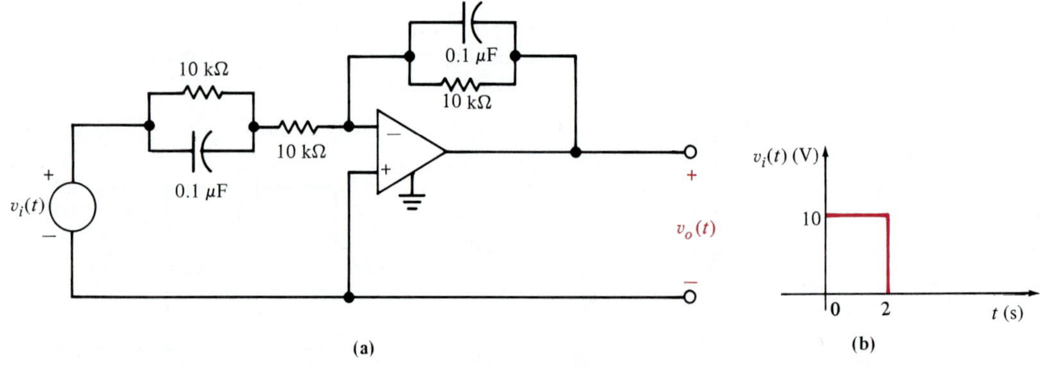

Figure P7.70

CHAPTER

RLC Circuits

In this chapter we extend our analysis of circuits that contain only a single storage element to the case where an inductor and a capacitor are present simultaneously. Although the *RLC* circuits are more complicated than those we have analyzed in Chapter 7, we will follow a development similar to that used earlier in obtaining a solution.

Our presentation will deal only with very simple circuits, since the analysis can quickly become very complicated for networks that contain more than one loop or one nonreference node. Once we have presented the mathematical development of the response equations, we will analyze several networks in which the parameters have been chosen to illustrate the different types of response. In addition, we will extend our PSPICE analysis to networks that contain both an inductor and a capacitor.

A more efficient method for providing a general mathematical solution for *RLC* networks will be presented in Chapter 17.

8.1 The Basic Circuit Equation

To begin our development, let us consider the two basic *RLC* circuits shown in Fig. 8.1. We assume that energy may be initially stored in both the inductor and capacitor. The node equation for the parallel *RLC* circuit is

$$\frac{v}{R} + \frac{1}{L}\int_{t_0}^{t} v(x)\,dx + i_L(t_0) + C\frac{dv}{dt} = i_S(t)$$

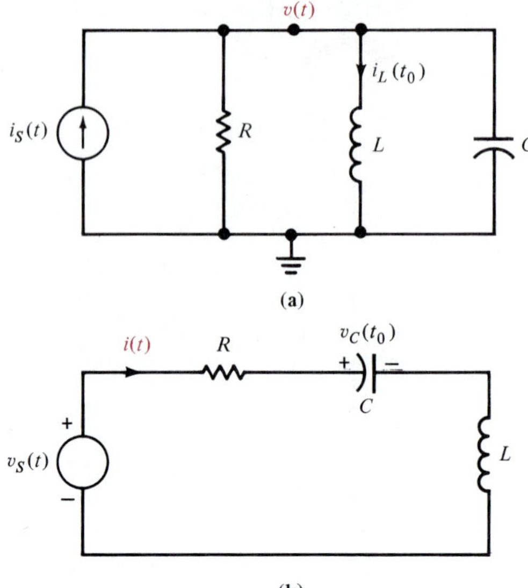

Figure 8.1 Parallel and series *RLC* circuits.

Similarly, the loop equation for the series *RLC* circuit is

$$Ri + \frac{1}{C}\int_{t_0}^{t} i(x)\,dx + v_C(t_0) + L\frac{di}{dt} = v_S(t)$$

Note that the equation for the node voltage in the parallel circuit is of the same form as that for the loop current in the series circuit. Therefore, the solution of these two circuits is dependent on solving one equation. If the two equations above are differentiated with respect to time, we obtain

$$C\frac{d^2v}{dt^2} + \frac{1}{R}\frac{dv}{dt} + \frac{v}{L} = \frac{di_S}{dt}$$

and

$$L\frac{d^2i}{dt^2} + R\frac{di}{dt} + \frac{i}{C} = \frac{dv_S}{dt}$$

Since both circuits lead to a second-order differential equation with constant coefficients, we will concentrate our analysis on this type of equation.

8.2 Mathematical Development of the Response Equations

In concert with our development of the solution of a first-order differential equation that results from the analysis of either an *RL* or an *RC* circuit as outlined in Chapter 7, we will now employ the same approach here to obtain the solution of a second-order differential equation that results from the analysis of *RLC* circuits. As a general rule, for this case we are confronted with an equation of the form

$$\frac{d^2x(t)}{dt^2} + a_1\frac{dx(t)}{dt} + a_2x(t) = f(t) \tag{8.1}$$

Once again we use the fact that if $x(t) = x_p(t)$ is a solution to Eq. (8.1), and if $x(t) = x_c(t)$ is a solution to the homogeneous equation

$$\frac{d^2x(t)}{dt^2} + a_1\frac{dx(t)}{dt} + a_2x(t) = 0$$

then

$$x(t) = x_p(t) + x_c(t)$$

is a solution to the original Eq. (8.1). If we again confine ourselves to a constant forcing function [i.e., $f(t) = A$], the development at the beginning of Chapter 7 shows that the solution of Eq. (8.1) will be of the form

$$x(t) = \frac{A}{a_2} + x_c(t) \tag{8.2}$$

Let us now turn our attention to the solution of the homogeneous equation

$$\frac{d^2x(t)}{dt^2} + a_1\frac{dx(t)}{dt} + a_2x(t) = 0$$

where a_1 and a_2 are constants. For simplicity we will rewrite the equation in the form

$$\frac{d^2x(t)}{dt^2} + 2\alpha\frac{dx(t)}{dt} + \omega_0^2x(t) = 0 \tag{8.3}$$

where we have made the following simple substitutions for the constants $a_1 = 2\alpha$ and $a_2 = \omega_0^2$. Following the development of a solution for the first-order homogeneous differential equation in Chapter 7, the solution of Eq. (8.3) must be a function whose first- and second-order derivatives have the same form, so that the left-hand side of Eq. (8.3) will become identically zero for all t. Again we assume that

$$x(t) = Ke^{st}$$

Substituting this expression into Eq. (8.3) yields

$$s^2Ke^{st} + 2\alpha sKe^{st} + \omega_0^2Ke^{st} = 0$$

Dividing both sides of the equation by Ke^{st} yields

$$s^2 + 2\alpha s + \omega_0^2 = 0 \tag{8.4}$$

This equation is commonly called the *characteristic equation*. α is called the exponential *damping coefficient* and ω_0 is referred to as the *undamped resonant frequency*. The importance of this terminology will become clear as we proceed with the development. If this equation is satisfied, our assumed solution $x(t) = Ke^{st}$ is correct. Employing the quadratic formula, we find that Eq. (8.4) is satisfied if

$$s = \frac{-2\alpha \pm \sqrt{4\alpha^2 - 4\omega_0^2}}{2} \tag{8.5}$$

$$= -\alpha \pm \sqrt{\alpha^2 - \omega_0^2}$$

Therefore, there are two values of s, s_1 and s_2, that satisfy Eq. (8.4).

$$\begin{aligned} s_1 &= -\alpha + \sqrt{\alpha^2 - \omega_0^2} \\ s_2 &= -\alpha - \sqrt{\alpha^2 - \omega_0^2} \end{aligned} \tag{8.6}$$

This means that $x_1(t) = K_1 e^{s_1 t}$ is a solution of Eq. (8.3) and that $x_2(t) = K_2 e^{s_2 t}$ is also a solution of Eq. (8.3); that is,

$$\frac{d^2}{dt^2}(K_1 e^{s_1 t}) + 2\alpha \frac{d}{dt}(K_1 e^{s_1 t}) + \omega_0^2 K_1 e^{s_1 t} = 0$$

and

$$\frac{d^2}{dt^2}(K_2 e^{s_2 t}) + 2\alpha \frac{d}{dt}(K_2 e^{s_2 t}) + \omega_0^2 K_2 e^{s_2 t} = 0$$

The addition of these two equations produces the equality

$$\frac{d^2}{dt^2}(K_1 e^{s_1 t} + K_2 e^{s_2 t}) + 2\alpha \frac{d}{dt}(K_1 e^{s_1 t} + K_2 e^{s_2 t}) + \omega_0^2(K_1 e^{s_1 t} + K_2 e^{s_2 t}) = 0$$

Note that the sum of the two solutions is also a solution. Therefore, in general, the complementary solution of Eq. (8.1) is of the form

$$x_c(t) = K_1 e^{s_1 t} + K_2 e^{s_2 t} \tag{8.7}$$

K_1 and K_2 are constants that can be evaluated via the initial conditions $x(0)$ and $dx(0)/dt$. For example, since

$$x(t) = K_1 e^{s_1 t} + K_2 e^{s_2 t}$$

then

$$x(0) = K_1 + K_2$$

and

$$\left.\frac{dx(t)}{dt}\right|_{t=0} = \frac{dx(0)}{dt} = s_1 K_1 + s_2 K_2$$

Hence $x(0)$ and $dx(0)/dt$ produce two simultaneous equations, which when solved yield the constants K_1 and K_2.

Close examination of Eqs. (8.6) and (8.7) indicates that the form of the solution of the homogeneous equation is dependent on the relative magnitude of the values of α and ω_0. For example, if $\alpha > \omega_0$, the roots of the characteristic equation, s_1 and s_2, also called the *natural frequencies* because they determine the natural (unforced) response of the network, are real and unequal; if $\alpha < \omega_0$, the roots are complex numbers; and finally if $\alpha = \omega_0$, the roots are real and equal. Each of these cases is very important; hence we will now examine each one in some detail.

Case 1, $\alpha > \omega_0$. This case is commonly called *overdamped*. The natural frequencies s_1 and s_2 are real and unequal, and therefore the natural response of the network described by the second-order differential equation is of the form

$$x_c(t) = K_1 e^{-(\alpha - \sqrt{\alpha^2 - \omega_0^2})t} + K_2 e^{-(\alpha + \sqrt{\alpha^2 - \omega_0^2})t} \tag{8.8}$$

where K_1 and K_2 are found from the initial conditions. This indicates that the natural response is the sum of two decaying exponentials.

Case 2, $\alpha < \omega_0$. This case is called *underdamped*. Since $\omega_0 > \alpha$, the roots of the characteristic equation given in Eq. (8.6) can be written as

$$s_1 = -\alpha + \sqrt{-(\omega_0^2 - \alpha^2)} = -\alpha + j\omega_n$$

$$s_2 = -\alpha - \sqrt{-(\omega_0^2 - \alpha^2)} = -\alpha - j\omega_n$$

where $j = \sqrt{-1}$ and $\omega_n = \sqrt{\omega_0^2 - \alpha^2}$. Thus the natural frequencies are complex numbers. The natural response is then

$$x_c(t) = K_1 e^{-(\alpha - j\omega_n)t} + K_2 e^{-(\alpha + j\omega_n)t}$$

This equation can be simplified in the following manner. First the equation is rewritten as

$$x_c(t) = e^{-\alpha t}(K_1 e^{j\omega_n t} + K_2 e^{-j\omega_n t})$$

Then by using Euler's identities,

$$e^{j\theta} = \cos\theta + j\sin\theta$$

and

$$e^{-j\theta} = \cos\theta - j\sin\theta$$

we obtain

$$\begin{aligned} x_c(t) &= e^{-\alpha t}[K_1(\cos\omega_n t + j\sin\omega_n t) + K_2(\cos\omega_n t - j\sin\omega_n t)] \\ &= e^{-\alpha t}[(K_1 + K_2)\cos\omega_n t + (jK_1 - jK_2)\sin\omega_n t] \\ &= e^{-\alpha t}(A_1\cos\omega_n t + A_2\sin\omega_n t) \end{aligned} \tag{8.9}$$

where A_1 and A_2, like K_1 and K_2, are constants, which are evaluated using the initial conditions $x(0)$ and $dx(0)/dt$. If $x_c(t)$ is real, K_1 and K_2 will be complex and $K_2 = K_1^*$. $A_1 = K_1 + K_2$ is therefore two times the real part of K_1, and $A_2 = jK_1 - jK_2$ is -2 times the imaginary part of K_1. A_1 and A_2 are real numbers. This illustrates that the natural response is an exponentially damped oscillatory response.

Case 3, $\alpha = \omega_0$. This case, called *critically damped*, results in

$$s_1 = s_2 = -\alpha$$

as shown in Eq. (8.6). Therefore, Eq. (8.7) reduces to

$$x_c(t) = K_3 e^{-\alpha t}$$

where $K_3 = K_1 + K_2$. However, this cannot be a solution to the second-order differential equation (8.3) because in general it is not possible to satisfy the two initial conditions $x(0)$ and $dx(0)/dt$ with the single constant K_3.

In the case where the characteristic equation has repeated roots, a solution can be obtained in the following manner. If $x_1(t)$ is known to be a solution of the second-order homogeneous equation, then via the substitution $x(t) = x_1(t)y(t)$ we can transform the given differential equation into a first-order equation in $dy(t)/dt$. Since this resulting equation is only a function of $y(t)$, it can be solved to find the general solution $x(t) = x_1(t)y(t)$.

For the present case, $s_1 = s_2 = -\alpha$, and hence the basic equation is

$$\frac{d^2x(t)}{dt^2} + 2\alpha\frac{dx(t)}{dt} + \alpha^2 x(t) = 0 \tag{8.10}$$

and one known solution is

$$x_1(t) = K_3 e^{-\alpha t}$$

By employing the substitution

$$x_2(t) = x_1(t)y(t) = K_3 e^{-\alpha t} y(t)$$

Eq. (8.10) becomes

$$\frac{d^2}{dt^2}[K_3 e^{-\alpha t} y(t)] + 2\alpha\frac{d}{dt}[K_3 e^{-\alpha t} y(t)] + \alpha^2 K_3 e^{-\alpha t} y(t) = 0$$

Evaluating the derivatives, we obtain

$$\frac{d}{dt}[K_3 e^{-\alpha t} y(t)] = -K_3 \alpha e^{-\alpha t} y(t) + K_3 e^{-\alpha t}\frac{dy(t)}{dt}$$

$$\frac{d^2}{dt^2}[K_3 e^{-\alpha t} y(t)] = K_3 \alpha^2 e^{-\alpha t} y(t) - 2K_3 \alpha e^{-\alpha t}\frac{dy(t)}{dt} + K_3 e^{-\alpha t}\frac{d^2y(t)}{dt^2}$$

Substituting these expressions into the equation above yields

$$K_3 e^{-\alpha t}\frac{d^2y(t)}{dt^2} = 0$$

Therefore,

$$\frac{d^2y(t)}{dt^2} = 0$$

and hence

$$y(t) = A_1 + A_2 t$$

Therefore, the general solution is

$$\begin{aligned} x_2(t) &= x_1(t)y(t) \\ &= K_3 e^{-\alpha t}(A_1 + A_2 t) \end{aligned}$$

which can be written as

$$x_2(t) = B_1 e^{-\alpha t} + B_2 t e^{-\alpha t} \tag{8.11}$$

where B_1 and B_2 are constants derived from the initial conditions.

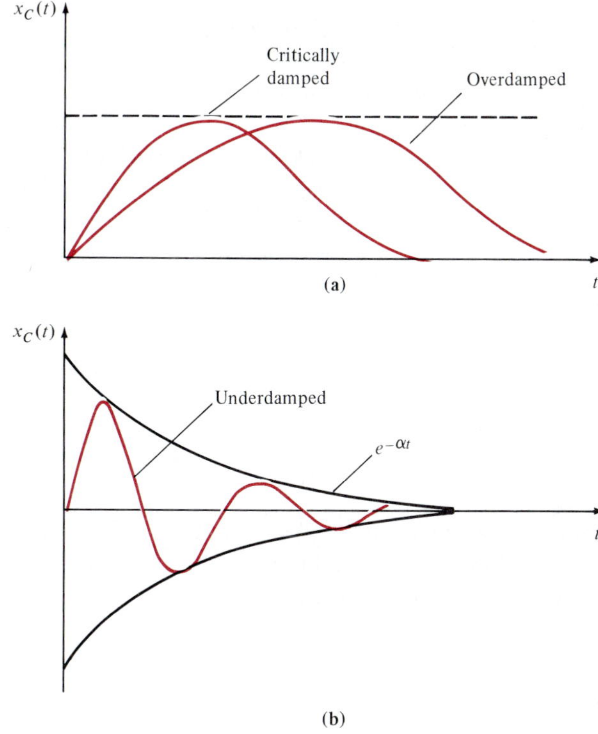

Figure 8.2 Comparison of overdamped, critically damped, and underdamped responses.

It is informative to sketch the natural response for the three cases we have discussed: overdamped, Eq. (8.8); underdamped, Eq. (8.9); and critically damped, Eq. (8.11). Figure 8.2 graphically illustrates the three cases for the situations in which $x_C(0) = 0$. Note that the critically damped response peaks and decays faster than the overdamped response. The underdamped response is an exponentially damped sinusoid whose rate of decay is dependent on the factor α. Actually, the terms $\pm e^{-\alpha t}$ define what is called the *envelope* of the response, and the damped oscillations (i.e., the oscillations of decreasing amplitude) exhibited by the waveform in Fig. 8.2b are called *ringing*.

DRILL EXERCISE

D8.1. A parallel *RLC* circuit has the following circuit parameters: $R = 1\ \Omega$, $L = 2$ H, and $C = 2$ F. Compute the damping ratio and the undamped natural frequency of this network.

Ans: $\alpha = 0.25$, $\omega_0 = 0.5$ rad/sec.

D8.2. A series *RLC* circuit consists of $R = 2\ \Omega$, $L = 1$ H, and a capacitor. Determine the type of response exhibited by the network if (a) $C = \frac{1}{2}$ F, (b) $C = 1$ F, and (c) $C = 2$ F.

Ans: (a) Underdamped, (b) critically damped, (c) overdamped.

8.3

The Network Response

We will now analyze a number of simple *RLC* networks that contain both nonzero initial conditions and constant forcing functions. Circuits that exhibit overdamped, underdamped, and critically damped responses will be considered. The approach will illustrate not only the differential equation techniques, but also the manner in which equivalent circuits, valid at $t = 0+$ and $t = \infty$, are used in the analysis. The following examples will serve to demonstrate the analysis techniques.

EXAMPLE 8.1

Consider the parallel *RLC* circuit shown in Fig. 8.3. The second-order differential equation that describes the voltage $v(t)$ is

$$\frac{d^2v}{dt^2} + \frac{1}{RC}\frac{dv}{dt} + \frac{v}{LC} = 0$$

A comparison of this equation with Eqs. (8.3) and (8.4) indicates that for the parallel *RLC* circuit the damping coefficient is $1/2RC$ and the resonant frequency is $1/\sqrt{LC}$. If the circuit parameters are $R = 2\ \Omega$, $C = \frac{1}{5}$ F, and $L = 5$ H, the equation becomes

$$\frac{d^2v}{dt^2} + 2.5\frac{dv}{dt} + v = 0$$

Let us assume that the initial conditions on the storage elements are $i_L(0) = -1$ A, and $v_C(0) = 4$ V.

The characteristic equation for the network is

$$s^2 + 2.5s + 1 = 0$$

and the roots are

$$s_1 = -2$$

$$s_2 = -0.5$$

Since the roots are real and unequal, the circuit is overdamped, and $v(t)$ is of the form

$$v(t) = K_1e^{-2t} + K_2e^{-0.5t}$$

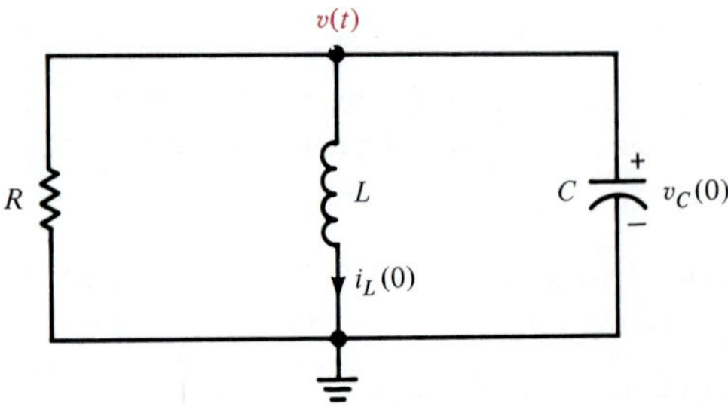

Figure 8.3 Parallel *RLC* circuit.

The initial conditions are now employed to determine the constants K_1 and K_2. Since $v(t) = v_C(t)$,

$$v_C(0) = v(0) = 4 = K_1 + K_2$$

The second equation needed to determine K_1 and K_2 is normally obtained from the expression

$$\frac{dv(t)}{dt} = -2K_1e^{-2t} - 0.5K_2e^{-0.5t}$$

However, the second initial condition is not $dv(0)/dt$. If this were the case, we would simply evaluate the equation above at $t = 0$. This would produce a second equation in the unknowns K_1 and K_2. We can, however, circumvent this problem by noting that the node equation for the circuit can be written as

$$C\frac{dv(t)}{dt} + \frac{v(t)}{R} + i_L(t) = 0$$

or

$$\frac{dv(t)}{dt} = \frac{-1}{RC}v(t) - \frac{i_L(t)}{C}$$

At $t = 0$,

$$\begin{aligned}\frac{dv(0)}{dt} &= \frac{-1}{RC}v(0) - \frac{1}{C}i_L(0)\\ &= -2.5(4) - 5(-1)\\ &= -5\end{aligned}$$

But since

$$\frac{dv(t)}{dt} = -2K_1e^{-2t} - 0.5K_2e^{-0.5t}$$

then when $t = 0$

$$-5 = -2K_1 - 0.5K_2$$

This equation, together with the equation

$$4 = K_1 + K_2$$

produces the constants $K_1 = 2$ and $K_2 = 2$. Therefore, the final equation for the voltage is

$$v(t) = 2e^{-2t} + 2e^{-0.5t}\text{ V}$$

Note that the voltage equation satisfies the initial condition $v(0) = 4$ V.

Before leaving this example, let us examine this problem from another viewpoint. Since the circuit is linear, the characteristic equation for any current or voltage in the network will be the same. Therefore, the current in the inductor must be of the form

$$i_L(t) = K_3 e^{-2t} + K_4 e^{-0.5t}$$

Using the initial conditions, we can evaluate the constants K_3 and K_4, and thus obtain an exact expression for $i_L(t)$.

$$i_L(0) = -1 = K_3 + K_4$$

and

$$\frac{di_L(0)}{dt} = -2K_3 - 0.5K_4$$

However, once again we do not know $di_L(0)/dt$. We do know that

$$v(0) = L\frac{di_L(0)}{dt}$$

and therefore,

$$\frac{di_L(0)}{dt} = \frac{v(0)}{L} = \frac{4}{5}$$

Hence

$$\tfrac{4}{5} = -2K_3 - 0.5K_4$$

This equation, together with the equation

$$-1 = K_3 + K_4$$

comprise the two simultaneous linearly independent equations in K_3 and K_4. Solving these two equations yields $K_3 = -\frac{1}{5}$ and $K_4 = -\frac{4}{5}$. Therefore, $i_L(t)$ is

$$i_L(t) = -\tfrac{1}{5}e^{-2t} - \tfrac{4}{5}e^{-0.5t} \text{ A}$$

The voltage across the inductor is related to the inductor current by the expression

$$\begin{aligned} v(t) &= L\frac{di_L(t)}{dt} \\ &= 5\frac{d}{dt}\left(-\frac{1}{5}e^{-2t} - \frac{4}{5}e^{-0.5t}\right) \\ &= 2e^{-2t} + 2e^{-0.5t} \text{ V} \end{aligned}$$

which is, of course, the expression we derived earlier. The response curve for this voltage $v(t)$ is shown in Fig. 8.4.

Note that in comparison with the *RL* and *RC* circuits we analyzed in Chapter 7, the response of this *RLC* circuit is controlled by two time constants. The first term has a time constant of $\frac{1}{2}$, and the second term has a time constant of 2. ■

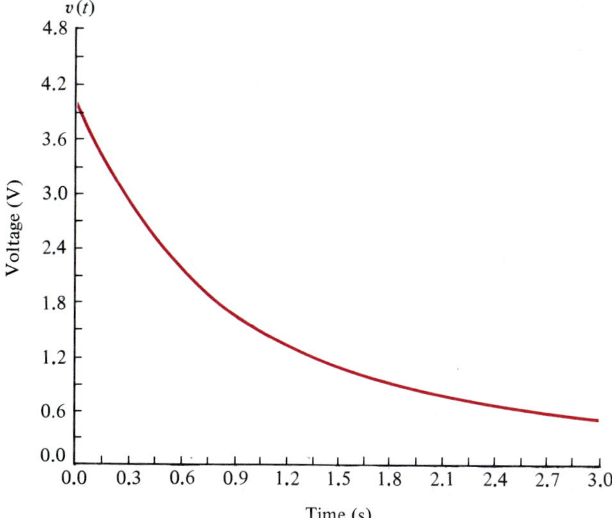

Figure 8.4 Overdamped response.

EXAMPLE 8.2

The series *RLC* circuit shown in Fig. 8.5 has the following parameters: $C = 0.04$ F, $L = 1$ H, $R = 6\ \Omega$, $i_L(0) = 4$ A, and $v_C(0) = -4$ V. The equation for the current in the circuit is given by the expression

$$\frac{d^2i}{dt^2} + \frac{R}{L}\frac{di}{dt} + \frac{i}{LC} = 0$$

A comparison of this equation with Eqs. (8.3) and (8.4) illustrates that for a series *RLC* circuit the damping coefficient is $R/2L$ and the resonant frequency is $1/\sqrt{LC}$. Substituting the circuit element values into the equation above yields

$$\frac{d^2i}{dt^2} + 6\frac{di}{dt} + 25i = 0$$

The characteristic equation is then

$$s^2 + 6s + 25 = 0$$

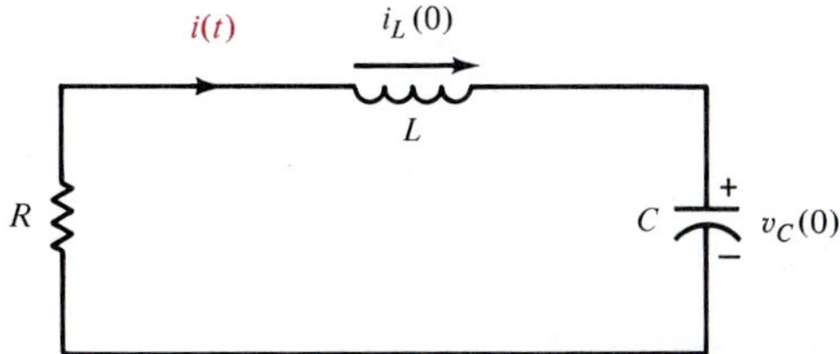

Figure 8.5 Series *RLC* circuit.

and the roots are

$$s_1 = -3 + j4$$

$$s_2 = -3 - j4$$

Since the roots are complex, the circuit is underdamped, and the expression for $i(t)$ is

$$i(t) = K_1 e^{-3t} \cos 4t + K_2 e^{-3t} \sin 4t$$

Using the initial conditions, we find that

$$i(0) = 4 = K_1$$

and

$$\frac{di}{dt} = -4K_1 e^{-3t} \sin 4t - 3K_1 e^{-3t} \cos 4t + 4K_2 e^{-3t} \cos 4t - 3K_2 e^{-3t} \sin 4t$$

and thus

$$\frac{di(0)}{dt} = -3K_1 + 4K_2$$

Although we do not know $di(0)/dt$, we can find it via KVL. From the circuit we note that

$$Ri(0) + L\frac{di(0)}{dt} + v_C(0) = 0$$

or

$$\frac{di(0)}{dt} = -\frac{R}{L}i(0) - \frac{v_C(0)}{L}$$

$$= -\frac{6}{1}(4) + \frac{4}{1}$$

$$= -20$$

Therefore,

$$-3K_1 + 4K_2 = -20$$

and since $K_1 = 4$, $K_2 = -2$. The expression then for $i(t)$ is

$$i(t) = 4e^{-3t} \cos 4t - 2e^{-3t} \sin 4t \text{ A}$$

Note that this expression satisfies the initial condition $i(0) = 4$.

The voltage across the capacitor could be determined via KVL using this current:

$$Ri(t) + L\frac{di(t)}{dt} + v_C(t) = 0$$

or

$$v_C(t) = -Ri(t) - L\frac{di(t)}{dt}$$

Substituting the expression above for $i(t)$ into this equation yields

$$v_C(t) = -4e^{-3t}\cos 4t + 22e^{-3t}\sin 4t \text{ V}$$

Note that this expression satisfies the initial condition $v_C(0) = -4$ V.

Once the characteristic equation was known, we could have simply assumed that $v_C(t)$ was of the form

$$v_C(t) = K_3e^{-3t}\cos 4t + K_4e^{-3t}\sin 4t$$

The constants K_3 and K_4 can be obtained via the initial conditions

$$v_C(0) = -4 = K_3$$

and

$$\frac{dv_C(t)}{dt} = -3K_3e^{-3t}\cos 4t - 4K_3e^{-3t}\sin 4t - 3K_4e^{-3t}\sin 4t + 4K_4e^{-3t}\cos 4t$$

This expression can be related to known initial conditions using the expression

$$i(t) = C\frac{dv_C(t)}{dt}$$

or

$$\frac{dv_C(t)}{dt} = \frac{i(t)}{C}$$

Setting the two expressions above equal to one another and evaluating the resultant equation at $t = 0$ yields

$$\frac{i(0)}{C} = -3K_3 + 4K_4$$

$$100 = -3K_3 + 4K_4$$

Since $K_3 = -4$, $K_4 = 22$, and therefore

$$v_C(t) = -4e^{-3t}\cos 4t + 22e^{-3t}\sin 4t \text{ V}$$

which is equivalent to the expression derived earlier. A plot of the function $v_C(t)$ is shown in Fig. 8.6. ■

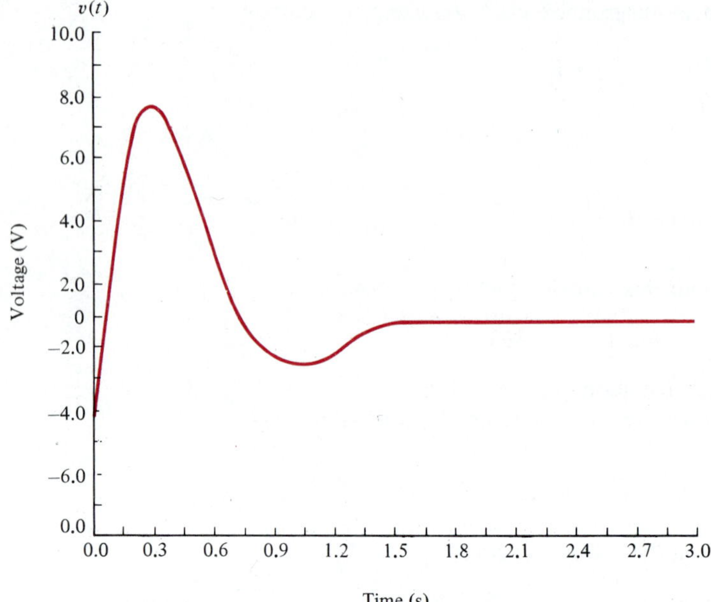

Figure 8.6 Underdamped response.

EXAMPLE 8.3

Let us examine the circuit in Fig. 8.7, which is slightly more complicated than the two we have considered earlier. The two equations that describe the network are

$$L\frac{di(t)}{dt} + R_1 i(t) + v(t) = 0$$

$$i(t) = C\frac{dv(t)}{dt} + \frac{v(t)}{R_2}$$

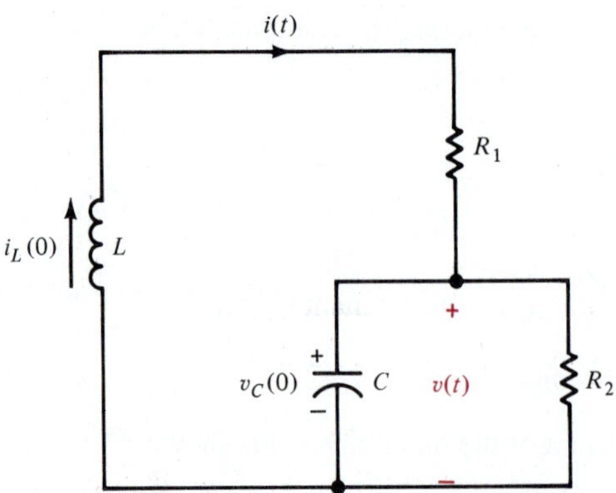

Figure 8.7 Series/parallel *RLC* circuit.

Substituting the second equation into the first yields

$$\frac{d^2v}{dt^2} + \left(\frac{1}{R_2C} + \frac{R_1}{L}\right)\frac{dv}{dt} + \frac{R_1 + R_2}{R_2LC}v = 0$$

If the circuit parameters and initial conditions are

$$R_1 = 10\ \Omega \qquad C = \tfrac{1}{8}\text{ F} \qquad v_C(0) = 1\text{ V}$$

$$R_2 = 8\ \Omega \qquad L = 2\text{ H} \qquad i_L(0) = \tfrac{1}{2}\text{ A}$$

the differential equation becomes

$$\frac{d^2v}{dt^2} + 6\frac{dv}{dt} + 9v = 0$$

The characteristic equation is then

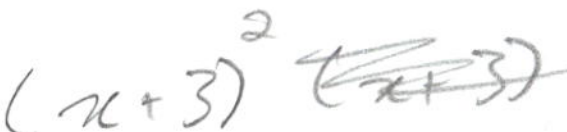

$$s^2 + 6s + 9 = 0$$

and hence the roots are

$$s_1 = -3$$

$$s_2 = -3$$

Since the roots are real and equal, the circuit is critically damped. The term $v(t)$ is then given by the expression

$$v(t) = K_1e^{-3t} + K_2te^{-3t}$$

Since $v(t) = v_C(t)$,

$$v(0) = v_C(0) = 1 = K_1$$

In addition,

$$\frac{dv(t)}{dt} = -3K_1e^{-3t} + K_2e^{-3t} - 3K_2te^{-3t}$$

However,

$$\frac{dv(t)}{dt} = \frac{i(t)}{C} - \frac{v(t)}{R_2C}$$

Setting these two expressions equal to one another and evaluating the resultant equation at $t = 0$ yields

$$\frac{1/2}{1/8} - \frac{1}{1} = -3K_1 + K_2$$

$$3 = -3K_1 + K_2$$

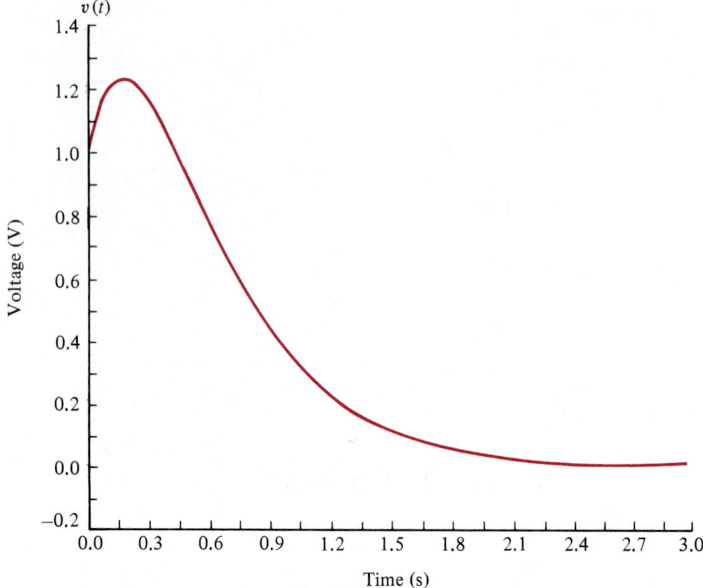

Figure 8.8 Critically damped response.

Since $K_1 = 1$, $K_2 = 6$ and the expression for $v(t)$ is

$$v(t) = e^{-3t} + 6te^{-3t} \text{ V}$$

Note that the expression satisfies the initial condition $v(0) = 1$.

As demonstrated in the preceding examples, once the roots of the characteristic equation are known, the current $i(t)$ could have been expressed as

$$i(t) = K_3e^{-3t} + K_4te^{-3t}$$

The constants can be evaluated as follows:

$$i(0) = \tfrac{1}{2} = K_3$$

and

$$\frac{di(t)}{dt} = -3K_3e^{-3t} + K_4e^{-3t} - 3K_4te^{-3t}$$

but

$$\frac{di(t)}{dt} = -\frac{R_1}{L}i(t) - \frac{v(t)}{L}$$

Equating the two expressions and evaluating the resultant equation at $t = 0$ yields

$$-3 = -3K_3 + K_4$$

Hence

$$K_4 = -\tfrac{3}{2}$$

Therefore,

$$i(t) = \tfrac{1}{2}e^{-3t} - \tfrac{3}{2}te^{-3t} \text{ A}$$

If this expression for the current is employed in the circuit equation,

$$v(t) = -L\frac{di(t)}{dt} - R_1 i(t)$$

we obtain

$$v(t) = e^{-3t} + 6te^{-3t} \text{ V}$$

which is identical to the expression derived earlier. This function is shown in Fig. 8.8. ■

DRILL EXERCISE

D8.3. The switch in the network in Fig. D8.3 opens at $t = 0$. Find $i(t)$ for $t > 0$.

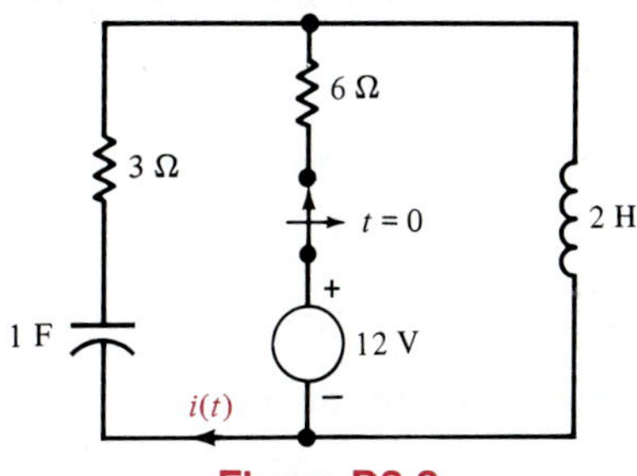

Figure D8.3

Ans: $i(t) = -2e^{-t/2} + 4e^{-t}$ A.

D8.4. The switch in the network in Fig. D8.4 moves from position 1 to position 2 at $t = 0$. Find $v_o(t)$ for $t > 0$.

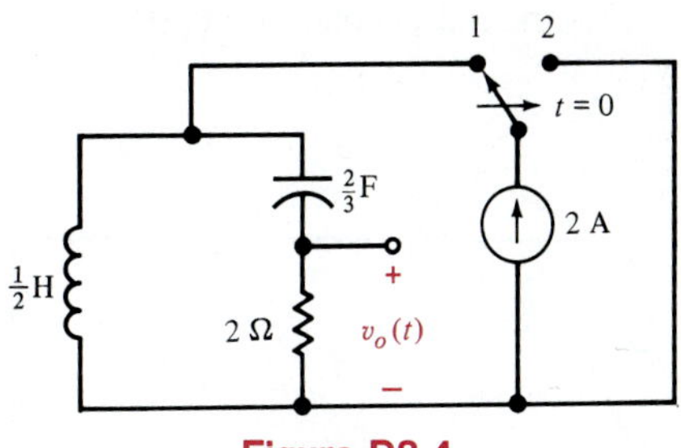

Figure D8.4

Ans: $v_o(t) = 2(e^{-t} - 3e^{-3t})$ V.

EXAMPLE 8.4

Consider the circuit shown in Fig. 8.9. This circuit is the same as that analyzed in Example 8.2, except that a constant forcing function is present. The circuit parameters are the same as those used in Example 8.2:

$$C = 0.04 \text{ F} \qquad i_L(0) = 4 \text{ A}$$

$$L = 1 \text{ H} \qquad v_C(0) = -4 \text{ V}$$

$$R = 6\,\Omega$$

We want to find an expression for $v_C(t)$ for $t > 0$.

From our earlier mathematical development we know that the general solution of this problem will consist of a particular solution plus a complementary solution. From Example 8.2 we know that the complementary solution is of the form $K_3e^{-3t}\cos 4t + K_4e^{-3t}\sin 4t$. The particular solution is a constant, since the input is a constant and therefore the general solution is

$$v_C(t) = K_3e^{-3t}\cos 4t + K_4e^{-3t}\sin 4t + K_5$$

An examination of the circuit shows that in the steady state the final value of $v_C(t)$ is 12 V, since in the steady-state condition, the inductor is a short circuit and the capacitor is an open circuit. Thus $K_5 = 12$. The steady-state value could also be immediately calculated from the differential equation. The form of the general solution is then

$$v_C(t) = K_3e^{-3t}\cos 4t + K_4e^{-3t}\sin 4t + 12$$

The initial conditions can now be used to evaluate the constants K_3 and K_4.

$$v_C(0) = -4 = K_3 + 12$$

$$-16 = K_3$$

Since the derivative of a constant is zero, the results of Example 8.2 show that

$$\frac{dv_C(0)}{dt} = \frac{i(0)}{C} = 100 = -3K_3 + 4K_4$$

and since $K_3 = -16$, $K_4 = 13$. Therefore, the general solution for $v_C(t)$ is

$$v_C(t) = 12 - 16e^{-3t}\cos 4t + 13e^{-3t}\sin 4t \text{ V}$$

Note that this equation satisfies the initial condition $v_C(0) = -4$, and the final condition $v_C(\infty) = 12$ V. ■

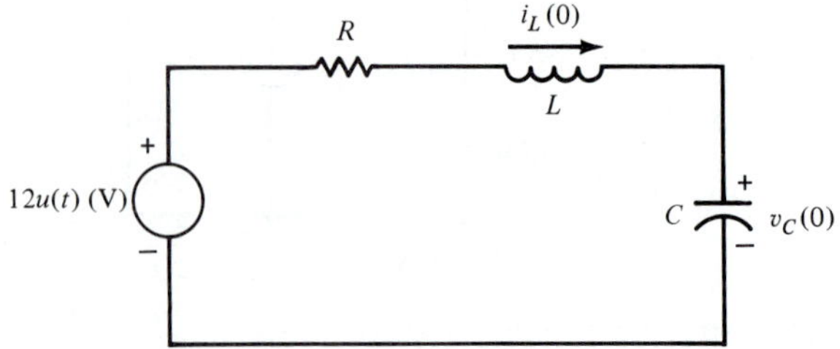

Figure 8.9 Series *RLC* circuit with a step function input.

EXAMPLE 8.5

Let us examine the circuit shown in Fig. 8.10. A close examination of this circuit will indicate that it is identical to that shown in Example 8.3 except that a constant forcing function is present. We assume the circuit is in steady state at $t = 0-$. The equations that describe the circuit for $t > 0$ are

$$L\frac{di(t)}{dt} + R_1 i(t) + v(t) = 24$$

$$i(t) = C\frac{dv(t)}{dt} + \frac{v(t)}{R_2}$$

Combining these equations, we obtain

$$\frac{d^2v(t)}{dt^2} + \left(\frac{1}{R_2C} + \frac{R_1}{L}\right)\frac{dv(t)}{dt} + \frac{R_1 + R_2}{R_2LC}v(t) = \frac{24}{LC}$$

If the circuit parameters are $R_1 = 10\ \Omega$, $R_2 = 2\ \Omega$, $L = 2$ H, and $C = \frac{1}{4}$ F, the differential equation for the output voltage reduces to

$$\frac{d^2v(t)}{dt^2} + 7\frac{dv(t)}{dt} + 12v(t) = 48$$

The characteristic equation is

$$s^2 + 7s + 12 = 0$$

and hence the roots are

$$s_1 = -3$$

$$s_2 = -4$$

The circuit response is overdamped, and therefore the general solution is of the form

$$v(t) = K_1e^{-3t} + K_2e^{-4t} + K_3$$

The steady-state value of the voltage, K_3, can be computed from Fig. 8.11a. Note that

$$v(\infty) = 4\text{ V} = K_3$$

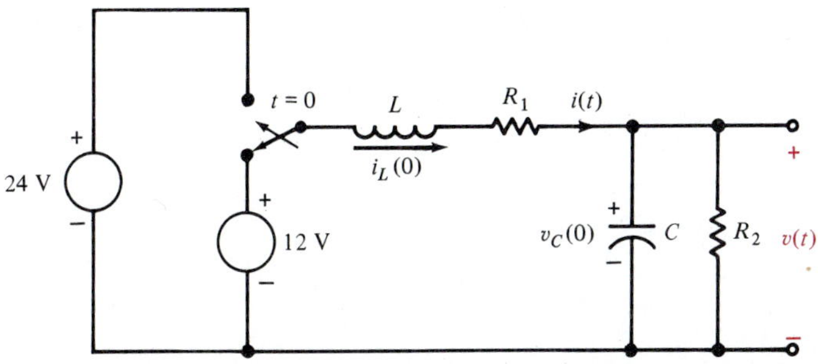

Figure 8.10 Series/parallel *RLC* circuit with a constant forcing function.

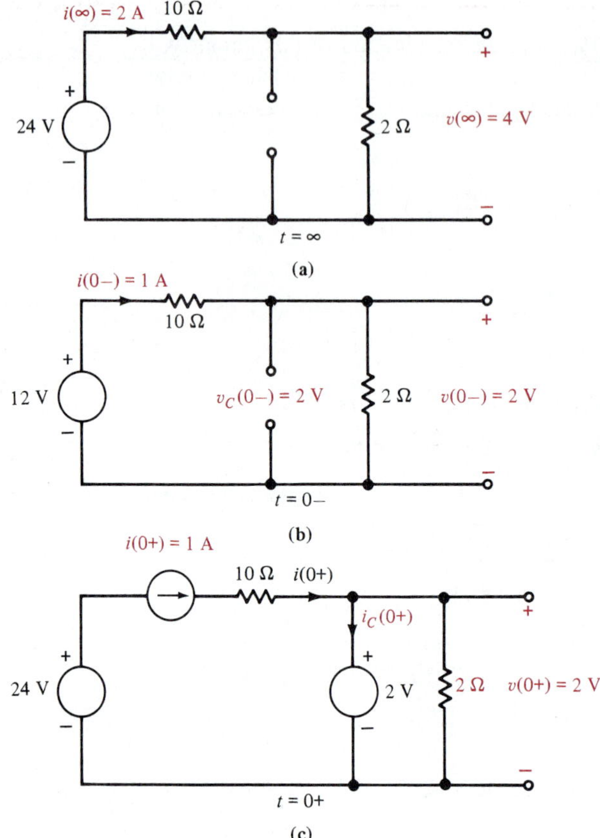

Figure 8.11 Equivalent circuits at $t = \infty$, $t = 0-$, and $t = 0+$ for the circuit in Fig. 8.10.

The initial conditions can be calculated from Fig. 8.11b and c, which are valid at $t = 0-$ and $t = 0+$, respectively. Note that $v(0+) = 2$ V and hence from the response equation

$$v(0+) = 2 \text{ V} = K_1 + K_2 + 4$$

$$-2 = K_1 + K_2$$

Figure 8.11c illustrates that $i(0+) = 1$. From the response equation we see that

$$\frac{dv(0)}{dt} = -3K_1 - 4K_2$$

and since

$$\frac{dv(0)}{dt} = \frac{i(0)}{C} - \frac{v(0)}{R_2C}$$

$$= 4 - 4$$

$$= 0$$

then

$$0 = -3K_1 - 4K_2$$

Solving the two equations for K_1 and K_2 yields $K_1 = -8$ and $K_2 = 6$. Therefore, the general solution for the voltage response is

$$v(t) = 4 - 8e^{-3t} + 6e^{-4t} \text{ V}$$

Note that this equation satisfies both the initial and final values of $v(t)$. ■

DRILL EXERCISE

D8.5. The switch in the network in Fig. D8.5 moves from position 1 to position 2 at $t = 0$. Compute $i_o(t)$ for $t > 0$ and use this current to determine $v_o(t)$ for $t > 0$.

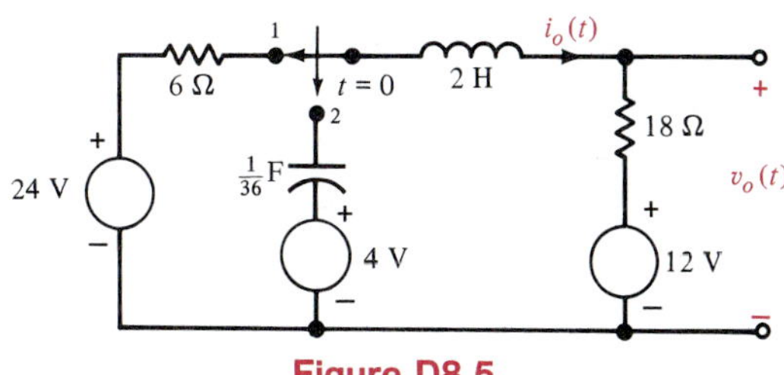

Figure D8.5

Ans: $i_o(t) = -\frac{1}{2}e^{-3t} + e^{-6t}$ A, $v_o(t) = 12 + 18i_o(t)$ V.

The examples clearly indicate that the general solution of even very simple *RLC* circuits can be quite complicated. Although we can continue our current mathematical analysis and present techniques for solving more general networks, better methods exist. For example, in Chapter 17 the Laplace transform technique will be applied to analyze *RLC* circuits with a variety of forcing functions, and the reader will find that the transform technique is not only easier to use but provides some additional insight in the analysis.

8.4 PSPICE Analysis of *RLC* Circuits

In the previous sections of this chapter, our analysis techniques have been mathematical in nature. Although the mathematical approach does indicate the type of response that the network will produce, it does not lead quickly to a solution. In contrast, a PSPICE program consisting of only a few statements is sufficient to yield results that easily display the circuit behavior. Now, using PSPICE, we reexamine some of the circuits that we treated in earlier examples, and illustrate the ease with which it can be applied in general.

EXAMPLE 8.6

The *RLC* circuit in Example 8.2 is redrawn for a PSPICE analysis in Fig. 8.12. Let us write a PSPICE program that will plot the capacitor voltage in intervals of 0.1 sec over a 3-sec range.

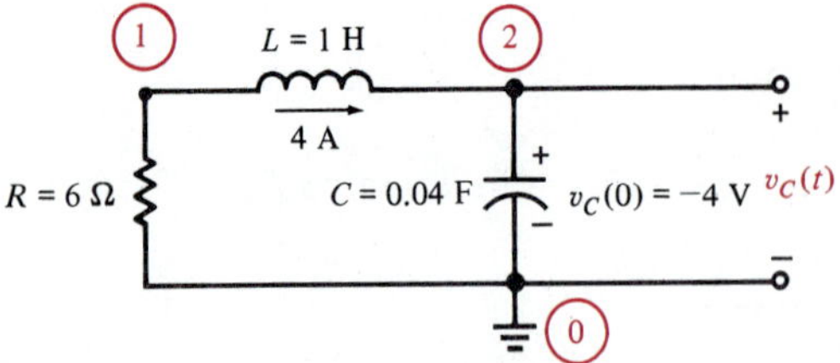

Figure 8.12 Circuit in Fig. 8.5 redrawn for PSPICE analysis.

The following program plots the capacitor voltage.

```
EXAMPLE 8.6  PLOT OF CAPACITOR VOLTAGE IN
*EXAMPLE 8.2
R     1     0     6
L     1     2     1     IC = 4
C     2     0     0.04  IC = -4
.TRAN       0.1   3     UIC
.PROBE
.END
```

A plot of the capacitor voltage is shown in Fig. 8.6. ■

EXAMPLE 8.7

The *RLC* circuit in Example 8.3 is redrawn in Fig. 8.13 for a PSPICE analysis. Let us write a PSPICE program that will plot the voltage $v(t)$ in intervals of 0.1 sec over a 3-sec range. With reference to Fig. 8.13 the PSPICE program is

```
EXAMPLE 8.7  PLOT OF V(T) IN EXAMPLE 8.3
L     0     1     2     IC = 0.5
R1    1     2     10
C     2     0     0.125       IC = 1
R2    2     0     8
.TRAN       0.1   3     UIC
.PROBE
.END
```

A plot of the voltage $v(t)$ is shown in Fig. 8.8. ■

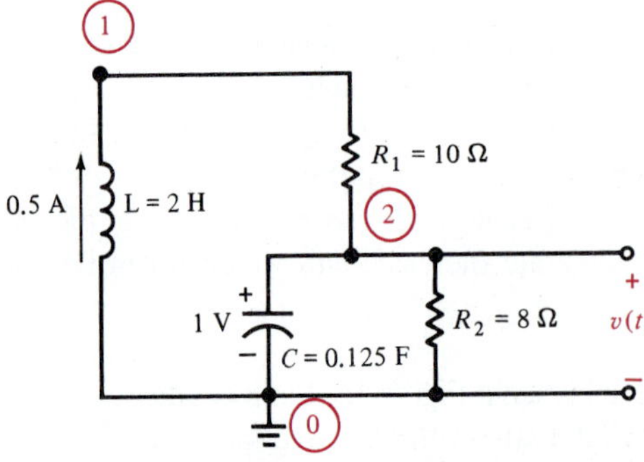

Figure 8.13 Circuit in Example 8.3 redrawn for PSPICE analysis.

EXAMPLE 8.8

Given the network in Fig. 8.14a, let us plot $v_o(t)$ and $i_o(t)$ for $t > 0$ over a 10-sec interval using a 100-ms step size.

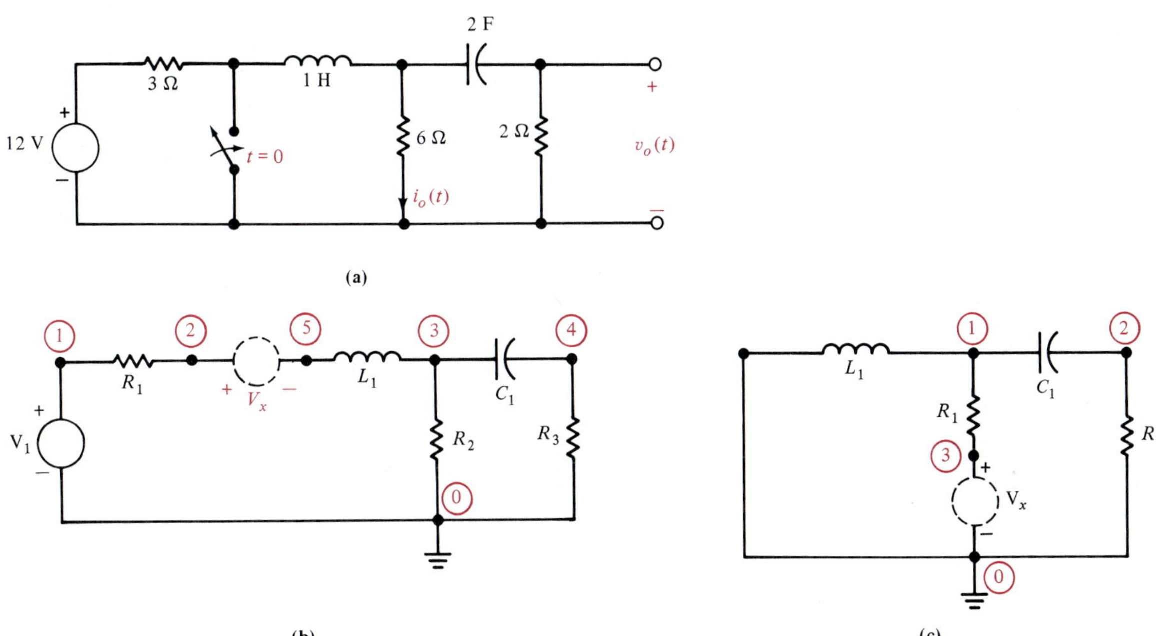

Figure 8.14 Networks used in Example 8.8.

The initial conditions are derived from the network in Fig. 8.14b. The PSPICE program for this circuit is

```
INITIAL NETWORK
V1   1   0   DC   12
VX   2   5   DC   0
R1   1   2   3
R2   3   0   6
R3   4   0   2
L1   5   3   1
C1   3   4   2
* USE DC ANALYSIS
.DC V1 12 12 1
.PRINT DC I(VX) V(3) V(4)
.END
```

PSPICE does not yield the voltage between nodes 3 and 4. However, $V(3)$ and $V(4)$ can be used to determine $V(3) - V(4) = 8$ V. I(VX) = 1.33 A. The network for $t > 0$ is shown in Fig. 8.14c. The program for this network is

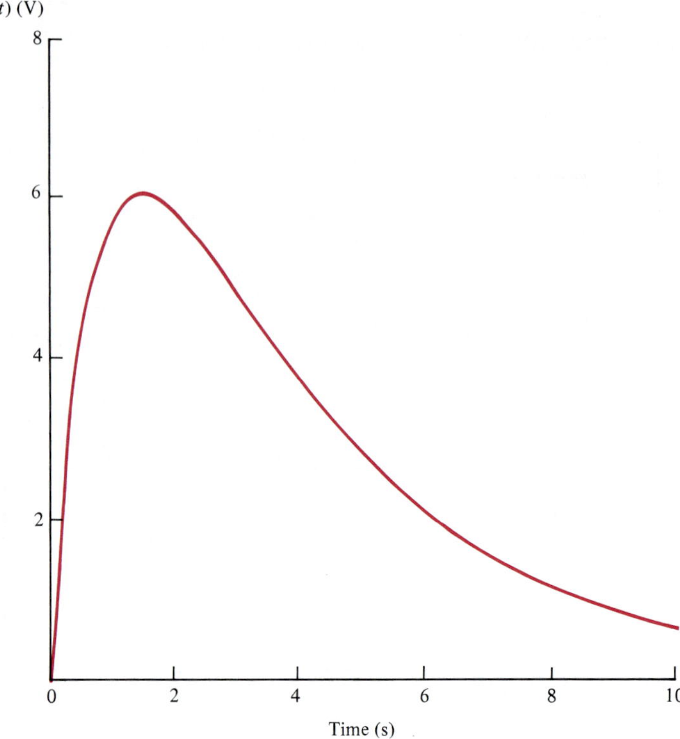

Figure 8.15a Output plot for Example 8.8.

```
FINAL NETWORK
R1   1   3   6
R2   2   0   2
L1   1   0   1    IC = 1.33A
C1   2   1   2    IC = 8V
VX   3   0   DC   0
* USE TRANSIENT ANALYSIS
.TRAN .1 10 UIC
.PROBE
.END
```

The plots for $v_o(t)$ and $i_o(t)$ are shown in Fig. 8.15a and b, respectively. ■

EXAMPLE 8.9

Consider the network shown in Fig. 8.16a. The network is labeled for a PSPICE analysis as shown in Fig. 8.16b. We wish to plot the output voltage $v_{out}(t)$ for $t > 0$ for a 10-sec interval with a 100-ms step size.

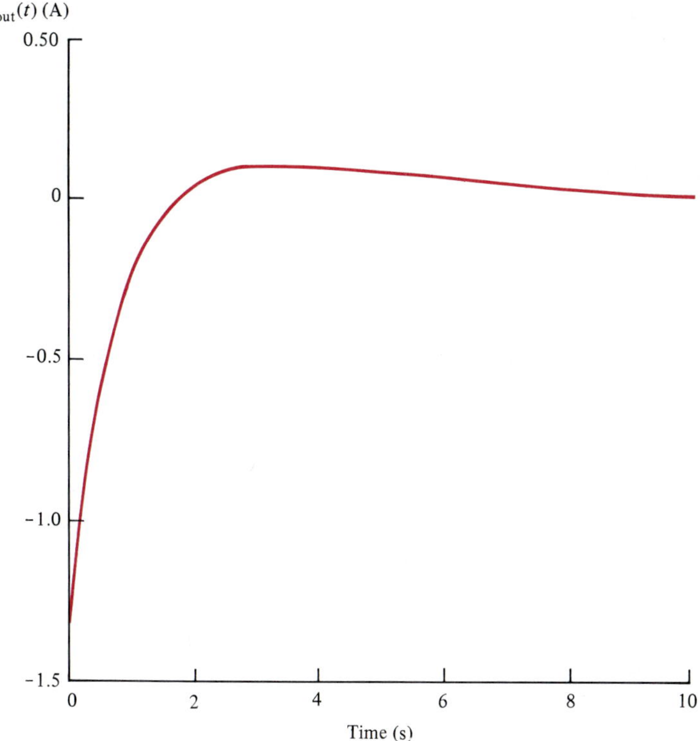

Figure 8.15b Output plot for Example 8.8.

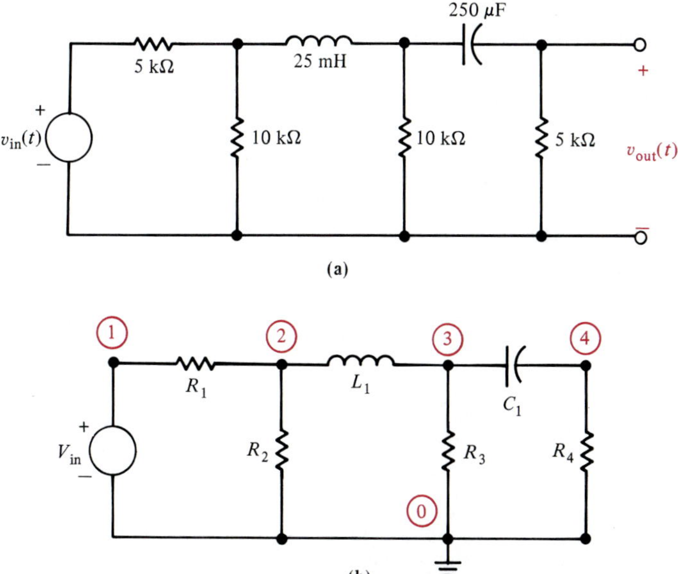

Figure 8.16 Circuits used in Example 8.9.

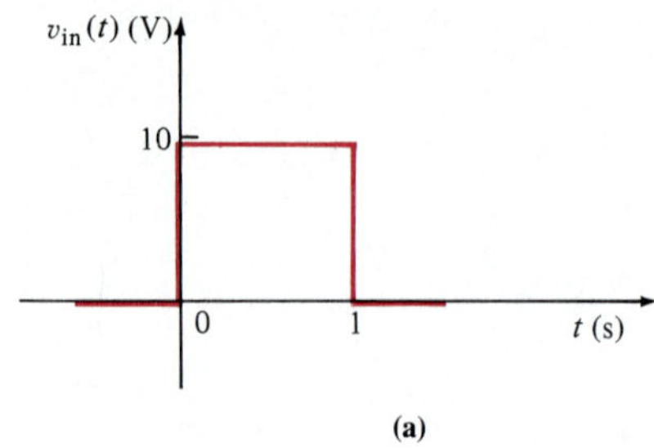

(a)

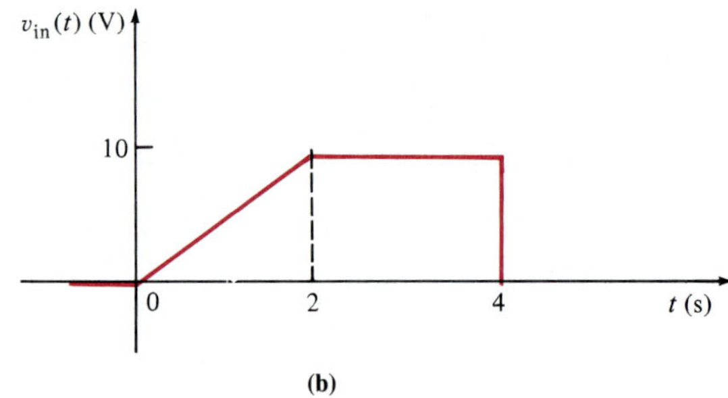

(b)

Figure 8.17 Inputs for Example 8.9.

The four inputs to the network are as follows:

1. The pulse shown in Fig. 8.17a.
2. The piecewise linear function shown in Fig. 8.17b.
3. The exponential function defined by the equation

$$\begin{aligned} v_{in}(t) &= 10(1 - e^{-t/4}) \text{ V} && 0 \leq t \leq 3 \text{ sec} \\ &= 10(1 - e^{-t/4}) - 10e^{-(t-3)} \text{ V} && 3 \text{ sec} \leq t \end{aligned}$$

4. The sinusoidal function is defined by the equation

$$v_{in}(t) = 10e^{-0.5t} \sin 2\pi t \text{ V}$$

The program for the network with the first input is

```
PULSE INPUT
VIN 1   0   PWL(0 10 1 10 1.001 0 10 0)
R1  1   2   5K
R2  2   0   10K
R3  3   0   10K
R4  4   0   5K
C1  3   4   250UF IC=0
L1  2   3   25MH  IC=0
```

```
* TRANSFER FUNCTIONS AND TRANSIENT ANALYSIS
.TF V(4) VIN
.TRAN .1 10 UIC
.PROBE
.END
```

The input and output voltages are shown in Fig. 8.18.

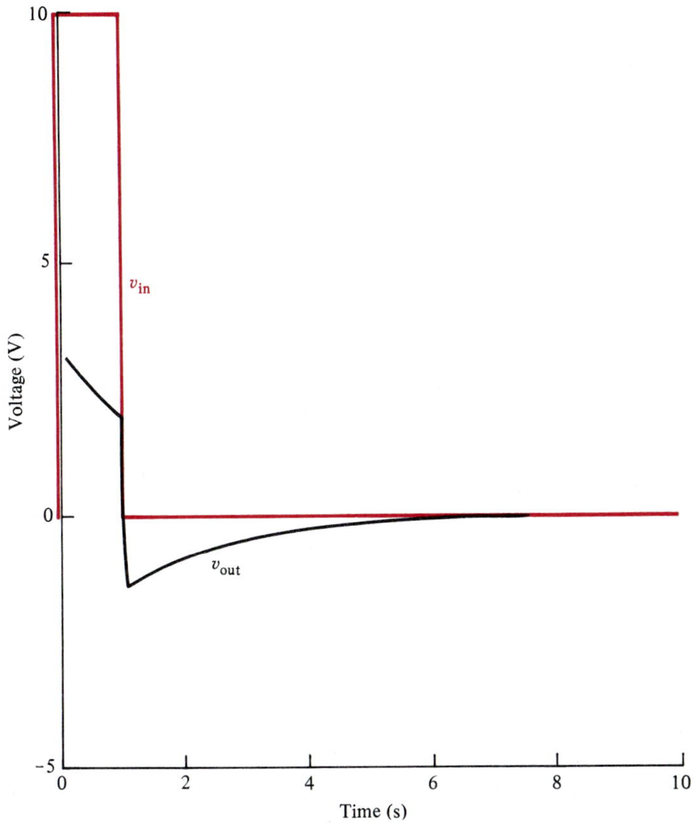

Figure 8.18 Input and output plots for the first input in Example 8.9.

The PSPICE statement for the second input is

```
VIN 1 0 PWL(0 0 2 10 4 10 4.001 0)
```

If the input voltage statement in the original program is replaced with this statement, the input and output voltages are as shown in Fig. 8.19.

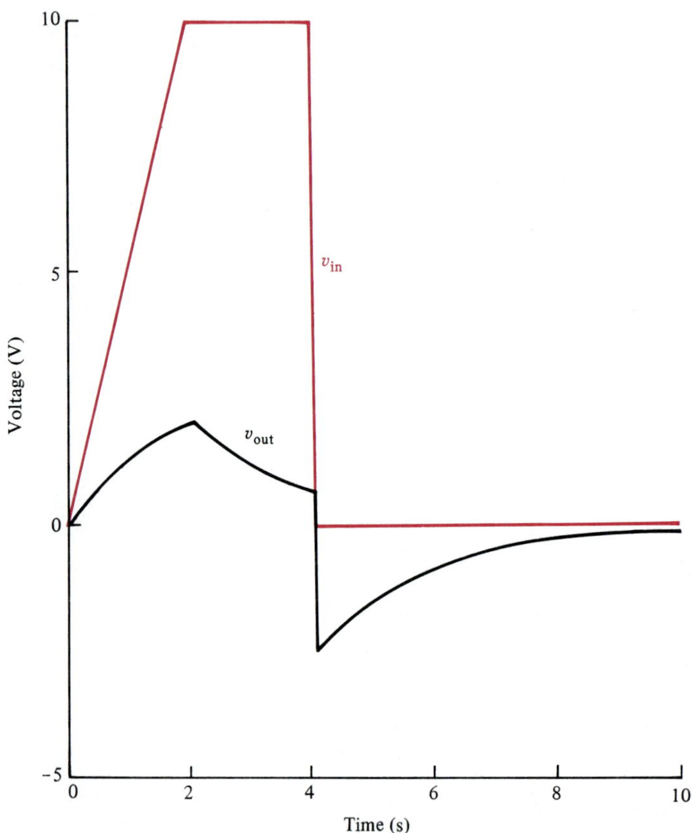

Figure 8.19 Input and output plots for the second input in Example 8.9.

The PSPICE statement for the third input is

```
VIN 1 0 EXP(0 10 0 4 3 1)
```

If the input voltage statement in the original program is replaced with this statement for the exponential source, the input and output voltages are as shown in Fig. 8.20.

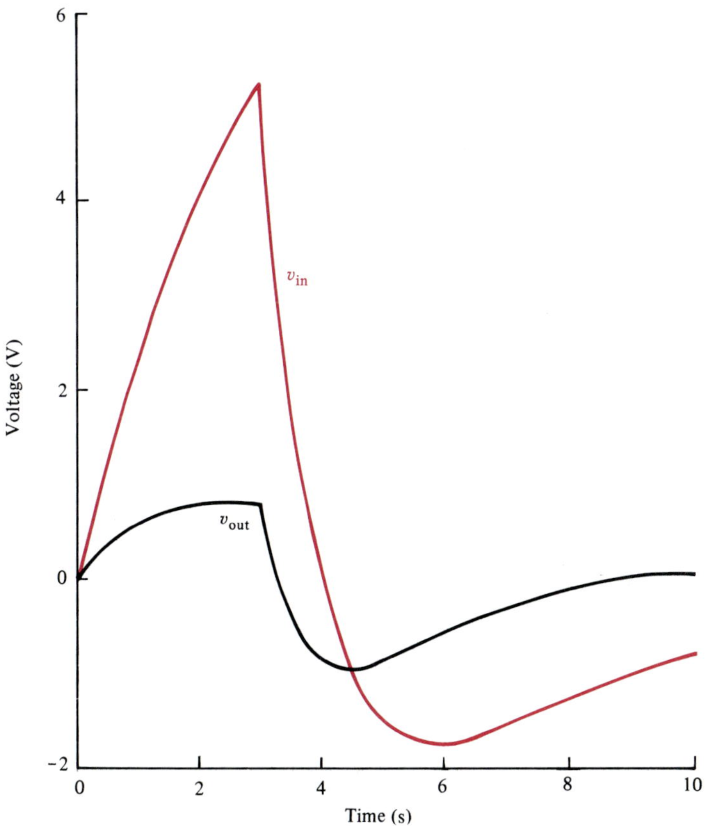

Figure 8.20 Input and output plots for the third input in Example 8.9.

The PSPICE statement for the fourth input is

```
VIN 1 0 SIN(0 10 1 0 .5)
```

Finally, if the input voltage statement in the original program is replaced with this statement for the sinusoidal source, the input and output voltages are as shown in Fig. 8.21. ■

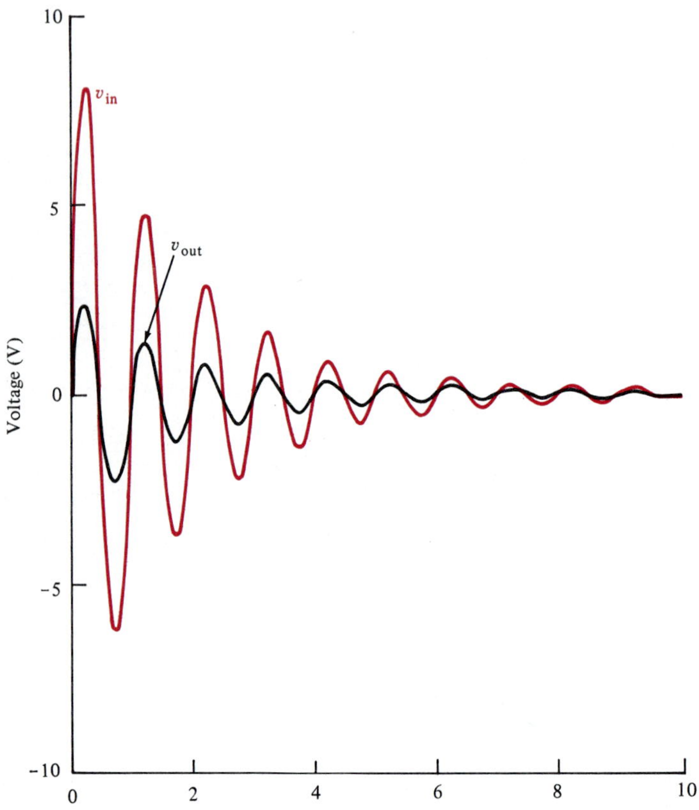

Figure 8.21 Input and output plots for the fourth input in Example 8.9.

DRILL EXERCISE

D8.6. For the network in Fig. D8.6, use PSPICE to plot the transient response of $i(t)$ from 0 to 0.5 sec in increments of 10 ms.

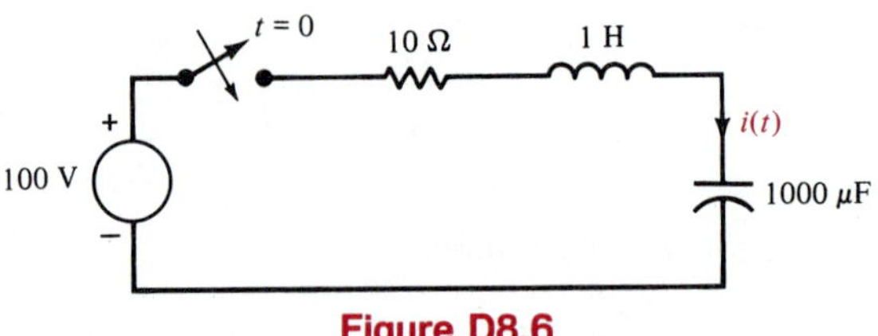

Figure D8.6

Ans:

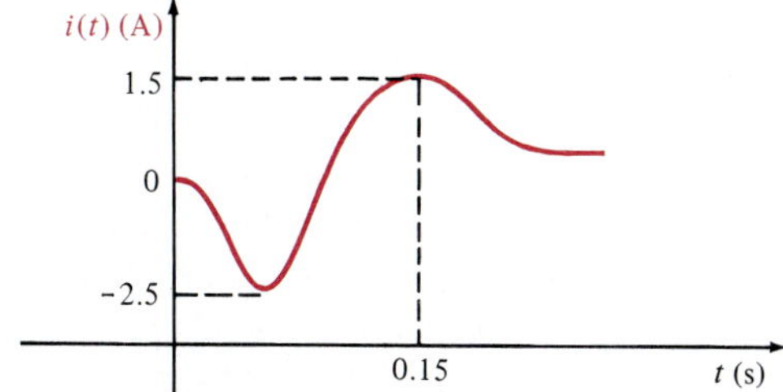

D8.7. Given the network in Fig. D8.7a, use PSPICE to plot $v_o(t)$, $0 \leq t \leq 10$ sec, using a 100-ms step size if the input is given in Fig. D8.7b.

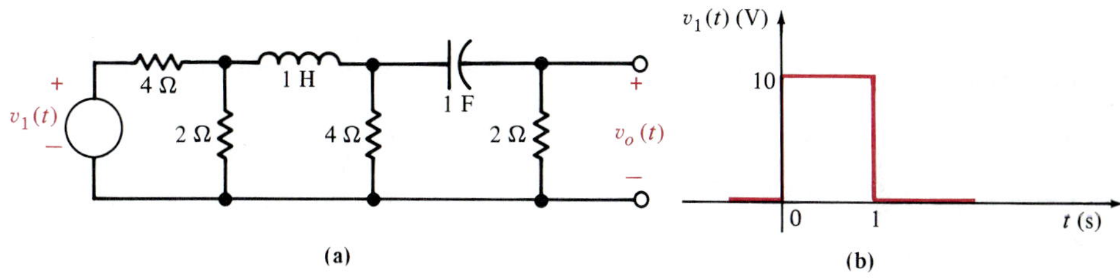

Figure D8.7

Ans:

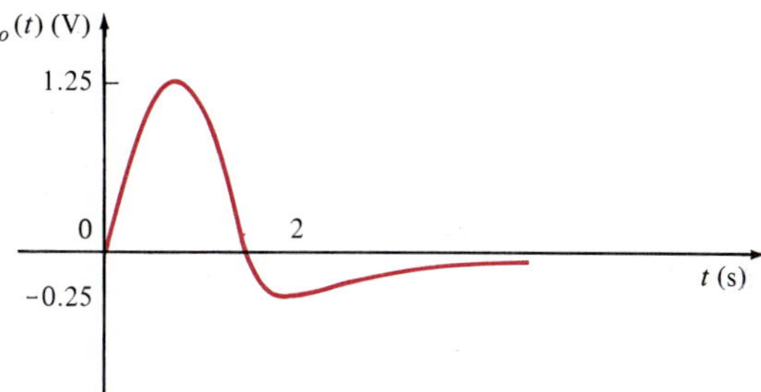

8.5 Summary

We have shown that the analysis of *RLC* circuits leads to a second-order differential equation. The roots of the circuit's characteristic equation control the type of response. These roots are a function of the circuit's damping ratio and undamped resonant frequency. If the roots are real and unequal, the response is overdamped; if the roots are complex conjugates, the response is underdamped; and if the roots are real and equal the response is critically damped.

Finally, the PSPICE analysis of *RLC* circuits is presented.

KEY POINTS

- The characteristic equation for an *RLC* circuit is $s^2 + 2\alpha s + \omega_0^2 = 0$, where α is the damping coefficient and ω_0 is the undamped resonant frequency.
- The two roots of the characteristic equation for an *RLC* circuit are real and unequal if $\alpha > \omega_0$; they are complex numbers if $\alpha < \omega_0$, and they are real and equal if $\alpha = \omega_0$.
- The response of an *RLC* circuit is said to be overdamped if the roots of the network's characteristic equation are real and unequal, underdamped if they are complex numbers, and critically damped if they are real and equal.
- The PSPICE analysis of networks containing both an inductor and a capacitor is performed in exactly the same manner as outlined in Chapter 7.

PROBLEMS

8.1. The voltage $v_1(t)$ in a network is defined by the equation

$$\frac{d^2v_1(t)}{dt^2} + 2\left[\frac{dv_1(t)}{dt}\right] + 5v_1(t) = 0$$

Find
(a) The characteristic equation of the network.
(b) The circuit's natural frequencies.
(c) The expression for $v_1(t)$.

8.2. The output voltage of a circuit is described by the differential equation

$$\frac{d^2v_o(t)}{dt^2} + 6\left[\frac{dv_o(t)}{dt}\right] + 10v_o(t) = 0$$

Find
(a) The characteristic equation of the circuit.
(b) The network's natural frequencies.
(c) The equation for $v_o(t)$.

8.3. The terminal current in a network is described by the equation

$$\frac{d^2i_o(t)}{dt^2} + 10\left[\frac{di_o(t)}{dt}\right] + 25i_o(t) = 0$$

Find
(a) The characteristic equation of the network.
(b) The network's natural frequencies.
(c) The equation for $i_o(t)$.

8.4. The network voltage $v_o(t)$ is described by the differential equation

$$\frac{d^2v_o(t)}{dt^2} + 6\left[\frac{dv_o(t)}{dt}\right] + 9v_o(t) = 0$$

Find
(a) The characteristic equation of the network.
(b) The network's natural frequencies.
(c) The expression for $v_o(t)$.

8.5. The differential equation that describes the current $i_o(t)$ in a network is

$$\frac{d^2i_o(t)}{dt^2} + 6\left[\frac{di_o(t)}{dt}\right] + 8i_o(t) = 0$$

Find
(a) The characteristic equation of the network.
(b) The network's natural frequencies.
(c) The expression for $i_o(t)$.

8.6. The differential equation for the voltage $v_1(t)$ in a network is

$$\frac{d^2v_1(t)}{dt^2} + 7\left[\frac{dv_1t}{dt}\right] + 12v_1(t) = 0$$

Find
(a) The characteristic equation for the network.
(b) The natural frequencies for the network.
(c) The expression for $v_1(t)$.

8.7. The parameters for a parallel *RLC* circuit are $R = 1\ \Omega$, $L = \frac{1}{5}$ H, and $C = \frac{1}{4}$ F. Determine the type of damping exhibited by the circuit.

8.8. A series *RLC* circuit contains a resistor $R = 2\ \Omega$ and a capacitor $C = \frac{1}{8}$ F. Select the value of the inductor so that the circuit is critically damped.

8.9. A current in an *RLC* network is described by the equation

$$\frac{d^2i(t)}{dt^2} + 6\left[\frac{di(t)}{dt}\right] + 10i(t) = 0$$

If the initial conditions are $i(0) = 1$ A and $i'(0) = 0$, find $i(t)$ for $t > 0$.

8.10. A branch current in a network is described by the equation

$$\frac{d^2i_1(t)}{dt^2} + 4\left[\frac{di_1(t)}{dt}\right] + \frac{17}{4}i_1(t) = 0$$

If the initial conditions are $i_1(0) = 1$ A and $i_1(\pi) = 0$, find $i_1(t)$ for $t > 0$.

8.11. The current in a circuit is described by the equation

$$\frac{d^2i_o(t)}{dt^2} + 4\left[\frac{di_o(t)}{dt}\right] + 4i_o(t) = 0$$

The initial conditions are $i_o(0) = 1$ A and $i_o'(0) = 0$. Find $i_o(t)$ for $t > 0$.

8.12. The terminal voltage in a network is described by the equation

$$\frac{d^2v_o(t)}{dt^2} + 8\left[\frac{dv_o(t)}{dt}\right] + 16v_o(t) = 0$$

If the initial conditions are $v_o(0) = 1$ and $v_o'(0) = 2$ V, find $v_o(t)$ for $t > 0$.

8.13. The output voltage of an *RLC* network is described by the equation

$$\frac{d^2v_o(t)}{dt^2} + 5\left[\frac{dv_o(t)}{dt}\right] + 6v_o(t) = 0$$

If the initial conditions are $v_o(0) = 1$ V and $v_o'(0) = 2$ V, find $v_o(t)$ for $t > 0$.

8.14. If the voltage $v_o(t)$ in a circuit is described by the equation

$$\frac{d^2v_o(t)}{dt^2} + 4\left[\frac{dv_o(t)}{dt}\right] + 4v_o(t) = 0$$

and the initial conditions are $v_o(0) = 2$ V and $v_o'(0) = 4$ V, find $v_o(t)$ for $t > 0$.

8.15. The current in an *RLC* network is given by the expression

$$\frac{d^2i_o(t)}{dt^2} + 4\left[\frac{di_o(t)}{dt}\right] + 3i_o(t) = 6$$

The initial conditions are $i_o(0) = 1$ A and $i_o'(0) = 0$. Find $i_o(t)$ for $t > 0$.

8.16. The output voltage in an *RLC* network is given by the expression

$$\frac{d^2v_o(t)}{dt^2} + 7\left[\frac{dv_o(t)}{dt}\right] + 10v_o(t) = 20e^{-3t}$$

If $v_o(0) = 4$ V and $v_o'(0) = 0$, find $v_o(t)$ for $t > 0$.

8.17. The voltage $v_1(t)$ in an *RLC* network is given by the expression

$$\frac{d^2v_1(t)}{dt^2} + 2\left[\frac{dv_1(t)}{dt}\right] + v_1(t) = 4$$

If $v_1(0) = 0$ and $v_1'(0) = 1$ *V*, find $v_1(t)$ for $t > 0$.

8.18. The current $i_1(t)$ is given by the expression

$$\frac{d^2i_1(t)}{dt^2} + 8\left[\frac{di_1(t)}{dt}\right] + 16i_1(t) = 8e^{-2t}$$

If $i_1(0) = 0$ and $i_1'(0) = 0$, find $i_1(t)$ for $t > 0$.

8.19. The voltage $v(t)$ is described by the equation

$$\frac{d^2v(t)}{dt^2} + 4\left[\frac{dv(t)}{dt}\right] + 8v(t) = 12$$

If $v(0) = 0$ and $v'(0) = 0$, find $v(t)$ for $t > 0$.

8.20. The current $i_A(t)$ is expressed by the equation

$$\frac{d^2i_A(t)}{dt^2} + 8\left[\frac{di_A(t)}{dt}\right] + 20i_A(t) = 10e^{-t}$$

If $i_A(0) = 0$ and $i_A'(0) = 0$, find $i_A(t)$ for $t > 0$.

8.21. For the underdamped circuit shown in Fig. P8.21, determine the voltage $v(t)$ if the initial conditions on the storage elements are $i_L(0) = 1$ A and $v_C(0) = 10$ V.

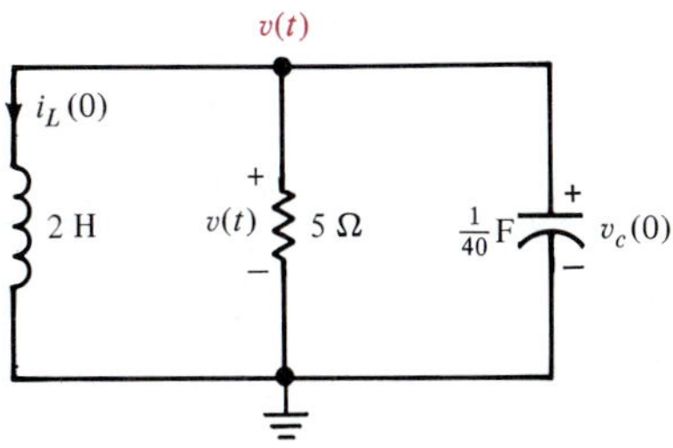

Figure P8.21

8.22. Given the circuit and the initial conditions of Problem 8.21, determine the current through the inductor.

8.23. In the critically damped circuit shown in Fig. P8.23, the initial conditions on the storage elements are $i_L(0) = 2$ A and $v_C(0) = 5$ V. Determine the voltage $v(t)$.

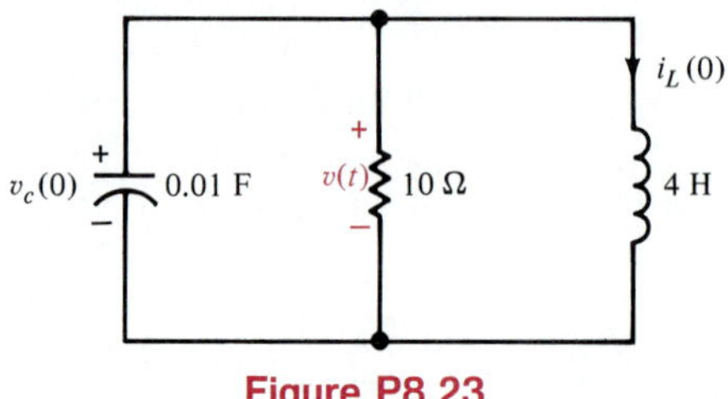

Figure P8.23

8.24. Given the circuit and the initial conditions from Problem 8.23, determine the current $i_L(t)$ that is flowing through the inductor.

8.25. Given the circuit in Fig. P8.25, find the equation for $i(t)$, $t > 0$.

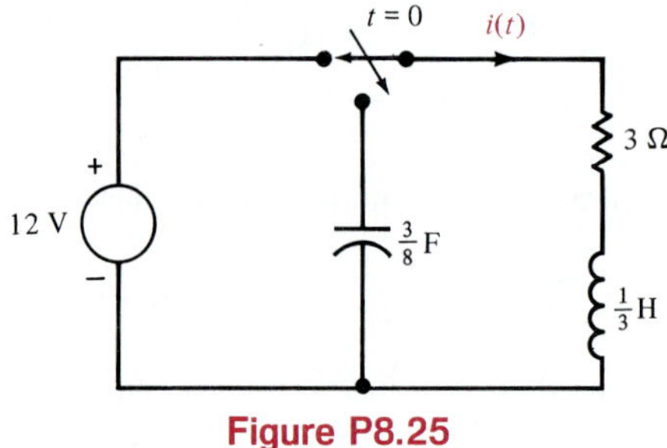

Figure P8.25

8.26. In the circuit shown in Fig. P8.26, switch action occurs at $t = 0$. Determine the voltage $v_o(t)$, $t > 0$.

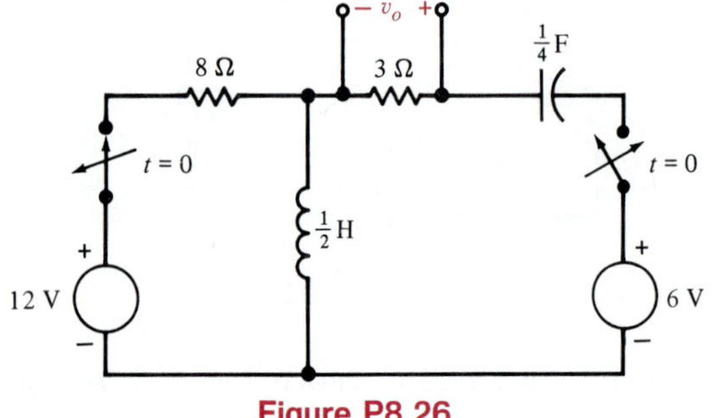

Figure P8.26

8.27. In the circuit shown in Fig. P8.27, find $v(t)$, $t > 0$.

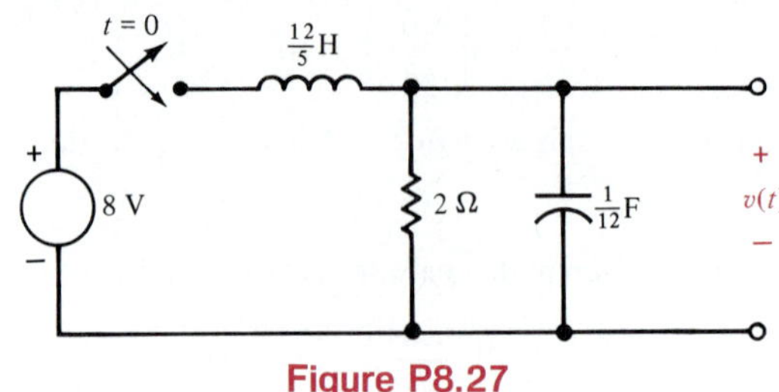

Figure P8.27

8.28. Given the network in Fig. P8.28, use PSPICE to plot $v_o(t)$ over a 10-sec interval starting at $t = 0$ using a 100-ms step size.

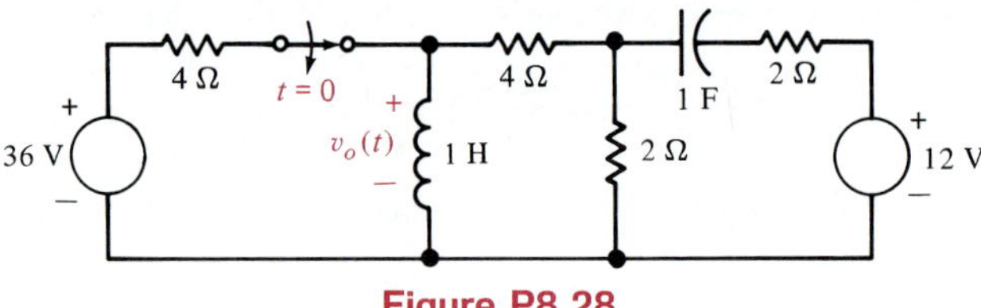

Figure P8.28

8.29. Given the network in Fig. P8.29, use PSPICE to plot $v_o(t)$ over a 10-sec interval starting at $t = 0$ using a 100-ms step size.

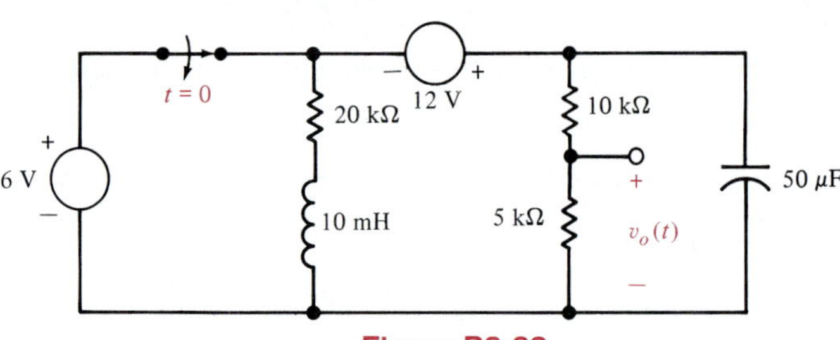

Figure P8.29

8.30. For the network shown in Fig. P8.30, use PSPICE to plot $v_o(t)$ over a 10-sec interval starting at $t = 0$ using a 100-ms step size.

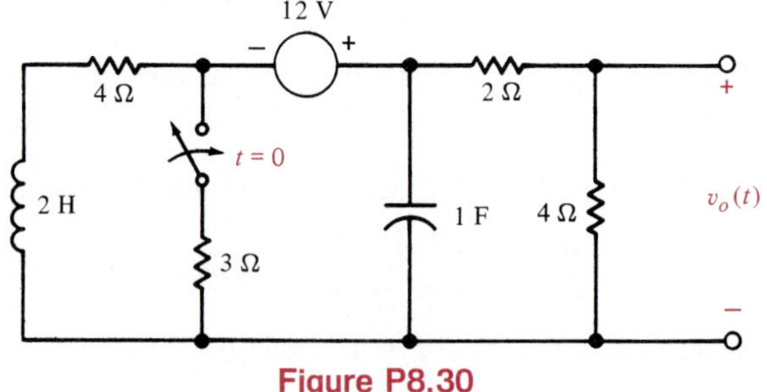

Figure P8.30

8.31. Given the network in Fig. P8.31, use PSPICE to plot $v_o(t)$ over a 10-sec interval starting at $t = 0$ using a 100-msec step size.

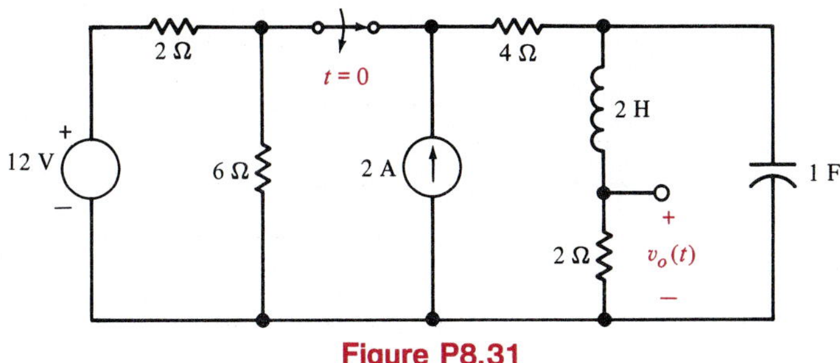

Figure P8.31

8.32. Given the network in Fig. P8.32a and the input voltage shown in Fig. P8.32b, use PSPICE to plot the voltage $v_o(t)$ over the interval $0 \leq t \leq 4$ sec using a 20-msec step size.

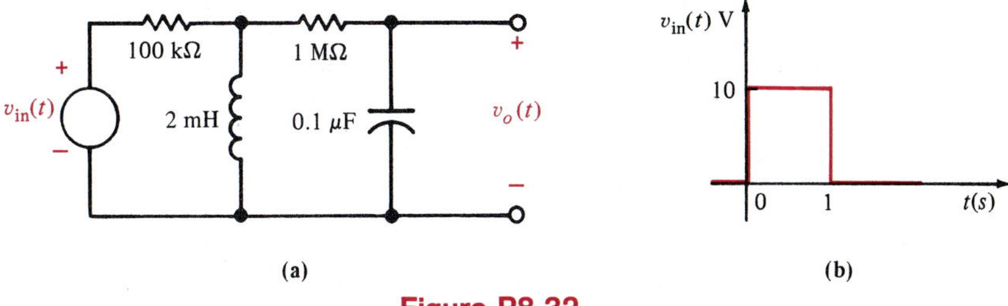

Figure P8.32

8.33. Given the network in Fig. P8.33a and the input voltage shown in Fig. P8.33b, plot $v_o(t)$ using PSPICE over the interval $0 \leq t \leq 10$ sec using a 100-msec step size.

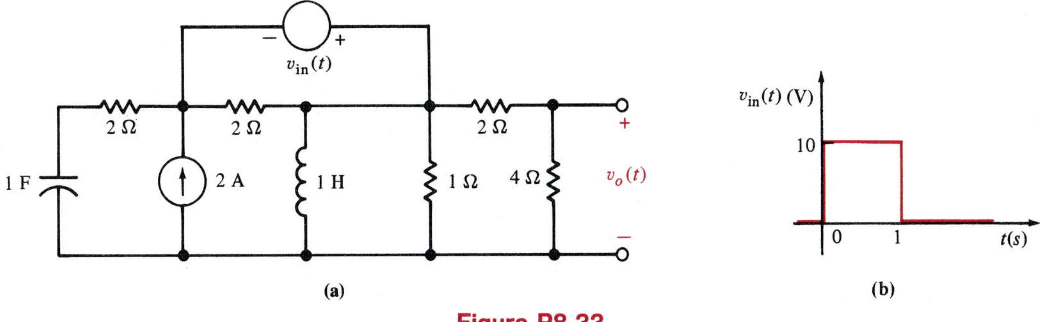

Figure P8.33

8.34. Given the network in Fig. P8.34 and the input

$$v_S(t) = 10(1 - e^{-t/4})\text{ V} \qquad 0 \le t \le 3\text{ sec}$$
$$= 10(1 - e^{-t/4}) - 10e^{-(t-3)}\text{ V} \qquad 3\text{ sec} < t$$

plot the voltage $v_o(t)$ over the interval $0 \le t \le 10$ sec using PSPICE with a 50-msec step size.

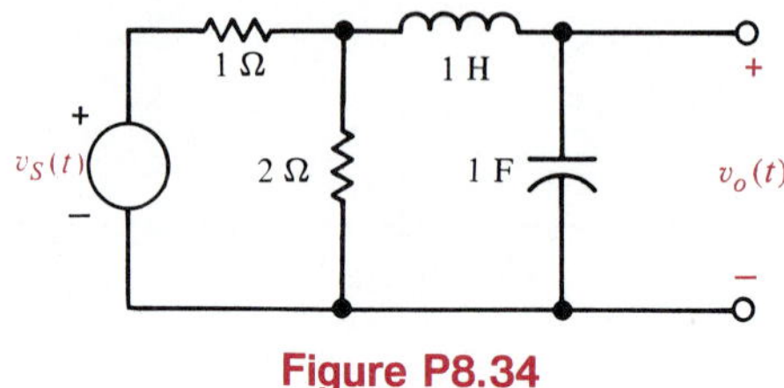

Figure P8.34

8.35. Given the network in Fig. P8.35a and the input in Fig. P8.35b, using PSPICE plot $v_o(t)$ in the interval $0 \le t \le 4$ sec using a 20-msec step size.

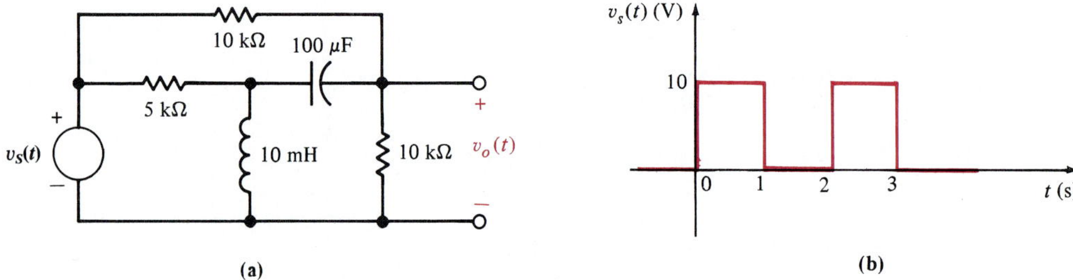

Figure P8.35

8.36. Repeat Problem 8.35 if the input voltage is

$$v_S(t) = 10(1 - e^{-t/2})\ V \qquad 0 \le t \le 3\text{ sec}$$
$$= 10(1 - e^{-t/2}) - 10e^{-(t-3)/4}\text{ A} \qquad 3\text{ sec} < t$$

Use a 30-msec step size over the interval $0 \le t \le 6$ sec.

CHAPTER

9

Sinusoids and Phasors

In the preceding chapters we have considered in some detail both the natural and forced response of a network. We found that the natural response was a characteristic of the network and was independent of the forcing function. The forced response, however, depends directly on the type of forcing function, which until now has generally been a constant. At this point we will diverge from this tack to consider an extremely important excitation: the *sinusoidal forcing function*. Nature is replete with examples of sinusoidal phenomena, and although this is important to us as we examine many different types of physical systems, one reason that we can appreciate at this point for studying this forcing function is that it is the dominant waveform in the electric power industry. The signal present at the ac outlets in our home, office, laboratory, and so on is sinusoidal. In addition, we will demonstrate in Chapter 18 that via Fourier analysis we can represent any periodic electrical signal by a sum of sinusoids.

In this chapter we concentrate on the steady-state forced response of networks with sinusoidal driving functions. We will ignore the initial conditions and the transient or natural response, which will eventually vanish for the type of circuits with which we will be dealing. We refer to this as an *ac steady-state analysis*. Before proceeding with this analysis, however, we must thoroughly understand the nature of the sinusoidal function.

9.1 Sinusoids

Let us begin our discussion of sinusoidal functions by considering the sine wave

$$x(\omega t) = X_M \sin \omega t \tag{9.1}$$

where $x(t)$ could represent either $v(t)$ or $i(t)$. X_M is the *amplitude* or *maximum value*, ω is the *radian* or *angular frequency*, and ωt is the *argument* of the sine function. A plot of the function in Eq. (9.1) as a function of its argument is shown in Fig. 9.1a. Obviously, the function repeats itself every 2π radians and therefore the *period* of the function is 2π radians. This condition is described mathematically as $x(\omega t + 2\pi) = x(\omega t)$ or in general for period T as

$$x[\omega(t + T)] = x(\omega t) \tag{9.2}$$

meaning that the function has the same value at time $t + T$ as it does at time t.

The waveform can also be plotted as a function of time as shown in Fig. 9.1b. Note that this function goes through one period every T seconds, or in other words, in 1 second it goes through $1/T$ periods or cycles. The number of cycles per second, called hertz, is the frequency f, where

$$f = \frac{1}{T} \tag{9.3}$$

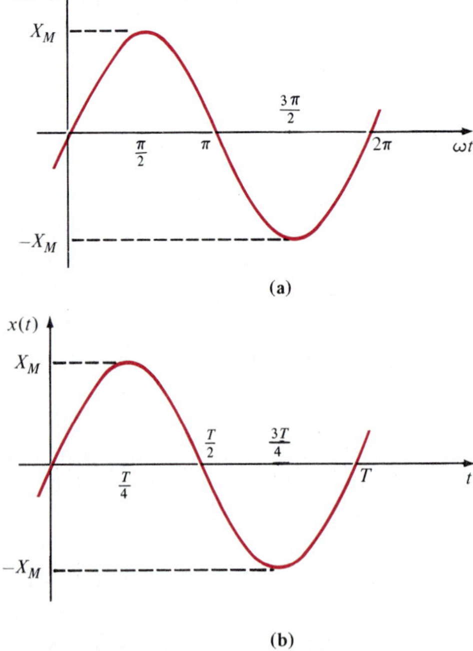

Figure 9.1 Plots of a sine wave as a function of both ωt and t.

Now since $\omega T = 2\pi$ as shown in Fig. 9.1a, we find that

$$\omega = \frac{2\pi}{T} = 2\pi f \tag{9.4}$$

which is of course the general relationship among period in seconds, frequency in hertz, and radian frequency.

Now that we have discussed some of the basic properties of a sine wave, let us consider the following general expression for a sinusoidal function:

$$x(t) = X_M \sin(\omega t + \theta) \tag{9.5}$$

In this case $(\omega t + \theta)$ is the argument of the sine function, and θ is called the *phase angle*. A plot of this function is shown in Fig. 9.2, together with the original function in Eq. (9.1) for comparison. Because of the presence of the phase angle, any point on the waveform $X_M \sin(\omega t + \theta)$ occurs θ radians earlier in time than the corresponding point on the waveform $X_M \sin \omega t$. Therefore, we say that $X_M \sin \omega t$ *lags* $X_M \sin(\omega t + \theta)$ by θ radians. In the more general situation, if

$$x_1(t) = X_{M_1} \sin(\omega t + \theta)$$

and

$$x_2(t) = X_{M_2} \sin(\omega t + \phi)$$

then $x_1(t)$ leads $x_2(t)$ by $\theta - \phi$ radians and $x_2(t)$ lags $x_1(t)$ by $\theta - \phi$ radians. If $\theta = \phi$, the waveforms are identical and the functions are said to be *in phase*. If $\theta \neq \phi$, the functions are *out of phase*.

The phase angle is normally expressed in degrees rather than radians, and therefore we will simply state at this point that we will use the two forms interchangeably, that is,

$$x(t) = X_M \sin\left(\omega t + \frac{\pi}{2}\right) = X_M \sin(\omega t + 90°) \tag{9.6}$$

In addition, it should be noted that adding to the argument integer multiples of either 2π radians or 360° does not change the original function. This can easily be shown mathematically, but is visibly evident when examining the waveform, as shown in Fig. 9.2.

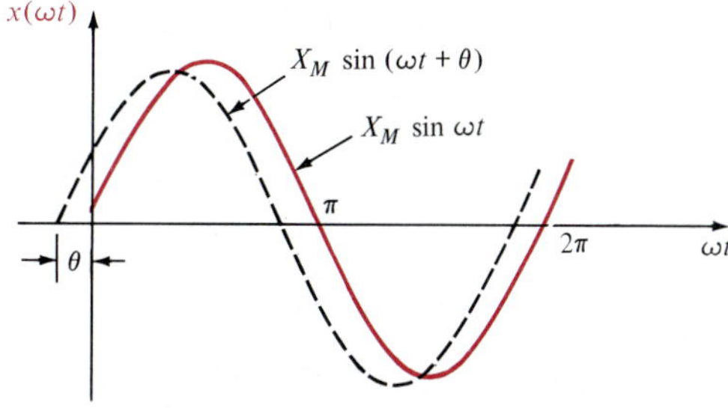

Figure 9.2 Graphical illustration of $X_M \sin(\omega t + \theta)$ leading $X_M \sin \omega t$ by θ radians.

Although our discussion has centered about the sine function, we could just as easily have used the cosine function, since the two waveforms differ only by a phase angle; that is,

$$\cos \omega t = \sin\left(\omega t + \frac{\pi}{2}\right) \tag{9.7}$$

$$\sin \omega t = \cos\left(\omega t - \frac{\pi}{2}\right) \tag{9.8}$$

It should be noted that when comparing one sinusoidal function with another *of the same frequency* to determine the phase difference, it is necessary to express both functions as either sines or cosines with positive amplitudes. Once in this format, the phase angle between the functions can be computed as outlined above. Two other trigonometric identities that normally prove useful in phase angle determination are

$$-\cos(\omega t) = \cos(\omega t \pm 180°) \tag{9.9}$$

$$-\sin(\omega t) = \sin(\omega t \pm 180°) \tag{9.10}$$

Finally, the angle-sum and angle-difference relationships for sines and cosines may be useful in the manipulation of sinusoidal functions.

These relations are

$$\begin{aligned}
\sin(\alpha + \beta) &= \sin\alpha\cos\beta + \cos\alpha\sin\beta \\
\cos(\alpha + \beta) &= \cos\alpha\cos\beta - \sin\alpha\sin\beta \\
\sin(\alpha - \beta) &= \sin\alpha\cos\beta - \cos\alpha\sin\beta \\
\cos(\alpha - \beta) &= \cos\alpha\cos\beta + \sin\alpha\sin\beta
\end{aligned} \tag{9.11}$$

Our interest in these relations will revolve about the cases where $\alpha = \omega t$ and $\beta = \theta$. Since θ is a constant, the equations above indicate that either a sinusoidal or cosinusoidal function with a phase angle can be written as the sum of a sine function and a cosine function with the same arguments. For example, using Eq. (9.11), Eq. (9.5) can be written as

$$\begin{aligned}
x(t) &= X_M \sin(\omega t + \theta) \\
&= X_M(\sin\omega t\cos\theta + \cos\omega t\sin\theta) \\
&= A\sin\omega t + B\cos\omega t
\end{aligned}$$

where

$$A = X_M \cos\theta$$

$$B = X_M \sin\theta$$

or

$$x(t) = \sqrt{A^2 + B^2}\sin(\omega t + \tan^{-1}(B/A))$$

EXAMPLE 9.1

Two voltages $v_1(t)$ and $v_2(t)$ are given by the equations

$$v_1(t) = 12\sin(377t + 45°)\text{ V}$$

$$v_2(t) = 6\sin(377t - 15°)\text{ V}$$

We wish to determine the frequency of the voltages and the phase angle between $v_1(t)$ and $v_2(t)$.

The frequency f in hertz (Hz) is given by the expression

$$f = \frac{\omega}{2\pi} = \frac{377}{2\pi} = 60 \text{ Hz}$$

The phase angle between the voltages is $45° - (-15°) = 60°$; that is, $v_1(t)$ leads $v_2(t)$ by 60° or $v_2(t)$ lags $v_1(t)$ by 60°. ■

EXAMPLE 9.2

We wish to plot the waveforms for the following functions: (a) $v(t) = 1 \cos (\omega t + 45°)$, (b) $v(t) = 1 \cos (\omega t + 225°)$, and (c) $v(t) = 1 \cos (\omega t - 315°)$.

In Fig. 9.3a is shown a plot of the function $v(t) = 1 \cos \omega t$. Figure 9.3b is a plot of

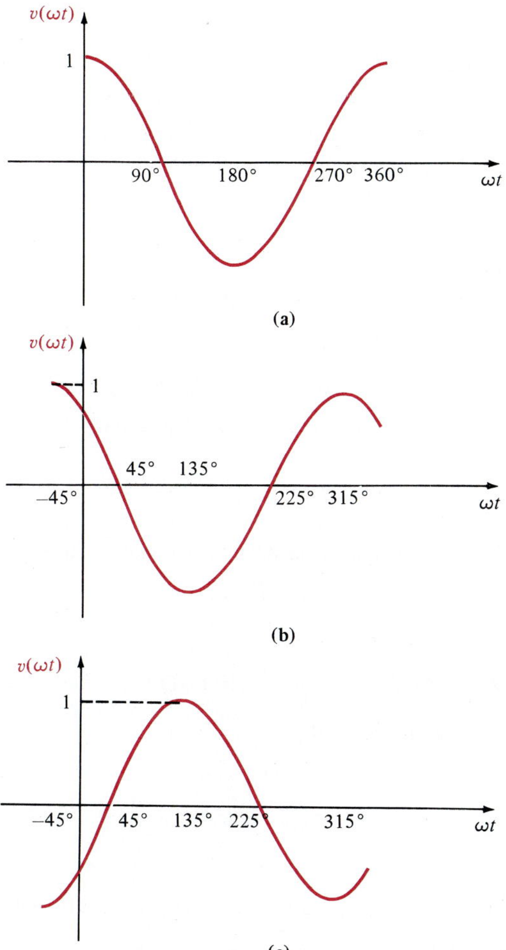

Figure 9.3 Cosine waveforms with various phase angles.

the function $v(t) = 1 \cos(\omega t + 45°)$. Figure 9.3c is a plot of the function $v(t) = 1 \cos(\omega t + 225°)$. Note that since

$$v(t) = 1 \cos(\omega t + 225°) = 1 \cos(\omega t + 45° + 180°)$$

this waveform is 180° out of phase with the waveform in Fig. 9.3b; that is, $\cos(\omega t + 225°) = -\cos(\omega t + 45°)$, and Fig. 9.3c is the negative of Fig. 9.3b. Finally, since the function

$$v(t) = 1 \cos(\omega t - 315°) = 1 \cos(\omega t - 315° + 360°) = 1 \cos(\omega t + 45°)$$

this function is identical to that shown in Fig. 9.3b. ■

EXAMPLE 9.3

Determine the phase angle between the two voltages $v_1(t) = 12 \sin(\omega t + 60°)$ V and $v_2(t) = -6 \cos(\omega t + 30°)$ V.

Using Eq. (9.9), $v_2(t)$ can be written as

$$v_2(t) = -6 \cos(\omega t + 30°) = 6 \cos(\omega t + 210°) \text{ V}$$

Then employing Eq. (9.7), we obtain

$$\begin{aligned} v_2(t) &= 6 \cos(\omega t + 210°) \\ &= 6 \sin(\omega t + 300°) \text{ V} \end{aligned}$$

Now that both voltages of the same frequency are expressed as sine waves with positive amplitudes, the phase angle between $v_1(t)$ and $v_2(t)$ is $60° - (300°) = -240°$ [i.e., $v_1(t)$ leads $v_2(t)$ by $+120°$]. ■

DRILL EXERCISE

D9.1. Given the voltage $v(t) = 120 \cos(314t + \pi/4)$ V, determine the frequency of the voltage in hertz and the phase angle in degrees.

Ans: $f = 50$ Hz, $\theta = 45°$.

D9.2. Express the following functions as sine functions with positive amplitudes.
(a) $A \cos(\omega t - 25°)$.
(b) $-B \cos(\omega t + 60°)$.
(c) $-C \sin(\omega t + 20°)$.

Ans: (a) $A \sin(\omega t + 65°)$, (b) $B \sin(\omega t + 330°)$, (c) $C \sin(\omega t + 200°)$.

D9.3. Three branch currents in a network are known to be

$$\begin{aligned} i_1(t) &= 2 \sin(377t + 45°) \text{ A} \\ i_2(t) &= 0.5 \cos(377t + 10°) \text{ A} \\ i_3(t) &= -0.25 \sin(377t + 60°) \text{ A} \end{aligned}$$

Determine the phase angles by which $i_1(t)$ leads $i_2(t)$ and $i_1(t)$ leads $i_3(t)$.

Ans: i_1 leads i_2 by $-55°$, i_1 leads i_3 by $-195°$.

9.2 Sinusoidal and Complex Forcing Functions

In the preceding chapters we applied a constant forcing function to a network and found that the steady-state response was also constant.

In a similar manner, if we apply a sinusoidal forcing function to a linear network, the steady-state voltages and currents in the network will also be sinusoidal. This should also be clear from the KVL and KCL equations. For example, if one branch voltage is a sinusoid of some frequency, the other branch voltages must be sinusoids of the same frequency if KVL is to apply around any closed path. This means, of course, that the forced solutions of the differential equations that describe a network with a sinusoidal forcing function are sinusoidal functions of time. For example, if we assume that our input function is a voltage $v(t)$ and our output response is a current $i(t)$ as shown in Fig. 9.4, then if $v(t) = A \sin (\omega t + \theta)$, $i(t)$ will be of the form $i(t) = B \sin (\omega t + \phi)$. The critical point here is that we know the form of the output response, and therefore the solution involves simply determining the values of the two parameters B and ϕ.

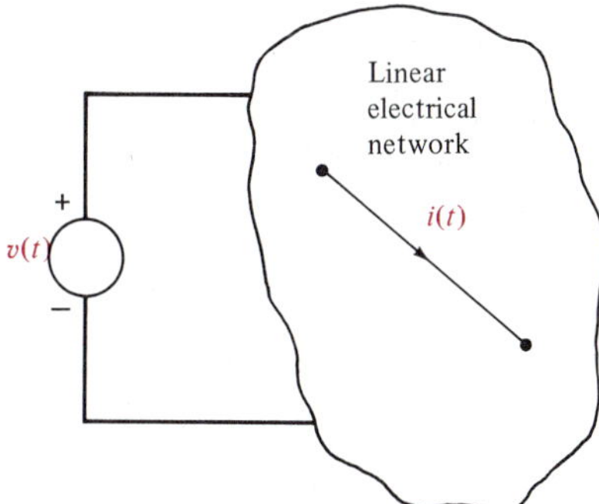

Figure 9.4 Current response to an applied voltage in an electrical network.

EXAMPLE 9.4

Consider the circuit in Fig. 9.5. The KVL equation for this circuit is

$$L \frac{di(t)}{dt} + Ri(t) = V_M \cos \omega t$$

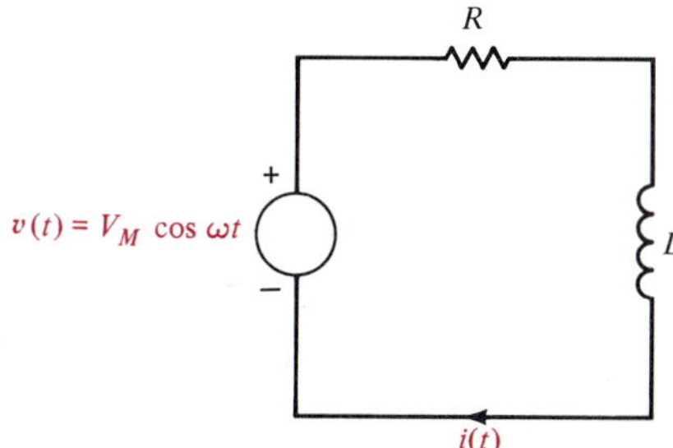

Figure 9.5 *RL* circuit.

Since the input forcing function is $V_M \cos \omega t$, we assume that the forced response component of the current $i(t)$ is of the form

$$i(t) = A \cos (\omega t + \phi)$$

which can be written using Eq. (9.11) as

$$\begin{aligned} i(t) &= A \cos \phi \cos \omega t - A \sin \phi \sin \omega t \\ &= A_1 \cos \omega t + A_2 \sin \omega t \end{aligned}$$

Note that this is, as we observed in Chapter 7, of the form of the forcing function $\cos \omega t$ and its derivative $\sin \omega t$. Substituting this form for $i(t)$ into the differential equation above yields

$$L \frac{d}{dt}(A_1 \cos \omega t + A_2 \sin \omega t) + R(A_1 \cos \omega t + A_2 \sin \omega t) = V_M \cos \omega t$$

Evaluating the indicated derivative produces

$$-A_1 \omega L \sin \omega t + A_2 \omega L \cos \omega t + RA_1 \cos \omega t + RA_2 \sin \omega t = V_M \cos \omega t$$

By equating coefficients of the sine and cosine functions, we obtain

$$\begin{aligned} -A_1 \omega L + RA_2 &= 0 \\ A_1 R + A_2 \omega L &= V_M \end{aligned}$$

that is, two simultaneous equations in the unknowns A_1 and A_2. Solving these two equations for A_1 and A_2 yields

$$A_1 = \frac{RV_M}{R^2 + \omega^2 L^2}$$

$$A_2 = \frac{\omega L V_M}{R^2 + \omega^2 L^2}$$

Therefore,

$$i(t) = \frac{RV_M}{R^2 + \omega^2 L^2} \cos \omega t + \frac{\omega L V_M}{R^2 + \omega^2 L^2} \sin \omega t$$

which using the last identity in Eq. (9.11) can be written as

$$i(t) = A \cos (\omega t + \phi)$$

where A and ϕ are determined as follows:

$$A \cos \phi = \frac{RV_M}{R^2 + \omega^2 L^2}$$

$$A \sin \phi = \frac{-\omega L V_M}{R^2 + \omega^2 L^2}$$

Hence

$$\tan \phi = \frac{A \sin \phi}{A \cos \phi} = -\frac{\omega L}{R}$$

and therefore

$$\phi = -\tan^{-1} \frac{\omega L}{R}$$

since

$$(A \cos \phi)^2 + (A \sin \phi)^2 = A^2(\cos^2 \phi + \sin^2 \phi) = A^2$$

$$A^2 = \frac{R^2 V_M^2}{(R^2 + \omega^2 L^2)^2} + \frac{(\omega L)^2 V_M^2}{(R^2 + \omega^2 L^2)^2} = \frac{V_M^2}{R^2 + \omega^2 L^2}$$

$$A = \frac{V_M}{\sqrt{R^2 + \omega^2 L^2}}$$

Hence the final expression for $i(t)$ is

$$i(t) = \frac{V_M}{\sqrt{R^2 + \omega^2 L^2}} \cos\left(\omega t - \tan^{-1} \frac{\omega L}{R}\right)$$

The analysis above indicates that ϕ is zero if $L = 0$ and hence $i(t)$ is in phase with $v(t)$. If $R = 0$, $\phi = 90°$ and the current lags the voltage by 90°. If R and L are both present, the current lags the voltage by some angle between 0° and 90°. ■

This example illustrates an important point—solving even a simple one-loop circuit containing one resistor and one inductor is very complicated when compared to the solution of a single-loop circuit containing only two resistors. Imagine for a moment how laborious it would be to solve a more complicated circuit using the procedure we have employed in Example 9.4. To circumvent this approach, we will establish a correspondence between sinusoidal time functions and complex numbers. We will then show that this relationship leads to a set of algebraic equations for currents and voltages in a network (e.g., loop currents or node voltages) in which the coefficients of the variables are complex numbers. Hence, once again we will find that determining the currents or voltages in a circuit can be accomplished by solving a set of algebraic equations; however, in this case, their solution is complicated by the fact that variables in the equations have complex, rather than real, coefficients.

The vehicle we will employ to establish a relationship between time-varying sinusoidal functions and complex numbers is Euler's equation, which for our purposes is written as

$$e^{j\omega t} = \cos \omega t + j \sin \omega t \tag{9.12}$$

This complex function has a real part and an imaginary part:

$$\begin{aligned} \text{Re}\,(e^{j\omega t}) &= \cos \omega t \\ \text{Im}\,(e^{j\omega t}) &= \sin \omega t \end{aligned} \tag{9.13}$$

where Re (·) and Im (·) represent the real part and the imaginary part, respectively, of the function in the parentheses.

Now suppose that we select as our forcing function in Fig. 9.4 the nonrealizable voltage

$$v(t) = V_M e^{j\omega t} \tag{9.14}$$

which because of Euler's identity can be written as

$$v(t) = V_M \cos \omega t + jV_M \sin \omega t \tag{9.15}$$

The real and imaginary parts of this function are each realizable. We think of this complex forcing function as two forcing functions, a real one and an imaginary one, and as a consequence of linearity, the principle of superposition applies and thus the current response can be written as

$$i(t) = I_M \cos(\omega t + \phi) + jI_M \sin(\omega t + \phi) \tag{9.16}$$

where $I_M \cos(\omega t + \phi)$ is the response due to $V_M \cos \omega t$ and $jI_M \sin(\omega t + \phi)$ is the response due to $jV_M \sin \omega t$. This expression for the current containing both a real and an imaginary term can be written via Euler's equation as

$$i(t) = I_M e^{j(\omega t + \phi)} \tag{9.17}$$

Because of the relationships above we find that rather than applying the forcing function $V_M \cos \omega t$ and calculating the response $I_M \cos(\omega t + \phi)$, we can apply the complex forcing function $V_M e^{j\omega t}$ and calculate the response $I_M e^{j(\omega t + \phi)}$, the real part of which is the desired response $I_M \cos(\omega t + \phi)$. Although this procedure may initially appear to be more complicated, it is not. It is via this technique that we will convert the differential equation to an algebraic equation which is much easier to solve.

EXAMPLE 9.5

Once again, let us determine the current in the *RL* circuit examined in Example 9.4. However, rather than applying $V_M \cos \omega t$ we will apply $V_M e^{j\omega t}$.

The forced response will be of the form

$$i(t) = I_M e^{j(\omega t + \phi)}$$

where only I_M and ϕ are unknown. Substituting $v(t)$ and $i(t)$ into the differential equation for the circuit, we obtain

$$RI_M e^{j(\omega t + \phi)} + L \frac{d}{dt}(I_M e^{j(\omega t + \phi)}) = V_M e^{j\omega t}$$

Taking the indicated derivative, we obtain

$$RI_M e^{j(\omega t + \phi)} + j\omega L I_M e^{j(\omega t + \phi)} = V_M e^{j\omega t}$$

Dividing each term of the equation by the common factor $e^{j\omega t}$ yields

$$RI_M e^{j\phi} + j\omega L I_M e^{j\phi} = V_M$$

which is an algebraic equation with complex coefficients. This equation can be written as

$$I_M e^{j\phi} = \frac{V_M}{R + j\omega L}$$

Converting the right-hand side of the equation to exponential or polar form produces the equation

$$I_M e^{j\phi} = \frac{V_M}{\sqrt{R^2 + \omega^2 L^2}} e^{j[-\tan^{-1}(\omega L/R)]}$$

(A quick refresher on complex numbers is given in Appendix C for readers who need to sharpen their skills in this area.) The form above clearly indicates that the magnitude and phase of the resulting current are

$$I_M = \frac{V_M}{\sqrt{R^2 + \omega^2 L^2}}$$

and

$$\phi = -\tan^{-1}\frac{\omega L}{R}$$

However, since our actual forcing function was $V_M \cos \omega t$ rather than $[V_M e^{j\omega t}$, our actual response is the real part of the complex response:

$$\begin{aligned} i(t) &= I_M \cos(\omega t + \phi) \\ &= \frac{V_M}{\sqrt{R^2 + \omega^2 L^2}} \cos\left(\omega t - \tan^{-1}\frac{\omega L}{R}\right) \end{aligned}$$

Note that this is identical to the response obtained in the previous example by solving the differential equation for the current $i(t)$. ■

9.3 Phasors

Once again let us assume that the forcing function for a linear network is of the form

$$v(t) = V_M e^{j\omega t}$$

Then every steady-state voltage or current in the network will have the same form and the same frequency ω; for example, a current $i(t)$ will be of the form $i(t) = I_M e^{j(\omega t + \phi)}$.

As we proceed in our subsequent circuit analyses, we will simply note the frequency and then drop the factor $e^{j\omega t}$ since it is common to every term in the describing equations. Dropping the term $e^{j\omega t}$ indicates that every voltage or current can be fully described by a magnitude and phase. For example, a voltage $v(t)$ can be written in exponential form as

$$v(t) = V_M \cos(\omega t + \theta) = \text{Re}\,[V_M e^{j(\omega t + \theta)}] \qquad (9.18)$$

or as a complex number

$$v(t) = \text{Re}\,(V_M \underline{/\theta}\, e^{j\omega t}) \qquad (9.19)$$

Since we are working with a complex forcing function, the real part of which is the desired answer, and each term in the equation will contain $e^{j\omega t}$, we can drop Re (·) and $e^{j\omega t}$ and work only with the complex number $V_M \underline{/\theta}$. This complex representation is commonly called a *phasor*. As a distinguishing feature, phasors will be written in boldface type. In a completely identical manner a voltage $v(t) = V_M \cos(\omega t + \theta) = \text{Re}[V_M e^{j(\omega t + \theta)}]$ and a current $i(t) = I_M \cos(\omega t + \phi) = \text{Re}[I_M e^{j(\omega t + \phi)}]$ are written in phasor notation as $\mathbf{V} = V_M \angle \theta$ and $\mathbf{I} = I_M \angle \phi$, respectively.

EXAMPLE 9.6

Again, we consider the *RL* circuit in Example 9.4. The differential equation is

$$L\frac{di(t)}{dt} + Ri(t) = V_M \cos \omega t$$

The forcing function can be replaced by a complex forcing function that is written as $\mathbf{V}e^{j\omega t}$ with phasor $\mathbf{V} = V_M \angle 0°$. Similarly, the forced response component of the current $i(t)$ can be replaced by a complex function that is written as $\mathbf{I}e^{j\omega t}$ with phasor $\mathbf{I} = I_M \angle \phi$. From our previous discussions we recall that the solution of the differential equation is the real part of this current.

Using the complex forcing function, we find that the differential equation becomes

$$L\frac{d}{dt}(\mathbf{I}e^{j\omega t}) + R\mathbf{I}e^{j\omega t} = \mathbf{V}e^{j\omega t}$$

$$j\omega L\mathbf{I}e^{j\omega t} + R\mathbf{I}e^{j\omega t} = \mathbf{V}e^{j\omega t}$$

Note that $e^{j\omega t}$ is a common factor and, as we have already indicated, can be eliminated leaving the phasors, that is,

$$j\omega L\mathbf{I} + R\mathbf{I} = \mathbf{V}$$

Therefore,

$$\mathbf{I} = \frac{\mathbf{V}}{R + j\omega L} = I_M \angle \phi = \frac{V_M}{\sqrt{R^2 + \omega^2 L^2}} \angle -\tan^{-1}\frac{\omega L}{R}$$

Thus

$$i(t) = \frac{V_M}{\sqrt{R^2 + \omega^2 L^2}} \cos\left(\omega t - \tan^{-1}\frac{\omega L}{R}\right)$$

which once again is the function we obtained earlier. ■

We define relations between phasors after the $e^{j\omega t}$ term has been eliminated as "phasor, or frequency domain, analysis." Thus we have transformed a set of differential equations with sinusoidal forcing functions in the time domain to a set of algebraic equations containing complex numbers in the frequency domain. In effect, we are now faced with solving a set of algebraic equations for the unknown phasors. The phasors are then simply transformed back to the time domain to yield the solution of the original set of differential equations. In addition, we note that the solution of sinusoidal steady-state circuits would be relatively simple if we could write the phasor equation directly from the circuit discription. In Section 9.4 we will lay the groundwork for doing just that.

Table 9.1 Phasor Representation

Time Domain	Frequency Domain
$A\cos(\omega t \pm \theta)$	$A\angle \pm \theta$
$A\sin(\omega t \pm \theta)$	$A\angle \pm \theta - 90°$

Note that in our discussions we have tacitly assumed that sinusoidal functions would be represented as phasors with a phase angle based on a cosine function. Therefore, if sine functions are used, we will simply employ the relationship in Eq. (9.7) to obtain the proper phase angle.

In summary, while $v(t)$ represents a voltage in the time domain, the phasor **V** represents the voltage in the frequency domain. The phasor contains only magnitude and phase information, and the frequency is implicit in this representation. The transformation from the time domain to the frequency domain, as well as the reverse transformation, is shown in Table 9.1. Recall that the phase angle is based on a cosine function, and therefore, if a sine function is involved, a 90° shift factor must be employed, as shown in the table.

The following examples illustrate the use of the phasor transformation.

EXAMPLE 9.7

Convert the time functions $v(t) = 24\cos(377t - 45°)$ and $i(t) = 12\sin(377t + 120°)$ to phasors.

Using the phasor transformation shown above, we have

$$\mathbf{V} = 24\angle -45°$$

$$\mathbf{I} = 12\angle 120° - 90° = 12\angle 30°$$

■

EXAMPLE 9.8

Convert the phasors $\mathbf{V} = 16\angle 20°$ and $\mathbf{I} = 10\angle -75°$ from the frequency domain to the time domain if the frequency is 60 Hz.

Employing the reverse transformation for phasors, we find that

$$v(t) = 16\cos(377t + 20°)$$

$$i(t) = 10\cos(377t - 75°)$$

■

DRILL EXERCISE

D9.4. Convert the following voltage functions to phasors.

$$v_1(t) = 12\cos(377t - 425°)\text{ V}$$

$$v_2(t) = 18\sin(2513t + 4.2°)\text{ V}$$

Ans: $\mathbf{V}_1 = 12\angle -425°$ V, $\mathbf{V}_2 = 18\angle -85.8°$ V.

D9.5. Convert the following phasors to the time domain if the frequency is 400 Hz.

$$\mathbf{V}_1 = 10\angle 20°$$

$$\mathbf{V}_2 = 12\angle -60°$$

Ans: $v_1(t) = 10\cos(800\pi t + 20°)$ V, $v_2(t) = 12\cos(800\pi t - 60°)$ V.

9.4 Phasor Relationships for Circuit Elements

As we proceed in our development of the techniques required to analyze circuits in the sinusoidal steady state, we are now in a position to establish the phasor relationships between voltage and current for the three passive elements R, L, and C.

In the case of a resistor as shown in Fig. 9.6a, the voltage–current relationship is known to be

$$v(t) = Ri(t) \tag{9.20}$$

Applying the complex voltage $V_M e^{j(\omega t+\theta_v)}$ results in the complex current $I_M e^{j(\omega t+\theta_i)}$, and therefore, Eq. (9.20) becomes

$$V_M e^{j(\omega t+\theta_v)} = RI_M e^{j(\omega t+\theta_i)}$$

which reduces to

$$V_M e^{j\theta_v} = RI_M e^{j\theta_i} \tag{9.21}$$

Equation (9.21) can be written in phasor form as

$$\mathbf{V} = R\mathbf{I} \tag{9.22}$$

where

$$\mathbf{V} = V_M e^{j\theta_v} = V_M \underline{/\theta_v} \qquad \text{and} \qquad \mathbf{I} = I_M e^{j\theta_i} = I_M \underline{/\theta_i}$$

From Eq. (9.21) we see that $\theta_v = \theta_i$ and thus the current and voltage for this circuit are *in phase*.

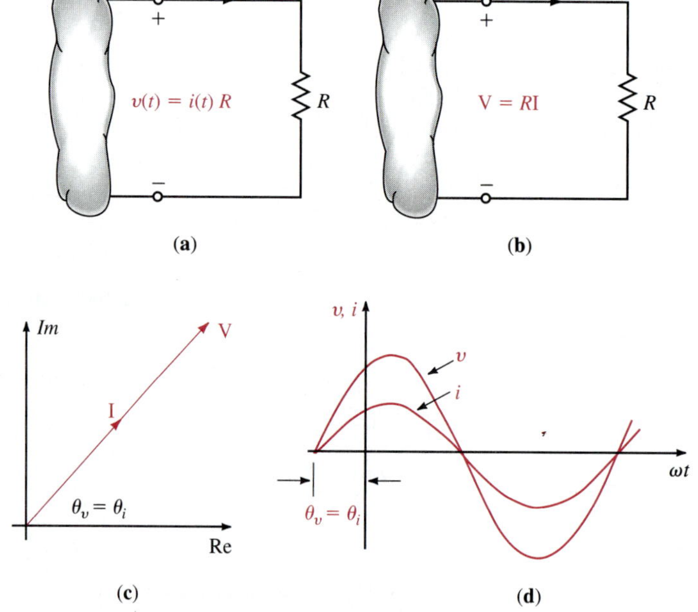

Figure 9.6 Voltage–current relationships for a resistor.

Historically, complex numbers have been represented as points on a graph in which the x-axis represents the real axis and the y-axis the imaginary axis. The line segment connecting the origin with the point provides a convenient representation of the magnitude and angle when the complex number is written in polar form. A review of Appendix C will indicate how these complex numbers or line segments can be added, subtracted, and so on. Since phasors are complex numbers it is convenient to represent the phasor voltage and current graphically as line segments. A plot of the line segments representing the phasors is called a *phasor diagram*. This pictorial representation of phasors provides immediate information on the relative magnitude of one phasor with another, the angle between two phasors, and the relative position of one phasor with respect to another (i.e., leading or lagging). A phasor diagram and the sinusoidal waveforms for the resistor are shown in Fig. 9.6c and d, respectively. A phasor diagram will be drawn for each of the other circuit elements in the remainder of this section.

EXAMPLE 9.9

If the voltage $v(t) = 24 \cos (377t + 75°)$ V is applied to a 6-Ω resistor as shown in Fig. 9.6a, we wish to determine the resultant current.

Since the phasor voltage is

$$\mathbf{V} = 24 \underline{/75°} \text{ V}$$

the phasor current from Eq. (9.22) is

$$\mathbf{I} = \frac{24 \underline{/75°}}{6} = 4 \underline{/75°} \text{ A}$$

which in the time domain is

$$i(t) = 4 \cos (377t + 75°) \text{ A}$$

DRILL EXERCISE

D9.6. The current in a 4-Ω resistor is known to be $\mathbf{I} = 12 \underline{/60°}$ A. Express the voltage across the resistor as a time function if the frequency of the current is 60 Hz.

Ans: $v(t) = 48 \cos (377t + 60°)$ V.

The voltage–current relationship for an inductor, as shown in Fig. 9.7a, is

$$v(t) = L \frac{di(t)}{dt} \tag{9.23}$$

Substituting the complex voltage and current into this equation yields

$$V_M e^{j(\omega t + \theta_v)} = L \frac{d}{dt} I_M e^{j(\omega t + \theta_i)}$$

which reduces to

$$V_M e^{j\theta_v} = j\omega L I_M e^{j\theta_i} \tag{9.24}$$

Equation (9.24) in phasor notation is

$$\mathbf{V} = j\omega L \mathbf{I} \tag{9.25}$$

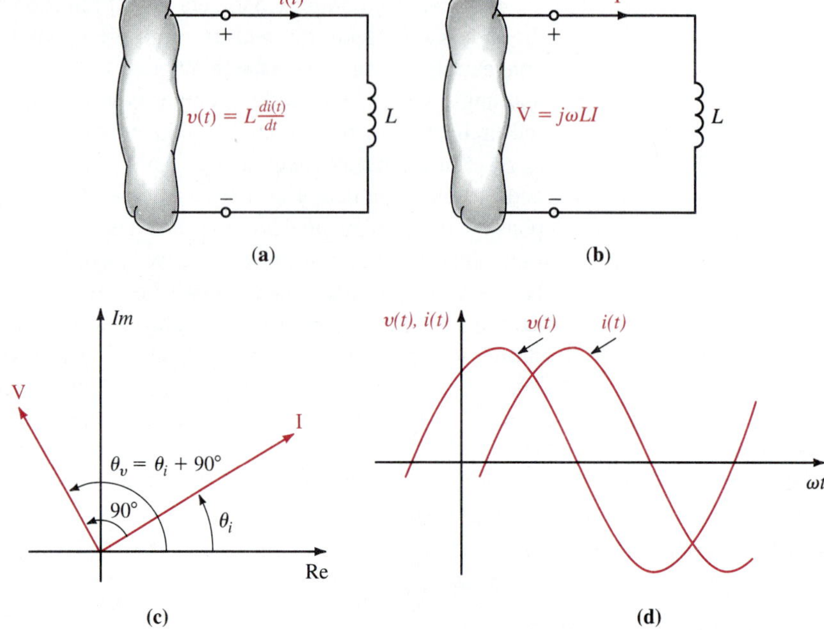

Figure 9.7 Voltage–current relationships for an inductor.

Note that the differential equation in the time domain (9.23) has been converted to an algebraic equation with complex coefficients in the frequency domain. This relationship is shown in Fig. 9.7b. Since the imaginary operator $j = 1e^{j90°} = 1\ \underline{/90°} = \sqrt{-1}$, Eq. (9.24) can be written as

$$V_M e^{j\theta_v} = \omega L I_M e^{j(\theta_i + 90°)} \tag{9.26}$$

Therefore, the voltage and current are *90° out of phase* and in particular the voltage leads the current by 90° or the current lags the voltage by 90°. The phasor diagram and the sinusoidal waveforms for the inductor circuit are shown in Fig. 9.7c and d, respectively.

EXAMPLE 9.10

The voltage $v(t) = 12 \cos (377t + 20°)$ V is applied to a 20-mH inductor as shown in Fig. 9.7a. Find the resultant current.

The phasor current is

$$\begin{aligned}\mathbf{I} &= \frac{\mathbf{V}}{j\omega L} = \frac{12\ \underline{/20°}}{\omega L\ \underline{/90°}}\\ &= \frac{12\ \underline{/20°}}{(377)(20 \times 10^{-3})\ \underline{/90°}}\\ &= 1.59\ \underline{/-70°}\text{ A}\end{aligned}$$

or

$$i(t) = 1.59 \cos (377t - 70°) \text{ A}$$

■

DRILL EXERCISE

D9.7. The current in a 0.05-H inductor is $\mathbf{I} = 4 \underline{/-30°}$ A. If the frequency of the current is 60 Hz, determine the voltage across the inductor.

Ans: $v_L(t) = 75.4 \cos(377t + 60°)$ V.

The voltage–current relationship for our last passive element, the capacitor, as shown in Fig. 9.8a, is

$$i(t) = C\frac{dv(t)}{dt} \tag{9.27}$$

Once again employing the complex voltage and current, we obtain

$$I_M e^{j(\omega t + \theta_i)} = C\frac{d}{dt} V_M e^{j(\omega t + \theta_v)}$$

which reduces to

$$I_M e^{j\theta_i} = j\omega C V_M e^{j\theta_v} \tag{9.28}$$

In phasor notation this equation becomes

$$\mathbf{I} = j\omega C\mathbf{V} \tag{9.29}$$

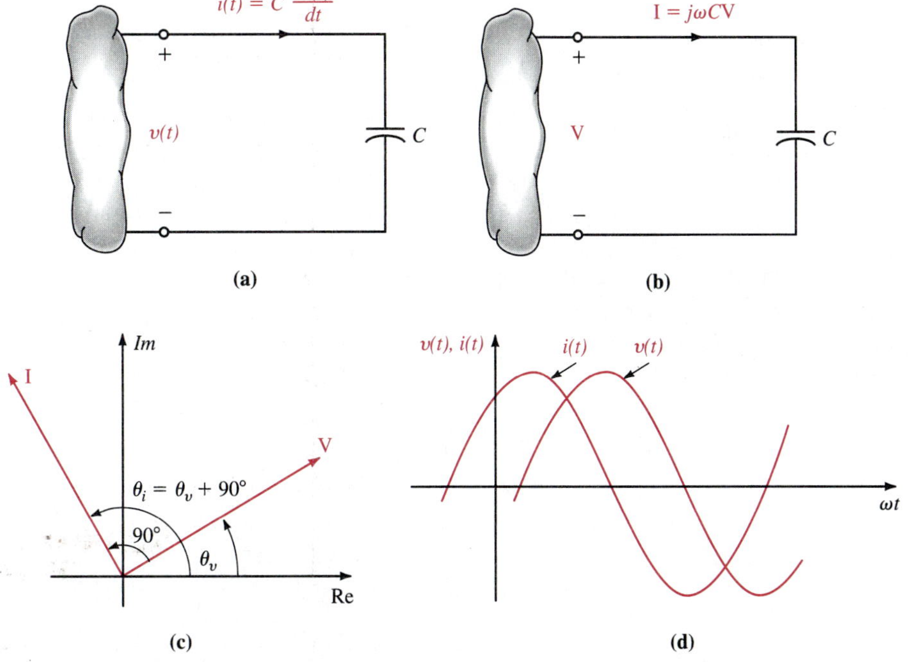

Figure 9.8 Voltage–current relationships for a capacitor.

Equation (9.27), a differential equation in the time domain, has been transformed into Eq. (9.29), an algebraic equation with complex coefficients in the frequency domain. The phasor relationship is shown in Fig. 9.8b. Substituting $j = 1e^{j90°}$ into Eq. (9.28) yields

$$I_M e^{j\theta_i} = \omega C V_M e^{j(\theta_v + 90°)} \tag{9.30}$$

Note that the voltage and current are *90° out of phase*. Equation (9.30) states that the current leads the voltage by 90° or the voltage lags the current by 90°. The phasor diagram and the sinusoidal waveforms for the capacitor circuit are shown in Fig. 9.8c and d, respectively.

EXAMPLE 9.11

The voltage $v(t) = 100 \cos (377t + 15°)$ V is applied to a 100-μF capacitor as shown in Fig. 9.8a. The resultant phasor current is

$$\begin{aligned}\mathbf{I} &= j\omega C(100 \underline{/15°}) \\ &= (377)(100 \times 10^{-6} \underline{/90°})(100 \underline{/15°}) \\ &= 3.77 \underline{/105°} \text{ A}\end{aligned}$$

Therefore, the current written as a time function is

$$i(t) = 3.77 \cos (377t + 105°) \text{ A}$$

■

DRILL EXERCISE

D9.8. The current in a 150-μF capacitor is $\mathbf{I} = 3.6 \underline{/-145°}$ A. If the frequency of the current is 60 Hz, determine the voltage across the capacitor.

Ans: $\mathbf{V}_C = 63.67 \underline{/-235°}$ V.

It is interesting to note that the phasor concept can be used to derive trigonometric identities. For example, consider the function $A \cos \omega t + B \sin \omega t$. Since

$$e^{j\omega t} = \cos \omega t + j \sin \omega t$$

and

$$-je^{j\omega t} = -j \cos \omega t + \sin \omega t$$

therefore,

$$\text{Re}\,(-je^{j\omega t}) = \sin \omega t$$

and hence

$$\begin{aligned}A \cos \omega t + B \sin \omega t &= \text{Re}\,(Ae^{j\omega t} - jBe^{j\omega t}) \\ &= \text{Re}\,[(A - jB)e^{j\omega t}] \\ &= \text{Re}\,(D \underline{/\theta}\, e^{j\omega t}) \\ &= D \cos (\omega t + \theta)\end{aligned}$$

where the phasor $A - jB$ is written in polar form as

$$A - jB = D \underline{/\theta}$$

where $D = \sqrt{A^2 + B^2}$ and $\theta = -\tan^{-1}(B/A)$.

EXAMPLE 9.12

We wish to convert the function

$$v(t) = 10 \cos (377t + 30°) + 5 \sin (377t - 20°)$$

to a single sinusoidal function using the phasor concept.

From the discussion above we note that

$$\begin{aligned} D \underline{/\theta} &= A - jB \\ &= 10 \underline{/30°} - j(5 \underline{/-20°}) \\ &= 10 \underline{/30°} - 5 \underline{/70°} \\ &= 6.96 \underline{/2.47°} \end{aligned}$$

and therefore

$$v(t) = 6.96 \cos (377t + 2.47°)$$

■

EXAMPLE 9.13

We wish to prove the trigonometric identity

$$\cos (\alpha + \beta) = \cos \alpha \cos \beta - \sin \alpha \sin \beta$$

using the phasor concept.

We can write the expression $\cos (\alpha + \beta)$ as

$$\begin{aligned} \cos (\alpha + \beta) &= \text{Re} [e^{j(\alpha + \beta)}] \\ &= \text{Re} (e^{j\alpha} e^{j\beta}) \\ &= \text{Re} [(\cos \alpha + j \sin \alpha)(\cos \beta + j \sin \beta)] \\ &= \cos \alpha \cos \beta - \sin \alpha \sin \beta \end{aligned}$$

■

This discussion illustrates that our newly found technique of phasors has many uses beyond ac circuit analysis.

9.5 Impedance and Admittance

We have examined each of the circuit elements in the frequency domain on an individual basis. We now wish to treat these passive circuit elements in a more general fashion. We now define the two-terminal input *impedance* **Z**, also referred to as the driving point impedance, in exactly the same manner in which we defined resistance earlier. Later we will examine another type of impedance, called transfer impedance.

Impedance is defined as the ratio of the phasor voltage **V** to the phasor current **I**:

$$\mathbf{Z} = \frac{\mathbf{V}}{\mathbf{I}} \tag{9.31}$$

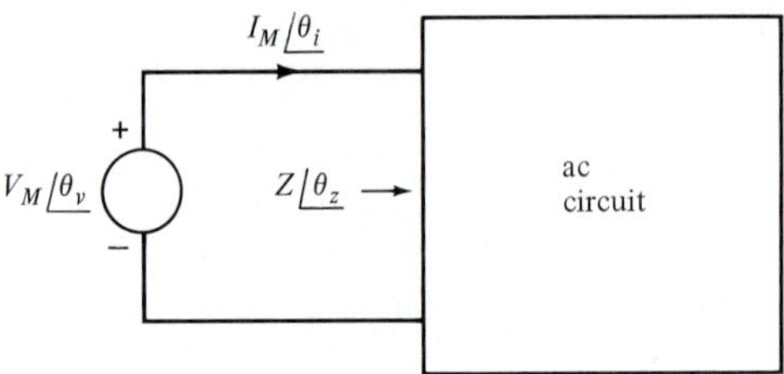

Figure 9.9 General impedance relationship.

at the two terminals of the element related to one another by the passive sign convention, as illustrated in Fig. 9.9. Since **V** and **I** are complex, the impedance **Z** is complex and

$$\mathbf{Z} = \frac{V_M \angle \theta_v}{I_M \angle \theta_i} = \frac{V_M}{I_M} \angle \theta_v - \theta_i = Z \angle \theta_z \tag{9.32}$$

Since **Z** is the ratio of **V** to **I**, the units of **Z** are ohms. Thus impedance in an ac circuit is analogous to resistance in a dc circuit. In rectangular form, impedance is expressed as

$$\mathbf{Z}(j\omega) = R(\omega) + jX(\omega) \tag{9.33}$$

where $R(\omega)$ is the real, or resistive, component and $X(\omega)$ is the imaginary, or reactive, component. In general, we simply refer to R as the resistance and X as the reactance. It is important to note that R and X are real functions of ω and therefore $\mathbf{Z}(j\omega)$ is frequency dependent. Equation (9.33) clearly indicates that **Z** is a complex number; however, it is not a phasor.

Equations (9.32) and (9.33) indicate that

$$Z \angle \theta_z = R + jX \tag{9.34}$$

Therefore,

$$\begin{aligned} Z &= \sqrt{R^2 + X^2} \\ \theta_z &= \tan^{-1} \frac{X}{R} \end{aligned} \tag{9.35}$$

where

$$R = Z \cos \theta_z$$

$$X = Z \sin \theta_z$$

For the individual passive elements the impedance is as shown in Table 9.2. However, just as it was advantageous to know how to determine the equivalent resistance of a series

Table 9.2 Passive Element Impedances

Passive Element	Impedance
R	$\mathbf{Z} = R$
L	$\mathbf{Z} = j\omega L = jX_L = \omega L \angle 90°,\ X_L = \omega L$
C	$\mathbf{Z} = \dfrac{1}{j\omega C} = jX_C = -\dfrac{1}{\omega C} \angle 90°,\ X_C = -\dfrac{1}{\omega C}$

and/or parallel combination of resistors in a dc circuit, we now want to learn how to determine the equivalent impedance in an ac circuit when the passive elements are interconnected. The determination of equivalent impedance is based on KCL and KVL. Therefore, we must see if these laws are valid in the frequency domain. Suppose, for example, that a circuit is driven by voltage sources of the form $V_{M_m} \cos(\omega t + \theta_m)$. Then every steady-state current in the network will be of the form $I_{M_k} \cos(\omega t + \phi_k)$. At any node in the circuit, KCL, written in the time domain, is

$$i_1(t) + i_2(t) + \cdots + i_n(t) = 0$$

or

$$I_{M_1} \cos(\omega t + \phi_1) + I_{M_2} \cos(\omega t + \phi_2) + \cdots + I_{M_n} \cos(\omega t + \phi_n) = 0$$

From our previous work we can immediately apply the phasor transformation to the equation above to obtain

$$\mathbf{I}_1 + \mathbf{I}_2 + \cdots + \mathbf{I}_n = 0$$

However, this equation is simply KCL in the frequency domain. In a similar manner we can show that KVL applies in the frequency domain. Using the fact that KCL and KVL are valid in the frequency domain, we can show, as was done in Chapter 2 for resistors, that impedances can be combined using the same rules that we established for resistor combinations, that is, if $\mathbf{Z}_1, \mathbf{Z}_2, \mathbf{Z}_3, \ldots, \mathbf{Z}_n$ are connected in series, the equivalent impedance $\mathbf{Z}_s$ is

$$\mathbf{Z}_s = \mathbf{Z}_1 + \mathbf{Z}_2 + \mathbf{Z}_3 + \cdots + \mathbf{Z}_n \tag{9.36}$$

and if $\mathbf{Z}_1, \mathbf{Z}_2, \mathbf{Z}_3, \ldots, \mathbf{Z}_n$ are connected in parallel, the equivalent impedance is

$$\frac{1}{\mathbf{Z}_p} = \frac{1}{\mathbf{Z}_1} + \frac{1}{\mathbf{Z}_2} + \frac{1}{\mathbf{Z}_3} + \cdots + \frac{1}{\mathbf{Z}_n} \tag{9.37}$$

EXAMPLE 9.14

Determine the equivalent impedance of the network shown in Fig. 9.10 if the frequency is $f = 60$ Hz. Then compute the current $i(t)$ if the voltage source is $v(t) = 50 \cos(\omega t + 30°)$ V. Finally, calculate the equivalent impedance if the frequency is $f = 400$ Hz.

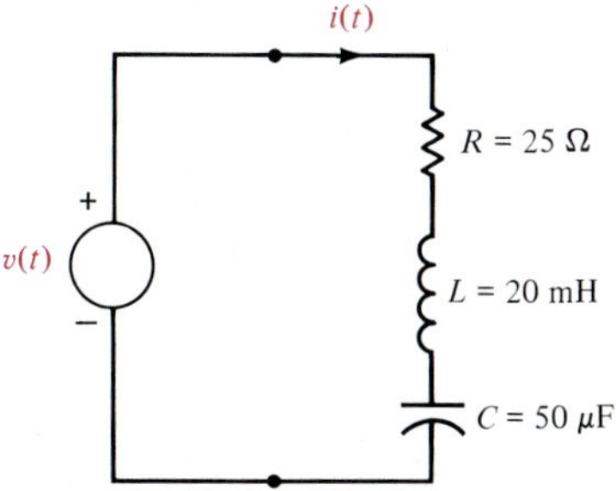

Figure 9.10 Series ac circuit.

The impedances of the individual elements at 60 Hz are

$$\mathbf{Z}_R = 25\ \Omega$$

$$\mathbf{Z}_L = j\omega L = j(2\pi \times 60)(20 \times 10^{-3}) = j7.54\ \Omega$$

$$\mathbf{Z}_C = \frac{-j}{\omega C} = \frac{-j}{(2\pi \times 60)(50 \times 10^{-6})} = -j53.05\ \Omega$$

Since the elements are in series,

$$\begin{aligned}\mathbf{Z} &= \mathbf{Z}_R + \mathbf{Z}_L + \mathbf{Z}_C \\ &= 25 - j45.51\ \Omega\end{aligned}$$

the current in the circuit is given by

$$\mathbf{I} = \frac{\mathbf{V}}{\mathbf{Z}} = \frac{50\ \underline{/30^\circ}}{25 - j45.51} = \frac{50\ \underline{/30^\circ}}{51.92\ \underline{/-61.22^\circ}} = 0.96\ \underline{/91.22^\circ}\ \text{A}$$

or in the time domain, $i(t) = 0.96 \cos(377t + 91.22^\circ)$ A.

If the frequency is 400 Hz, the impedance of each element is

$$\mathbf{Z}_R = 25\ \Omega$$

$$\mathbf{Z}_L = j\omega L = j50.27\ \Omega$$

$$\mathbf{Z}_C = \frac{-j}{\omega C} = -j7.96\ \Omega$$

The total impedance is then

$$\mathbf{Z} = 25 + j42.31 = 49.14\ \underline{/59.42^\circ}\ \Omega$$

It is important to note that at the frequency $f = 60$ Hz, the reactance of the circuit is capacitive, that is, if the impedance is written as $R + jX$, $X < 0$; however, at $f = 400$ Hz the reactance is inductive since $X > 0$. ■

EXAMPLE 9.15

Determine the equivalent impedance of the circuit in Fig. 9.11 at 60 Hz. Compute the voltage $\mathbf{V}_S$ if the current $\mathbf{I}$ is known to be $\mathbf{I} = 0.5\ \underline{/-22.98^\circ}$ A.

The individual impedances are

$$\mathbf{Z}_R = 20\ \Omega$$

$$\mathbf{Z}_L = j\omega L = j(2\pi \times 60)(40 \times 10^{-3}) = j15.08\ \Omega$$

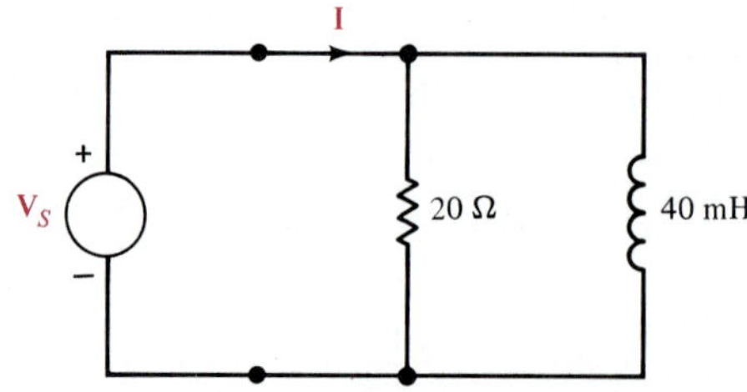

Figure 9.11 Parallel ac circuit.

The equivalent impedance of the parallel elements is

$$\mathbf{Z} = \frac{\mathbf{Z}_R\mathbf{Z}_L}{\mathbf{Z}_R + \mathbf{Z}_L}$$

Using the values for $\mathbf{Z}_R$ and $\mathbf{Z}_L$, we obtain

$$\mathbf{Z} = \frac{(20)(j15.08)}{20 + j15.08}$$

$$= \frac{301.60 \underline{/90^\circ}}{25.05 \underline{/37.02^\circ}} = 7.25 + j9.61 \ \Omega$$

The reader should note carefully that the resistive component of the impedance is *not* the 20-Ω resistor. Recall that $\mathbf{Z}(j\omega) = R(\omega) + jx(\omega)$, and therefore, in general, the resistive component of an equivalent impedance is a function of ω.

If the current $\mathbf{I} = 0.5 \underline{/-22.98^\circ}$ A, the voltage $\mathbf{V}_S$ is

$$\mathbf{V}_S = \mathbf{IZ}$$

$$= (0.5 \underline{/-22.98^\circ})(12.04 \underline{/52.98^\circ})$$

$$= 6.02 \underline{/30^\circ} \text{ V}$$

DRILL EXERCISE

D9.9. Compute the voltage $v(t)$ in the network in Fig. D9.9.

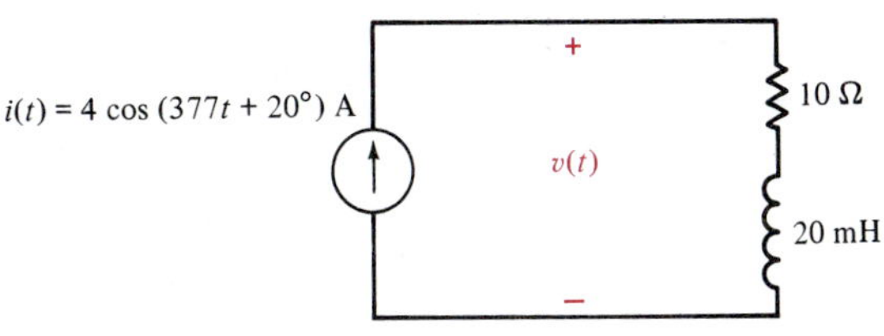

Figure D9.9

Ans: $v(t) = 50.1 \cos(377t + 57^\circ)$ V.

D9.10. Find the current $i(t)$ in the network in Fig. D9.10.

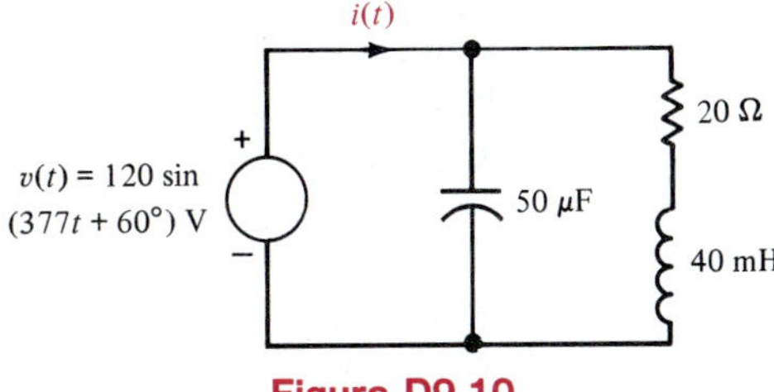

Figure D9.10

Ans: $i(t) = 3.88 \cos(377t - 39.2^\circ)$ A.

Another quantity that is very useful in the analysis of ac circuits is the two-terminal input *admittance*, which is the reciprocal of impedance; that is,

$$\mathbf{Y} = \frac{1}{\mathbf{Z}} = \frac{\mathbf{I}}{\mathbf{V}} \tag{9.38}$$

The units of $\mathbf{Y}$ are siemens, and this quantity is analogous to conductance in resistive dc circuits. Since $\mathbf{Z}$ is a complex number, $\mathbf{Y}$ is also a complex number.

$$\mathbf{Y} = Y_M \underline{/\theta_y} \tag{9.39}$$

which is written in rectangular form as

$$\mathbf{Y} = G + jB \tag{9.40}$$

where G and B are called *conductance* and *susceptance*, respectively. Because of the relationship between $\mathbf{Y}$ and $\mathbf{Z}$, we can express the components of one quantity as a function of the components of the other.

$$G + jB = \frac{1}{R + jX} \tag{9.41}$$

Rationalizing the right-hand side of this equation yields

$$G + jB = \frac{R - jX}{R^2 + X^2}$$

and therefore,

$$G = \frac{R}{R^2 + X^2}, \qquad B = \frac{-X}{R^2 + X^2} \tag{9.42}$$

In a similar manner we can show that

$$R = \frac{G}{G^2 + B^2} \qquad X = \frac{-B}{G^2 + B^2} \tag{9.43}$$

It is very important to note that in general R and G are *not* reciprocals of one another. The same is true for X and B. The purely resistive case is an exception. In the purely reactive case the quantities are negative reciprocals of one another.

The admittance of the individual passive elements is

$$\begin{aligned} \mathbf{Y}_R &= \frac{1}{R} = G \\ \mathbf{Y}_L &= \frac{1}{j\omega L} = -\frac{1}{\omega L} \underline{/90^\circ} \\ \mathbf{Y}_C &= j\omega C = \omega C \underline{/90^\circ} \end{aligned} \tag{9.44}$$

Once again, since KCL and KVL are valid in the frequency domain, we can show, using the same approach outlined in Chapter 2 for conductance in resistive circuits, that the rules for combining admittances are the same as those for combining conductances; that is, if $\mathbf{Y}_1, \mathbf{Y}_2, \mathbf{Y}_3, \ldots, \mathbf{Y}_n$ are connected in parallel, the equivalent admittance is

$$\mathbf{Y}_p = \mathbf{Y}_1 + \mathbf{Y}_2 + \cdots + \mathbf{Y}_n \tag{9.45}$$

and if $\mathbf{Y}_1, \mathbf{Y}_2, \cdots, \mathbf{Y}_n$ are connected in series, the equivalent admittance is

$$\frac{1}{\mathbf{Y}_s} = \frac{1}{\mathbf{Y}_1} + \frac{1}{\mathbf{Y}_2} + \cdots + \frac{1}{\mathbf{Y}_n} \tag{9.46}$$

EXAMPLE 9.16

If the equivalent impedance of a network is $\mathbf{Z} = 10 \underline{/30^\circ}\ \Omega$, compute the equivalent admittance and draw the equivalent circuits.

Converting $\mathbf{Z}$ to rectangular form gives us

$$\begin{aligned}\mathbf{Z} &= 10 \underline{/30^\circ} \\ &= 8.66 + j5.0\ \Omega\end{aligned}$$

The admittance is then

$$\begin{aligned}\mathbf{Y} &= \frac{1}{\mathbf{Z}} \\ &= 0.1 \underline{/-30^\circ} \\ &= 0.0866 - j0.05\ \text{S}\end{aligned}$$

The equivalent circuits are shown in Fig. 9.12. A quick check of the calculations can be made by converting $\mathbf{Y}_R$ to $\mathbf{Z}_R$ and $\mathbf{Y}_L$ to $\mathbf{Z}_L$, and then determining the impedance of the parallel combination; for example,

$$\mathbf{Z}_R = \frac{1}{\mathbf{Y}_R} = \frac{1}{0.0866} = 11.55\ \Omega$$

$$\mathbf{Z}_L = \frac{1}{\mathbf{Y}_L} = \frac{1}{-j0.05} = j20\ \Omega$$

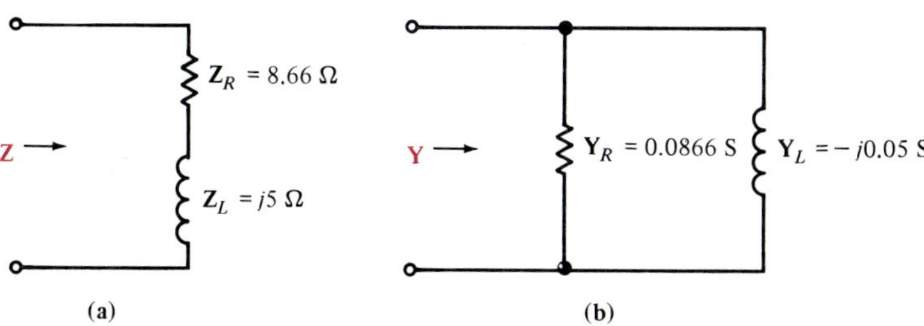

Figure 9.12 Equivalent circuits for impedance and admittance.

and then

$$\mathbf{Z} = \frac{\mathbf{Z}_R\mathbf{Z}_L}{\mathbf{Z}_R + \mathbf{Z}_L}$$

$$= \frac{(11.55)(j20)}{11.55 + j20}$$

$$= \frac{231 \angle 90^\circ}{23.1 \angle 60^\circ} = 10 \angle 30^\circ \ \Omega$$

which is our original impedance. ■

EXAMPLE 9.17

Calculate the equivalent admittance $\mathbf{Y}_p$ for the network in Fig. 9.13 and use it to determine the current $\mathbf{I}$ if $\mathbf{V}_S = 60 \angle 45^\circ$ V.

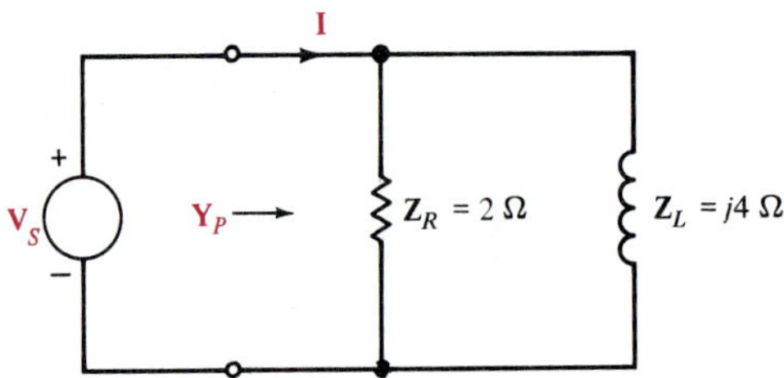

Figure 9.13 Example parallel circuit.

From Fig. 9.13 we note that

$$\mathbf{Y}_R = \frac{1}{\mathbf{Z}_R} = \frac{1}{2} \text{ S}$$

$$\mathbf{Y}_L = \frac{1}{\mathbf{Z}_L} = \frac{-j}{4} \text{ S}$$

Therefore,

$$\mathbf{Y}_p = \frac{1}{2} - j\frac{1}{4} \text{ S}$$

and hence

$$\mathbf{I} = \mathbf{Y}_p\mathbf{V}_S$$

$$= \left(\frac{1}{2} - j\frac{1}{4}\right)(60 \angle 45^\circ)$$

$$= 33.6 \angle 18.43^\circ \text{ A}$$

A quick check of this result can be made using the equivalent impedance

$$\begin{aligned}\mathbf{Z}_p &= \frac{(2)(j4)}{2 + j4} \\ &= \frac{8j}{2 + j4} \\ &= \frac{8 \angle 90^\circ}{4.47 \angle 63.43^\circ} \\ &= 1.79 \angle 26.57^\circ \ \Omega\end{aligned}$$

$$\begin{aligned}\mathbf{I} &= \frac{\mathbf{V}_S}{\mathbf{Z}_p} \\ &= \frac{60 \angle 45^\circ}{1.79 \angle 26.57^\circ} \\ &= 33.6 \angle 18.43^\circ \ \text{A}\end{aligned}$$

which, of course, checks with our previous result. ■

Calculate the equivalent impedance for the circuit in Fig. 9.14 and use this value to determine the voltage $\mathbf{V}_S$ if the current is $\mathbf{I} = 10 \angle 30^\circ$ A.

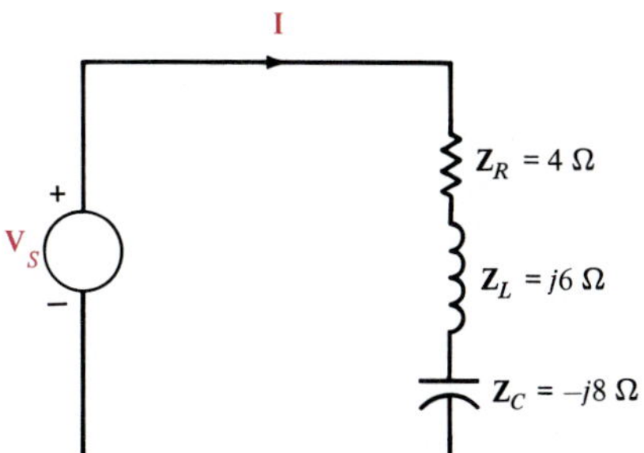

Figure 9.14 Example series circuit.

As shown in Fig. 9.14,

$$\begin{aligned}\mathbf{Z}_S &= \mathbf{Z}_R + \mathbf{Z}_L + \mathbf{Z}_C \\ &= 4 - j2 \ \Omega\end{aligned}$$

Then

$$\begin{aligned}\mathbf{V}_S &= \mathbf{I}\mathbf{Z}_S \\ &= (4 - j2)(10 \angle 30^\circ) \\ &= (4.47 \angle -26.57^\circ)(10 \angle 30^\circ) \\ &= 44.7 \angle 3.43^\circ \ \text{V}\end{aligned}$$

■

DRILL EXERCISE

D9.11. Find the current **I** in the network in Fig. D9.11.

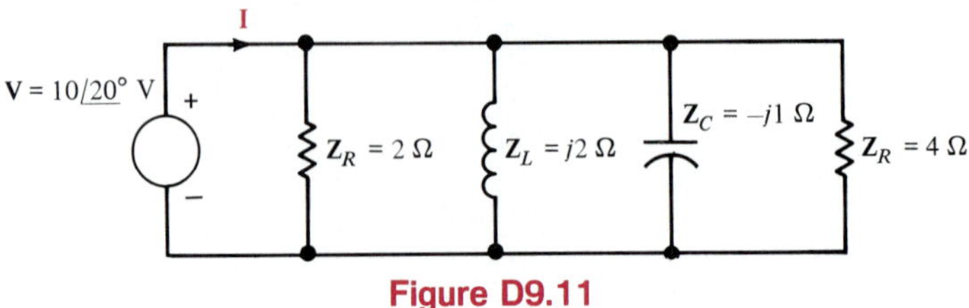

Figure D9.11

Ans: $\mathbf{I} = 9\ \underline{/53.7^\circ}$ A.

D9.12. Determine the voltage across the current source in the network in Fig. D9.12.

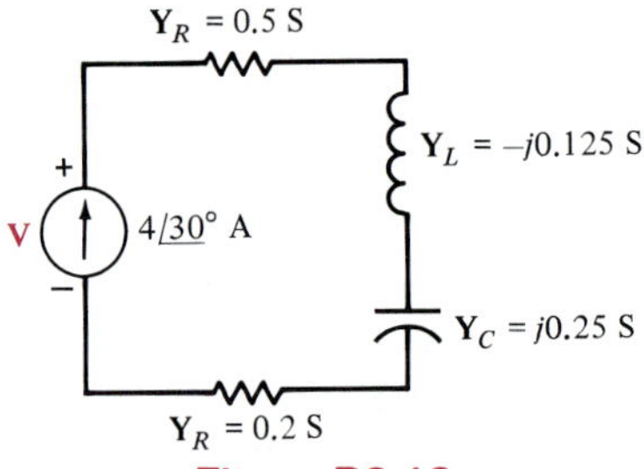

Figure D9.12

Ans: $\mathbf{V} = 32.4\ \underline{/59.7^\circ}$ V.

As a prelude to our analysis of more general ac circuits, let us examine the techniques for computing the impedance or admittance of circuits in which numerous passive elements are interconnected. The following example illustrates that our technique is based simply on the repeated application of Eqs. (9.36), (9.37), (9.45), and (9.46), and is analogous to our earlier computations of equivalent resistance.

EXAMPLE 9.19

Consider the network shown in Fig. 9.15a. The impedance of each element is given in the figure. We wish to calculate the equivalent impedance of the network $\mathbf{Z}_{eq}$ at terminals *A-B*.

The equivalent impedance $\mathbf{Z}_{eq}$ could be calculated in a variety of ways; we could use only impedances, or only admittances, or a combination of the two. We will use two approaches to illustrate the various techniques involved. We begin by noting that the circuit in Fig. 9.15a can be represented by the circuit in Fig. 9.15b. Using strictly an impedance approach, we note that

$$\mathbf{Z}_4 = \frac{(j4)(-j2)}{j4 - j2}$$

$$= -j4\ \Omega$$

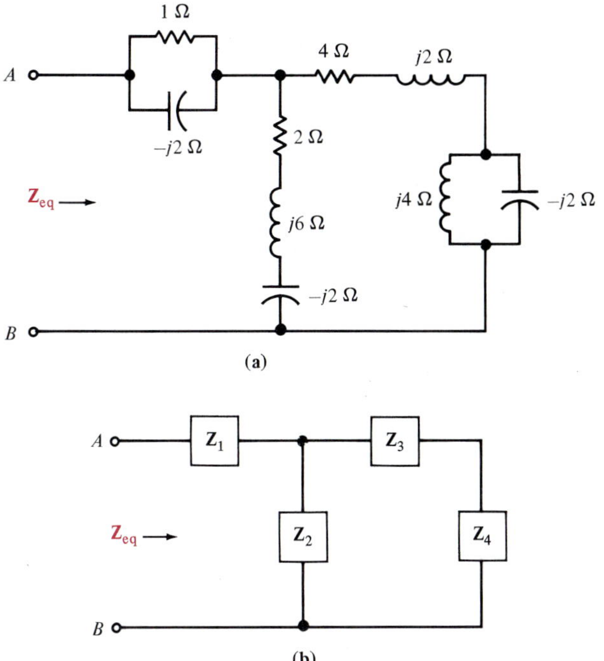

Figure 9.15 Example circuits for determining equivalent impedance.

Since $\mathbf{Z}_3 = 4 + j2\ \Omega$, then $\mathbf{Z}_{34}$, which is the combined impedance of $\mathbf{Z}_3$ and $\mathbf{Z}_4$, is

$$\begin{aligned}\mathbf{Z}_{34} &= 4 + j2 - j4 \\ &= 4 - j2\ \Omega\end{aligned}$$

The figure indicates that

$$\begin{aligned}\mathbf{Z}_2 &= 2 + j6 - j2 \\ &= 2 + j4\ \Omega\end{aligned}$$

and therefore $\mathbf{Z}_{234}$, which is the combined impedance of $\mathbf{Z}_2$, $\mathbf{Z}_3$, and $\mathbf{Z}_4$, is

$$\begin{aligned}\mathbf{Z}_{234} &= \frac{(2 + j4)(4 - j2)}{(2 + j4) + (4 - j2)} \\ &= \frac{16 + j12}{6 + j2}\ \Omega\end{aligned}$$

If we multiply numerator and denominator by $6 - j2$ and perform the indicated algebra, we obtain

$$\mathbf{Z}_{234} = 3 + j1\ \Omega$$

From the figure

$$\mathbf{Z}_1 = \frac{(1)(-j2)}{1 - j2}$$

$$= \frac{4}{5} - j\frac{2}{5}\ \Omega$$

Therefore, if $\mathbf{Z}_{1234}$ is the combined impedance of $\mathbf{Z}_1$, $\mathbf{Z}_2$, $\mathbf{Z}_3$, and $\mathbf{Z}_4$, then

$$\mathbf{Z}_{eq} = \mathbf{Z}_{1234} = 3 + j1 + \frac{4}{5} - j\frac{2}{5}$$

$$= 3.8 + j0.6\ \Omega$$

We could obtain the same result by using both impedances and admittances, as illustrated below.

$$\mathbf{Y}_4 = \mathbf{Y}_L + \mathbf{Y}_C$$

$$= \frac{1}{j4} + \frac{1}{-j2}$$

$$= j\frac{1}{4}\ \mathrm{S}$$

Therefore,

$$\mathbf{Z}_4 = -j4\ \Omega$$

Now

$$\mathbf{Z}_{34} = \mathbf{Z}_3 + \mathbf{Z}_4$$

$$= (4 + j2) + (-j4)$$

$$= 4 - j2\ \Omega$$

and hence

$$\mathbf{Y}_{34} = \frac{1}{\mathbf{Z}_{34}}$$

$$= \frac{1}{4 - j2}$$

$$= 0.20 + j0.10\ \mathrm{S}$$

Since

$$\mathbf{Z}_2 = 2 + j6 - j2$$

$$= 2 + j4\ \Omega$$

then

$$\mathbf{Y}_2 = \frac{1}{2 + j4}$$

$$= 0.10 - j0.20 \text{ S}$$

$$\mathbf{Y}_{234} = \mathbf{Y}_2 + \mathbf{Y}_{34}$$

$$= 0.30 - j0.10 \text{ S}$$

The reader should note carefully our approach—we are adding impedances in series and adding admittances in parallel.

From $\mathbf{Y}_{234}$ we can compute $\mathbf{Z}_{234}$ as

$$\mathbf{Z}_{234} = \frac{1}{\mathbf{Y}_{234}}$$

$$= \frac{1}{0.30 - j0.10}$$

$$= 3 + j1 \ \Omega$$

Now

$$\mathbf{Y}_1 = \mathbf{Y}_R + \mathbf{Y}_C$$

$$= \frac{1}{1} + \frac{1}{-j2}$$

$$= 1 + j\frac{1}{2} \text{ S}$$

and then

$$\mathbf{Z}_1 = \frac{1}{1 + j\dfrac{1}{2}}$$

$$= 0.8 - j0.4 \ \Omega$$

Therefore,

$$\mathbf{Z}_{eq} = \mathbf{Z}_1 + \mathbf{Z}_{234}$$

$$= 0.8 - j0.4 + 3 + j1$$

$$= 3.8 + j0.6 \ \Omega$$

which is exactly what we obtained using an impedance approach. ■

DRILL EXERCISE

D9.13. Compute the impedance $\mathbf{Z}_T$ in the network in Fig. D9.13.

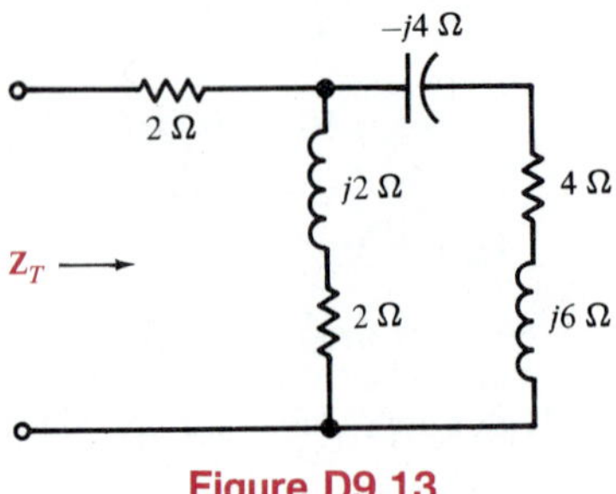

Figure D9.13

Ans: $\mathbf{Z}_T = 3.38 + j1.08\ \Omega$.

The wye-to-delta and delta-to-wye transformations presented earlier for resistance are also valid for impedance in the frequency domain. Therefore, the impedances shown in Fig. 9.16 are related by the following equations:

$$\mathbf{Z}_a = \frac{\mathbf{Z}_1\mathbf{Z}_2}{\mathbf{Z}_1 + \mathbf{Z}_2 + \mathbf{Z}_3}$$

$$\mathbf{Z}_b = \frac{\mathbf{Z}_1\mathbf{Z}_3}{\mathbf{Z}_1 + \mathbf{Z}_2 + \mathbf{Z}_3} \qquad (9.47)$$

$$\mathbf{Z}_c = \frac{\mathbf{Z}_2\mathbf{Z}_3}{\mathbf{Z}_1 + \mathbf{Z}_2 + \mathbf{Z}_3}$$

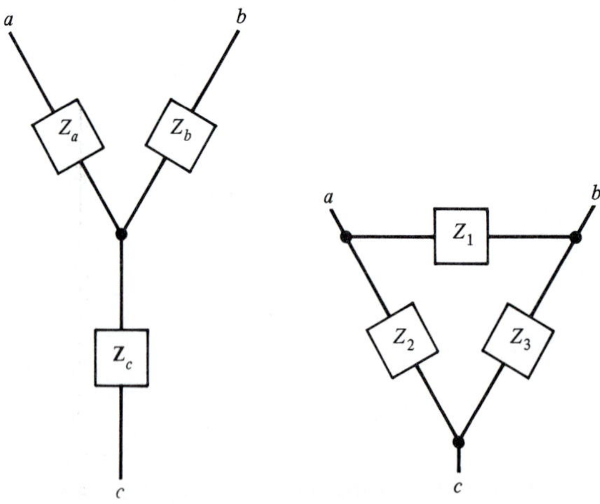

Figure 9.16 Wye and delta network configurations.

and

$$\mathbf{Z}_1 = \frac{\mathbf{Z}_a\mathbf{Z}_b + \mathbf{Z}_b\mathbf{Z}_c + \mathbf{Z}_c\mathbf{Z}_a}{\mathbf{Z}_c}$$
$$\mathbf{Z}_2 = \frac{\mathbf{Z}_a\mathbf{Z}_b + \mathbf{Z}_b\mathbf{Z}_c + \mathbf{Z}_c\mathbf{Z}_a}{\mathbf{Z}_b} \tag{9.48}$$
$$\mathbf{Z}_3 = \frac{\mathbf{Z}_a\mathbf{Z}_b + \mathbf{Z}_b\mathbf{Z}_c + \mathbf{Z}_c\mathbf{Z}_a}{\mathbf{Z}_a}$$

These equations are general relationships and therefore apply to any set of impedances connected in a wye or delta configuration.

EXAMPLE 9.20

Let us determine the impedance $\mathbf{Z}_{eq}$ at the terminals *A*-*B* of the network in Fig. 9.17a.

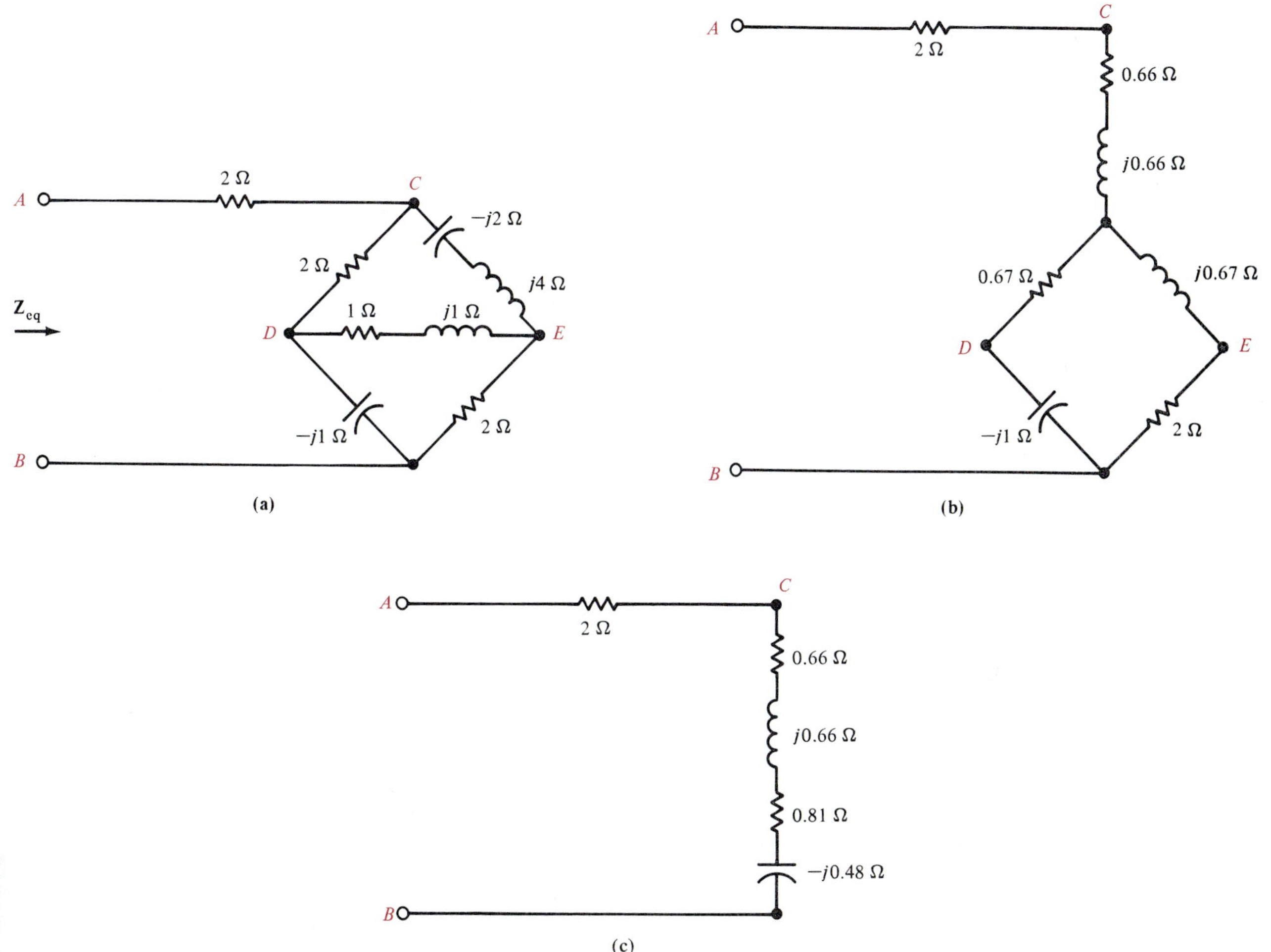

Figure 9.17 Circuits used in Example 9.20.

To simplify the network, we must convert one of the back-to-back deltas into a wye. If we select the delta defined by the points C, D, and E, the impedances of the corresponding wye are calculated from Eq. (9.47) as

$$\frac{(2)(-j2+j4)}{2-j2+j4+1+j1} = \frac{2(j2)}{3+j3} = 0.66 + j0.66\ \Omega$$

$$\frac{(2)(1+j1)}{3+j3} = 0.67\ \Omega$$

$$\frac{(-j2+j4)(1+j1)}{3+j3} = j0.67\ \Omega$$

and shown in Fig. 9.17b. If we now combine the two impedances in parallel, we obtain

$$\frac{(0.67-j1)(2+j0.67)}{0.67-j1+2+j0.67} = 0.81 - j0.48\ \Omega$$

which reduces the network to that shown in Fig. 9.17c. Hence the equivalent impedance is

$$\begin{aligned} \mathbf{Z}_{eq} &= 2 + 0.66 + j0.66 + 0.81 - j0.48 \\ &= 3.47 + j0.18\ \Omega \end{aligned}$$

■

DRILL EXERCISE

D9.14. Determine $\mathbf{Z}_{eq}$ at the terminals A-B of the network shown in Fig. D9.14.

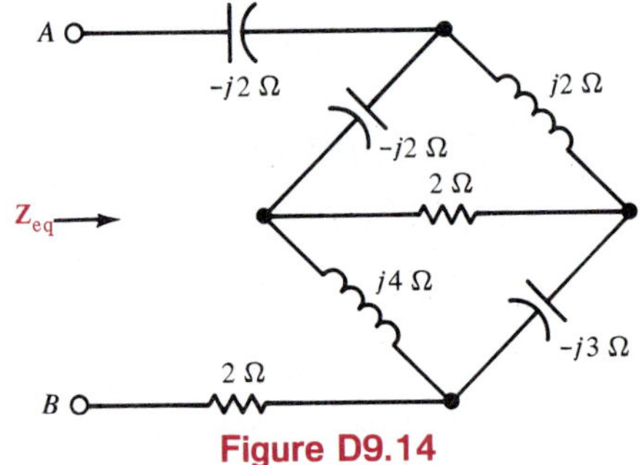

Figure D9.14

Ans: $\mathbf{Z}_{eq} = 4 - j4\ \Omega$.

EXAMPLE 9.21

We wish to compute the voltage $\mathbf{V}_o$ in the network in Fig. 9.18.

Using the results of Example 9.20 we find that the source current is

$$\begin{aligned} \mathbf{I}_S &= \frac{12\ \underline{/0^\circ}}{3.47 + j0.18} \\ &= 3.45\ \underline{/-2.97^\circ}\ \text{A} \end{aligned}$$

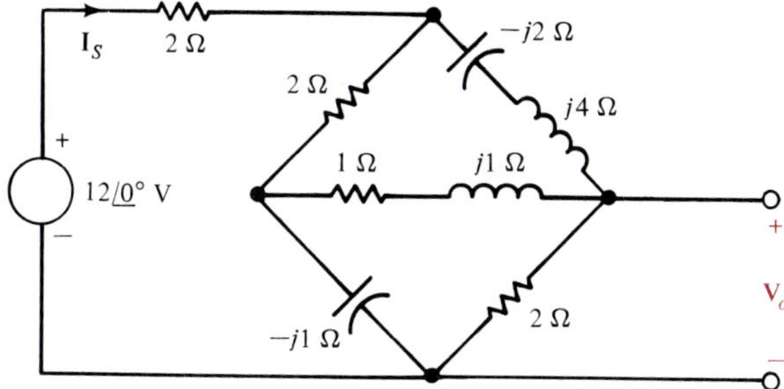

Figure 9.18 Circuit used in Example 9.21.

Employing current division, the current in the 2-Ω resistor connected from E to B in Fig. 9.17b is

$$\mathbf{I}_{2\Omega} = \frac{(3.45\ \underline{/-2.97^\circ})(0.67 - j1)}{2.67 - j0.33}$$
$$= 1.54\ \underline{/-52.1^\circ}\ \text{A}$$

Therefore,

$$\mathbf{V}_o = (1.54\ \underline{/-52.1^\circ})(2)$$
$$= 3.08\ \underline{/-52.1^\circ}\ \text{V}$$

Impedance and admittance are functions of frequency, and therefore their values change as the frequency changes. These changes in **Z** and **Y** have a resultant effect on the current–voltage relationships in a network. This impact of changes in frequency on circuit parameters can be easily seen via a phasor diagram. The following examples will serve to illustrate these points.

EXAMPLE 9.22

Consider the circuit shown in Fig. 9.19. The pertinent variables are labeled on the figure. For convenience in forming a phasor diagram we select **V** as a reference phasor and arbitrarily assign it a 0° phase angle. We will, therefore, measure all currents with respect to this phasor. We suffer no loss of generality by assigning **V** a 0° phase angle, since if

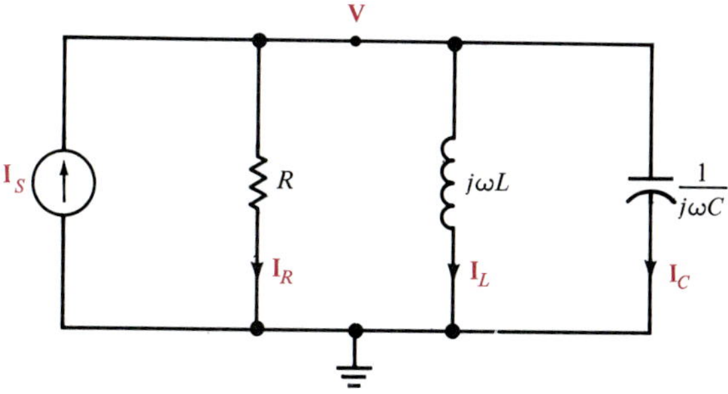

Figure 9.19 Example parallel circuit.

it is actually, for example, 30°, we will simply rotate the entire phasor diagram by 30° because all the currents are measured with respect to this phasor.

At the upper node in the circuit KCL is

$$\mathbf{I}_S = \mathbf{I}_R + \mathbf{I}_L + \mathbf{I}_C = \frac{\mathbf{V}}{R} + \frac{\mathbf{V}}{j\omega L} + \frac{\mathbf{V}}{1/j\omega C}$$

Since $\mathbf{V} = V_M \underline{/0^\circ}$, then

$$\mathbf{I}_S = \frac{V_M \underline{/0^\circ}}{R} + \frac{V_M \underline{/-90^\circ}}{\omega L} + V_M \omega C \underline{/90^\circ}$$

The phasor diagram that illustrates the phase relationship between $\mathbf{V}$, $\mathbf{I}_R$, $\mathbf{I}_L$, and $\mathbf{I}_C$ is shown in Fig. 9.20a. For small values of ω such that the magnitude of $\mathbf{I}_L$ is greater than that of $\mathbf{I}_C$, the phasor diagram for the currents is shown in Fig. 9.20b. In the case of large values of ω, that is, those for which $\mathbf{I}_C$ is greater than $\mathbf{I}_L$, the phasor diagram for the currents is shown in Fig. 9.20c. Note that as ω increases, the phasor $\mathbf{I}_S$ moves from $\mathbf{I}_{S_1}$ to $\mathbf{I}_{S_n}$ along a locus of points specified by the dashed line shown in Fig. 9.20d.

Note that $\mathbf{I}_S$ is in phase with $\mathbf{V}$ when $\mathbf{I}_C = \mathbf{I}_L$ or, in other words, when $\omega L = 1/\omega C$. Hence the node voltage $\mathbf{V}$ is in phase with the current source $\mathbf{I}_S$ when

$$\omega = \frac{1}{\sqrt{LC}}$$

This can also be seen from the KCL equation

$$\mathbf{I}_S = \left[\frac{1}{R} + j\left(\omega C - \frac{1}{\omega L}\right)\right]\mathbf{V}$$

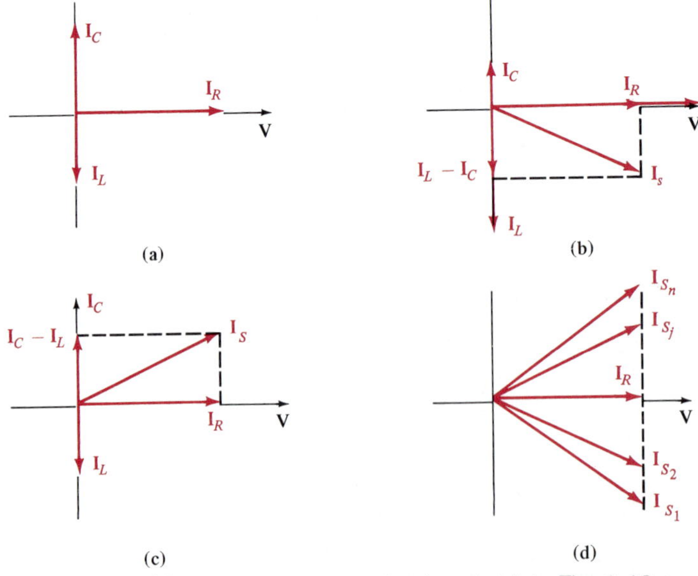

Figure 9.20 Phasor diagrams for the circuit in Fig. 9.19.

EXAMPLE 9.23

Let us examine the series circuit shown in Fig. 9.21a. KVL for this circuit is of the form

$$\mathbf{V}_S = \mathbf{V}_R + \mathbf{V}_L + \mathbf{V}_C$$

$$= \mathbf{I}R + \omega L\mathbf{I} \angle 90^\circ + \frac{\mathbf{I}}{\omega C} \angle -90^\circ$$

If we select $\mathbf{I}$ as a reference phasor so that $\mathbf{I} = I_M \angle 0^\circ$, then if $\omega L I_M > I_M/\omega C$, the phasor diagram will be of the form shown in Fig. 9.21b. Specifically, if $\omega = 377$ rad/sec (i.e., $f = 60$ Hz), then $\omega L = 6$ and $1/\omega C = 2$. Under these conditions the phasor diagram is as shown in Fig. 9.21c. If, however, we select $\mathbf{V}_S$ as reference with, for example,

$$v_S(t) = 12\sqrt{2} \cos(377t + 90^\circ) \text{ V}$$

then

$$\mathbf{I} = \frac{\mathbf{V}}{\mathbf{Z}} = \frac{12\sqrt{2} \angle 90^\circ}{4 + j6 - j2}$$

$$= \frac{12\sqrt{2} \angle 90^\circ}{4\sqrt{2} \angle 45^\circ}$$

$$= 3 \angle 45^\circ \text{ A}$$

and the entire phasor diagram, as shown in Fig. 9.21b and c, is rotated 45° as shown in Fig. 9.21d. ■

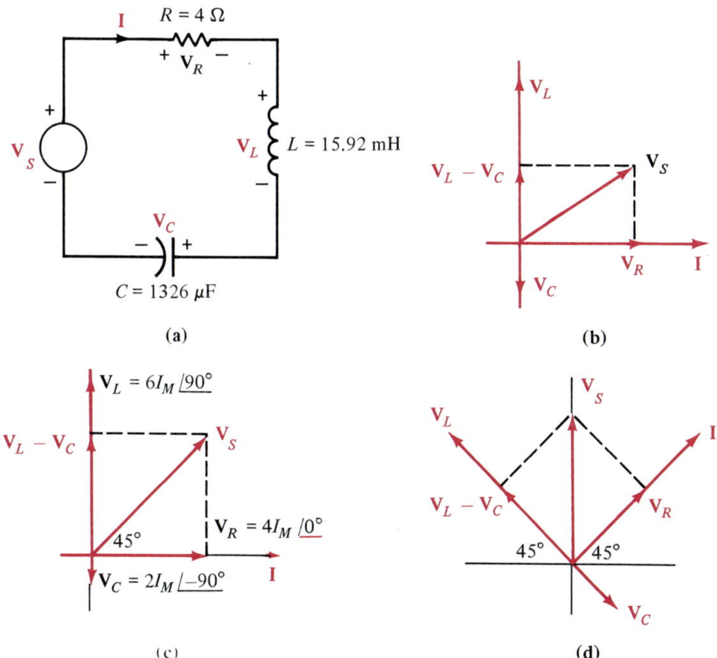

Figure 9.21 Series circuit and certain specific phasor diagrams.

DRILL EXERCISE

D9.15. Draw a phasor diagram illustrating all currents and voltages for the network in Fig. D9.15.

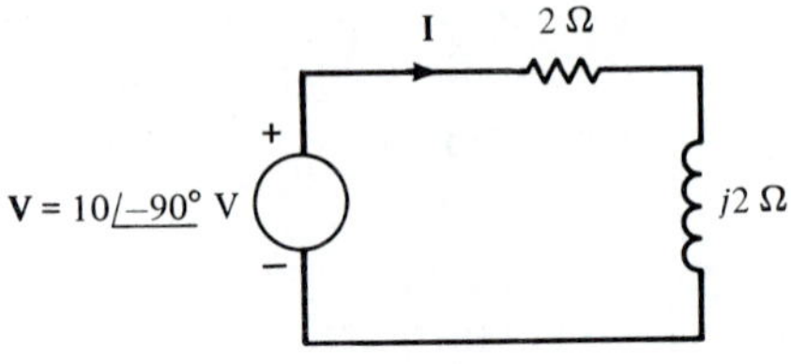

Figure D9.15

Ans:

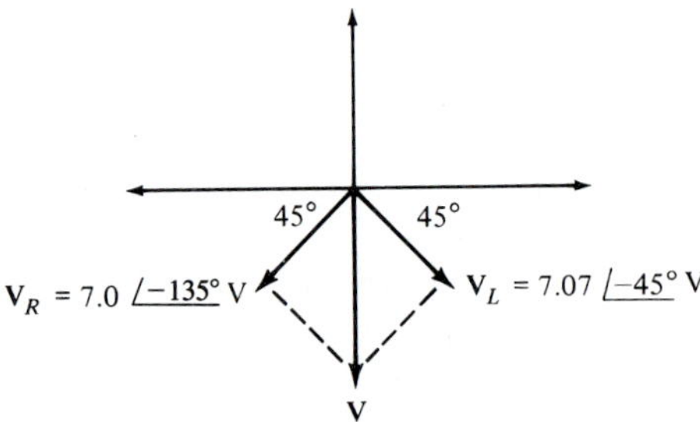

D9.16. Draw a phasor diagram illustrating all currents and voltages for the network in Fig. D9.16.

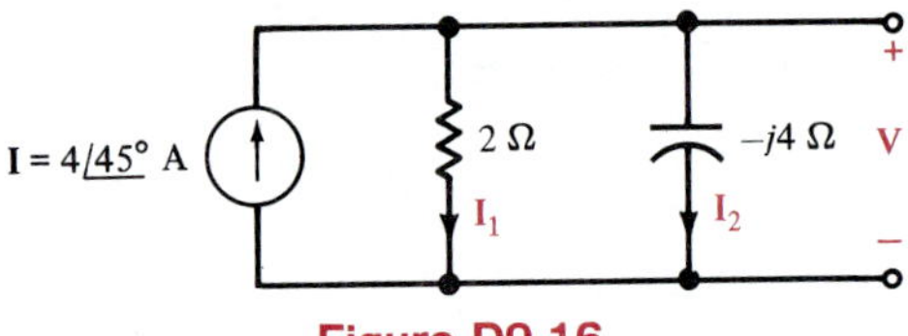

Figure D9.16

Ans:

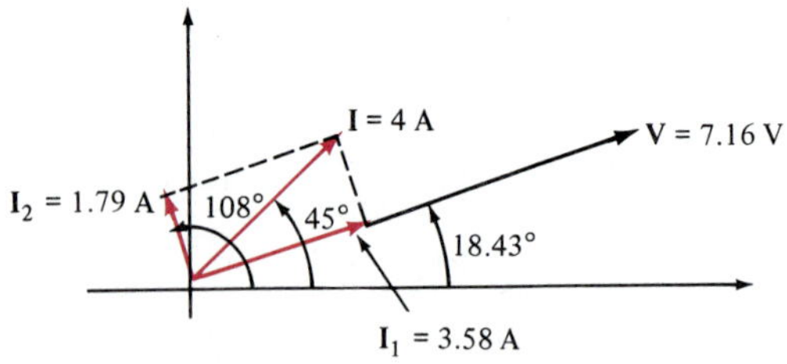

9.6 Basic Analysis Using Kirchhoff's Laws

We have shown that Kirchhoff's laws apply in the frequency domain, and therefore they can be used to compute steady-state voltages and currents in ac circuits. This approach involves expressing these voltages and currents as phasors, and once this is done, the ac steady-state analysis employing phasor equations is performed in an identical fashion to that used in the dc analysis of resistive circuits. Complex number algebra is the tool that is used for the mathematical manipulation of the phasor equations, which, of course, have complex coefficients. We will begin by illustrating that the techniques we have applied in the solution of dc resistive circuits are valid in ac circuit analysis also—the only difference being that in steady-state ac circuit analysis the algebraic phasor equations have complex coefficients. The PSPICE circuit analysis program can also be applied to analyze the ac circuits which we will present here and in Chapter 10.

EXAMPLE 9.24

We wish to calculate all the voltages and currents in the circuit shown in Fig. 9.22. Our approach will be as follows. We will calculate the total impedance seen by the source $\mathbf{V}_S$. Then we will use this to determine $\mathbf{I}_1$. Knowing $\mathbf{I}_1$, we can compute $\mathbf{V}_1$ using KVL. Knowing $\mathbf{V}_1$, we can compute $\mathbf{I}_2$ and $\mathbf{I}_3$, and so on.

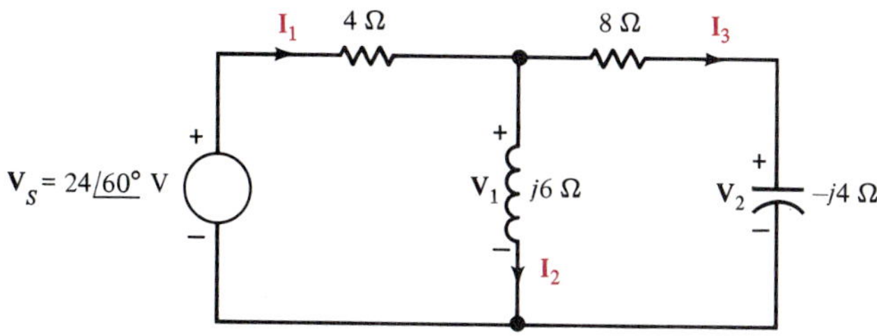

Figure 9.22 Example ac circuit.

The total impedance seen by the source $\mathbf{V}_S$ is

$$\mathbf{Z}_{eq} = 4 + \frac{(j6)(8 - j4)}{j6 + 8 - j4}$$

$$= 4 + \frac{24 + j48}{8 + j2}$$

$$= 4 + 4.24 + j4.94$$

$$= 9.61 \angle 30.94^\circ \ \Omega$$

Then

$$\mathbf{I}_1 = \frac{\mathbf{V}_S}{\mathbf{Z}_{eq}} = \frac{24 \angle 60^\circ}{9.61 \angle 30.94^\circ}$$

$$= 2.5 \angle 29.06^\circ \text{ A}$$

$\mathbf{V}_1$ can be determined using KVL:

$$\mathbf{V}_1 = \mathbf{V}_S - 4\mathbf{I}_1$$

$$= 24 \angle 60^\circ - 10 \angle 29.06^\circ$$

$$= 3.26 + j15.92 = 16.25 \angle 78.43^\circ \text{ V}$$

Note that $\mathbf{V}_1$ could also be computed via voltage division:

$$\mathbf{V}_1 = \frac{\mathbf{V}_S \dfrac{(j6)(8 - j4)}{j6 + 8 - j4}}{4 + \dfrac{(j6)(8 - j4)}{j6 + 8 - j4}} \text{ V}$$

which from our previous calculation is

$$\mathbf{V}_1 = \frac{(24 \underline{/60^\circ})(6.51 \underline{/49.39^\circ})}{9.61 \underline{/30.49^\circ}}$$

$$= 16.25 \underline{/78.43^\circ} \text{ V}$$

Knowing $\mathbf{V}_1$, we can calculate both $\mathbf{I}_2$ and $\mathbf{I}_3$:

$$\mathbf{I}_2 = \frac{\mathbf{V}_1}{j6} = \frac{16.25 \underline{/78.43^\circ}}{6 \underline{/90^\circ}}$$

$$= 2.71 \underline{/-11.56^\circ} \text{ A}$$

and

$$\mathbf{I}_3 = \frac{\mathbf{V}_1}{8 - j4}$$

$$= 1.82 \underline{/105^\circ} \text{ A}$$

Note that $\mathbf{I}_2$ and $\mathbf{I}_3$ could have been calculated by current division. For example, $\mathbf{I}_2$ could be determined by

$$\mathbf{I}_2 = \frac{\mathbf{I}_1(8 - j4)}{8 - j4 + j6}$$

$$= \frac{(2.5 \underline{/29.06^\circ})(8.94 \underline{/-26.57^\circ})}{8 + j2}$$

$$= 2.71 \underline{/-11.56^\circ} \text{ A}$$

Finally, $\mathbf{V}_2$ can be computed as

$$\mathbf{V}_2 = \mathbf{I}_3(-j4)$$

$$= 7.28 \underline{/15^\circ} \text{ V}$$

This value could also have been computed by voltage division. The phasor diagram for the currents $\mathbf{I}_1$, $\mathbf{I}_2$, and $\mathbf{I}_3$ is shown in Fig. 9.23 and is an illustration of KCL. ■

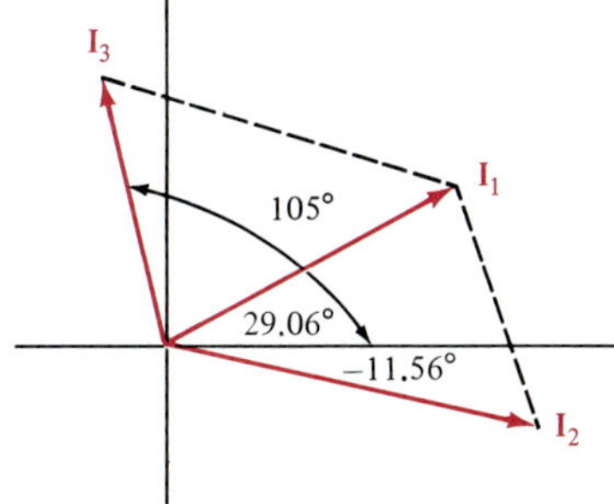

Figure 9.23 Phasor diagram for the currents in Example 9.24.

EXAMPLE 9.25

If the current in the 4-Ω resistor in Fig. 9.24 is known to be $\mathbf{I}_4 = 3\ \underline{/45^\circ}$ A, calculate the value of the source voltage $\mathbf{V}_S$.

Since $\mathbf{I}_4 = 3\ \underline{/45^\circ}$ A, then

$$\begin{aligned}\mathbf{V}_2 &= 4\mathbf{I}_4\\ &= 12\ \underline{/45^\circ}\text{ V}\end{aligned}$$

Hence

$$\begin{aligned}\mathbf{I}_5 &= \frac{\mathbf{V}_2}{8 + j5 - j2}\\ &= 1.41\ \underline{/24.44^\circ} = 1.28 + j0.58\text{ A}\end{aligned}$$

Employing KCL, we get

$$\begin{aligned}\mathbf{I}_3 &= \mathbf{I}_4 + \mathbf{I}_5\\ &= 2.12 + j2.12 + 1.28 + j0.58\\ &= 3.40 + j2.70 = 4.34\ \underline{/38.45^\circ}\text{ A}\end{aligned}$$

$\mathbf{V}_1$ can then be calculated using KVL:

$$\begin{aligned}\mathbf{V}_1 &= \mathbf{I}_3(5 + j2) + \mathbf{V}_2\\ &= (4.34\ \underline{/38.45^\circ})(5.39\ \underline{/21.8^\circ}) + 12\ \underline{/45^\circ}\\ &= 20.10 + j28.80 = 35.12\ \underline{/55.09^\circ}\text{ V}\end{aligned}$$

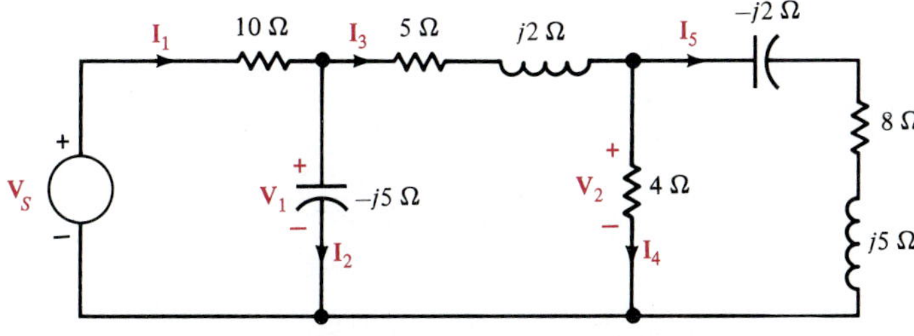

Figure 9.24 Example ac circuit.

$\mathbf{I}_2$ can now be calculated as

$$\mathbf{I}_2 = \frac{\mathbf{V}_1}{-j5}$$

$$= 7.02 \underline{/145.09^\circ} \text{ A}$$

$\mathbf{I}_1$ can be computed via KCL:

$$\mathbf{I}_1 = \mathbf{I}_2 + \mathbf{I}_3$$

$$= 7.02 \underline{/145.09^\circ} + 4.34 \underline{/38.45^\circ}$$

$$= -2.36 + j6.72 = 7.12 \underline{/109.35^\circ} \text{ A}$$

$\mathbf{V}_S$ is then computed using KVL:

$$\mathbf{V}_S = 10\mathbf{I}_1 + \mathbf{V}_1$$

$$= 71.2 \underline{/109.35^\circ} + 35.12 \underline{/55.09^\circ}$$

$$= -3.5 + j96 = 96.06 \underline{/92.09^\circ} \text{ V}$$

Note that these are exactly the same kind of calculations made earlier in the analysis of resistive dc circuits without the use of complex numbers.

As a quick check on our calculations, let us examine KVL around the entire outer loop:

$$\mathbf{V}_S - 10\mathbf{I}_1 - (5 + j2)\mathbf{I}_3 - (8 + j3)\mathbf{I}_5 = 0$$

$$-3.5 + j96 + 23.60 - j67.20 - 11.61 - j20.31 - 8.49 - j8.49 = 0$$

$$0 = 0$$

which indicates that our calculations are correct. ■

DRILL EXERCISE

D9.17. Find the currents $\mathbf{I}_1$, $\mathbf{I}_2$, and $\mathbf{I}_3$ in the network in Fig. D9.17.

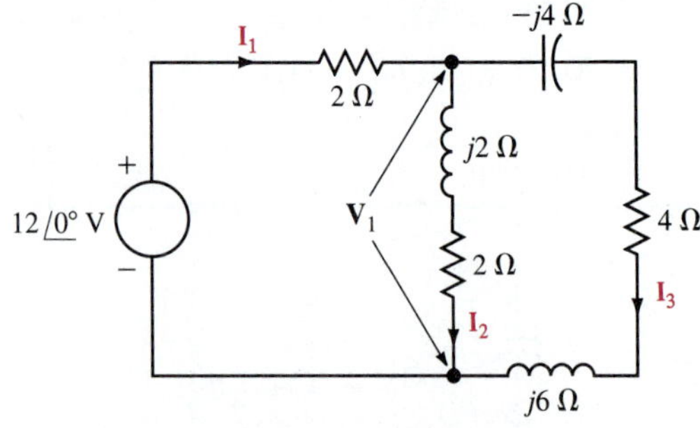

Figure D9.17

Ans: $\mathbf{I}_1 = 3.38 \underline{/-17.7^\circ}$ A, $\mathbf{I}_2 = 2.1 \underline{/-24.7^\circ}$ A, $\mathbf{I}_3 = 1.33 \underline{/-6.3^\circ}$ A.

D9.18. In the network in Fig. D9.18, $\mathbf{V}_o$ is known to be $8 \underline{/45°}$ V. Compute $\mathbf{V}_S$.

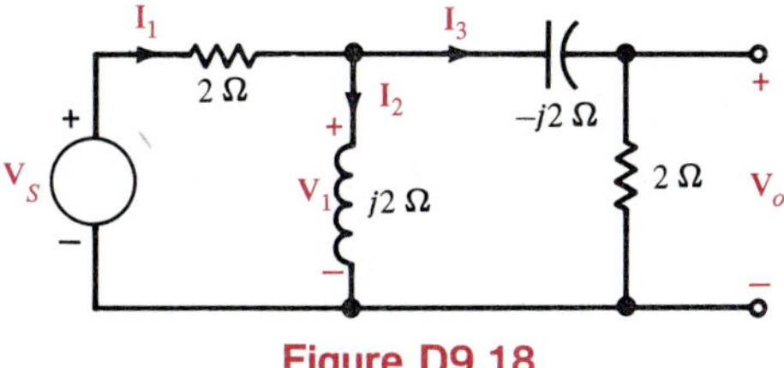

Figure D9.18

Ans: $\mathbf{V}_S = 17.88 \underline{/-18.45°}$ V.

9.7 Use of Phasors in Operational Amplifier Circuits

The phasor equations that govern the operation of the op-amp are identical to those in the time domain. Therefore, the analysis techniques that we have presented in earlier chapters are also valid in ac op-amp circuits.

EXAMPLE 9.26

We want to derive the input–output relationship in the frequency domain for the op-amp shown in Fig. 9.25, and then determine the output voltage when the circuit is a differentiator (i.e., $\mathbf{Z}_i = 1/j\omega C_i$ and $\mathbf{Z}_F = R_F$) or an integrator (i.e., $Z_i = R_i$ and $\mathbf{Z}_F = 1/j\omega C_F$).

Using the op-amp characteristics described in Chapter 3, we find that the KCL phasor equation at the op-amp input is

$$\frac{\mathbf{V}_i}{\mathbf{Z}_i} + \frac{\mathbf{V}_o}{\mathbf{Z}_F} = 0$$

Therefore,

$$\frac{\mathbf{V}_o}{\mathbf{V}_i} = -\frac{\mathbf{Z}_F}{\mathbf{Z}_i}$$

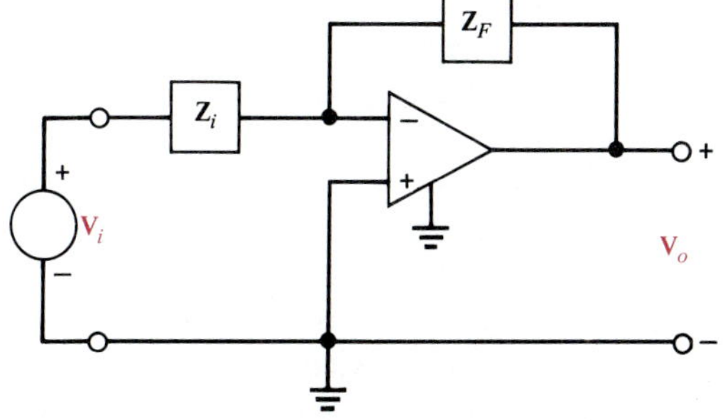

Figure 9.25 Ac operational amplifier circuit.

For a differentiator configuration, $\mathbf{Z}_F = R_F$ and $\mathbf{Z}_i = 1/j\omega C_i$, and hence

$$\mathbf{V}_o = j\omega C_i R_F \mathbf{V}_i$$

In the integrator case $\mathbf{Z}_F = 1/j\omega C_F$ and $\mathbf{Z}_i = R_i$, and therefore,

$$\mathbf{V}_o = \frac{-\mathbf{V}_i}{j\omega C_F R_i}$$

■

These relationships are very useful in phasor circuits containing differentiators and integrators. In addition, the reader will soon learn that differentiation in the time domain is equivalent to multiplying by $j\omega$ in the frequency domain, and integration in the time domain is equivalent to dividing by $j\omega$ in the frequency domain.

9.8 Summary

In this chapter it was shown that our solution approach for ac circuits involves an analysis in the frequency domain. A set of differential equations with sinusoidal forcing functions in the time domain was transformed into a set of algebraic equations with complex coefficients in the frequency domain. The phasor method was introduced and phasor relationships were established for the circuit elements. Algebraic equations in the frequency domain involving the unknown circuit quantities were solved using the phasor technique.

Impedance and admittance were introduced and used in conjunction with phasors to solve ac circuits containing a single source. The solution technique is based on the following facts: (1) Kirchhoff's laws hold for phasor currents and voltages, and (2) for an impedance $\mathbf{Z}$, its voltage and current are related by $\mathbf{V} = \mathbf{IZ}$.

KEY POINTS

- The sinusoidal function $x(t) = X_M \sin(\omega t + \theta)$ has an amplitude of X_M, a radian frequency of ω, a period of $2\pi/\omega$, and a phase angle of θ.
- If $x_1(t) = X_{M_1} \sin(\omega t + \theta)$ and $x_2(t) = X_{M_2} \sin(\omega t + \phi)$, $x_1(t)$ leads $x_2(t)$ by $\theta - \phi$ radians and $x_2(t)$ lags $x_1(t)$ by $\theta - \phi$ radians.
- When comparing one sinusoidal function with another of the same frequency to determine the phase difference, it is necessary to express both functions as either sines or cosines with positive amplitudes.
- The sinusoidal voltage $v(t) = V_M \cos(\omega t + \theta)$ can be written in exponential form as $v(t) = \text{Re}\,[V_M e^{j(\omega t+\theta)}]$ and in phasor form as $\mathbf{V} = V_M \angle\theta$.
- If θ_v and θ_i represent the phase angles of the voltage across and the current through a circuit element, then $\theta_i = \theta_v$ if the element is a resistor, θ_i lags θ_v by 90° if the element is an inductor, θ_i leads θ_v by 90° if the element is a capacitor.
- Impedance, $\mathbf{Z}$, is defined as the ratio of the phasor voltage, $\mathbf{V}$, to the phasor current, $\mathbf{I}$, where $\mathbf{Z} = R$ for a resistor, $\mathbf{Z} = j\omega L$ for an inductor, and $\mathbf{Z} = 1/j\omega C$ for a capacitor.

- **Z** and **Y** are functions of frequency, and therefore, their values change as frequency changes.
- KCL and KVL apply in the frequency domain.

PROBLEMS

9.1. Determine the relative position of the two sine waves.

$$v_1(t) = 12 \sin (377t - 45°)$$

$$v_2(t) = 6 \sin (377t + 675°)$$

9.2. Two voltages are given by the equations

$$v_1(t) = 100 \sin (377t - 210°) \text{ V}$$

$$v_2(t) = -50 \sin (377t - 285°) \text{ V}$$

Find the phase angle between $v_1(t)$ and $v_2(t)$.

9.3. Given the following currents

$$i_1(t) = 4 \sin (377t - 10°) \text{ A}$$

$$i_2(t) = -2 \cos (377t - 195°) \text{ A}$$

$$i_3(t) = -1 \sin (377t - 250°) \text{ A}$$

Compute the phase angle between each pair of currents.

9.4. Determine the phase angles by which $v_1(t)$ leads $i_1(t)$ and $v_1(t)$ leads $i_2(t)$, where

$$v_1(t) = 4 \sin (377t + 25°)$$

$$i_1(t) = 0.05 \cos (377t - 10°)$$

$$i_2(t) = -0.1 \sin (377t + 75°)$$

9.5. Express the following phasors as cosine functions with a frequency of 60 Hz.
(a) $\mathbf{V}_1 = 24 \underline{/-45°}$ V.
(b) $\mathbf{V}_2 = 10 \underline{/120°}$ V.

9.6. Convert the following voltages to phasors in the frequency domain.
(a) $v_1(t) = 12 \cos (2\pi 400t + 60°)$ V.
(b) $v_2(t) = 6 \sin (2\pi 400t - 20°)$ V.

9.7. Calculate the current in the resistor in Fig. P9.7 if the voltage input is
(a) $v_1(t) = 10 \cos (377t + 180°)$ V.
(b) $v_2(t) = 12 \sin (377t + 45°)$ V.
Give the answers in both the time and frequency domains.

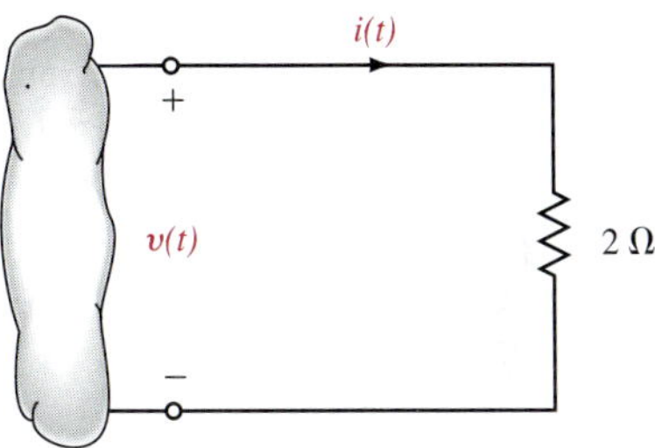

Figure P9.7

9.8. Calculate the current in the capacitor shown in Fig. P9.8 if the voltage input is
(a) $v_1(t) = 16 \cos (377t - 22°)$ V.
(b) $v_2(t) = 8 \sin (377t + 64°)$ V.
Give the answers in both the time and frequency domains.

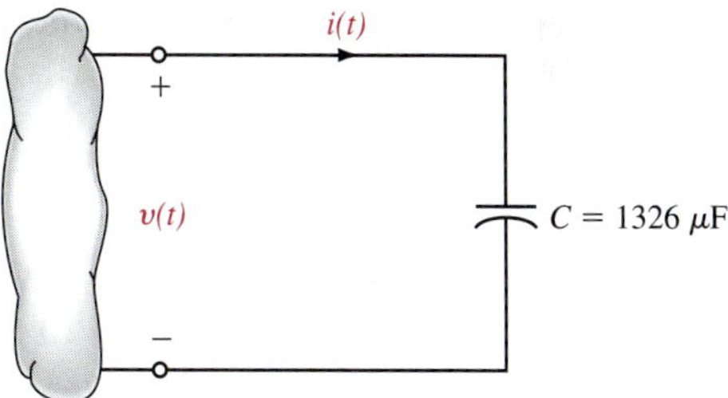

Figure P9.8

9.9. Calculate the current in the inductor shown in Fig. P9.9 if the voltage input is
(a) $v_1(t) = 24 \cos (377t + 12°)$ V.
(b) $v_2(t) = 18 \sin (377t - 48°)$ V.
Give the answers in both the time and frequency domains.

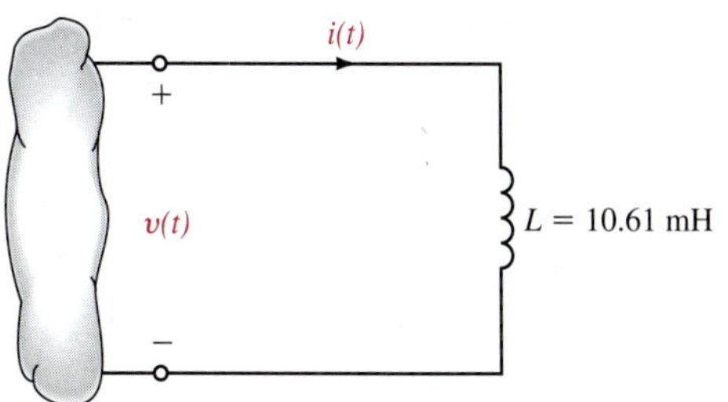

Figure P9.9

9.10. Calculate the equivalent impedance at terminals *A-B* in the circuit shown in Fig. P9.10.

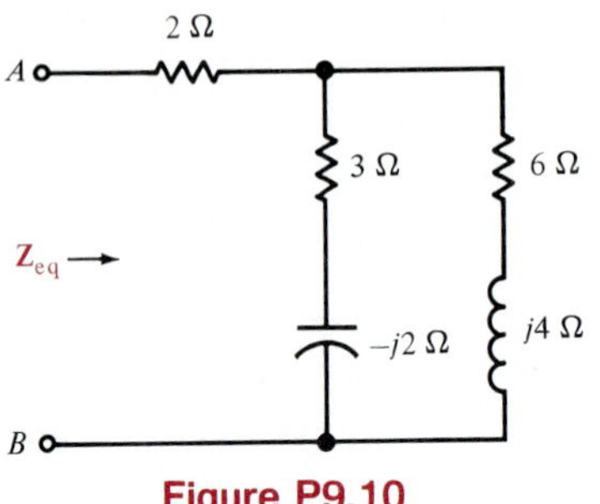

Figure P9.10

9.11. Find $\mathbf{Z}_T$ in the network in Fig. P9.11.

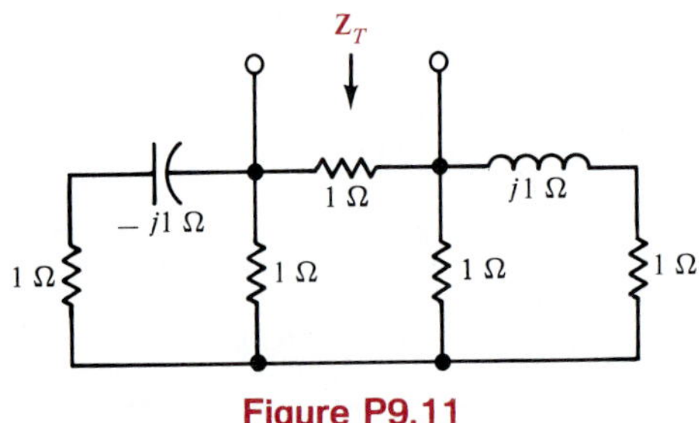

Figure P9.11

9.12. In the network in Fig. P9.12, find $\mathbf{Z}(j\omega)$ at a frequency of 60 Hz.

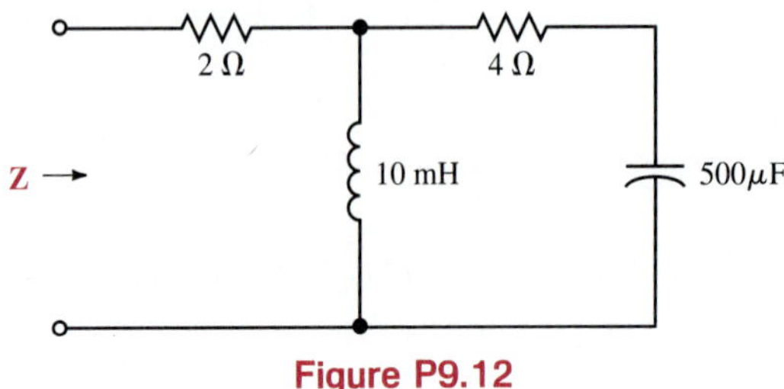

Figure P9.12

9.13. In Europe, the electric power grid operates at a frequency of 50 Hz. Find $\mathbf{Z}(j\omega)$ in Problem 9.12 at $f = 50$ Hz.

9.14. Find $\mathbf{Z}(j\omega)$ in the network in Fig. P9.14.

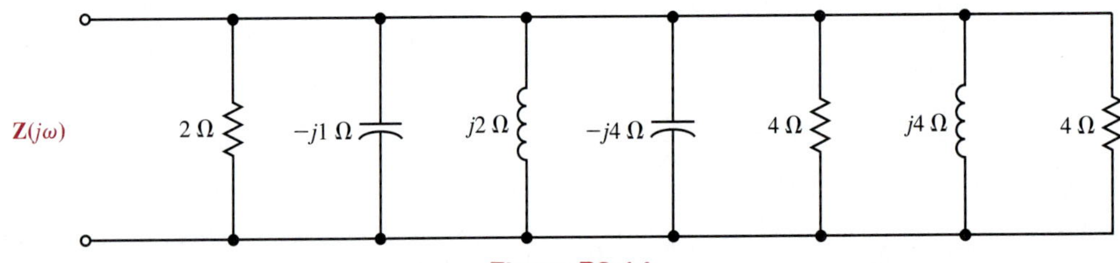

Figure P9.14

9.15. Find $\mathbf{Y}(j\omega)$ in the network in Fig. P9.15.

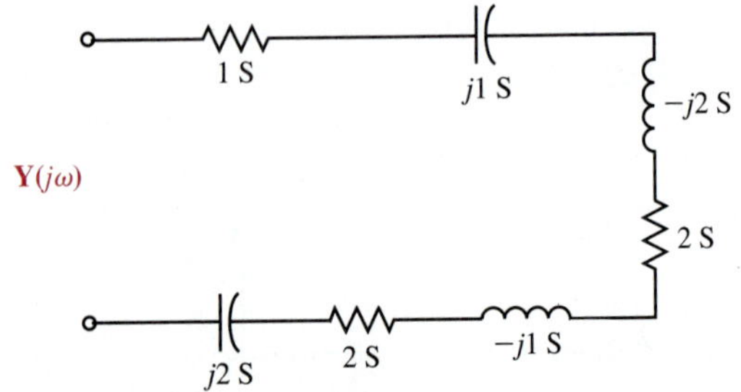

Figure P9.15

9.16. Calculate $\mathbf{Y}_{eq}$ as shown in Fig. P9.16.

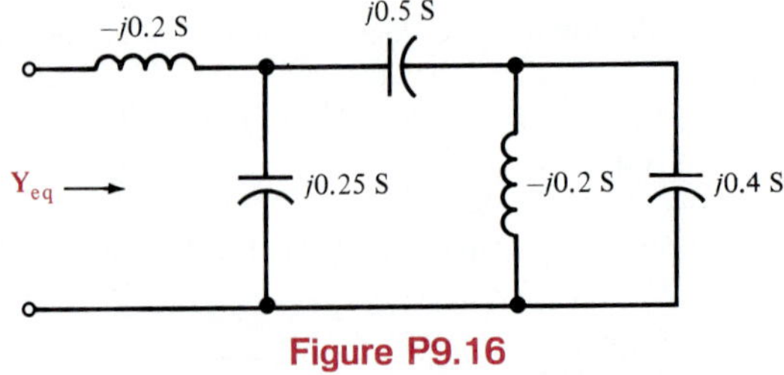

Figure P9.16

9.17. Find $\mathbf{Z}(j\omega)$ in the network in Fig. P9.17.

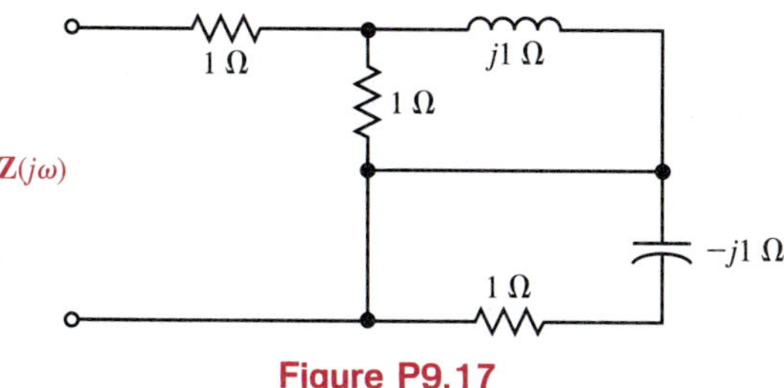

Figure P9.17

9.18. Find $\mathbf{Z}(j\omega)$ in the network in Fig. P9.18.

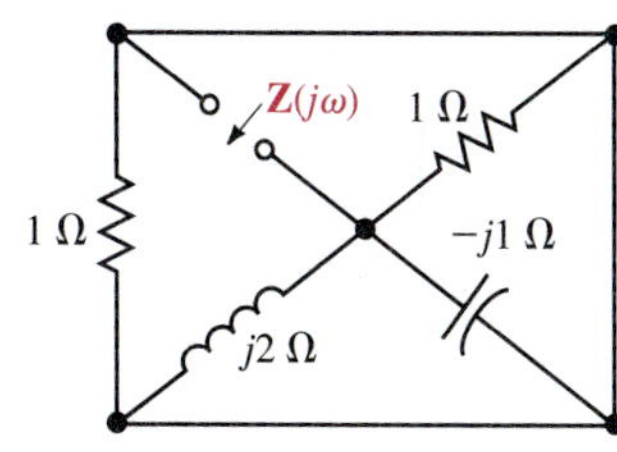

Figure P9.18

9.19. Find $\mathbf{Z}_T(j\omega)$ in the network in Fig. P9.19.

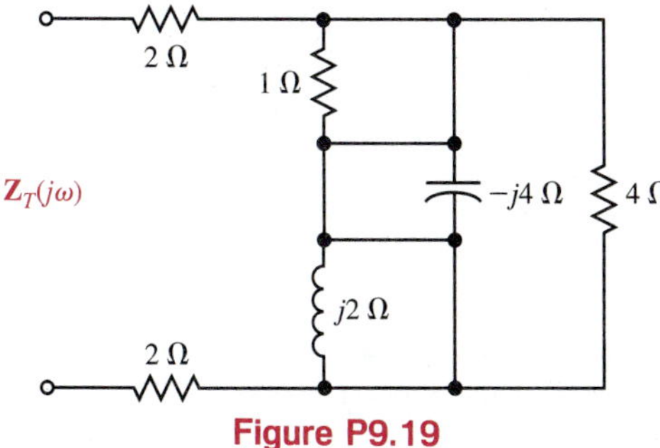

Figure P9.19

9.20. Find $\mathbf{Z}(j\omega)$ in the network in Fig. P9.20.

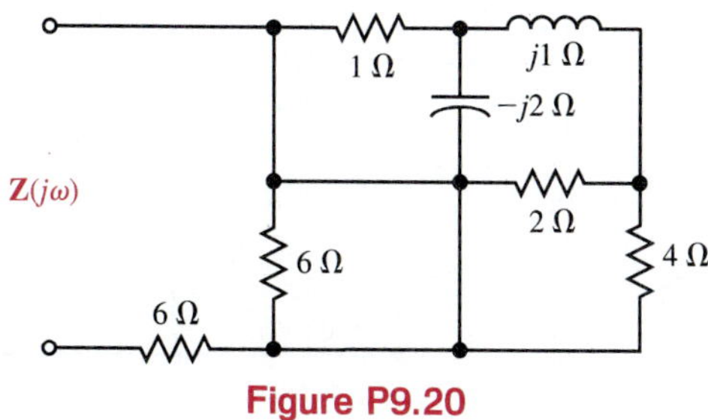

Figure P9.20

9.21. Find $\mathbf{Z}(j\omega)$ in the network in Fig. P9.21.

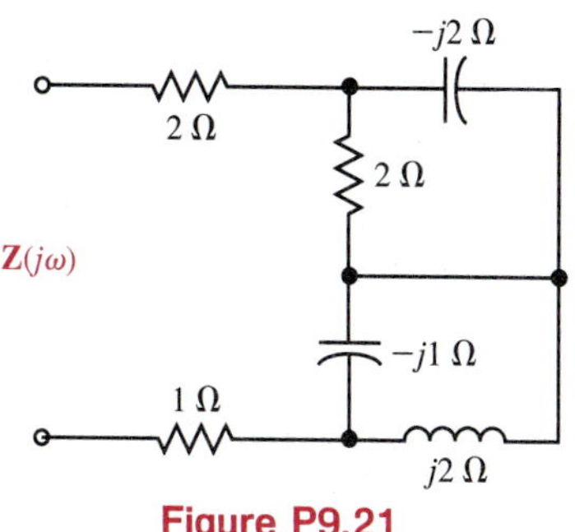

Figure P9.21

9.22. Find $\mathbf{Z}(j\omega)$ in the network in Fig. P9.22.

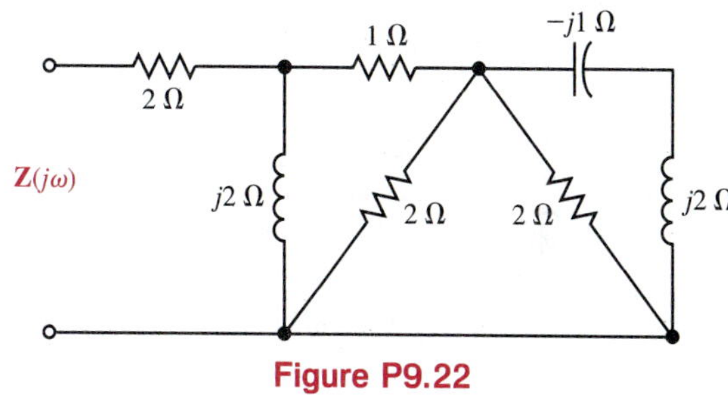

Figure P9.22

9.23. Find $\mathbf{Z}_T$ in the network in Fig. P9.23.

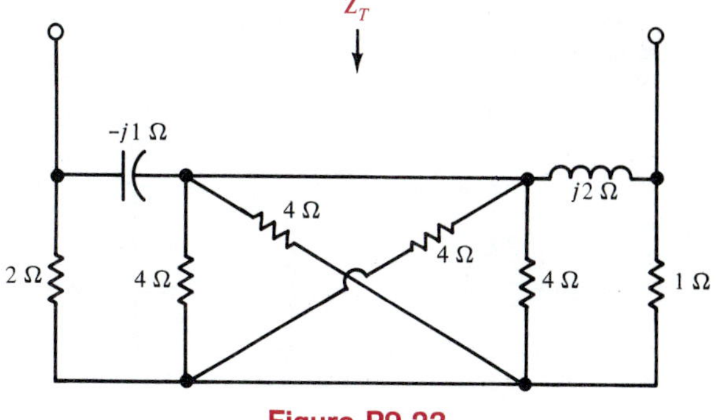

Figure P9.23

9.24. In the circuit shown in Fig. P9.24, determine the value of the inductance such that the current is in phase with the source voltage.

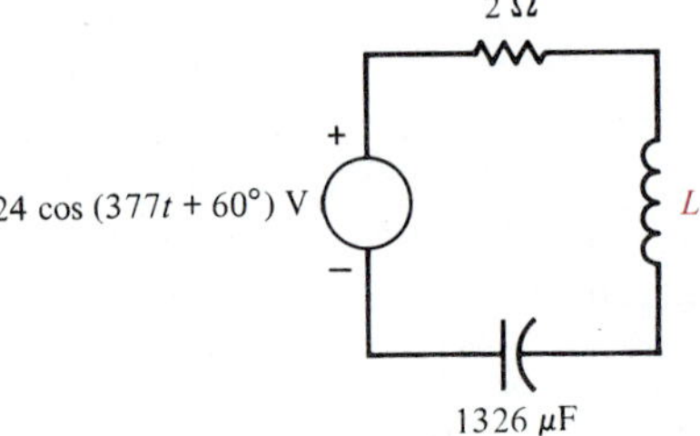

Figure P9.24

9.25. Find $\mathbf{I}_1$ and $\mathbf{V}_0$ in the network in Fig. P9.25.

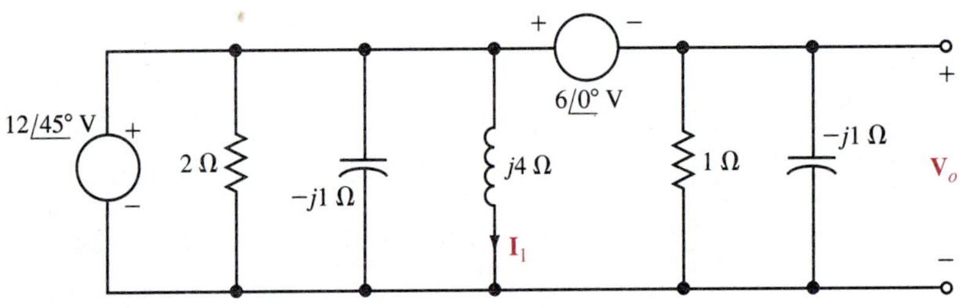

Figure P9.25

9.26. Find $\mathbf{V}_0$ in the circuit in Fig. P9.26.

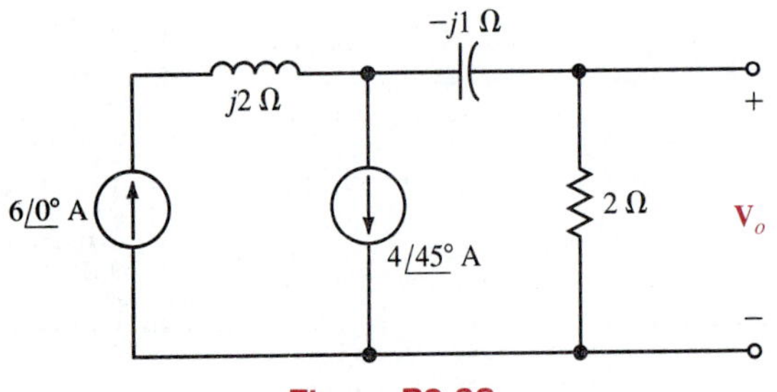

Figure P9.26

9.27. Using $\mathbf{V}_S$ as a reference, plot the phasor diagram for all voltages and currents in Fig. P9.27.

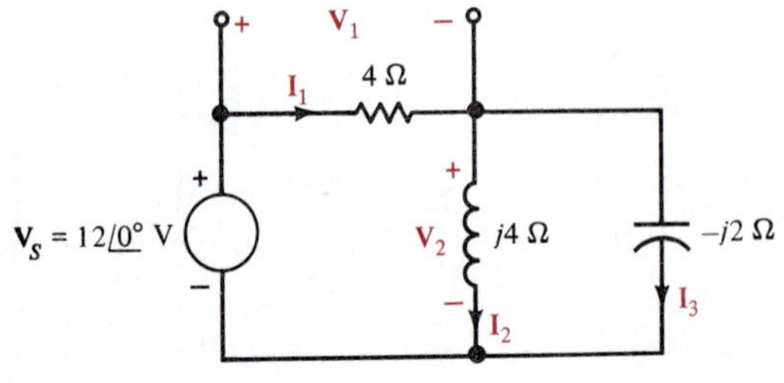

Figure P9.27

9.28. In the network in Fig. P9.28, find $\mathbf{V}_0$ and plot the phasor diagram for the currents.

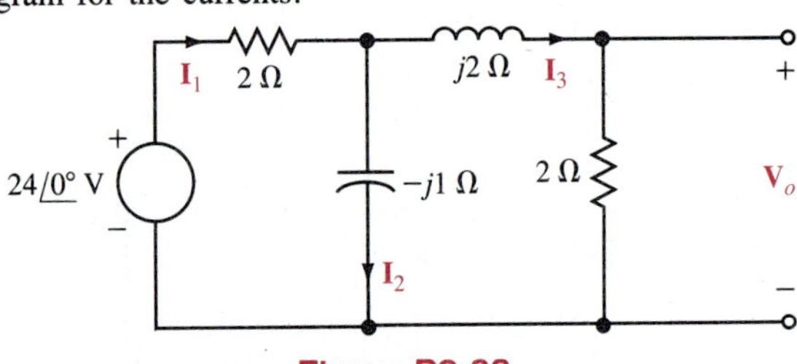

Figure P9.28

9.29. Given the network in Fig. P9.29, plot the phasor diagrams for all voltages and currents.

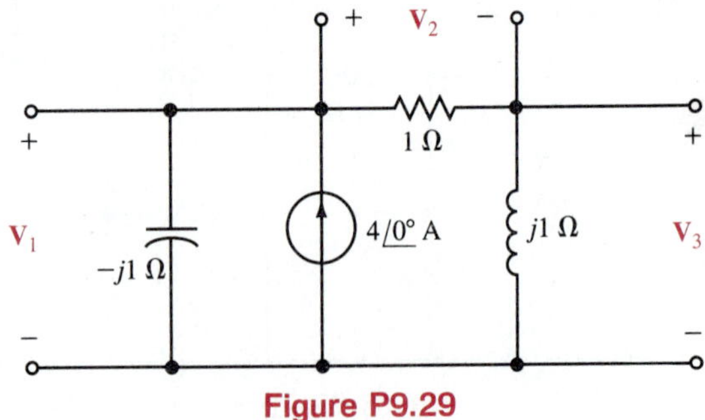

Figure P9.29

9.30. Given the network in Fig. P9.30, plot the phasor diagrams for (a) the currents $\mathbf{I}_1$, $\mathbf{I}_2$, and $\mathbf{I}_3$ and (b) the currents $\mathbf{I}_3$, $\mathbf{I}_4$, and $\mathbf{I}_5$.

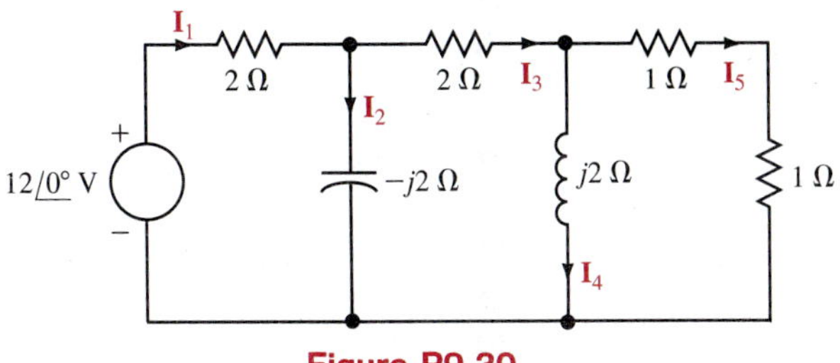

Figure P9.30

9.31. Calculate all the currents in the circuit shown in Fig. P9.31.

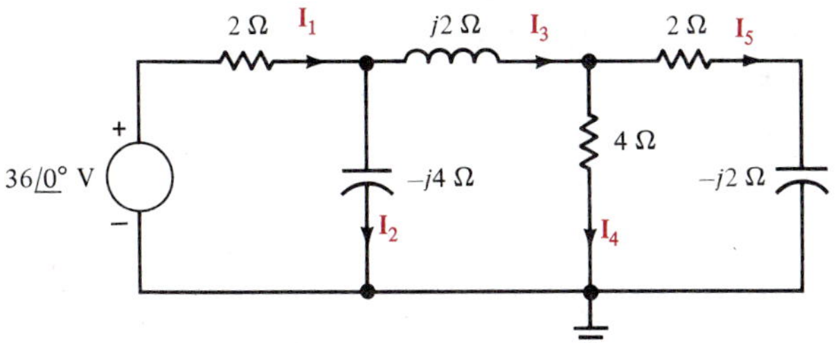

Figure P9.31

9.32. Find $\mathbf{I}_o$ in the network in Fig. P9.32.

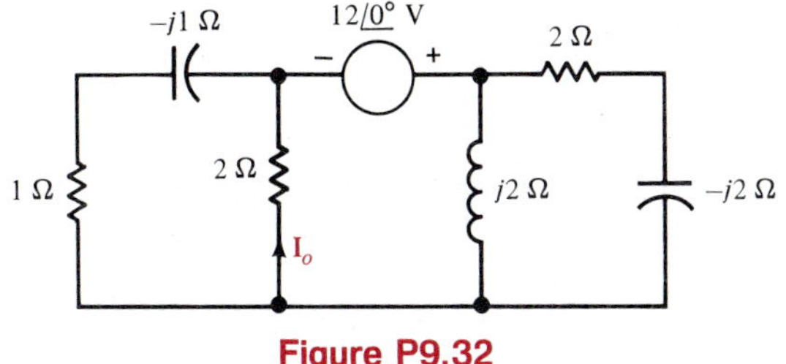

Figure P9.32

9.33. Find $\mathbf{I}_o$ in the network in Fig. P9.33.

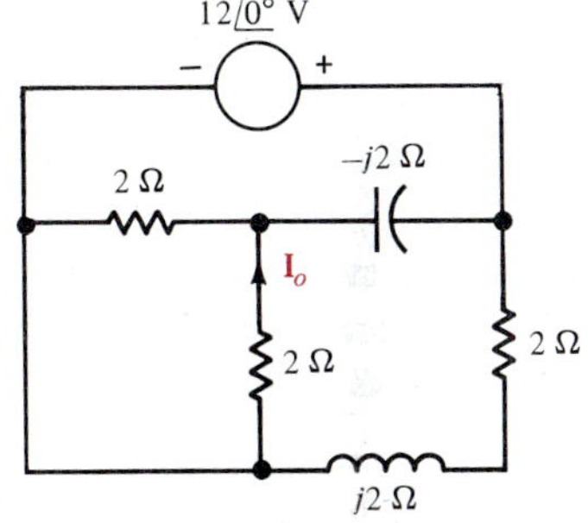

Figure P9.33

9.34. Find $\mathbf{V}_o$ in the network in Fig. P9.34.

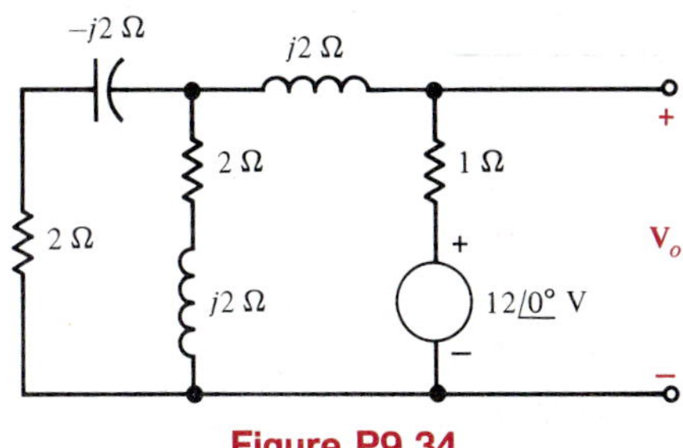

Figure P9.34

9.35. Find $\mathbf{I}_o$ in the network in Fig. P9.35.

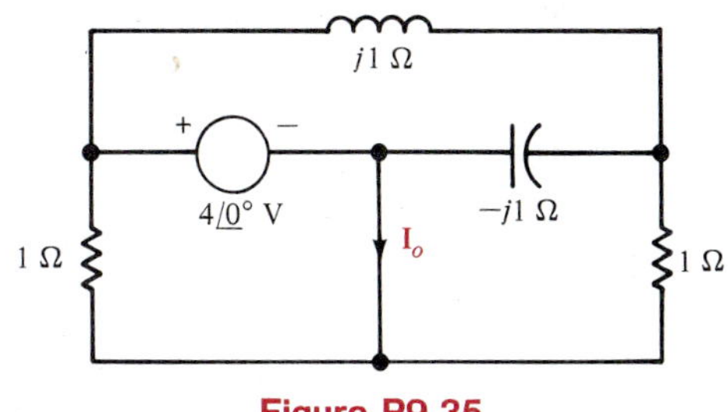

Figure P9.35

9.36. Find $\mathbf{I}_o$ in the network in Fig. P9.36.

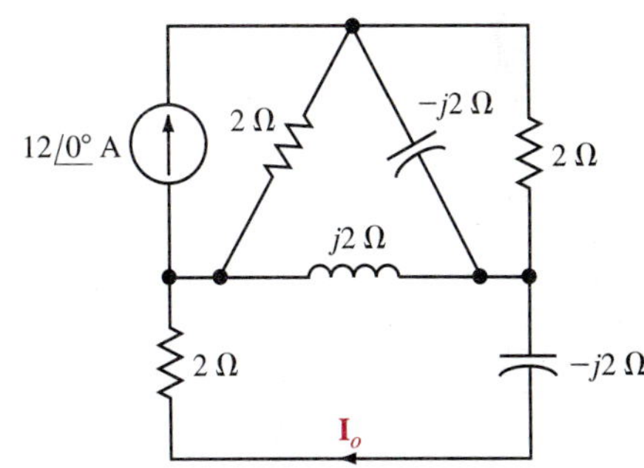

Figure P9.36

9.37. Find $\mathbf{I}_o$ in the network in Fig. P9.37.

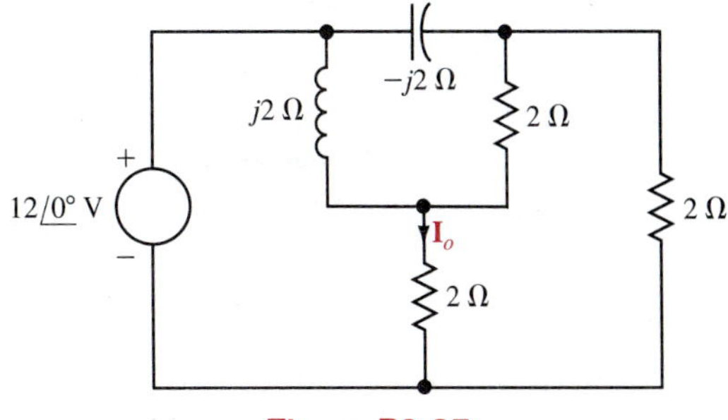

Figure P9.37

9.38. Find $\mathbf{I}_o$ in the network in Fig. P9.38.

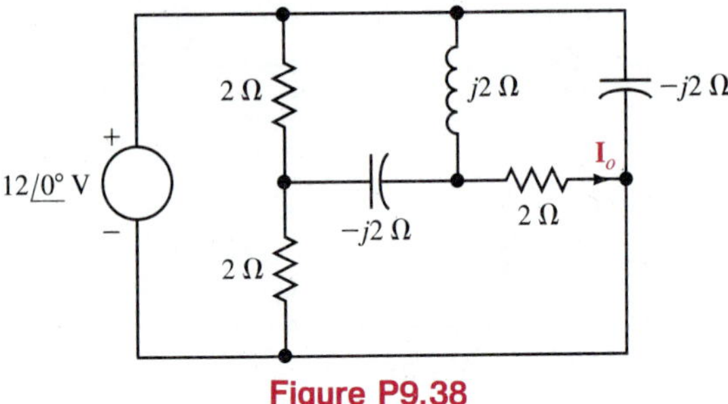

Figure P9.38

9.39. Find $\mathbf{I}_o$ in the network in Fig. P9.39.

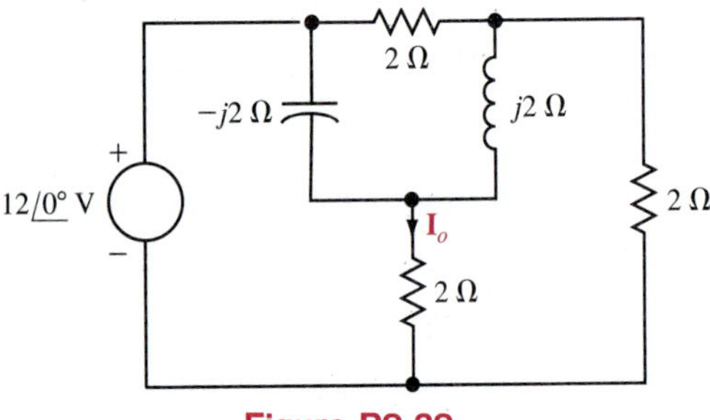

Figure P9.39

9.40. In the network in Fig. P9.40, if $\mathbf{V}_x = 4\ \underline{/45°}$ V, find $\mathbf{I}_o$.

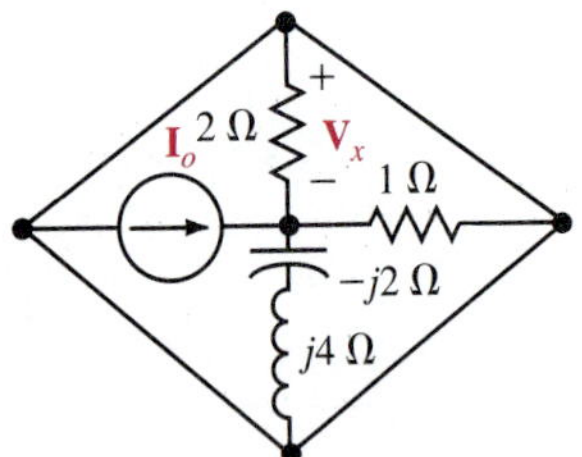

Figure P9.40

9.41. Find $\mathbf{I}_o$ in the network in Fig. P9.41.

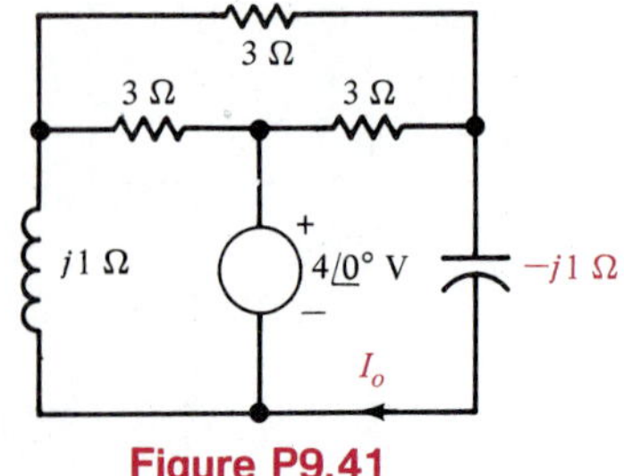

Figure P9.41

9.42. Find $\mathbf{V}_o$ in the network in Fig. P9.42.

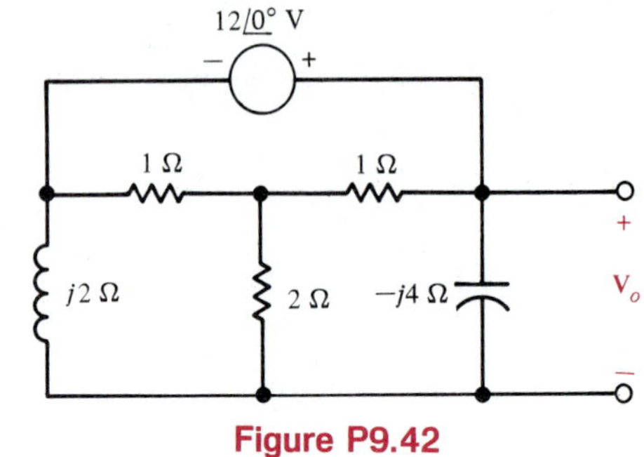

Figure P9.42

9.43. Find $\mathbf{V}_o$ in the network in Fig. P9.43.

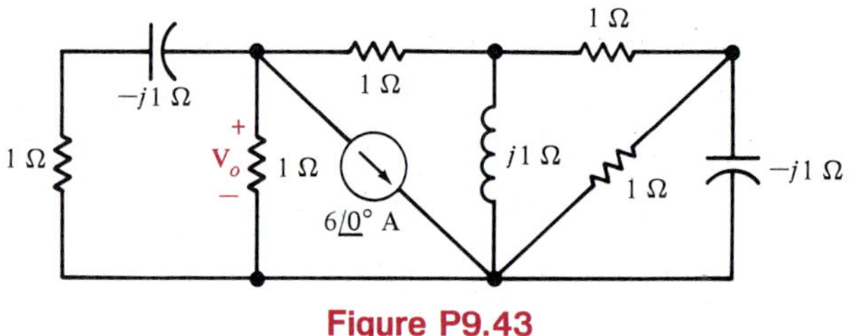

Figure P9.43

9.44. Find $\mathbf{V}_o$ in the network in Fig. P9.44.

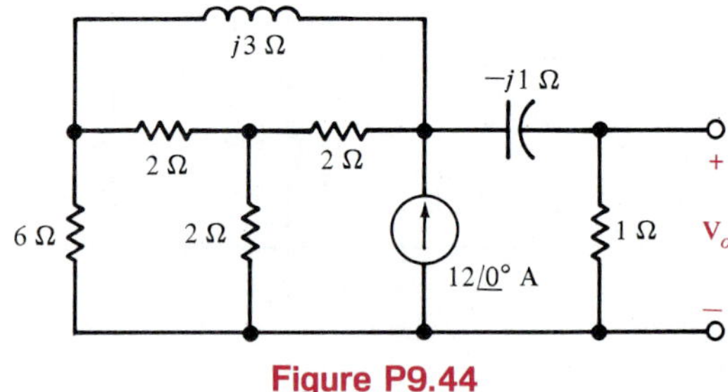

Figure P9.44

9.45. In the circuit in Fig. P9.45, if $\mathbf{V}_o = 4\ \underline{/45°}$ V, find $\mathbf{I}_1$.

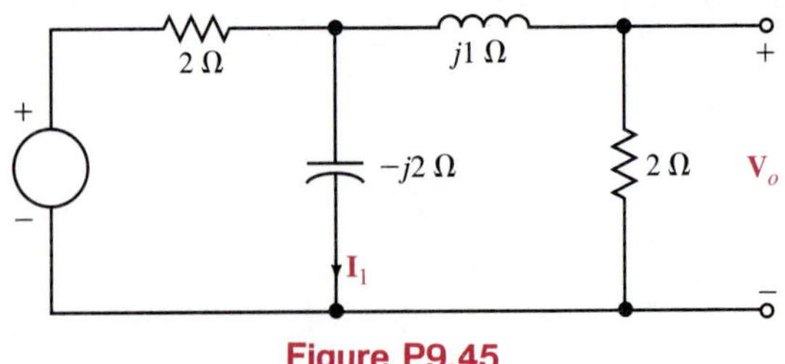

Figure P9.45

9.46. Find $\mathbf{V}_x$ in the network in Fig. P9.46.

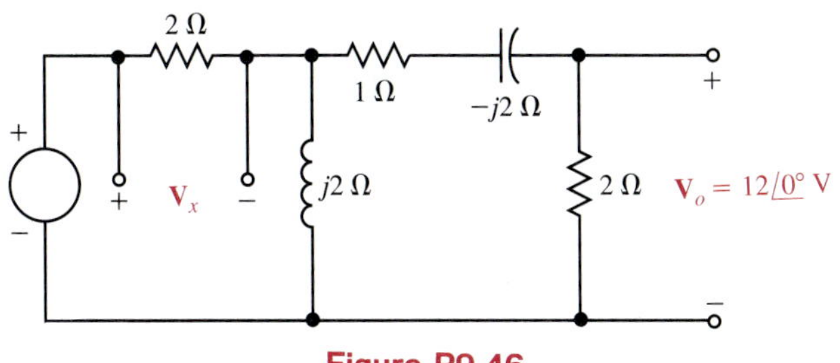

Figure P9.46

9.47. Find $\mathbf{I}_x$ in the circuit in Fig. P9.47.

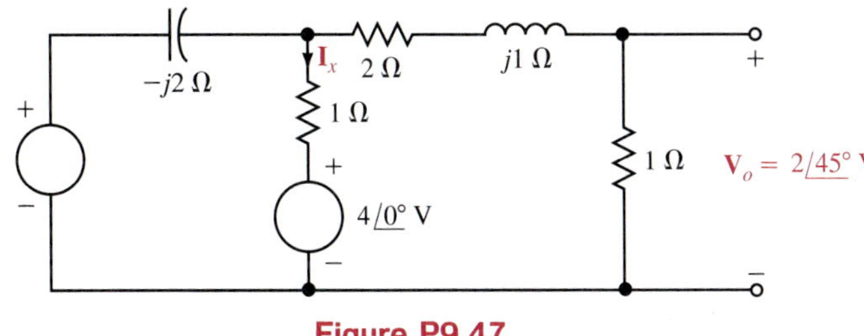

Figure P9.47

9.48. Find $\mathbf{V}_S$ in the network in Fig. P9.48 if $\mathbf{I}_1 = 2\ \underline{/0^\circ}$ A.

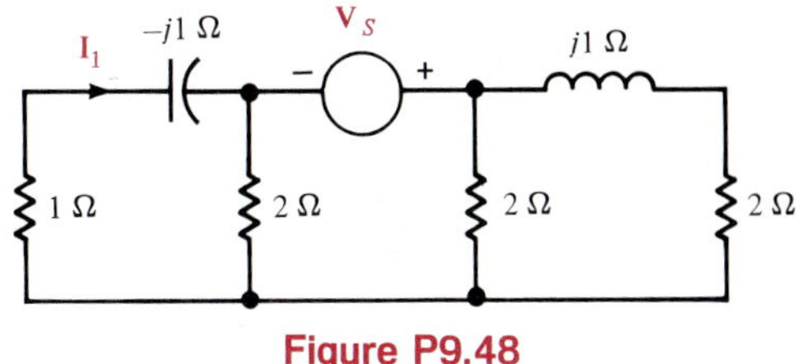

Figure P9.48

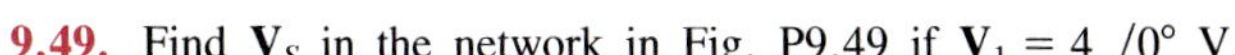

9.49. Find $\mathbf{V}_S$ in the network in Fig. P9.49 if $\mathbf{V}_1 = 4\ \underline{/0^\circ}$ V.

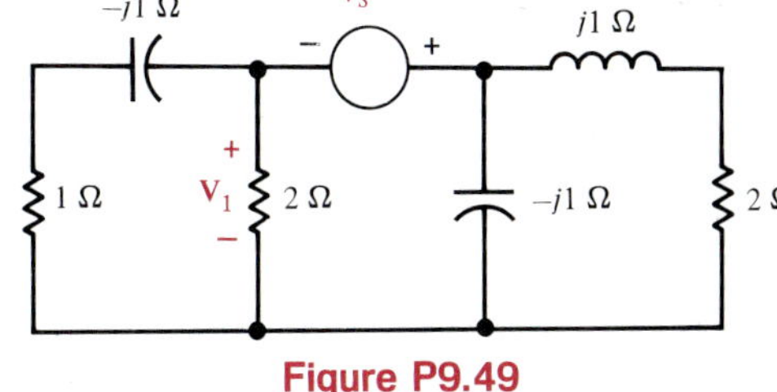

Figure P9.49

9.50. In the network in Fig. P9.50, if $\mathbf{I}_x = 5\ \underline{/45^\circ}$ A, find $\mathbf{V}_o$.

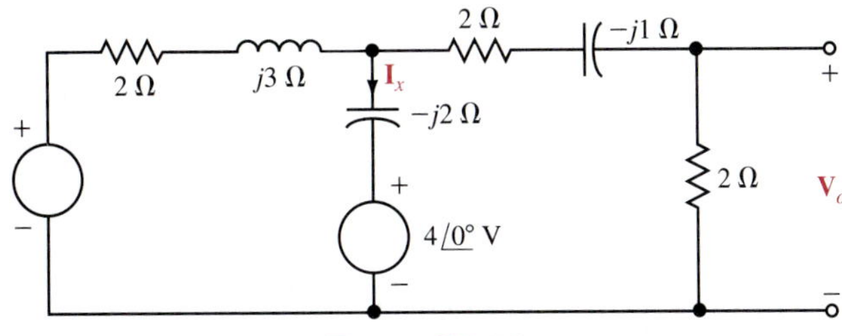

Figure P9.50

9.51. In the circuit in Fig. P9.51, if $\mathbf{V}_x = 6\ \underline{/30^\circ}$ V, find $\mathbf{V}_o$.

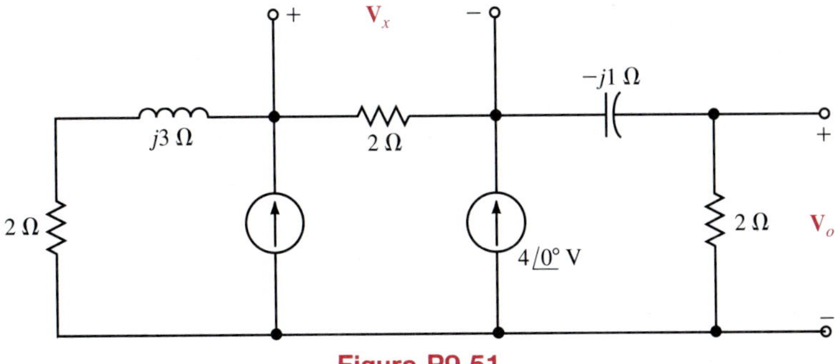

Figure P9.51

9.52. Find $\mathbf{V}_S$ in the network in Fig. P9.52 if $\mathbf{I}_o = 2 \angle 0^\circ$ A.

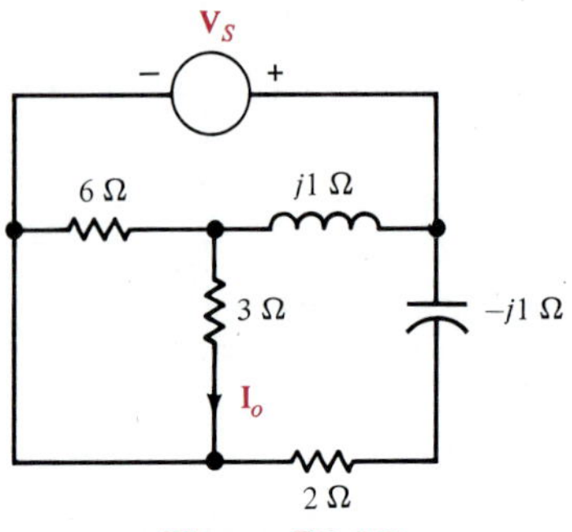

Figure P9.52

9.53. If $\mathbf{I}_o = 4 \angle 0^\circ$ A, find $\mathbf{V}_S$ in Fig. P9.53.

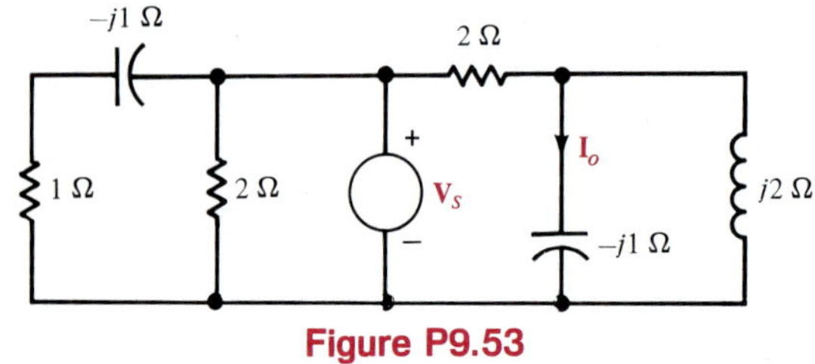

Figure P9.53

9.54. Given $\mathbf{V}_S = 12 \angle 0^\circ$ V, find $\mathbf{I}_S$ in Fig. P9.54.

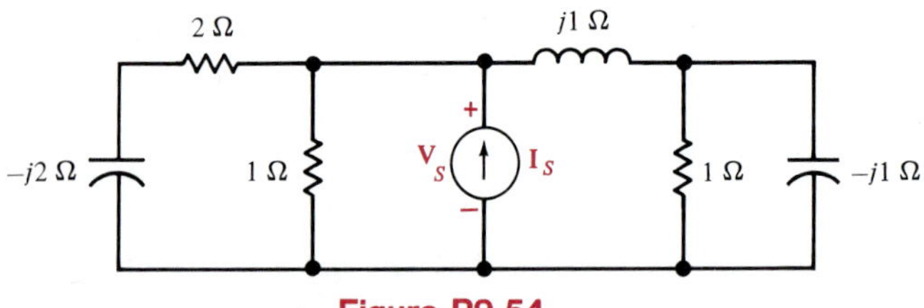

Figure P9.54

9.55. Find $\mathbf{I}_S$ in the network in Fig. P9.55 if $\mathbf{V}_1 = 8 \angle 0^\circ$ V.

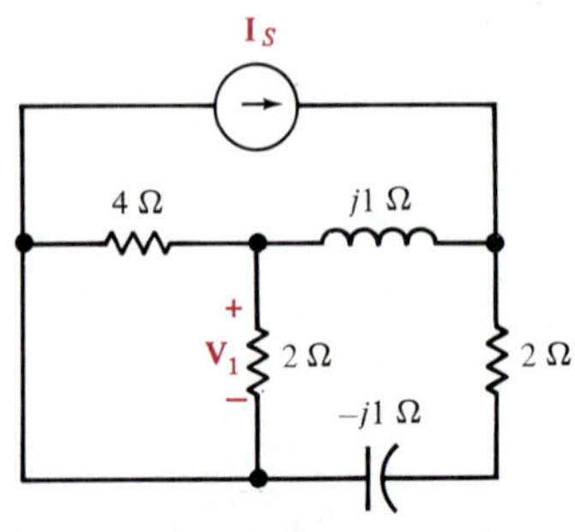

Figure P9.55

9.56. Given $\mathbf{I}$, find $\mathbf{V}_S$ in Fig. P9.56.

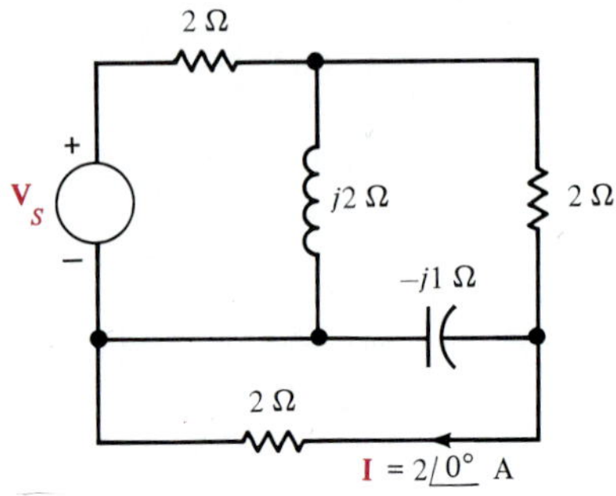

Figure P9.56

9.57. If $\mathbf{V}_1 = 10 \angle 0^\circ$ V, find $\mathbf{V}_S$ in Fig. P9.57.

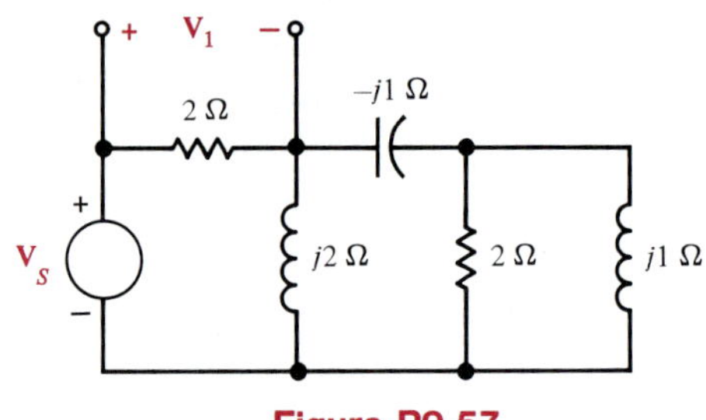

Figure P9.57

9.58. If $\mathbf{V}_1 = 4 \angle 0^\circ$ V, find $\mathbf{I}_o$ in Fig. P9.58.

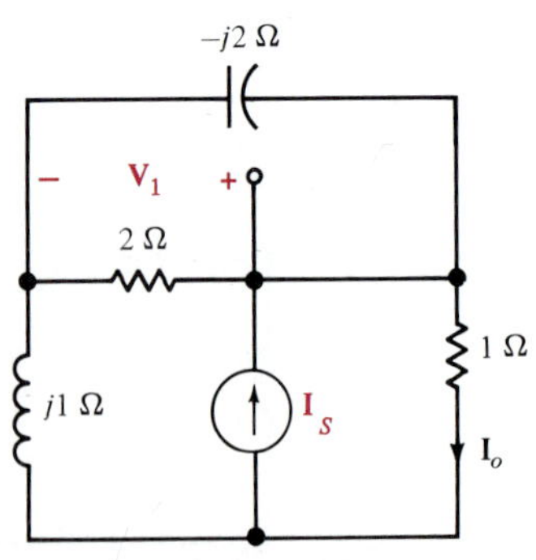

Figure P9.58

9.59. In the network in Fig. P9.59, $\mathbf{V}_o$ is known to be $4 \angle 45°$ V; find $\mathbf{Z}$.

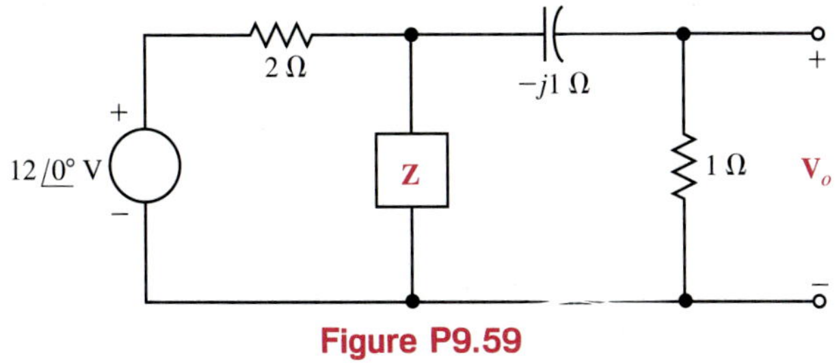

Figure P9.59

9.60. If $\mathbf{V}_o = 2 \angle 45°$ V in the network in Fig. P9.60, find $\mathbf{Z}$.

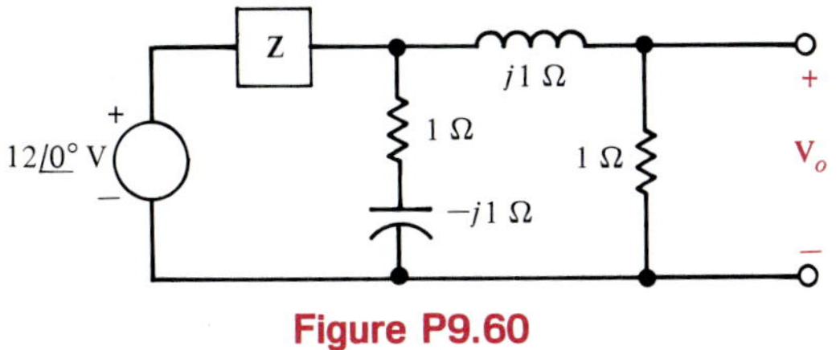

Figure P9.60

9.61. In the network in Fig. P9.61, $\mathbf{V}_1 = 2 \angle 45°$ V. Find $\mathbf{Z}$.

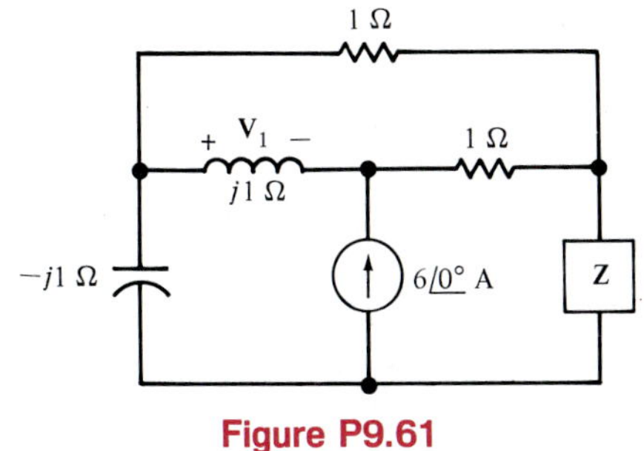

Figure P9.61

9.62. Find $\mathbf{Z}$ in the network in Fig. P9.62, if $\mathbf{V}_o = 6 \angle 45°$ V.

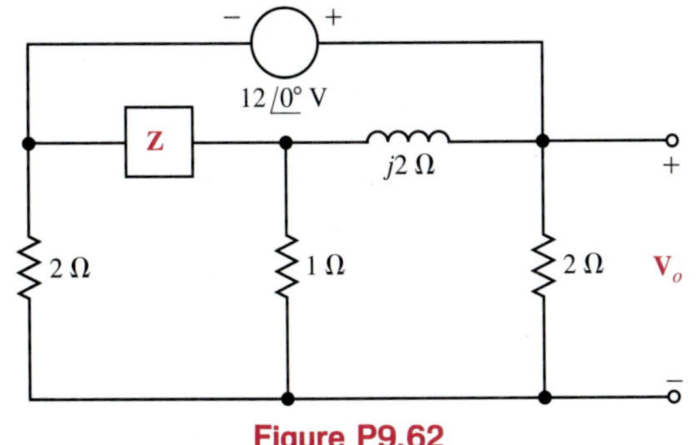

Figure P9.62

9.63. Derive the input–output relationship $\mathbf{V}_o/\mathbf{V}_i$ in the frequency domain for the op-amp circuit in Fig. P9.63.

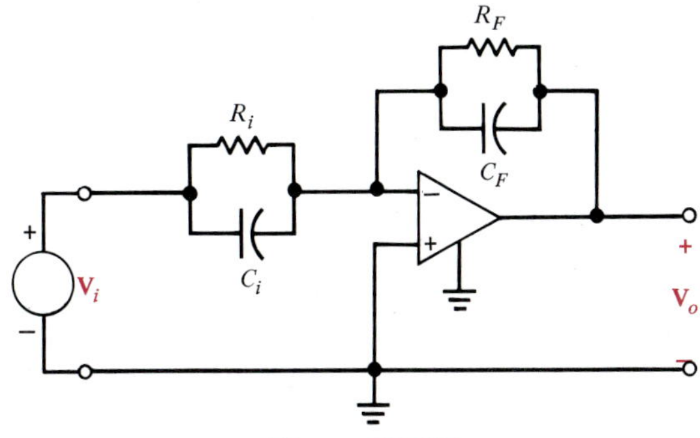

Figure P9.63

9.64. Determine the expression for the output voltage in the frequency domain for the circuit shown in Fig. P9.64.

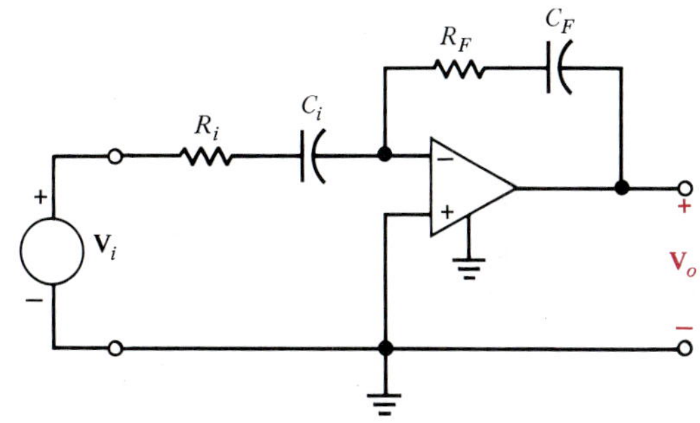

Figure P9.64

9.65

(a) Determine the input–output relationship in the frequency domain for the op-amp circuit shown in Fig. P9.65.

(b) Compute the transfer function $\mathbf{V}_o/\mathbf{V}_i$ if $\mathbf{Z}_i = R_i + 1/j\omega C_i$ and $\mathbf{Z}_F = 1/j\omega C_F$.

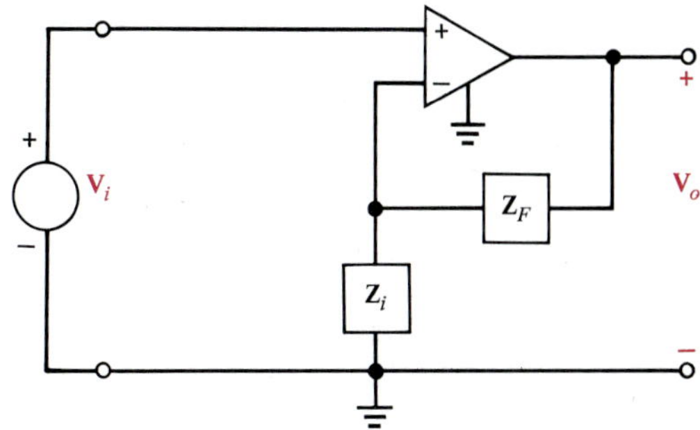

Figure P9.65

CHAPTER

10

Sinusoidal Steady-State Analysis

We have demonstrated in Chapter 9 the use of KCL, KVL, and the relationship $\mathbf{V} = \mathbf{IZ}$ in the solution of ac steady-state circuit problems. Since the network theorems and other circuit analysis techniques, which we successfully employed to solve resistive dc circuits, are applicable to linear circuits, it would appear that these circuit analysis tools could be applied to networks containing R, L, and C elements. In this chapter we demonstrate the applicability of these various techniques to the steady-state analysis of ac circuits. Since we have discussed the analysis techniques in earlier chapters, we concentrate here on illustrating their use via numerous examples. As we proceed, it will become obvious to the reader that the difference between the material in this chapter and that in preceding chapters is our use of phasors, impedance, and admittance, which lead to equations involving complex numbers.

Once again our purpose in employing the network theorems is to simplify the problem. Although the theorems can be readily applied in many cases, they do not always lead to a simpler solution than that obtained by a node or mesh analysis. For example, when applying superposition to a problem with two independent sources, we may actually find that the application of superposition makes the problem twice as complicated as the original one. Therefore, as we proceed with the various techniques, the reader is encouraged to examine each problem from a broad perspective in an attempt to determine the best method(s) of attack.

The examples employed in this chapter were specifically chosen to demonstrate the applicability of the various circuit analysis techniques to a variety of problems. In addition, an attempt has been made to help readers understand the various ways in which a circuit analysis problem can be approached by using more than one method to solve a particular problem. In this chapter ac circuits will also be analyzed using PSPICE.

10.1 Linearity

In order to illustrate the principle of linearity and its use in the solution of ac steady-state circuit problems, consider the following example.

EXAMPLE 10.1

The input voltage to the circuit shown in Fig. 10.1 is $v_S(t) = 12 \cos(377t + 30°)$ V. We wish to determine the current $\mathbf{I}_4$ using linearity. Therefore, we will assume that $\mathbf{I}_4 = 1 \underline{/0°}$ A, compute the required value of $\mathbf{V}_S$, and then use the actual value of $\mathbf{V}_S$ to determine the actual value of $\mathbf{I}_4$.

If $\mathbf{I}_4 = 1 \underline{/0°}$ A, then $\mathbf{V}_2 = (2)(1 \underline{/0°}) = 2 \underline{/0°}$ V. Hence

$$\mathbf{I}_5 = \frac{2 \underline{/0°}}{2 \underline{/-90°}} = 1 \underline{/90°} \text{ A}$$

Applying KCL, we have

$$\begin{aligned}\mathbf{I}_3 &= \mathbf{I}_4 + \mathbf{I}_5 \\ &= 1 + j1 \text{ A}\end{aligned}$$

Now applying KVL gives us

$$\begin{aligned}\mathbf{V}_1 &= (1 + j1)(j4) + \mathbf{V}_2 \\ &= -2 + j4 \text{ V}\end{aligned}$$

$\mathbf{I}_2$ is then

$$\mathbf{I}_2 = \frac{-2 + j4}{-j3} = \frac{-4}{3} - j\frac{2}{3} \text{ A}$$

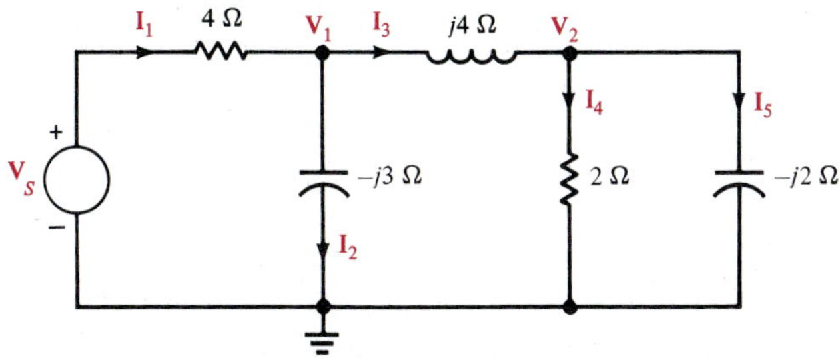

Figure 10.1 Example circuit used to illustrate linearity.

Again using KCL, we obtain

$$\mathbf{I}_1 = \mathbf{I}_2 + \mathbf{I}_3$$

$$= \frac{-4}{3} - \frac{j2}{3} + 1 + j1 = \frac{-1}{3} + j\frac{1}{3}\text{ A}$$

KVL around the loop at the far left of the circuit gives us

$$\mathbf{V}_S = 4(\mathbf{I}_1) + \mathbf{V}_1$$

$$= -\frac{4}{3} + j\frac{4}{3} - 2 + j4$$

$$= -\frac{10}{3} + j\frac{16}{3} = 6.29 \underline{/122^\circ}\text{ V}$$

Now since $\mathbf{V}_S = 6.29 \underline{/122^\circ}$ V produces an $\mathbf{I}_4 = 1 \underline{/0^\circ}$ A, then $\mathbf{V}_S = 12 \underline{/30^\circ}$ V produces an $\mathbf{I}_4$ of

$$\mathbf{I}_4 = \frac{(12 \underline{/30^\circ})(1 \underline{/0^\circ})}{6.29 \underline{/122^\circ}} = 1.91 \underline{/-92^\circ}\text{ A}$$

Therefore,

$$i_4(t) = 1.91 \cos (377t - 92^\circ)\text{ A}$$

■

DRILL EXERCISE

D10.1. Compute $\mathbf{V}_o$ in the network in Fig. D10.1 using linearity by first assuming that $\mathbf{V}_o = 1 \underline{/0^\circ}$ V.

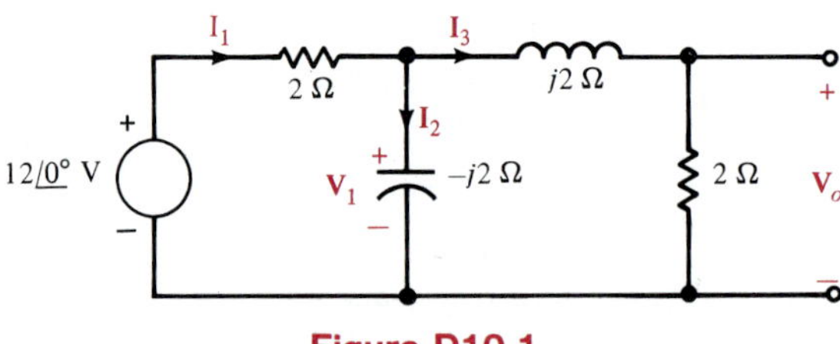

Figure D10.1

Ans: $\mathbf{V}_o = 5.36 \underline{/-63.43^\circ}$ V.

10.2 Nodal Analysis

The following examples illustrate that nodal analysis is performed in exactly the same manner as it was in dc resistive circuits. In this section nodal analysis will be employed in the frequency domain using the phasor technique. At this point our experience indicates that nodal analysis will be a feasible technique when the network contains a number of voltage sources. In this case the use of supernodes will typically simplify the solution procedure.

EXAMPLE 10.2

We wish to calculate the node voltages in the circuit in Fig. 10.2.

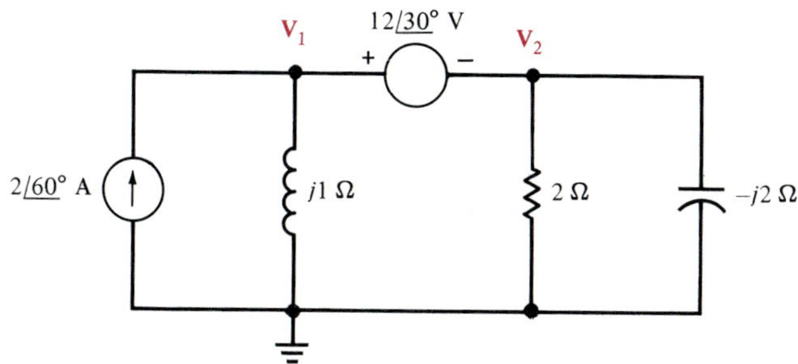

Figure 10.2 Example circuit used to illustrate nodal analysis.

To begin, we note that $\mathbf{V}_1$ and $\mathbf{V}_2$ form a supernode (i.e., $\mathbf{V}_1 = \mathbf{V}_2 + 12 \angle 30°$). Therefore, only one equation, together with this constraint equation, is needed to determine the node voltages.

Applying KCL at the supernode yields the equation

$$\frac{\mathbf{V}_1}{j1} + \frac{\mathbf{V}_1 - 12 \angle 30°}{2} + \frac{\mathbf{V}_1 - 12 \angle 30°}{-j2} = 2 \angle 60°$$

$$\mathbf{V}_1\left(\frac{1}{j1} + \frac{1}{2} + \frac{1}{-j2}\right) = 2 \angle 60° + \frac{12 \angle 30°}{2} - \frac{12 \angle 30°}{j2}$$

Simplifying this equation, we obtain

$$(0.71 \angle -45°)\,\mathbf{V}_1 = 3.2 + j9.93$$

or

$$\mathbf{V}_1 = 14.69 \angle 117.14° \text{ V}$$

Then using the constraint equation gives us

$$\begin{aligned}\mathbf{V}_2 &= \mathbf{V}_1 - 12 \angle 30° \\ &= -6.70 + j13.07 - (10.4 + j6) \\ &= -17.1 + j7.07 = 18.50 \angle 157.54° \text{ V}\end{aligned}$$

■

DRILL EXERCISE

D10.2. Use nodal analysis to find $\mathbf{V}_o$ in the network in Fig. D10.2.

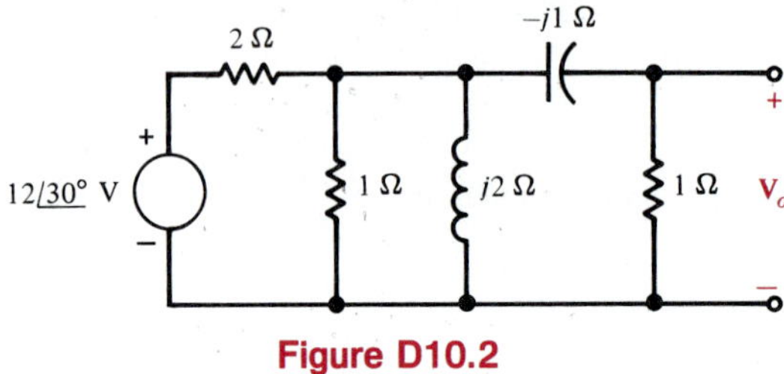

Figure D10.2

Ans: $\mathbf{V}_o = 2.12 \angle 75^\circ$ V.

EXAMPLE 10.3

Let us compute the current $\mathbf{I}_o$ in the network in Fig. 10.3.

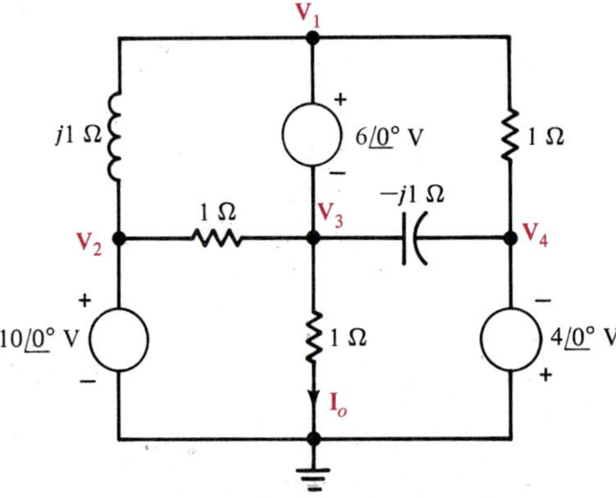

Figure 10.3 Example circuit used to illustrate nodal analysis.

The network has three supernodes which enclose each voltage source. The constraint equations for the supernodes are

$$\mathbf{V}_1 = \mathbf{V}_3 + 6 \angle 0^\circ \text{ V}$$

$$\mathbf{V}_2 = 10 \angle 0^\circ \text{ V}$$

$$\mathbf{V}_4 = -4 \angle 0^\circ \text{ V}$$

We need one additional linear independent equation to determine the node voltage $\mathbf{V}_3$, which will in turn yield $\mathbf{I}_o$. The KCL equation at the supernode containing the $6 \angle 0^\circ$ V voltage source is

$$\frac{\mathbf{V}_1 - \mathbf{V}_2}{j1} + \frac{\mathbf{V}_1 - \mathbf{V}_4}{1} + \frac{\mathbf{V}_3 - \mathbf{V}_2}{1} + \frac{\mathbf{V}_3}{1} + \frac{\mathbf{V}_3 - \mathbf{V}_4}{-j1} = 0$$

which reduces to

$$(1 - j)\mathbf{V}_1 - (1 - j)\mathbf{V}_2 - (1 + j)\mathbf{V}_4 + (2 + j)\mathbf{V}_3 = 0$$

Employing the constraint equations, we obtain

$$(1 - j)(\mathbf{V}_3 + 6 \underline{/0^\circ}) + (2 + j)\mathbf{V}_3 = (1 - j)10 + (1 + j)(-4 \underline{/0^\circ})$$

Solving this equation for $\mathbf{V}_3$ yields

$$\mathbf{V}_3 = \tfrac{8}{3} \underline{/-90^\circ} \text{ V}$$

and therefore

$$\mathbf{I}_o = \tfrac{8}{3} \underline{/-90^\circ} \text{ A}$$

EXAMPLE 10.4

We wish to determine $\mathbf{V}_o$ in the network in Fig. 10.4.

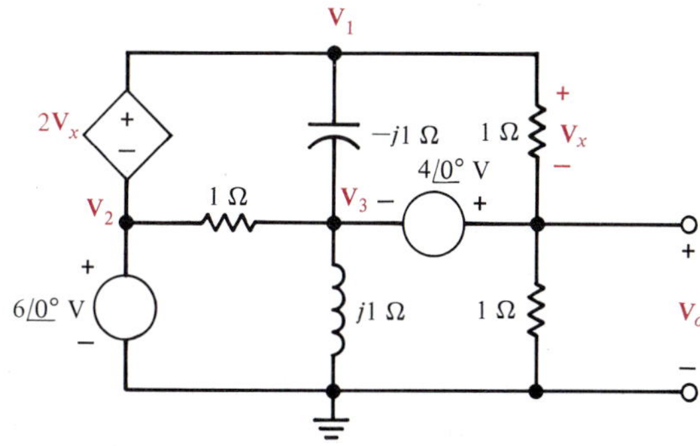

Figure 10.4 Example circuit used in nodal analysis.

The network contains two supernodes, one containing the sources $6 \underline{/0^\circ}$ V and $2\mathbf{V}_x$ and the other containing the $4 \underline{/0^\circ}$ V source. The constraint equations for the network are

$$\mathbf{V}_2 = 6 \underline{/0^\circ}$$

$$\mathbf{V}_3 = \mathbf{V}_o - 4 \underline{/0^\circ}$$

and around the outer loop

$$6 \underline{/0^\circ} + 2\mathbf{V}_x - \mathbf{V}_x - \mathbf{V}_o = 0$$

or

$$\mathbf{V}_x = \mathbf{V}_o - 6 \underline{/0^\circ}$$

We have three linearly independent equations. The fourth equation is obtained by employing KCL at the supernode containing the $4 \underline{/0^\circ}$ V source. That equation is

$$\frac{\mathbf{V}_o - (6 \underline{/0^\circ} + 2\mathbf{V}_x)}{1} + \frac{\mathbf{V}_3 - (6 \underline{/0^\circ} + 2\mathbf{V}_x)}{-j1} + \frac{\mathbf{V}_3 - 6 \underline{/0^\circ}}{1} + \frac{\mathbf{V}_3}{j1} + \frac{\mathbf{V}_o}{1} = 0$$

Substituting the equations for $\mathbf{V}_3$ and $\mathbf{V}_x$ into the equation above yields

$$\mathbf{V}_o = 3.22 \underline{/7.13^\circ} \text{ V}$$

DRILL EXERCISE

D10.3. Use nodal analysis to determine the node voltages in the network in Fig. D10.3.

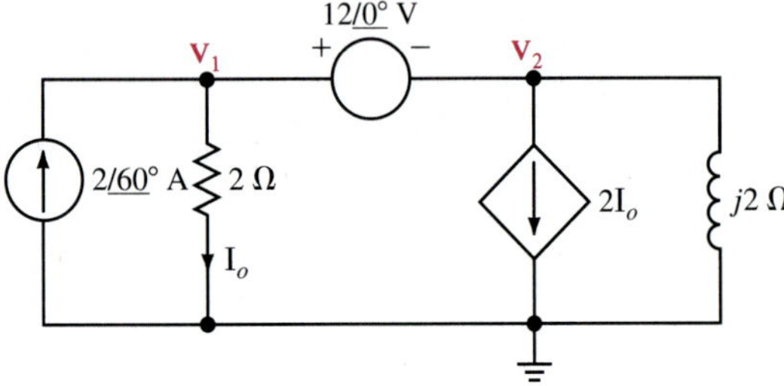

Figure D10.3

Ans: $\mathbf{V}_1 = 2.77 \underline{/-58.4^\circ}$ V, $\mathbf{V}_2 = -10.81 \underline{/12.6^\circ}$ V.

EXAMPLE 10.5

Calculate the output voltage for the op-amp circuit shown in Fig. 10.5a. Use the equivalent circuit for the op-amp in Fig. 3.33b and assume that $R_i = \infty$ and $R_o = 0$.

The original network containing the equivalent op-amp circuit is shown in Fig. 10.5b. The dependent source creates a supernode, and therefore only two equations are needed to compute the output voltage. These equations are

$$(\mathbf{V}_1 - \mathbf{V}_S)j\omega C_1 + \frac{\mathbf{V}_1}{R_1} + (\mathbf{V}_1 - \mathbf{V})j\omega C_2 + (\mathbf{V}_1 + A\mathbf{V})j\omega C_3 = 0$$

$$(\mathbf{V} - \mathbf{V}_1)j\omega C_2 + \frac{\mathbf{V}}{R_2} = 0$$

or

$$\mathbf{V}_1\left(j\omega C_1 + \frac{1}{R_1} + j\omega C_2 + j\omega C_3\right) - \mathbf{V}(j\omega C_2 - Aj\omega C_3) = j\omega C_1 \mathbf{V}_S$$

$$-\mathbf{V}_1(j\omega C_2) + \mathbf{V}\left(j\omega C_2 + \frac{1}{R_2}\right) = 0$$

Solving for the voltage $\mathbf{V}$, we obtain

$$\mathbf{V} = \frac{-\omega^2 C_1 C_2 \mathbf{V}_S}{[1/R_1 + j\omega(C_1 + C_2 + C_3)](j\omega C_2 + 1/R_2) - j\omega C_2(j\omega C_2 - j\omega C_3 A)}$$

Simplifying this function yields

$$\mathbf{V} = \frac{-\omega^2 C_1 C_2 R_1 R_2 \mathbf{V}_S}{1 + j\omega[R_2 C_2 + R_1(C_1 + C_2 + C_3)] - \omega^2 R_1 R_2 C_2(C_1 + C_3 + C_3 A)}$$

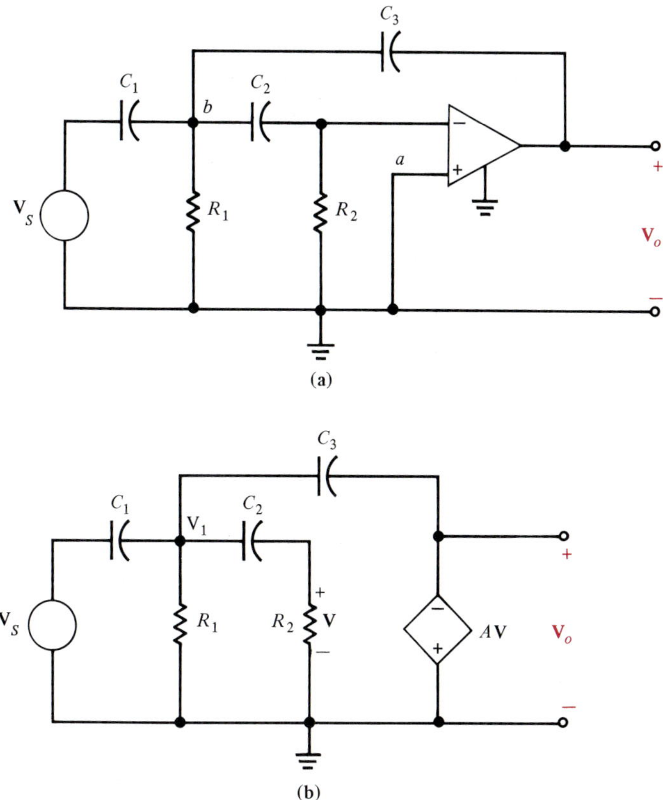

Figure 10.5 Operational amplifier circuit used to illustrate nodal analysis.

The output voltage $\mathbf{V}_o$ is then

$$\mathbf{V}_o = \frac{+A\omega^2 C_1 C_2 R_1 R_2 \mathbf{V}_S}{1 + j\omega[R_2C_2 + R_1(C_1 + C_2 + C_3)] - \omega^2 R_1 R_2 C_2 (C_1 + C_3 + C_3 A)}$$

Note that as $A \rightarrow \infty$ the input–output relationship becomes

$$\mathbf{V}_o = -\frac{C_1}{C_3}\mathbf{V}_S$$

It is informative to compare this result with our previous discussions. The way this circuit is configured, an error signal is developed between points a and b. From the discussion in Chapter 3, the feedback works to minimize the error signal and as $A \rightarrow \infty$, the error signal is reduced to zero. Since a is at zero potential, there is no ac voltage across the elements R_1, C_2, and R_2 and hence they have no effect. Therefore, the network is in essence that shown in Fig. 3.35a, where R_2 is replaced with $1/j\omega C_3$ and R_1 is replaced with $1/j\omega C_1$. ■

EXAMPLE 10.6

The circuit shown in Fig. 10.6 represents a portion of an equivalent circuit for a transistor amplifier. We wish to calculate the input admittance assuming that the circuit is energized on the left side by a source having voltage $\mathbf{V}$ and current $\mathbf{I}_1$. The two node equations are

$$-\mathbf{I}_1 + j\omega C_1\mathbf{V} + j\omega C_2(\mathbf{V} - \mathbf{V}_o) = 0$$

$$j\omega C_2(\mathbf{V}_o - \mathbf{V}) + g_m\mathbf{V} + \frac{\mathbf{V}_o}{R_L} = 0$$

or

$$(j\omega C_1 + j\omega C_2)\mathbf{V} - j\omega C_2\mathbf{V}_o = \mathbf{I}_1$$

$$(g_m - j\omega C_2)\mathbf{V} + \left(j\omega C_2 + \frac{1}{R_L}\right)\mathbf{V}_o = 0$$

Solving these equations for $\mathbf{V}$ yields

$$\mathbf{V} = \frac{\mathbf{I}_1(j\omega C_2 + 1/R_L)}{(j\omega C_1 + j\omega C_2)(j\omega C_2 + 1/R_L) + j\omega C_2(g_m - j\omega C_2)}$$

The input admittance is

$$\mathbf{Y}_i = \frac{\mathbf{I}_1}{\mathbf{V}} = j\omega(C_1 + C_2) + \frac{j\omega C_2(g_m - j\omega C_2)}{j\omega C_2 + 1/R_L}$$

In electronics we find that for an important range of frequencies $1/R_L \gg \omega C_2$ and $g_m \gg \omega C_2$. Under these conditions the equation above reduces to

$$\mathbf{Y}_i = j\omega C_1 + j\omega C_2 + j\omega C_2 g_m R_L$$

$$= j\omega[C_1 + C_2(1 + g_m R_L)]$$

The quantity $g_m R_L$ is generally large, and therefore the equation above indicates that the input capacitance of the amplifier is significantly increased due to the capacitor C_2, which links the input and the output of the amplifier. This increase in the input capacitance of an amplifier stage is known in electronics as the *Miller effect*. ■

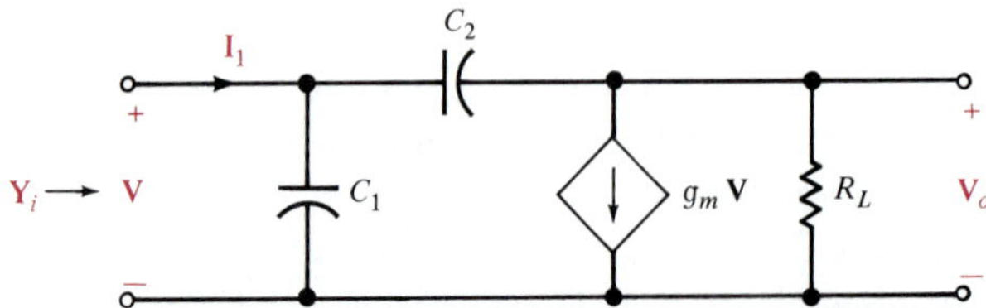

Figure 10.6 Portion of a transistor amplifier used to illustrate nodal analysis.

10.3 Mesh Analysis

The following examples illustrate the use of mesh analysis in ac steady-state circuits. Our experience indicates that mesh analysis will be most feasible when the network contains a number of current sources. In this case the analysis can be simplified by defining the mesh or loop currents to flow through the current sources.

EXAMPLE 10.7

Let us determine the voltage $\mathbf{V}_o$ in the circuit in Fig. 10.7.

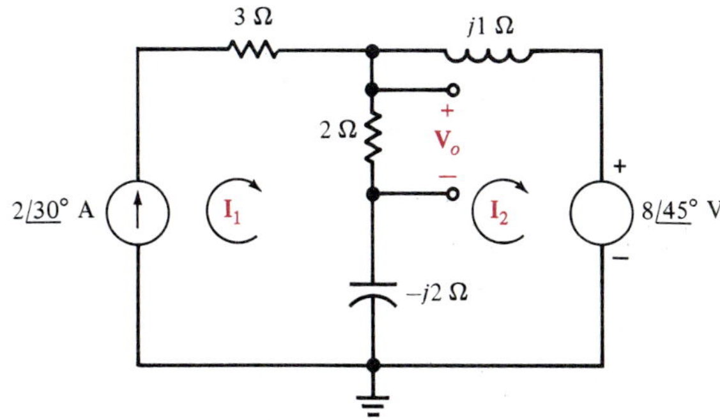

Figure 10.7 Circuit used to illustrate mesh equations.

Although there are two meshes, we need only one equation to determine all the mesh currents, because of the presence of the current source. KVL for the right mesh is

$$(\mathbf{I}_2 - 2\,\underline{/30^\circ})(2 - j2) + \mathbf{I}_2(j1) = -8\,\underline{/45^\circ}$$

or

$$\mathbf{I}_2 = \frac{(2\,\underline{/-30^\circ})(2 - j2) - 8\,\underline{/45^\circ}}{2 - j1} = 3.18\,\underline{/-64.96^\circ}\text{ A}$$

To solve for $\mathbf{V}_o$, we need the current through the 2-Ω resistor.

$$\begin{aligned}\mathbf{I}_1 - \mathbf{I}_2 &= 2\,\underline{/30^\circ} - 3.18\,\underline{/-64.96^\circ}\\ &= 0.38 + j3.88\text{ A}\end{aligned}$$

and hence

$$\begin{aligned}\mathbf{V}_o &= 2(\mathbf{I}_1 - \mathbf{I}_2)\\ &= 0.76 + j7.76\text{ V}\end{aligned}$$

■

The following example illustrates the ease with which a fairly complicated network can be analyzed by selecting the proper method.

EXAMPLE 10.8

Consider the network in Fig. 10.8a. We wish to compute the voltage $\mathbf{V}_o$.

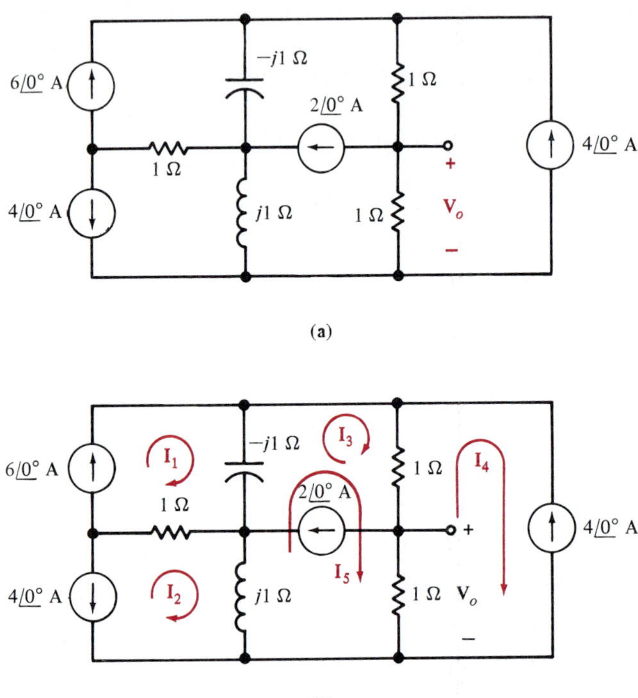

Figure 10.8 Circuits used in Example 10.8.

Because of the presence of the current sources, the currents $\mathbf{I}_1$, $\mathbf{I}_2$, $\mathbf{I}_3$, and $\mathbf{I}_4$ are defined to flow through a current source as shown in Fig. 10.8b. Therefore, the constraint equations are

$$\mathbf{I}_1 = 6\,\underline{/0^\circ}$$

$$\mathbf{I}_2 = -4\underline{/0^\circ}$$

$$\mathbf{I}_3 = 2\,\underline{/0^\circ}$$

$$\mathbf{I}_4 = -4\,\underline{/0^\circ}$$

Finally, the current $\mathbf{I}_5$ is defined to flow through a path that does not contain a current source. Such a path is shown in Fig. 10.8b. The KVL equation for the path through which $\mathbf{I}_5$ flows is

$$j1(\mathbf{I}_5 - \mathbf{I}_2) - j1(\mathbf{I}_5 + \mathbf{I}_3 - \mathbf{I}_1) + 1(\mathbf{I}_5 + \mathbf{I}_3 - \mathbf{I}_4) + 1(\mathbf{I}_5 - \mathbf{I}_4) = 0$$

or

$$2\mathbf{I}_5 = j\mathbf{I}_2 - (1 - j)\mathbf{I}_3 + 2\mathbf{I}_4 - j\mathbf{I}_1$$

Using the values defined by the constraint equations, we obtain

$$\mathbf{I}_5 = -(5 + 4j) \text{ A}$$

and then

$$\begin{aligned}\mathbf{V}_o &= 1(\mathbf{I}_5 - \mathbf{I}_4) \\ &= -(1 + 4j) \text{ V}\end{aligned}$$

Imagine for a moment the problem of determining $\mathbf{V}_o$ using nodal analysis. ■

Consider now the application of mesh analysis to a network containing a dependent source.

EXAMPLE 10.9

Let us determine the voltage $\mathbf{V}_o$ in the network in Fig. 10.9a.

Because there are two current sources present in the network, mesh analysis will be employed to obtain a solution. The mesh currents are defined in Fig. 10.9b. The constraint equations for the network are

$$\mathbf{I}_2 = -4 \underline{/0^\circ}$$

$$\mathbf{I}_x = \mathbf{I}_4 - \mathbf{I}_2 = \mathbf{I}_4 + 4 \underline{/0^\circ}$$

$$\mathbf{I}_3 = 2\mathbf{I}_x = 2\mathbf{I}_4 + 8 \underline{/0^\circ}$$

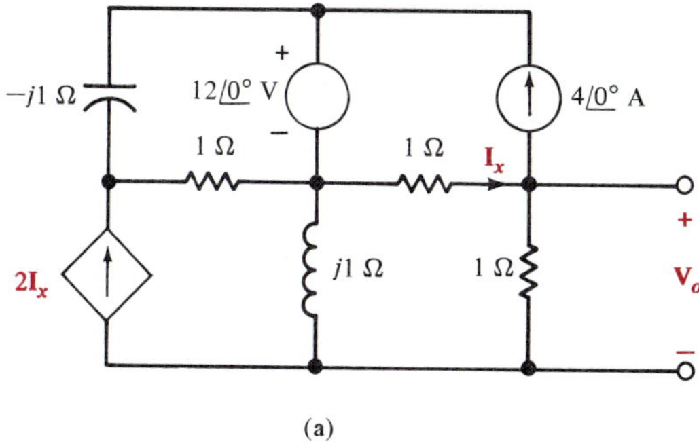

(a)

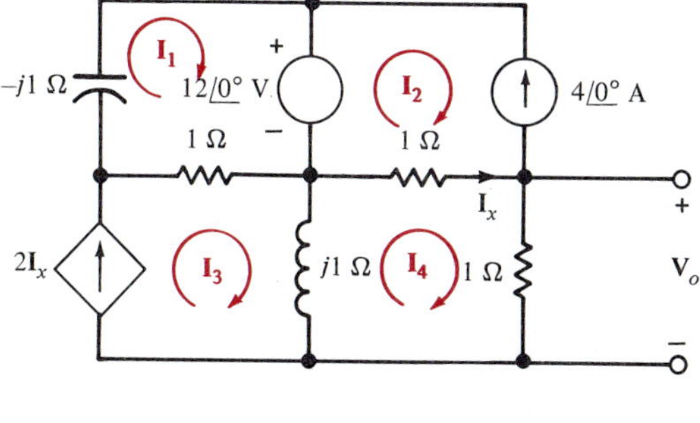

(b)

Figure 10.9 Circuits used in Example 10.9.

The KVL equations for mesh 1 and mesh 4 are

$$-j1\mathbf{I}_1 + 1(\mathbf{I}_1 - \mathbf{I}_3) = -12 \angle 0^\circ$$

$$j1(\mathbf{I}_4 - \mathbf{I}_3) + 1(\mathbf{I}_4 - \mathbf{I}_2) + \mathbf{I}_4 = 0$$

Note that if the constraint equations are substituted into the second KVL equation, the only unknown in the equation is $\mathbf{I}_4$. This substitution yields

$$\mathbf{I}_4 = -4 \angle -36.87^\circ \text{ A}$$

and hence

$$\mathbf{V}_o = -4 \angle -36.87^\circ \text{ V}$$

DRILL EXERCISES

D10.4. Use mesh equations to find $\mathbf{V}_o$ in the network in Fig. D10.4.

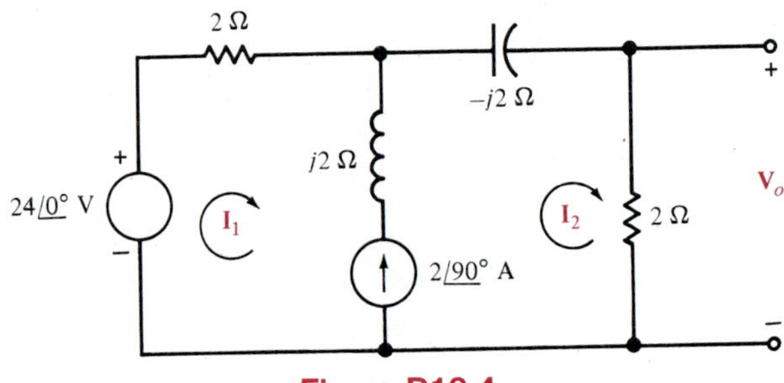

Figure D10.4

Ans: $\mathbf{V}_o = 10.88 \angle 36^\circ$ V.

D10.5. Find $\mathbf{V}_o$ in the network in Fig. D10.5 using mesh equations.

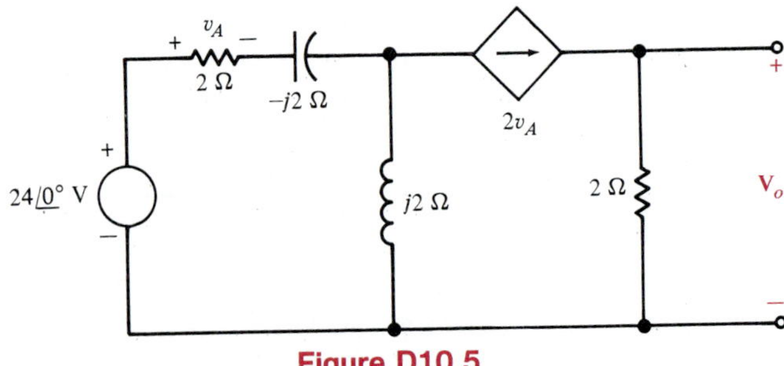

Figure D10.5

Ans: $\mathbf{V}_o = 23.28 \angle 75.96^\circ$ V.

10.4 Superposition

We will now illustrate the principle of superposition in ac steady-state circuit analysis. Our vehicle for illustrating this principle will be the circuit analyzed in Example 10.7 and shown in Fig. 10.7.

EXAMPLE 10.10

Let us use superposition to determine $\mathbf{V}_o$ in the circuit in Fig. 10.7.

We first replace the voltage source with a short circuit and compute $\mathbf{V}_o'$, which is the component of $\mathbf{V}_o$ due to the current source, as shown in Fig. 10.10a. Using current division, we obtain

$$\mathbf{V}_o' = \frac{(2\,\underline{/30^\circ})(j1)}{2 - j2 + j1}(2)$$
$$= 1.79\,\underline{/146.57^\circ}\text{ V}$$

Now replacing the current source with an open circuit, we can determine $\mathbf{V}_o''$, which is the component of $\mathbf{V}_o$ due to the voltage source, as shown in Fig. 10.10b. Using voltage division, we have

$$\mathbf{V}_o'' = \frac{8\,\underline{/45^\circ}}{2 - j2 + j1}(2)$$
$$= 7.14\,\underline{/71.57^\circ}\text{ V}$$

Finally,

$$\mathbf{V}_o = \mathbf{V}_o' + \mathbf{V}_o''$$
$$= 1.79\,\underline{/146.57^\circ} + 7.14\,\underline{/71.57^\circ}$$
$$= 0.76 + j7.76\text{ V}$$

■

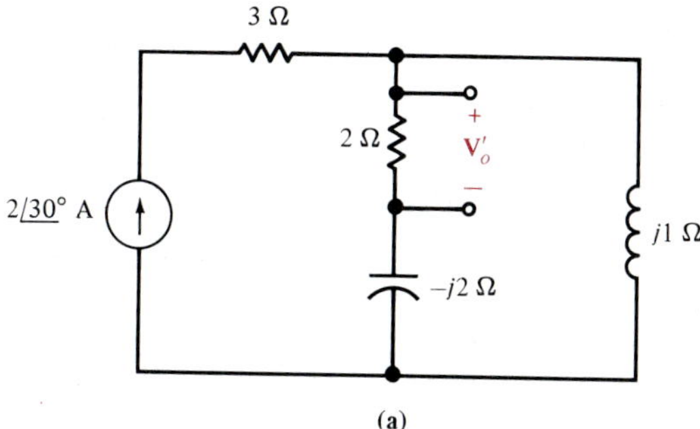

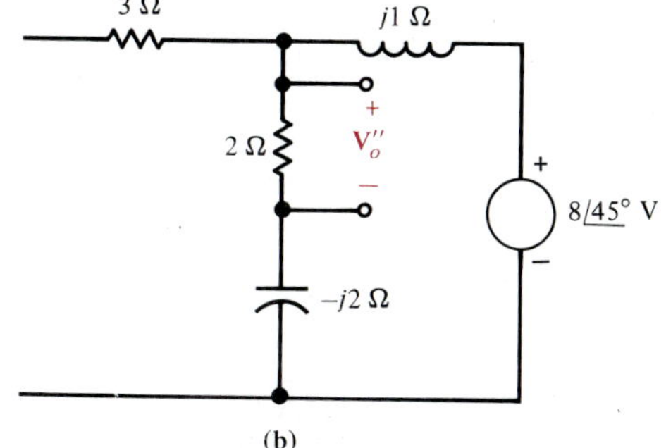

Figure 10.10 Circuits used to determine $\mathbf{V}_o$ in Fig. 10.6 via superposition.

DRILL EXERCISE

D10.6. Using superposition, find $\mathbf{V}_o$ in the network in Fig. D10.6.

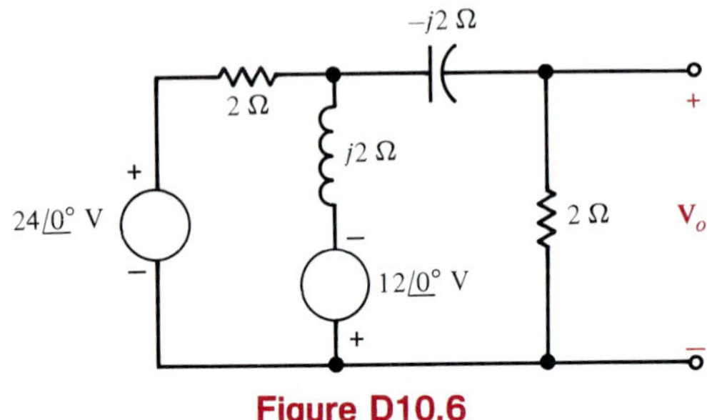

Figure D10.6

Ans: $\mathbf{V}_o = 12 \angle 90° \text{ V}$.

D10.7. Find $\mathbf{V}_o$ in the network in Fig. D10.7 using superposition.

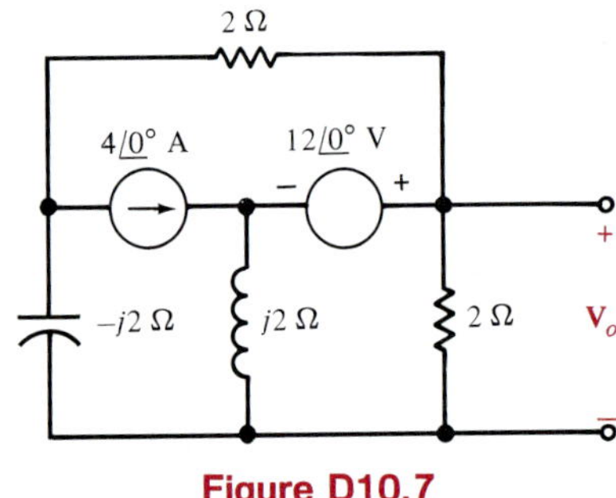

Figure D10.7

Ans: $\mathbf{V}_o = 5.66 \angle -45° \text{ V}$.

An important special case of polyphase circuits is the single-phase three-wire system shown in Fig. 10.11a. Its importance stems from the fact that it is the normal system found in households. Note that the voltage sources are equal (i.e., $\mathbf{V}_{an} = \mathbf{V}_{nb} = \mathbf{V}$), so that the magnitudes are equal and the phases are equal (single phase), and therefore the line-to-line voltage $\mathbf{V}_{ab} = 2\mathbf{V}_{an} = 2\mathbf{V}_{nb} = 2\mathbf{V}$. Typically, lights or small appliances are connected from one line to *neutral n*, and large appliances (e.g., hot water heaters) are connected line to line. Lights operate at about 120 V and the hot water heater operates at approximately 240 V.

Let us now attach two identical loads to the single-phase three-wire voltage system using perfect conductors as shown in Fig. 10.11b. From the figure we note that

$$\mathbf{I}_{aA} = \frac{\mathbf{V}}{\mathbf{Z}_L}$$

and

$$\mathbf{I}_{bB} = -\frac{\mathbf{V}}{\mathbf{Z}_L}$$

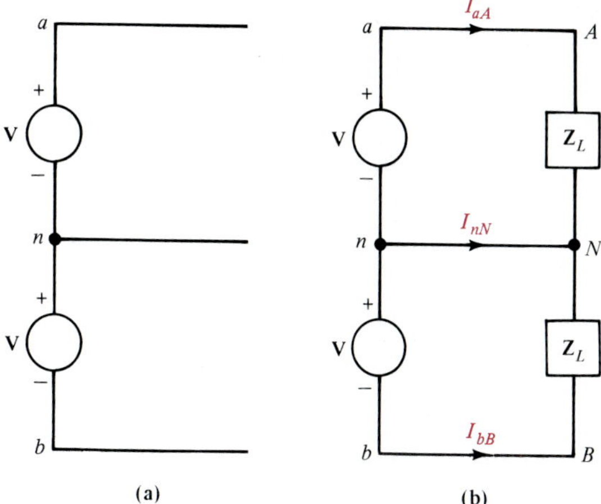

Figure 10.11 Single-phase three-wire system.

KCL at point N is

$$\mathbf{I}_{nN} = -(\mathbf{I}_{aA} + \mathbf{I}_{bB})$$
$$= -\left(\frac{\mathbf{V}}{\mathbf{Z}_L} - \frac{\mathbf{V}}{\mathbf{Z}_L}\right)$$
$$= 0$$

Note that there is no current in the neutral wire, and therefore it could be removed without affecting the remainder of the system; that is, all the voltages and currents would be unchanged. One is naturally led to wonder just how far the simplicity exhibited by this system will extend. For example, what would happen if each line had a line impedance, if the neutral conductor had an impedance associated with it, and if there were a load tied from line to line? To explore these questions, consider the circuit in Fig. 10.12a. Although we could examine this circuit using many of the techniques we have employed in previous chapters, the symmetry of the network suggests that perhaps superposition may lead us to some conclusions without having to resort to a brute-force assault. Employing superposition, we consider the two circuits in Fig. 10.12b and c. The currents in Fig. 10.12b are labeled arbitrarily. Because of the symmetrical relationship between Fig. 10.12b and c, the currents in Fig. 10.12c correspond directly to those in Fig. 10.12b. If we add the two *phasor* currents in each branch, we find that the neutral current is again zero. A neutral current of zero is a direct result of the symmetrical nature of the network. If either the line impedances $\mathbf{Z}_{\text{line}}$ or the load impedances $\mathbf{Z}_L$ are unequal, the neutral current will be nonzero. We will make direct use of these concepts when we study three-phase networks in Chapter 12.

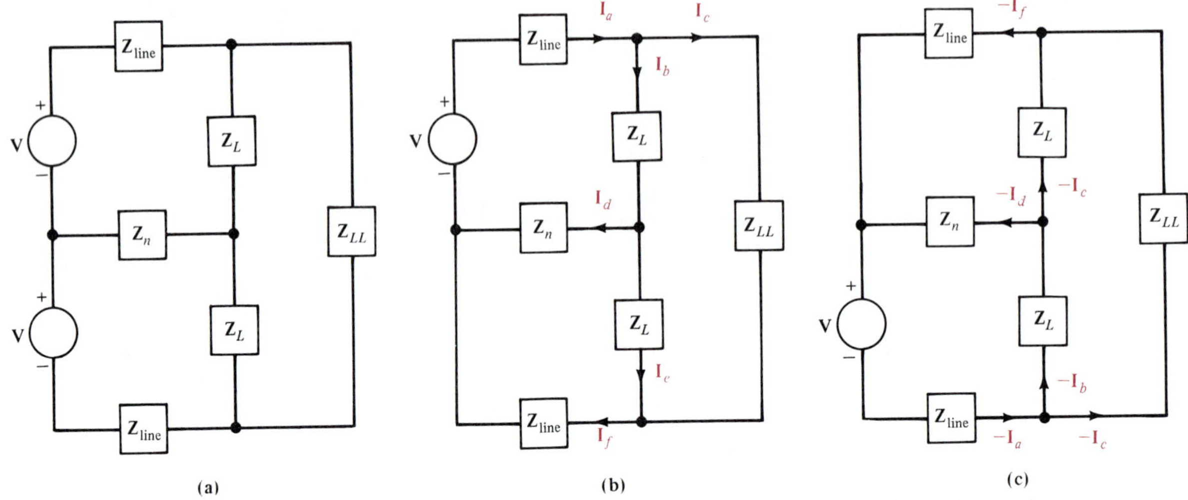

Figure 10.12 Circuits for analyzing a single-phase three-wire system.

10.5 Source Transformation

We have found that source transformation can be a very useful tool in simplifying a network. If the network has the proper topology (i.e., a ladder network which resembles cascaded window panes), the successive application of source transformation can be a feasible method. Source transformation can also be used in combination with other techniques for network simplification. The application of source transformation to ac steady-state circuit analysis will be illustrated by the following example.

EXAMPLE 10.11

We wish to determine $\mathbf{I}_4$ in the circuit in Fig. 10.1 using source transformation.

First we convert the voltage source and series 4-Ω resistor to a current source and parallel 4-Ω resistor as shown in Fig. 10.13a. The current source and parallel resistor and capacitor are now converted into an equivalent circuit consisting of a voltage source in series with an impedance as shown in Fig. 10.13b, where

$$\mathbf{Z}_1 = \frac{(4)(-j3)}{4 - j3}$$

$$= \frac{-j12}{4 - j3}\ \Omega$$

Combining $\mathbf{Z}_1$ with the inductor yields $\mathbf{Z}_2$, where

$$\mathbf{Z}_2 = \mathbf{Z}_1 + j4 = \frac{12 + j4}{4 - j3}\ \Omega$$

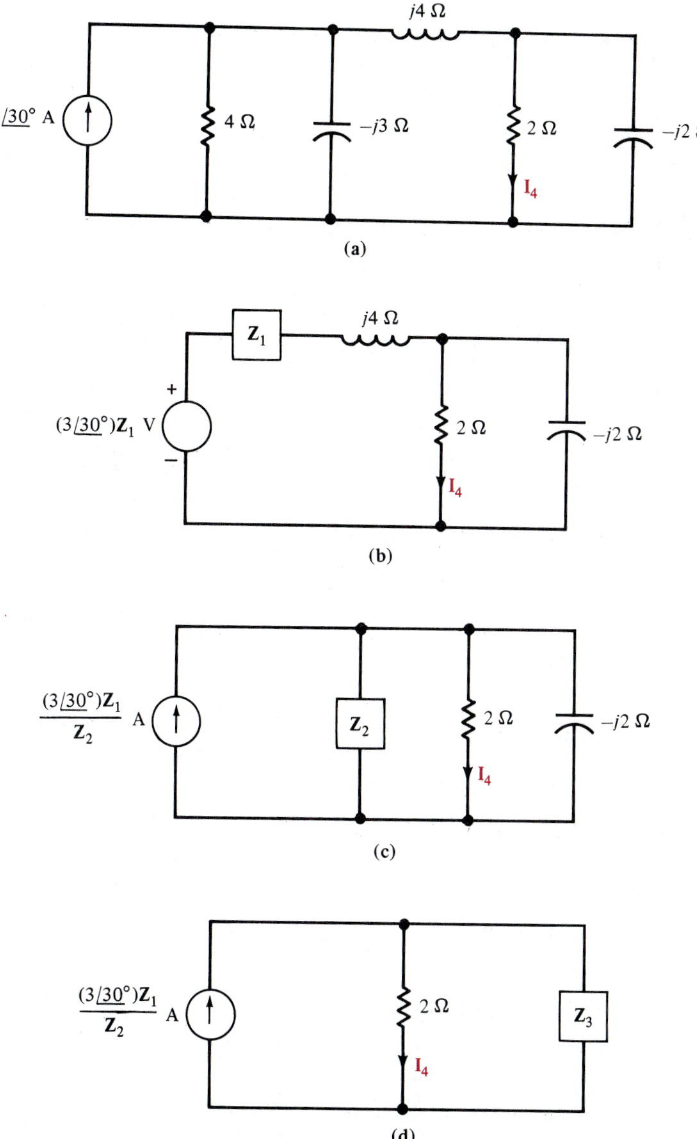

Figure 10.13 Circuits used in solving Example 10.1 via source transformation.

Converting the circuit in Fig. 10.13b back to a current source in parallel with $\mathbf{Z}_2$ produces the circuit in Fig. 10.13c. Combining $\mathbf{Z}_2$ with the parallel capacitor produces the circuit in Fig. 10.13d, where

$$\mathbf{Z}_3 = \frac{(-j2)(\mathbf{Z}_2)}{-j2 + \mathbf{Z}_2}$$

$$= \frac{4 - j12}{3 - j2}\ \Omega$$

The value of the current source is

$$\frac{(3\,\underline{/30^\circ})\mathbf{Z}_1}{\mathbf{Z}_2} = \frac{(3\,\underline{/30^\circ})[-j12/(4-j3)]}{\dfrac{12+j4}{4-j3}} = 2.85\,\underline{/-78.43^\circ}\text{ A}$$

Now employing current division

$$\mathbf{I}_4 = \frac{(2.85\,\underline{/-78.43^\circ})[(4-j12)/(3-j2)]}{2+\dfrac{4-j12}{3-j2}}$$

$$= 1.91\,\underline{/-92^\circ}\text{ A}$$

DRILL EXERCISE

D10.8. Find $\mathbf{V}_o$ in the network in Fig. D10.6 using source transformation.

Ans: $\mathbf{V}_o = 12\,\underline{/90^\circ}$ V.

10.6

Thévenin's and Norton's Theorems

In the following examples we will demonstrate the use of Thévenin's and Norton's theorems in the solution of a variety of problems including those in which these theorems are combined with other analysis techniques. We will illustrate the use of Thévenin's and Norton's theorems in the frequency domain and show that the open-circuit voltage and short-circuit current are phasors, and therefore their ratio is a complex quantity which we call the Thévenin equivalent impedance.

Thévenin's theorem is extremely useful in reducing a portion of a network to a single source in series with an impedance. As a gigantic example of this simplification, suppose that we are making calculations for the proper design of the power distribution in a locale such as the university's electric power network. In order to analyze the local system, the entire U.S. power grid, to which the local power network is connected, can be represented by a Thévenin equivalent circuit; that is, the total electric power system of the United States is replaced by two elements in order to study the campus power system. These Thévenin equivalent parameters are readily available from the local power companies. The power companies routinely calculate these values when evaluating their system using a computer simulation of their network.

EXAMPLE 10.12

We will determine the voltage $\mathbf{V}_o$ in the circuit in Fig. 10.7 using Thévenin's theorem.

The open-circuit voltage $\mathbf{V}_{oc}$ is obtained from the circuit in Fig. 10.14a. Since the current in the outer loop is $2\,\underline{/30^\circ}$ A, then

$$\mathbf{V}_{oc} = (2\,\underline{/30^\circ})(j1) + 8\,\underline{/45^\circ}$$

$$= 8.73\,\underline{/57.75^\circ}\text{ V}$$

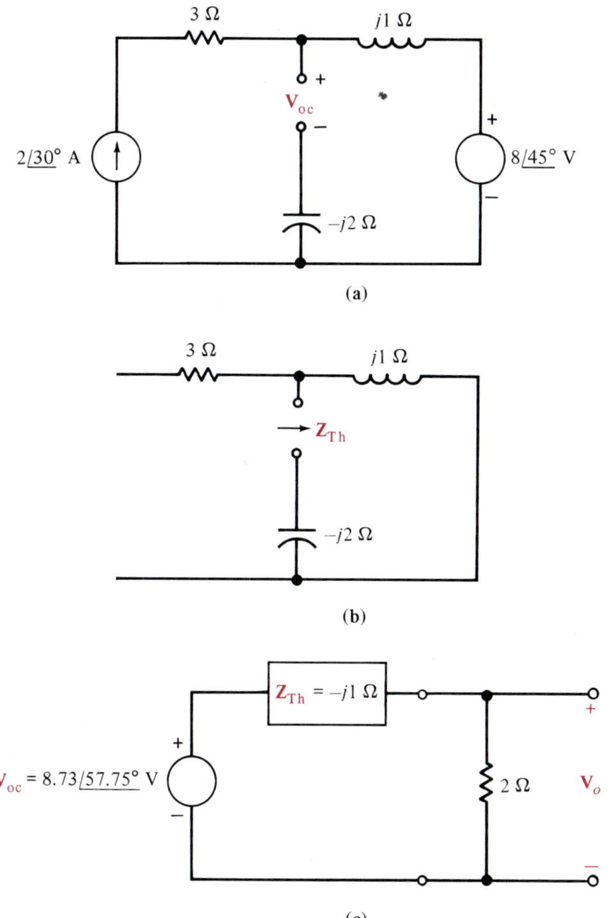

Figure 10.14 Circuits used to compute $\mathbf{V}_o$ in Fig. 10.7 using Thévenin's theorem.

The Thévenin equivalent impedance is found by turning off the independent sources; that is, replacing current sources with an open circuit and voltage sources with short circuits, and computing the impedance at the terminals, where $\mathbf{V}_{oc}$ is determined as shown in Fig. 10.14b. Note that

$$\mathbf{Z}_{Th} = -j2 + j1 = -j1\ \Omega$$

We could have determined the short-circuit current and then calculated $\mathbf{Z}_{Th}$ from $\mathbf{V}_{oc}/\mathbf{I}_{sc}$, but that would have been much more difficult in this case.

The Thévenin equivalent circuit is then shown in Fig. 10.14c together with the 2-Ω load. The voltage $\mathbf{V}_o$ is then calculated as

$$\begin{aligned}\mathbf{V}_o &= \frac{(8.73\ \underline{/57.75°})(2)}{2 - j1} \\ &= 0.76 + j7.76\ \text{V}\end{aligned}$$

■

EXAMPLE 10.13

Let us determine the voltage $\mathbf{V}_o$ in the circuit in Example 10.8 using Thévenin's theorem.

The open-circuit voltage is computed from the network in Fig. 10.15a. Applying KCL at the nodes in the network indicates that

$$\mathbf{I}_1 = 4 \angle 0^\circ - 2 \angle 0^\circ = 2 \angle 0^\circ \text{ A}$$

$$\mathbf{I}_2 = 6 \angle 0^\circ + \mathbf{I}_1 = 8 \angle 0^\circ \text{ A}$$

$$\mathbf{I}_3 = 6 \angle 0^\circ + 4 \angle 0^\circ = 10 \angle 0^\circ \text{ A}$$

and

$$\mathbf{I}_4 = \mathbf{I}_2 + 2 \angle 0^\circ - \mathbf{I}_3 = 0$$

KVL around the loop containing $\mathbf{V}_{oc}$ and only passive elements is

$$+\mathbf{V}_{oc} + (1)(2 \angle 0^\circ) - (-j1)\mathbf{I}_2 - j1\mathbf{I}_4 = 0$$

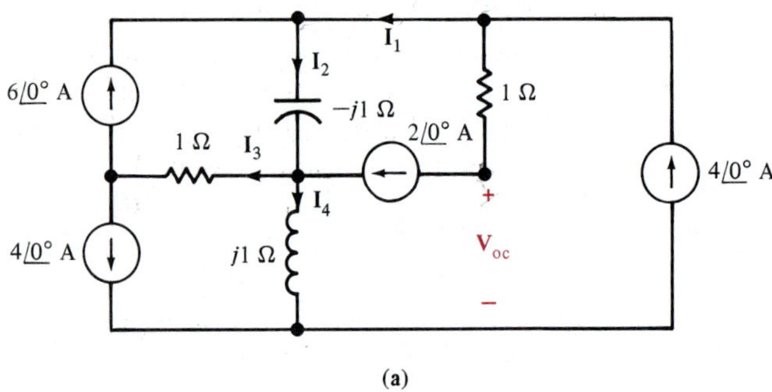

(a)

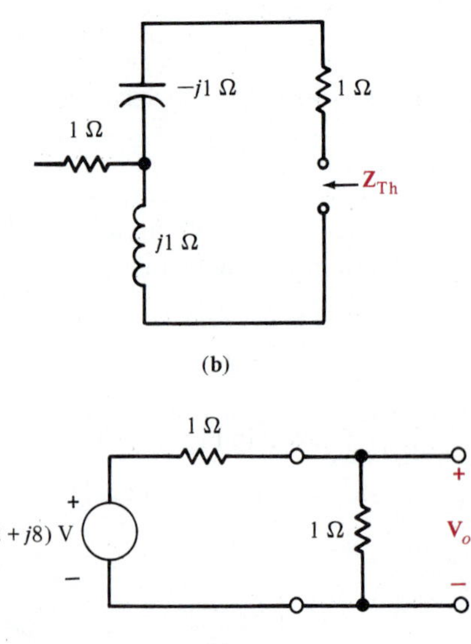

(c)

Figure 10.15 Circuits used in Example 10.13.

Therefore,

$$\mathbf{V}_{oc} = -(2 + j8) \text{ V}$$

$\mathbf{Z}_{Th}$ is determined from the network in Fig. 10.15b as $\mathbf{Z}_{Th} = 1\ \Omega$. The Thévenin equivalent network connected to the output load resistor is shown in Fig. 10.15c and the output voltage is

$$\mathbf{V}_o = \tfrac{1}{2}(-2 - j8)$$
$$= -(1 + j4) \text{ V}$$

which is the value obtained in Example 10.8. ■

EXAMPLE 10.14

Let us use Thévenin's theorem to determine the voltage $\mathbf{V}_o$ in the network in Example 10.9.

The open-circuit voltage is determined from the network in Fig. 10.16a. Note that $\mathbf{I}'_x = 4\,\underline{/0^\circ}$ A and since $2\mathbf{I}'_x$ flows through the inductor, the open-circuit voltage $\mathbf{V}_{oc}$ is

$$\mathbf{V}_{oc} = -1(4\,\underline{/0^\circ}) + j1(2\mathbf{I}'_x)$$
$$= -4 + j8 \text{ V}$$

To determine the Thévenin equivalent impedance, we turn off the independent sources, apply a test voltage source to the output terminals, and compute the current leaving the test source. As shown in Fig. 10.16b, since $\mathbf{I}''_x$ flows in the test source, KCL requires that the current in the inductor be $\mathbf{I}''_x$ also. KVL around the mesh containing the test source indicates that

$$j1\mathbf{I}''_x - 1\mathbf{I}''_x - \mathbf{V}_{test} = 0$$

Therefore,

$$\mathbf{I}''_x = \frac{-\mathbf{V}_{test}}{1 - j}$$

Then

$$\mathbf{Z}_{Th} = \frac{\mathbf{V}_{test}}{-\mathbf{I}''_x}$$
$$= 1 - j\,\Omega$$

If the Thévenin equivalent network is now connected to the load as shown in Fig. 10.16c, the output voltage $\mathbf{V}_o$ is found to be

$$\mathbf{V}_o = \frac{-4 + j8}{2 - j1}(1)$$
$$= -4\,\underline{/-36.87^\circ} \text{ V}$$

which is identical to that obtained in Example 10.9. ■

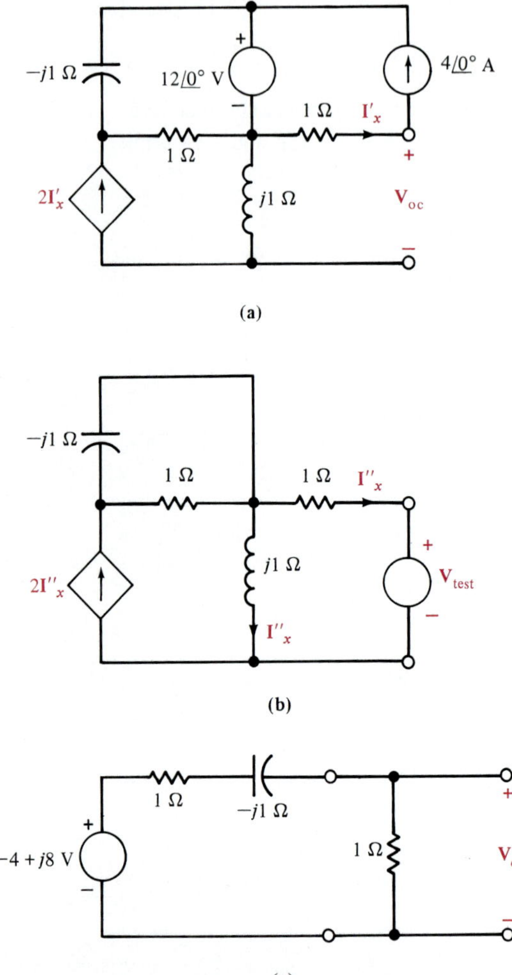

Figure 10.16 Circuits used in Example 10.14.

The following example illustrates the use of Norton's theorem. Once again the problem has been solved in an earlier example, and therefore we can check our results and compare the methods.

EXAMPLE 10.15

Let us employ Norton's theorem to find the current in the 2-Ω resistor in Fig. 10.2, which flows from the node labeled $\mathbf{V}_2$ to ground.

The short-circuit current is computed from the circuit in Fig. 10.17a. Because the short circuit places the voltage source directly across the inductor, the inductor current is

$$\mathbf{I}_L = \frac{12 \angle 30^\circ}{j1} = 12 \angle -60^\circ \text{ A}$$

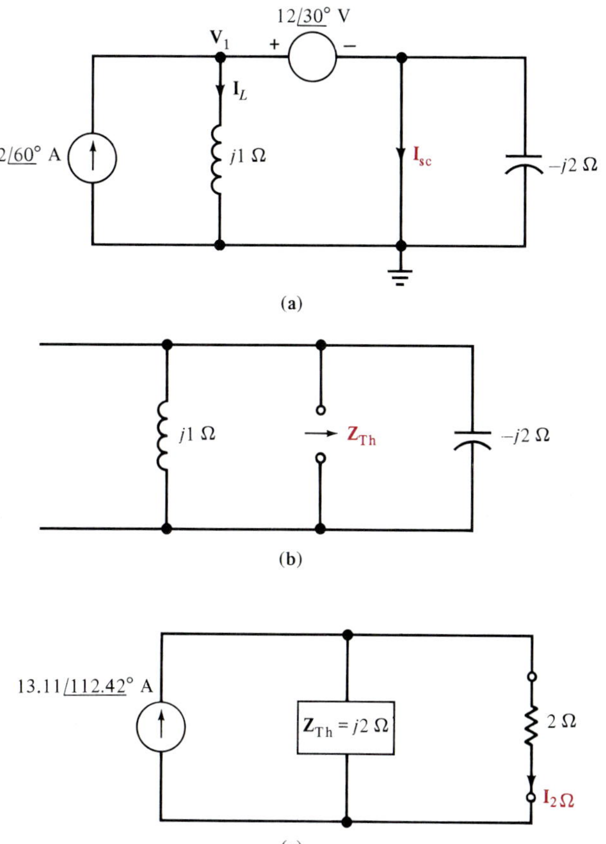

Figure 10.17 Circuits used to compute the current in the 2-Ω resistor in Fig. 10.2 using Norton's theorem.

KCL applied at node 1 then indicates that

$$\begin{aligned}\mathbf{I}_{sc} &= 2 \underline{/60^\circ} - 12 \underline{/-60^\circ} \\ &= 13.11 \underline{/112.42^\circ} \text{ A}\end{aligned}$$

The Thévenin equivalent impedance is calculated from Fig. 10.17b as

$$\mathbf{Z}_{Th} = \frac{(j1)(-j2)}{j1 - j2} = j2\ \Omega$$

Therefore, the current in the 2-Ω resistor in Fig. 10.2 is computed by employing the Norton equivalent circuit as shown in Fig. 10.17c.

$$\begin{aligned}\mathbf{I}_{2\Omega} &= \frac{(13.11 \underline{/112.42^\circ})(j2)}{2 + j2} \\ &= \frac{18.50 \underline{/157.54^\circ}}{2} \text{ A}\end{aligned}$$

Since the voltage $\mathbf{V}_2$ in Fig. 10.2 was computed to be $\mathbf{V}_2 = 18.50 \underline{/157.54^\circ}$ V in Example 10.2, the value of $\mathbf{I}_{2\Omega}$ checks with previous results. ■

DRILL EXERCISE

D10.9. Use Thévenin's theorem to compute $\mathbf{V}_o$ in the network in Fig. D10.6.

Ans: $\mathbf{V}_o = 12 \underline{/90^\circ}$ V.

D10.10. Use Norton's theorem to find $\mathbf{V}_o$ in the network in Fig. D10.6.

Ans: $\mathbf{V}_o = 12 \underline{/90^\circ}$ V.

D10.11. Use Thévenin's theorem to find $\mathbf{V}_o$ in the network in Fig. D10.7.

Ans: $\mathbf{V}_o = 5.66 \underline{/-45^\circ}$ V.

D10.12. Find $\mathbf{V}_o$ in the network in Fig. D10.4 using Thévenin's theorem.

Ans: $\mathbf{V}_o = 10.86 \underline{/36.53^\circ}$ V.

EXAMPLE 10.16

We wish to find the output voltage $\mathbf{V}_o$ in the network shown in Fig. 10.18a.

The network is fairly complicated, and therefore, it would appear that a direct attack might be time consuming. Since it is the output voltage at the right of the network that is required, we can begin by simplifying the left portion of the network. Note that if we form a Thévenin equivalent for the portion of the network to the left of the dependent source, we can reduce the circuit to one which has two unknown node voltages, one of which is the output voltage $\mathbf{V}_o$.

The portion of the network to the left of the current source is shown in Fig. 10.18b. The open-circuit voltage is

$$\mathbf{V}_{oc} = \frac{(12 \underline{/0^\circ})(-j1)}{1 - j1}$$

$$= 8.49 \underline{/-45^\circ} \text{ V}$$

The Thévenin equivalent impedance derived from Fig. 10.18c is

$$\mathbf{Z}_{Th} = 1 + \frac{-j1}{1 - j1}$$

$$= 1.5 - j0.5\ \Omega$$

The simplified network employing the Thévenin equivalent is shown in Fig. 10.18d. This result could have readily been obtained by repeated application of source transformation.

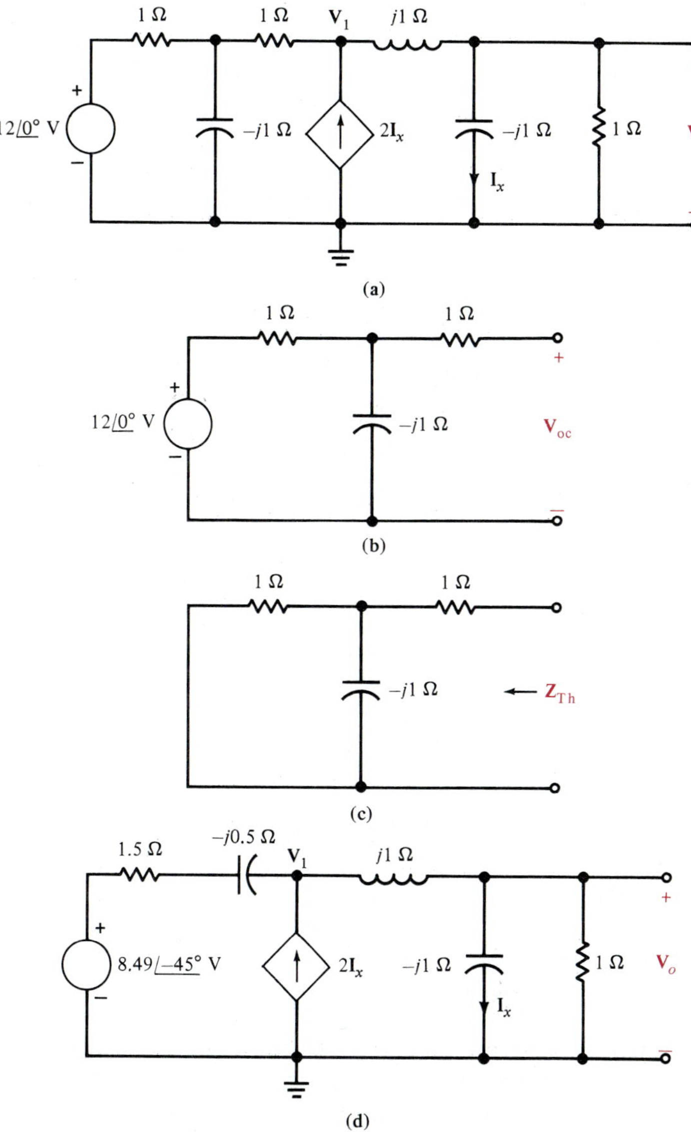

Figure 10.18 Example network.

The node equations for the equivalent network are

$$\frac{\mathbf{V}_1 - 8.49\ \underline{/-45^\circ}}{1.5 - j0.5} + \frac{\mathbf{V}_1 - \mathbf{V}_o}{j1} = 2\mathbf{I}_x$$

$$\frac{\mathbf{V}_o - \mathbf{V}_1}{j1} + \frac{\mathbf{V}_o}{-j1} + \frac{\mathbf{V}_o}{1} = 0$$

$$\mathbf{I}_x = \frac{\mathbf{V}_o}{-j1}$$

Solving these equations yields

$$\mathbf{V}_o = 6.00 \underline{/0^\circ} \text{ V}$$

■

10.7 PSPICE Analysis Techniques

The ac-analysis portion of PSPICE is similar to the dc-analysis part except that a steady-state frequency-domain solution is obtained. The circuit solved by PSPICE is a frequency-domain circuit. Since the analysis is in the frequency domain, all ac sources must be sinusoidal and have the same frequency. The computer calculates the impedances of all branches from the input values of the elements (and the frequency), and the programmer furnishes the magnitudes and phase angles of all sources. Dependent sources are also phasors and are always a constant times a branch voltage or current.

We have already presented, in Chapters 4 and 7, all the necessary tools for a PSPICE analysis in this case, with the exception of the proper solution control statement. For an ac analysis the solution control statement is

```
.AC XXX NO FSTART FSTOP
```

XXX is DEC for decade variation, OCT for octave variation, or LIN for linear variation. NO is number of points per decade, octave, or just total points, depending on whether DEC, OCT, or LIN was specified. FSTART is the starting frequency, and FSTOP is the final frequency both in hertz. Note that the AC statement is not meaningful if no independent ac sources are specified.

For the ac analysis, five additional output variables can be used by replacing V by

VR	real part
VI	imaginary part
VM	magnitude
VP	phase
VDB	$20 \log_{10}$ (magnitude)

Note that in this chapter we will use only the linear variation. In fact, we will compute the unknown voltages and currents at only one frequency; that is, the source frequency will not be varied. Decade and octave variations are explained and used in Chapter 14.

A typical print statement used in ac analysis is

```
.PRINT AC VM(1) VP(1) IM(1) IP(1) VM(3,4)
VP(3,4)
```

Note that in contrast to a dc analysis, we can compute the voltage or current between two nonreference nodes.

The following examples will serve to illustrate the use of PSPICE in ac-circuit analysis.

EXAMPLE 10.17

Let us use PSPICE to determine the voltage $\mathbf{V}_o$ in the network in Fig. 10.19.

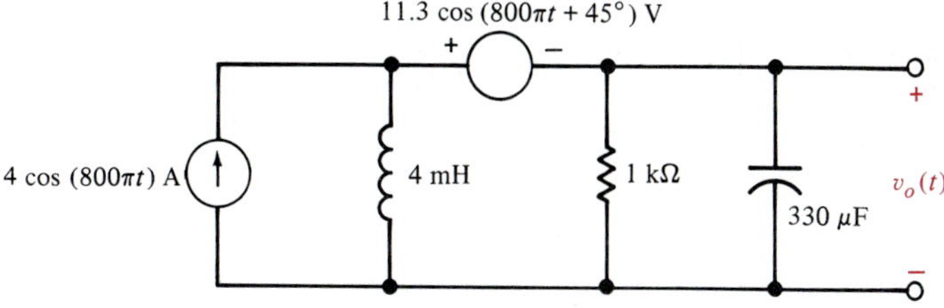

Figure 10.19 Circuit used in Example 10.17.

Note that the frequency of the sources is 400 Hz. The PSPICE program that computes the output voltage is listed below.

```
EXAMPLE 10.17
V1   1   2   AC   11.3   45
I1   0   1   AC    4   0
R1   2   0   1K
C1   2   0   330U
L1   1   0   4MH
*AC ANALYSIS
.AC LIN 1 400 400
.PRINT AC VM(2) VP(2)
.END
```

The computer output listing includes

```
   FREQ              VM(2)              VP(2)
4.000E+02         4.524E+00         -7.599E+01
```

■

EXAMPLE 10.18

Let us use PSPICE to determine the output voltage in the network in Fig. 10.20a.

Note that the source frequency is 60 Hz. The PSPICE program for computing the output voltage $\mathbf{V}_o$ is

```
EXAMPLE 10.18
VC   1   0   AC   12   0
VX   1   5   AC   0
HX   3   0   VX   2
R1   5   2   2K
R2   2   3   10K
C1   4   0   200UF
L1   2   4   20MH
* ANALYSIS
.AC LIN 1 60 60
.PRINT AC VM(2,3) VP(2,3)
.END
```

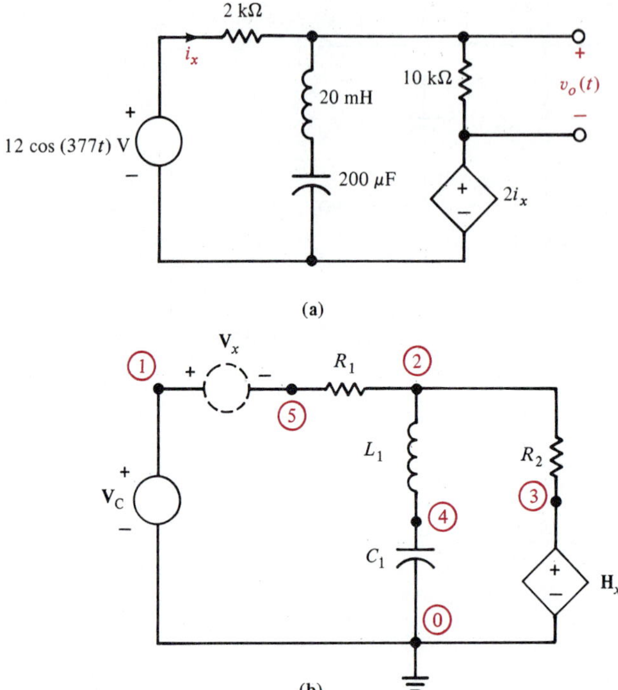

Figure 10.20 Circuits employed in Example 10.18.

The computer output listing includes the following:

```
    FREQ            VM(2,3)          VP(2,3)
  6.000E+01        3.637E-02       -1.091E+02
```

EXAMPLE 10.19

Let us consider once again the problem in Example 10.16. The network in Fig. 10.18a is redrawn for a PSPICE analysis in Fig. 10.21. Note that in Example 10.16 the network is

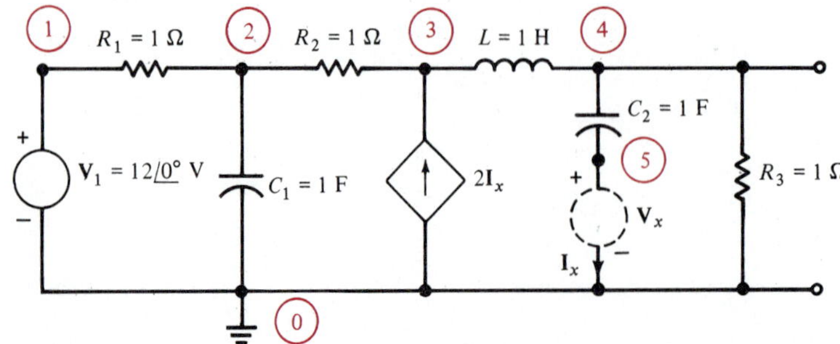

Figure 10.21 Network in Fig. 10.18a redrawn for PSPICE analysis.

specified in the frequency domain and the frequency was not explicitly stated. For PSPICE analysis we must state the frequency and the parameter values. We are free to select any frequency and circuit parameters that are consistent with the impedance values given. A convenient value is $f = 1/2\pi$ hertz, which yields $\omega = 1$ rad/sec.

The PSPICE program for computing the output voltage is

```
EXAMPLE 10.16 AC ANALYSIS
V1   1   0   AC   12   0
R1   1   2   1
C1   2   0   1
R2   2   3   1
F1   0   3   VX   2
L1   3   4   1
C2   4   5   1
VX   5   0   AC   0    0
R3   4   0   1
.AC LIN 1    0.159155   0.159155
.PRINT   AC  VM(4)      VP(4)
.END
```

The computer output for the program is

```
    FREQ              VM(4)              VP(4)
1.592E-01          6.000E+00          2.04-05
```

■

Since our analysis approach in this chapter has been confined to a single frequency, the PSPICE examples have maintained that format. However, the use of the plot routines for analysis over a range of frequencies is presented in Chapter 14.

DRILL EXERCISE

D10.13. Given the network in Fig. D10.13, compute $\mathbf{V}_o$ and $\mathbf{I}_o$ using PSPICE if the frequency is $f = 1/2\pi$ Hz.

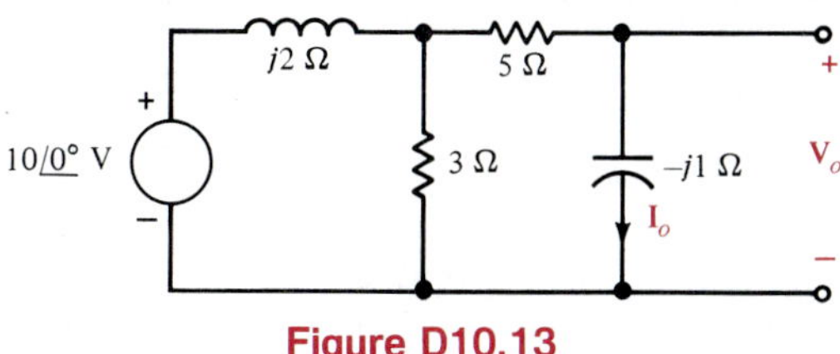

Figure D10.13

Ans: $\mathbf{V}_o = 1.402\ \underline{/-127.4°}$ V, $\mathbf{I}_o = 1.402\ \underline{/-37.4°}$ A.

D10.14. Using PSPICE, compute $\mathbf{V}_o$ in the network in Fig. D10.14.

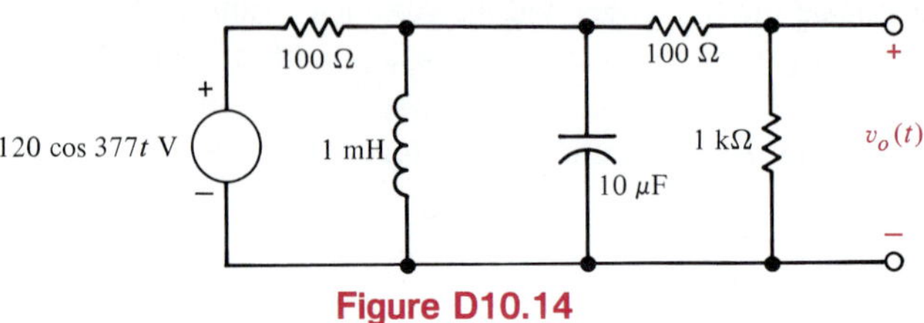

Figure D10.14

Ans: $\mathbf{V}_o = 0.4118 \angle 89.76^\circ$ V.

10.8 Summary

This chapter has illustrated that through the use of phasors, impedance, and admittance, we can determine the steady-state response of an ac-circuit to a sinusoidal input using such techniques as nodal analysis, loop analysis, superposition, source transformation, and Thévenin's and Norton's theorems.

We have also demonstrated the use of PSPICE in the solution of ac steady-state circuit problems. Finally, it has been shown that the ac-analysis portion of the PSPICE circuit-analysis program is a simple and efficient technique for solving ac circuit analysis problems.

KEY POINTS

- The principle of linearity can be applied to ac steady-state circuit problems.
- All the solution techniques applied to dc circuits—nodal analysis, loop analysis, superposition, source transformation, Thévenin's theorem, and Norton's theorem—are all applicable in the solution of ac steady-state circuit problems.
- With the addition of a solution control statement and five output variables, the PSPICE techniques presented in earlier chapters can be applied to the solution of ac steady-state circuit problems.

PROBLEMS

10.1. Given the network in Fig. P10.1, use linearity and the assumption that $\mathbf{V}_o = 1 \angle 0^\circ$ V to determine the actual value of $\mathbf{V}_o$ if $\mathbf{V}_S = 24 \angle 0^\circ$ V.

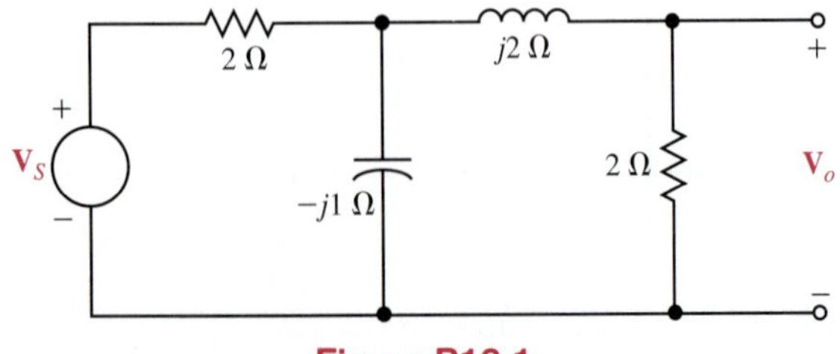

Figure P10.1

10.2. Given the circuit in Fig. P10.2, use linearity and the assumption that $\mathbf{I}_o = 1 \angle 0^\circ$ A to find the actual value of $\mathbf{I}_o$ if $\mathbf{V}_S = 12 \angle 0^\circ$ V.

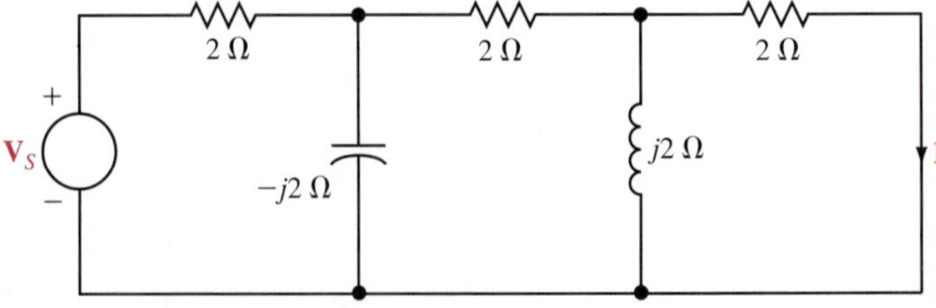

Figure P10.2

10.3. Given the network in Fig. P10.3, use linearity and the assumption that $\mathbf{I}_o = 1\,\underline{/0°}$ A to find the actual value of $\mathbf{I}_o$ if $\mathbf{I}_S = 12\,\underline{/0°}$ A.

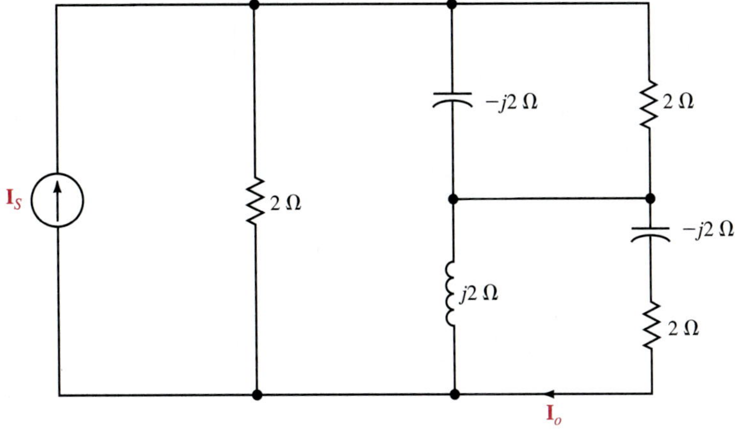

Figure P10.3

10.4. Find $\mathbf{V}_o$ in the network in Fig. P10.4.

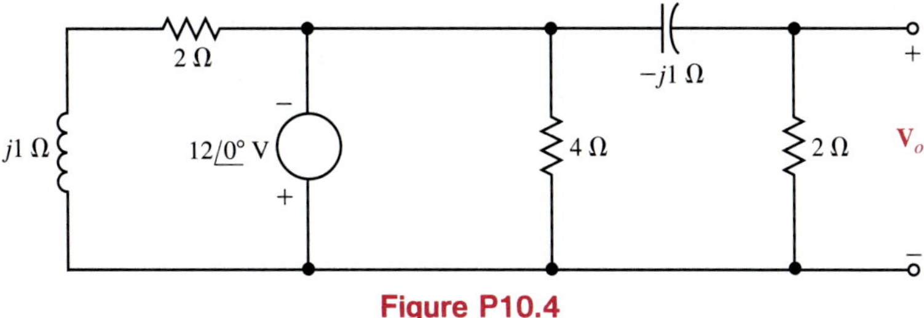

Figure P10.4

10.5. Determine $\mathbf{V}_o$ in the circuit in Fig. P10.5.

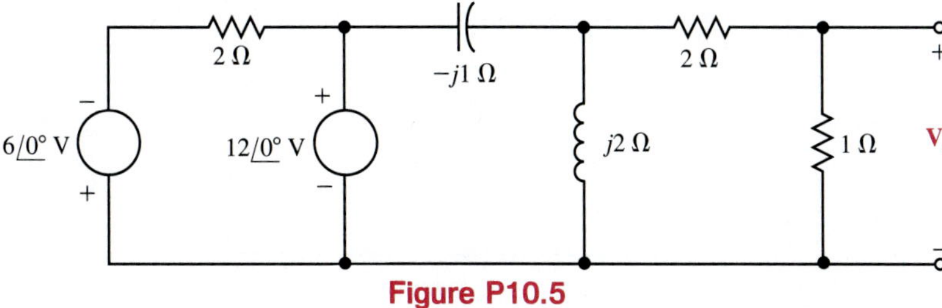

Figure P10.5

10.6. Find $\mathbf{V}_o$ in the network in Fig. P10.6.

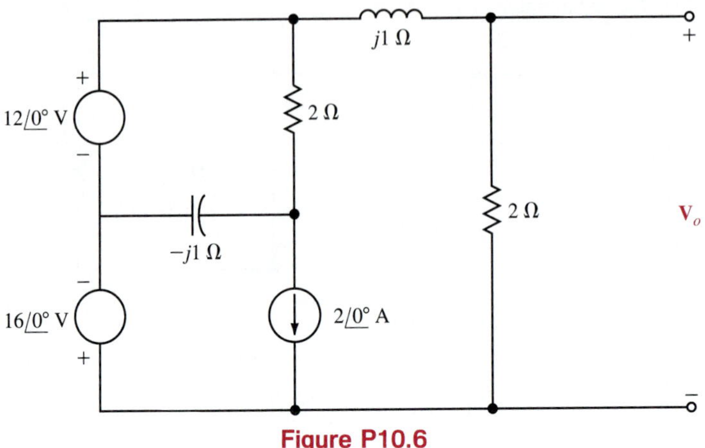

Figure P10.6

10.7. Use nodal analysis to find $\mathbf{V}_o$ in the circuit in Fig. P10.7.

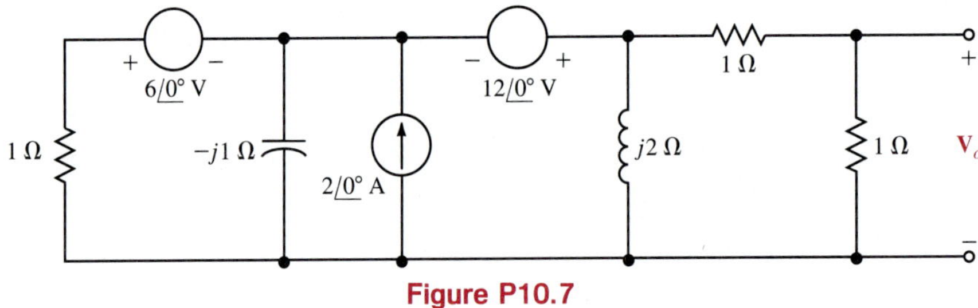

Figure P10.7

10.8. Use nodal analysis to determine $\mathbf{I}_o$ in the network in Fig. P10.8.

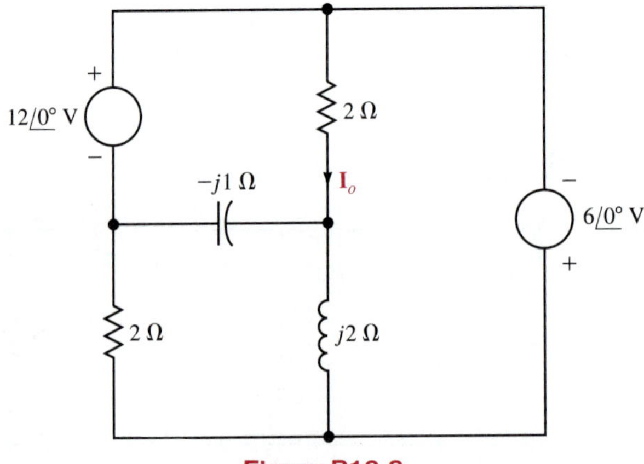

Figure P10.8

10.9. Find $\mathbf{V}_o$ in the circuit in Fig. P10.9 using nodal analysis.

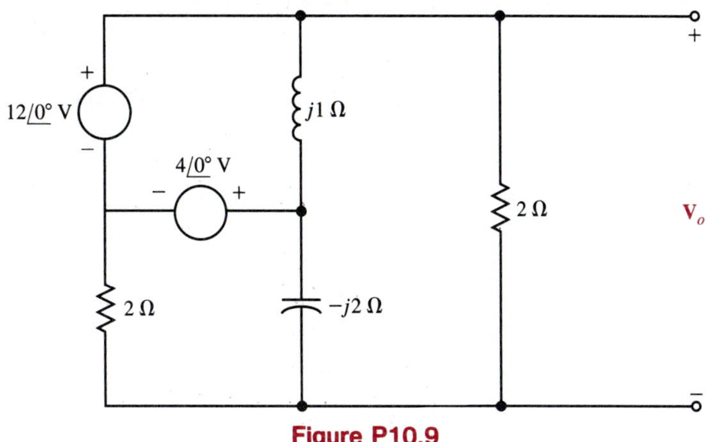

Figure P10.9

10.10. Find $\mathbf{I}_o$ in the network in Fig. P10.10 using nodal analysis.

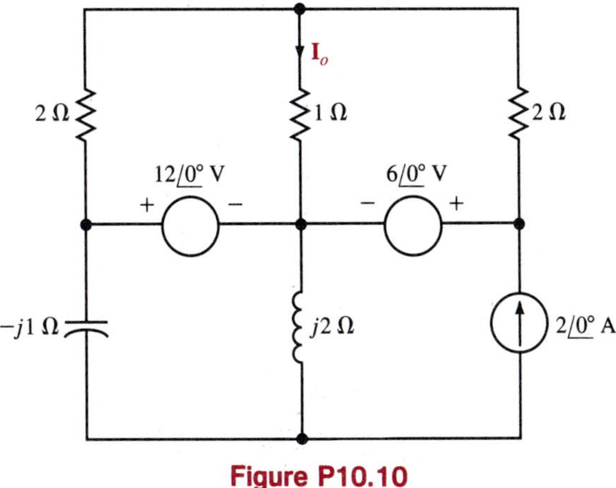

Figure P10.10

10.11. Find $\mathbf{V}_o$ in the network in Fig. P10.11 using nodal analysis.

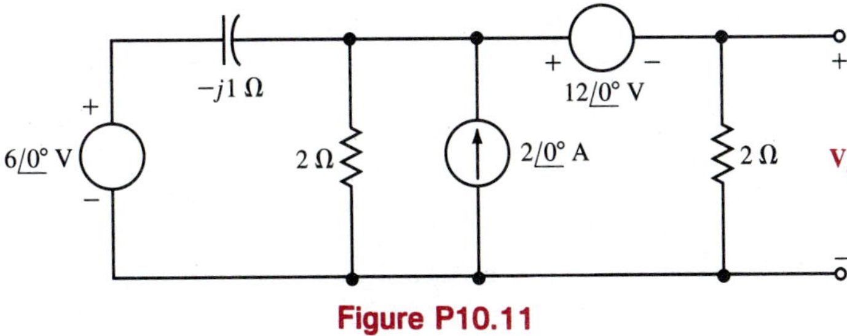

Figure P10.11

10.12. Use nodal analysis to find $\mathbf{I}_o$ in the circuit in Fig. P10.12.

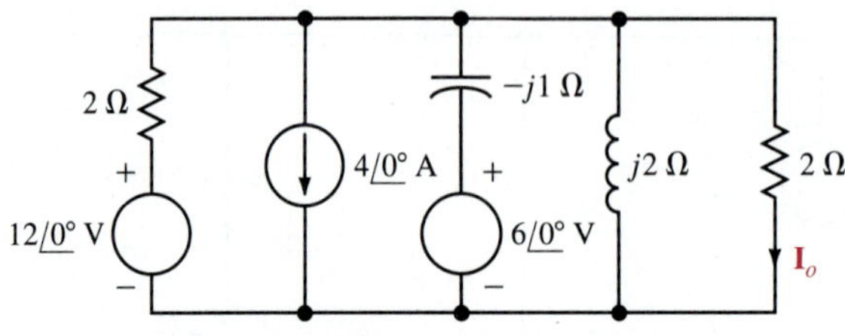

Figure P10.12

10.13. Using nodal analysis, find $\mathbf{I}_o$ in the circuit in Fig. P10.13.

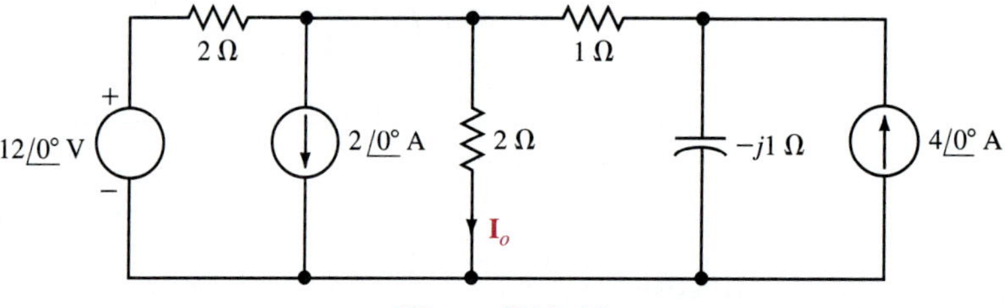

Figure P10.13

10.14. Find $\mathbf{V}_o$ in the network in Fig. P10.14 using nodal analysis.

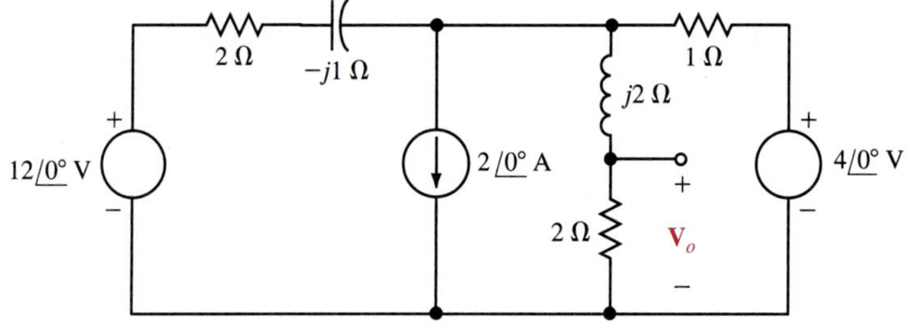

Figure P10.14

10.15. Find $\mathbf{I}_o$ in the circuit in Fig. P10.15 using nodal analysis.

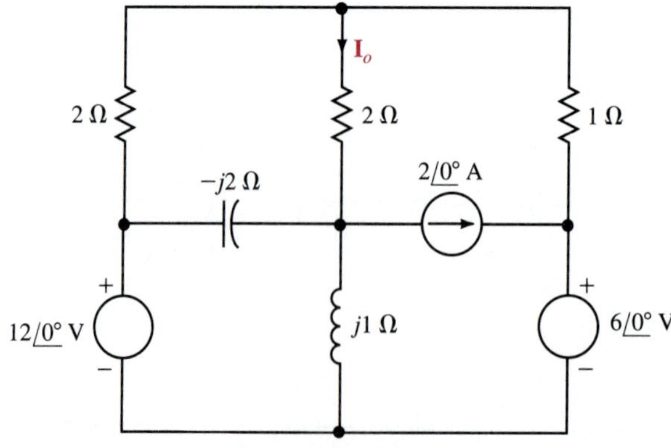

Figure P10.15

10.16. Find $\mathbf{V}_o$ in the network in Fig. P10.16 using nodal analysis.

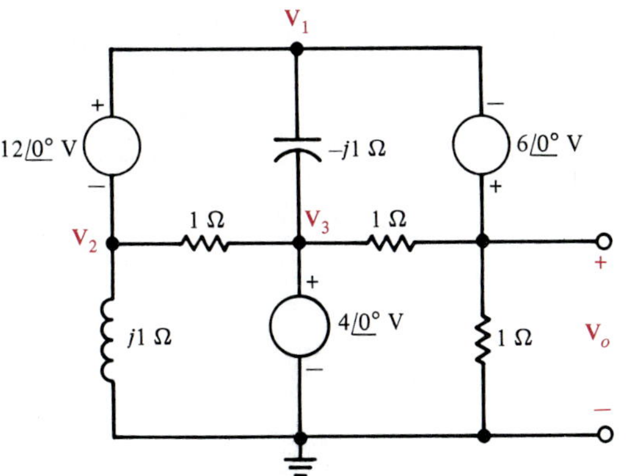

Figure P10.16

10.17. Find $\mathbf{I}_o$ in the network in Fig. P10.17 using nodal analysis.

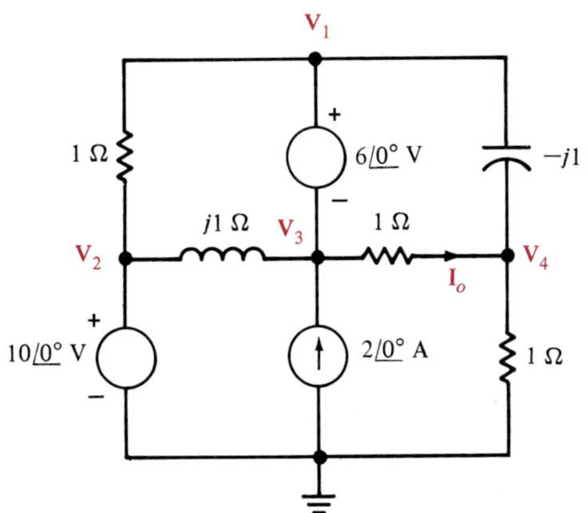

Figure P10.17

10.18. Find the voltage across the inductor in the circuit shown in Fig. P10.18 using nodal analysis.

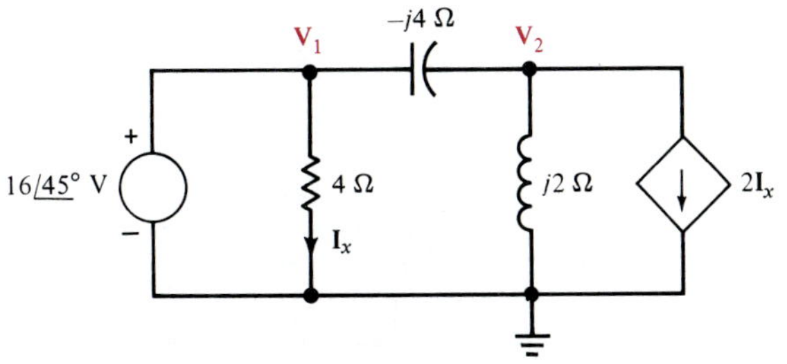

Figure P10.18

10.19. The low-frequency equivalent circuit for a common-emitter transistor amplifier is shown in Fig. P10.19. Compute the voltage gain $\mathbf{V}_o/\mathbf{V}_S$.

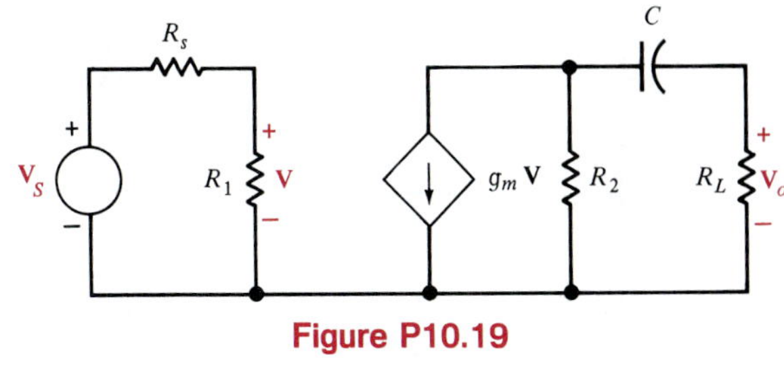

Figure P10.19

10.20. Use nodal analysis to find $\mathbf{V}_o$ in the circuit in Fig. P10.20.

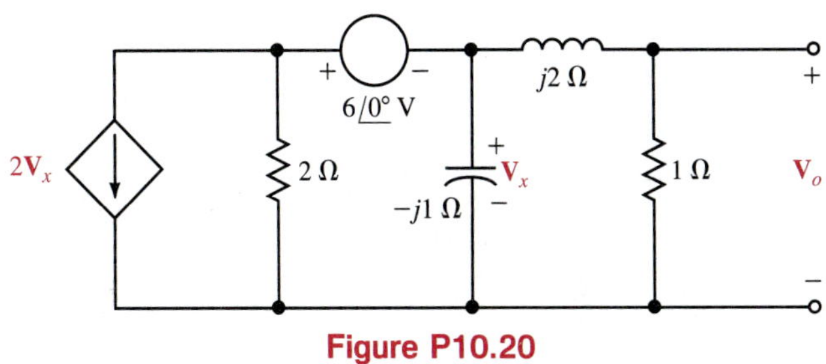

Figure P10.20

10.21. Use nodal analysis to find $\mathbf{V}_o$ in the network in Fig. P10.21.

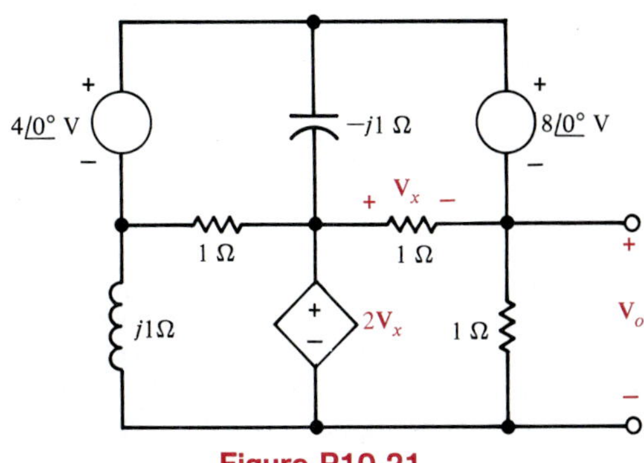

Figure P10.21

10.22. Find $\mathbf{I}_o$ in the network in Fig. P10.22.

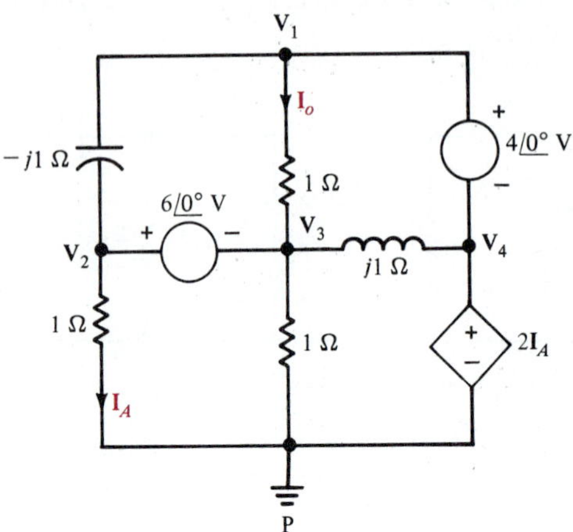

Figure P10.22

10.23. Find the voltage gain $\mathbf{V}_o/\mathbf{V}_S$ for the low-pass filter shown in Fig. P10.23 (and described in Chapter 14) using nodal analysis. Assume that the op-amp is ideal so that the op-amp together with resistors R_3 and R_4 can be represented by a gain of $A = (1 + R_4/R_3)$, using the representation of Fig. 3.33b with $R_i = \infty$ and $R_o = 0$.

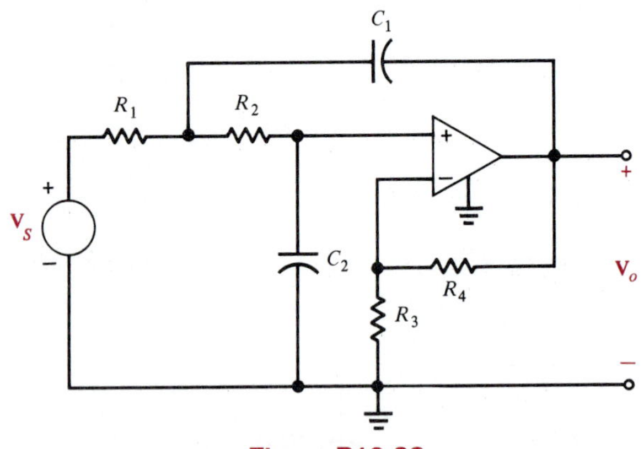

Figure P10.23

10.24. Find the voltage gain $\mathbf{V}_o/\mathbf{V}_S$ for the high-pass filter shown in Fig. P10.24 (and described in Chapter 14) using nodal analysis. Assume that the op-amp is ideal so that the op-amp together with resistors R_3 and R_4 can be represented by a gain of $A = (1 + R_4/R_3)$, using the representation in Fig. 3.33b with $R_i = \infty$ and $R_o = 0$.

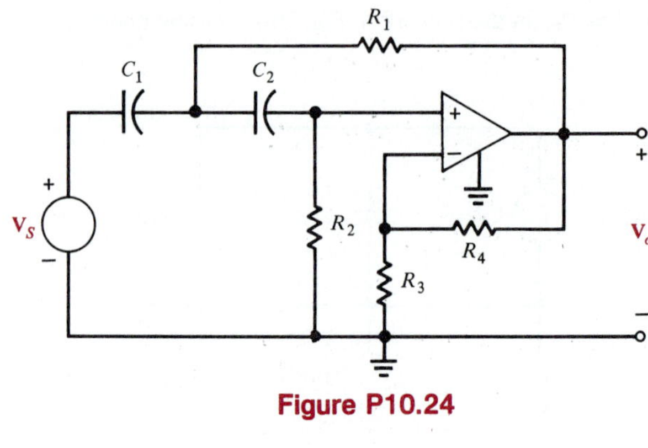

Figure P10.24

10.25. Use mesh analysis to find $\mathbf{V}_o$ in the network in Fig. P10.25.

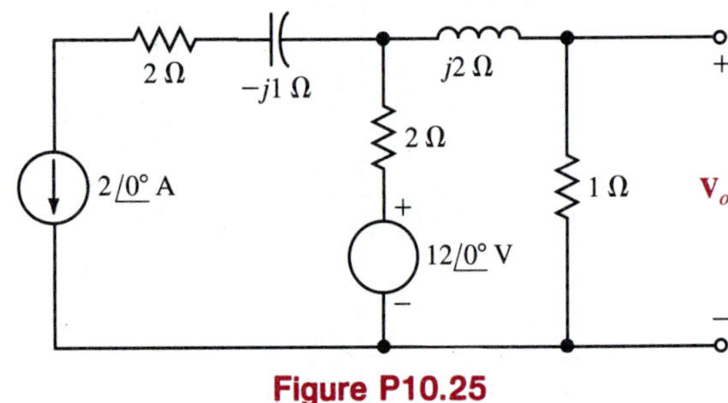

Figure P10.25

10.26. Use mesh analysis to find $\mathbf{V}_o$ in the circuit shown in Fig. P10.26.

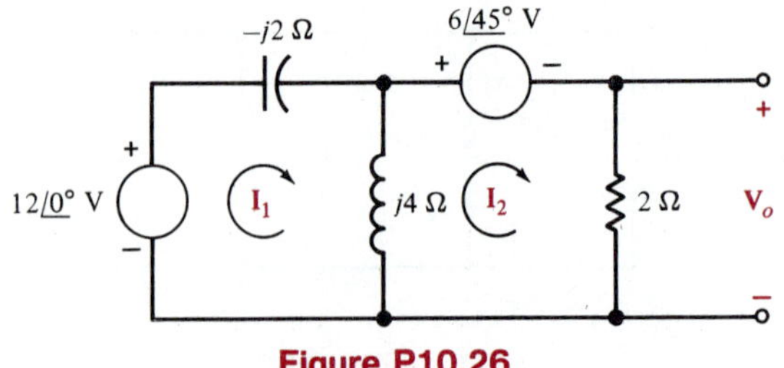

Figure P10.26

10.27. Use mesh analysis to find $\mathbf{V}_o$ in the circuit shown in Fig. P10.27.

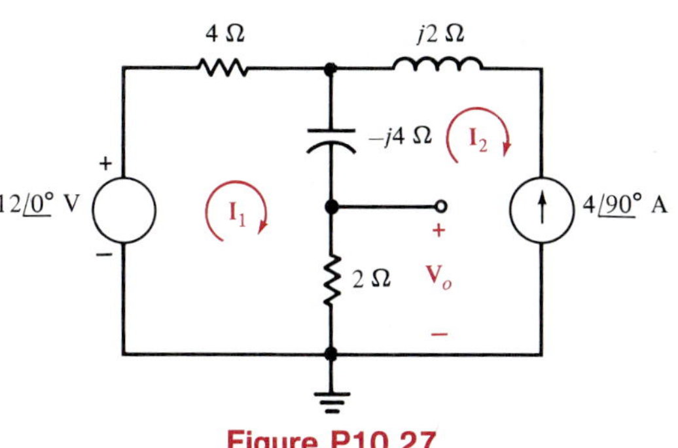

Figure P10.27

10.28. Find $\mathbf{V}_o$ in the circuit in Fig. P10.28 using mesh analysis.

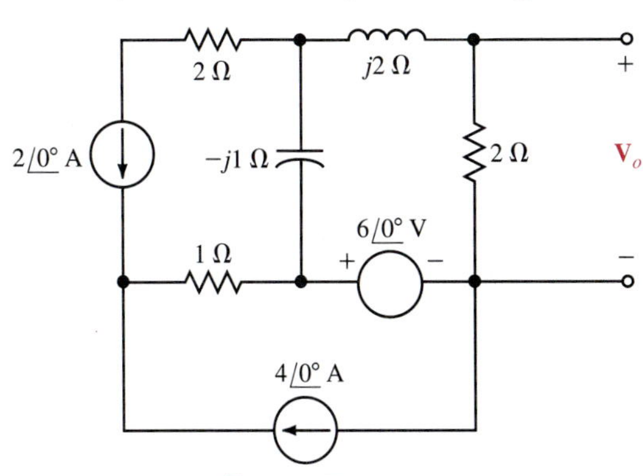

Figure P10.28

10.29. Use mesh analysis to determine $\mathbf{V}_o$ in the circuit in Fig. P10.29.

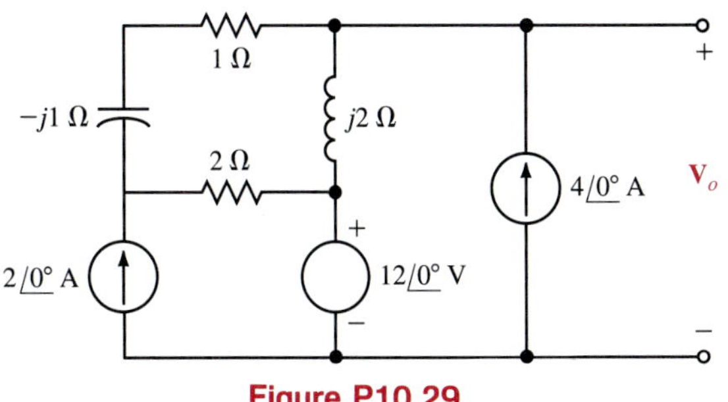

Figure P10.29

10.30. Using loop analysis, find $\mathbf{I}_o$ in the network in Fig. P10.30.

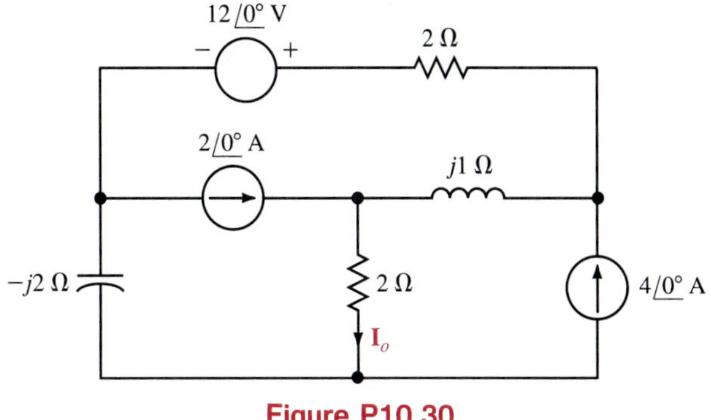

Figure P10.30

10.31. Use mesh analysis to find $\mathbf{V}_o$ in the circuit in Fig. P10.31.

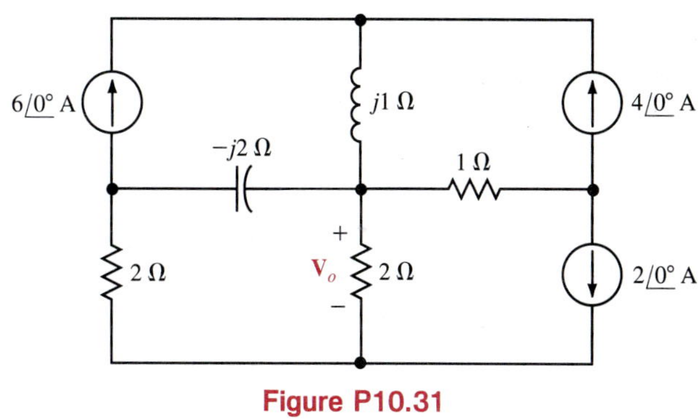

Figure P10.31

10.32. Using loop analysis, find $\mathbf{V}_o$ in the network in Fig. P10.32.

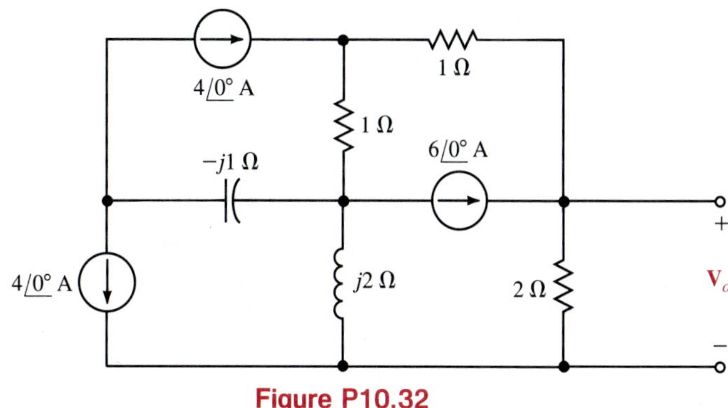

Figure P10.32

10.33. Using loop analysis, determine $\mathbf{V}_o$ in the network in Fig. P10.33.

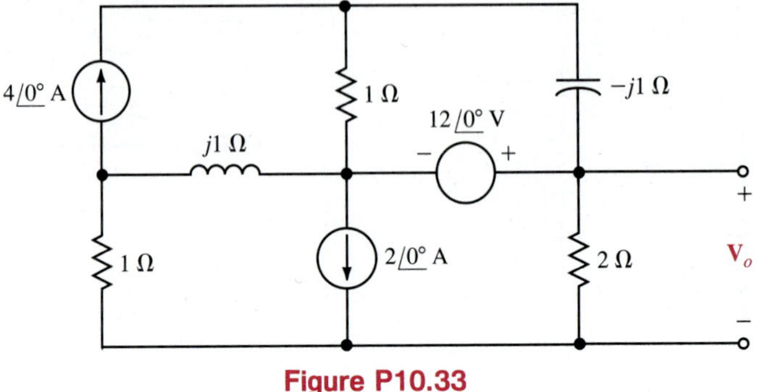

Figure P10.33

10.34. Use mesh analysis to find $\mathbf{V}_o$ in the circuit in Fig. P10.34.

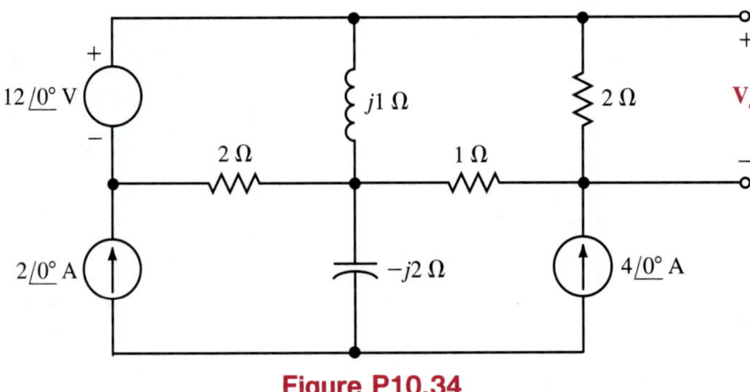

Figure P10.34

10.35. Use loop analysis to find $\mathbf{V}_o$ in the network in Fig. P10.35.

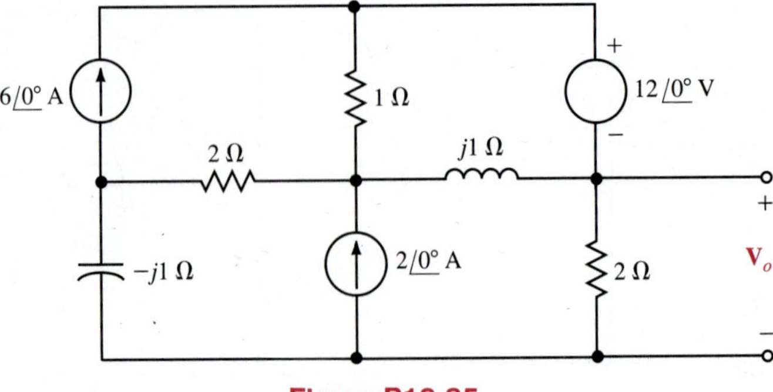

Figure P10.35

10.36. Find $\mathbf{V}_o$ in the network in Fig. P10.36.

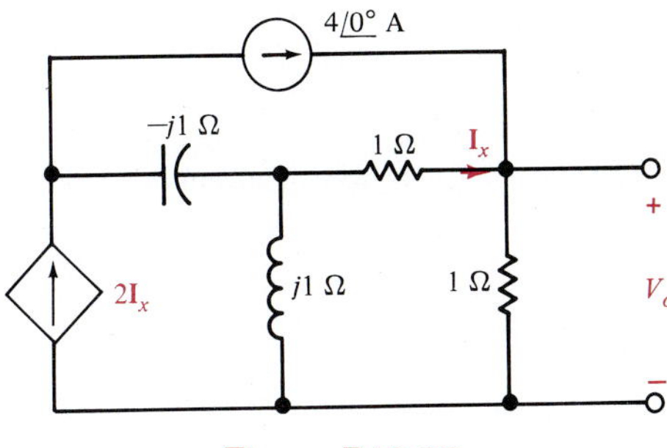

Figure P10.36

10.37. Find $\mathbf{V}_o$ in the network in Fig. P10.37.

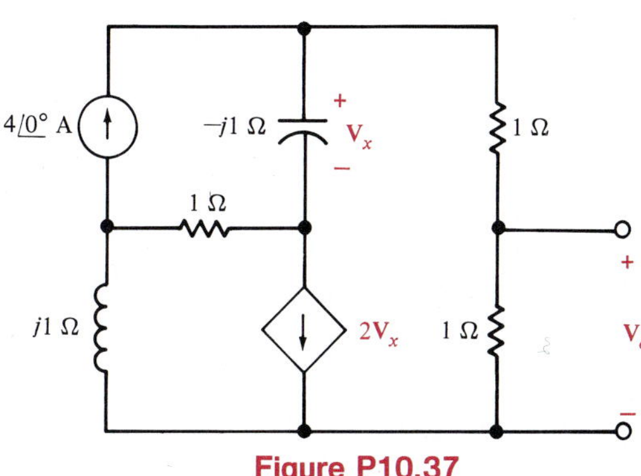

Figure P10.37

10.38. Using superposition, find $\mathbf{V}_o$ in the circuit in Fig. P10.38.

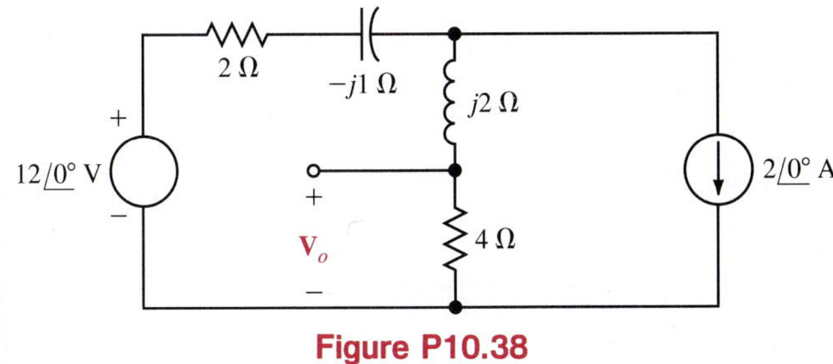

Figure P10.38

10.39. Use superposition to find $\mathbf{V}_o$ in the network in Fig. P10.39.

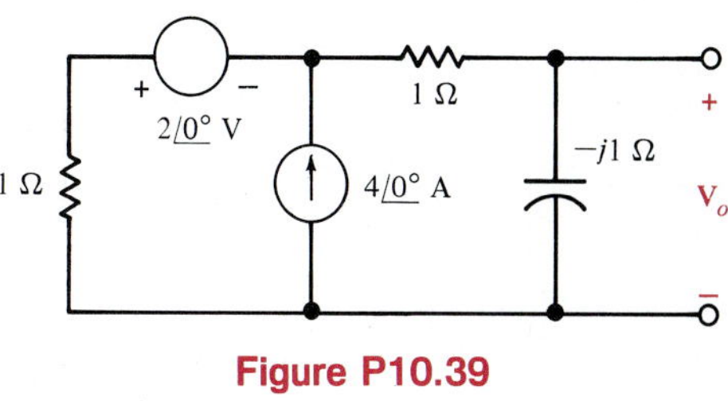

Figure P10.39

10.40. Use superposition to determine $\mathbf{V}_o$ in the circuit in Fig. P10.40.

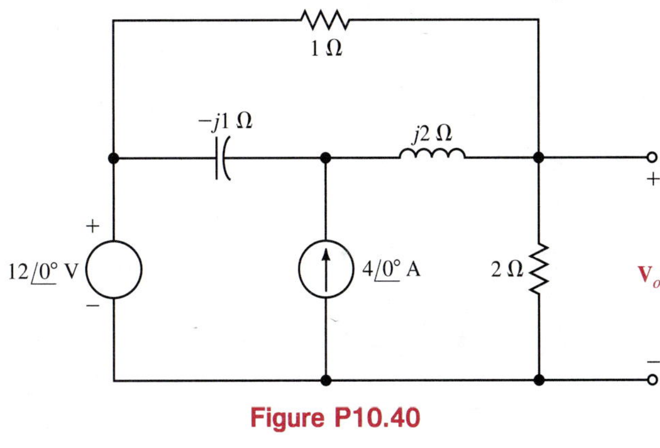

Figure P10.40

10.41. Find $\mathbf{V}_o$ in the network in Fig. P10.41 using superposition.

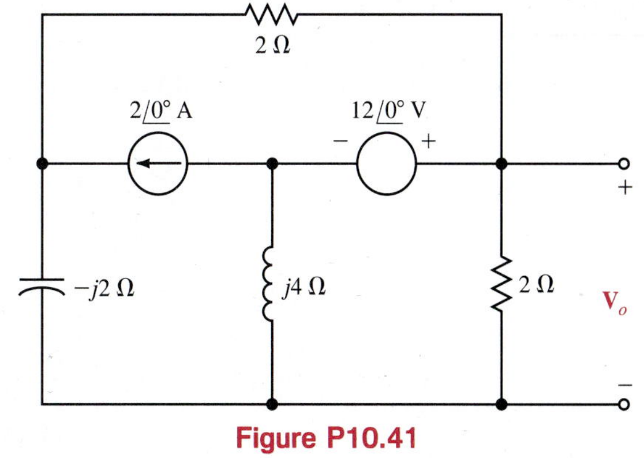

Figure P10.41

10.42. Find $\mathbf{V}_o$ in the network in Fig. P10.42 using superposition.

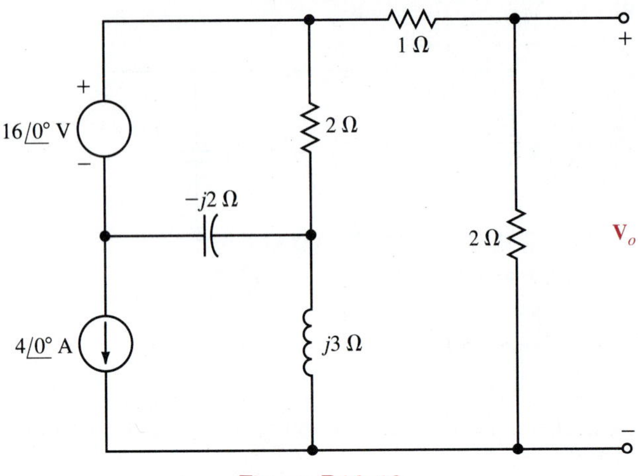

Figure P10.42

10.43. Use source transformation to determine $\mathbf{V}_o$ in the network in Fig. P10.11.

10.44. Using source transformation, find $\mathbf{I}_o$ in the network in Fig. P10.12.

10.45. Solve Problem 10.13 using source transformation.

10.46. Solve Problem 10.14 using source transformation.

10.47. Solve Problem 10.26 using Thévenin's theorem.

10.48. Solve Problem 10.27 using Thévenin's theorem.

10.49. Use Thévenin's theorem to find $\mathbf{V}_o$ in the network in Fig. P10.49.

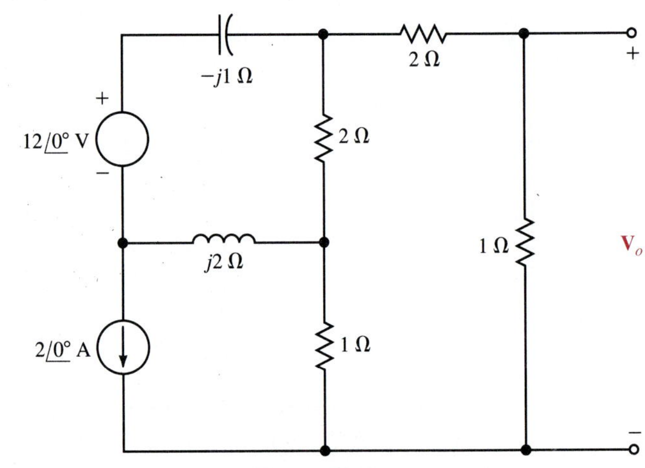

Figure P10.49

10.50. Apply Thévenin's theorem twice to find $\mathbf{V}_o$ in the circuit in Fig. P10.50.

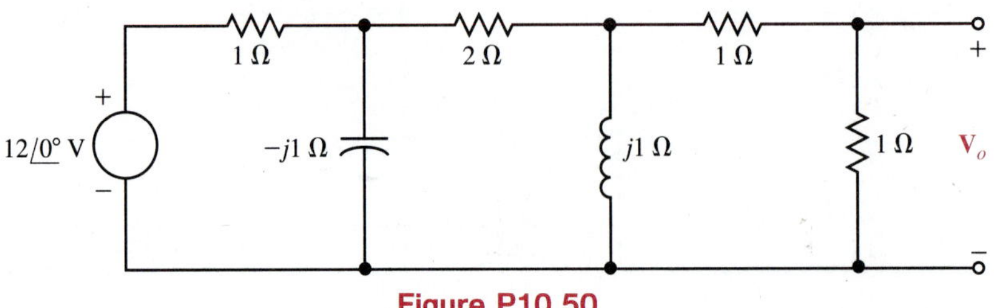

Figure P10.50

10.51. Use Thévenin's theorem to find $\mathbf{V}_o$ in the network in Fig. P10.51.

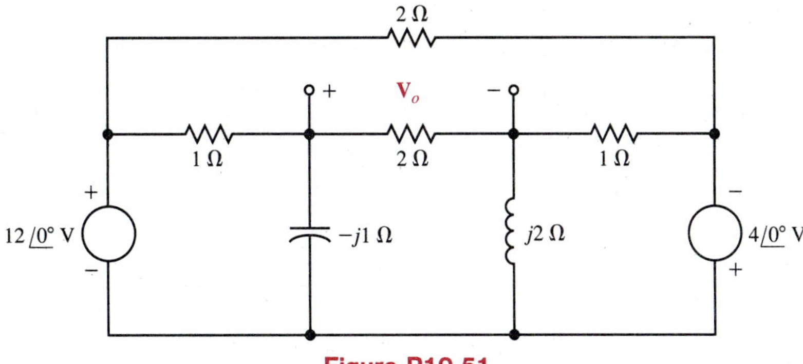

Figure P10.51

10.52. Given the circuit in Fig. P10.52, find the Thévenin equivalent circuit at the terminals *A-B*.

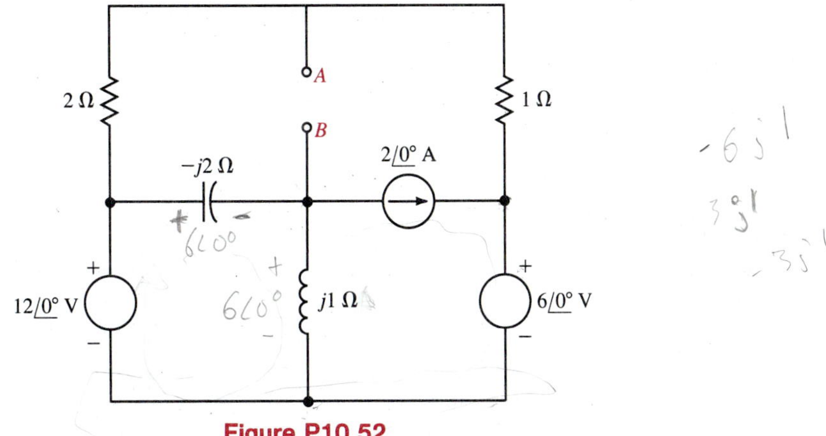

Figure P10.52

10.53. Given the network in Fig. P10.53, find the Thévenin equivalent of the network at the terminals *A-B*.

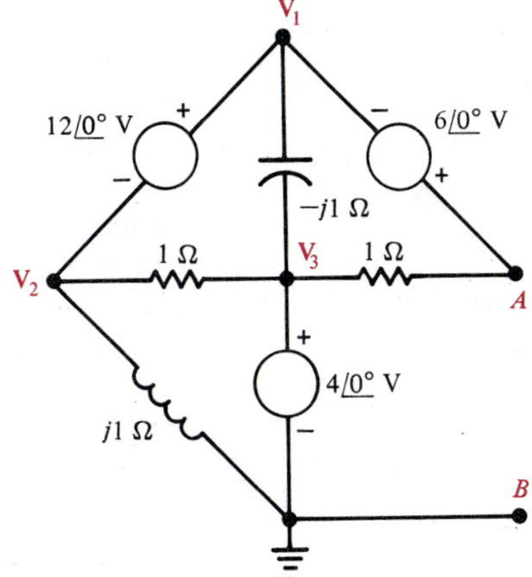

Figure P10.53

10.54. Solve Problem 10.18 using Thévenin's theorem.

10.55. Solve Problem 10.20 using Thévenin's theorem.

10.56. Solve Problem 10.21 using Thévenin's theorem.

10.57. Solve Problem 10.22 using Thévenin's theorem.

10.58. Find the Thévenin equivalent for the network in Fig. P10.58 at the terminals *A-B*.

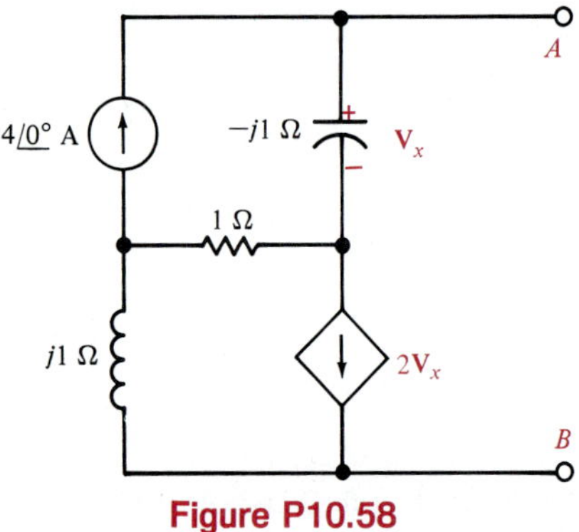

Figure P10.58

10.59. Find the current in the inductor in the circuit shown in Fig. P10.59 using Norton's theorem.

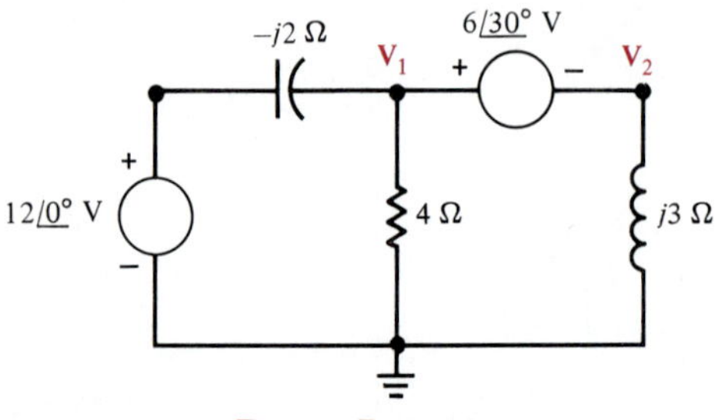

Figure P10.59

10.60. Find $\mathbf{I}_o$ in the network in Fig. P10.60 using Norton's theorem.

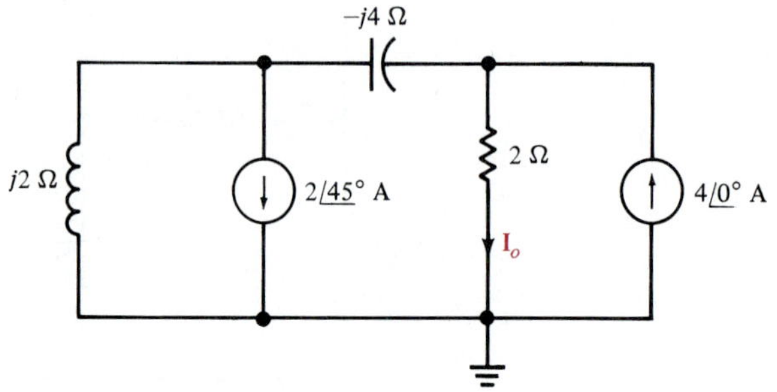

Figure P10.60

10.61. Find $\mathbf{V}_x$ in the circuit in Fig. P10.61 using Norton's theorem.

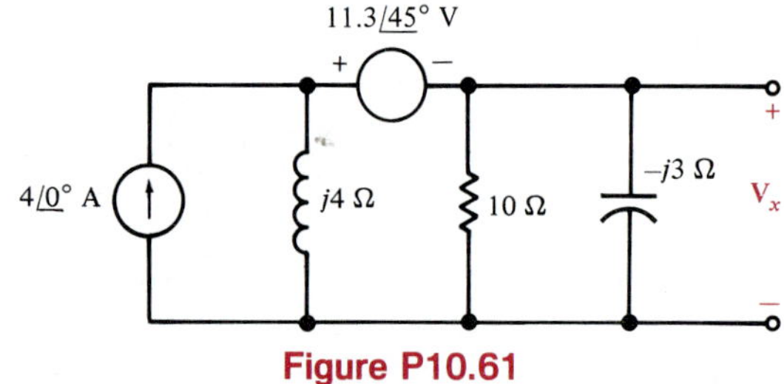

Figure P10.61

10.62. Find the Thévenin equivalent for the network in Fig. P10.62 at the terminals *A-B*.

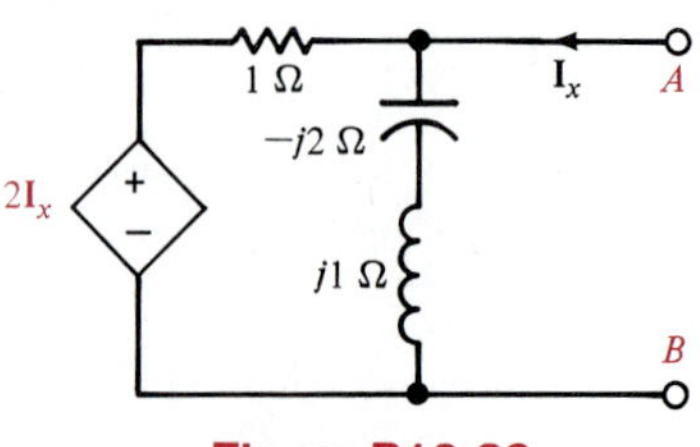

Figure P10.62

10.63. Find the Thévenin equivalent of the network in Fig. P10.63 at the terminals *A-B*.

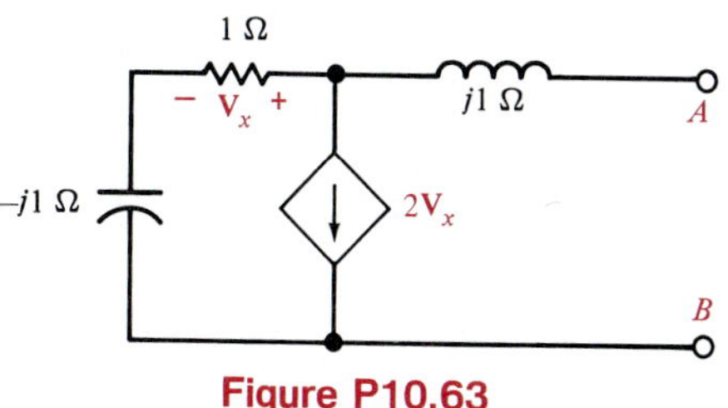

Figure P10.63

10.64. Find the Thévenin equivalent of the network in Fig. P10.64 at the terminals *A-B*.

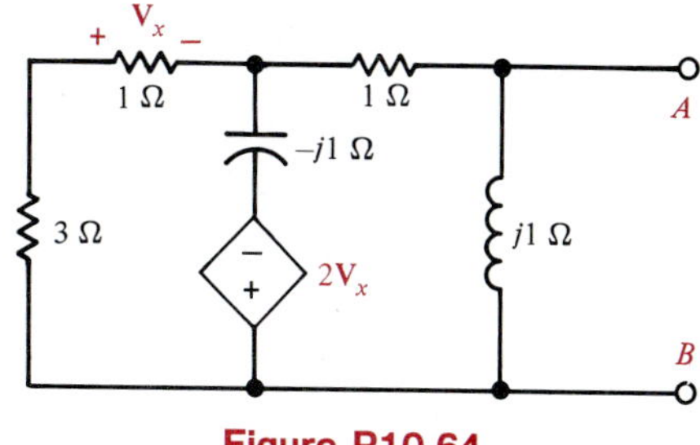

Figure P10.64

10.65. The network in Fig. P10.65 operates at $f = 400$ Hz. Use PSPICE to find the current $\mathbf{I}_o$.

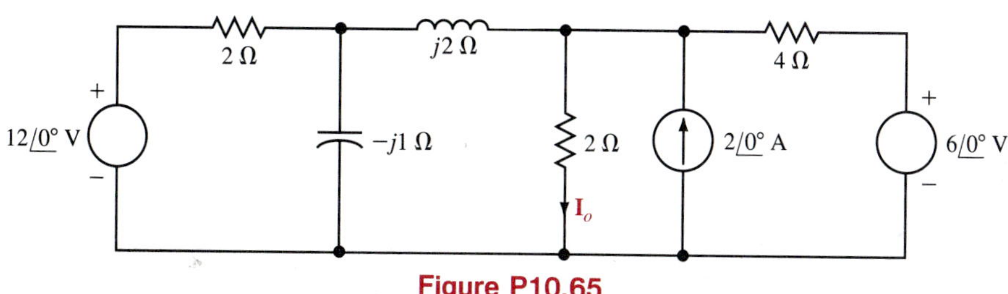

Figure P10.65

10.66. The circuit in Fig. P10.66 operates at $f = 60$ Hz. Use PSPICE to find the voltage $\mathbf{V}_o$.

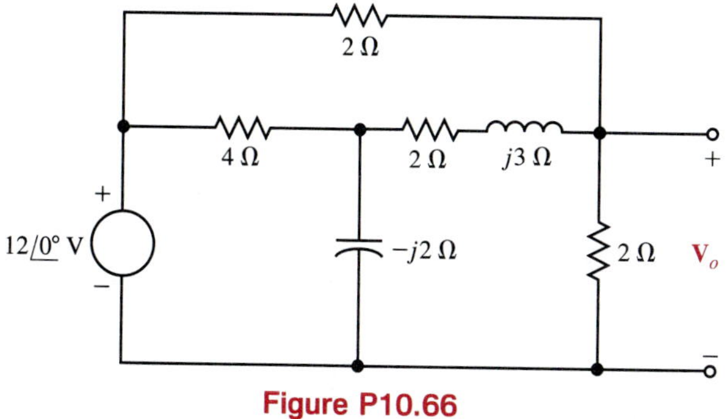

Figure P10.66

10.67. The network in Fig. P10.67 operates at $f = 400$ Hz. Use PSPICE to find the current $\mathbf{I}_o$.

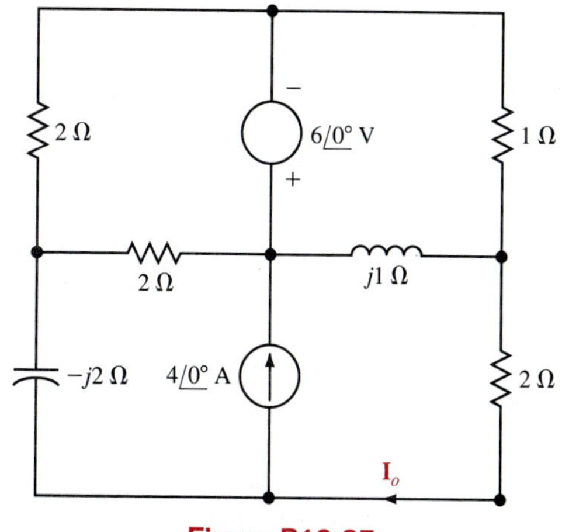

Figure P10.67

10.68. The network in Fig. P10.68 operates at $f = 400$ Hz. Use PSPICE to find the current $\mathbf{I}_o$.

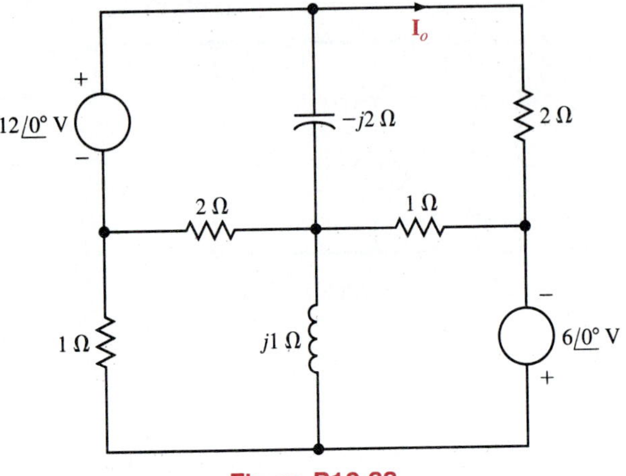

Figure P10.68

10.69. The network in Fig. P10.69 operates at $f = 60$ Hz. Use PSPICE to find the voltage $\mathbf{V}_o$.

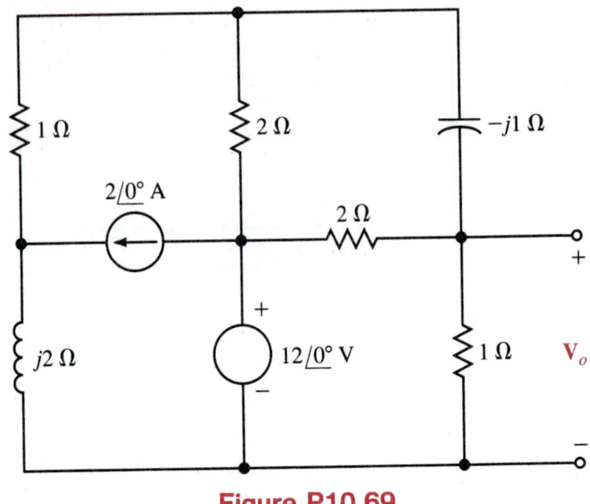

Figure P10.69

10.70. Use PSPICE to find $\mathbf{V}_o$ in the network in Fig. P10.70.

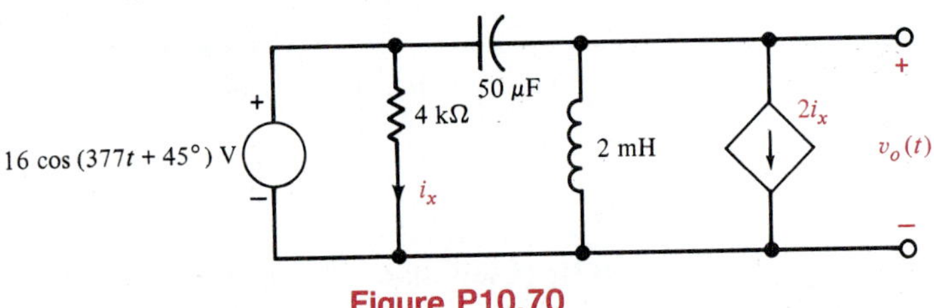

Figure P10.70

10.71. Use PSPICE to calculate both the small-signal transfer function $\mathbf{I}_o/\mathbf{I}_S$ and the output current in the network in Fig. P10.71.

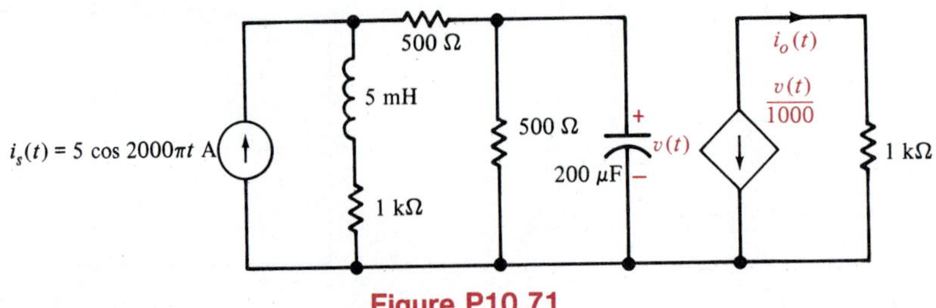

Figure P10.71

10.72. Use PSPICE to calculate both the dc small-signal transfer function $\mathbf{V}_o/\mathbf{V}_S$ and the output voltage $\mathbf{V}_o$ in the network in Fig. P10.72.

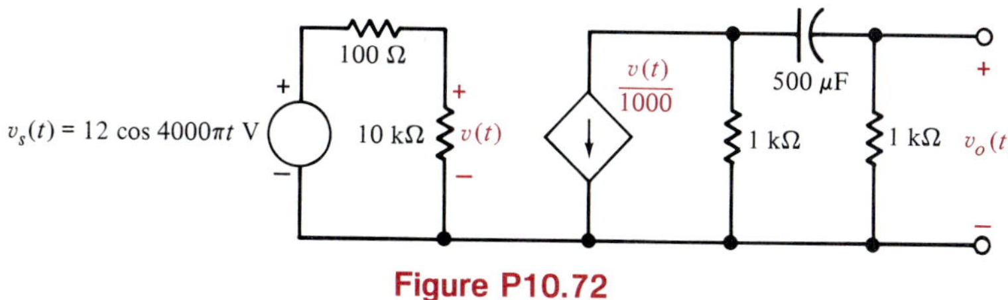

Figure P10.72

10.73. The network in Fig. P10.73 operates at $f = 60$ Hz. Use PSPICE to find the current $\mathbf{I}_o$.

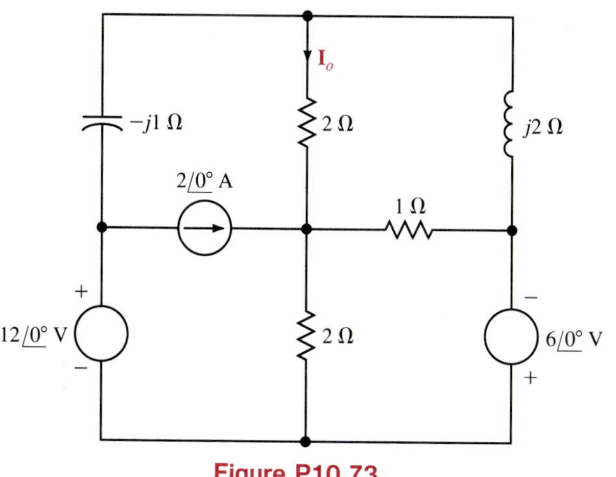

Figure P10.73

10.74. The network in Fig. P10.74 operates at $f = 60$ Hz. Use PSPICE to find the voltage $\mathbf{V}_o$.

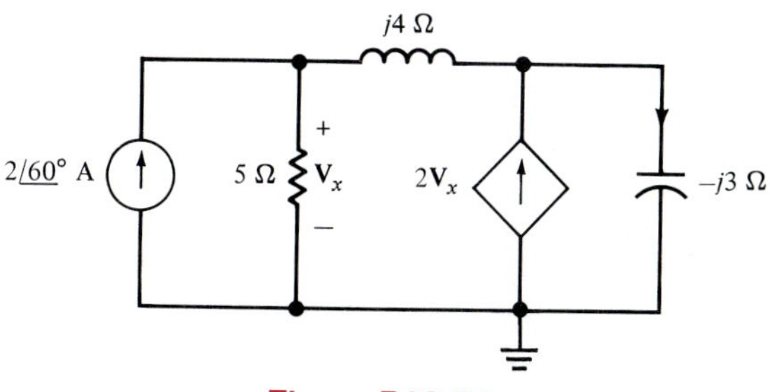

Figure P10.74

10.75. The network in Fig. P10.75 operates at 60 Hz. Find the currents $\mathbf{I}_o$ and $\mathbf{I}_x$ using PSPICE.

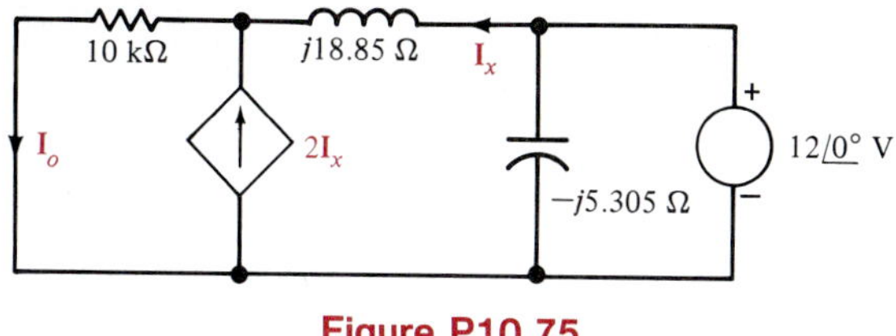

Figure P10.75

10.76. The circuit in Fig. P10.76 operates at $f = 60$ Hz. Use PSPICE to compute the voltage $\mathbf{V}_o$.

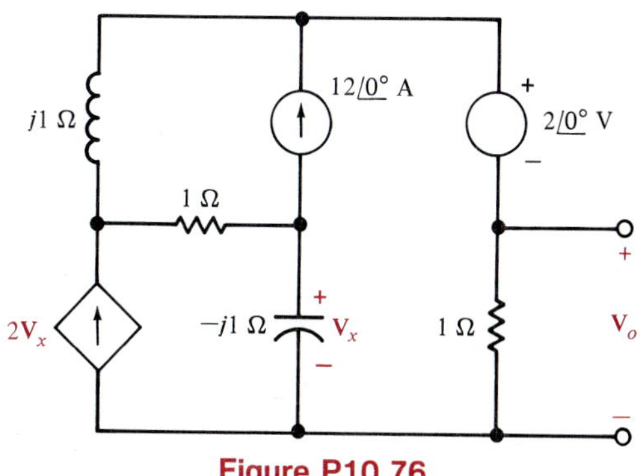

Figure P10.76

CHAPTER

Steady-State Power Analysis

We have been primarily concerned in the preceding chapters with determining the voltage or current at some point in a network. Of equal importance to us in many situations is the power that is supplied or absorbed by some element. Typically, electrical and electronic devices have peak power or maximum instantaneous power ratings that cannot be exceeded without damaging the devices.

In electrical and electronic systems, power comes in all sizes. The power absorbed by some device on an integrated-circuit chip may be in picowatts, whereas the power supplied by a large generating station may be in gigawatts. Note that the range between these two examples is phenomenally large (10^{21}).

In our previous work we defined instantaneous power to be the product of voltage and current. Average power, obtained by averaging the instantaneous power, is the average rate at which energy is absorbed or supplied. In the dc case, where both current and voltage are constant, the instantaneous power is equal to the average power. However, as we will demonstrate, this is not the case when the currents and voltages are sinusoidal functions of time.

In this chapter we explore the many ramifications of power in ac circuits. We examine instantaneous power, average power, maximum power transfer, average power for periodic nonsinusoidal waveforms, the power factor, complex power, and power measurement. Finally, safety considerations will be introduced and discussed through a variety of examples.

11.1

Instantaneous Power

By employing the sign convention adopted in the earlier chapters, we can compute the instantaneous power supplied or absorbed by any device as the product of the instantaneous voltage across the device and the instantaneous current through it.

Consider the circuit shown in Fig. 11.1. In general, the steady-state voltage and current for the network can be written as

$$v(t) = V_M \cos(\omega t + \theta_v) \tag{11.1}$$

$$i(t) = I_M \cos(\omega t + \theta_i) \tag{11.2}$$

The instantaneous power is then

$$\begin{aligned} p(t) &= v(t)i(t) \\ &= V_M I_M \cos(\omega t + \theta_v)\cos(\omega t + \theta_i) \end{aligned} \tag{11.3}$$

Employing the following trigonometric identity,

$$\cos\phi_1 \cos\phi_2 = \tfrac{1}{2}[\cos(\phi_1 - \phi_2) + \cos(\phi_1 + \phi_2)] \tag{11.4}$$

we find that the instantaneous power can be written as

$$p(t) = \frac{V_M I_M}{2}[\cos(\theta_v - \theta_i) + \cos(2\omega t + \theta_v + \theta_i)] \tag{11.5}$$

Note that the instantaneous power consists of two terms. The first term is a constant (i.e., it is time independent), and the second term is a cosine wave of twice the excitation frequency. We will examine this equation in more detail in Section 11.2.

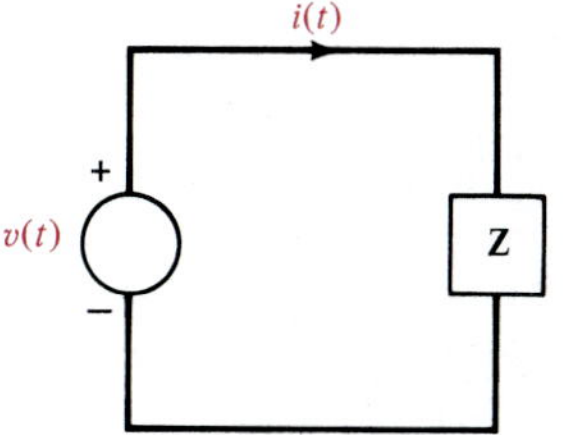

Figure 11.1 Simple ac network.

EXAMPLE 11.1

The circuit in Fig. 11.1 has the following parameters: $v(t) = 4\cos(\omega t + 60°)$ V and $\mathbf{Z} = 2\,\underline{/30°}\ \Omega$. We wish to determine equations for the current and the instantaneous power as a function of time, and plot these functions with the voltage on a single graph for comparison.

Since

$$\begin{aligned} \mathbf{I} &= \frac{4\,\underline{/60°}}{2\,\underline{/30°}} \\ &= 2\,\underline{/30°}\text{ A} \end{aligned}$$

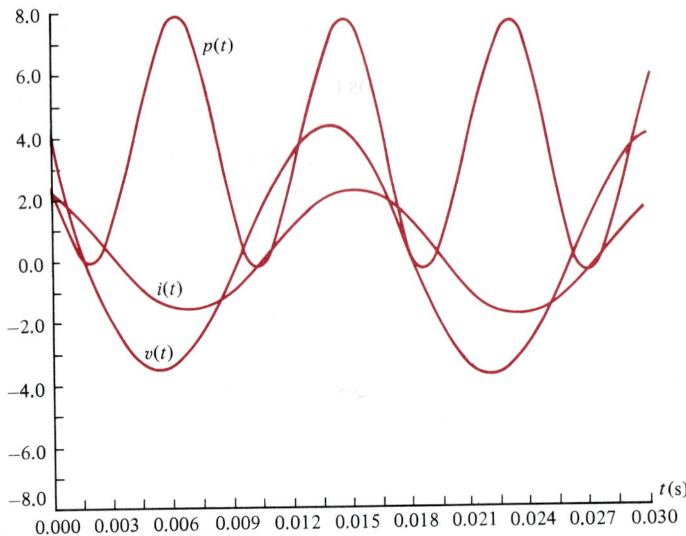

Figure 11.2 Plots of $v(t)$, $i(t)$, and $p(t)$ for the circuit in Example 11.1 using $f = 60$ Hz.

then

$$i(t) = 2 \cos (\omega t + 30°) \text{ A}$$

From Eq. (11.5),

$$\begin{aligned} p(t) &= 4[\cos (30°) + \cos (2\omega t + 90°)] \\ &= 3.46 + 4 \cos (2\omega t + 90°) \text{ W} \end{aligned}$$

A plot of this function, together with plots of the voltage and current, are shown in Fig. 11.2. As can be seen in the figure, the instantaneous power has an average value, and the frequency is twice that of the voltage or current. ■

11.2 Average Power

The average value of any periodic waveform (e.g., a sinusoidal function) can be computed by integrating the function over a complete period and dividing this result by the period. Therefore, if the voltage and current are given by Eqs. (11.1) and (11.2), respectively, the average power is

$$\begin{aligned} P &= \frac{1}{T} \int_{t_0}^{t_0+T} p(t)\, dt \\ &= \frac{1}{T} \int_{t_0}^{t_0+T} V_m I_M \cos (\omega t + \theta_v) \cos (\omega t + \theta_i)\, dt \end{aligned} \tag{11.6}$$

where t_0 is arbitrary, $T = 2\pi/\omega$ is the period of the voltage or current, and P is measured in watts. Actually, we may average the waveform over any integral number of periods so that Eq. (11.6) can also be written as

$$P = \frac{1}{nT}\int_{t_0}^{t_0+nT} V_M I_M \cos(\omega t + \theta_v)\cos(\omega t + \theta_i)\, dt \tag{11.7}$$

where n is a positive integer.

Employing Eq. (11.5) in the expression (11.6), we obtain

$$P = \frac{1}{T}\int_{t_0}^{t_0+T} \frac{V_M I_M}{2}[\cos(\theta_v - \theta_i) + \cos(2\omega t + \theta_v + \theta_i)]\, dt \tag{11.8}$$

We could, of course, plod through the indicated integration; however, with a little forethought we can determine the result by inspection. The first term is independent of t, and therefore a constant in the integration. Integrating the constant over the period and dividing by the period simply results in the original constant. The second term is a cosine wave. It is well known that the average value of a cosine wave over one complete period or an integral number of periods is zero, and therefore the second term in Eq. (11.8) vanishes. In view of this discussion, Eq. (11.8) reduces to

$$P = \tfrac{1}{2} V_M I_M \cos(\theta_v - \theta_i) \tag{11.9}$$

Note that since $\cos(-\theta) = \cos(\theta)$, the argument for the cosine function can be either $\theta_v - \theta_i$ or $\theta_i - \theta_v$. In addition, note that $\theta_v - \theta_i$ is the angle of the circuit impedance as shown in Fig. 11.1. Therefore, for a purely resistive circuit,

$$P = \tfrac{1}{2} V_M I_M \tag{11.10}$$

and for a purely reactive circuit,

$$\begin{aligned} P &= \tfrac{1}{2} V_M I_M \cos(90°) \\ &= 0 \end{aligned}$$

Because purely reactive impedances absorb no average power, they are often called *lossless elements*. The purely reactive network operates in a mode in which it stores energy over one part of the period and releases it over another.

EXAMPLE 11.2

We wish to determine the average power absorbed by the impedance shown in Fig. 11.3.

From the figure we note that

$$\mathbf{I} = \frac{\mathbf{V}}{\mathbf{Z}} = \frac{V_M \angle \theta_v}{2 + j2} = \frac{10 \angle 60°}{2.83 \angle 45°} = 3.53 \angle 15° \text{ A}$$

Therefore,

$$I_M = 3.53 \text{ A} \qquad \text{and} \qquad \theta_i = 15°$$

Hence

$$\begin{aligned} P &= \tfrac{1}{2} V_M I_M \cos(\theta_v - \theta_i) \\ &= \tfrac{1}{2}(10)(3.53)\cos(60° - 15°) \\ &= 12.5 \text{ W} \end{aligned}$$

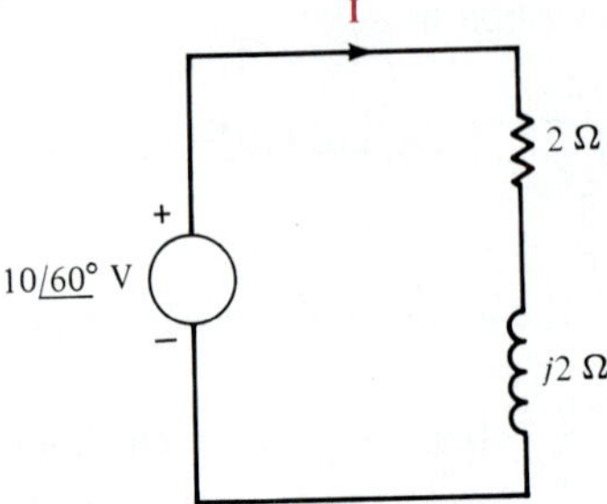

Figure 11.3 Example *RL* circuit.

Since the inductor absorbs no power, we can employ Eq. (11.10) provided that V_M in that equation is the voltage across the resistor. Using voltage division, we obtain

$$\mathbf{V}_R = \frac{(10\ \underline{/60^\circ})(2)}{2 + j2} = 7.07\ \underline{/15^\circ}\ \text{V}$$

and therefore,

$$\begin{aligned} P &= \tfrac{1}{2}(7.07)(3.53) \\ &= 12.5\ \text{W} \end{aligned}$$

In addition, using Ohm's law, we could also employ the expressions

$$P = \frac{1}{2}\frac{V_M^2}{R}$$

or

$$P = \tfrac{1}{2}I_M^2 R$$

where once again we must be careful that the V_M and I_M in these equations refer to the voltage across the resistor and the current through it, respectively. ■

EXAMPLE 11.3

For the circuit shown in Fig. 11.4 we wish to determine both the total average power absorbed and the total average power supplied.

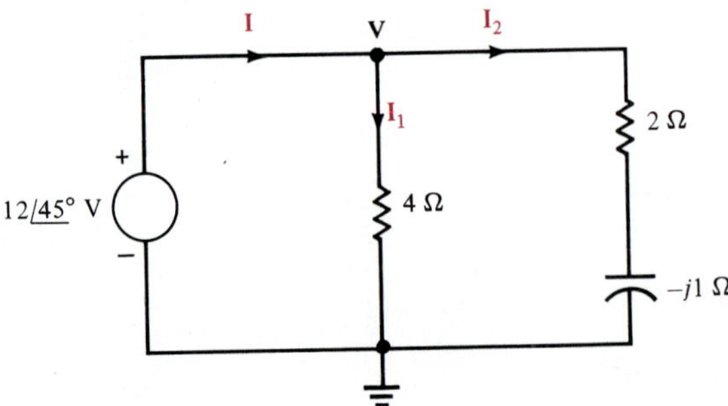

Figure 11.4 Example circuit for illustrating a power balance.

From the figure we note that

$$\mathbf{I}_1 = \frac{12 \underline{/45^\circ}}{4} = 3 \underline{/45^\circ} \text{ A}$$

$$\mathbf{I}_2 = \frac{12 \underline{/45^\circ}}{2 - j1} = \frac{12 \underline{/45^\circ}}{2.24 \underline{/-26.57^\circ}} = 5.37 \underline{/71.57^\circ} \text{ A}$$

and therefore,

$$\begin{aligned} \mathbf{I} &= \mathbf{I}_1 + \mathbf{I}_2 \\ &= 3 \underline{/45^\circ} + 5.37 \underline{/71.57^\circ} \\ &= 8.16 \underline{/62.08^\circ} \text{ A} \end{aligned}$$

The average power absorbed in the 4-Ω resistor is

$$P_4 = \tfrac{1}{2} V_M I_M = \tfrac{1}{2}(12)(3) = 18 \text{ W}$$

The average power absorbed in the 2-Ω resistor is

$$P_2 = \tfrac{1}{2} I_M^2 R = \tfrac{1}{2}(5.37)^2(2) = 28.8 \text{ W}$$

Therefore, the total average power absorbed is

$$P_A = 18 + 28.8 = 46.8 \text{ W}$$

Note that we could have calculated the power absorbed in the 2-Ω resistor using $\frac{1}{2}V_M^2/R$ if we had first calculated the voltage across the 2-Ω resistor.

The total average power supplied by the source is

$$\begin{aligned} P_S &= +\tfrac{1}{2} V_M I_M \cos(\theta_v - \theta_i) \\ &= +\tfrac{1}{2}(12)(8.16) \cos(45^\circ - 62.08^\circ) \\ &= +46.8 \text{ W} \end{aligned}$$

Thus the total average power supplied is, of course, equal to the total average power absorbed. ■

DRILL EXERCISE

D11.1. Find the average power absorbed by each resistor in the network in Fig. D11.1.

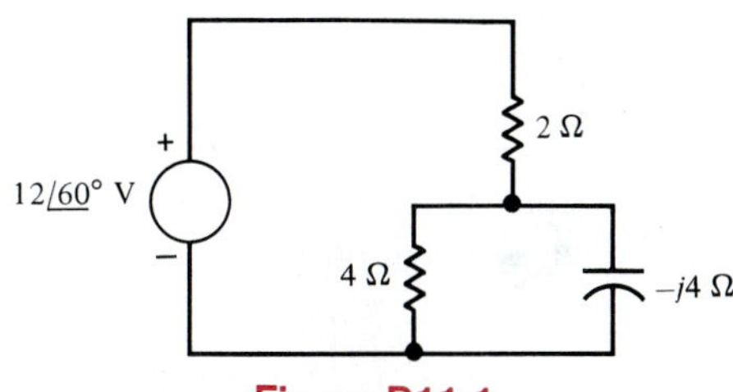

Figure D11.1

Ans: $P_{2\Omega} = 7.20$ W, $P_{4\Omega} = 7.20$ W.

D11.2. Given the network in Fig. D11.2, find the average power absorbed by each passive circuit element and the total average power supplied by the current source.

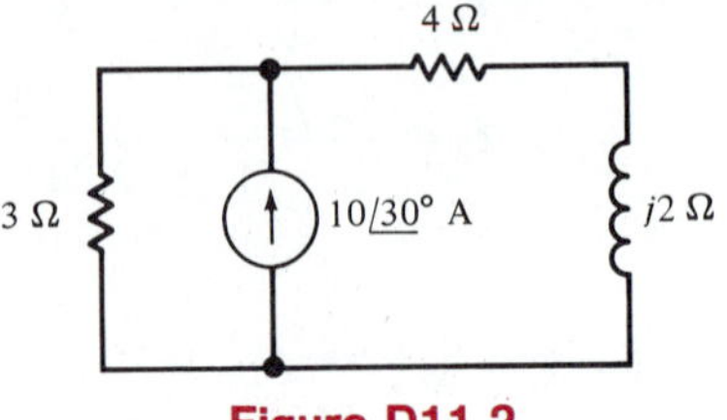

Figure D11.2

Ans: $P_{3\Omega} = 56.55$ W, $P_{4\Omega} = 33.95$ W, $P_L = 0$, $P_{cs} = -90.50$ W.

When determining average power, if more than one source is present in a network, we can use any of our network analysis techniques to find the necessary voltage and/or current to compute the power. However, we must remember that in general we cannot apply superposition to power. We should, however, mention that there is a special case in which superposition does apply to power. If the current is of the form

$$i(t) = I_1 \cos(\omega_1 t + \theta_1) + I_2 \cos(\omega_2 t + \theta_2)$$

and if $n\omega_1 = m\omega_2$, where n and m are two different integers, then $n\omega_1 = 2\pi/(T/n)$ and $m\omega_2 = 2\pi/(T/m)$, so that $\cos \omega_1 t$ has n full periods in time T and $\cos \omega_2 t$ has m full periods in time T. These sinusoids are said to be harmonically related. This subject will be discussed in detail in Chapter 18.

The average power that is absorbed by resistor R, in a time interval T, is given by

$$P = \frac{1}{T}\int_0^T [I_1 \cos(\omega_1 t + \theta_1) + I_2 \cos(\omega_2 t + \theta_2)]^2 R\, dt$$

where both ω_1 and ω_2 are periodic in T. The equation above can be written as

$$\begin{aligned} P &= \frac{1}{T}\int_0^T [I_1^2 \cos^2(\omega_1 t + \theta_1) + I_2^2 \cos^2(\omega_2 t + \theta_2) \\ &\quad + 2I_1 I_2 \cos(\omega_1 t + \theta_1)\cos(\omega_2 t + \theta_2)]R\, dt \\ &= \frac{I_1^2}{2}R + \frac{I_2^2}{2}R + \frac{1}{T}\int_0^T \{I_1 I_2 \cos[(\omega_1 + \omega_2)t + \theta_1 + \theta_2] \\ &\quad + I_1 I_2 \cos[(\omega_1 - \omega_2)t + \theta_1 - \theta_2]\}R\, dt \end{aligned}$$

Since $(\omega_1 + \omega_2) = (n + m)(2\pi/T)$ and $(\omega_1 - \omega_2) = (n - m)(2\pi/T)$, the integral will be zero. Therefore,

$$P = \frac{I_1^2}{2}R + \frac{I_2^2}{2}R$$

Superposition also applies if one, and only one, of the sources is dc.

EXAMPLE 11.4

Consider the network shown in Fig. 11.5. We wish to determine the total average power absorbed and supplied by each element.

From the figure we note that

$$\mathbf{I}_2 = \frac{12 \underline{/30^\circ}}{2} = 6 \underline{/30^\circ} \text{ A}$$

and

$$\mathbf{I}_3 = \frac{12 \underline{/30^\circ} - 6 \underline{/0^\circ}}{j1} = \frac{4.39 + j6}{j1} = 7.43 \underline{/-36.19^\circ} \text{ A}$$

The power absorbed by the 2-Ω resistor is

$$P_2 = \tfrac{1}{2} V_M I_M = \tfrac{1}{2}(12)(6) = 36 \text{ W}$$

According to the direction of $\mathbf{I}_3$ the 6 $\underline{/0^\circ}$-V source is absorbing power. The power it absorbs is given by

$$\begin{aligned} P_{6\underline{/0^\circ}} &= \tfrac{1}{2} V_M I_M \cos(\theta_v - \theta_i) \\ &= \tfrac{1}{2}(6)(7.43) \cos[0^\circ - (-36.19^\circ)] \\ &= 18 \text{ W} \end{aligned}$$

At this point an obvious question arises: How do we know if the 6 $\underline{/0^\circ}$-V source is supplying power to the remainder of the network or absorbing it? The answer to this question is actually straightforward. If we employ our passive sign convention that was adopted in the earlier chapters, that is, if the current reference direction enters the positive terminal of the source and the answer is positive, the source is absorbing power. If the answer is negative, the source is supplying power to the remainder of the circuit. A generator sign convention could have been used and under this condition the interpretation of the sign of the answer would be reversed. Note that once the sign convention is adopted and used, the sign for average power will be negative only if the angle difference is greater than 90° (i.e., $|\theta_v - \theta_i| > 90^\circ$).

To obtain the power supplied to the network, we compute $\mathbf{I}_1$ as

$$\begin{aligned} \mathbf{I}_1 &= \mathbf{I}_2 + \mathbf{I}_3 \\ &= 6 \underline{/30^\circ} + 7.43 \underline{/-36.19^\circ} \\ &= 11.29 \underline{/-7.07^\circ} \text{ A} \end{aligned}$$

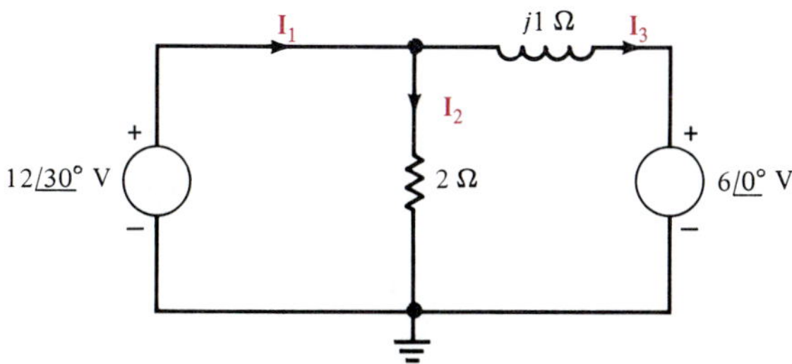

Figure 11.5 Example *RL* circuit with two sources.

Therefore, the power supplied by the 12 $\underline{/30°}$-V source using the generator sign convention is

$$P_S = +\tfrac{1}{2}(12)(11.29)\cos(30° + 7.07°)$$
$$= +54 \text{ W}$$

and hence the power absorbed is equal to the power supplied. ■

EXAMPLE 11.5

The current in a 2-Ω resistor is of the form

$$i(t) = 4\cos(377t + 30°) + 2\cos(754t + 60°)$$

We wish to find the average power absorbed by the resistor.

Since $\omega_1 = 377$ and $\omega_2 = 754$ (i.e., $\omega_2 = 2\omega_1$), then

$$P = \tfrac{1}{2}(4^2 + 2^2)2 = 20 \text{ W}$$ ■

DRILL EXERCISE

D11.3. Determine the total average power absorbed and supplied by each element in the network in Fig. D11.3.

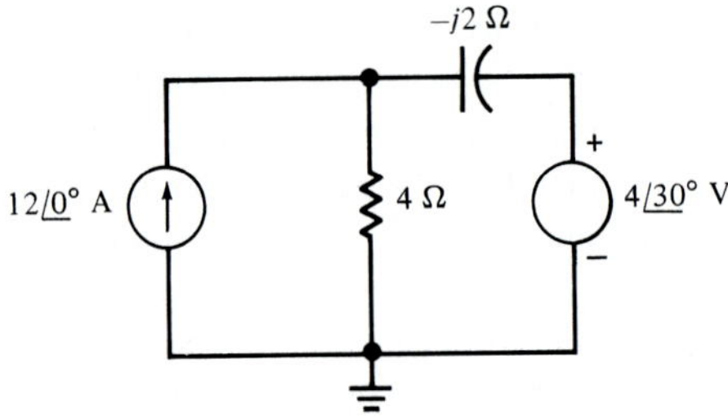

Figure D11.3

Ans: $P_{cs} = -69.3$ W, $\mathbf{P}_{V_S} = 19.8$ W, $P_{4\Omega} = 49.5$ W, $P_C = 0$.

D11.4. Given the network in Fig. D11.4, determine the total average power absorbed or supplied by each element.

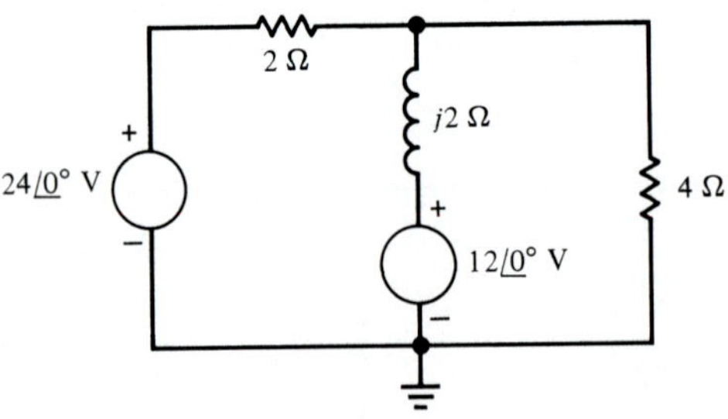

Figure D11.4

Ans: $P_{24\underline{/0°}} = -55.4$ W, $P_{12\underline{/0°}} = 5.5$ W, $P_{2\Omega} = 22.1$ W, $P_{4\Omega} = 27.8$ W, $P_L = 0$.

11.3 Maximum Average Power Transfer

In our study of resistive networks we addressed the problem of maximum power transfer to a resistive load. We showed that if the network excluding the load was represented by a Thévenin equivalent circuit, maximum power transfer would result if the value of the load resistor was equal to the Thévenin equivalent resistance (i.e., $R_L = R_{\text{Th}}$). We will now reexamine this issue within the present context to determine the load impedance for the network shown in Fig. 11.6 that will result in maximum average power being absorbed by the load impedance $\mathbf{Z}_L$.

The equation for average power at the load is

$$P_L = \tfrac{1}{2}V_L I_L \cos(\theta_{V_L} - \theta_{i_L}) \tag{11.11}$$

The phasor current and voltage at the load are given by the expressions

$$\mathbf{I}_L = \frac{\mathbf{V}_{\text{oc}}}{\mathbf{Z}_{\text{Th}} + \mathbf{Z}_L} \tag{11.12}$$

$$\mathbf{V}_L = \frac{\mathbf{V}_{\text{oc}}\mathbf{Z}_L}{\mathbf{Z}_{\text{Th}} + \mathbf{Z}_L} \tag{11.13}$$

where

$$\mathbf{Z}_{\text{Th}} = R_{\text{Th}} + jX_{\text{Th}} \tag{11.14}$$

and

$$\mathbf{Z}_L = R_L + jX_L \tag{11.15}$$

The magnitude of the phasor current and voltage are given by the expressions

$$I_L = \frac{V_{\text{oc}}}{[(R_{\text{Th}} + R_L)^2 + (X_{\text{Th}} + X_L)^2]^{1/2}} \tag{11.16}$$

$$V_L = \frac{V_{\text{oc}}(R_L^2 + X_L^2)^{1/2}}{[(R_{\text{Th}} + R_L)^2 + (X_{\text{Th}} + X_L)^2]^{1/2}} \tag{11.17}$$

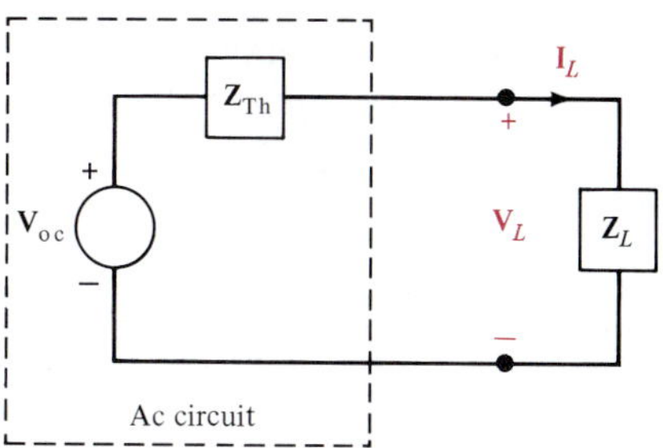

Figure 11.6 Circuit used to examine maximum power transfer.

The phase angles for the phasor current and voltage are contained in the quantity $(\theta_{V_L} - \theta_{i_L})$. Note also that $\theta_{V_L} - \theta_{i_L} = \theta_{\mathbf{Z}_L}$ and in addition

$$\cos \theta_{\mathbf{Z}_L} = \frac{R_L}{[R_L^2 + X_L^2]^{1/2}} \tag{11.18}$$

Substituting Eqs. (11.16) to (11.18) into Eq. (11.11) yields

$$P_L = \frac{1}{2} \frac{V_{oc}^2 R_L}{(R_{Th} + R_L)^2 + (X_{Th} + X_L)^2} \tag{11.19}$$

which could, of course, be obtained directly from Eq. (11.16) using $\frac{1}{2}I_L^2 R_L$. Once again, a little forethought will save us some work. From the standpoint of maximizing P_L, V_{oc} is a constant. The quantity $(X_{Th} + X_L)$ absorbs no power and therefore any nonzero value of this quantity only serves to reduce P_L. Hence we can eliminate this term by selecting $X_L = -X_{Th}$. Our problem then reduces to maximizing

$$P_L = \frac{1}{2} \frac{V_{oc}^2 R_L}{(R_L + R_{Th})^2} \tag{11.20}$$

However, this is the same quantity we maximized in the purely resistive case by selecting $R_L = R_{Th}$. Therefore, for maximum average power transfer to the load shown in Fig. 11.6, $\mathbf{Z}_L$ should be chosen so that

$$\mathbf{Z}_L = R_L + jX_L = R_{Th} - jX_{Th} = \mathbf{Z}_{Th}^* \tag{11.21}$$

Finally, if the load impedance is purely resistive (i.e., $X_L = 0$), the condition for maximum average power transfer can be derived via the expression

$$\frac{dP_L}{dR_L} = 0$$

where P_L is the expression in Eq. (11.19) with $X_L = 0$. The value of R_L that maximizes P_L under the condition $X_L = 0$ is

$$R_L = \sqrt{R_{Th}^2 + X_{Th}^2} \tag{11.22}$$

EXAMPLE 11.6

Given the circuit in Fig. 11.7a we wish to find the value of $\mathbf{Z}_L$ for maximum average power transfer. In addition, we wish to find the value of the maximum average power delivered to the load.

In order to solve the problem, we form a Thévenin equivalent at the load. The circuit in Fig. 11.7b is used to compute the open-circuit voltage

$$\mathbf{V}_{oc} = \frac{4 \underline{/0^\circ}\ (2)}{6 + j1}(4) = 5.28 \underline{/-9.46^\circ} \text{ V}$$

The Thévenin equivalent impedance can be derived from the circuit in Fig. 11.7c. As shown in the figure,

$$\mathbf{Z}_{Th} = \frac{4(2 + j1)}{6 + j1} = 1.40 + j0.43\ \Omega$$

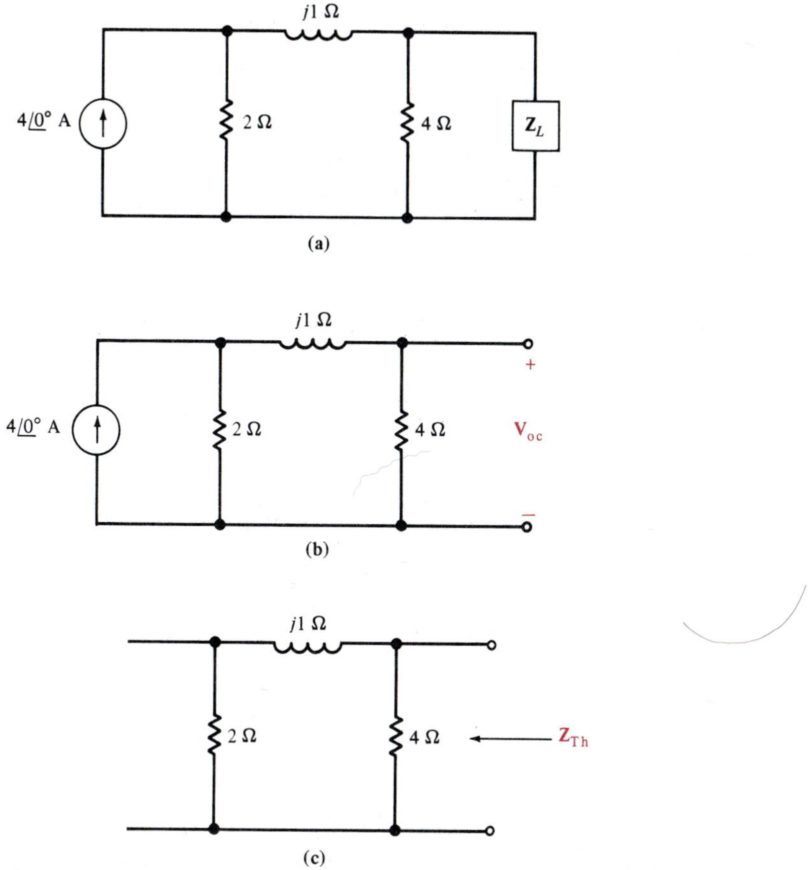

Figure 11.7 Circuits for illustrating maximum average power transfer.

Therefore, $\mathbf{Z}_L$ for maximum average power transfer is

$$\mathbf{Z}_L = 1.40 - j0.43\ \Omega$$

With $\mathbf{Z}_L$ as given above, the current in the load is

$$\mathbf{I} = \frac{5.28\ \underline{/-9.46^\circ}}{2.8} = 1.89\ \underline{/-9.46^\circ}\ \text{A}$$

Therefore, the maximum average power transferred to the load is

$$P_L = \tfrac{1}{2}I_M^2 R_L = \tfrac{1}{2}(1.89)^2(1.4) = 2.50\ \text{W}$$

EXAMPLE 11.7

For the circuit shown in Fig. 11.8a, we wish to find the value of $\mathbf{Z}_L$ for maximum average power transfer. In addition, let us determine the value of the maximum average power delivered to the load.

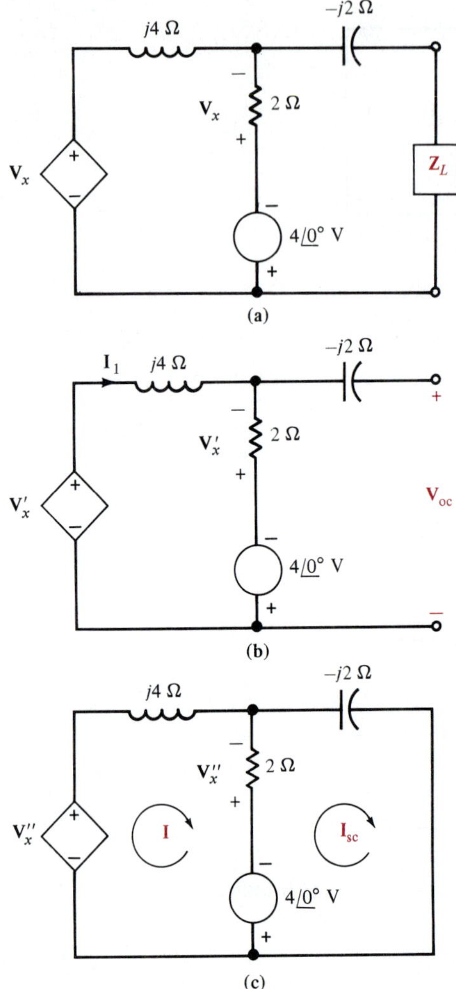

Figure 11.8 Circuits for illustrating maximum average power transfer.

We will first reduce the circuit, with the exception of the load, to a Thévenin equivalent circuit. The open-circuit voltage can be computed from Fig. 11.8b. The equations for the circuit are

$$\mathbf{V}_x' + 4 = (2 + j4)\mathbf{I}_1$$

$$\mathbf{V}_x' = -2\mathbf{I}_1$$

Solving for $\mathbf{I}_1$, we obtain

$$\mathbf{I}_1 = \frac{1 \angle -45^\circ}{\sqrt{2}}$$

The open-circuit voltage is then

$$\begin{aligned}\mathbf{V}_{oc} &= 2\mathbf{I}_1 - 4\,\underline{/0^\circ} \\ &= \sqrt{2}\,\underline{/-45^\circ} - 4\,\underline{/0^\circ} \\ &= -3 - j1 \\ &= -3.16\,\underline{/18.43^\circ}\text{ V}\end{aligned}$$

The short-circuit current can be derived from Fig. 11.8c. The equations for this circuit are

$$\begin{aligned}\mathbf{V}_x'' + 4 &= (2 + j4)\mathbf{I} - 2\mathbf{I}_{sc} \\ -4 &= -2\mathbf{I} + (2 - j2)\mathbf{I}_{sc} \\ \mathbf{V}_x'' &= -2(\mathbf{I} - \mathbf{I}_{sc})\end{aligned}$$

Solving these equations for $\mathbf{I}_{sc}$ yields

$$\mathbf{I}_{sc} = -(1 + j2)\text{ A}$$

The Thévenin equivalent impedance is then

$$\mathbf{Z}_{Th} = \frac{\mathbf{V}_{oc}}{\mathbf{I}_{sc}} = \frac{3 + j1}{1 + j2} = 1 - j1\ \Omega$$

Therefore, for maximum average power transfer the load impedance should be

$$\mathbf{Z}_L = 1 + j1\ \Omega$$

The current in this load $\mathbf{Z}_L$ is then

$$\mathbf{I}_L = \frac{\mathbf{V}_{oc}}{\mathbf{Z}_{Th} + \mathbf{Z}_L} = \frac{-3 - j1}{2} = -1.58\,\underline{/18.43^\circ}\text{ A}$$

Hence the maximum average power transferred to the load is

$$\begin{aligned}P_L &= \tfrac{1}{2}(1.58)^2(1) \\ &= 1.25\text{ W}\end{aligned}$$

■

DRILL EXERCISE

D11.5. Given the network in Fig. D11.5, find $\mathbf{Z}_L$ for maximum average power transfer and the maximum average power transferred to the load.

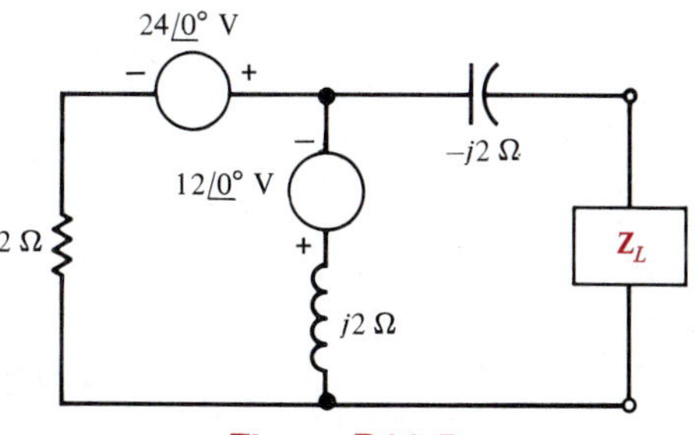

Figure D11.5

Ans: $\mathbf{Z}_L = 1 + j1\Omega$, $P_L = 44.94$ W.

D11.6. Find $\mathbf{Z}_L$ for maximum average power transfer and the maximum average power transferred to the load in the network in Fig. D11.6.

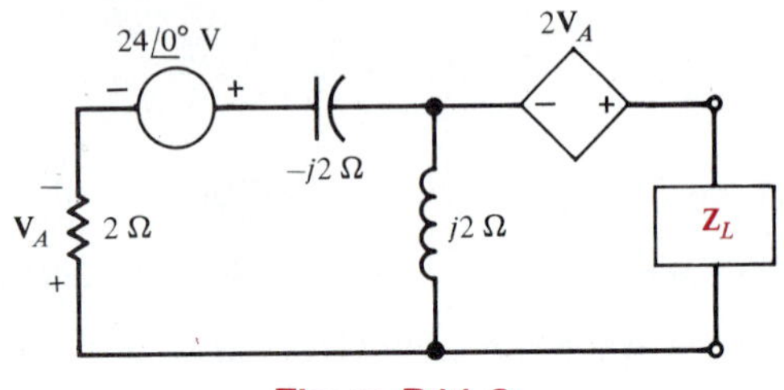

Figure D11.6

Ans: $\mathbf{Z}_L = 2 + j2\ \Omega$, $P_L = 180$ W.

11.4 Effective or rms Values

In the preceding sections of this chapter we have shown that the average power absorbed by a resistive load is directly dependent on the type, or types, of sources that are delivering power to the load. For example, if the source was dc, the average power absorbed was I^2R, and if the source was sinusoidal, the average power was $\frac{1}{2}I_M^2R$. Although these two types of waveforms are extremely important, they are by no means the only waveforms we will encounter in circuit analysis. Therefore, a technique by which we can compare the *effectiveness* of different sources in delivering power to a resistive load would be quite useful.

In order to accomplish this comparison, we define what is called the *effective value of a periodic waveform*, representing either voltage or current. Although either quantity could be used, we will employ current in the definition. Hence we define the effective value of a periodic current as a constant, or dc value, which delivers the same average power to a resistor R. Let us call the constant current I_{eff}. Then the average power delivered to a resistor as a result of this current is

$$P = I_{\text{eff}}^2 R$$

Similarly, the average power delivered to a resistor by a periodic current $i(t)$ is

$$P = \frac{1}{T}\int_{t_0}^{t_0+T} i^2(t)R\, dt$$

Equating these two expressions, we find that

$$I_{\text{eff}} = \sqrt{\frac{1}{T}\int_{t_0}^{t_0+T} i^2(t)\, dt} \tag{11.23}$$

Note that this effective value is found by first determining the *square* of the current, then computing the average or *mean* value, and finally taking the square *root*. Thus in ''reading'' the mathematical Eq. (11.23), we are determining the root mean square, which we abbreviate as rms, and therefore I_{eff} is called I_{rms}.

Since dc is a constant, the rms value of dc is simply the constant value. Let us now determine the rms value of other waveforms. The most important waveform is the sinusoid, and therefore, we address this particular one in the following example.

EXAMPLE 11.8

We wish to compute the rms value of the waveform $i(t) = I_M \cos(\omega t - \theta)$, which has a period of $T = 2\pi/\omega$.

Substituting these expressions into Eq. (11.23) yields

$$I_{\text{rms}} = \left[\frac{1}{T}\int_0^T I_M^2 \cos^2(\omega t - \theta)\, dt\right]^{1/2}$$

Using the trigonometric identity

$$\cos^2\phi = \tfrac{1}{2} + \tfrac{1}{2}\cos 2\phi,$$

we find that the equation above can be expressed as

$$I_{\text{rms}} = I_M\left\{\frac{\omega}{2\pi}\int_0^{2\pi/\omega} \left[\tfrac{1}{2} + \tfrac{1}{2}\cos(2\omega t - 2\theta)\right] dt\right\}^{1/2}$$

Since we know that the average or mean value of a cosine wave is zero,

$$I_{\text{rms}} = I_M\left(\frac{\omega}{2\pi}\int_0^{2\pi/\omega} \frac{1}{2}\, dt\right)^{1/2}$$

$$= I_M\left[\frac{\omega}{2\pi}\left(\frac{t}{2}\right)\Bigg|_0^{2\pi/\omega}\right]^{1/2} = \frac{I_M}{\sqrt{2}} \tag{11.24}$$

Therefore, the rms value of a sinusoid is equal to the maximum value divided by the $\sqrt{2}$. Hence a sinusoidal current with a maximum value of I_M delivers the same average power to a resistor R as a dc current with a value of $I_M/\sqrt{2}$. ■

Upon using the rms values for voltage and current, the average power can be written in general as

$$P = V_{\text{rms}} I_{\text{rms}} \cos(\theta_v - \theta_i) \tag{11.25}$$

The power absorbed by a resistor R is

$$P = I_{\text{rms}}^2 R = \frac{V_{\text{rms}}^2}{R} \tag{11.26}$$

In dealing with voltages and currents in numerous electrical applications, it is important to know whether the values quoted are maximum, average, rms, or what. For example, the normal 120-V ac electrical outlets have an rms value of 120 V, an average value of 0 V, and a maximum value of $120\sqrt{2}$ V.

Finally, if the current in a resistor R is composed of a sum of harmonically related sinusoids, the power absorbed by the resistor can be expressed as

$$P = (I_{1\text{rms}}^2 + I_{2\text{rms}}^2 + \cdots + I_{n\text{rms}}^2)R \tag{11.27}$$

where the rms value of the total current is

$$I_{\text{rms}} = \sqrt{I_{1\text{rms}}^2 + I_{2\text{rms}}^2 + \cdots + I_{n\text{rms}}^2} \tag{11.28}$$

Each component represents a current of different frequency.

EXAMPLE 11.9

We wish to compute the rms value of the voltage waveform shown in Fig. 11.9.

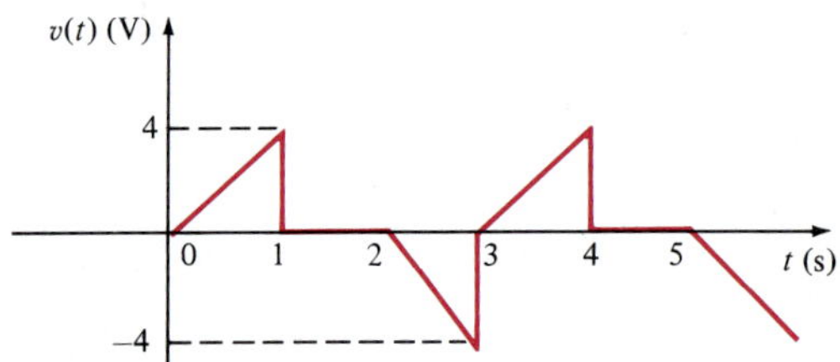

Figure 11.9 Waveform used to illustrate rms values.

The waveform is periodic with period $T = 3$ sec. The equation for the voltage in the time frame $0 \leq t \leq 3$ sec is

$$v(t) = \begin{cases} 4t \text{ V} & 0 < t \leq 1 \text{ sec} \\ 0 \text{ V} & 1 < t \leq 2 \text{ sec} \\ -4t + 8 \text{ V} & 2 < t \leq 3 \text{ sec} \end{cases}$$

The rms value is

$$\begin{aligned} V_{\text{rms}} &= \left\{\frac{1}{3}\left[\int_0^1 (4t)^2\,dt + \int_1^2 (0)^2\,dt + \int_2^3 (8-4t)^2\,dt\right]\right\}^{1/2} \\ &= \left[\frac{1}{3}\left(\frac{16t^3}{3}\bigg|_0^1 + \left(64t - \frac{64t^2}{2} + \frac{16t^3}{3}\right)\bigg|_2^3\right)\right]^{1/2} \\ &= 1.89 \text{ V} \end{aligned}$$

EXAMPLE 11.10

Determine the rms value of the current waveform in Fig. 11.10 and use this value to compute the average power delivered to a 2-Ω resistor through which this current is flowing.

The current waveform is periodic with a period of $T = 4$ sec. The rms value is

$$\begin{aligned} I_{\text{rms}} &= \left\{\frac{1}{4}\left[\int_0^2 (4)^2\,dt + \int_2^4 (-4)^2\,dt\right]\right\}^{1/2} \\ &= \left[\frac{1}{4}\left(16t\bigg|_0^2 + 16t\bigg|_2^4\right)\right]^{1/2} \\ &= 4 \text{ A} \end{aligned}$$

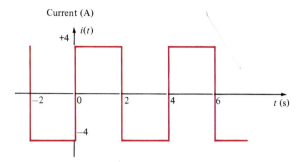

Figure 11.10 Waveform used to illustrate rms values.

The average power delivered to a 2-Ω resistor with this current is

$$P = I_{\text{rms}}^2 R = (4)^2(2) = 32 \text{ W}$$

■

EXAMPLE 11.11

We wish to find the rms value of the current:

$$i(t) = 12 \sin 377t + 6 \sin (754t + 30°) \text{ A}$$

Since the frequencies of the two sinusoidal waves are different,

$$I_{\text{rms}}^2 = I_{1\text{rms}}^2 + I_{2\text{rms}}^2$$

$$= \left(\frac{12}{\sqrt{2}}\right)^2 + \left(\frac{6}{\sqrt{2}}\right)^2$$

Therefore,

$$I_{\text{rms}} = 9.49 \text{ A}$$

■

DRILL EXERCISE

D11.7. Compute the rms value of the voltage waveform shown in Fig. D11.7.

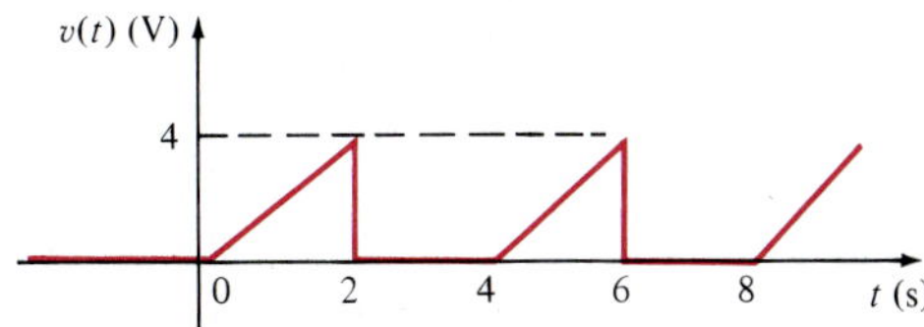

Figure D11.7

Ans: $V_{\text{rms}} = 1.633$ V.

D11.8. The current waveform in Fig. D11.8 is flowing through a 4-Ω resistor. Compute the average power delivered to the resistor.

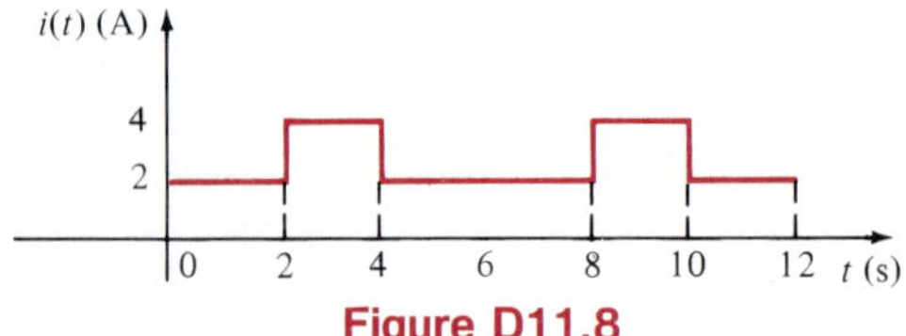

Figure D11.8

Ans: $P = 32$ W.

D11.9. The current waveform in Fig. D11.9 is flowing through a 10-Ω resistor. Determine the average power delivered to the resistor.

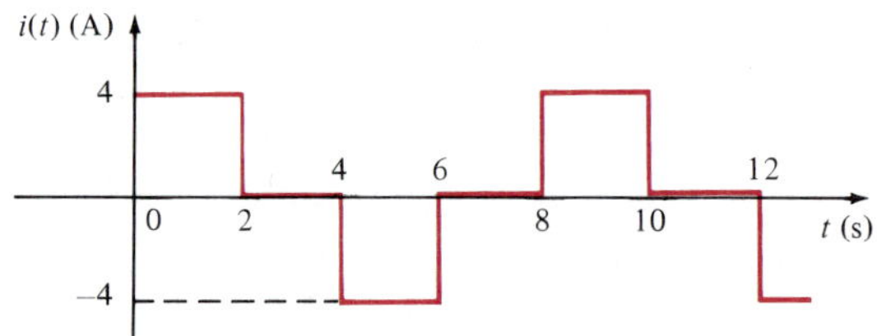

Figure D11.9

Ans: $P = 80$ W.

D11.10. The current $i(t)$ is flowing in a 4-Ω resistor. Determine the power absorbed by the resistor if

$$i(t) = 4 \sin (377t - 30°) + 6 \cos (754t - 45°) \text{ A}$$

Ans: $P = 104$ W.

11.5 The Power Factor

The power factor is a very important quantity. Its importance stems in part from the economic impact it has on industrial users of large amounts of power. In this section we carefully define this term and then illustrate its significance via some practical examples.

In Section 11.4 we showed that a load operating in the ac steady state is delivered an average power of

$$P = V_{rms} I_{rms} \cos (\theta_v - \theta_i)$$

We will now further define the terms in this important equation. The product $V_{rms}I_{rms}$ is referred to as the *apparent power*. Although the term $\cos (\theta_v - \theta_i)$ is a dimensionless quantity, and the units of P are watts, apparent power is normally stated in volt-amperes (VA) or kilovolt-amperes (kVA) in order to distinguish it from average power.

We now define the *power factor* (pf) as the ratio of the average power to the apparent power; that is,

$$\text{pf} = \frac{P}{V_{\text{rms}} I_{\text{rms}}} = \cos(\theta_v - \theta_i) \tag{11.29}$$

where

$$\cos(\theta_v - \theta_i) = \cos\theta_{\mathbf{Z}_L} \tag{11.30}$$

The angle $\theta_v - \theta_i = \theta_{\mathbf{Z}_L}$ is the phase angle of the load impedance and is often referred to as the *power factor angle*. The two extreme positions for this angle correspond to a purely resistive load where $\theta_{\mathbf{Z}_L}=0$, the pf is 1, and the purely reactive load where $\theta_{\mathbf{Z}_L}=\pm 90°$ and the pf is 0. It is, of course, possible to have a unity pf for a load containing R, L, and C elements if the values of the circuit elements are such that a zero phase angle is obtained at the particular operating frequency.

There is, of course, a whole range of power factor angles between $\pm 90°$ and 0°. If the load is an equivalent RC combination, then the pf angle lies between the limits $-90° < \theta_{\mathbf{Z}_L} < 0°$. On the other hand, if the load is an equivalent RL combination, then the pf angle lies between the limits $0 < \theta_{\mathbf{Z}_L} < 90°$. Obviously, confusion in identifying the type of load could result, due to the fact that $\cos\theta_{\mathbf{Z}_L} = \cos(-\theta_{\mathbf{Z}_L})$. To circumvent this problem, the pf is said to be either *leading* or *lagging*, where these two terms *refer to the phase of the current with respect to the voltage*. Since the current leads the voltage in an RC load, the load has a leading pf. In a similar manner, an RL load has a lagging pf; therefore, load impedances of $\mathbf{Z}_L = 1 - j1$ and $\mathbf{Z}_L = 2 + j1$ have power factors of $\cos(-45°) = 0.707$ leading and $\cos(26.59°) = 0.894$ lagging, respectively.

EXAMPLE 11.12

An industrial load consumes 88 kW at a pf of 0.707 lagging from a 480-V rms line. The transmission line resistance from the power company's transformer to the plant is 0.08 Ω. Let us determine the power that must be supplied by the power company (a) under present conditions and (b) if the pf is somehow changed to 0.90 lagging.

(a) The equivalent circuit for these conditions is shown in Fig. 11.11. Using Eq. (11.29), we obtain the magnitude of the rms current into the plant:

$$\begin{aligned} I_{\text{rms}} &= \frac{P_L}{(\text{pf})(V_{\text{rms}})} \\ &= \frac{(88)(10^3)}{(0.707)(480)} \\ &= 259.3 \text{ A rms} \end{aligned}$$

The power that must be supplied by the power company is

$$\begin{aligned} P_S &= P_L + (0.08)I_{\text{rms}}^2 \\ &= 88{,}000 + (0.08)(259.3)^2 \\ &= 93.38 \text{ kW} \end{aligned}$$

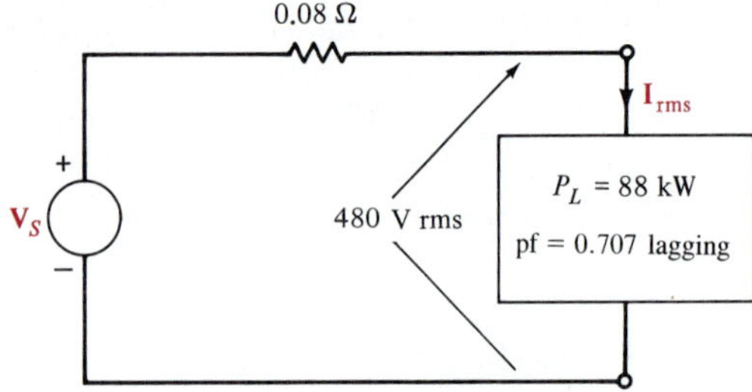

Figure 11.11 Example circuit for examining changes in power factor.

(b) Suppose now that the pf is somehow changed to 0.90 lagging but the voltage remains constant at 480 V. The rms load current for this condition is

$$
\begin{aligned}
I_{rms} &= \frac{P_L}{(\text{pf})(V_{rms})} \\
&= \frac{(88)(10^3)}{(0.90)(480)} \\
&= 203.7 \text{ A rms}
\end{aligned}
$$

Under these conditions the power company must generate

$$
\begin{aligned}
P_S &= P_L + (0.08)I_{rms}^2 \\
&= 88{,}000 + (0.08)(203.7)^2 \\
&= 91.32 \text{ kW}
\end{aligned}
$$

Note carefully the difference between the two cases. A simple change in the pf of the load from 0.707 lagging to 0.90 lagging has had an interesting effect. Note that in the first case the power company must generate 93.38 kW in order to supply the plant with 88 kW of power because the low power factor means that the line losses will be high—5.38 kW. However, in the second case the power company need only generate 91.32 kW in order to supply the plant with its required power, and the corresponding line losses are only 3.32 kW. ■

The example clearly indicates the economic impact of the load's power factor. A low power factor at the load means that the utility generators must be capable of carrying more current at constant voltage, and they must also supply power for higher I_{rms}^2R line losses than would be required if the load's power factor were high. Since line losses represent energy expended in heat and benefit no one, the utility will insist that a plant maintain a high pf, typically 0.90 lagging, and adjust their rate schedule to penalize plants that do not conform to this requirement. In the next section we will demonstrate a simple and economical technique for achieving this power factor correction.

DRILL EXERCISE

D11.11. An industrial load consumes 100 kW at 0.707 pf lagging. The 60-Hz line voltage at the load is $480\,\underline{/0^\circ}$ V rms. The transmission-line resistance between the power company's transformer and the load is 0.1 Ω. Determine the power savings that could be obtained if the pf is changed to 0.94 lagging.

Ans: Power saved is 3.771 kW.

11.6 Complex Power

In our study of ac steady-state power, it is convenient to introduce another quantity, which is commonly called *complex power*. To develop the relationship between this quantity and others we have presented in the preceding sections, consider the circuit shown in Fig. 11.12.

The complex power is defined to be

$$\mathbf{S} = \mathbf{V}_{\text{rms}}\mathbf{I}^*_{\text{rms}} \tag{11.31}$$

where $\mathbf{I}^*_{\text{rms}}$ refers to the complex conjugate of $\mathbf{I}_{\text{rms}}$; that is, if $\mathbf{I}_{\text{rms}} = I_{\text{rms}}\,\underline{/\theta_i} = I_R + jI_I$, then $\mathbf{I}^*_{\text{rms}} = I_{\text{rms}}\,\underline{/-\theta_i} = I_R - jI_I$. Complex power is then

$$\mathbf{S} = V_{\text{rms}}\,\underline{/\theta_v}\,I_{\text{rms}}\,\underline{/-\theta_i} = V_{\text{rms}}I_{\text{rms}}\,\underline{/\theta_v - \theta_i} \tag{11.32}$$

or

$$\mathbf{S} = V_{\text{rms}}I_{\text{rms}}\cos(\theta_v - \theta_i) + jV_{\text{rms}}I_{\text{rms}}\sin(\theta_v - \theta_i) \tag{11.33}$$

where, of course, $\theta_v - \theta_i = \theta_{\mathbf{Z}}$. We note from Eq. (11.33) that the real part of the complex power is simply the *real* or *average power*. The imaginary part of **S** we call the *reactive* or *quadrature power*. Therefore, complex power can be expressed in the form

$$\mathbf{S} = P + jQ \tag{11.34}$$

where

$$P = \text{Re}(\mathbf{S}) = V_{\text{rms}}I_{\text{rms}}\cos(\theta_v - \theta_i) \tag{11.35}$$

$$Q = \text{Im}(\mathbf{S}) = V_{\text{rms}}I_{\text{rms}}\sin(\theta_v - \theta_i) \tag{11.36}$$

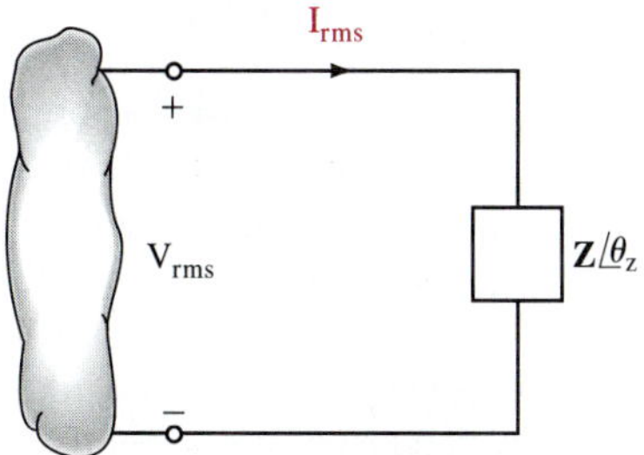

Figure 11.12 Circuit used to explain power relationships.

and as shown in Eq. (11.33), the magnitude of the complex power is what we have called the *apparent power*, and the phase angle for complex power is simply the power factor angle. Complex power like apparent power is measured in volt-amperes, real power is measured in watts, and in order to distinguish Q from the other quantities, which in fact have the same dimensions, it is measured in volt-amperes reactive, or var.

In addition to the relationships expressed above, we note that since

$$\cos(\theta_v - \theta_i) = \frac{\text{Re}(\mathbf{Z})}{|\mathbf{Z}|}$$

$$\sin(\theta_v - \theta_i) = \frac{\text{Im}(\mathbf{Z})}{|\mathbf{Z}|}$$

and

$$I_{\text{rms}} = \frac{V_{\text{rms}}}{|\mathbf{Z}|}$$

Eqs. (11.35) and (11.36) can be written as

$$P = I_{\text{rms}}^2 \,\text{Re}(\mathbf{Z}) \tag{11.37}$$

$$Q = I_{\text{rms}}^2 \,\text{Im}(\mathbf{Z}) \tag{11.38}$$

and therefore, Eq. (11.34) can be expressed as

$$\mathbf{S} = I_{\text{rms}}^2 \mathbf{Z} \tag{11.39}$$

The diagrams in Fig. 11.13 serve to explain further the relationships among the various quantities of power. As shown in Fig. 11.13a, the phasor current can be split into two components: one that is in phase with $\mathbf{V}_{\text{rms}}$ and one that is 90° out of phase with $\mathbf{V}_{\text{rms}}$. Equations (11.35) and (11.36) illustrate that the in-phase component produces the real power, and the 90° component, called the *quadrature component*, produces the reactive or quadrature power. In addition, Eqs. (11.34) to (11.36) indicate that

$$\tan(\theta_v - \theta_i) = \frac{Q}{P} \tag{11.40}$$

which relates the pf angle to P and Q in what is called the *power triangle*.

The relationships among $\mathbf{S}$, P, and Q can be expressed via the diagrams shown in Fig. 11.13b and c. In Fig. 11.13b we note the following conditions. If Q is positive, the load is inductive, the power factor is lagging, and the complex number $\mathbf{S}$ lies in the first quadrant. If Q is negative, the load is capacitive, the power factor is leading, and the complex number $\mathbf{S}$ lies in the fourth quadrant. If Q is zero, the load is resistive, the power factor is unity, and the complex number $\mathbf{S}$ lies along the positive real axis. Figure 11.13c illustrates the relationships expressed by Eqs. (11.37) to (11.39) for an inductive load.

Finally, it is important to state that complex power, like energy, is conserved; that is, the total complex power delivered to any number of individual loads is equal to the sum of the complex powers delivered to each individual load, regardless of how the loads are interconnected.

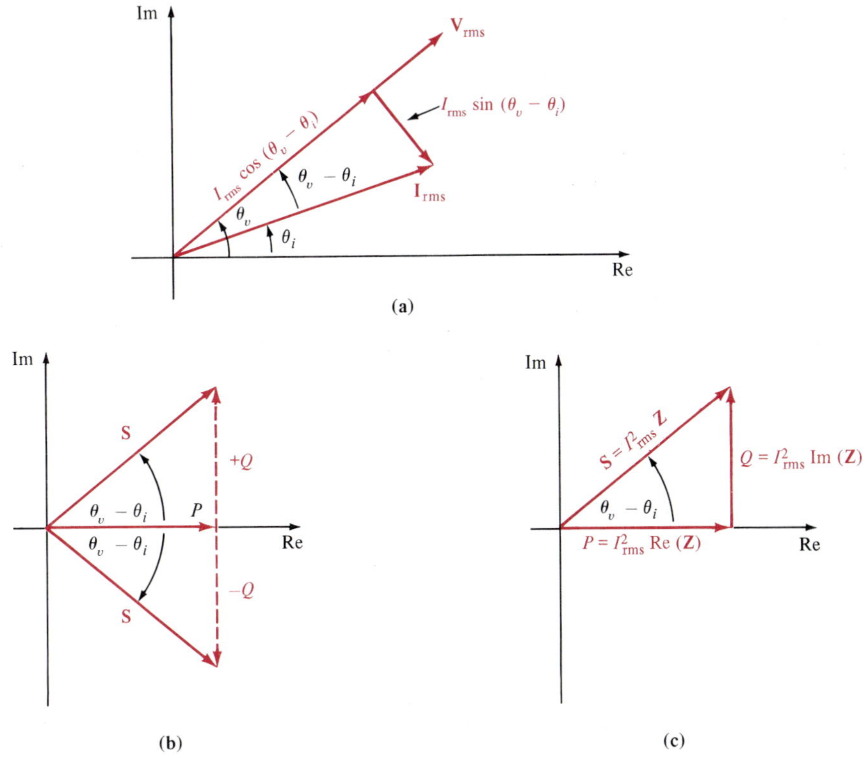

Figure 11.13 Diagram for illustrating power relationships.

EXAMPLE 11.13

A load operates at 20 kW, 0.8 pf lagging. The load voltage is $220\,\underline{/0^\circ}$ V rms at 60 Hz. The impedance of the line is $0.09 + j0.3\ \Omega$. We wish to determine the voltage and power factor at the input to the line.

The circuit diagram for this problem is shown in Fig. 11.14. The magnitude of the rms load current is

$$I_L = \frac{(20)(10^3)}{(220)(0.8)}$$

$$= 113.64 \text{ A rms}$$

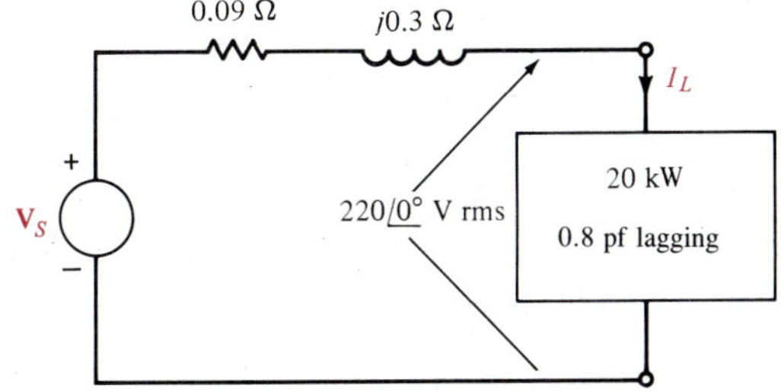

Figure 11.14 Example circuit for power analysis.

This current lags the voltage by $\theta = \cos^{-1} 0.8 = 36.87°$. The reactive power at the load is

$$\begin{aligned} Q_L &= P_L \tan \theta \\ &= (20)(10^3)(0.75) \\ &= 15{,}000 \text{ var} \end{aligned}$$

The real power loss in the line is

$$\begin{aligned} P_{\text{line}} &= (113.64)^2(0.09) \\ &= 1162.26 \text{ W} \end{aligned}$$

The reactive power loss in the line is

$$\begin{aligned} Q_{\text{line}} &= (113.64)^2(0.3) \\ &= 3874.21 \text{ var} \end{aligned}$$

The power supplied at the generator must equal that consumed in the system, and hence

$$\begin{aligned} P_S &= P_L + P_{\text{line}} \\ &= 20{,}000 + 1162.26 \\ &= 21{,}162.26 \text{ W} \end{aligned}$$

and

$$\begin{aligned} Q_S &= Q_L + Q_{\text{line}} \\ &= 15{,}000 + 3874.21 \\ &= 18{,}874.21 \text{ var} \end{aligned}$$

Therefore, the complex power at the generator is

$$\begin{aligned} \mathbf{S}_S &= P_S + jQ_S \\ &= 21{,}162.26 + j18{,}874.21 \\ &= 28{,}356.25 \underline{/41.73°} \text{ VA} \end{aligned}$$

Hence the generator voltage is

$$\begin{aligned} V_S &= \frac{|\mathbf{S}_S|}{I_L} = \frac{28{,}356.25}{113.64} \\ &= 249.53 \text{ V rms} \end{aligned}$$

and the generator power factor is

$$\cos(41.73°) = 0.75 \text{ lagging}$$

We could have solved this problem using KVL. For example, we calculated the load current as

$$\mathbf{I}_L = 113.64 \underline{/-36.87°} \text{ A rms}$$

Hence the voltage drop in the transmission line is

$$\mathbf{V}_{\text{line}} = (113.64 \underline{/-36.87^\circ})(0.09 + j0.3)$$

$$= 35.59 \underline{/36.43^\circ} \text{ V rms}$$

Therefore, the generator voltage is

$$\mathbf{V}_S = 220 \underline{/0^\circ} + 35.59 \underline{/36.43^\circ}$$

$$= 249.53 \underline{/4.86^\circ} \text{ V rms}$$

Hence the generator voltage is 249.53 V rms. In addition,

$$\theta_v - \theta_i = 4.86^\circ - (-36.87^\circ) = 41.73^\circ$$

and therefore,

$$\text{pf} = \cos(41.73^\circ) = 0.75 \text{ lagging}$$

■

EXAMPLE 11.14

Two networks A and B are connected by two conductors having a net impedance of $\mathbf{Z} = 0 + j1\Omega$ as shown in Fig. 11.15. The voltages at the terminals of the networks are $\mathbf{V}_A = 120 \underline{/30^\circ}$ V rms and $\mathbf{V}_B = 120 \underline{/0^\circ}$ V rms. We wish to determine the average power flow between the networks and identify which is the source and which is the load.

As shown in Fig. 11.15,

$$\mathbf{I} = \frac{\mathbf{V}_A - \mathbf{V}_B}{\mathbf{Z}}$$

$$= \frac{120 \underline{/30^\circ} - 120 \underline{/0^\circ}}{j1}$$

$$= 62.12 \underline{/15^\circ} \text{ A rms}$$

The power delivered by network A is

$$P_A = |\mathbf{V}_A|\,|\mathbf{I}| \cos(\theta_{\mathbf{V}_A} - \theta_{\mathbf{I}})$$

$$= (120)(62.12) \cos(30^\circ - 15^\circ)$$

$$= 7200.4 \text{ W}$$

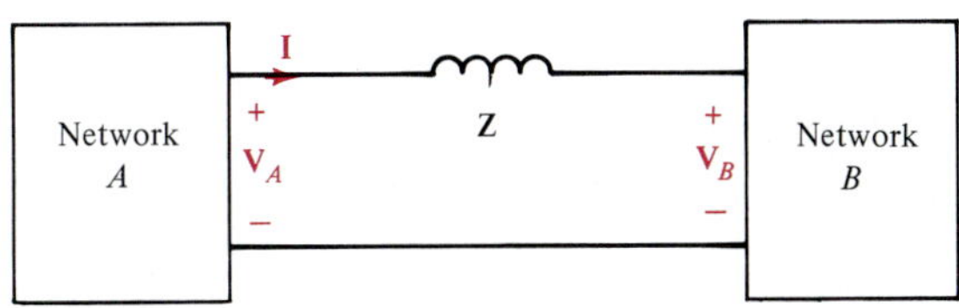

Figure 11.15 Network used in Example 11.14.

The power absorbed by network B is

$$\begin{aligned} P_B &= |\mathbf{V}_B|\,|\mathbf{I}| \cos(\theta_{\mathbf{V}_B} - \theta_{\mathbf{I}}) \\ &= (120)(62.12)\cos(0^\circ - 15^\circ) \\ &= 7200.4 \text{ W} \end{aligned}$$

If the power flow had actually been from network B to network A, the resultant signs on P_A and P_B would have been negative. ■

DRILL EXERCISE

D11.12. An industrial load requires 40 kW at 0.84 pf lagging. The load voltage is 220 $\angle 0^\circ$ V rms at 60 Hz. The transmission-line impedance is $0.1 + j0.25\ \Omega$. Determine the real and reactive power losses in the line and the real and reactive power required at the input to the transmission line.

Ans: $P_{\text{line}} = 4.685$ kW, $Q_{\text{line}} = 11.713$ kvar, $P_S = 44.685$ kW, $Q_S = 37.55$ kvar.

D11.13. A load requires 60 kW at 0.85 pf lagging. The 60-Hz line voltage at the load is 220 $\angle 0^\circ$ V rms. If the transmission-line impedance is $0.12 + j0.18\ \Omega$, determine the line voltage and power factor at the input.

Ans: $\mathbf{V}_{\text{in}} = 284.6 \angle 5.8^\circ$ V rms, $\text{pf}_{\text{in}} = 0.792$ lagging.

Industrial plants that require large amounts of power have a wide variety of loads. However, by nature the loads normally have a lagging power factor. In view of the results obtained in Example 11.12, we are naturally led to ask if there is any convenient technique for raising the power factor of a load. Since a typical load may be a bank of induction motors or other expensive machinery, the technique for raising the pf should be an economical one in order to be feasible.

To answer the question we pose, let us examine the circuit shown in Fig. 11.16. The circuit illustrates a typical industrial load. In parallel with this load we have placed a capacitor. The original complex power for the load $\mathbf{Z}_L$, which we will denote as $\mathbf{S}_{\text{old}}$, is

$$\mathbf{S}_{\text{old}} = P_{\text{old}} + jQ_{\text{old}} = |\mathbf{S}_{\text{old}}| \angle \theta_{\text{old}}$$

The new complex power that results from adding a capacitor is

$$\mathbf{S}_{\text{new}} = P_{\text{old}} + jQ_{\text{new}} = |\mathbf{S}_{\text{new}}| \angle \theta_{\text{new}}$$

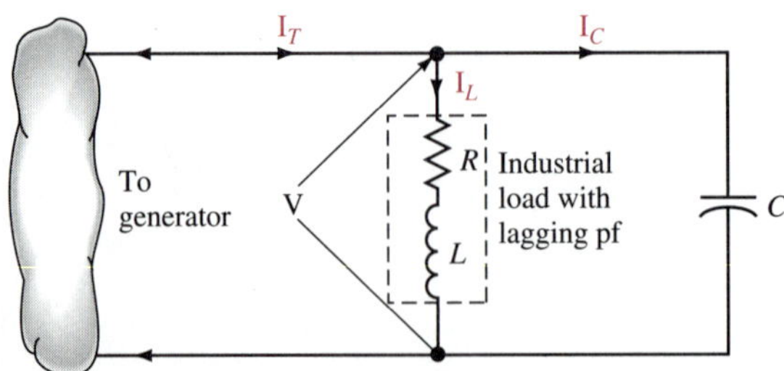

Figure 11.16 Circuit for power factor correction.

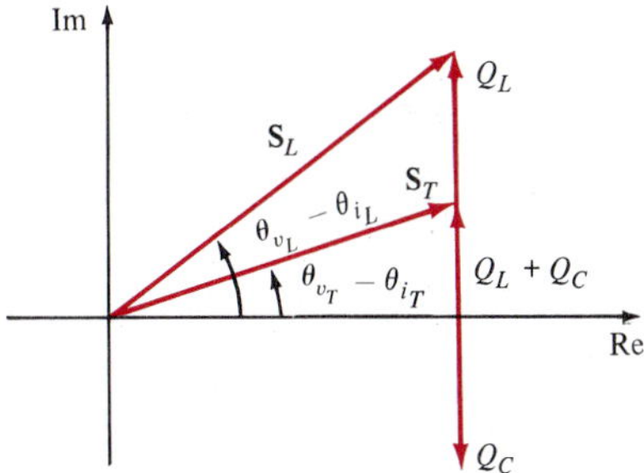

Figure 11.17 Illustration of the technique for power factor correction.

where θ_{new} is specified by the required power factor. The difference between the new and old complex powers is caused by the addition of the capacitor. Hence

$$\mathbf{S}_{\text{cap}} = \mathbf{S}_{\text{new}} - \mathbf{S}_{\text{old}} \tag{11.41}$$

and since the capacitor is purely reactive,

$$\mathbf{S}_{\text{cap}} = +jQ_C = -j\omega C V_{\text{rms}}^2 \tag{11.42}$$

Equation (11.42) can be used to find the required value of C in order to achieve the new specified power factor. The procedure we have described is illustrated in Fig. 11.17, where $\theta_{\text{old}} = \theta_{v_L} - \theta_{i_L}$ and $\theta_{\text{new}} = \theta_{v_T} - \theta_{i_T}$. Hence we can obtain a particular power factor for the total load simply by a judicious selection of a capacitor and placing it in parallel with the original load. In general, we want the power factor to be large, and therefore the power factor angle must be small (i.e., the larger the desired power factor, the smaller the angle $\theta_{v_T} - \theta_{i_T}$).

EXAMPLE 11.15

An industrial load consisting of a bank of induction motors consumes 50 kW at a pf of 0.8 lagging from a 220 $\underline{/0°}$-V rms, 60-Hz line. We wish to raise the pf to 0.95 lagging by placing a bank of capacitors in parallel with the load.

The circuit diagram for this problem is shown in Fig. 11.18. $P_L = 50$ kW and since $\cos^{-1} 0.8 = 36.87°$, $\theta_{\text{old}} = 36.87°$. Therefore,

$$Q_{\text{old}} = P_{\text{old}} \tan \theta_{\text{old}} = (50)(10^3)(0.75) = 37.5 \text{ kvar}$$

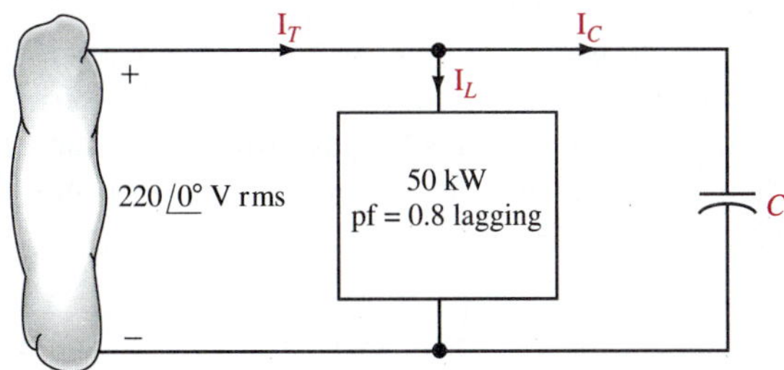

Figure 11.18 Example circuit for power factor correction.

Hence

$$\mathbf{S}_{\text{old}} = 50{,}000 + j37{,}500$$

Since the required power factor is 0.95, $\theta_{\text{new}} = 18.19°$. Then

$$\begin{aligned}\mathbf{S}_{\text{new}} &= 50{,}000 + j50{,}000 \tan 18.19° \\ &= 50{,}000 + j16{,}430\end{aligned}$$

and therefore,

$$\begin{aligned}\mathbf{S}_{\text{cap}} &= \mathbf{S}_{\text{new}} - \mathbf{S}_{\text{old}} \\ &= -j21{,}070\end{aligned}$$

Hence

$$\begin{aligned}C &= \frac{21{,}070}{(377)(220)^2} \\ &= 1155\ \mu\text{F}\end{aligned}$$

By using a capacitor of this magnitude in parallel with the industrial load, we create, from the utility's perspective, a load pf of 0.95 lagging. However, the parameters of the actual load remain unchanged. ■

DRILL EXERCISE

D11.14. Compute the value of the capacitor necessary to change the power factor in Drill Exercise D11.11 to 0.95 lagging.

Ans: $C = 773\ \mu\text{F}$.

11.7 Power Measurements

In the preceding sections we have illustrated techniques for computing power. We will now show how power is actually measured in an electrical network. An instrument used to measure average power is the wattmeter. This instrument contains a low-impedance current coil (which ideally has zero impedance) that is connected in series with the load, and a high-impedance voltage coil (which ideally has infinite impedance) that is connected across the load. If the voltage and current are periodic and the wattmeter is connected as shown in Fig. 11.19a, it will read

$$P = \frac{1}{T}\int_0^T v(t)i(t)\,dt$$

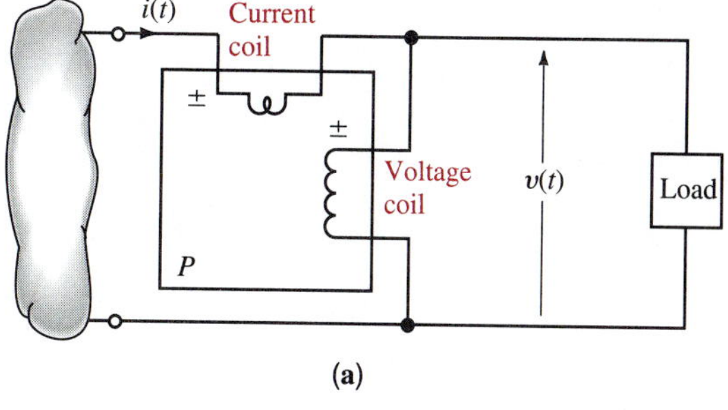

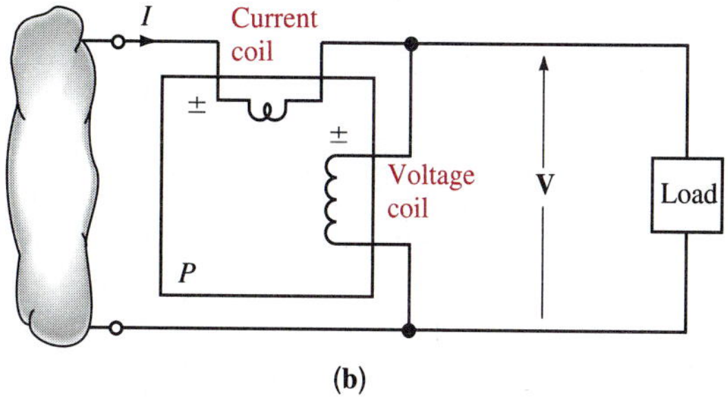

Figure 11.19 Wattmeter connections for power measurement.

where $v(t)$ and $i(t)$ are defined on the figure. Note that $i(t)$ is referenced entering the $\pm$ terminal of the current coil and $v(t)$ is referenced positive at the $\pm$ terminal of the voltage coil. In the frequency domain the equivalent circuit is shown in Fig. 11.19b, where the current and voltage are referenced the same as in the time domain and the wattmeter reading will be

$$P = \text{Re}\,(\mathbf{VI}^*) = |\mathbf{V}|\,|\mathbf{I}|\cos(\theta_v - \theta_i)$$

If $v(t)$ and $i(t)$ or $\mathbf{V}$ and $\mathbf{I}$ are correctly chosen, the reading will be average power. In Fig. 11.19 the connections will produce a reading of power delivered to the load. Since the two coils are completely isolated from one another, they could be connected anywhere in the circuit and the reading may or may not have meaning.

Note that if one of the coils on the wattmeter is reversed, the equations for the power are the negative of what they were before the coil reversal due to the change in the variable reference as related to the $\pm$ terminal. Due to the physical construction of wattmeters, the $\pm$ terminal of the potential coil should always be connected to the same line as the current coil, as shown in Fig. 11.20a. If it becomes necessary to reverse a winding to produce an upscale reading, either coil can be reversed. For example, reversing the current coil results in the network shown in Fig. 11.20b. Note that if the $\pm$ terminal of the potential coil is connected to the line containing the current coil and the meter is reading upscale, the power, P, is flowing through the wattmeter from circuit A to circuit B.

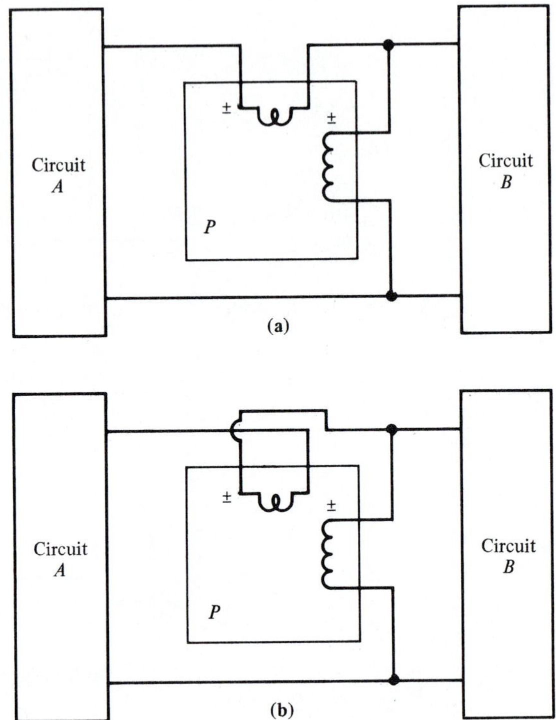

Figure 11.20 Wattmeter connections.

EXAMPLE 11.16

Given the network shown in Fig. 11.21, we wish to determine the wattmeter reading.

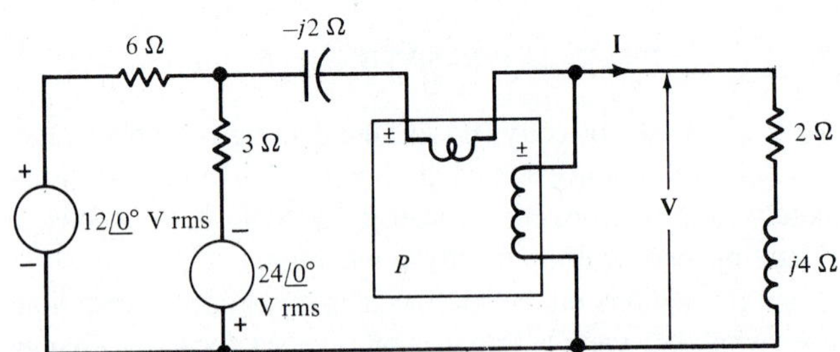

Figure 11.21 Use of the wattmeter in power measurement.

Using any one of a number of techniques (e.g., Thévenin's theorem), we can show that

$$\mathbf{I} = 2.68\ \underline{/153.43^\circ}\ \text{A rms}$$

$$\mathbf{V} = 11.99\ \underline{/216.86^\circ}\ \text{V rms}$$

and therefore the wattmeter reads

$$P = \text{Re}\,(\mathbf{VI}^*) = |\mathbf{V}||\mathbf{I}| \cos (\theta_v - \theta_i)$$
$$= (11.99)(2.68) \cos (216.86° - 153.43°)$$
$$= 14.37 \text{ W}$$

the average power delivered to the impedance $(2 + j4)\ \Omega$. ■

DRILL EXERCISE

D11.15. Determine the wattmeter reading in the network in Fig. D11.15.

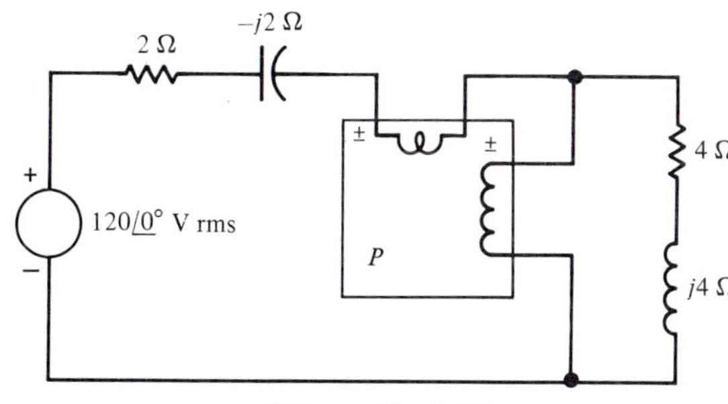

Figure D11.15

Ans: $P = 1440.24$ W.

11.8 Safety Considerations

Although this book is concerned primarily with the theory of circuit analysis, we recognize that most students, by this point in their study, will have begun to relate the theory to the electrical devices and systems that they encounter in the world around them. Thus it seems advisable to depart briefly from the theoretical and spend some time discussing the very practical and important subject of safety. Electrical safety is a very broad and diverse topic that would require several volumes for a comprehensive treatment. Instead, we will limit our discussion to a few introductory concepts and illustrate them with examples.

It would be difficult to imagine that anyone in our society could have reached adolescence without having experienced some form of electrical shock. Whether that shock was from a harmless electrostatic discharge or from accidental contact with an energized electrical circuit, the response was probably the same—an immediate and involuntary muscular reaction. In either case, the cause of the reaction is current flowing through the body. The severity of the shock depends on several factors, the most important of which are the magnitude, the duration, and the pathway of the current through the body.

The effect of electrical shock varies widely from person to person. Figure 11.22 shows the general reactions that occur as a result of 60-Hz ac current flow through the body from hand to hand, with the heart in the conduction pathway. Observe that there is an intermediate range of current, from about 0.1 to 0.2 A, which is most likely to be fatal. Current

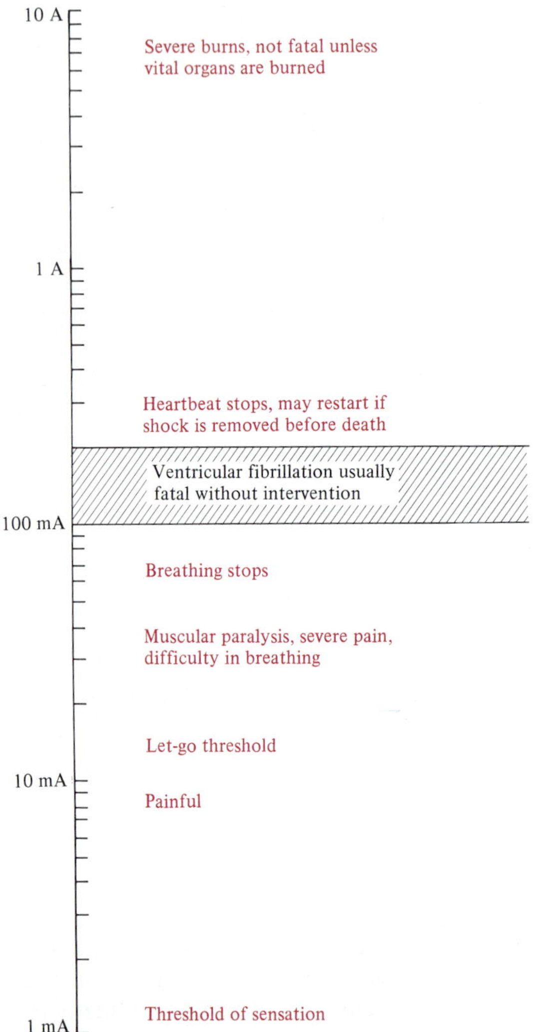

Figure 11.22 Effects of electrical shock. (From C. F. Dalziel and W. R. Lee, "Lethal Electric Currents," *IEEE Spectrum*, Feb. 1969, pp. 44–50, and C. F. Dalziel, "Electric Shock Hazard," *IEEE Spectrum*, Feb. 1972, pp. 41–50.)

levels in this range are apt to produce ventricular fibrillation, a disruption of the orderly contractions of the heart muscle. Recovery of the heartbeat generally does not occur without immediate medical intervention. Current levels above that fatal range tend to cause the heart muscle to contract severely, and if the shock is removed soon enough, the heart may resume beating on its own.

The voltage required to produce a given current depends on the quality of the contact to the body and the impedance of the body between the points of contact. The electrostatic voltage such as might be produced by sliding across a car seat on a dry winter day may be on the order of 20,000 to 40,000 V, and the current surge upon touching the door handle, on the order of 40 A. However, the pathway for the current flow is mainly over the body surface and its duration is for only a few microseconds. Although that shock could be disastrous for some electronic components, it causes nothing more than mild discomfort and aggravation to a human being.

Electrical appliances found about the home typically require 120 or 240 V rms for operation. Although the voltage level is small compared with that of the electrostatic shock, the potential for harm to the individual and to property is much greater. Accidental contact is more apt to result in current flow either from hand to hand or from hand to foot—either of which will subject the heart to shock. Moreover, the relatively slowly changing (low-frequency) 60-Hz current tends to penetrate more deeply into the body as opposed to remaining on the surface as a rapidly changing (high-frequency) current would tend to do. In addition, the energy source has the capability of sustaining a current flow without depletion. Thus subsequent discussion will concentrate primarily on hazards associated with the 60-Hz ac power system.

The single-phase three-wire system introduced in Chapter 10 is commonly, although not exclusively, used for electrical power distribution in residences. Two important aspects of this, or any, system that relate to safety were not mentioned earlier: circuit fusing and grounding.

Each branch circuit, regardless of the type loads it serves, is protected from excessive current flow by circuit breakers or fuses. Receptacle circuits are generally limited to 20 amps and lighting circuits to 15 amps. Clearly, these cannot protect persons from lethal shock. The primary purpose of these current-limiting devices is to protect equipment.

The neutral conductor of the power system is connected to ground (earth) at a multitude of points throughout the system and, in particular, at the service entrance to the residence. The connection to earth may be by way of a driven ground rod or by contact to a cold water pipe of a buried metallic water system. The 120-V branch circuits radiating from the distribution panel (fuse box) generally consist of three conductors rather than only two as was shown back in Chapter 10, Fig. 10.11. The third conductor is the ground wire, as shown in Fig. 11.23.

The ground conductor may appear to be redundant, since it plays no role in the normal operation of a load that might be connected to the receptacle. Its role is illustrated by the following example.

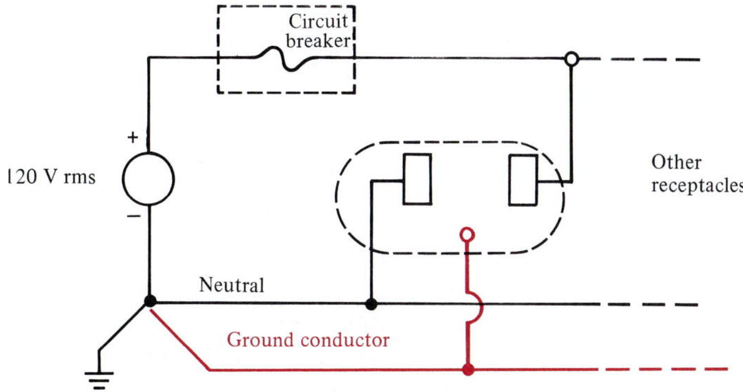

Figure 11.23 A household receptacle.

EXAMPLE 11.17

Joe College has a workshop in his basement where he uses a variety of power tools such as drills, saws, and sanders. The basement floor is concrete, and being below ground level, it is usually damp. Damp concrete is a relatively good conductor. Unknown to Joe, the insulation on a wire in his electric drill has been nicked and the wire is in contact with (or shorted to) the metal case of the drill, as shown in Fig. 11.24.

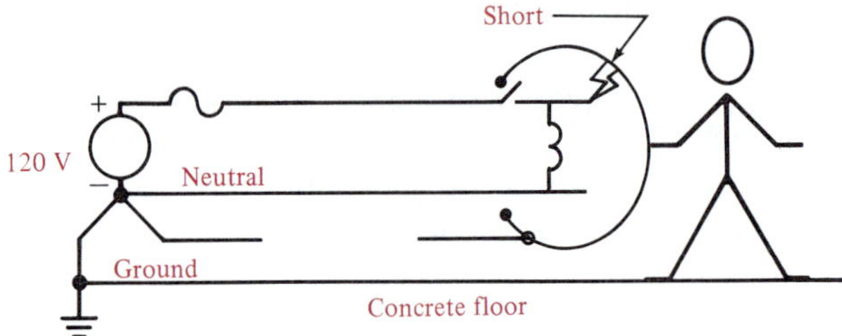

Figure 11.24 Faulty circuit.

Without the ground conductor connected to the metal case of the tool, Joe would receive a severe, perhaps fatal, shock when he attempts to use the drill. The voltage between his hand and his feet would be 120 V, and the current through his body would be limited by the resistance of his body and of the concrete floor. Typically, the circuit breakers would not operate. However, if the ground conductor is present and properly connected to the drill case, the case will remain at ground potential, the 120-V conductor becomes shorted to ground, the circuit breaker operates, and Joe lives to drill another hole. ■

It was mentioned earlier that the circuit breaker or fuse cannot provide effective protection against shock. There is, however, a special type of device called a ground-fault interrupter (GFI) which can provide protection for personnel. This device detects current flow outside the normal circuit. Consider the circuit of Fig. 11.24. In the absence of a fault condition the current in the neutral conductor must be the same as that in the line conductor. If a fault occurs, the neutral and line currents will differ by the current flowing to ground through the fault. The GFI detects that imbalance of currents in the neutral and line and opens the circuit in response. Its principle of operation is illustrated by the following example.

EXAMPLE 11.18

Consider the action of the magnetic circuit in Fig. 11.25. Under normal operating conditions, i_1 and i_2 are equal, and if the coils in the neutral and line conductors are identical, as we learned in basic physics, the magnetic flux in the case will be zero. Consequently, no voltage will be induced in the sensing coil.

If a fault should occur at the load, current will flow in the ground conductor and perhaps in the earth; thus i_1 and i_2 will no longer be equal, the magnetic flux will not be zero, and a voltage will be induced in the sensing coil. That voltage can be used to activate a circuit breaker. This is the essence of the GFI device. ■

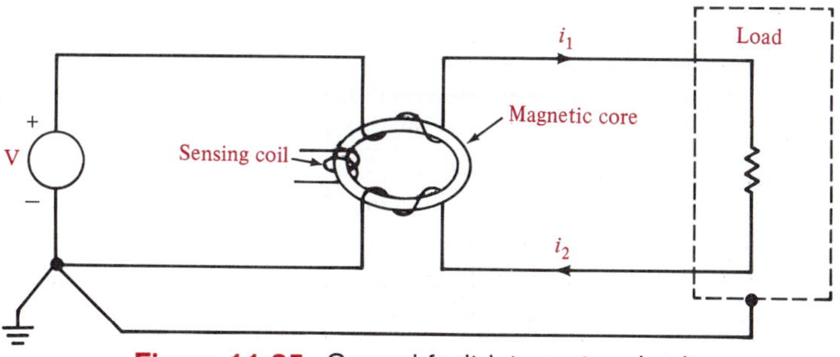

Figure 11.25 Ground-fault interrupter circuit.

Ground-fault interrupters are available in the form of circuit breakers and also as receptacles. They are now required in branch circuits that serve outlets in areas such as bathrooms, basements, garages, and outdoor sites. The devices will operate at ground-fault currents on the order of a few milliamperes. Unfortunately, the GFI is a relatively new device and electrical code requirements are generally not retroactive. Thus few older residences have them.

Requirements for the installation and maintenance of electrical systems are meticulously defined by various codes that have been established to provide protection of personnel and property. Installation, alteration, or repair of electrical devices and systems should be undertaken only by qualified persons. The subject matter that we study in circuit analysis does not provide that qualification.

The following examples illustrate the potential hazards that can be encountered in a variety of everyday situations. We begin by revisiting a situation described in a previous example.

EXAMPLE 11.19

Suppose that a man is working on the roof of a mobile home with a hand drill. It is early in the day, the man is barefoot, and dew covers the mobile home. The ground prong on the electrical plug of the drill has been removed. Will the man be shocked if the "hot" electrical line shorts to the case of the drill?

To analyze this problem, we must construct a model that adequately represents the situation described. In his book *Medical Instrumentation* (Houghton Mifflin Company, Boston, 1978), John G. Webster suggests the following values for resistance of the human body: $R_{\text{skin}}(\text{dry}) = 15\text{ k}\Omega$, $R_{\text{skin}}(\text{wet}) = 150\ \Omega$, $R_{\text{limb}}(\text{arm or leg}) = 100\ \Omega$, and $R_{\text{trunk}} = 200\ \Omega$.

The network model is shown in Fig. 11.26. Note that since the ground line is open circuited, a closed path exists from the hot wire through the short, the human body, the mobile home, and the ground. For the conditions stated above, we assume that the surface contact resistances R_{sc_1} and R_{sc_2} and 150 Ω each. The body resistance, R_{body}, consisting of arm, trunk, and leg, is 400 Ω. The mobile home resistance is assumed to be zero, and the ground resistance, R_{gnd}, from the mobile home ground to the actual source ground is assumed to be 1 Ω. Therefore, the magnitude of the current through the body from hand to

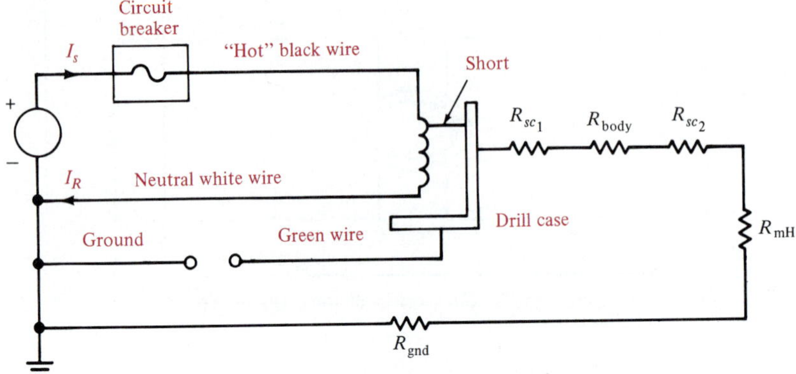

Figure 11.26 Model for Example 11.19.

foot would be

$$\mathbf{I}_{body} = \frac{120}{R_{sc_1} + R_{body} + R_{sc_2} + R_{gnd}}$$

$$= \frac{120}{701}$$

$$= 171 \text{ mA}$$

A current of this magnitude can easily cause heart failure.

It is important to note that additional protection would be provided if the circuit breaker were a ground-fault interrupter. ■

EXAMPLE 11.20

Two boys are playing basketball in their backyard. In order to cool off, they decide to jump into their pool. The pool has a vinyl lining, so the water is electrically insulated from the earth. Unknown to the boys, there is a ground fault in one of the pool lights. One boy jumps in and while standing in the pool with water up to his chest, reaches up to pull in the other boy, who is holding onto a grounded hand rail as shown in Fig. 11.27a. What is the impact of this action?

The action in Fig. 11.27a is modeled as shown in Fig. 11.27b. Note that since a ground fault has occurred, there exists a current path through the two boys. Assuming that the fault, pool, and railing resistances are approximately zero, the magnitude of the current through the two boys would be

$$\mathbf{I} = \frac{120}{(3R_{arm}) + 3(R_{wet\ contact}) + R_{trunk}}$$

$$= \frac{120}{950}$$

$$= 126 \text{ mA}$$

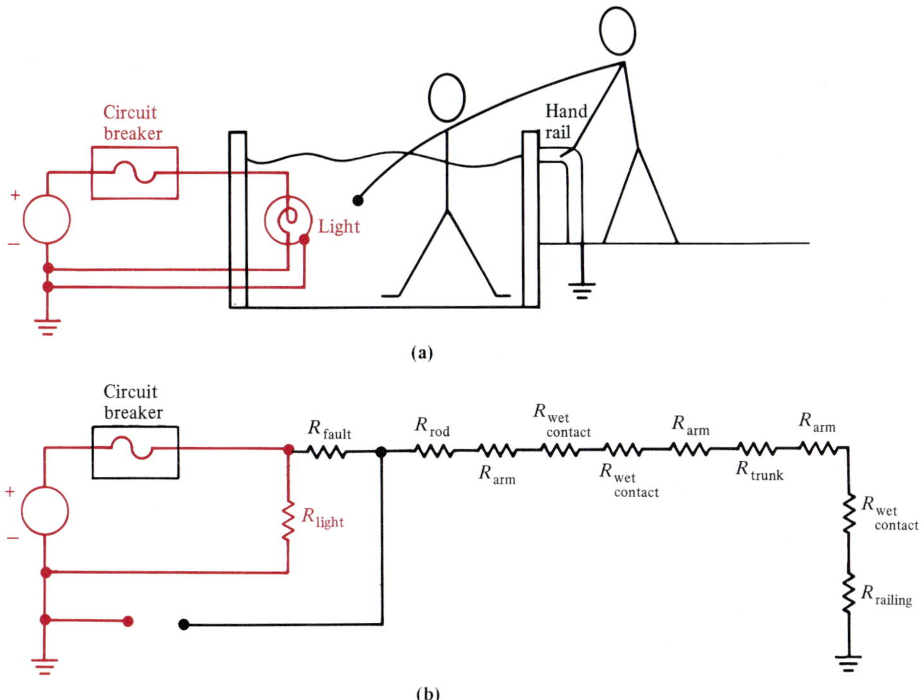

Figure 11.27 Diagrams used in Example 11.20.

This current level would cause severe shock in both boys. The boy outside the pool could experience heart failure. ■

EXAMPLE 11.21

A patient in a medical laboratory has a muscle stimulator attached to her left forearm. Her heart rate is being monitored by an EKG machine with two differential electrodes over the heart and the ground electrode attached to her right ankle. This activity is illustrated in Fig. 11.28a. The stimulator acts as a current source that drives 150 mA through the muscle from the active electrode to the passive electrode. If the laboratory technician mistakenly decides to connect the passive electrode of the stimulator to the ground electrode of the EKG system to achieve a common ground, is there any risk?

When the passive electrode of the stimulator is connected to the ground electrode of the EKG system, the equivalent network in Fig. 11.28b illustrates the two paths for the stimulator current. Using current division, the body current is

$$\mathbf{I}_{\text{body}} = \frac{(150)(10^{-3})(100)}{100 + 100 + 200 + 100}$$

$$= 30 \text{ mA}$$

Therefore, a dangerously high level of current will flow from the stimulator through the body to the EKG ground. ■

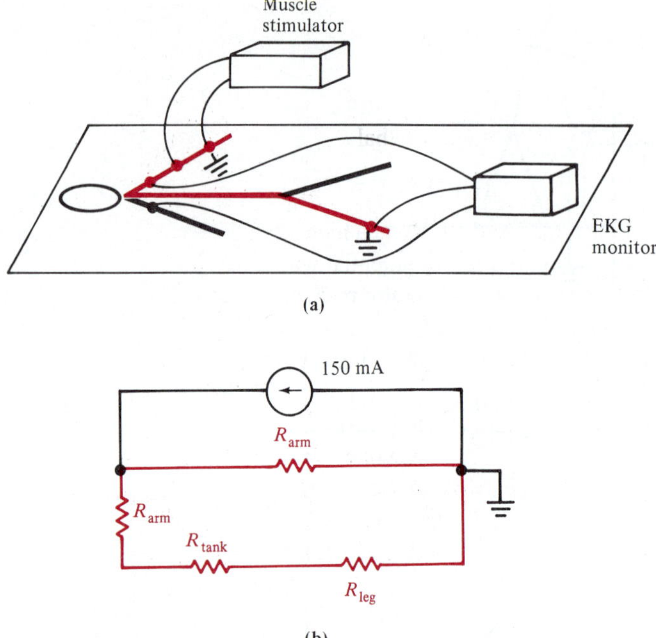

Figure 11.28 Diagrams used in Example 11.21.

EXAMPLE 11.22

A cardiac care patient with a pacing electrode has ignored the hospital rules and is listening to a cheap stereo. The stereo has an amplified 60-Hz hum that is very annoying. The patient decides to dismantle the stereo partially in an attempt to eliminate the hum. In the process, while he is holding one of the speaker wires, the other touches the pacing electrode. What are the risks in this situation?

Let us suppose that the patient's skin is damp and that the 60-Hz voltage across the speaker wires is only 10 mV. Then the circuit model in this case would be shown in Fig. 11.29. The current through the heart would be

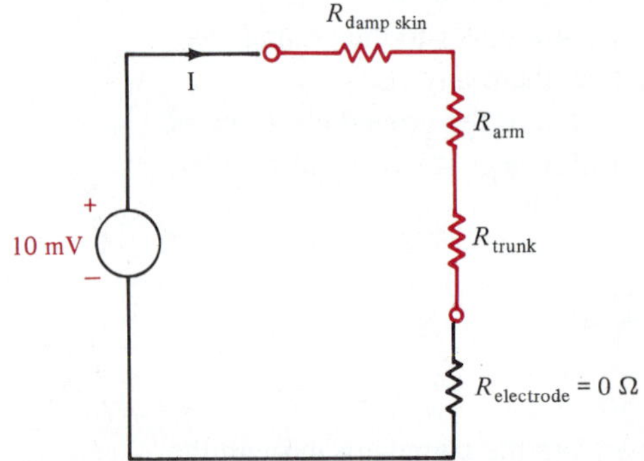

Figure 11.29 Circuit model for Example 11.22.

$$\mathbf{I} = \frac{(10)(10^{-3})}{150 + 100 + 200}$$

$$= 22.2\ \mu\text{A}$$

It is known that 10 μA delivered directly to the heart is potentially lethal. ■

EXAMPLE 11.23

While maneuvering in a muddy area, a crane operator accidentally touched a high-voltage line with the boom of the crane as illustrated in Fig. 11.30a. The line potential was 7200 V. The neutral conductor was grounded at the pole. When the crane operator realized what had happened, he jumped from the crane and ran in the direction of the pole, which was approximately 10 m away. He was electrocuted as he ran. Can we explain this very tragic accident?

The conditions depicted in Fig. 11.30a can be modeled as shown in Fig. 11.30b. The crane was at 7200 V with respect to earth. Therefore, a gradient of 720 V/m existed along the earth between the crane and the power pole. This earth between the crane and the pole is modeled as a resistance. If the man's stride was about 1 m, the difference in potential between his feet was approximately 720 V. A man standing in the same area with his feet together was unharmed. ■

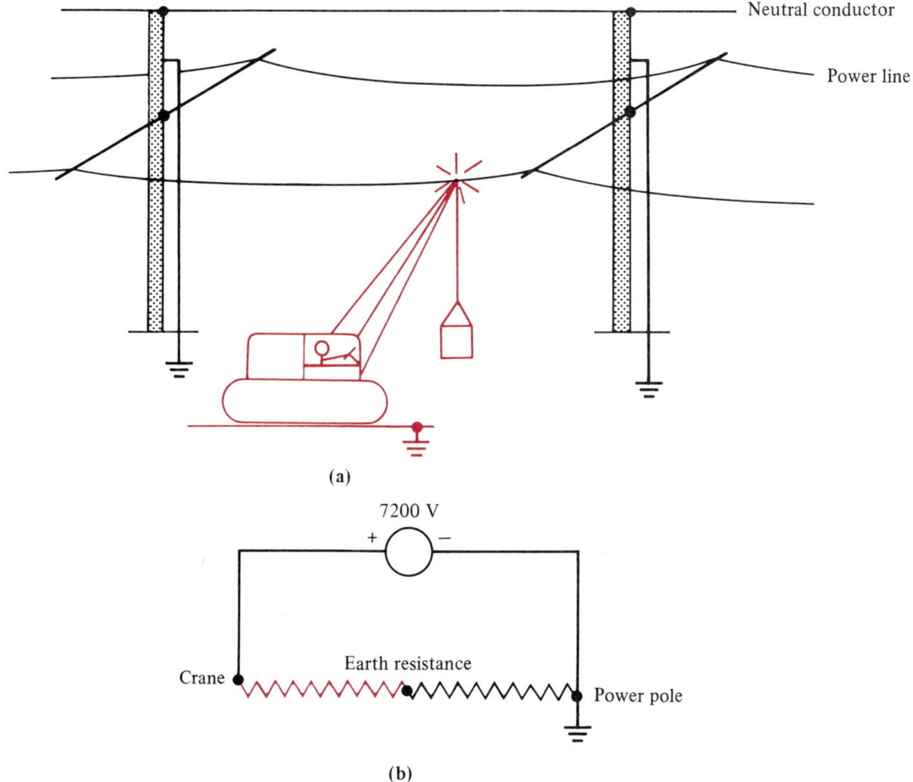

Figure 11.30 Illustrations used in Example 11.23.

The examples of this section have been provided in an attempt to illustrate some of the potential dangers that exist when working or playing around electric power. In the worst case, failure to prevent an electrical accident can result in death. However, even nonlethal electrical contacts can cause such things as burns or falls. Therefore, we must always be alert to ensure not only our own safety, but also that of others who work and play with us.

The following guidelines will help us minimize the chances of injury.

1. Avoid working on energized electrical systems.
2. Always assume that an electrical system is energized unless you can absolutely verify that it is not.
3. Never make repairs or alterations which are not in compliance with the provisions of the prevailing code.
4. Do not work on potentially hazardous electrical systems alone.
5. If another person is ''frozen'' to an energized electrical circuit, deenergize the circuit if possible. If that cannot be done, use nonconductive material such as dry wooden boards, sticks, belts, and articles of clothing to separate the body from the contact. Act quickly but take care to protect yourself.
6. When handling long metallic equipment such as ladders, antennas, and so on, outdoors, be continuously aware of overhead power lines and avoid any possibility of contact with them.

DRILL EXERCISE

D11.16. A woman is driving her car in a violent rainstorm. While she is waiting at an intersection, a power line falls across her car and makes contact. The power line voltage is 7200 V.

(a) Assuming that the resistance of the car is negligible, what is the potential current through her body if while holding the door handle with a dry hand, she steps out onto the wet ground?

(b) If she remained in the car, what would happen?

Ans: (a) $\mathbf{I} = 463$ mA, extremely dangerous; (b) she should be safe.

Safety when working with electric power must always be a primary consideration. Regardless of how efficient or expedient an electrical network is for a particular application, it is worthless if it is also hazardous to human life.

In addition to the numerous deaths that occur each year due to electrical accidents, fire damage that results from improper use of electrical wiring and distribution equipment amounts to millions of dollars per year.

To prevent the loss of life and damage to property, very detailed procedures and specifications have been established for the construction and operation of electrical systems to ensure their safe operation. The *National Electrical Code* ANSI C1 (ANSI—American National Standards Institute) is the primary guide. There are other codes, however: for example, the national electric safety code, ANSI C2, which deals with safety requirements for public utilities. The Underwriters' Laboratory (UL) tests all types of devices and systems to ensure that they are safe for use by the general public. We find the UL label on all types of electrical equipment that is used in the home, such as appliances and extension cords.

Electric energy plays a very central role in our lives. It is extremely important to our general health and well-being. However, if not properly used, it can be lethal.

11.9 Summary

The basic power relationships which apply in ac steady-state circuits have been presented. Instantaneous power and average power were defined. Techniques for maximum average power transfer, which is analogous to maximum power transfer for dc circuits, was presented for various load conditions.

The effective, or rms, value of a periodic waveform was introduced as a means of measuring the effectiveness of a source in delivering power to a resistive load.

The power factor angle was introduced, together with a scheme for correcting it if necessary. Complex power and its relationship to real and reactive power were also presented.

Techniques for making power measurements were presented, and finally, safety considerations were introduced and discussed through a variety of examples.

KEY POINTS

- If the current and voltage are sinusoidal functions of time, the instantaneous power is equal to a time-independent average value plus a sinusoidal term which has a frequency twice that of the voltage or current.
- Capacitors and inductors are lossless elements and absorb no average power.
- When multiple sources are present in a network, superposition cannot be used in computing power unless each source is of a different frequency.
- To obtain the maximum average power transfer to a load, the load impedance should be chosen equal to the conjugate of the Thévenin equivalent impedance representing the remainder of the network.
- The effective value of a periodic waveform is found by determining the root-mean-square value of the waveform.
- The rms value of a sinusoidal function is equal to the maximum value of the sinusoid divided by $\sqrt{2}$.
- Apparent power is defined as the product $V_{\text{rms}} I_{\text{rms}}$.
- The power factor is defined as the ratio of the average power to the apparent power and is said to be leading when the phase of the current leads the voltage, and lagging when the phase of the current lags the voltage.
- The power factor of a load with a lagging power factor can be corrected by placing a capacitor in parallel with the load.
- Complex power, $\mathbf{S}$, is defined as the product $\mathbf{V}_{\text{rms}} \mathbf{I}^*_{\text{rms}}$.
- The complex power $\mathbf{S}$ can be written as $\mathbf{S} = P + jQ$, where P is the real or average power and Q is the imaginary or quadrature power.

PROBLEMS

11.1. Find the average power absorbed by the network shown in Fig. P11.1.

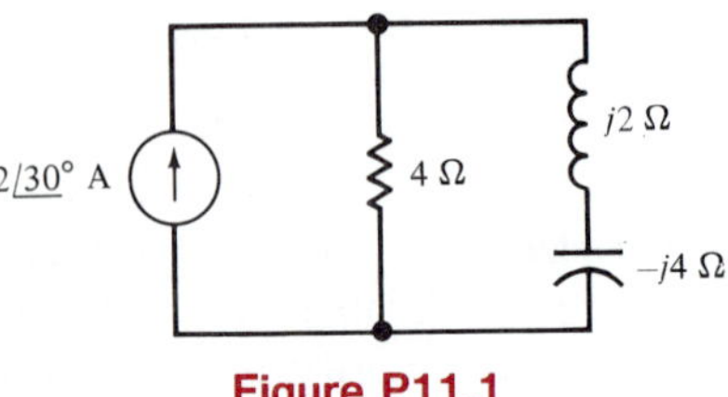

Figure P11.1

11.2. Given the network in Fig. P11.2, find the power supplied and the average power absorbed by each element.

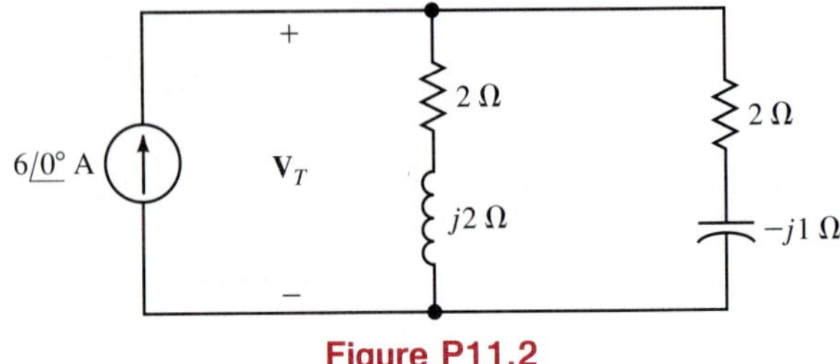

Figure P11.2

11.3. Given the circuit in Fig. P11.3, find the average power supplied and the average power absorbed by each element.

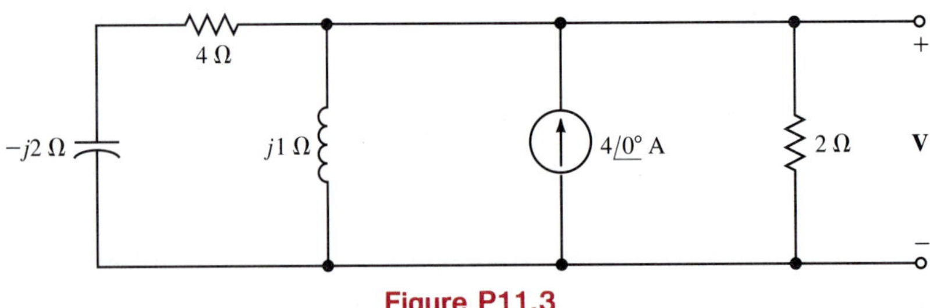

Figure P11.3

11.4. Given the network in Fig. P11.4, find the total average power supplied and the average power absorbed in the 4-Ω resistor.

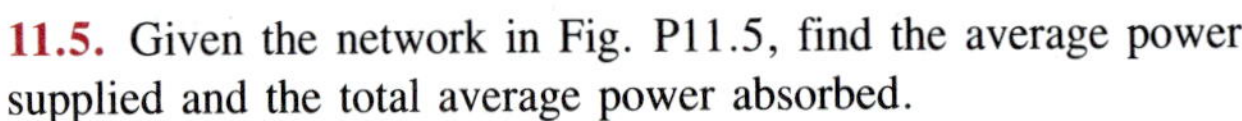

11.5. Given the network in Fig. P11.5, find the average power supplied and the total average power absorbed.

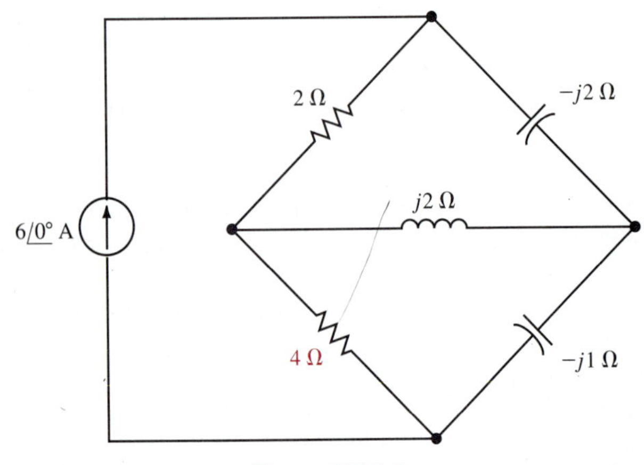

Figure P11.4

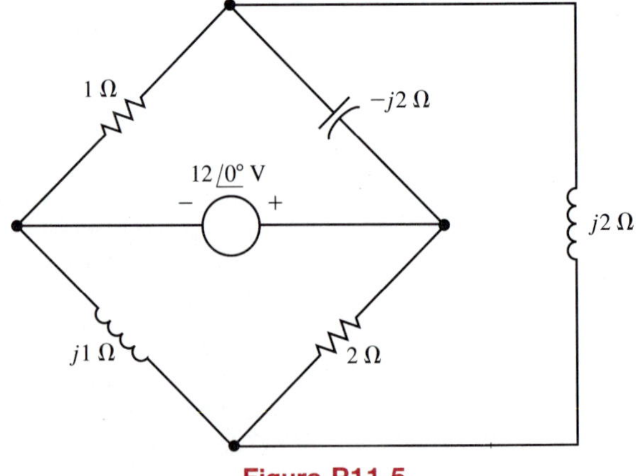

Figure P11.5

11.6. Given the network in Fig. P11.6, determine which elements are supplying power, which ones are absorbing power, and how much power is being supplied and absorbed?

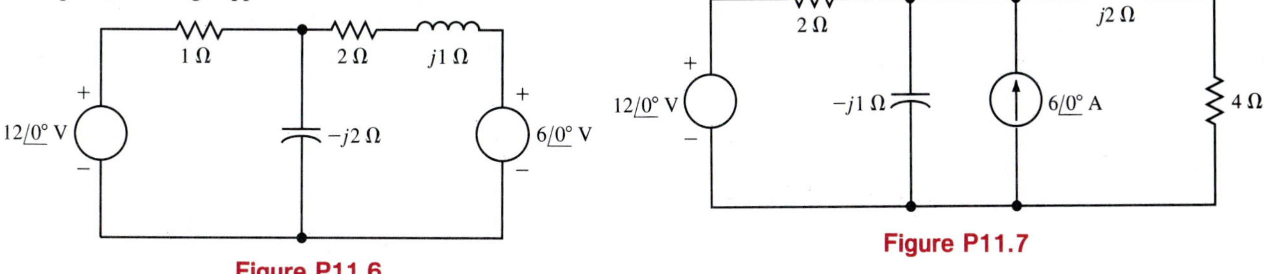

Figure P11.6

11.7. Given the circuit in Fig. P11.7, find the average power absorbed by the network.

Figure P11.7

11.8. Given the network in Fig. P11.8, find the average power supplied to the circuit.

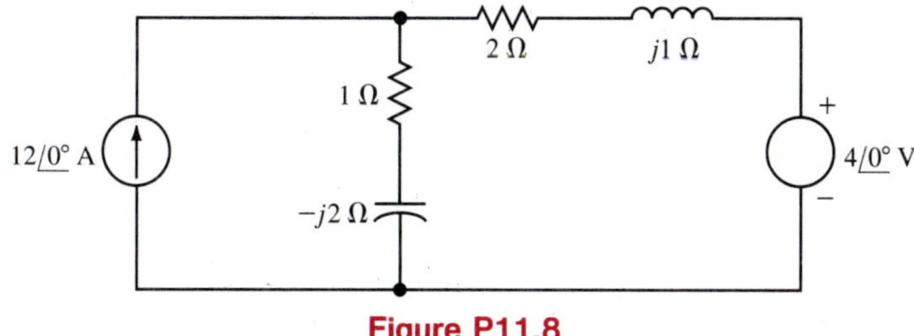

Figure P11.8

11.9. Given the circuit in Fig. P11.9, determine the amount of average power supplied to the network.

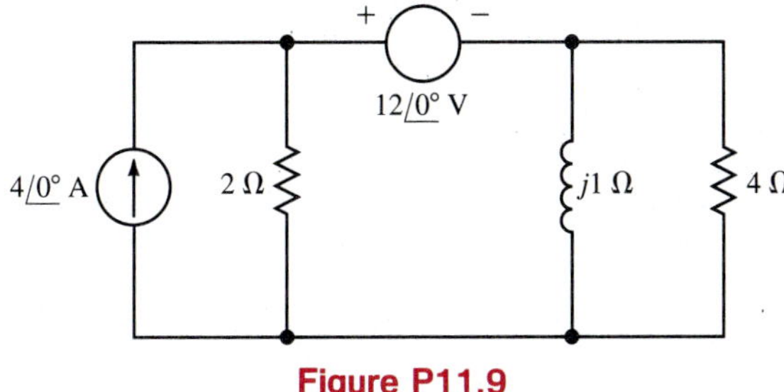

Figure P11.9

11.10. Determine the average power absorbed by the 4-Ω resistor in the network shown in Fig. P11.10.

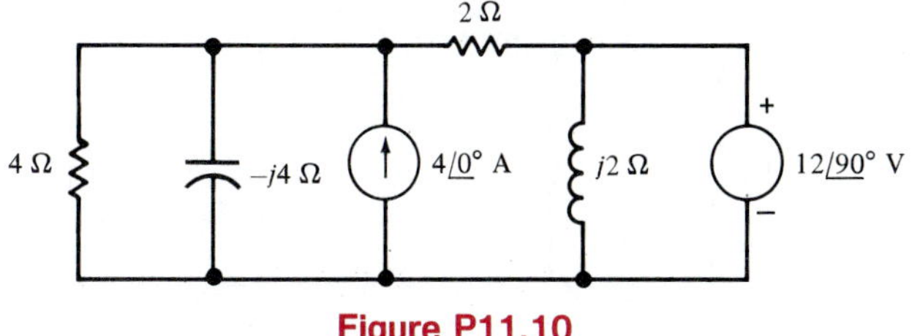

Figure P11.10

11.11. Find the average power absorbed by the 2-Ω resistor in the circuit shown in Fig. P11.11.

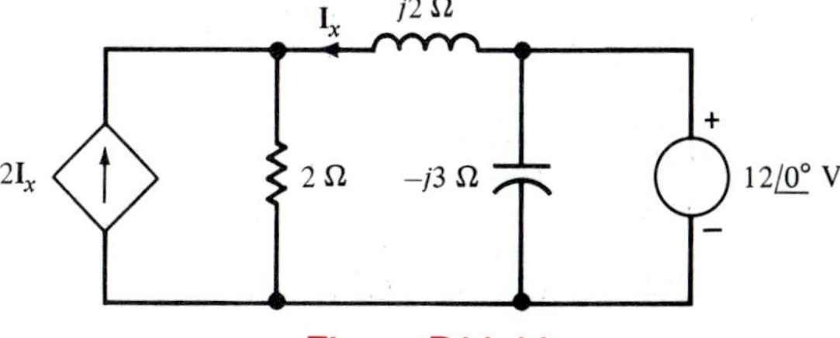

Figure P11.11

11.12. Given the network in Fig. P11.12, determine which elements are supplying power, which ones are absorbing power, and how much power is being supplied and absorbed.

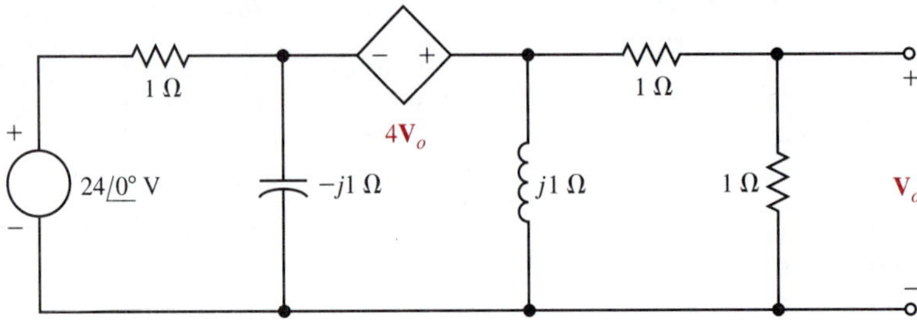

Figure P11.12

11.13. Determine the average power absorbed by the 2-kΩ output resistor in Fig. P11.13.

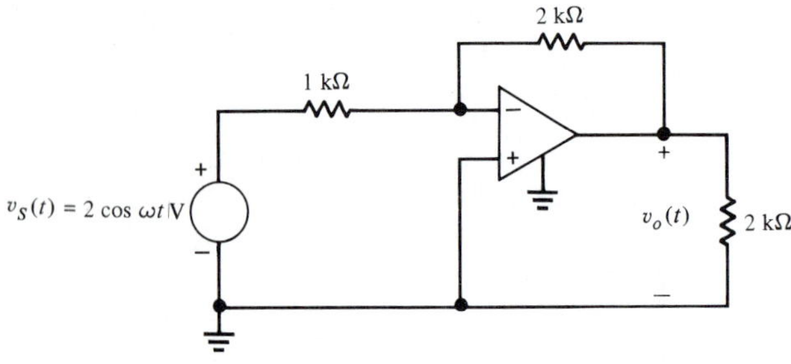

Figure P11.13

11.14. Determine the average power absorbed by the 4-kΩ resistor in Fig. P11.14.

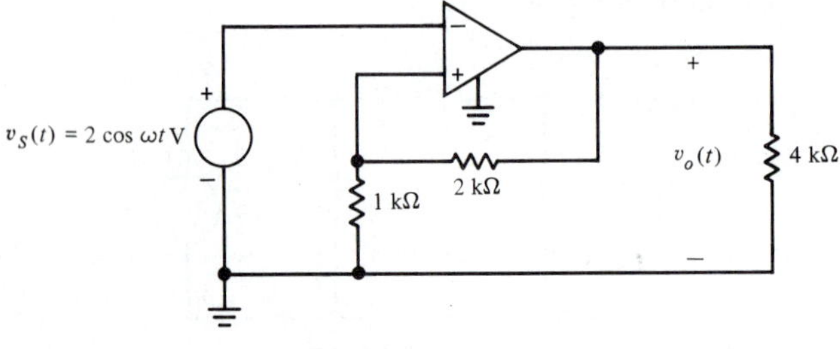

Figure P11.14

11.15. Determine the impedance $\mathbf{Z}_L$ for maximum average power transfer and the value of the maximum average power transferred to $\mathbf{Z}_L$ for the circuit shown in Fig. P11.15.

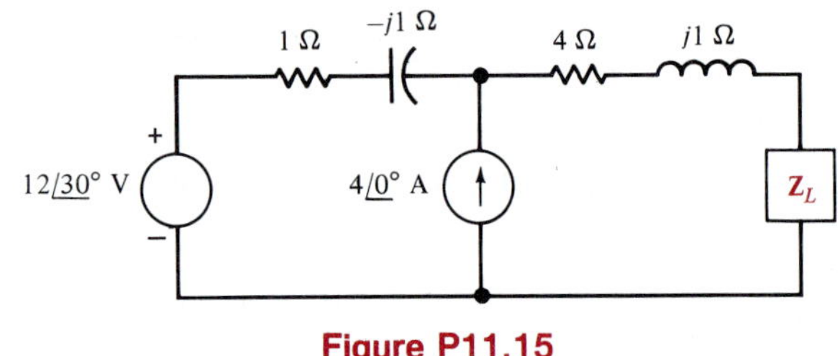

Figure P11.15

11.16. Determine the impedance $\mathbf{Z}_L$ for maximum average power transfer and the value of the maximum average power absorbed by the load in the network shown in Fig. P11.16.

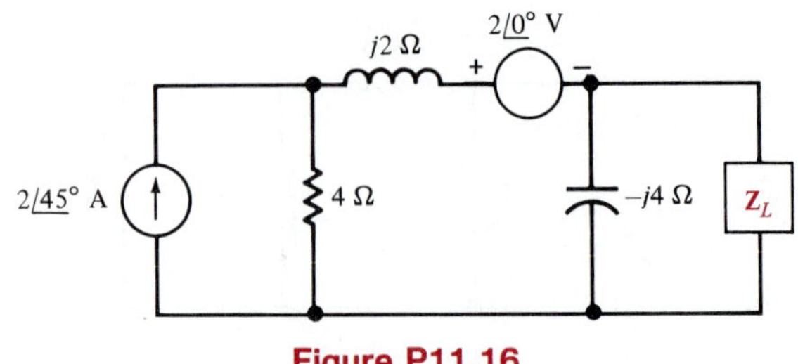

Figure P11.16

11.17. In the network in Fig. P11.17, find $\mathbf{Z}_L$ for maximum average power transfer and the maximum average power transferred.

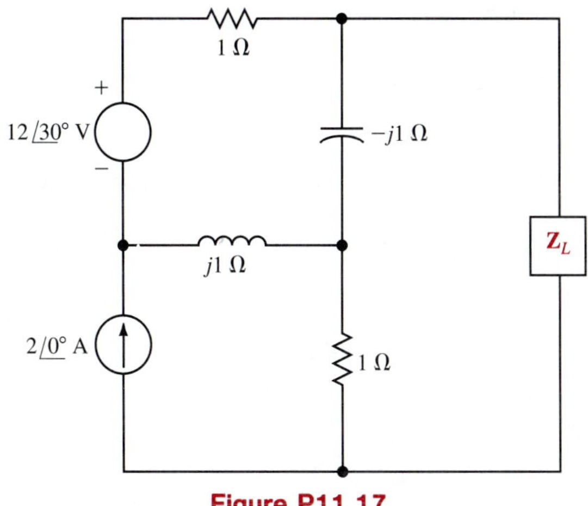

Figure P11.17

11.18. Repeat Problem 11.17 for the network in Fig. P11.18.

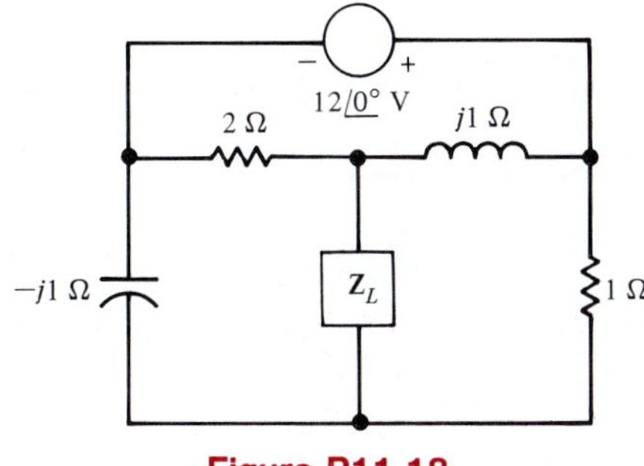

Figure P11.18

11.19. Repeat Problem 11.17 for the network in Fig. P11.19.

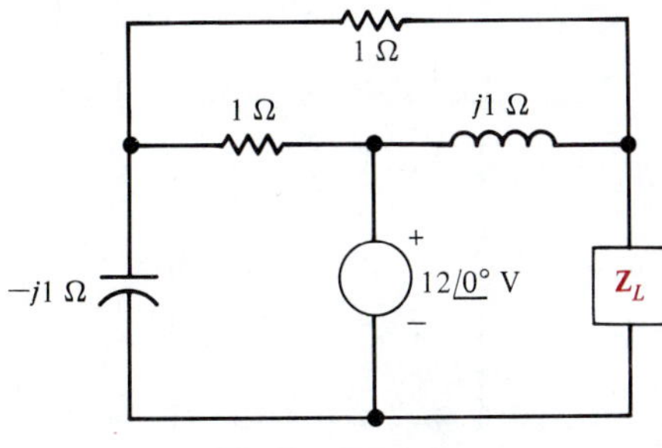

Figure P11.19

11.20. Determine the impedance $\mathbf{Z}_L$ for maximum average power transfer and the value of the maximum average power absorbed by the load in the network shown in Fig. P11.20.

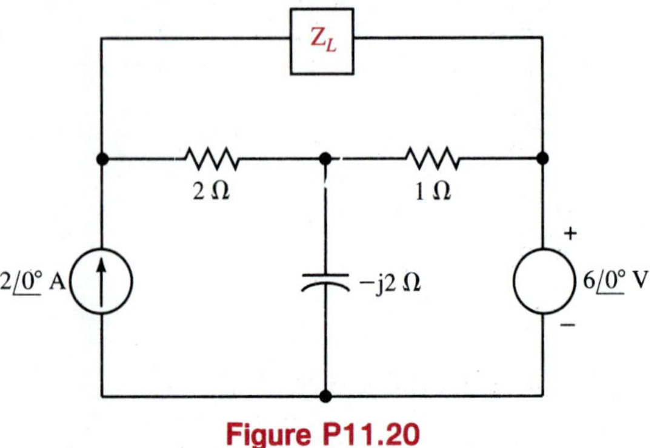

Figure P11.20

11.21. Determine the impedance $\mathbf{Z}_L$ for maximum average power transfer and the value of the maximum average power absorbed by the load in the network shown in Fig. P11.21.

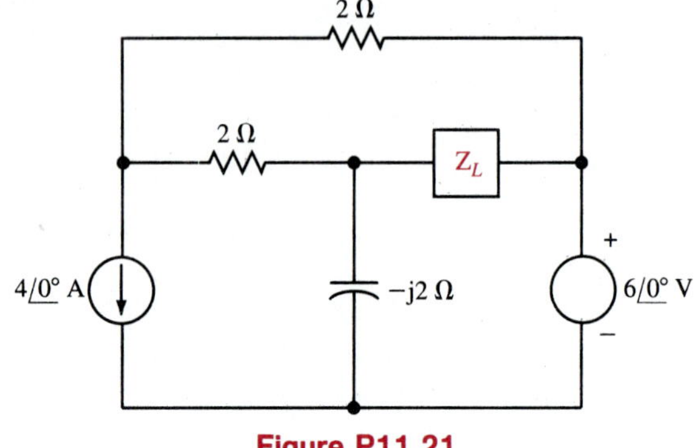

Figure P11.21

11.22. Determine the impedance $\mathbf{Z}_L$ for maximum average power transfer and the value of the maximum average power absorbed by the load in the network shown in Fig. P11.22.

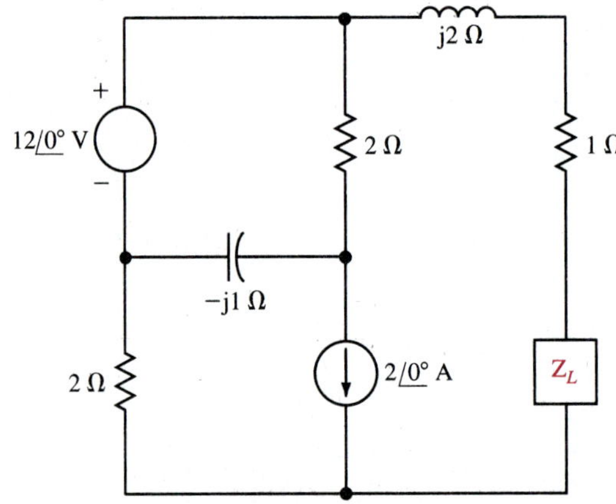

Figure P11.22

11.23. Determine the impedance $\mathbf{Z}_L$ for maximum average power transfer and the value of the maximum average power absorbed by the load in the network shown in Fig. P11.23.

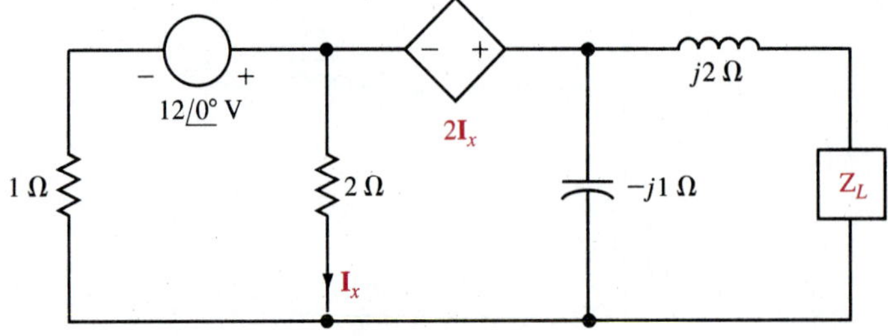

Figure P11.23

11.24. Determine the impedance $\mathbf{Z}_L$ for maximum average power transfer and the value of the maximum average power absorbed by the load in the network shown in Fig. P11.24.

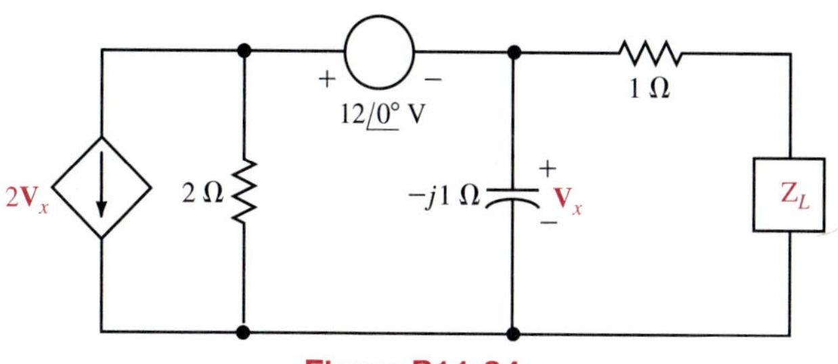

Figure P11.24

11.25. Given the network in Fig. P11.25, find $\mathbf{Z}_L$ for maximum average power transfer and the maximum average power transferred.

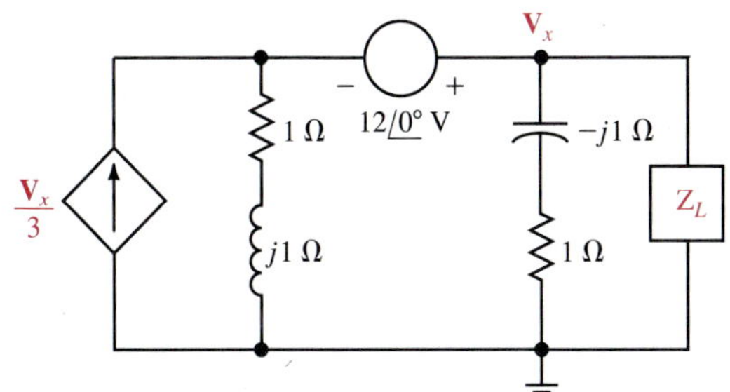

Figure P11.25

11.26. Find the impedance $\mathbf{Z}_L$ for maximum average power transfer and the value of the maximum average power transferred to $\mathbf{Z}_L$ for the circuit shown in Fig. P11.26.

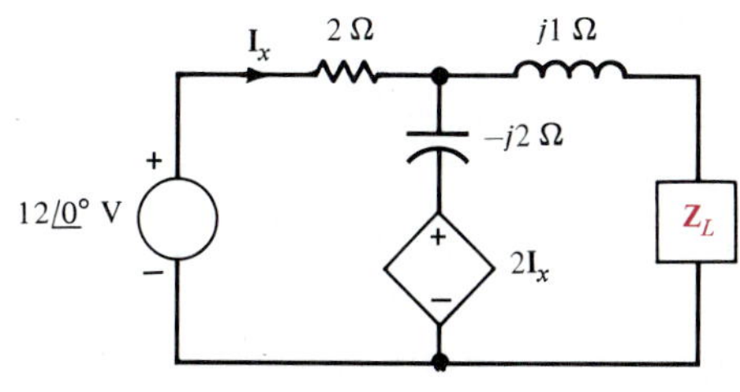

Figure P11.26

11.27. Repeat Problem 11.26 for the network in Fig. P11.27.

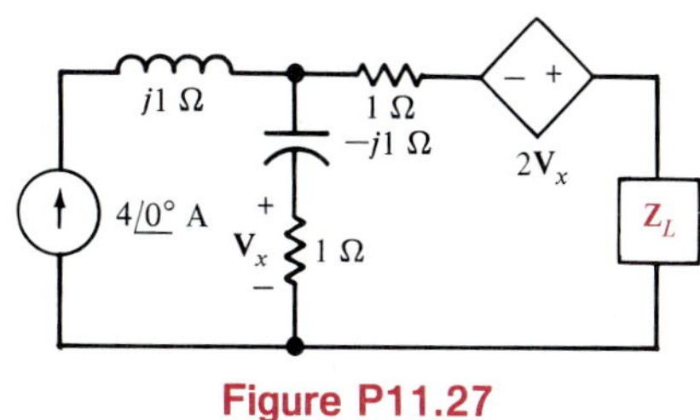

Figure P11.27

11.28. Repeat Problem 11.26 for the network in Fig. P11.28.

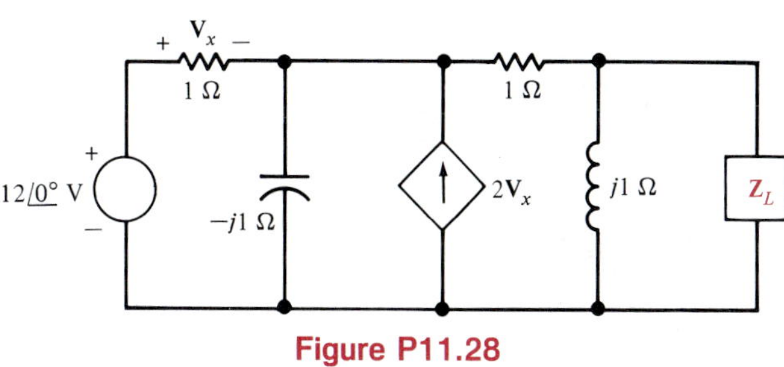

Figure P11.28

11.29. Find the rms value of the waveform shown in Fig. P11.29.

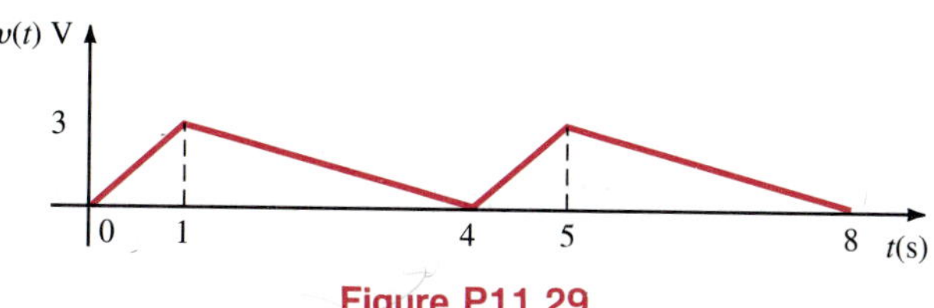

Figure P11.29

11.30. Calculate the rms value of the waveform in Fig. P11.30.

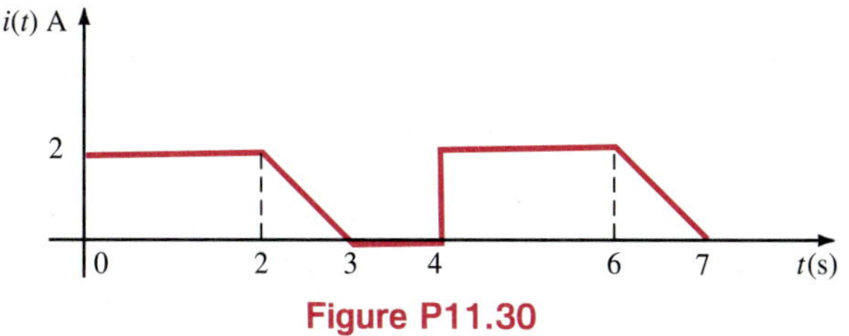

Figure P11.30

11.31. Calculate the rms value of the waveform shown in Fig. P11.31.

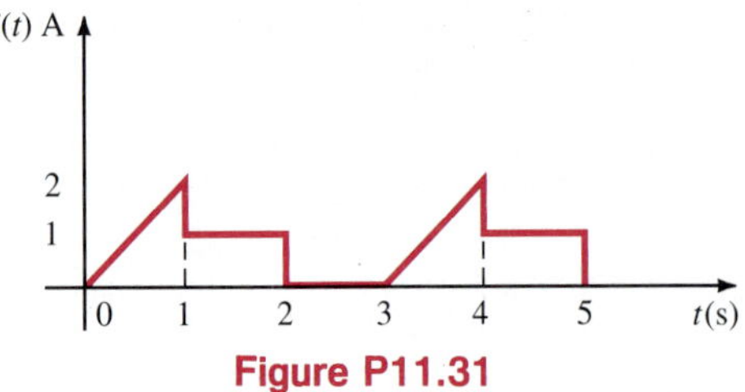

Figure P11.31

11.32. Calculate the rms value of the waveform shown in Fig. P11.32.

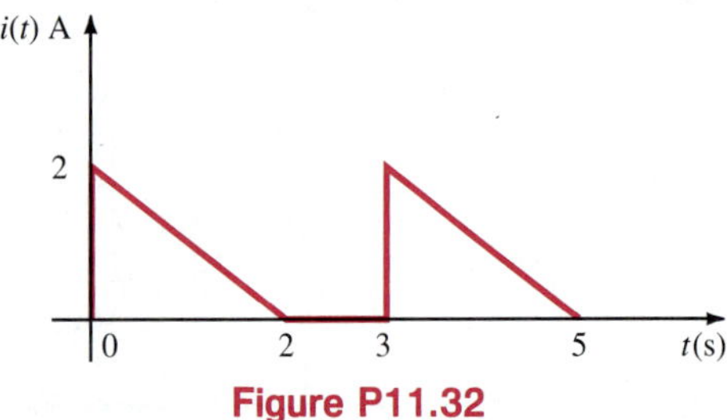

Figure P11.32

11.33. Calculate the rms value of the waveform shown in Fig. P11.33.

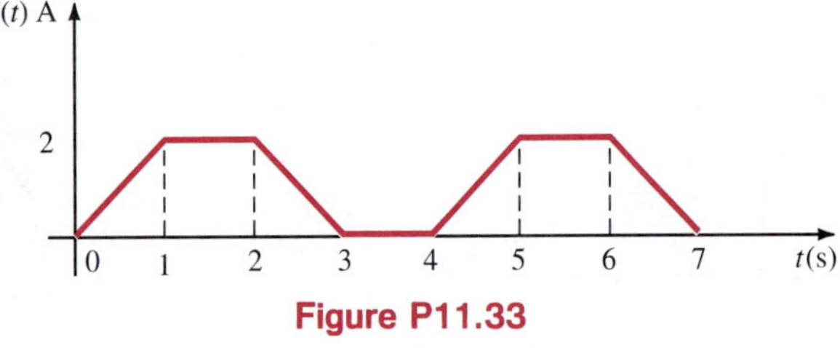

Figure P11.33

11.34. Calculate the rms value of the waveform shown in Fig. P11.34.

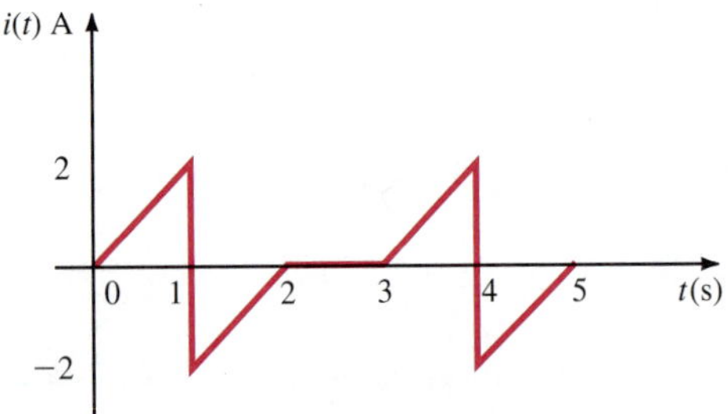

Figure P11.34

11.35. Find the rms value of the waveform shown in Fig. P11.35.

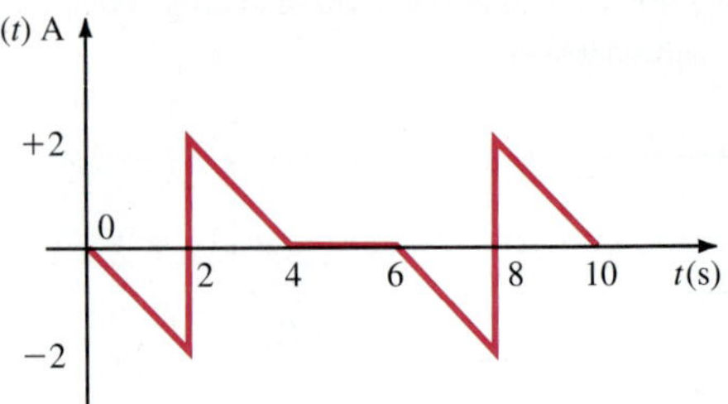

Figure P11.35

11.36. Calculate the rms value of the waveform shown in Fig. P11.36.

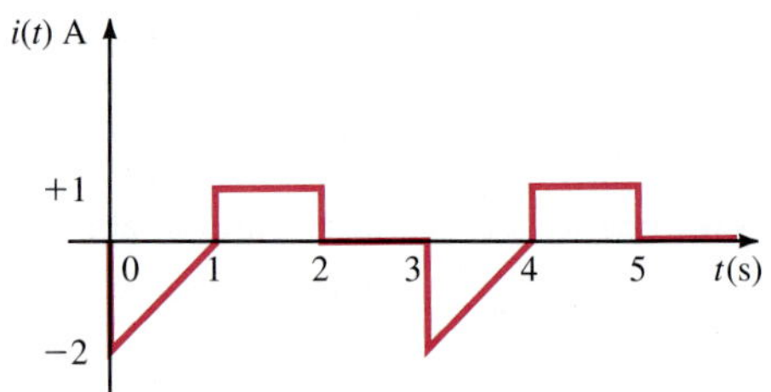

Figure P11.36

11.37. Compute the rms value of the waveform in Fig. P11.37.

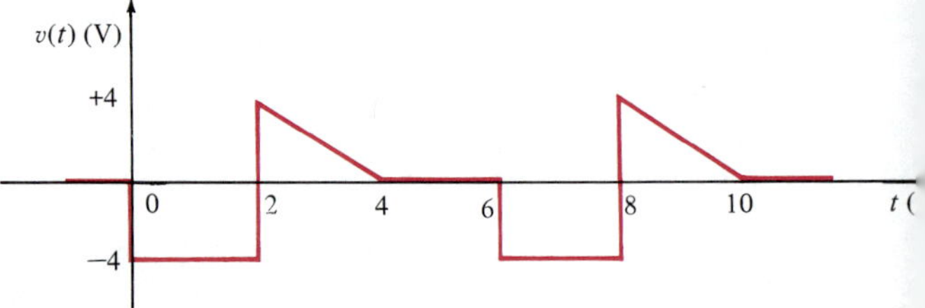

Figure P11.37

11.38. Find the rms value of the exponential waveform shown in Fig. P11.38.

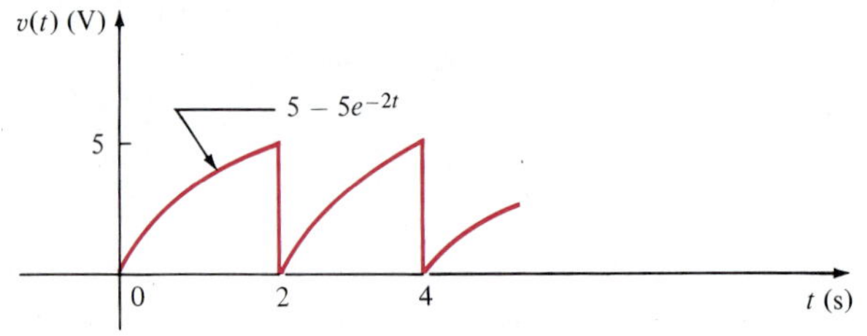

Figure P11.38

11.39. Determine the rms value of the rectified sine wave shown in Fig. P11.39. Use the fact that $\sin^2 \theta = \frac{1}{2} - \frac{1}{2} \cos 2\theta$.

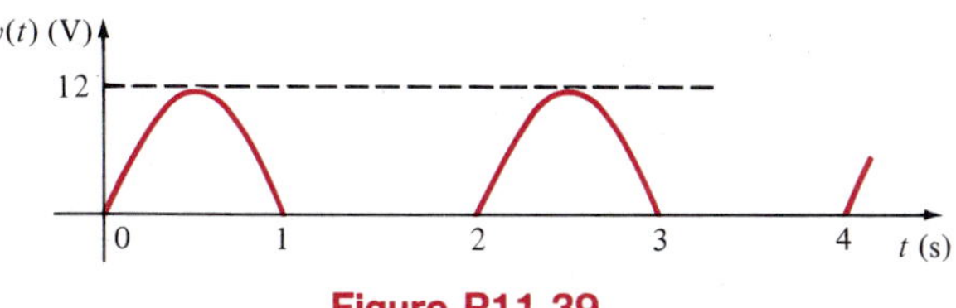

Figure P11.39

11.40. Compute the effective value of the voltage that is given by the expression

$$v(t) = 4 \sin (\omega t + 30°) + 6 \sin (2\omega t + 90°) \text{ V}$$

11.41. For the following sinusoidal voltages, write the phasor voltage with rms magnitude in polar form.
(a) $v_1(t) = 50\sqrt{2} \cos (377t + 20°)$ V.
(b) $v_2(t) = 32/\sqrt{2} \cos (377t - 30°)$ V.
(c) $v_3(t) = 20 \sin (2\pi t + 10°)$ V.

11.42. An industrial load consumes 100 kW at 0.8 pf lagging. If an ammeter in the transmission line indicates that the load current is 284 A rms, find the load voltage.

11.43. An industrial plant with an inductive load consumes 10 kW of power from a 220-V rms line. If the load power factor is 0.8, what is the angle by which the load voltage leads the load current?

11.44. An industrial load that consumes 80 kW is supplied by the power company through a transmission line with 0.1 Ω resistance, with 84 kW. If the voltage at the load is 440 V rms, find the power factor at the load.

11.45. The power company must generate 100 kW in order to supply an industrial load with 94 kW through a transmission line with 0.09 Ω resistance. If the load power factor is 0.83 lagging, find the load voltage.

11.46. The power company supplies 80 kW to an industrial load. The load draws 220 A rms from the transmission line. If the load voltage is 440 V rms and the load power factor is 0.8 lagging, find the losses in the transmission line.

11.47. An industrial load operates at 40 kW, 0.84 pf lagging, and a voltage of $220 \angle 0°$ V rms. The real and reactive power losses in the transmission line are 2 kW and 8 kvar, respectively. Find the impedance of the transmission line and the complex power at the generator.

11.48. The input voltage for an industrial load which consumes 24 kW at 0.86 pf lagging is $220 \angle 0°$ V rms. If the real and reactive power losses in the transmission-line feeders are 1.6 kW and 2.4 kvar, respectively, find the voltage and power factor at the input to the transmission line.

11.49. An industrial load operates at 30 kW, 0.8 pf lagging. The load voltage is $220 \angle 0°$ V rms. The real and reactive power losses in the transmission-line feeder are 1.8 kW and 2.4 kvar, respectively. Find the impedance of the transmission line and the input voltage to the line.

11.50. Determine the real power, reactive power, the complex power, and the power factor for a load having the following characteristics.
(a) $\mathbf{I} = 2 \angle 40°$ A rms, $\mathbf{V} = 450 \angle 70°$ V rms.
(b) $\mathbf{I} = 1.5 \angle -20°$ A rms, $\mathbf{Z} = 5000 \angle 15°$ Ω.
(c) $\mathbf{V} = 200 \angle +35°$ V rms, $\mathbf{Z} = 1500 \angle -15°$ Ω.

11.51. A transmission line with impedance $0.08 + j0.25$ Ω is used to deliver power to a load. The load is inductive and the load voltage is $220 \angle 0°$ V rms at 60 Hz. If the load requires 12 kW and the real power loss in the line is 560 W, determine the power factor angle of the load.

11.52. Determine the complex power input to the transmission line of Problem 11.51, and from this value compute the input voltage and power factor.

11.53. Given the network in Fig. P11.53, determine the input voltage $\mathbf{V}_S$.

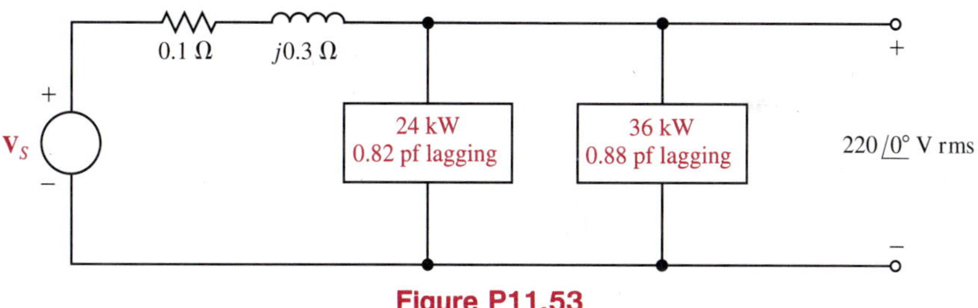

Figure P11.53

11.54. Find the source voltage in the network shown in Fig. P11.54.

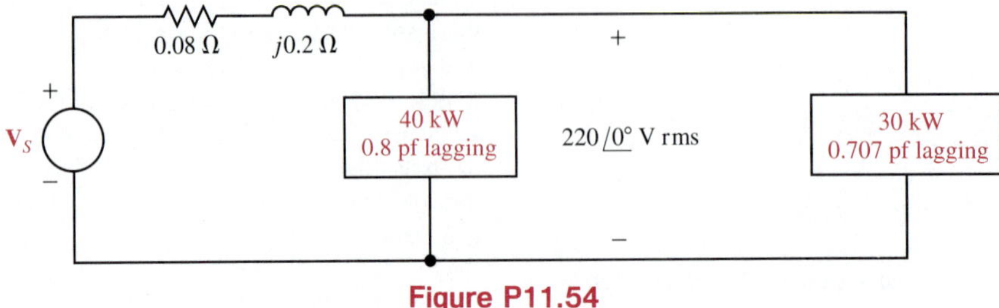

Figure P11.54

11.55. Use Kirchhoff's laws to compute the source voltage of the network shown in Fig. P11.55.

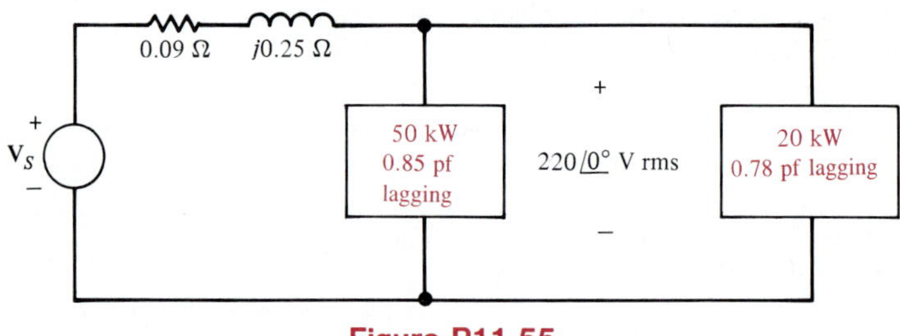

Figure P11.55

11.56. Find the input source voltage and the power factor of the source for the network shown in Fig. P11.56.

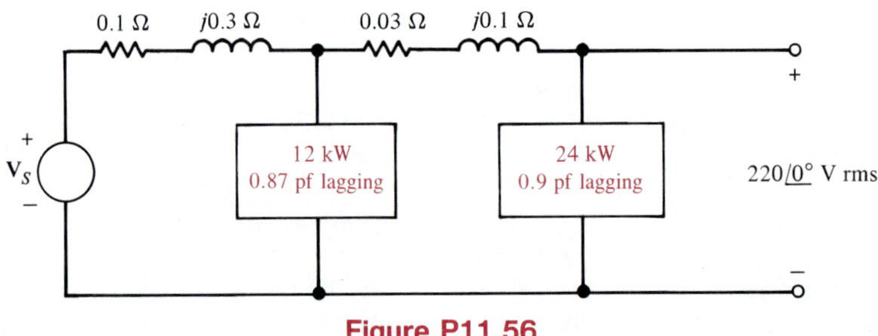

Figure P11.56

11.57. Given the network in Fig. P11.57, compute the input source voltage and the input power factor.

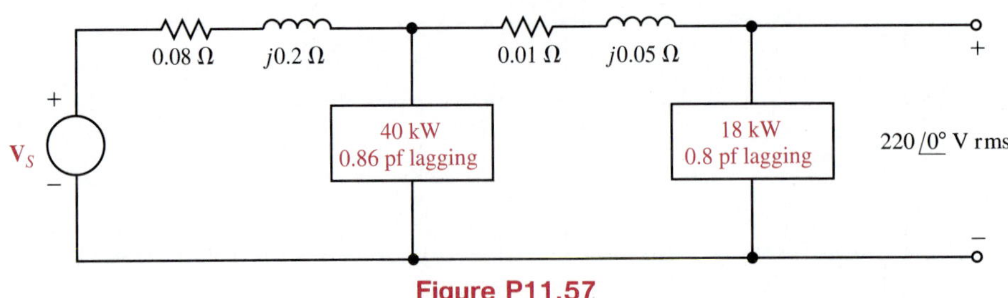

Figure P11.57

11.58. What value of capacitance must be placed in parallel with the 18-kW load in Problem 11.57 in order to raise the power factor of this load to 0.9 lagging?

11.59. An industrial load is supplied through a transmission line that has a line impedance of $0.1 + j0.2\ \Omega$. The 60-Hz line voltage at the load is $480\ \underline{/0^\circ}$ V rms. The load consumes 124 kW at 0.75 pf lagging. What value of capacitance when placed in parallel with the load will change the power factor to 0.9 lagging?

11.60. The 60-Hz line voltage for a 60-kW, 0.76-pf lagging industrial load is $440\ \underline{/0^\circ}$ V rms. Find the value of capacitance which when placed in parallel with the load will raise the power factor to 0.9 lagging.

11.61. An industrial load consumes 44 kW at 0.82 pf lagging from a $220\ \underline{/0^\circ}$-V-rms, 60-Hz line. A bank of capacitors totaling 900 μF is available. If these capacitors are placed in parallel with the load, what is the new power factor of the total load?

11.62. A bank of induction motors consumes 36 kW at 0.78 pf lagging from a 60-Hz, $220\ \underline{/0^\circ}$-V-rms line. If 500 μF of capacitors is placed in parallel with the load, find the new power factor of the combined load.

11.63. Given the network in Fig. P11.63, determine the wattmeter reading.

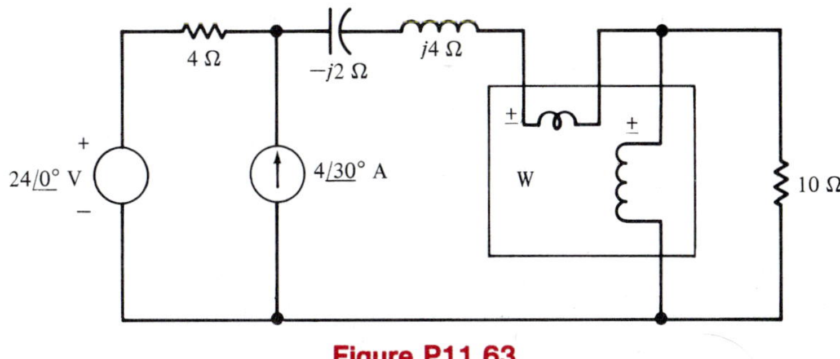

Figure P11.63

11.64. Given the network in Fig. P11.64, compute the wattmeter reading.

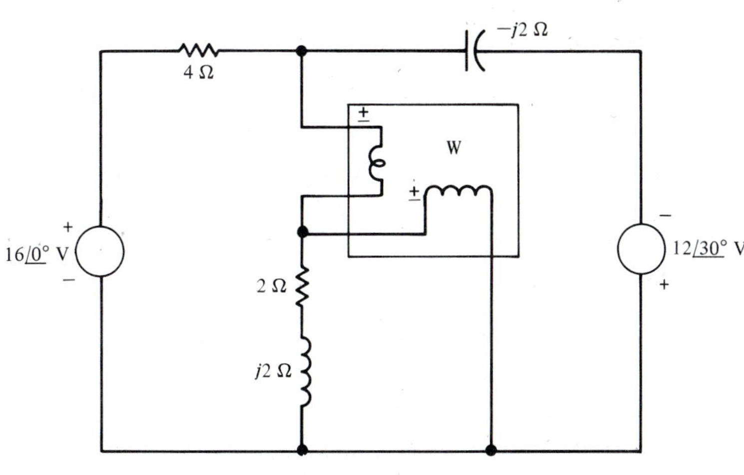

Figure P11.64

11.65. In order to test a light socket, a woman, while standing on cushions that insulate her from the ground, sticks her finger into the socket as shown in Fig. P11.65. The tip of her finger makes contact with one side of the line and the side of her finger makes contact with the other side of the line. Assuming that any portion of a limb has a resistance of 100 Ω, is there any current in the body? Is there any current in the vicinity of the heart?

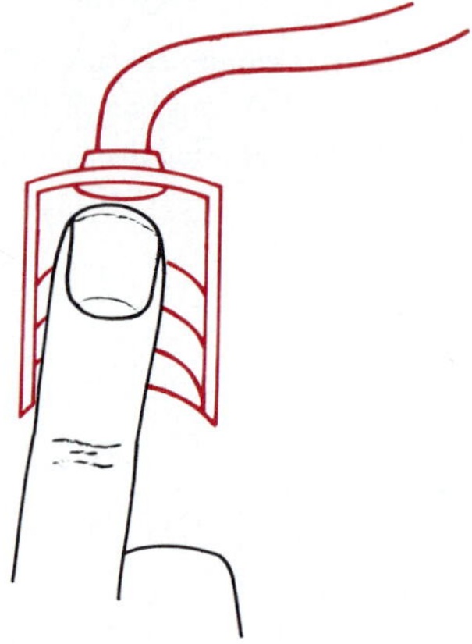

Figure P11.65

11.66. An inexperienced mechanic is installing a 12-V battery in a car. The negative terminal has been connected. He is currently tightening the bolts on the positive terminal. With a tight grip on the wrench, he turns it so that the gold ring on his finger makes contact with the frame of the car. This situation is modeled in Fig. P11.66, where we assume that the resistance of the wrench is negligible and the resistance of the contact is as follows:

$$R_1 = R_{\text{bolt to wrench}} = 0.01\ \Omega$$

$$R_2 = R_{\text{wrench to ring}} = 0.01\Omega$$

$$R_3 = R_{\text{ring to frame}} = 0.01\ \Omega$$

What power is quickly dissipated in the gold ring, and what is the impact of this power dissipation?

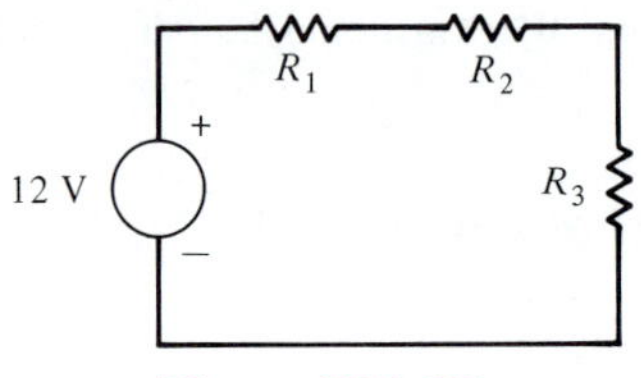

Figure P11.66

11.67. A man and his son are flying a kite. The kite becomes entangled in a 7200-V power line close to a power pole. The man crawls up the pole to remove the kite. While trying to remove the kite, the man accidentally touches the 7200-V line. Assuming the power pole is well grounded, what is the potential current through the man's body?

CHAPTER

12

Polyphase Circuits

12.1 Three-Phase Circuits

In this chapter we add a new dimension to our study of ac steady-state circuits. Up to this point we have dealt with what we refer to as single-phase circuits. Now we extend our analysis techniques to polyphase circuits or, more specifically, three-phase circuits, that is, circuits containing three voltage sources that are one-third of a cycle apart in time. There are a number of important reasons why we study three-phase circuits.

It is more advantageous and economical to generate and transmit electric power in the polyphase mode rather than with single-phase systems. As a result, most electric power is transmitted in polyphase circuits. In the United States the power system frequency is 60 Hz, whereas in other parts of the world 50 Hz is common.

Power transmission is most efficiently accomplished at very high voltage. Since this voltage can be extremely high in comparison to the level at which it is normally used (e.g., in the household), there is a need to raise and lower the voltage. This can be accomplished in ac systems using transformers, which we will study in Chapter 13.

As the name implies, three-phase circuits are those in which the forcing function is a three-phase system of voltages. If the three sinusoidal voltages have the same magnitude and frequency and each voltage is 120° out of phase with the other two, the voltages are said to be *balanced*. If the loads are such that the currents produced by the voltages are also balanced, the entire circuit is referred to as a *balanced three-phase circuit*.

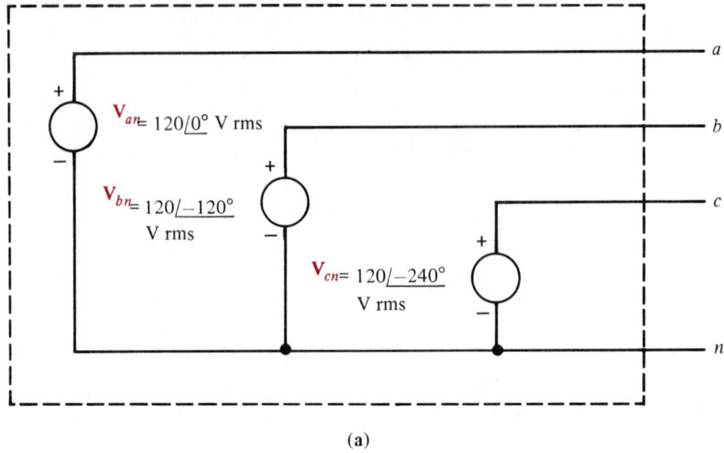

(a)

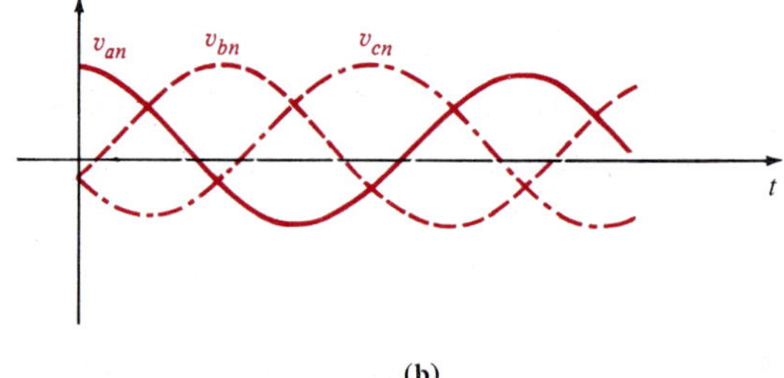

(b)

Figure 12.1 Balanced three-phase voltages.

A balanced set of three-phase voltages can be represented in the frequency domain as shown in Fig. 12.1a where we have assumed that their magnitudes are 120 V rms. From the figure we note that

$$\begin{aligned}\mathbf{V}_{an} &= 120\,\underline{/0°} \text{ V rms}\\ \mathbf{V}_{bn} &= 120\,\underline{/-120°} \text{ V rms}\\ \mathbf{V}_{cn} &= 120\,\underline{/-240°} \text{ V rms}\\ &= 120\,\underline{/120°} \text{ V rms}\end{aligned} \tag{12.1}$$

Note that our double-subscript notation is exactly the same as that employed in the earlier chapters; that is, $\mathbf{V}_{an}$ means the voltage at point a with respect to the point n. We will also employ the double-subscript notation for currents; that is, $\mathbf{I}_{an}$ is used to represent the current from a to n. However, we must be very careful in this case to describe the precise path, since in a circuit there will be more than one path between the two points. For example, in the case of a single loop the two possible currents in the two paths will be 180° out of phase with each other.

The phasor voltages above can be expressed in the time domain as

$$\begin{aligned} v_{an}(t) &= 120\sqrt{2}\cos \omega t \text{ V} \\ v_{bn}(t) &= 120\sqrt{2}\cos (\omega t - 120°) \text{ V} \\ v_{cn}(t) &= 120\sqrt{2}\cos (\omega t + 120°) \text{ V} \end{aligned} \tag{12.2}$$

These time functions are shown in Fig. 12.1b.

Finally, let us examine the instantaneous power generated by a three-phase system. Assume that the voltages in Fig. 12.1 are

$$\begin{aligned} v_{an}(t) &= V_m \cos \omega t \text{ V} \\ v_{bn}(t) &= V_m \cos (\omega t - 120°) \text{ V} \\ v_{cn}(t) &= V_m \cos (\omega t - 240°) \text{ V} \end{aligned} \tag{12.3}$$

If the load is balanced, the currents produced by the sources are

$$\begin{aligned} i_a(t) &= I_m \cos (\omega t - \theta) \text{ A} \\ i_b(t) &= I_m \cos (\omega t - \theta - 120°) \text{ A} \\ i_c(t) &= I_m \cos (\omega t - \theta - 240°) \text{ A} \end{aligned} \tag{12.4}$$

The instantaneous power produced by the system is

$$\begin{aligned} p(t) &= p_a(t) + p_b(t) + p_c(t) \\ &= V_m I_m[\cos \omega t \cos (\omega t - \theta) + \cos (\omega t - 120°) \cos (\omega t - \theta - 120°) \\ &\qquad + \cos (\omega t - 240°) \cos (\omega t - \theta - 240°)] \end{aligned} \tag{12.5}$$

Using the trigonometric identity

$$\cos \alpha \cos \beta = \tfrac{1}{2}[\cos (\alpha - \beta) + \cos (\alpha + \beta)] \tag{12.6}$$

Eq. (12.5) becomes

$$\begin{aligned} p(t) = \frac{V_m I_m}{2} &[\cos \theta + \cos (2\omega t - \theta) + \cos \theta \\ &+ \cos (2\omega t - \theta - 240°) + \cos \theta + \cos (2\omega t - \theta - 480°)] \end{aligned} \tag{12.7}$$

which can be written as

$$\begin{aligned} p(t) = \frac{V_m I_m}{2} &[3 \cos \theta + \cos (2\omega t - \theta) \\ &+ \cos (2\omega t - \theta - 120°) + \cos (2\omega t - \theta + 120°)] \end{aligned} \tag{12.8}$$

There exists a trigonometric identity which allows us to simplify the expression above. The identity, which we will prove later using phasors, is

$$\cos \phi + \cos (\phi - 120°) + \cos (\phi + 120°) = 0 \tag{12.9}$$

If we employ this identity, the expression for the power becomes

$$p(t) = 3\frac{V_m I_m}{2}\cos\theta \text{ W} \tag{12.10}$$

Note that this equation indicates that the instantaneous power is always constant in time, rather than pulsating as in the single-phase case. Therefore, power delivery from a three-phase voltage source is very smooth, which is another important reason why power is generated in three-phase form.

12.2 Three-Phase Connections

By far the most important polyphase voltage source is the balanced three-phase source. This source, as illustrated by Fig. 12.2, has the following properties. The phase voltages, that is, the voltage from each line a, b, and c to the neutral n, are given by

$$\begin{aligned}\mathbf{V}_{an} &= V_p \underline{/0^\circ}\\ \mathbf{V}_{bn} &= V_p \underline{/-120^\circ}\\ \mathbf{V}_{cn} &= V_p \underline{/+120^\circ}\end{aligned} \tag{12.11}$$

The phasor diagram for these voltages is shown in Fig. 12.3a. The phase sequence of this set is said to be *abc*, meaning that $\mathbf{V}_{bn}$ *lags* $\mathbf{V}_{an}$ by 120°. The other possibility is for $\mathbf{V}_{bn}$ to *lead* $\mathbf{V}_{an}$ by 120° and $\mathbf{V}_{cn}$ to lead $\mathbf{V}_{bn}$ by 120°. This sequence, which is described as *acb*, is shown in Fig. 12.3b and is given by

$$\begin{aligned}\mathbf{V}_{an} &= V_p \underline{/0^\circ}\\ \mathbf{V}_{bn} &= V_p \underline{/+120^\circ}\\ \mathbf{V}_{cn} &= V_p \underline{/-120^\circ}\end{aligned} \tag{12.12}$$

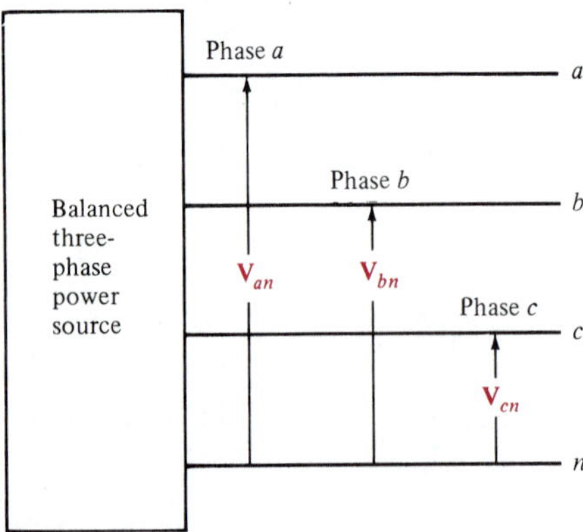

Figure 12.2 Balanced three-phase voltage source.

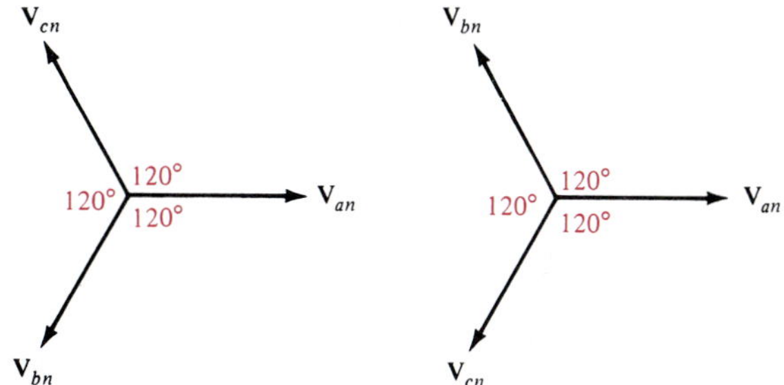

Figure 12.3 Phasor diagram for a balanced three-phase voltage source: (a) sequence *abc*; (b) sequence *acb*.

We will standardize our notation so that we always label the voltages $\mathbf{V}_{an}$, $\mathbf{V}_{bn}$, and $\mathbf{V}_{cn}$ and observe them in the order *abc*. Furthermore, we will normally assume with no loss of generality that $\underline{/\mathbf{V}_{an}} = 0°$.

An important property of the balanced voltage set is that

$$\mathbf{V}_{an} + \mathbf{V}_{bn} + \mathbf{V}_{cn} = 0 \tag{12.13}$$

This property can easily be seen by resolving the voltage phasors into components along the real and imaginary axes. It can also be demonstrated via Eq. (12.9).

From the standpoint of the user who connects a load to the balanced three-phase voltage source, it is not important how the voltages are generated. It is important to note, however, that if the load currents generated by connecting a load to the power source shown in Fig. 12.2 are also *balanced*, there are two possible equivalent configurations for the load. The equivalent load can be considered as being connected in either a *wye* (Y) or a *delta* (Δ) configuration. The balanced wye configuration is shown in Fig. 12.4a and equivalently in Fig. 12.4b. The delta configuration is shown in Fig. 12.5a and equiva-

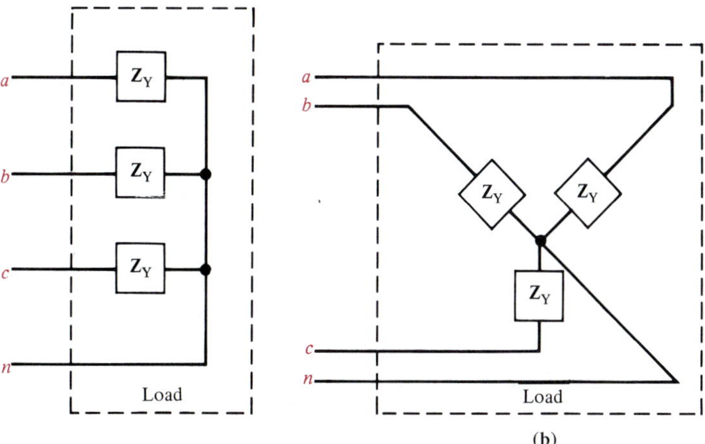

Figure 12.4 Wye (Y)-connected loads.

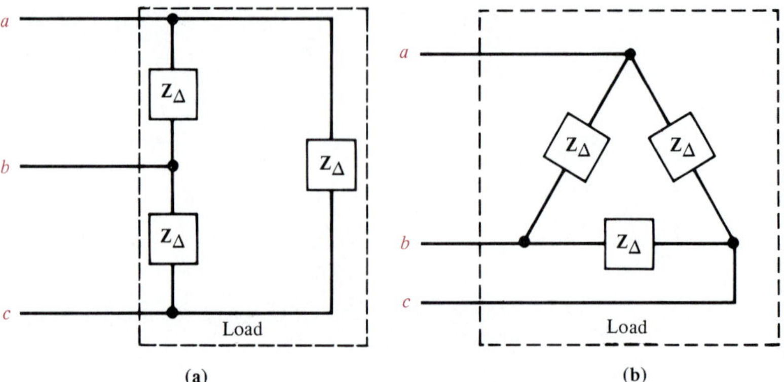

Figure 12.5 Delta (Δ)-connected loads.

lently in Fig. 12.5b. Note that in the case of the delta connection, there is no neutral line. The actual function of the neutral line in the wye connection will be examined and it will be shown that in a balanced system the neutral line carries no current and, for purposes of analysis, may be omitted.

Balanced Wye–Wye Connection

Suppose now that the source and load are both connected in a wye, as shown in Fig. 12.6. The phase voltages with positive phase sequence are

$$\begin{aligned}\mathbf{V}_{an} &= V_p \angle 0^\circ \\ \mathbf{V}_{bn} &= V_p \angle -120^\circ \\ \mathbf{V}_{cn} &= V_p \angle +120^\circ\end{aligned} \tag{12.14}$$

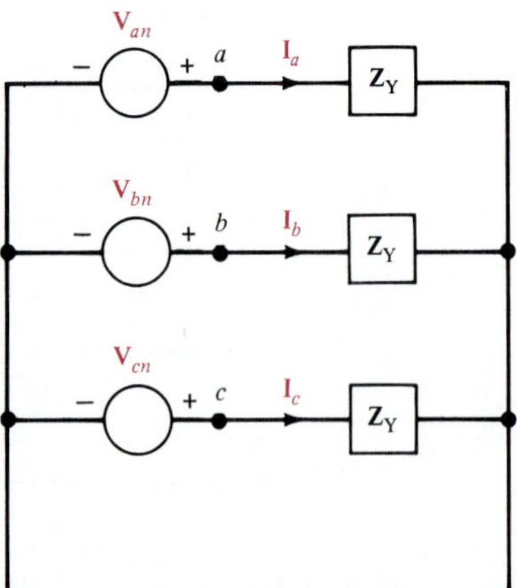

Figure 12.6 Balanced three-phase wye–wye connection.

where V_p, the phase voltage, is the magnitude of the phasor voltage from the neutral to any line. The *line-to-line* or, simply, *line voltages* can be calculated using KVL; for example,

$$\begin{aligned}\mathbf{V}_{ab} &= \mathbf{V}_{an} - \mathbf{V}_{bn}\\ &= V_p \underline{/0^\circ} - V_p \underline{/-120^\circ}\\ &= V_p - V_p\left[-\frac{1}{2} - j\frac{\sqrt{3}}{2}\right]\\ &= V_p\left[\frac{3}{2} + j\frac{\sqrt{3}}{2}\right]\\ &= \sqrt{3}\, V_p \underline{/30^\circ}\end{aligned}$$

The phasor addition is shown in Fig. 12.7a. In a similar manner, we obtain the set of line-to-line voltages as

$$\begin{aligned}\mathbf{V}_{ab} &= \sqrt{3}\, V_p \underline{/30^\circ}\\ \mathbf{V}_{bc} &= \sqrt{3}\, V_p \underline{/-90^\circ} \qquad (12.15)\\ \mathbf{V}_{ca} &= \sqrt{3}\, V_p \underline{/-210^\circ}\end{aligned}$$

All the line voltages together with the phase voltages are shown in Fig. 12.7b. We will denote the magnitude of the line voltages as V_L, and therefore, for a balanced system,

$$V_L = \sqrt{3}\, V_p \qquad (12.16)$$

Hence in a wye-connected system, the line voltage is equal to $\sqrt{3}$ times the phase voltage.

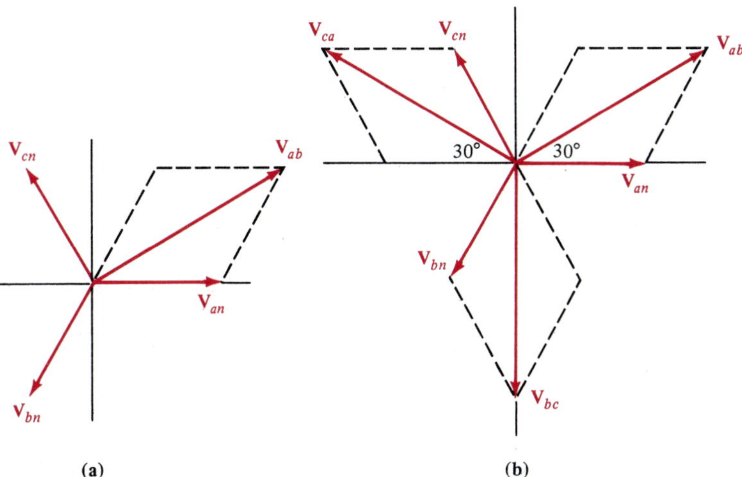

Figure 12.7 Phasor representation of phase and line voltages in a balanced wye–wye system.

As shown in Fig. 12.6, the line current for the a phase is

$$\mathbf{I}_a = \frac{\mathbf{V}_{an}}{\mathbf{Z}_Y} = \frac{V_p \angle 0^\circ}{\mathbf{Z}_Y} \tag{12.17}$$

where $\mathbf{I}_b$ and $\mathbf{I}_c$ have the same magnitude but lag $\mathbf{I}_a$ by 120° and 240°, respectively.

The neutral current $\mathbf{I}_n$ is then

$$\mathbf{I}_n = (\mathbf{I}_a + \mathbf{I}_b + \mathbf{I}_c) = 0 \tag{12.18}$$

Since there is no current in the neutral, this conductor could contain any impedance or it could be an open or a short circuit, without changing the results found above.

As illustrated by the wye–wye connection in Fig. 12.6 the current in the line connecting the source to the load is the same as the phase current flowing through the impedance $\mathbf{Z}_Y$. Therefore, in a wye–wye connection,

$$I_L = I_Y \tag{12.19}$$

where I_L is the magnitude of the line current and I_Y is the magnitude of the current in a wye-connected load.

It is important to note that although we have a three-phase system composed of three sources and three loads, we can analyze only one phase and use the phase sequence to obtain the voltages and currents in the other phases. This is, of course, a direct result of the balanced condition. We may even have impedances present in the lines; however, as long as the system remains balanced, we need analyze only one phase. If the line impedances in lines a, b, and c are equal, the system will be balanced. Recall that the balance of the system is unaffected by whatever appears in the neutral line and since the neutral line impedance is arbitrary, we assume that it is zero (i.e., a short circuit).

EXAMPLE 12.1

An abc-sequence three-phase voltage source connected in a balanced wye has a line voltage of $\mathbf{V}_{ab} = 208 \angle -30^\circ$ V rms. We wish to determine the phase voltages.

The magnitude of the phase voltage is given by the expression

$$V_p = \frac{208}{\sqrt{3}}$$

$$V_p = 120 \text{ V rms}$$

The phase relationships between the line and phase voltages are shown in Fig. 12.7. From this figure we note that

$$\mathbf{V}_{an} = 120 \angle -60^\circ \text{ V rms}$$

$$\mathbf{V}_{bn} = 120 \angle -180^\circ \text{ V rms}$$

$$\mathbf{V}_{cn} = 120 \angle +60^\circ \text{ V rms}$$

The magnitudes of these voltages are quite common and one often hears that the electric service in a building, for example, is three-phase 208/120 V. ■

EXAMPLE 12.2

A three-phase wye-connected load is supplied by an *abc*-sequence balanced three-phase wye-connected source with a phase voltage of 120 V rms. If the line impedance and load impedance per phase are $1 + j1\ \Omega$ and $20 + j10\ \Omega$, respectively, we wish to determine the value of the line currents and the load voltages.

The phase voltages are

$$\mathbf{V}_{an} = 120\ \underline{/0^\circ}\ \text{V rms}$$

$$\mathbf{V}_{bn} = 120\ \underline{/-120^\circ}\ \text{V rms}$$

$$\mathbf{V}_{cn} = 120\ \underline{/+120^\circ}\ \text{V rms}$$

The per phase circuit diagram is shown in Fig. 12.8. The line current for the *a* phase is

$$\mathbf{I}_{aA} = \frac{120\ \underline{/0^\circ}}{21 + j11}$$

$$= 5.06\ \underline{/-27.65^\circ}\ \text{A rms}$$

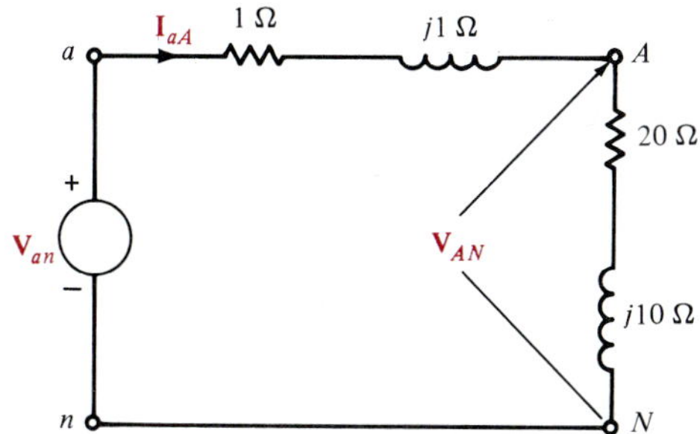

Figure 12.8 Per phase circuit diagram for the problem in Example 12.2.

The load voltage for the *a* phase, which we call $\mathbf{V}_{AN}$, is

$$\mathbf{V}_{AN} = (5.06\ \underline{/-27.65^\circ})(20 + j10)$$

$$= 113.15\ \underline{/-1.08^\circ}\ \text{V rms}$$

The corresponding line currents and load voltages for the *b* and *c* phases are

$$\mathbf{I}_{bB} = 5.06\ \underline{/-147.65^\circ}\ \text{A rms} \qquad \mathbf{V}_{BN} = 113.15\ \underline{/-121.08^\circ}\ \text{V rms}$$

$$\mathbf{I}_{cC} = 5.06\ \underline{/-267.65^\circ}\ \text{A rms} \qquad \mathbf{V}_{CN} = 113.15\ \underline{/-241.08^\circ}\ \text{V rms}$$

To reemphasize and clarify our terminology, phase voltage, V_p, is the magnitude of the phasor voltage from the neutral to any line, while line voltage, V_L, is the magnitude of the phasor voltage between any two lines. Thus the values of V_L and V_p will depend on the point at which they are calculated in the system. ■

DRILL EXERCISE

D12.1. The voltage for the *a* phase of an *abc*-phase-sequence balanced wye-connected source is $\mathbf{V}_{an} = 120\ \underline{/90^\circ}$ V rms. Determine the line voltages for this source.

Ans: $\mathbf{V}_{ab} = 208\ \underline{/120^\circ}$ V rms, $\mathbf{V}_{bc} = 208\ \underline{/0^\circ}$ V rms, $\mathbf{V}_{ca} = 208\ \underline{/-120^\circ}$ V rms.

D12.2. An *abc*-phase-sequence three-phase voltage source connected in a balanced wye has a line voltage of $\mathbf{V}_{ab} = 208\ \underline{/0^\circ}$ V rms. Determine the phase voltages of the source.

Ans: $\mathbf{V}_{an} = 120\ \underline{/-30^\circ}$ V rms, $\mathbf{V}_{bn} = 120\ \underline{/-150^\circ}$ V rms, $\mathbf{V}_{cn} = 120\ \underline{/-270^\circ}$ V rms.

D12.3. A three-phase wye load is supplied by an *abc*-sequence balanced three-phase wye-connected source through a transmission line with an impedance of $1 + j1$ ohms per phase. The load impedance is $8 + j3$ ohms per phase. If the load voltage for the *a* phase is $104.02\ \underline{/26.6^\circ}$ V rms (i.e., $V_p = 104.02$ V rms at the load end), determine the phase voltages of the source.

Ans: $\mathbf{V}_{an} = 120\ \underline{/30^\circ}$ V rms, $\mathbf{V}_{bn} = 120\ \underline{/-90^\circ}$ V rms, $\mathbf{V}_{cn} = 120\ \underline{/-210^\circ}$ V rms.

Balanced Wye–Delta Connection

Another important three-phase circuit is the balanced wye–delta system, that is, a wye-connected source and a delta-connected load, as shown in Fig. 12.9. From Fig. 12.9 we note that for this connection the line-to-line voltage is equal to the voltage across each load impedance in the delta-connected load. Therefore, if the phase voltages of the source are

$$
\begin{aligned}
\mathbf{V}_{an} &= V_p\ \underline{/0^\circ} \\
\mathbf{V}_{bn} &= V_p\ \underline{/-120^\circ} \\
\mathbf{V}_{cn} &= V_p\ \underline{/+120^\circ}
\end{aligned}
\tag{12.20}
$$

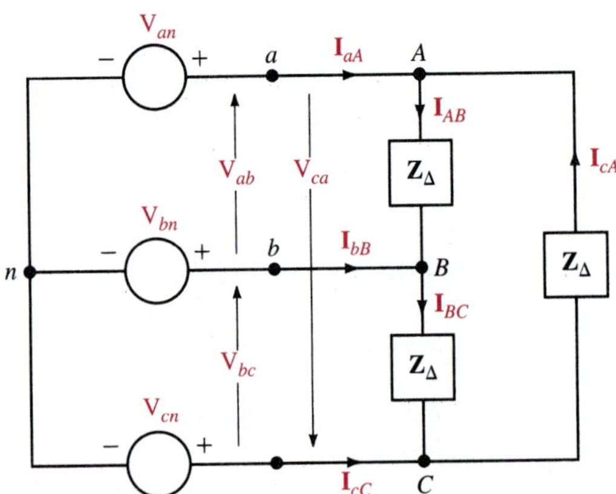

Figure 12.9 Balanced three-phase wye–delta system.

then the line voltages are

$$\begin{aligned} \mathbf{V}_{ab} &= \sqrt{3}\, V_p \,\underline{/30^\circ} = V_L \,\underline{/30^\circ} = \mathbf{V}_{AB} \\ \mathbf{V}_{bc} &= \sqrt{3}\, V_p \,\underline{/-90^\circ} = V_L \,\underline{/-90^\circ} = \mathbf{V}_{BC} \\ \mathbf{V}_{ca} &= \sqrt{3}\, V_p \,\underline{/-210^\circ} = V_L \,\underline{/-210^\circ} = \mathbf{V}_{CA} \end{aligned} \tag{12.21}$$

where V_L is the magnitude of the line voltage at the delta-connected load and at the source.

From Fig. 12.9 we note that if $\mathbf{Z}_\Delta = Z_\Delta \,\underline{/\theta}$, the phase currents at the load are

$$\mathbf{I}_{AB} = \frac{\mathbf{V}_{AB}}{\mathbf{Z}_\Delta} = I_\Delta \,\underline{/30^\circ - \theta} \tag{12.22}$$

where $\mathbf{I}_{BC}$ and $\mathbf{I}_{CA}$ have the same magnitude but lag $\mathbf{I}_{AB}$ by 120° and 240°, respectively.

The magnitude of the phase currents in the Δ is

$$I_\Delta = \frac{V_L}{Z_\Delta} \tag{12.23}$$

KCL can now be employed in conjunction with the phase currents to determine the line currents. For example,

$$\begin{aligned} \mathbf{I}_{aA} &= \mathbf{I}_{AB} + \mathbf{I}_{AC} \\ &= \mathbf{I}_{AB} - \mathbf{I}_{CA} \end{aligned}$$

Using the same approach as that employed earlier to determine the relationship between the line voltages and the phase voltages in a wye–wye connection, we can show that since $\underline{/\mathbf{V}_{an}} = 0^\circ$,

$$\mathbf{I}_{aA} = \sqrt{3}\, I_\Delta \,\underline{/-\theta} \tag{12.24}$$

where $\mathbf{I}_{bB}$ and $\mathbf{I}_{cC}$ have the same magnitude but lag $\mathbf{I}_{aA}$ by 120° and 240°, respectively, and therefore the relation between the magnitudes of the line currents and the delta currents is

$$I_L = \sqrt{3}\, I_\Delta \tag{12.25}$$

The phasor diagram in Fig. 12.10 illustrates all the important relationships among the currents and voltages for a delta-connected load.

EXAMPLE 12.3

An *abc*-phase-sequence three-phase voltage source connected in a balanced wye supplies power to a balanced delta-connected load. The load current $\mathbf{I}_{AB} = 4 \,\underline{/20^\circ}$ A rms. We wish to determine the line currents.

From the relationships among the Δ and line currents as specified in Eqs. (12.22) and (12.24) and illustrated in Fig. 12.10, we note that if

$$\mathbf{I}_{AB} = 4 \,\underline{/20^\circ} \text{ A rms}$$

then

$$\begin{aligned} \mathbf{I}_{aA} &= 4\sqrt{3} \,\underline{/-10^\circ} \text{ A rms} \\ \mathbf{I}_{bB} &= 4\sqrt{3} \,\underline{/-130^\circ} \text{ A rms} \\ \mathbf{I}_{cC} &= 4\sqrt{3} \,\underline{/+110^\circ} \text{ A rms} \end{aligned}$$

■

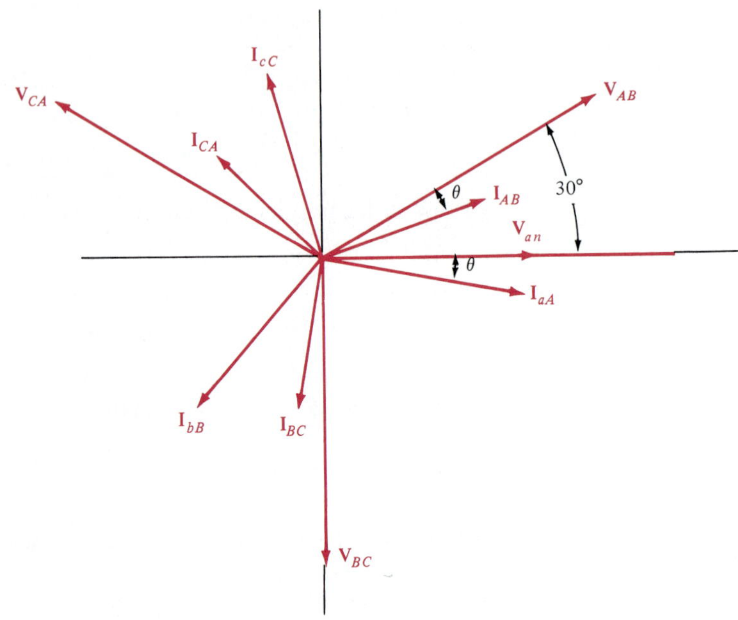

Figure 12.10 Current and voltage relationships for a delta-connected load.

EXAMPLE 12.4

A balanced delta-connected load contains a 10-Ω resistor in series with a 20-mH inductor in each phase. The voltage source is an *abc*-sequence three-phase 60-Hz, balanced wye with a voltage $\mathbf{V}_{an} = 120\ \underline{/30°}$ V rms. We wish to determine all Δ currents and line currents.

The impedance per phase in the delta load is $\mathbf{Z}_\Delta = 10 + j7.54\ \Omega$. The line voltage $\mathbf{V}_{ab} = 120\sqrt{3}\ \underline{/60°}$ V rms. Since there is no line impedance, $\mathbf{V}_{AB} = \mathbf{V}_{ab} = 120\sqrt{3}\ \underline{/60°}$ V rms. Hence

$$\mathbf{I}_{AB} = \frac{120\sqrt{3}\ \underline{/60°}}{10 + j7.54}$$

$$= 16.60\ \underline{/+22.98°}\text{ A rms}$$

Then from the relationship specified in Eqs. (12.22) and (12.24), we have

$$\mathbf{I}_{aA} = 16.60\sqrt{3}\ \underline{/-7.02°}$$

$$= 28.75\ \underline{/-7.02°}\text{ A rms}$$

Therefore, the remaining phase and line currents are

$$\mathbf{I}_{BC} = 16.60\ \underline{/-97.02°}\text{ A rms} \qquad \mathbf{I}_{bB} = 28.75\ \underline{/-127.02°}\text{ A rms}$$

$$\mathbf{I}_{CA} = 16.60\ \underline{/+142.98°}\text{ A rms} \qquad \mathbf{I}_{cC} = 28.75\ \underline{/112.98°}\text{ A rms}$$

■

DRILL EXERCISE

D12.4. An *abc*-sequence three-phase voltage source connected in a balanced wye supplies power to a balanced delta-connected load. The line current for the *a* phase is $\mathbf{I}_{aA} = 12 \angle 40^\circ$ A rms. Find the phase currents in the delta-connected load.

Ans: $\mathbf{I}_{AB} = 6.93 \angle 70^\circ$ A rms, $\mathbf{I}_{BC} = 6.93 \angle 50^\circ$ A rms, $\mathbf{I}_{CA} = 6.96 \angle -170^\circ$ A rms.

D12.5. An *abc*-sequence balanced three-phase wye-connected source supplies power to a balanced delta-connected load. The load impedance per phase is $12 + j8\ \Omega$. If the current $\mathbf{I}_{AB}$ in one phase of the delta is $14.42 \angle 86.31^\circ$ A rms, determine the line currents and phase voltages at the source.

Ans: $\mathbf{I}_a = 24.98 \angle 56.31^\circ$ A rms, $\mathbf{I}_b = \mathbf{I}_a \angle -120^\circ$, $\mathbf{I}_c = \mathbf{I}_a \angle -240^\circ$, $\mathbf{V}_{an} = 120 \angle 90^\circ$ V rms, $\mathbf{V}_{bn} = \mathbf{V}_{an} \angle -120^\circ$, $\mathbf{V}_{cn} = \mathbf{V}_{an} \angle -240^\circ$.

Delta-Connected Source

Up to this point we have concentrated our discussion on circuits that have wye-connected sources. However, our analysis of the wye wye and wye–delta connections provides us with the information necessary to handle a delta-connected source.

Consider the delta-connected source shown in Fig. 12.11a. Note that the sources are connected line to line. We found earlier that the relationship between line-to-line and line-to-neutral voltages was given by Eq. (12.15) and illustrated in Fig. 12.7 for an *abc* phase sequence of voltages. Therefore, if the delta sources are

$$\begin{aligned} \mathbf{V}_{ab} &= V_L \angle 0^\circ \\ \mathbf{V}_{bc} &= V_L \angle -120^\circ \\ \mathbf{V}_{ca} &= V_L \angle +120^\circ \end{aligned} \tag{12.26}$$

where V_L is the magnitude of the phase voltage, the equivalent wye sources shown in Fig. 12.11b are

$$\begin{aligned} \mathbf{V}_{an} &= \frac{V_L}{\sqrt{3}} \angle -30^\circ = V_p \angle -30^\circ \\ \mathbf{V}_{bn} &= \frac{V_L}{\sqrt{3}} \angle -150^\circ = V_p \angle -150^\circ \\ \mathbf{V}_{cn} &= \frac{V_L}{\sqrt{3}} \angle -270^\circ = V_p \angle +90^\circ \end{aligned} \tag{12.27}$$

where V_p is the magnitude of the phase voltage of an equivalent wye-connected source. In addition, if the line currents are known to be

$$\begin{aligned} \mathbf{I}_a &= I_L \angle \theta \\ \mathbf{I}_b &= I_L \angle \theta - 120^\circ \\ \mathbf{I}_c &= I_L \angle \theta + 120^\circ \end{aligned} \tag{12.28}$$

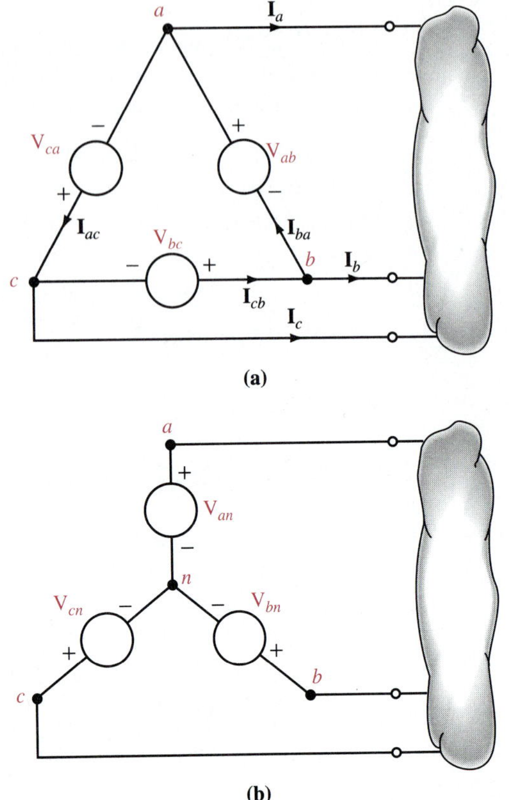

Figure 12.11 Sources connected in delta and wye.

the phase current $\mathbf{I}_{ba}$ in the delta source is

$$\mathbf{I}_{ba} = \frac{I_L}{\sqrt{3}} \underline{/\theta + 30^\circ} = I_\Delta \underline{/\theta + 30^\circ} \tag{12.29}$$

where $\mathbf{I}_{cb}$ and $\mathbf{I}_{ac}$ have the same magnitude, $\mathbf{I}_\Delta$, but lag $\mathbf{I}_{ba}$ by 120° and 240°, respectively. Therefore, if we encounter a network containing a delta-connected source, we can easily convert the source from delta to wye so that all the techniques we have discussed previously can be applied in an analysis.

EXAMPLE 12.5

Consider the network shown in Fig. 12.12a. We wish to determine the line currents, the magnitude of the line voltage at the load, and the magnitude of the phase currents in the delta source.

The single-phase diagram for the network is shown in Fig. 12.12b. The line current $\mathbf{I}_{aA}$ is

$$\mathbf{I}_{aA} = \frac{(208/\sqrt{3}) \underline{/-30^\circ}}{12.1 + j4.2}$$

$$= 9.38 \underline{/-49.14^\circ} \text{ A rms}$$

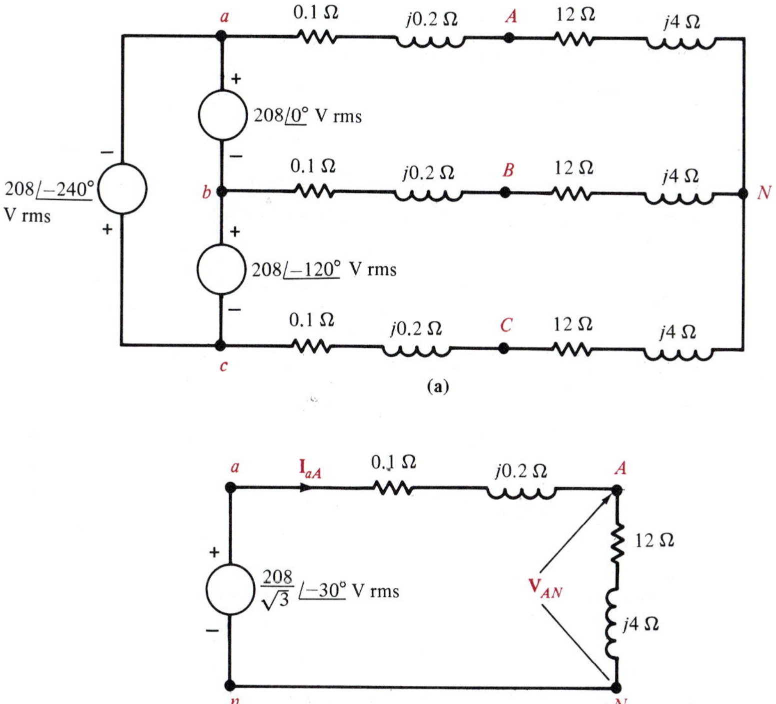

Figure 12.12 Delta–wye network and an equivalent single-phase (*a*-phase) diagram.

and thus $\mathbf{I}_{bB} = 9.38\ \underline{/-169.14°}$ A rms and $\mathbf{I}_{cC} = 9.38\ \underline{/70.86°}$ A rms. The voltage $\mathbf{V}_{AN}$ is then

$$\mathbf{V}_{AN} = (9.38\ \underline{/-49.14°})(12 + j4)$$
$$= 118.66\ \underline{/-30.71°} \text{ V rms}$$

Therefore, the magnitude of the line voltage at the load is

$$V_L = \sqrt{3}\,(118.66)$$
$$= 205.53 \text{ V rms}$$

The current $\mathbf{I}_{ba}$ is

$$\mathbf{I}_{ba} = \frac{9.38}{\sqrt{3}}\ \underline{/-49.14° + 30°}$$
$$= 5.42\ \underline{/-19.14°} \text{ A rms}$$

Therefore, the magnitude of the phase current in the delta source is $|\mathbf{I}_{ba}| = I_\Delta = 5.42$ A rms.

The phase voltage at the source is $V_p = 208/\sqrt{3} = 120$ V rms, while the phase voltage at the load is $V_p = 205.53/\sqrt{3} = 118.67$ V rms. Clearly, we must be careful with our notation and specify where the phase or line voltage is taken. ■

DRILL EXERCISE

D12.6. Consider the network shown in Fig. D12.6. Compute the magnitude of the line voltages at the load and the magnitude of the phase currents in the delta-connected source.

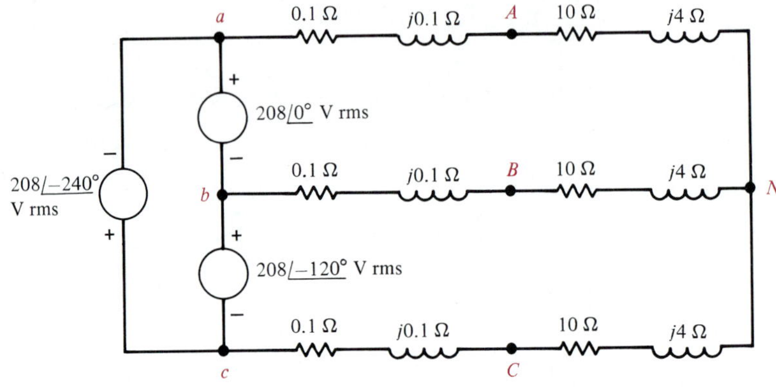

Figure D12.6

Ans: $I_\Delta = 6.36$ A rms, $V_L = 241$ V rms.

Wye⇌Delta Transformations

We have stated earlier that for a balanced system, the equivalent load configuration may be either wye or delta. Equations (9.47) and (9.48) illustrate the general relationship between the impedances in the wye and delta configuration shown in Fig. 9.16. However, for the balanced case where $\mathbf{Z}_a = \mathbf{Z}_b = \mathbf{Z}_c$ and $\mathbf{Z}_1 = \mathbf{Z}_2 = \mathbf{Z}_3$, Eqs. (9.47) and (9.48) reduce to

$$\mathbf{Z}_Y = \tfrac{1}{3}\mathbf{Z}_\Delta \tag{12.30}$$

and

$$\mathbf{Z}_\Delta = 3\mathbf{Z}_Y \tag{12.31}$$

EXAMPLE 12.6

A three-phase load impedance consists of a balanced wye in parallel with a balanced delta, as shown in Fig. 12.13a. We wish to determine the equivalent delta load.

The equivalent delta for the given Y is

$$\begin{aligned}\mathbf{Z}_\Delta &= 3\mathbf{Z}_Y \\ &= 12 + j3\ \Omega\end{aligned}$$

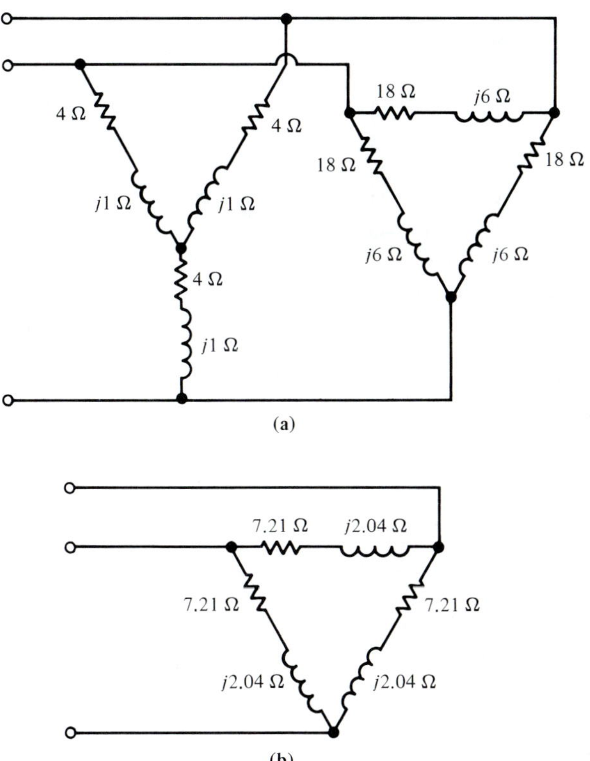

Figure 12.13 Three-phase load consisting of a balanced wye in parallel with a balanced delta converted to an equivalent delta load: (a) original load; (b) equivalent load.

This delta-connected impedance is in parallel with the original one, and therefore, the impedance per phase of the combined delta is

$$\mathbf{Z}'_\Delta = \frac{(12 + j3)(18 + j6)}{12 + j3 + 18 + j6}$$

$$= 7.21 + j2.04\ \Omega$$

which is shown as the equivalent load in Fig. 12.13b. ■

EXAMPLE 12.7

Consider the network shown in Fig. 12.14a. We wish to determine all the load currents.

The circuit diagram for the a phase is shown in Fig. 12.14b. Since the parallel combination of 30 Ω and 20 Ω is 12 Ω,

$$\mathbf{I}_{aA} = \frac{120\ \underline{/0^\circ}}{16}$$

$$= 7.5\ \underline{/0^\circ}\ \text{A rms}$$

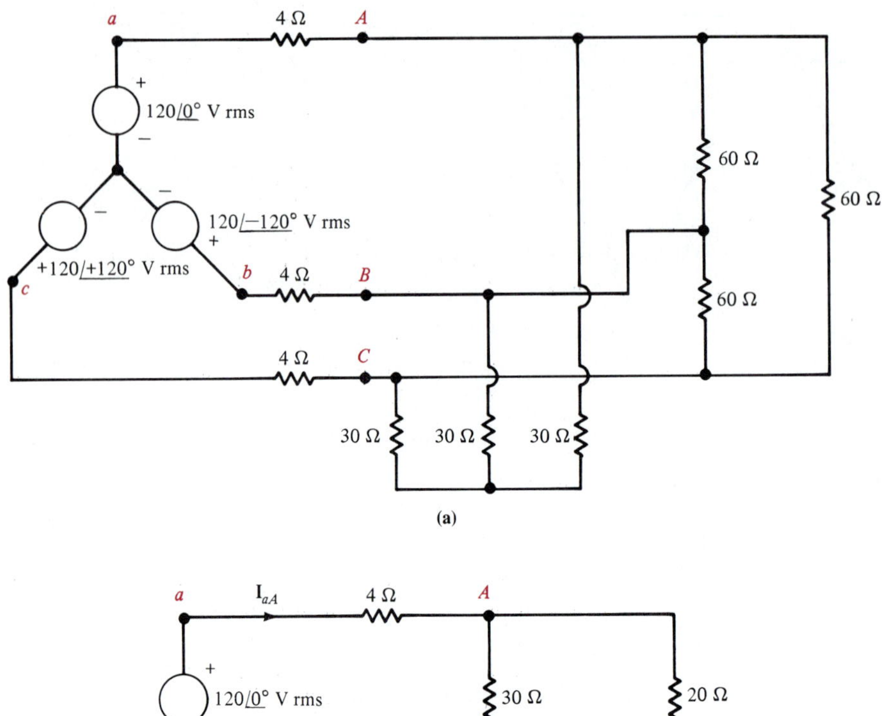

Figure 12.14 Networks used in Example 12.7: (a) original network; (b) *a*-phase equivalent network.

and hence

$$\mathbf{V}_{AN} = (7.5 \underline{/0^\circ})(12)$$
$$= 90 \underline{/0^\circ} \text{ V rms}$$

In the original wye-connected load,

$$\mathbf{I}_{AN} = \frac{90 \underline{/0^\circ}}{30}$$
$$= 3 \underline{/0^\circ} \text{ A rms}$$

and therefore $\mathbf{I}_{BN} = 3 \underline{/-120^\circ}$ A rms and $\mathbf{I}_{CN} = 3 \underline{/+120^\circ}$ A rms.

For the original delta-connected load

$$\mathbf{V}_{AB} = 90\sqrt{3} \underline{/0^\circ + 30^\circ}$$
$$= 155.88 \underline{/30^\circ} \text{ V rms}$$

and hence

$$\mathbf{I}_{AB} = \frac{155.88 \underline{/30^\circ}}{60}$$

$$= 2.60 \underline{/30^\circ} \text{ A rms}$$

Therefore, $\mathbf{I}_{BC} = 2.60 \underline{/-90^\circ}$ A rms and $\mathbf{I}_{CA} = 2.60 \underline{/+150^\circ}$ A rms.

As shown in Fig. 12.14a, the line current must be equal to the sum of the load currents. Hence

$$\mathbf{I}_{aA} = \mathbf{I}_{AN} + \mathbf{I}_{AB} + \mathbf{I}_{AC}$$

$$= \mathbf{I}_{AN} + \mathbf{I}_{AB} - \mathbf{I}_{CA}$$

$$= 3 \underline{/0^\circ} + 2.6 \underline{/30^\circ} - 2.6 \underline{/150^\circ}$$

$$= 7.5 \underline{/0^\circ} \text{ A rms}$$

which is a simple check on our calculations. ■

EXAMPLE 12.8

A balanced three-phase system has a load consisting of a balanced wye in parallel with a balanced delta. The impedance per phase for the wye is $10 + j6\ \Omega$ and for the delta is $24 + j9\ \Omega$. The source is a balanced wye with an *abc* phase sequence, and the line voltage $\mathbf{V}_{ab} = 208 \underline{/30^\circ}$ V rms. If the line impedance per phase is $1 + j0.5\ \Omega$, we want to determine the line currents and the load phase voltages when the load is converted to an equivalent wye.

Converting the delta load to an equivalent wye load, we obtain

$$\mathbf{Z}_{Y1} = \tfrac{1}{3}\mathbf{Z}_{\Delta 1} = 8 + j3\ \Omega$$

The per phase equivalent circuit is shown in Fig. 12.15. Since the line voltage at the source is $\mathbf{V}_{ab} = 208 \underline{/30^\circ}$ V rms, the phase voltage at the source is $\mathbf{V}_{an} = 120 \underline{/0^\circ}$ V rms.

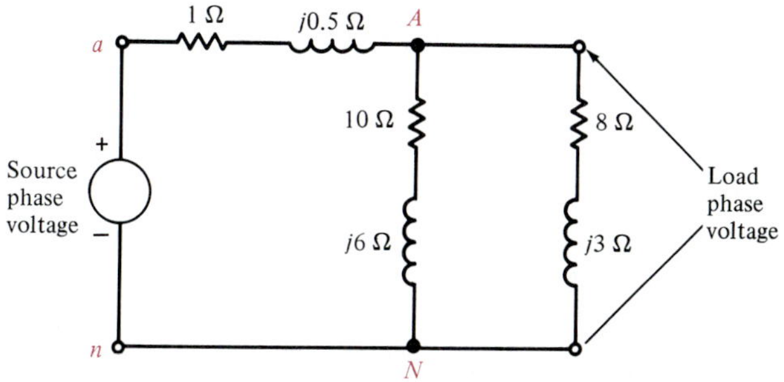

Figure 12.15 Per phase equivalent circuit for Example 12.8.

Recall that since the loads are balanced, the neutral points can be connected. The equivalent wye load impedance is

$$\mathbf{Z}_Y = \frac{\mathbf{Z}_{Y1}\mathbf{Z}_{Y2}}{\mathbf{Z}_{Y1} + \mathbf{Z}_{Y2}}$$

$$= \frac{(10 + j6)(8 + j3)}{10 + j6 + 8 + j3} = 4.95 \underline{/24.95^\circ}$$

$$= 4.49 + j2.09\ \Omega$$

The line current $\mathbf{I}_a$ is then

$$\mathbf{I}_a = \frac{\mathbf{V}_{an}}{\mathbf{Z}_{\text{line}} + \mathbf{Z}_Y}$$

$$= \frac{120 \underline{/0^\circ}}{1 + j0.5 + 4.49 + j2.09}$$

$$= 19.77 \underline{/-25.26^\circ}\ \text{A rms}$$

The phase voltage at the load is then $V_p = |\mathbf{V}_{AN}|$, where

$$\mathbf{V}_{AN} = \mathbf{I}_a\mathbf{Z}_Y$$

$$= (19.77 \underline{/-25.26^\circ})(4.95 \underline{/24.95^\circ})$$

$$= 97.86 \underline{/-0.31^\circ}\ \text{V rms}$$

Therefore, the line currents and load phase voltages are

$$\mathbf{I}_{aA} = 19.77 \underline{/-25.26^\circ}\ \text{A rms} \qquad \mathbf{V}_{AN} = 97.86 \underline{/-0.31^\circ}\ \text{V rms}$$

$$\mathbf{I}_{bB} = 19.77 \underline{/-145.26^\circ}\ \text{A rms} \qquad \mathbf{V}_{BN} = 97.86 \underline{/-120.31^\circ}\ \text{V rms}$$

$$\mathbf{I}_{cC} = 19.77 \underline{/+94.74^\circ}\ \text{A rms} \qquad \mathbf{V}_{CN} = 97.86 \underline{/+119.69^\circ}\ \text{V rms}$$

Thus at the load $V_p = 97.86$ V rms and $V_L = \sqrt{3}\, V_p = 169.50$ V rms. ■

DRILL EXERCISE

D12.7. In a balanced three-phase system the load consists of a balanced wye in parallel with a balanced delta. The impedance per phase for the wye is $8 + j4\ \Omega$ and for the delta is $18 + j6\ \Omega$. The source is an *abc*-phase-sequence balanced wye and $\mathbf{V}_{an} = 120 \underline{/60^\circ}$ V rms. If the line impedance per phase is $1 + j1\ \Omega$, determine the magnitude of the phase currents in each load.

Ans: $I_\Delta = 1.68$ A rms.

Power Relationships

Whether the load is connected in a wye or a delta, the real and reactive power per phase are

$$P_p = V_p I_L \cos\theta$$
$$Q_p = V_p I_L \sin\theta \qquad (12.32)$$

where θ is the angle between the phase voltage and the line current, or

$$P_p = \frac{V_L I_L}{\sqrt{3}} \cos \theta$$
$$Q_p = \frac{V_L I_L}{\sqrt{3}} \sin \theta \tag{12.33}$$

The total real and reactive power for all three phases is then

$$P_T = \sqrt{3}\, V_L I_L \cos \theta$$
$$Q_T = \sqrt{3}\, V_L I_L \sin \theta \tag{12.34}$$

and therefore, the magnitude of the complex power (apparent power) is

$$\begin{aligned} |\mathbf{S}_T| &= \sqrt{P_T^2 + Q_T^2} \\ &= \sqrt{3}\, V_L I_L \end{aligned} \tag{12.35}$$

and

$$\angle \mathbf{S}_T = \theta$$

EXAMPLE 12.9

A three-phase balanced wye–delta system has a line voltage of 208 V rms. The total real power absorbed by the load is 1200 W. If the power factor angle of the load is 20° lagging, we wish to determine the magnitude of the line current and the value of the load impedance per phase in the delta.

The line current can be obtained from Eq. (12.33). Since the real power per phase is 400 W,

$$400 = \frac{208 I_L}{\sqrt{3}} \cos 20°$$

$$I_L = 3.5 \text{ A rms}$$

The magnitude of the current in each leg of the delta-connected load is

$$\begin{aligned} I_\Delta &= \frac{I_L}{\sqrt{3}} \\ &= 2.04 \text{ A rms} \end{aligned}$$

Therefore, the magnitude of the delta impedance in each phase of the load is

$$\begin{aligned} |\mathbf{Z}_\Delta| &= \frac{\mathbf{V}_L}{I_\Delta} \\ &= \frac{208}{2.04} \\ &= 101.77\ \Omega \end{aligned}$$

Since the power factor angle is 20° lagging, the load impedance is

$$\mathbf{Z}_\Delta = 101.77 \underline{/20^\circ}$$
$$= 95.63 + j34.81\ \Omega$$

■

EXAMPLE 12.10

For the circuit in Example 12.2 we wish to determine the real and reactive power per phase at the load and the total real power, reactive power, and the complex power at the source.

From the data in Example 12.2 the complex power per phase at the load is

$$\mathbf{S}_{\text{load}} = \mathbf{VI}^*$$
$$= (113.15 \underline{/-1.08^\circ})(5.06 \underline{/27.65^\circ})$$
$$= 572.54 \underline{/26.57^\circ}$$
$$= 512.07 + j256.1 \text{ VA}$$

Therefore, the real and reactive power per phase at the load are 512.07 W and 256.1 var, respectively.

The complex power per phase at the source is

$$\mathbf{S}_{\text{source}} = \mathbf{VI}^*$$
$$= (120 \underline{/0^\circ})(5.06 \underline{/27.65^\circ})$$
$$= 607.2 \underline{/27.65^\circ}$$
$$= 537.90 + j281.8 \text{ VA}$$

and therefore, total real power, reactive power, and apparent power at the source are 1613.7 W, 845.4 var, and 1821.6 VA, respectively. ■

EXAMPLE 12.11

A balanced three-phase source serves three loads as follows:

Load 1: 24 kW at 0.6 lagging power factor
Load 2: 10 kW at unity power factor
Load 3: 12 kVA at 0.8 leading power factor

If the line voltage at the loads is 208 V rms at 60 Hz, we wish to determine the line current and the combined power factor of the loads.

From the data we find that

$$\mathbf{S}_1 = 24{,}000 + j32{,}000$$
$$\mathbf{S}_2 = 10{,}000 + j0$$
$$\mathbf{S}_3 = 12{,}000 \underline{/-36.9^\circ} = 9600 - j7200$$

Therefore,

$$\mathbf{S}_{\text{load}} = 43{,}600 + j24{,}800$$

$$= 50{,}160 \underline{/29.63^\circ} \text{ VA}$$

$$I_L = \frac{|\mathbf{S}_{\text{load}}|}{\sqrt{3}\, V_L}$$

$$= \frac{50{,}160}{208\sqrt{3}}$$

$$I_L = 139.23 \text{ A rms}$$

and the combined power factor is

$$\text{pf}_{\text{load}} = \cos 29.63^\circ$$

$$= 0.869 \text{ lagging}$$

EXAMPLE 12.12

Given the three-phase system in Example 12.11, let us determine the line voltage and power factor at the source if the line impedance is $\mathbf{Z}_{\text{line}} = 0.05 + j0.02\ \Omega$.

The complex power absorbed by the line impedances is

$$\mathbf{S}_{\text{line}} = 3(\mathbf{R}_{\text{line}} I_L^2 + j\mathbf{X}_{\text{line}} I_L^2)$$

$$= 3077 + j1231 \text{ VA}$$

The complex power delivered by the source is then

$$\mathbf{S}_S = \mathbf{S}_{\text{load}} + \mathbf{S}_{\text{line}}$$

$$= 43{,}600 + j24{,}800 + 3077 + j1231$$

$$= 53{,}445 \underline{/29.15^\circ} \text{ VA}$$

The line voltage at the source is then

$$V_{L_S} = \frac{|\mathbf{S}_S|}{\sqrt{3}\, I_L}$$

$$= 215.43 \text{ V rms}$$

and the power factor at the source is

$$\text{pf}_S = \cos 29.15^\circ$$

$$= 0.873 \text{ lagging}$$

EXAMPLE 12.13

Two balanced three-phase systems, X and Y, are interconnected with lines having impedance $\mathbf{Z}_{\text{line}} = 1 + j2\ \Omega$. The line voltages are $\mathbf{V}_{ab} = 12 \underline{/0^\circ}$ kV rms and $\mathbf{V}_{AB} = 12 \underline{/5^\circ}$ kV rms as shown in Fig. 12.16a. We wish to determine which system is the source, which is the load, and the average power supplied by the source and absorbed by the load.

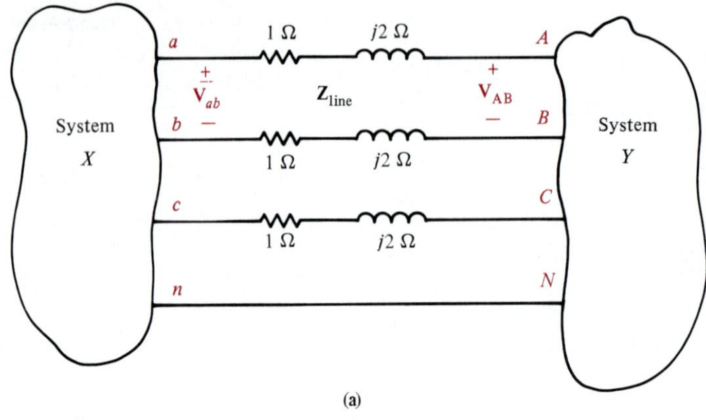

(a)

$\mathbf{I}_{aA}$ a — 1 Ω — j2 Ω — A

$\mathbf{V}_{an} = \frac{12}{\sqrt{3}} \underline{/-30^\circ}$ kV rms $\qquad \mathbf{V}_{AN} = \frac{12}{\sqrt{3}} \underline{/-25^\circ}$ kV rms

n $\qquad$ N

(b)

Figure 12.16 Circuits used in Example 12.13: (a) original three-phase system; (b) per phase circuit.

When we draw the per phase circuit for the system as shown in Fig. 12.16b, the analysis will be essentially the same as that of Example 11.14.

The network in Fig. 12.16b indicates that

$$\mathbf{I}_{aA} = \frac{\mathbf{V}_{an} - \mathbf{V}_{AN}}{\mathbf{Z}_{\text{line}}}$$

$$= \frac{\frac{12{,}000}{\sqrt{3}} \underline{/-30^\circ} - \frac{12{,}000}{\sqrt{3}} \underline{/-25^\circ}}{\sqrt{5} \underline{/63.43^\circ}}$$

$$= 270.30 \underline{/-180.93^\circ} \text{ A rms}$$

The average power absorbed by system Y is

$$P_Y = \sqrt{3}\, |\mathbf{V}_{AB}||\mathbf{I}_{aA}| \cos(\theta_{\mathbf{V}_{AN}} - \theta_{\mathbf{I}_{aA}})$$

$$= \sqrt{3}\,(12{,}000)(270.30) \cos(-25^\circ + 180.93^\circ)$$

$$= -5.130 \text{ MW}$$

Note that system Y is not the load, but rather the source and supplies 5.130 MW.

System X absorbs the following average power:

$$P_X = \sqrt{3}\,|\mathbf{V}_{ab}||\mathbf{I}_{Aa}| \cos(\theta_{\mathbf{V}_{an}} - \theta_{\mathbf{I}_{Aa}})$$

where

$$\mathbf{I}_{Aa} = -\mathbf{I}_{aA} = 270.30 \underline{/-0.93^\circ} \text{ A rms}$$

Therefore,

$$P_X = \sqrt{3}\,(12{,}000)(270.30) \cos(-30^\circ + 0.93^\circ)$$
$$= 4.910 \text{ MW}$$

and hence system X is the load.

The difference in the power supplied by system Y and that absorbed by system X is, of course, the power absorbed by the resistance of the three lines. ■

The preceding example illustrates an interesting point. Note that the phase difference between the two ends of the power line determines the direction of the power flow. Since the numerous power companies throughout the United States are tied together to form the U.S. power grid, the phase difference across the interconnecting transmission lines reflects the manner in which power is transferred between power companies.

DRILL EXERCISE

D12.8. A three-phase balanced wye–wye system has a line voltage of 208 V rms. The total real power absorbed by the load is 12 kW at 0.8 pf lagging. Determine the per phase impedance of the load.

Ans: $\mathbf{Z} = 2.88 \underline{/36.87^\circ}\ \Omega$.

D12.9. For the balanced wye–wye system described in Drill Exercise D12.3, determine the real and reactive power and the complex power at both the source and the load.

Ans: $\mathbf{S}_{\text{load}} = 1186.27 + j444.94$ VA, $\mathbf{S}_{\text{source}} = 1335.65 + j593.55$ VA.

D12.10. A 480-V rms line feeds two balanced three-phase loads. If the two loads are rated as follows,

Load 1: 5 kVA at 0.8 pf lagging

Load 2: 10 kVA at 0.9 pf lagging

determine the magnitude of the line current from the 480-V rms source.

Ans: $I_L = 17.97$ A rms.

12.3 PSPICE Analysis of Three-Phase Circuits

All the material required to apply PSPICE to three-phase circuits has been covered in previous chapters. Therefore, our approach in this section will be to demonstrate the use of PSPICE through several examples. In contrast to the earlier sections of this chapter, we will analyze the entire three-phase network and not rely on the balance condition to treat only a single phase. In addition, we will illustrate how to solve two problems with PSPICE that naturally occur in a delta-connected system. As we have indicated earlier, when only impedance is specified, we will, for convenience, assume in our analysis that $\omega = 1$ rad/sec and thus $f = 1/2\pi$ Hz.

EXAMPLE 12.14

A balanced wye–wye three-phase system which is labeled for a PSPICE analysis is shown in Fig. 12.17. The following PSPICE program will calculate the line currents, the phase load voltages, and the line-to-line load voltages. $f = 1/2\pi$ Hz. The program is

```
EXAMPLE 12.14
VAN 1 0 AC 120 0
VBN 5 0 AC 120 -120
VCN 6 0 AC 120 -240
RAN 1 2 10
RBN 5 4 10
RCN 6 7 10
LAN 2 3 2
LBN 4 3 2
LCN 7 3 2
.PRINT AC IM(VAN) IP(VAN) VM(1,3) VP(1,3)
.PRINT AC IM(VBN) IP(VBN) VM(5,3) VP(5,3)
.PRINT AC IM(VCN) IP(VCN) VM(6,3) VP(6,3)
.PRINT AC VM(1,5) VP(1,5) VM(5,6) VP(5,6)
+ VM(6,1) VP(6,1)
.AC LIN 1 .159155 .159155
.END
```

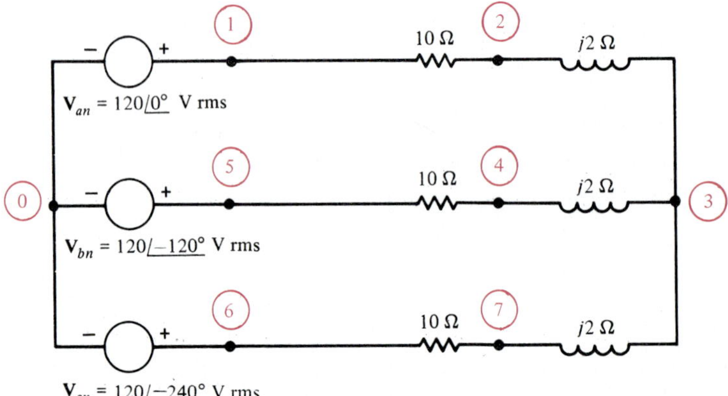

Figure 12.17 Balanced wye–wye three-phase system labeled for PSPICE analysis.

The computer output contains

$$\mathbf{I}(\text{VAN}) = 11.77 \angle 168.7^\circ \text{ A rms}$$
$$\mathbf{V}(1,3) = 120 \angle 0^\circ \text{ V rms}$$
$$\mathbf{V}(1,5) = 207.8 \angle 30^\circ \text{ V rms}$$

Recalling that **I**(VAN) is the current from node 1 to node 0 through the source $\mathbf{V}_{an}$, we see that the line current flowing from nodes 1 to 2 is $-\mathbf{I}(\text{VAN}) = 11.77 \angle -11.3^\circ$ A rms. ■

EXAMPLE 12.15

A balanced wye–delta system labeled for a PSPICE analysis is shown in Fig. 12.18. The following PSPICE program will calculate the source currents, load currents, and load voltages. $f = 1/2\pi$ Hz.

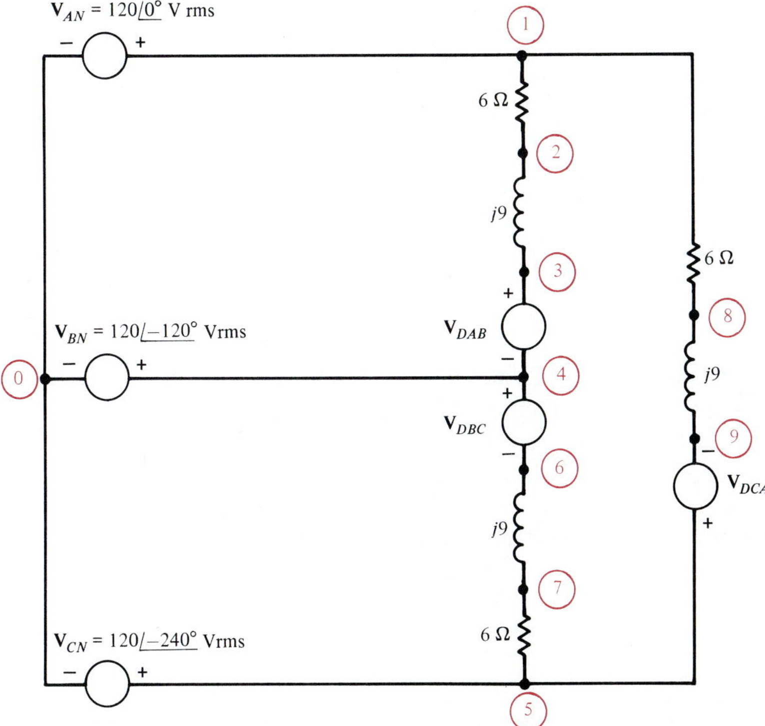

Figure 12.18 Balanced wye–delta three-phase system labeled for PSPICE analysis.

The computer program is

```
EXAMPLE 12.15
VAN 1 0 AC 120 0
VBN 4 0 AC 120 -120
VCN 5 0 AC 120 -240
RAB 1 2 6
LAB 2 3 9
VDAB 3 4 AC 0
RBC 7 5 6
LBC 6 7 9
VDBC 4 6 AC 0
RCA 1 8 6
LCA 8 9 9
VDCA 5 9 AC 0
.PRINT AC IM(VAN) IP(VAN) IM(VBN) IP(VBN)
+ IM(VCN) IP(VCN)
.PRINT AC IM(VDAB) IP(VDAB) VM(1,4) VP(1,4)
.PRINT AC IM(VDBC) IP(VDBC) VM(4,5) VP(4,5)
.PRINT AC IM(VDCA) IP(VDCA) VM(5,1) VP(5,1)
.AC LIN 1 .159155 .159155
.END
```

The computer output contains

```
I(VAN) = 33.28 /123.7° A rms
I(VDAB) = 19.22 /-26.31° A rms
V(1,4) = 207.8 /30° V rms
```

■

EXAMPLE 12.16

A balanced three-phase delta-connected system is shown in Fig. 12.19a. We wish to use PSPICE to calculate the currents and voltages in this network. From a PSPICE standpoint this circuit has two problems: There is a loop of voltage sources, and there is no convenient ground node. Therefore, the circuit has been modified as shown in Fig. 12.19b. The 1-pΩ resistors placed in series around the voltage sources alleviate the problem of a voltage source loop. The 1-TΩ resistors permit us to introduce the ground node required by PSPICE. The position and value of these resistors are selected to maintain a balanced system and have a minimum impact on the operation of the original network.

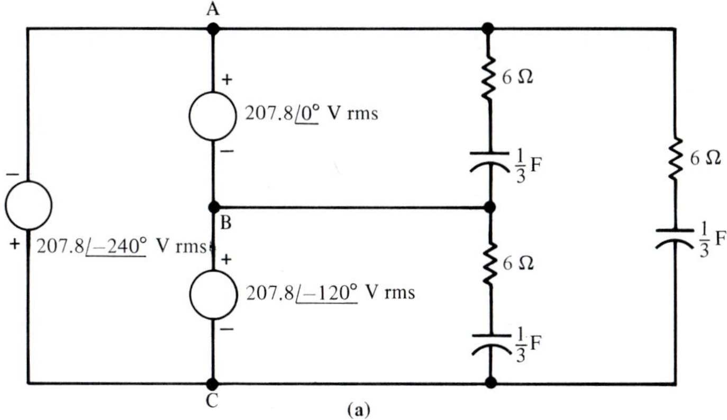

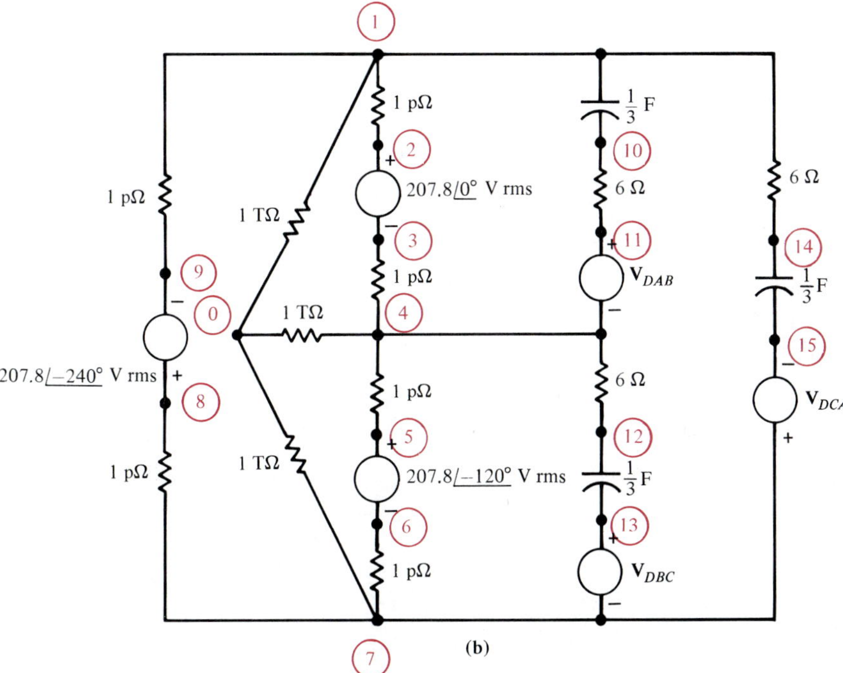

Figure 12.19 Three-phase delta-connected system analyzed using PSPICE.

The PSPICE program, written for the network in Fig. 12.19b, which calculates the source currents, load currents, and load voltages, follows. $f = 1/2\pi$ Hz.

```
EXAMPLE 12.16
VAB 2 3 AC 207.8 0
RSAB1 1 2 1P
RSAB2 3 4 1P
VBC 5 6 AC 207.8 -120
RSBC1 4 5 1P
RSBC2 6 7 1P
VCA 8 9 AC 207.8 -240
RSCA1 7 8 1P
RSCA2 9 1 1P
RSHZA 1 0 1T
RSHZB 4 0 1T
RSHZC 7 0 1T
RAB 10 11 6
CAB 1 10 .33333
VDAB 11 4 AC 0
RBC 4 12 6
CCB 12 13 .33333
VDBC 13 7 AC 0
RCA 1 14 6
CCA 14 15 .33333
VDCA 7 15 AC 0
.PRINT AC IM(VAB) IP(VAB) VM(1,4) VP(1,4)
.PRINT AC IM(VBC) IP(VBC) VM(4,7) VP(4,7)
.PRINT AC IM(VCA) IP(VCA) VM(7,1) VP(7,1)
.PRINT AC IM(VDAB) IP(VDAB)
.PRINT AC IM(VDBC) IP(VDBC)
.PRINT AC IM(VDCA) IP(VDCA)
.AC LIN 1 .159155 .159155
.END
```

The computer output contains

$$\mathbf{I}(VAB) = 30.98 \underline{/-153.4^\circ} \text{ A rms}$$
$$\mathbf{V}(1,4) = 207.8 \underline{/0^\circ} \text{ V rms}$$
$$\mathbf{I}(VDAB) = 30.98 \underline{/26.57^\circ} \text{ A rms}$$

■

DRILL EXERCISE

D12.11. Given the network in Fig. D12.11, use PSPICE to determine the line current, line voltage, and load voltage. $f = 1/2\pi$ Hz.

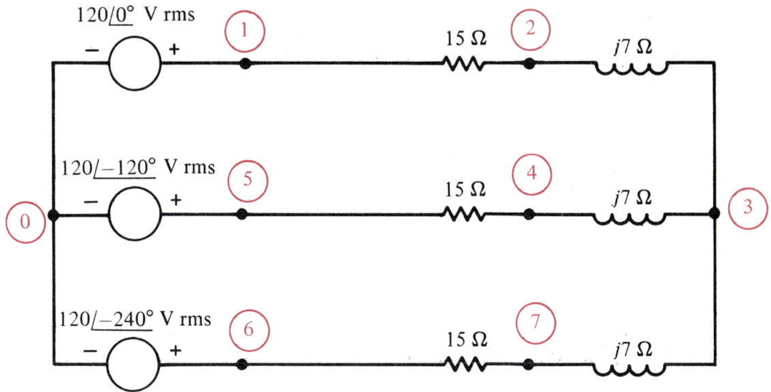

Figure D12.11

Ans: $\mathbf{I}_a = 7.249\ \underline{/-25°}$ A rms, $\mathbf{V}(1, 3) = 120\ \underline{/0°}$ V rms, $\mathbf{V}(1, 5) = 207.8\ \underline{/30°}$ V rms.

D12.12. The network in Fig. D12.12 is labeled for a PSPICE analysis. Write a PSPICE program for calculating the source currents, load currents, load voltage, and the voltage drop across the line impedance. $f = 1/2\pi$ Hz.

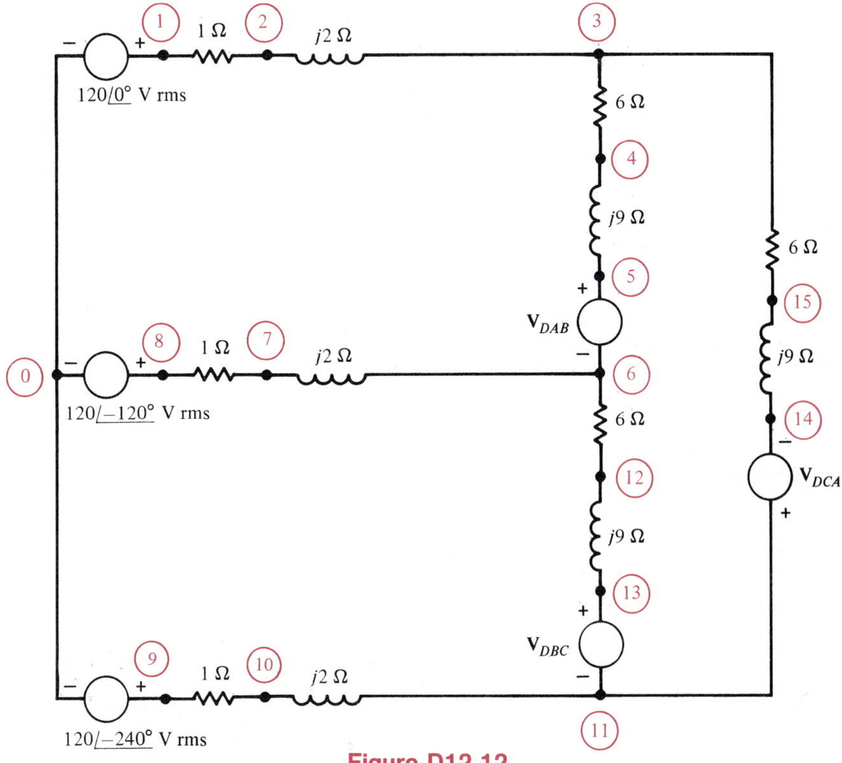

Figure D12.12

Ans: $\mathbf{I}_S = 20.58\ \underline{/121°}$ A rms, $\mathbf{I}_L = 11.88\ \underline{/-29.04°}$ A rms, $\mathbf{V}_L = 128.5\ \underline{/27.27°}$ V rms, $\mathbf{V}_{\text{line}} = 46.02\ \underline{/4.4°}$ V rms.

D12.13. The network in Fig. D12.13 is labeled for a PSPICE analysis. Write a PSPICE program for calculating the line currents, load currents, load voltages, and source currents. What is the purpose of the 1-TΩ resistors in the network? $f = 1/2\pi$ Hz.

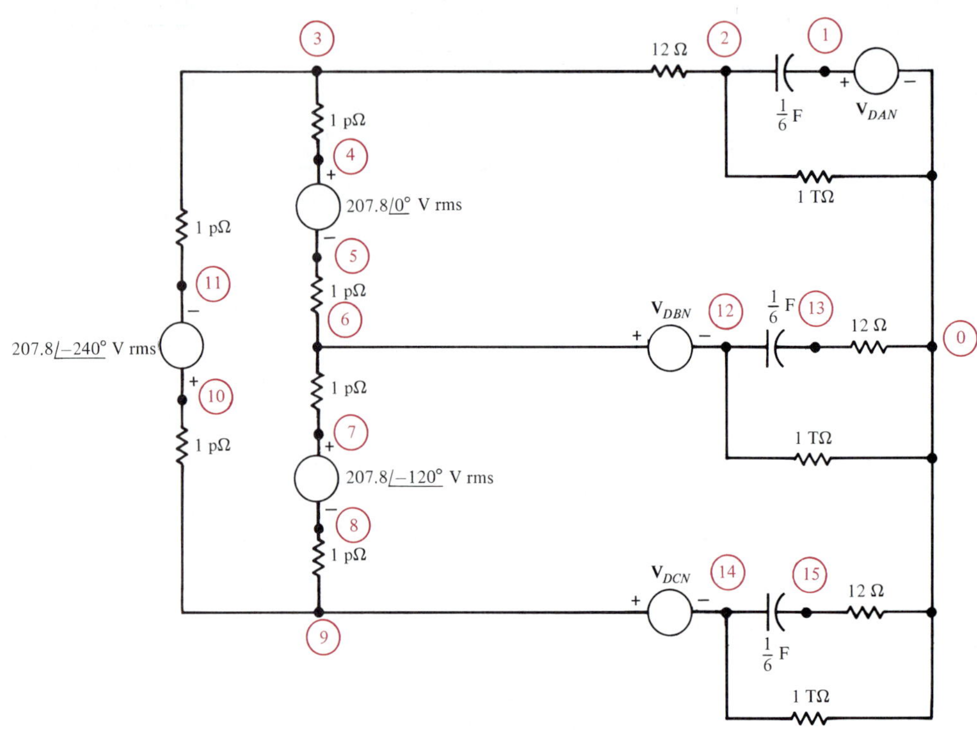

Figure D12.13

Ans: $\mathbf{I}_S = 5.162\ \underline{/-153.4^\circ}$ A rms, $\mathbf{I}_{\text{line}} = 8.942\ \underline{/-3.4^\circ}$ A rms, $\mathbf{I}_L = 8.942\ \underline{/-3.4^\circ}$ A rms, $\mathbf{V}_L = 120\ \underline{/-30^\circ}$ V rms.

12.4 Three-Phase Measurement

Power Measurement

The measurement of three-phase power would seem to be a simple task, since it would appear that in order to accomplish the measurement we would simply replicate the single-phase measurement for each phase. For a balanced system this task would become even simpler since we need only measure the power in one phase and multiply the wattmeter reading by 3. However, as we pointed out earlier, the neutral terminal in the wye-connected load may be inaccessible, and of course, in a delta-connected load there is no neutral point. Our measurement technique must, therefore, deal only with external lines to the load.

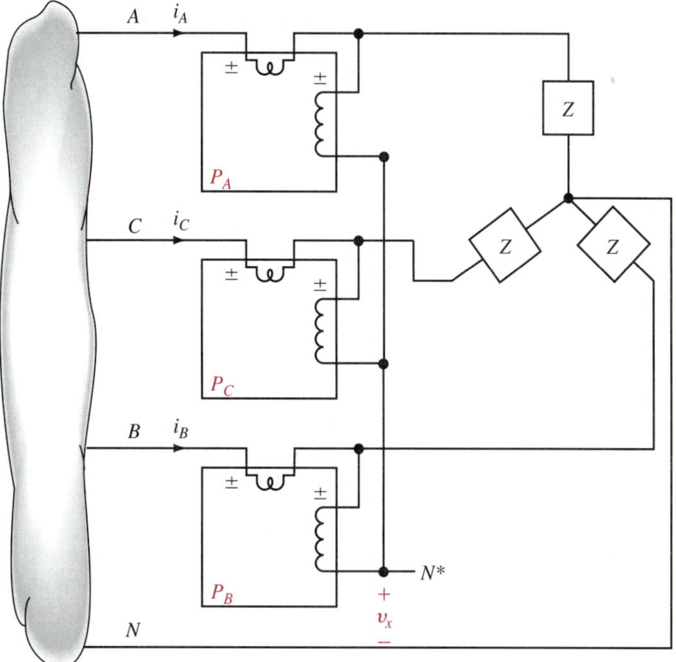

Figure 12.20 Wattmeter connections for power measurement.

The method we will describe is applicable for both wye and delta connections; however, we will present the measurement method using a balanced wye-connected load. Consider the circuit shown in Fig. 12.20. The current coil of each wattmeter is connected in series with the line current, and the voltage coil of each wattmeter is connected between a line and what we will call the virtual neutral N^*, which is nothing more than an arbitrary point.

The average power measured by wattmeter A is

$$P_A = \frac{1}{T} \int_0^T v_{AN^*} i_A \, dt \tag{12.36}$$

where T is the period of all voltages and currents in the system. In a similar manner,

$$P_B = \frac{1}{T} \int_0^T v_{BN^*} i_B \, dt \tag{12.37}$$

$$P_C = \frac{1}{T} \int_0^T v_{CN^*} i_C \, dt \tag{12.38}$$

The sum of all the wattmeter measurements is

$$P = \frac{1}{T} \int_0^T (v_{AN^*} i_A + v_{BN^*} i_B + v_{CN^*} i_C) \, dt \tag{12.39}$$

As shown in Fig. 12.20, the voltages v_{AN*}, v_{BN*}, and v_{CN*} can be expressed as

$$\begin{aligned} v_{AN*} &= v_{AN} - v_X \\ v_{BN*} &= v_{BN} - v_X \\ v_{CN*} &= v_{CN} - v_X \end{aligned} \tag{12.40}$$

Substituting Eq. (12.40) into Eq. (12.39) yields

$$P = \frac{1}{T}\int_0^T (v_{AN}i_A + v_{BN}i_B + v_{CN}i_C)\,dt - \frac{1}{T}\int_0^T v_X(i_A + i_B + i_C)\,dt \tag{12.41}$$

However,

$$i_A + i_B + i_C = 0 \tag{12.42}$$

and hence the expression for the power reduces to

$$P = \frac{1}{T}\int_0^T v_{AN}i_A\,dt + \frac{1}{T}\int_0^T v_{BN}i_B\,dt + \frac{1}{T}\int_0^T v_{CN}i_C\,dt \tag{12.43}$$

which we recognize to be the total power absorbed by the three-phase wye load. Therefore, using the three wattmeters shown in Fig. 12.20, we can measure the power absorbed by a three-phase load, and in addition, this method applies whether the system is balanced or unbalanced, and whether the load is wye or delta connected.

If we connect the virtual neutral to line C, the voltage coil on wattmeter C will have zero volts across it; the wattmeter will read zero watts, and can thus be removed. Hence the two wattmeters A and B will then measure all the power in the three-phase load. This configuration is shown in Fig. 12.21 and is known as the *two-wattmeter method* for power

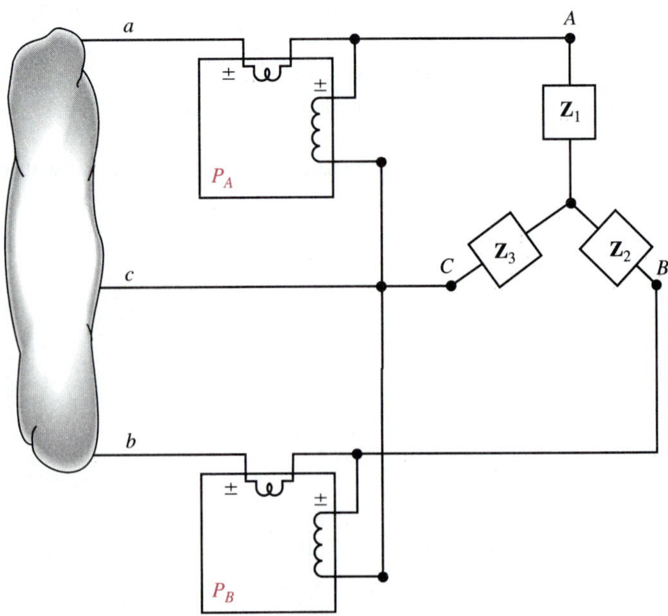

Figure 12.21 Two-wattmeter method for power measurement.

measurement. The two-wattmeter method can always be used when the load is balanced, but in the unbalanced case, is valid only for a three-wire load, that is, a load with no neutral conductor. In general, if there are n wires from the source to the load, $n - 1$ wattmeters are required. As shown in Fig. 12.21, the total power measured by the two-wattmeter method is

$$\begin{aligned} P_T = P_A + P_B &= |\mathbf{V}_{AC}||\mathbf{I}_{aA}| \cos (\underline{/\mathbf{V}_{AC}} - \underline{/\mathbf{I}_{aA}}) \\ &\quad + |\mathbf{V}_{BC}||\mathbf{I}_{bB}| \cos (\underline{/\mathbf{V}_{BC}} - \underline{/\mathbf{I}_{bB}}) \end{aligned} \tag{12.44}$$

where, for example, $\underline{/\mathbf{V}_{AC}}$ represents the phase angle of the voltage $\mathbf{V}_{AC}$.

EXAMPLE 12.17

A balanced wye–delta system has an *abc*-phase-sequence source with $\mathbf{V}_{an} = 120 \underline{/0^\circ}$ V rms. The balanced load has a phase impedance of $10 + j5\ \Omega$. We wish to find the power absorbed by the load using the two-wattmeter method.

If $\mathbf{V}_{an} = 120 \underline{/0^\circ}$ V rms, then

$$\mathbf{V}_{AB} = 208 \underline{/30^\circ} \text{ V rms}$$

$$\mathbf{V}_{BC} = 208 \underline{/-90^\circ} \text{ V rms}$$

$$\mathbf{V}_{CA} = 208 \underline{/-210^\circ} \text{ V rms}$$

and hence $\mathbf{V}_{AC} = 208 \underline{/-30^\circ}$ V rms. Since the phase impedance is $10 + j5 = 11.18 \underline{/26.57^\circ}\ \Omega$, the magnitude of the delta current is

$$I_\Delta = \frac{208}{11.18} = 18.60 \text{ A rms}$$

The average power per phase is, therefore,

$$\begin{aligned} P_p &= (18.60)^2(10) \\ &= 3461 \text{ W} \end{aligned}$$

and hence the total power is

$$P_T = 10{,}383 \text{ W}$$

Using the two-wattmeter method, we note from Eq. (12.24) that

$$\mathbf{I}_{aA} = 18.60 \sqrt{3} \underline{/-26.57^\circ} \text{ A rms}$$

$$\mathbf{I}_{bB} = 18.60 \sqrt{3} \underline{/-120 - 26.57^\circ} \text{ A rms}$$

Therefore, using Eq. (12.44), we have

$$\begin{aligned} P_T &= (208)(32.22) \cos (-30^\circ + 26.57^\circ) \\ &\quad + (208)(32.22) \cos (-90^\circ + 120^\circ + 26.57^\circ) \\ &= 10{,}383 \text{ W} \end{aligned}$$

■

DRILL EXERCISE

D12.14. Given the data in Drill Exercise D12.5, if the two-wattmeter method is used to measure power, find the reading of each wattmeter.

Ans: $P_A = 5185$ W, $P_B = 2303$ W.

Power Factor Measurement

The two-wattmeter method can also be used to compute the power factor angle if the load is balanced. To illustrate how this can be accomplished, consider the wye-connected load shown in Fig. 12.21, where $\mathbf{Z} = Z_Y \underline{/\theta°}$. If the source is an *abc*-sequence balanced wye with $\underline{/\mathbf{V}_{an}} = 0°$, then as shown in Eq. (12.44),

$$P_A = |\mathbf{V}_{AC}||\mathbf{I}_{aA}| \cos (\underline{/\mathbf{V}_{AC}} - \underline{/\mathbf{I}_{aA}})$$

From Eq. (12.13), $|\mathbf{V}_{AC}| = V_L$ and $\underline{/\mathbf{V}_{AC}} = \underline{/\mathbf{V}_{CA}} + 180° = -30°$. Similarly, from Eq. (12.18), $|\mathbf{I}_{aA}| = I_L$ and $\underline{/\mathbf{I}_{aA}} = -\theta$. Therefore,

$$P_A = V_L I_L \cos (-30° + \theta)$$

In a similar manner we can show that

$$P_B = V_L I_L \cos (30° + \theta)$$

Then the ratio of the two wattmeter readings is

$$\frac{P_A}{P_B} = \frac{\cos (\theta - 30°)}{\cos (\theta + 30°)}$$

If we now employ the trigonometric identities in Eq. (9.11) and recall that $\cos 30° = \sqrt{3}/2$ and $\sin 30° = 1/2$, it is straightforward to show that the equation above can be reduced to

$$\tan \theta = \frac{(P_A - P_B)\sqrt{3}}{P_A + P_B}$$

and since $P_T = P_A + P_B$,

$$\theta = \tan^{-1} \frac{(P_A - P_B)\sqrt{3}}{P_T} \tag{12.45}$$

The equations above indicate that if $P_A = P_B$, the load is resistive; if $P_A > P_B$, the load is inductive; and if $P_A < P_B$, the load is capacitive. Finally, this technique is valid whether the load is wye or delta connected.

EXAMPLE 12.18

In a balanced wye–delta system two wattmeters are connected to measure the total power. We wish to determine the power factor of the load if the wattmeter readings are $P_A = 1200$ W and $P_B = 480$ W.

Using Eq. (12.45), we have

$$\theta = \tan^{-1} \frac{(1200 - 480)\sqrt{3}}{1680}$$

$$= 36.59°$$

Therefore,

$$\cos\theta = \text{pf} = 0.80 \text{ lagging}$$

EXAMPLE 12.19

If the source in Example 12.18 has a line voltage of 208 V rms, we wish to find the delta load impedance.

The total power $P_T = P_A + P_B = 1680$ W; therefore, the power per phase is $P_p = 1680/3 = 560$ W. Hence using the expression

$$V_L I_\Delta \cos\theta = P_p$$

$$I_\Delta = \frac{560}{(208)(0.8)}$$

$$= 3.37 \text{ A rms}$$

Therefore,

$$\mathbf{Z} = \frac{208}{3.37} \angle 36.59° = 61.81 \angle 36.59°$$

$$= 49.63 + j36.84\ \Omega$$

EXAMPLE 12.20

In a balanced wye–delta system two wattmeters are used to measure total power. Wattmeter A reads 800 W and wattmeter B reads 400 W after the current coil terminals are reversed. If the line voltage is 208 V rms, we wish to determine the total average power, the power factor, and the impedance of the load.

From the data given $P_A = 800$ and $P_B = -400$; therefore, the total power P_T is

$$P_T = P_A + P_B$$

$$= 400 \text{ W}$$

The power factor is computed from

$$\theta = \tan^{-1} \frac{[800 - (-400)]\sqrt{3}}{400}$$

$$\theta = 79.11°$$

and therefore,

$$\cos\theta = \text{pf} = 0.19 \text{ lagging}$$

The delta impedance is determined from

$$V_L I_\Delta \cos\theta = P_p$$

$$(208)I_\Delta(0.19) = \frac{400}{3}$$

$$I_\Delta = 3.37 \text{ A rms}$$

Therefore,

$$\mathbf{Z}_\Delta = \frac{208}{3.37} \underline{/79.11^\circ} = 61.65 \underline{/79.11^\circ}\ \Omega$$

■

DRILL EXERCISE

D12.15. Two wattmeters are used to measure the total power in the load of a balanced wye–wye system. The line voltage is 208 V rms and the wattmeter readings are $P_A = 1600$ W and $P_B = 840$ W. Compute the impedance per phase of the load.

Ans: $\mathbf{Z} = 15.58 \underline{/28.35^\circ}\ \Omega$.

D12.16. Two wattmeters are used to measure the total power in the load of a balanced wye–wye system. The line voltage is 208 V rms. Wattmeter A reads 1280 W and wattmeter B reads 540 W when the current coil terminals are reversed. Determine the power factor of the load and the impedance per phase of the load.

Ans: pf = 0.23 lagging, $\mathbf{Z} = 13.43 \underline{/76.79^\circ}\ \Omega$.

12.5

Power Factor Correction

In Section 11.6 we illustrated a simple technique for raising the power factor of a load. The method involved judiciously selecting a capacitor and placing it in parallel with the load. In a balanced three-phase system, power factor correction is performed in exactly the same manner. It is important to note, however, that the $\mathbf{S}_{cap}$ specified in Eq. (11.42) is provided by three capacitors, and in addition, V_{rms} in the equation is the voltage across each capacitor. The following example illustrates the technique.

EXAMPLE 12.21

In the balanced three-phase system shown in Fig. 12.22, the line voltage is 34.5 kV rms at 60 Hz. We wish to find the values of the capacitors C such that the total load has a power factor of 0.94 leading.

Following the development outlined in Section 11.6 for single-phase power factor correction, we obtain

$$\mathbf{S}_{old} = 24 \underline{/\cos^{-1} 0.78} \text{ MVA}$$

$$= 18.72 + j15.02 \text{ MVA}$$

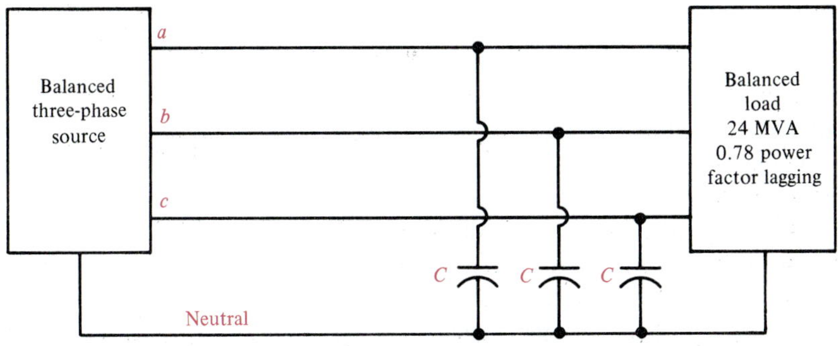

Figure 12.22 Network used in Example 12.21.

and

$$\theta_{\text{new}} = -\cos^{-1} 0.94$$
$$= -19.95°$$

Therefore,

$$\mathbf{S}_{\text{new}} = 18.72 + j18.72 \tan(-19.95°)$$
$$= 18.72 - j6.80 \text{ MVA}$$

and

$$\mathbf{S}_{\text{cap}} = \mathbf{S}_{\text{new}} - \mathbf{S}_{\text{old}}$$
$$= -j21.82 \text{ MVA}$$

However,

$$-j\omega C\, V_{\text{rms}}^2 = -j21.82 \text{ MVA}$$

and since the line voltage is 34.5 kV rms, then

$$(377)\left(\frac{34.5}{\sqrt{3}}\right)^2 C = \frac{21.82}{3}$$

and hence

$$C = 48.6\ \mu\text{F}$$

DRILL EXERCISE

D12.17. Find C in Example 12.21 such that the total load has a power factor of 0.90 lagging.

Ans: $C = 13.26\ \mu$F.

Finally, the reader will recall that our entire discussion in this chapter has focused on balanced systems. It is extremely important, however, to point out that in the unbalanced three-phase system the problem is much more complicated because of the mutual inductive coupling between phases in power apparatus.

12.6 Summary

We have shown that an important advantage of the balanced three-phase system is that it provides very smooth power delivery. Because of the balanced condition it is possible to analyze a circuit on a "per phase" basis, thereby providing a significant computational shortcut to a solution. The relationships between the various voltages and currents for the balanced system in both a wye and a delta configuration were found to be very simple. The ability to perform wye-to-delta and delta-to-wye transformations was shown to be an important ingredient in our computational approach.

Techniques for power measurement in a three-phase environment were presented. The two-wattmeter method for the measurement of three-phase power was derived, and the use of this method for power factor measurement was presented. Finally, power factor correction for balanced three-phase systems was discussed.

KEY POINTS

- A balanced three-phase voltage source has three sinusoidal voltages of the same magnitude and frequency, and each voltage is 120° out of phase with the others.
- If the load currents generated by connecting a load to a balanced three-phase voltage source are also balanced, the load is connected in either a balanced wye or a balanced delta configuration.
- A positive-phase-sequence balanced voltage source is one in which $\mathbf{V}_{bn}$ lags $\mathbf{V}_{an}$ by 120° and $\mathbf{V}_{cn}$ lags $\mathbf{V}_{bn}$ by 120°.
- There is no current in the neutral line of a balanced wye–wye system.
- Balanced three-phase ac circuits may be analyzed on a per phase basis.
- The two-wattmeter method is a technique for measuring real power in a three-phase system using only two wattmeters.
- The power factor angle of a load in a balanced three-phase system can be computed using two wattmeters.
- Power factor correction in a balanced three-phase environment is performed in the same manner as in the single-phase case.

PROBLEMS

12.1. In a three-phase balanced wye–wye system, the source is an *abc*-sequence set of voltages with $\mathbf{V}_{an} = 120 \angle 60°$ V rms. The per phase impedance of the load is $12 + j16\ \Omega$. If the line impedance per phase is $0.8 + j1.4\ \Omega$, find the line currents and the load voltages.

12.2. An *abc*-sequence set of voltages feeds a balanced three-phase wye–wye system. The line and load impedances are $0.6 + j1\ \Omega$ and $18 + j14\ \Omega$, respectively. If the load voltage on the a phase is $\mathbf{V}_{AN} = 114.47 \angle 18.99°$ V rms, determine the voltages at the line input.

12.3. In a balanced three-phase wye–wye system, the source is an *abc*-sequence set of voltages. The load voltage on the a phase is $\mathbf{V}_{AN} = 108.58 \angle 79.81°$ V rms, $\mathbf{Z}_{\text{line}} = 1 + j1.4\ \Omega$, and $\mathbf{Z}_{\text{load}} = 10 + j13\ \Omega$. Determine the input sequence of voltages.

12.4. A balanced *abc* sequence of voltages feeds a balanced three-phase wye–wye system. The line and load impedances are $0.6 + j0.9\ \Omega$ and $8 + j12\ \Omega$, respectively. The load voltage on the a phase is $\mathbf{V}_{AN} = 116.63 \angle 10°$ V rms. Find the line voltage $\mathbf{V}_{ab}$.

12.5. In a balanced three-phase wye–wye system, the source is an *abc*-sequence set of voltages. $\mathbf{Z}_{\text{line}} = 1 + j1.8\ \Omega$, $\mathbf{Z}_{\text{load}} = 14 + j12\ \Omega$, and the load voltage on the *a* phase is $\mathbf{V}_{AN} = 398.1 \underline{/17.99^\circ}$ V rms. Find the line voltage $\mathbf{V}_{ab}$.

12.6. An *abc*-sequence balanced three-phase source feeds a balanced wye–wye system. $\mathbf{Z}_{\text{line}} = 0.8 + j0.2\ \Omega$, $\mathbf{Z}_{\text{load}} = 12 + j6\ \Omega$, and $\underline{/\mathbf{V}_{AN}} = 30^\circ$. The total power absorbed by the load is 2 kW. Determine the total power loss in the lines.

12.7. An *abc*-phase-sequence balanced three-phase source feeds a balanced load. The system is wye–wye connected. The load impedance is $10 + j6\ \Omega$, the line impedance is $1 + j0.5\ \Omega$, and $\underline{/\mathbf{V}_{AN}} = 60^\circ$. The total power loss in the lines is 470.44 W. Find $\mathbf{V}_{AN}$ and the magnitude of the source voltage.

12.8. An *abc*-phase-sequence source supplies power in a balanced three-phase wye–wye system. The load impedance is $20 + j16\ \Omega$, the line impedance is $2 + j1\ \Omega$, and the total power loss in the lines is 600 W. Find the line current if the load is suddenly short circuited.

12.9. In a balanced three-phase wye–wye system the magnitude of the line current is known to be 6.18 A rms. The load power factor is 0.83 lagging, and the total power absorbed by the load is 1718.66 W. If the line impedance is $1 + j1\ \Omega$, compute the load impedance and the magnitude of the phase voltage at the source.

12.10. In a balanced three-phase wye–wye system, the source is an *abc*-sequence set of voltages and $\mathbf{V}_{an} = 120 \underline{/50^\circ}$ V rms. The load voltage on the *a* phase is $110.65 \underline{/29.03^\circ}$ V rms and the load impedance is $16 + j20\ \Omega$. Find the line impedance.

12.11. An *abc*-sequence set of voltages feeds a balanced three-phase wye–wye system. If the input voltage $\mathbf{V}_{an} = 440 \underline{/10^\circ}$ V rms, the load voltage on the *a* phase is $\mathbf{V}_{AN} = 398.32 \underline{/8.72^\circ}$ V rms, and $\mathbf{Z}_{\text{load}}$ is $20 + j24\ \Omega$, find the line impedance.

12.12. In a balanced three-phase wye–wye system, the source is an *abc*-sequence set of voltages and $\mathbf{V}_{an} = 120 \underline{/40^\circ}$ V rms. If the *a*-phase line current and line impedance are known to be $7.10 \underline{/-10.28^\circ}$ A rms and $0.8 + j1\ \Omega$, respectively, find the load impedance.

12.13. An *abc*-sequence set of voltages feeds a balanced three-phase wye–wye system. If the line current in the *a* phase is $16.78 \underline{/20.98^\circ}$ A rms, the line impedance is $1.2 + j1.8\ \Omega$, and the input voltage $\mathbf{V}_{an} = 440 \underline{/70^\circ}$ V rms, find the load impedance.

12.14. In a balanced three-phase wye–wye system, the source is an *abc*-sequence set of voltages and $\mathbf{V}_{an} = 120 \underline{/20^\circ}$ V rms. If the line impedance and load voltage are $0.5 + j0.8\ \Omega$ and $\mathbf{V}_{AN} = 112.95 \underline{/19.71^\circ}$ V rms, find the load impedance.

12.15. An *abc*-sequence set of voltages feeds a balanced three-phase wye–wye system. If $\mathbf{V}_{an} = 440 \underline{/30^\circ}$ V rms, $\mathbf{V}_{AN} = 413.28 \underline{/29.78^\circ}$ V rms, and $\mathbf{Z}_{\text{line}} = 1.2 + j1.5\ \Omega$, find the load impedance.

12.16. An *abc*-phase-sequence three-phase balanced wye-connected source supplies a balanced delta-connected load. The impedance per phase in the delta load is $12 + j9\ \Omega$. The line voltage at the source is $\mathbf{V}_{ab} = 120\sqrt{3} \underline{/40^\circ}$ V rms. If the line impedance is zero, find the line currents in the balanced wye–delta system.

12.17. An *abc*-phase-sequence three-phase balanced wye-connected 60-Hz source supplies a balanced delta-connected load. The phase impedance in the load consists of a 20-Ω resistor in series with a 50-mH inductor, and the phase voltage at the source is $\mathbf{V}_{an} = 120 \underline{/20^\circ}$ V rms. If the line impedance is zero, find the line currents in the system.

12.18. An *abc*-phase-sequence three-phase balanced wye-connected source supplies power to a balanced delta-connected load. The impedance per phase in the load is $14 + j12\ \Omega$. If the source voltage for the *a* phase is $\mathbf{V}_{an} = 120 \underline{/80^\circ}$ V rms, and the line impedance is zero, find the phase currents in the wye-connected source.

12.19. An *abc*-phase-sequence three-phase balanced wye-connected source supplies a balanced delta-connected load. The impedance per phase of the delta load is $10 + j8\ \Omega$. If the line impedance is zero and the line current in the *a* phase is known to be $\mathbf{I}_{aA} = 28.10 \underline{/-28.66^\circ}$ A rms, find the load voltage $\mathbf{V}_{AB}$.

12.20. An *abc*-phase-sequence three-phase balanced wye-connected source supplies power to a balanced delta-connected load. The impedance per phase of the delta load is $14 + j11\ \Omega$. If the line impedance is zero and the line current in the *a* phase is $\mathbf{I}_{aA} = 20.22 \underline{/31.84^\circ}$ A rms, find the voltages of the balanced source.

12.21. In a balanced three-phase wye–delta system, the source has an *abc* phase sequence and $\mathbf{V}_{an} = 120 \underline{/0^\circ}$ V rms. If the line current is $\mathbf{I}_{aA} = 4.8 \underline{/20^\circ}$ A rms, find the load impedance per phase in the delta.

12.22. In a balanced three-phase wye–delta system the source has an *abc* phase sequence and $\mathbf{V}_{an} = 120 \underline{/40^\circ}$ V rms. The line and load impedances are $0.5 + j0.4\ \Omega$ and $24 + j18\ \Omega$, respectively. Find the delta currents in the load.

12.23. Given the network in Fig. P12.23, compute the line currents and the magnitude of the phase voltage at the load.

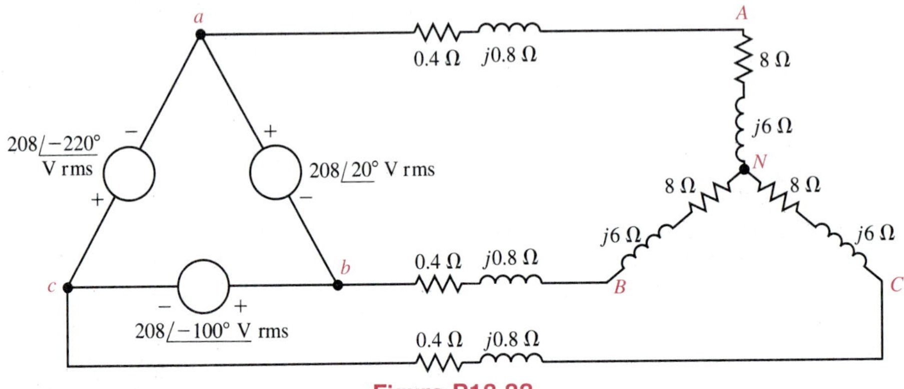

Figure P12.23

12.24. In a three-phase balanced system a delta-connected source supplies power to a wye-connected load. If the line impedance is $0.2 + j0.4\ \Omega$, the load impedance $6 + j4\ \Omega$, and the source phase voltage $\mathbf{V}_{ab} = 208\ \underline{/40^\circ}$ V rms, find the magnitude of the line voltage at the load.

12.25. In a three-phase balanced system a delta-connected source supplies power to a wye-connected load. The per phase values of the line and load impedance are $0.5 + j0.5\ \Omega$ and $9 + j6\ \Omega$, respectively. If the source phase voltage is $\mathbf{V}_{ab} = 208\ \underline{/60^\circ}$ V rms, find the magnitude of the currents in the delta sources.

12.26. In a three-phase balanced delta–delta system, the source has an *abc* phase sequence. The line and load impedances are $0.3 + j0.2\ \Omega$ and $9 + j6\ \Omega$, respectively. If the load current in the delta is $\mathbf{I}_{AB} = 15\ \underline{/40^\circ}$ A rms, find the phase voltages of the source.

12.27. In a balanced three-phase delta–delta system, the source has an *abc* phase sequence. The phase angle for the source voltage is $\underline{/\mathbf{V}_{ab}} = 40^\circ$ and $\mathbf{I}_{ba} = 4\ \underline{/15^\circ}$ A rms. If the total power absorbed by the load is 1400 W, find the load impedance.

12.28. A three-phase load impedance consists of a balanced wye in parallel with a balanced delta. What is the equivalent wye load and what is the equivalent delta load if the phase impedance of the wye and delta are $6 + j3\ \Omega$ and $15 + j12\ \Omega$, respectively?

12.29. In a balanced three-phase system, the *abc*-phase-sequence source is wye connected and $\mathbf{V}_{an} = 120\ \underline{/20^\circ}$ V rms. The load consists of two balanced wyes with phase impedances of $8 + j6\ \Omega$ and $12 + j8\ \Omega$. If the line impedance is zero, find the line currents and the phase current in each load.

12.30. In a balanced three-phase system, the source is a balanced wye with an *abc* phase sequence and $\mathbf{V}_{ab} = 208\ \underline{/60^\circ}$ V rms. The load consists of a balanced wye with a phase impedance of $8 + j5\ \Omega$ in parallel with a balanced delta with a phase impedance of $21 + j12\ \Omega$. If the line impedance is $1.2 + j1\ \Omega$, find the phase currents in the balanced wye load.

12.31. In a balanced three-phase system, the source is a balanced wye with an *abc* phase sequence and $\mathbf{V}_{an} = 120\ \underline{/0^\circ}$ V rms. The load is a balanced wye with a phase impedance of $6 + j4\ \Omega$ in parallel with a balanced delta having a phase impedance of $12 + j6\ \Omega$. If the line impedance is $1 + j1\ \Omega$, find the magnitude of the phase voltage of the delta.

12.32. In a balanced three-phase system, the source is a balanced wye with an *abc* phase sequence and $\mathbf{V}_{ab} = 208\ \underline{/50^\circ}$ V rms. The load is a balanced wye in parallel with a balanced delta. The phase impedance of the wye is $5 + j3\ \Omega$ and the phase impedance of the delta is $18 + j12\ \Omega$. If the line impedance is $1 + j0.8\ \Omega$, find the line currents and the phase currents in the loads.

12.33. In a balanced three-phase system the source, which has an *abc* phase sequence, is connected in delta and $\mathbf{V}_{ab} = 208\ \underline{/55^\circ}$ V rms. There are two loads connected in parallel. Load 1 is connected in wye and the phase impedance is $4 + j3\ \Omega$. Load 2 is connected in wye and the phase impedance is $8 + j6\ \Omega$. Compute the delta currents in the source if the line impedance connecting the source to the loads is $0.2 + j0.1\ \Omega$.

12.34. In a balanced three-phase system the source has an *abc* phase sequence and is connected in delta. There are two parallel wye-connected loads. The phase impedance of load 1 and load 2 is $4 + j4\ \Omega$ and $10 + j4\ \Omega$, respectively. The line impedance connecting the source to the loads is $0.3 + j0.2\ \Omega$. If the current in the a phase of load 1 is $\mathbf{I}_{AN_1} = 10 \underline{/20^\circ}$ A rms, find the delta currents in the source.

12.35. In a balanced three-phase system the source has an abc phase sequence and is connected in delta. There are two loads connected in parallel. The line connecting the source to the loads has an impedance of $0.2 + j0.1\ \Omega$. Load 1 is connected in wye and the phase impedance is $4 + j2\ \Omega$. Load 2 is connected in delta and the phase impedance is $12 + j9\ \Omega$. The current $\mathbf{I}_{AB}$ in the delta load is $16 \underline{/45^\circ}$ A rms. Find the phase voltages of the source.

12.36. In a balanced three-phase system the source has an abc phase sequence and is connected in delta. There are two loads connected in parallel. Load 1 is connected in wye and has a phase impedance of $6 + j2\ \Omega$. Load 2 is connected in delta and has a phase impedance of $9 + j3\ \Omega$. The line impedance is $0.4 + j0.3\ \Omega$. Determine the phase voltages of the source if the current in the a phase of load 1 is $\mathbf{I}_{AN_1} = 12 \underline{/30^\circ}$ A rms.

12.37. A three-phase abc-sequence wye-connected source supplies 14 kVA with a power factor of 0.75 lagging to a parallel combination of a wye load and a delta load. If the wye load consumes 9 kVA at a power factor of 0.6 lagging and has a phase-a current of $10 \underline{/-30^\circ}$ A rms, determine the phase impedance of the delta load.

12.38. An abc-sequence wye-connected source having a phase-a voltage of $120 \underline{/0^\circ}$ V rms is attached to a wye-connected load having an impedance of $80 \underline{/70^\circ}\ \Omega$. If the line impedance is $4 \underline{/20^\circ}\ \Omega$, determine the total complex power produced by the voltage sources and the real and reactive power dissipated by the load.

12.39. A three-phase abc-sequence wye-connected source with $\mathbf{V}_{an} = 220 \underline{/0^\circ}$ V rms supplies power to a wye-connected load that consumes 50 kW of power in each phase at a pf of 0.8 lagging. Three capacitors are found that have an impedance of $-j2.0\ \Omega$, and they are connected in parallel with the previous load in a wye configuration. Determine the power factor of the combined load as seen by the source.

12.40. If the three capacitors in the network in Problem 12.39 are connected in a delta configuration, determine the power factor of the combined load as seen by the source.

12.41. Two industrial plants represent balanced three-phase loads. The plants receive their power from a balanced three-phase source with a line voltage of 4.6 kV rms. Plant 1 is rated at 300 kVA, 0.8 pf lagging and plant 2 is rated at 350 kVA, 0.84 pf lagging. Determine the power line current.

12.42. A three-phase balanced wye–wye system has a line voltage of 208 V rms. The line current is 6 A rms and the total real power absorbed by the load is 1800 W. Determine the load impedance per phase.

12.43. The magnitude of the complex power (apparent power) supplied by a three-phase balanced wye–wye system is 3600 VA. The line voltage is 208 V rms. If the line impedance is negligible and the power factor angle of the load is 25°, determine the load impedance.

12.44. In a balanced three-phase system, the source has an abc phase sequence, is Y connected, and $\mathbf{V}_{an} = 120 \underline{/20^\circ}$ V rms. The source feeds two loads, both of which are Y connected. The impedance of load 1 is $8 + j6\ \Omega$. The complex power for the a phase of load 2 is $600 \underline{/36^\circ}$ VA. Find the line current for the a phase and the total complex power of the source.

12.45. A cluster of loads are served by a balanced three-phase source with a line voltage of 4160 V rms. Load 1 is 240 kVA at 0.8 pf lagging and load 2 is 160 kVA at 0.92 pf lagging. A third load is unknown except that it has a power factor of unity. If the line current is measured and found to be 62 A rms, find the complex power of the unknown load.

12.46. The following loads are served by a balanced three-phase source:

Load 1: 18 kVA at 0.8 pf lagging
Load 2: 8 kVA at 0.8 pf leading
Load 3: 12 kVA at 0.75 pf lagging

The load voltage is 208 V rms at 60 Hz. If the line impedance is negligible, find the power factor of the source.

12.47. A small shopping center contains three stores that represent three balanced three-phase loads. The power lines to the shopping center represent a three-phase source with a line voltage of 13.8 kV rms. The three loads are

Load 1: 500 kVA at 0.8 pf lagging
Load 2: 400 kVA at 0.85 pf lagging
Load 3: 300 kVA at 0.90 pf lagging

Find the power line current.

12.48. A balanced three-phase source serves the following loads:

Load 1: 18 kVA at 0.8 pf lagging
Load 2: 10 kVA at 0.7 pf leading
Load 3: 12 kW at unity pf
Load 4: 16 kVA at 0.6 pf lagging

The load voltage is 208 V rms at 60 Hz and the line impedance is $0.02 + j0.04\ \Omega$. Find the line voltage and power factor at the source.

12.49. A balanced three-phase source supplies power to three loads. The loads are

Load 1: 18 kW at 0.8 pf lagging

Load 2: 10 kVA at 0.6 pf leading

Load 3: unknown

If the line voltage at the loads is 208 V rms, the line current at the source is 116.39 A rms, and the combined power factor at the load is 0.86 lagging, find the unknown load.

12.50. A balanced three-phase source supplies power to three loads. The loads are

Load 1: 30 kVA at 0.8 pf lagging

Load 2: 24 kW at 0.6 pf leading

Load 3: unknown

The line voltage at the load and line current at the source are 208 V rms and 166.8 A rms, respectively. If the combined power factor at the load is unity, find the unknown load.

12.51. A balanced three-phase source supplies power to three loads. The loads are

Load 1: 20 kVA at 0.6 pf lagging

Load 2: 12 kW at 0.75 pf lagging

Load 3: unknown

If the line voltage at the load is 208 V rms, the line current at the source is 98.60 A rms, and the combined power factor at the load is 0.88 pf lagging, find the unknown load.

12.52. Calculate the line current, load current, and load voltages in the network in Fig. P12.52 using PSPICE. $f = 1/2\pi$ Hz.

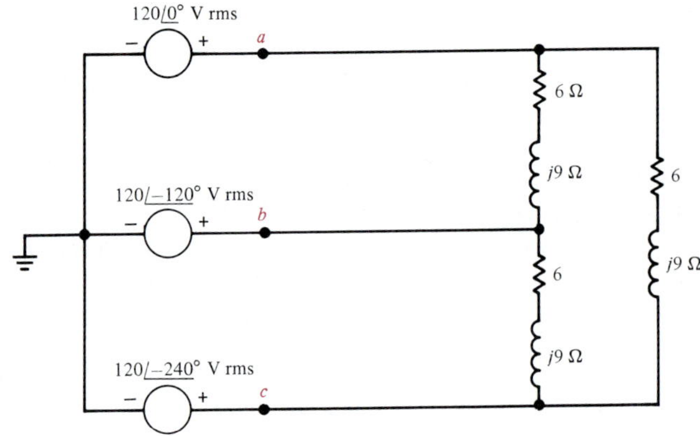

Figure P12.52

12.53. Find the source current, load current, and voltage across the line impedances in the network in Fig. P12.53 using PSPICE. $f = 1/2\pi$ Hz.

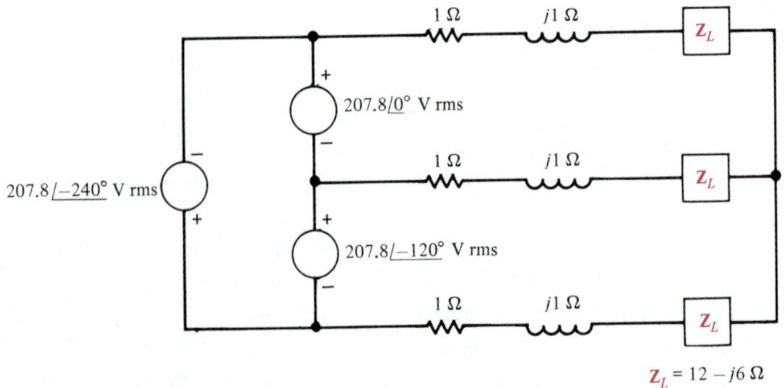

Figure P12.53

12.54. Use PSPICE to calculate the source current, both load currents, and both load voltages in the circuit in Fig. P12.54. $f = 1/2\pi$ Hz.

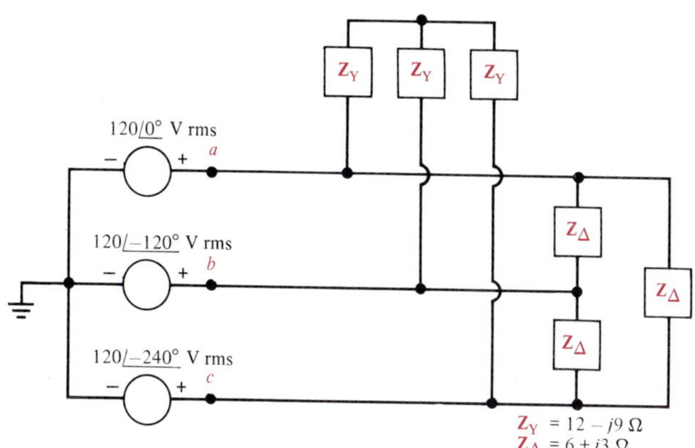

Figure P12.54

12.55. Given the circuit in Fig. P12.55, use PSPICE to calculate the source current, line-to-line voltage, power loss in the line impedance, and both load voltages. $f = 1/2\pi$ Hz.

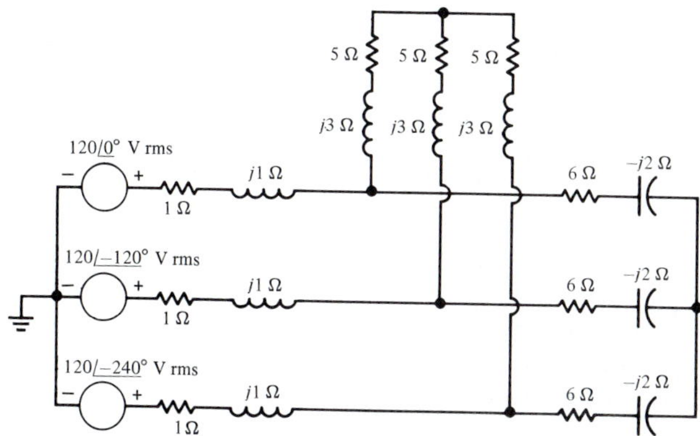

Figure P12.55

12.56. In a balanced three-phase system the source has an *abc* phase sequence, is connected in wye, and $\mathbf{V}_{an} = 120\,\underline{/45^\circ}$ V rms. There are two parallel loads connected in wye. Load 1 has a phase impedance of $4 + j3\ \Omega$ and load 2 has a phase impedance of $12 + j8\ \Omega$. Find the total power absorbed by the loads using the two-wattmeter method.

12.57. Two wattmeters are employed to measure the total power in a balanced three-phase wye–delta system. The phase voltage of the source is 120 V rms. Without reversing the meter terminals, the meter readings are $P_A = 1600$ W and $P_B = 800$ W. Determine the impedance per phase of the delta.

12.58. A balanced three-phase system has an *abc*-phase-sequence wye-connected source and $\mathbf{V}_{an} = 120\,\underline{/60^\circ}$ V rms. Two loads are connected in parallel. Load 1 is connected in wye, and the phase impedance is $6 + j2\ \Omega$. Load 2 is connected in delta, and the phase impedance is $8 + j3\ \Omega$. Find the total power absorbed by the loads using the two-wattmeter method.

12.59. A balanced wye–wye three-phase system has an *abc*-sequence source with $\mathbf{V}_{an} = 90\,\underline{/30^\circ}$ V rms. The balanced load has a per phase impedance of $24 + j16\ \Omega$. Calculate the readings of the wattmeters if the two-wattmeter method is used to measure the total three-phase power.

12.60. The balanced three-phase source in Fig. P12.60 has an *abc* phase sequence. Find the reading of each wattmeter in the network.

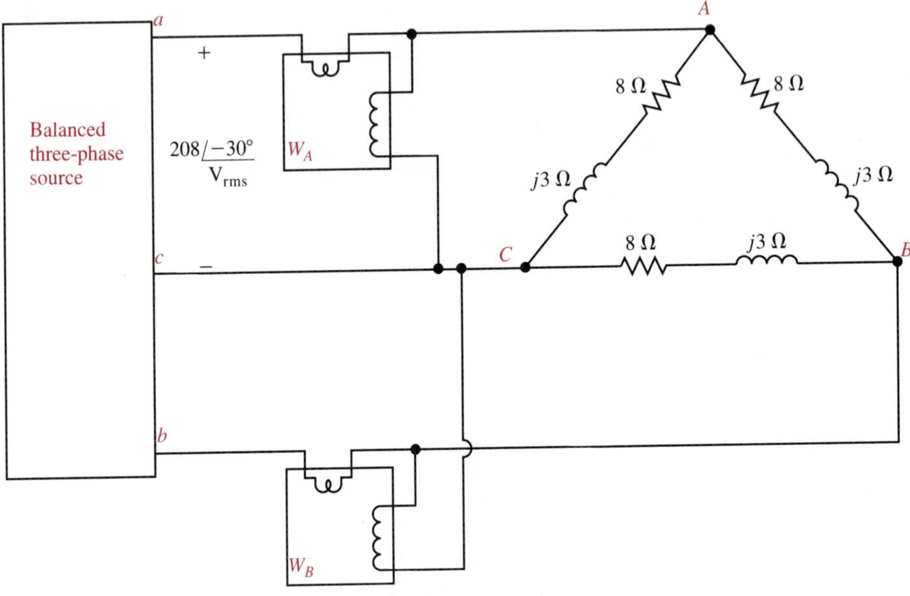

Figure P12.60

12.61. In a balanced wye–delta three-phase system two wattmeters are used to measure total power. The line voltage is 208 V rms. If wattmeter *A* reads 1200 W and wattmeter *B* reads 400 W when the current terminals are reversed, determine the power factor and the load impedance per phase.

12.62. A balanced wye–delta three-phase system employs two wattmeters to measure the total power. If the wattmeters read P_A = 800 W and P_B = 400 W without reversing the wattmeter terminals, and the line voltage is 208 V rms, find the value of the load impedance.

12.63. Given the network in Example 12.21, if the balanced three-phase load is 10 MVA at 0.8 pf lagging and the line voltage is 4.6 kV rms, find *C* so that the total load has a pf of 0.90 leading.

12.64. Find *C* in Problem 12.63 so that the total load has a power factor of 0.95 lagging.

12.65. Find *C* in the network in Fig. P12.65 such that the total load has a power factor of 0.92 leading.

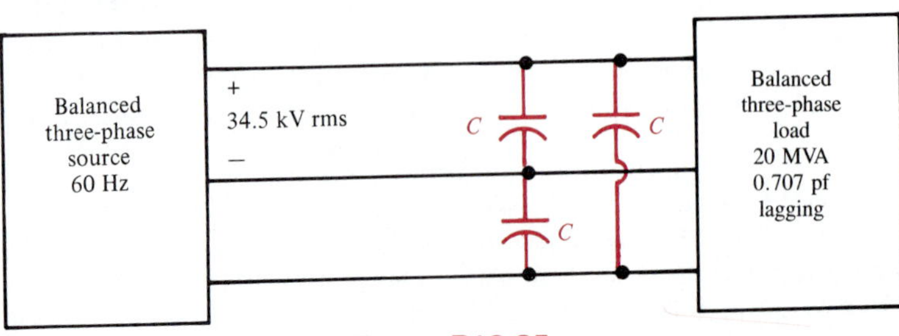

Figure P12.65

12.66. Find the value of C in Problem 12.65 such that the total load has a power factor of 0.92 lagging.

12.67. Find C in the network in Fig. P12.67 such that the total load has a power factor of 0.9 lagging.

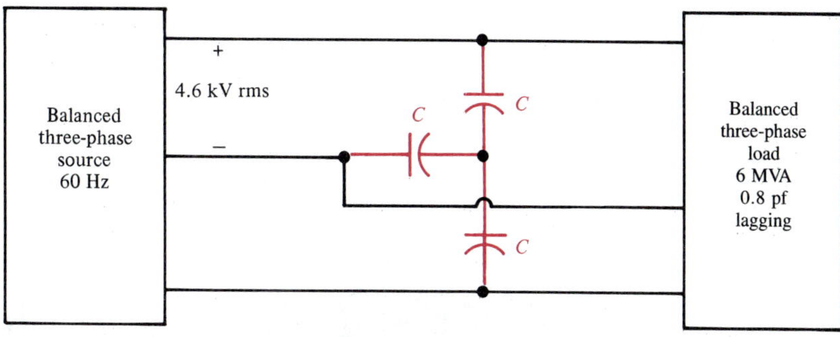

Figure P12.67

12.68. Find C in the network in Fig. P12.67 so that the total load has a power factor of 0.9 leading.

CHAPTER

13

Magnetically Coupled Networks

We now introduce a new four-terminal element, which is called a *transformer*. This circuit element consists of two inductors that are placed in close proximity to one another. Because of their close proximity, they share a common magnetic flux, and therefore, the inductor coils are said to be mutually coupled.

We begin with a general description of two coupled coils and then show how circuit equations can be written for networks containing coupled inductors. We will next consider coils that are coupled with good magnetic material and then derive an approximation for ideal coupling, called the *ideal transformer*. The PSPICE circuit analysis program will be applied to magnetically coupled networks and circuits containing ideal transformers. Finally, autotransformers and three-phase transformers are presented.

Transformers are very important because we encounter them in a variety of applications. For example, in communication systems transformers are used for impedance matching between sources and loads or transmission lines. Transformers provide dc isolation and isolation for safety. They are also used to provide matching between a grounded source and a balanced transmission line.

In power systems, transformers are used to step the voltage up or down. In fact, transformers are built into special wall plugs that are used to step down the voltage for the purpose of recharging batteries for calculators and hand tools.

13.1 Mutual Inductance

To begin our description of two coupled coils, we employ *Faraday's law*, which can be stated as follows: The induced voltage in a coil is proportional to the time rate of change of flux and the number of turns, N, in the coil. The two coupled coils are shown in Fig. 13.1 together with the following flux components.

ϕ_{l_1}	The flux in coil 1, which does not link coil 2, that is produced by the current in coil 1
ϕ_{l_2}	The flux in coil 2, which does not link coil 1, that is produced by the current in coil 2
ϕ_{12}	The flux in coil 1 produced by the current in coil 2
ϕ_{21}	The flux in coil 2 produced by the current in coil 1
$\phi_{11} = \phi_{l_1} + \phi_{21}$	The flux in coil 1 produced by the current in coil 1
$\phi_{22} = \phi_{l_2} + \phi_{12}$	The flux in coil 2 produced by the current in coil 2
ϕ_1	The total flux in coil 1
ϕ_2	The total flux in coil 2

In order to write the equations that describe the coupled coils, we define the voltages and currents, using the passive sign convention, at each pair of terminals as shown in Fig. 13.1.

Mathematically, Faraday's law can be written as

$$v_1(t) = N_1 \frac{d\phi_1}{dt} \tag{13.1}$$

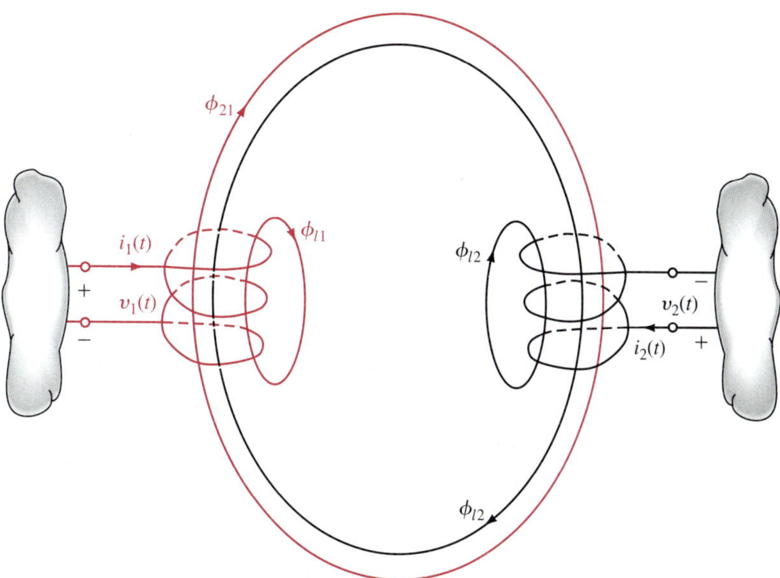

Figure 13.1 Flux relationships for mutually coupled coils.

The flux ϕ_1 will be equal to ϕ_{11}, the flux in coil 1 caused by current in coil 1, plus or minus the flux in coil 1 caused by current in coil 2; that is,

$$\phi_1 = \phi_{11} + \phi_{12} \tag{13.2}$$

If the current in coil 2 is such that the fluxes add, then the plus sign is used; if the current in coil 2 is such that the fluxes oppose one another, a minus sign is used. The equation for the voltage can now be written

$$\begin{aligned} v_1(t) &= N_1 \frac{d\phi_1}{dt} \\ &= N_1 \frac{d\phi_{11}}{dt} + N_1 \frac{d\phi_{12}}{dt} \end{aligned} \tag{13.3}$$

From basic physics we know that

$$\begin{aligned} \phi_{11} &= N_1 i_1 \mathcal{P}_{11} \\ \phi_{12} &= N_2 i_2 \mathcal{P}_{12} \end{aligned} \tag{13.4}$$

where the $\mathcal{P}$'s are constants (permeances) that depend on the magnetic paths taken by the flux components. The voltage equation can then be written

$$v_1(t) = N_1^2 \mathcal{P}_{11} \frac{di_1}{dt} + N_1 N_2 \mathcal{P}_{12} \frac{di_2}{dt} \tag{13.5}$$

The constant $N_1^2 \mathcal{P}_{11} = L_{11}$ (the same L that we used before) is now called the *self-inductance,* and the constant $N_1 N_2 \mathcal{P}_{12} = L_{12}$ is called the *mutual inductance*. Therefore,

$$v_1(t) = L_{11} \frac{di_1}{dt} + L_{12} \frac{di_2}{dt} \tag{13.6}$$

Using the same technique, we can write

$$v_2(t) = N_2^2 \mathcal{P}_{22} \frac{di_2}{dt} + N_1 N_2 \mathcal{P}_{21} \frac{di_1}{dt} \tag{13.7}$$

which can be written

$$v_2(t) = L_{22} \frac{di_2}{dt} + L_{21} \frac{di_1}{dt} \tag{13.8}$$

If the media through which the magnetic flux passes is linear, then $\mathcal{P}_{12} = \mathcal{P}_{21}$. Hence $L_{12} = L_{21} = M$. For convenience, let us define $L_1 = L_{11}$ and $L_2 = L_{22}$.

We now need to examine the physical details of coupled coils. In basic physics we learned the *right-hand rule,* which states that if we curl the fingers of our right hand around the coil in the direction of the current, the flux produced by the current is in the direction of our thumb.

In order to indicate the physical relationship of the coils and, therefore, simplify the sign convention for the mutual terms, we employ what is commonly called the *dot convention*. Dots are placed beside each coil so that if currents are entering both dotted

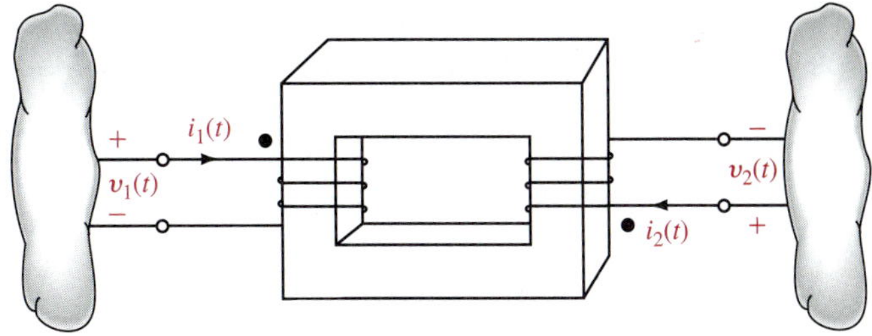

Figure 13.2 Illustration of the dot convention for coupled coils.

terminals or leaving both dotted terminals, the fluxes produced by these currents will add. In order to place the dots on a pair of coupled coils, we arbitrarily select one terminal of either coil and place a dot there. Using the right-hand rule, we determine the direction of the flux produced by this coil when current is entering the dotted terminal. We then examine the other coil to determine which terminal the current would have to enter to produce a flux that would add to the flux produced by the first coil. Place a dot on this terminal. The dots have been placed on the two coupled circuits in Fig. 13.2; verify that they are correct.

When the equations for the terminal voltages are written, the dots can be used to define the sign of the mutually induced voltages. If the currents $i_1(t)$ and $i_2(t)$ are both entering or leaving dots, the sign of the mutual voltage $M(di_2/dt)$ will be the same in an equation as the self-induced voltage $L_1(di_1/dt)$. If one current enters a dot and the other current leaves a dot, the mutual induced voltage and self-induced voltage terms will have opposite signs.

EXAMPLE 13.1

Determine the expressions for $v_1(t)$ and $v_2(t)$ in the circuits shown in Fig. 13.3.

For the circuit in Fig. 13.3 the voltage equations for the variables as assigned on the figure are

$$v_1(t) = L_1 \frac{di_1}{dt} + M \frac{di_2}{dt}$$

$$v_2(t) = L_2 \frac{di_2}{dt} + M \frac{di_1}{dt}$$

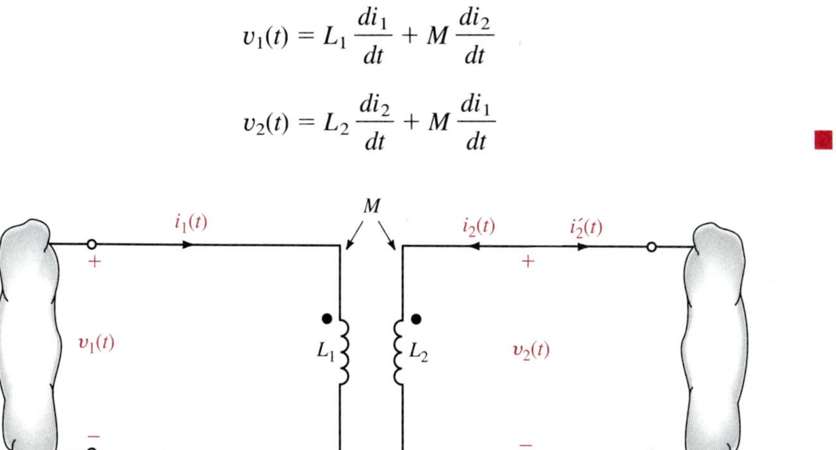

Figure 13.3 Two mutually coupled inductor circuits.

EXAMPLE 13.2

Determine the equations for $v_1(t)$ and $v_2(t)$ in the circuits shown in Fig 13.3 if we replace $i_2(t)$ with $i_2'(t)$ (i.e., secondary current leaving the dot).

The equations are

$$v_1(t) = L_1 \frac{di_1}{dt} - M \frac{di_2'}{dt}$$

$$v_2(t) = -L_2 \frac{di_2'}{dt} + M \frac{di_1}{dt}$$

■

DRILL EXERCISE

D13.1. Write the equations for the $v_1(t)$ and $v_2(t)$ in the circuit in Fig. D13.1.

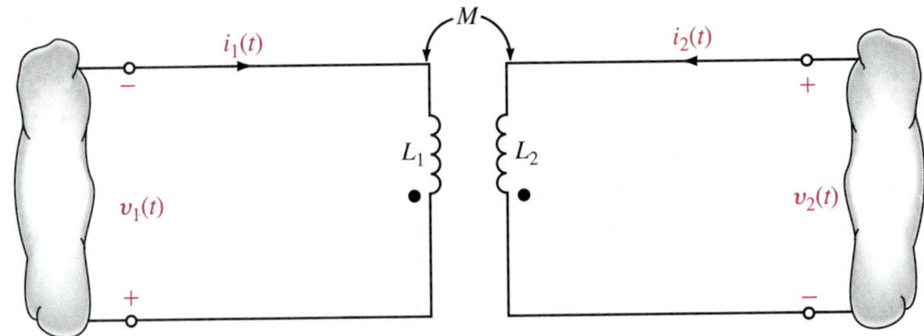

Figure D13.1

Ans: $v_1(t) = -L_1 \frac{di_1}{dt} - M \frac{di_2}{dt}$, $v_2(t) = L_2 \frac{di_2}{dt} + M \frac{di_1}{dt}$

Assume the coupled circuit in Fig. 13.3 is excited with a sinusoidal source. The voltages will be of the form $\mathbf{V}_1 e^{j\omega t}$ and $\mathbf{V}_2 e^{j\omega t}$, and the currents will be of the form $\mathbf{I}_1 e^{j\omega t}$ and $\mathbf{I}_2 e^{j\omega t}$, where $\mathbf{V}_1$, $\mathbf{V}_2$, $\mathbf{I}_1$, and $\mathbf{I}_2$ are phasors. Substituting these voltages and currents into Eqs. (13.6) and (13.8), we obtain

$$\begin{aligned} \mathbf{V}_1 &= j\omega L_1 \mathbf{I}_1 + j\omega M \mathbf{I}_2 \\ \mathbf{V}_2 &= j\omega L_2 \mathbf{I}_2 + j\omega M \mathbf{I}_1 \end{aligned} \tag{13.9}$$

The model of the coupled circuit in the frequency domain is identical to that in the time domain except for the way the elements and variables are labeled. The sign on the mutual terms is handled in the same manner as is done in the time domain.

EXAMPLE 13.3

The two mutually coupled coils in Fig. 13.4a can be interconnected in four possible ways. We wish to determine the equivalent inductance of each of the four possible interconnections.

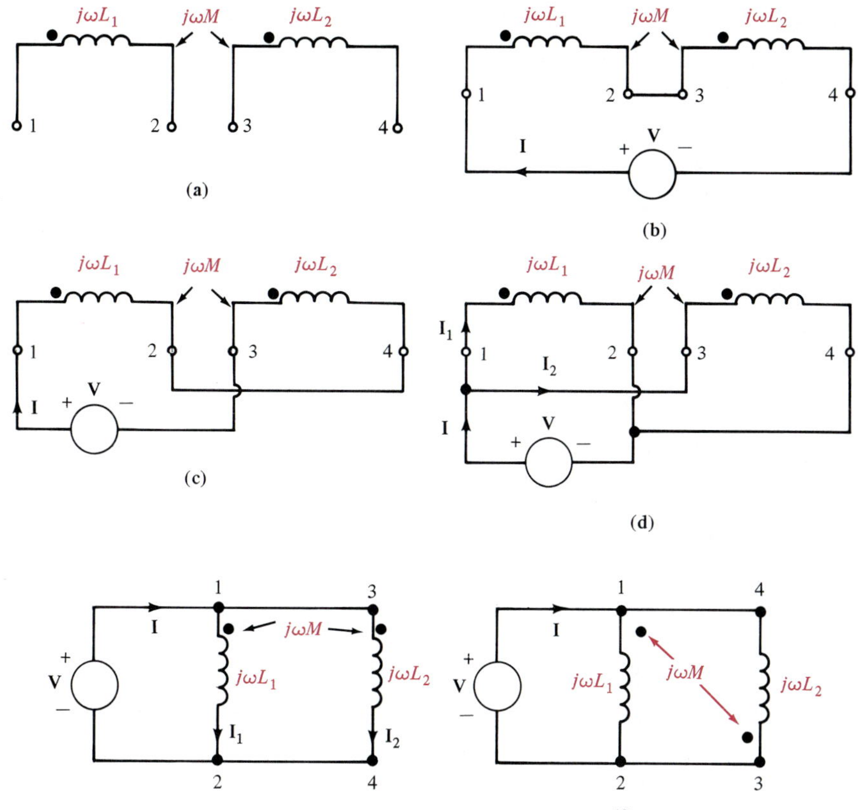

Figure 13.4 Circuits used in Example 13.3.

Case 1 is shown in Fig. 13.4b. In this case

$$\begin{aligned} \mathbf{V} &= j\omega L_1\mathbf{I} + j\omega M\mathbf{I} + j\omega L_2\mathbf{I} + j\omega M\mathbf{I} \\ &= j\omega L_{\text{eq}}\mathbf{I} \end{aligned}$$

where $L_{\text{eq}} = L_1 + L_2 + 2M$.

Case 2 is shown in Fig. 13.4c. Using KVL, we obtain

$$\begin{aligned} \mathbf{V} &= j\omega L_1\mathbf{I} - j\omega M\mathbf{I} + j\omega L_2\mathbf{I} - j\omega M\mathbf{I} \\ &= j\omega L_{\text{eq}}\mathbf{I} \end{aligned}$$

where $L_{\text{eq}} = L_1 + L_2 - 2M$.

Case 3 is shown in Fig. 13.4d and redrawn in Fig. 13.4e. The two KVL equations are

$$\mathbf{V} = j\omega L_1\mathbf{I}_1 + j\omega M\mathbf{I}_2$$

$$\mathbf{V} = j\omega M\mathbf{I}_1 + j\omega L_2\mathbf{I}_2$$

Solving these equations for $\mathbf{I}_1$ and $\mathbf{I}_2$ yields

$$\mathbf{I}_1 = \frac{\mathbf{V}(L_2 - M)}{j\omega(L_1 L_2 - M^2)}$$

$$\mathbf{I}_2 = \frac{\mathbf{V}(L_1 - M)}{j\omega(L_1 L_2 - M^2)}$$

Using KCL gives us

$$\mathbf{I} = \mathbf{I}_1 + \mathbf{I}_2 = \frac{\mathbf{V}(L_1 + L_2 - 2M)}{j\omega(L_1 L_2 - M^2)} = \frac{\mathbf{V}}{j\omega L_{\text{eq}}}$$

where

$$L_{\text{eq}} = \frac{L_1 L_2 - M^2}{L_1 + L_2 - 2M}$$

Case 4 is shown in Fig. 13.4f. The voltage equations in this case will be the same as those in Case 3 except that the signs of the mutual terms will be negative. Therefore,

$$L_{\text{eq}} = \frac{L_1 L_2 - M^2}{L_1 + L_2 + 2M}$$

EXAMPLE 13.4

We wish to determine the output voltage $\mathbf{V}_o$ in the circuit in Fig. 13.5.

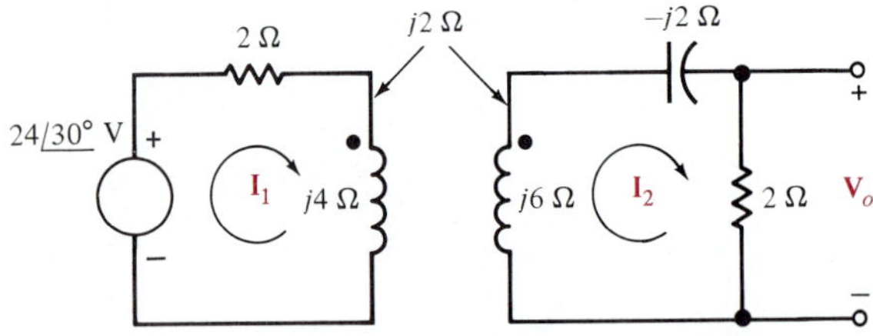

Figure 13.5 Example of a magnetically coupled circuit.

The two KVL equations for the network are

$$(2 + j4)\mathbf{I}_1 - j2\mathbf{I}_2 = 24 \angle 30^\circ$$

$$-j2\mathbf{I}_1 + (2 + j6 - j2)\mathbf{I}_2 = 0$$

Solving the equations yields

$$\mathbf{I}_2 = 2.68 \angle 3.43^\circ \text{ A}$$

Therefore,

$$\mathbf{V}_o = 2\mathbf{I}_2$$

$$= 5.37 \angle 3.43^\circ \text{ V}$$

Let us now consider a more complicated example involving mutual inductance.

EXAMPLE 13.5

Consider the circuit in Fig. 13.6. We wish to write the mesh equations for this network. Because of the multiple currents that are present in the coupled inductors, we must be very careful in writing the circuit equations.

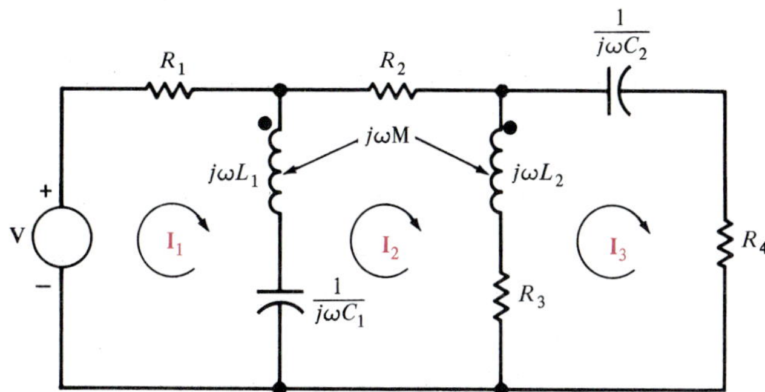

Figure 13.6 Example of a magnetically coupled circuit.

The mesh equations for the phasor network are

$$\mathbf{I}_1 R_1 + j\omega L_1(\mathbf{I}_1 - \mathbf{I}_2) + j\omega M(\mathbf{I}_2 - \mathbf{I}_3) + \frac{1}{j\omega C_1}(\mathbf{I}_1 - \mathbf{I}_2) = \mathbf{V}$$

$$\frac{1}{j\omega C_1}(\mathbf{I}_2 - \mathbf{I}_1) + j\omega L_1(\mathbf{I}_2 - \mathbf{I}_1) + j\omega M(\mathbf{I}_3 - \mathbf{I}_2) + R_2\mathbf{I}_2 + j\omega L_2(\mathbf{I}_2 - \mathbf{I}_3)$$

$$- j\omega M(\mathbf{I}_2 - \mathbf{I}_1) + R_3(\mathbf{I}_2 - \mathbf{I}_3) = 0$$

$$R_3(\mathbf{I}_3 - \mathbf{I}_2) + j\omega L_2(\mathbf{I}_3 - \mathbf{I}_2) + j\omega M(\mathbf{I}_2 - \mathbf{I}_1) + \frac{1}{j\omega C_2}\mathbf{I}_3 + R_4\mathbf{I}_3 = 0$$

which can be rewritten in the form

$$\left(R_1 + j\omega L_1 + \frac{1}{j\omega C_1}\right)\mathbf{I}_1 - \left(j\omega L_1 + \frac{1}{j\omega C_1} - j\omega M\right)\mathbf{I}_2 - j\omega M\mathbf{I}_3 = \mathbf{V}$$

$$- \left(j\omega L_1 + \frac{1}{j\omega C_1} - j\omega M\right)\mathbf{I}_1$$

$$+ \left(\frac{1}{j\omega C_1} + j\omega L_1 + R_2 + j\omega L_2 + R_3 - j2\omega M\right)\mathbf{I}_2$$

$$- (j\omega L_2 + R_3 - j\omega M)\mathbf{I}_3 = 0$$

$$- j\omega M\mathbf{I}_1 - (R_3 + j\omega L_2 - j\omega M)\mathbf{I}_2 + \left(R_3 + j\omega L_2 + \frac{1}{j\omega C_2} + R_4\right)\mathbf{I}_3 = 0$$

Note the symmetrical form of these equations. ■

DRILL EXERCISE

D13.2. Find the currents $\mathbf{I}_1$ and $\mathbf{I}_2$ and the output voltage $\mathbf{V}_o$ in the network in Fig. D13.2.

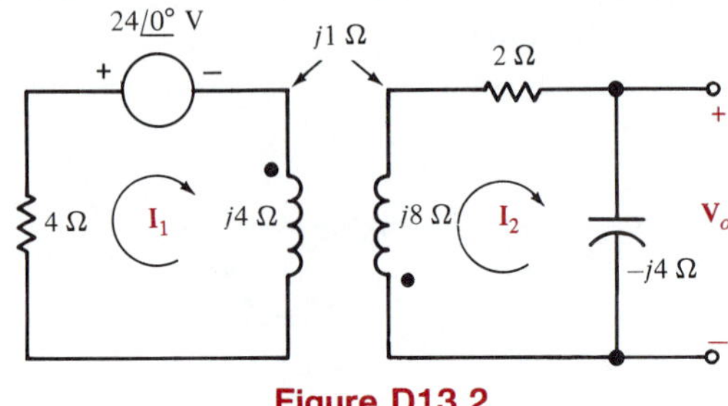

Figure D13.2

Ans: $\mathbf{I}_1 = -4.29\angle -42.8°$ A, $\mathbf{I}_2 = 0.96\angle -16.26°$ A, $\mathbf{V}_o = 3.84\angle -106.26°$ V.

D13.3. Write the KVL equations in standard form for the network in Fig. D13.3.

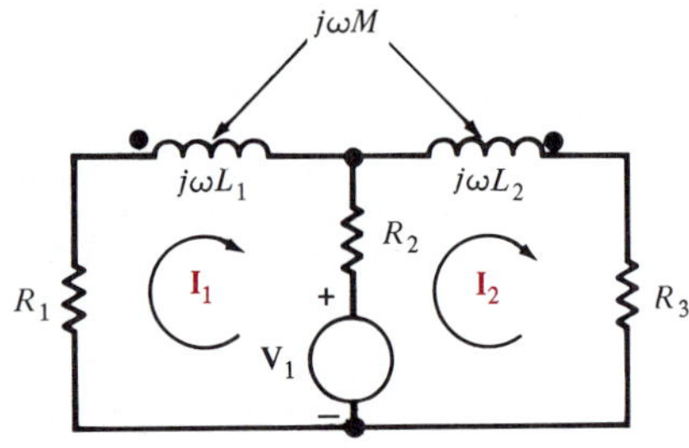

Figure D13.3

Ans: $(R_1 + j\omega L_1 + R_2)\mathbf{I}_1 - (R_2 + j\omega M)\mathbf{I}_2 = -\mathbf{V}_1$,
$-(R_2 + j\omega M)\mathbf{I}_1 + (R_2 + j\omega L_2 + R_3)\mathbf{I}_2 = \mathbf{V}_1$.

D13.4. Write the mesh equations for the network in Fig. D13.4.

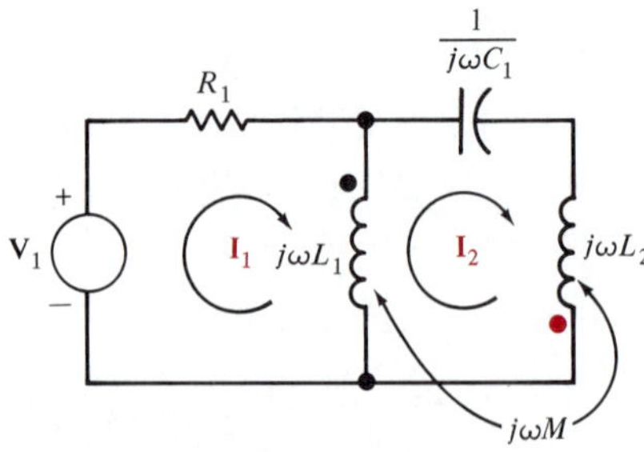

Figure D13.4

Ans: $(R_1 + j\omega L_1)\mathbf{I}_1 - (j\omega L_1 + j\omega M)\mathbf{I}_2 = \mathbf{V}_1$, $-(j\omega L_1 + j\omega M)\mathbf{I}_1 +$
$(j\omega L_1 + 1/j\omega C + j\omega L_2 + 2j\omega M)\mathbf{I}_2 = 0$.

13.2 Energy Analysis

We now perform an energy analysis on a pair of mutually coupled inductors, which will yield some interesting relationships for the circuit elements. Our analysis will involve the performance of an experiment on the network shown in Fig. 13.7. Before beginning the experiment, we set all voltages and currents in the circuit equal to zero. Once the circuit is quiescent, we begin by letting the current $i_1(t)$ increase from zero to some value I_1 with the right-side terminals open circuited. Since the right-side terminals are open, $i_2(t) = 0$, and therefore the power entering these terminals is zero. The instantaneous power entering the left-side terminals is

$$p(t) = v_1(t)i_1(t) = \left[L_1 \frac{di_1(t)}{dt}\right]i_1(t)$$

The energy stored within the coupled circuit at t_1 when $i_1(t) = I_1$ is then

$$\int_0^{t_1} v_1(t)i_1(t)\,dt = \int_0^{I_1} L_1 i_1(t)\,di_1(t) = \tfrac{1}{2}L_1 I_1^2$$

Continuing our experiment, starting at time t_1, we let the current $i_2(t)$ increase from zero to some value I_2 at time t_2 while holding $i_1(t)$ constant at I_1. The energy delivered through the right-side terminals is

$$\int_{t_1}^{t_2} v_2(t)i_2(t)\,dt = \int_0^{I_2} L_2 i_2(t)\,di_2(t) = \tfrac{1}{2}L_2 I_2^2$$

However, during the interval t_1 to t_2 the voltage $v_1(t)$ is

$$v_1(t) = L_1 \frac{di_1(t)}{dt} + M\frac{di_2(t)}{dt}$$

Since $i_1(t)$ is a constant I_1, the energy delivered through the left-side terminals is

$$\int_{t_1}^{t_2} v_1(t)i_1(t)\,dt = \int_{t_1}^{t_2} M\frac{di_2(t)}{dt} I_1\,dt = MI_1\int_0^{I_2} di_2(t)$$
$$= MI_1I_2$$

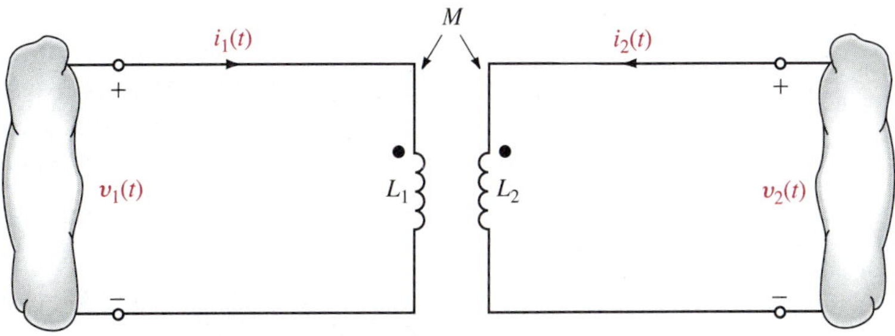

Figure 13.7 Magnetically coupled circuit.

Therefore, the total energy stored in the network for $t > t_2$ is

$$w = \tfrac{1}{2}L_1I_1^2 + \tfrac{1}{2}L_2I_2^2 + MI_1I_2 \tag{13.10}$$

We could, of course, repeat our entire experiment with either the dot on L_1 or L_2, but not both, reversed, and in this case the sign on the mutual inductance term would be negative, producing

$$w = \tfrac{1}{2}L_1I_1^2 + \tfrac{1}{2}L_2I_2^2 - MI_1I_2$$

It is very important for the reader to realize that in our derivation of the equation above, by means of the experiment, the values I_1 and I_2 could have been any values at any *time*; therefore, the energy stored in the magnetically coupled inductors at any instant of time is given by the expression

$$w(t) = \tfrac{1}{2}L_1[i_1(t)]^2 + \tfrac{1}{2}L_2[i_2(t)]^2 \pm Mi_1(t)i_2(t) \tag{13.11}$$

The two coupled inductors represent a passive network, and therefore, the energy stored within this network must be nonnegative for any values of the inductances and currents. The equation for the instantaneous energy stored in the magnetic circuit can be written as

$$w(t) = \tfrac{1}{2}L_1i_1^2 + \tfrac{1}{2}L_2i_2^2 \pm Mi_1i_2$$

Adding and subtracting the term $\frac{1}{2}(M^2/L_2)i_1^2$ and rearranging the equation yields

$$w(t) = \frac{1}{2}\left(L_1 - \frac{M^2}{L_2}\right)i_1^2 + \frac{1}{2}L_2\left(i_2 + \frac{M}{L_2}i_1\right)^2$$

From this expression we recognize that the instantaneous energy stored will be nonnegative if

$$M \le \sqrt{L_1L_2} \tag{13.12}$$

Note that this equation specifies an upper limit on the value of the mutual inductance.

We define the coefficient of coupling between the two inductors L_1 and L_2 as

$$k = \frac{M}{\sqrt{L_1L_2}} \tag{13.13}$$

and we note from Eq. (13.12) that its range of values is

$$0 \le k \le 1 \tag{13.14}$$

This coefficient is an indication of how much flux in one coil is linked with the other coil; that is, if all the flux in one coil reaches the other coil, then we have 100% coupling and $k = 1$. For large values of k (i.e., $k > 0.5$), the inductors are said to be tightly coupled, and for small values of k (i.e., $k \le 0.5$), the coils are said to be loosely coupled. The previous equations indicate that the value for the mutual inductance is confined to the range

$$0 \le M \le \sqrt{L_1L_2} \tag{13.15}$$

and that the upper limit is the geometric mean of the inductances L_1 and L_2.

EXAMPLE 13.6

The coupled circuit in Fig. 13.8a has a coefficient of coupling of 1 (i.e., $k = 1$). We wish to determine the energy stored in the mutually coupled inductors at time $t = 5$ ms. $L_1 = 2.653$ mH and $L_2 = 10.61$ mH.

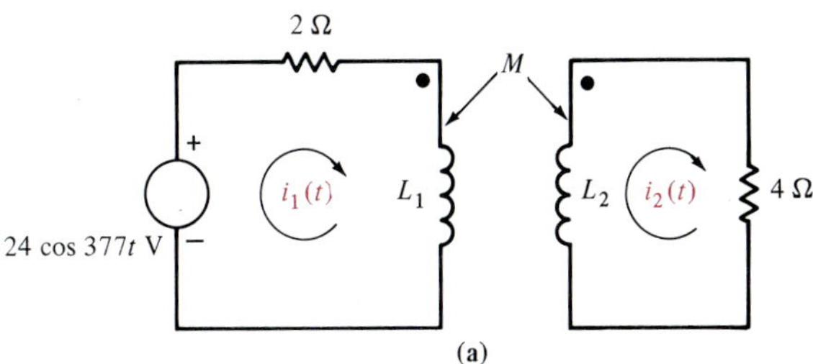

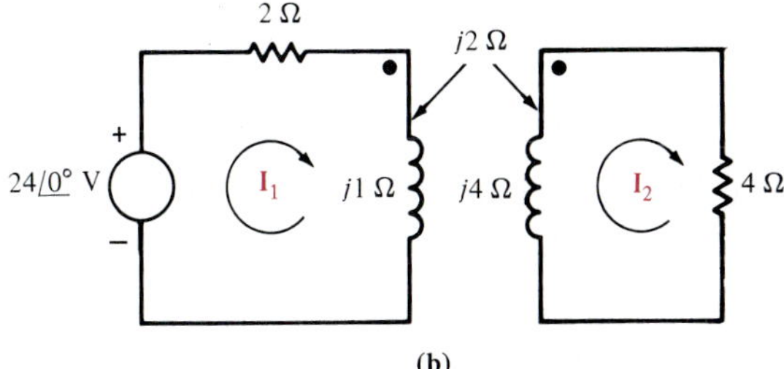

Figure 13.8 Example of a magnetically coupled circuit drawn in the time and frequency domains.

From the data the mutual inductance is

$$M = \sqrt{L_1 L_2} = 5.31 \text{ mH}$$

The frequency-domain equivalent circuit is shown in Fig. 13.8b, where the impedance values for X_{L_1}, X_{L_2}, and X_M are 1, 4, and 2, respectively. The mesh equations for the network are then

$$(2 + j1)\mathbf{I}_1 - 2j\mathbf{I}_2 = 24 \underline{/0^\circ}$$

$$-j2\mathbf{I}_1 + (4 + 4j)\mathbf{I}_2 = 0$$

Solving these equations for the two mesh currents yields

$$\mathbf{I}_1 = 9.41 \underline{/-11.31^\circ} \text{ A} \qquad \text{and} \qquad \mathbf{I}_2 = 3.33 \underline{/+33.69^\circ} \text{ A}$$

and therefore,

$$i_1(t) = 9.41 \cos (377t - 11.31^\circ) \text{ A}$$

$$i_2(t) = 3.33 \cos (377t + 33.69^\circ) \text{ A}$$

At $t = 5$ ms, $377t = 1.885$ rad or 108°, and therefore,

$$i_1(t = 5 \text{ ms}) = 9.41 \cos(108° - 11.31°) = -1.10 \text{ A}$$

$$i_2(t = 5 \text{ ms}) = 3.33 \cos(108° + 33.69°) = -2.61 \text{ A}$$

Therefore, the energy stored in the coupled inductors at $t = 5$ ms is

$$\begin{aligned} w(t)\big|_{t=0.005 \text{ sec}} &= \tfrac{1}{2}(2.653)(10^{-3})(-1.10)^2 + \tfrac{1}{2}(10.61)(10^{-3})(-2.61)^2 \\ &\quad - (5.31)(10^{-3})(-1.10)(-2.61) \\ &= (1.55)(10^{-3}) + (36.14)(10^{-3}) - (15.25)(10^{-3}) \\ &= 22.44 \text{ mJ} \end{aligned}$$

■

DRILL EXERCISE

D13.5. Two coils in a network are positioned such that there is 100% coupling between them. If the inductance of one coil is 10 mH and the mutual inductance is 6 mH, compute the inductance of the other coil.

Ans: $L_2 = 3.6$ mH.

D13.6. The network in Fig. D13.6 operates at 60 Hz. Compute the energy stored in the mutually coupled inductors at time $t = 10$ ms.

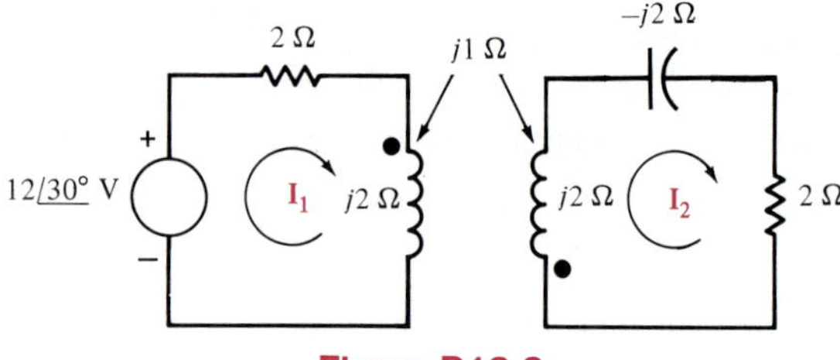

Figure D13.6

Ans: $w(10 \text{ ms}) = 39$ mJ.

13.3 The Linear Transformer

A transformer is a device that contains two or more coils that are coupled magnetically. A typical transformer network is shown in Fig. 13.9. The source is connected to what is called the *primary* of the transformer, and the load is connected to the *secondary*. Thus R_1 and L_1 refer to the resistance and self-inductance of the primary, and R_2 and L_2 refer to the secondary's resistance and self-inductance. The transformer is said to be *linear* if the magnetic permeability (μ) of the paths through which the fluxes flow is constant. Without the use of high μ material, the coefficient of coupling, k, is typically very small. Transformers of this type find wide application in such products as radio and TV receivers.

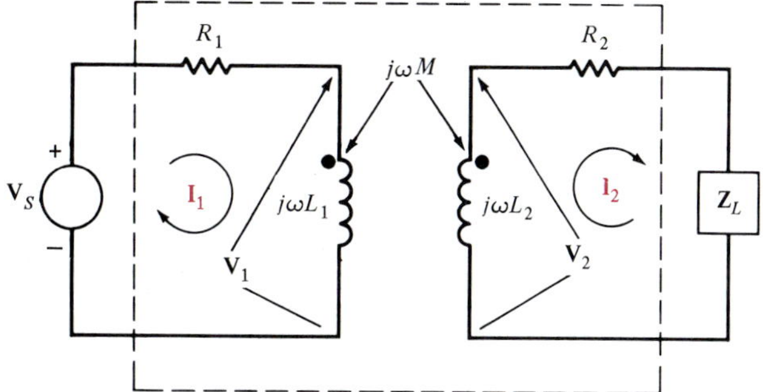

Figure 13.9 Transformer network.

There are a number of ways in which we can model the magnetically coupled inductor portion of the transformer. The model already developed is shown in Fig. 13.10a. All the models are developed so that the equations relating $\mathbf{V}_1$, $\mathbf{V}_2$, $\mathbf{I}_1$, and $\mathbf{I}_2$ in Fig. 13.10a are satisfied. These equations are

$$\begin{aligned} \mathbf{V}_1 &= j\omega L_1 \mathbf{I}_1 + j\omega M \mathbf{I}_2 \\ \mathbf{V}_2 &= j\omega L_2 \mathbf{I}_2 + j\omega M \mathbf{I}_1 \end{aligned} \tag{13.16}$$

One technique involves the use of dependent sources as shown in Fig. 13.10b. The circuit equations for this model are

$$\begin{aligned} \mathbf{V}_1 &= j\omega L_1 \mathbf{I}_1 + j\omega M \mathbf{I}_2 \\ \mathbf{V}_2 &= j\omega M \mathbf{I}_1 + j\omega L_2 \mathbf{I}_2 \end{aligned} \tag{13.17}$$

Another network involves replacing the transformer with an equivalent T network. This equivalent circuit is valid only if the four-terminal network can be replaced by the three-terminal network shown in Fig. 13.10c. These circuit equations are also

$$\begin{aligned} \mathbf{V}_1 &= j\omega L_1 \mathbf{I}_1 + j\omega M \mathbf{I}_2 \\ \mathbf{V}_2 &= j\omega M \mathbf{I}_1 + j\omega L_2 \mathbf{I}_2 \end{aligned} \tag{13.18}$$

If we simply examine these equations as they are given, they appear to represent two mesh equations in which the common element between the two meshes is an inductor of value M. In order for the total inductance in the first mesh to be L_1 and the inductance of the common element to be M, as indicated in the first equation above, an inductor of value $L_1 - M$ must be placed in the first mesh of the equivalent circuit. A similar argument for the second mesh indicates that an inductor of value $L_2 - M$ must be placed in that mesh of the equivalent circuit. The equivalent circuit resulting from this analysis is shown in Fig. 13.10d. Note that this network satisfies the circuit Eqs. (13.18).

If the relationship between the currents and the dots are such that the mutual terms in Eq. (13.18) are negative, we may simply replace M by $-M$ in the equivalent network in Fig. 13.10. This situation is illustrated in Example 13.7. From a mathematical modeling standpoint a negative inductance presents no problems; however, we must remember that such a physical element does not actually exist.

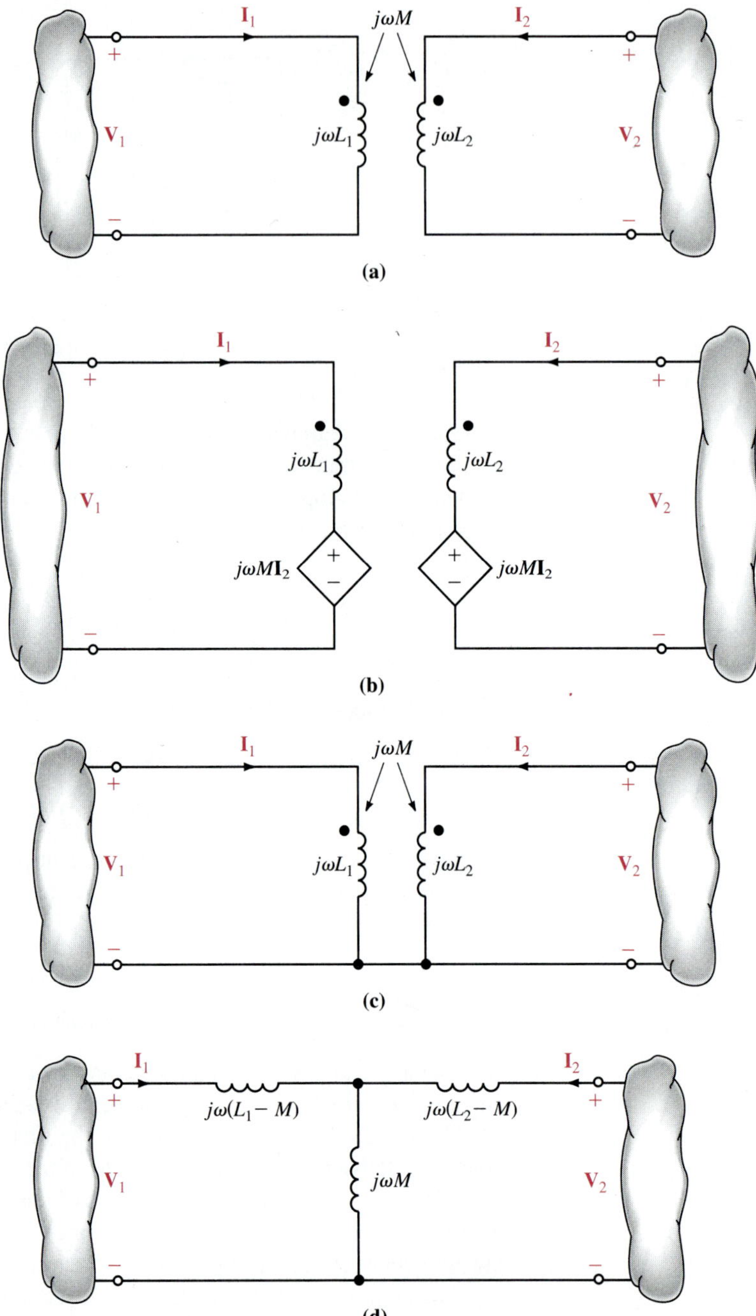

Figure 13.10 Transformer equivalent circuits.

EXAMPLE 13.7

The coefficient of coupling for the linear transformer in the network in Fig. 13.11a is $k = 0.1$. We wish to determine the equivalent T network for the transformer and redraw the circuit using this equivalent model.

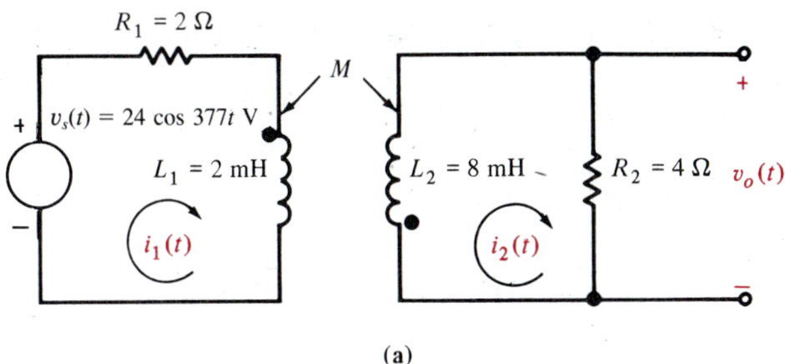

(a)

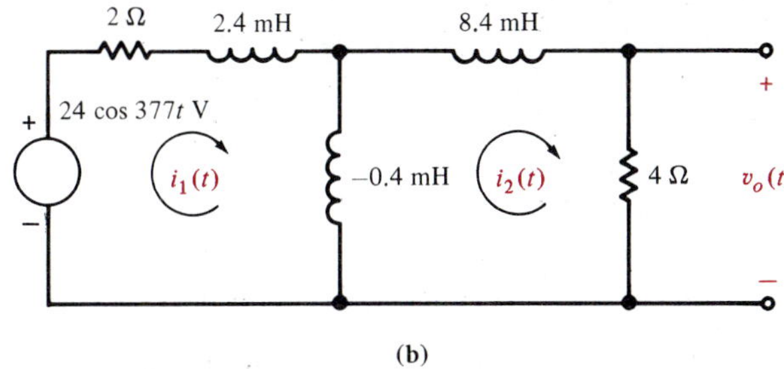

(b)

Figure 13.11 Example transformer circuit together with its T-equivalent circuit.

Using the relationship $k = M/\sqrt{L_1L_2}$, we find that

$$\begin{aligned} M &= k\sqrt{L_1L_2} \\ &= 0.1\sqrt{(16)(10^{-6})} \\ &= 0.4 \text{ mH} \end{aligned}$$

Since both currents are into the dot, the equations for the original network are

$$\mathbf{V}_S = R_1\mathbf{I}_1 + j\omega L_1\mathbf{I}_1 + j\omega M\mathbf{I}_2$$

$$0 = j\omega M\mathbf{I}_1 + j\omega L_2\mathbf{I}_2 + R_2\mathbf{I}_2$$

Note that the mutual term is positive in these equations. If in the T network we use the same direction for the currents, the mutual term in the first equation will be positive only if the inductor in the center of the T network is negative. Therefore, the center inductor is chosen to have a value of $-M$, and the other two inductors have values $L_1 + M$ and $L_2 + M$, as shown in Fig. 13.11b. Write the mesh equations for the network in Fig. 13.11b to verify that they are the same as the equations above. ■

With reference to Fig. 13.9, let us compute the input impedance to the transformer as seen by the source. The network equations are

$$\mathbf{V}_S = \mathbf{I}_1(R_1 + j\omega L_1) - j\omega M\mathbf{I}_2$$
$$0 = -j\omega M\mathbf{I}_1 + (R_2 + j\omega L_2 + \mathbf{Z}_L)\mathbf{I}_2$$

Solving the second equation for $\mathbf{I}_2$ and substituting it into the first equation yields

$$\mathbf{V}_S = \left(R_1 + j\omega L_1 + \frac{\omega^2 M^2}{R_2 + j\omega L_2 + \mathbf{Z}_L}\right)\mathbf{I}_1$$

Therefore, the input impedance is

$$\mathbf{Z}_i = \frac{\mathbf{V}_S}{\mathbf{I}_1} = R_1 + j\omega L_1 + \frac{\omega^2 M^2}{R_2 + j\omega L_2 + \mathbf{Z}_L} \tag{13.19}$$

As we look from the source into the network to determine $\mathbf{Z}_i$, we see the impedance of the primary (i.e., $R_1 + j\omega L_1$) plus an impedance that the secondary of the transformer reflects, due to mutual coupling, into the primary. This *reflected impedance* is

$$\mathbf{Z}_R = \frac{\omega^2 M^2}{R_2 + j\omega L_2 + \mathbf{Z}_L} \tag{13.20}$$

Note that this reflected impedance is independent of the dot locations.

If $\mathbf{Z}_L$ in Eq. (13.20) is written as

$$\mathbf{Z}_L = R_L + jX_L \tag{13.21}$$

then

$$\mathbf{Z}_R = \frac{\omega^2 M^2}{R_2 + R_L + j(\omega L_2 + X_L)}$$

which can be written as

$$\mathbf{Z}_R = \frac{\omega^2 M^2[(R_2 + R_L) - j(\omega L_2 + X_L)]}{(R_2 + R_L)^2 + (\omega L_2 + X_L)^2} \tag{13.22}$$

This equation illustrates that if X_L is an inductive reactance, or if X_L is a capacitive reactance with $\omega L_2 > X_L$, then the reflected reactance is capacitive. In general, the reflected reactance is opposite in sign to that of the total reactance in the secondary. If $\omega L_2 + X_L = 0$ (i.e., the secondary is in resonance), $\mathbf{Z}_R$ is purely resistive and

$$\mathbf{Z}_R = \frac{\omega^2 M^2}{R_2 + R_L} \tag{13.23}$$

EXAMPLE 13.8

For the network shown in Fig. 13.12 we wish to determine the input impedance.

Following the development that led to Eq. (13.19), we find that the input impedance is

$$\begin{aligned}\mathbf{Z}_i &= 12 + j10 + \frac{(1)^2}{16 + j8 - j4 + 4 + j6} \\ &= 12.04 + j9.98 \\ &= 15.64 \underline{/39.65^\circ}\ \Omega\end{aligned}$$

■

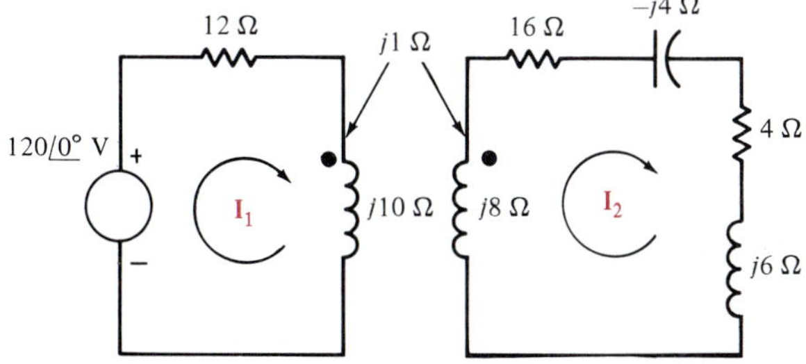

Figure 13.12 Example transformer circuit.

DRILL EXERCISE

D13.7. Given the network in Fig. D13.7, find the input impedance of the network and the current in the voltage source.

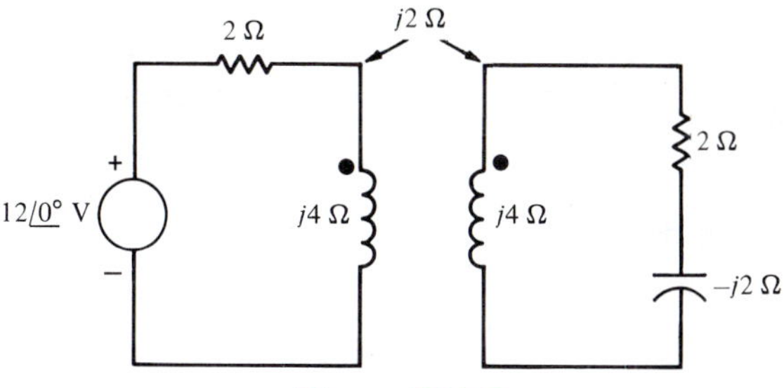

Figure D13.7

Ans: $\mathbf{Z}_i = 3 + j3\ \Omega$, $\mathbf{I}_S = 2 - j2$ A.

13.4 PSPICE Analysis of Magnetically Coupled Circuits

The PSPICE circuit analysis program can be employed to solve magnetically coupled circuits. To use PSPICE in this case we must present one additional statement that defines the coupling. The statement is

```
KXXXXXXX LYYYYYYY LZZZZZZZ value
```

LYYYYYYY and LZZZZZZZ are the names of the coupled inductors and value is the coefficient of coupling KXXXXXXX that must satisfy the relationship

$$0 < \text{KXXXXXXX} \leq 1$$

The ''dot'' convention applies in this analysis and the ''dot'' should be on the first node of each inductor.

The following examples illustrate the use of PSPICE for coupled circuits.

EXAMPLE 13.9

Let us consider once again the network in Example 13.4, which is redrawn in Fig. 13.13 for PSPICE analysis. Let us assume that the frequency is $f = 1/2\pi$ Hz. The coupling coefficient is then

$$k = \frac{2}{\sqrt{(4)(6)}} = 0.408248$$

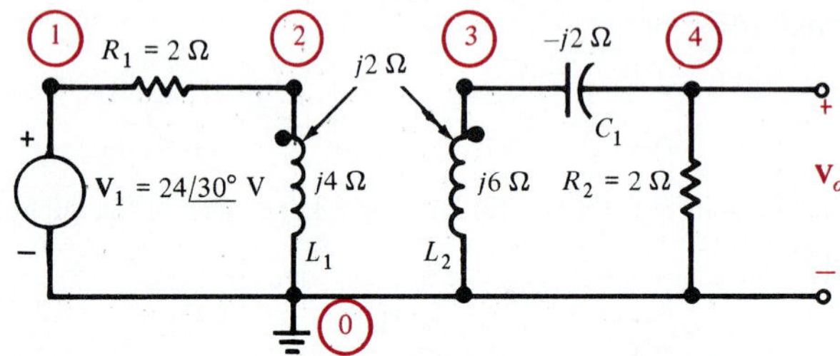

Figure 13.13 Figure 13.5 prepared for PSPICE analysis.

The PSPICE program for determining the output voltage is

```
EXAMPLE 13.9
V1   1    0    AC    24    30
R1   1    2    2
L1   2    0    4
L2   3    0    6
K1   L1   L2   0.408248
C1   3    4    0.5
R2   4    0    2
.AC LIN   1    0.159155   0.159155
.PRINT    AC   VM(4)      VP(4)
.END
```

The computer output is

```
FREQ          VM(4)        VP(4)
1.592E-01    5.367E+00    3.435E+00
```

■

EXAMPLE 13.10

The network in Fig. 13.14a is redrawn in Fig. 13.14b for PSPICE analysis. Let us write a program that will calculate the current in R_2 assuming that the frequency is $f = 1/2\pi$ Hz. The coefficient of coupling is

$$k = \frac{1}{\sqrt{(2)(3)}} = 0.408248$$

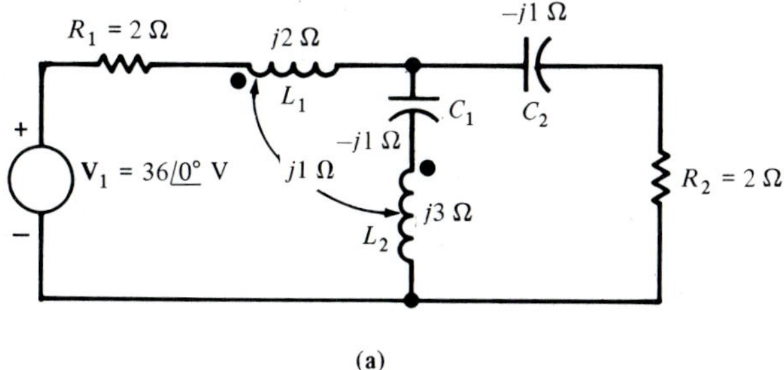

(a)

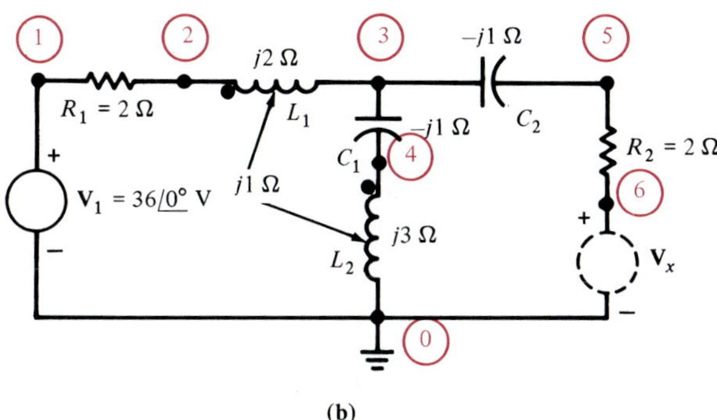

(b)

Figure 13.14 PSPICE analysis example.

The program is listed below.

```
EXAMPLE 13.10
V1    1     0     AC    36    0
R1    1     2     2
L1    2     3     2
C1    3     4     1
L2    4     0     3
K1    L1    L2    0.408248
C2    3     5     1
R2    5     6     2
VX    6     0     AC    0     0
.AC LIN   1     0.159155    0.159155
.PRINT    AC    IM(VX)      IP(VX)
.END
```

The computer output is

```
FREQ          IM(VX)        IP(VX)
1.592E-01     6.900E+00     2.657E+01
```

■

DRILL EXERCISE

D13.8. Given the network in Fig. D13.8, with $f = 1/2\pi$ Hz and the coefficient of coupling = 0.9, compute $\mathbf{V}_o$ by means of a PSPICE program.

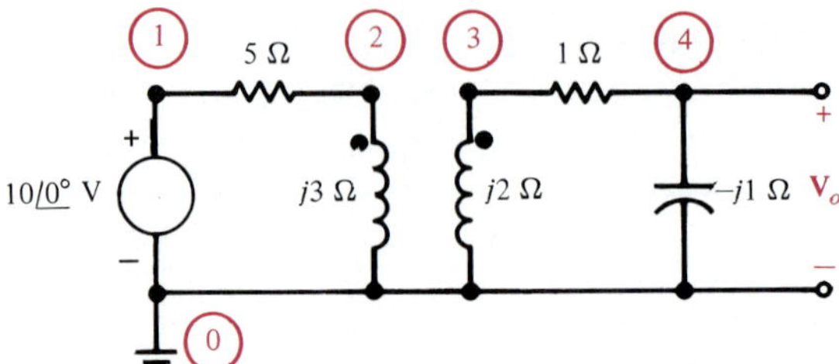

Figure D13.8

Ans: $\mathbf{V}_o = 2.092 \angle -49.39^\circ$ V.

D13.9. Given the network in Fig. D13.9 with $f = 1/2\pi$ Hz and the coefficient of coupling = 0.7071, write a PSPICE program to compute $\mathbf{V}_o$.

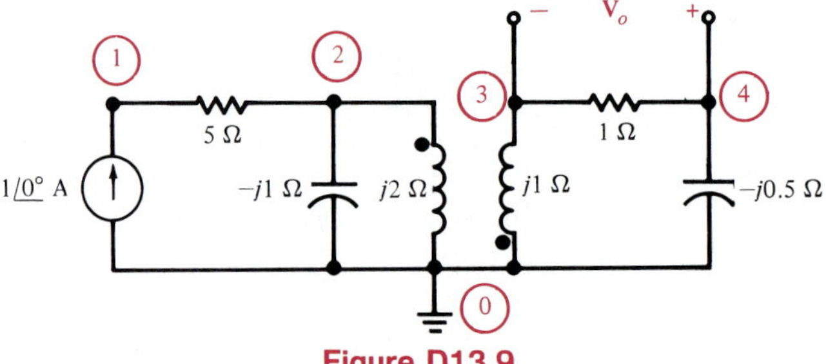

Figure D13.9

Ans: $\mathbf{V}_o = 0.8944 \angle -63.44^\circ$ V.

13.5 The Ideal Transformer

Consider the situation illustrated in Fig. 13.15, showing two coils of wire wound on a single closed magnetic core. The magnetic core concentrates the flux so that all the flux links all the turns of both coils. In the ideal case we also neglect wire resistance. Let us now examine the coupling equations under the condition that the same flux goes through each winding and therefore,

$$v_1(t) = N_1 \frac{d\phi_1}{dt} = N_1 \frac{d\phi}{dt}$$

and

$$v_2(t) = N_2 \frac{d\phi_2}{dt} = N_2 \frac{d\phi}{dt}$$

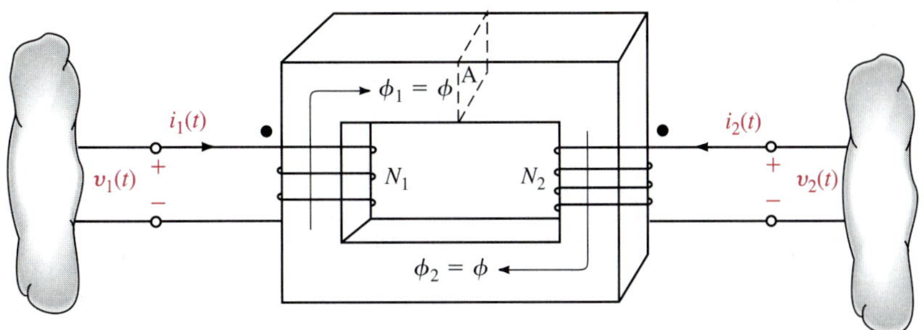

Figure 13.15 Transformer employing a magnetic core.

and therefore,

$$\frac{v_1}{v_2} = \frac{N_1}{N_2} \frac{\dfrac{d\phi}{dt}}{\dfrac{d\phi}{dt}} = \frac{N_1}{N_2} \tag{13.24}$$

Another relationship can be developed between the currents $i_1(t)$ and $i_2(t)$ and the number of turns in each coil. To develop this relationship, we employ, from electromagnetic field theory, Ampère's law, which is written in mathematical form as

$$\oint H \cdot dl = i_{\text{enclosed}} = N_1 i_1 + N_2 i_2 \tag{13.25}$$

where H is the magnetic field intensity and the integral is over the closed path traveled by the flux around the transformer core. For the ideal core material, $\mu = \infty$, $H = 0$. Therefore,

$$N_1 i_1 + N_2 i_2 = 0 \tag{13.26}$$

or

$$\frac{i_1}{i_2} = -\frac{N_2}{N_1} \tag{13.27}$$

Note that if we divide Equation (13.26) by N_1 and multiply it by v_1, we obtain

$$v_1 i_1 + \frac{N_2}{N_1} v_1 i_2 = 0$$

However, since $v_1 = (N_1/N_2)v_2$,

$$v_1 i_1 + v_2 i_2 = 0$$

and hence the total power into the device is zero, which means that an ideal transformer is lossless.

Therefore, to summarize the dot convention for an ideal transformer,

$$v_1 = \frac{N_1}{N_2} v_2 \tag{13.28a}$$

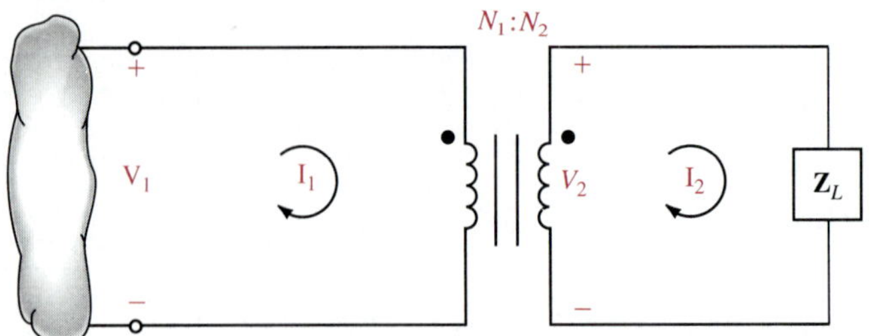

Figure 13.16 Ideal transformer circuit used to illustrate input impedance.

where both voltages are referenced positive at the dots and

$$N_1 i_1 + N_2 i_2 = 0 \tag{13.28b}$$

where both currents are defined entering the dots.

Consider now the circuit shown in Fig. 13.16, where the symbol used for the transformer indicates that it is an iron-core transformer. Because of the relationship between the dots and both the assigned currents and voltages, the phasor voltages $\mathbf{V}_1$ and $\mathbf{V}_2$ are related by the expression

$$\frac{\mathbf{V}_1}{\mathbf{V}_2} = \frac{N_1}{N_2}$$

and the phasor currents, from Eq. (13.28b), are related by

$$\frac{\mathbf{I}_1}{\mathbf{I}_2} = \frac{N_2}{N_1}$$

The sign in Eq. (13.28b) is reversed, since the direction of $\mathbf{I}_2$ is reversed. These two equations above can be rewritten as

$$\begin{aligned} \mathbf{V}_1 &= \frac{N_1}{N_2}\mathbf{V}_2 \\ \mathbf{I}_1 &= \frac{N_2}{N_1}\mathbf{I}_2 \end{aligned} \tag{13.29}$$

Also note that

$$\begin{aligned} \mathbf{S}_1 = \mathbf{V}_1\mathbf{I}_1^* &= \left(\frac{N_1}{N_2}\mathbf{V}_2\right)\left(\frac{N_2}{N_1}\mathbf{I}_2\right)^* \\ &= \mathbf{V}_2\mathbf{I}_2^* = \mathbf{S}_2 \end{aligned}$$

From the figure we note that $\mathbf{Z}_L = \mathbf{V}_2/\mathbf{I}_2$, and therefore the input impedance

$$\mathbf{Z}_1 = \frac{\mathbf{V}_1}{\mathbf{I}_1} = \left(\frac{N_1}{N_2}\right)^2 \mathbf{Z}_L \tag{13.30}$$

If we now define the turns ratio as

$$n = \frac{N_2}{N_1} \tag{13.31}$$

then the defining equations for the *ideal transformer* are

$$\begin{aligned} \mathbf{V}_1 &= \frac{\mathbf{V}_2}{n} \\ \mathbf{I}_1 &= n\mathbf{I}_2 \\ \mathbf{S}_1 &= \mathbf{S}_2 \\ \mathbf{Z}_1 &= \frac{\mathbf{Z}_L}{n^2} \end{aligned} \tag{13.32}$$

Equations (13.32) define the important relationships for an ideal transformer. Care must be exercised in using these relationships because the signs on the voltages and currents are dependent on the assigned references and how they are related to the dots as illustrated earlier.

EXAMPLE 13.11

We wish to determine the input impedance seen by the source in the circuit in Fig. 13.17 if $\mathbf{Z}_L = 10\ \underline{/30^\circ}\ \Omega$.

$$n = \frac{N_2}{N_1} = \frac{4}{1} = 4$$

and the reflected impedance is

$$\mathbf{Z}_1 = \frac{10\ \underline{/30^\circ}}{(4)^2}$$

$$= 0.54 + j0.313\ \Omega$$

Therefore, the input impedance seen by the source is

$$\mathbf{Z}_i = 4 - j2 + 0.54 + j0.313$$

$$= 4.54 - j1.69\ \Omega$$

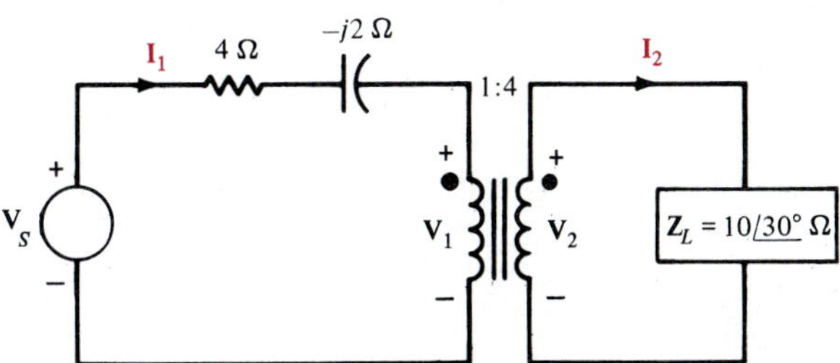

Figure 13.17 Example circuit containing an ideal transformer.

EXAMPLE 13.12

Given the circuit shown in Fig. 13.18, we wish to determine all indicated voltages and currents.

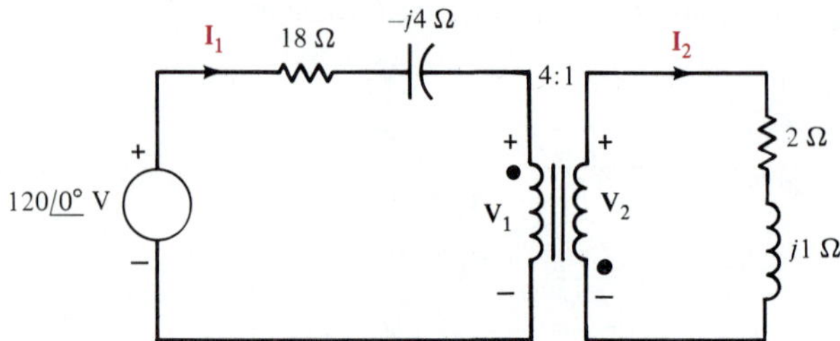

Figure 13.18 Ideal transformer circuit.

Because of the relationships between the dots and the currents and voltages, the transformer equations are

$$\mathbf{V}_1 = -\frac{\mathbf{V}_2}{n} \qquad \text{and} \qquad \mathbf{I}_1 = -n\mathbf{I}_2$$

where $n = \frac{1}{4}$. The reflected impedance at the input to the transformer is

$$\mathbf{Z}_1 = 16(2 + j1) = 32 + j16\ \Omega$$

Therefore, the current in the source is

$$\mathbf{I}_1 = \frac{120\ \underline{/0^\circ}}{18 - j4 + 32 + j16} = 2.23\ \underline{/-13.5^\circ}\ \text{A}$$

The voltage across the input to the transformer is then

$$\begin{aligned}\mathbf{V}_1 &= \mathbf{I}_1\mathbf{Z}_1\\ &= (2.33\ \underline{/-13.5^\circ})(32 + j16)\\ &= 83.50\ \underline{/13.07^\circ}\ \text{V}\end{aligned}$$

Hence $\mathbf{V}_2$ is

$$\begin{aligned}\mathbf{V}_2 &= -n\mathbf{V}_1\\ &= -\tfrac{1}{4}(83.50\ \underline{/13.07^\circ})\\ &= 20.88\ \underline{/193.07^\circ}\ \text{V}\end{aligned}$$

The current $\mathbf{I}_2$ is

$$\begin{aligned}\mathbf{I}_2 &= -\frac{\mathbf{I}_1}{n}\\ &= -4(2.33\ \underline{/-13.5^\circ})\\ &= 9.32\ \underline{/166.50^\circ}\ \text{A}\end{aligned}$$

■

DRILL EXERCISE

D13.10. Compute the current $\mathbf{I}_1$ in the network in Fig. D13.10.

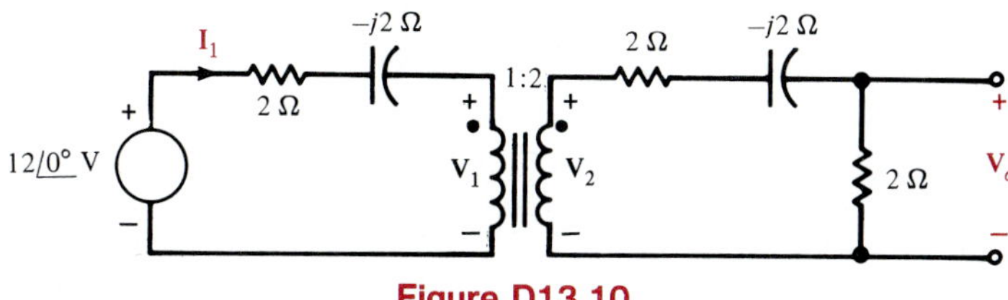

Figure D13.10

Ans: $\mathbf{I}_1 = 3.07\ \underline{/39.81°}$ A.

D13.11. Find $\mathbf{V}_o$ in the network in Fig. D13.10.

Ans: $\mathbf{V}_o = 3.07\ \underline{/39.82°}$ V.

D13.12. In the network in Fig. D13.12 the voltage $\mathbf{V}_o = 10\ \underline{/0°}$ V. Find the input voltage $\mathbf{V}_S$.

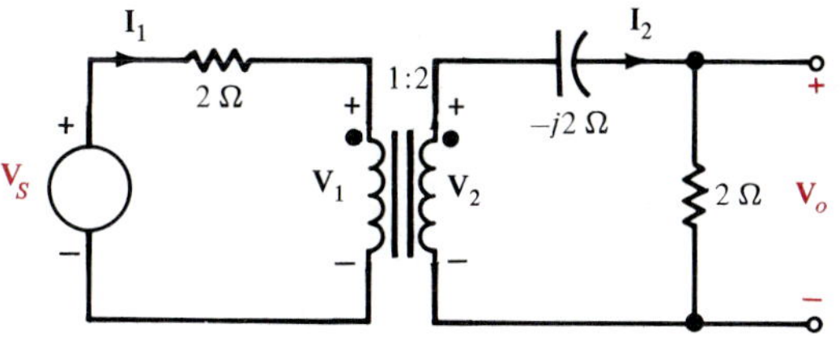

Figure D13.12

Ans: $\mathbf{V}_S = 25.5\ \underline{/-11.31°}$ V.

Another technique for simplifying the analysis of circuits containing an ideal transformer involves the use of either Thévenin's or Norton's theorems to obtain an equivalent circuit that replaces the transformer and either the primary or secondary circuit. This technique usually requires more effort, however, than the approach presented thus far. Let us demonstrate this approach by employing Thévenin's theorem to derive an equivalent circuit for the transformer and primary circuit of the network shown in Fig. 13.19a. The equations for the transformer in view of the direction of the currents and voltages and the position of the dots are

$$\mathbf{I}_1 = n\mathbf{I}_2$$

$$\mathbf{V}_1 = \frac{\mathbf{V}_2}{n}$$

Forming a Thévenin equivalent at the secondary terminals $2 - 2'$ as shown in Fig. 13.19b, we note that $\mathbf{I}_2 = 0$ and therefore from Eq. (13.32) $\mathbf{I}_1 = 0$. Hence

$$\mathbf{V}_{oc} = \mathbf{V}_2 = n\mathbf{V}_1 = n\mathbf{V}_{S_1}$$

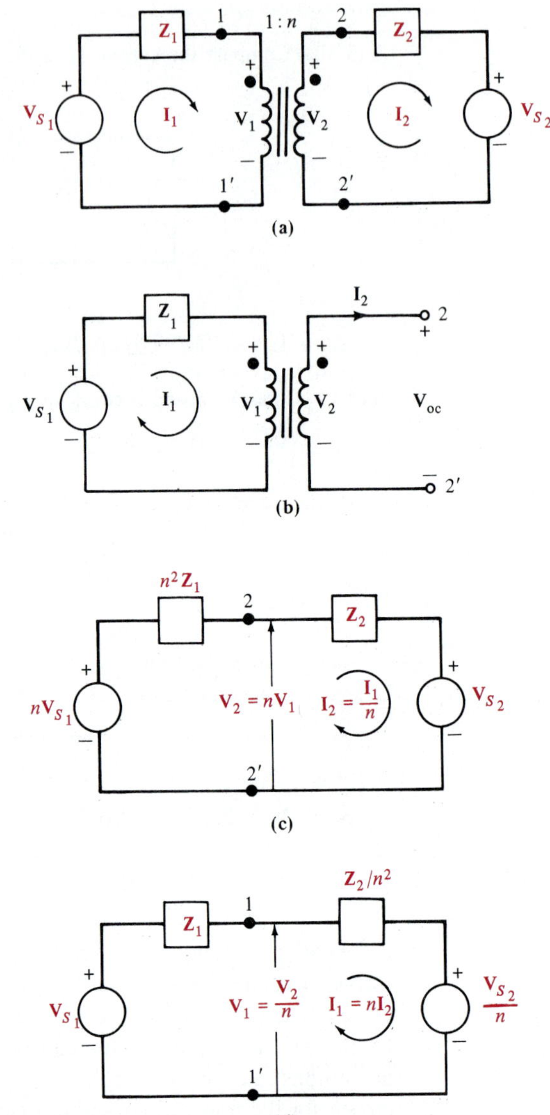

Figure 13.19 Circuit containing an ideal transformer and some of its equivalent networks.

The Thévenin equivalent impedance obtained by looking into the open-circuit terminals with $\mathbf{V}_{S_1}$ replaced by a short circuit is $\mathbf{Z}_1$, which when reflected into the secondary by the turns ratio is

$$\mathbf{Z}_{Th} = n^2\mathbf{Z}_1$$

Therefore, one of the resulting equivalent circuits for the network in Fig. 13.19a is as shown in Fig. 13.19c. In a similar manner we can show that replacing the transformer and its secondary circuit by an equivalent circuit results in the network shown in Fig. 13.19d.

It can be shown in general that when developing an equivalent circuit for the transformer and its primary circuit, each primary voltage is multiplied by n, each primary current is divided by n, and each primary impedance is multiplied by n^2. Similarly, when developing an equivalent circuit for the transformer and its secondary circuit, each secondary voltage is divided by n, each secondary current is multiplied by n, and each secondary impedance is divided by n^2. Powers are the same, whether calculated on the primary or secondary side.

The reader should recall from our previous analysis that if either dot on the transformer is reversed, then n is replaced by $-n$ in the equivalent circuits. In addition, it should be noted that the development of these equivalent circuits is predicated on the assumption that removing the transformer will divide the network into two parts; that is, there are no connections between the primary and secondary other than through the transformer. If any external connections exist, the equivalent circuit technique cannot in general be used. Finally, it should be pointed out that if the primary or secondary circuits are more complicated than those shown in Fig. 13.19a, Thévenin's theorem may be applied to reduce the network to that shown in Fig. 13.19a. Also, we can simply reflect the complicated circuit component by component from one side of the transformer to the other.

EXAMPLE 13.13

Given the circuit in Fig. 13.20a, we wish to draw the two networks obtained by replacing the transformer and the primary, and the transformer and the secondary, with equivalent circuits.

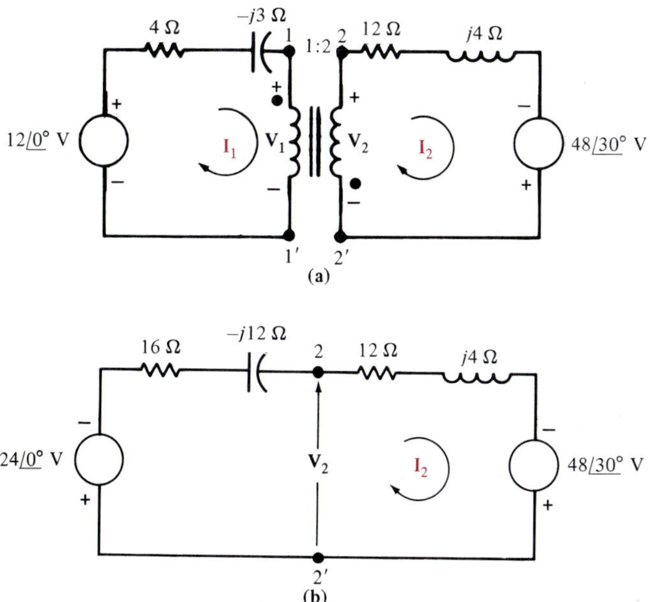

Figure 13.20 Example circuit and two equivalent circuits.

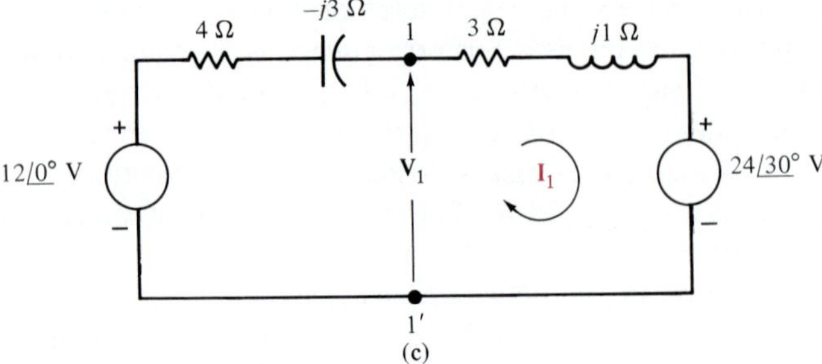

Figure 13.20 (continued)

Due to the relationship between the assigned currents and voltages and the location of the dots, the network containing an equivalent circuit for the primary, and the network containing an equivalent circuit for the secondary, are shown in Fig. 13.20b and c, respectively. The reader should note carefully the polarity of the voltage sources in the equivalent networks. ■

EXAMPLE 13.14

We wish to derive an equivalent network for the circuit in Fig. 13.21a and use it to determine the current $\mathbf{I}_1$.

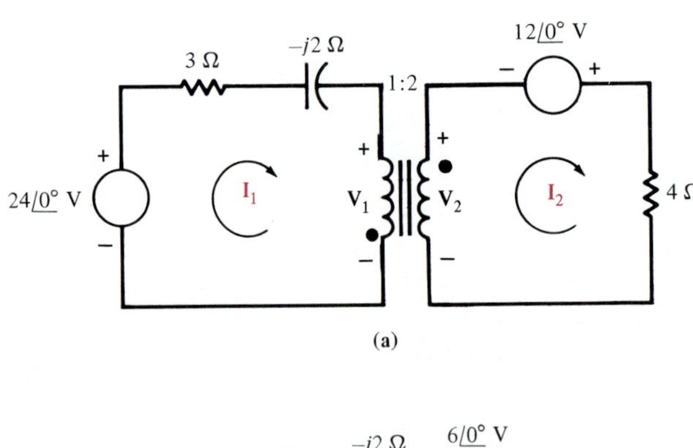

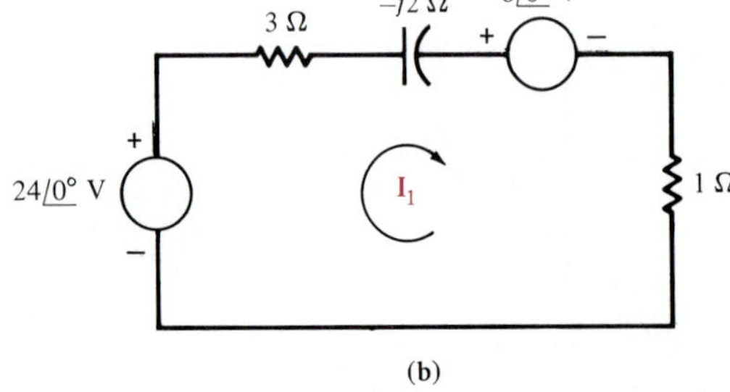

Figure 13.21 Example circuit containing an ideal transformer and an equivalent circuit.

For the given relationship between the assigned currents and voltages and the location of the dots, the network containing an equivalent circuit for the transformer and the secondary is shown in Fig. 13.21b. The current $\mathbf{I}_1$ is therefore

$$\mathbf{I}_1 = \frac{24 \underline{/0^\circ} - 6 \underline{/0^\circ}}{3 - j2 + 1}$$

$$= \frac{18 \underline{/0^\circ}}{4 - j2}$$

$$= 4.02 \underline{/26.57^\circ} \text{ A}$$

EXAMPLE 13.15

Let us determine the output voltage $\mathbf{V}_o$ in the circuit in Fig. 13.22a.

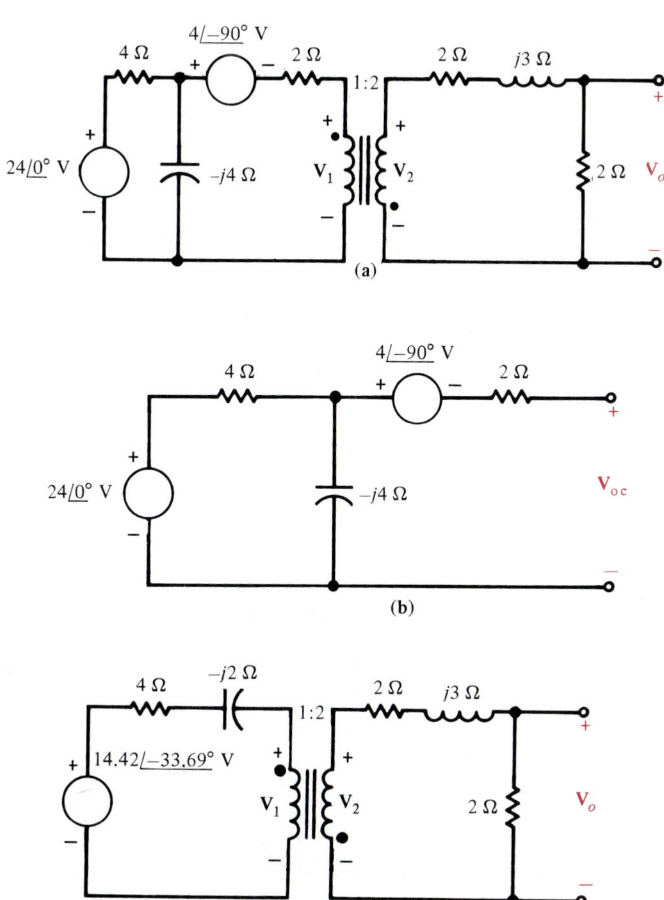

Figure 13.22 Example network and other circuits used to derive an equivalent network.

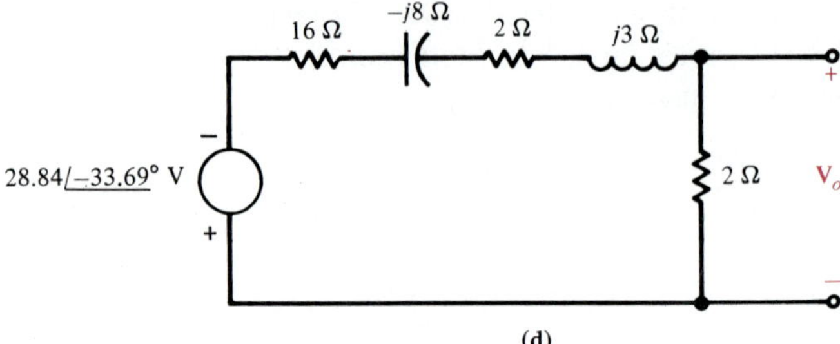

Figure 13.22 (continued)

We begin our attack by forming a Thévenin equivalent for the primary circuit. From Fig. 13.22b we can show that the open-circuit voltage is

$$\mathbf{V}_{oc} = \frac{24\,\underline{/0^\circ}}{4 - j4}(-j4) - 4\,\underline{/-90^\circ}$$

$$= 12 - j8 = 14.42\,\underline{/-33.69^\circ}\text{ V}$$

The Thévenin equivalent impedance looking into the open-circuit terminals with the voltage sources replaced by short circuits is

$$\mathbf{Z}_{Th} = \frac{(4)(-j4)}{4 - j4} + 2$$

$$= 4 - j2\ \Omega$$

The circuit in Fig. 13.22a thus reduces to that shown in Fig. 13.22c. Forming an equivalent circuit for the transformer and primary results in the network shown in Fig. 13.22d. Therefore, the voltage $\mathbf{V}_o$ is

$$\mathbf{V}_o = \frac{-28.84\,\underline{/-33.69^\circ}}{20 - j5}(2)$$

$$= 2.80\,\underline{/+160.35^\circ}\text{ V}$$

DRILL EXERCISE

D13.13. Given the network in Fig. D13.13, form an equivalent circuit for the transformer and secondary, and use the resultant network to compute $\mathbf{I}_1$.

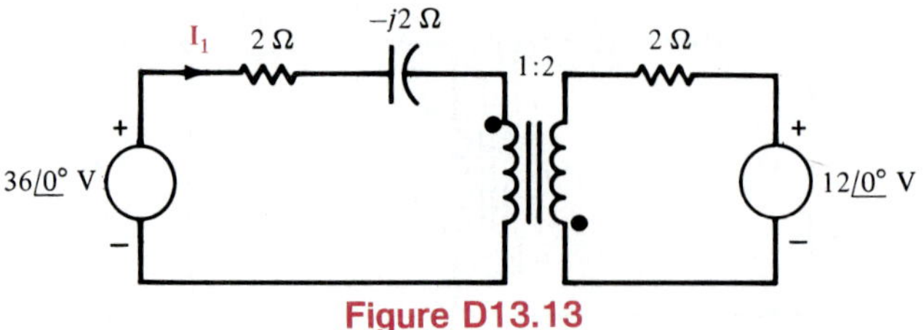

Figure D13.13

Ans: $\mathbf{I}_1 = 13.12\,\underline{/38.66^\circ}$ A.

D13.14. Given the network in Fig. D13.14, form an equivalent circuit for the transformer and primary, and use the resultant network to find $\mathbf{V}_o$.

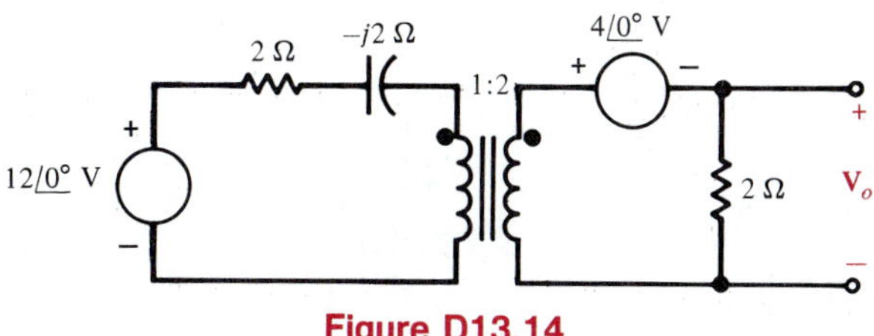

Figure D13.14

Ans: $\mathbf{V}_o = 3.12 \angle 38.66° \text{ V}$.

D13.15. Find $\mathbf{V}_o$ in the network in Fig. D13.15.

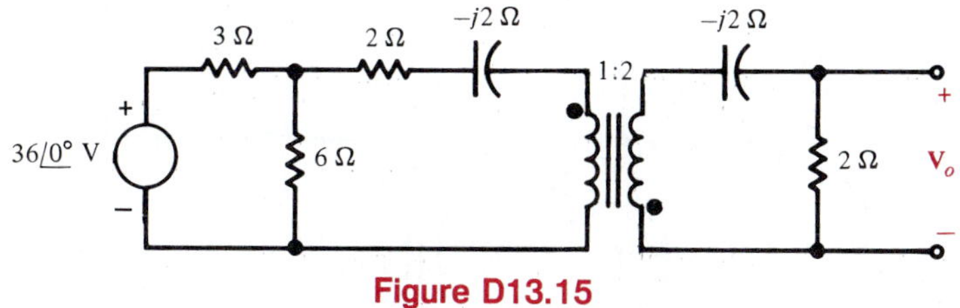

Figure D13.15

Ans: $\mathbf{V}_o = 4.66 \angle 29.05° \text{ V}$.

As a final point, consider the following example.

EXAMPLE 13.16

Given the circuit in Fig. 13.23a, we wish to find the value of the load R_o for maximum power transfer, and the value of the maximum power delivered to this load.

In order to determine the value of R_o for maximum power transfer, we will form a Thévenin equivalent at the terminals of R_o. The open-circuit voltage is derived from Fig. 13.23b as follows. The secondary resistance reflected into the primary is 16 Ω, and therefore,

$$\begin{aligned}\mathbf{I}_1 &= \frac{48 \angle 0°}{6 + 2 + 16} \\ &= 2 \angle 0° \text{ A}\end{aligned}$$

Therefore,

$$\begin{aligned}\mathbf{V}_1 &= (2 \angle 0°)(16) \\ &= 32 \angle 0° \text{ V}\end{aligned}$$

and by means of the turns ratio

$$\mathbf{I}_2 = 4 \angle 0° \text{ A}$$

$$\mathbf{V}_2 = 16 \angle 0° \text{ V}$$

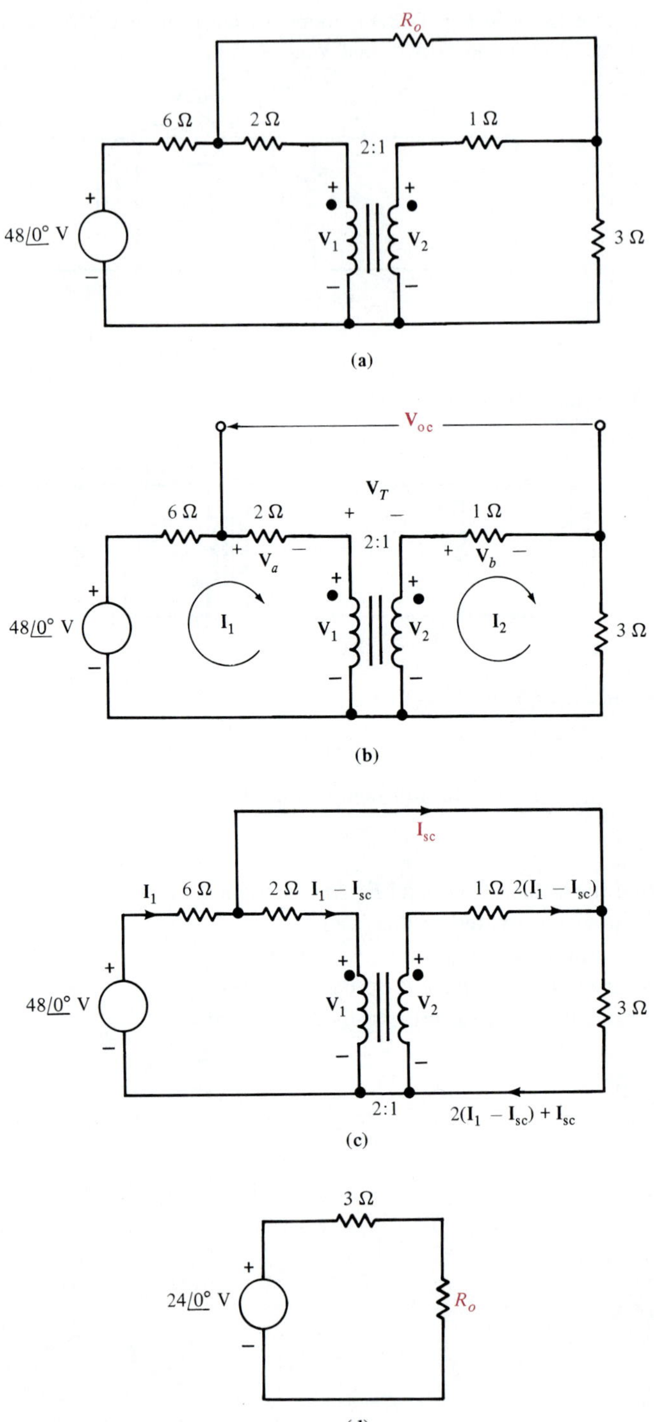

Figure 13.23 Maximum power transfer example involving an ideal transformer.

Now

$$\mathbf{V}_{oc} = \mathbf{V}_a + \mathbf{V}_T + \mathbf{V}_b = \mathbf{V}_a + \mathbf{V}_1 - \mathbf{V}_2 + \mathbf{V}_b = 24 \angle 0° \text{ V}$$

The short-circuit current is obtained from Fig. 13.23c. The equations for this network are

$$48 \angle 0° = 6\mathbf{I}_1 + 2(\mathbf{I}_1 - \mathbf{I}_{sc}) + \mathbf{V}_1$$

$$48 \angle 0° = 6\mathbf{I}_1 + 3[2(\mathbf{I}_1 - \mathbf{I}_{sc}) + \mathbf{I}_{sc}]$$

$$\mathbf{V}_2 = (1)[2(\mathbf{I}_1 - \mathbf{I}_{sc})] + 3[2(\mathbf{I}_1 - \mathbf{I}_{sc}) + \mathbf{I}_{sc}]$$

$$\mathbf{V}_1 = 2\mathbf{V}_2$$

These equations reduce to

$$48 \angle 0° = 12\mathbf{I}_1 - 3\mathbf{I}_{sc}$$

$$48 \angle 0° = 24\mathbf{I}_1 - 12\mathbf{I}_{sc}$$

and therefore,

$$\begin{bmatrix} \mathbf{I}_1 \\ \mathbf{I}_{sc} \end{bmatrix} = \frac{-1}{72} \begin{bmatrix} -12 & 3 \\ -24 & 12 \end{bmatrix} \begin{bmatrix} 48 \angle 0° \\ 48 \angle 0° \end{bmatrix}$$

$$= \frac{-1}{72} \begin{bmatrix} (-9)\, 48 \angle 0° \\ (-12)\, 48 \angle 0° \end{bmatrix} = \begin{bmatrix} 6.55 \angle 0° \text{ A} \\ 8 \angle 0° \text{ A} \end{bmatrix}$$

and

$$\mathbf{Z}_{Th} = \frac{24 \angle 0°}{8 \angle 0°} = 3 \ \Omega$$

Therefore, the equivalent circuit is shown in Fig. 13.23d, and $R_o = 3\ \Omega$ for maximum power transfer. The maximum power delivered to R_o is

$$P_{max} = \left[\frac{24}{2(3)}\right]^2 (3) = 48 \text{ W}$$

■

13.6 PSPICE Analysis of Ideal Transformer Circuits

The PSPICE circuit analysis program is a useful tool in solving ideal transformer circuits. To employ the power of PSPICE in this case, however, we must model the ideal transformer in terms of the circuit elements we have introduced thus far. The mathematical equations that govern the operation of the ideal transformer shown in Fig. 13.24a are

$$\mathbf{V}_1 = \frac{1}{n}\mathbf{V}_2$$

$$\mathbf{I}_2 = -\frac{1}{n}\mathbf{I}_1$$

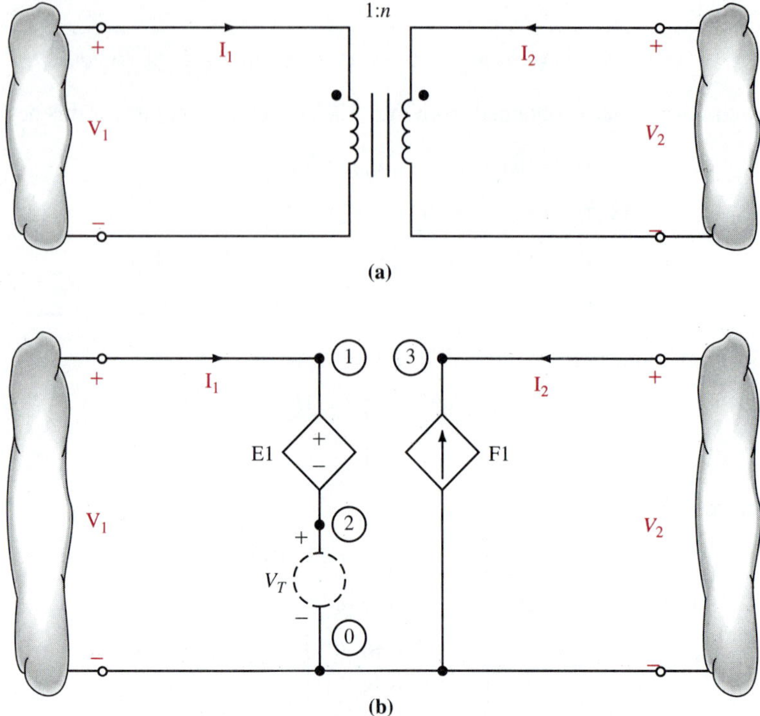

Figure 13.24 (a) Ideal transformer; (b) PSPICE model.

Note that both the voltage equation and the current equation exhibit a *dependent* relationship. Therefore, the equations can be modeled as shown in Fig. 13.24b, and the PSPICE program for this model is

```
E1    1    2    3     0     1/n
VT    2    0    0
F1    0    3    VT    1/n
```

where VT is a dummy test source used to measure $\mathbf{I}_1$. If either one of the dots is reversed, then the term $1/n$ in the PSPICE program should have a negative sign. Furthermore, recall that in a PSPICE analysis there must be a dc path to ground from every node. Therefore, it may be necessary in some cases to insert a very large resistor (e.g., 10 MΩ) across the current source from node ③ to node ⓪ to provide this path without affecting the results of the analysis.

EXAMPLE 13.17

Let us solve Drill Exercises D13.10 and D13.11 using PSPICE.

Employing our model for the ideal transformer, the network in Fig. D13.10 is labeled for a PSPICE analysis as shown in Fig. 13.25, where once again, for convenience, we assume that $f = 1/2\pi$ hertz. The PSPICE program and the output results are listed below.

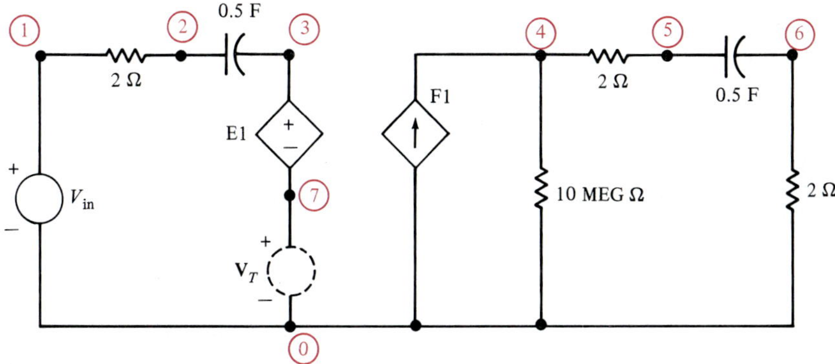

Figure 13.25 Network used in Example 13.17.

```
DRILL EXERCISES D13.10 AND D13.11 USING
*DEPENDENT SOURCES.
VIN 1 0 AC 12
R1 1 2 2
C1 2 3 0.5
E1 3 7 4 0 0.5
VTEST 7 0 0
F1 0 4 VTEST 0.5
* Add this to create a DC path to ground.
* PSpice will generate an error if there
* is not a DC path to ground from every node.
RDCPATH 4 0 10MEG
R2 4 5 2
C2 5 6 0.5
R3 6 0 2
* F = 1/2PI
.AC LIN 1 0.159155 0.159155
.PRINT AC VM(6) VP(6) IM(VTEST) IP(VTEST)
.END

AC ANALYSIS RESULTS

   FREQ         VM(6)        VP(6)        IM(VTEST)    IP(VTEST)
1.592E-01    3.073E+00    3.981E+01    3.073E+00    3.981E+01
```

DRILL EXERCISE

D13.16. Solve the problem presented in Example 13.15 using PSPICE.

Ans: See Example 13.15.

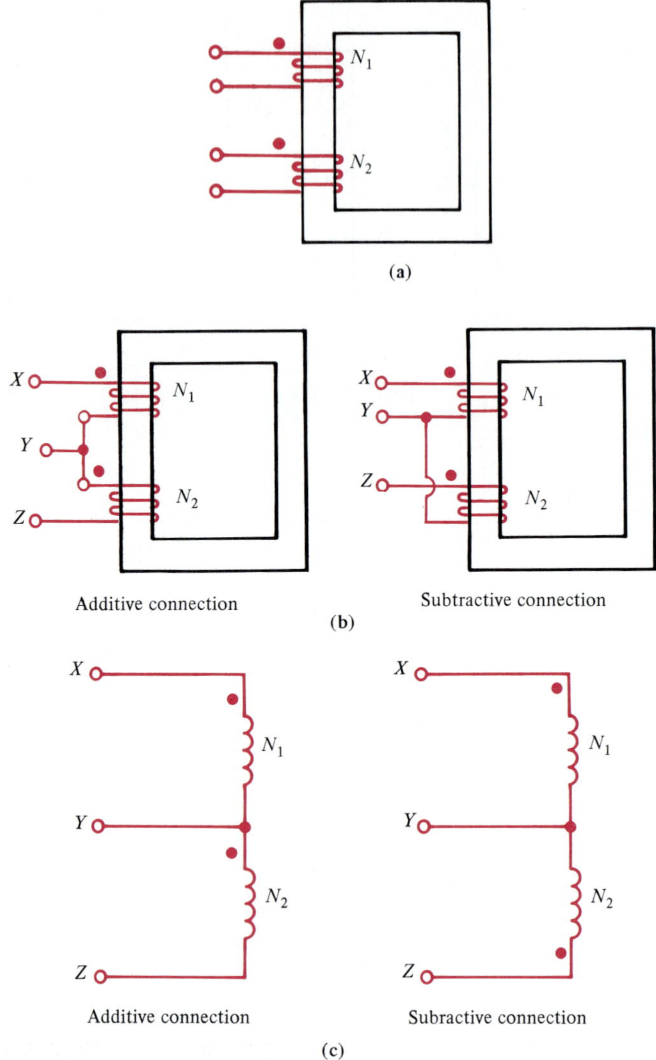

Figure 13.26 Autotransformer: (a) normal two-winding transformer with adjacent windings; (b) two-winding transformer interconnected to create a single-winding, three-terminal autotransformer; (c) symbolic representation of (b).

13.7 Ideal Autotransformers

The two winding transformers we have presented thus far provide electrical isolation between primary and secondary windings as shown in Fig. 13.26a. It is possible, however, to interconnect primary and secondary windings serially, creating a three-terminal device as shown in Fig. 13.26b and represented in Fig. 13.26c. As we shall see, this arrangement offers certain practical advantages over the isolated case. Note that the three-terminal arrangement is essentially one continuous winding with an internal tap point (terminal Y). Such a device is commonly available and is called an ''autotransformer.''

The tap point can be adjusted to provide a variable voltage at the output. The autotransformer can be used in any application that requires a normal transformer, provided that electrical isolation is not required. It is particularly useful where a variable output voltage supply is needed (e.g., in a laboratory environment).

Let us examine the step-down autotransformer connection in Fig. 13.27a. By transformer action

$$\frac{\mathbf{V}_{XY}}{\mathbf{V}_{YZ}} = \frac{N_1}{N_2} \qquad \text{and} \qquad N_1\mathbf{I}_1 = N_2\mathbf{I}_{ZY}$$

Using KVL gives us

$$\mathbf{V}_2 = \mathbf{V}_{YZ}$$

$$\mathbf{V}_1 = \mathbf{V}_{XY} + \mathbf{V}_{YZ} = \left(\frac{N_1}{N_2} + 1\right)\mathbf{V}_2$$

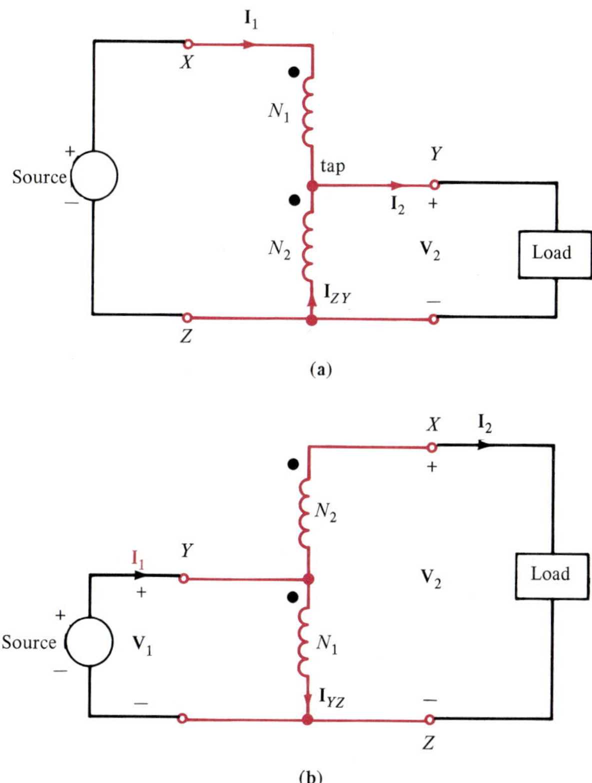

Figure 13.27 Autotransformer circuits: (a) step-down connection; (b) step-up connection.

and

$$\mathbf{I}_2 = \mathbf{I}_1 + \mathbf{I}_{ZY} = \left(1 + \frac{N_1}{N_2}\right)\mathbf{I}_1$$

Then

$$\frac{\mathbf{V}_{XY}\mathbf{I}_1}{\mathbf{V}_2\mathbf{I}_2} = \frac{\left(\dfrac{N_1}{N_2}\mathbf{V}_2\right)\left(\dfrac{N_2}{N_1 + N_2}\right)\mathbf{I}_2}{\mathbf{V}_2\mathbf{I}_2} = \frac{N_1}{N_1 + N_2} \tag{13.33}$$

What is the significance of this expression? We know that the power rating of winding N_1 must be the same as the rating of winding N_2. However, Eq. (13.33) illustrates that the power rating of winding N_1 (i.e., the power rating of the transformer) is only a fraction $[N_1/(N_1 + N_2)]$ of the power that is required by the load.

The following examples illustrate this point.

EXAMPLE 13.18

A 5-kVA 220 V/110 V transformer is to be connected to serve a 110-V load from a 330-V source.

(a) Draw the proper circuit diagram.

(b) Determine the maximum load that can be served without overloading the transformer.

(a) The circuit in Fig. 13.27a is appropriate if the 220-V winding is assigned to be "N_1" and the 110-V winding is assigned as "N_2."

(b) Since $N_1/N_2 = 220/110$, $N_1 = 2N_2$ and

$$\frac{N_1}{N_1 + N_2} = \frac{2N_2}{2N_2 + N_2} = \frac{2}{3}$$

Therefore,

$$\frac{\mathbf{V}_{XY}\mathbf{I}_1}{\mathbf{V}_2\mathbf{I}_2} = \frac{5 \text{ kVA}}{S_2} = \frac{2}{3}$$

and

$$\mathbf{S}_2 = \tfrac{3}{2}(\mathbf{S}) = 7.5 \text{ kVA}$$

or the load served can be one and one-half times the power rating of the transformer. ■

EXAMPLE 13.19

The transformer in Example 13.18 is to be used to serve a 330-V load from a 220-V source.

(a) Draw the appropriate circuit diagram.

(b) Calculate all currents if each winding operates at rated conditions.

(c) Calculate the apparent power of the load.

(a) The proper circuit diagram is shown in Fig. 13.27b, with the 220-V winding assigned as "N_1" and the 110-V winding assigned as "N_2."

(b) $\mathbf{I}_{YZ} = 5000/220 = 22.73$ A and $\mathbf{I}_2 = 5000/110 = 45.45$ A, and therefore, $\mathbf{I}_1 = \mathbf{I}_2 + \mathbf{I}_{YZ} = 68.18$ A. Note: At this point it may not be clear how we can ignore the complex nature of currents and voltages in adding them. However, Example 13.20 will examine this issue.

(c) $\mathbf{S}_2 = \mathbf{V}_2\mathbf{I}_2 = 15$ kVA. Note that this figure is three times winding ratings! ■

EXAMPLE 13.20

Under the conditions described in Example 13.19, let us determine all phasor voltages and currents if $\mathbf{V}_2 = 330 \angle 0°$ V and the load power factor is 0.8 lagging.

Suppose that $\mathbf{V}_1 = 220 \angle \alpha$. Then $\mathbf{V}_{XY} = (110/220)220 \angle \alpha = 110 \angle \alpha$ V. Now $\mathbf{V}_2 = \mathbf{V}_1 + \mathbf{V}_{XY} = 220 \angle \alpha + 110 \angle \alpha = 330 \angle \alpha$ V. However, since $\mathbf{V}_2 = 330 \angle 0°$ V, $\alpha = 0°$. Since the load power factor is lagging, $\theta_2 = -\cos^{-1}(0.8) = -36.9°$, and therefore $\mathbf{I}_2 = 45.45 \angle -36.9°$ A. Also, $\mathbf{I}_{YZ} = (N_2/N_1)\mathbf{I}_2 = (110/220)(45.45 \angle -36.9°) = 22.72 \angle -36.9°$ A. Hence $\mathbf{I}_1 = \mathbf{I}_2 + \mathbf{I}_{YZ} = 45.45 \angle -36.9° + 22.72 \angle -36.9° = 68.18 \angle -36.9°$ A. The analysis indicates that similar results would have been achieved for any angle α or power factor; that is, $\mathbf{V}_1$, $\mathbf{V}_{XY}$, and $\mathbf{V}_2$ are in phase *in general*, as are $\mathbf{I}_1$, $\mathbf{I}_2$, and $\mathbf{I}_{YZ}$. ■

The examples have illustrated that two-winding transformers are capable of more power delivery when connected as an autotransformer. In the two-winding transformer, power is transferred inductively while in the autotransformer power is transferred both inductively and conductively.

DRILL EXERCISE

D13.17. A 120/12-V two-winding transformer is available. What voltages can be obtained at the output if the transformer is connected as an autotransformer with a $120 \angle 0°$-V input?

Ans: $132 \angle 0°$ V and $108 \angle 0°$ V.

13.8 Three-Phase Transformers

Transformers play a key role in three-phase ac power transmission. They are used to step up the voltage from the generating station to the high-voltage power transmission network and to step down the voltage from the transmission network to the loads. The three-phase transformation may be accomplished using a *bank* of single-phase transformers or a three-phase transformer. If a bank of single-phase transformers is used, it is important to ensure that all of the transformers have similar characteristics, so as to maintain a balanced system.

There are four possible balanced ways in which the three-phase transformer can be connected: wye–wye, delta–delta, wye–delta, and delta–wye, as shown in Fig. 13.28. Commonly, "wye" is symbolized as "Y" and "delta" as "Δ."

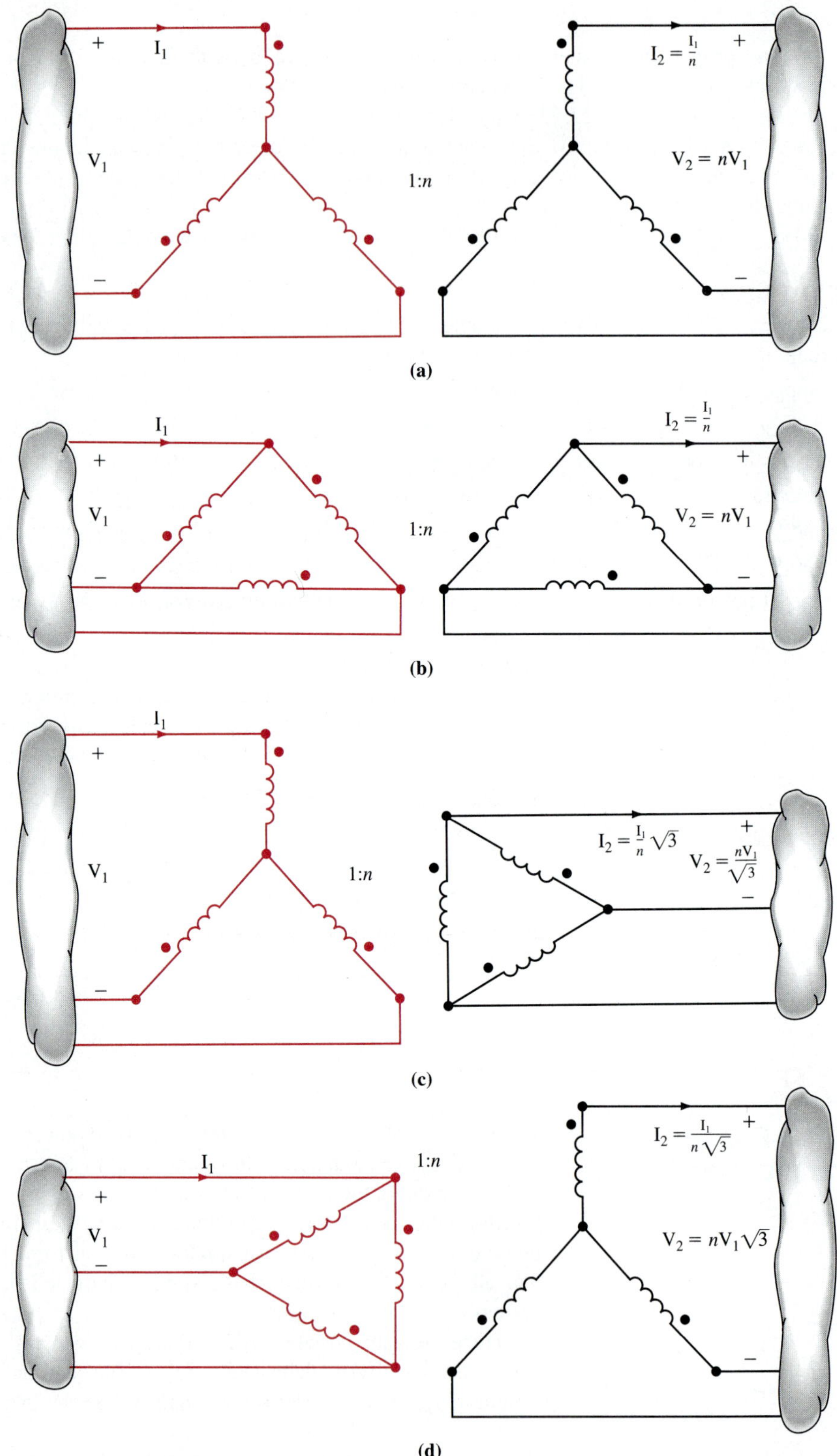

Figure 13.28 Three-phase transformer connection.

Transformer primaries are frequently connected in delta. The reason for using this delta configuration is that it allows the dominant third-harmonic component, which typically exists in the primary current, to circulate in the primary loop without being induced in the secondary.

The delta–delta configuration has a unique property. If one of the transformers is removed for any reason (e.g., repair or maintenance), the other two transformers, which now form an *open delta*, can still provide balanced three-phase voltages at a reduced load.

Whenever the secondary is connected in wye and the neutral line is used, both single-phase (line-to-neutral) and three-phase (line-to-line) loads can be supplied.

In wye–wye and delta–delta configurations, the line currents and voltages differ by the turns ratio n. In the wye–delta and delta–wye configurations, an additional $\sqrt{3}$ factor comes into play.

EXAMPLE 13.21

A three-phase 100-kVA 0.8-pf load is supplied by a three-phase feeder whose per phase impedance is $0.01 + j0.02\ \Omega$. At its sending end the feeder is connected to the low-voltage side of a three-phase 12-kVY: 208-V Δ transformer that we assume to be ideal. If the line voltage at the load is 200 V, we wish to find

(a) The line current at the load.

(b) The line voltage at the sending end of the feeder.

(c) The line voltage and current on the high-voltage side of the transformer.

The network is shown in Fig. 13.29.

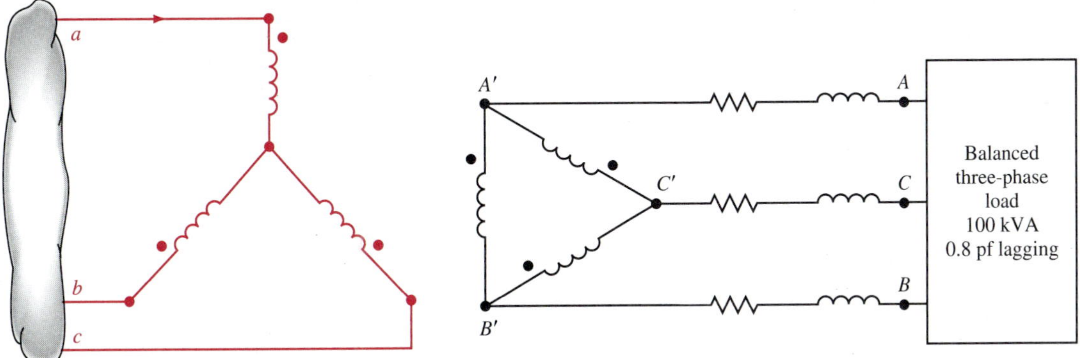

Figure 13.29 Network used in Example 13.21.

The line voltage at the load is

$$V_{AB} = V_L = 200\ \text{V}$$

and therefore,

$$V_{AN} = \frac{V_{AB}}{\sqrt{3}} = 115.5\ \text{V}$$

Recall that

$$\mathbf{S} = \sqrt{3}\, V_L I_L$$

and hence

$$I_L = \frac{100{,}000}{200\sqrt{3}}$$

$$= 288.7 \text{ A}$$

If we now assume that $\mathbf{V}_{AN} = 115.5 \underline{/0^\circ}$ V and note that

$$\theta = \cos^{-1}(\text{pf}) = \cos^{-1}(0.8) = \pm 36.9^\circ$$

since the power factor is lagging,

$$\mathbf{I}_{A'A} = 288.7 \underline{/-36.9^\circ} \text{ A}$$

Then at the feeder sending end,

$$\mathbf{V}_{A'N'} = (0.01 + j0.02)\mathbf{I}_{A'A} + 115.5 \underline{/0^\circ}$$

$$= 121.3 \underline{/1.36^\circ} \text{ V}$$

Thus the line voltage on the low-voltage side of the transformer is

$$V_L = V_{A'N'}\sqrt{3} = 210.1 \text{ V}$$

Three-phase power transformers are rated in *line* voltage, regardless of internal wye or delta connections. Since the transformer is assumed ideal, the line voltage on the high-voltage side of the transformer is

$$V_{LHV} = \left(\frac{12{,}000}{208}\right)(210.1)$$

$$= 12.12 \text{ kV}$$

Similarly,

$$I_{LHV} = \left(\frac{208}{12{,}000}\right)(288.7)$$

$$= 5.004 \text{ A}$$

It is interesting to note that the transformer per phase turns ratio is $(12{,}000/\sqrt{3})/208$. However, the phase voltage on the high-voltage side of the transformer is line to neutral because of the wye connection. Therefore, if the line voltage is determined using the per phase turns ratio, we obtain

$$V_{LHV} = \sqrt{3}\left(\frac{12{,}000/\sqrt{3}}{208}\right) V_{LLV}$$

and the $\sqrt{3}$ factor cancels. Similar considerations affect the current. ■

DRILL EXERCISE

D13.18. A balanced three-phase transformer is rated at 1000 MVA 25 kVΔ: 500 kVY. The balanced three-phase load connected to the high-voltage terminals of the transformer consumes 750 MW, 0.8 pf lagging, at 500 kV. Determine the magnitude of the line currents at the source and the load.

Ans: $I_S = 21.65$ kA, $I_L = 1083$ A.

13.9 Safety Considerations

Transistors are used extensively in modern electronic equipment to provide a low-voltage power supply. As examples, in computer systems a common voltage level is 5 V dc, portable radios use 9 V dc, and military and airplane equipment operates at 28 V dc. When transformers are used to connect these low-voltage transistor circuits to the power line, there is generally less danger of shock within the system because the transformer provides electrical isolation from the line voltage. However, from a safety standpoint, a transformer, although helpful in many situations, is not an absolute solution. We must always be vigilant when working with any electrical equipment to minimize the dangers of electrical shock.

In power electronics equipment or power systems the danger is severe. The problem in these cases is that of high voltage from a low-impedance source and we must constantly remember that the line voltage in our homes can be lethal.

Consider now the following example that illustrates a hidden danger that could surprise even the experienced professional, with devastating consequences.

EXAMPLE 13.22

Two adjacent homes, *A* and *B*, are fed from different transformers as shown in Fig. 13.30a. A surge on the line feeding house *B* has caused the circuit breaker *X-Y* to open. House *B* is now left without power. In an attempt to help his neighbor, the resident of house *A* volunteers to connect a long extension cord between a wall plug in house *A* and a wall plug in house *B*, as shown in Fig. 13.30b. Later, the lineman from the utility company comes to reconnect the circuit breaker. Unaware of the extension cord connection, the lineman believes that there is no voltage between points *X* and *Z*. However, because of the electrical connection between the two homes, 7200 V exists between the two points, and the lineman could be seriously injured or even killed if he comes in contact with this high voltage. ■

13.10 Summary

Magnetically coupled circuits have been presented and the circuit equations that describe these elements discussed. Mutual inductance has been defined and a dot convention adopted for indicating the physical relationship of the coils in order to simplify the sign

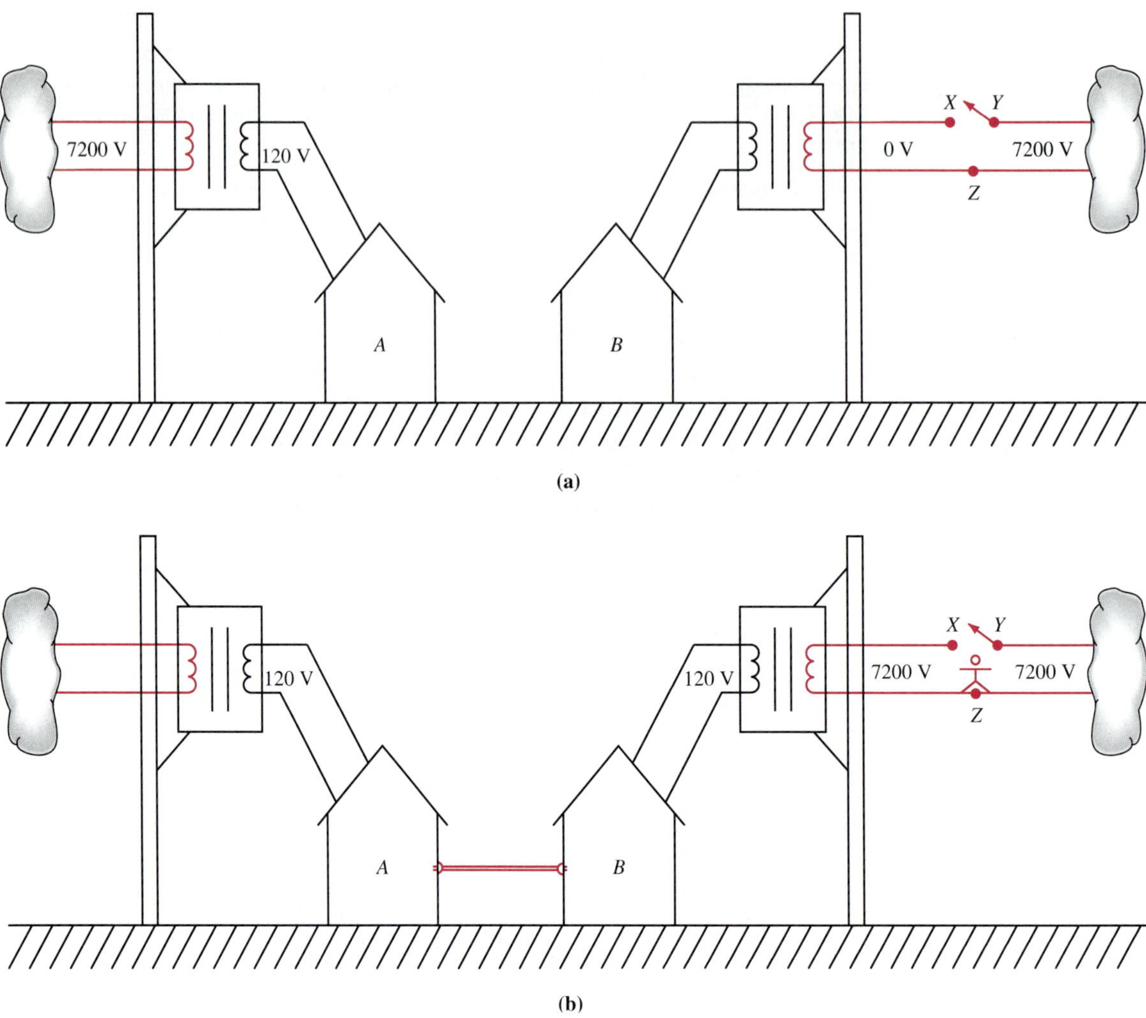

Figure 13.30 Diagrams used in Example 13.22.

convention for the mutual terms in the circuit equations. An energy analysis for mutually coupled coils has been performed, which has led to a definition of the coefficient of coupling between coils.

The linear transformer in which no magnetic material is used to couple the coils was discussed. Then coils coupled with good magnetic material were described and presented as an ideal transformer. Finally, Thévenin's theorem was employed to derive equivalent circuits for the transformer and either its primary or secondary circuit to simplify the analysis of circuits containing ideal transformers.

The PSPICE circuit analysis program was used to analyze circuits containing mutual inductance and ideal transformers.

Autotransformers and three-phase transformers were also presented. Finally, some safety considerations were discussed.

KEY POINTS

- A transformer is a device that contains two or more coils that are coupled magnetically.
- Inductor coils are said to be mutually coupled if they share a common magnetic flux.
- The dot convention is used to determine if the flux produced by current flowing through one inductor coil will add to or oppose the flux produced by another inductor coil.
- A transformer is said to be linear if the magnetic path material between the two coils has constant permeability.
- An ideal transformer is one in which the core permeability and winding conductivities are assumed to be infinite.
- The PSPICE material presented throughout the text can be applied to mutually coupled circuits and circuits that contain ideal transformers.
- The use of an autotransformer can be a cost-effective approach in some transformer applications.
- Three-phase transformers are commonly used in four connections: wye–wye, delta–delta, wye–delta, and delta–wye.

PROBLEMS

13.1. Find $\mathbf{V}_o$ in the network in Fig. P13.1.

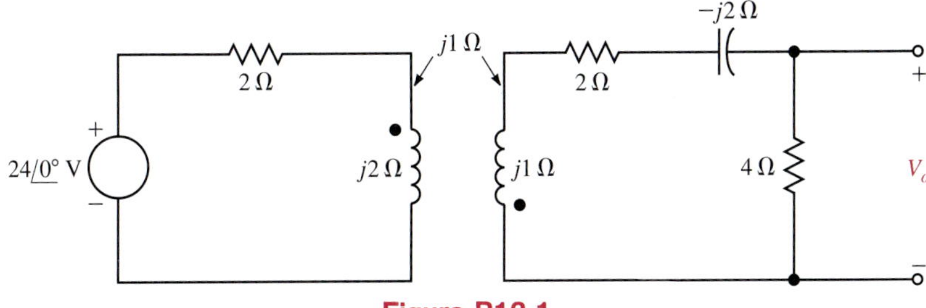

Figure P13.1

13.2. Find $\mathbf{V}_A$ in the network shown in Fig. P13.2.

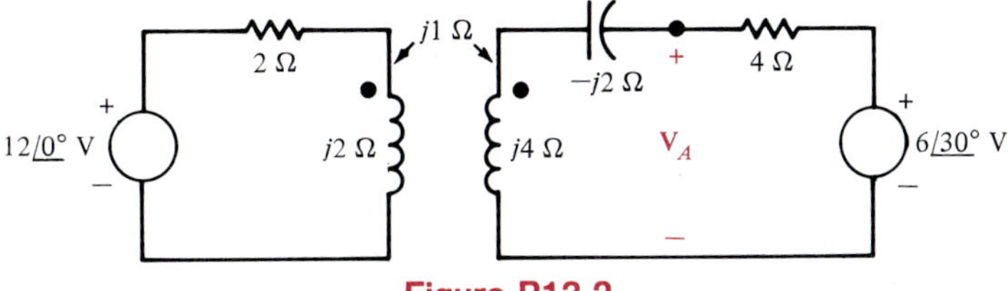

Figure P13.2

13.3. Given the network in Fig. P13.3,
(a) Find the equations for $v_a(t)$ and $v_b(t)$.
(b) Find the equations for $v_c(t)$ and $v_d(t)$.

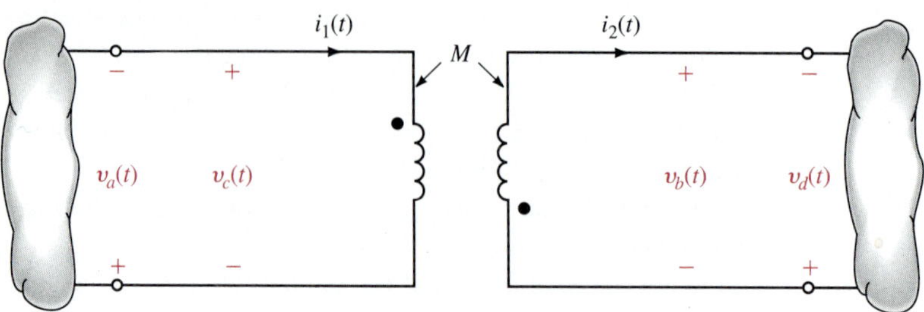

Figure P13.3

13.4. Write the mesh equations for the network shown in Fig. P13.4.

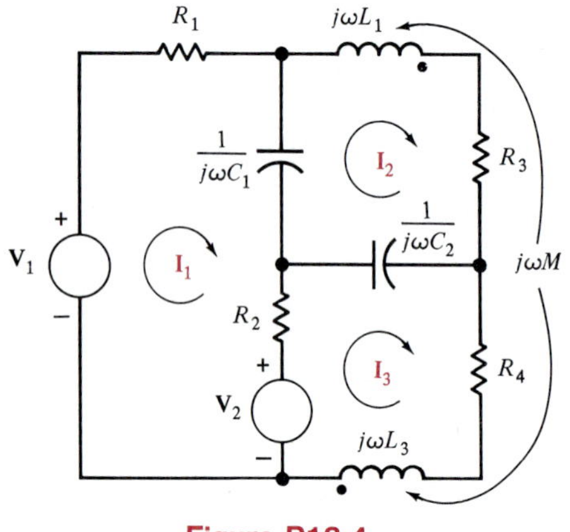

Figure P13.4

13.5. Write the mesh equations in standard form for the network in Fig. P13.5.

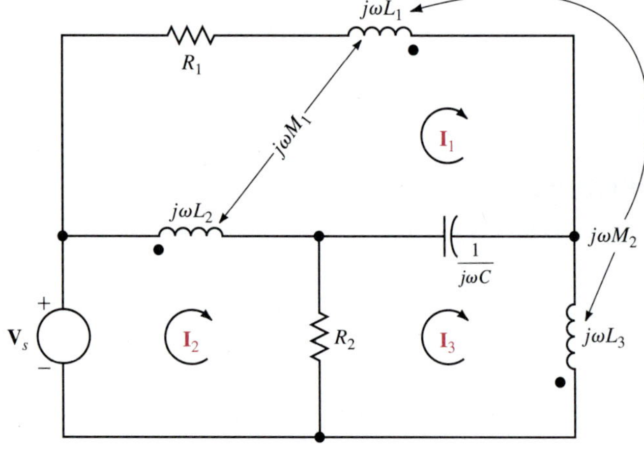

Figure P13.5

13.6. Find $\mathbf{V}_o$ in the network in Fig. P13.6.

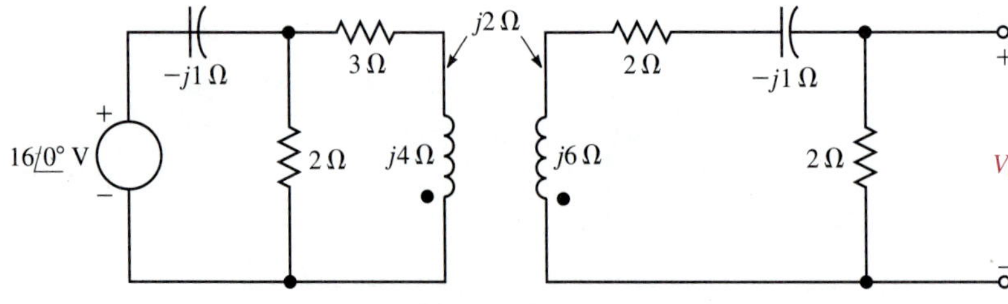

Figure P13.6

13.7. Find $\mathbf{V}_o$ in the circuit in Fig. P13.7.

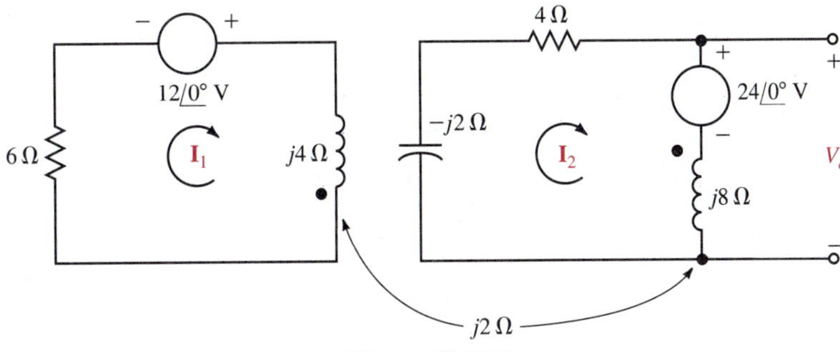

Figure P13.7

13.8. Find $\mathbf{V}_o$ in the network in Fig. P13.8.

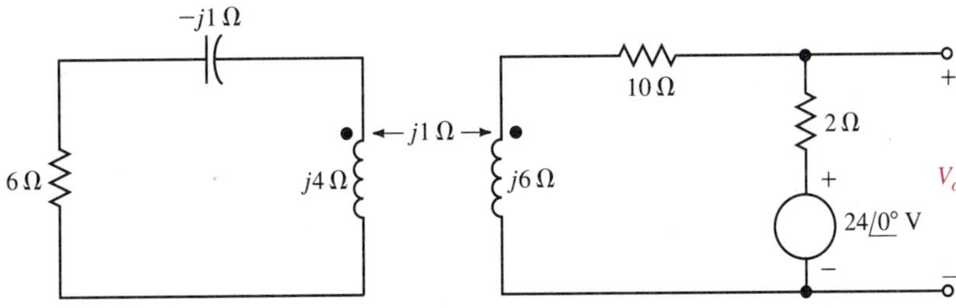

Figure P13.8

13.9. Find $\mathbf{V}_o$ in the circuit in Fig. P13.9.

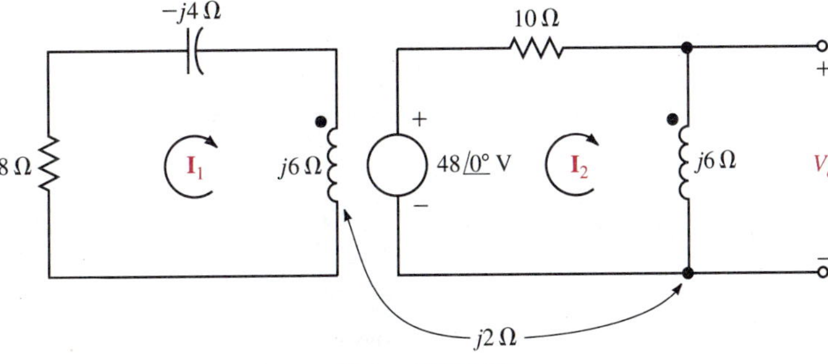

Figure P13.9

13.10. Find $\mathbf{V}_o$ in the network in Fig. P13.10.

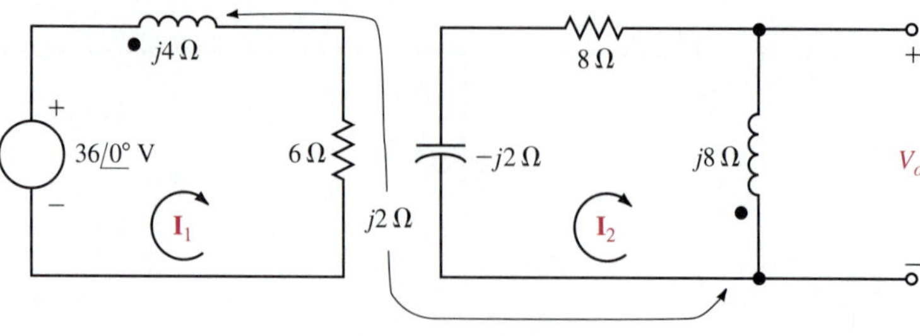

Figure P13.10

13.11. Find $\mathbf{V}_o$ in the circuit in Fig. P13.11.

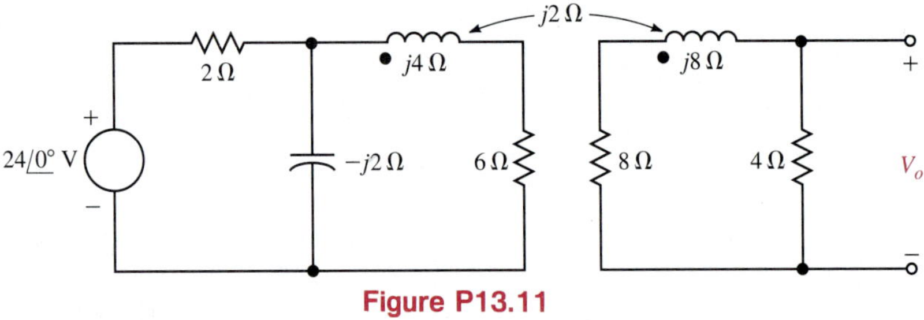

Figure P13.11

13.12. Find $\mathbf{V}_o$ in the network in Fig. P13.12.

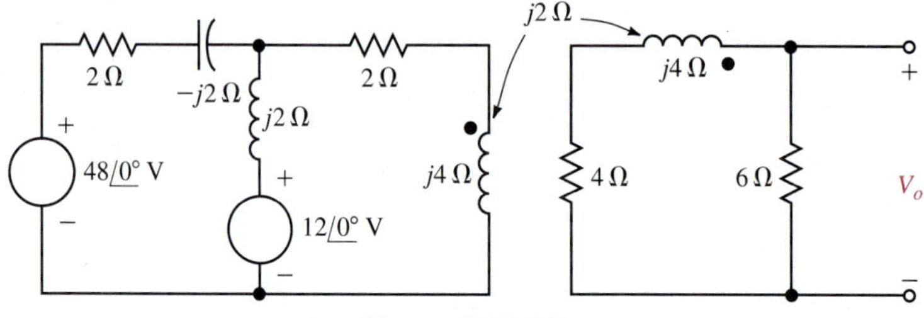

Figure P13.12

13.13. Find $\mathbf{I}_o$ in the circuit in Fig. P13.13.

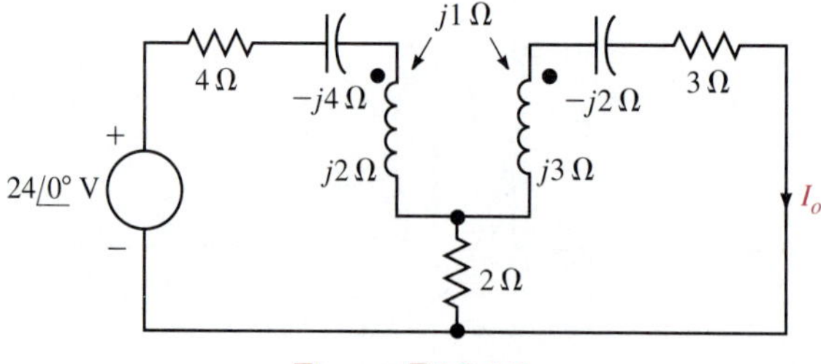

Figure P13.13

13.14. The currents in the network shown in Fig. P13.14 are

$$i_1(t) = 16 \cos (377t - 45°) \text{ mA}$$

$$i_2(t) = 3 \cos (377t - 45°) \text{ mA}$$

If the inductance values are $L_1 = 2$ H, $L_2 = 8$ H, and $M = 3$ H, determine the voltages $v_1(t)$ and $v_2(t)$.

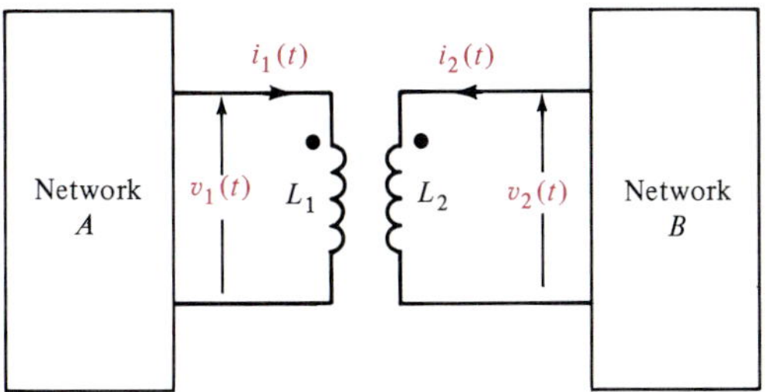

Figure P13.14

13.15. Determine the energy stored in the coupled inductors in Problem 13.14 at $t = 3$ ms.

13.16. Given the network shown in Fig. P13.16, determine the value of the capacitor C that will cause the reflected impedance to the primary to be purely resistive.

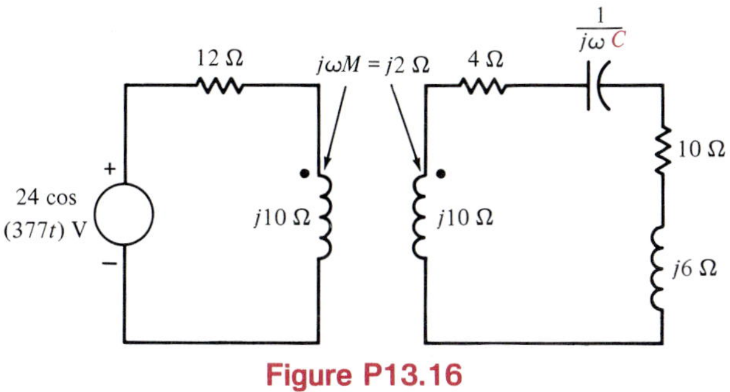

Figure P13.16

13.17. Analyze the network in Fig. P13.17 and determine if a value of $\mathbf{X}_c$ can be found such that the output voltage is equal to twice the input voltage.

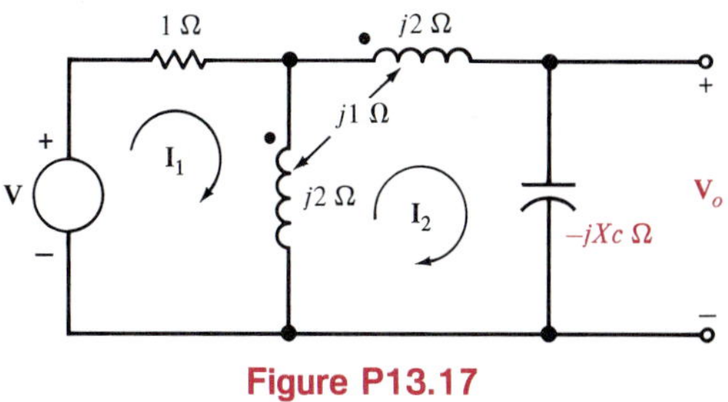

Figure P13.17

13.18. Compute the input impedance of the network in Fig. P13.18.

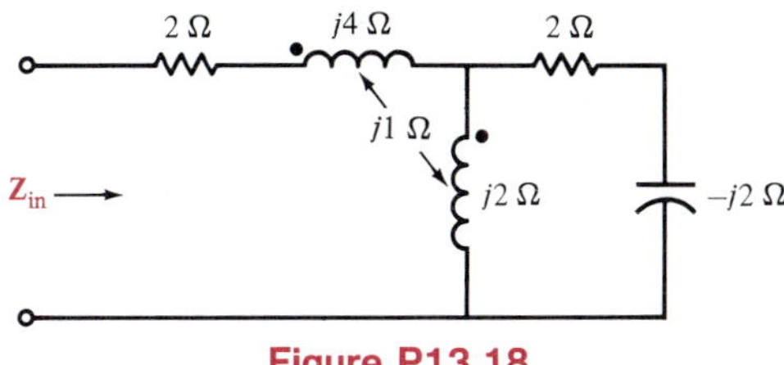

Figure P13.18

13.19. Write a PSPICE program to calculate $\mathbf{V}_o$ in the network shown in Fig. P13.19, where $f = 1/2\pi$ Hz.

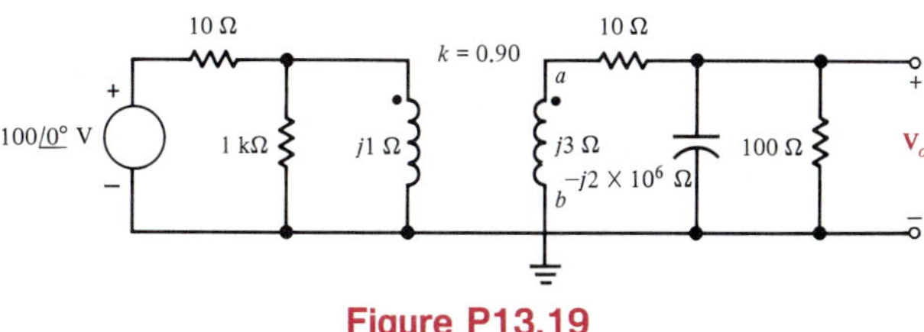

Figure P13.19

13.20. Repeat Problem 13.19 with the dot moved from position *a* to position *b*.

13.21. Use PSPICE to find $\mathbf{V}_o$ in the network in Fig. P13.21. The source frequency is 60 Hz.

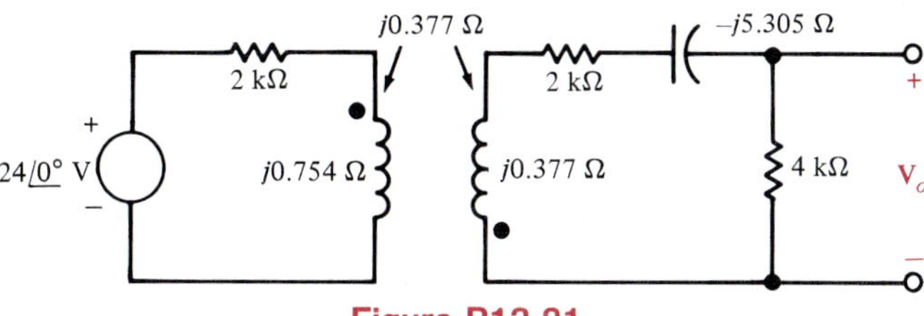

Figure P13.21

13.22. Use PSPICE to find $\mathbf{V}_o$ in the network in Fig. P13.22. The source frequency is 400 Hz and $k_{12} = 0.8$.

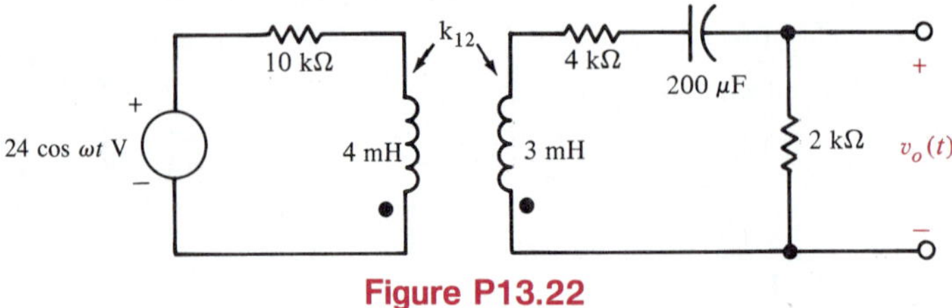

Figure P13.22

13.23. Given the network in Fig. P13.23, find the currents $i_1(t)$, $i_2(t)$, and $i_o(t)$ using PSPICE.

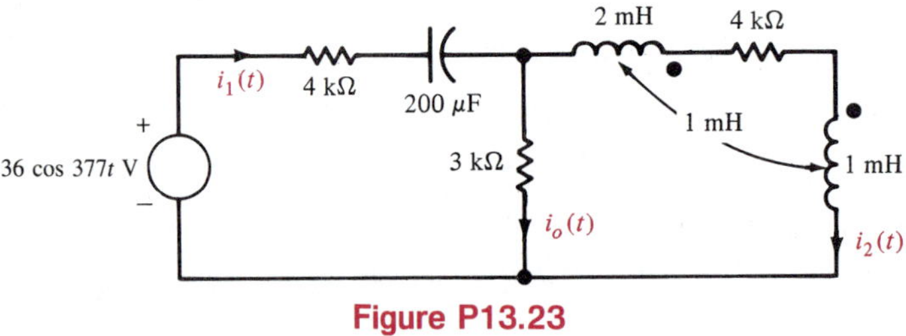

Figure P13.23

13.24. Repeat Problem 13.23 if the source frequency is 400 Hz.

13.25. Use PSPICE to find $v_0(t)$ in the network in Fig. P13.25. The source frequency is 400 Hz.

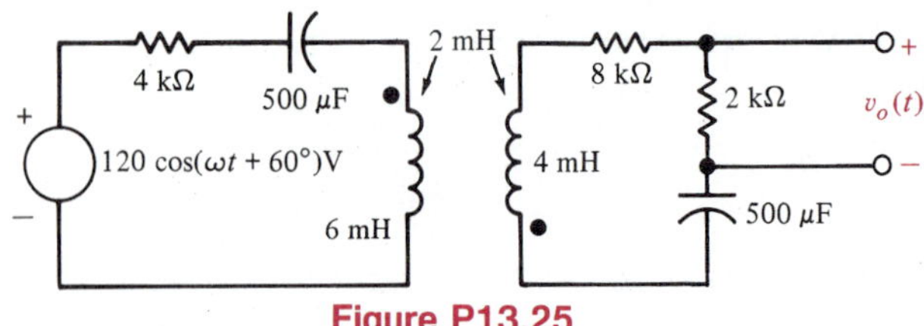

Figure P13.25

13.26. Use PSPICE to find the currents $\mathbf{I}_1$, $\mathbf{I}_2$, and $\mathbf{I}_3$ in the network in Fig. P13.26. The frequency of the sources is 60 Hz. $k = 0.6$

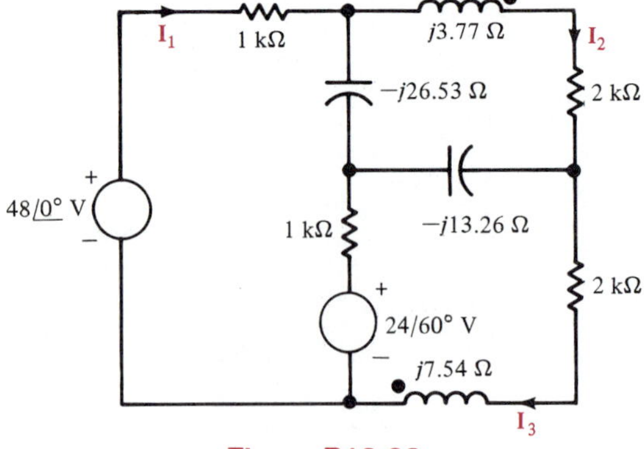

Figure P13.26

13.27. Given the network in Fig. P13.27, determine $\mathbf{V}_o$ using PSPICE. $f = 1/2\pi$ Hz. Let

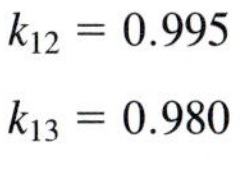

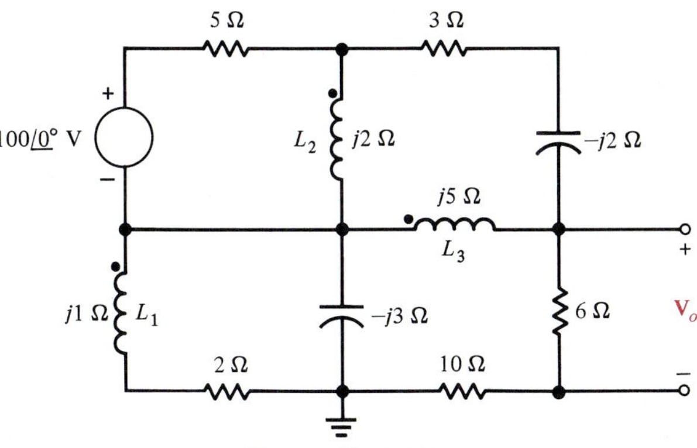

Figure P13.27

13.29. If the voltage source $\mathbf{V}_S$ in the circuit of Problem 13.28 is 50 $\angle 0°$ V, determine $\mathbf{V}_o$.

13.30. Determine the input impedance seen by the source in the network shown in Fig. P13.30.

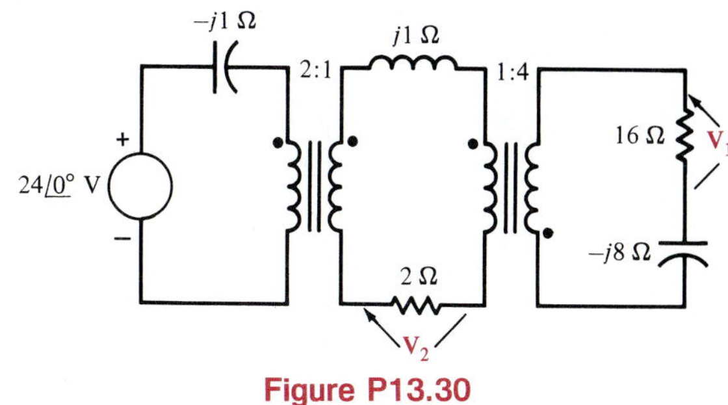

Figure P13.30

13.28. Given that $\mathbf{V}_o = 48 \angle 30°$ V in the circuit shown in Fig. P13.28, determine $\mathbf{V}_S$.

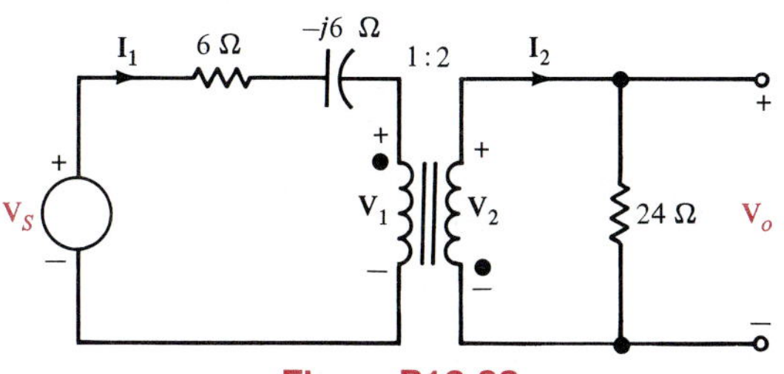

Figure P13.28

13.31. Find all currents and voltages in the network in Fig. P13.31.

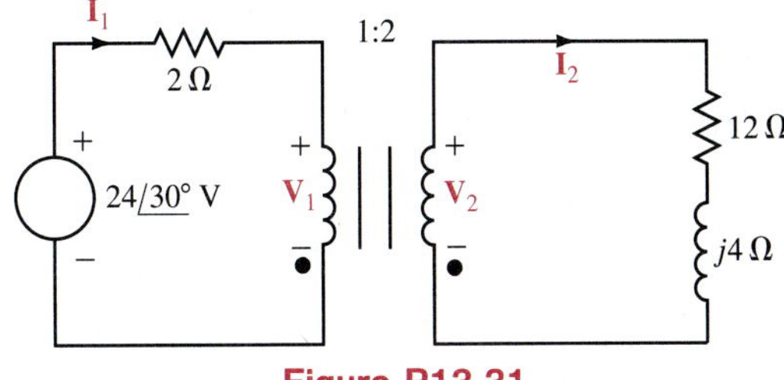

Figure P13.31

13.32. Find $\mathbf{I}_o$ in the network in Fig. P13.32.

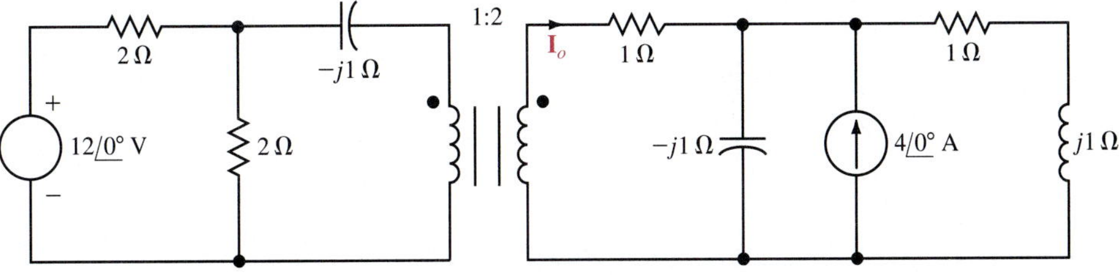

Figure P13.32

13.33. Find the voltage $\mathbf{V}_o$ in the network in Fig. P13.33.

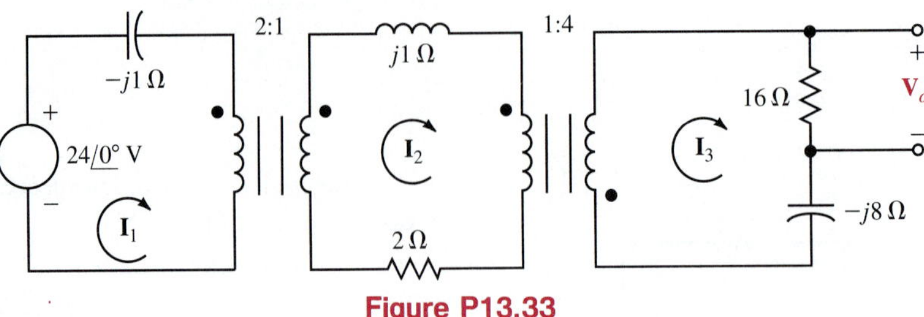

Figure P13.33

13.34. Find $\mathbf{V}_o$ in the circuit in Fig. P13.34.

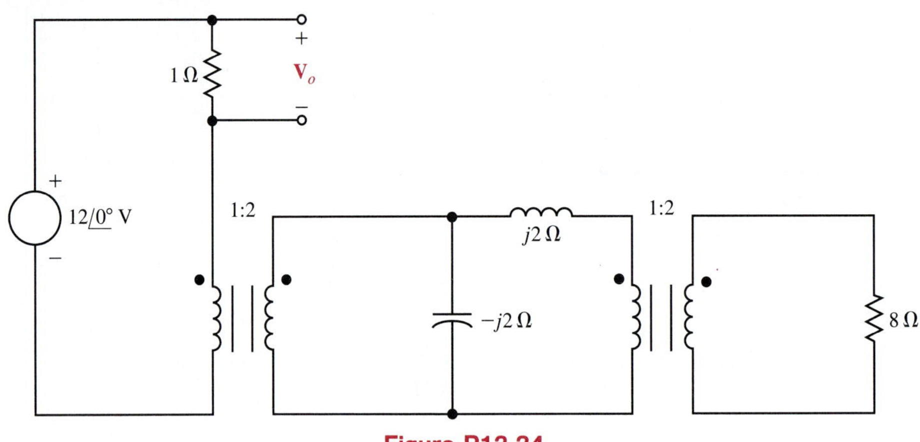

Figure P13.34

13.35. Find the current $\mathbf{I}$ in the network in Fig. P13.35.

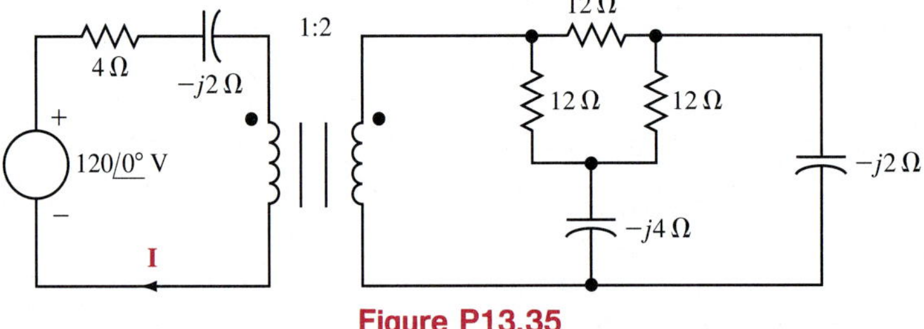

Figure P13.35

13.36. Find the current $\mathbf{I}$ in the network in Fig. P13.36.

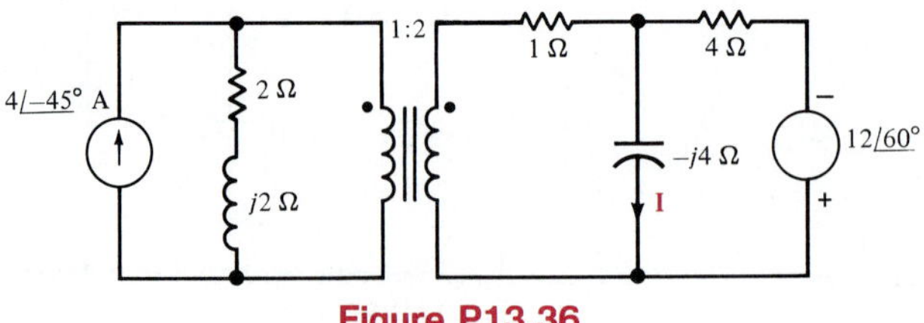

Figure P13.36

13.37. Find $\mathbf{V}_o$ in the network in Fig. P13.37.

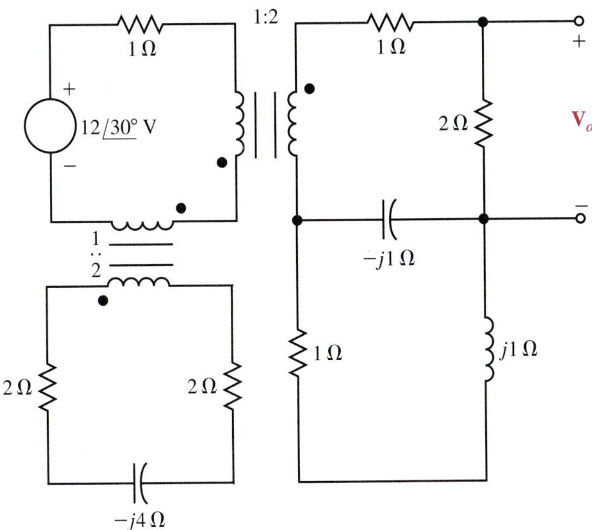

Figure P13.37

13.38. Find $\mathbf{V}_o$ in the circuit in Fig. P13.38.

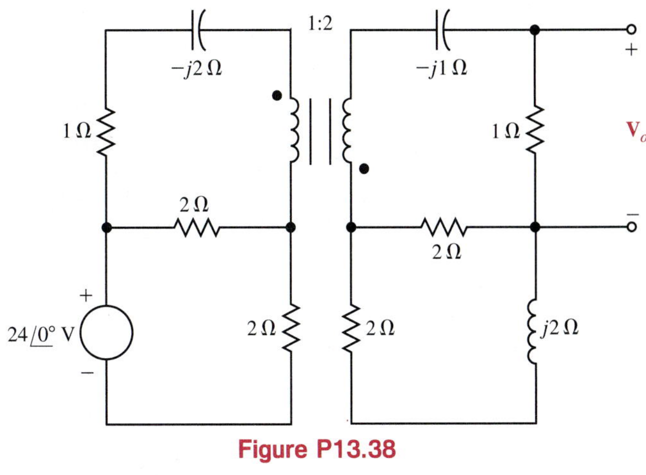

Figure P13.38

13.39. In the circuit in Fig. P13.39, if $\mathbf{I}_x = 6\ \underline{/45°}$ A, find $\mathbf{V}_o$.

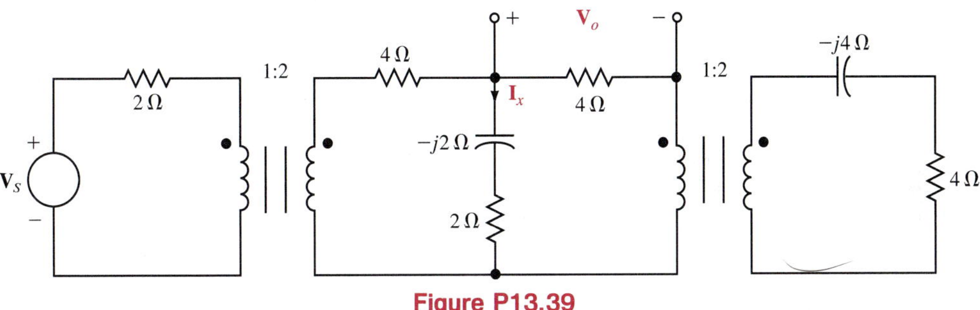

Figure P13.39

13.40. In the network in Fig. P13.40, if $\mathbf{I}_1 = 4\ \underline{/0°}$ A, find $\mathbf{V}_S$.

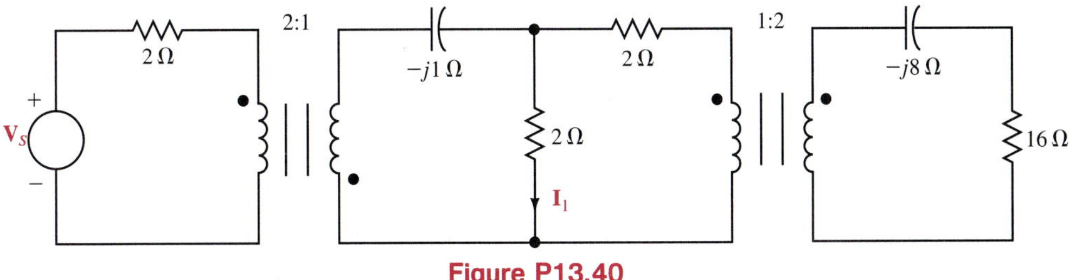

Figure P13.40

13.41. Find $\mathbf{V}_o$ in the network in Fig. P13.41.

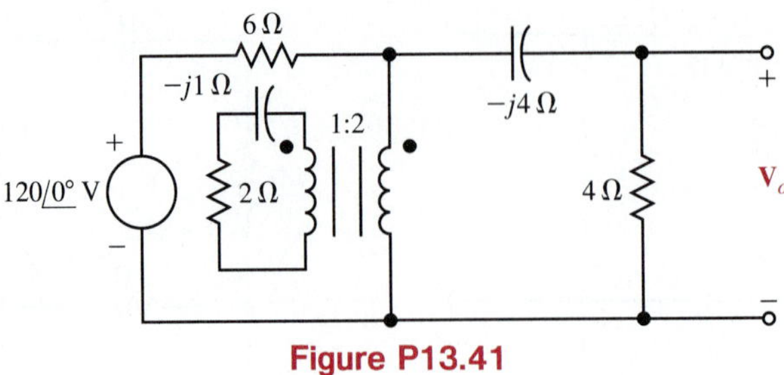

Figure P13.41

13.42. Find $\mathbf{I}_o$ in the circuit in Fig. P13.42.

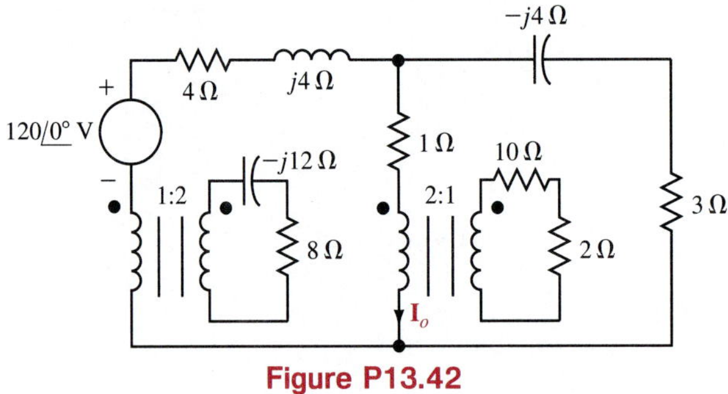

Figure P13.42

13.43. Find $\mathbf{V}_o$ in the network in Fig. P13.43.

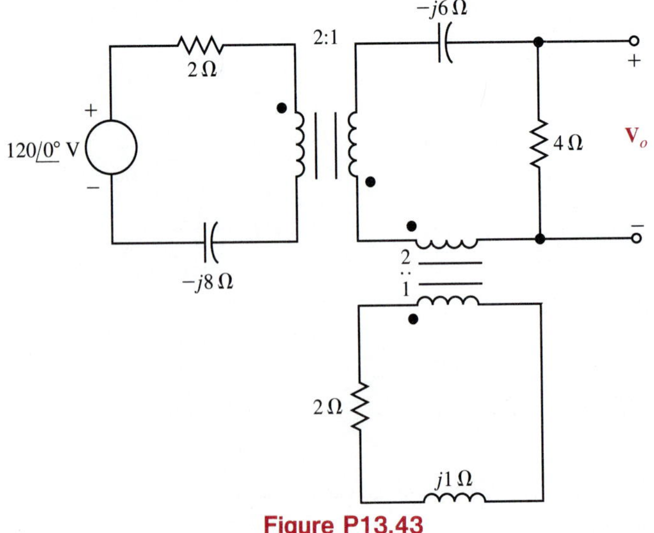

Figure P13.43

13.44. Find $\mathbf{V}_o$ in the circuit in Fig. P13.44.

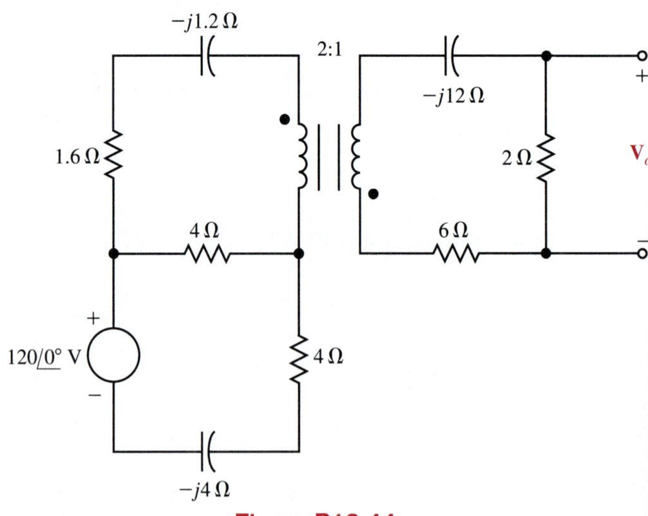

Figure P13.44

13.45. Find $\mathbf{I}_o$ in the network in Fig. P13.45.

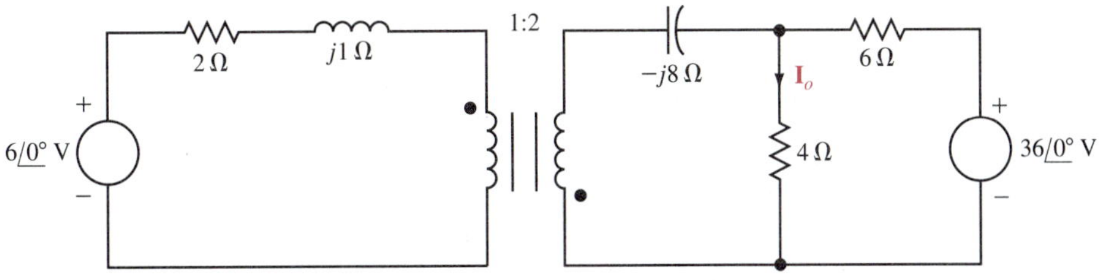

Figure P13.45

13.46. Find $\mathbf{V}_o$ in the network in Fig. P13.46.

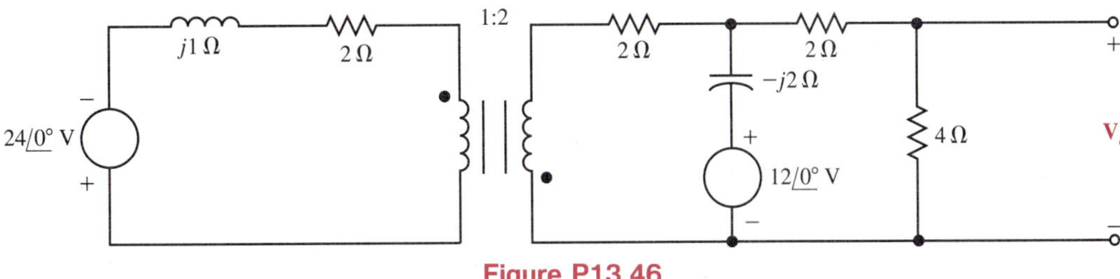

Figure P13.46

13.47. The output stage of an amplifier is to be matched to the impedance of a speaker as shown in Fig. P13.47. If the impedance of the speaker is 8 Ω and the amplifier requires a load impedance of 3.2 kΩ, determine the turns ratio of the ideal transformer.

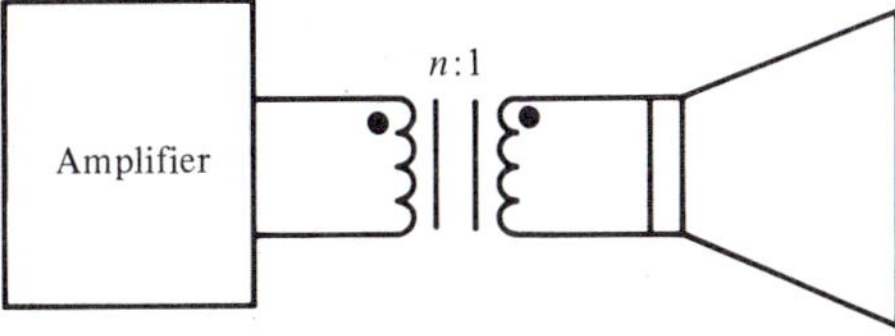

Figure P13.47

13.48. In the network shown in Fig. P13.48, $\mathbf{V}_L = 120 \angle 0°$ V rms. $\mathbf{Z}_L$ absorbs 500 watts at 0.85 pf lagging. Find $\mathbf{V}_S$.

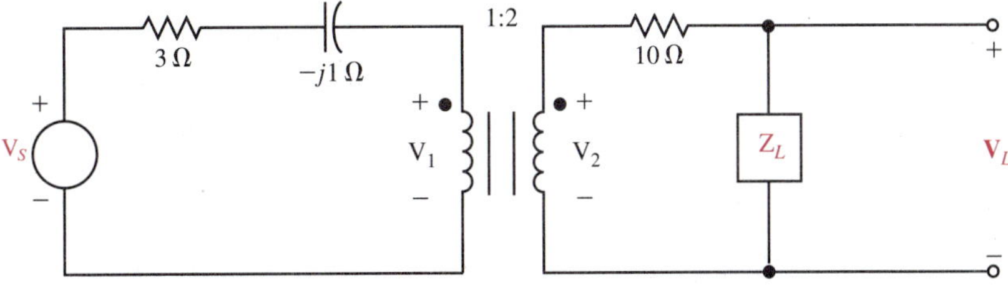

Figure P13.48

13.49. Find the input voltage $\mathbf{V}_S$ and the input power factor in the network in Fig. P13.49.

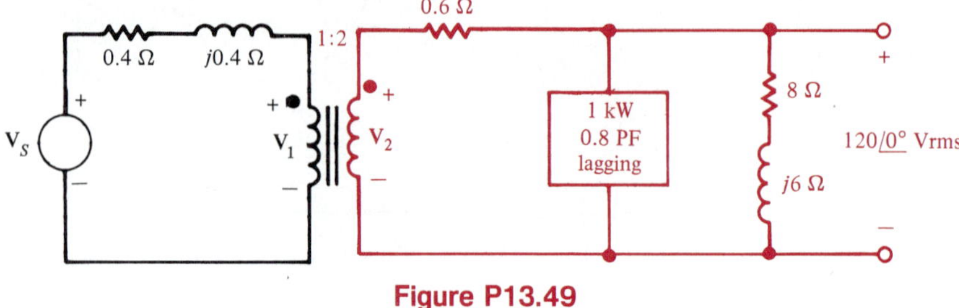

Figure P13.49

13.50. Find $\mathbf{I}_S$ in the circuit in Fig. P13.50.

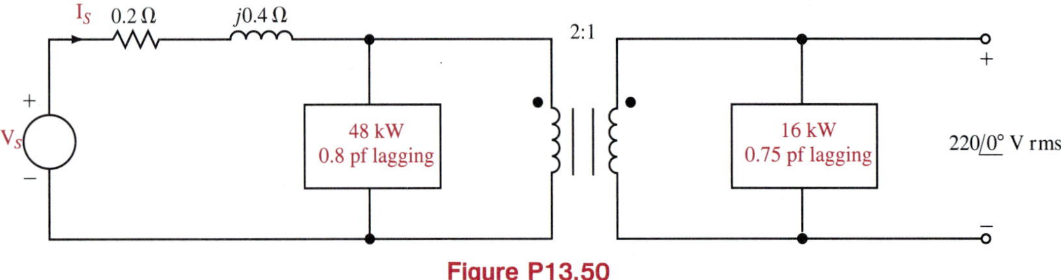

Figure P13.50

13.51. Find $\mathbf{V}_S$ in the network in Fig. P13.51.

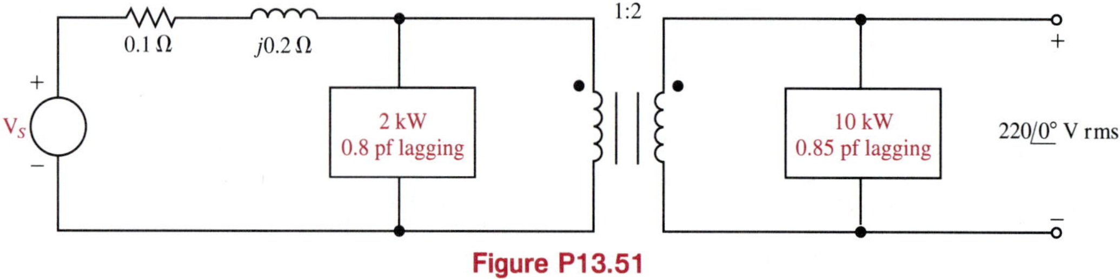

Figure P13.51

13.52. Determine the wattmeter reading in the network in Fig. P13.52.

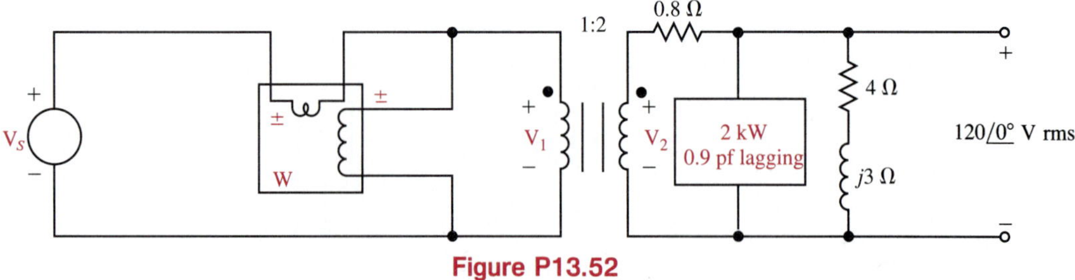

Figure P13.52

13.53. Determine the input impedance of the circuit shown in Fig. P13.53.

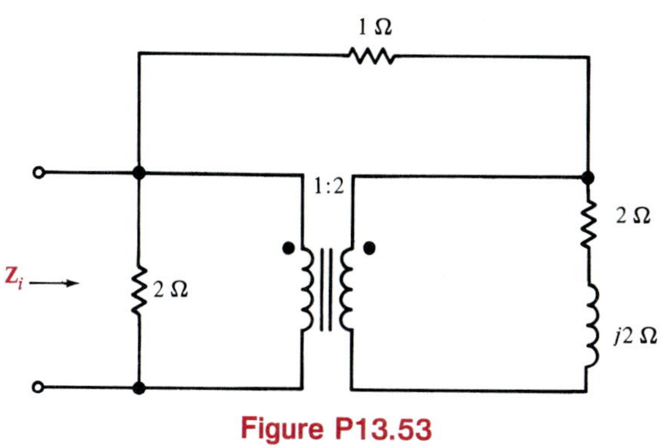

Figure P13.53

13.54. Find the value of the load resistance R_o in the network shown in Fig. P13.54 for maximum power transfer.

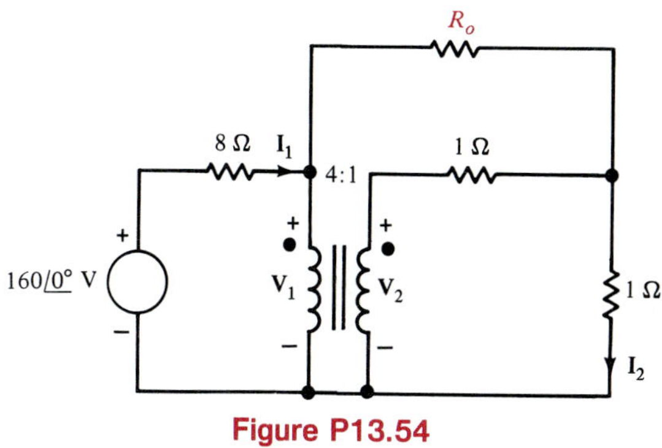

Figure P13.54

13.55. In the network in Fig. P13.55, I_A is known to be $10\,\angle 30°$ A. Use PSPICE to find the voltage V_o.

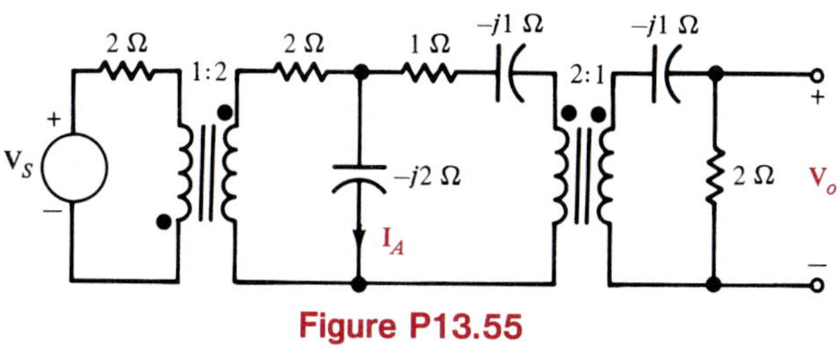

Figure P13.55

13.56. Use PSPICE to find **I** in the network in Fig. P13.56.

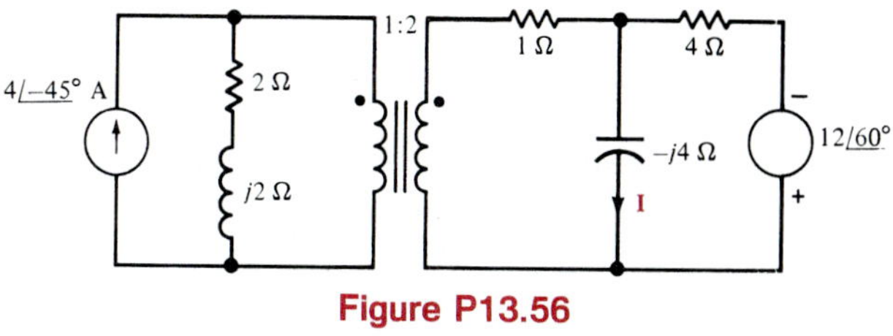

Figure P13.56

13.57. Use PSPICE to find **I** in the network in Fig. P13.57.

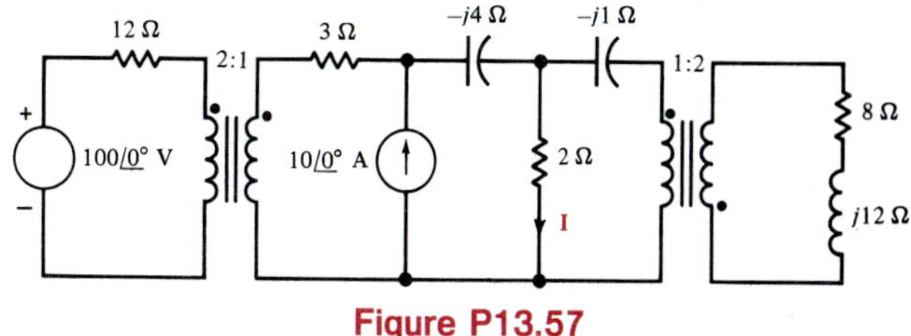

Figure P13.57

13.58. Form an equivalent circuit for the transformer and primary in the network shown in Fig. P13.58, and use this circuit and PSPICE to find the current I_2.

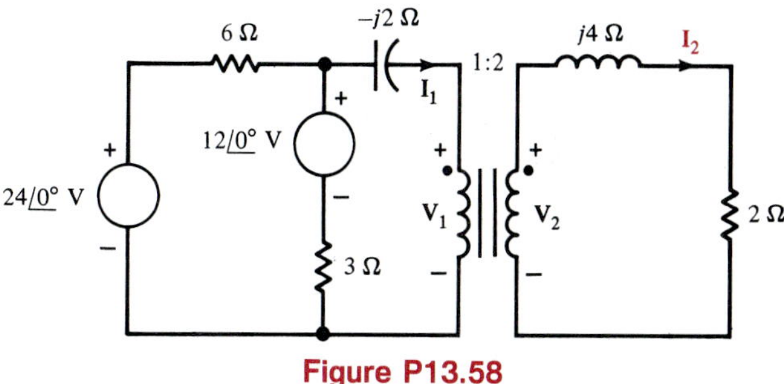

Figure P13.58

13.59. A 36-kVA, 2400/240-V, two-winding transformer is to be connected as an autotransformer to supply 2160 V to a load. Draw the transformer connection scheme and determine the kVA rating of the autotransformer.

13.60. A 2300/230-V two-winding transformer rated at 50 kVA is connected as an autotransformer to step up the voltage from 2300 V to 2530 V. What is the kVA output of this autotransformer?

13.61. A 440/220-V two-winding transformer rated as 15 kVA is connected as an autotransformer to step down the voltage from 660 V to 220 V. Determine the volt-ampere rating of the autotransformer and compare it with that of the original transformer.

13.62. A 440/110-V two-winding transformer rated as 20 kVA is connected as shown in Fig. P13.62. Determine the voltage $\mathbf{V}_2$ and the kVA rating of the transformer in the configuration shown.

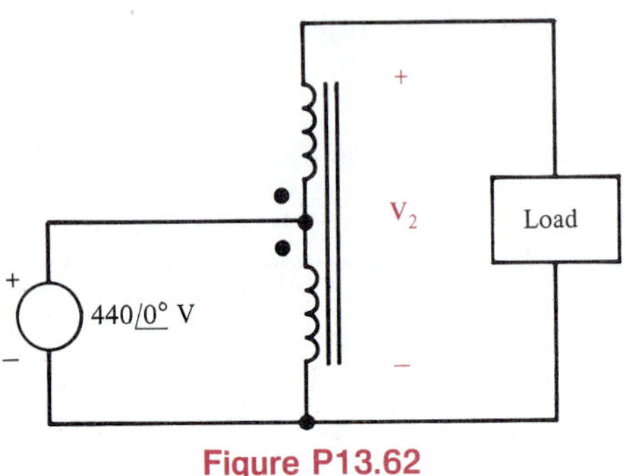

Figure P13.62

13.63. A bank of single-phase transformers is connected in a high-voltage wye to a low-voltage delta configuration. Each transformer is rated at 60 kVA, 13.2 kV/240 V rms. If the transformers serve a balanced three-phase load that consumes 66.67 kVA at 0.75 pf lagging, what are the magnitudes of the line currents at the source and the load?

13.64. A balanced three-phase transformer is rated at 240Δ/208Y rms. The transformer serves a balanced three-phase load that consumes 12.5 kVA as 0.8 pf lagging. The magnitude of the line voltage at the load is 200 V rms and the line impedance is $0.1 + j0.2\ \Omega$. Find the magnitude of the line current at the load and the magnitude of the line voltage at the primary of the transformer.

13.65. A balanced three-phase transformer 13.2kY/240Δ V rms is used in the network in Fig. P13.65 to supply a load. The magnitude of the line voltage at the load is 230 V rms. Find the magnitude of the line current at the load and the magnitude of the line voltage at the primary of the transformer.

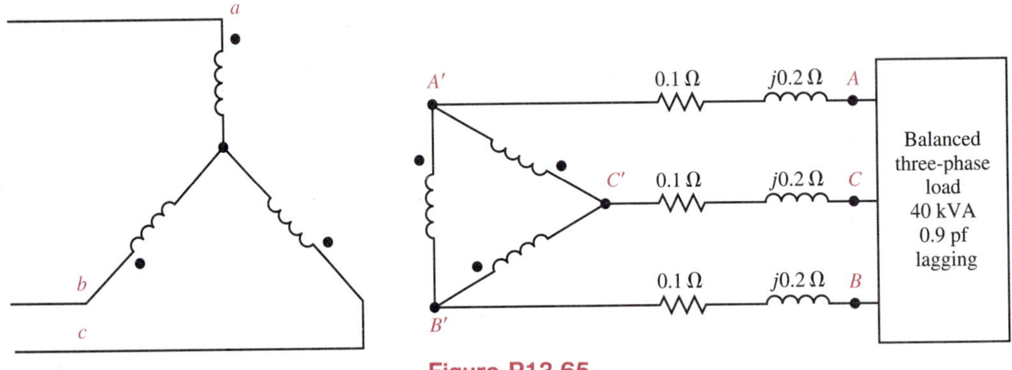

Figure P13.65

CHAPTER

14

Frequency Characteristics

In this chapter we examine the frequency characteristics of electrical networks. Of particular interest to us will be the effect of the source frequency on a network. Techniques for determining the frequency response of a network are presented and standard plots that display the network's performance as a function of frequency are discussed.

The concept of resonance is examined in some detail. The various parameters used to define the frequency selectivity of a network, such as bandwidth, cutoff frequency, and quality factor, are discussed. Network scaling for both magnitude and frequency is also presented.

Networks with special filtering properties are examined. Specifically, low-pass, high-pass, band-pass, and band-elimination filters are discussed. Techniques for designing active filters (containing op-amps) are presented. PSPICE is used to plot a network's frequency response and thus display the network's frequency-selective properties.

Networks that exhibit particular filtering characteristics have wide application in numerous types of communication and control systems where it is necessary to pass certain frequencies and reject others.

14.1 General Network Characteristics

In the earlier chapters we have investigated the response of networks to a number of different types of excitation. For example, we analyzed networks in which the forcing function was a constant (dc) or a sinusoid (ac). In the latter case the steady-state response of the network to a sinusoidal forcing function was obtained by determining the response due to the function $\mathbf{X}e^{j\omega t}$, where $\mathbf{X}e^{j\omega t}$ represented either current or voltage, and then taking the real part of the resultant function. The steady-state response was $y(t) = \mathbf{Y}e^{j\omega t}$. We demonstrated in Chapter 9 that the application of the forcing function $\mathbf{X}e^{j\omega t}$, which would appear to complicate the analysis, actually simplified the task, because it reduced the differential equation for the steady-state response to a set of algebraic equations with complex coefficients. We now change emphasis and examine the performance of a network as the frequency changes (i.e., frequency will become our variable). Networks in communication and control systems are often designed to be frequency selective; in other words, they pass certain frequencies and reject others. For example, in a stereo system if the networks are designed so that each frequency in the audio range is amplified the same amount, high-fidelity sound reproduction is achieved. If some frequencies are amplified more than others, distortion occurs, and the output is not an exact duplicate of the input.

To simplify our notation, we can represent a linear network by a block diagram as shown in Fig. 14.1a. The block represents the network, $x(t)$ represents the input or excitation, and $y(t)$ represents the output or response. Both $x(t)$ and $y(t)$ may be either a voltage or a current. As a further simplification we will substitute s for $j\omega$ as shown in Fig. 14.1b. At this point, we simply make the substitution for convenience, eliminating the complex coefficients in the algebraic manipulations. We will call this variable, s, complex frequency. At this point it is always equal to $j\omega$. However, in Chapter 16 it will be allowed to take on the value $\sigma + j\omega$ and will be given a more general meaning and use.

Since the network in Fig. 14.1 is linear, the input $x(t) = \mathbf{X}e^{j\omega t}$ will produce a steady-state response $y(t)\mathbf{Y}e^{j\omega t} = \mathbf{H}(j\omega)\mathbf{X}e^{j\omega t}$, where $\mathbf{H}(j\omega)$ will depend on the frequency ω. Using the s notation $y(t) = \mathbf{Y}e^{st} = \mathbf{H}(s)\mathbf{X}e^{st}$. The ratio of reponse to input is called the *network function* and is

$$\mathbf{H}(s) = \frac{\mathbf{Y}}{\mathbf{X}} \tag{14.1}$$

which is dependent on the complex frequency s and is independent of time.

Since both the input and output can generally be either voltage or current, there are four possible network functions. For example, if the input is current and the output is

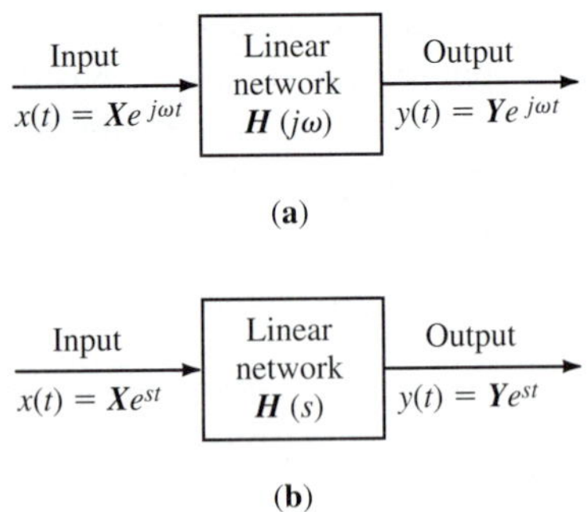

Figure 14.1 Representation of a linear network.

voltage, the network function is an impedance, more specifically a *transfer* impedance $\mathbf{Z}(s)$. Similarly, the other network functions are a transfer admittance $\mathbf{Y}(s)$, a voltage gain $\mathbf{G_V}(s)$, and a current gain $\mathbf{G_I}(s)$. These functions will be examined in more detail in Chapter 15 when we study two-port networks. We should also note that in addition to the *transfer functions* that we have just defined, there are *driving-point functions*. Driving-point functions are admittance or impedance, which are defined at a single pair of terminals. For example, the input impedance of a network is a driving-point impedance.

EXAMPLE 14.1

We wish to determine the transfer admittance $[\mathbf{I}_2(s)/\mathbf{V}_1(s)]$ and the voltage gain of the network shown in Fig. 14.2.

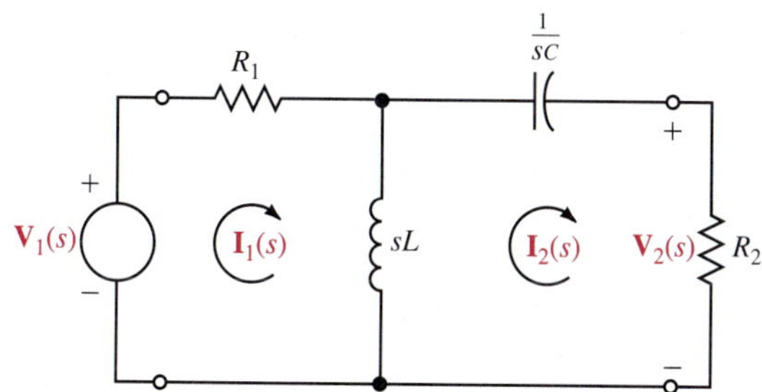

Figure 14.2 Circuit employed in Example 14.1.

The mesh equations for the network are

$$(R_1 + sL)\mathbf{I}_1(s) - sL\mathbf{I}_2(s) = \mathbf{V}_1(s)$$

$$-sL\mathbf{I}_1(s) + \left(R_2 + sL + \frac{1}{sC}\right)\mathbf{I}_2(s) = 0$$

$$\mathbf{V}_2(s) = \mathbf{I}_2(s)R_2$$

Solving the equations for $\mathbf{I}_2(s)$ yields

$$\mathbf{I}_2(s) = \frac{sL\mathbf{V}_1(s)}{(R_1 + sL)(R_2 + sL + 1/sC) - s^2L^2}$$

Therefore, the transfer admittance $[\mathbf{I}_2(s)/\mathbf{V}_1(s)]$ is

$$\mathbf{Y}_T(s) = \frac{\mathbf{I}_2(s)}{\mathbf{V}_1(s)} = \frac{LCs^2}{(R_1 + R_2)LCs^2 + (L + R_1R_2C)s + R_1}$$

and the voltage gain is

$$\mathbf{G_V}(s) = \frac{\mathbf{V}_2(s)}{\mathbf{V}_1(s)} = \frac{LCR_2s^2}{(R_1 + R_2)LCs^2 + (L + R_1R_2C)s + R_1}$$ ■

DRILL EXERCISE

D14.1. Find the input impedance $\mathbf{Z}_i(s)$ for the network in Fig. D14.1.

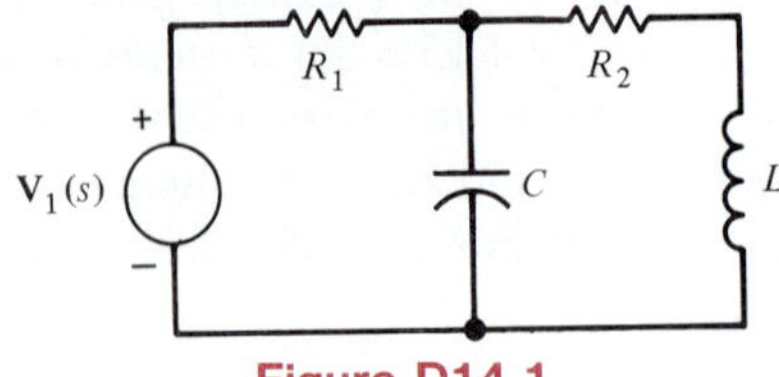

Figure D14.1

Ans: $\mathbf{Z}_i(s) = \dfrac{(R_1 + R_2) + s(CR_1R_2 + L) + s^2LCR_1}{1 + sCR_2 + s^2LC}$.

D14.2. Find the current gain $\mathbf{G}_\mathrm{I}(s)$ for the network in Fig. D14.2.

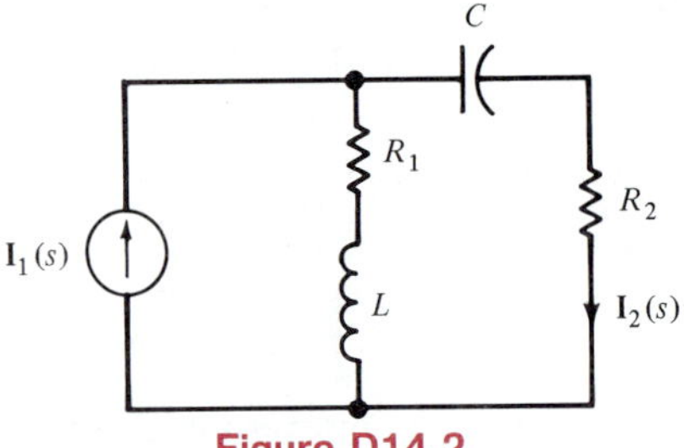

Figure D14.2

Ans: $\mathbf{G}_\mathrm{I}(s) = \dfrac{sC(R_1 + sL)}{1 + sC(R_1 + R_2) + s^2LC}$.

Poles and Zeros

In general, the network functions can be expressed as the ratio of two polynomials in s. Example 14.1 is an illustration of this fact. In addition, we note that since the values of our circuit elements, or controlled sources, are real numbers, the coefficients of the two polynomials will be real. Therefore, we will express a network function in the form

$$\mathbf{H}(s) = \frac{N(s)}{D(s)} = \frac{a_m s^m + a_{m-1}s^{m-1} + \cdots + a_1 s + a_0}{b_n s^n + b_{n-1}s^{n-1} + \cdots + b_1 s + b_0} \tag{14.2}$$

where $N(s)$ is the numerator polynomial of degree m and $D(s)$ is the denominator polynomial of degree n. Equation (14.2) can also be written in the form

$$\mathbf{H}(s) = \frac{K_0(s - z_1)(s - z_2)\cdots(s - z_m)}{(s - p_1)(s - p_2)\cdots(s - p_n)} \tag{14.3}$$

where K_0 is a constant, $z_1, \ldots, z_m$ are the roots of $N(s)$, and $p_1, \ldots, p_n$ are the roots of $D(s)$. Note that if $s = z_1$ or $z_2, \ldots, z_m$, then $\mathbf{H}(s)$ becomes zero and hence $z_1, \ldots, z_m$

are called *zeros* of the function. Similarly, if $s = p_1$ or $p_2, \ldots, p_n$, then $\mathbf{H}(s)$ becomes infinite and therefore $p_1, \ldots, p_n$ are called *poles* of the function. The zeros or poles may actually be complex. However, if they are complex, they must occur in conjugate pairs since the coefficients of the polynomial are real. The representation of the network function specified in Eq. (14.3) is extremely important and is generally employed to represent any linear time-invariant system. The importance of this form stems from the fact that the dynamic properties of a system can be gleaned from an examination of the system poles.

14.2 Sinusoidal Frequency Analysis

Although there are specific cases in which a network operates at only one frequency (e.g., a power system network), in general we are interested in the behavior of a network as a function of frequency. In a sinusoidal steady-state analysis, the network function can be expressed as

$$\mathbf{H}(j\omega) = M(\omega)e^{j\phi(\omega)} \tag{14.4}$$

where $M(\omega) = |\mathbf{H}(j\omega)|$ and $\phi(\omega)$ is the phase. A plot of these two functions, which are commonly called the *magnitude* and *phase characteristics*, display the manner in which the response varies with the input frequency ω. We will now illustrate the manner in which to perform a frequency-domain analysis by simply evaluating the function at various frequencies within the range of interest or by using what is called a *Bode plot* (named after Hendrik W. Bode).

Basic Frequency Response Plots

If the network function is evaluated at some point $s = j\omega$, its value will be a complex quantity having a magnitude and phase. Therefore, in the case of a transfer function, two plots are used to describe the behavior of the network function: the amplitude plot and the phase plot. If the network function is a driving-point function, then separate graphs for the resistance and reactance or for the conductance and susceptance are normally used. In either case, however, we can evaluate the function $\mathbf{H}(j\omega)$ as a function of frequency by simply substituting values for ω into $\mathbf{H}(j\omega)$, which will yield both a magnitude and phase at each frequency.

Suppose that the network function is of the form

$$\mathbf{H}(s) = \frac{K_0(s - z_1)(s - z_2)}{(s - p_1)(s - p_2)} \tag{14.5}$$

If we evaluate this function at some point $s = j\omega_0$, then

$$\mathbf{H}(s = j\omega_0) = \frac{K_0(j\omega_0 - z_1)(j\omega_0 - z_2)}{(j\omega_0 - p_1)(j\omega_0 - p_2)} \tag{14.6}$$

and each term of the form $(j\omega_0 - z_i)$ and $(j\omega_0 - p_j)$ is a complex number having a magnitude and phase. Therefore, Eq. (14.6) can be written as

$$\mathbf{H}(j\omega_0) = \frac{K_0(Z_1 e^{j\theta_1})(Z_2 e^{j\theta_2})}{(P_1 e^{j\phi_1})(P_2 e^{j\phi_2})} \tag{14.7}$$

or

$$\mathbf{H}(j\omega_0) = \frac{K_0 Z_1 Z_2}{P_1 P_2} e^{j(\theta_1 + \theta_2 - \phi_1 - \phi_2)} \tag{14.8}$$

where, for example, $(j\omega_0 - z_1) = Z_1 e^{j\theta_1}$.

In general, the magnitude and phase of the network function can be expressed as

$$|\mathbf{H}(j\omega_0)| = \frac{K_0 \prod_{i=1}^{m} Z_i}{\prod_{j=1}^{n} P_j} \tag{14.9}$$

and

$$\underline{/\mathbf{H}(j\omega_0)} = \sum_{i=1}^{m} \theta_i - \sum_{j=1}^{n} \phi_j \tag{14.10}$$

where m is the number of finite zeros of $\mathbf{H}(j\omega)$ and n is the number of finite poles. If we evaluate these functions for all values of ω from 0 to ∞, we will obtain the magnitude and phase characteristics, or equivalently, the frequency response of $\mathbf{H}(j\omega)$.

EXAMPLE 14.2

We wish to determine the frequency response of a network with the transfer function

$$\mathbf{H}(s) = \frac{2s}{s^2 + 2s + 17}$$

$\mathbf{H}(s)$ can be written as

$$\mathbf{H}(s) = \frac{2s}{(s + 1 + j4)(s + 1 - j4)}$$

and hence

$$\mathbf{H}(j\omega) = \frac{2\omega \underline{/90^\circ}}{[1 + j(\omega + 4)][1 + j(\omega - 4)]}$$

The frequency response, which consists of the magnitude and phase of $\mathbf{H}(j\omega)$ as a function of frequency, can be obtained simply by evaluating $\mathbf{H}(j\omega)$ for various values of ω. The magnitude and phase of $\mathbf{H}(j\omega)$ for several values of ω are

ω	0	1	2	3	4	$\sqrt{17}$	5	6	7	8
$M(\omega)$	0	0.12	0.29	0.60	0.99	1.0	0.78	0.53	0.40	0.32
$\phi(\omega)$	90°	83°	73°	53°	7°	0°	− 39°	− 58°	− 66°	− 71°

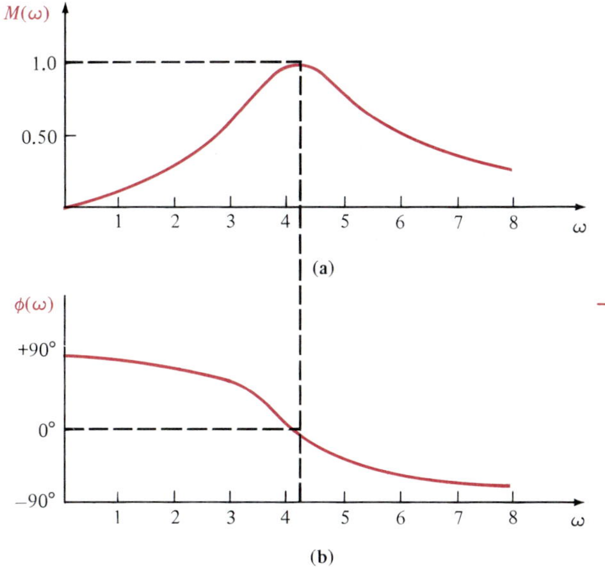

Figure 14.3 Graphs employed in Example 14.2.

Sketches of the amplitude and phase characteristics for the network function are shown in Fig. 14.3a and b. ■

Example 14.2 illustrates some rather subtle points that are true in general. Note that in the vicinity of a zero the network response approaches a minimum, and in the vicinity of a pole the network response approaches a maximum. The importance of this relationship stems from the fact that if we want to design a network filter to pass certain frequencies and reject others, we design the filter so that it has poles in the passband frequency range and zeros in the band-rejection frequency range.

DRILL EXERCISE

D14.3. Given the following network function $\mathbf{H}(j\omega)$, tabulate the values for $M(\omega)$ and $\phi(\omega)$ versus ω for $0 \leq \omega \leq 10$ rad/sec in increments of 1 rad/sec.

$$\mathbf{H}(j\omega) = \frac{j\omega + 2}{(j\omega + 6)(j\omega + 8)}$$

Ans:

ω	0	1	2	3	...
$M(\omega)$	0.042	0.0455	0.0543	0.063	...
$\phi(\omega)$	0°	10°	12.6°	9.2°	...

Frequency Response Using a Bode Plot

If the network characteristics are plotted on a semilog scale, that is, a linear scale for the ordinate and a logarithmic scale for the abscissa, they are known as *Bode plots*. In this case we plot $20 \log_{10} M(\omega)$ versus $\log_{10} (\omega)$ instead of $M(\omega)$ versus ω. The advantage of this technique is that rather than plotting the characteristic point by point, we can employ straight-line approximations to obtain the characteristic very efficiently. The ordinate for the magnitude plot is the decibel (dB). This unit was originally employed to measure the ratio of two powers; that is,

$$\text{number of dB} = 10 \log_{10} \frac{P_2}{P_1} \tag{14.11}$$

If the powers are absorbed by two equal resistors, then

$$\begin{aligned} \text{number of dB} &= 10 \log_{10} \frac{|\mathbf{V}_2|^2/R}{|\mathbf{V}_1|^2/R} = 10 \log_{10} \frac{|\mathbf{I}_2|^2 R}{|\mathbf{I}_1|^2 R} \\ &= 20 \log_{10} \left| \frac{\mathbf{V}_2}{\mathbf{V}_1} \right| = 20 \log_{10} \left| \frac{\mathbf{I}_2}{\mathbf{I}_1} \right| \end{aligned} \tag{14.12}$$

The term ''dB'' has become so popular that it now is used for voltage and current ratios, as illustrated in Eq. (14.12), without regard to the impedance employed in each case.

In the sinusoidal steady-state case, $\mathbf{H}(j\omega)$ in Eq. (14.3) can be expressed in general as

$$\mathbf{H}(j\omega) = \frac{K_0(j\omega)^{\pm N}(1 + j\omega\tau_1)(1 + j\omega\tau_2)[1 + 2\zeta_3(j\omega\tau_3) + (j\omega\tau_3)^2]\cdots}{(1 + j\omega\tau_a)[1 + 2\zeta_b(j\omega\tau_b) + (j\omega\tau_b)^2]\cdots} \tag{14.13}$$

Note that this equation contains the following typical factors:

1. A frequency-independent factor, K_0
2. Poles or zeros at the origin of the form $j\omega$, that is, $(j\omega)^{+N}$ for zeros and $(j\omega)^{-N}$ for poles
3. Poles or zeros of the form $(1 + j\omega\tau)$
4. Quadratic poles or zeros of the form $1 + 2\zeta(j\omega\tau) + (j\omega\tau)^2$

Taking the logarithm of the magnitude of the function $\mathbf{H}(j\omega)$ in Eq. (14.13) yields

$$\begin{aligned} 20 \log_{10} |\mathbf{H}(j\omega)| &= 20 \log_{10} K_0 \pm 20N \log_{10} |j\omega| \\ &\quad + 20 \log_{10} |1 + j\omega\tau_1| \\ &\quad + 20 \log_{10} |1 + j\omega\tau_2| \\ &\quad + 20 \log_{10} |1 + 2\zeta_3(j\omega\tau_3) + (j\omega\tau_3)^2| \\ &\quad + \cdots - 20 \log_{10} |1 + j\omega\tau_a| - 20 \log_{10} |1 \\ &\quad + 2\zeta_b(j\omega\tau_b) + (j\omega\tau_b)^2| \cdots \end{aligned} \tag{14.14}$$

Note that we have used the fact that the log of the product of two or more terms is equal to the sum of the logs of the individual terms, the log of the quotient of two terms is equal to the difference of the logs of the individual terms, and the fact that $\log_{10} A^n = n \log_{10} A$.

The phase angle for $\mathbf{H}(j\omega)$ is

$$\underline{/\mathbf{H}(j\omega)} = 0 \pm N(90°) + \tan^{-1} \omega\tau_1 + \tan^{-1} \omega\tau_2 + \tan^{-1}\left(\frac{2\zeta_3\omega\tau_3}{1 - \omega^2\tau_3^2}\right) + \cdots - \tan^{-1} \omega\tau_a - \tan^{-1}\left(\frac{2\zeta_b\omega\tau_b}{1 - \omega^2\tau_b^2}\right) \cdots \tag{14.15}$$

As Eqs. (14.14) and (14.15) indicate, we will simply plot each factor individually on a common graph and then sum them algebraically to obtain the total characteristic.

The term $20 \log_{10} K_0$ represents a constant magnitude with zero phase shift, as shown in Fig. 14.4a. Poles or zeros at the origin are of the form $(j\omega)^{\pm N}$, where + is used for a zero and − is used for a pole. The magnitude of this function is $\pm 20N \log_{10} \omega$, which is a straight line on semilog paper with a slope of $\pm 20N$ dB/decade; that is, the value will change by 20 each time the frequency is multiplied by 10, and the phase of this function is a constant $\pm N(90°)$. The magnitude and phase characteristics for poles and zeros at the origin are shown in Fig. 14.4b and c, respectively.

Linear approximations can be employed when a simple pole or zero of the form $(1 + j\omega\tau)$ is present in the network function. For $\omega\tau \ll 1$, $(1 + j\omega\tau) \approx 1$, and therefore, $20 \log_{10} |(1 + j\omega\tau)| = 20 \log_{10} 1 = 0$ dB. Similarly, if $\omega\tau \gg 1$, then $(1 + j\omega\tau) \approx j\omega\tau$, and hence $20 \log_{10} |(1 + j\omega\tau)| = 20 \log_{10} \omega\tau$. Therefore, for $\omega\tau \ll 1$ the response is 0 dB and for $\omega\tau \gg 1$ the response has a slope that is the same as that of a simple pole or zero at the origin. The intersection of these two asymptotes, one for $\omega\tau \ll 1$ and one for $\omega\tau \gg 1$, is the point where $\omega\tau = 1$ or $\omega = 1/\tau$, which is called the *break frequency*. At this break frequency, where $\omega = 1/\tau$, $20 \log_{10} |(1 + j1)| = 20 \log_{10}(2)^{1/2} = 3$ dB. Therefore, the actual curve deviates from the asymptotes by 3 dB at the break frequency. It can be shown that at one-half and twice the break frequency, the deviations are 1 dB. The phase angle associated with a simple pole or zero is $\phi = \tan^{-1} \omega\tau$, which is a simple arc-tangent curve. Therefore, the phase shift is 45° at the break frequency and 26.6° and 63.4° at one-half and twice the break frequency, respectively. The actual magnitude curve for a pole of this form is shown in Fig. 14.5a. For a zero the magnitude curve and the asymptote for $\omega\tau \gg 1$ have a positive slope, and the phase curve extends from 0° to +90° as shown in Fig. 14.5b. If multiple poles or zeros of the form $(1 + j\omega\tau)^N$ are present, then the slope of the high-frequency asymptote is multiplied by N, the deviation between the actual curve and the asymptote at the break frequency is $3N$ dB, and the phase curve extends from 0 to $N(90°)$ and is $N(45°)$ at the break frequency.

Quadratic poles or zeros are of the form $1 + 2\zeta(j\omega\tau) + (j\omega\tau)^2$. This term is a function not only of ω, but the dimensionless term ζ, which is called the *damping ratio*. If $\zeta > 1$ or $\zeta = 1$, the roots are real and unequal or real and equal, respectively, and these two cases have already been addressed. If $\zeta < 1$, the roots are complex conjugates, and it

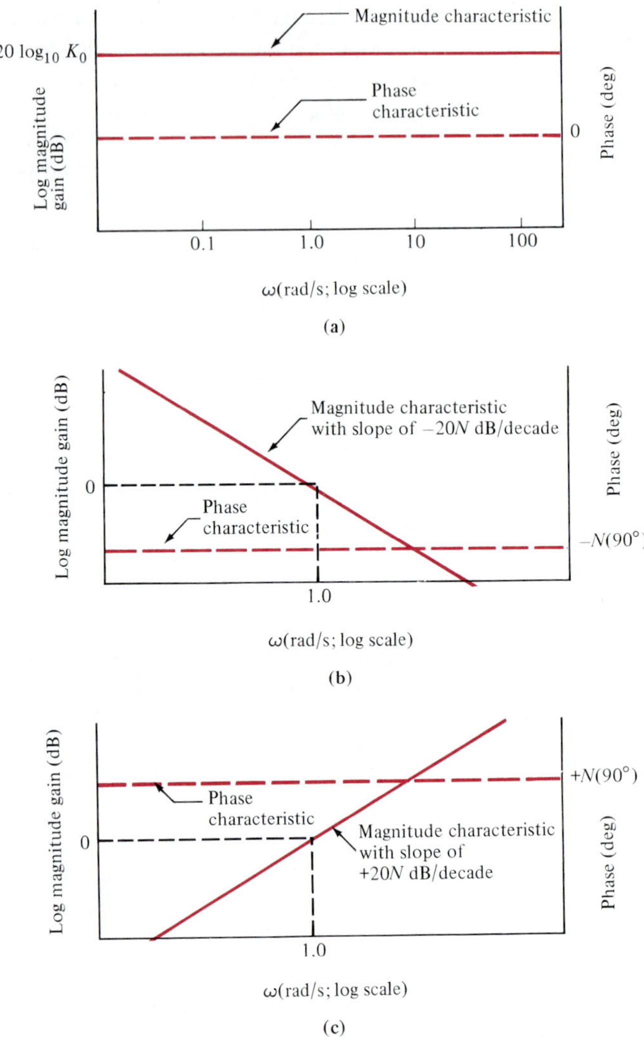

Figure 14.4 Magnitude and phase characteristics for a constant term and poles and zeros at the origin.

is this case which we will examine now. Following the argument above for a simple pole or zero, the log magnitude of the quadratic factor is 0 dB for $\omega\tau \ll 1$. For $\omega\tau \gg 1$,

$$20 \log_{10} |1 - (\omega\tau)^2 + 2j\zeta(\omega\tau)| \approx 20 \log_{10} |(\omega\tau)^2| = 40 \log_{10} |\omega\tau|$$

and therefore, for $\omega\tau \gg 1$, the slope of the log magnitude curve is +40 dB/decade for a quadratic zero and −40 dB/decade for a quadratic pole. Between the two extremes, $\omega\tau \ll 1$ and $\omega\tau \gg 1$, the behavior of the function is dependent on the damping ratio ζ. Figure 14.6a illustrates the manner in which the log magnitude curve for a quadratic *pole* changes as a function of the damping ratio. The phase shift for the quadratic factor is $\tan^{-1} 2\zeta\omega\tau/[1 - (\omega\tau)^2]$. The phase plot for quadratic *poles* is shown in Fig. 14.6b. Note

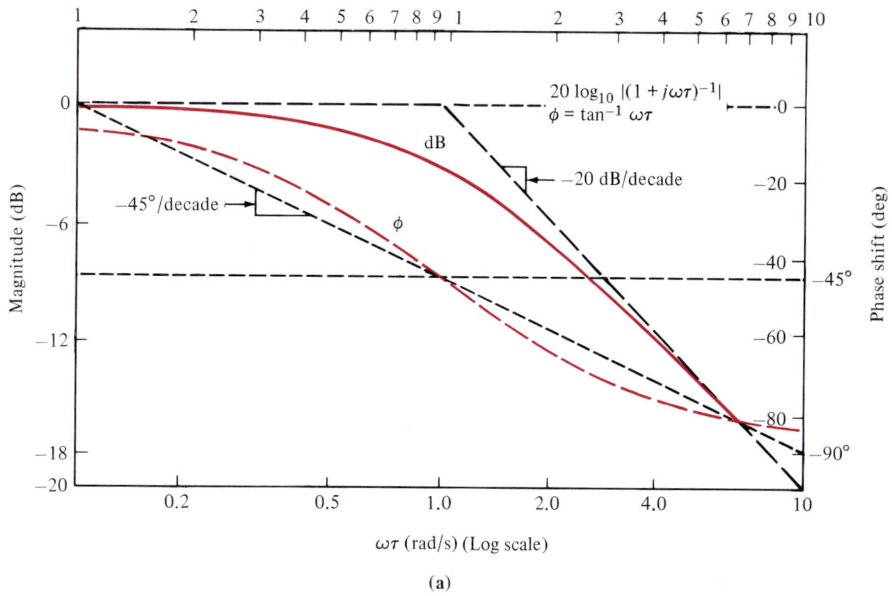

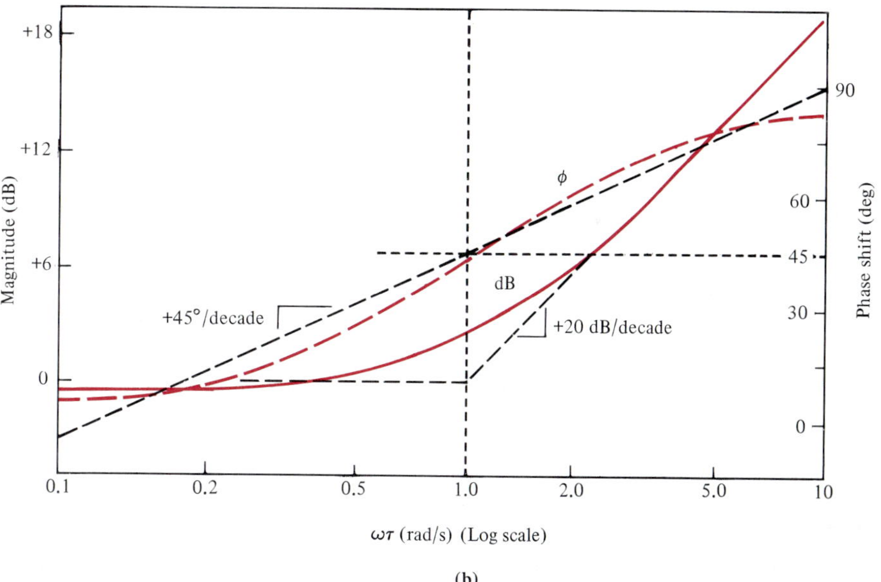

Figure 14.5 Magnitude and phase plot (a) for a simple pole, and (b) for a simple zero.

that in this case the phase changes from 0° at frequencies for which $\omega\tau \ll 1$ to $-180°$ at frequencies for which $\omega\tau \gg 1$. For quadratic zeros the magnitude and phase curves are inverted; that is, the log magnitude curve has a slope of +40 dB/decade for $\omega\tau \gg 1$, and the phase curve is 0° for $\omega\tau \ll 1$ and $+180°$ for $\omega\tau \gg 1$.

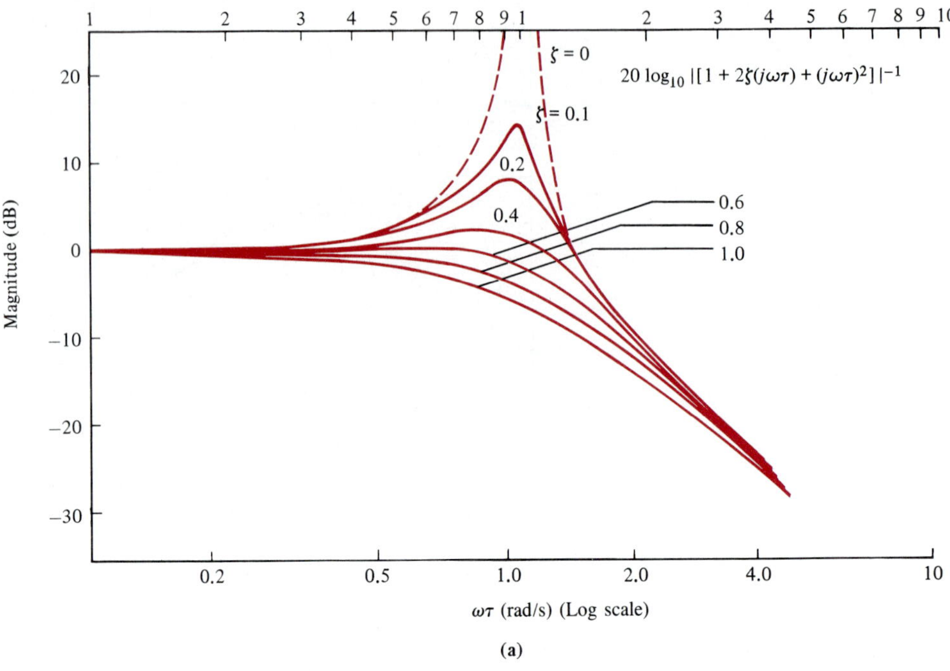

(a)

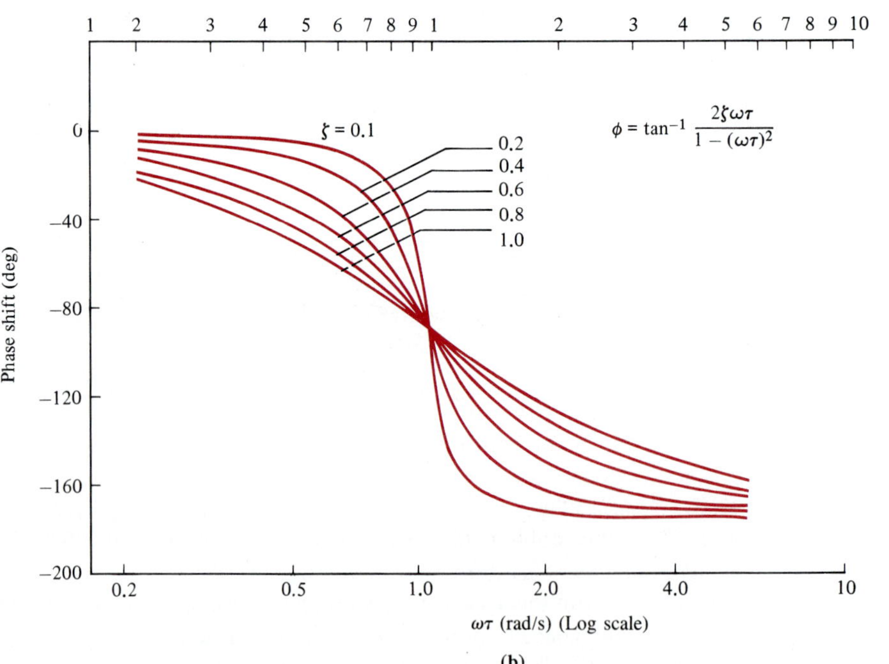

(b)

Figure 14.6 Magnitude and phase characteristics for quadratic poles.

EXAMPLE 14.3

We want to generate the magnitude and phase plots for the transfer function

$$\mathbf{G_V}(j\omega) = \frac{10(0.1j\omega + 1)}{(j\omega + 1)(0.02j\omega + 1)}$$

Note that this function is in standard form, since every term is of the form $(j\omega\tau + 1)$. In order to determine the composite magnitude and phase characteristics, we will plot the individual asymptotic terms and then add them as specified in Eqs. (14.14) and (14.15). Let us consider the magnitude plot first. Since $K_0 = 10$, $20 \log_{10} 10 = 20$ dB, which is a constant independent of frequency, as shown in Fig. 14.7a. The zero of the transfer function contributes a term of the form $+20 \log_{10} |1 + 0.1j\omega|$, which is 0 dB for $0.1\omega \ll 1$, has a slope of +20 dB/decade for $0.1\omega \gg 1$, and has a break frequency at $\omega = 10$ rad/sec. The poles have break frequencies at $\omega = 1$ and $\omega = 50$ rad/sec. The pole with break frequency at $\omega = 1$ rad/sec contributes a term of the form $-20 \log_{10} |1 + j\omega|$, which is 0 dB for $\omega \ll 1$ and has a slope of −20 dB/decade for $\omega \gg 1$. A similar argument can be made for the pole that has a break frequency at $\omega = 50$ rad/sec. These factors are all plotted individually in Fig. 14.7a.

Consider now the individual phase curves. The term K_0 is not a function of ω and does not contribute to the phase of the transfer function. The phase curve for the zero is $+\tan^{-1} 0.1\,\omega$, which is an arctangent curve that extends from 0° for $0.1\omega \ll 1$ to +90° for $0.1\omega \gg 1$ and has a phase of +45° at the break frequency. The phase curves for the two poles are $-\tan^{-1} \omega$ and $-\tan^{-1} 0.02\omega$. The term $-\tan^{-1} \omega$ is 0° for $\omega \ll 1$, −90° for $\omega \gg 1$, and −45° at the break frequency $\omega = 1$. The phase curve for the remaining pole is plotted in a similar fashion. All the individual phase curves are shown in Fig. 14.7a.

As specified in Eqs. (14.14) and (14.15), the composite magnitude and phase of the transfer function are obtained simply by adding the individual terms. The composite curves are plotted in Fig. 14.7b. Note that the actual magnitude curve (solid line) differs from the straight-line approximation (dashed line) by 3 dB at the break frequencies and 1 dB at one-half and twice the break frequencies. ■

EXAMPLE 14.4

Let us draw the Bode plot for the following transfer function:

$$\mathbf{G_V}(j\omega) = \frac{25(j\omega + 1)}{(j\omega)^2(0.1j\omega + 1)}$$

Once again all the individual terms for both magnitude and phase are plotted in Fig. 14.8a. The straight line with a slope of −40 dB/decade is generated by the double pole at the origin. This line is a plot of $-40 \log_{10} \omega$ versus ω and therefore passes through 0 dB at $\omega = 1$ rad/sec. The phase for the double pole is a constant −180° for all frequencies. The remainder of the terms are plotted as illustrated in Example 14.3.

The composite plots are shown in Fig. 14.8b. Once again they are obtained simply by adding the individual terms in Fig. 14.8a. Note that for frequencies for which $\omega \ll 1$, the slope of the magnitude curve is −40 dB/decade. At $\omega = 1$ rad/sec, which is the break frequency of the zero, the magnitude curve changes slope to −20 dB/decade. At $\omega = 10$

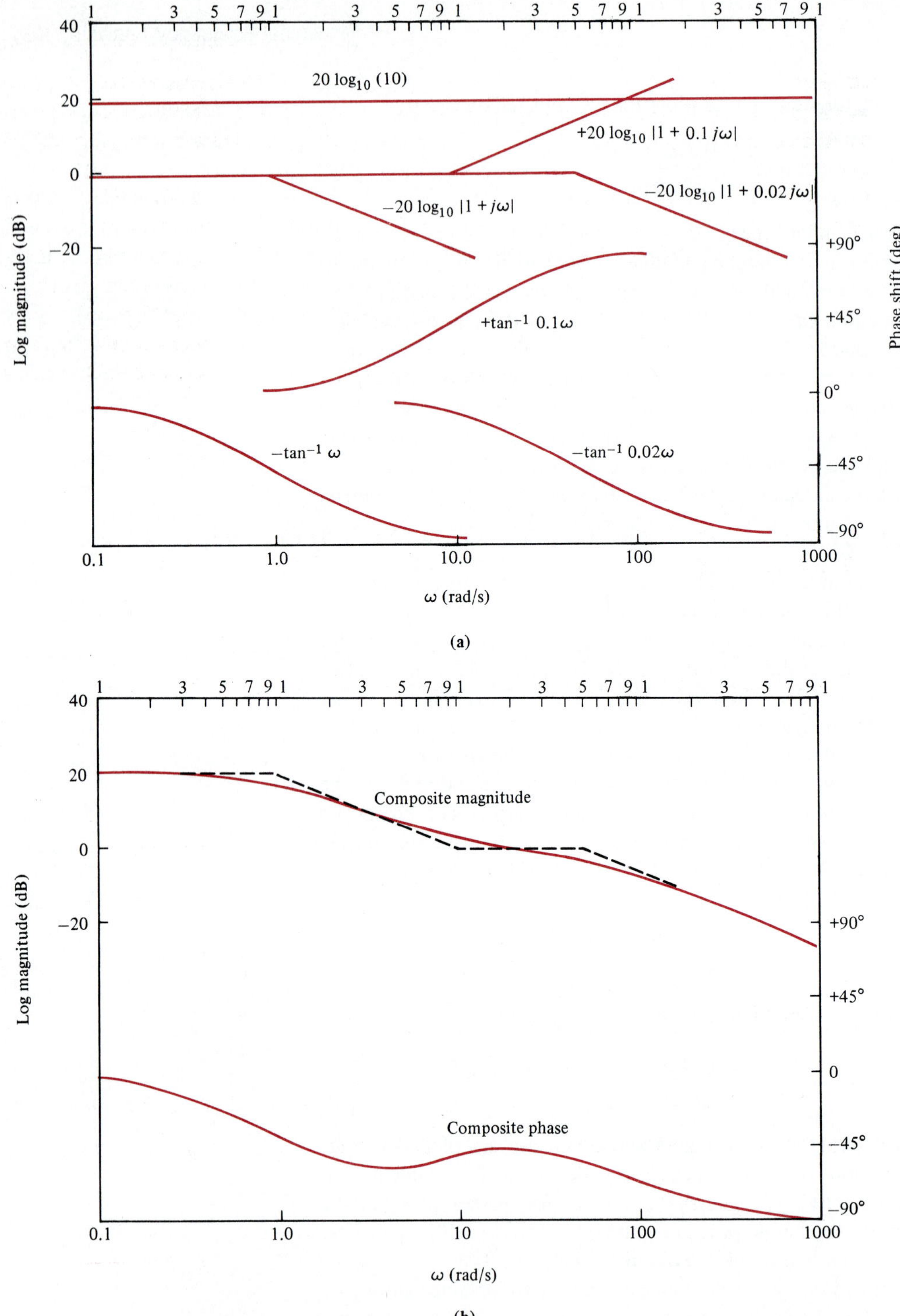

Figure 14.7 (a) Magnitude and phase components for the poles and zeros of the transfer function in Example 14.3; (b) Bode plot for the transfer function in Example 14.3.

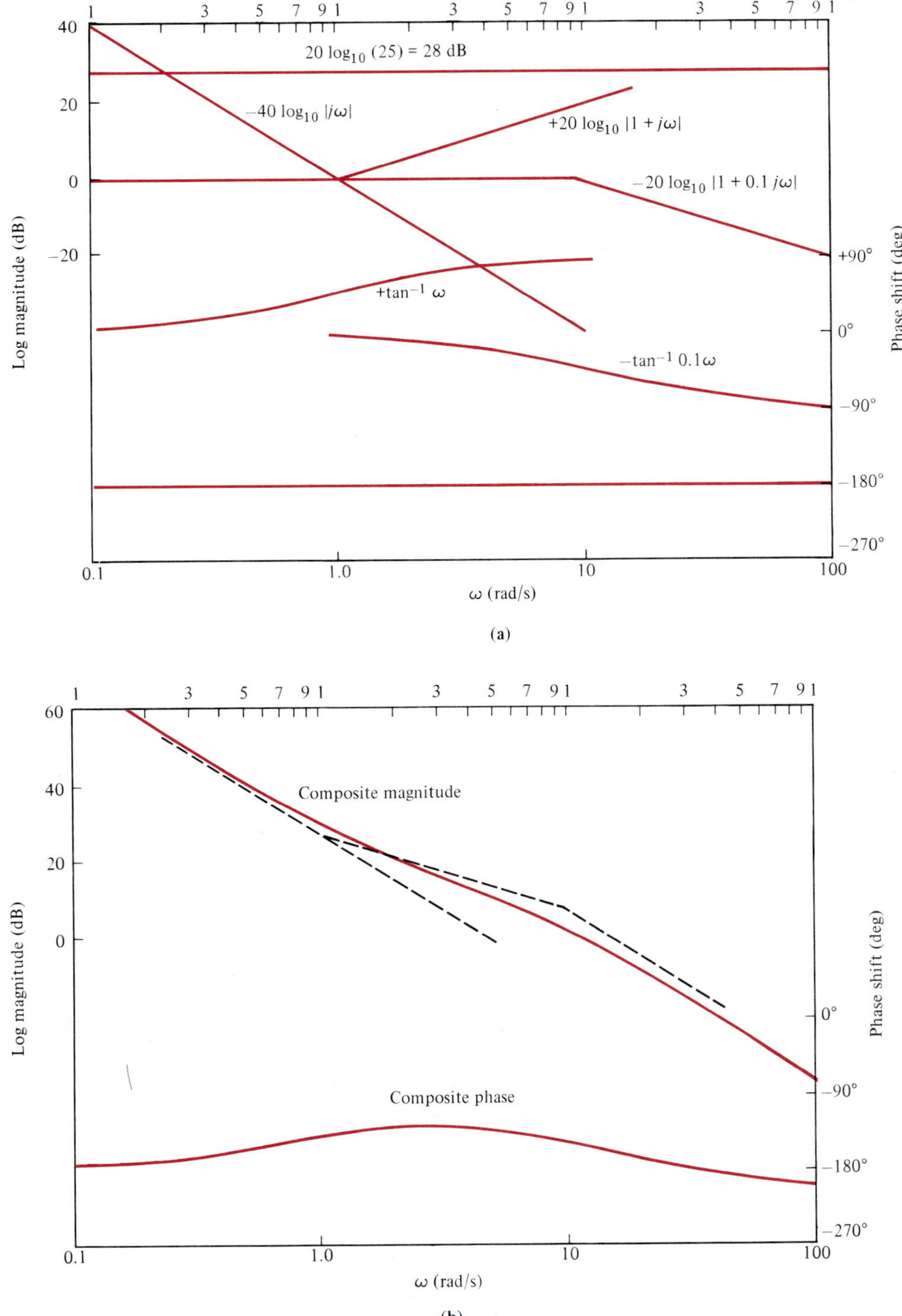

Figure 14.8 (a) Magnitude and phase components for the poles and zeros of the transfer function in Example 14.4; (b) Bode plot for the transfer function in Example 14.4.

rad/sec, which is the break frequency of the pole, the slope of the magnitude curve changes back to −40 dB/decade.

The composite phase curve starts at −180° due to the double pole at the origin. Since the first break frequency encountered is a zero, the phase curve shifts toward −90°. However, before the composite phase reaches −90°, the pole with break frequency ω = 10 rad/sec begins to shift the composite curve back toward −180°. ■

Example 14.4 illustrates the manner in which to plot directly terms of the form $K_0/(j\omega)^N$. For terms of this form, the initial slope of $-20N$ dB/decade will intersect the 0-dB axis at a frequency of $(K_0)^{1/N}$ rad/sec; that is, $-20 \log_{10} |K_0/(j\omega)^N| = 0$ dB implies that $K_0/(\omega)^N = 1$, and therefore, $\omega = (K_0)^{1/N}$ rad/sec. Note that the projected slope of the magnitude curve in Example 14.4 intersects the 0-dB axis at $\omega = (25)^{1/2} = 5$ rad/sec.

Similarly, it can be shown that for terms of the form $K_0(j\omega)^N$, the initial slope of $+20N$ dB/decade will intersect the 0-dB axis at a frequency of $\omega = (1/K_0)^{1/N}$ rad/sec; that is, $+20 \log_{10} |K_0 j\omega)^N| = 0$ dB implies that $K_0(\omega)^N = 1$, and therefore $\omega = (1/K_0)^{1/N}$ rad/sec.

By applying the concepts we have just demonstrated, we can normally plot the log magnitude characteristic of a transfer function directly in one step.

EXAMPLE 14.5

We wish to generate the Bode plot for the following transfer function:

$$\mathbf{G_V}(j\omega) = \frac{100(j\omega + 1)}{j\omega(j\omega + 10)}$$

This function can be written in standard form as

$$\mathbf{G_V}(j\omega) = \frac{10(j\omega + 1)}{j\omega(0.1j\omega + 1)}$$

The Bode plot for $\mathbf{G_V}(j\omega)$ is shown in Fig. 14.9. The magnitude curve can be plotted in one step as follows. Since the transfer function has a frequency-independent gain factor of 10 and a pole at the origin, the composite magnitude curve will have an initial slope of −20 dB/decade, which intersects the 0-dB axis at $\omega = 10$ rad/sec. The initial slope of −20 dB/decade changes at the first break frequency, which, in this case, is a zero with break frequency $\omega = 1$ rad/sec. Since the magnitude characteristic for the zero is 0 dB for $\omega \ll 1$ rad/sec and +20 dB/decade for $\omega \gg 1$ rad/sec, the composite magnitude curve changes slope from −20 dB/decade to 0 dB/decade (− 20 dB/decade + 20 dB/decade) at $\omega = 1$ rad/sec. The composite magnitude characteristic, which now has a slope of 0 dB/decade for $\omega > 1$ rad/sec, continues until the next break frequency is encountered, which in this case is a pole with break frequency $\omega = 10$ rad/sec. Since the magnitude characteristic for the pole is 0 dB/decade for $\omega < 10$ rad/sec and −20 dB/decade for $\omega > 10$ rad/sec, the composite magnitude characteristic changes slope from 0 dB/decade to −20 dB/decade (0 dB/decade − 20 dB/decade) at $\omega = 10$ rad/sec. This final slope of −20 dB/decade continues for all values of $\omega > 10$ rad/sec. Note that the composite

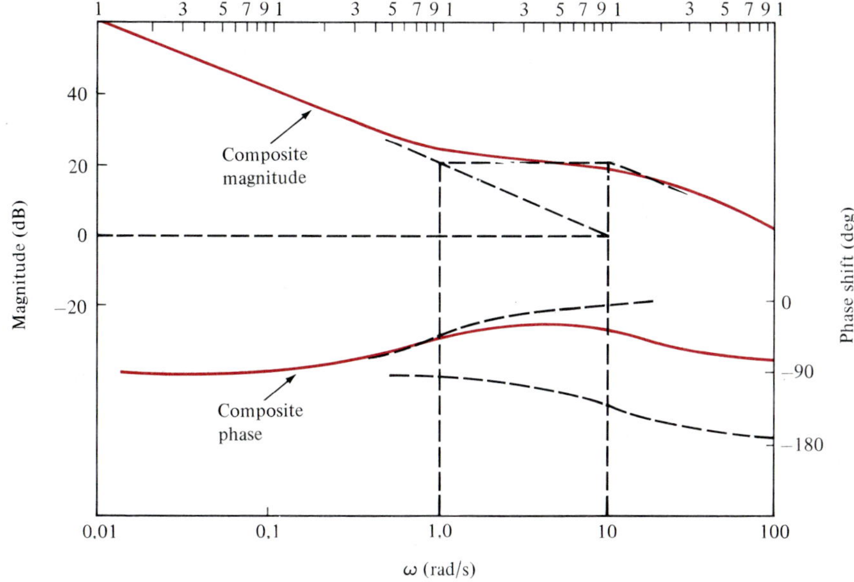

Figure 14.9 Bode plot for the function in Example 14.5.

magnitude curve in Fig. 14.9 has been adjusted for the 3-dB difference between the actual curve and the straight-line asymptotes at the break frequencies. At low frequencies the phase due to the pole at the origin is $-90°$. This curve, together with the two arctangent curves, one for the zero with break frequency $\omega = 1$ rad/sec, and one for the pole with a break frequency of $\omega = 10$ rad/sec, are summed to yield the composite phase curve as shown in Fig. 14.9. ■

DRILL EXERCISE

D14.4. *Sketch* the magnitude characteristic of the Bode plot, labeling all critical slopes and points for the function

$$\mathbf{G}(j\omega) = \frac{10^4(j\omega + 2)}{(j\omega + 10)(j\omega + 100)}$$

Ans:

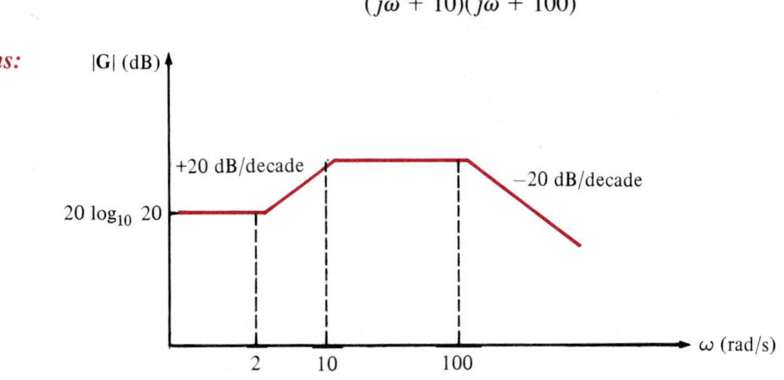

D14.5. *Sketch* the magnitude characteristic of the Bode plot, labeling all critical slopes and points for the function

$$\mathbf{G}(j\omega) = \frac{10}{j\omega(0.05j\omega + 1)}$$

Ans:

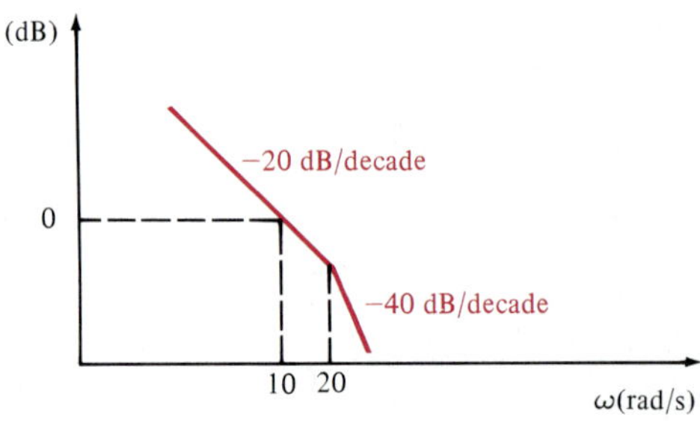

D14.6. *Sketch* the magnitude characteristic of the Bode plot, labeling all critical slopes and points for the function

$$\mathbf{G}(j\omega) = \frac{100(0.02j\omega + 1)}{(j\omega)^2}$$

Ans:

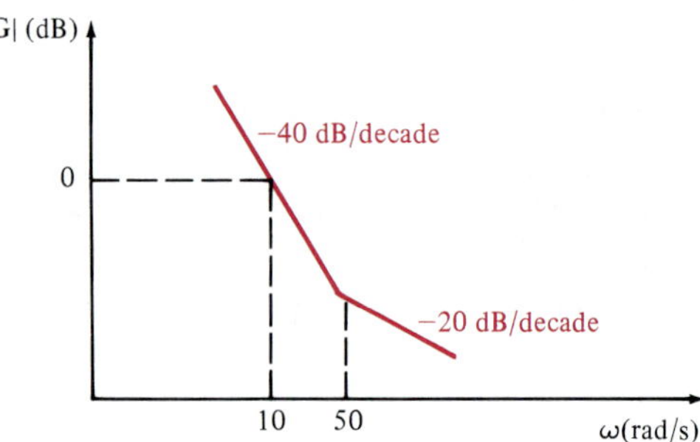

D14.7. *Sketch* the magnitude characteristic of the Bode plot, labeling all critical slopes and points for the function

$$\mathbf{G}(j\omega) = \frac{10j\omega}{(j\omega + 1)(j\omega + 10)}$$

Ans:

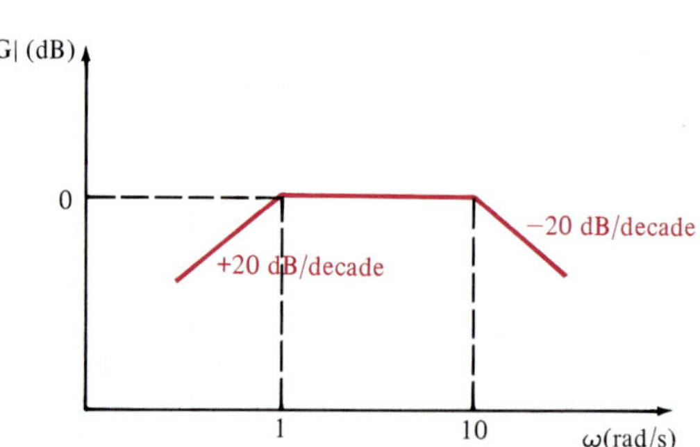

EXAMPLE 14.6

We wish to generate the Bode plot for the following transfer function:

$$\mathbf{G_V}(j\omega) = \frac{25j\omega}{(j\omega + 0.5)[(j\omega)^2 + 4j\omega + 100]}$$

Expressing this function in standard form, we obtain

$$\mathbf{G_V}(j\omega) = \frac{0.5j\omega}{(2j\omega + 1)[(j\omega/10)^2 + j\omega/25 + 1]}$$

The Bode plot is shown in Fig. 14.10. The initial low-frequency slope due to the zero at the origin is +20 dB/decade, and this slope intersects the 0-dB line at $\omega = 1/K_0 = 2$ rad/sec. At $\omega = 0.5$ rad/sec the slope changes from +20 dB/decade to 0 dB/decade due to the presence of the pole with a break frequency at $\omega = 0.5$ rad/sec. The quadratic term has a center frequency of $\omega = 10$ rad/sec (i.e., $\tau = 1/10$). Since

$$2\zeta\tau = \frac{1}{25}$$

and

$$\tau = 0.1$$

then

$$\zeta = 0.2$$

Plotting the curve in Fig. 14.6a with a damping ratio of $\zeta = 0.2$ at the center frequency $\omega = 10$ rad/sec completes the composite magnitude curve for the transfer function.

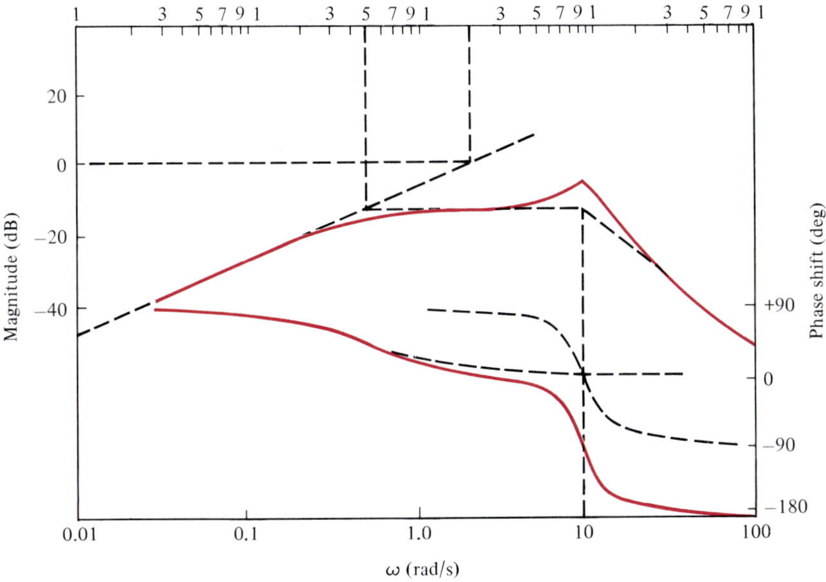

Figure 14.10 Bode plot for the transfer function in Example 14.6.

The initial low-frequency phase curve is $+90°$, due to the zero at the origin. This curve, together with the phase curve for the simple pole and the phase curve for the quadratic term, as defined in Fig. 14.6b, are combined to yield the composite phase curve. ■

DRILL EXERCISE

D14.8. Given the following function $\mathbf{G}(j\omega)$, sketch the magnitude characteristic of the Bode plot, labeling all critical slopes and points.

$$\mathbf{G}(j\omega) = \frac{0.2(j\omega + 1)}{j\omega[(j\omega/12)^2 + j\omega/36 + 1]}$$

Ans:

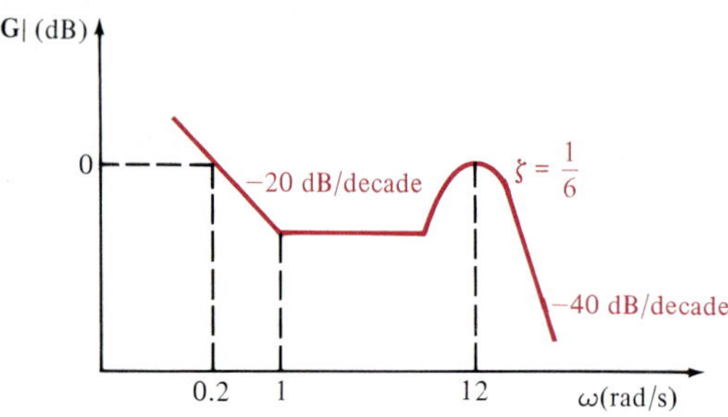

EXAMPLE 14.7

Given the asymptotic magnitude characteristic shown in Fig. 14.11, we wish to determine the transfer function $\mathbf{G_V}(j\omega)$.

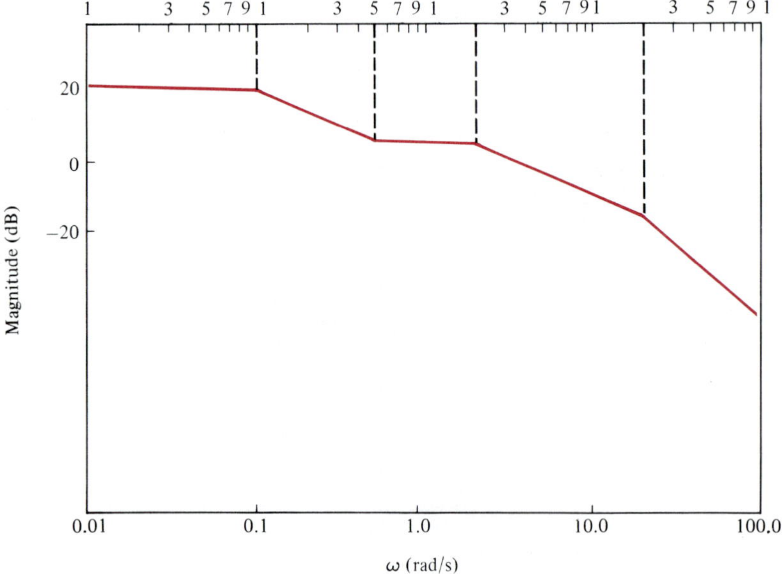

Figure 14.11 Straight-line magnitude plot employed in Example 14.7.

Since the initial slope is 0 dB/decade, and the level of the characteristic is 20 dB, the factor K_0 can be obtained from the expression

$$20 \text{ dB} = 20 \log_{10} K_0$$

and hence

$$K_0 = 10$$

The -20-dB/decade slope starting at $\omega = 0.1$ rad/sec indicates that the first pole has a break frequency at $\omega = 0.1$ rad/sec, and therefore one of the factors in the denominator is $(10j\omega + 1)$. The slope changes by $+20$ dB/decade at $\omega = 0.5$ rad/sec, indicating that there is a zero present with a break frequency at $\omega = 0.5$ rad/sec, and therefore the numerator has a factor of $(2j\omega + 1)$. Two additional poles are present with break frequencies at $\omega = 2$ rad/sec and $\omega = 20$ rad/sec. Therefore, the composite transfer function is

$$\mathbf{G}_V(j\omega) = \frac{10(2j\omega + 1)}{(10j\omega + 1)(0.5j\omega + 1)(0.05j\omega + 1)}$$

The reader should note carefully the ramifications this example has with regard to network design. ■

DRILL EXERCISE

D14.9. Find the transfer function $\mathbf{G}(j\omega)$ if the straight-line magnitude characteristic approximation for this function is as shown in Fig. D14.9.

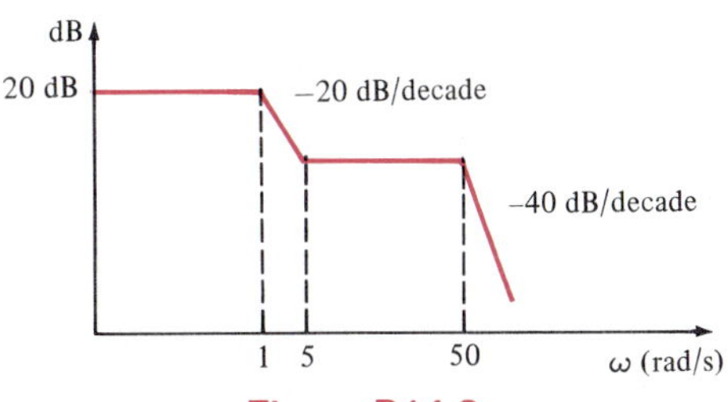

Figure D14.9

Ans: $\mathbf{G}(j\omega) = \dfrac{10\left(\dfrac{j\omega}{5} + 1\right)}{(j\omega + 1)\left(\dfrac{j\omega}{50} + 1\right)^2}$.

D14.10. Determine the transfer function $\mathbf{G}(j\omega)$ if the straight-line magnitude characteristic approximation for this function is as shown in Fig. D14.10.

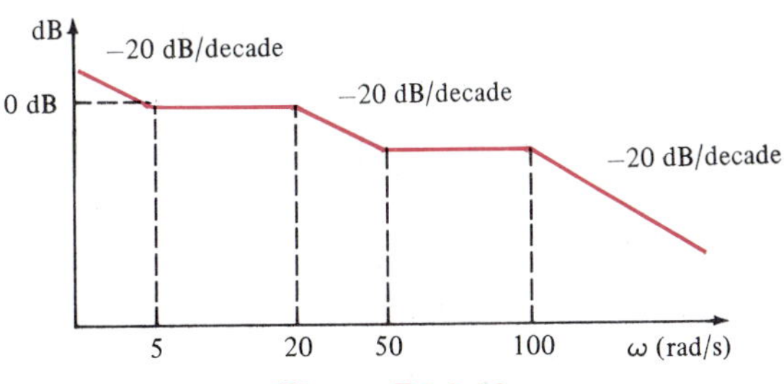

Figure D14.10

Ans: $\mathbf{G}(j\omega) = \dfrac{5\left(\dfrac{j\omega}{5}+1\right)\left(\dfrac{j\omega}{50}+1\right)}{j\omega\left(\dfrac{j\omega}{20}+1\right)\left(\dfrac{j\omega}{100}+1\right)}$.

14.3 Resonant Circuits

Two circuits with extremely important frequency characteristics are shown in Fig. 14.12. The input impedance for the series RLC circuit is

$$\mathbf{Z}(j\omega) = R + j\omega L + \frac{1}{j\omega C} \tag{14.16}$$

and the input admittance for the parallel RLC circuit is

$$\mathbf{Y}(j\omega) = G + j\omega C + \frac{1}{j\omega L} \tag{14.17}$$

Note that these two equations have the same general form. The imaginary terms in both of the equations above will be zero if

$$\omega L = \frac{1}{\omega C}$$

The value of ω that satisfies this equation is

$$\omega_0 = \frac{1}{\sqrt{LC}} \tag{14.18}$$

and at this value of ω the impedance of the series circuit becomes

$$\mathbf{Z}(j\omega_0) = R \tag{14.19}$$

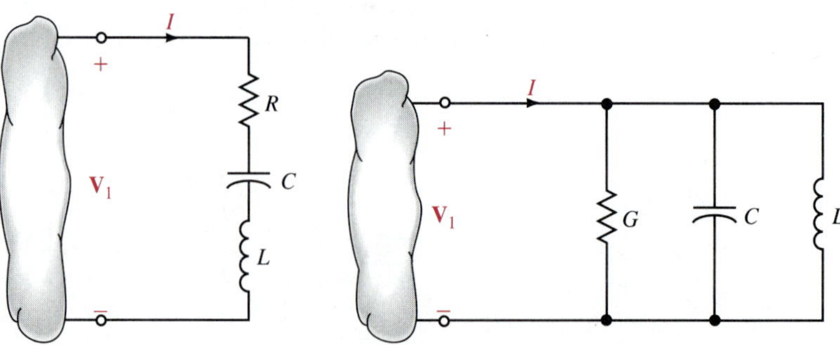

Figure 14.12 Series and parallel *RLC* circuits.

and the admittance of the parallel circuit is

$$\mathbf{Y}(j\omega_0) = G \tag{14.20}$$

This frequency ω_0, at which the impedance of the series circuit or the admittance of the parallel circuit is purely real, is called the *resonant frequency*, and the circuits themselves, at this frequency, are said to be *in resonance*. Resonance is a very important consideration in engineering design. For example, engineers designing the attitude control system for the Saturn vehicles had to ensure that the control system frequency did not excite the body bending (resonant) frequencies of the vehicle. Excitation of the bending frequencies would cause oscillation that, if continued unchecked, would result in a buildup of stress until the vehicle would finally break apart.

At resonance the voltage and current are in phase, and therefore, the phase angle is zero and the power factor is unity. In the series case, at resonance the impedance is a minimum, and therefore the current is maximum for a given voltage. Figure 14.13 illus-

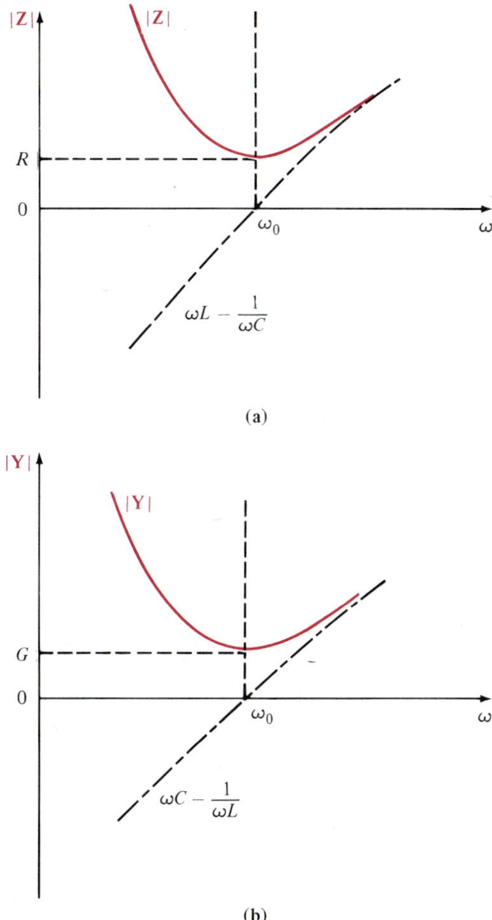

Figure 14.13 Frequency response of (a) a series, and (b) a parallel *RLC* circuit.

trates the frequency response of both the series and parallel *RLC* circuits. Note that at low frequencies the impedance of the series circuit is dominated by the capacitive term and the admittance of the parallel circuit is dominated by the inductive term. At high frequencies the impedance of the series circuit is dominated by the inductive term, and the admittance of the parallel circuit is dominated by the capacitive term.

Resonance can be viewed from another perspective—that of the phasor diagram. Once again we will consider the series and parallel cases together in order to illustrate the similarities between them. In the series case the current is common to every element, and in the parallel case the voltage is a common variable. Therefore, the current in the series circuit and the voltage in the parallel circuit are employed as references. Phasor diagrams for both circuits are shown in Fig. 14.14 for the three frequency values $\omega < \omega_0$, $\omega = \omega_0$, and $\omega > \omega_0$.

In the series case when $\omega < \omega_0$, $\mathbf{V}_C > \mathbf{V}_L$, θ_Z is negative and the voltage $\mathbf{V}_1$ lags the current. If $\omega = \omega_0$, $\mathbf{V}_L = \mathbf{V}_C$, θ_Z is zero, and the voltage $\mathbf{V}_1$ is in phase with the current. If $\omega > \omega_0$, $\mathbf{V}_L > \mathbf{V}_C$, θ_Z is positive, and the voltage $\mathbf{V}_1$ leads the current. Similar statements can be made for the parallel case in Fig. 14.14b. Because of the close relationship between series and parallel resonance, as illustrated by the preceding material, we will concentrate most of our discussion on the series case in the following developments.

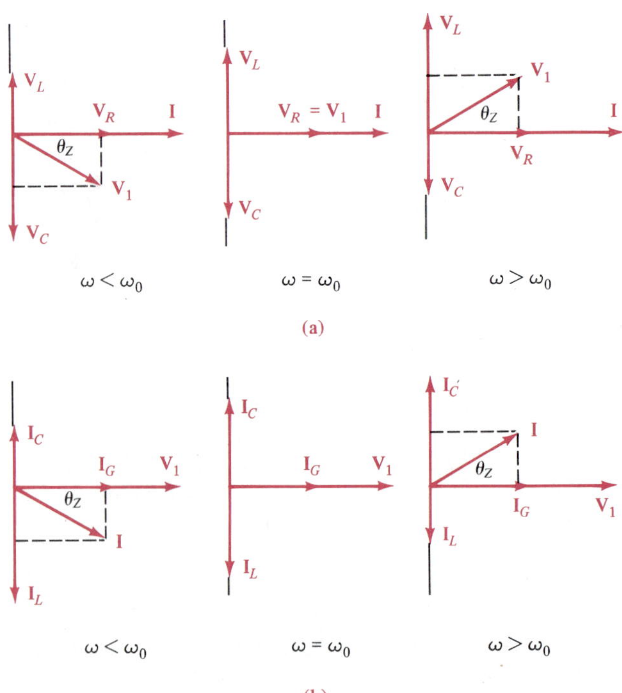

Figure 14.14 Phasor diagrams for (a) a series *RLC* circuit, and (b) a parallel *GLC* circuit.

For the series circuit we define what is commonly called the *quality factor* Q as

$$Q = \frac{\omega_0 L}{R} = \frac{1}{\omega_0 CR} = \frac{1}{R}\sqrt{\frac{L}{C}} \tag{14.21}$$

Q is a very important factor in resonant circuits, and its ramifications will be illustrated throughout the remainder of this section.

EXAMPLE 14.8

Consider the network shown in Fig. 14.15. Let us determine the resonant frequency, the voltages across each element at resonance, and the value of the quality factor.

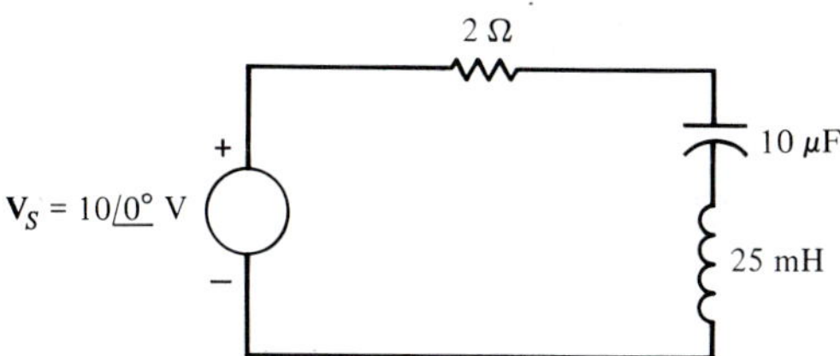

Figure 14.15 Series circuit.

The resonant frequency is obtained from the expression

$$\begin{aligned} \omega_0 &= \frac{1}{\sqrt{LC}} \\ &= \frac{1}{\sqrt{(25)(10^{-3})(10)(10^{-6})}} \\ &= 2000 \text{ rad/sec} \end{aligned}$$

At this resonant frequency

$$\mathbf{I} = \frac{\mathbf{V}}{\mathbf{Z}} = \frac{\mathbf{V}}{R} = 5 \underline{/0°} \text{ A}$$

Therefore,

$$\begin{aligned} \mathbf{V}_R &= (5 \underline{/0°})(2) = 10 \underline{/0°} \text{ V} \\ \mathbf{V}_L &= j\omega_0 L\mathbf{I} = 250 \underline{/90°} \text{ V} \\ \mathbf{V}_C &= \frac{\mathbf{I}}{j\omega_0 C} = 250 \underline{/-90°} \text{ V} \end{aligned}$$

Note the magnitude of the voltages across the inductor and capacitor with respect to the input voltage. Note also that these voltages are equal and are 180° out of phase with one another. Therefore, the phasor diagram for this condition is shown in Fig. 14.14a for $\omega = \omega_0$. The quality factor Q derived from Eq. (14.21) is

$$Q = \frac{\omega_0 L}{R} = \frac{(2)(10^3)(25)(10^{-3})}{2} = 25$$

It is interesting to note that the voltages across the inductor and capacitor can be written in terms of Q as

$$|\mathbf{V}_L| = \omega_0 L|\mathbf{I}| = \frac{\omega_0 L}{R}\mathbf{V}_S = Q\mathbf{V}_S$$

and

$$|\mathbf{V}_C| = \frac{|\mathbf{I}|}{\omega_0 C} = \frac{1}{\omega_0 CR}\mathbf{V}_S = Q\mathbf{V}_S$$

This analysis indicates that for a given current there is a resonant voltage rise across the inductor and capacitor which is equal to the product of Q and the applied voltage. ■

DRILL EXERCISE

D14.11. Given the network in Fig. D14.11, find the value of C that will place the circuit in resonance at 1800 rad/sec.

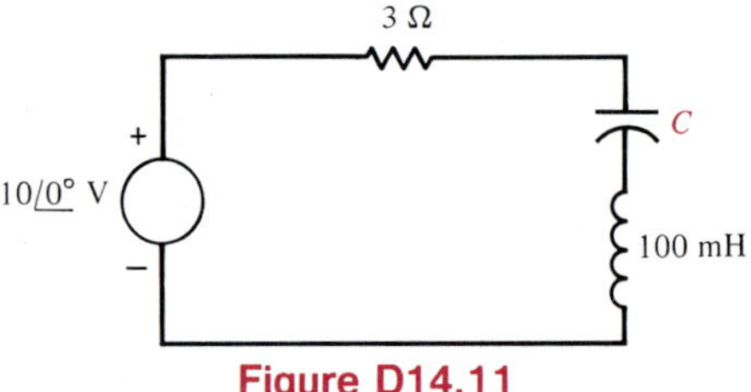

Figure D14.11

Ans: $C = 3.09\ \mu\text{F}$.

D14.12. Given the network in D14.11, determine the Q of the network and the magnitude of the voltage across the capacitor.

Ans: $Q = 60$, $|\mathbf{V}_C| = 600$ V.

The impedance of the circuit in Fig. 14.12a is given by the Eq. (14.16), which can be expressed as an admittance,

$$\mathbf{Y}(j\omega) = \frac{1}{R[1 + j(1/R)(\omega L - 1/\omega C)]}$$

$$= \frac{1}{R[1 + j(\omega L/R - 1/\omega CR)]}$$

$$= \frac{1}{R[1 + jQ(\omega L/RQ - 1/\omega CRQ)]} \tag{14.22}$$

Using the fact that $Q = \omega_0 L/R = 1/\omega_0 CR$, Eq. (14.22) becomes

$$\mathbf{Y}(j\omega) = \frac{1}{R[1 + jQ(\omega/\omega_0 - \omega_0/\omega)]} \tag{14.23}$$

Since $\mathbf{I} = \mathbf{Y}\mathbf{V}_1$ and the voltage across the resistor is $\mathbf{V}_R = \mathbf{I}R$, then

$$\frac{\mathbf{V}_R}{\mathbf{V}_1} = \mathbf{G}_\mathbf{V}(j\omega) = \frac{1}{1 + jQ(\omega/\omega_0 - \omega_0/\omega)} \tag{14.24}$$

and the magnitude and phase are

$$M(\omega) = \frac{1}{[1 + Q^2(\omega/\omega_0 - \omega_0/\omega)^2]^{1/2}} \tag{14.25}$$

and

$$\phi(\omega) = -\tan^{-1} Q\left(\frac{\omega}{\omega_0} - \frac{\omega_0}{\omega}\right) \tag{14.26}$$

The sketches for these functions are shown in Fig. 14.16. Note that the circuit has the form of a band-pass filter. The bandwidth as shown is the difference between the two half-power frequencies. Since power is proportional to the square of the magnitude, these two frequencies may be derived by setting the magnitude $M(\omega) = 1/\sqrt{2}$; that is,

$$\left|\frac{1}{1 + jQ(\omega/\omega_0 - \omega_0/\omega)}\right| = \frac{1}{\sqrt{2}}$$

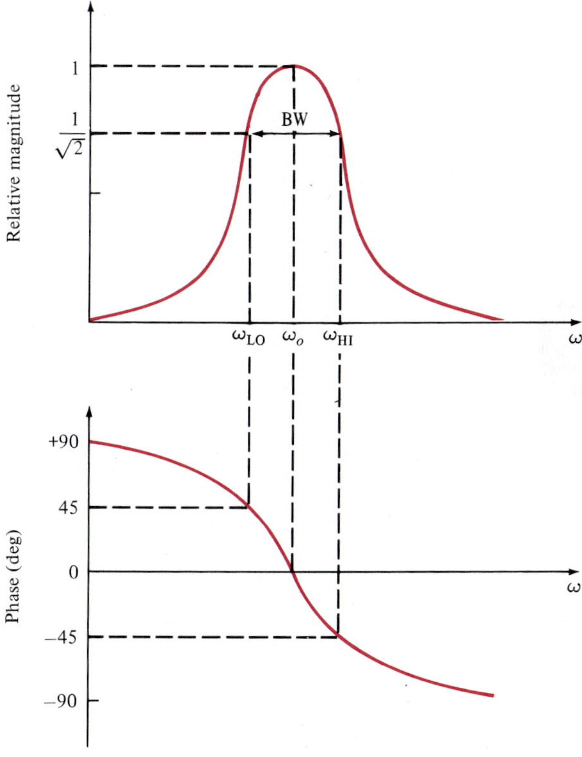

Figure 14.16 Magnitude and phase curves for Eq. (14.24).

Therefore,

$$Q\left(\frac{\omega}{\omega_0} - \frac{\omega_0}{\omega}\right) = \pm 1 \tag{14.27}$$

Solving this equation, we obtain four frequencies,

$$\omega = \pm \frac{\omega_0}{2Q} \pm \omega_0 \sqrt{\left(\frac{1}{2Q}\right)^2 + 1} \tag{14.28}$$

Taking only the positive values, we obtain

$$\omega_{LO} = \omega_0\left[-\frac{1}{2Q} + \sqrt{\left(\frac{1}{2Q}\right)^2 + 1}\right] \tag{14.29}$$

$$\omega_{HI} = \omega_0\left[\frac{1}{2Q} + \sqrt{\left(\frac{1}{2Q}\right)^2 + 1}\right]$$

Subtracting these two equations yields the bandwidth as shown in Fig. 14.16:

$$\text{BW} = \omega_{HI} - \omega_{LO} = \frac{\omega_0}{Q} \tag{14.31}$$

and multiplying the two equations yields

$$\omega_0^2 = \omega_{LO}\omega_{HI} \tag{14.32}$$

which illustrates that the resonant frequency is the geometric mean of the two half-power frequencies. Recall that the half-power frequencies are the points at which the log-magnitude curve is down 3 dB from its maximum value. Therefore, the difference between the 3-dB frequencies, which is of course the bandwidth, is often called the 3-dB bandwidth.

DRILL EXERCISE

D14.13. For the network in Fig. D14.11, compute the two half-power frequencies and the bandwidth of the network.

Ans: $\omega_{HI} = 1815$ rad/sec, $\omega_{LO} = 1785$ rad/sec, BW = 30 rad/sec.

Equation (14.21) indicates the dependence of Q on R. A high-Q series circuit has a small value of R, and as we will illustrate later, a high-Q parallel circuit has a relatively large value of R.

Equation (14.31) illustrates that the bandwidth is inversely proportional to Q. Therefore, the frequency selectivity of the circuit is determined by the value of Q. A high-Q circuit has a small bandwith, and therefore, the circuit is very selective. The manner in which Q affects the frequency selectivity of the network is graphically illustrated in Fig. 14.17. Hence, if we pass a signal with a wide frequency range through a high-Q circuit, only the frequency components within the bandwidth of the network will not be attenuated; that is, the network acts like a band-pass filter.

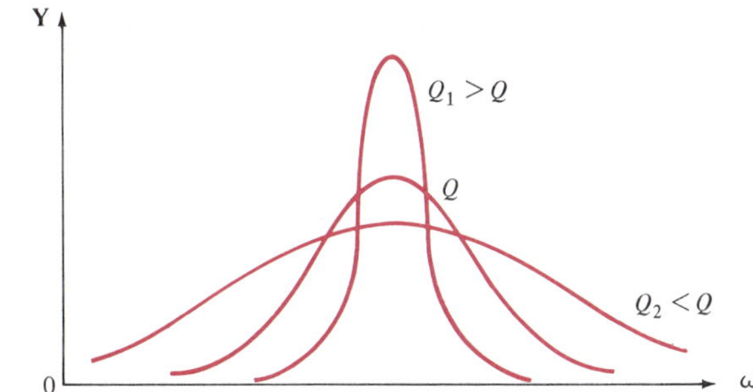

Figure 14.17 Network frequency response as a function of Q.

Q has a more general meaning which we can explore via an energy analysis of the series circuit. Recall from Chapter 6 that an inductor stores energy in its magnetic field and a capacitor stores energy in its electric field. When a network is in resonance there is a continuous exchange of energy between the magnetic field of the inductor and the electric field of the capacitor. During each half-cycle the energy stored in the inductor's magnetic field will vary from zero to a maximum value and back to zero again. The capacitor operates in a similar manner. The energy exchange takes place in the following way. During one quarter-cycle the capacitor absorbs energy as quickly as the inductor gives it up, and during the following one quarter-cycle the inductor absorbs energy as fast as it is released by the capacitor. Although the energy stored in each element is continuously varying, the total energy stored in the resonant circuit is constant and therefore not changing with time.

In order to define Q in terms of energy, suppose that the excitation is at the frequency ω_0 and that

$$i(t) = I_M \cos \omega_0 t$$

The energies stored in the inductor and capacitor are

$$w_L(t) = \tfrac{1}{2} L I_M^2 \cos^2 \omega_0 t$$

and

$$w_C(t) = \frac{1}{2} \frac{I_M^2}{\omega_0^2 C} \sin^2 \omega_0 t$$

The total energy stored is then

$$\begin{aligned} w_S(t) &= w_L(t) + w_C(t) \\ &= \tfrac{1}{2} I_M^2 \left(L \cos^2 \omega_0 t + \frac{1}{\omega_0^2 C} \sin^2 \omega_0 t \right) \end{aligned}$$

However, since $\omega_0 = 1/\sqrt{LC}$, the equation above reduces to

$$w_S(t) = \tfrac{1}{2} L I_M^2 \equiv W_S \tag{14.33}$$

which is the maximum energy stored at resonance.

The energy dissipated per cycle can be derived by multiplying the average power absorbed by the resistor by the period of one cycle:

$$W_D = (\tfrac{1}{2}I_M^2 R)\frac{2\pi}{\omega_0} \tag{14.34}$$

Then

$$\begin{aligned}\frac{W_S}{W_D} &= \frac{\frac{1}{2}LI_M^2}{\frac{1}{2}I_M^2 R(2\pi/\omega_0)} \\ &= \frac{\omega_0 L}{R}\frac{1}{2\pi}\end{aligned} \tag{14.35}$$

Using Eq. (14.22), we find that this equation becomes

$$\frac{W_S}{W_D} = \frac{Q}{2\pi}$$

or

$$Q = 2\pi\frac{W_S}{W_D} \tag{14.36}$$

where we recall that W_S is the maximum energy stored at resonance and W_D is the energy dissipated per cycle. The importance of this definition of Q stems from the fact that this expression is applicable to acoustic, electrical, and mechanical systems and therefore is generally considered to be the basic definition of Q.

EXAMPLE 14.9

Given a series circuit with $R = 2\ \Omega$, $L = 2$ mH, and $C = 5\ \mu$F, we wish to determine the resonant frequency, the quality factor, and the bandwidth for the circuit. Then we will determine the change in Q and the BW if R is changed from 2 to 0.2 Ω.

Using Eq. (14.18), we have

$$\begin{aligned}\omega_0 &= \frac{1}{\sqrt{LC}} = \frac{1}{[(2)(10^{-3})(5)(10^{-6})]^{1/2}} \\ &= 10^4 \text{ rad/sec}\end{aligned}$$

and therefore, the resonant frequency is $10^4/2\pi = 1592$ Hz. The quality factor is

$$\begin{aligned}Q &= \frac{\omega_0 L}{R} = \frac{(10^4)(2)(10^{-3})}{2} \\ &= 10\end{aligned}$$

and the bandwidth is

$$\begin{aligned}\text{BW} &= \frac{\omega_0}{Q} = \frac{10^4}{10} \\ &= 10^3 \text{ rad/sec}\end{aligned}$$

If R is changed to $R = 0.2\ \Omega$, the new value of Q is 100, and therefore the new BW is 10^2 rad/sec. ■

DRILL EXERCISE

D14.14. A series circuit is composed of $R = 2\ \Omega$, $L = 40$ mH, and $C = 100\ \mu$F. Determine the bandwidth of this circuit about its resonant frequency.

Ans: BW = 50 rad/sec, $\omega_0 = 500$ rad/sec.

D14.15. A series *RLC* circuit has the following properties. $R = 4\ \Omega$, $\omega_0 = 4000$ rad/sec, and the BW = 100 rad/sec. Determine the values of L and C.

Ans: $L = 40$ mH, $C = 1.56\ \mu$F.

EXAMPLE 14.10

We wish to determine the parameters R, L, and C so that the circuit shown in Fig. 14.18 operates as a band-pass filter with an ω_0 of 1000 rad/sec and a bandwidth of 100 rad/sec.

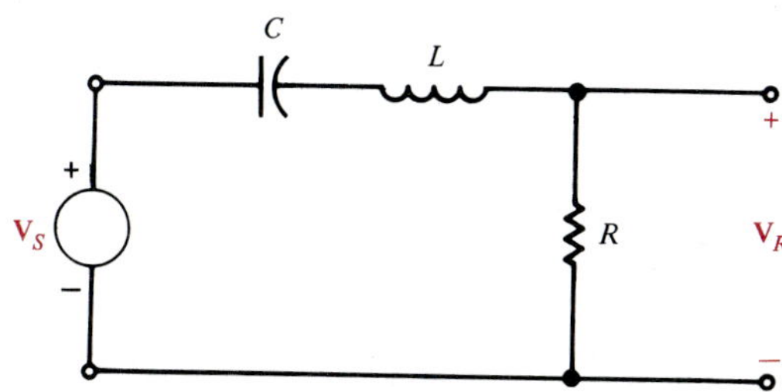

Figure 14.18 Series *RLC* circuit.

The voltage gain for the network is

$$\mathbf{G_V}(j\omega) = \frac{(R/L)j\omega}{(j\omega)^2 + (R/L)j\omega + 1/LC}$$

Hence

$$\omega_0 = \frac{1}{\sqrt{LC}}$$

and since $\omega_0 = 10^3$,

$$\frac{1}{LC} = 10^6$$

The bandwidth is

$$\text{BW} = \frac{\omega_0}{Q}$$

Then

$$Q = \frac{\omega_0}{\text{BW}} = \frac{1000}{100}$$

$$= 10$$

However,

$$Q = \frac{\omega_0 L}{R}$$

Therefore,

$$\frac{1000L}{R} = 10$$

Note that we have two equations in the three unknown circuit parameters R, L, and C. Hence if we select $C = 1\ \mu\text{F}$, then

$$L = \frac{1}{10^6 C} = 1\text{ H}$$

and

$$\frac{1000(1)}{R} = 10$$

yields

$$R = 100\ \Omega$$

Therefore, the parameters $R = 100\ \Omega$, $L = 1$ H, and $C = 1\ \mu\text{F}$ will produce the proper filter characteristics. ■

In Example 14.8 we found that the voltage across the capacitor or inductor in the series resonant circuit could be quite high. In fact it was equal to Q times the magnitude of the source voltage. With this in mind, let us reexamine this network as shown in Fig. 14.19. The output voltage for the network is

$$\mathbf{V}_o = \left(\frac{1/j\omega C}{R + j\omega L + 1/j\omega C}\right)\mathbf{V}_S$$

which can be written as

$$\mathbf{V}_o = \frac{\mathbf{V}_S}{1 - \omega^2 LC + j\omega CR}$$

The magnitude of this voltage can be expressed as

$$|\mathbf{V}_o| = \frac{\mathbf{V}_S}{\sqrt{(1 - \omega^2 LC)^2 + (\omega CR)^2}} \tag{14.37}$$

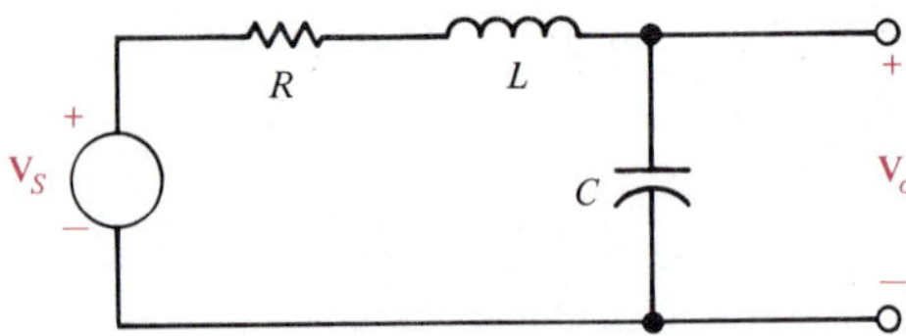

Figure 14.19 Series resonant circuit.

In view of the previous discussion, we might assume that the maximum value of the output voltage would occur at the resonant frequency ω_0. Let's see if this assumption is correct. The frequency at which $|\mathbf{V}_o|$ is maximum is the nonzero value of ω, which satisfies the equation

$$\frac{d|\mathbf{V}_o|}{d\omega} = 0 \tag{14.38}$$

If we perform the indicated operation and solve for the nonzero ω_{max}, we obtain

$$\omega_{\text{max}} = \sqrt{\frac{1}{LC} - \frac{1}{2}\left(\frac{R}{L}\right)^2} \tag{14.39}$$

By employing the relationships $\omega_0^2 = 1/LC$ and $Q = \omega_0 L/R$, the expression for ω_{max} can be written as

$$\begin{aligned} \omega_{\text{max}} &= \sqrt{\omega_0^2 - \frac{1}{2}\left(\frac{\omega_0}{Q}\right)^2} \\ &= \omega_0 \sqrt{1 - \frac{1}{2Q^2}} \end{aligned} \tag{14.40}$$

Clearly, $\omega_{\text{max}} \neq \omega_0$; however, ω_0 closely approximates ω_{max} if the Q is high. In addition, if we substitute Eq. (14.40) into Eq. (14.37) and use the relationships $\omega_0^2 = 1/LC$ and $\omega_0^2 C^2 R^2 = 1/Q^2$, we find that

$$|\mathbf{V}_o|_{\text{max}} = \frac{QV_S}{\sqrt{1 - 1/4Q^2}} \tag{14.41}$$

Again, we see that $|\mathbf{V}_o|_{\text{max}} \approx QV_S$ if the network has a high Q.

EXAMPLE 14.11

Given the network in Fig. 14.19, we wish to determine ω_0 and ω_{max} for $R = 50\ \Omega$ and $R = 1\ \Omega$ if $L = 50$ mH and $C = 5\ \mu$F.

The network parameters yield

$$\begin{aligned} \omega_0 &= \frac{1}{\sqrt{LC}} \\ &= \frac{1}{\sqrt{(5)(10^{-2})(5)(10^{-6})}} \\ &= 2000 \text{ rad/sec} \end{aligned}$$

If $R = 50\ \Omega$, then

$$\begin{aligned} Q &= \frac{\omega_0 L}{R} \\ &= \frac{(2000)(0.05)}{50} \\ &= 2 \end{aligned}$$

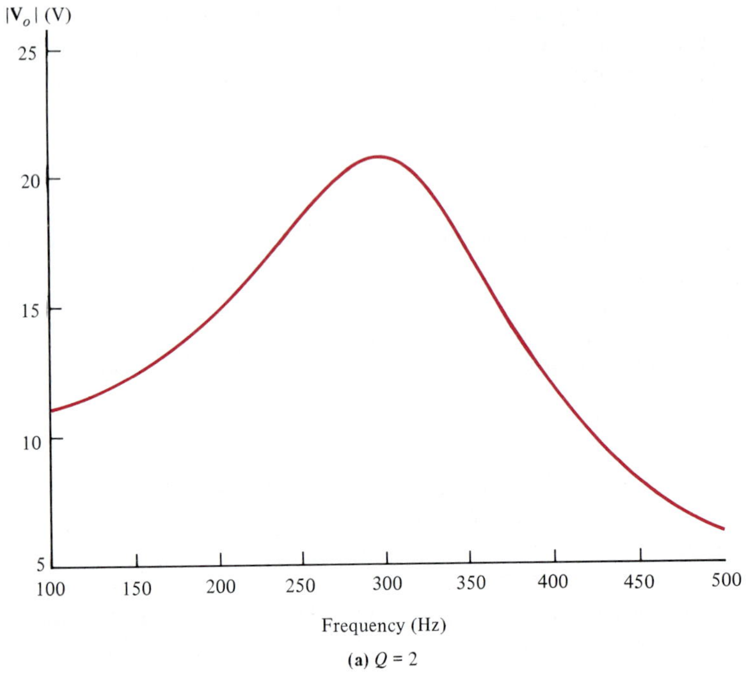

(a) $Q = 2$

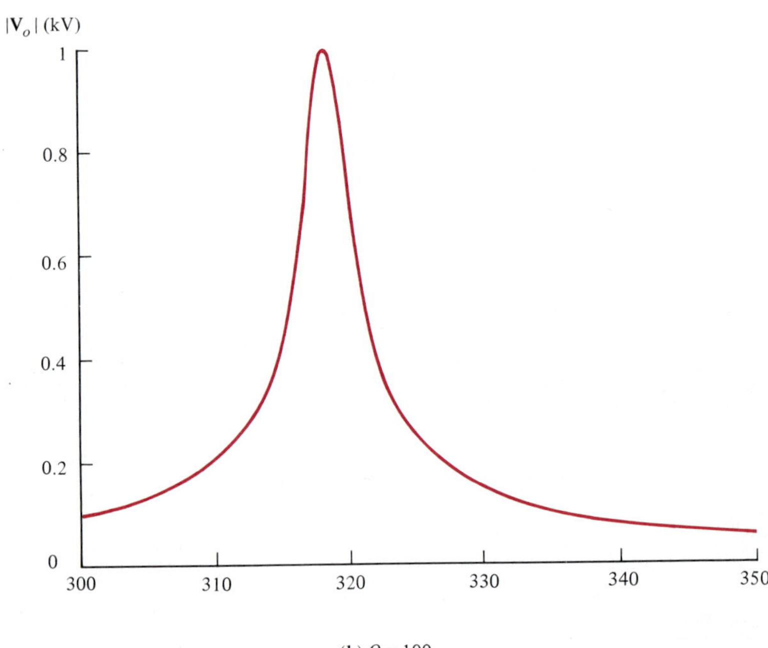

(b) $Q = 100$

Figure 14.20 Frequency response plots for the network in Fig. 14.19 with (a) $R = 50\ \Omega$ and (b) $R = 1\ \Omega$.

and

$$\omega_{max} = \omega_0 \sqrt{1 - \frac{1}{2Q^2}}$$

$$= 2000 \sqrt{1 - \frac{1}{8}}$$

$$= 1871 \text{ rad/sec}$$

If $R = 1\ \Omega$, then $Q = 100$ and $\omega_{max} = 2000$ rad/sec.

A plot of $|\mathbf{V}_o|$ versus frequency for the network with $R = 50\ \Omega$ and $R = 1\ \Omega$ is shown in Fig. 14.20a and b, respectively. Note that when the Q of the network is small, the frequency response is not selective and $\omega_0 \neq \omega_{max}$. However, if the Q is large, the frequency response is very selective and $\omega_0 \cong \omega_{max}$. ■

In our presentation of resonance thus far, we have focused most of our discussion on the series resonant circuit. We should recall, however, that the equations for the impedance of the series circuit and the admittance of the parallel circuit are similar. Therefore, the networks possess similar properties, as we will illustrate in the following examples.

Consider the network shown in Fig. 14.21. The source current $\mathbf{I}_S$ can be expressed as

$$\mathbf{I}_S = \mathbf{I}_G + \mathbf{I}_C + \mathbf{I}_L$$

$$= \mathbf{V}_S G + j\omega C \mathbf{V}_S + \frac{\mathbf{V}_S}{j\omega L}$$

$$= \mathbf{V}_S \left[G + j\left(\omega C - \frac{1}{\omega L} \right) \right]$$

When the network is in resonance,

$$\mathbf{I}_S = G\mathbf{V}_S$$

that is, all the source current flows through the conductance G. Does this mean that there is no current in L or C? Definitely not! $\mathbf{I}_C$ and $\mathbf{I}_L$ are equal in magnitude but 180° out of phase with one another. Therefore, $\mathbf{I}_x$, as shown in Fig. 14.21, is zero. In addition, if $G = 0$, the source current is zero. What is actually taking place, however, is an energy exchange between the electric field of the capacitor and the magnetic field of the inductor. As one increases, the other decreases, and vice versa.

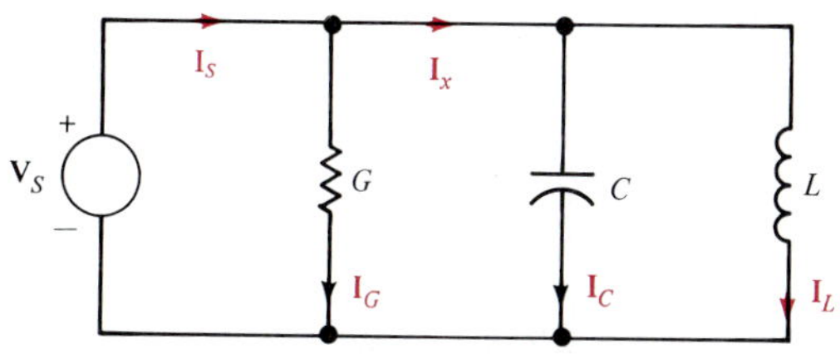

Figure 14.21 Parallel *RLC* circuit.

EXAMPLE 14.12

The network in Fig. 14.21 has the following parameters:

$$\mathbf{V}_S = 120 \underline{/0^\circ} \text{ V}, G = 0.01 \text{ S}, C = 600 \ \mu\text{F, and } L = 120 \text{ mH}$$

If the source operates at the resonant frequency of the network, compute all the branch currents.

The resonant frequency for the network is

$$\begin{aligned}\omega_0 &= \frac{1}{\sqrt{LC}} \\ &= \frac{1}{\sqrt{(120)(10^{-3})(600)(10^{-6})}} \\ &= 117.85 \text{ rad/sec}\end{aligned}$$

At this frequency

$$\mathbf{Y}_C = j\omega_0 C = j7.07 \times 10^{-2} \text{ S}$$

and

$$\mathbf{Y}_L = -j\left(\frac{1}{\omega_0 L}\right) = -j7.07 \times 10^{-2} \text{ S}$$

The branch currents are then

$$\begin{aligned}\mathbf{I}_G &= G\mathbf{V}_S = 1.2 \underline{/0^\circ} \text{ A} \\ \mathbf{I}_C &= \mathbf{Y}_C\mathbf{V}_S = 8.49 \underline{/90^\circ} \text{ A} \\ \mathbf{I}_L &= \mathbf{Y}_L\mathbf{V}_S = 8.49 \underline{/-90^\circ} \text{ A}\end{aligned}$$

and

$$\mathbf{I}_S = \mathbf{I}_G + \mathbf{I}_C + \mathbf{I}_L = \mathbf{I}_G = 1.2 \underline{/0^\circ} \text{ A}$$

As the analysis indicates, the source supplies only the losses in the resistive element. In addition, the source voltage and current are in phase, and therefore, the power factor is unity. ■

EXAMPLE 14.13

Given the parallel *RLC* circuit in Fig. 14.22,

(a) Derive the expression for the resonant frequency, the half-power frequencies, the bandwidth, and the quality factor for the transfer characteristic $\mathbf{V}_{\text{out}}/\mathbf{I}_{\text{in}}$ in terms of the circuit parameters R, L, and C.

(b) Compute the quantities in part (a) if $R = 1 \text{ k}\Omega$, $L = 10$ mH, and $C = 100 \ \mu$F.

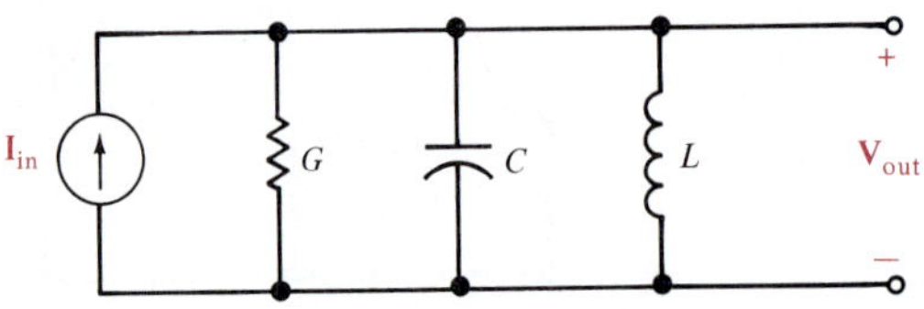

Figure 14.22 Circuit used in Example 14.13.

(a) The output voltage can be written as

$$\mathbf{V}_{\text{out}} = \frac{\mathbf{I}_{\text{in}}}{\mathbf{Y}_T}$$

and therefore, the magnitude of the transfer characteristic can be expressed as

$$\left|\frac{\mathbf{V}_{\text{out}}}{\mathbf{I}_{\text{in}}}\right| = \frac{1}{\sqrt{(1/R^2) + (\omega C - 1/\omega L)^2}}$$

The transfer characteristic is a maximum at the resonant frequency

$$\omega_0 = \frac{1}{\sqrt{LC}} \tag{14.42}$$

and at this frequency

$$\left|\frac{\mathbf{V}_{\text{out}}}{\mathbf{I}_{\text{in}}}\right|_{\max} = R \tag{14.43}$$

As demonstrated earlier, at the half-power frequencies the magnitude is equal to $1/\sqrt{2}$ of its maximum value, and hence the half-power frequencies can be obtained from the expression

$$\frac{1}{\sqrt{(1/R^2) + (\omega C - 1/\omega L)^2}} = \frac{R}{\sqrt{2}}$$

Solving this equation and taking only the positive values of ω yields

$$\omega_{\text{LO}} = -\frac{1}{2RC} + \sqrt{\frac{1}{(2RC)^2} + \frac{1}{LC}} \tag{14.44}$$

and

$$\omega_{\text{HI}} = \frac{1}{2RC} + \sqrt{\frac{1}{(2RC)^2} + \frac{1}{LC}} \tag{14.45}$$

Subtracting these two half-power frequencies yields the bandwidth

$$\begin{aligned} \text{BW} &= \omega_{\text{HI}} - \omega_{\text{LO}} \\ &= \frac{1}{RC} \end{aligned} \tag{14.46}$$

Therefore, the quality factor is

$$\begin{aligned} Q &= \frac{\omega_0}{\text{BW}} \\ &= \frac{RC}{\sqrt{LC}} \\ &= R\sqrt{\frac{C}{L}} \end{aligned} \tag{14.47}$$

Using Eqs. (14.42), (14.46), and (14.47), we can write Eqs. (14.44) and (14.45) as

$$\omega_{LO} = \omega_0 \left[\frac{-1}{2Q} + \sqrt{\frac{1}{(2Q)^2} + 1} \right] \tag{14.48}$$

$$\omega_{HI} = \omega_0 \left[\frac{1}{2Q} + \sqrt{\frac{1}{(2Q)^2} + 1} \right] \tag{14.49}$$

(b) Using the values given for the circuit components, we find that

$$\omega_0 = \frac{1}{\sqrt{(10^{-2})(10^{-4})}} = 10^3 \text{ rad/sec}$$

The half-power frequencies are

$$\omega_{LO} = \frac{-1}{(2)(10^3)(10^{-4})} + \sqrt{\frac{1}{(2)(10^{-1})} + 10^6}$$

$$= 995 \text{ rad/sec}$$

and

$$\omega_{HI} = 1005 \text{ rad/sec}$$

Therefore, the bandwidth is

$$BW = \omega_{HI} - \omega_{LO} = 10 \text{ rad/sec}$$

and

$$Q = 10^3 \sqrt{\frac{10^{-4}}{10^{-2}}}$$

$$= 100$$

DRILL EXERCISE

D14.16. A parallel *RLC* circuit has the following parameters: $R = 2$ kΩ, $L = 20$ mH, and $C = 150$ μF. Determine the resonant frequency, the Q, and the bandwidth of the circuit.

Ans: $\omega_0 = 577$ rad/sec, $Q = 173$, and BW = 3.33 rad/sec.

D14.17. A parallel *RLC* circuit has the following parameters: $R = 6$ kΩ, BW = 1000 rad/sec, and $Q = 120$. Determine the values of L, C, and ω_0.

Ans: $L = 417.5$ μH, $C = 0.167$ μF, and $\omega_0 = 119{,}760$ rad/sec.

EXAMPLE 14.14

A stereo receiver is tuned to 98 MHz on the FM band. The tuning knob controls a variable capacitor in a parallel resonant circuit. If the inductance of the circuit is 0.1 μH and the Q is 120, determine the values of C and G.

Using the expression for the resonant frequency, we obtain

$$C = \frac{1}{\omega_0^2 L}$$
$$= \frac{1}{(2\pi \times 98 \times 10^6)^2(0.1 \times 10^{-6})}$$
$$= 26.4 \text{ pF}$$

The conductance is

$$G = \frac{1}{\omega_0 LQ}$$
$$= \frac{1}{(2\pi \times 98 \times 10^6)(10^{-7})(120)}$$
$$= 135\ \mu\text{S}$$

■

EXAMPLE 14.15

Given the data in Example 14.14, suppose that another FM station in the vicinity is broadcasting at 98.1 MHz. Let us determine the relative value of the voltage across the resonant circuit at this frequency compared with that at 98 MHz, assuming that the current produced by both signals have the same amplitude.

At the resonant frequency of 98 MHz the voltage across the circuit is

$$\mathbf{V} = \frac{\mathbf{I}}{\mathbf{Y}} = \frac{\mathbf{I}}{(135)(10^{-6})} = 7407\mathbf{I}$$

At 98.1 MHz the voltage is

$$\mathbf{V} = \frac{\mathbf{I}}{\sqrt{G^2 + (\omega C - 1/\omega L)^2}}$$

where

$$G = 135\ \mu\text{S}$$
$$\omega C = (2\pi \times 98.1 \times 10^6)(26.4 \times 10^{-12}) = 16{,}272 \times 10^{-6} \text{ S}$$
$$\omega L = (2\pi \times 98.1 \times 10^6)(10^{-7}) = 61.6381\ \Omega$$

and therefore,

$$\frac{1}{\omega L} = 16{,}224 \times 10^{-6} \text{ S}$$

Hence

$$\mathbf{V} = \frac{\mathbf{I}}{(14{,}650)(10^{-6})} = 6979\mathbf{I}$$

This analysis indicates that the magnitude of the voltage across the resonant circuit produced by the 98-MHz signal is almost identical to that produced by the 98.1-MHz signal. Therefore, the network is not very frequency selective. ■

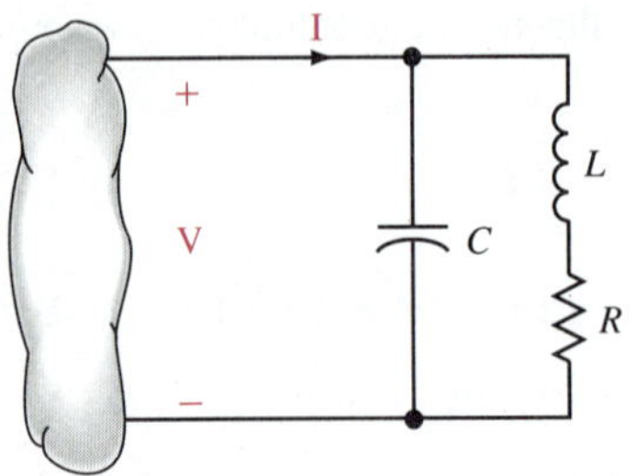

Figure 14.23 Practical parallel resonant circuit.

In general, the resistance of the winding of an inductor cannot be neglected, and hence a more practical parallel resonant circuit is the one shown in Fig. 14.23. The input admittance of this circuit is

$$\mathbf{Y}(j\omega) = j\omega C + \frac{1}{R + j\omega L}$$

$$= j\omega C + \frac{R - j\omega L}{R^2 + \omega^2 L^2}$$

$$= \frac{R}{R^2 + \omega^2 L^2} + j\left(\omega C - \frac{\omega L}{R^2 + \omega^2 L^2}\right)$$

The resonant frequency at which the admittance is purely real is

$$\omega_r C - \frac{\omega_r L}{R^2 + \omega_r^2 L^2} = 0$$

$$\omega_r = \sqrt{\frac{1}{LC} - \frac{R^2}{L^2}} \tag{14.50}$$

EXAMPLE 14.16

Given the tank circuit in Fig. 14.24, let us determine ω_0 and ω_r for $R = 50\ \Omega$ and $R = 5\ \Omega$.

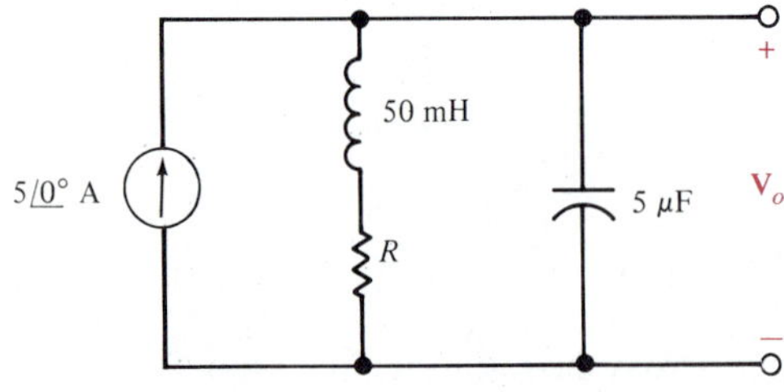

Figure 14.24 Tank circuit used in Example 14.16.

Using the network parameter values, we obtain

$$\omega_0 = \frac{1}{\sqrt{LC}}$$

$$= \frac{1}{\sqrt{(0.05)(5)(10^{-6})}}$$

$$= 2000 \text{ rad/sec}$$

$$= 318.3 \text{ Hz}$$

If $R = 50\ \Omega$, then

$$\omega_r = \sqrt{\frac{1}{LC} - \frac{R^2}{L^2}}$$

$$= \sqrt{\frac{1}{(0.05)(5)(10^{-6})} - \left(\frac{50}{0.05}\right)^2}$$

$$= 1732 \text{ rad/sec}$$

$$= 275.7 \text{ Hz}$$

If $R = 5\ \Omega$, then

$$\omega_r = \sqrt{\frac{1}{(0.05)(5)(10^{-6})} - \left(\frac{5}{0.05}\right)^2}$$

$$= 1997 \text{ rad/sec}$$

$$= 317.9 \text{ Hz}$$

Note that as $R \to 0$, $\omega_r \to \omega_0$. This fact is also illustrated in the frequency response curves in Fig. 14.25a and b, where we have plotted $|\mathbf{V}_o|$ versus frequency for $R = 50\ \Omega$ and $R = 5\ \Omega$, respectively. ■

Let us now try to relate some of the things we have learned about resonance to the Bode plots we presented earlier. The admittance for the series resonant circuit is

$$\mathbf{Y}(j\omega) = \frac{1}{R + j\omega L + 1/j\omega C}$$

$$= \frac{j\omega C}{(j\omega)^2 LC + j\omega CR + 1} \tag{14.51}$$

The standard form for the quadratic factor is

$$(j\omega\tau)^2 + 2\zeta\omega\tau j + 1$$

where $\tau = 1/\omega_0$, and hence in general the quadratic factor can be written as

$$\frac{(j\omega)^2}{\omega_0^2} + \frac{2\zeta\omega}{\omega_0} j + 1 \tag{14.52}$$

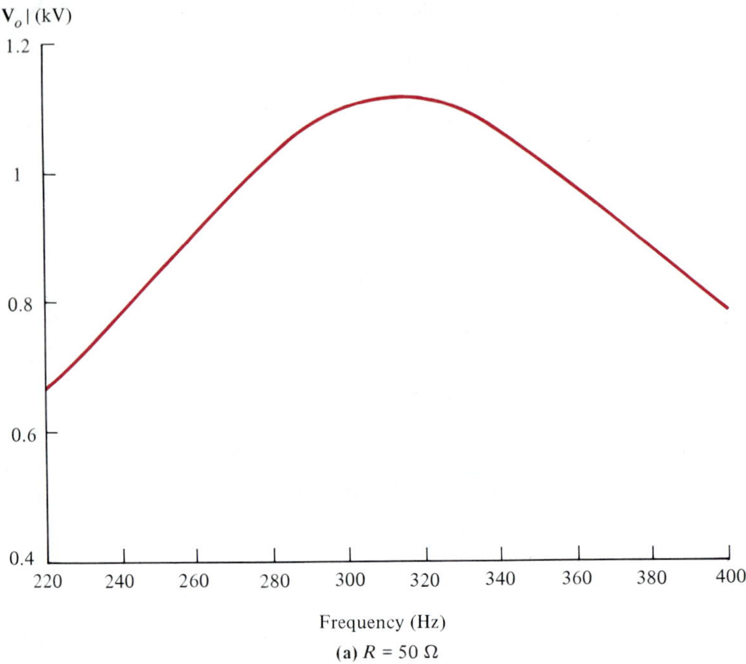

(a) $R = 50\ \Omega$

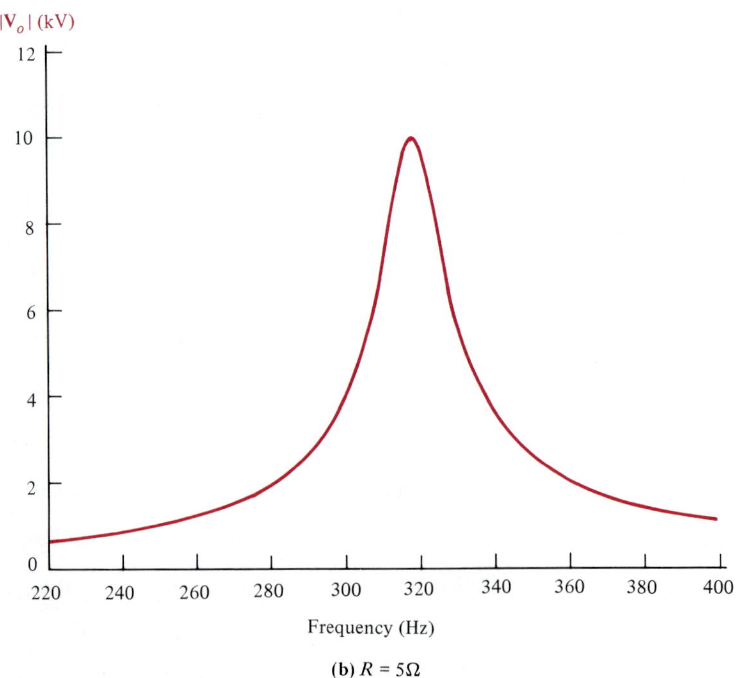

(b) $R = 5\Omega$

Figure 14.25 Frequency response curves for Example 14.16.

If we now compare this form of the quadratic factor with the denominator of $\mathbf{Y}(j\omega)$, we find that

$$\omega_0^2 = \frac{1}{LC}$$

$$\frac{2\zeta}{\omega_0} = CR$$

and therefore

$$\zeta = \frac{R}{2}\sqrt{\frac{C}{L}}$$

However, from Eq. (14.21),

$$Q = \frac{1}{R}\sqrt{\frac{L}{C}}$$

and hence

$$Q = \frac{1}{2\zeta} \tag{14.53}$$

To illustrate the significance of this equation, consider the Bode plot for the function $\mathbf{Y}(j\omega)$. The plot has an initial slope of +20 dB/decade due to the zero at the origin. If $\zeta > 1$, the poles represented by the quadratic factor in the denominator will simply roll off the frequency response as illustrated in Fig. 14.6a, and at high frequencies the slope of the composite characteristic will be −20 dB/decade. Note from Eq. (14.53) that if $\zeta > 1$, the Q of the circuit is very small. However, if $0 < \zeta < 1$, the frequency response will peak as shown in Fig. 14.6a, and the sharpness of the peak will be controlled by ζ. If ζ is very small, the peak of the frequency response is very narrow, the Q of the network is very large, and the circuit is very selective in filtering the input signal. Equation (14.53) together with Fig. 14.17 illustrate the connections among the frequency response, the Q, and the ζ of a network.

14.4 Scaling

Throughout this book we have employed a host of examples to illustrate the concepts being discussed. In many cases the actual values of the parameters were unrealistic in a practical sense, even though they may have simplified the presentation. In this section we illustrate how to *scale* the circuits to make them more realistic.

There are two ways to scale a circuit: *magnitude or impedance scaling* and *frequency scaling*. To magnitude-scale a circuit, we simply multiply the impedance of each element by a scale factor K_M. Therefore, a resistor R becomes $K_M R$. Multiplying the impedance of an inductor $j\omega L$ by K_M yields a new inductor $K_M L$, and multiplying the impedance of a capacitor $1/j\omega C$ by K_M yields a new capacitor C/K_M. Therefore, in magnitude scaling,

$$R' \rightarrow K_M R$$
$$L' \rightarrow K_M L \tag{14.54}$$
$$C' \rightarrow \frac{C}{K_M}$$

since

$$\omega'_0 = \frac{1}{\sqrt{LC}} = \frac{1}{\sqrt{K_M LC/K_M}} = \omega_0$$

and Q' is

$$Q' = \frac{\omega_0 L}{R} = \frac{\omega_0 K_M L}{K_M R} = Q$$

The resonant frequency, the quality factor, and therefore the bandwidth are unaffected by magnitude scaling.

In frequency scaling the scale factor is denoted as K_F. The resistor is frequency independent and, therefore, unaffected by this scaling. The new inductor L', which has the same impedance at the scaled frequency ω'_1, must satisfy the equation

$$j\omega_1 L = j\omega'_1 L'$$

where $\omega'_1 = K_F \omega_1$. Therefore,

$$j\omega_1 L = jK_F \omega_1 L'$$

Hence the new inductor value is

$$L' = \frac{L}{K_F}$$

Using a similar argument, we find that

$$C' = \frac{C}{K_F}$$

Therefore, to frequency-scale by a factor K_F,

$$R' \rightarrow R$$
$$L' \rightarrow \frac{L}{K_F} \tag{14.55}$$
$$C' \rightarrow \frac{C}{K_F}$$

Note that

$$\omega'_0 = \frac{1}{\sqrt{(L/K_F)(C/K_F)}} = K_F \omega_0$$

and

$$Q' = \frac{K_F \omega_0 L}{R K_F} = Q$$

and therefore,

$$\text{BW}' = K_F(\text{BW})$$

Hence the resonant frequency and bandwidth of the circuit are affected by frequency scaling.

EXAMPLE 14.17

If the values of the circuit parameters in Fig. 14.23 are $R = 2\ \Omega$, $L = 1$ H, and $C = \frac{1}{2}$ F, let us determine the values of the elements if the circuit is magnitude scaled by a factor $K_M = 10^2$ and frequency scaled by a factor $K_F = 10^6$.

The magnitude scaling yields

$$R' = 2K_M = 200\ \Omega$$

$$L' = (1)K_M = 100\ \text{H}$$

$$C' = \frac{1}{2}\frac{1}{K_M} = \frac{1}{200}\ \text{F}$$

Applying frequency scaling to these values yields the final results:

$$R'' = 200\ \Omega$$

$$L'' = \frac{100}{K_F} = 100\ \mu\text{H}$$

$$C'' = \frac{1}{200}\frac{1}{K_F} = 0.005\ \mu\text{F}$$

■

DRILL EXERCISE

D14.18. An *RLC* network has the following parameter values: $R = 10\ \Omega$, $L = 1$ H, and $C = 2$ F. Determine the values of the circuit elements if the circuit is magnitude scaled by a factor of 100 and frequency scaled by a factor of 10,000.

Ans: $R = 1\ \text{k}\Omega$, $L = 10$ mH, $C = 2\ \mu$F.

14.5 Frequency Response Using PSPICE

The PSPICE circuit analysis program can be used to demonstrate network characteristics by means of the plotting routine. The following examples illustrate its use.

EXAMPLE 14.18

Let us examine the network in Fig. 14.26a, which has been labeled for PSPICE analysis. We wish to write a PSPICE program that will plot the output voltage using 20 points over the frequency range 100 to 200 Hz.

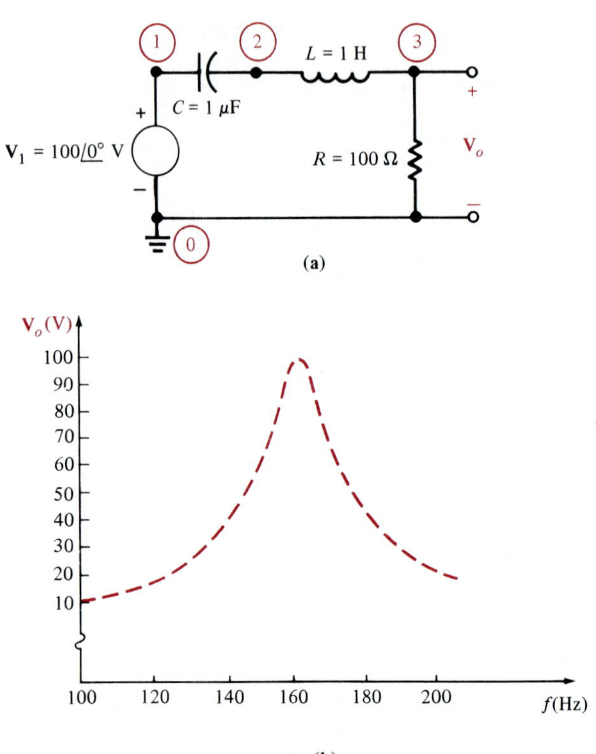

Figure 14.26 PSPICE analysis of a series *RLC* circuit.

The PSPICE program is listed below.

```
EXAMPLE 14.18
V1    1    0    AC    100   0
C     1    2    1UF
R     3    0    100
L     2    3    1H
.AC   LIN  20   100   200
.PROBE
.END
```

The output plot is shown in Fig. 14.26b. Note that the resonant frequency of the circuit, calculated to be 160 Hz, is demonstrated on the plot. ■

EXAMPLE 14.19

Consider the network shown in Fig. 14.27 which has been labeled for a PSPICE analysis. We wish to write a PSPICE program to plot the output voltage over the frequency range from 1 Hz to 10 MHz using 5 points per decade.

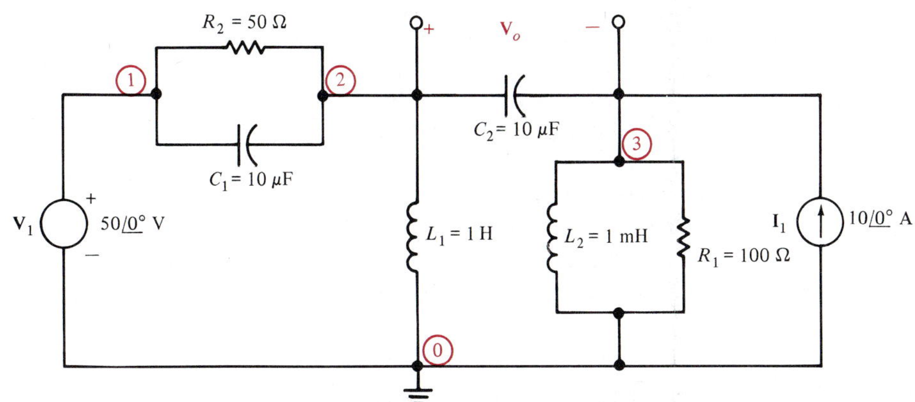

Figure 14.27 Network used in Example 14.19.

The PSPICE program is listed below.

```
EXAMPLE 14.19
*INDEPENDENT VOLTAGE SOURCE
V1   1   0   AC   50   0
*INDEPENDENT CURRENT SOURCE
I1   0   3   AC   10   0
*RESISTOR VALUES
R1   3   0   100
R2   1   2   50
*CAPACITOR VALUES
C1   1   2   10U
C2   2   3   10U
*INDUCTOR VALUES
L1   2   0   1
L2   3   0   1M
.AC   DEC   5   1   10MEG
.PROBE
.END
```

The frequency response plot is shown in Fig. 14.28. Note that although the network exhibits a dominant resonant frequency, the frequency response has more than one maximum because of the multiplicity of inductors and capacitors. ■

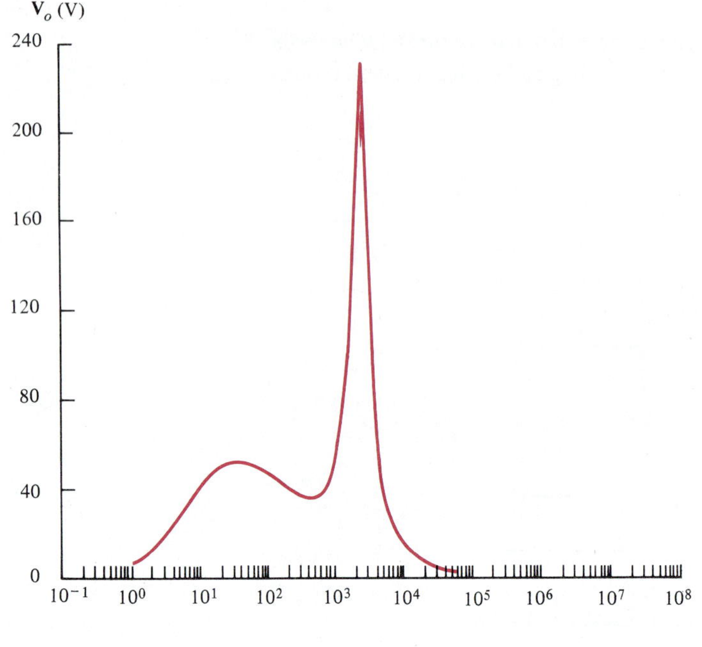

Figure 14.28 Frequency response plot for the network in Example 14.19.

DRILL EXERCISE

D14.19. Write a PSPICE program and use it to plot the frequency response of the voltage $\mathbf{V}_C$ in the network in Fig. D14.19 over the frequency range 100 to 1000 Hz using 50 points.

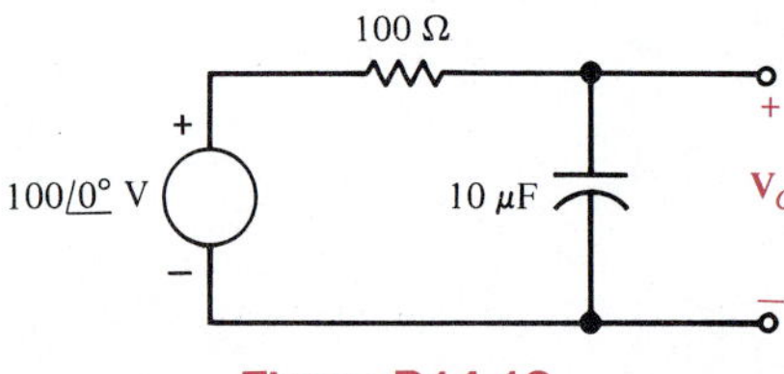

Figure D14.19

Ans:

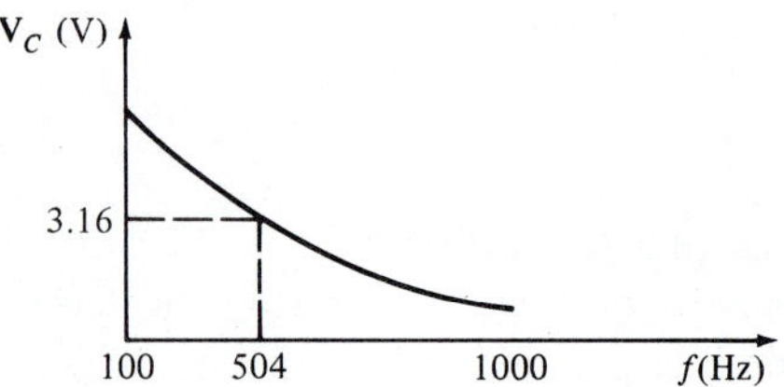

D14.20. Write a PSPICE program and use it to plot the frequency response of $\mathbf{V}_o$ in the network in Fig. D14.20 using 10 points per decade over the interval from 10 Hz to 10 kHz.

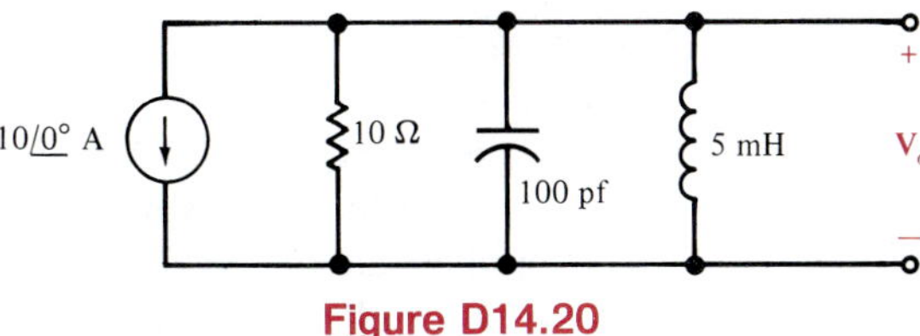

Figure D14.20

Ans:

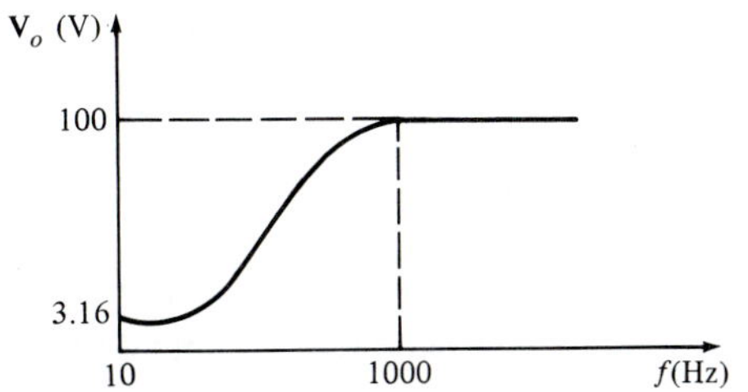

D14.21. Write a PSPICE program and use it to plot the voltage across the 1-Ω resistor in the network in Fig. D14.21 using 2 points per octave over the frequency range 1 Hz to 400 kHz.

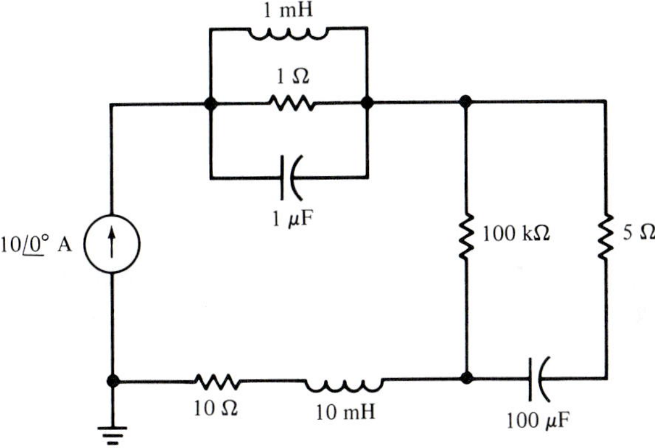

Figure D14.21

Ans:

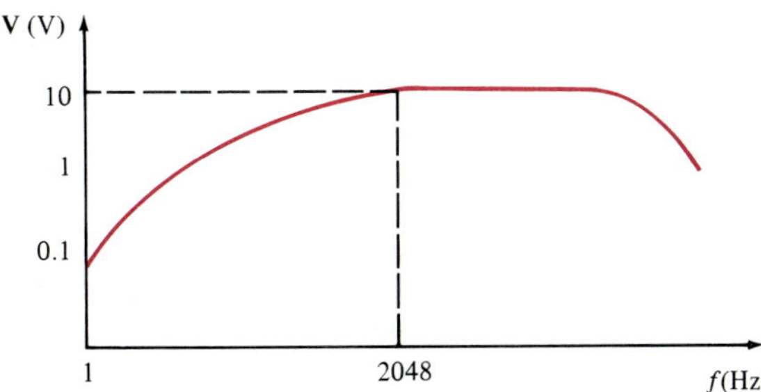

14.6

Filter Networks

Passive Filters

A filter network is generally designed to pass signals with a specific frequency range and reject or attenuate signals whose frequency spectrum is outside this passband. The most common filters are *low-pass* filters, which pass low frequencies and reject high frequencies; *high-pass* filters, which pass high frequencies and block low frequencies; *band-pass* filters, which pass some particular band of frequencies and reject all frequencies outside the range; and *band-rejection* filters, which are specifically designed to reject a particular band of frequencies and pass all other frequencies.

The ideal frequency characteristic for a low-pass filter is shown in Fig. 14.29a. Also shown is a typical or physically realizable characteristic. Ideally, we would like the low-pass filter to pass all frequencies to some frequency ω_0 and pass no frequency above that value; however, it is not possible to design such a filter with linear circuit elements. Hence we must be content to employ filters that we can actually build in the laboratory, and these filters have frequency characteristics that are simply not ideal.

A simple low-pass filter network is shown in Fig. 14.29b. The voltage gain for the network is

$$\mathbf{G_V}(j\omega) = \frac{1}{1 + j\omega RC} \tag{14.56}$$

which can be written as

$$\mathbf{G_V}(j\omega) = \frac{1}{1 + j\omega\tau} \tag{14.57}$$

where $\tau = RC$, the time constant. The amplitude characteristic is

$$M(\omega) = \frac{1}{[1 + (\omega\tau)^2]^{1/2}} \tag{14.58}$$

and the phase characteristic is

$$\phi(\omega) = -\tan^{-1}\omega\tau \tag{14.59}$$

Note that at the break frequency, $\omega = 1/\tau$ and the amplitude is

$$M\left(\omega = \frac{1}{\tau}\right) = \frac{1}{\sqrt{2}} \tag{14.60}$$

The break frequency is also commonly called the *half-power frequency*. This name is derived from the fact that if the voltage or current is $1/\sqrt{2}$ of its maximum value, then the power, which is proportional to the square of the voltage or current, is one-half its maximum value.

The magnitude, in decibels, and phase curves for this simple low-pass circuit are shown in Fig. 14.29c. Note that the magnitude curve is flat for low frequencies and rolls

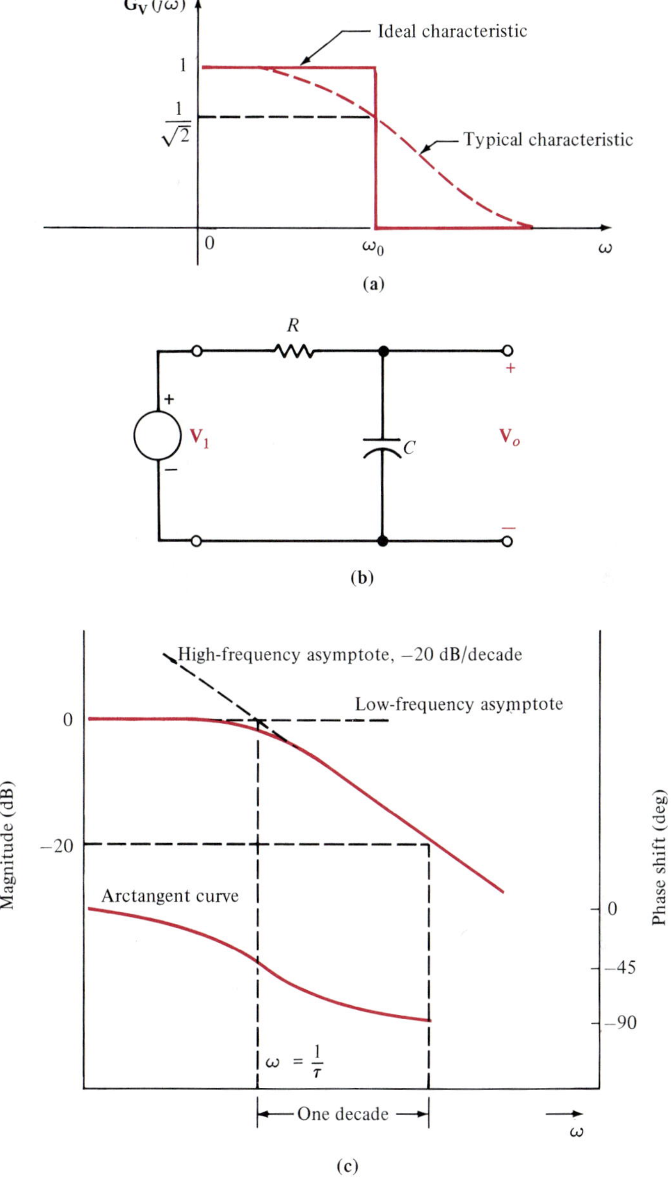

Figure 14.29 Low-pass filter circuit and its frequency characteristics.

off at high frequencies. The phase shifts from 0° at low frequencies to −90° at high frequencies.

The ideal frequency characteristic for a high-pass filter is shown in Fig. 14.30a, together with a typical characteristic that we could achieve with linear circuit components.

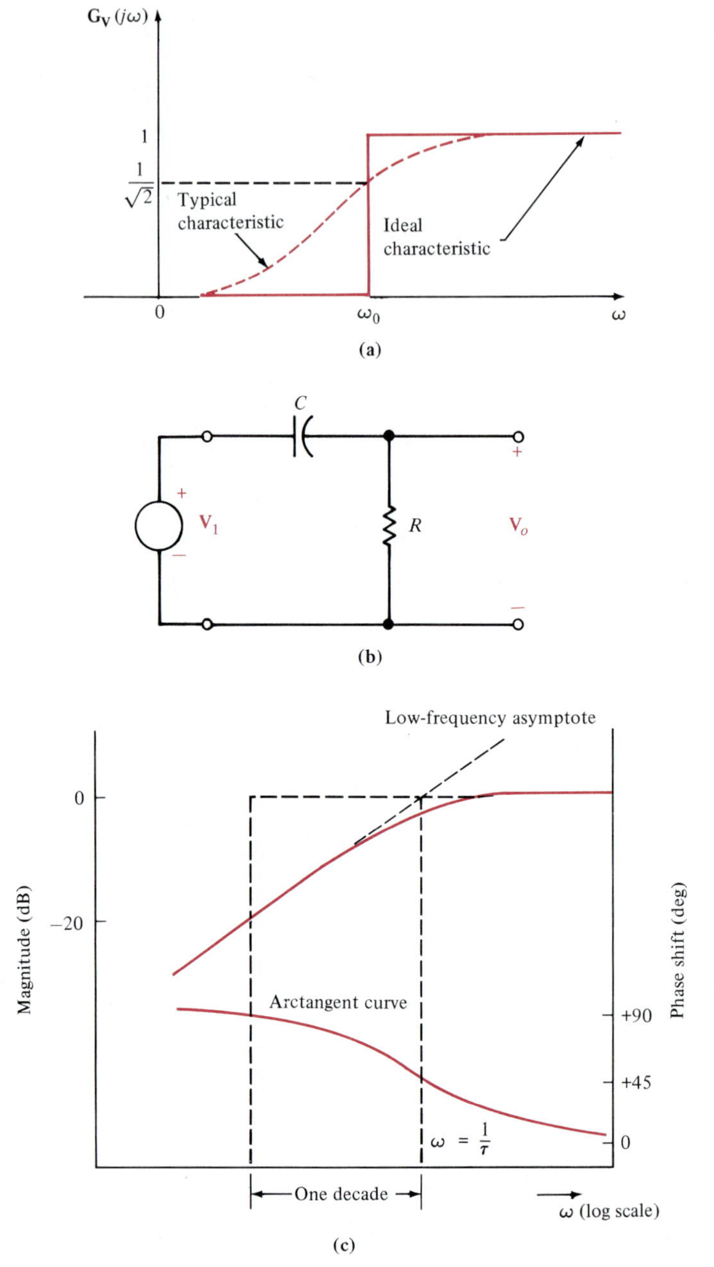

Figure 14.30 High-pass filter circuit and frequency characteristics.

Ideally, the high-pass filter passes all frequencies above some frequency ω_0 and no frequencies below that value.

A simple high-pass filter network is shown in Fig. 14.30b. This is the same network as shown in Fig. 14.29b, except that the output voltage is taken across the resistor. The voltage gain for this network is

$$\mathbf{G_V}(j\omega) = \frac{j\omega\tau}{1 + j\omega\tau} \tag{14.61}$$

where once again $\tau = RC$. The magnitude of this function is

$$M(\omega) = \frac{\omega\tau}{[1 + (\omega\tau)^2]^{1/2}} \tag{14.62}$$

and the phase is

$$\phi(\omega) = \frac{\pi}{2} - \tan^{-1} \omega\tau \tag{14.63}$$

The half-power frequency is $\omega = 1/\tau$, and the phase at this frequency is 45°.

The magnitude and phase curves for this high-pass filter are shown in Fig. 14.30c. At low frequencies the magnitude curve has a slope of +20 dB/decade due to the term $\omega\tau$ in the numerator of Eq. (14.62). Then at the break frequency the curve begins to flatten out. The phase curve is derived from Eq. (14.63).

Ideal and typical amplitude characteristics for simple band-pass and band-rejection filters are shown in Fig. 14.31a and b, respectively. Simple networks that are capable of realizing the typical characteristics of each filter are shown below the characteristics in Fig. 14.31c and d. ω_0 is the center frequency of the pass or rejection band and the frequency at which the maximum or minimum amplitude occurs. ω_{LO} and ω_{HI} are the lower and upper break frequencies or *cutoff frequencies*, where the amplitude is $1/\sqrt{2}$ of the maximum value. The width of the pass or rejection band is called the *bandwidth,* and hence

$$\text{BW} = \omega_{HI} - \omega_{LO} \tag{14.64}$$

To illustrate these points, let us consider the band-pass filter. The voltage transfer function is

$$\mathbf{G_V}(j\omega) = \frac{R}{R + j(\omega L - 1/\omega C)}$$

and therefore, the amplitude characteristic is

$$M(\omega) = \frac{RC\omega}{\sqrt{(RC\omega)^2 + (\omega^2 LC - 1)^2}}$$

At low frequencies

$$M(\omega) \approx \frac{RC\omega}{1} \approx 0$$

At high frequencies

$$M(\omega) \simeq \frac{RC\omega}{\omega^2 LC} \approx \frac{R}{\omega L} \approx 0$$

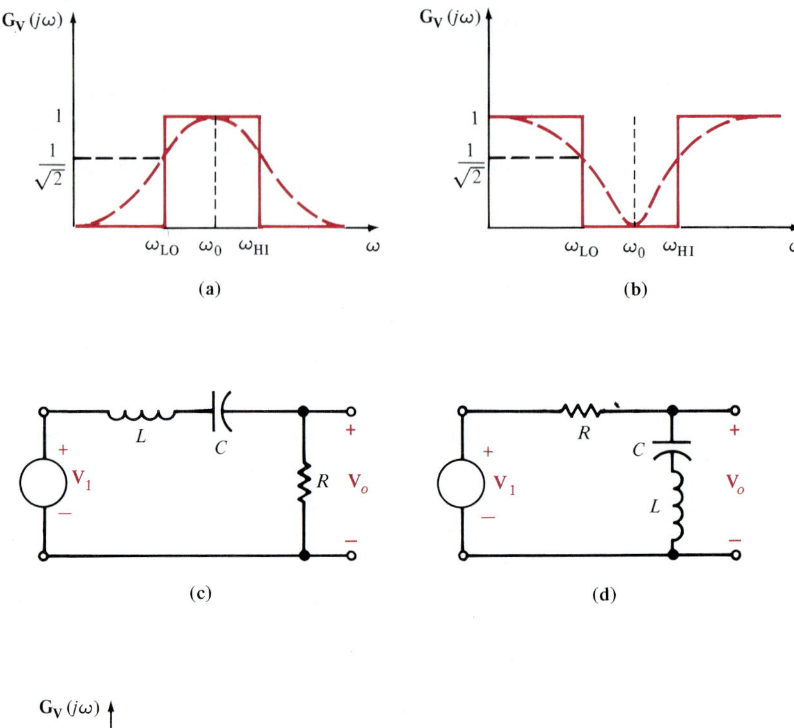

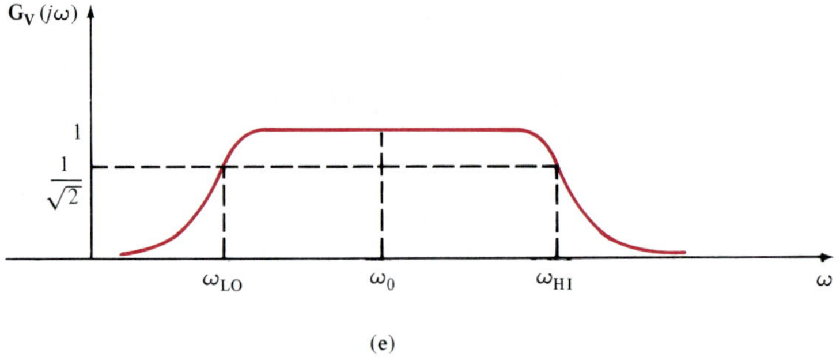

Figure 14.31 Band-pass and band-rejection filters and characteristics.

In the midfrequency range $(RC\omega)^2 \gg (\omega^2LC - 1)^2$, and thus $M(\omega) \approx 1$. Therefore, the frequency characteristic for this filter is shown in Fig. 14.31e. The center frequency is $\omega_0 = 1/\sqrt{LC}$. At the lower cutoff frequency

$$\omega^2LC - 1 = -RC\omega$$

or

$$\omega^2 + \frac{R\omega}{L} - \omega_0^2 = 0$$

Solving this expression for ω_{LO}, we obtain

$$\omega_{LO} = \frac{-(R/L) + \sqrt{(R/L)^2 + 4\omega_0^2}}{2}$$

At the upper cutoff frequency

$$\omega^2 LC - 1 = +RC\omega$$

or

$$\omega^2 - \frac{R}{L}\omega - \omega_0^2 = 0$$

Solving this expression for ω_{HI}, we obtain

$$\omega_{HI} = \frac{+(R/L) + \sqrt{(R/L)^2 + 4\omega_0^2}}{2}$$

Therefore, the bandwidth of the filter is

$$BW = \omega_{HI} - \omega_{LO} = \frac{R}{L}$$

EXAMPLE 14.20

Consider the network in Fig. 14.32a, which is a low-pass filter prepared for PSPICE analysis. Let us write a PSPICE program to plot the output voltage using 10 points per decade over the frequency range from 1 Hz to 1 kHz.

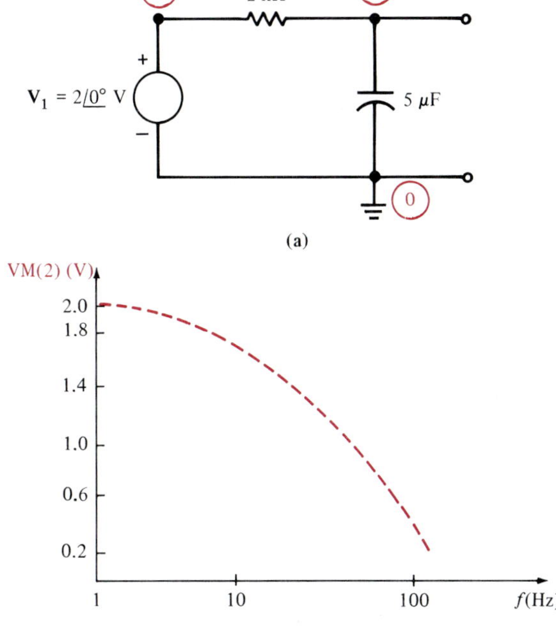

Figure 14.32 PSPICE analysis of a low-pass filter.

The program is listed below.

```
EXAMPLE 14.20
V1    1    0    AC   2   0
R     1    2    2K
C     2    0    5UF
.AC   DEC  10   1      1000
.PROBE
.END
```

The output plot is shown in Fig. 14.32b. ■

DRILL EXERCISE

D14.22. Given the filter network shown in Fig. D14.22, sketch the magnitude characteristic of the Bode plot for $\mathbf{G_V}(j\omega)$.

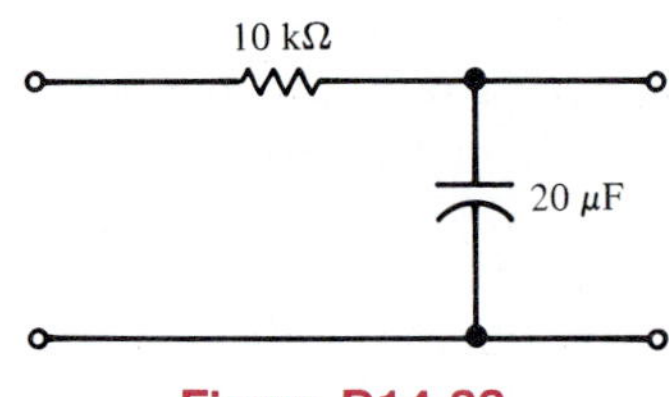

Figure D14.22

Ans:

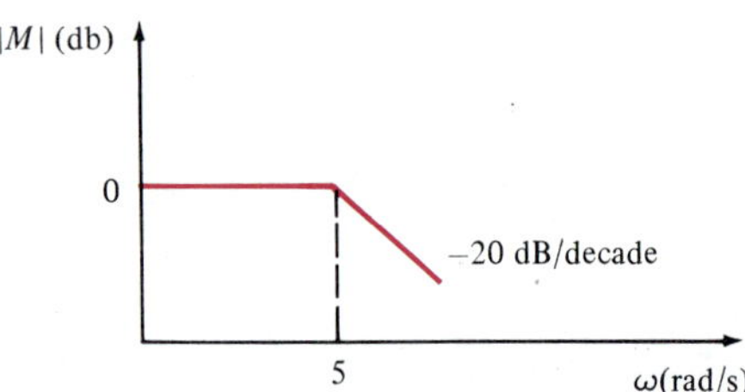

D14.23. Given the filter network in Fig. D14.23 sketch the magnitude characteristic of the Bode plot for $\mathbf{G_V}(j\omega)$.

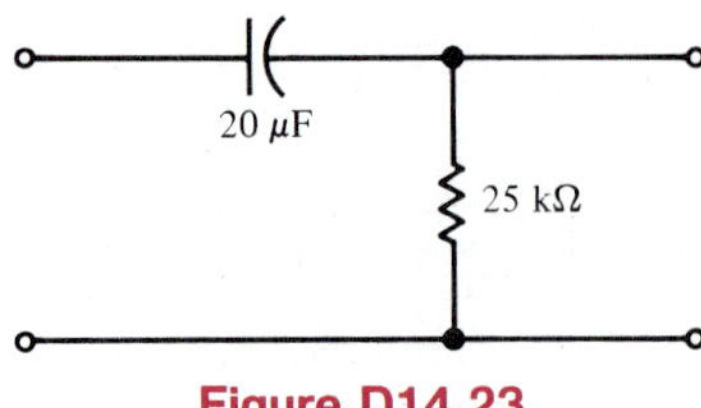

Figure D14.23

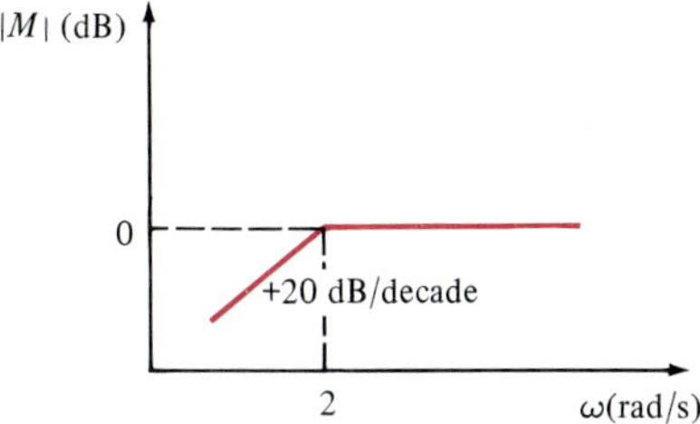

D14.24. A band-pass filter network is shown in Fig. D14.24. Sketch the magnitude characteristic of the Bode plot for $\mathbf{G_V}(j\omega)$.

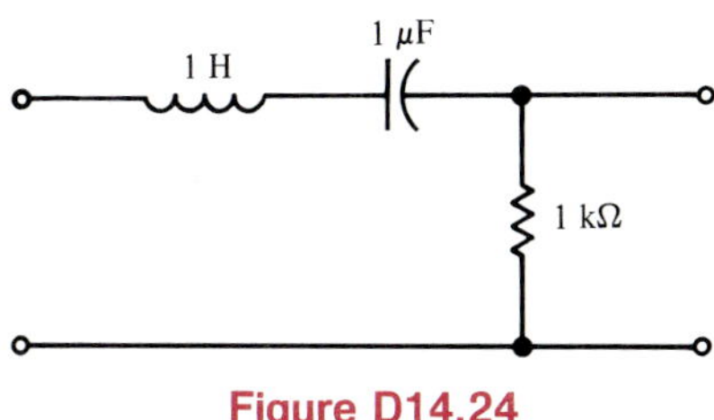

Figure D14.24

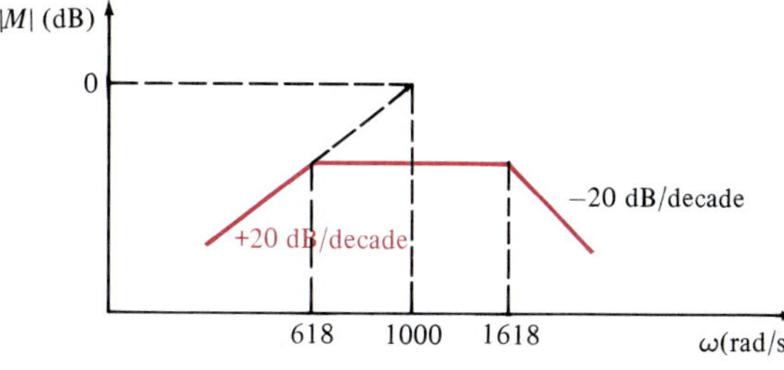

Active Filters

In the preceding section we saw that the four major classes of filters (*low-pass, high-pass, band-pass,* and *band-rejection*) are realizable with simple, passive element circuits. However, passive filters do have some serious drawbacks. One obvious problem is the inability to generate a network with a gain > 1 since a purely passive network cannot add energy to a signal. Another serious drawback of passive filters is the need in many topologies for inductive elements. Inductors are generally expensive and are not usually available in precise values. In addition, inductors usually come in odd shapes (toroids, bobbins, E-cores, etc.) and are not easily handled by existing automated printed circuit board assembly machines. By applying operational amplifiers in linear feedback circuits, one can generate all of the primary filter types using only resistors, capacitors, and the

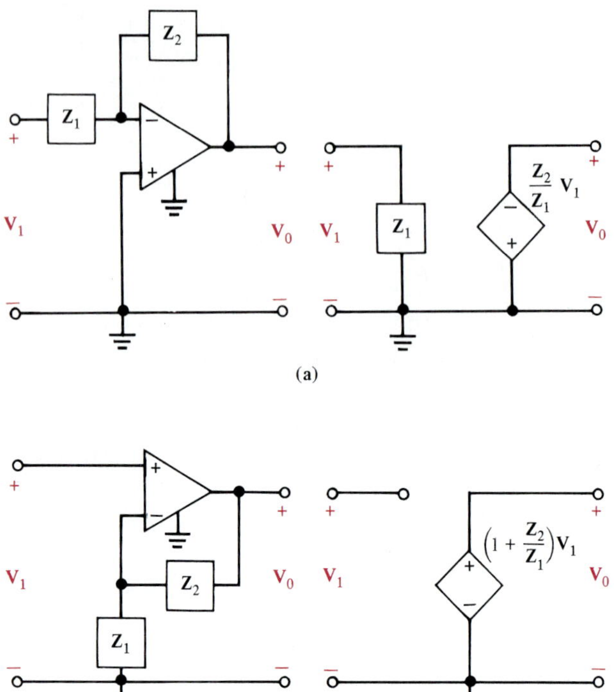

Figure 14.33 Equivalent circuits for the (a) inverting and (b) noninverting operational amplifier circuits.

op-amp integrated circuits themselves. The only exceptions to this are applications that involve very high frequency and/or high power.

The equivalent circuits for the operational amplifiers derived in Chapter 3 are valid in the sinusoidal steady-state case also, when we replace the attendant resistors with impedances. The equivalent circuits for the basic inverting and noninverting op-amp circuits are shown in Fig. 14.33a and b, respectively. Particular filter characteristics are obtained by judiciously selecting the impedances $\mathbf{Z}_1$ and $\mathbf{Z}_2$.

EXAMPLE 14.21

Let us determine the filter characteristics of the network shown in Fig. 14.34.

The impedances as illustrated in Fig. 14.33a are

$$\mathbf{Z}_1 = R_1$$

and

$$\mathbf{Z}_2 = \frac{R_2/j\omega C}{R_2 + 1/j\omega C} = \frac{R_2}{j\omega R_2 C + 1}$$

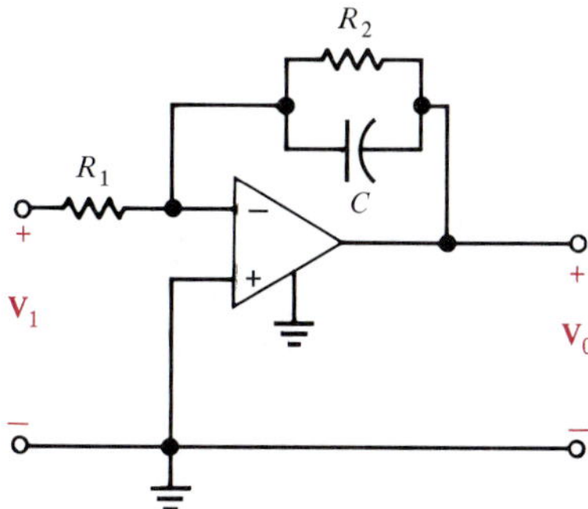

Figure 14.34 Operational amplifier filter circuit.

Therefore, the voltage gain of the network is

$$\mathbf{G_V}(j\omega) = \frac{\mathbf{V}_o(j\omega)}{\mathbf{V}_1(j\omega)} = \frac{-R_2/R_1}{j\omega R_2 C + 1}$$

Note that the transfer function is that of a low-pass filter. ■

DRILL EXERCISE

D14.25. Use PSPICE to generate the Bode plot for the network in Fig. 14.34 over the frequency range from 100 Hz to 10 kHz. The parameter values are $R_1 = 500\ \Omega$, $R_2 = 1\ \text{k}\Omega$, and $C_1 = 0.1\ \mu\text{F}$, and the op-amp parameters are $R_i = \infty$, $R_o = 0$, and $A = 10^5$. In addition, find the cutoff frequency.

Ans:

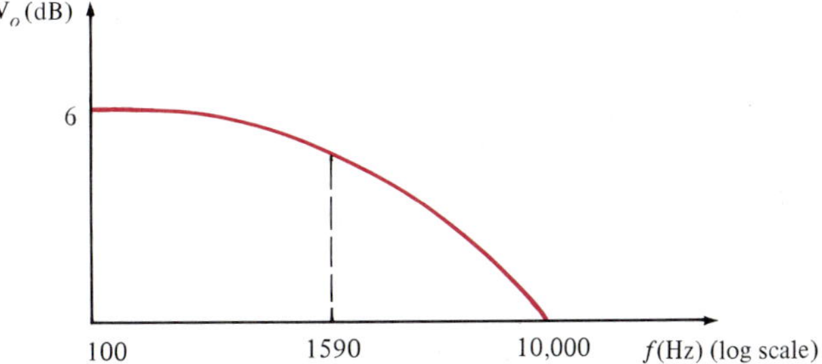

EXAMPLE 14.22

We will show that the amplitude characteristic for the filter network in Fig. 14.35a is as shown in Fig. 14.35b.

Comparing this network with that in Fig. 14.33b, we see that

$$\mathbf{Z}_1 = \frac{1}{j\omega C_1}$$

and

$$\mathbf{Z}_2 = \frac{R}{j\omega R C_2 + 1}$$

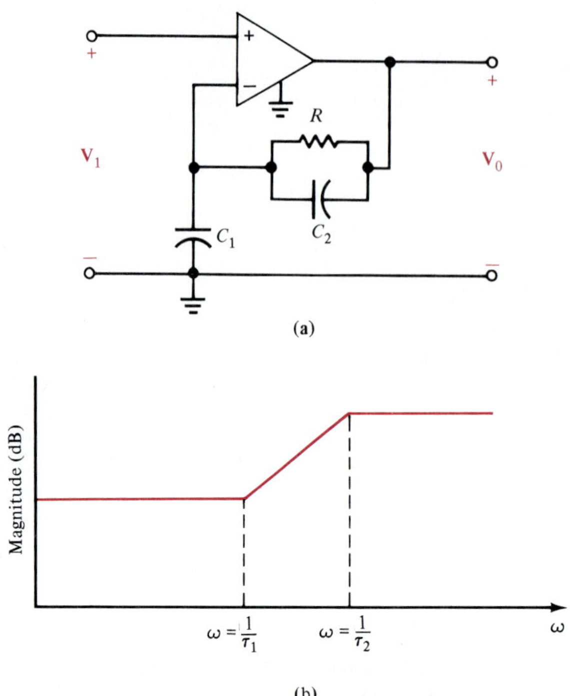

Figure 14.35 Operational amplifier circuit and its amplitude characteristic.

Therefore, the voltage gain for the network as a function of frequency is

$$\mathbf{G_V}(j\omega) = \frac{\mathbf{V}_o(j\omega)}{\mathbf{V}_1(j\omega)} = 1 + \frac{R/(j\omega RC_2 + 1)}{1/j\omega C_1}$$

$$= \frac{j\omega(RC_1 + RC_2) + 1}{j\omega RC_2 + 1}$$

$$= \frac{j\omega\tau_1 + 1}{j\omega\tau_2 + 1}$$

where $\tau_1 = R(C_1 + C_2)$ and $\tau_2 = RC_2$. Since $\tau_1 > \tau_2$, the amplitude characteristic is of the form shown in Fig. 14.35b. Note that the low frequencies have a gain of 1; however, the high frequencies are amplified. The exact amount of amplification is determined through selection of the circuit parameters. ■

DRILL EXERCISE

D14.26. Given the filter network shown in Fig. D14.26, determine the transfer function $\mathbf{G_V}(j\omega)$, sketch the magnitude characteristic of the Bode plot for $\mathbf{G_V}(j\omega)$, and identify the filter characteristics of the network.

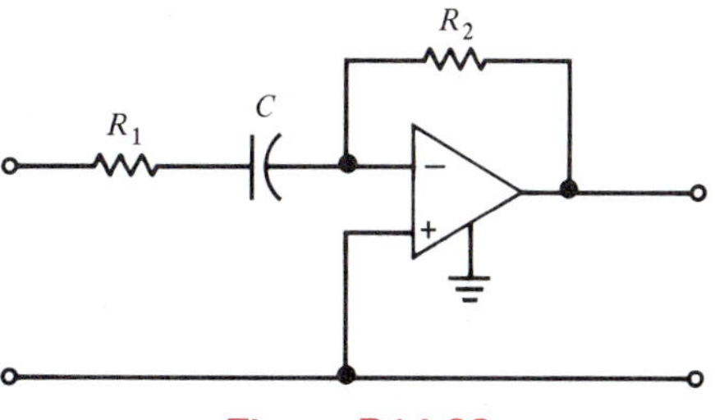

Figure D14.26

Ans: $\mathbf{G}_{\mathbf{V}}(j\omega) = \dfrac{-j\omega CR_2}{1 + j\omega CR_1}$; this is a high-pass filter.

$|M|$ dB

$20\log_{10} CR_2$

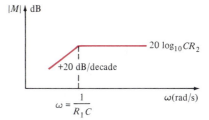

+20 dB/decade

$\omega = \dfrac{1}{R_1 C}$

ω(rad/s)

All the circuits considered so far in this section have been first-order filters. In other words, they all had no more than one pole and/or one zero. In many applications it is desired to generate a circuit with a frequency selectivity greater than that afforded by first-order circuits. The next logical step is to consider the class of second-order filters. For most active filter applications, if an order greater than two is desired, one usually takes two or more active filter circuits and places them in series so that the total response is the desired higher-order response. This is done because first- and second-order filters are well understood and easily obtained with single op-amp circuits.

In general, second-order filters will have a transfer function with a denominator containing quadratic poles of the form $s^2 + As + B$. For high-pass and low-pass circuits, $B = \omega_c^2$ and $A = 2\zeta\omega_c$. For these circuits, ω_c is the cutoff frequency, and ζ is the damping ratio discussed earlier. Using the VDB output operator with the ac analysis portion of PSPICE, one can easily find the cutoff frequency by finding the appropriate shift in the decibel response.

For band-pass circuits, $B = \omega_0^2$ and $A = \omega_0/Q$, where ω_0 is the center frequency and Q is the quality factor for the circuit. Notice that $Q = 1/2\zeta$. Q is a measure of the selectivity of these circuits. The bandwidth is ω_0/Q, as discussed previously.

The transfer function of the second-order low-pass active filter can generally be written as

$$\mathbf{H}(s) = \frac{H_0\omega_c^2}{s^2 + 2\zeta\omega_c s + \omega_c^2} \tag{14.65}$$

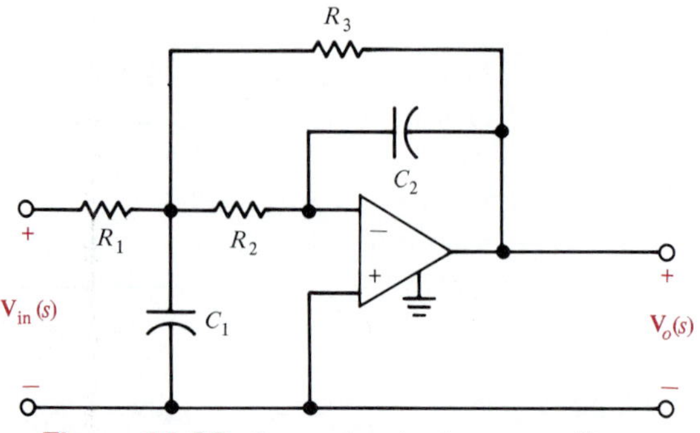

Figure 14.36 Second-order low-pass filter.

where H_0 is the dc gain. A circuit that exhibits this transfer function is illustrated in Fig. 14.36, and has the following transfer function:

$$\mathbf{H}(s) = \frac{\mathbf{V}_o(s)}{\mathbf{V}_{in}(s)} = \frac{-\left(\dfrac{R_3}{R_1}\right)\left(\dfrac{1}{R_3R_2C_1C_2}\right)}{s^2 + s\left(\dfrac{1}{R_1C_1} + \dfrac{1}{R_2C_1} + \dfrac{1}{R_3C_1}\right) + \dfrac{1}{R_3R_2C_1C_2}} \qquad (14.66)$$

EXAMPLE 14.23

We wish to determine the damping ratio, cutoff frequency, and dc gain H_o for the network in Fig. 14.36 if $R_1 = R_2 = R_3 = 5\ \text{k}\Omega$ and $C_1 = C_2 = 0.1\ \mu\text{F}$.

Comparing Eqs. (14.65) and (14.66), we find that

$$\omega_c = \frac{1}{\sqrt{R_3R_2C_1C_2}}$$

$$2\zeta\omega_c = \frac{1}{C_1}\left(\frac{1}{R_1} + \frac{1}{R_2} + \frac{1}{R_3}\right)$$

and therefore,

$$\zeta = \frac{1}{2}\sqrt{\frac{C_2}{C_1}}\left(\frac{1}{R_1} + \frac{1}{R_2} + \frac{1}{R_3}\right)\sqrt{R_2R_3}$$

In addition, we note that

$$H_o = -\frac{R_3}{R_1}$$

Substituting the given parameter values into the equation above yields

$$\omega_c = 2000\ \text{rad/sec} = 318\ \text{Hz}$$

$$\zeta = 1.5$$

and

$$H_o = -1$$

EXAMPLE 14.24

We wish to vary the capacitors C_1 and C_2 in Example 14.23 to achieve damping ratios of 1.0, 0.75, 0.50, and 0.25 while maintaining ω_c constant at 2000 rad/sec.

As shown in the cutoff-frequency equation in Example 14.23, if ω_c is to remain constant at 2000 rad/sec, the product of C_1 and C_2 must remain constant. Using the capacitance values in Example 14.23, we have

$$C_1C_2 = (10)^{-14}$$

or

$$C_2 = \frac{(10)^{-14}}{C_1}$$

Substituting this expression into the equation for the damping ratio yields

$$\zeta = \frac{\sqrt{10^{-14}}}{\sqrt{C_1}\,\sqrt{C_1}}\left[\frac{1}{2}\left(\frac{1}{R_1}+\frac{1}{R_2}+\frac{1}{R_3}\right)\right]\sqrt{R_2R_3}$$

$$= \frac{(0.15)(10^{-6})}{C_1}$$

or

$$C_1 = \frac{(0.15)(10^{-6})}{\zeta}$$

Therefore, for $\zeta = 1.0$, 0.75, 0.50, and 0.25, the corresponding values for C_1 are 0.15, 0.20, 0.30, and 0.6 μF, respectively. The values of C_2 that correspond to these values of C_1 are 67, 50, 33, and 17 nF, respectively. ■

This example illustrates that we can adjust the network parameters to achieve a specified transient response while maintaining the cutoff frequency of the filter constant. In fact, as a general rule we design filters with specific characteristics through proper manipulation of the network parameters.

EXAMPLE 14.25

We will now use PSPICE to demonstrate that the transient response of the circuits generated in Example 14.24 will exhibit increasing overshoot and ringing as ζ is decreased. We will apply a -1-V step function to the input of the network and employ the op-amp model with $R_i = \infty\ \Omega$, $R_o = 0\ \Omega$, and $A = 10^5$.

The following PSPICE program will produce the transient response for the four cases of damping. The values for C_1 and C_2 are given for a damping ratio of 0.5.

```
TRANSIENT RESPONSE FOR EXAMPLE 14.24
R1   1 2 5K
R2   2 3 5K
R3   2 4 5K
```

```
* Change these two values to set ζ
C1   2 0 0.3UF
C2   3 4 33NF
VIN 1 0 PWL(0 0 1MS 0 1.001MS -1V 50MS -1V)
*Op-Amp dependent generator A = 1E5
EOUT 4 0 3 0 -1E5
.TRAN 0.2MS 10MS
.PROBE
.END
```

The results are shown in Fig. 14.37. The curves indicate that a $\zeta = 0.75$ might be a good design compromise between rapid step response and minimum overshoot.

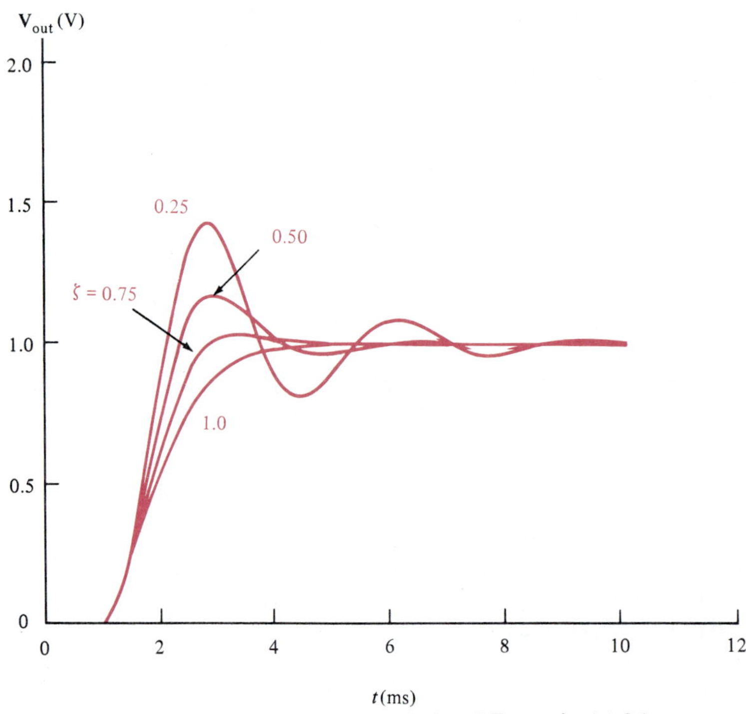

Figure 14.37 Transient analysis of Example 14.24.

DRILL EXERCISE

D14.27. Verify that Eq. 14.66 is the transfer function for the network in Fig. 14.36.

When a particular circuit or device is either routinely employed or replicated within a circuit, it is often convenient to define what PSPICE refers to as a *subcircuit*. The subcircuit, which defines a group of circuit elements, is automatically inserted wherever the subcircuit is referenced. Subcircuits may contain any number of elements, even other subcircuits. The general form for the subcircuit statements is

```
.SUBCKT SUBNAM N1 < N2 N3 . . . >
      .
      .
      .
      .
      .
      .
.ENDS <SUBNAM>
```

where SUBNAM is the subcircuit name and N_1, N_2, N_3, ··· are the nodes used to connect the subcircuit into the network. The .ENDS <SUBNAM> statement is the last statement in the subcircuit, and <SUBNAM> is needed only when subcircuits are nested. The elements that define the subcircuit are listed between the .SUBCKT and .ENDS statements.

Subcircuits are called with the statement

```
XYYYYYYY N1 < N2 N3 . . . > SUBNAM
```

where N_1, N_2, N_3, . . . define the nodes in the network to which the subcircuit is connected. Note that nodes referenced in the .SUBCKT statement are not necessarily numbered the same as those defined in the subcircuit call. In contrast to the *local* nodes used to define the subcircuit, the nodes used in the subcircuit call are *global* except in cases where subcircuits are nested. The reference node is always global and need not be defined in the .SUBCKT statement.

EXAMPLE 14.26

A low-pass filter is shown in Fig. 14.38a. Let us use PSPICE to plot the frequency response of the filter over the range 1 to 10,000 Hz using 10 points per decade.

We will use a subcircuit to define the op-amp using the model shown in Fig. 14.38b. The local nodes used for the subcircuit are also shown in Fig. 14.38b. The global nodes used for the original network are shown in Fig. 14.38a. The PSPICE program for plotting the frequency response is listed below.

```
EXAMPLE 14.26
VS   1   0   AC   12 0
R1   1   2   100
R2   2   3   100
R3   5   0   5K
R4   4   5   50K
C1   2   3   10U
C2   3   0   10U
X1   3   5   4   OPAMP
```

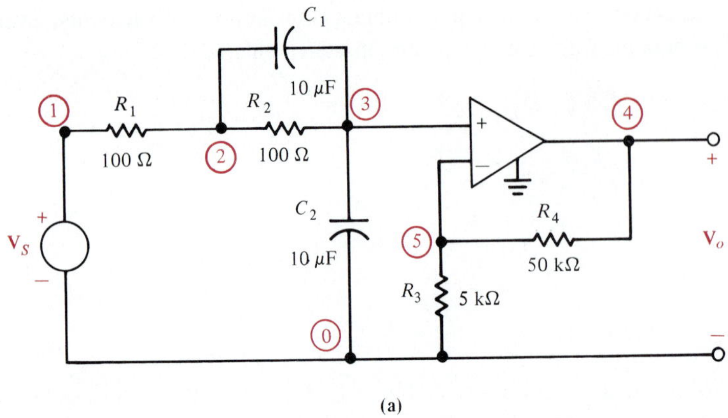

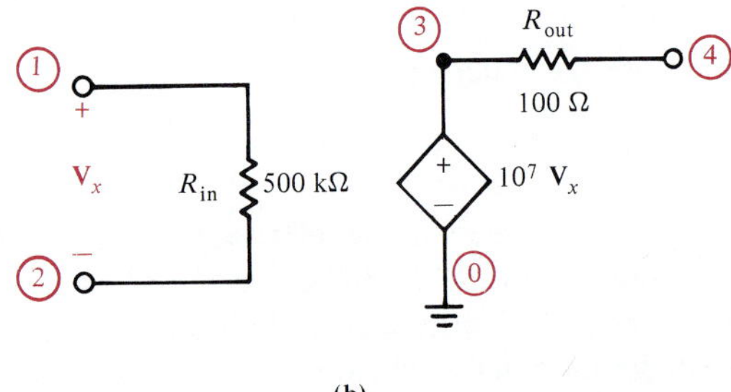

Figure 14.38 Circuits used in Example 14.26: (a) low-pass filter; (b) op-amp subcircuit.

```
*ANALYSIS
.AC DEC 10 1 10K
.PROBE
* OPAMP SUBCIRCUIT
.SUBCKT OPAMP 1 2 4
RIN    1 2 500K
ROUT   3 4 100
EOUT   3 0 1 2 1E7
.ENDS
*
.END
```

The frequency response plot is shown in Fig. 14.39. ■

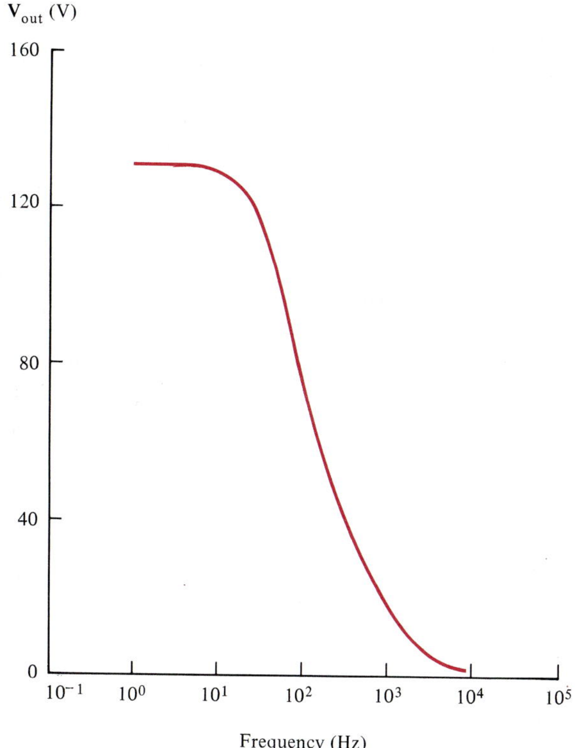

Figure 14.39 Frequency response plot for the network in Example 14.26.

EXAMPLE 14.27

A high-pass filter network is shown in Fig. 14.40a. Let us use PSPICE to plot the frequency response of the filter over the range from 1 to 100 kHz using 10 points per decade.

The subcircuit for the op-amp is shown in Fig. 14.40b, and the PSPICE program for plotting the frequency response is shown below.

```
EXAMPLE 14.27
VS   1   0   AC    12   0
R1   1   2   100
R2   2   3   100
R3   2   5   100
R4   4   5   100
C1   2   3   50U
C2   3   4   100U
X1   2   3   OPAMP
X2   4   5   OPAMP
*ANALYSIS
.AC DEC 10 1 100K
.PROBE
```

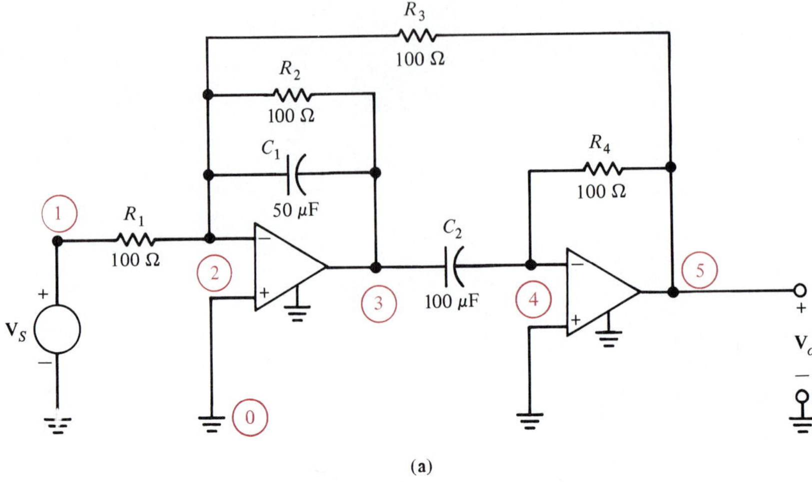

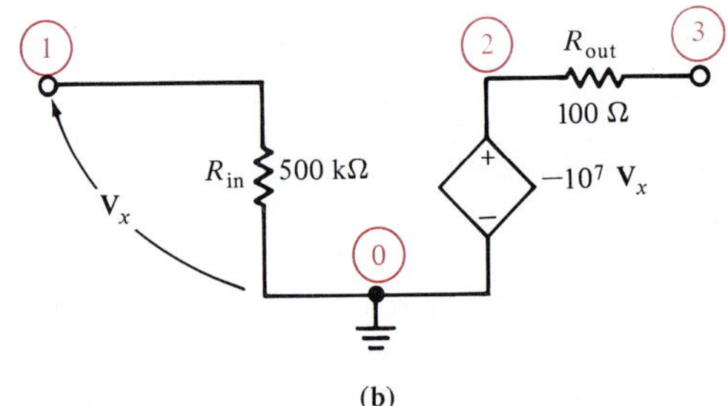

Figure 14.40 Circuits used in Example 14.27: (a) high-pass filter; (b) op-amp subcircuit.

```
*OPAMP
.SUBCKT OPAMP 1 3
RIN  1 0 500K
ROUT 2 3 100
EOUT 2 0 1 0-1E7
.ENDS
*
.END
```

The frequency response plot is shown in Fig. 14.41. ■

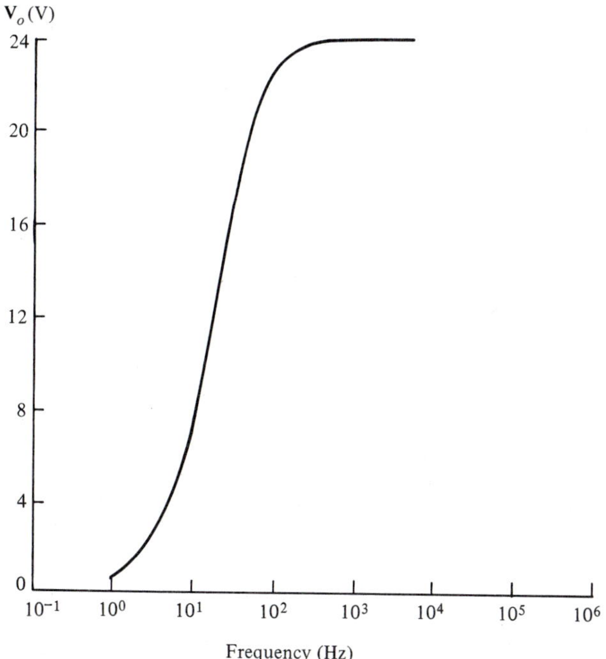

Figure 14.41 Frequency response plot for the network in Example 14.27.

The general transfer function for the second-order band-pass filter is

$$\frac{\mathbf{V}_o(s)}{\mathbf{V}_S(s)} = \frac{sH_o}{s^2 + \dfrac{\omega_o}{Q}s + \omega_0^2} \tag{14.67}$$

As discussed earlier, ω_0 is the center frequency of the band-pass characteristic and Q is the quality factor. Recall that for low-pass filters, $\mathbf{H}_o$ was the passband or dc gain. For a band-pass filter, the gain is maximum at the center frequency, ω_0. To find this maximum gain we substitute $s = j\omega_0$ in the expression above to obtain

$$\begin{aligned}\frac{\mathbf{V}_o(j\omega_0)}{\mathbf{V}_S(j\omega_0)} &= \frac{j\omega_0 H_o}{-\omega_0^2 + j\omega_0(\omega_0 0/Q) + \omega_0^2} \\ &= \frac{QH_o}{\omega_0}\end{aligned} \tag{14.68}$$

In addition, the difference between the high and low half-power frequencies (i.e., $\omega_{HI} - \omega_{LO}$) is, of course, the bandwidth

$$\omega_{HI} - \omega_{LO} = \text{BW} = \frac{\omega_0}{Q} \tag{14.69}$$

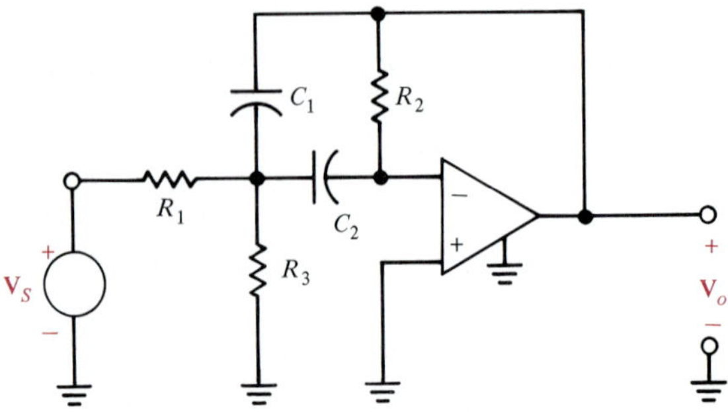

Figure 14.42 Second-order band-pass filter.

Q is a measure of the selectivity of the band-pass filter, and as the equation indicates, as Q is increased, the bandwith is decreased.

An op-amp implementation of a band-pass filter is shown in Fig. 14.42. The transfer function for this network is

$$\frac{\mathbf{V}_o(s)}{\mathbf{V}_S(s)} = \frac{-\left(\dfrac{1}{R_1C_1}\right)s}{s^2 + \left(\dfrac{1}{R_2C_1} + \dfrac{1}{R_2C_2}\right)s + \dfrac{1 + R_1/R_3}{R_1R_2C_1C_2}} \tag{14.70}$$

Comparing this expression to the more general expression for the band-pass filter yields the following definitions:

$$\omega_0 = \left(\frac{1 + R_1/R_3}{R_1R_2C_1C_2}\right)^{1/2}$$

$$\frac{Q}{\omega_0} = \frac{R_2C_1C_2}{C_1 + C_2} \tag{14.71}$$

$$\frac{Q}{\omega_0}\omega_0 = \frac{R_2C_1C_2}{C_1 + C_2}\left(\frac{1 + R_1/R_3}{R_1R_2C_1C_2}\right)^{1/2}$$

These expressions can be simplifed to yield

$$Q = \frac{(1 + R_1/R_3)^{1/2}}{1 + C_1/C_2}\left(\frac{R_2C_1}{R_1C_2}\right)^{1/2} \tag{14.72}$$

$$\text{BW} = \frac{\omega_0}{Q} = \frac{1}{R_2}\left(\frac{1}{C_1} + \frac{1}{C_2}\right) \tag{14.73}$$

and

$$\left.\frac{\mathbf{V}_o}{\mathbf{V}_S}\right|_{\omega=\omega_0} = \frac{QH_o}{\omega_0} = -\frac{R_2}{R_1}\left(\frac{1}{1 + C_1/C_2}\right) \tag{14.74}$$

EXAMPLE 14.28

We wish to find a new expression for Eqs. (14.71) to (14.74) under the condition that $C_1 = C_2 = C$.

Using the condition the equations reduce to

$$\omega_0 = \frac{1}{C}\sqrt{\frac{1 + R_1/R_3}{R_1 R_2}}$$

$$Q = \frac{1}{2}\sqrt{\frac{R_2}{R_1}}\sqrt{1 + \frac{R_1}{R_3}}$$

$$\text{BW} = \frac{2}{R_2 C}$$

and

$$\left.\frac{\mathbf{V}_o}{\mathbf{V}_S}\right|_{\omega=\omega_0} = -\frac{R_2}{2R_1}$$

■

EXAMPLE 14.29

Let us use the equations in Example 14.28 to design a band-pass filter of the form shown in Fig. 14.42 with a BW = 2000 rad/s, $(\mathbf{V}_0/\mathbf{V}_S)(\omega_0) = -5$, and $Q = 3$. Use $C = 0.1\ \mu\text{F}$, and determine the center frequency of the filter.

Using the filter equations, we find that

$$\text{BW} = \frac{2}{R_2 C}$$

$$2000 = \frac{2}{R_2(10)^{-7}}$$

$$R_2 = 10\ \text{k}\Omega$$

$$\frac{\mathbf{V}_o}{\mathbf{V}_S}(\omega_0) = -\frac{R_2}{2R_1}$$

$$-5 = -\frac{10{,}000}{2R_1}$$

$$R_1 = 1\ \text{k}\Omega$$

and

$$Q = \frac{1}{2}\sqrt{\frac{R_2}{R_1}}\sqrt{1 + \frac{R_1}{R_3}}$$

$$3 = \frac{1}{2}\sqrt{\frac{10{,}000}{1000}}\sqrt{1 + \frac{1000}{R_3}}$$

or

$$R_3 = 385\ \Omega$$

Therefore, $R_1 = 1\ \text{k}\Omega$, $R_2 = 10\ \text{k}\Omega$, $R_3 = 385\ \Omega$, and $C = 0.1\ \mu\text{F}$ completely define the band-pass filter shown in Fig. 14.42. The center frequency of the filter is

$$\omega_0 = \frac{1}{C}\sqrt{\frac{1 + R_1/R_3}{R_1 R_2}}$$

$$= \frac{1}{10^{-7}}\sqrt{\frac{1 + (1000/385)}{(1000)(10{,}000)}}$$

$$= 6000 \text{ rad/sec}$$

EXAMPLE 14.30

We wish to use PSPICE to plot the Bode plot for the filter designed in Example 14.29. We will employ the op-amp model, in which $R_i = \infty$, $R_o = 0$, and $A = 10^5$, and plot over the frequency range from 600 to 60 kHz using 30 points per decade.

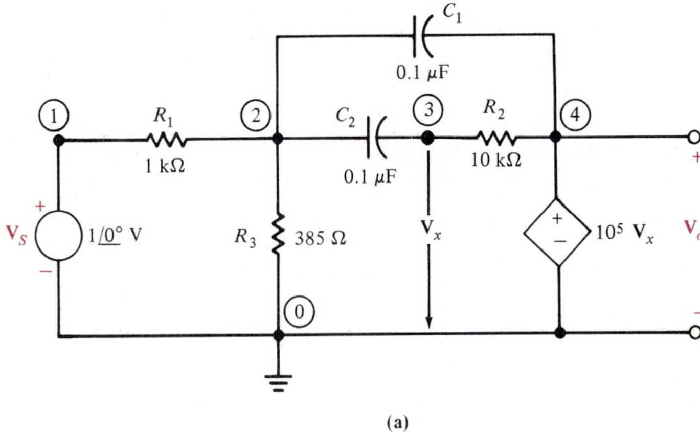

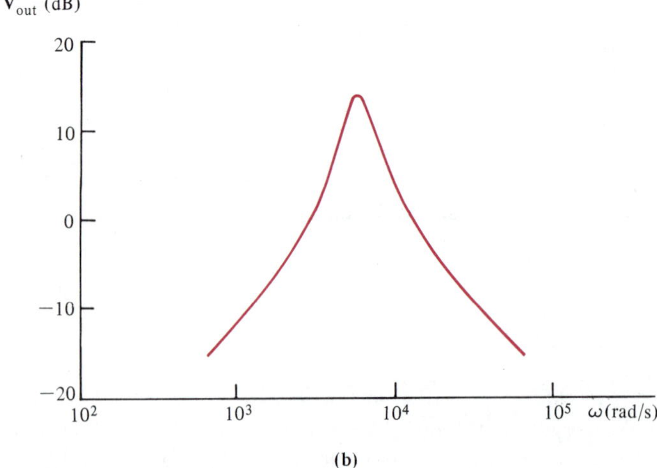

Figure 14.43 Figures employed in Example 14.30: (a) band-pass filter equivalent circuits; (b) Bode plot.

The equivalent circuit for the filter is shown in Fig. 14.43a. The PSPICE program is listed below.

```
SECOND-ORDER BAND-PASS FILTER EXAMPLE
VIN 1 0 AC 1 0
R1 1 2 1K
C1 2 4 0.1UF
C2 2 3 0.1UF
R2 3 4 10K
R3 2 0 385
E1 4 0 0 3 1E5
.AC DEC 30 600 60K
.PROBE
.END
```

The Bode plot is shown in Fig. 14.43b. As can be seen from the plot, the center frequency is 6 krad/sec and BW = 2 krad/sec. ■

DRILL EXERCISE

D14.28. Verify that Eq. 14.70 is the transfer function for the band-pass filter in Fig. 14.42.

It is instructive to examine briefly the frequency response limitations of an op-amp. Thus far, our modeling of the op-amp has been limited to finite gain, input resistance, and output resistance. The astute reader may have noticed that this model has infinite frequency response (i.e., there are no poles or zeros in the forward gain, A). As developing engineers, we should be skeptical when an infinite response is associated with a real, physical system. Of course, real operational amplifiers do have a finite frequency response, and their forward gain is more realistically represented as

$$A(j\omega) = \frac{A_o[1 + j(f/f_{z_1})][1 + j(f/f_{z_2})] \cdots}{[1 + j(f/f_{p_1})][1 + j(f/f_{p_2})] \cdots}$$

where $1 + j(f/f_{p_1}) = 1 + j\omega\tau_{p_1}$, where $\omega = 2\pi f$ and $\tau_{p_1} = 1/2\pi f_{p_1}$.

There are, however, a large class of op-amps that are specifically designed to have a single-pole forward gain response, which ensures stable and reliable operation in a wide range of applications. The single-pole gain of the amplifier can be written as

$$A = \frac{A_o}{1 + j(f/f_p)}$$

Typical values for f_p are in the range from 5 to 30 Hz. These data prompt us to ask if such a low-frequency pole does not severely limit the usefulness of the amplifier. The following example will illustrate that the frequency response is actually limited by a factor known as the gain–bandwidth product, GBW, which for a single-pole op-amp is GBW = $A_o f_p$.

EXAMPLE 14.31

Consider the unity-gain op-amp buffer circuit shown in Fig. 14.44a. The equivalent circuit for the amplifier is shown in Fig. 14.44b, where we have assured that $R_i = \infty$, $R_o = 0$, and $\mathbf{A} = A_o/[1 + j(f/f_p)]$. KVL yields

$$\mathbf{V}_e = \mathbf{V}_S - \mathbf{A}\mathbf{V}_e$$

or

$$(\mathbf{A} + 1)\mathbf{V}_e = \mathbf{V}_S$$

However,

$$\mathbf{V}_e = \mathbf{V}_S - \mathbf{V}_o$$

and therefore,

$$(\mathbf{A} + 1)\mathbf{V}_o = \mathbf{A}\mathbf{V}_S$$

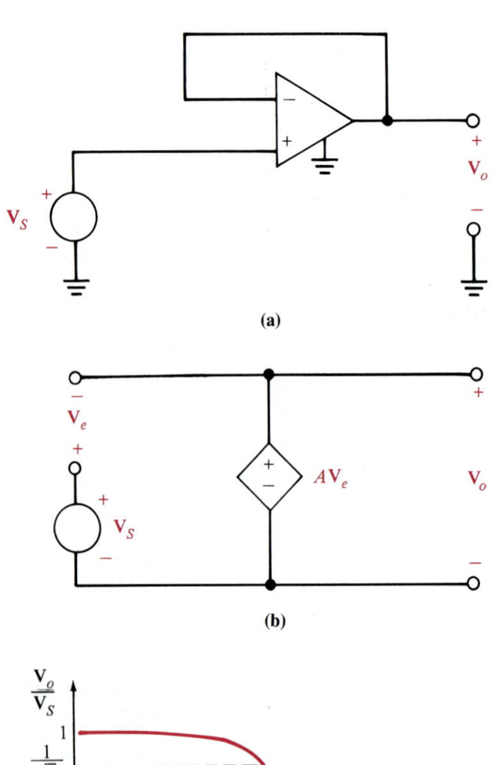

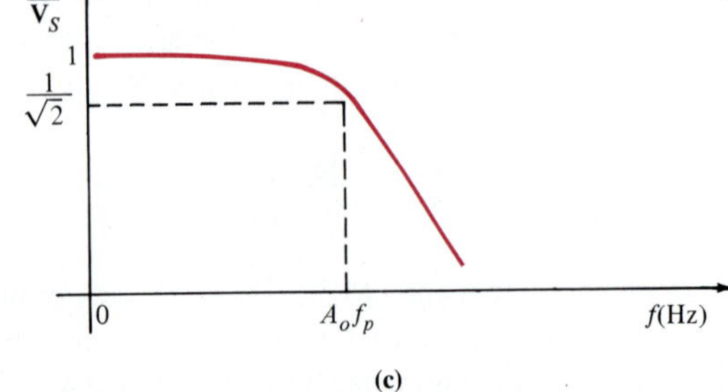

Figure 14.44 Figures used in Example 14.31: (a) buffer amplifier; (b) buffer amplifier equivalent circuit; (c) unity-gain buffer amplifier frequency response.

and

$$\frac{\mathbf{V}_o}{\mathbf{V}_S} = \frac{\mathbf{A}}{1 + \mathbf{A}}$$

Substituting for **A** yields

$$\frac{\mathbf{V}_o}{\mathbf{V}_S} = \frac{A_o/[1 + j(f/f_p)]}{1 + A_o/[1 + j(f/f_p)]}$$

$$= \frac{A_o}{1 + j(f/f_p) + A_o}$$

Since f_p is small, we assume that $f/f_p \gg 1$ and therefore,

$$\frac{\mathbf{V}_o}{\mathbf{V}_S} \simeq \frac{A_o}{A_o + j(f/f_p)}$$

$$\simeq \frac{1}{1 + j(f/f_p A_o)}$$

Thus at low frequencies $\mathbf{V}_o/\mathbf{V}_S = 1$ and at very high frequencies ($f \gg A_o f_p$), the gain is $\mathbf{V}_o/\mathbf{V}_S \equiv 1/j(f/f_p A_o) = -j(A_o f_p/f)$. Therefore, the op-amp looks like a low-pass filter with a cutoff frequency equal to $f_c = A_o f_p$ (i.e., $\omega_c = 2\pi A_o f_p$), as shown in Fig. 14.44c. ■

The equivalent circuit for a single-pole op-amp is shown in Fig. 14.45. The low-frequency pole is set at $f_p = 1/2\pi RC$. For example, if $R = 100$ kΩ and $C = 159$ nF, the op-amp would have a pole at 10 Hz. In the equivalent network the first dependent source models the dc gain. The second dependent source buffers the output, so that the pole frequency will not be affected by the output load impedance.

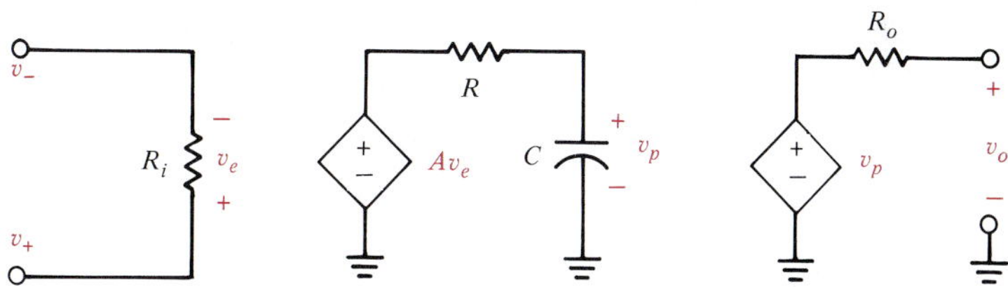

Figure 14.45 Single-pole op-amp equivalent circuit.

EXAMPLE 14.32

Given the high-pass filter network shown in Fig. D14.26 with $R_1 = 1$ kΩ, $R_2 = 10$ kΩ, and $C = 1.59$ nF, we wish to determine the cutoff frequency f_c, and use PSPICE to generate the Bode plot for both the simple and single-pole op-amp models. We will use $A = (2)10^5$, $R_i = \infty$, and $R_o = 0$ for the simple op-amp model. For the single-pole model we will use $f_p = 5$ Hz.

The transfer function for the filter is

$$\frac{\mathbf{V}_o}{\mathbf{V}_S}(s) = \frac{R_2 Cs}{R_1 Cs + 1}$$

Therefore,

$$\omega_c = \frac{1}{R_1 C}$$

or

$$f_c = \frac{1}{2\pi R_1 C}$$

$$= \frac{1}{2\pi(10)^3(1.59)(10)^{-9}}$$

$$= 100 \text{ kHz}$$

The PSPICE equivalent circuit for the filter using the simple op-amp model is shown in Fig. 14.46a. The PSPICE program, which produces a Bode plot over the frequency range from 1 kHz to 1 MHz using 20 points per decade, is listed below.

```
HIGH-PASS FILTER WITH SIMPLE OP-AMP MODEL
VIN 1 0 AC 1 0
R1 1 2 1K
C1 2 3 1.59NF
R2 3 4 10K
E1 4 0 0 3 2E5
.AC DEC 20 1K 1000K
.PROBE
.END
```

The PSPICE equivalent circuit for the filter with the single-pole op-amp model is shown in Fig. 14.46b, where we have used $R_{p_1} = 100$ kΩ and $C_{p_1} = 318.3$ nF to satisfy the condition $f_p = 5$ Hz. The PSPICE program for generating the Bode plot for this network is listed below.

```
HIGH-PASS FILTER WITH SINGLE-POLE OP-AMP MODEL
VIN 1 0 AC 1 0
R1 1 2 1K
C1 2 3 1.59NF
R2 3 4 10K
E1 5 0 0 3 2E5
RP1 5 6 100K
CP1 6 0 318.3NF
E2 4 0 6 0 1
.AC DEC 20 1K 1000K
.PROBE
.END
```

The Bode plots for the filter with both op-amp models are shown in Fig. 14.47. Note that the op-amp with a single internal pole rolls off the high-frequency characteristic, resulting in a composite frequency response similar to that of a band-pass filter. ■

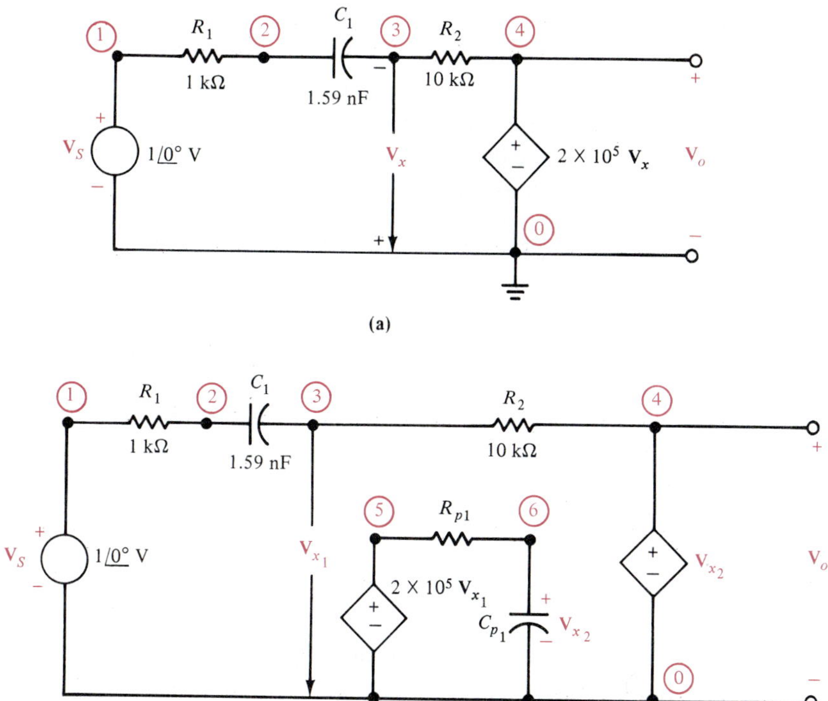

Figure 14.46 Circuits used in Example 14.32: (a) high-pass filter with simple op-amp model; (b) high-pass filter with single-pole op-amp model.

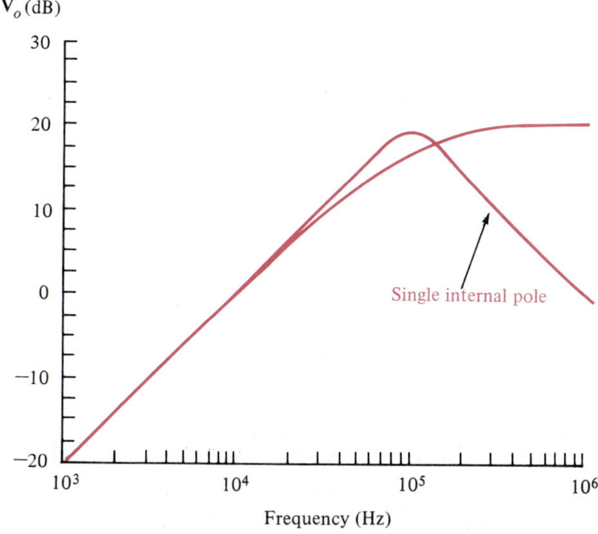

Figure 14.47 Bode plots for a high-pass filter using different op-amp models.

14.7 Summary

In this chapter we have examined the frequency characteristics of networks. The network function has been introduced and the frequency response of a network has been analyzed using both the pole–zero and Bode plots. These plots are very important because they indicate the behavior of the network as a function of frequency.

Resonant circuits have been presented and discussed. At resonance the input voltage and current to a network are in phase and the power factor is unity. The bandwidth, half-power frequencies, and quality factor for resonant circuits have been introduced and network scaling has been presented as an aid to network design with practical circuit element values.

Passive and active filters were introduced. Low-pass, high-pass, band-pass, and band-rejection filters were examined. Procedures for the analysis and design of active filters were presented and discussed.

Finally, PSPICE was used to analyze filters and plot their frequency and transient responses.

KEY POINTS

- There are four common network or transfer functions: $\mathbf{Z}(j\omega)$ is the ratio of the output voltage to the input current, $\mathbf{Y}(j\omega)$ is the ratio of the output current to the input voltage, $\mathbf{G_V}(j\omega)$ is the ratio of the output voltage to the input voltage, and $\mathbf{G_I}(j\omega)$ is the ratio of the output current to the input current.
- Values of the general frequency variable "s" that cause the magnitude of a network function to become zero are called "zeros" of that function; likewise, "s" values that drive the magnitude to infinity are called "poles."
- Bode plots are semilog plots of the magnitude and phase of transfer functions as a function of frequency.
- Straight-line approximations can be used to sketch quickly the magnitude characteristics of a Bode plot. The error between the actual curve and the straight-line approximation may be calculated, if necessary.
- Resonant frequency is defined as the frequency at which the impedance of a series *RLC* circuit or the admittance of a parallel *RLC* circuit is purely real.
- The quality factor, Q, is a measure of the sharpness of the resonance peak. The higher the Q, the sharper the peak.
- The break, cutoff, or half-power frequencies are the frequencies at which the magnitude characteristic of a filter is $1/\sqrt{2}$ of its maximum value.
- Unrealistic parameter values of passive circuit elements can be both magnitude and frequency scaled to produce realistic circuit element values.
- The four most common types of filter networks are low-pass, high-pass, band-pass, and band-rejection.
- The bandwidth of a band-pass or band-rejection filter is defined as the difference in frequency between the half-power points.
- The design of certain filters, for example, low-pass, high-pass, and so on, with specific characteristics can be quickly accomplished from known transfer functions.

PROBLEMS

14.1. Find the transfer impedance $\mathbf{V}_o(s)/\mathbf{I}_S(s)$ for the network shown in Fig. P14.1.

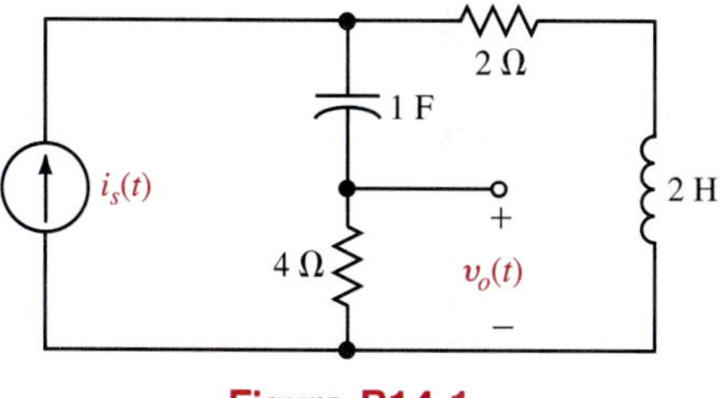

Figure P14.1

14.2. Determine the voltage transfer function $\mathbf{V}_o(s)/\mathbf{V}_i(s)$ for the network shown in Fig. P14.2.

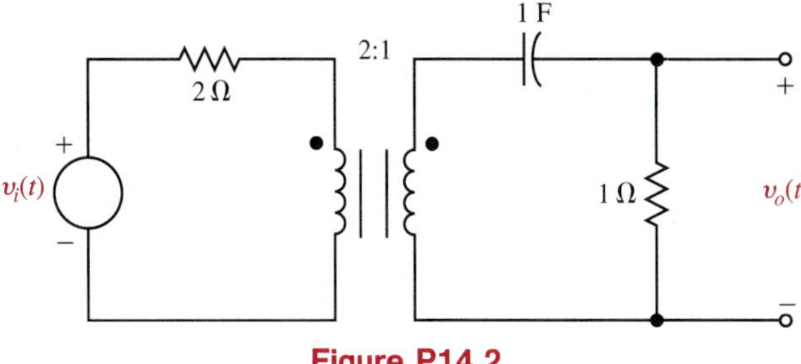

Figure P14.2

14.3. Find the driving point impedance as a function of s for the network shown in Fig. P14.3.

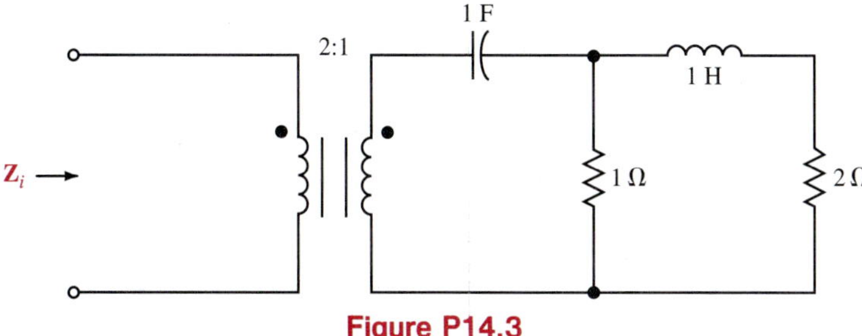

Figure P14.3

14.4. Find the current gain $\mathbf{I}_o(s)/\mathbf{I}_S(s)$ for the circuit shown in Fig. P14.4.

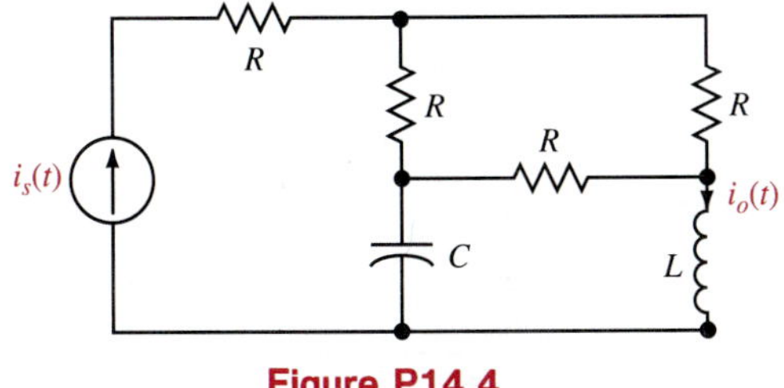

Figure P14.4

14.5. Find the driving point impedance at the input terminals of the circuit in Fig. P14.5.

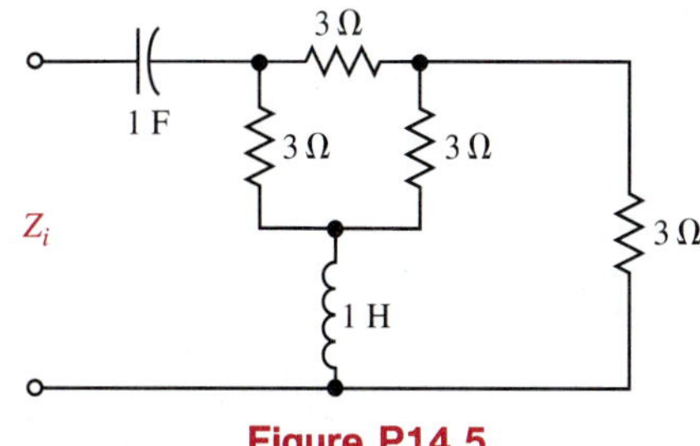

Figure P14.5

14.6. Draw the Bode plot for the network function

$$\mathbf{H}(j\omega) = \frac{j\omega 2 + 1}{j\omega 10 + 1}$$

14.7. Draw the Bode plot for the network function

$$\mathbf{H}(j\omega) = \frac{10(10j\omega + 1)}{(100j\omega + 1)(j\omega + 1)}$$

14.8. Draw the Bode plot for the network function

$$\mathbf{H}(j\omega) = \frac{j\omega}{(j\omega + 1)(0.1j\omega + 1)}$$

14.9. Sketch the magnitude characteristic of the Bode plot for the transfer function

$$\mathbf{H}(j\omega) = \frac{10}{j\omega(0.1j\omega + 1)}$$

14.10. Sketch the magnitude characteristic of the Bode plot for the transfer function

$$\mathbf{H}(j\omega) = \frac{20(j\omega + 1)}{j\omega(0.1j\omega + 1)(0.01j\omega + 1)}$$

14.11. Sketch the magnitude characteristic of the Bode plot for the transfer function

$$\mathbf{H}(j\omega) = \frac{640(j\omega + 1)(0.02j\omega + 1)}{(j\omega)^2(j\omega + 10)}$$

14.12. Sketch the magnitude characteristic of the Bode plot for the transfer function

$$\mathbf{G}(j\omega) = \frac{400(j\omega + 2)(j\omega + 50)}{-\omega^2(j\omega + 200)^2}$$

14.13. Sketch the magnitude characteristic of the Bode plot for the transfer function

$$\mathbf{H}(j\omega) = \frac{100(j\omega)}{(j\omega + 1)(j\omega + 10)(j\omega + 50)}$$

14.14. Sketch the magnitude characteristic of the Bode plot for the transfer function

$$\mathbf{H}(j\omega) = \frac{10^5(5j\omega + 1)^2}{(j\omega)^2(j\omega + 10)(j\omega + 50)^2}$$

14.15. Sketch the magnitude characteristic of the Bode plot for the transfer function

$$\mathbf{H}(j\omega) = \frac{10^2(j\omega)^2}{(j\omega + 1)(j\omega + 10)^2(j\omega + 100)}$$

14.16. Sketch the magnitude characteristic of the Bode plot for the transfer function

$$\mathbf{G}(j\omega) = \frac{-\omega^2}{(j\omega + 1)^3}$$

14.17. Sketch the magnitude characteristic of the Bode plot for the transfer function

$$\mathbf{G}(j\omega) = \frac{-\omega^2 10^4}{(j\omega + 1)^2(j\omega + 10)(j\omega + 100)^2}$$

14.18. Sketch the magnitude characteristic of the Bode plot for the transfer function

$$\mathbf{H}(j\omega) = \frac{81(j\omega + 0.1)}{(j\omega)(-\omega^2 + 3.6j\omega + 81)}$$

14.19. Sketch the magnitude characteristic of the Bode plot for the transfer function

$$\mathbf{H}(j\omega) = \frac{6.4(j\omega)}{(j\omega + 1)(-\omega^2 + 8j\omega + 64)}$$

14.20. Find $\mathbf{H}(j\omega)$ if its magnitude characteristic is shown in Fig. P14.20.

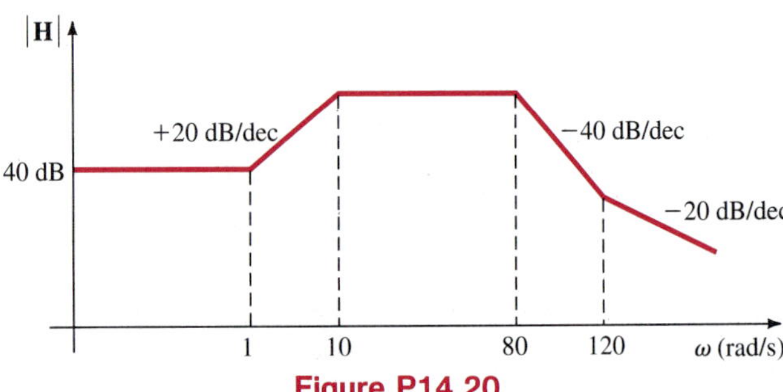

Figure P14.20

14.21. Find $\mathbf{H}(j\omega)$ if its amplitude characteristic is shown in Fig. P14.21.

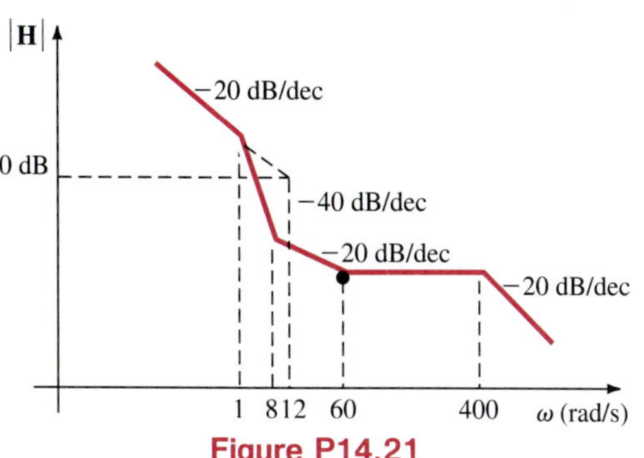

Figure P14.21

14.22. Find $\mathbf{H}(j\omega)$ if its amplitude characteristic is shown in Fig. P14.22.

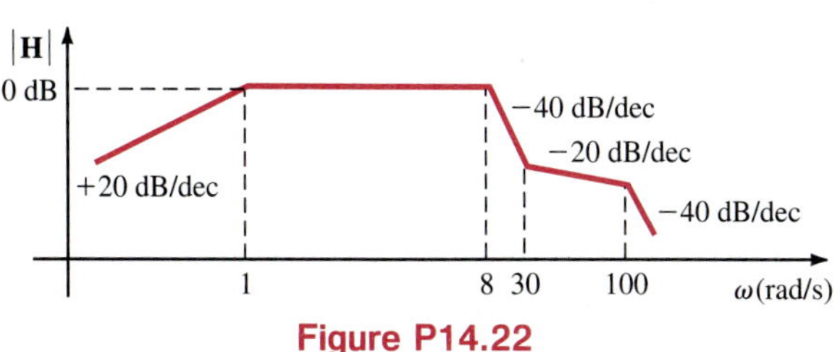

Figure P14.22

14.23. Find $\mathbf{H}(j\omega)$ if its amplitude characteristic is shown in Fig. P14.23.

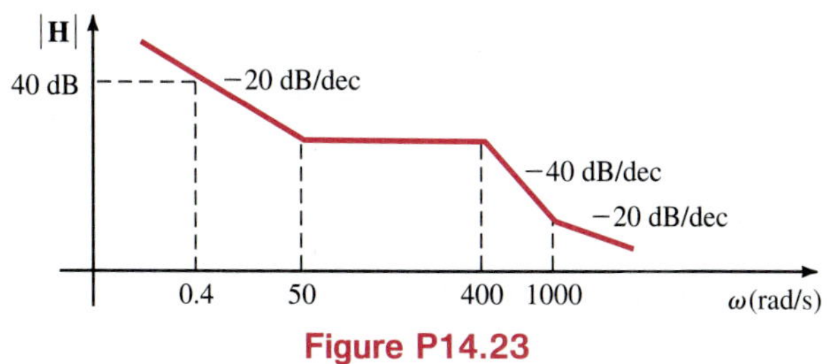

Figure P14.23

14.24. Find $\mathbf{H}(j\omega)$ if its magnitude characteristic is shown in Fig. P14.24.

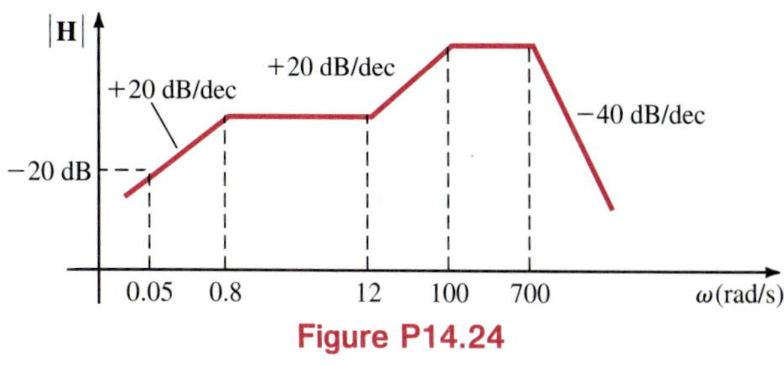

Figure P14.24

14.25. Given the magnitude characteristic in Fig. P14.25, find $\mathbf{H}(j\omega)$.

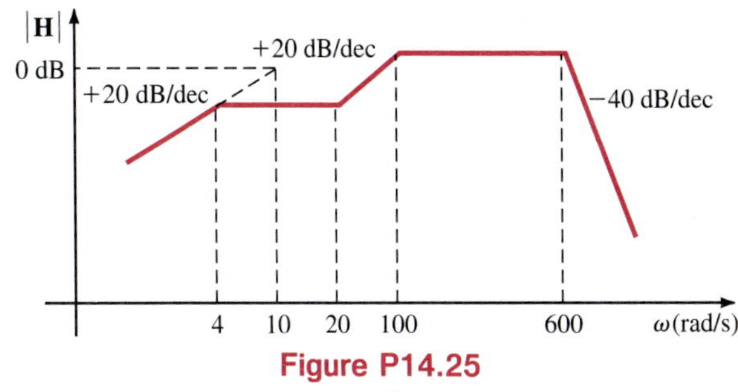

Figure P14.25

14.26. Given the magnitude characteristic in Fig. P14.26, find $\mathbf{G}(j\omega)$.

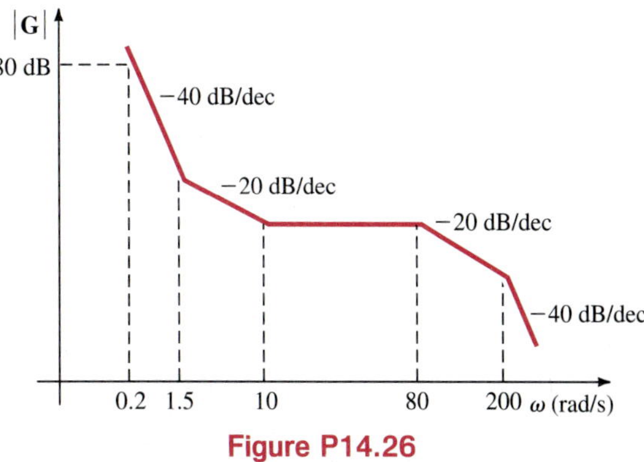

Figure P14.26

14.27. Find $\mathbf{G}(j\omega)$ if the amplitude characteristic for this function is shown in Fig. P14.27.

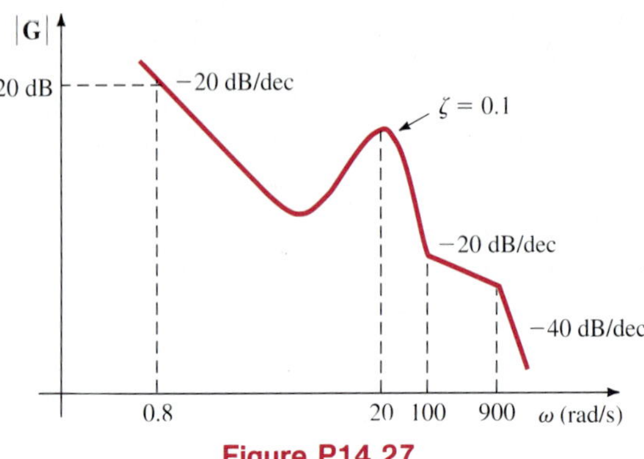

Figure P14.27

14.28. Given the magnitude characteristic of the Bode plot shown in Fig. P14.28, find $\mathbf{G}(j\omega)$.

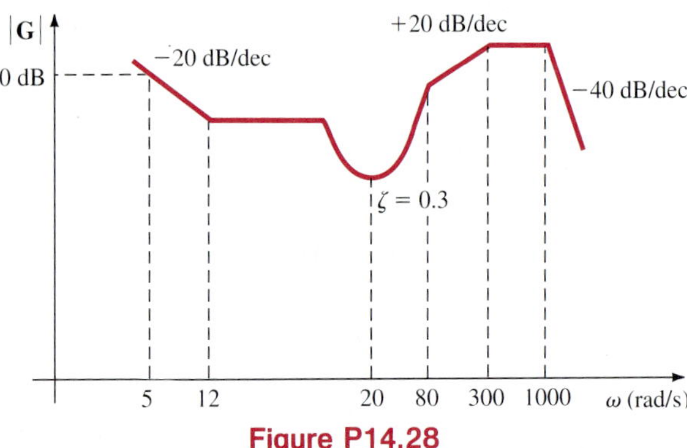

Figure P14.28

14.29. Given the series RLC circuit in Fig. P14.29, if $R = 10\ \Omega$, find the values of L and C such that the network will have a resonant frequency of 100 kHz and a bandwidth of 1 kHz.

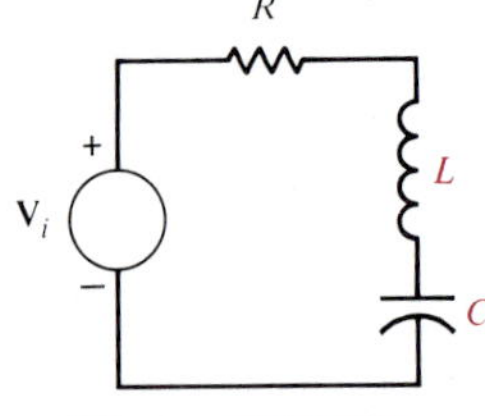

Figure P14.29

14.30. Given the series RLC circuit in Fig. P14.30,

(a) Derive the expression for the half-power frequencies, the resonant frequency, the bandwidth, and the quality factor for the transfer characteristic $\mathbf{I}/\mathbf{V}_{in}$ in terms of R, L, C.

(b) Compute the quantities in part (a) if $R = 10\ \Omega$, $L = 100$ mH, and $C = 10\ \mu$F.

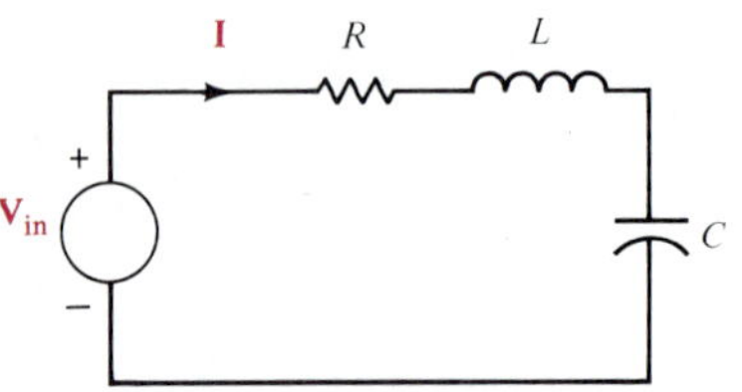

Figure P14.30

14.31. The series RLC circuit in Fig. P14.31 is driven by a variable-frequency source. If the resonant frequency of the network is selected as $\omega_0 = 1200$ rad/sec, find the value of C. In addition, compute the current at resonance and at $\omega_0/4$ and $4\omega_0$.

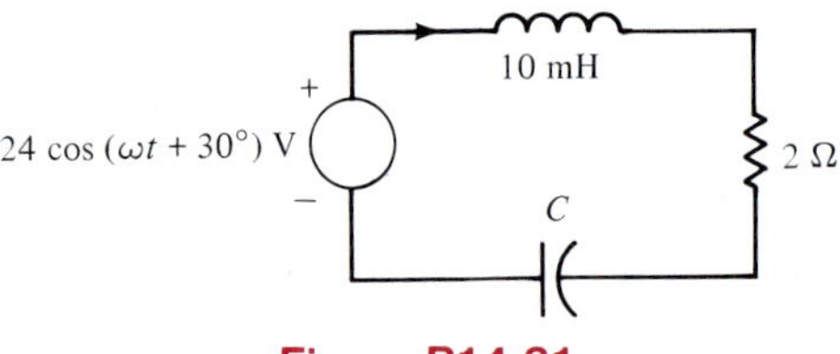

Figure P14.31

14.32. Given the network in Fig. P14.32, find ω_0, Q, ω_{max}, and $|\mathbf{V}_o|_{max}$.

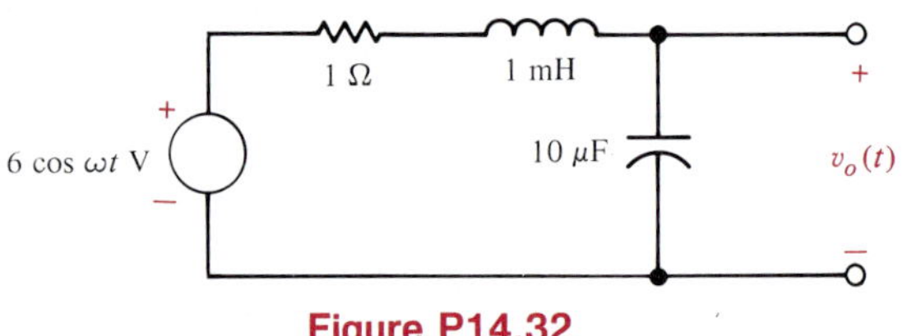

Figure P14.32

14.33. Repeat Problem 14.32 if the value of R is changed to 0.1 Ω.

14.34. Given the network in Fig. P14.34, find $|\mathbf{V}_o|_{max}$.

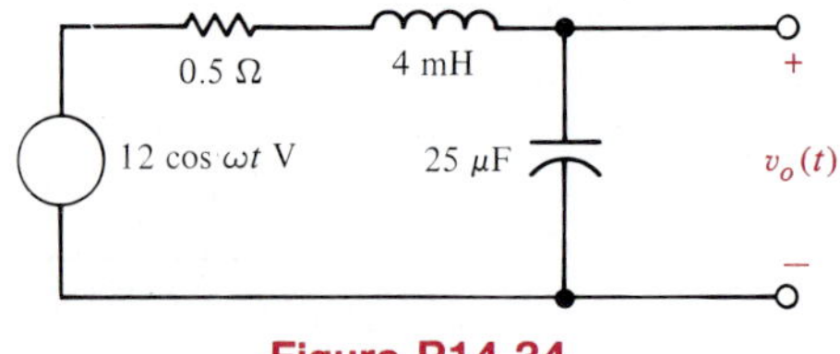

Figure P14.34

14.35. In Problem 14.34, if the resistance is changed to 1 Ω and the capacitance is changed to 50 μF, calculate ω_0, ω_{max}, and $|\mathbf{V}_o|_{max}$.

14.36. Determine ω_o and Q for the network in Fig. P14.36.

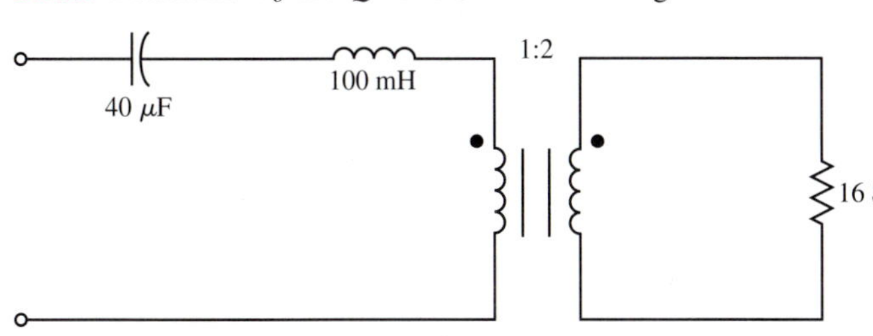

Figure P14.36

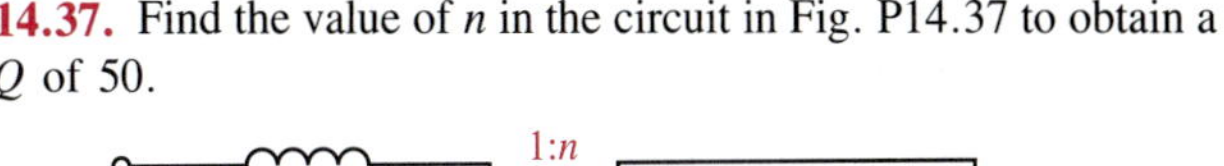

14.37. Find the value of n in the circuit in Fig. P14.37 to obtain a Q of 50.

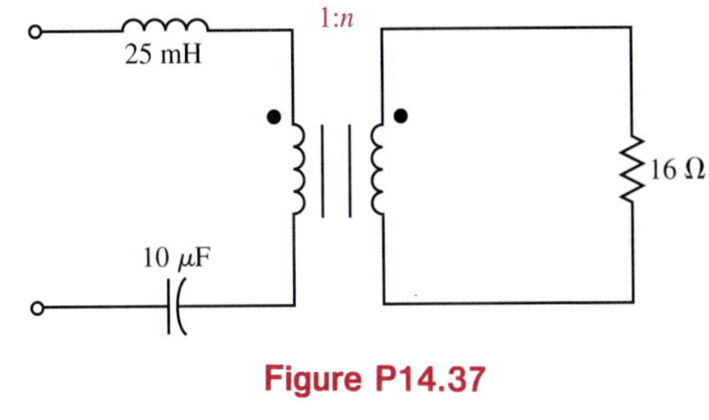

Figure P14.37

14.38. Find ω_o and Q for the circuit in Fig. P14.38.

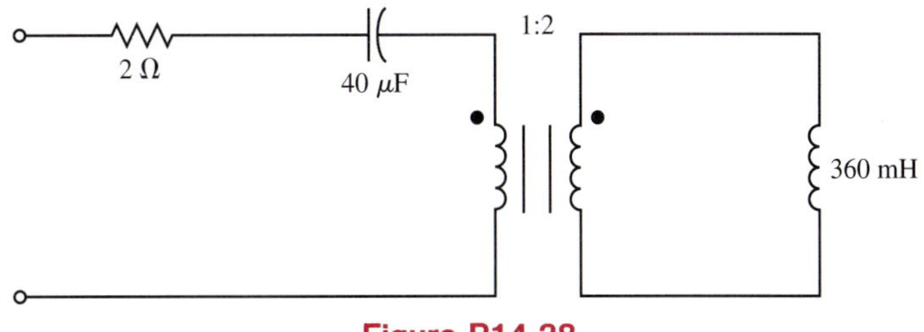

Figure P14.38

14.39. In the network in Fig. P14.39, $\omega_o = 1000$ rad/sec and at that frequency the impedance seen by the source is 4 Ω. If $L =$ 20 mH, determine the Q of the network and the bandwidth.

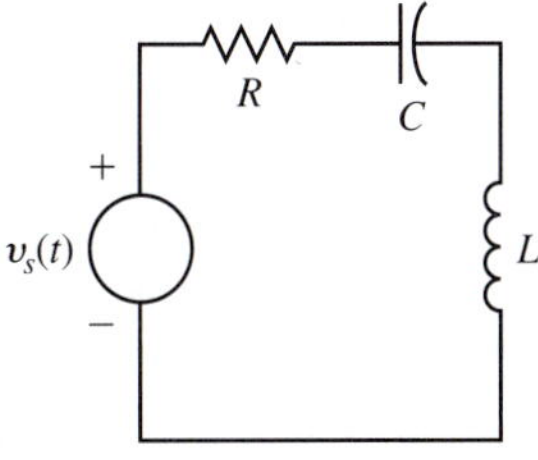

Figure P14.39

14.40. A series *RLC* circuit is driven by a signal generator. The resonant frequency of the network is known to be 1600 rad/sec, and at that frequency the impedance seen by the signal generator is 3 Ω. If $C = 20\ \mu$F, find L, Q, and the bandwidth.

14.41. In the network in Fig. P14.41, the inductor value is 30 mH, and the circuit is driven by a variable-frequency source. If the magnitude of the current at resonance is 12 A, $\omega_o =$ 1000 rad/sec, and $L = 10$ mH, find C, Q, and the bandwidth of the circuit.

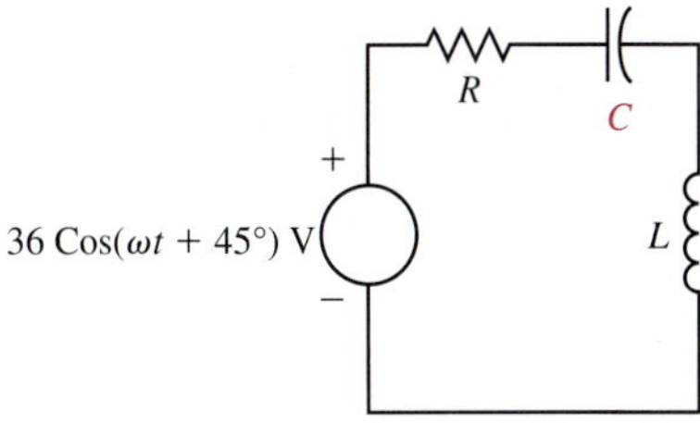

Figure P14.41

14.42. In the network in Fig. P14.42, the source operates at the network's resonant frequency. Find ω_o and Q for the circuit and the output voltage $\mathbf{V}_o$.

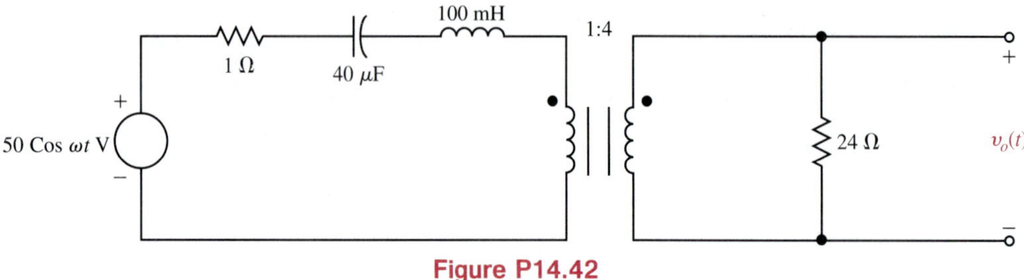

Figure P14.42

14.43. A variable frequency voltage source drives the network in Fig. P14.43. Determine the resonant frequency, Q, BW, and the average power dissipated by the network at resonance.

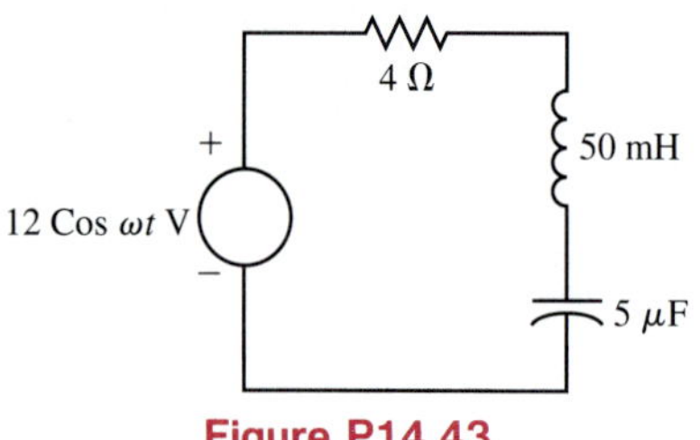

Figure P14.43

14.44. The resonant frequency for the network in Fig. P14.44 is 10,000 rad/sec, and the BW is 1000 rad/sec. The inductor value is 20 mH, and its internal resistance is 4 Ω. Determine the values of R and C and the average power dissipated by the network when the source is tuned to the resonant frequency.

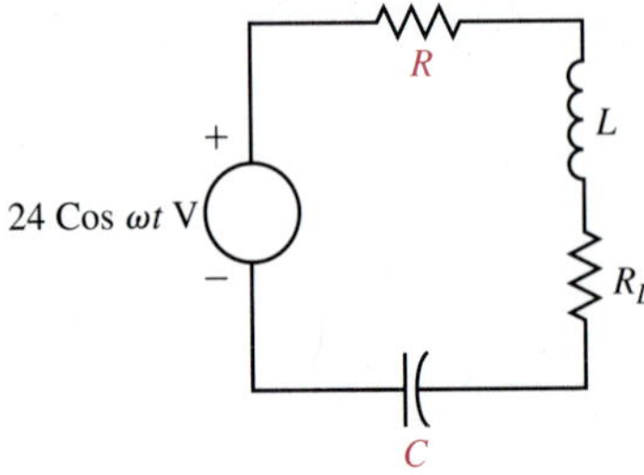

Figure P14.44

14.45. Consider the network in Fig. P14.45. If $R = 2$ kΩ, $L = 20$ mH, $C = 50$ μF, and $R_S = \infty$, determine the resonant frequency ω_0, the Q of the network, and the bandwidth of the network. What impact does an R_S of 10 kΩ have on the quantities determined?

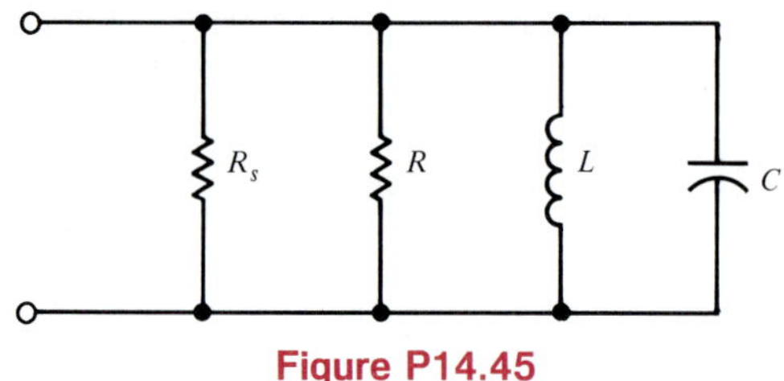

Figure P14.45

14.46. The source in the network in Fig. P14.46 is $i_S(t) = \cos 1000t + \cos 1500t$ A. $R = 200$ Ω and $C = 500$ μF. If $\omega_0 = 1000$ rad/sec, find L, Q, and the BW. Compute the output voltage $v_o(t)$ and discuss the magnitude of the output voltage at the two input frequencies.

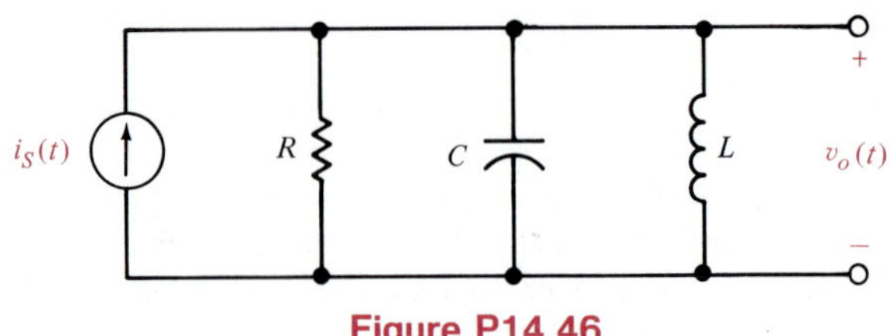

Figure P14.46

14.47. Find the resonant frequency and the quality factor for the network in Fig. P14.47.

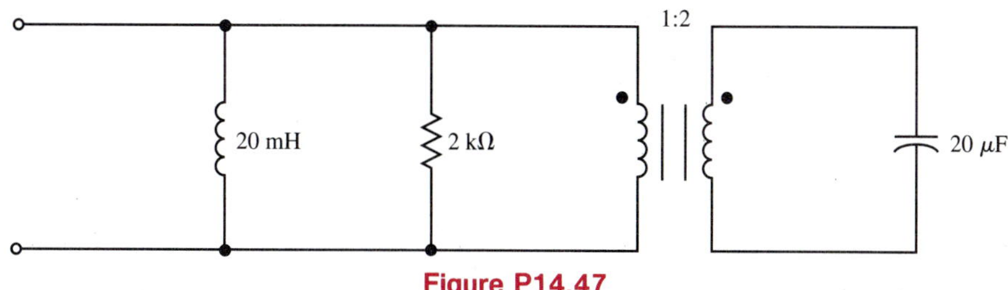

Figure P14.47

14.48. A parallel *RLC* circuit, which is driven by a variable-frequency 2-A current source, has the following values: $R = 1$ kΩ, $L = 100$ mH, and $C = 10$ μF. Find the bandwidth of the network, the half-power frequencies, and the voltage across the network at the half-power frequencies.

14.49. Determine the parameters of a parallel resonant circuit which has the following properties: $\omega_0 = 2$ Mrad/sec, BW = 20 krad/sec, and an impedance at resonance of 2000 Ω.

14.50. A parallel *RLC* circuit, which is driven by a variable-frequency 10-A source, has the following parameters: $R = 500$ Ω, $L = 0.5$ mH, and $C = 20$ μF. Find the resonant frequency, the Q, the average power dissipated at the resonant frequency, the BW, and the average power dissipated at the half-power frequencies.

14.51. Given the circuit in Fig. P14.51, compute ω_o, Q, BW, and the output voltage when the source frequency is the resonant frequency of the network.

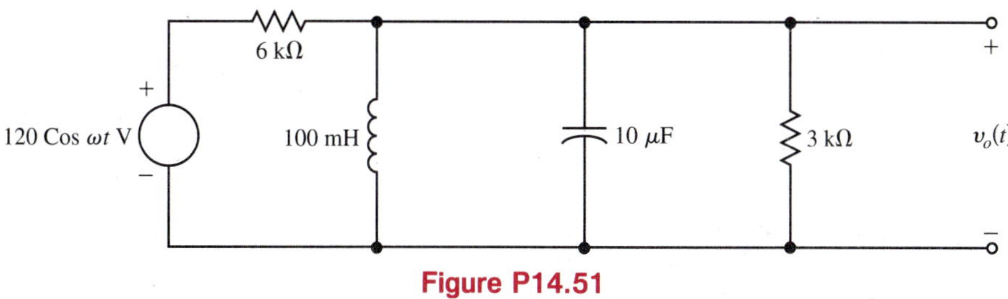

Figure P14.51

14.52. Determine the expression for the resonant frequency of the network shown in Fig. P14.52.

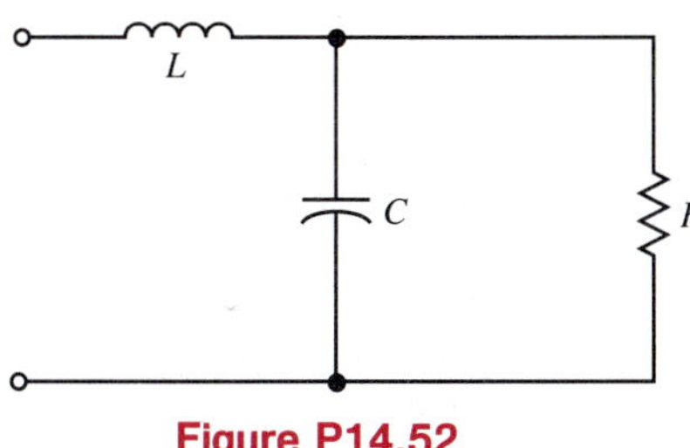

Figure P14.52

14.53. Given the network in Fig. P14.53, determine the minimum value of $\mathbf{Z}_1$ that will cause the circuit to be in resonance.

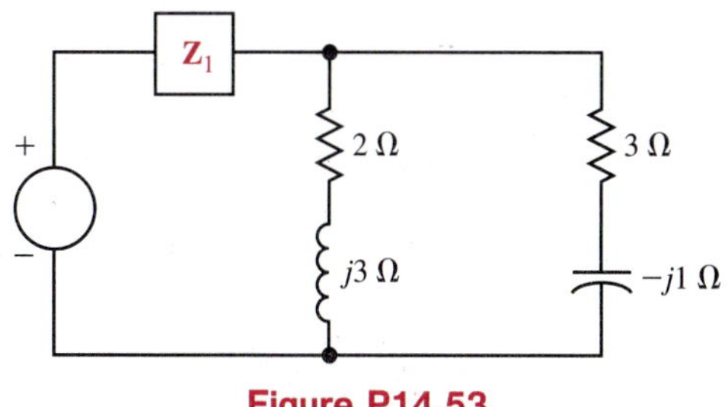

Figure P14.53

14.54. Plot $\mathbf{I}_x$ in the network in Fig. P14.54, using PSPICE. Use 5 points per decade over the frequency range 0.1 to 10 MHz.

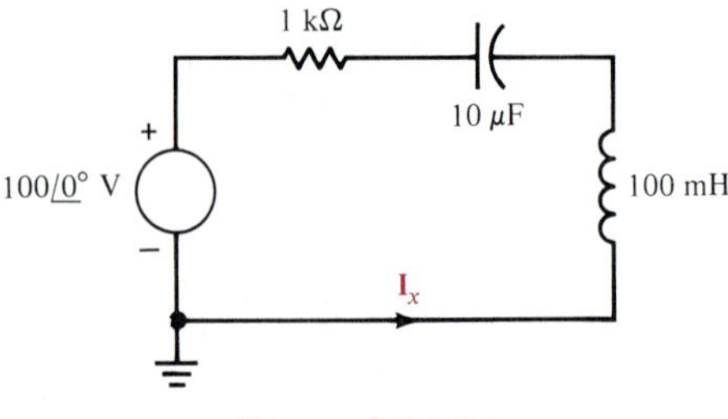

Figure P14.54

14.55. Given the network in Fig. P14.55, use 5 points per decade to plot the frequency response of **V** over the range 0.1 Hz to 1 MHz, using PSPICE.

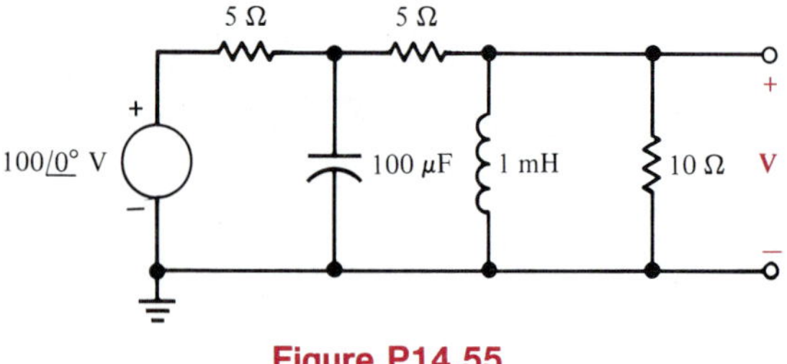

Figure P14.55

14.56. Given the network in Fig. P14.56, plot $\mathbf{V}_o$ using PSPICE. Use 5 points per decade over the range 0.1 Hz to 1 MHz.

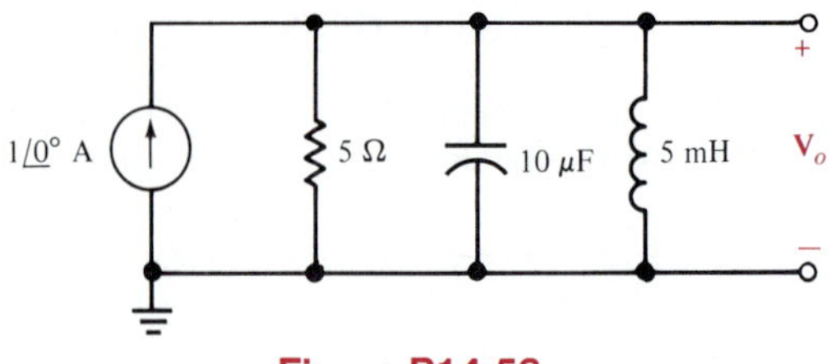

Figure P14.56

14.57. Repeat Problem 14.56 with $C = 100\ \mu$F.

14.58. Determine the new parameters of the network shown in Fig. P14.58 if $\mathbf{Z}_{\text{new}} = 1000\mathbf{Z}_{\text{old}}$.

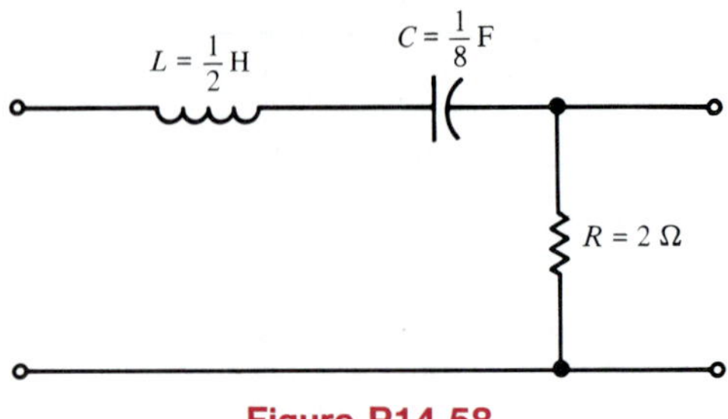

Figure P14.58

14.59. Determine the new parameters of the network in Problem 14.58 if $\omega_{\text{new}} = 10^5\omega_{\text{old}}$.

14.60. Determine what type of filter the network shown in Fig. P14.60 represents by determining the voltage transfer function.

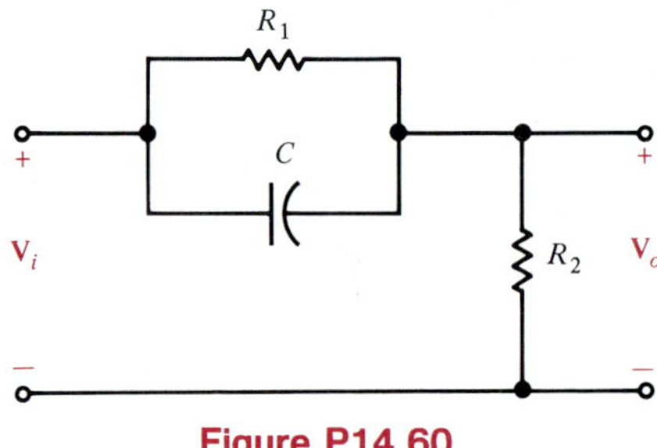

Figure P14.60

14.61. Given the network in Fig. P14.61, sketch the magnitude characteristic of the transfer function

$$\mathbf{G}_\mathbf{V}(j\omega) = \frac{\mathbf{V}_o}{\mathbf{V}_1}(j\omega)$$

Identify the type of filter.

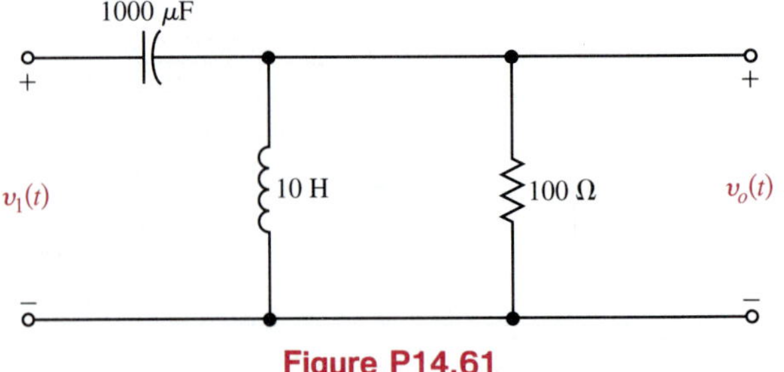

Figure P14.61

14.62. Find the voltage transfer function

$$\frac{\mathbf{V}_o}{\mathbf{V}_1}(j\omega)$$

for the network shown in Fig. P14.62. What type of filter is this?

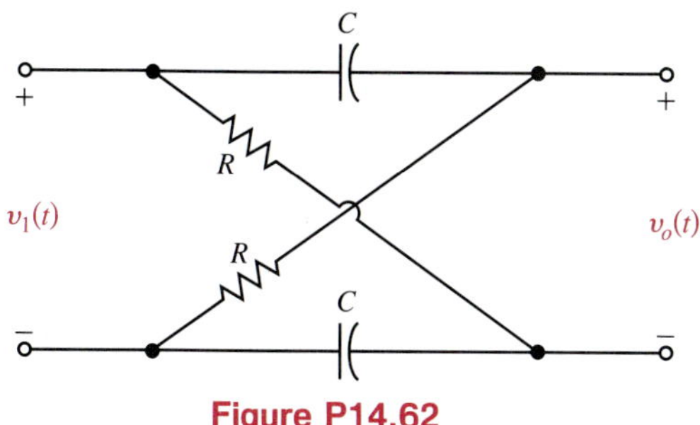

Figure P14.62

14.63. Given the network shown in Fig. P14.63,
(a) Determine the voltage transfer function.
(b) Determine what type of filter the network represents.

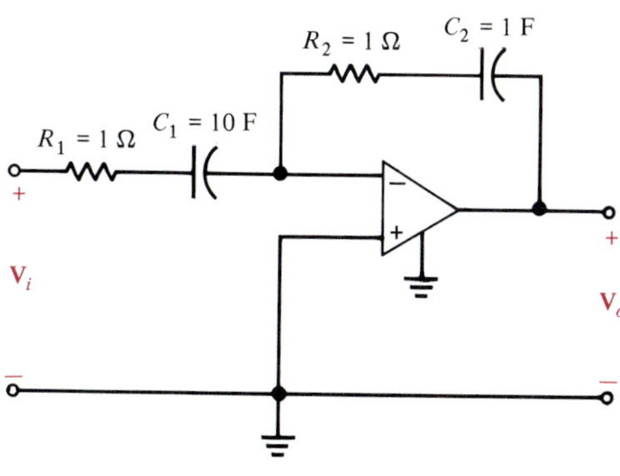

Figure P14.63

14.64. Determine the voltage transfer function and its magnitude characteristic for the network shown in Fig. P14.64 and identify the filter properties.

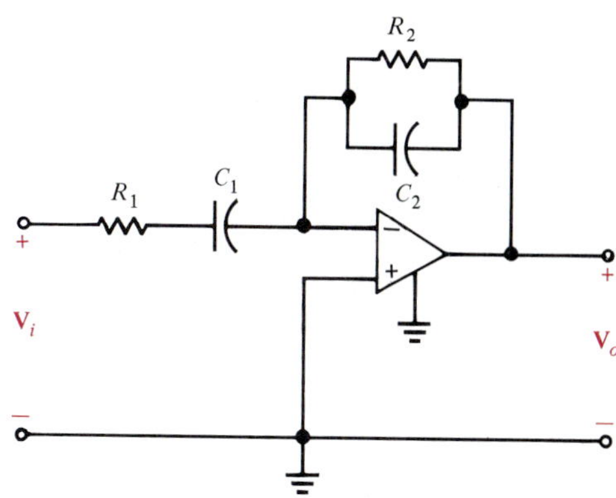

Figure P14.64

14.65. Given the network in Fig. P14.65, find the transfer function

$$\frac{\mathbf{V}_o}{\mathbf{V}_1}(j\omega)$$

and determine what type of filter the network represents.

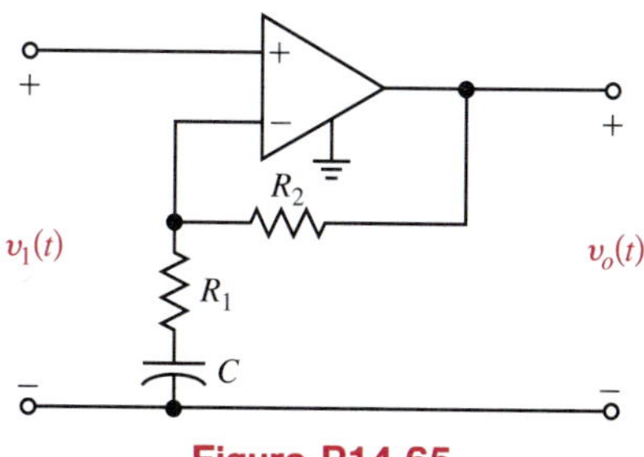

Figure P14.65

14.66. Given the network in Fig. P14.66, and employing the voltage follower analyzed in Chapter 3, determine the voltage transfer function, and its magnitude characteristic. What type of filter does the network represent?

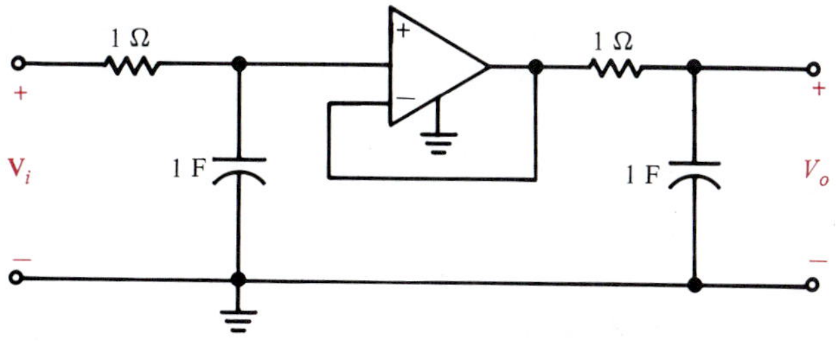

Figure P14.66

14.67. Use PSPICE to plot the frequency response of the active filter shown in Fig. P14.67. Plot over the range from 1 Hz to 10 kHz using 10 points per decade. For the op-amp model use R_{in} = 500 kΩ, R_{out} = 100 Ω, and $A = 10^7$. What type of filter is this?

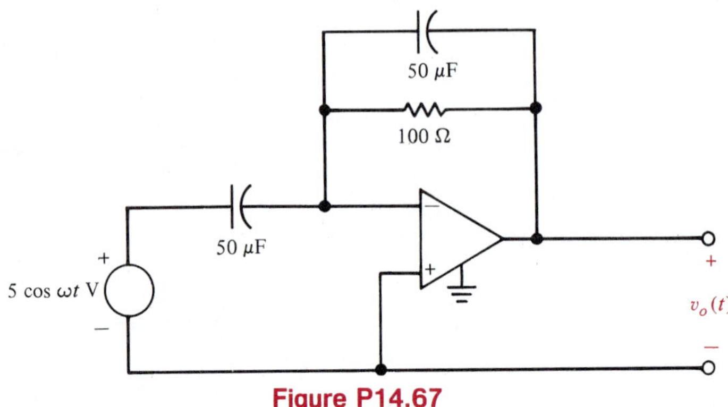

Figure P14.67

14.68. Use PSPICE to plot the frequency response of the active filter in Fig. P14.68. Plot over the range from 1 Hz to 100 kHz using 10 points per decade. In modeling the op-amps, use R_{in} = 500 kΩ, R_{out} = 100 Ω, and $A = 10^7$. What type of filter is this?

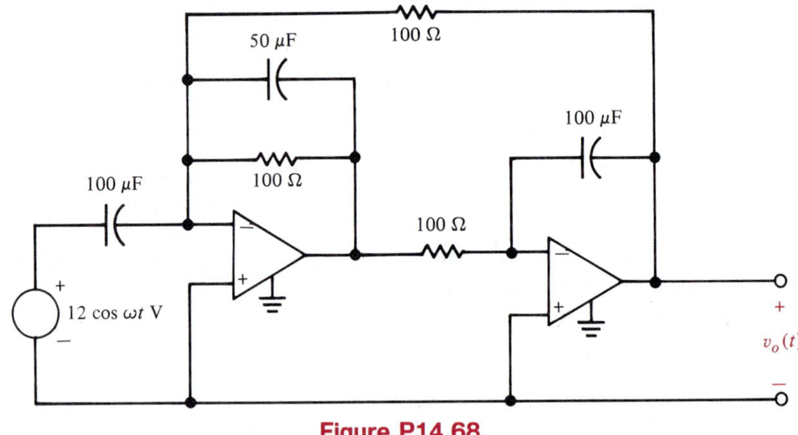

Figure P14.68

14.69. If two first-order low-pass filters are placed in series, the result is a second-order filter. Find H_o, ζ, and ω_c for the network in Fig. P14.69.

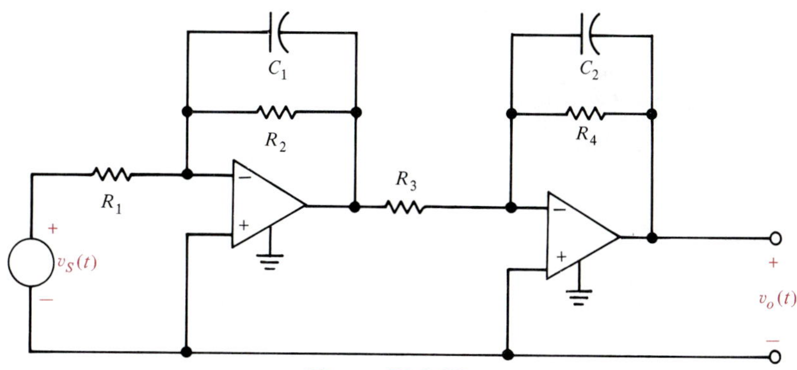

Figure P14.69

14.70. Given the second-order low-pass filter in Fig. P14.69, design a filter that has $H_o = 100$ and $f_c = 5$ kHz. Set $R_1 = R_3 = 1$ kΩ, and let $R_2 = R_4$ and $C_1 = C_2$. Verify the results with PSPICE by generating a Bode plot for the network over the range 500 Hz to 50 kHz using 20 points per decade. Use an op-amp model with $R_i = \infty$, $R_o = 0$, and $A = (2)10^5$.

14.71. The second-order low-pass filter shown in Fig. P14.71 has the transfer function

$$\frac{\mathbf{V}_o}{\mathbf{V}_1}(s) = \frac{\dfrac{-R_3}{R_1}\left(\dfrac{1}{R_2R_3C_1C_2}\right)}{s^2 + \dfrac{s}{C_1}\left(\dfrac{1}{R_1} + \dfrac{1}{R_2} + \dfrac{1}{R_3}\right) + \dfrac{1}{R_2R_3C_1C_2}}$$

Design a filter with $H_o = -10$ and $f_c = 5$ kHz, assuming that $C_1 = C_2 = 10$ nF and $R_1 = 1$ kΩ.

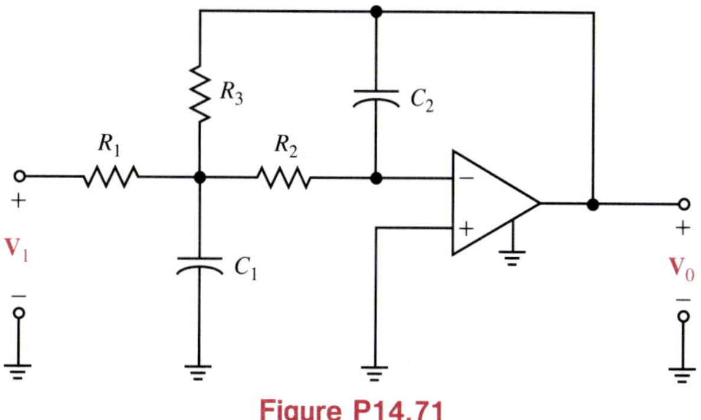

Figure P14.71

14.72. Design a second-order high-pass active filter with $\omega_c = 10$ krad/sec, $H_o = -5$, and $\zeta = 1$. Set $C_2 = C_3 = C$ and $R_1 = 1$ kΩ.

14.73. Using the equations for a general second-order band-pass filter, design a filter with a center frequency gain of -5, $\omega_0 = 50$ krad/sec, and a BW $= 10$ krad/sec. Let $C_1 = C_2 = C$ and $R_1 = 1$ kΩ. What is the Q of this circuit? Use PSPICE to generate a Bode plot for the filter over the range 100 Hz to 100 kHz using 30 points per decade. Use an op-amp model with $R_i = \infty$, $R_o = 0$, and $A = (2)10^5$.

CHAPTER

15

Two-Port Networks

We say that the linear network in Fig. 15.1a has a single *port*, that is, a single pair of terminals. The pair of terminals *A-B* that constitute this port could represent a single element (e.g., *R*, *L*, or *C*), or it could be some interconnection of these elements. The linear network in Fig. 15.1b is called a two-port. As a general rule the terminals *A-B* represent the input port and the terminals *C-D* represent the output port.

We study two-ports and the parameters that describe them for a number of reasons. For example, most circuits or systems have at least two ports. We may put an input signal into one port and obtain an output signal from another. The parameters of the two-port completely describe its behavior in terms of the voltage and current at each port. Thus knowing the parameters of a two-port network permits us to describe its operation when it is connected into a larger network. Two-port networks are also important in modeling electronic devices and system components. For example, in electronics, two-port networks are employed to model such things as transistors and op-amps. Other examples of electrical components modeled by two-ports are transformers and transmission lines.

In general, we describe the two-port as a network consisting of *R*, *L*, and *C* elements, transformers, op-amps, dependent sources, but no independent sources. The network has an input port and an output port, and as is the case with an op-amp, one terminal may be common to both ports.

Four popular types of two-port parameters are examined: admittance, impedance, hybrid, and transmission. We demonstrate the usefulness of each set of parameters, show how they are related to one another, and finally illustrate how two-port networks can be interconnected in parallel, series, or cascade.

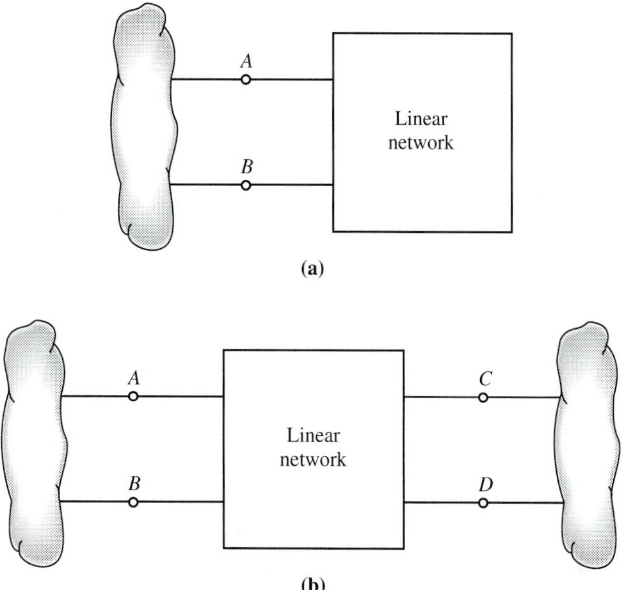

Figure 15.1 (a) Single-port network; (b) two-port network.

15.1 Admittance Parameters

In the two-port network shown in Fig. 15.2, it is customary to label the voltages and currents as shown; that is, the upper terminals are positive with respect to the lower terminals and the currents are into the two-port at the upper terminals and, because KCL must be satisfied at each port, the current is out of the two-port at the lower terminals. Since the network is linear and contains no independent sources, the principle of superposition can be applied to determine the current $\mathbf{I}_1$, which can be written as the sum of two components, one due to $\mathbf{V}_1$ and one due to $\mathbf{V}_2$. Using this principle, we can write

$$\mathbf{I}_1 = \mathbf{y}_{11}\mathbf{V}_1 + \mathbf{y}_{12}\mathbf{V}_2$$

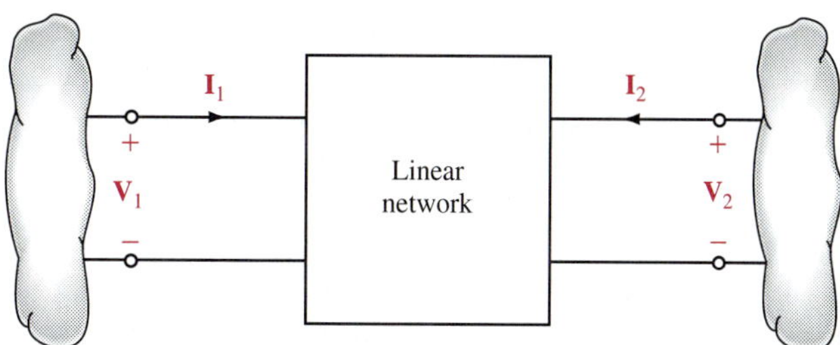

Figure 15.2 Generalized two-port network.

where $\mathbf{y}_{11}$ and $\mathbf{y}_{12}$ are essentially constants of proportionality with units of siemens. In a similar manner $\mathbf{I}_2$ can be written as

$$\mathbf{I}_2 = \mathbf{y}_{21}\mathbf{V}_1 + \mathbf{y}_{22}\mathbf{V}_2$$

Therefore, the two equations that describe the two-port are

$$\begin{aligned} \mathbf{I}_1 &= \mathbf{y}_{11}\mathbf{V}_1 + \mathbf{y}_{12}\mathbf{V}_2 \\ \mathbf{I}_2 &= \mathbf{y}_{21}\mathbf{V}_1 + \mathbf{y}_{22}\mathbf{V}_2 \end{aligned} \tag{15.1}$$

or in matrix form,

$$\begin{bmatrix} \mathbf{I}_1 \\ \mathbf{I}_2 \end{bmatrix} = \begin{bmatrix} \mathbf{y}_{11} & \mathbf{y}_{12} \\ \mathbf{y}_{21} & \mathbf{y}_{22} \end{bmatrix} \begin{bmatrix} \mathbf{V}_1 \\ \mathbf{V}_2 \end{bmatrix}$$

Note that subscript 1 refers to the input port and subscript 2 refers to the output port, and the equations describe what we will call the *Y parameters* for a network. If these parameters $\mathbf{y}_{11}$, $\mathbf{y}_{12}$, $\mathbf{y}_{21}$, and $\mathbf{y}_{22}$ are known, the input/output operation of the two-port is completely defined.

From Eqs. (15.1) we can determine the Y parameters in the following manner. Note from the equations that $\mathbf{y}_{11}$ is equal to $\mathbf{I}_1$ divided by $\mathbf{V}_1$ with the output short-circuited (i.e., $\mathbf{V}_2 = 0$).

$$\mathbf{y}_{11} = \left.\frac{\mathbf{I}_1}{\mathbf{V}_1}\right|_{\mathbf{V}_2=0} \tag{15.2}$$

Since $\mathbf{y}_{11}$ is an admittance at the input measured in siemens with the output short-circuited, it is called the *short-circuit input admittance*. The equations indicate that the other Y parameters can be determined in a similar manner:

$$\begin{aligned} \mathbf{y}_{12} &= \left.\frac{\mathbf{I}_1}{\mathbf{V}_2}\right|_{\mathbf{V}_1=0} \\ \mathbf{y}_{21} &= \left.\frac{\mathbf{I}_2}{\mathbf{V}_1}\right|_{\mathbf{V}_2=0} \\ \mathbf{y}_{22} &= \left.\frac{\mathbf{I}_2}{\mathbf{V}_2}\right|_{\mathbf{V}_1=0} \end{aligned} \tag{15.3}$$

$\mathbf{y}_{12}$ and $\mathbf{y}_{21}$ are called the *short-circuit transfer admittances,* and $\mathbf{y}_{22}$ is called the *short-circuit output admittance*. As a group the Y parameters are referred to as the *short-circuit admittance parameters*. Note that by applying the definitions above, these parameters could be determined experimentally for a two-port whose actual configuration is unknown.

EXAMPLE 15.1

Let us determine the Y parameters for the network shown in Fig. 15.3a.

From our definitions the admittance $\mathbf{y}_{11}$ is equal to $\mathbf{I}_1$ divided by $\mathbf{V}_1$, with $\mathbf{V}_2 = 0$ as shown in Fig. 15.3b. Therefore, with the output shorted, the two 6-Ω resistors are in

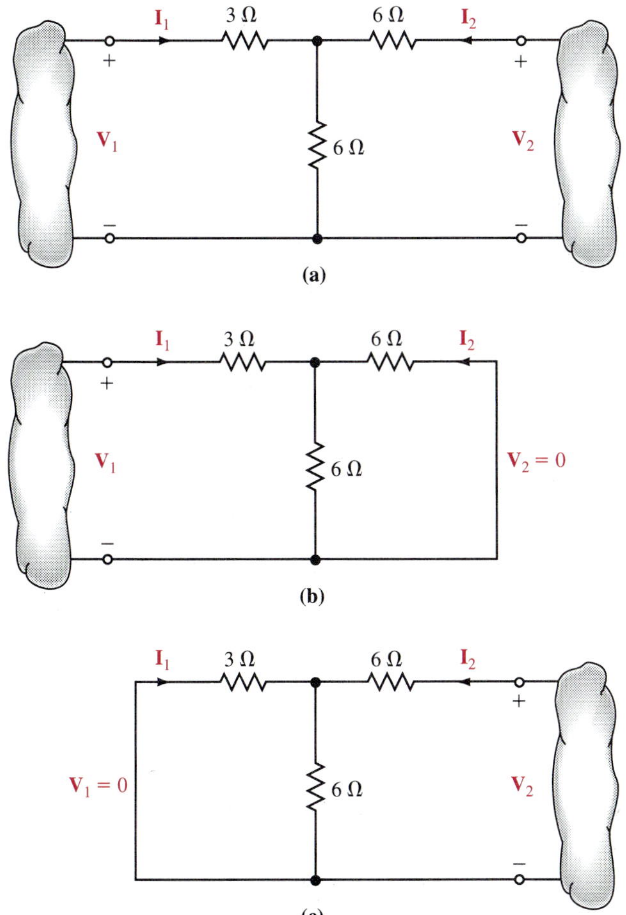

Figure 15.3 Two-port and the configurations used to compute the Y parameters.

parallel and $\mathbf{V}_1 = 6\mathbf{I}_1$; hence

$$\mathbf{y}_{11} = \left.\frac{\mathbf{I}_1}{\mathbf{V}_1}\right|_{\mathbf{V}_2=0} = \frac{\mathbf{I}_1}{6\mathbf{I}_1} = \frac{1}{6}\text{ S}$$

The parameter $\mathbf{y}_{12}$ is evaluated using the circuit in Fig. 15.3c.

$$\mathbf{I}_2 = \frac{\mathbf{V}_2}{6 + \dfrac{(3)(6)}{3+6}}$$

Employing current division, we obtain

$$-\mathbf{I}_1 = \mathbf{I}_2\left(\frac{6}{3+6}\right)$$

Hence

$$\mathbf{y}_{12} = -\tfrac{1}{12}\,\text{S}$$

In a similar manner, $\mathbf{y}_{21}$ can be determined from Fig. 15.3b.

$$\mathbf{I}_1 = \frac{\mathbf{V}_1}{3 + \dfrac{(6)(6)}{6 + 6}}$$

and

$$-\mathbf{I}_2 = \mathbf{I}_1\left(\frac{6}{6 + 6}\right)$$

Therefore,

$$\mathbf{y}_{21} = -\tfrac{1}{12}\,\text{S}$$

Finally, $\mathbf{y}_{22}$ is obtained from Fig. 15.3c as

$$\mathbf{I}_2 = \frac{\mathbf{V}_2}{6 + \dfrac{(6)(3)}{6 + 3}}$$

and hence

$$\mathbf{y}_{22} = \tfrac{1}{8}\,\text{S}$$

From the values computed above we find that the equations that describe the two-port operation by means of admittance parameters are

$$\mathbf{I}_1 = \tfrac{1}{6}\mathbf{V}_1 - \tfrac{1}{12}\mathbf{V}_2$$

$$\mathbf{I}_2 = -\tfrac{1}{12}\mathbf{V}_1 + \tfrac{1}{8}\mathbf{V}_2$$

or in matrix form,

$$\begin{bmatrix} \mathbf{I}_1 \\ \mathbf{I}_2 \end{bmatrix} = \begin{bmatrix} \frac{1}{6} & -\frac{1}{12} \\ -\frac{1}{12} & \frac{1}{8} \end{bmatrix} \begin{bmatrix} \mathbf{V}_1 \\ \mathbf{V}_2 \end{bmatrix}$$

■

EXAMPLE 15.2

We wish to determine the Y parameters for the two-port shown in Fig. 15.4a. Once these parameters are known, we will determine the current in a 4-Ω load which is connected to the output port when a 2-A current source is applied at the input port.

From Fig. 15.4b, we note that

$$\mathbf{I}_1 = \mathbf{V}_1(\tfrac{1}{1} + \tfrac{1}{2})$$

Therefore,

$$\mathbf{y}_{11} = \tfrac{3}{2}\,\text{S}$$

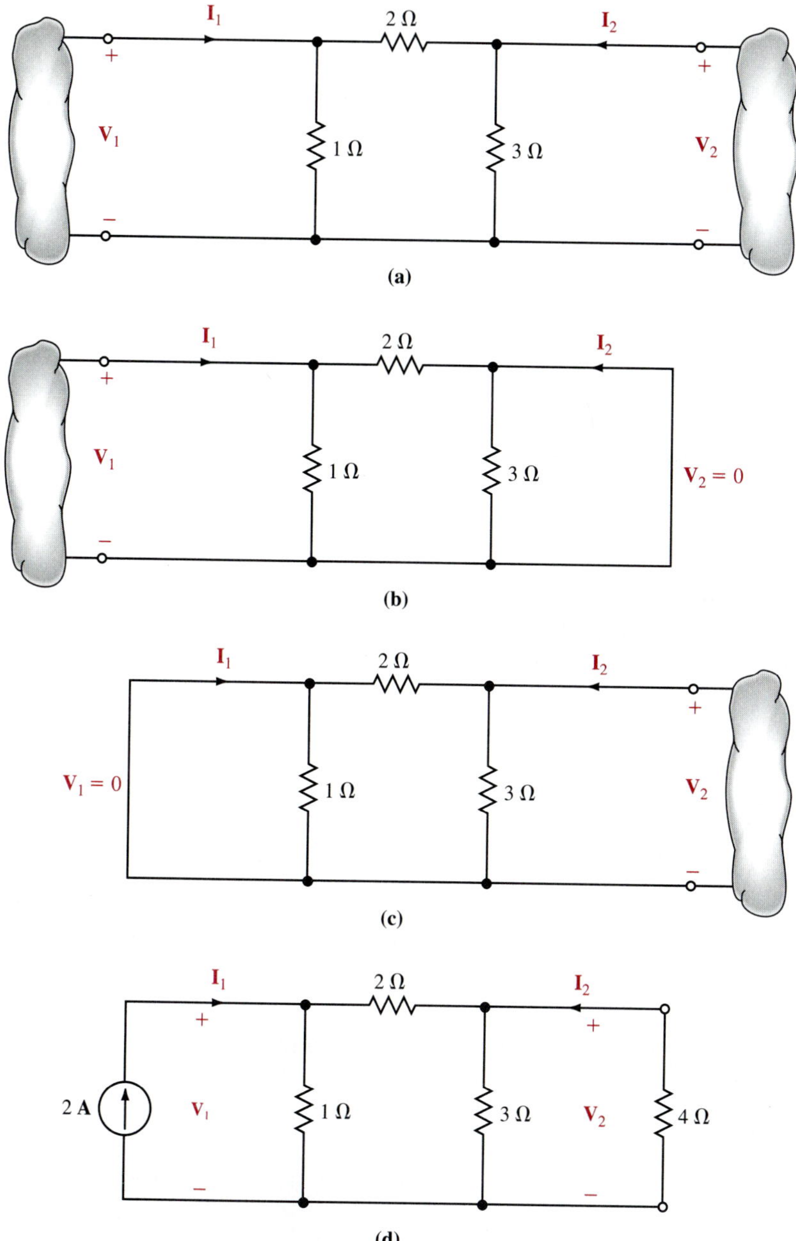

Figure 15.4 Networks employed in Example 15.2.

As shown in Fig. 15.4c,

$$\mathbf{I}_1 = -\frac{\mathbf{V}_2}{2}$$

and hence

$$\mathbf{y}_{12} = -\tfrac{1}{2}\text{ S}$$

Also, $\mathbf{y}_{21}$ is computed from Fig. 15.4b using the equation

$$\mathbf{I}_2 = -\frac{\mathbf{V}_1}{2}$$

and therefore,

$$\mathbf{y}_{21} = -\tfrac{1}{2}\text{ S}$$

Finally, $\mathbf{y}_{22}$ can be derived from Fig. 15.4c using

$$\mathbf{I}_2 = \mathbf{V}_2(\tfrac{1}{3} + \tfrac{1}{2})$$

and

$$\mathbf{y}_{22} = \tfrac{5}{6}\text{ S}$$

Therefore, the equations that describe the two-port itself are

$$\mathbf{I}_1 = \tfrac{3}{2}\mathbf{V}_1 - \tfrac{1}{2}\mathbf{V}_2$$

$$\mathbf{I}_2 = -\tfrac{1}{2}\mathbf{V}_1 + \tfrac{5}{6}\mathbf{V}_2$$

These equations can now be employed to determine the operation of the two-port for some given set of terminal conditions. The terminal conditions we will examine are shown in Fig. 15.4d. From this figure we note that

$$\mathbf{I}_1 = 2\text{ A} \qquad \text{and} \qquad \mathbf{V}_2 = -4\mathbf{I}_2$$

Combining these with the two-port equations above yields

$$2 = \tfrac{3}{2}\mathbf{V}_1 - \tfrac{1}{2}\mathbf{V}_2$$

$$0 = -\tfrac{1}{2}\mathbf{V}_1 + \tfrac{13}{12}\mathbf{V}_2$$

or in matrix form

$$\begin{bmatrix} \frac{3}{2} & -\frac{1}{2} \\ -\frac{1}{2} & \frac{13}{12} \end{bmatrix} \begin{bmatrix} \mathbf{V}_1 \\ \mathbf{V}_2 \end{bmatrix} = \begin{bmatrix} 2 \\ 0 \end{bmatrix}$$

Note carefully that these equations are simply the nodal equations for the network in Fig. 15.4d. Solving the equations, we obtain $\mathbf{V}_2 = \frac{8}{11}$ V and therefore $\mathbf{I}_2 = -\frac{2}{11}$ A. ■

DRILL EXERCISE

D15.1. Find the Y parameters for the two-port network shown in Fig. D15.1.

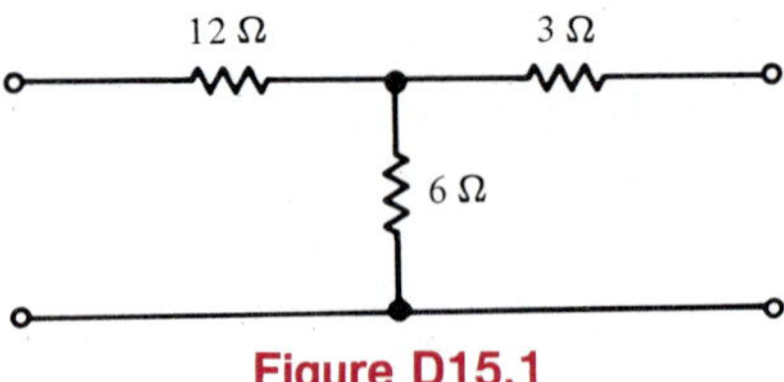

Figure D15.1

Ans: $\mathbf{y}_{11} = \frac{1}{14}$ S, $\mathbf{y}_{12} = \mathbf{y}_{21} = -\frac{1}{21}$ S, $\mathbf{y}_{22} = \frac{1}{7}$ S.

D15.2. Find the Y parameters for the two-port network shown in Fig. D15.2.

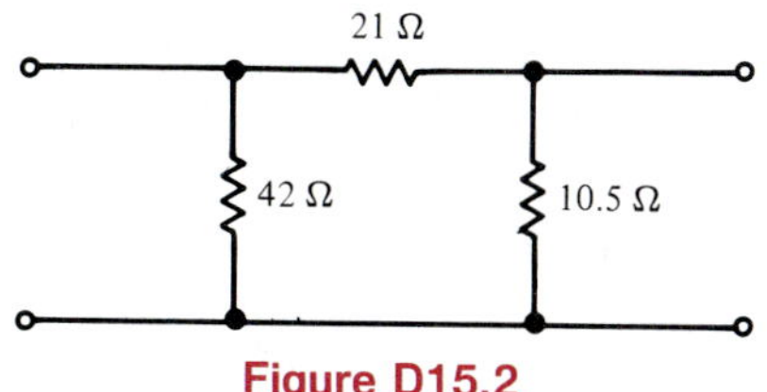

Figure D15.2

Ans: $\mathbf{y}_{11} = \frac{1}{14}$ S, $\mathbf{y}_{12} = \mathbf{y}_{21} = -\frac{1}{21}$ S, $\mathbf{y}_{22} = \frac{1}{7}$ S.

D15.3. If a 10-A source is connected to the input of the two-port in Fig. D15.2, find the current in a 5-Ω resistor connected to the output port.

Ans: $\mathbf{I}_2 = -4.29$ A.

15.2 Impedance Parameters

Once again if we assume that the two-port is a linear network that contains no independent sources, then by means of superposition we can write the input and output voltages as the sum of two components, one due to $\mathbf{I}_1$ and one due to $\mathbf{I}_2$:

$$\begin{aligned} \mathbf{V}_1 &= \mathbf{z}_{11}\mathbf{I}_1 + \mathbf{z}_{12}\mathbf{I}_2 \\ \mathbf{V}_2 &= \mathbf{z}_{21}\mathbf{I}_1 + \mathbf{z}_{22}\mathbf{I}_2 \end{aligned} \tag{15.4}$$

These equations, which describe the two-port, can also be written in matrix form as

$$\begin{bmatrix} \mathbf{V}_1 \\ \mathbf{V}_2 \end{bmatrix} = \begin{bmatrix} \mathbf{z}_{11} & \mathbf{z}_{12} \\ \mathbf{z}_{21} & \mathbf{z}_{22} \end{bmatrix} \begin{bmatrix} \mathbf{I}_1 \\ \mathbf{I}_2 \end{bmatrix} \tag{15.5}$$

Like the Y parameters, these *Z parameters* can be derived as follows:

$$\begin{aligned} \mathbf{z}_{11} &= \left.\frac{\mathbf{V}_1}{\mathbf{I}_1}\right|_{\mathbf{I}_2=0} \\ \mathbf{z}_{12} &= \left.\frac{\mathbf{V}_1}{\mathbf{I}_2}\right|_{\mathbf{I}_1=0} \\ \mathbf{z}_{21} &= \left.\frac{\mathbf{V}_2}{\mathbf{I}_1}\right|_{\mathbf{I}_2=0} \\ \mathbf{z}_{22} &= \left.\frac{\mathbf{V}_2}{\mathbf{I}_2}\right|_{\mathbf{I}_1=0} \end{aligned} \tag{15.6}$$

In the equations above, setting $\mathbf{I}_1$ or $\mathbf{I}_2 = 0$ is equivalent to open-circuiting the input or output port. Therefore, the Z parameters are called the *open-circuit impedance parameters*. $\mathbf{z}_{11}$ is called the *open-circuit input impedance*, $\mathbf{z}_{22}$ is called the *open-circuit output impedance*, and $\mathbf{z}_{12}$ and $\mathbf{z}_{21}$ are termed *open-circuit transfer impedances*.

EXAMPLE 15.3

We wish to find the Z parameters for the network in Fig. 15.5a. Once the parameters are known, we will use them to find the current in a 4-Ω resistor that is connected to the output terminals when a 12 $\underline{/0^\circ}$-V source with an internal impedance of $1 + j0\ \Omega$ is connected to the input.

From Fig. 15.5a we note that

$$\mathbf{z}_{11} = 2 - j4\ \Omega$$

$$\mathbf{z}_{12} = -j4\ \Omega$$

$$\mathbf{z}_{21} = -j4\ \Omega$$

$$\mathbf{z}_{22} = -j4 + j2 = -j2\ \Omega$$

The equations for the two-port are, therefore,

$$\mathbf{V}_1 = (2 - j4)\mathbf{I}_1 - j4\mathbf{I}_2$$

$$\mathbf{V}_2 = -j4\mathbf{I}_1 - j2\mathbf{I}_2$$

The terminal conditions for the network shown in Fig. 15.5b are

$$\mathbf{V}_1 = 12\ \underline{/0^\circ} - (1)\mathbf{I}_1$$

$$\mathbf{V}_2 = -4\mathbf{I}_2$$

Combining these with the two-port equations yields

$$12\ \underline{/0^\circ} = (3 - j4)\mathbf{I}_1 - j4\mathbf{I}_2$$

$$0 = -j4\mathbf{I}_1 + (4 - j2)\mathbf{I}_2$$

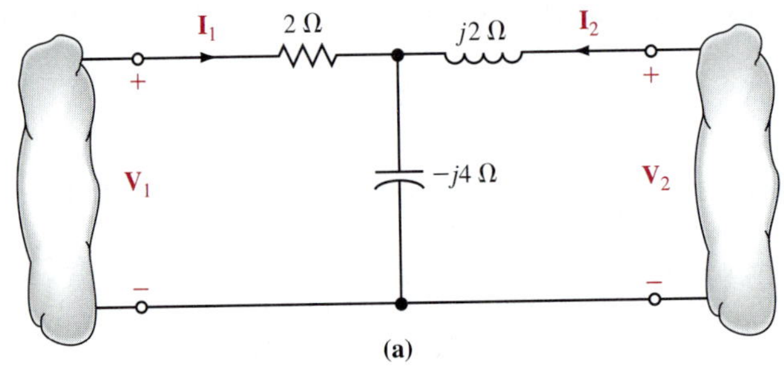

(a)

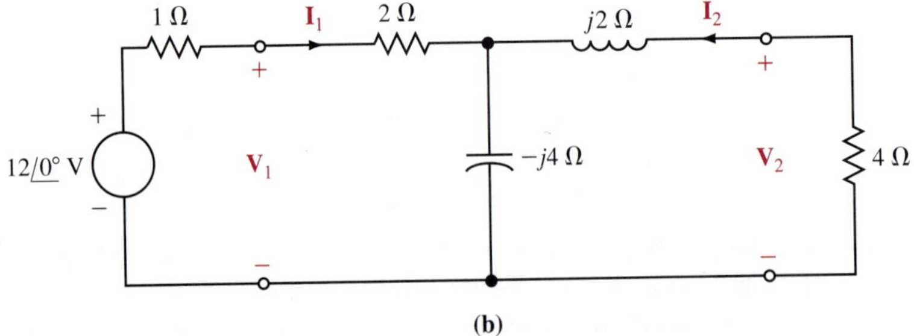

(b)

Figure 15.5 Circuits employed in Example 15.3.

It is interesting to note that these equations are the mesh equations for the network. If we solve the equations for $\mathbf{I}_2$, we obtain $\mathbf{I}_2 = 1.61 \underline{/137.73°}$ A, which is the current in the 4-Ω load. ■

EXAMPLE 15.4

Let us determine the Z parameters for the two-port network shown in Fig. 15.6a.

$\mathbf{z}_{11}$ is derived from Fig. 15.6b via the equation

$$\mathbf{I}_1 = \frac{\mathbf{V}_1}{R_1} + \beta \mathbf{V}_1$$

and hence

$$\mathbf{z}_{11} = \frac{R_1}{1 + \beta R_1}$$

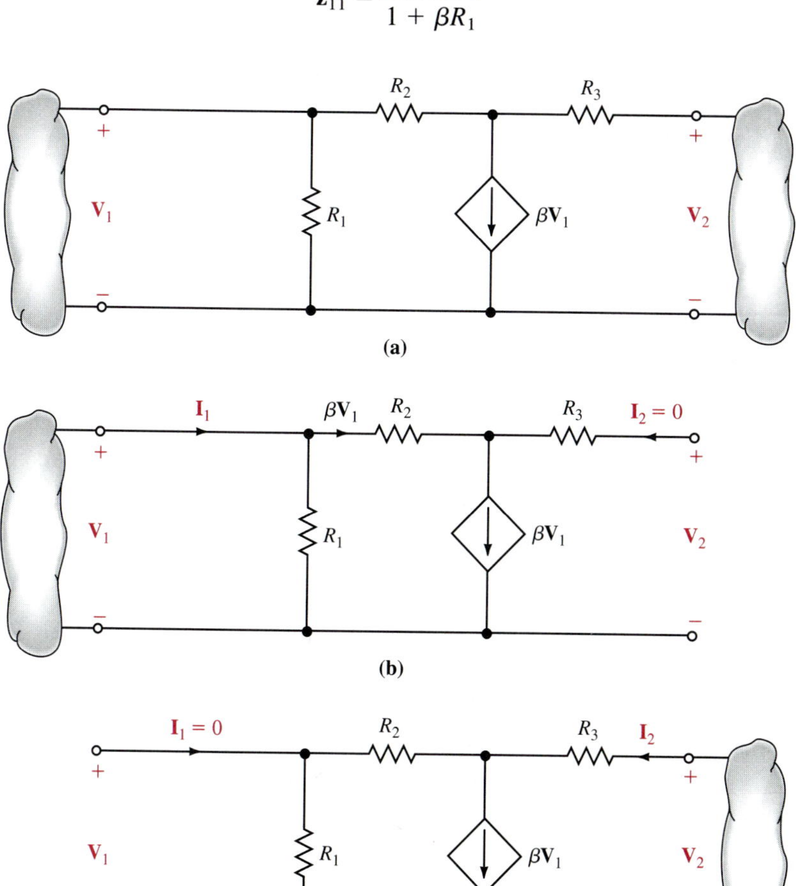

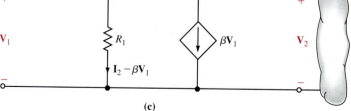

Figure 15.6 Circuits employed in Example 15.4.

$\mathbf{z}_{12}$ is obtained from Fig. 15.6c. Note that with $\mathbf{I}_1 = 0$, the current in resistors R_1 and R_2 is $\mathbf{I}_2 - \beta\mathbf{V}_1$. Therefore,

$$\mathbf{V}_1 = R_1(\mathbf{I}_2 - \beta\mathbf{V}_1)$$

and hence

$$\mathbf{z}_{12} = \frac{R_1}{1 + \beta R_1}$$

The parameter $\mathbf{z}_{21}$ is derived from Fig. 15.6b. Since $\mathbf{I}_2 = 0$,

$$\mathbf{I}_1 = \frac{\mathbf{V}_1}{R_1} + \beta\mathbf{V}_1$$

In addition, KVL implies that

$$\mathbf{V}_1 - \beta\mathbf{V}_1 R_2 = \mathbf{V}_2$$

From these two equations we obtain

$$\mathbf{z}_{21} = \frac{R_1(1 - \beta R_2)}{1 + \beta R_1}$$

Then $\mathbf{z}_{22}$ can be obtained from Fig. 15.6c. Note that KVL around the outer loop is

$$\mathbf{V}_2 = \mathbf{I}_2 R_3 + (\mathbf{I}_2 - \beta\mathbf{V}_1)R_2 + \mathbf{V}_1$$

but

$$\mathbf{V}_1 = R_1(\mathbf{I}_2 - \beta\mathbf{V}_1)$$

Solving the two equations yields

$$\mathbf{z}_{22} = \frac{R_1 + R_2 + R_3 + \beta R_1 R_3}{1 + \beta R_1}$$

Therefore, the two-port equations for the network are

$$\mathbf{V}_1 = \frac{R_1}{1 + \beta R_1}\mathbf{I}_1 + \frac{R_1}{1 + \beta R_1}\mathbf{I}_2$$

$$\mathbf{V}_2 = \frac{R_1(1 - \beta R_2)}{1 + \beta R_1}\mathbf{I}_1 + \frac{R_1 + R_2 + R_3 + \beta R_1 R_3}{1 + \beta R_1}\mathbf{I}_2$$

■

At this point it is important for the reader to note that in all of our two-port analyses, the application of KVL and KCL will produce the equations needed to determine the parameters.

DRILL EXERCISE

D15.4. Find the Z parameters for the network in Fig. D15.1. Then compute the current in a 4-Ω load if a 24 $\underline{/0°}$-V source is connected at the input port.

Ans: $\mathbf{I}_2 = -0.73 \underline{/0°}$ A.

D15.5. Find the Z parameters for the two-port in Fig. D15.5.

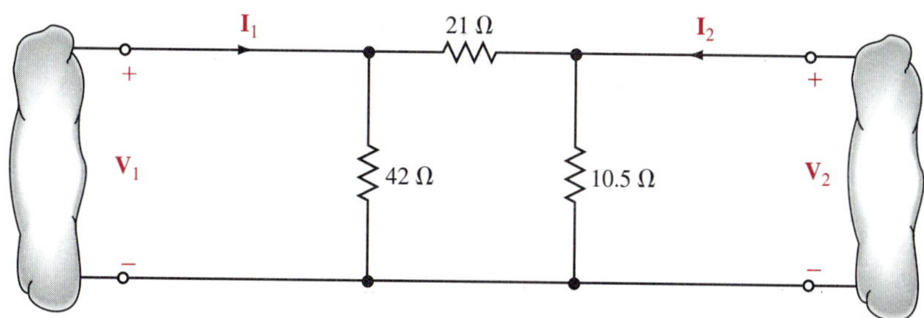

Figure D15.5

Ans: $\mathbf{z}_{11} = 18\ \Omega$, $\mathbf{z}_{12} = \mathbf{z}_{21} = 6\ \Omega$, $\mathbf{z}_{22} = 9\ \Omega$.

15.3 Hybrid Parameters

Under the assumptions used to develop the Y and Z parameters, we can obtain what are commonly called the *hybrid parameters*. In the pair of equations that define these parameters, $\mathbf{V}_1$ and $\mathbf{I}_2$ are the independent variables. Therefore, the two-port equations in terms of the hybrid parameters are

$$\begin{aligned} \mathbf{V}_1 &= \mathbf{h}_{11}\mathbf{I}_1 + \mathbf{h}_{12}\mathbf{V}_2 \\ \mathbf{I}_2 &= \mathbf{h}_{21}\mathbf{I}_1 + \mathbf{h}_{22}\mathbf{V}_2 \end{aligned} \tag{15.7}$$

or in matrix form,

$$\begin{bmatrix} \mathbf{V}_1 \\ \mathbf{I}_2 \end{bmatrix} = \begin{bmatrix} \mathbf{h}_{11} & \mathbf{h}_{12} \\ \mathbf{h}_{21} & \mathbf{h}_{22} \end{bmatrix} \begin{bmatrix} \mathbf{I}_1 \\ \mathbf{V}_2 \end{bmatrix} \tag{15.8}$$

These parameters are especially important in transistor circuit analysis. The parameters are determined via the following equations:

$$\begin{aligned} \mathbf{h}_{11} &= \left.\frac{\mathbf{V}_1}{\mathbf{I}_1}\right|_{\mathbf{V}_2=0} \\ \mathbf{h}_{12} &= \left.\frac{\mathbf{V}_1}{\mathbf{V}_2}\right|_{\mathbf{I}_1=0} \\ \mathbf{h}_{21} &= \left.\frac{\mathbf{I}_2}{\mathbf{I}_1}\right|_{\mathbf{V}_2=0} \\ \mathbf{h}_{22} &= \left.\frac{\mathbf{I}_2}{\mathbf{V}_2}\right|_{\mathbf{I}_1=0} \end{aligned} \tag{15.9}$$

The parameters $\mathbf{h}_{11}$, $\mathbf{h}_{12}$, $\mathbf{h}_{21}$, and $\mathbf{h}_{22}$ represent the *short-circuit input impedance*, the *open-circuit reverse voltage gain*, the *short-circuit forward current gain*, and the *open-*

circuit output admittance, respectively. Because of this mix of parameters, they are called *hybrid parameters*. In transistor circuit analysis the parameters $\mathbf{h}_{11}$, $\mathbf{h}_{12}$, $\mathbf{h}_{21}$, and $\mathbf{h}_{22}$ are normally labeled $\mathbf{h}_i$, $\mathbf{h}_r$, $\mathbf{h}_f$, and $\mathbf{h}_o$.

EXAMPLE 15.5

If the hybrid parameters for the two-port shown in Fig. 15.7 are known, we wish to find an expression for the input impedance $\mathbf{Z}_{\text{in}}$ of the network.

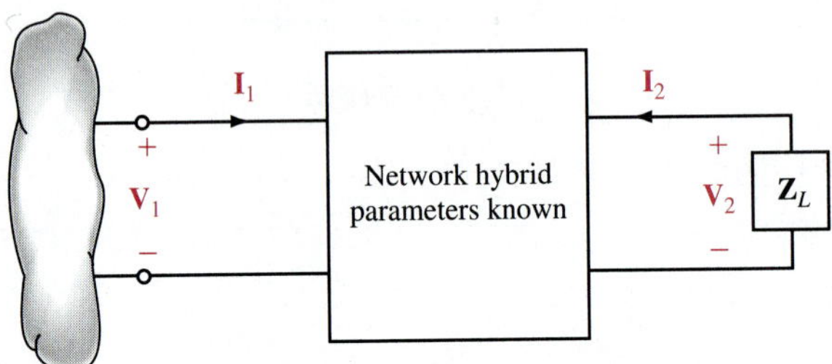

Figure 15.7 Network used in Example 15.5.

The equations for the two-port are

$$\mathbf{V}_1 = \mathbf{h}_{11}\mathbf{I}_1 + \mathbf{h}_{12}\mathbf{V}_2$$

$$\mathbf{I}_2 = \mathbf{h}_{21}\mathbf{I}_1 + \mathbf{h}_{22}\mathbf{V}_2$$

However,

$$\mathbf{V}_2 = -\mathbf{I}_2\mathbf{Z}_L$$

Therefore,

$$\mathbf{V}_1 = \mathbf{h}_{11}\mathbf{I}_1 - \mathbf{h}_{12}\mathbf{Z}_L\mathbf{I}_2$$

$$0 = \mathbf{h}_{21}\mathbf{I}_1 - (1 + \mathbf{h}_{22}\mathbf{Z}_L)\mathbf{I}_2$$

Solving the two equations for $\mathbf{I}_1$ and then forming the ratio $\mathbf{V}_1/\mathbf{I}_1$ yields

$$\mathbf{Z}_{\text{in}} = \frac{\mathbf{V}_1}{\mathbf{I}_1} = \mathbf{h}_{11} - \frac{\mathbf{h}_{12}\mathbf{h}_{21}\mathbf{Z}_L}{1 + \mathbf{h}_{22}\mathbf{Z}_L}$$

EXAMPLE 15.6

Let us determine the hybrid parameters for the network shown in Fig. 15.8a.

$\mathbf{h}_{11}$ is determined using Fig. 15.8b. From KVL

$$\mathbf{V}_1 - \mathbf{I}_1 R_1 = \mathbf{V}$$

and using KCL we can write

$$\mathbf{I}_1 = \frac{\mathbf{V}}{R_2} + j\omega C\mathbf{V} + \alpha\mathbf{I}_1$$

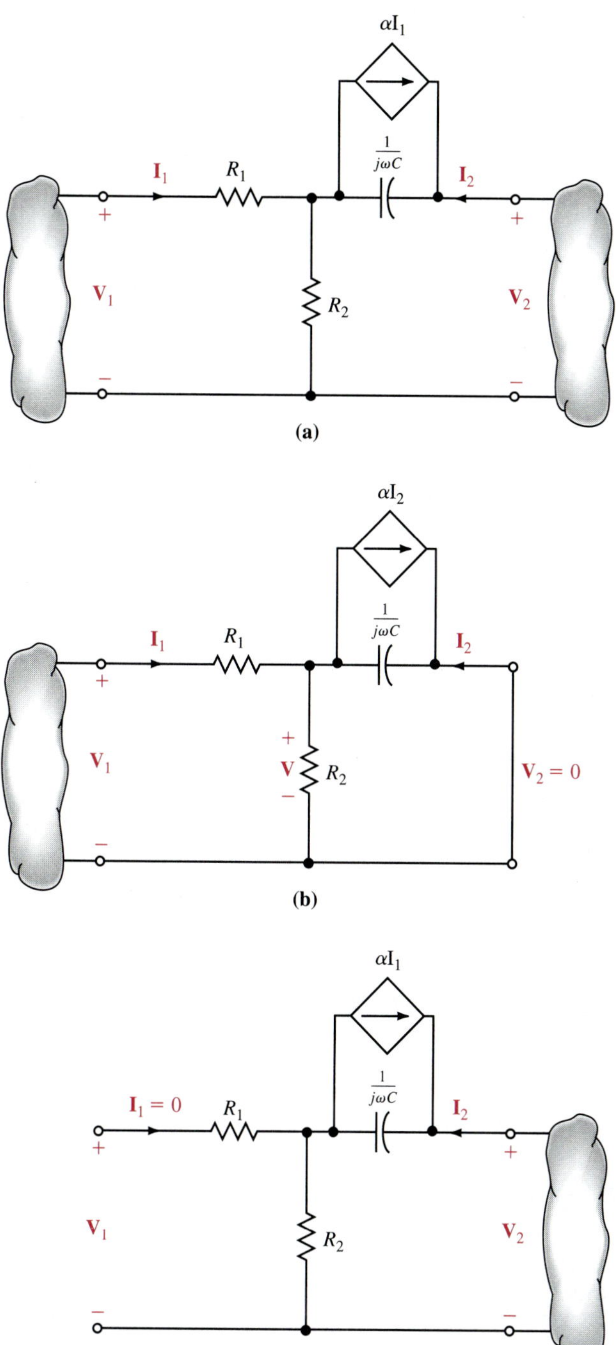

Figure 15.8 Circuits employed in Example 15.6.

Solving these two equations, we obtain

$$\mathbf{h}_{11} = R_1 + \frac{1-\alpha}{1/R_2 + j\omega C}$$

The parameter $\mathbf{h}_{12}$ is derived from Fig. 15.8c. Since $\mathbf{I}_1 = 0$, the voltage across R_2 is $\mathbf{V}_1$, and therefore,

$$\mathbf{V}_1 = \frac{\mathbf{V}_2}{R_2 + 1/j\omega C} R_2$$

from which we find that

$$\mathbf{h}_{12} = \frac{R_2}{R_2 + 1/j\omega C}$$

Figure 15.8b is used to compute $\mathbf{h}_{21}$. The two KCL equations for this network are

$$\mathbf{I}_1 = \frac{\mathbf{V}}{R_2} + j\omega C\mathbf{V} + \alpha\mathbf{I}_1$$

$$\mathbf{I}_2 = -(j\omega C\mathbf{V} + \alpha\mathbf{I}_1)$$

Solving these equations, we can show that

$$\mathbf{h}_{21} = -\frac{\alpha + j\omega CR_2}{1 + j\omega CR_2}$$

Finally, $\mathbf{h}_{22}$ is obtained from Fig. 15.8c. The KVL equation for this network is

$$\mathbf{V}_2 = \mathbf{I}_2\left(R_2 + \frac{1}{j\omega C}\right)$$

and hence

$$\mathbf{h}_{22} = \frac{1}{R_2 + 1/j\omega C}$$

Therefore, the two-port equations in terms of the hybrid parameters for the network in Fig. 15.8a are

$$\mathbf{V}_1 = \left(R_1 + \frac{1-\alpha}{1/R_2 + j\omega C}\right)\mathbf{I}_1 + \frac{R_2}{R_2 + 1/j\omega C}\mathbf{V}_2$$

$$\mathbf{I}_2 = -\frac{\alpha + j\omega CR_2}{1 + j\omega CR_2}\mathbf{I}_1 + \frac{1}{R_2 + 1/j\omega C}\mathbf{V}_2$$

■

EXAMPLE 15.7

An equivalent circuit for the op-amp in Fig. 15.9a is shown in Fig. 15.9b. We will determine the hybrid parameters for this network.

Parameter $\mathbf{h}_{11}$ is derived from Fig. 15.9c. With the output shorted, $\mathbf{h}_{11}$ is a function of only R_i, R_1, and R_2 and

$$\mathbf{h}_{11} = R_i + \frac{R_1 R_2}{R_1 + R_2}$$

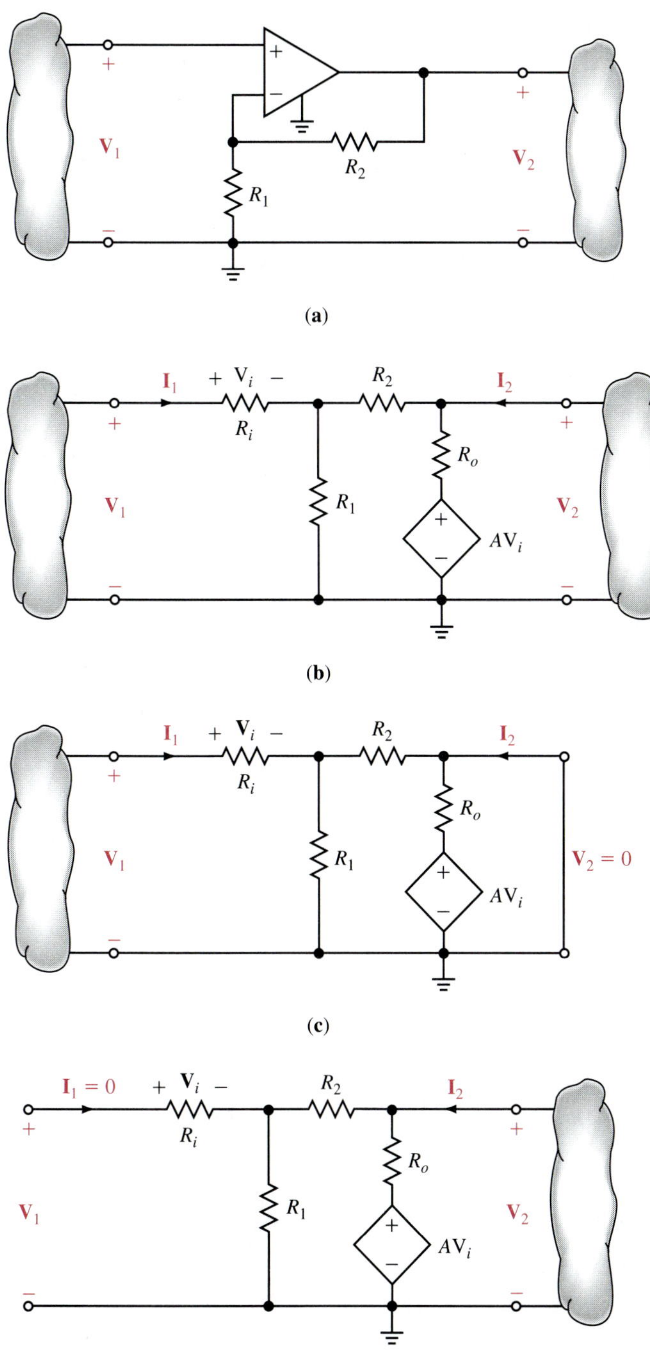

Figure 15.9 Circuits employed in Example 15.7.

Figure 15.9d is used to derive $\mathbf{h}_{12}$. Since $\mathbf{I}_1 = 0$, $\mathbf{V}_i = 0$ and the relationship between $\mathbf{V}_1$ and $\mathbf{V}_2$ is a simple voltage divider.

$$\mathbf{V}_1 = \frac{\mathbf{V}_2 R_1}{R_1 + R_2}$$

Therefore,

$$\mathbf{h}_{12} = \frac{R_1}{R_1 + R_2}$$

KVL and KCL can be applied to Fig. 15.9c to determine $\mathbf{h}_{21}$. The two equations that relate $\mathbf{I}_2$ to $\mathbf{I}_1$ are

$$\mathbf{V}_i = \mathbf{I}_1 R_i$$

$$\mathbf{I}_2 = \frac{-A\mathbf{V}_i}{R_o} - \frac{\mathbf{I}_1 R_1}{R_1 + R_2}$$

Therefore,

$$\mathbf{h}_{21} = -\left(\frac{AR_i}{R_o} + \frac{R_1}{R_1 + R_2}\right)$$

Finally, the relationship between $\mathbf{I}_2$ and $\mathbf{V}_2$ in Fig. 15.9d is

$$\frac{\mathbf{V}_2}{\mathbf{I}_2} = \frac{R_o(R_1 + R_2)}{R_o + R_1 + R_2}$$

and therefore,

$$\mathbf{h}_{22} = \frac{R_o + R_1 + R_2}{R_o(R_1 + R_2)}$$

The network equations are, therefore,

$$\mathbf{V}_1 = \left(R_i + \frac{R_1 R_2}{R_1 + R_2}\right)\mathbf{I}_1 + \frac{R_1}{R_1 + R_2}\mathbf{V}_2$$

$$\mathbf{I}_2 = -\left(\frac{AR_i}{R_o} + \frac{R_1}{R_1 + R_2}\right)\mathbf{I}_1 + \frac{R_o + R_1 + R_2}{R_o(R_1 + R_2)}\mathbf{V}_2$$

■

DRILL EXERCISE

D15.6. Find the hybrid parameters for the network shown in Fig. D15.1.

Ans: $\mathbf{h}_{11} = 14\ \Omega$, $\mathbf{h}_{12} = \frac{2}{3}$, $\mathbf{h}_{21} = -\frac{2}{3}$, $\mathbf{h}_{22} = \frac{1}{9}$ S.

D15.7. If a 4-Ω load is connected to the output port of the network examined in Drill Exercise D15.6, determine the input impedance of the two-port with the load connected.

Ans: $\mathbf{Z}_i = 15.23\ \Omega$.

D15.8. Compute the hybrid parameters for the network in Fig. D15.2.

Ans: $\mathbf{h}_{11} = 4\ \Omega$, $\mathbf{h}_{12} = \frac{2}{3}$, $\mathbf{h}_{21} = -\frac{2}{3}$, $\mathbf{h}_{22} = \frac{1}{9}$ S.

D15.9. Consider the network in Fig. D15.9. The two-port is a hybrid model for a basic transistor. Determine the voltage gain of the entire network, $\mathbf{V}_2/\mathbf{V}_S$, if a source $\mathbf{V}_S$ with internal resistance R_1 is applied at the input to the two-port and a load R_L is connected at the output port.

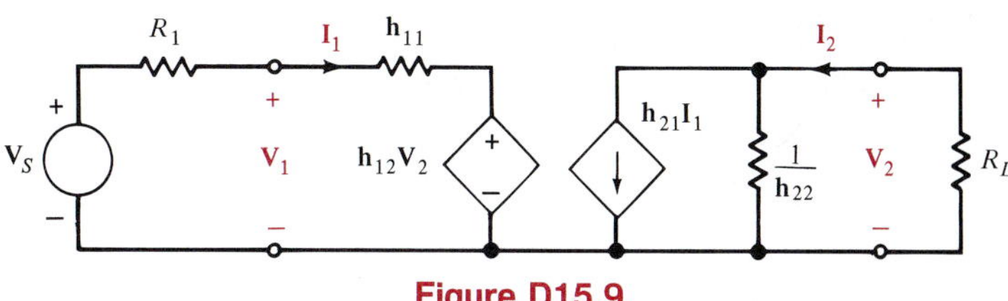

Figure D15.9

Ans: $\dfrac{\mathbf{V}_2}{\mathbf{V}_S} = \dfrac{-\mathbf{h}_{21}}{(R_1 + \mathbf{h}_{11})(\mathbf{h}_{22} + 1/R_L) - \mathbf{h}_{21}\mathbf{h}_{12}}$.

15.4 Transmission Parameters

The final parameters we will discuss are called the *transmission parameters*. They are defined by the equations

$$\mathbf{V}_1 = \mathbf{A}\mathbf{V}_2 - \mathbf{B}\mathbf{I}_2$$
$$\mathbf{I}_1 = \mathbf{C}\mathbf{V}_2 - \mathbf{D}\mathbf{I}_2 \tag{15.10}$$

or in matrix form,

$$\begin{bmatrix} \mathbf{V}_1 \\ \mathbf{I}_1 \end{bmatrix} = \begin{bmatrix} \mathbf{A} & \mathbf{B} \\ \mathbf{C} & \mathbf{D} \end{bmatrix} \begin{bmatrix} \mathbf{V}_2 \\ -\mathbf{I}_2 \end{bmatrix} \tag{15.11}$$

These parameters are very useful in the analysis of circuits connected in cascade, as we will demonstrate later. The parameters are determined via the following equations:

$$\mathbf{A} = \left.\frac{\mathbf{V}_1}{\mathbf{V}_2}\right|_{\mathbf{I}_2=0}$$
$$\mathbf{B} = \left.\frac{\mathbf{V}_1}{-\mathbf{I}_2}\right|_{\mathbf{V}_2=0}$$
$$\mathbf{C} = \left.\frac{\mathbf{I}_1}{\mathbf{V}_2}\right|_{\mathbf{I}_2=0}$$
$$\mathbf{D} = \left.\frac{\mathbf{I}_1}{-\mathbf{I}_2}\right|_{\mathbf{V}_2=0} \tag{15.12}$$

A, **B**, **C**, and **D** represent the *open-circuit voltage ratio*, the *negative short-circuit transfer impedance*, the *open-circuit transfer admittance*, and the *negative short-circuit current ratio*, respectively. For obvious reasons the transmission parameters are commonly referred to as the *ABCD parameters*.

EXAMPLE 15.8

Let us determine the transmission parameters for the network in Fig. 15.10a.

The parameter **A** is determined from Fig. 15.10b. Since $\mathbf{I}_2 = 0$, $\mathbf{V}_1$ and $\mathbf{V}_2$ are related via the two KVL equations

$$\mathbf{V}_2 = \gamma\mathbf{I}_1 + \mathbf{I}_1 R_2$$

$$\mathbf{V}_1 = \mathbf{I}_1(R_1 + R_2)$$

Solving these equations for $\mathbf{V}_1/\mathbf{V}_2$ yields

$$\mathbf{A} = \frac{R_1 + R_2}{\gamma + R_2}$$

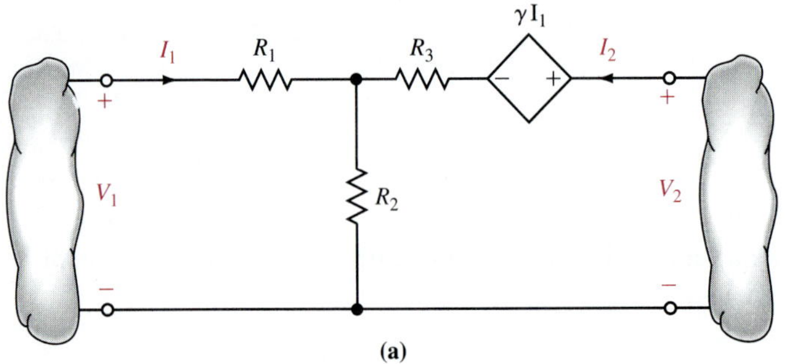

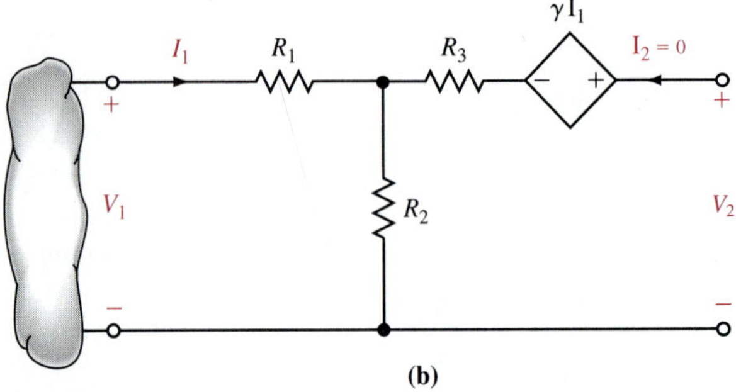

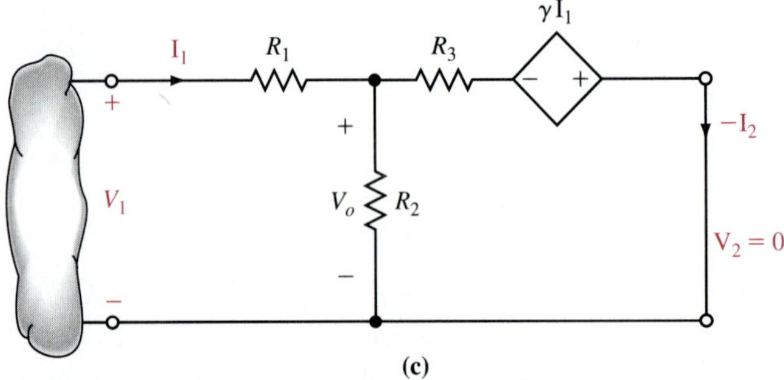

Figure 15.10 Circuits used in Example 15.8.

Parameter **B** is obtained from Fig. 15.10c. In this case it is convenient to assume that $-\mathbf{I}_2 = 1$ A and solve for $\mathbf{V}_1$. Under this condition

$$\mathbf{V}_o = R_3 - \gamma \mathbf{I}_1$$

and

$$\mathbf{I}_1 = 1 + \frac{\mathbf{V}_o}{R_2}$$

Therefore,

$$\mathbf{I}_1 = \frac{R_2 + R_3}{R_2 + \gamma}$$

Then

$$\begin{aligned}\mathbf{V}_1 &= \mathbf{I}_1 R_1 + \mathbf{V}_o \\ &= \frac{R_2 + R_3}{R_2 + \gamma} R_1 + R_3 - \gamma \frac{R_2 + R_3}{R_2 + \gamma} \\ &= R_3 + \frac{(R_1 - \gamma)(R_2 + R_3)}{R_2 + \gamma}\end{aligned}$$

and hence

$$\mathbf{B} = R_3 + \frac{(R_1 - \gamma)(R_2 + R_3)}{R_2 + \gamma}$$

In Fig. 15.10b if we assume that $\mathbf{I}_1 = 1$ A, then $\mathbf{V}_2$ is

$$\mathbf{V}_2 = R_2 + \gamma$$

and therefore

$$\mathbf{C} = \frac{1}{R_2 + \gamma}$$

Figure 15.10c is employed to determine parameter **D**. In our analysis above, in which we derived parameter **B**, we found that if we assumed that $-\mathbf{I}_2 = 1$ A, then

$$\mathbf{I}_1 = \frac{R_2 + R_3}{R_2 + \gamma}$$

and therefore,

$$\mathbf{D} = \frac{R_2 + R_3}{R_2 + \gamma}$$

The two-port equations for the network are, therefore,

$$\mathbf{V}_1 = \frac{R_1 + R_2}{R_2 + \gamma} \mathbf{V}_2 - \left[R_3 + \frac{(R_1 - \gamma)(R_2 + R_3)}{R_2 + \gamma} \right] \mathbf{I}_2$$

$$\mathbf{I}_1 = \frac{1}{R_2 + \gamma} \mathbf{V}_2 - \frac{R_2 + R_3}{R_2 + \gamma} \mathbf{I}_2$$

■

EXAMPLE 15.9

We will now determine the transmission parameters for the network in Fig. 15.11a. The most convenient way in which to attack this type of problem is to assume a value for one of the variables in a parameter and solve the network for the other variable using KVL and KCL. To simplify our analysis, we have also defined variables $\mathbf{V}_A$, $\mathbf{V}_B$, and $\mathbf{I}_o$ as shown in Fig. 15.11b and c.

Parameter $\mathbf{A}$ is derived from Fig. 15.11b. If we assume that $\mathbf{V}_2 = 1$ V, then $\mathbf{V}_B = 1$ V and $\mathbf{I}_o = j\omega$ A, and hence

$$\mathbf{V}_A = 2\mathbf{I}_o + \mathbf{V}_B = 2j\omega + 1 \text{ V}$$

Then

$$\begin{aligned}\mathbf{I}_1 &= j\omega\mathbf{V}_A + \mathbf{I}_o \\ &= -2\omega^2 + 2j\omega \text{ A}\end{aligned}$$

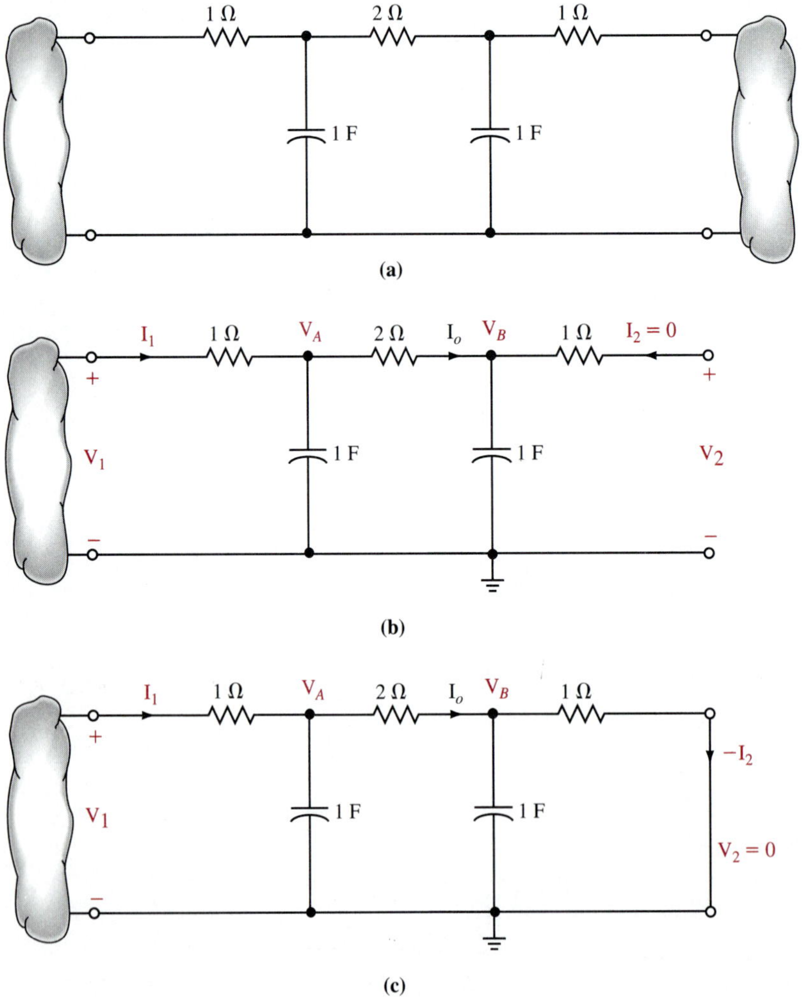

Figure 15.11 Circuits used in Example 15.9.

Finally,

$$\begin{aligned}\mathbf{V}_1 &= (1)\mathbf{I}_1 + \mathbf{V}_A \\ &= -2\omega^2 + 4j\omega + 1 \text{ V}\end{aligned}$$

and therefore,

$$\mathbf{A} = -2\omega^2 + 4j\omega + 1$$

Figure 15.11c is used to derive parameter **B**. Assuming that $-\mathbf{I}_2 = 1$ A, then $\mathbf{V}_B = 1$ V, $\mathbf{I}_o = 1 + j\omega$ A, and

$$\mathbf{V}_A = 2\mathbf{I}_o + \mathbf{V}_B = 3 + j2\omega \text{ V}$$

Then

$$\begin{aligned}\mathbf{I}_1 &= \mathbf{I}_o + j\omega\mathbf{V}_A \\ &= (1 + j\omega) + j\omega(3 + 2\omega) \\ &= 1 + 4j\omega - 2\omega^2 \text{ A}\end{aligned}$$

Finally,

$$\begin{aligned}\mathbf{V}_1 &= (1)\mathbf{I}_1 + \mathbf{V}_A \\ &= 4 + 6j\omega - 2\omega^2 \text{ V}\end{aligned}$$

Therefore,

$$\mathbf{B} = 4 + 6j\omega - 2\omega^2$$

From the analysis above used to derive parameter **A**, we find that

$$\mathbf{C} = -2\omega^2 + 2j\omega$$

In a similar manner it follows from the derivation of parameter **B** that

$$\mathbf{D} = -2\omega^2 + 4j\omega + 1$$

Therefore, the two-port equations are

$$\begin{aligned}\mathbf{V}_1 &= (-2\omega^2 + 4j\omega + 1)\mathbf{V}_2 - (-2\omega^2 + 6j\omega + 4)\mathbf{I}_2 \\ \mathbf{I}_1 &= (-2\omega^2 + 2j\omega)\mathbf{V}_2 - (-2\omega^2 + 4j\omega + 1)\mathbf{I}_2\end{aligned}$$

■

DRILL EXERCISE

D15.10. Find the transmission parameters for the network shown in Fig. D15.1.

Ans: $\mathbf{A} = 3$, $\mathbf{B} = 21\ \Omega$, $\mathbf{C} = \frac{1}{6}$ S, $\mathbf{D} = \frac{3}{2}$.

D15.11. Compute the transmission parameters for the two-port in Fig. D15.2.

Ans: $\mathbf{A} = 3$, $\mathbf{B} = 21\ \Omega$, $\mathbf{C} = \frac{1}{6}$ S, $\mathbf{D} = \frac{3}{2}$.

One final note concerning the derivation of two-port parameters is in order. It is entirely possible that some or all of the two-port parameters for a particular network may not exist. For example, consider the network in Fig. 15.12. The parameter $\mathbf{y}_{11}$ is determined by short-circuiting the output and computing the ratio $\mathbf{I}_1/\mathbf{V}_1$. However, shorting the output implies that $2\mathbf{I}_o = 0$, and therefore $\mathbf{I}_o = 0$. However, $\mathbf{I}_o$ cannot be zero if a nonzero voltage $\mathbf{V}_1$ is applied at the input. Therefore, this network does not have a set of Y parameters.

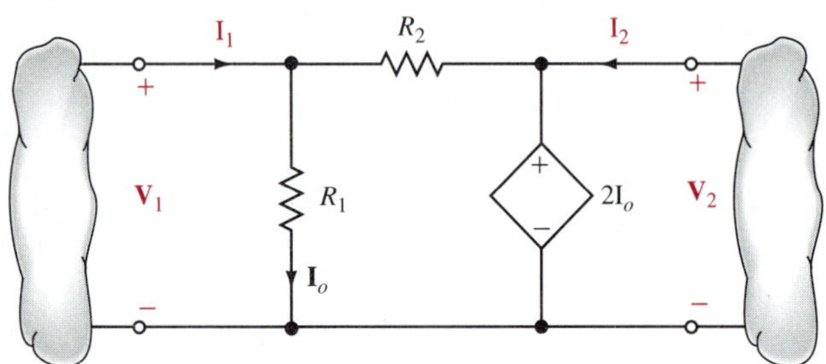

Figure 15.12 Example network that does not have a set of Y parameters unless the source has an internal impedance.

15.5 Equivalent Circuits

A number of standard equivalent circuits can be employed to represent a two-port network. These equivalent circuits have the same terminal characteristics as the original network, although they may not be physically realizable because they may possibly contain negative impedances. The equivalent circuits that are perhaps the most popular are shown in Fig. 15.13. Since there are four coefficients in the two-port equations, each of the circuits contains four elements. The reader can easily verify the validity of the circuits in Fig. 15.13 by writing the circuit equations that characterize each.

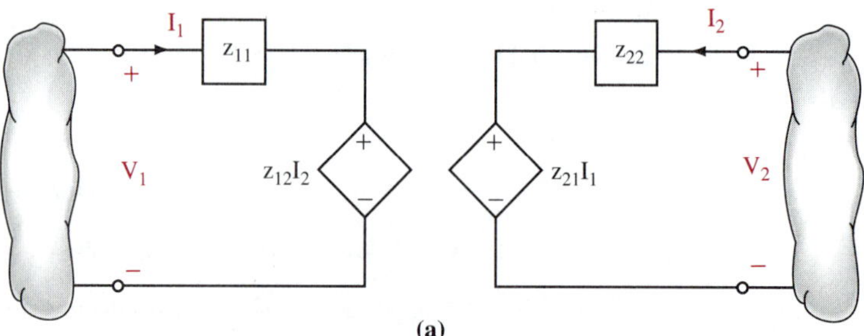

Figure 15.13 Some standard equivalent circuits for two-ports.

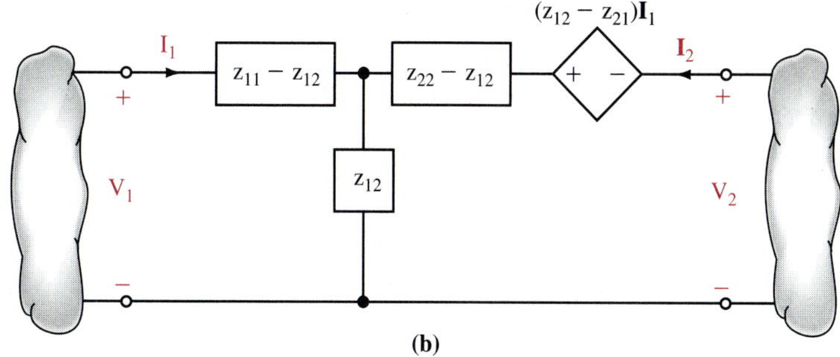

(b)

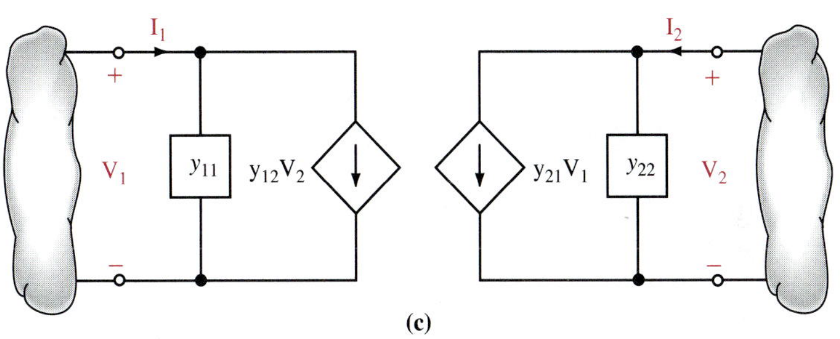

(c)

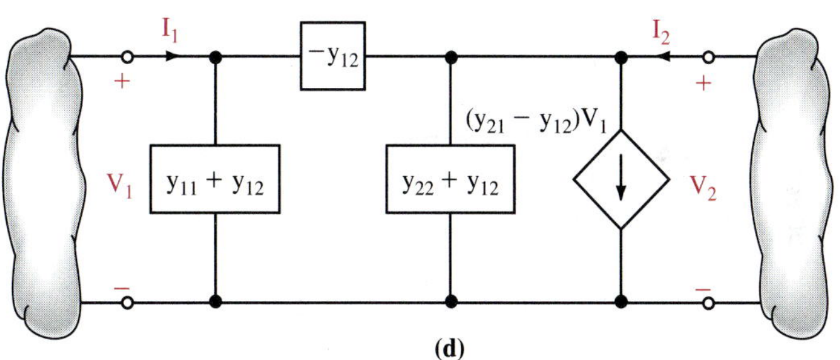

(d)

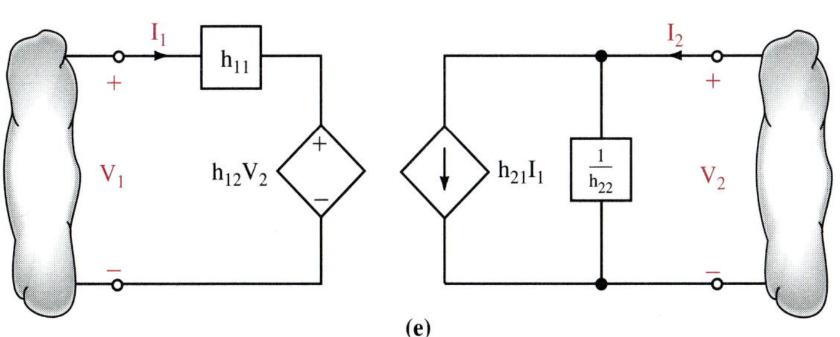

(e)

Figure 15.13 (continued)

ter C ersions

If all the two-port parameters for a network exist, it is possible to relate one set of parameters to another since the parameters interrelate the variables $\mathbf{V}_1$, $\mathbf{I}_1$, $\mathbf{V}_2$, and $\mathbf{I}_2$. The following examples illustrate the manner in which to determine one set of parameters from another.

EXAMPLE 15.10

Let us show how to determine the hybrid parameters if the Z parameters are known.

The two-port equations involving the Z parameters are

$$\mathbf{V}_1 = \mathbf{z}_{11}\mathbf{I}_1 + \mathbf{z}_{12}\mathbf{I}_2$$

$$\mathbf{V}_2 = \mathbf{z}_{21}\mathbf{I}_1 + \mathbf{z}_{22}\mathbf{I}_2$$

If we solve the second Z-parameter equation for $\mathbf{I}_2$, it will be directly in the form of the second hybrid-parameter equation. Therefore,

$$\mathbf{I}_2 = \frac{-\mathbf{z}_{21}}{\mathbf{z}_{22}}\mathbf{I}_1 + \frac{1}{\mathbf{z}_{22}}\mathbf{V}_2$$

Substituting this equation for $\mathbf{I}_2$ into the first Z-parameter equation, we obtain

$$\mathbf{V}_1 = \frac{\mathbf{z}_{11}\mathbf{z}_{22} - \mathbf{z}_{12}\mathbf{z}_{21}}{\mathbf{z}_{22}}\mathbf{I}_1 + \frac{\mathbf{z}_{12}}{\mathbf{z}_{22}}\mathbf{V}_2$$

Comparing the equations for $\mathbf{I}_2$ and $\mathbf{V}_1$ in terms of the Z parameters with the defining equations for the hybrid parameters shows that

$$\mathbf{h}_{11} = \frac{\Delta_Z}{\mathbf{z}_{22}} \qquad \mathbf{h}_{12} = \frac{\mathbf{z}_{12}}{\mathbf{z}_{22}}$$

$$\mathbf{h}_{21} = \frac{-\mathbf{z}_{21}}{\mathbf{z}_{22}} \qquad \mathbf{h}_{22} = \frac{1}{\mathbf{z}_{22}}$$

where $\Delta_Z = \mathbf{z}_{11}\mathbf{z}_{22} - \mathbf{z}_{12}\mathbf{z}_{21}$. ■

EXAMPLE 15.11

Let us determine the Z parameters from the Y parameters. We can easily illustrate the conversion using matrix analysis because of the nature of the two sets of equations.

In matrix form we can write

$$\begin{bmatrix} \mathbf{z}_{11} & \mathbf{z}_{12} \\ \mathbf{z}_{21} & \mathbf{z}_{22} \end{bmatrix} \begin{bmatrix} \mathbf{I}_1 \\ \mathbf{I}_2 \end{bmatrix} = \begin{bmatrix} \mathbf{V}_1 \\ \mathbf{V}_2 \end{bmatrix}$$

and

$$\begin{bmatrix} \mathbf{y}_{11} & \mathbf{y}_{12} \\ \mathbf{y}_{21} & \mathbf{y}_{22} \end{bmatrix} \begin{bmatrix} \mathbf{V}_1 \\ \mathbf{V}_2 \end{bmatrix} = \begin{bmatrix} \mathbf{I}_1 \\ \mathbf{I}_2 \end{bmatrix}$$

From the second equation we note that

$$\begin{bmatrix} \mathbf{V}_1 \\ \mathbf{V}_2 \end{bmatrix} = \begin{bmatrix} \mathbf{y}_{11} & \mathbf{y}_{12} \\ \mathbf{y}_{21} & \mathbf{y}_{22} \end{bmatrix}^{-1} \begin{bmatrix} \mathbf{I}_1 \\ \mathbf{I}_2 \end{bmatrix}$$

and hence we can determine the Z parameters by inverting the matrix for the Y parameters (i.e., $[\mathbf{Z}] = [\mathbf{Y}]^{-1}$). The determinant for the matrix is

$$\Delta_Y = \mathbf{y}_{11}\mathbf{y}_{22} - \mathbf{y}_{12}\mathbf{y}_{21}$$

The adjoint of the matrix is

$$\begin{bmatrix} \mathbf{y}_{22} & -\mathbf{y}_{12} \\ -\mathbf{y}_{21} & \mathbf{y}_{11} \end{bmatrix}$$

Therefore, the inverse is

$$\frac{1}{\Delta_Y}\begin{bmatrix} \mathbf{y}_{22} & -\mathbf{y}_{12} \\ -\mathbf{y}_{21} & \mathbf{y}_{11} \end{bmatrix}$$

and hence the Z parameters in terms of the Y parameters are

$$\mathbf{z}_{11} = \frac{\mathbf{y}_{22}}{\Delta_Y} \qquad \mathbf{z}_{12} = \frac{-\mathbf{y}_{12}}{\Delta_Y}$$

$$\mathbf{z}_{21} = \frac{-\mathbf{y}_{21}}{\Delta_Y} \qquad \mathbf{z}_{22} = \frac{\mathbf{y}_{11}}{\Delta_Y}$$

■

Table 15.1 lists all the conversion formulas that relate one set of two-port parameters to another. Note that Δ_Z, Δ_Y, Δ_H, and Δ_T refer to the determinants of the matrices for the Z, Y, hybrid, and ABCD parameters, respectively. Therefore, given one set of parameters for a network, we can use Table 15.1 to find others.

DRILL EXERCISE

D15.12. Determine the Y parameters for a two-port if the Z parameters are

$$\mathbf{Z} = \begin{bmatrix} 18 & 6 \\ 6 & 9 \end{bmatrix}$$

Ans: $\mathbf{y}_{11} = \frac{1}{14}$ S, $\mathbf{y}_{12} = \mathbf{y}_{21} = -\frac{1}{21}$ S, $\mathbf{y}_{22} = \frac{1}{7}$ S.

D15.13. Given the Z parameters for a two-port in Drill Exercise D15.12, determine the hybrid parameters for this network.

Ans: $\mathbf{h}_{11} = 14\ \Omega$, $\mathbf{h}_{12} = \frac{2}{3}$, $\mathbf{h}_{21} = -\frac{2}{3}$, $\mathbf{h}_{22} = \frac{1}{9}$ S.

Table 15.1 Two-Port Parameter Conversion Formulas

$\begin{bmatrix} \mathbf{z}_{11} & \mathbf{z}_{12} \\ \mathbf{z}_{21} & \mathbf{z}_{22} \end{bmatrix}$	$\begin{bmatrix} \frac{\mathbf{y}_{22}}{\Delta_Y} & \frac{-\mathbf{y}_{12}}{\Delta_Y} \\ \frac{-\mathbf{y}_{21}}{\Delta_Y} & \frac{\mathbf{y}_{11}}{\Delta_Y} \end{bmatrix}$	$\begin{bmatrix} \frac{\mathbf{A}}{\mathbf{C}} & \frac{\Delta_T}{\mathbf{C}} \\ \frac{1}{\mathbf{C}} & \frac{\mathbf{D}}{\mathbf{C}} \end{bmatrix}$	$\begin{bmatrix} \frac{\Delta_H}{\mathbf{h}_{22}} & \frac{\mathbf{h}_{12}}{\mathbf{h}_{22}} \\ \frac{-\mathbf{h}_{21}}{\mathbf{h}_{22}} & \frac{1}{\mathbf{h}_{22}} \end{bmatrix}$
$\begin{bmatrix} \frac{\mathbf{z}_{22}}{\Delta_Z} & \frac{-\mathbf{z}_{12}}{\Delta_Z} \\ \frac{-\mathbf{z}_{21}}{\Delta_Z} & \frac{\mathbf{z}_{11}}{\Delta_Z} \end{bmatrix}$	$\begin{bmatrix} \mathbf{y}_{11} & \mathbf{y}_{12} \\ \mathbf{y}_{21} & \mathbf{y}_{22} \end{bmatrix}$	$\begin{bmatrix} \frac{\mathbf{D}}{\mathbf{B}} & \frac{-\Delta_T}{\mathbf{B}} \\ -\frac{1}{\mathbf{B}} & \frac{\mathbf{A}}{\mathbf{B}} \end{bmatrix}$	$\begin{bmatrix} \frac{1}{\mathbf{h}_{11}} & \frac{-\mathbf{h}_{12}}{\mathbf{h}_{11}} \\ \frac{\mathbf{h}_{21}}{\mathbf{h}_{11}} & \frac{\Delta_H}{\mathbf{h}_{11}} \end{bmatrix}$
$\begin{bmatrix} \frac{\mathbf{z}_{11}}{\mathbf{z}_{21}} & \frac{\Delta_Z}{\mathbf{z}_{21}} \\ \frac{1}{\mathbf{z}_{21}} & \frac{\mathbf{z}_{22}}{\mathbf{z}_{21}} \end{bmatrix}$	$\begin{bmatrix} \frac{-\mathbf{y}_{22}}{\mathbf{y}_{21}} & \frac{-1}{\mathbf{y}_{21}} \\ \frac{-\Delta_Y}{\mathbf{y}_{21}} & \frac{-\mathbf{y}_{11}}{\mathbf{y}_{21}} \end{bmatrix}$	$\begin{bmatrix} \mathbf{A} & \mathbf{B} \\ \mathbf{C} & \mathbf{D} \end{bmatrix}$	$\begin{bmatrix} \frac{-\Delta_H}{\mathbf{h}_{21}} & \frac{-\mathbf{h}_{11}}{\mathbf{h}_{21}} \\ \frac{-\mathbf{h}_{22}}{\mathbf{h}_{21}} & \frac{-1}{\mathbf{h}_{21}} \end{bmatrix}$
$\begin{bmatrix} \frac{\Delta_Z}{\mathbf{z}_{22}} & \frac{\mathbf{z}_{12}}{\mathbf{z}_{22}} \\ \frac{-\mathbf{z}_{21}}{\mathbf{z}_{22}} & \frac{1}{\mathbf{z}_{22}} \end{bmatrix}$	$\begin{bmatrix} \frac{1}{\mathbf{y}_{11}} & \frac{-\mathbf{y}_{12}}{\mathbf{y}_{11}} \\ \frac{\mathbf{y}_{21}}{\mathbf{y}_{11}} & \frac{\Delta_Y}{\mathbf{y}_{11}} \end{bmatrix}$	$\begin{bmatrix} \frac{\mathbf{B}}{\mathbf{D}} & \frac{\Delta_T}{\mathbf{D}} \\ -\frac{1}{\mathbf{D}} & \frac{\mathbf{C}}{\mathbf{D}} \end{bmatrix}$	$\begin{bmatrix} \mathbf{h}_{11} & \mathbf{h}_{12} \\ \mathbf{h}_{21} & \mathbf{h}_{22} \end{bmatrix}$

15.7 Interconnection of Two-Ports

In this section we will illustrate techniques for treating a network as a combination of subnetworks. We will, therefore, analyze a two-port network as an interconnection of simpler two-ports. Although two-ports can be interconnected in a variety of ways, we will treat only three types of connections: parallel, series, and cascade.

Parallel Interconnection

Suppose that a two-port N is composed of two two-ports N_a and N_b which are interconnected as shown in Fig. 15.14a. The defining equations for networks N_a and N_b are

$$\begin{aligned} \mathbf{I}_{1a} &= \mathbf{y}_{11a}\mathbf{V}_{1a} + \mathbf{y}_{12a}\mathbf{V}_{2a} \\ \mathbf{I}_{2a} &= \mathbf{y}_{21a}\mathbf{V}_{1a} + \mathbf{y}_{22a}\mathbf{V}_{2a} \end{aligned} \tag{15.13}$$

$$\begin{aligned} \mathbf{I}_{1b} &= \mathbf{y}_{11b}\mathbf{V}_{1b} + \mathbf{y}_{12b}\mathbf{V}_{2b} \\ \mathbf{I}_{2b} &= \mathbf{y}_{21b}\mathbf{V}_{1b} + \mathbf{y}_{22b}\mathbf{V}_{2b} \end{aligned} \tag{15.14}$$

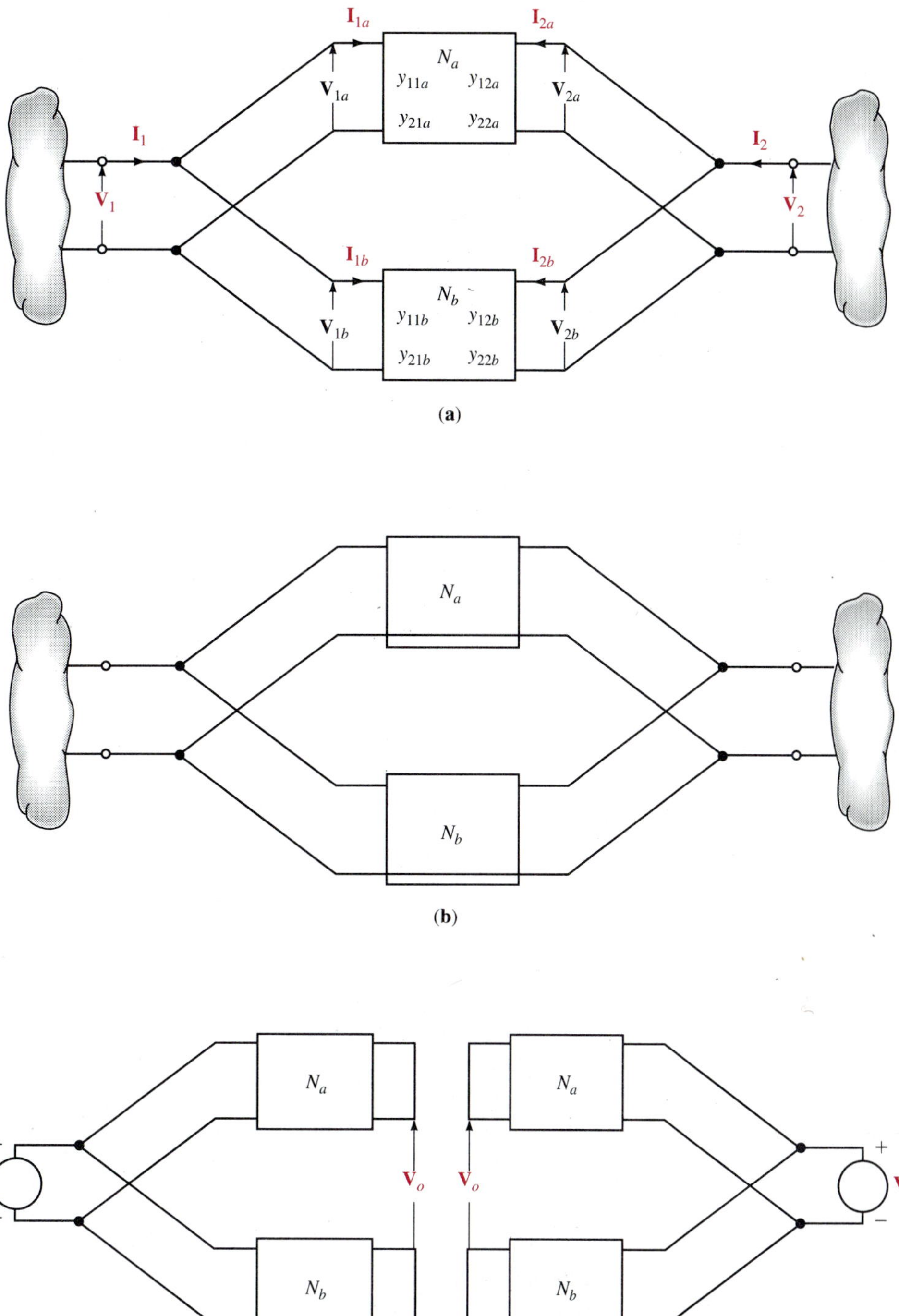

Figure 15.14 Some circuits describing the parallel interconnection of two-ports.

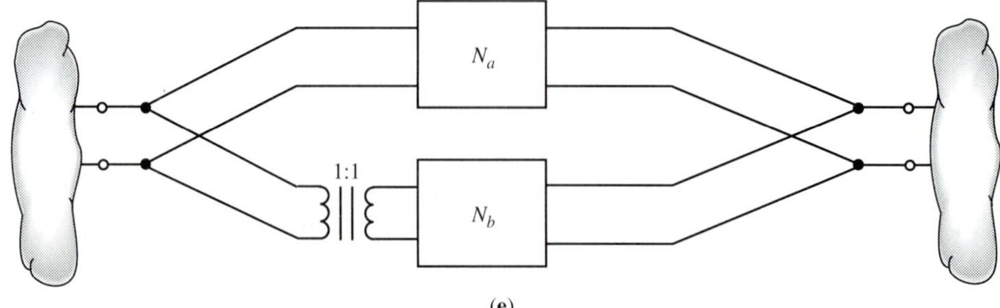

(e)

Figure 15.14 (continued)

Provided that the terminal characteristics of the two networks N_a and N_b are not altered by the interconnection illustrated in Fig. 15.14a, then

$$\begin{aligned} \mathbf{V}_1 &= \mathbf{V}_{1a} = \mathbf{V}_{1b} \\ \mathbf{V}_2 &= \mathbf{V}_{2a} = \mathbf{V}_{2b} \\ \mathbf{I}_1 &= \mathbf{I}_{1a} + \mathbf{I}_{1b} \\ \mathbf{I}_2 &= \mathbf{I}_{2a} + \mathbf{I}_{2b} \end{aligned} \tag{15.15}$$

$$\begin{aligned} \mathbf{I}_1 &= (\mathbf{y}_{11a} + \mathbf{y}_{11b})\mathbf{V}_1 + (\mathbf{y}_{12a} + \mathbf{y}_{12b})\mathbf{V}_2 \\ \mathbf{I}_2 &= (\mathbf{y}_{21a} + \mathbf{y}_{21b})\mathbf{V}_1 + (\mathbf{y}_{22a} + \mathbf{y}_{22b})\mathbf{V}_2 \end{aligned} \tag{15.16}$$

Therefore, the Y parameters for the total network are

$$\begin{bmatrix} \mathbf{y}_{11} & \mathbf{y}_{12} \\ \mathbf{y}_{21} & \mathbf{y}_{22} \end{bmatrix} = \begin{bmatrix} \mathbf{y}_{11a} + \mathbf{y}_{11b} & \mathbf{y}_{12a} + \mathbf{y}_{12b} \\ \mathbf{y}_{21a} + \mathbf{y}_{21b} & \mathbf{y}_{22a} + \mathbf{y}_{22b} \end{bmatrix} \tag{15.17}$$

and hence to determine the Y parameters for the total network, we simply add the Y parameters of the two networks N_a and N_b. Equation (15.16) is always valid if the two networks are of the form shown in Fig. 15.14b. In the general case, Eq. (15.16) is valid if the networks satisfy the Brune tests (named after O. Brune, who developed these procedures). The Brune tests require that the voltage $\mathbf{V}_o$ be zero when the networks N_a and N_b are interconnected as shown in Fig. 15.14c and d. If networks N_a and N_b fail these tests, a one-to-one ideal transformer can be employed to ensure that Eqs. (15.16) are satisfied. One possible location of the transformer is shown in Fig. 15.14e; however, the transformer may actually be located at any one of the four ports shown in Fig. 15.14a. An example in which an ideal transformer is required to provide the necessary isolation is shown in Fig. 15.15. Note that if the two two-ports are connected in parallel without the use of an ideal transformer, impedances $\mathbf{z}_7$ and $\mathbf{z}_8$ are short-circuited by the interconnection, and hence Eq. (15.16) does not apply.

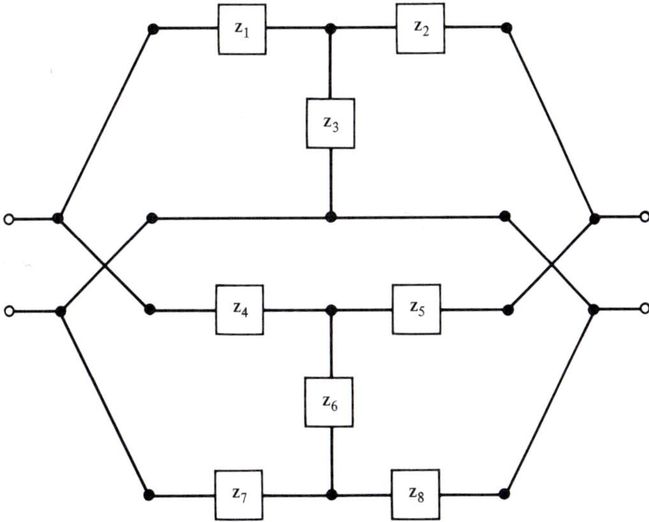

Figure 15.15 Example where parallel connection of two-ports cannot be performed without the use of an ideal transformer.

EXAMPLE 15.12

We wish to determine the Y parameters for the network shown in Fig. 15.16a by considering it to be a parallel combination of two networks as shown in Figure 15.16b. The capacitive network will be referred to as N_a and the resistive network will be referred to as N_b.

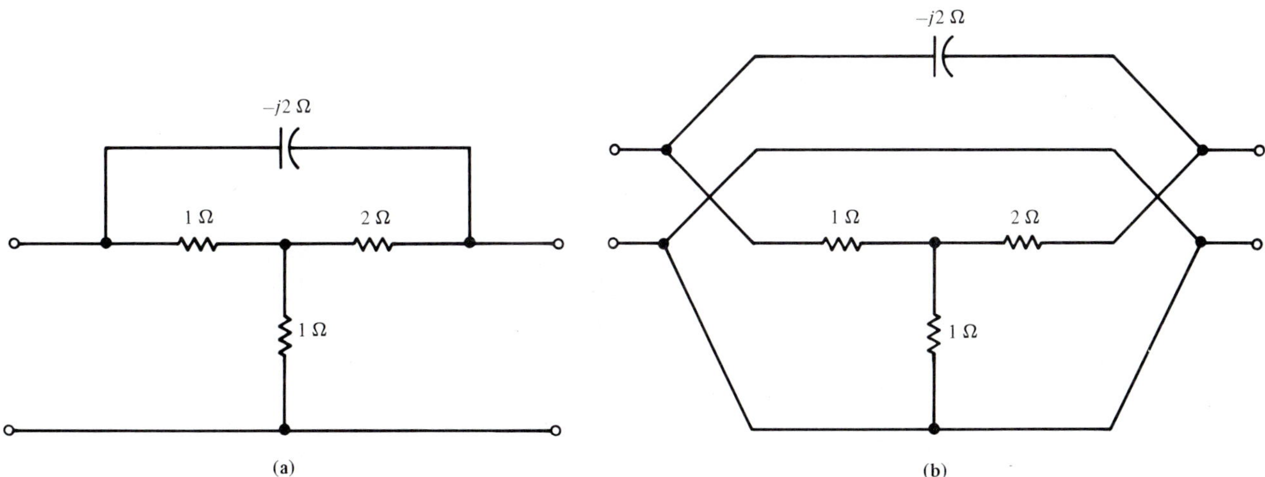

Figure 15.16 Network composed of the parallel combination of two subnetworks.

The Y parameters for N_a are

$$\mathbf{y}_{11a} = j\tfrac{1}{2}\text{ S} \qquad \mathbf{y}_{12a} = -j\tfrac{1}{2}\text{ S}$$

$$\mathbf{y}_{21a} = -j\tfrac{1}{2}\text{ S} \qquad \mathbf{y}_{22a} = j\tfrac{1}{2}\text{ S}$$

and the Y parameters for N_b are

$$\mathbf{y}_{11b} = \tfrac{3}{5}\text{ S} \qquad \mathbf{y}_{12b} = -\tfrac{1}{5}\text{ S}$$

$$\mathbf{y}_{21b} = -\tfrac{1}{5}\text{ S} \qquad \mathbf{y}_{22b} = \tfrac{2}{5}\text{ S}$$

Hence the Y parameters for the network in Fig. 15.16 are

$$\mathbf{y}_{11} = \tfrac{3}{5} + j\tfrac{1}{2}\text{ S} \qquad \mathbf{y}_{12} = -(\tfrac{1}{5} + j\tfrac{1}{2})\text{ S}$$

$$\mathbf{y}_{21} = -(\tfrac{1}{5} + j\tfrac{1}{2})\text{ S} \qquad \mathbf{y}_{22} = \tfrac{2}{5} + j\tfrac{1}{2}\text{ S}$$

To gain an appreciation for the simplicity of this approach, the reader need only try to find the Y parameters for the network in Fig. 15.16a directly. ■

EXAMPLE 15.13

Consider the amplifier equivalent circuit shown in Fig. 15.17a. We will derive the Y parameters for this network via the parallel interconnection scheme shown in Fig. 15.17b.

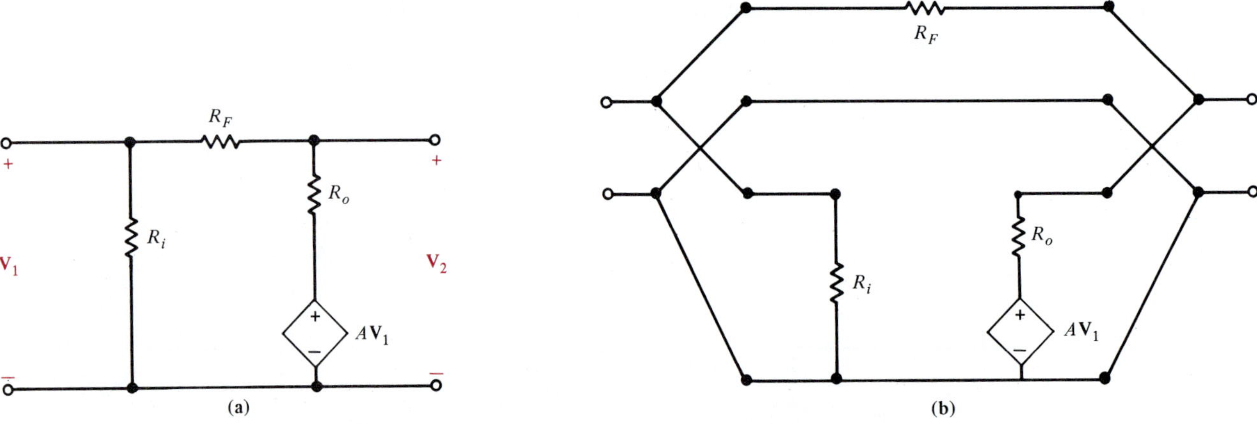

Figure 15.17 Amplifier circuit composed of two parallel networks.

We will refer to the upper network in Fig. 15.17b as N_a and the lower network as N_b. The Y parameters for N_a are

$$\mathbf{y}_{11a} = \frac{1}{R_F} \qquad \mathbf{y}_{12a} = -\frac{1}{R_F}$$

$$\mathbf{y}_{21a} = -\frac{1}{R_F} \qquad \mathbf{y}_{22a} = \frac{1}{R_F}$$

and the Y parameters for N_b are

$$\mathbf{y}_{11b} = \frac{1}{R_i} \qquad \mathbf{y}_{12b} = 0$$

$$\mathbf{y}_{21b} = \frac{-A}{R_o} \qquad \mathbf{y}_{22b} = \frac{1}{R_o}$$

Hence the Y parameters for the network in Fig. 15.17a are

$$\mathbf{y}_{11} = \frac{1}{R_F} + \frac{1}{R_i} \qquad \mathbf{y}_{12} = -\frac{1}{R_F}$$

$$\mathbf{y}_{21} = -\left(\frac{1}{R_F} + \frac{A}{R_o}\right) \qquad \mathbf{y}_{22} = \frac{1}{R_F} + \frac{1}{R_o}$$

■

DRILL EXERCISE

D15.14. Find the Y parameters of the network in Fig. D15.2 by considering the network to be a parallel interconnection of two two-ports as shown in Fig. D15.14.

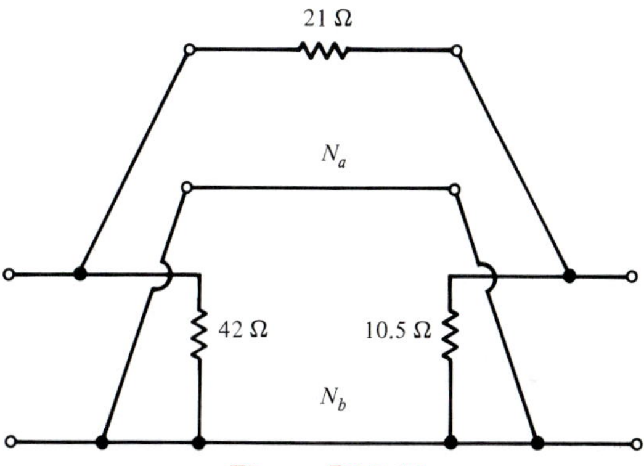

Figure D15.14

Ans: $\mathbf{y}_{11} = \frac{1}{14}$ S, $\mathbf{y}_{12} = \mathbf{y}_{21} = -\frac{1}{21}$ S, $\mathbf{y}_{22} = \frac{1}{7}$ S.

Series Interconnection

Consider the two-port N, which is composed of the series connection of N_a and N_b, as shown in Fig. 15.18a. The defining equations for the individual two-port networks N_a and N_b are

$$\begin{aligned} \mathbf{V}_{1a} &= \mathbf{z}_{11a}\mathbf{I}_{1a} + \mathbf{z}_{12a}\mathbf{I}_{2a} \\ \mathbf{V}_{2a} &= \mathbf{z}_{21a}\mathbf{I}_{1a} + \mathbf{z}_{22a}\mathbf{I}_{2a} \end{aligned} \tag{15.18}$$

$$\begin{aligned} \mathbf{V}_{1b} &= \mathbf{z}_{11b}\mathbf{I}_{1b} + \mathbf{z}_{12b}\mathbf{I}_{2b} \\ \mathbf{V}_{2b} &= \mathbf{z}_{21b}\mathbf{I}_{1b} + \mathbf{z}_{22b}\mathbf{I}_{2b} \end{aligned} \tag{15.19}$$

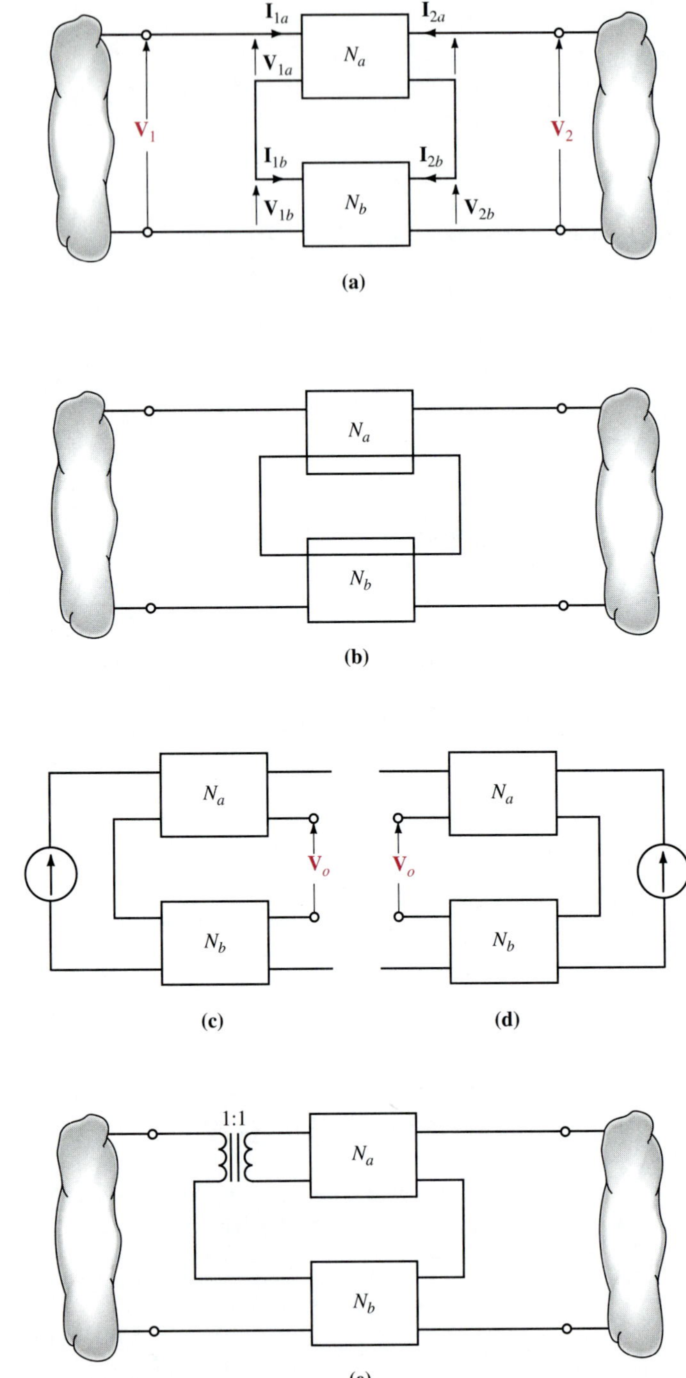

Figure 15.18 Some circuits describing the series interconnection of two-ports.

Once again, as long as the terminal characteristics of the two networks N_a and N_b are not altered by the series interconnection, then

$$\begin{aligned} \mathbf{I}_1 &= \mathbf{I}_{1a} = \mathbf{I}_{1b} \\ \mathbf{I}_2 &= \mathbf{I}_{2a} = \mathbf{I}_{2b} \end{aligned} \tag{15.20}$$

and

$$\begin{aligned} \mathbf{V}_1 &= \mathbf{V}_{1a} + \mathbf{V}_{1b} = (\mathbf{z}_{11a} + \mathbf{z}_{11b})\mathbf{I}_1 + (\mathbf{z}_{12a} + \mathbf{z}_{12b})\mathbf{I}_2 \\ \mathbf{V}_2 &= \mathbf{V}_{2a} + \mathbf{V}_{2b} = (\mathbf{z}_{21a} + \mathbf{z}_{21b})\mathbf{I}_1 + (\mathbf{z}_{22a} + \mathbf{z}_{22b})\mathbf{I}_2 \end{aligned} \tag{15.21}$$

Therefore, the Z parameters for the total network are

$$\begin{bmatrix} \mathbf{z}_{11} & \mathbf{z}_{12} \\ \mathbf{z}_{21} & \mathbf{z}_{22} \end{bmatrix} = \begin{bmatrix} \mathbf{z}_{11a} + \mathbf{z}_{11b} & \mathbf{z}_{12a} + \mathbf{z}_{12b} \\ \mathbf{z}_{21a} + \mathbf{z}_{21b} & \mathbf{z}_{22a} + \mathbf{z}_{22b} \end{bmatrix} \tag{15.22}$$

Therefore, the Z parameters for the total network are equal to the sum of the Z parameters for the networks N_a and N_b. This procedure is always valid if the networks N_a and N_b are of the form shown in Fig. 15.18b. The Brune tests for the series interconnection of two-ports requires that $\mathbf{V}_o$ be zero when the networks are interconnected as shown in Fig. 15.18c and d. If these conditions are not satisfied, a one-to-one ideal transformer can be employed in the manner described for the parallel interconnection case and illustrated in Fig. 15.18e. For example, a transformer would be required if the two networks in Fig. 15.15 were interconnected in series, since $\mathbf{z}_4$ and $\mathbf{z}_5$ would be short-circuited.

EXAMPLE 15.14

Let us determine the Z parameters for the network shown in Fig. 15.16a. The circuit is redrawn in Fig. 15.19, illustrating a series interconnection. The upper network will be referred to as N_a and the lower network as N_b.

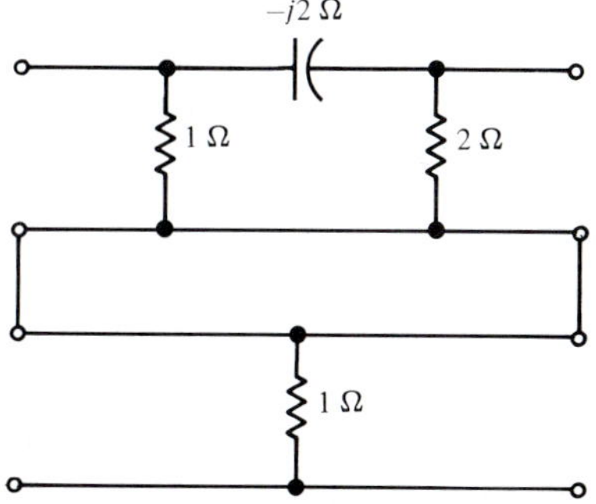

Figure 15.19 Network in Fig. 15.16a redrawn as a series interconnection of two networks.

The Z parameters for N_a are

$$\mathbf{z}_{11a} = \frac{2 - 2j}{3 - 2j}\ \Omega \qquad \mathbf{z}_{12a} = \frac{2}{3 - 2j}\ \Omega$$

$$\mathbf{z}_{21a} = \frac{2}{3 - 2j}\ \Omega \qquad \mathbf{z}_{22a} = \frac{2 - 4j}{3 - 2j}\ \Omega$$

and the Z parameters for N_b are

$$\mathbf{z}_{11b} = \mathbf{z}_{12b} = \mathbf{z}_{21b} = \mathbf{z}_{22b} = 1\ \Omega$$

Hence the Z parameters for the total network are

$$\mathbf{z}_{11} = \frac{5 - 4j}{3 - 2j}\ \Omega \qquad \mathbf{z}_{12} = \frac{5 - 2j}{3 - 2j}\ \Omega$$

$$\mathbf{z}_{21} = \frac{5 - 2j}{3 - 2j}\ \Omega \qquad \mathbf{z}_{22} = \frac{5 - 6j}{3 - 2j}\ \Omega$$

We could easily check these results against those obtained in Example 15.12 by applying the conversion formulas in Table 15.1. ■

DRILL EXERCISE

D15.15. Find the Z parameters of the network in Fig. D15.1 by considering the circuit to be a series interconnection of two two-ports as shown in Fig. D15.15.

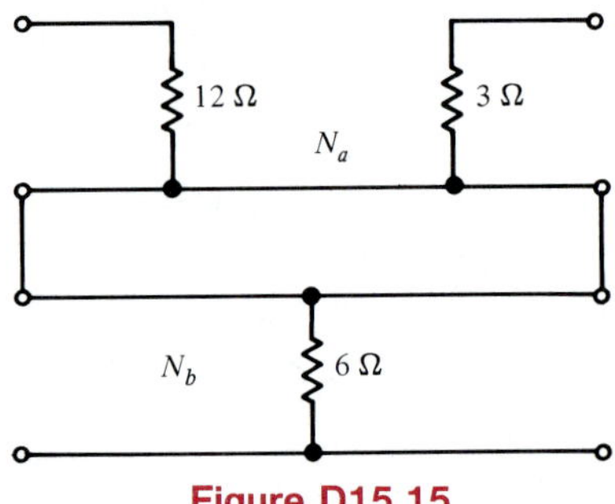

Figure D15.15

Ans: $\mathbf{z}_{11} = 18\ \Omega$, $\mathbf{z}_{12} = \mathbf{z}_{21} = 6\ \Omega$, $\mathbf{z}_{22} = 9\ \Omega$.

Cascade Interconnection

A two-port N is composed of the cascade interconnection of N_a and N_b as shown in Fig. 15.20, where the parameter equations for N_a and N_b are

$$\begin{bmatrix} \mathbf{V}_{1a} \\ \mathbf{I}_{1a} \end{bmatrix} = \begin{bmatrix} \mathbf{A}_a & \mathbf{B}_a \\ \mathbf{C}_a & \mathbf{D}_a \end{bmatrix} \begin{bmatrix} \mathbf{V}_{2a} \\ -\mathbf{I}_{2a} \end{bmatrix} \tag{15.23}$$

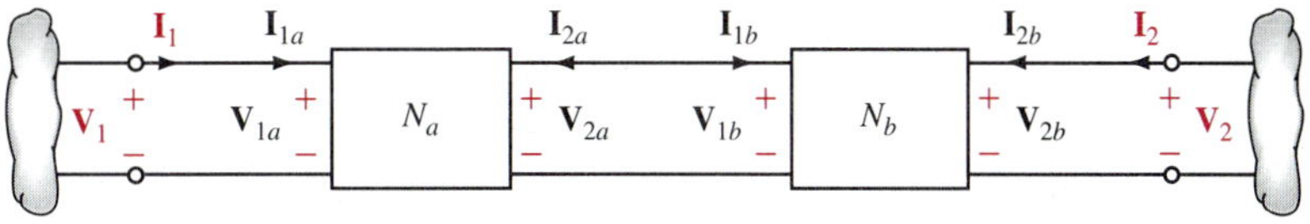

Figure 15.20 Cascade interconnection of networks.

$$\begin{bmatrix} \mathbf{V}_{1b} \\ \mathbf{I}_{1b} \end{bmatrix} = \begin{bmatrix} \mathbf{A}_b & \mathbf{B}_b \\ \mathbf{C}_b & \mathbf{D}_b \end{bmatrix} \begin{bmatrix} \mathbf{V}_{2b} \\ -\mathbf{I}_{2b} \end{bmatrix} \tag{15.24}$$

From Fig. 15.20, however, we note that

$$\begin{bmatrix} \mathbf{V}_1 \\ \mathbf{I}_1 \end{bmatrix} = \begin{bmatrix} \mathbf{V}_{1a} \\ \mathbf{I}_{1a} \end{bmatrix}, \quad \begin{bmatrix} \mathbf{V}_{2a} \\ -\mathbf{I}_{2a} \end{bmatrix} = \begin{bmatrix} \mathbf{V}_{1b} \\ \mathbf{I}_{1b} \end{bmatrix}, \quad \begin{bmatrix} \mathbf{V}_{2b} \\ -\mathbf{I}_{2b} \end{bmatrix} = \begin{bmatrix} \mathbf{V}_2 \\ -\mathbf{I}_2 \end{bmatrix}$$

and therefore, the equations for the total network are

$$\begin{bmatrix} \mathbf{V}_1 \\ \mathbf{I}_1 \end{bmatrix} = \begin{bmatrix} \mathbf{A}_a & \mathbf{B}_a \\ \mathbf{C}_a & \mathbf{D}_a \end{bmatrix} \begin{bmatrix} \mathbf{A}_b & \mathbf{B}_b \\ \mathbf{C}_b & \mathbf{D}_b \end{bmatrix} \begin{bmatrix} \mathbf{V}_2 \\ -\mathbf{I}_2 \end{bmatrix} \tag{15.25}$$

Hence the transmission parameters for the total network are derived by matrix multiplication as indicated above. The order of the matrix multiplication is important and is performed in the order in which the networks are interconnected.

EXAMPLE 15.15

Let us reexamine the circuit that we considered in Example 15.9 and derive its two-port parameters by considering it to be a cascade connection of two networks as shown in Fig. 15.21.

The ABCD parameters for the identical T networks can be easily calculated as

$$\mathbf{A} = 1 + j\omega \qquad \mathbf{B} = 2 + j\omega$$

$$\mathbf{C} = j\omega \qquad \mathbf{D} = 1 + j\omega$$

Therefore, the transmission parameters for the total network are

$$\begin{bmatrix} \mathbf{A} & \mathbf{B} \\ \mathbf{C} & \mathbf{D} \end{bmatrix} = \begin{bmatrix} 1 + j\omega & 2 + j\omega \\ j\omega & 1 + j\omega \end{bmatrix} \begin{bmatrix} 1 + j\omega & 2 + j\omega \\ j\omega & 1 + j\omega \end{bmatrix}$$

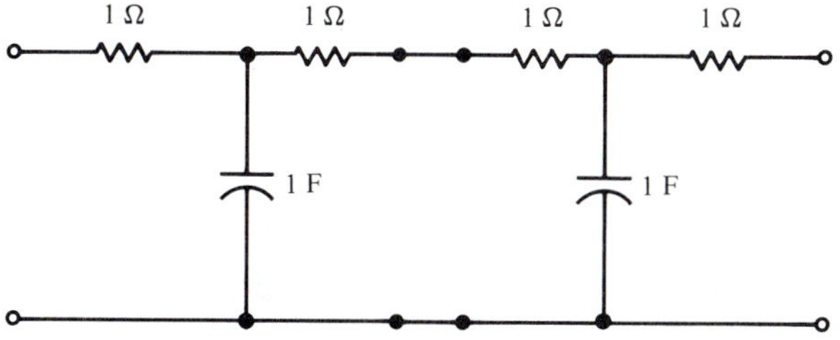

Figure 15.21 Two-port in Fig. 15.11 redrawn as a cascade connection of two networks.

Performing the matrix multiplication, we obtain

$$\begin{bmatrix} \mathbf{A} & \mathbf{B} \\ \mathbf{C} & \mathbf{D} \end{bmatrix} = \begin{bmatrix} 1 + 4j\omega - 2\omega^2 & 4 + 6j\omega - 2\omega^2 \\ 2j\omega - 2\omega^2 & 1 + 4j\omega - 2\omega^2 \end{bmatrix}$$

DRILL EXERCISE

D15.16. Find the transmission parameters of the network in Fig. D15.1 by considering the circuit to be a cascade interconnection of three two-ports as shown in Fig. D15.16.

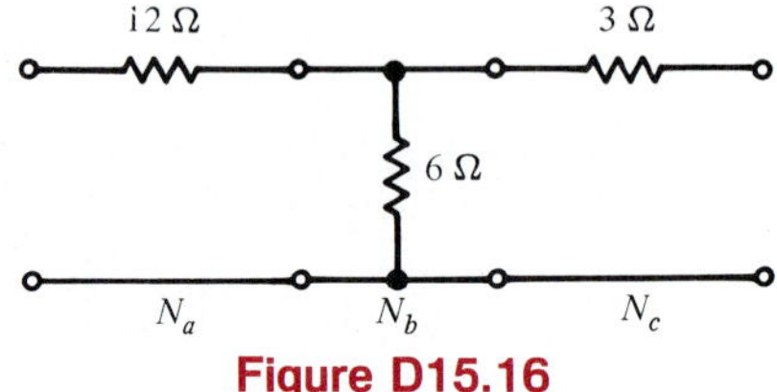

Figure D15.16

Ans: $\mathbf{A} = 3$, $\mathbf{B} = 21\ \Omega$, $\mathbf{C} = \frac{1}{6}$ S, $\mathbf{D} = \frac{3}{2}$.

15.8

T–Π Equivalent Networks

Two very important two-ports are the T and Π networks shown in Fig. 15.22. Because we encounter these two geometrical forms often in two-port analyses, it is instructive to determine the conditions under which these two networks are equivalent. In order to

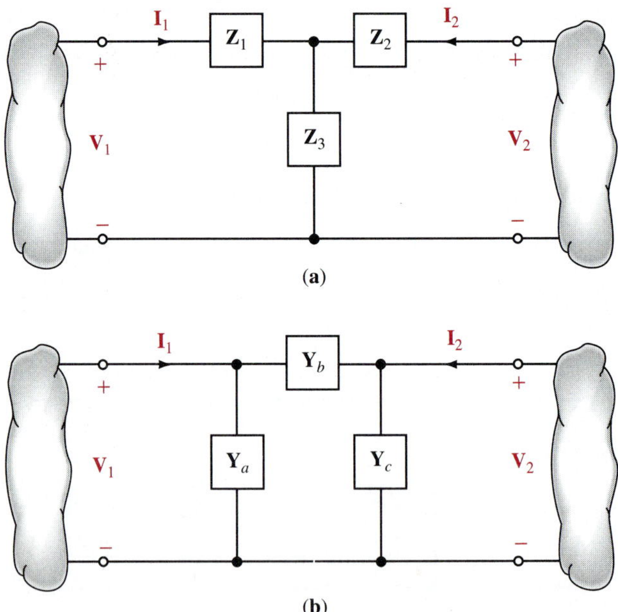

Figure 15.22 T and Π two-port networks.

determine the equivalence relationship, we will examine Z-parameter equations for the T network and the Y-parameter equations for the Π network.

For the T network the equations are

$$\mathbf{V}_1 = (\mathbf{Z}_1 + \mathbf{Z}_3)\mathbf{I}_1 + \mathbf{Z}_3\mathbf{I}_2$$
$$\mathbf{V}_2 = \mathbf{Z}_3\mathbf{I}_1 + (\mathbf{Z}_2 + \mathbf{Z}_3)\mathbf{I}_2$$

and for the Π network the equations are

$$\mathbf{I}_1 = (\mathbf{Y}_a + \mathbf{Y}_b)\mathbf{V}_1 - \mathbf{Y}_b\mathbf{V}_2$$
$$\mathbf{I}_2 = -\mathbf{Y}_b\mathbf{V}_1 + (\mathbf{Y}_b + \mathbf{Y}_c)\mathbf{V}_2$$

Solving the equations for the T network in terms of $\mathbf{I}_1$ and $\mathbf{I}_2$, we obtain

$$\mathbf{I}_1 = \left(\frac{\mathbf{Z}_2 + \mathbf{Z}_3}{\mathbf{D}_1}\right)\mathbf{V}_1 - \frac{\mathbf{Z}_3\mathbf{V}_2}{\mathbf{D}_1}$$
$$\mathbf{I}_2 = -\frac{\mathbf{Z}_3\mathbf{V}_1}{\mathbf{D}_1} + \left(\frac{\mathbf{Z}_1 + \mathbf{Z}_3}{\mathbf{D}_1}\right)\mathbf{V}_2$$

where $\mathbf{D}_1 = \mathbf{Z}_1\mathbf{Z}_2 + \mathbf{Z}_2\mathbf{Z}_3 + \mathbf{Z}_1\mathbf{Z}_3$. Comparing these equations with those for the Π network, we find that

$$\mathbf{Y}_a = \frac{\mathbf{Z}_2}{\mathbf{D}_1}$$
$$\mathbf{Y}_b = \frac{\mathbf{Z}_3}{\mathbf{D}_1} \qquad (15.26)$$
$$\mathbf{Y}_c = \frac{\mathbf{Z}_1}{\mathbf{D}_1}$$

or in terms of the impedances of the Π network,

$$\mathbf{Z}_a = \frac{\mathbf{D}_1}{\mathbf{Z}_2}$$
$$\mathbf{Z}_b = \frac{\mathbf{D}_1}{\mathbf{Z}_3} \qquad (15.27)$$
$$\mathbf{Z}_c = \frac{\mathbf{D}_1}{\mathbf{Z}_1}$$

If we reverse this procedure and solve the equations for the Π network in terms of $\mathbf{V}_1$ and $\mathbf{V}_2$ and then compare the resultant equations with those for the T network, we find that

$$\mathbf{Z}_1 = \frac{\mathbf{Y}_c}{\mathbf{D}_2}$$
$$\mathbf{Z}_2 = \frac{\mathbf{Y}_a}{\mathbf{D}_2} \qquad (15.28)$$
$$\mathbf{Z}_3 = \frac{\mathbf{Y}_b}{\mathbf{D}_2}$$

where $\mathbf{D}_2 = \mathbf{Y}_a\mathbf{Y}_b + \mathbf{Y}_b\mathbf{Y}_c + \mathbf{Y}_a\mathbf{Y}_c$. Equation (15.28) can also be written in the form

$$\mathbf{Z}_1 = \frac{\mathbf{Z}_a\mathbf{Z}_b}{\mathbf{Z}_a + \mathbf{Z}_b + \mathbf{Z}_c}$$
$$\mathbf{Z}_2 = \frac{\mathbf{Z}_b\mathbf{Z}_c}{\mathbf{Z}_a + \mathbf{Z}_b + \mathbf{Z}_c} \qquad (15.29)$$
$$\mathbf{Z}_3 = \frac{\mathbf{Z}_a\mathbf{Z}_c}{\mathbf{Z}_a + \mathbf{Z}_b + \mathbf{Z}_c}$$

Do Eqs. (15.27) and (15.29) look familiar? They should! They are simply the wye ⇆ delta transformations employed in Chapter 2, since the T is a wye-connected network and the Π is a delta-connected network.

EXAMPLE 15.16

Let us determine both the equivalent-Π and equivalent-T networks for the two-port shown in Fig. 15.23a.

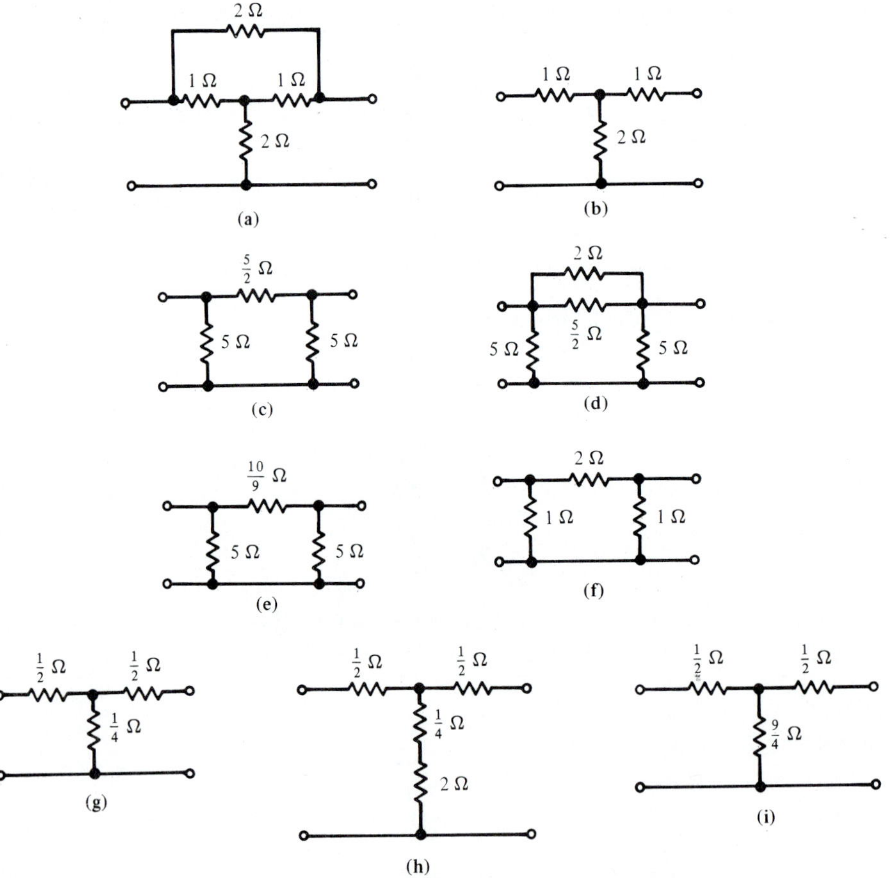

Figure 15.23 T–Π equivalent two-port analysis.

Using the T-to-Π transformation equations, the T portion of the two-port shown in Fig. 15.23b is converted to a Π network in Fig. 15.23c. Combining the 2-Ω resistor in the original network with the Π network in Fig. 15.23c yields the network in Fig. 15.23d, which can immediately be converted to the final Π-equivalent network in Fig. 15.23e.

In a similar manner the Π portion of the original two-port shown in Fig. 15.23f is converted to the T network in Fig. 15.23g. Combining the remaining 2-Ω resistor in the original network with the T network in Fig. 15.23g yields the network in Fig. 15.23h, which can be converted immediately to the final T-equivalent network in Fig. 15.23i.

■

DRILL EXERCISE

D15.17. Show that the two-ports in Figs. D15.1 and D15.2 are equivalent.

15.9 Two-Ports Embedded Within a Network

We may encounter situations in which a two-port is an integral part of a larger network. For example, the two-port may be some type of "black box" where all that is known concerning it is a set of two-port parameters. We will now try to illustrate that all the techniques we have described earlier can readily be applied in these situations also.

Consider the following example of a two-port embedded within a larger network.

EXAMPLE 15.17

Consider the network in Fig. 15.24a. The hybrid parameters of the two-port are known to be

$$\begin{bmatrix} \mathbf{h}_{11} & \mathbf{h}_{12} \\ \mathbf{h}_{21} & \mathbf{h}_{22} \end{bmatrix} = \begin{bmatrix} 100 & 0 \\ 1 & 10^{-3} \end{bmatrix}$$

We wish to determine the output voltage $\mathbf{V}_o$ of the entire network.

Our approach to the problem will be as follows. We will first find the input impedance of the two-port, add it to the 100-Ω resistor, and reflect this impedance to the left side of the transformer. We will then solve the two-loop circuit in which the transformer and everything to the right of it is represented by the reflected impedance. The voltage and current on the left side of the transformer will then be reflected to the right side. The latter voltage and current will be used in conjunction with the two-port parameters to determine the current in the load and therefore $\mathbf{V}_o$.

Using the results of Example 15.5, we find that the input impedance of the two-port is

$$\mathbf{Z}_{\text{in}} = \mathbf{h}_{11} - \frac{\mathbf{h}_{12}\mathbf{h}_{21}\mathbf{Z}_L}{1 + \mathbf{h}_{22}\mathbf{Z}_L}$$

where the hybrid parameters are given and $\mathbf{Z}_L = 100 + j100\ \Omega$. $\mathbf{Z}_{\text{in}}$ is, therefore, 100 Ω, and hence when this impedance is added to the 100 Ω in Fig. 15.24a and reflected to the

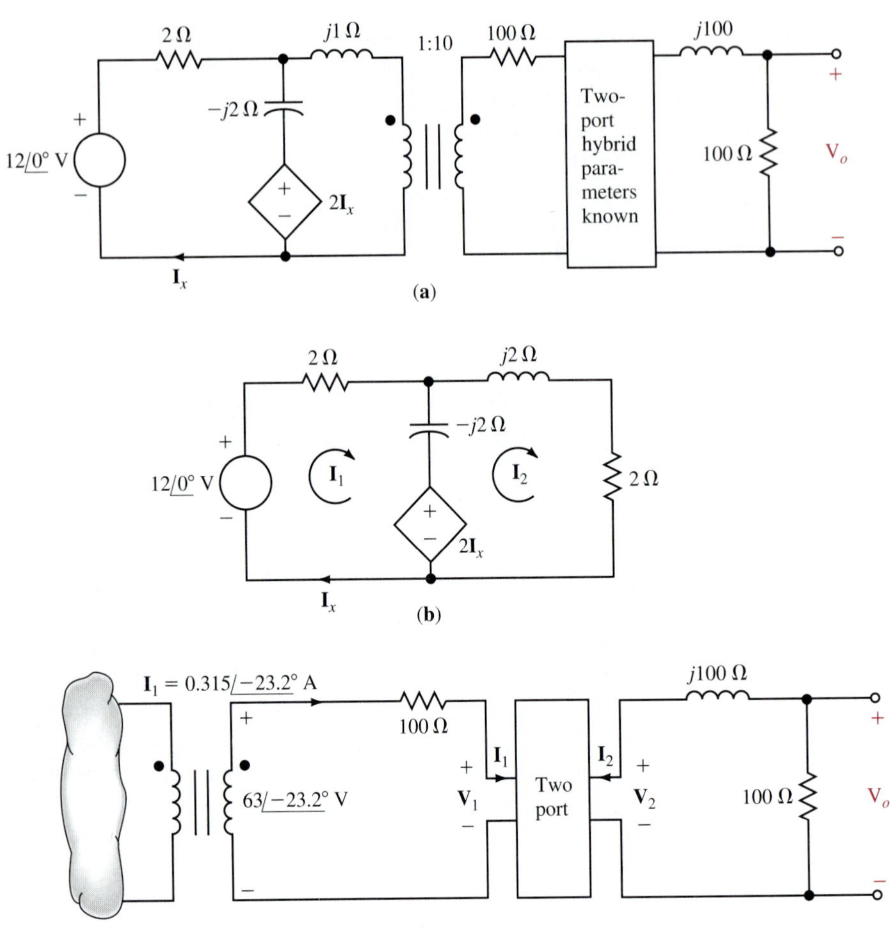

Figure 15.24 Circuits used in Example 15.17.

left side of the transformer, we obtain the network in Fig. 15.24b. The mesh equations for this network are

$$\mathbf{I}_1 = \mathbf{I}_x$$

$$(2 - j2)\mathbf{I}_1 - (-j2)\mathbf{I}_2 = 12\,\underline{/0^\circ} - 2\mathbf{I}_1$$

$$-(-j2)\mathbf{I}_1 + (2 - j1)\mathbf{I}_2 = 2\mathbf{I}_1$$

Solving these equations yields $\mathbf{I}_2 = 3.15\,\underline{/-23.2^\circ}$ A, and therefore, the voltage on the left side of the transformer is $6.30\,\underline{/-23.2^\circ}$ V. Reflecting voltage and current to the right side of the transformer yields the results shown in Fig. 15.24c. Note that $\mathbf{I}_1 = 0.315\,\underline{/-23.2^\circ}$ A and that $\mathbf{V}_2 = -\mathbf{I}_2(100 + j100)$. The equation for the two-port that relates the input and output currents is

$$\mathbf{I}_2 = \mathbf{h}_{21}\mathbf{I}_1 + \mathbf{h}_{22}\mathbf{V}_2$$

$$= \mathbf{h}_{21}\mathbf{I}_1 - \mathbf{I}_2(100 + j100)\ 10^{-3}$$

$$= \frac{1(\mathbf{I}_1)}{1 + (100 + j100)}\ 10^{-3}$$

Therefore,

$$\mathbf{V}_o = -100\mathbf{I}_2$$

$$= -28.52\ \underline{/-28.4^\circ}\ \text{V}$$

DRILL EXERCISE

D15.18. Given the network in Fig. D15.18, determine $\mathbf{V}_o$ if the hybrid parameters for the two-port are

$$\begin{bmatrix} \mathbf{h}_{11} & \mathbf{h}_{12} \\ \mathbf{h}_{21} & \mathbf{h}_{22} \end{bmatrix} = \begin{bmatrix} 14 & \frac{2}{3} \\ -\frac{2}{3} & \frac{1}{9} \end{bmatrix}$$

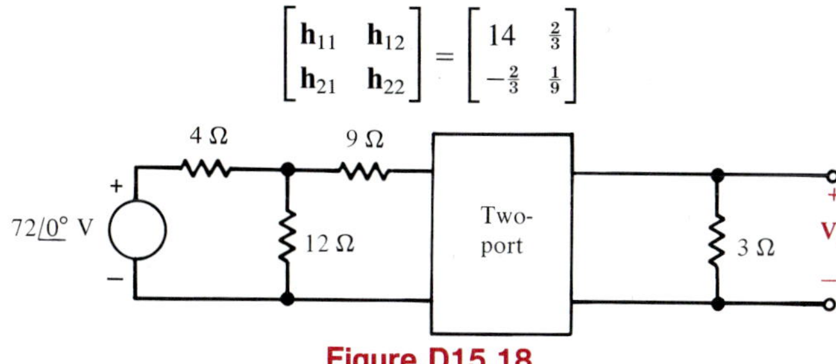

Figure D15.18

Ans: $\mathbf{V}_o = 3\ \underline{/0^\circ}$ V.

15.10 Summary

In this chapter two-port networks have been described using admittance, impedance, hybrid, and transmission parameters. Equivalent circuits for two-port networks have been described and conversion formulas for changing from one set of parameters to another have been presented.

Finally, we have shown that a two-port may be treated as an interconnection of simpler networks. The interconnections described are parallel, series, and cascade. Finally, we have shown that the T- and Π-equivalent networks are the same as the wye and delta networks described in Chapter 2.

KEY POINTS

- Four of the most common parameters used to describe a two-port network are the admittance, impedance, hybrid, and transmission parameters.

- Some or all of the two-port parameters for a network may not exist.
- If all the two-port parameters for a network exist, a set of conversion formulas can be used to relate one set of two-port parameters to another.
- The Brune tests must be satisfied in order to interconnect two-ports.
- When interconnecting two-ports, the Y parameters are added for a parallel connection, the Z parameters are added for a series connection, and the transmission parameters in matrix form are multiplied together for a cascade connection.
- T-to-Π transformations are shown to be the same as wye ⇆ delta transformations.

PROBLEMS

15.1. Given the two networks in Fig. P15.1, find the Y parameters for the circuit in (a) and the Z parameters for the circuit in (b).

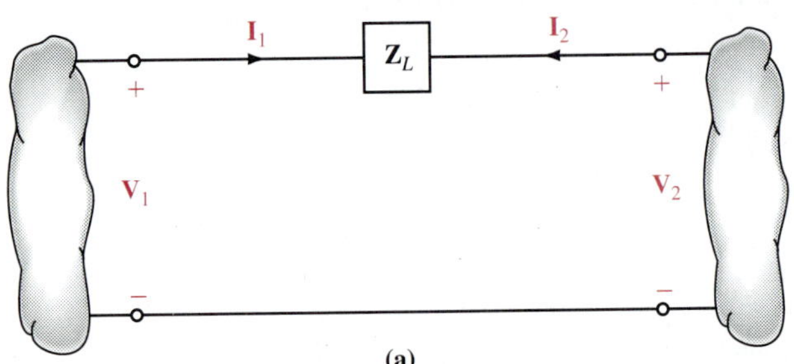

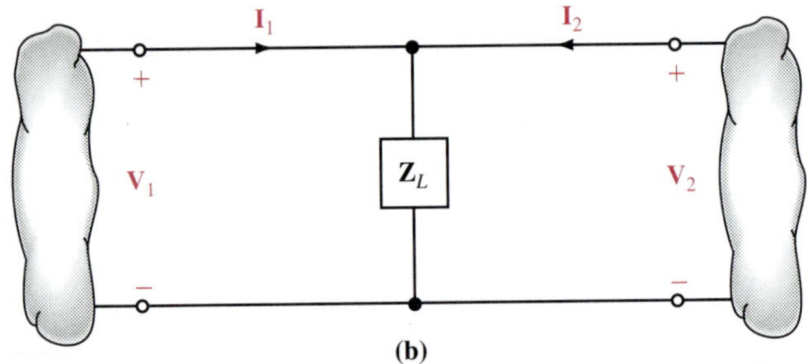

Figure P15.1

15.2. Find the Y parameters for the two-port network shown in Fig. P15.2.

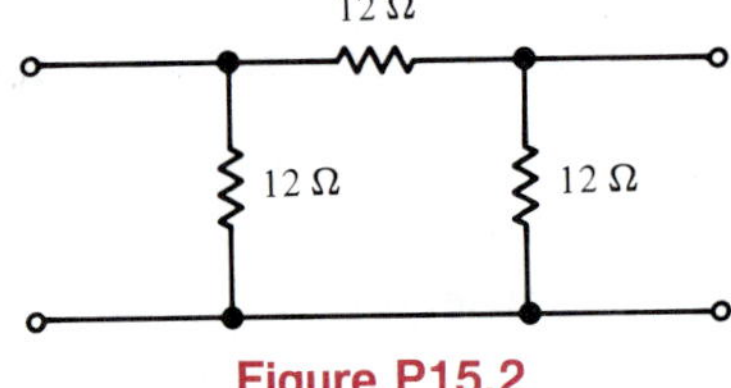

Figure P15.2

15.3. Find the Z parameters for the two-port in Problem 15.2.

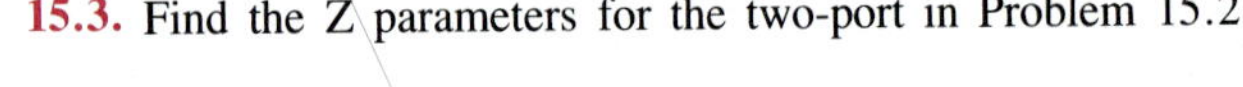

15.4. If a 12-A source is connected at the input port of the network shown in Fig. P15.2, find the current in a 4-Ω load resistor.

15.5. Find the Y parameters for the two-port network in Fig. P15.5.

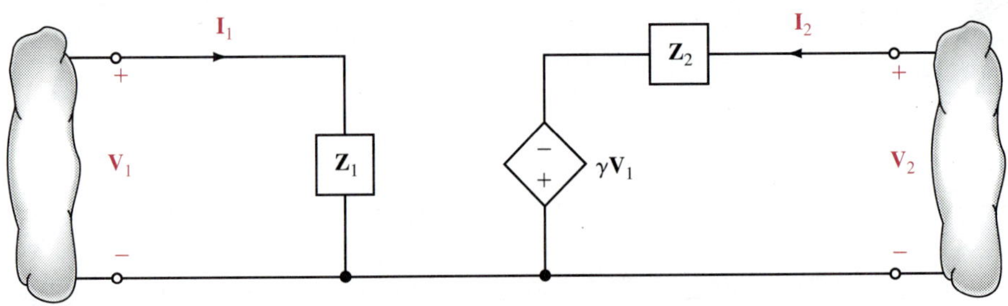

Figure P15.5

15.6. Find the Z parameters for the network in Fig. P15.5.

15.7. Find the Z parameters for the network shown in Fig. P15.7.

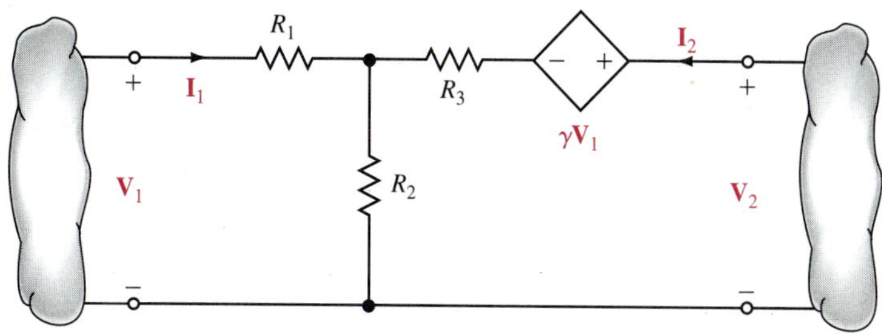

Figure P15.7

15.8. Find the Y parameters for the network in Fig. P15.8.

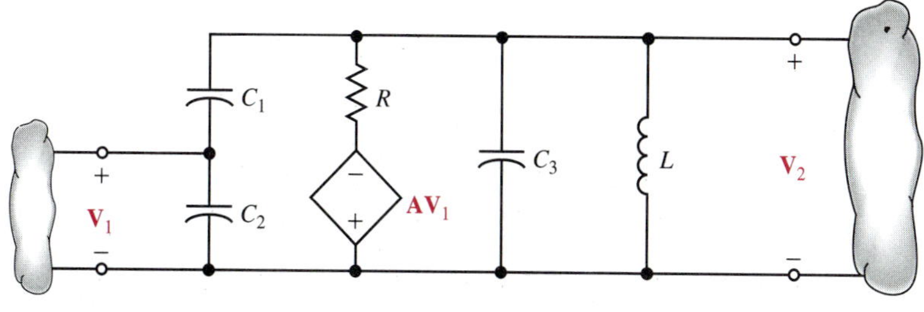

Figure P15.8

15.9. Given the network in Fig. P15.9, (a) find the Z parameters for the transformer, (b) write the terminal equation at each end of the two-port, and (c) use the information obtained to find $\mathbf{V}_2$.

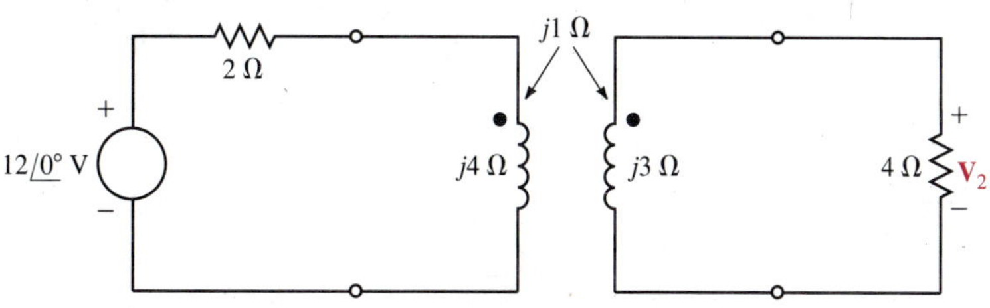

Figure P15.9

15.10. Compute the Z parameters of the network shown in Fig. P15.10.

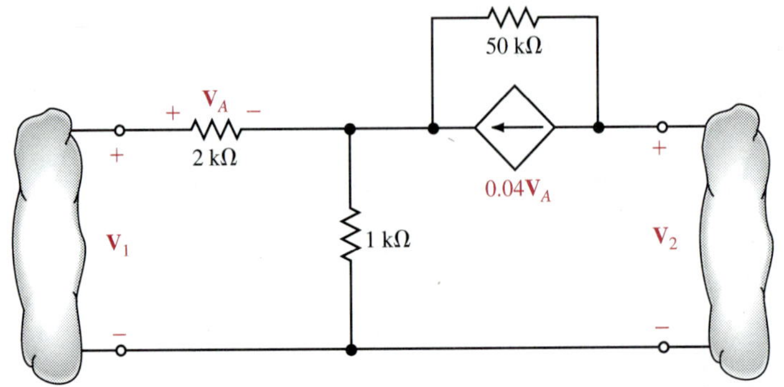

Figure P15.10

15.11. Compute the input impedance of the network in Fig. P15.10 if a 4-kΩ load is connected to the output port.

15.12. Compute the output impedance of the two-port network in Fig. P15.10.

15.13. Determine the voltage gain of the two-port network in Fig. P15.10 when a 4-kΩ load resistor is connected to the output port.

15.14. Find the Z parameters for the two-port network shown in Fig. P15.14.

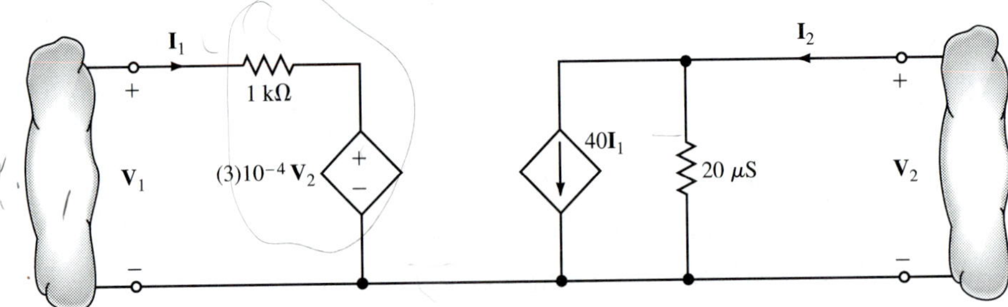

Figure P15.14

15.15. Find the voltage gain of the two-port in Fig. P15.14 if a 12-kΩ load is connected to the output port.

15.16. Find the input impedance of the network in Fig. P15.14.

15.17. Find the Z parameters of the two-port in Fig. P15.17.

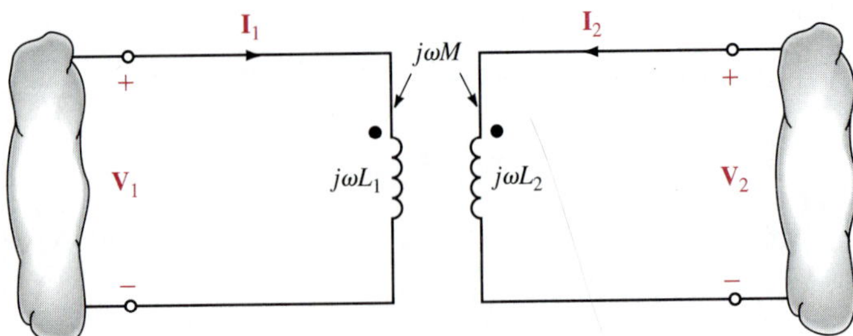

Figure P15.17

15.18. Determine the Z parameters for the two-port network in Fig. P15.18.

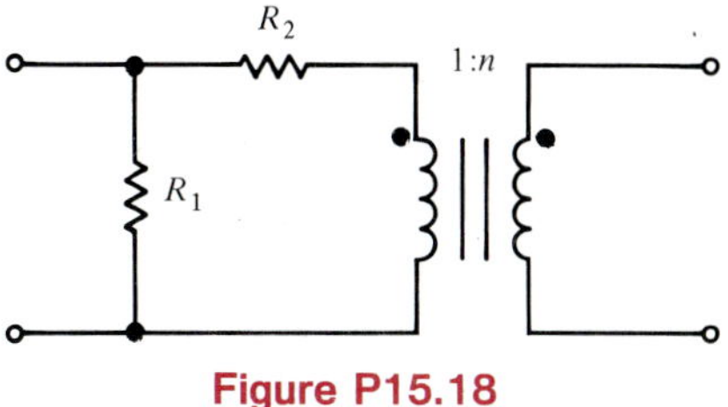

Figure P15.18

15.19. Find the ABCD parameters for the networks in Fig. P15.1.

15.20. Find the ABCD parameters for the circuit in Fig. P15.5.

15.21. Find the transmission parameters for the network in Fig. P15.21.

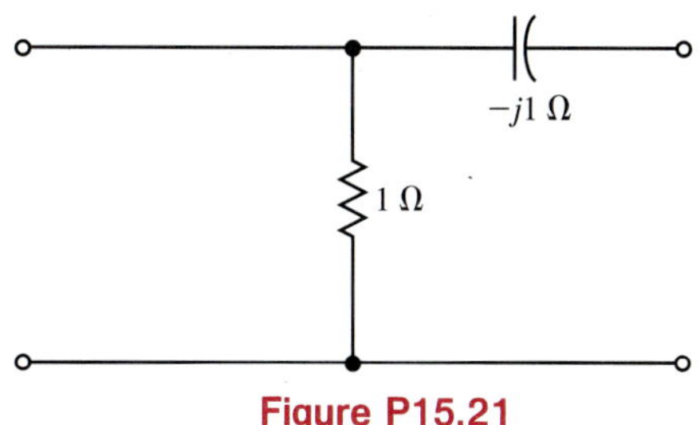

Figure P15.21

15.22. Find the transmission parameters for the circuit in Fig. P15.22.

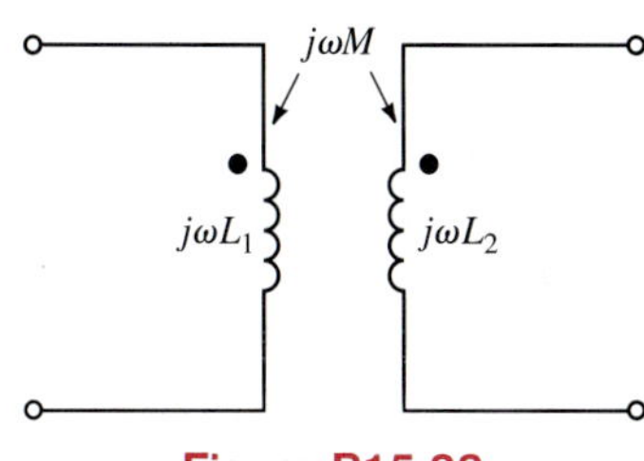

Figure P15.22

15.23. Find the transmission parameters for the two-port network in Fig. P15.23.

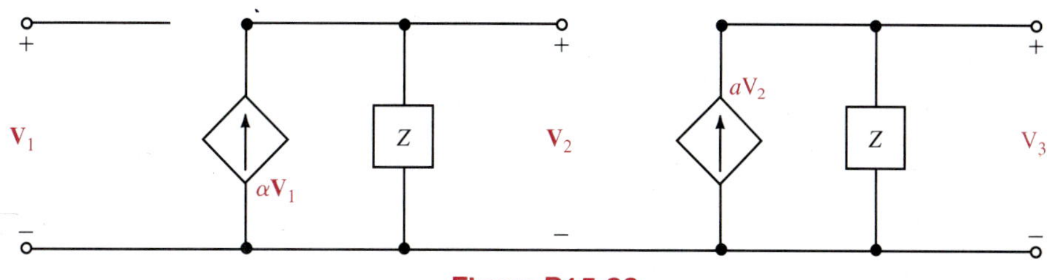

Figure P15.23

15.24. Find the transmission parameters for the network in Fig. P15.24 and use the information to find $\mathbf{Z}_i$.

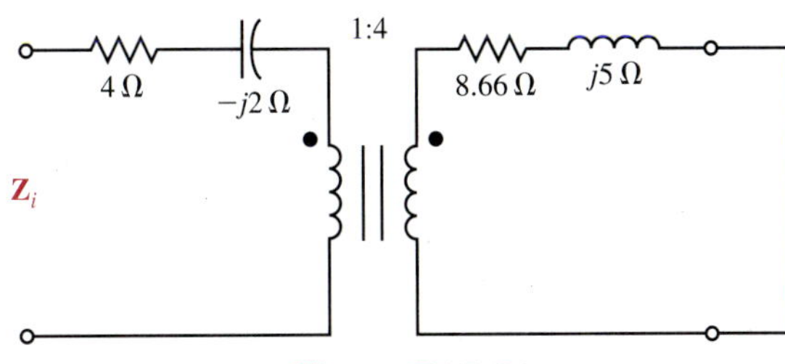

Figure P15.24

15.25. Given the network in Fig. P15.25, find the transmission parameters for the two-port and then find $\mathbf{I}_o$ using the terminal conditions.

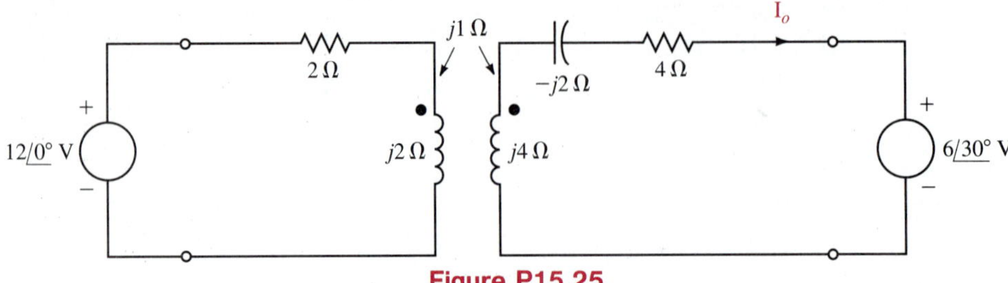

Figure P15.25

15.26. Find the hybrid parameters for the network in Problem 15.2.

15.27. Determine the input impedance of the network shown in Fig. P15.27, in terms of the hybrid parameters of the two-port and two-port load $\mathbf{Z}_L$.

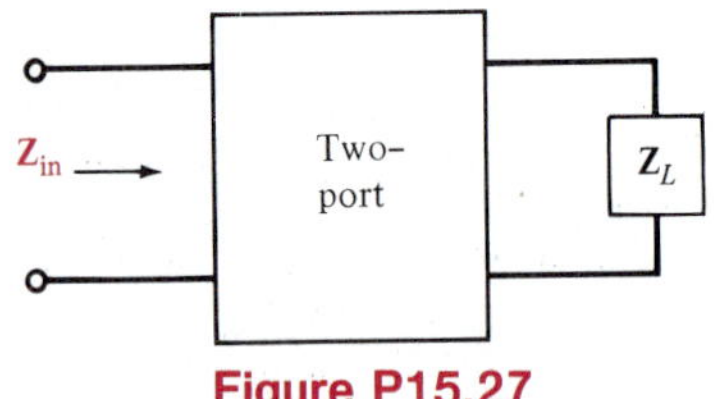

Figure P15.27

15.28. Following are the hybrid parameters for a network.

$$\begin{bmatrix} \mathbf{h}_{11} & \mathbf{h}_{12} \\ \mathbf{h}_{21} & \mathbf{h}_{22} \end{bmatrix} = \begin{bmatrix} \frac{11}{5} & \frac{2}{5} \\ -\frac{2}{5} & \frac{1}{5} \end{bmatrix}$$

Determine the Y parameters for the network.

15.29. Find the voltage gain $\mathbf{V}_2/\mathbf{V}_1$ for the network in Fig. P15.29 using the Z parameters.

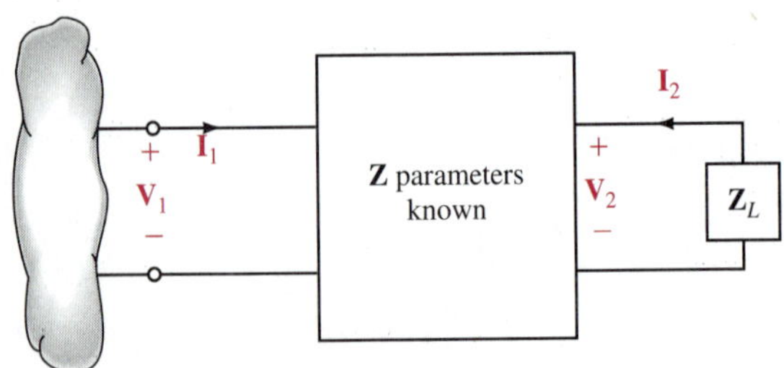

Figure P15.29

15.30. Find the Z parameters in terms of the ABCD parameters.

15.31. The G parameters are defined as follows:

$$\mathbf{I}_1 = g_{11}\mathbf{V}_1 + g_{12}\mathbf{I}_2$$
$$\mathbf{V}_2 = g_{21}\mathbf{V}_1 + g_{22}\mathbf{I}_2$$

Find the Z parameters in terms of the G parameters.

15.32. If the **h** parameters for a network are

$$\begin{bmatrix} \mathbf{h}_{11} & \mathbf{h}_{12} \\ \mathbf{h}_{21} & \mathbf{h}_{22} \end{bmatrix} = \begin{bmatrix} \frac{11}{5} & \frac{2}{5} \\ -\frac{2}{5} & \frac{1}{5} \end{bmatrix}$$

Find the ABCD parameters.

15.33. If the Y parameters for a network are known to be

$$\begin{bmatrix} \mathbf{y}_{11} & \mathbf{y}_{12} \\ \mathbf{y}_{21} & \mathbf{y}_{22} \end{bmatrix} = \begin{bmatrix} \frac{5}{11} & -\frac{2}{11} \\ -\frac{2}{11} & \frac{3}{11} \end{bmatrix}$$

Find the Z parameters.

15.34. Find the ABCD parameters for the network in Fig. P15.23.

15.35. Find the ABCD parameters for the circuit in Fig. P15.35.

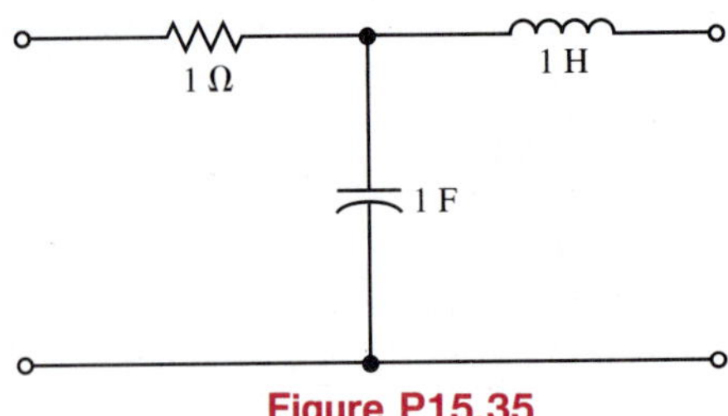

Figure P15.35

15.36. Find the transmission parameters for the two-port in Fig. P15.36.

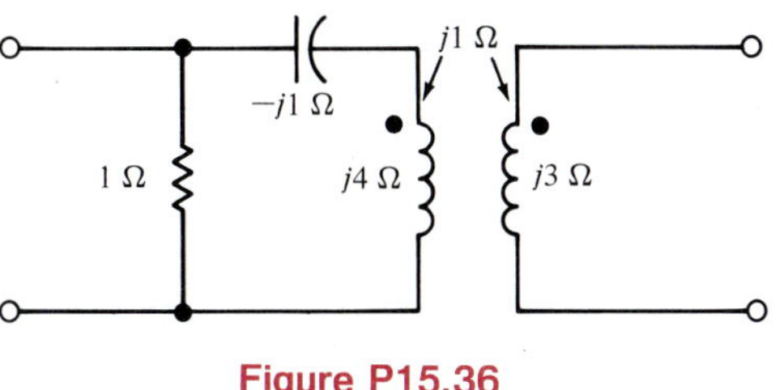

Figure P15.36

15.37. Find the transmission parameters for the two-port in Fig. P15.37.

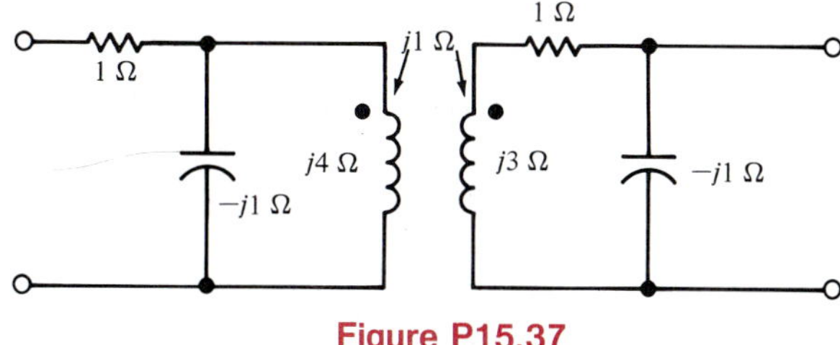

Figure P15.37

15.38. Find the transmission parameters of the two-port in Fig. P15.38 and then use the terminal conditions to compute $\mathbf{I}_o$.

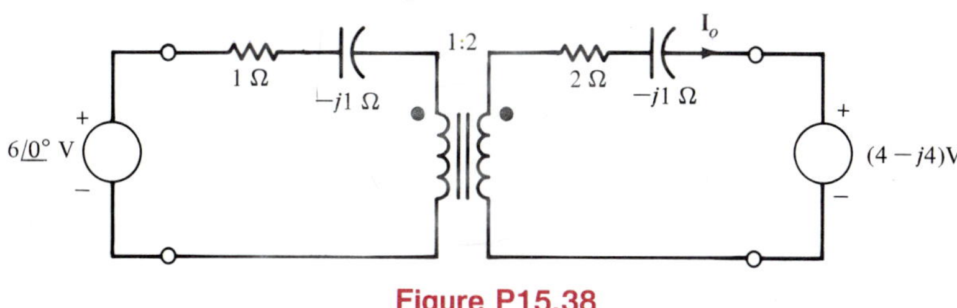

Figure P15.38

15.39. Find the Y parameters of the two-port in Fig. P15.39. Find the input admittance of the network when the capacitor is connected to the output port.

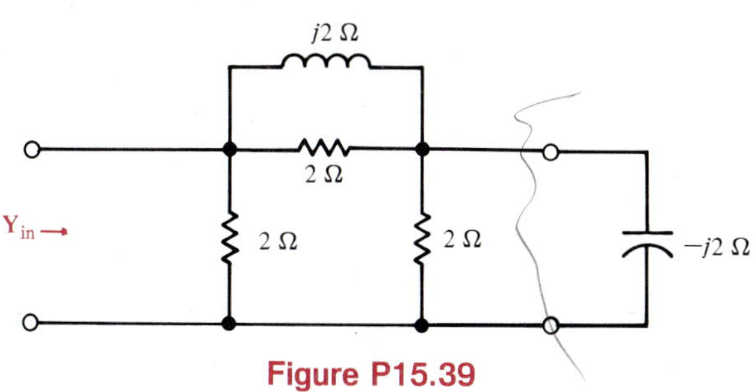

Figure P15.39

15.40. Find the Y parameters for the two-port network in Fig. P15.40.

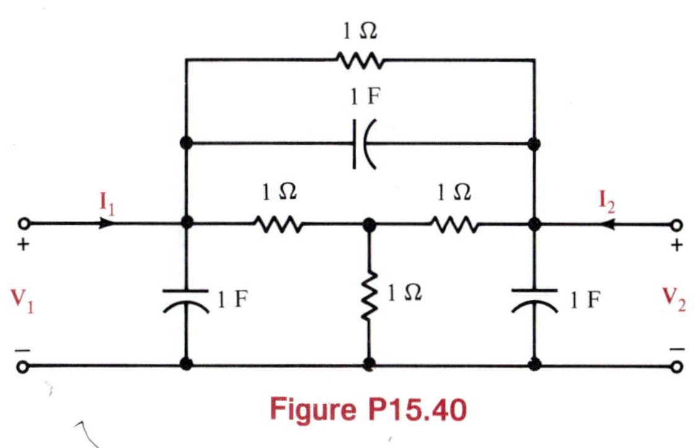

Figure P15.40

15.41. Find the Y parameters for the two-port in Fig. P15.41.

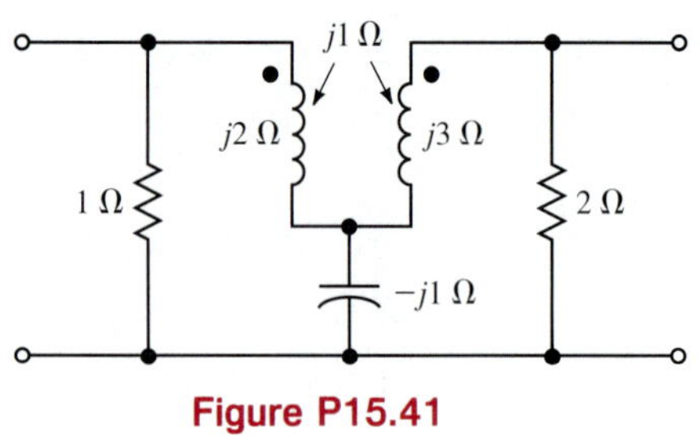

Figure P15.41

15.42. Find the Y parameters for the network in Fig. P15.42.

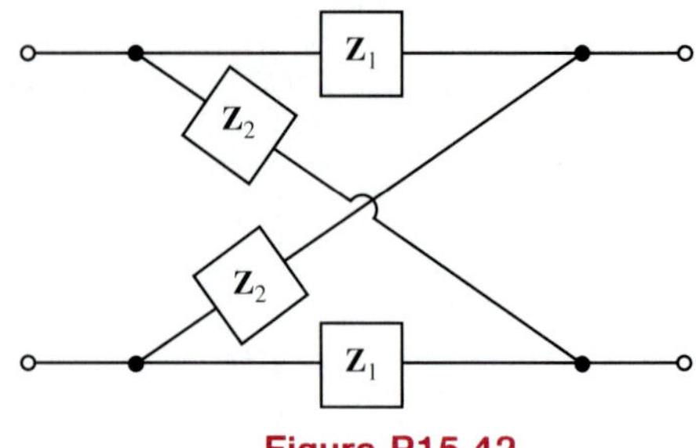

Figure P15.42

15.43. Determine the input impedance in the network shown in Fig. P15.43 assuming that the transmission parameters for the two identical two-ports are known.

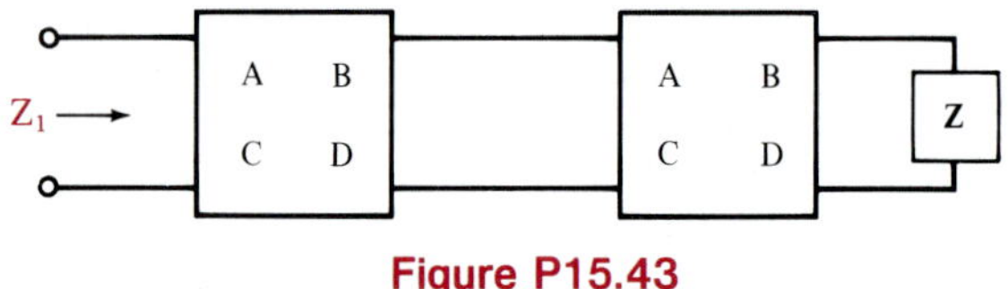

Figure P15.43

15.44. Find the transmission parameters of the two-port in Fig. P15.44 and then use the terminal conditions to find $\mathbf{I}_o$.

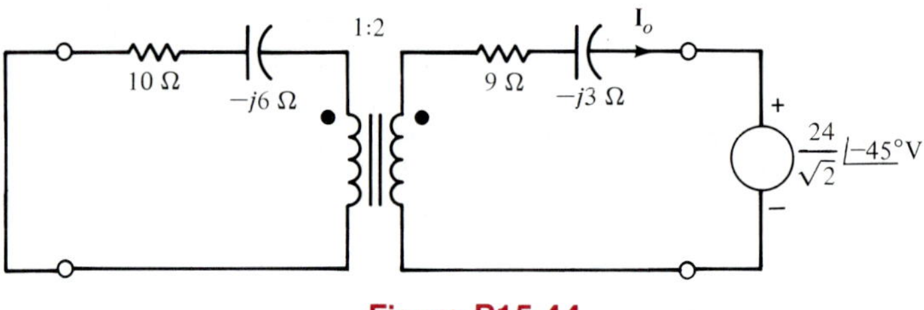

Figure P15.44

15.45. Determine the Y parameters for the network shown in Fig. P15.45.

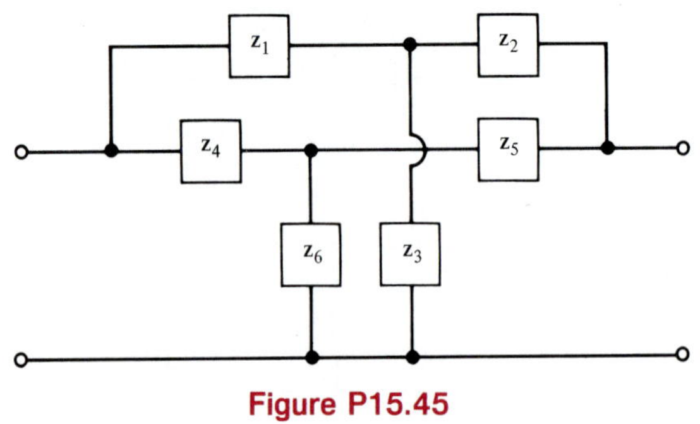

Figure P15.45

15.46. Find the Y parameters of the two-port in Fig. P15.46.

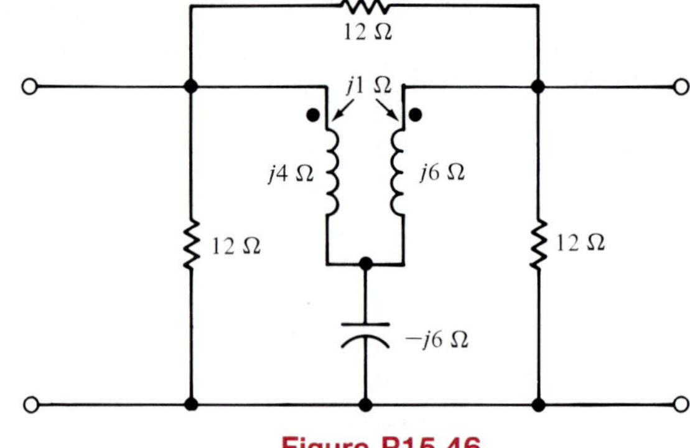

Figure P15.46

15.47. Find the Z parameters for the two-port in Fig. P15.47 and then determine $\mathbf{I}_o$ for the specified terminal conditions.

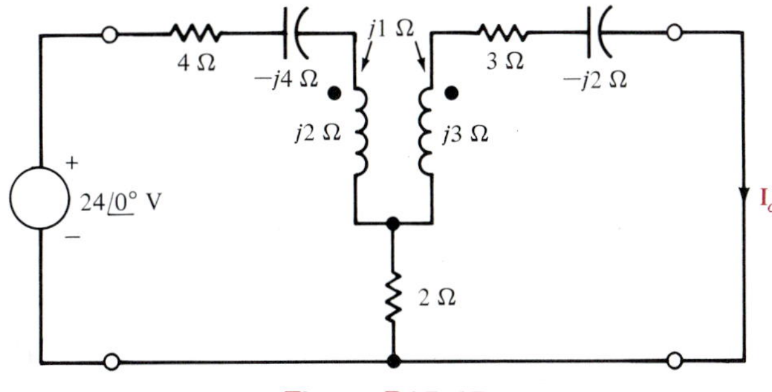

Figure P15.47

15.48. Determine the output voltage $\mathbf{V}_o$ in the network in Fig. P15.48 if the Z parameters for the two-port are

$$Z = \begin{bmatrix} 3 & 2 \\ 2 & 3 \end{bmatrix}$$

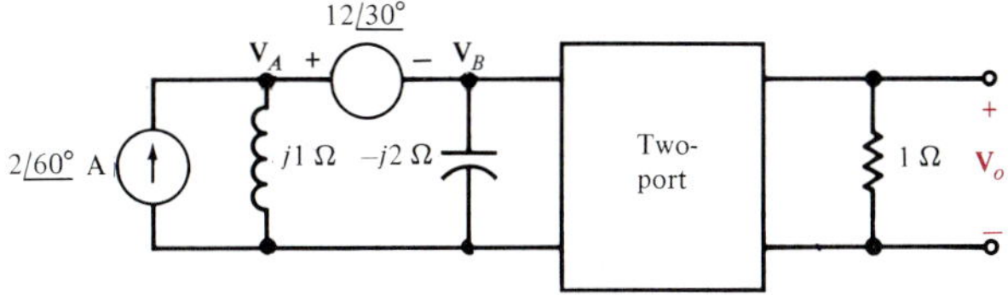

Figure P15.48

15.49. Determine the output voltage $\mathbf{V}_o$ in the network in Fig. P15.49 if the Z parameters of the two-port are

$$Z = \begin{bmatrix} 5 & 4 \\ 4 & 12 \end{bmatrix}$$

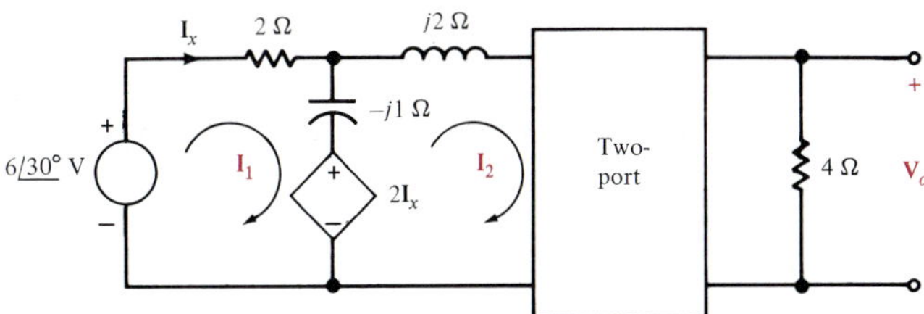

Figure P15.49

CHAPTER

16

The Laplace Transform

We will now introduce the Laplace transform. This is an extremely important technique in that for a given set of initial conditions, it will yield the total response of the circuit consisting of both the natural and forced responses in one operation.

Our use of the Laplace transform to solve circuit problems is analogous to that of using phasors in sinusoidal steady-state analysis. Using the Laplace transform we transform the circuit problem from the time domain to the frequency domain, solve the problem using algebra in the frequency domain, and then convert the solution in the complex frequency domain back to the time domain. Therefore, as we shall see, the Laplace transform is an integral transform that converts a set of linear simultaneous integrodifferential equations to a set of simultaneous algebraic equations.

Our approach is to define the Laplace transform, derive some of the transform pairs, consider some of the important properties of the transform, illustrate the inverse transform operation, introduce the convolution integral, and finally apply the transform to the solution of linear constant-coefficient integrodifferential equations.

16.1 Definition

The Laplace transform of a function $f(t)$ is defined by the equation

$$\mathscr{L}[f(t)] = \mathbf{F}(s) = \int_0^\infty f(t)e^{-st}\,dt \tag{16.1}$$

where s is the complex frequency

$$s = \sigma + j\omega \tag{16.2}$$

and the function $f(t)$ is assumed to possess the property that

$$f(t) = 0 \qquad \text{for } t < 0$$

Note that the Laplace transform is unilateral ($0 \leq t < \infty$), in contrast to the Fourier transform (see Chapter 18), which is bilateral ($-\infty < t < \infty$). In our analysis of circuits using the Laplace transform, we will focus our attention on the time interval $t \geq 0$. It is important to note that it is the initial conditions that account for the operation of the circuit prior to $t = 0$, and therefore our analyses will describe the circuit operation for $t \geq 0$.

In order for a function $f(t)$ to possess a Laplace transform, it must satisfy the condition

$$\int_0^\infty e^{-\sigma t}|f(t)|\, dt < \infty \tag{16.3}$$

for some real value of σ. Because of the convergence factor $e^{-\sigma t}$, there are a number of important functions that have Laplace transforms, even though Fourier transforms for these functions do not exist. All of the inputs we will apply to circuits possess Laplace transforms. Functions that do not have Laplace transforms (e.g., e^{t^2}) are of no interest to us in circuit analysis.

The inverse Laplace transform, which is analogous to the inverse Fourier transform, is defined by the relationship

$$\mathscr{L}^{-1}[\mathbf{F}(s)] = f(t) = \frac{1}{2\pi j}\int_{\sigma_1 - j\infty}^{\sigma_1 + j\infty} \mathbf{F}(s)e^{st}\, ds \tag{16.4}$$

where σ_1 is real and $\sigma_1 > \sigma$ in Eq. (16.3). The evaluation of this integral is based on complex variable theory, and therefore we will circumvent its use by developing and using a set of Laplace transform pairs.

16.2 Some Important Transform Pairs

We will now develop a number of basic transform pairs that are very useful in circuit analysis.

There are two singularity functions that are very important in circuit analysis: (1) the unit step function, $u(t)$, discussed in Chapter 7, and (2) the unit impulse or delta function, $\delta(t)$. They are called *singularity functions* because they are either not finite or they do not possess finite derivatives everywhere. They are mathematical models for signals that we employ in circuit analysis.

The unit impulse function can be represented in the limit by the rectangular pulse shown in Fig. 16.1a as $a \rightarrow 0$. The function is defined by the following:

$$\delta(t - t_0) = 0 \qquad t \neq t_0$$

$$\int_{t_0-\epsilon}^{t_0+\epsilon} \delta(t - t_0)\, dt = 1 \qquad \epsilon > 0$$

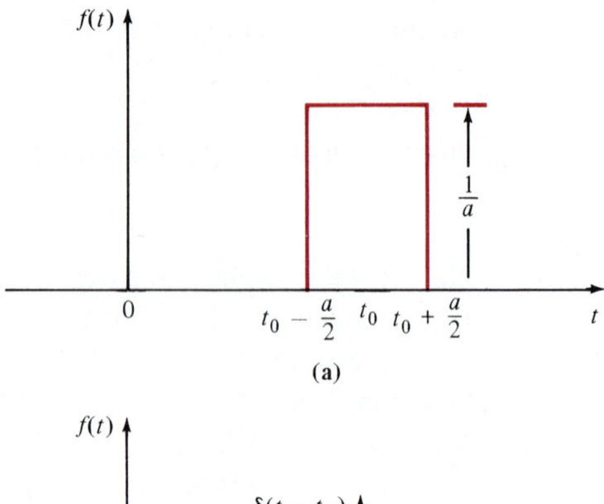

(a)

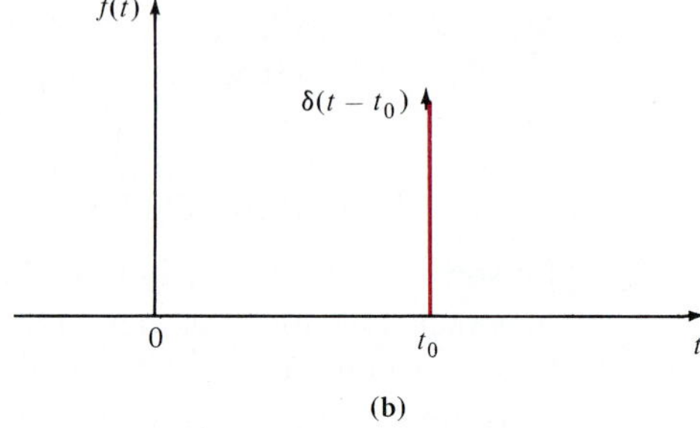

(b)

Figure 16.1 Representations of the unit impulse.

The unit impulse is zero except at $t = t_0$, where it is undefined, but it has unit area (sometimes referred to as *strength*). We represent the unit impulse function on a plot as shown in Fig. 16.1b.

An important property of the unit impulse function is what is often called the *sampling property,* which is exhibited by the following integral:

$$\int_{t_1}^{t_2} f(t)\,\delta(t - t_0)\,dt = \begin{cases} f(t_0) & t_1 < t_0 < t_2 \\ 0 & t_0 < t_1, \quad t_0 > t_2 \end{cases}$$

for a finite t_0 and any $f(t)$ continuous at t_0. Note that the unit impulse function simply samples the value of $f(t)$ at $t = t_0$.

EXAMPLE 16.1

The Laplace transform of the impulse function is

$$\mathbf{F}(s) = \int_0^{\infty} \delta(t - t_0)e^{-st}\,dt$$

Using the sampling property of the delta function, we obtain

$$\mathcal{L}[\delta(t - t_0)] = e^{-t_0 s}$$

In the limit as $t_0 \to 0$, $e^{-t_0 s} \to 1$, and therefore

$$\mathcal{L}[\delta(t)] = \mathbf{F}(s) = 1$$

■

EXAMPLE 16.2

The Laplace transform of the function

$$f(t) = te^{-at}\,\delta(t-1)$$

is

$$\mathbf{F}(s) = \int_0^\infty te^{-at}\,\delta(t-1)e^{-st}\,dt$$

$$= \int_0^\infty \delta(t-1)te^{-(s+a)t}\,dt$$

Once again employing the sampling property of the impulse, we obtain

$$\mathbf{F}(s) = 1e^{-(s+a)} = \frac{e^{-s}}{e^{a}}$$

■

EXAMPLE 16.3

The Laplace transform of the unit step function defined in Chapter 7 is

$$\mathbf{F}(s) = \int_0^\infty u(t)e^{-st}\,dt$$

$$= \int_0^\infty 1e^{-st}\,dt$$

$$= -\frac{1}{s}e^{-st}\Big|_0^\infty$$

$$= \frac{1}{s} \qquad \sigma > 0$$

Therefore,

$$\mathcal{L}[u(t)] = \mathbf{F}(s) = \frac{1}{s}$$

■

EXAMPLE 16.4

The Laplace transform of the time-shifted unit step function shown in Fig. 16.2 is

$$\mathbf{F}(s) = \int_0^\infty u(t-a)e^{-st}\,dt$$

Note that

$$u(t-a) = \begin{cases} 1 & a < t < \infty \\ 0 & t < a \end{cases}$$

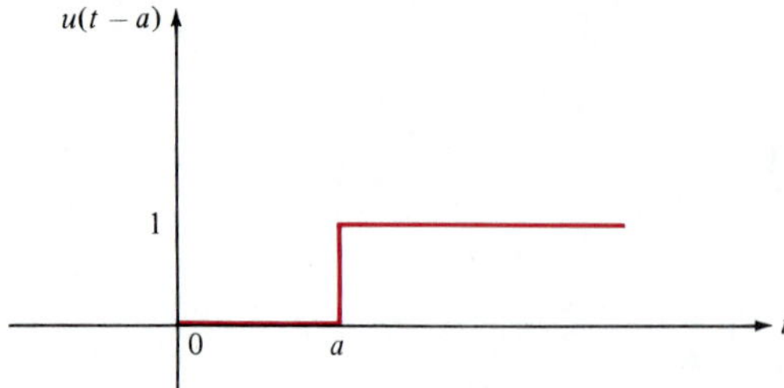

Figure 16.2 Time-shifted unit step function.

Therefore,

$$\mathbf{F}(s) = \int_a^\infty e^{-st}\,dt$$

$$= \frac{e^{-as}}{s}$$

■

EXAMPLE 16.5

The Laplace transform for the cosine function is

$$\mathbf{F}(s) = \int_0^\infty \cos \omega t \, e^{-st}\,dt$$

$$= \int_0^\infty \frac{e^{j\omega t} + e^{-j\omega t}}{2} e^{-st}\,dt$$

$$= \int_0^\infty \frac{e^{-(s-j\omega)t} + e^{-(s+j\omega)t}}{2}\,dt$$

$$= \frac{1}{2}\left(\frac{1}{s - j\omega} + \frac{1}{s + j\omega}\right) \qquad \sigma > 0$$

$$= \frac{s}{s^2 + \omega^2}$$

■

A short table of useful Laplace transform pairs is shown in Table 16.1.

DRILL EXERCISE

D16.1. If $f(t) = e^{-at} \cos \omega t \; \delta(t - 2)$, find $\mathbf{F}(s)$.

Ans: $\mathbf{F}(s) = e^{-2(s+a)} \cos 2\omega$.

D16.2. Find $\mathbf{F}(s)$ if $f(t) = tu(t - 1)$.

Ans: $\mathbf{F}(s) = \dfrac{e^{-s}}{s} + \dfrac{e^{-s}}{s^2}$.

D16.3. Show that if $f(t) = \sin \omega t$, then $\mathbf{F}(s) = \omega/(s^2 + \omega^2)$.

Table 16.1 Short Table of Laplace Transform Pairs

$f(t)$	$\mathbf{F}(s)$
$\delta(t)$	1
$u(t)$	$\frac{1}{s}$
e^{-at}	$\frac{1}{s+a}$
t	$\frac{1}{s^2}$
$\frac{t^n}{n!}$	$\frac{1}{s^{n+1}}$
te^{-at}	$\frac{1}{(s+a)^2}$
$\frac{t^n e^{-at}}{n!}$	$\frac{1}{(s+a)^{n+1}}$
$\sin bt$	$\frac{b}{s^2+b^2}$
$\cos bt$	$\frac{s}{s^2+b^2}$
$e^{-at}\sin bt$	$\frac{b}{(s+a)^2+b^2}$
$e^{-at}\cos bt$	$\frac{s+a}{(s+a)^2+b^2}$

16.3 Some Useful Properties of the Transform

We now present a number of important properties of the Laplace transform and illustrate their usefulness by examples.

Theorem 1

$$\mathscr{L}[Af(t)] = AF(s) \tag{16.5}$$

The transform is defined as

$$\mathscr{L}[Af(t)] = \int_0^\infty Af(t)e^{-st}\,dt$$

Since A is not a function of time,

$$\mathcal{L}[Af(t)] = A\int_0^\infty f(t)e^{-st}\,dt$$
$$= A\mathbf{F}(s)$$

EXAMPLE 16.6

The Laplace transform of the function

$$f(t) = e^{-ax}\sin y\cos\omega t$$

can be found using the results of Example 16.5 as

$$\mathbf{F}(s) = e^{-ax}\sin y\,\frac{s}{s^2+\omega^2}$$

■

Theorem 2

$$\mathcal{L}[f_1(t) \pm f_2(t)] = \mathbf{F}_1(s) \pm \mathbf{F}_2(s) \tag{16.6}$$

by definition

$$\mathcal{L}[f_1(t) \pm f_2(t)] = \int_0^\infty [f_1(t) \pm f_2(t)]e^{-st}\,dt$$
$$= \int_0^\infty f_1(t)e^{-st}\,dt \pm \int_0^\infty f_2(t)e^{-st}\,dt$$
$$= \mathbf{F}_1(s) \pm \mathbf{F}_2(s)$$

EXAMPLE 16.7

The Laplace transform of the function

$$f(t) = e^{-t} + e^{-2t}$$

is

$$\mathbf{F}(s) = \int_0^\infty (e^{-t} + e^{-2t})e^{-st}\,dt$$
$$= \int_0^\infty e^{-(s+1)t}\,dt + \int_0^\infty e^{-(s+2)t}\,dt$$
$$= \frac{1}{s+1} + \frac{1}{s+2}$$

where $\mathcal{L}[e^{-t}] = 1/(s+1)$ and $\mathcal{L}[e^{-2t}] = 1/(s+2)$. ■

Theorem 3

$$\mathscr{L}[f(at)] = \frac{1}{a}\mathbf{F}\left(\frac{s}{a}\right) \qquad a > 0 \tag{16.7}$$

The Laplace transform of f(at) is

$$\mathscr{L}[f(at)] = \int_0^\infty f(at)e^{-st}\,dt$$

Now let $\lambda = at$ and $d\lambda = a\,dt$. Then

$$\begin{aligned}\mathscr{L}[f(at)] &= \int_0^\infty f(\lambda)e^{-(\lambda/a)s}\,\frac{d\lambda}{a}\\ &= \frac{1}{a}\int_0^\infty f(\lambda)e^{-(s/a)\lambda}\,d\lambda\\ &= \frac{1}{a}\mathbf{F}\left(\frac{s}{a}\right) \qquad a > 0\end{aligned}$$

EXAMPLE 16.8

We wish to find the Laplace transform of $f(t) = \cos \omega(t/2)$.

We have found that $\mathscr{L}[\cos \omega t] = s/(s^2 + \omega^2)$. Therefore, using Theorem 3 yields

$$\begin{aligned}\cos \omega\left(\frac{t}{2}\right) &= \frac{4s}{(2s)^2 + \omega^2}\\ &= \frac{s}{s^2 + \omega^2/4}\end{aligned}$$

which should be fairly obvious, since $\mathscr{L}[\cos \alpha t] = s/(s^2 + \alpha^2)$, where in this case $\alpha = \omega/2$. ■

Theorem 4

$$\mathscr{L}[f(t - t_0)u(t - t_0)] = e^{-t_0 s}\mathbf{F}(s) \qquad t_0 \geq 0 \tag{16.8}$$

This theorem, commonly known as the shifting theorem, is illustrated as follows:

$$\begin{aligned}\mathscr{L}[f(t - t_0)u(t - t_0)] &= \int_0^\infty f(t - t_0)u(t - t_0)e^{-st}\,dt\\ &= \int_{t_0}^\infty f(t - t_0)e^{-st}\,dt\end{aligned}$$

If we now let $\lambda = t - t_0$ *and* $d\lambda = dt$, *then*

$$\mathcal{L}[f(t-t_0)u(t-t_0)] = \int_0^\infty f(\lambda)e^{-s(\lambda+t_0)}\,d\lambda$$

$$= e^{-t_0 s}\int_0^\infty f(\lambda)e^{-s\lambda}\,d\lambda$$

$$= e^{-t_0 s}\mathbf{F}(s) \qquad t_0 \geq 0$$

EXAMPLE 16.9

If $f(t) = (t-1)u(t-1)$, then by employing Theorem 4, we find that

$$\mathcal{L}[(t-1)u(t-1)] = e^{-s}\mathcal{L}[t] = \frac{e^{-s}}{s^2}$$

■

Theorem 5

$$\mathcal{L}[f(t)u(t-t_0)] = e^{-t_0 s}\mathcal{L}[f(t+t_0)] \tag{16.9}$$

By definition,

$$\mathcal{L}[f(t)u(t-t_0)] = \int_0^\infty f(t)u(t-t_0)e^{-st}\,dt$$

Now let $\lambda = t - t_0$ *and* $d\lambda = dt$, *and therefore,*

$$\int_0^\infty f(t)u(t-t_0)e^{-st}\,dt = \int_{-t_0}^\infty f(\lambda+t_0)u(\lambda)e^{-s(\lambda+t_0)}\,d\lambda$$

$$= \int_0^\infty f(\lambda+t_0)e^{-s(\lambda+t_0)}\,d\lambda$$

$$= e^{-t_0 s}\mathcal{L}[f(t+t_0)]$$

EXAMPLE 16.10

If $f(t) = tu(t-1)$, then $\mathcal{L}[f(t)]$ can be found from Theorem 5 as

$$\mathcal{L}[tu(t-1)] = e^{-s}\mathcal{L}[t+1]$$

$$= e^{-s}\left(\frac{1}{s^2} + \frac{1}{s}\right)$$

■

Theorem 6

$$\mathcal{L}[e^{-at}f(t)] = \mathbf{F}(s+a) \tag{16.10}$$

By definition,

$$\mathcal{L}[e^{-at}f(t)] = \int_0^\infty e^{-at}f(t)e^{-st}\,dt$$

$$= \int_0^\infty f(t)e^{-(s+a)}\,dt$$

$$= \mathbf{F}(s+a)$$

EXAMPLE 16.11

Since the Laplace transform of cos ωt is known to be

$$\mathscr{L}[\cos \omega t] = \frac{s}{s^2 + \omega^2}$$

then

$$\mathscr{L}[e^{-at} \cos \omega t] = \frac{s + a}{(s + a)^2 + \omega^2}$$

■

Theorem 7

$$\mathscr{L}\left[\frac{d^n f(t)}{dt^n}\right] = s^n \mathbf{F}(s) - s^{n-1} f(0) - s^{n-2} f'_{(0)} \cdots s^0 f^{n-1}_{(0)} \qquad (16.11)$$

Let us begin by examining $\mathscr{L}[df(t)/dt]$. *By definition,*

$$\mathscr{L}\left[\frac{df(t)}{dt}\right] = \int_0^\infty \frac{df(t)}{dt} e^{-st}\, dt$$

Using integration by parts gives us

$$u = e^{-st} \qquad dv = \frac{df(t)}{dt} dt = df(t)$$

$$du = -se^{-st}\, dt \qquad v = f(t)$$

Hence

$$\mathscr{L}\left[\frac{df(t)}{dt}\right] = f(t)e^{-st}\Big|_0^\infty + s\int_0^\infty f(t)e^{-st}\, dt$$

If we assume that the Laplace transform of the function $f(t)$ *exists so that*

$$\lim_{t\to\infty} e^{-st} f(t) = 0$$

then

$$\mathscr{L}\left[\frac{df(t)}{dt}\right] = -f(0) + s\mathbf{F}(s)$$

Using this result, we can write

$$\begin{aligned}\mathscr{L}\left[\frac{d^2 f(t)}{dt^2}\right] &= s\mathscr{L}[f'(t)] - f'(0) \\ &= s[s\mathbf{F}(s) - f(0)] - f'(0) \\ &= s^2\mathbf{F}(s) - sf(0) - f'(0)\end{aligned}$$

By continuing in this manner we can demonstrate the original statement.

We will illustrate the usefulness of this theorem later in this chapter when we employ it to solve differential equations.

Theorem 8

$$\mathcal{L}[tf(t)] = \frac{-d\mathbf{F}(s)}{ds} \tag{16.12}$$

By definition,

$$\mathbf{F}(s) = \int_0^\infty f(t)e^{-st}\,dt$$

Then

$$\begin{aligned}\frac{d\mathbf{F}(s)}{ds} &= \int_0^\infty f(t)(-te^{-st})\,dt \\ &= -\int_0^\infty tf(t)e^{-st}\,dt \\ &= -\mathcal{L}[tf(t)]\end{aligned}$$

EXAMPLE 16.12

Let us demonstrate the use of Theorem 8. To begin, we know that

$$\mathcal{L}[u(t)] = \frac{1}{s}$$

Then

$$\mathcal{L}[tu(t)] = \frac{-d}{ds}\left(\frac{1}{s}\right) = \frac{1}{s^2}$$

Continuing, we note that

$$\mathcal{L}[t^2u(t)] = -\frac{d}{ds}\left(\frac{1}{s^2}\right) = \frac{2}{s^3}$$

and

$$\mathcal{L}[t^3u(t)] = \frac{-d}{ds}\left(\frac{2}{s^3}\right) = \frac{3!}{s^4}$$

and in general,

$$\mathcal{L}[t^nu(t)] = \frac{n!}{s^{n+1}}$$

■

Theorem 9

$$\mathcal{L}\left[\frac{f(t)}{t}\right] = \int_s^\infty \mathbf{F}(\lambda)\,d\lambda \tag{16.13}$$

By definition,

$$\int_0^\infty f(t)e^{-\lambda t}\,dt = \mathbf{F}(\lambda)$$

Therefore,

$$\int_s^\infty \int_0^\infty f(t)e^{-\lambda t}\,dt\,d\lambda = \int_s^\infty \mathbf{F}(\lambda)\,d\lambda$$

Since $f(t)$ is Laplace transformable, we can change the order of integration so that

$$\int_0^\infty f(t)\int_s^\infty e^{-\lambda t}\,d\lambda\,dt = \int_s^\infty \mathbf{F}(\lambda)\,d\lambda$$

and hence

$$\int_0^\infty \frac{f(t)}{t}e^{-st}\,dt = \int_s^\infty \mathbf{F}(\lambda)\,d\lambda$$

and therefore,

$$\mathcal{L}\left[\frac{f(t)}{t}\right] = \int_s^\infty \mathbf{F}(\lambda)\,d\lambda$$

EXAMPLE 16.13

If $f(t) = te^{-at}$, then

$$\mathbf{F}(\lambda) = \frac{1}{(\lambda + a)^2}$$

Therefore,

$$\int_s^\infty \mathbf{F}(\lambda)\,d\lambda = \int_s^\infty \frac{1}{(\lambda + a)^2}\,d\lambda = \left.\frac{-1}{\lambda + a}\right|_s^\infty = \frac{1}{s + a}$$

Hence

$$f_1(t) = \frac{f(t)}{t} = \frac{te^{-at}}{t} = e^{-at} \quad \text{and} \quad \mathbf{F}_1(s) = \frac{1}{s + a}$$

Theorem 10

$$\mathcal{L}\left[\int_0^t f(\lambda)\,d\lambda\right] = \frac{1}{s}\mathbf{F}(s) \tag{16.14}$$

We begin with the expression

$$\mathcal{L}\left[\int_0^t f(\lambda)\,d\lambda\right] = \int_0^\infty \int_0^t f(\lambda)\,d\lambda e^{-st}\,dt$$

Integrating by parts yields

$$u = \int_0^t f(\lambda)\, d\lambda \qquad dv = e^{-st}\, dt$$

$$du = f(t)\, dt \qquad v = \frac{-1}{s} e^{-st}$$

Therefore,

$$\mathcal{L}\left[\int_0^t f(\lambda)\, d\lambda\right] = \frac{-e^{-st}}{s} \int_0^t f(\lambda)\, d\lambda \Big|_0^\infty + \frac{1}{s} \int_0^\infty f(t) e^{-st}\, dt$$

$$= \frac{1}{s} \mathbf{F}(s)$$

We employ this theorem later in the chapter when we examine integrodifferential equations. The theorems we have presented are listed in Table 16.2 for quick reference.

Table 16.2 Some Useful Properties of the Laplace Transform

$f(t)$	$\mathbf{F}(s)$
$Af(t)$	$A\mathbf{F}(s)$
$f_1(t) \pm f_2(t)$	$\mathbf{F}_1(s) \pm \mathbf{F}_2(s)$
$f(at)$	$\frac{1}{a}\mathbf{F}\left(\frac{s}{a}\right), \quad a > 0$
$f(t - t_0)u(t - t_0),\ t_0 \geq 0$	$e^{-t_0 s}\mathbf{F}(s)$
$f(t)u(t - t_0)$	$e^{-t_0 s}\mathcal{L}[f(t + t_0)]$
$e^{-at}f(t)$	$\mathbf{F}(s + a)$
$\frac{d^n f(t)}{dt^n}$	$s^n\mathbf{F}(s) - s^{n-1}f(0) - s^{n-2}f^1(0) \cdots s^0 f^{n-1}(0)$
$tf(t)$	$-\frac{d\mathbf{F}(s)}{ds}$
$\frac{f(t)}{t}$	$\int_s^\infty \mathbf{F}(\lambda)\, d\lambda$
$\int_0^t f(\lambda)\, d\lambda$	$\frac{1}{s}\mathbf{F}(s)$
$\int_0^t f_1(\lambda) f_2(t - \lambda)\, d\lambda$	$\mathbf{F}_1(s)\mathbf{F}_2(s)$

DRILL EXERCISE

D16.4. Find $\mathbf{F}(s)$ if $f(t) = \frac{1}{2}(t - 4e^{-2t})$.

Ans: $\mathbf{F}(s) = \dfrac{1}{2s^2} - \dfrac{2}{s + 2}$.

D16.5. If $f(t) = te^{-(t-1)}u(t-1) - e^{-(t-1)}u(t-1)$, determine $\mathbf{F}(s)$.

Ans: $\mathbf{F}(s) = \dfrac{e^{-s}}{(s+1)^2}$.

D16.6. Find $\mathbf{F}(s)$ if $f(t) = e^{-4t}(t - e^{-t})$.

Ans: $\mathbf{F}(s) = \dfrac{1}{(s+4)^2} - \dfrac{1}{s+5}$.

D16.7. Find the Laplace transform of the function $te^{-4x}/(a^2+4)$.

Ans: $\mathbf{F}(s) = \dfrac{e^{-4x}}{s^2(a^2+4)}$.

D16.8. If $f(t) = \cos \omega t u(t-1)$, find $\mathbf{F}(s)$.

Ans: $\mathbf{F}(s) = e^{-s}\left(\dfrac{s\cos\omega}{s^2+\omega^2} - \dfrac{\omega\sin\omega}{s^2+\omega^2}\right)$.

D16.9. Use Theorem 8 to demonstrate that $\mathscr{L}[te^{-at}] = 1/(s+a)^2$.

16.4 The Gate Function

The *gate function* is defined as the difference of two unit step functions of the form

$$f(t) = u(t-\tau) - u(t-\tau-T_0) \tag{16.15}$$

The function is shown in Fig. 16.3. Note that the function is a rectangular pulse of unit height and width T_0 which starts at $t = \tau$. The importance of this gate is the fact that any

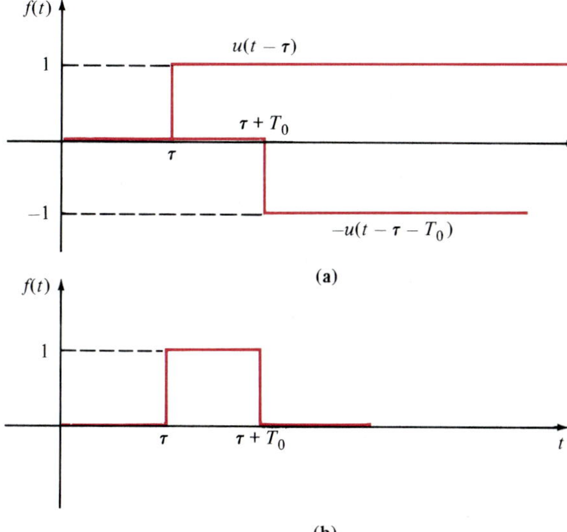

Figure 16.3 Gate function generated from two unit step functions.

function $f(t)$ which is multiplied by the gate is equal to $f(t)$ within the gate interval $\tau < t < \tau + T_0$ and is zero everywhere else. Therefore, we can use the gate function to express a complex waveform by simply treating every different portion of the waveform separately using a gate. The following example illustrates the approach.

EXAMPLE 16.14

We wish to determine the Laplace transform of the sawtooth function shown in Fig. 16.4.

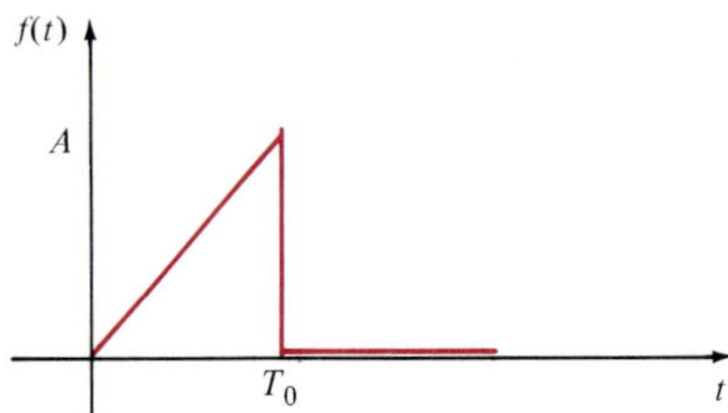

Figure 16.4 Sawtooth waveform.

Using the gate function, we can express the sawtooth as

$$\begin{aligned} f(t) &= \frac{A}{T_0} t[u(t) - u(t - T_0)] \\ &= \frac{A}{T_0} tu(t) - \frac{A}{T_0} tu(t - T_0) \\ &= \frac{A}{T_0} tu(t) - \frac{A}{T_0}(t - T_0 + T_0)u(t - T_0) \\ &= \frac{A}{T_0} tu(t) - \frac{A}{T_0}(t - T_0)u(t - T_0) - Au(t - T_0) \end{aligned}$$

Then $\mathbf{F}(s)$ is

$$\begin{aligned} \mathbf{F}(s) &= \frac{A}{T_0}\left(\frac{1}{s^2} - \frac{1}{s^2} e^{-T_0 s} - \frac{T_0}{s} e^{-T_0 s}\right) \\ &= \frac{A}{T_0 s^2}[1 - (1 + T_0 s)e^{-T_0 s}] \end{aligned}$$

■

Another technique for expressing a complex waveform involves the use of a combination of steps and ramps of the form $Au(t - t_0)$ and $B(t - t_0)u(t - t_0)$, respectively. The Laplace transforms of these functions are simply

$$\mathcal{L}[Au(t - t_0)] = \frac{A}{s} e^{-t_0 s}$$

and

$$\mathcal{L}[B(t - t_0)u(t - t_0)] = \frac{B}{s^2}e^{-t_0 s}$$

For each discontinuity in the waveform we will have a step in the expression and for each change in slope we will have a ramp in the expression. The following example illustrates the technique.

EXAMPLE 16.15

Let us determine the Laplace transform of the waveform shown in Fig. 16.5.

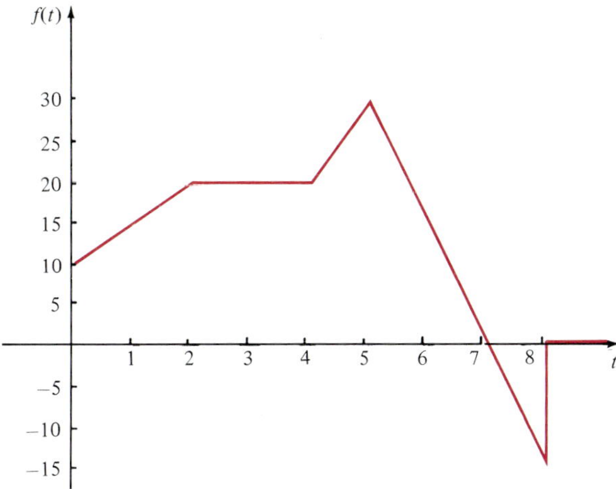

Figure 16.5 Piecewise linear function used in Example 16.15.

Starting at time $t = 0$, there is a discontinuity of 10; therefore, the function contains the term $10u(t)$. The slope changes from 0 to 5 at $t = 0$ and therefore the expression contains the term $5tu(t)$. At time $t = 2$ the slope changes from 5 to 0, and therefore the expression contains the term $-5(t - 2)u(t - 2)$. At time $t = 4$ the slope changes from 0 to 10, and hence the expression will contain the term $10(t - 4)u(t - 4)$. Continuing in this manner, we obtain

$$f(t) = 10u(t) + 5tu(t) - 5(t - 2)u(t - 2) + 10(t - 4)u(t - 4) - 25(t - 5)u(t - 5) + 15(t - 8)u(t - 8) + 15u(t - 8)$$

Therefore,

$$\mathbf{F}(s) = \frac{10}{s} + \frac{5}{s^2} - \frac{5}{s^2}e^{-2s} + \frac{10}{s^2}e^{-4s} - \frac{25}{s^2}e^{-5s} + \frac{15}{s^2}e^{-8s} + \frac{15}{s}e^{-8s}$$ ■

DRILL EXERCISE

D16.10. Find the Laplace transform of the waveform in Fig. D16.10.

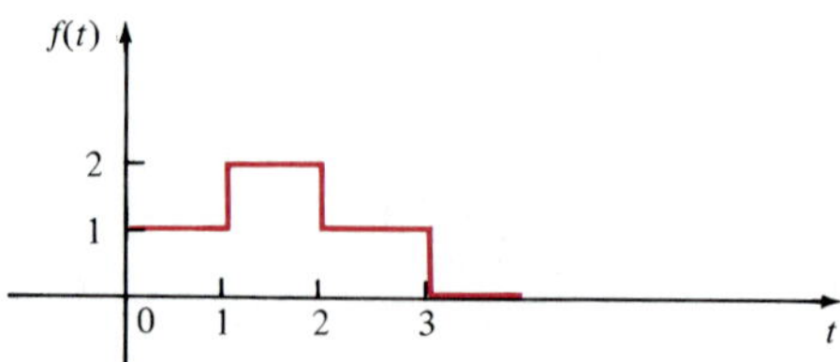

Figure D16.10

Ans: $\mathbf{F}(s) = \frac{1}{s} + \frac{e^{-s}}{s} - \frac{e^{-2s}}{s} - \frac{e^{-3s}}{s}$.

D16.11. Determine the Laplace transform of the waveform in Fig. D16.11.

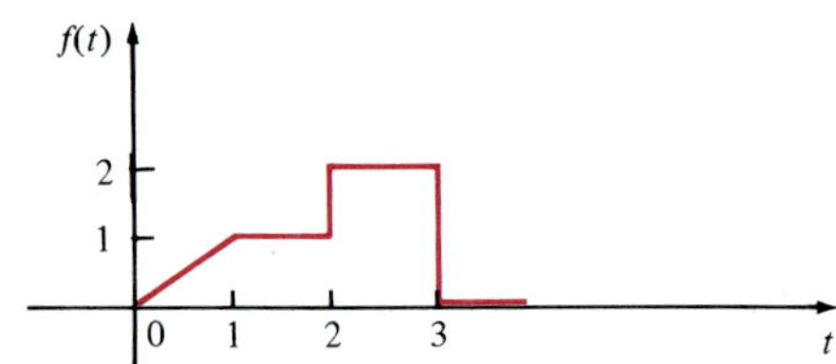

Figure D16.11

Ans: $\mathbf{F}(s) = \frac{1}{s^2} - \frac{e^{-s}}{s^2} + \frac{e^{-2s}}{s} - \frac{2e^{-3s}}{s}$.

16.5 Periodic Functions

If $f(t)$ is a periodic function with period T for $t \geq 0$ and is zero for $t < 0$, then $f(t)$ can be expressed as

$$f(t) = f_1(t) + f_2(t) + f_3(t) + \cdots \qquad (16.16)$$

where $f_1(t)$ is the function in the interval $0 \leq t < T$ and is zero elsewhere, $f_2(t)$ is the function in the interval $T \leq t < 2T$ and is zero elsewhere, and so on. However, since $f_2(t)$ in the interval $T \leq t < 2T$ is identical to $f_1(t)$ in the interval $0 \leq t < T$, and $f_3(t)$ in the interval $2T \leq t < 3T$ is identical to $f_1(t)$ in the interval $0 \leq t < T$, and so on, then $f(t)$ can be written as

$$f(t) = f_1(t) + f_1(t - T)u(t - T) + f_1(t - 2T)u(t - 2T) + \cdots \qquad (16.17)$$

Hence, using Theorem 4, we can write

$$\begin{aligned} \mathcal{L}[f(t)] &= \mathbf{F}_1(s) + \mathbf{F}_1(s)e^{-Ts} + \mathbf{F}_1(s)e^{-2Ts} + \cdots \\ &= \mathbf{F}_1(s)(1 + e^{-Ts} + e^{-2Ts} + e^{-3Ts} + \cdots) \end{aligned} \qquad (16.18)$$

If $|x| < 1$, then

$$\frac{1}{1-x} = 1 + x + x^2 + x^3 + \cdots$$

since $x = e^{-sT}$; $|x| = |e^{-sT}| < 1$ provided that $T > 0$. Therefore,

$$\mathscr{L}[f(t)] = \frac{\mathbf{F}_1(s)}{1 - e^{-Ts}} \tag{16.19}$$

where $\mathbf{F}_1(s) = \mathscr{L}[f_1(t)]$, the Laplace transform of the first period of $f(t)$, and $\mathbf{F}_1(s) = \int_0^T f(t)e^{-st}\,dt$.

EXAMPLE 16.16

The Laplace transform of the periodic waveform shown in Fig. 16.6 is

$$\mathscr{L}[f(t)] = \frac{\mathbf{F}_1(s)}{1 - e^{-Ts}}$$

where from Example 16.14,

$$\mathbf{F}_1(s) = \frac{A}{T_0 s^2}[1 - (1 + T_0 s)e^{-T_0 s}]$$

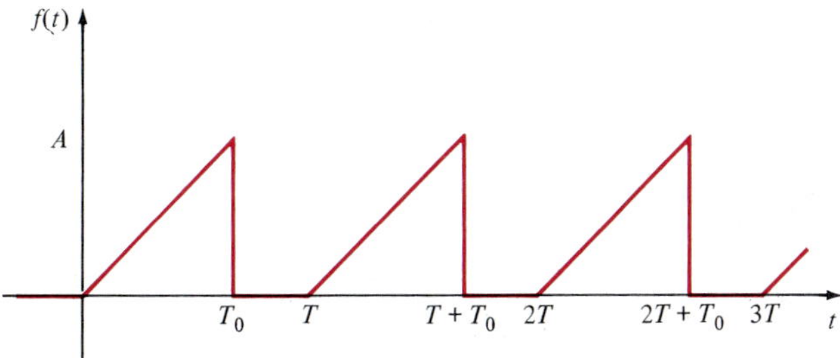

Figure 16.6 Periodic sawtooth waveform. ■

DRILL EXERCISE

D16.12. Determine the Laplace transform of the periodic waveform in Fig. D16.12.

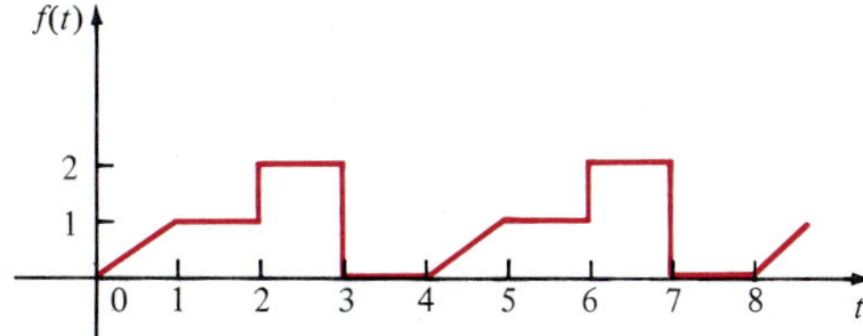

Figure D16.12

Ans: $\mathbf{F}(s) = \frac{1}{1 - e^{-4s}}\left(\frac{1}{s^2} - \frac{e^{-s}}{s^2} + \frac{e^{-2s}}{s} - \frac{2e^{-3s}}{s}\right)$.

D16.13. Find the Laplace transform of the periodic waveform shown in Fig. D16.13.

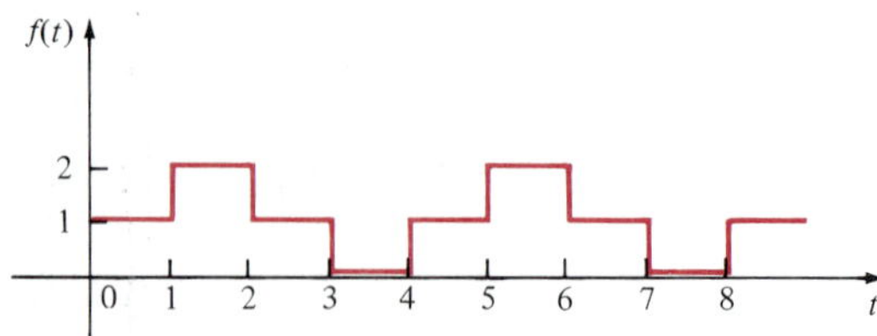

Figure D16.13

Ans: $\mathbf{F}(s) = \frac{1}{1 - e^{-4s}}\left(\frac{1}{s} + \frac{e^{-s}}{s} - \frac{e^{-2s}}{s} - \frac{e^{-3s}}{s}\right).$

16.6 Performing the Inverse Transform

As we begin our discussion of this topic, let us outline the procedure we will use in applying the Laplace transform to circuit analysis. First, we will transform the problem from the time domain to the complex frequency domain. Next, we will solve the circuit equations algebraically in the complex frequency domain. Finally, we will transform the solution from the frequency domain back to the time domain. It is this latter operation that we discuss now.

The algebraic solution of the circuit equations in the complex frequency domain results in a rational function of s of the form

$$\mathbf{F}(s) = \frac{\mathbf{P}(s)}{\mathbf{Q}(s)} = \frac{a_m s^m + a_{m-1}s^{m-1} + \cdots + a_1 s + a_0}{b_n s^n + b_{n-1}s^{n-1} + \cdots + b_1 s + b_0} \tag{16.20}$$

The roots of the polynomial $\mathbf{P}(s)$ are called the *zeros* of the function $\mathbf{F}(s)$ because at these values of s, $\mathbf{F}(s) = 0$. Similarly, the roots of the polynomial $\mathbf{Q}(s)$ are called *poles* of $\mathbf{F}(s)$, since at these values of s, $\mathbf{F}(s)$ becomes infinite.

If $\mathbf{F}(s)$ is a proper rational function of s, then $n > m$. However, if this is not the case, we simply divide $\mathbf{P}(s)$ by $\mathbf{Q}(s)$ to obtain a quotient and a remainder; that is,

$$\frac{\mathbf{P}(s)}{\mathbf{Q}(s)} = C_{m-n}s^{m-n} + \cdots + C_2 s^2 + C_1 s + C_0 + \frac{\mathbf{P}_1(s)}{\mathbf{Q}(s)} \tag{16.21}$$

Now $\mathbf{P}_1(s)/\mathbf{Q}(s)$ is a proper rational function of s. Let us examine the possible forms of the roots of $\mathbf{Q}(s)$.

1. If the roots are simple, $\mathbf{P}_1(s)/\mathbf{Q}(s)$ can be expressed in partial fraction form as

$$\frac{\mathbf{P}_1(s)}{\mathbf{Q}(s)} = \frac{K_1}{s + p_1} + \frac{K_2}{s + p_2} + \cdots + \frac{K_n}{s + p_n} \tag{16.22}$$

2. If $\mathbf{Q}(s)$ has simple complex roots, they will appear in complex-conjugate pairs and the partial fraction expansion of $\mathbf{P}_1(s)/\mathbf{Q}(s)$ for each pair of complex-conjugate roots will be of the form

$$\frac{\mathbf{P}_1(s)}{\mathbf{Q}_1(s)(s+\alpha-j\beta)(s+\alpha+j\beta)} = \frac{K_1}{s+\alpha-j\beta} + \frac{K_1^*}{s+\alpha+j\beta} + \cdots \tag{16.23}$$

where $\mathbf{Q}(s) = \mathbf{Q}_1(s)(s+\alpha-j\beta)(s+\alpha+j\beta)$ and K_1^* is the complex conjugate of K_1.

3. If $\mathbf{Q}(s)$ has a root of multiplicity r, the partial fraction expansion for each such root will be of the form

$$\frac{\mathbf{P}_1(s)}{\mathbf{Q}_1(s)(s+p_1)^r} = \frac{K_{11}}{(s+p_1)} + \frac{K_{12}}{(s+p_1)^2} + \cdots + \frac{K_{1r}}{(s+p_1)^r} + \cdots \tag{16.24}$$

The importance of these partial fraction expansions stems from the fact that once the function $\mathbf{F}(s)$ is expressed in this form, the individual inverse Laplace transforms can be obtained from known and tabulated transform pairs. The sum of these inverse Laplace transforms then yields the desired time function, $f(t) = \mathcal{L}^{-1}[\mathbf{F}(s)]$.

Simple Poles

Let us assume that all the poles of $\mathbf{F}(s)$ are simple, so that the partial fraction expansion of $\mathbf{F}(s)$ is of the form

$$\mathbf{F}(s) = \frac{\mathbf{P}(s)}{\mathbf{Q}(s)} = \frac{K_1}{s+p_1} + \frac{K_2}{s+p_2} + \cdots + \frac{K_n}{s+p_n} \tag{16.25}$$

Then the constant K_i can be computed by multiplying both sides of this equation by $(s + p_i)$ and evaluating the equation at $s = -p_i$; that is,

$$\left.\frac{(s+p_i)\mathbf{P}(s)}{\mathbf{Q}(s)}\right|_{s=-p_i} = 0 + \cdots + 0 + K_i + 0 + \cdots + 0 \qquad i = 1, 2, \ldots, n \tag{16.26}$$

Once all of the K_i terms are known, the time function $f(t) = \mathcal{L}^{-1}[\mathbf{F}(s)]$ can be obtained using the Laplace transform pair

$$\mathcal{L}^{-1}\left[\frac{1}{s+a}\right] = e^{-at} \tag{16.27}$$

EXAMPLE 16.17

Given that

$$\mathbf{F}(s) = \frac{12(s+1)(s+3)}{s(s+2)(s+4)(s+5)}$$

let us find the function $f(t) = \mathcal{L}^{-1}[\mathbf{F}(s)]$.

Expressing $\mathbf{F}(s)$ in a partial fraction expansion, we obtain

$$\frac{12(s+1)(s+3)}{s(s+2)(s+4)(s+5)} = \frac{K_0}{s} + \frac{K_1}{s+2} + \frac{K_2}{s+4} + \frac{K_3}{s+5}$$

To determine K_0, we multiply both sides of the equation by s to obtain the equation

$$\frac{12(s+1)(s+3)}{(s+2)(s+4)(s+5)} = K_0 + \frac{K_1 s}{s+2} + \frac{K_2 s}{s+4} + \frac{K_3 s}{s+5}$$

Evaluating the equation at $s = 0$ yields

$$\frac{(12)(1)(3)}{(2)(4)(5)} = K_0 + 0 + 0 + 0$$

or

$$K_0 = \frac{36}{40}$$

Similarly,

$$(s+2)\mathbf{F}(s)\Big|_{s=-2} = \frac{12(s+1)(s+3)}{s(s+4)(s+5)}\Big|_{s=-2} = K_1$$

or

$$K_1 = 1$$

Using the same approach, we find that $K_2 = \frac{36}{8}$ and $K_3 = -\frac{32}{5}$. Hence $\mathbf{F}(s)$ can be written as

$$\mathbf{F}(s) = \frac{36/40}{s} + \frac{1}{s+2} + \frac{36/8}{s+4} - \frac{32/5}{s+5}$$

Then $f(t) = \mathscr{L}^{-1}[\mathbf{F}(s)]$ is

$$f(t) = \left(\frac{36}{40} + 1e^{-2t} + \frac{36}{8}e^{-4t} - \frac{32}{5}e^{-5t}\right)u(t)$$

■

DRILL EXERCISE

D16.14. Find $f(t)$ if $\mathbf{F}(s) = 10(s+6)/(s+1)(s+3)$.

Ans: $f(t) = (25e^{-t} - 15e^{-3t})u(t)$.

D16.15. If $\mathbf{F}(s) = 12(s+2)/s(s+1)$, find $f(t)$.

Ans: $f(t) = (24 - 12e^{-t})u(t)$.

Complex-Conjugate Poles

Let us assume that $\mathbf{F}(s)$ has one pair of complex-conjugate poles. The partial fraction expansion of $\mathbf{F}(s)$ can then be written as

$$\mathbf{F}(s) = \frac{\mathbf{P}_1(s)}{\mathbf{Q}_1(s)(s+\alpha-j\beta)(s+\alpha+j\beta)} = \frac{K_1}{s+\alpha-j\beta} + \frac{K_1^*}{s+\alpha+j\beta} + \cdots \tag{16.28}$$

The constant K_1 can then be determined using the procedure employed for simple poles; that is,

$$(s+\alpha-j\beta)\mathbf{F}(s)\bigg|_{s=-\alpha+j\beta} = K_1 \tag{16.29}$$

In this case K_1 is in general a complex number that can be expressed as $|K_1|\ \underline{/\theta}$. Then $K_1^* = |K_1|\ \underline{/-\theta}$. Hence, the partial fraction expansion can be expressed in the form

$$\begin{aligned}\mathbf{F}(s) &= \frac{|K_1|\ \underline{/\theta}}{s+\alpha-j\beta} + \frac{|K_1|\ \underline{/-\theta}}{s+\alpha+j\beta} + \cdots \\ &= \frac{|K_1|e^{j\theta}}{s+\alpha-j\beta} + \frac{|K_1|e^{-j\theta}}{s+\alpha+j\beta} + \cdots\end{aligned} \tag{16.30}$$

The corresponding time function is then of the form

$$\begin{aligned}f(t) = \mathscr{L}^{-1}[\mathbf{F}(s)] &= |K_1|e^{j\theta}e^{-(\alpha-j\beta)t} + |K_1|e^{-j\theta}e^{-(\alpha+j\beta)t} + \cdots \\ &= |K_1|e^{-\alpha t}[e^{j(\beta t+\theta)} + e^{-j(\beta t+\theta)}] + \cdots \\ &= 2|K_1|e^{-\alpha t}\cos(\beta t+\theta) + \cdots\end{aligned} \tag{16.31}$$

EXAMPLE 16.18

Let us determine the time function $f(t)$ for the function

$$\mathbf{F}(s) = \frac{10(s+1)}{s(s^2+2s+2)}$$

The function $\mathbf{F}(s)$ can be expressed as a partial fraction expansion of the form

$$\mathbf{F}(s) = \frac{10(s+1)}{s(s+1-j1)(s+1+j1)} = \frac{K_0}{s} + \frac{K_1}{s+1-j1} + \frac{K_1^*}{s+1+j1}$$

Then

$$s\mathbf{F}(s)\bigg|_{s=0} = K_0$$

$$\frac{(10)(1)}{(1+j1)(1-j1)} = 5 = K_0$$

In a similar manner,

$$(s + 1 - j1)\mathbf{F}(s)\Big|_{s=-1+j1} = K_1$$

$$\frac{(10)(-1 + j1 + 1)}{(-1 + j1)(-1 + j1 + 1 + j1)} = K_1$$

$$\frac{5}{\sqrt{2}} \angle -135^\circ = K_1$$

Then, of course,

$$K_1^* = \frac{5}{\sqrt{2}} \angle 135^\circ$$

Hence

$$\mathbf{F}(s) = \frac{5}{s} + \frac{(5/\sqrt{2}) \angle -135^\circ}{s + 1 - j1} + \frac{(5/\sqrt{2}) \angle 135^\circ}{s + 1 + j1}$$

and therefore,

$$f(t) = [5 + 5\sqrt{2}e^{-t} \cos (t - 135^\circ)]u(t)$$

■

DRILL EXERCISE

D16.16. Determine $f(t)$ if $\mathbf{F}(s) = s/(s^2 + 4s + 8)$.

Ans: $f(t) = 1.41e^{-2t} \cos (2t + 45^\circ)u(t)$.

D16.17. If $\mathbf{F}(s) = (s + 1)/(s^2 + 6s + 13)$, find $f(t)$.

Ans: $f(t) = 1.41e^{-3t} \cos (2t + 45^\circ)u(t)$.

Multiple Poles

Let us suppose that $\mathbf{F}(s)$ has a pole of multiplicity r. Then $\mathbf{F}(s)$ can be written in a partial fraction expansion of the form

$$\begin{aligned}\mathbf{F}(s) &= \frac{\mathbf{P}_1(s)}{\mathbf{Q}_1(s)(s + p_1)^r} \\ &= \frac{K_{11}}{s + p_1} + \frac{K_{12}}{(s + p_1)^2} + \cdots + \frac{K_{1r}}{(s + p_1)^r} + \cdots \end{aligned} \tag{16.32}$$

Employing the approach for a simple pole, we can evaluate K_{1r} as

$$(s + p_1)^r\mathbf{F}(s)\Big|_{s=-p_1} = K_{1r} \tag{16.33}$$

In order to evaluate K_{1r-1} we multiply $\mathbf{F}(s)$ by $(s+p_1)^r$ as we did to determine K_{1r}; however, prior to evaluating the equation at $s=-p_1$, we take the derivative with respect to s. The proof that this will yield K_{1r-1} can be obtained by multiplying both sides of Eq. (16.32) by $(s+p_1)^r$ and then taking the derivative with respect to s. Now when we evaluate the equation at $s=-p_1$, the only term remaining on the right side of the equation is K_{1r-1}, and therefore,

$$\frac{d}{ds}[(s+p_1)^r\mathbf{F}(s)]\bigg|_{s=-p_1} = K_{1r-1} \tag{16.34}$$

K_{1r-2} can be computed in a similar fashion, and in that case the equation is

$$\frac{d^2}{ds^2}[(s+p_1)^r\mathbf{F}(s)]\bigg|_{s=-p_1} = (2!)K_{1r-2} \tag{16.35}$$

The general expression for this case is

$$K_{1j} = \frac{1}{(r-j)!}\frac{d^{r-j}}{ds^{r-j}}[(s+p_1)^r\mathbf{F}(s)]\bigg|_{s=-p_1} \tag{16.36}$$

Let us illustrate this procedure with an example.

EXAMPLE 16.19

Given the following function $\mathbf{F}(s)$, let us determine the corresponding time function $f(t)=\mathcal{L}^{-1}[\mathbf{F}(s)]$.

$$\mathbf{F}(s) = \frac{10(s+3)}{(s+1)^3(s+2)}$$

Expressing $\mathbf{F}(s)$ as a partial fraction expansion, we obtain

$$\mathbf{F}(s) = \frac{10(s+3)}{(s+1)^3(s+2)} = \frac{K_{11}}{s+1} + \frac{K_{12}}{(s+1)^2} + \frac{K_{13}}{(s+1)^3} + \frac{K_2}{s+2}$$

Then

$$(s+1)^3\mathbf{F}(s)\bigg|_{s=-1} = K_{13}$$

$$20 = K_{13}$$

K_{12} is now determined by the equation

$$\frac{d}{ds}[(s+1)^3\mathbf{F}(s)]\bigg|_{s=-1} = K_{12}$$

$$\frac{-10}{(s+2)^2}\bigg|_{s=-1} = -10 = K_{12}$$

In a similar fashion K_{11} is computed from the equation

$$\frac{d^2}{ds^2}[(s+1)^3\mathbf{F}(s)]\bigg|_{s=-1} = 2K_{11}$$

$$\frac{20}{(s+2)^3}\bigg|_{s=-1} = 20 = 2K_{11}$$

Therefore,

$$10 = K_{11}$$

In addition,

$$(s+2)\mathbf{F}(s)\bigg|_{s=-2} = K_2$$

$$-10 = K_2$$

Hence $\mathbf{F}(s)$ can be expressed as

$$\mathbf{F}(s) = \frac{10}{s+1} - \frac{10}{(s+1)^2} + \frac{20}{(s+1)^3} - \frac{10}{s+2}$$

Now we employ the transform pair

$$\mathcal{L}^{-1}\left[\frac{1}{(s+a)^{n+1}}\right] = \frac{t^n}{n!}e^{-at}$$

and hence

$$f(t) = (10e^{-t} - 10te^{-t} + 10t^2e^{-t} - 10e^{-2t})u(t)$$

■

DRILL EXERCISE

D16.18. Determine $f(t)$ if $\mathbf{F}(s) = s/(s+1)^2$.

Ans: $f(t) = (e^{-t} - te^{-t})u(t)$.

D16.19. If $\mathbf{F}(s) = (s+2)/s^2(s+1)$, find $f(t)$.

Ans: $f(t) = (-1 + 2t + e^{-t})u(t)$.

16.7 Convolution Integral

Convolution is a very important concept and has wide application in circuit and systems analysis. We will first illustrate the connection that exists between the convolution integral and the Laplace transform. We then indicate the manner in which the convolution integral is applied in circuit analysis.

Theorem 11

$$\textit{If } f(t) = \int_0^t f_1(t-\lambda)f_2(\lambda)\, d\lambda = \int_0^t f_1(\lambda)f_2(t-\lambda)\, d\lambda \tag{16.37}$$

and $\mathcal{L}[f(t)] = \mathbf{F}(s)$, $\mathcal{L}[f_1(t)] = \mathbf{F}_1(s)$, *and* $\mathcal{L}[f_2(t)] = \mathbf{F}_2(s)$, *then*

$$\mathbf{F}(s) = \mathbf{F}_1(s)\mathbf{F}_2(s) \tag{16.38}$$

Our demonstration begins with the definition

$$\mathcal{L}[f(t)] = \int_0^\infty \left[\int_0^t f_1(t-\lambda)f_2(\lambda)\, d\lambda\right] e^{-st}\, dt$$

We now force the function into the proper format by introducing into the integral within the brackets the unit step function $u(t-\lambda)$. We can do this because

$$u(t-\lambda) = \begin{cases} 1 & \text{for } \lambda < t \\ 0 & \text{for } \lambda > t \end{cases} \tag{16.39}$$

The first condition in Eq. (16.39) ensures that the insertion of the unit step function has no impact within the limits of integration. The second condition in Eq. (16.39) allows us to change the upper limit of integration from t to ∞. Therefore,

$$\mathcal{L}[f(t)] = \int_0^\infty \left[\int_0^\infty f_1(t-\lambda)u(t-\lambda)f_2(\lambda)\, d\lambda\right] e^{-st}\, dt$$

which can be written as

$$\mathcal{L}[f(t)] = \int_0^\infty f_2(\lambda)\left[\int_0^\infty f_1(t-\lambda)u(t-\lambda)e^{-st}\, dt\right] d\lambda$$

Note that the integral within the brackets is the shifting theorem illustrated in Example 16.9. Hence the equation can be written as

$$\begin{aligned} \mathcal{L}[f(t)] &= \int_0^\infty f_2(\lambda)\mathbf{F}_1(s)e^{-s\lambda}\, d\lambda \\ &= \mathbf{F}_1(s)\int_0^\infty f_2(\lambda)e^{-s\lambda}\, d\lambda \\ &= \mathbf{F}_1(s)\mathbf{F}_2(s) \end{aligned}$$

Note that convolution in the time domain corresponds to multiplication in the frequency domain.

Let us now illustrate the use of Theorem 11 in the evaluation of an inverse Laplace transform.

EXAMPLE 16.20

Let us determine the inverse Laplace transform of the function $\mathbf{F}(s)$, where

$$\mathbf{F}(s) = \left(\frac{s}{s+1}\right)^2$$

Using convolution, we let

$$\mathbf{F}_1(s) = \mathbf{F}_2(s) = \frac{s}{s+1} = 1 - \frac{1}{s+1}$$

Therefore,

$$f_1(t) = f_2(t) = \delta(t) - e^{-t}$$

Hence

$$\begin{aligned} f(t) &= \int_0^t f_1(\lambda) f_2(t-\lambda)\, d\lambda \\ &= \int_0^t [\delta(\lambda) - e^{-\lambda}][\delta(t-\lambda) - e^{-(t-\lambda)}]\, d\lambda \\ &= \int_0^t \delta(\lambda)\delta(t-\lambda)\, d\lambda - \int_0^t \delta(\lambda) e^{-(t-\lambda)}\, d\lambda \\ &\quad - \int_0^t \delta(t-\lambda) e^{-\lambda}\, d\lambda + \int_0^t e^{-\lambda} e^{-(t-\lambda)}\, d\lambda \end{aligned}$$

Employing the sampling property of the impulse function, we obtain

$$f(t) = \delta(t) - (2e^{-t} - te^{-t})u(t)$$

which is the inverse transform of $\mathbf{F}(s)$.

For comparison, let us determine $f(t)$ from $\mathbf{F}(s)$ using the partial-fraction expansion method described earlier. $\mathbf{F}(s)$ can be written as

$$\mathbf{F}(s) = \left(\frac{s}{s+1}\right)^2 = 1 - \frac{2s+1}{(s+1)^2}$$

The second term can then be expanded as

$$\frac{2s+1}{(s+1)^2} = \frac{K_{11}}{s+1} + \frac{K_{12}}{(s+1)^2}$$

Evaluating the constants, we obtain $K_{11} = 2$ and $K_{12} = -1$. Therefore,

$$\mathbf{F}(s) = 1 - \frac{2}{s+1} + \frac{1}{(s+1)^2}$$

and hence

$$f(t) = \delta(t) - (2e^{-t} - te^{-t})u(t)$$ ■

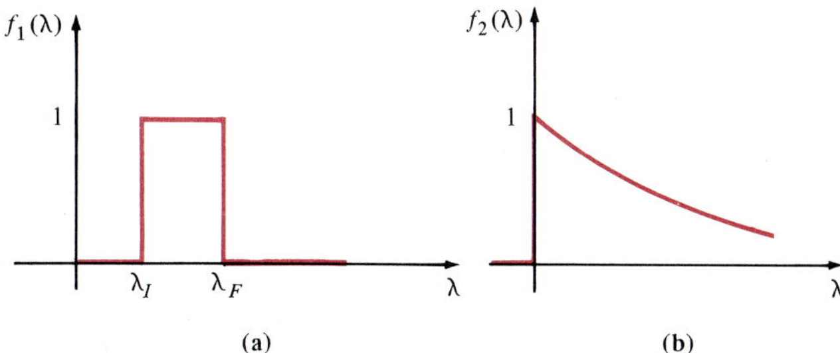

Figure 16.7 Two functions $f_1(\lambda)$ and $f_2(\lambda)$.

Although we can employ convolution to derive an inverse Laplace transform, the example, although quite simple, illustrates that this is a very poor approach. If the function $\mathbf{F}(s)$ is very complicated, the mathematics can become unwieldy. Convolution is, however, a very powerful and useful tool. To understand its usefulness, let us interpret the operation of the convolution integral by examining it from a graphical standpoint.

Suppose that the functions $f_1(\lambda)$ and $f_2(\lambda)$ are as shown in Fig. 16.7. If $f_2(\lambda) = e^{-\alpha\lambda}u(\lambda)$, the time-shifted function

$$f_2(\lambda - t) = e^{-\alpha(\lambda-t)}u(\lambda - t)$$

is as shown in Fig. 16.8a. If we now change the sign of the argument so that the function is

$$f_2(t - \lambda) = e^{-\alpha(t-\lambda)}u(t - \lambda),$$

then the function is folded or reflected about the point $\lambda = t$ as shown in Fig. 16.8b. The integrand of the convolution integral, $f_1(\lambda)f_2(t - \lambda)$, will be nonzero only when the two functions overlap. The overlap, or product of the two functions, for several values of t is shown shaded in Fig. 16.9. The convolution integral for various values of t is equal to the shaded area shown in Fig. 16.9. A plot of these areas as t varies is shown in Fig. 16.10. If we had shifted and folded the function $f_1(\lambda)$, multiplied it by $f_2(\lambda)$ and integrated, we would again obtain the curve in Fig. 16.10.

This graphical explanation of convolution has hopefully provided some additional insight. It also serves, however, to illustrate that *convolution is a very important simulation tool*. For example, if we know the impulse response of a network, we can use

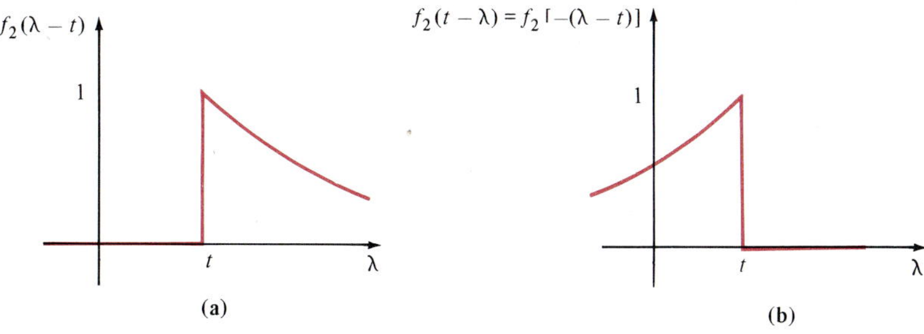

Figure 16.8 Shifting and folding operation on $f_2(\lambda)$.

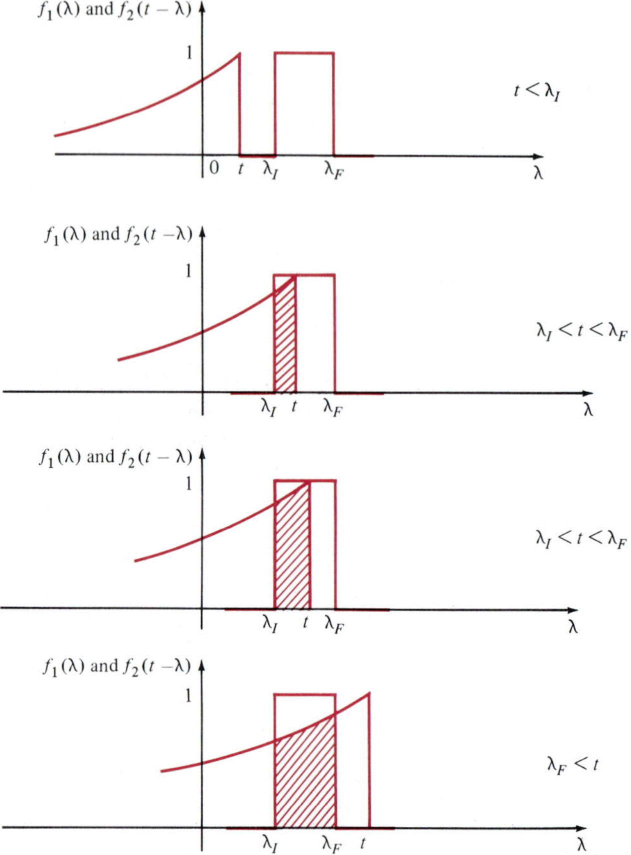

Figure 16.9 Product of the functions $f_1(\lambda)$ and $f_2(t-\lambda)$ shown shaded in the figures for different values of t.

convolution to determine the network's response to an input that may be available only as an experimental curve obtained in the laboratory. Thus convolution permits us to obtain the network response to inputs that cannot be written as analytical functions, but can be simulated on a digital computer.

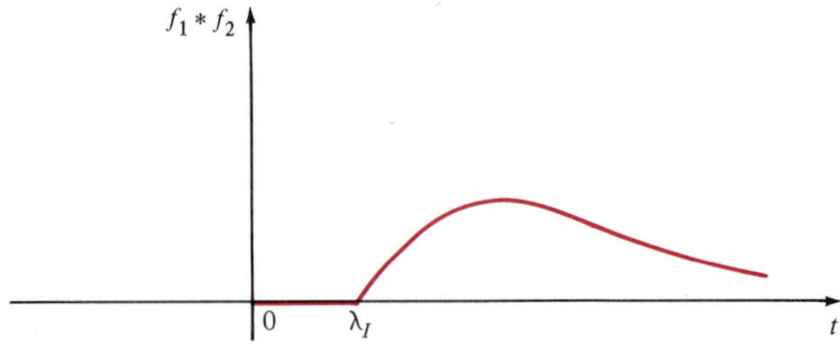

Figure 16.10 Plot of the convolution integral.

16.8 Initial-Value and Final-Value Theorems

Suppose that we wish to determine the initial or final value of a circuit response in the time domain from the Laplace transform of the function in the s-domain without performing the inverse transform. If we determine the function $f(t) = \mathcal{L}^{-1}[\mathbf{F}(s)]$, we can find the initial value by evaluating $f(t)$ as $t \to 0$ and the final value by evaluating $f(t)$ as $t \to \infty$. It would be very convenient, however, if we could simply determine the initial and final values from $\mathbf{F}(s)$ without having to perform the inverse transform. The initial- and final-value theorems allow us to do just that.

The *initial-value theorem* states that

$$\lim_{t \to 0} f(t) = \lim_{s \to \infty} s\mathbf{F}(s) \tag{16.40}$$

provided that $f(t)$ and its first derivative are transformable.

The proof of this theorem employs the Laplace transform of the function $df(t)/dt$.

$$\int_0^\infty \frac{df(t)}{dt} e^{-st}\, dt = s\mathbf{F}(s) - f(0)$$

Taking the limit of both sides as $s \to \infty$, we find that

$$\lim_{s \to \infty} \int_0^\infty \frac{df(t)}{dt} e^{-st}\, dt = \lim_{s \to \infty} [s\mathbf{F}(s) - f(0)]$$

and since

$$\int_0^\infty \frac{df(t)}{dt} \lim_{s \to \infty} e^{-st}\, dt = 0$$

then

$$f(0) = \lim_{s \to \infty} s\mathbf{F}(s)$$

which is, of course,

$$\lim_{t \to 0} f(t) = \lim_{s \to \infty} s\mathbf{F}(s)$$

The *final-value theorem* states that

$$\lim_{t \to \infty} f(t) = \lim_{s \to 0} s\mathbf{F}(s) \tag{16.41}$$

provided that $f(t)$ and its first derivative are transformable and that $f(\infty)$ exists. This latter requirement means that the poles of $\mathbf{F}(s)$ must have negative real parts with the exception that there can be a simple pole at $s = 0$.

The proof of this theorem also involves the Laplace transform of the function $df(t)/dt$.

$$\int_0^\infty \frac{df(t)}{dt} e^{-st}\, dt = s\mathbf{F}(s) - f(0)$$

Taking the limit of both sides as $s \to 0$ gives us

$$\lim_{s \to 0} \int_0^\infty \frac{df(t)}{dt} e^{-st}\, dt = \lim_{s \to 0} [s\mathbf{F}(s) - f(0)]$$

Therefore,

$$\int_0^\infty \frac{df(t)}{dt}\, dt = \lim_{s \to 0} [s\mathbf{F}(s) - f(0)]$$

and

$$f(\infty) - f(0) = \lim_{s \to 0} s\mathbf{F}(s) - f(0)$$

and hence

$$f(\infty) = \lim_{t \to \infty} f(t) = \lim_{s \to 0} s\mathbf{F}(s)$$

EXAMPLE 16.21

Let us determine the initial and final values for the function in Example 16.18.

The function is

$$\mathbf{F}(s) = \frac{10(s + 1)}{s(s^2 + 2s + 2)}$$

and the corresponding time function is

$$f(t) = 5 + 5\sqrt{2}\, e^{-t} \cos(t - 135°)$$

Applying the initial-value theorem, we have

$$\begin{aligned} f(0) &= \lim_{s \to \infty} s\mathbf{F}(s) \\ &= \lim_{s \to \infty} \frac{10(s + 1)}{s^2 + 2s + 2} \\ &= 0 \end{aligned}$$

The poles of $\mathbf{F}(s)$ are $s = 0$ and $s = -1 \pm j1$, so the final-value theorem is applicable. Thus,

$$\begin{aligned} f(\infty) &= \lim_{s \to 0} s\mathbf{F}(s) \\ &= \lim_{s \to 0} \frac{10(s + 1)}{s^2 + 2s + 2} \\ &= 5 \end{aligned}$$

Note that these values could be obtained directly from the time function $f(t)$. ■

DRILL EXERCISE

D16.20. Find the initial and final values of the function $f(t)$ if $\mathbf{F}(s) = \mathcal{L}[f(t)]$ is given by the expression

$$\mathbf{F}(s) = \frac{(s+1)^2}{s(s+2)(s^2+2s+2)}$$

Ans: $f(0) = 0$ and $f(\infty) = \frac{1}{4}$.

16.9 Applications to Differential Equations

Let us now begin to apply the power of the Laplace transform. In this section we demonstrate its use in the solution of linear, constant-coefficient integrodifferential equations. In Chapter 17 we illustrate the use of the Laplace transform to circuit analysis. In our approach we employ the transform pairs we have presented and discussed to convert the integrodifferential equation in the time domain to an algebraic equation in the s-domain. The solution in the s-domain will then be converted back to the time domain using the transform pairs. The solution we obtain will be the complete solution since the initial conditions are automatically included in the transform process.

EXAMPLE 16.22

Given the following differential equation, let us find $y(t)$.

$$\frac{d^2y(t)}{dt^2} + \frac{5dy(t)}{dt} + 6y(t) = e^{-t}$$

where $y(0) = 1$ and $y'(0) = 1$.

Transforming the equation, we obtain

$$s^2\mathbf{Y}(s) - sy(0) - y'(0) + 5[s\mathbf{Y}(s) - y(0)] + 6\mathbf{Y}(s) = \frac{1}{s+1}$$

or

$$\mathbf{Y}(s)[s^2 + 5s + 6] = \frac{1}{s+1} + s + 1 + 5$$

Therefore,

$$\mathbf{Y}(s) = \frac{1}{(s+1)(s^2+5s+6)} + \frac{s+6}{s^2+5s+6}$$

The first term on the right-hand side of the equation is the response for zero initial conditions. The second term on the right-hand side of the equation modifies the complementary term to give the required initial conditions. Note that the particular integral term is not affected by this second term. Using the techniques described earlier to determine the inverse transform, we obtain

$$y(t) = (\tfrac{1}{2}e^{-t} + 3e^{-2t} - \tfrac{5}{2}e^{-3t})u(t)$$

■

Our approach can be applied equally well to a set of simultaneous linear constant-coefficient differential equations. The following example illustrates this use.

EXAMPLE 16.23

Let us use the Laplace transform technique to solve the simultaneous equations

$$\frac{dx(t)}{dt} + \frac{dy(t)}{dt} + 2y(t) = 4u(t)$$

$$2\frac{dx(t)}{dt} + x(t) + \frac{dy(t)}{dt} = 2u(t)$$

where $y(0) = 1$ and $x(0) = -1$.

Transforming the equations, we obtain

$$s\mathbf{X}(s) - x(0) + s\mathbf{Y}(s) - y(0) + 2\mathbf{Y}(s) = \frac{4}{s}$$

$$2s\mathbf{X}(s) - 2x(0) + \mathbf{X}(s) + s\mathbf{Y}(s) - y(0) = \frac{2}{s}$$

These equations reduce to

$$s\mathbf{X}(s) + (s + 2)\mathbf{Y}(s) = \frac{4}{s}$$

$$(2s + 1)\mathbf{X}(s) + s\mathbf{Y}(s) = \frac{2}{s} - 1$$

Solving the algebraic equations, we find that

$$\mathbf{X}(s) = \frac{-(s^2 + 4s - 4)}{s(s^2 + 5s + 2)}$$

$$\mathbf{Y}(s) = \frac{s^2 + 6s + 4}{s(s^2 + 5s + 2)}$$

from which we obtain

$$x(t) = (2 - 3.07e^{-0.44t} - +0.08e^{-4.56t})u(t)$$

$$y(t) = (2 - 0.86e^{-0.44t} - 0.14e^{-4.56t})u(t)$$

■

EXAMPLE 16.24

Finally, let us employ the Laplace transform to solve the equation

$$\frac{dy(t)}{dt} + 2y(t) + \int_0^t y(\lambda)e^{-2(t-\lambda)}\,d\lambda = 10u(t) \qquad y(0) = 0$$

Applying the transform, we obtain

$$s\mathbf{Y}(s) + 2\mathbf{Y}(s) + \frac{\mathbf{Y}(s)}{s+2} = \frac{10}{s}$$

$$\mathbf{Y}(s)\left(s + 2 + \frac{1}{s+2}\right) = \frac{10}{s}$$

$$\mathbf{Y}(s) = \frac{10(s+2)}{s(s^2+4s+5)}$$

Expressing the function in a partial-fraction expansion, we obtain

$$\frac{10(s+2)}{s(s+2-j1)(s+2+j1)} = \frac{K_0}{s} + \frac{\mathrm{K}_1}{s+2-j1} + \frac{K_1^*}{s+2+j1}$$

$$\left.\frac{10(s+2)}{s^2+4s+5}\right|_{s=0} = K_0$$

$$4 = K_0$$

In a similar manner,

$$\left.\frac{10(s+2)}{s(s+2+j1)}\right|_{s=-2+j1} = K_1$$

$$2.236 \underline{/-153.43^\circ} = K_1$$

Therefore,

$$2.236 \underline{/153.43^\circ} = K_1^*$$

The partial-fraction expansion of $\mathbf{Y}(s)$ is then

$$\mathbf{Y}(s) = \frac{4}{s} + \frac{2.236 \underline{/-153.43^\circ}}{s+2-j1} + \frac{2.236 \underline{/153.43^\circ}}{s+2+j1}$$

and therefore,

$$y(t) = [4 + 4.472e^{-2t}\cos(t - 153.43^\circ)]u(t)$$ ■

DRILL EXERCISE

D16.21. Solve the differential equation

$$\frac{d^2y(t)}{dt^2} + 11\frac{dy(t)}{dt} + 30y(t) = 4u(t) \qquad y(0) = y'(0) = 0$$

Ans: $y(t) = (\frac{4}{30} - \frac{4}{5}e^{-5t} + \frac{2}{3}e^{-6t})u(t)$.

D16.22. Determine $y(t)$ if

$$\frac{dy(t)}{dt} + \int_0^t y(\lambda)e^{-2(t-\lambda)}\,d\lambda = u(t) \qquad y(0) = 0$$

Ans: $y(t) = (2 - 2e^{-t} - te^{-t})u(t)$.

16.10 Summary

In this chapter we have introduced a very powerful analysis technique called the Laplace transform. We have shown that by using this technique we can transform a set of linear, constant coefficient integrodifferential equations in the time domain into a set of algebraic equations in the s-domain. The equations in the s-domain can be solved algebraically and then the solution is transformed back to the time domain. The transformation from the s-domain to the time domain was accomplished by using a set of known transform pairs. The convolution integral was introduced and its use in the inversion of Laplace transforms was demonstrated. Finally, the transform technique was applied to the solution of integrodifferential equations.

KEY POINTS

- The Laplace transform allows us to convert a problem from the time domain to the frequency domain, solve the problem using algebra in the frequency domain, and then convert the solution in the frequency domain back to the time domain.
- The Laplace transform is defined as

$$\mathcal{L}[f(t)] = \mathbf{F}(s) = \int_0^{\infty} f(t)e^{-st}\,dt$$

- A set of transform pairs exist that permit us to transform a function from the time domain to the frequency domain, and vice versa.
- Ten important properties of the Laplace transform have been presented that facilitate its use.
- The gate function aids us in determining the Laplace transform of a waveform in the time domain.
- The partial-fraction expansion method can be applied to determine the inverse Laplace transform of functions containing simple, multiple, and complex-conjugate poles.
- The convolution integral can be used to determine the inverse Laplace transform.
- The initial- and final-value theorems can be applied to determine initial and final values of time-domain functions when their Laplace transforms are known.
- The Laplace transform can be used to solve integrodifferential equations. The complete solution is obtained since the initial conditions are automatically included in the transform process.

PROBLEMS

16.1. Demonstrate the following equalities:

(a) $\mathcal{L}[e^{-at}] = \dfrac{1}{s+a}$.

(b) $\mathcal{L}[\sin \omega t] = \dfrac{\omega}{s^2+\omega^2}$.

(c) $\mathcal{L}[t] = \dfrac{1}{s^2}$.

16.2. Find the $\mathcal{L}[f(t)]$ if $f(t)$ is given by the expression

$$f(t) = \frac{e^{-(a+t)}}{4x}$$

16.3. Find the $\mathcal{L}[f(t)]$ if $f(t)$ is

$$f(t) = a + a^2 t + ae^{-at} + a^2 \cos at$$

16.4. Find $\mathbf{F}(s)$ if $f(t) = e^{-at} \sin \omega t u(t-1)$.

16.5. Find $\mathbf{F}(s)$ if $f(t) = te^{-at}u(t-4)$.

16.6. Use Theorem 5 to find $\mathscr{L}[f(t)]$ if

$$f(t) = e^{-at}u(t-1)$$

16.7. Use Theorem 6 to find $\mathscr{L}[f(t)]$ if

$$f(t) = te^{-at}u(t-1)$$

16.8. Use the shifting theorem to determine $\mathscr{L}[f(t)]$, where

$$f(t) = [t - 1 + e^{-(t-1)}]u(t-1)$$

16.9. Use the shifting theorem to determine $\mathscr{L}[f(t)]$, where

$$f(t) = [e^{-(t-2)} - e^{-2(t-2)}]u(t-2)$$

16.10. If $f(t) = t \cos \omega t u(t-1)$, find $\mathbf{F}(s)$.

16.11. Find $\mathbf{F}(s)$ if $f(t)$ is given by the waveform in Fig. P16.11.

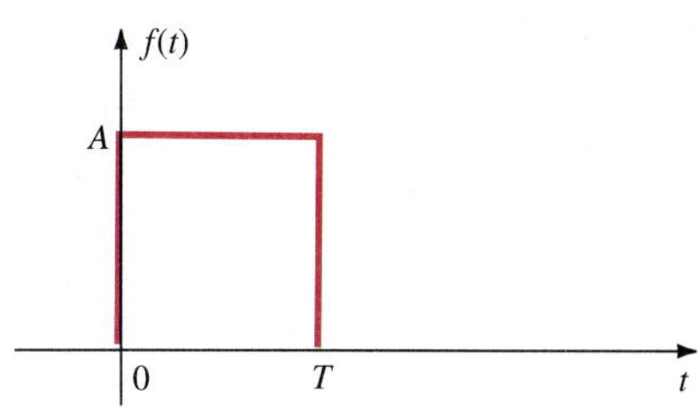

Figure P16.11

16.12. Find $\mathbf{F}(s)$ if $f(t)$ is expressed by the waveform in Fig. P16.12.

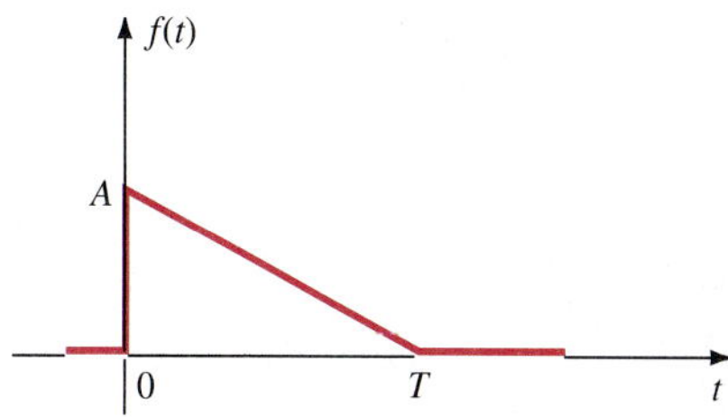

Figure P16.12

16.13. Find $\mathbf{F}(s)$ if $f(t)$ is given by the waveform in Fig. P16.13.

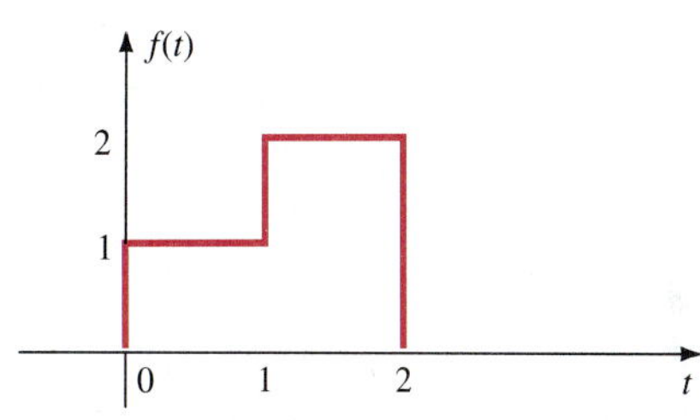

Figure P16.13

16.14. Find the Laplace transform of the waveform shown in Fig. P16.14.

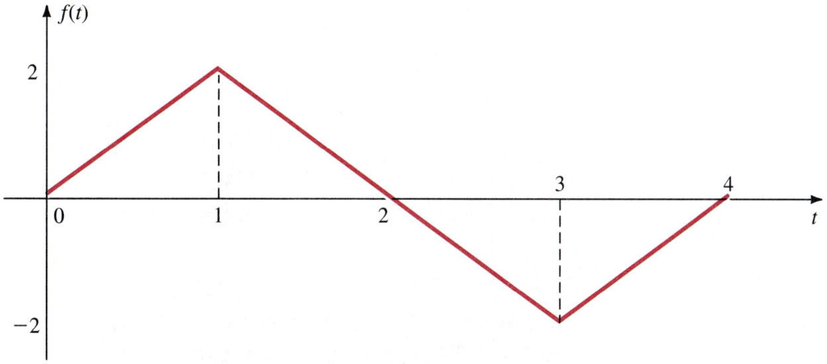

Figure P16.14

16.15. Find the Laplace transform of the periodic waveform shown in Fig. P16.15.

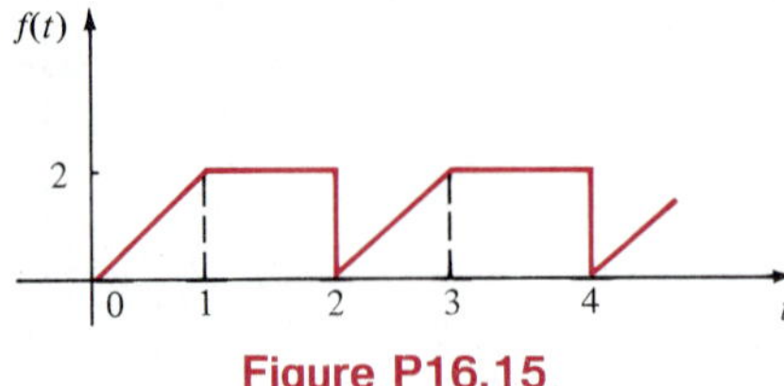

Figure P16.15

16.16. Find the Laplace transform of the waveform shown in Fig. P16.16.

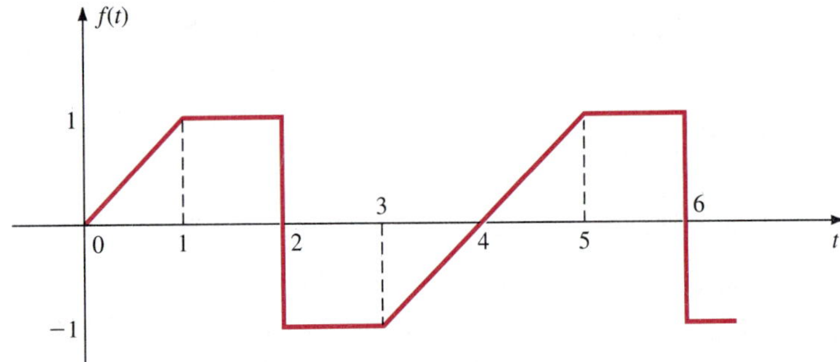

Figure P16.16

16.17. Find the Laplace transform of the waveform shown in Fig. P16.17.

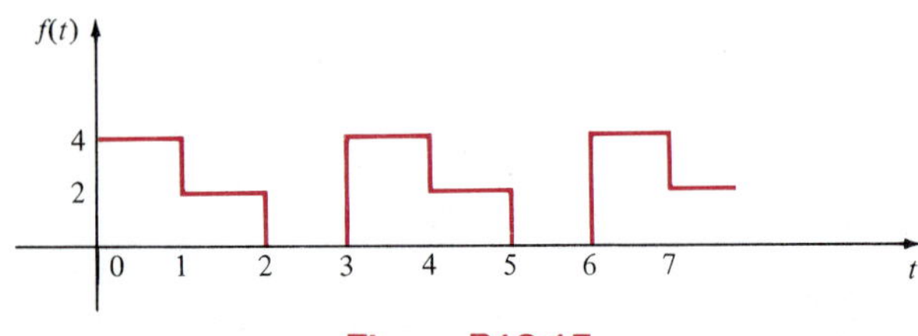

Figure P16.17

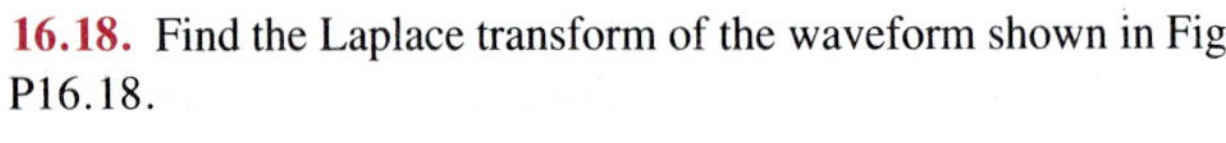
16.18. Find the Laplace transform of the waveform shown in Fig. P16.18.

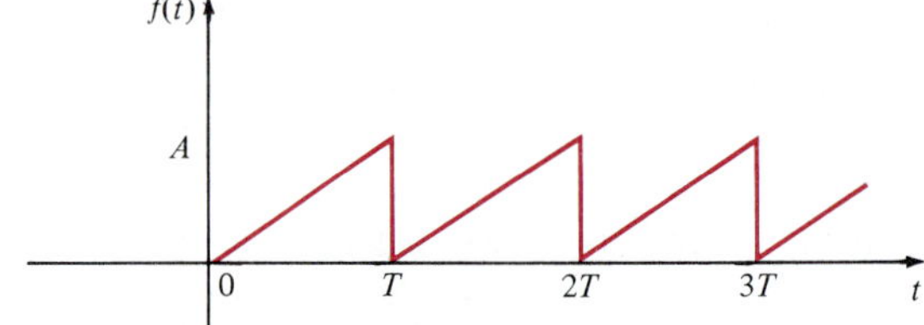

Figure P16.18

16.19. Find the Laplace transform of the waveform shown in Fig. P16.19.

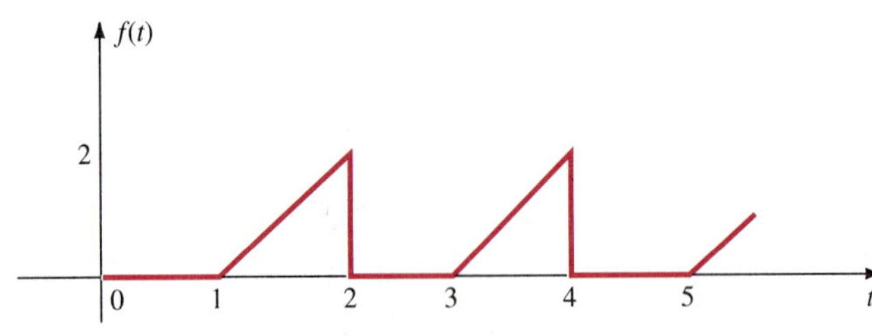

Figure P16.19

16.20. Find the Laplace transform of the waveform shown in Fig. P16.20.

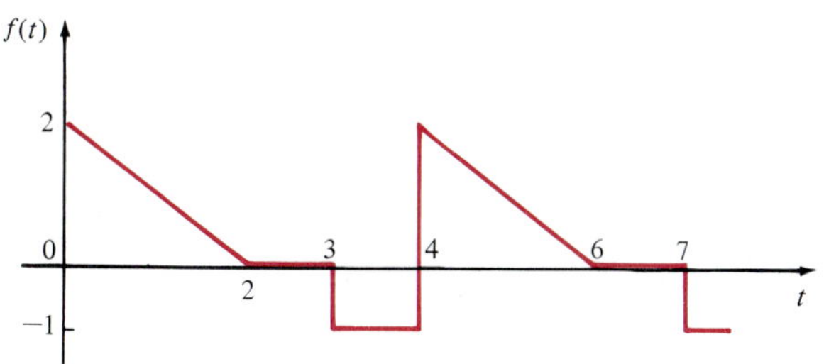

Figure P16.20

16.21. Find the Laplace transform of the waveform shown in Fig. P16.21.

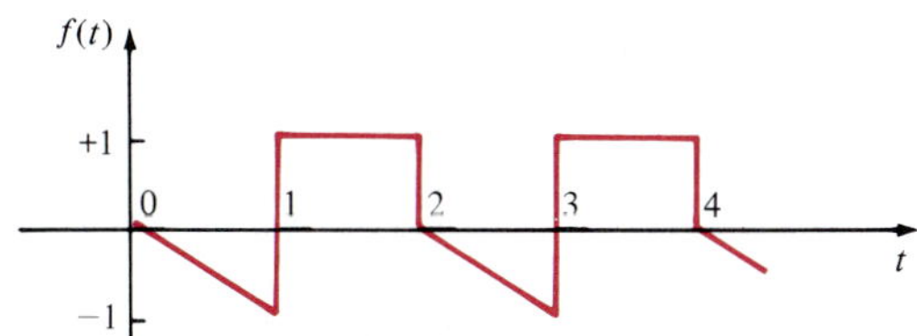

Figure P16.21

16.22. Find the Laplace transform of the waveform shown in Fig. P16.22.

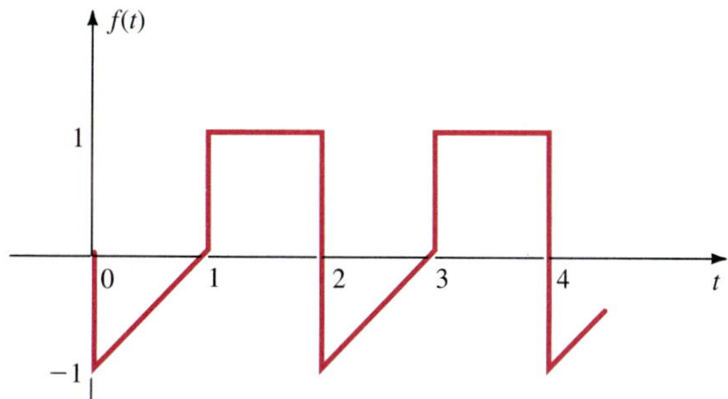

Figure P16.22

16.23. Find the Laplace transform of the waveform shown in Fig. P16.23.

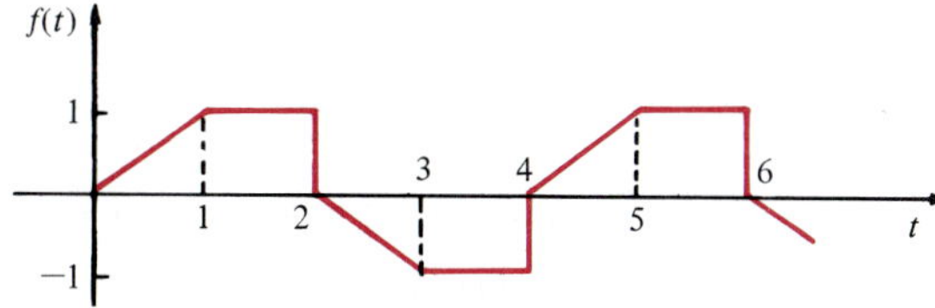

Figure P16.23

16.24. Find the Laplace transform of the waveform shown in Fig. P16.24.

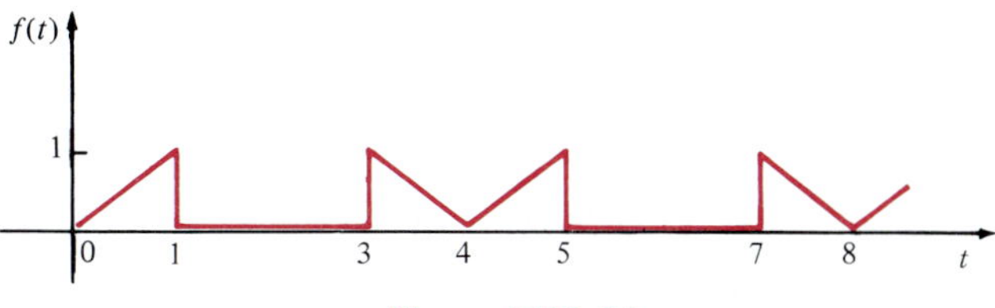

Figure P16.24

16.25. Find the Laplace transform of the waveform shown in Fig. P16.25.

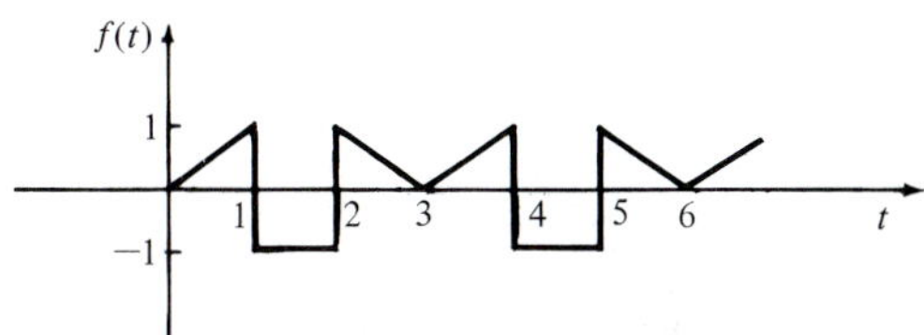

Figure P16.25

16.26. Determine the Laplace transform of the periodic waveform shown in Fig. P16.26.

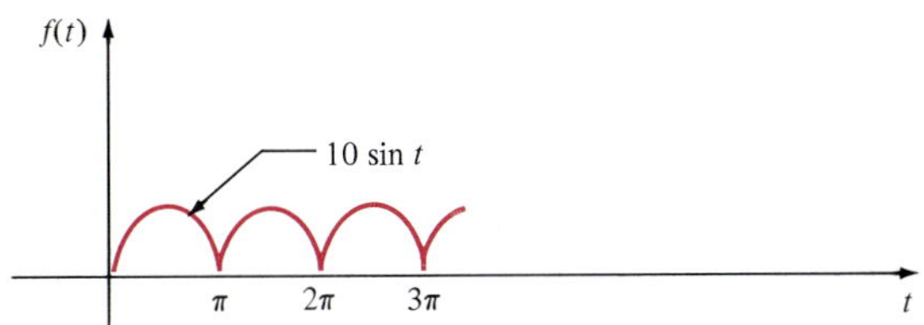

Figure P16.26

16.27. Determine the Laplace transform of the periodic waveform shown in Fig. P16.27.

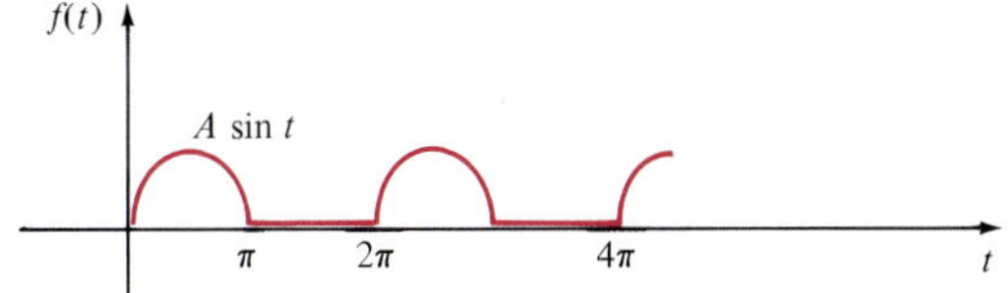

Figure P16.27

16.28. Given the following functions $\mathbf{F}(s)$, find $f(t)$.

(a) $\mathbf{F}(s) = \dfrac{s + 10}{(s + 4)(s + 6)}$.

(b) $\mathbf{F}(s) = \dfrac{24}{(s + 2)(s + 8)}$.

16.29. Given the following functions $\mathbf{F}(s)$, find $f(t)$.

(a) $\mathbf{F}(s) = \dfrac{s}{(s+8)(s+12)}$.

(b) $\mathbf{F}(s) = \dfrac{s^2+3s+2}{s^2+2s+1}$.

16.30. Given the following functions $\mathbf{F}(s)$, find $f(t)$.

(a) $\mathbf{F}(s) = \dfrac{s^2+7s+12}{(s+2)(s+4)(s+6)}$.

(b) $\mathbf{F}(s) = \dfrac{(s+3)(s+6)}{s(s^2+10s+24)}$.

16.31. Given the following functions $\mathbf{F}(s)$, find $f(t)$.

(a) $\mathbf{F}(s) = \dfrac{s^3+2s^2+s}{s^2(s^2+5s+4)}$.

(b) $\mathbf{F}(s) = \dfrac{s^2+4s+3}{(s^2+2s+1)(s^2+7s+12)}$.

16.32. Given the following functions $\mathbf{F}(s)$, find $f(t)$.

(a) $\mathbf{F}(s) = \dfrac{s^2+4s+5}{(s+1)(s+4)}$.

(b) $\mathbf{F}(s) = \dfrac{s+10}{s^2}$.

16.33. Given the following functions $\mathbf{F}(s)$, find $f(t)$.

(a) $\mathbf{F}(s) = \dfrac{s+8}{s^2(s+6)}$.

(b) $\mathbf{F}(s) = \dfrac{1}{s^2(s+1)^2}$.

16.34. Given the following functions $\mathbf{F}(s)$, find $f(t)$.

(a) $\mathbf{F}(s) = \dfrac{s+4}{(s+2)^2}$.

(b) $\mathbf{F}(s) = \dfrac{s+6}{s(s+1)^2}$.

16.35. Given the following functions $\mathbf{F}(s)$, find $f(t)$.

(a) $\mathbf{F}(s) = \dfrac{s^2}{(s+1)^2(s+2)}$.

(b) $\mathbf{F}(s) = \dfrac{s^2+9s+20}{s(s+4)^3(s+5)}$.

16.36. Given the following functions $\mathbf{F}(s)$, find $f(t)$.

(a) $\mathbf{F}(s) = \dfrac{s^2+6s+8}{s^2(s+2)(s^2+10s+16)}$.

(b) $\mathbf{F}(s) = \dfrac{s^2+9s+18}{s(s+4)(s^2+10s+24)}$.

16.37. Given the following functions $\mathbf{F}(s)$, find $f(t)$.

(a) $\mathbf{F}(s) = \dfrac{s(s+6)}{(s+3)(s^2+6s+18)}$.

(b) $\mathbf{F}(s) = \dfrac{(s+4)(s+8)}{s(s^2+8s+32)}$.

16.38. Given the following functions $\mathbf{F}(s)$, find $f(t)$.

(a) $\mathbf{F}(s) = \dfrac{s+2}{(s^2+4s+5)(s^2+4s+8)}$.

(b) $\mathbf{F}(s) = \dfrac{s(s+2)}{s^2+2s+2}$.

16.39. Given the following functions $\mathbf{F}(s)$, find $f(t)$.

(a) $\mathbf{F}(s) = \dfrac{(s+1)(s+3)}{(s+2)(s^2+4s+8)}$.

(b) $\mathbf{F}(s) = \dfrac{(s+2)^2}{s^2+4s+8}$.

16.40. Find $f(t)$ if $\mathbf{F}(s)$ is given by the expression

$$\mathbf{F}(s) = \frac{s+1}{s(s+2)(s^2+2s+2)}$$

16.41. Find $f(t)$ if $\mathbf{F}(s)$ is given by the expression

$$\mathbf{F}(s) = \frac{s(s+1)}{(s+2)^3(s+3)}$$

16.42. Find $f(t)$ if $\mathbf{F}(s)$ is given by the expression

$$\mathbf{F}(s) = \frac{12(s+2)}{s^2(s+1)(s^2+4s+8)}$$

16.43. Find $f(t)$ if $\mathbf{F}(s)$ is given by the following functions:

(a) $\mathbf{F}(s) = \dfrac{(s^2+2s+3)e^{-2s}}{s(s+1)(s+2)}$.

(b) $\mathbf{F}(s) = \dfrac{(s+2)e^{-4s}}{s^2(s+1)}$.

16.44. Find $f(t)$ if $\mathbf{F}(s)$ is given by the following functions:

(a) $\mathbf{F}(s) = \dfrac{2(s+1)e^{-s}}{(s+2)(s+3)}$.

(b) $\mathbf{F}(s) = \dfrac{10(s+2)e^{-2s}}{(s+1)(s+4)}$.

16.45. Find $f(t)$ if $\mathbf{F}(s)$ is given by the following function:

$$\mathbf{F}(s) = \frac{(s+1)e^{-s}}{s(s+2)(s^2+2s+2)}$$

16.46. Find the inverse Laplace transform of the function

$$\mathbf{F}(s) = \frac{10s(s+2)e^{-4s}}{(s+1)^2(s^2+2s+2)}$$

16.47. Find $f(t)$ if $\mathbf{F}(s)$ is given by the expression

$$\mathbf{F}(s) = \frac{1-e^{-s}}{s^2(s+1)}$$

16.48. Find $f(t)$ if $\mathbf{F}(s)$ is given by the expression

$$\mathbf{F}(s) = \frac{s^2e^{-2s}}{(s^2+1)(s+1)(s^2+2s+2)}$$

16.49. Find $f(t)$ using convolution if $\mathbf{F}(s)$ is

$$\mathbf{F}(s) = \frac{1}{(s+1)(s+2)}$$

16.50. Use convolution to find $f(t)$ if

$$\mathbf{F}(s) = \frac{1}{(s+1)(s+2)^2}$$

16.51. Determine the initial and final values of $f(t)$ if $\mathbf{F}(s)$ is given by the expressions

(a) $\mathbf{F}(s) = \dfrac{2(s+2)}{s(s+1)}$

(b) $\mathbf{F}(s) = \dfrac{2(s^2+2s+6)}{(s+1)(s+2)(s+3)}$

(c) $\mathbf{F}(s) = \dfrac{2s^2}{(s+1)(s^2+2s+2)}$

16.52. Find the initial and final values of the time function $f(t)$ if $\mathbf{F}(s)$ is given as

(a) $\mathbf{F}(s) = \dfrac{10(s+2)}{(s+1)(s+3)}$.

(b) $\mathbf{F}(s) = \dfrac{s^2+2s+4}{(s+6)(s^3+4s^2+8s+10)}$.

(c) $\mathbf{F}(s) = \dfrac{2s}{s^2+2s+2}$.

16.53. Find the final values of the time function $f(t)$ given that

(a) $\mathbf{F}(s) = \dfrac{10(s+1)}{(s+2)(s-3)}$.

(b) $\mathbf{F}(s) = \dfrac{10}{s^2+4}$.

16.54. Solve the following differential equations using Laplace transforms.

(a) $\dfrac{dx(t)}{dt} + 3x(t) = e^{-t}, \qquad x(0) = 1.$

(b) $\dfrac{dx(t)}{dt} + 4x(t) = 2u(t), \qquad x(0) = 2.$

16.55. Solve the following differential equations using Laplace transforms.

(a) $\dfrac{d^2y(t)}{dt^2} + \dfrac{2dy(t)}{dt} + y(t) = e^{-2t},$
$y(0) = y'(0) = 0.$

(b) $\dfrac{d^2y(t)}{dt^2} + \dfrac{4dy(t)}{dt} + 4y(t) = u(t),$
$y(0) = 0, \ y'(0) = 1.$

16.56. Use Laplace transforms to find $y(t)$ if

$$\frac{dy(t)}{dt} + 5y(t) + 4\int_0^t y(x)\,dx = u(t), \qquad y(0) = 0$$

16.57. Solve the following integrodifferential equation using Laplace transforms.

$$\frac{dy(t)}{dt} + 2y(t) + \int_0^t y(\lambda)\,d\lambda = 1 - e^{-2t}, \qquad y(0) = 0, \quad t > 0$$

16.58. Use Laplace transforms to solve the following integrodifferential equation.

$$\frac{dy(t)}{dt} + 2y(t) + \int_0^t y(\lambda)e^{-2(t-\lambda)}\,d\lambda = 4u(t),$$

$$y(0) = 1, \quad t > 0$$

16.59. Find $y(t)$ if

$$y(t) + \int_0^t y(\lambda)(t-\lambda)\,d\lambda = e^{-t}$$

CHAPTER

17

Application of the Laplace Transform to Circuit Analysis

We have demonstrated in Chapter 16 how the Laplace transform can be used to solve linear constant-coefficient differential equations. Since all of the linear networks with which we are dealing can be described by linear constant-coefficient differential equations, use of the Laplace transform for circuit analysis would appear to be a feasible approach. The terminal characteristics for each circuit element can be described in the s-domain by transforming the appropriate time-domain equations. Kirchhoff's laws when applied to a circuit produce a set of integrodifferential equations in terms of the terminal characteristics of the network elements which when transformed yield a set of algebraic equations in the s-domain. Therefore, a complex frequency-domain analysis in which the passive network elements are represented by their transform impedance or admittance; and sources, whether independent or dependent, are represented in terms of their transform variables, can be performed. Such an analysis in the s-domain is algebraic and all of the techniques derived in dc analysis will apply. Therefore, the analysis is similar to the dc analysis of resistive networks and all of the network analysis techniques and network theorems that were applied in dc analysis are valid in the s-domain (e.g., node analysis, loop analysis, superposition, source transformation, Thevenin's theorem, Norton's theorem, and combinations of impedance or admittance). Therefore, our approach will be to transform each of the circuit elements, draw an s-domain equivalent circuit, and then using the transformed network, solve the circuit equations algebraically in the s-domain. Finally, a table of transform pairs can be employed to obtain a complete (transient plus steady-state) solution. Our approach in this chapter is to employ the circuit element models and demonstrate, using a variety of examples, many of the concepts and techniques that have been presented earlier.

17.1 Circuit Element Models

Let us examine the terminal characteristics of the network elements in the s-domain. The voltage–current relationship for a resistor in the time domain using the passive sign convention is

$$v(t) = Ri(t) \tag{17.1}$$

Using the Laplace transform, this relationship in the s-domain is

$$\mathbf{V}(s) = R\mathbf{I}(s) \tag{17.2}$$

Therefore, the time-domain and complex-frequency-domain representations of this element are as shown in Fig. 17.1a.

The time-domain relationships for a capacitor using the passive sign convention are

$$v(t) = \frac{1}{C}\int_0^t i(x)\,dx + v(0) \tag{17.3}$$

$$i(t) = C\frac{dv(t)}{dt} \tag{17.4}$$

The s-domain equations for the capacitor are then

$$\mathbf{V}(s) = \frac{\mathbf{I}(s)}{sC} + \frac{v(0)}{s} \tag{17.5}$$

$$\mathbf{I}(s) = sC\mathbf{V}(s) - Cv(0) \tag{17.6}$$

and hence the s-domain representation of this element is as shown in Fig. 17.1b.

For the inductor, the voltage–current relationships using the passive sign convention are

$$v(t) = L\frac{di(t)}{dt} \tag{17.7}$$

$$i(t) = \frac{1}{L}\int_0^t v(x)\,dx + i(0) \tag{17.8}$$

The relationships in the s-domain are then

$$\mathbf{V}(s) = sL\mathbf{I}(s) - Li(0) \tag{17.9}$$

$$\mathbf{I}(s) = \frac{\mathbf{V}(s)}{sL} + \frac{i(0)}{s} \tag{17.10}$$

The s-domain representation of this element is shown in Fig. 17.1c.

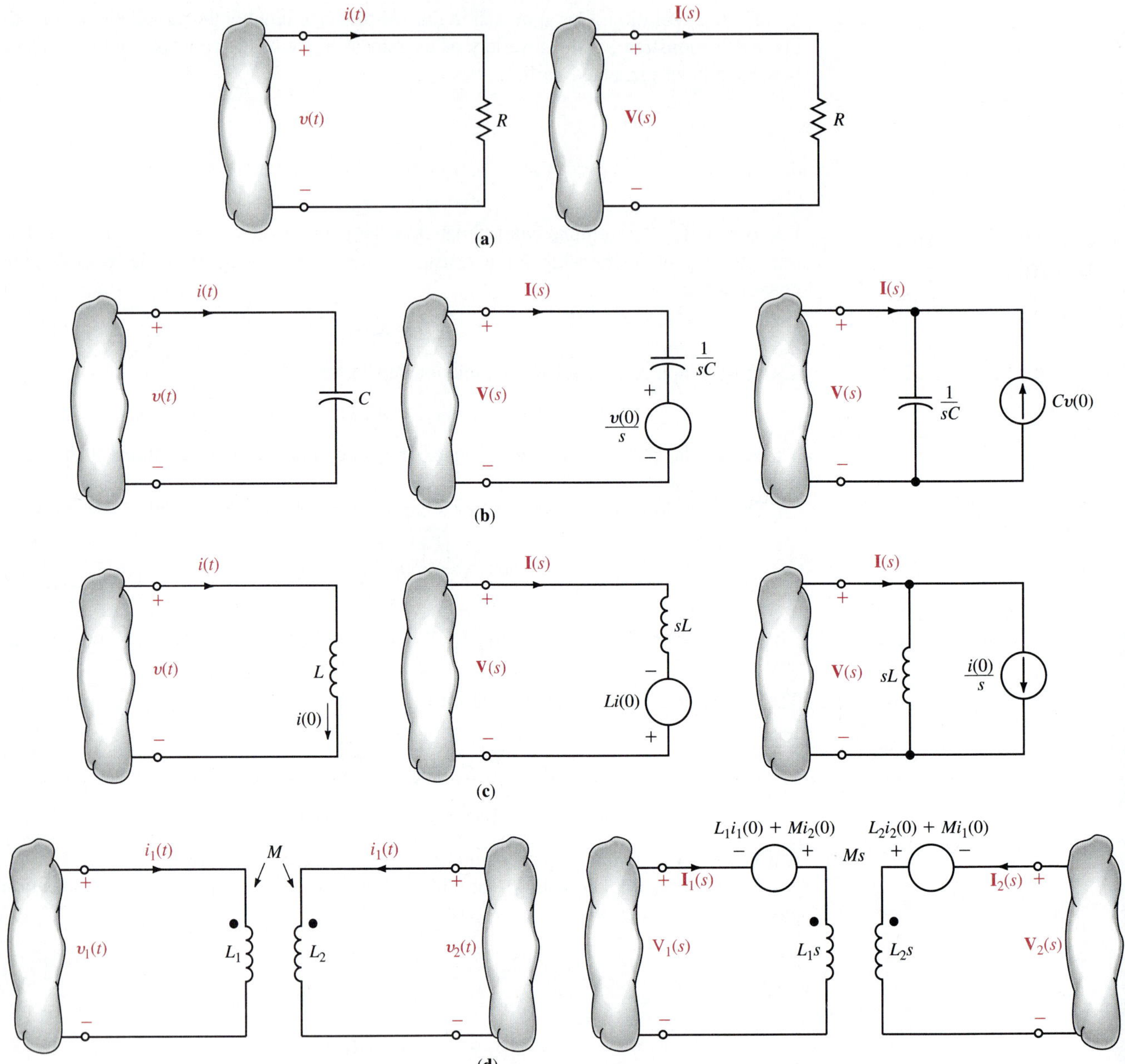

Figure 17.1 Time-domain and *s*-domain representations of circuit elements.

Using the passive sign convention, we find that the voltage–current relationships for the coupled inductors shown in Fig. 17.1d are

$$v_1(t) = L_1 \frac{di_1(t)}{dt} + M \frac{di_2(t)}{dt}$$
$$v_2(t) = L_2 \frac{di_2(t)}{dt} + M \frac{di_1(t)}{dt} \qquad (17.11)$$

The relationships in the s-domain are then

$$\mathbf{V}_1(s) = L_1 s\mathbf{I}_1(s) - L_1 i_1(0) + Ms\mathbf{I}_2(s) - Mi_2(0)$$
$$\mathbf{V}_2(s) = L_2 s\mathbf{I}_2(s) - L_2 i_2(0) + Ms\mathbf{I}_1(s) - Mi_1(0) \qquad (17.12)$$

Independent and dependent voltage and current sources can also be represented by their transforms; that is,

$$\mathbf{V}_1(s) = \mathcal{L}[v_1(t)]$$
$$\mathbf{I}_2(s) = \mathcal{L}[i_2(t)] \qquad (17.13)$$

and if $v_1(t) = Ai_2(t)$, which represents a current-controlled voltage source, then

$$\mathbf{V}_1(s) = A\mathbf{I}_2(s) \qquad (17.14)$$

The reader should note carefully the direction of the current sources and the polarity of the voltage sources in the transformed network which result from the initial conditions. If the polarity of the initial voltage or direction of the initial current is reversed, the sources in the transformed circuit that results from the initial condition are also reversed.

17.2 Analysis Techniques

Now that we have the s-domain representation for the circuit elements, we are in a position to analyze networks using a transformed circuit.

EXAMPLE 17.1

Given the circuits in Fig. 17.2a and b, we wish to write the mesh equations in the s-domain for the network in Fig. 17.2a and the node equations in the s-domain for the network in Fig. 17.2b.

The transformed circuit using the impedance values for the network parameters in Fig. 17.2a is shown in Fig. 17.2c. The mesh equations for this network are

$$\left(R_1 + \frac{1}{sC_1} + \frac{1}{sC_2} + sL_1\right)\mathbf{I}_1(s) - \left(\frac{1}{sC_2} + sL_1\right)\mathbf{I}_2(s)$$
$$= \mathbf{V}_A(s) - \frac{v_1(0)}{s} + \frac{v_2(0)}{s} - L_1 i_1(0)$$

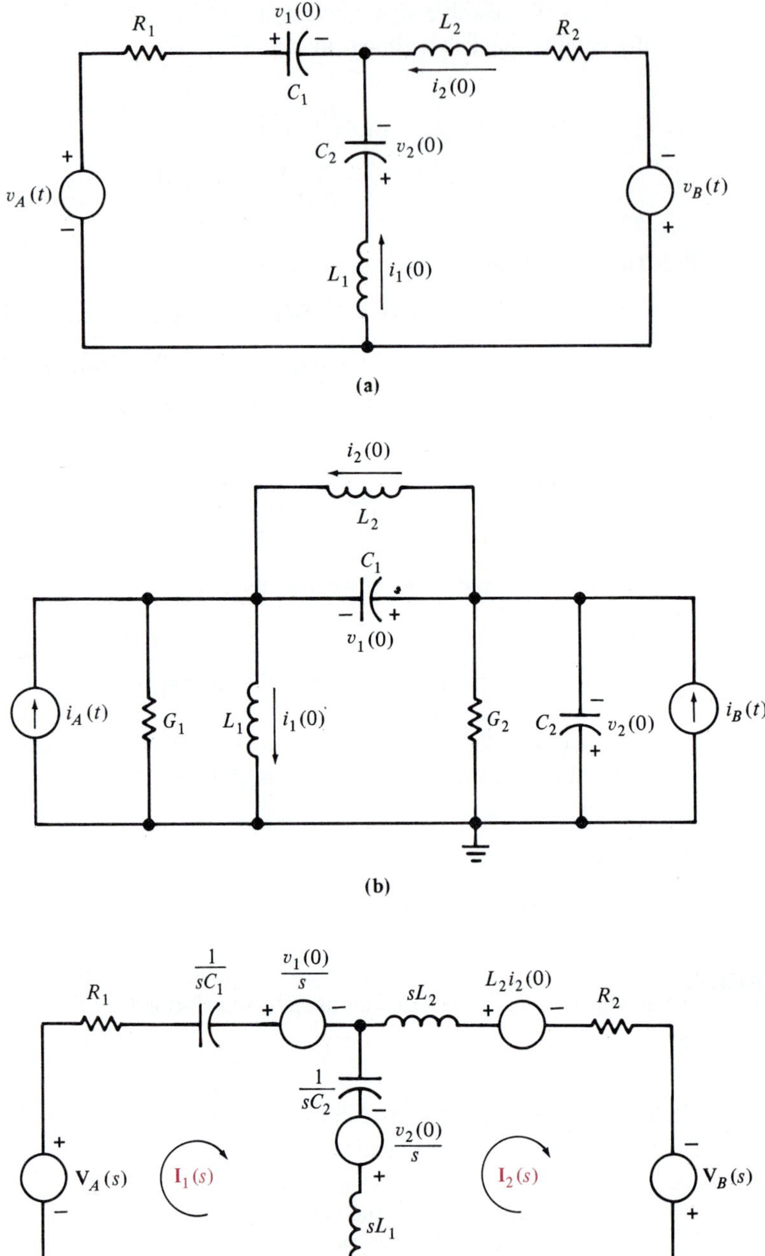

Figure 17.2 Circuits used in Example 17.1.

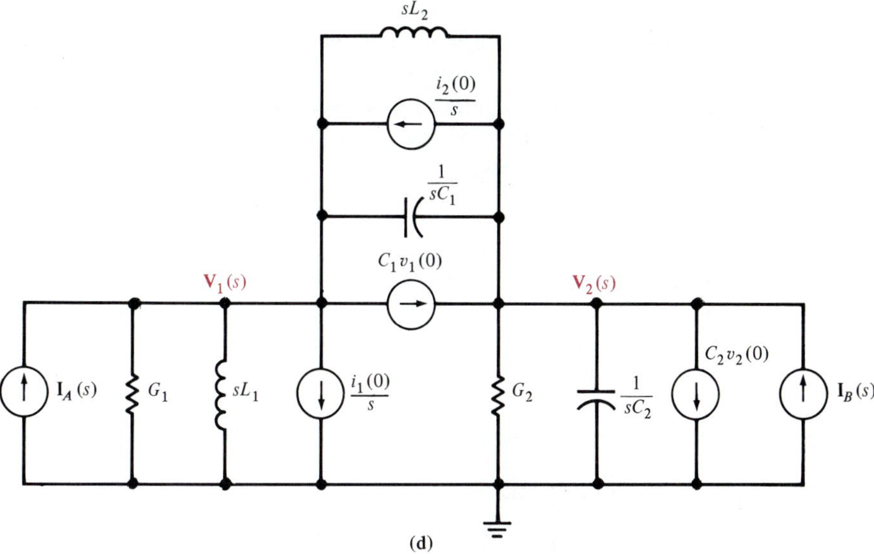

Figure 17.2 (continued)

$$-\left(\frac{1}{sC_2} + sL_1\right)\mathbf{I}_1(s) + \left(\frac{1}{sC_2} + sL_1 + sL_2 + R_2\right)\mathbf{I}_2(s)$$

$$= L_1 i_1(0) - \frac{v_2(0)}{s} - L_2 i_2(0) + \mathbf{V}_B(s)$$

The transformed circuit using the impedance values for the network parameters in Fig. 17.2b is shown in Fig. 17.2d. The node equations for this network are

$$\left(G_1 + \frac{1}{sL_1} + sC_1 + \frac{1}{sL_2}\right)\mathbf{V}_1(s) - \left(\frac{1}{sL_2} + sC_1\right)\mathbf{V}_2(s)$$

$$= \mathbf{I}_A(s) - \frac{i_1(0)}{s} + \frac{i_2(0)}{s} - C_1 v_1(0)$$

$$-\left(\frac{1}{sL_2} + sC_1\right)\mathbf{V}_1(s) + \left(\frac{1}{sL_2} + sC_1 + G_2 + sC_2\right)\mathbf{V}_2(s)$$

$$= C_1 v_1(0) - \frac{i_2(0)}{s} - C_2 v_2(0) + \mathbf{I}_B(s)$$

■

Example 17.1 attempts to illustrate the manner in which to employ the two s-domain representations of the inductor and capacitor circuit elements when initial conditions are present. In the following examples we illustrate the use of a number of analysis techniques in obtaining the complete response of a transformed network. The circuits analyzed have been specifically chosen to demonstrate the use of the Laplace transform to circuits with a variety of passive and active elements.

EXAMPLE 17.2

Consider the network shown in Fig. 17.3a. Let us determine the node voltages $v_1(t)$ and $v_2(t)$ using both nodal analysis and superposition.

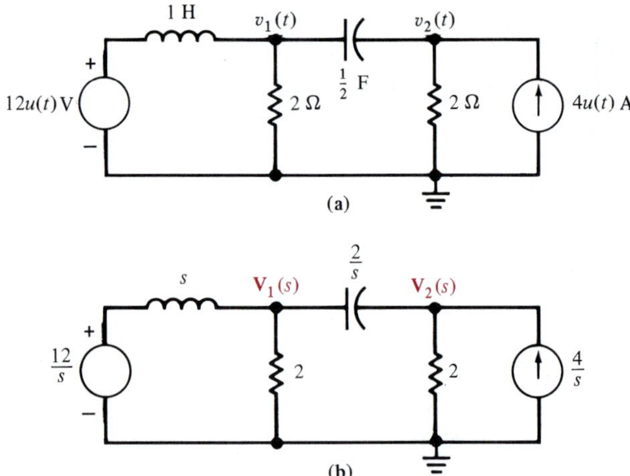

Figure 17.3 Circuits used in Example 17.2.

The transformed network is shown in Fig. 17.3b. The node equations for the network are

$$\mathbf{V}_1(s)\left(\frac{1}{s}+\frac{1}{2}+\frac{s}{2}\right)-\mathbf{V}_2(s)\left(\frac{s}{2}\right)=\frac{12}{s^2}$$

$$-\mathbf{V}_1(s)\left(\frac{s}{2}\right)+\mathbf{V}_2(s)\left(\frac{s}{2}+\frac{1}{2}\right)=\frac{4}{s}$$

The equations in matrix form are

$$\begin{bmatrix} \dfrac{s^2+s+2}{2s} & \dfrac{-s}{2} \\ \dfrac{-s}{2} & \dfrac{s+1}{2} \end{bmatrix}\begin{bmatrix} \mathbf{V}_1(s) \\ \mathbf{V}_2(s) \end{bmatrix}=\begin{bmatrix} \dfrac{12}{s^2} \\ \dfrac{4}{s} \end{bmatrix}$$

Solving for the node voltages, we obtain

$$\begin{bmatrix} \mathbf{V}_1(s) \\ \mathbf{V}_2(s) \end{bmatrix} = \begin{bmatrix} \dfrac{s^2+s+2}{2s} & \dfrac{-s}{2} \\ \dfrac{-s}{2} & \dfrac{s+1}{2} \end{bmatrix}^{-1} \begin{bmatrix} \dfrac{12}{s^2} \\ \dfrac{4}{s} \end{bmatrix}$$

$$= \frac{4s}{2s^2+3s+2} \begin{bmatrix} \dfrac{s+1}{2} & \dfrac{s}{2} \\ \dfrac{s}{2} & \dfrac{s^2+s+2}{2s} \end{bmatrix} \begin{bmatrix} \dfrac{12}{s^2} \\ \dfrac{4}{s} \end{bmatrix}$$

$$= \begin{bmatrix} \dfrac{4(s^2+3s+3)}{s(s^2+3s/2+1)} \\ \dfrac{4(s^2+4s+2)}{s(s^2+3s/2+1)} \end{bmatrix}$$

Let us now determine $v_1(t) = \mathcal{L}^{-1}[\mathbf{V}_1(s)]$ and $v_2(t) = \mathcal{L}^{-1}[\mathbf{V}_2(s)]$. $\mathbf{V}_1(s)$ can be written as

$$\mathbf{V}_1(s) = \frac{4(s^2+3s+3)}{s(s^2+\frac{3}{2}s+1)} = \frac{4(s^2+3s+3)}{s[s+\frac{3}{4}+j(\sqrt{7}/4)][s+\frac{3}{4}-j(\sqrt{7}/4)]}$$

Therefore,

$$\frac{4(s^2+3s+3)}{s[s+\frac{3}{4}-j(\sqrt{7}/4)][s+\frac{3}{4}+j(\sqrt{7}/4)]} = \frac{K_0}{s} + \frac{K_1}{s+\frac{3}{4}-j(\sqrt{7}/4)} + \frac{K_1^*}{s+\frac{3}{4}+j(\sqrt{7}/4)}$$

Evaluating the constants K_0 and K_1, we obtain

$$\left.\frac{4(s^2+3s+3)}{s^2+\frac{3}{2}s+1}\right|_{s=0} = K_0$$

$$12 = K_0$$

and

$$\left.\frac{4(s^2+3s+3)}{s[s+\frac{3}{4}+j(\sqrt{7}/4)]}\right|_{s=-3/4+j(\sqrt{7}/4)} = K_1$$

$$4\underline{/180^\circ} = K_1$$

Therefore,

$$v_1(t) = \left[12 + 8e^{-(3/4)t}\cos\left(\frac{\sqrt{7}}{4}t + 180^\circ\right)\right]u(t)\ \text{V}$$

In a similar manner,

$$\mathbf{V}_2(s) = \frac{4(s^2 + 4s + 2)}{s(s^2 + \frac{3}{2}s + 1)} = \frac{K_0}{s} + \frac{K_1}{s + \frac{3}{4} - j(\sqrt{7}/4)} + \frac{K_1^*}{s + \frac{3}{4} + j(\sqrt{7}/4)}$$

Evaluating the constants K_0 and K_1, we obtain

$$\left.\frac{4(s^2 + 4s + 2)}{s^2 + \frac{3}{2}s + 1}\right|_{s=0} = K_0$$

$$8 = K_0$$

and

$$\left.\frac{4(s^2 + 4s + 2)}{s[s + \frac{3}{4} + j(\sqrt{7}/4)}\right|_{s=-3/4+j(\sqrt{7}/4)} = K_1$$

$$5.66\underline{/-110.7^\circ} = K_1$$

Therefore,

$$v_2(t) = \left[8 + 11.32e^{-(3/4)t}\cos\left(\frac{\sqrt{7}}{4}t - 110.7^\circ\right)\right]u(t)\text{ V}$$

It is interesting to note that as $t \to \infty$, the sources in the circuit appear to be dc sources. In the dc case the inductor acts like a short circuit and the capacitor acts like an open circuit. Under these conditions the voltage $v_1(\infty) = 12$ V, and the voltage $v_2(\infty) = 8$ V. It is reassuring to realize that the equations for $v_1(t)$ and $v_2(t)$ yield these results also. ■

EXAMPLE 17.3

Consider the network shown in Fig. 17.4a. We wish to determine the output voltage $v_o(t)$. As we begin to attack the problem we note two things. First, because the source $12u(t)$ is connected between $v_1(t)$ and $v_2(t)$, we have a supernode. Second, if $v_2(t)$ is known, $v_o(t)$ can be easily obtained by voltage division. Hence we will use nodal analysis in conjunction with voltage division to obtain a solution.

The transformed network is shown in Fig. 17.4b. KCL for the supernode is

$$\frac{\mathbf{V}_1(s)}{2} + \mathbf{V}_1(s)\frac{s}{2} - 2\mathbf{I}(s) + \frac{\mathbf{V}_2(s)}{s+1} = 0$$

However,

$$\mathbf{I}(s) = -\frac{\mathbf{V}_1(s)}{2}$$

and

$$\mathbf{V}_1(s) = \mathbf{V}_2(s) - \frac{12}{s}$$

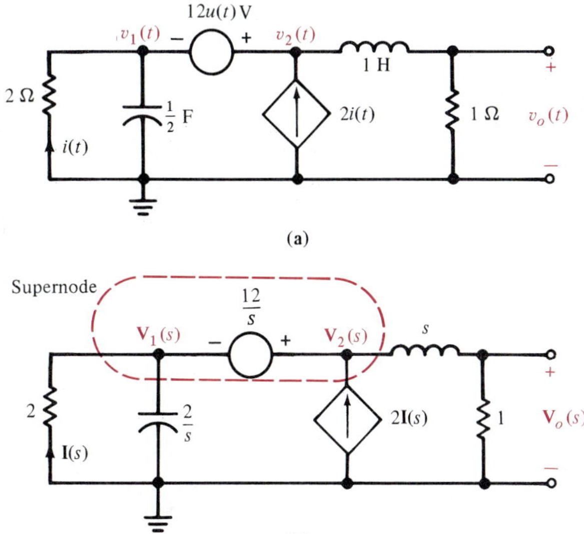

Figure 17.4 Circuits used in Example 17.3.

Substituting the last two equations into the first equation yields

$$\left[\mathbf{V}_2(s) - \frac{12}{s}\right]\frac{s+3}{2} + \frac{\mathbf{V}_2(s)}{s+1} = 0$$

or

$$\mathbf{V}_2(s) = \frac{12(s+1)(s+3)}{s(s^2+4s+5)}$$

Employing a voltage divider, we obtain

$$\begin{aligned}\mathbf{V}_o(s) &= \mathbf{V}_2(s)\frac{1}{s+1} \\ &= \frac{12(s+3)}{s(s^2+4s+5)} \\ &= \frac{12(s+3)}{s(s+2-j1)(s+2+j1)}\end{aligned}$$

which can be written as

$$\frac{12(s+3)}{s(s+2-j1)(s+2+j1)} = \frac{K_0}{s} + \frac{K_1}{s+2-j1} + \frac{K_1^*}{s+2+j1}$$

Evaluating the constants, we obtain

$$\left.\frac{12(s+3)}{s^2+4s+5}\right|_{s=0} = K_0$$

$$\frac{36}{5} = K_0$$

and

$$\left.\frac{12(s+3)}{s(s+2+j1)}\right|_{s=-2+j1} = K_1$$

$$3.79\underline{/161.57^\circ} = K_1$$

Therefore,

$$v_o(t) = [7.2 + 7.58e^{-2t}\cos(t + 161.57^\circ)]u(t) \text{ V}$$

EXAMPLE 17.4

Let us examine the network in Fig. 17.5a. We wish to determine the output voltage $v_o(t)$. We will solve this problem using mesh equations.

The transformed network is shown in Fig. 17.5b. KVL for the right-hand loop is

$$\frac{12}{s} - [\mathbf{I}_2(s) - \mathbf{I}_1(s)]s - \frac{\mathbf{I}_2(s)}{s} - 2\mathbf{I}_2(s) = 0$$

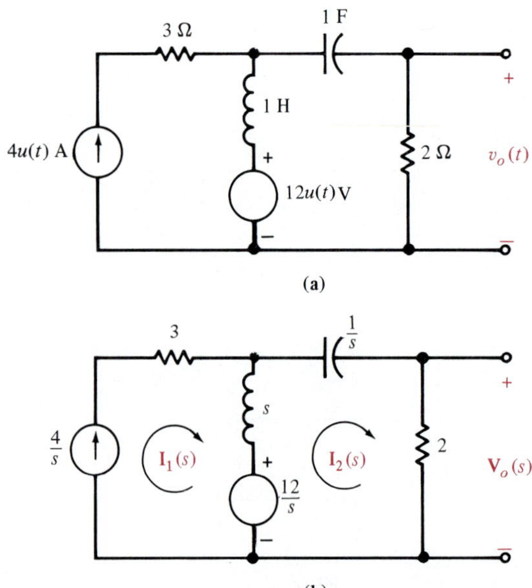

Figure 17.5 Circuits used in Example 17.4.

However, $\mathbf{I}_1(s) = 4/s$, and hence

$$\mathbf{I}_2(s) = \frac{4(s+3)}{(s+1)^2}$$

Therefore,

$$\mathbf{V}_o(s) = \frac{8(s+3)}{(s+1)^2}$$

$\mathbf{V}_o(s)$ can be written as

$$\mathbf{V}_o(s) = \frac{8(s+3)}{(s+1)^2} = \frac{K_{11}}{(s+1)^2} + \frac{K_{12}}{s+1}$$

Evaluating the constants, we obtain

$$8(s+3)\Big|_{s=-1} = K_{11}$$
$$16 = K_{11}$$

and

$$\frac{d}{ds}[8(s+3)]\Big|_{s=-1} = K_{12}$$

$$8 = K_{12}$$

Therefore,

$$v_o(t) = (16te^{-t} + 8e^{-t})u(t) \text{ V}$$

■

EXAMPLE 17.5

Let us determine the voltage $v_o(t)$ in the network in Fig. 17.6a using Thévenin's theorem.

The transformed network is shown in Fig. 17.6b. The open-circuit voltage can be computed from the network in Fig. 17.6c. Note that

$$\mathbf{V}_{oc}(s) + \mathbf{V}'_x(s) = \frac{\mathbf{V}'_x(s)}{2}$$

Therefore,

$$\mathbf{V}'_x(s) = -2\mathbf{V}_{oc}(s)$$

KCL at the supernode including the two independent sources is

$$\frac{\mathbf{V}_{oc}(s) + (4/s)}{s} + \frac{\mathbf{V}_{oc}(s) + (4/s) - \frac{1}{2}(-2\mathbf{V}_{oc}(s))}{1}$$
$$+ \frac{\mathbf{V}_{oc}(s) + (8/s) - \frac{1}{2}(-2\mathbf{V}_{oc}(s))}{1/s} + \frac{\mathbf{V}_{oc}(s) - \frac{1}{2}(-2\mathbf{V}_{oc}(s))}{1} = 0$$

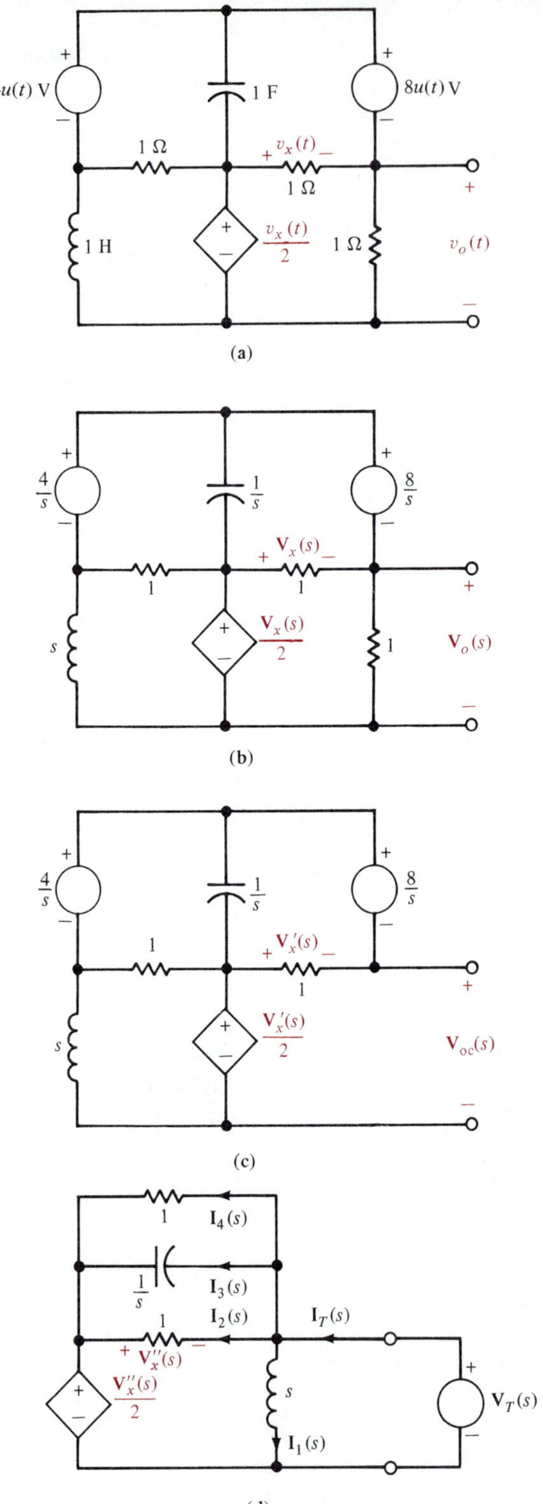

Figure 17.6 Circuits used in Example 17.5.

Solving this equation for $\mathbf{V}_{oc}(s)$ yields

$$\mathbf{V}_{oc}(s)\left(\frac{1}{s} + 2 + 2s + 2\right) = -\left(\frac{4}{s^2} + \frac{4}{s} + 8\right)$$

or

$$\mathbf{V}_{oc}(s) = \frac{-(8s^2 + 4s + 4)}{s(2s^2 + 4s + 1)}$$

In order to determine the Thévenin equivalent impedance we will zero the independent sources and apply a test source $\mathbf{V}_T(s)$ at the open-circuit terminals as shown in Fig. 17.6d. KVL illustrates that

$$\mathbf{V}_T(s) + \mathbf{V}_x''(s) - \frac{\mathbf{V}_x''(s)}{2} = 0$$

so that

$$-\frac{\mathbf{V}_x''(s)}{2} = \mathbf{V}_T(s)$$

Then

$$\begin{aligned}\mathbf{I}_T(s) &= \mathbf{I}_1(s) + \mathbf{I}_2(s) + \mathbf{I}_3(s) + \mathbf{I}_4(s)\\ &= \frac{\mathbf{V}_T(s)}{s} + \frac{\mathbf{V}_T(s) - (-\mathbf{V}_T(s))}{1} + \frac{\mathbf{V}_T(s) - (-\mathbf{V}_T(s))}{1/s} + \frac{\mathbf{V}_T(s) - (-\mathbf{V}_T(s))}{1}\\ &= \mathbf{V}_T(s)\left(\frac{1}{s} + 2 + 2s + 2\right)\end{aligned}$$

And hence

$$\mathbf{Z}_{Th}(s) = \frac{\mathbf{V}_T(s)}{\mathbf{I}_T(s)} = \frac{s}{2s^2 + 4s + 1}$$

Then

$$\begin{aligned}\mathbf{V}_o(s) &= \frac{\mathbf{V}_{oc}(s)}{\mathbf{Z}_{Th}(s) + 1}(1)\\ &= \frac{-(8s^2 + 4s + 4)}{s(2s^2 + 5s + 1)}\end{aligned}$$

This function can be expressed in the form

$$\frac{-2(2s^2 + 2s + 1)}{s(s^2 + \frac{5}{2}s + \frac{1}{2})} = \frac{K_0}{s} + \frac{K_1}{s + 0.22} + \frac{K_2}{s + 2.28}$$

Solving for K_0, K_1, K_2 in the manner illustrated earlier yields

$$K_0 = -4$$
$$K_1 = 2.9$$

and

$$K_2 = -2.9$$

Therefore,

$$v_o(t) = \left[-4 + 2.9e^{-0.22t} - 2.9e^{-2.28t} \right] u(t)$$

EXAMPLE 17.6

We wish to find the Y parameters for the two-port network shown in Fig. 17.7a. Then use the two-port equations to find $v_2(t)$ if a $1u(t)$-A current source is connected at the input.

The Y parameters for the network in Fig. 17.7a can be found from the Y parameters of the networks in Fig. 17.7b and c. By connecting the networks in Fig. 17.7b and c in parallel, we generate the original network in Fig. 17.7a. Therefore, we will compute the Y parameters for the two-ports in Fig. 17.7b and c and add them to determine the Y parameters for the network in Fig. 17.7a.

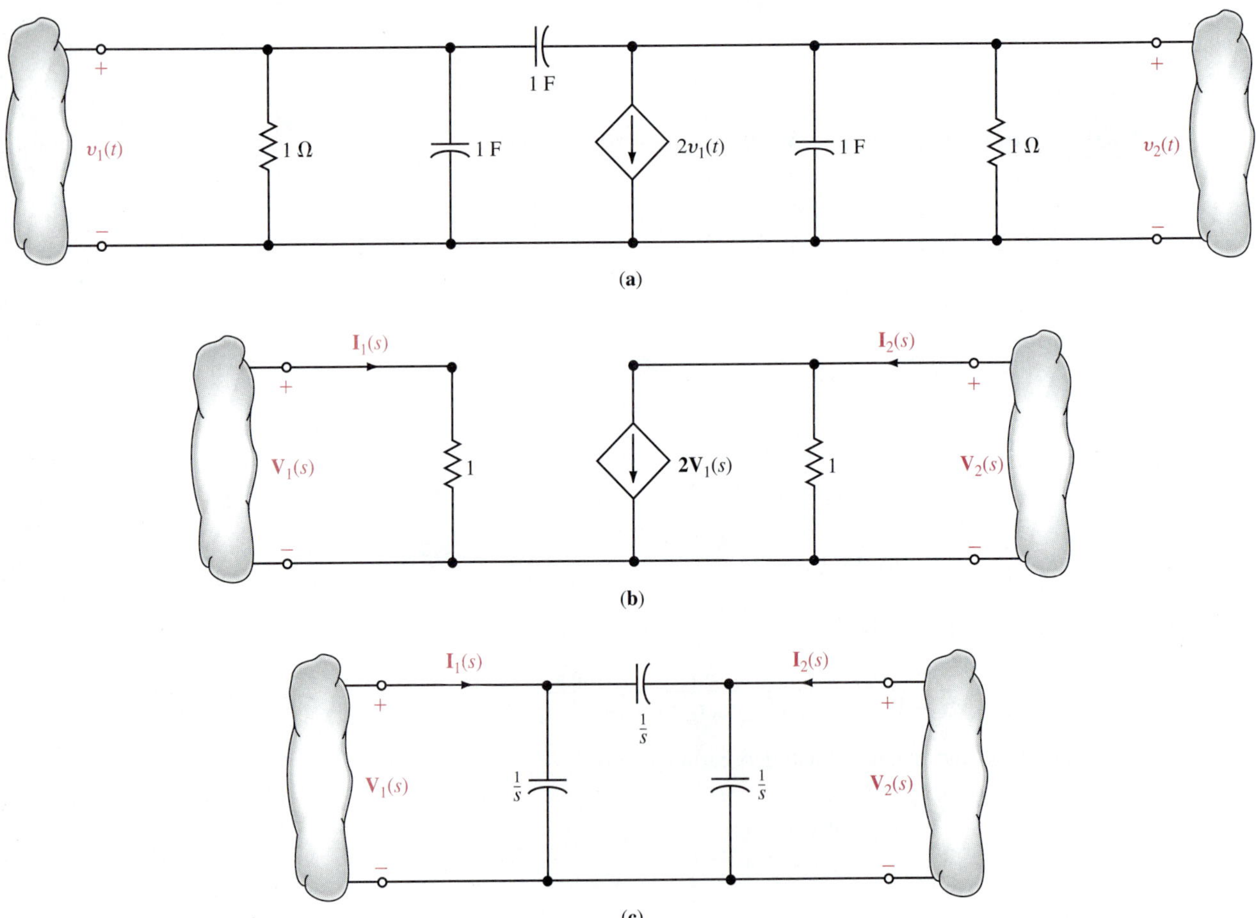

Figure 17.7 Circuits used in Example 17.6.

Using the techniques in Section 15.1, we find that the parameters for the network in Fig. 17.7b are

$$\mathbf{y}_{11}(s) = \left.\frac{\mathbf{I}_1(s)}{\mathbf{V}_1(s)}\right|_{\mathbf{V}_2(s)=0} = 1$$

$$\mathbf{y}_{12}(s) = \left.\frac{\mathbf{I}_1(s)}{\mathbf{V}_2(s)}\right|_{\mathbf{V}_1(s)=0} = 0$$

$$\mathbf{y}_{21}(s) = \left.\frac{\mathbf{I}_2(s)}{\mathbf{V}_1(s)}\right|_{\mathbf{V}_2(s)=0} = 2$$

$$\mathbf{y}_{22}(s) = \left.\frac{\mathbf{I}_2(s)}{\mathbf{V}_2(s)}\right|_{\mathbf{V}_1(s)=0} = 1$$

Therefore,

$$\mathbf{Y}_1(s) = \begin{bmatrix} 1 & 0 \\ 2 & 1 \end{bmatrix}$$

In a similar manner we find that the Y parameters for the network in Fig. 17.7c are

$$\mathbf{Y}_2(s) = \begin{bmatrix} 2s & -s \\ -s & 2s \end{bmatrix}$$

Therefore the Y parameters for the total network in Fig. 17.7a are

$$\mathbf{Y}_T(s) = \begin{bmatrix} 1 + 2s & -s \\ 2 - s & 1 + 2s \end{bmatrix}$$

If a $1u(t)$-A current source is applied at the input of the network in Fig. 17.7a, the equations for the transformed network are

$$\begin{bmatrix} 2s + 1 & -s \\ 2 - s & 2s + 1 \end{bmatrix} \begin{bmatrix} \mathbf{V}_1(s) \\ \mathbf{V}_2(s) \end{bmatrix} = \begin{bmatrix} \frac{1}{s} \\ 0 \end{bmatrix}$$

and hence

$$\begin{bmatrix} \mathbf{V}_1(s) \\ \mathbf{V}_2(s) \end{bmatrix} = \frac{1}{3s^2 + 6s + 1} \begin{bmatrix} 2s + 1 & s \\ s - 2 & 2s + 1 \end{bmatrix} \begin{bmatrix} \frac{1}{s} \\ 0 \end{bmatrix}$$

and

$$\mathbf{V}_2(s) = \frac{s - 2}{s(3s^2 + 6s + 1)}$$

which can be written as

$$\mathbf{V}_2(s) = \frac{1}{3}\left[\frac{s-2}{s(s+1.82)(s+0.18)}\right]$$

$$= \frac{1}{3}\left(\frac{K_0}{s} + \frac{K_1}{s+1.82} + \frac{K_2}{s+0.18}\right)$$

Using the procedures illustrated earlier, we find that $K_0 = -6$, $K_1 = -1.28$, and $K_2 = 7.38$. Therefore,

$$\tfrac{1}{3}(-6 - 1.28e^{-1.82t} + 7.38e^{-0.18t}u(t)$$

DRILL EXERCISE

D17.1. Find $i_o(t)$ in the network in Fig. D17.1 using node equations.

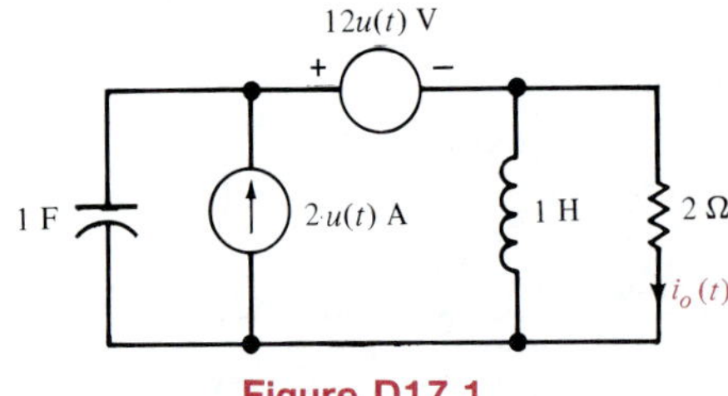

Figure D17.1

Ans: $i_o(t) = 6.53e^{-t/4} \cos [(\sqrt{15}/4)t - 156.72°]u(t)$ A.

D17.2. Find $v_o(t)$ in the network in Fig. D17.2 using loop equations.

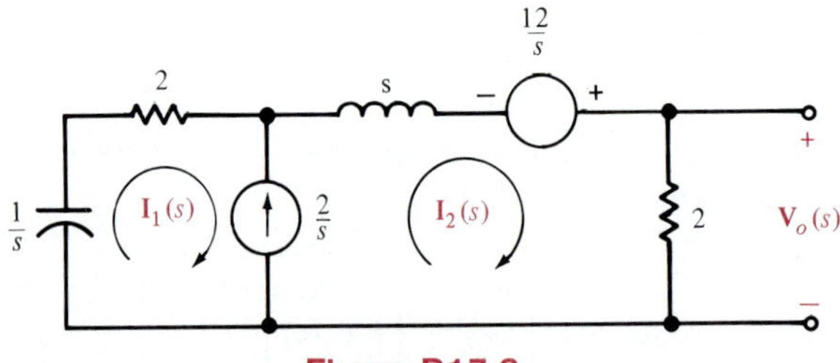

Figure D17.2

Ans: $v_o(t) = (4 - 8.93e^{-3.73t} + 4.93e^{-0.27t})u(t)$ V.

D17.3. Solve Drill Exercise D17.2 using Thévenin's theorem.

We will now illustrate the use of the Laplace transform in the transient analysis of circuits. We will analyze networks such as those considered in Chapters 7 and 8. Our approach will first be to determine the initial conditions for the capacitors and inductors in the network, and then we will employ the element models that were specified at the beginning of this chapter together with the circuit analysis techniques to obtain a solution. The following examples will demonstrate the approach.

EXAMPLE 17.7

In this example we analyze the circuit redrawn in Fig. 17.8a. At $t = 0$ the initial voltage across the capacitor is 1 V, and the initial current drawn through the inductor is 1 A. The circuit for $t > 0$ is shown in Fig. 17.8b with the initial conditions. The transformed network is shown in Fig. 17.8c.

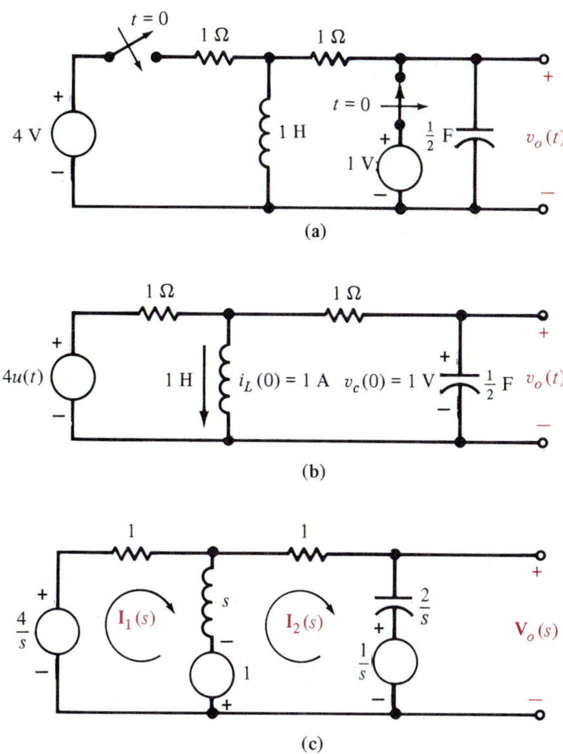

Figure 17.8 Circuits employed in Example 17.7.

The mesh equations for the transformed network are

$$(s + 1)\mathbf{I}_1(s) - s\mathbf{I}_2(s) = \frac{4}{s} + 1$$

$$-s\mathbf{I}_1(s) + \left(s + \frac{2}{s} + 1\right)\mathbf{I}_2(s) = \frac{-1}{s} - 1$$

which can be written in matrix form as

$$\begin{bmatrix} s + 1 & -s \\ -s & \dfrac{s^2 + s + 2}{s} \end{bmatrix} \begin{bmatrix} \mathbf{I}_1(s) \\ \mathbf{I}_2(s) \end{bmatrix} = \begin{bmatrix} \dfrac{s + 4}{s} \\ \dfrac{-(s + 1)}{s} \end{bmatrix}$$

Solving for the currents, we obtain

$$\begin{bmatrix} \mathbf{I}_1(s) \\ \mathbf{I}_2(s) \end{bmatrix} = \begin{bmatrix} s+1 & -s \\ -s & \dfrac{s^2+s+2}{s} \end{bmatrix}^{-1} \begin{bmatrix} \dfrac{s+4}{s} \\ \dfrac{-(s+1)}{s} \end{bmatrix}$$

$$= \frac{s}{2s^2+3s+2} \begin{bmatrix} \dfrac{s^2+s+2}{s} & s \\ s & s+1 \end{bmatrix} \begin{bmatrix} \dfrac{s+4}{s} \\ \dfrac{-(s+1)}{s} \end{bmatrix}$$

$$= \begin{bmatrix} \dfrac{4s^2+6s+8}{s(2s^2+3s+2)} \\ \dfrac{2s-1}{2s^2+3s+2} \end{bmatrix}$$

The output voltage is then

$$\mathbf{V}_o(s) = \frac{2}{s}\mathbf{I}_2(s) + \frac{1}{s}$$

$$= \frac{2}{s}\left(\frac{2s-1}{2s^2+3s+2}\right) + \frac{1}{s}$$

$$= \frac{s+\frac{7}{2}}{s^2+\frac{3}{2}s+1}$$

This function can be written in a partial-fraction expansion as

$$\frac{s+\frac{7}{2}}{s^2+\frac{3}{2}s+1} = \frac{K_1}{s+\frac{3}{4}-j(\sqrt{7}/4)} + \frac{K_1^*}{s+\frac{3}{4}+j(\sqrt{7}/4)}$$

Evaluating the constants, we obtain

$$\left.\frac{s+\frac{7}{2}}{s+\frac{3}{4}+j(\sqrt{7}/4)}\right|_{s=-(3/4)+j(\sqrt{7}/4)} = K_1$$

$$2.14\,\underline{/-76.5^\circ} = K_1$$

Therefore,

$$v_o(t) = \left[4.29e^{-(3/4)t}\cos\left(\frac{\sqrt{7}}{4}t - 76.5^\circ\right)\right]u(t)\text{ V}$$

■

EXAMPLE 17.8

We wish to find the output voltage $v_o(t)$ for $t > 0$ in the network in Fig. 17.9a.

In steady state prior to $t = 0$, the current $i_1(0)$ is

$$i_1(0) = \frac{12}{2+4} = 2\text{ A}$$

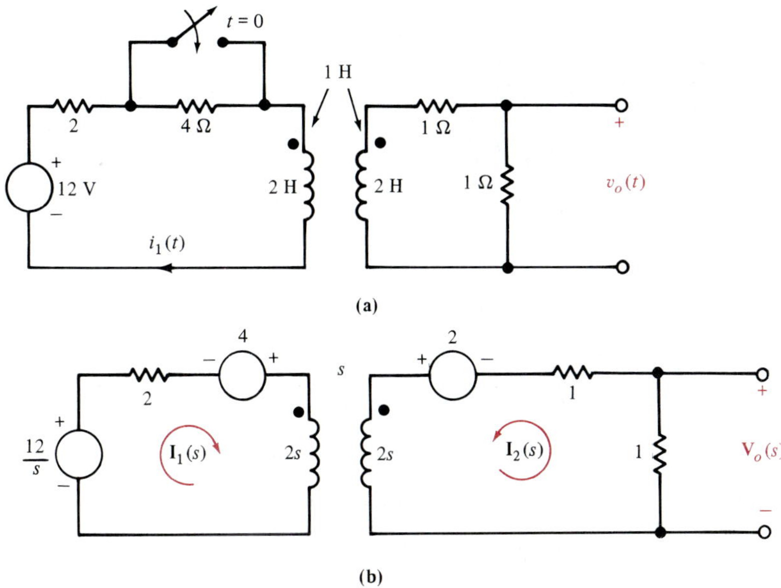

Figure 17.9 Circuits used in Example 17.8.

Hence

$$L_1 i_1(0) = 4$$

$$M i_1(0) = 2$$

Using these data, we obtain the transformed network as shown in Fig. 17.9b. The mesh equations for this network are

$$(2s + 2)\mathbf{I}_1(s) + s\mathbf{I}_2(s) = \frac{12}{s} + 4$$

$$s\mathbf{I}_1(s) + (2s + 2)\mathbf{I}_2(s) = 2$$

Solving these equations for the mesh currents, we obtain

$$\begin{bmatrix} \mathbf{I}_1(s) \\ \mathbf{I}_2(s) \end{bmatrix} = \frac{1}{3s^2 + 8s + 4} \begin{bmatrix} 2s + 2 & -s \\ -s & 2s + 2 \end{bmatrix} \begin{bmatrix} \dfrac{4s + 12}{s} \\ 2 \end{bmatrix}$$

which yields

$$\mathbf{I}_2(s) = \frac{-8}{3s^2 + 8s + 4}$$

and

$$\mathbf{V}_o(s) = -1\mathbf{I}_2(s) = \frac{\frac{8}{3}}{s^2 + \frac{8}{3}s + \frac{4}{3}}$$

This equation can be written as

$$\frac{\frac{8}{3}}{s^2 + \frac{8}{3}s + \frac{4}{3}} = \frac{K_1}{s + 2} + \frac{K_2}{s + \frac{2}{3}}$$

Solving for the constants yields $K_1 = -2$ and $K_2 = 2$. Therefore,

$$v_o(t) = 2(e^{-(2/3)t} - e^{-2t})u(t) \text{ V}$$

■

DRILL EXERCISE

D17.4. Solve Problem 7.1 using Laplace transforms.

Ans: $v_c(t) = 4e^{-1.2t}$ V.

D17.5. Solve Drill Exercise D7.6 using Laplace transforms.

Ans: $v_o(t) = \frac{24}{5} + \frac{1}{5}e^{-(5/8)t}$ V.

D17.6. Solve Drill Exercise D7.7 using Laplace transforms.

Ans: $v_o(t) = 6 - \frac{10}{3}e^{-2t}$ V.

D17.7. Find $v_o(t)$ for $t > 0$ in the network in Fig. D17.7.

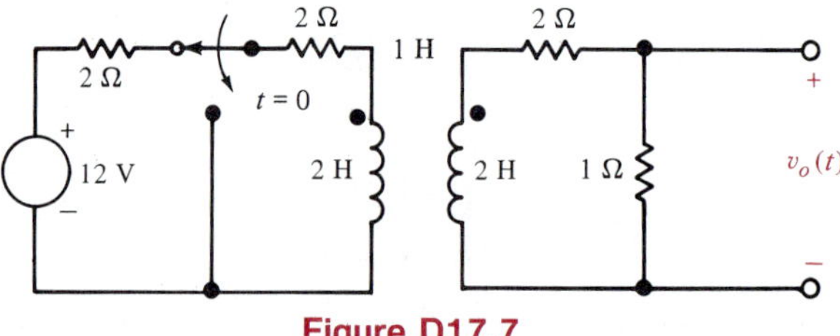

Figure D17.7

Ans: $v_o(t) = 1.13(e^{-2.55t} - e^{-0.78t})u(t)$ V.

In Chapter 16 we illustrated techniques for determining the Laplace transform of a complex waveform. By using these methods in conjunction with our analysis techniques, we can obtain the response of a network to any number of unusual input forcing functions. The following examples are provided to demonstrate the techniques.

EXAMPLE 17.9

In this example we analyze the circuit in Example 7.7 using Laplace transforms. The circuit and the input are redrawn in Fig. 17.10a. We wish to find the expression for the output voltage $v_o(t)$ for $t > 0$.

KCL can be used to compute the node voltage $\mathbf{V}_1(s)$ and then a simple voltage divider can be employed to find $\mathbf{V}_o(s)$. KCL yields

$$\frac{\mathbf{V}_1(s) - \mathbf{V}(s)}{6000} + \frac{\mathbf{V}_1(s)}{1/10^{-4}s} + \frac{\mathbf{V}_1(s)}{12{,}000} = 0$$

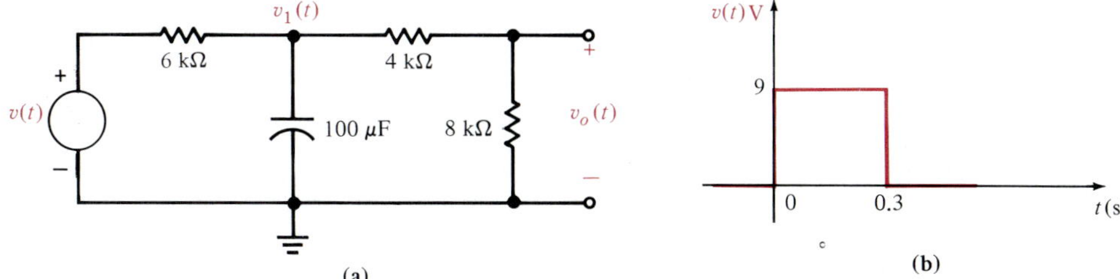

Figure 17.10 Network (a) and input (b) used in Example 17.9.

Solving this equation for $\mathbf{V}_1(s)$, we obtain

$$\mathbf{V}_1(s) = \frac{\mathbf{V}(s)}{6000(10^{-4}s + 2.5 \times 10^{-4})}$$

Note that

$$\mathbf{V}_o(s) = \frac{8000}{4000 + 8000}\mathbf{V}_1(s)$$

Therefore,

$$\mathbf{V}_o(s) = \frac{2\mathbf{V}(s)}{1.8s + 4.5}$$

$$= \frac{1}{0.9}\frac{\mathbf{V}(s)}{s + 2.5}$$

The expression for $\mathbf{V}(s) = \mathscr{L}[v(t)]$, where $v(t)$ is as shown in Fig. 17.10b, is

$$\mathbf{V}(s) = 9\left(\frac{1}{s} - \frac{e^{-0.3s}}{s}\right)$$

Substituting this equation into the expression for $\mathbf{V}_o(s)$ yields

$$\mathbf{V}_o(s) = \left(\frac{1}{0.9}\right)\frac{1}{s + 2.5}(9)\frac{1 - e^{-0.3s}}{s}$$

$$= \frac{10(1 - e^{-0.3s})}{s(s + 2.5)}$$

Since

$$\frac{10}{s(s + 2.5)} = \frac{4}{s} - \frac{4}{s + 2.5}$$

then

$$\mathbf{V}_o(s) = \left(\frac{4}{s} - \frac{4}{s + 2.5}\right)(1 - e^{-0.3s})$$

and therefore

$$v_o(t) = (4 - 4e^{-2.5t})u(t) - (4 - 4e^{-2.5(t-0.3)})u(t - 0.3) \text{ V}$$

This expression is of course identical to that obtained in Example 7.7. ■

EXAMPLE 17.10

Suppose that the input to the network of Example 17.9 is shown in Fig. 17.11. Let us find the network output for this new input.

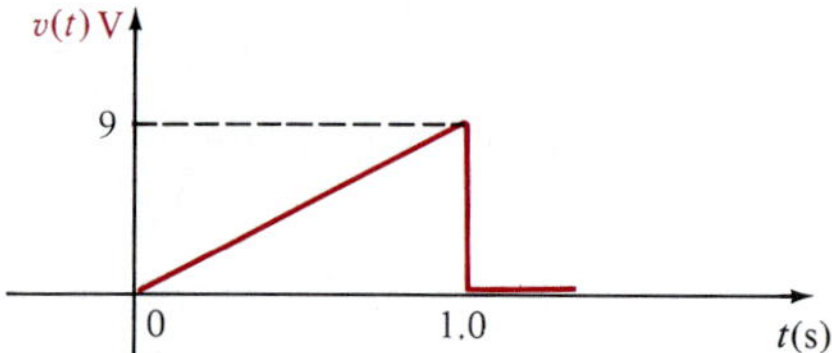

Figure 17.11 Input for the problem in Example 17.10.

As shown in Example 16.14, the Laplace transform of this sawtooth waveform is

$$\mathbf{V}(s) = \frac{9}{s^2}[1 - (1 + s)e^{-s}]$$

and hence the output is

$$\begin{aligned}\mathbf{V}_o(s) &= \left(\frac{1}{0.9}\right)\frac{1}{s + 2.5}\left(\frac{9}{s^2}\right)[1 - (1 + s)e^{-s}] \\ &= \frac{10}{s^2(s + 2.5)}[1 - (1 + s)e^{-s}] \\ &= \frac{10}{s^2(s + 2.5)} - \frac{10(s + 1)}{s^2(s + 2.5)}e^{-s}\end{aligned}$$

Expanding each of the terms above in a partial-fraction expansion, we obtain

$$\frac{10}{s^2(s + 2.5)} = \frac{4}{s^2} - \frac{1.6}{s} + \frac{1.6}{s + 2.5}$$

and

$$\frac{10(s + 1)}{s^2(s + 2.5)} = \frac{4}{s^2} + \frac{2.4}{s} - \frac{2.4}{s + 2.5}$$

Therefore,

$$v_o(t) = (4t - 1.6 + 1.6e^{-2.5t})u(t) - [4(t - 1) + 2.4 - 2.4e^{-2.5(t-1)}]u(t - 1) \text{ V}$$ ■

DRILL EXERCISE

D17.8. Find the expression for the voltage $v_o(t)$ for $t > 0$ in the network in Fig. D17.8a if the input is as shown in Fig. D17.8b.

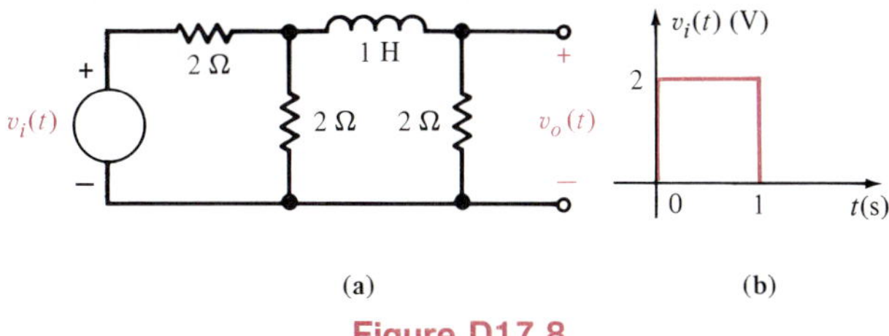

Figure D17.8

Ans: $v_o(t) = \frac{2}{3}[(1 - e^{-3t})u(t) - (1 - e^{-3(t-1)})u(t-1)]$ V.

D17.9. Find the expression for the voltage $v_o(t)$ for $t > 0$ in the network in Fig. D17.9a if the input is as shown in Fig. D17.9b.

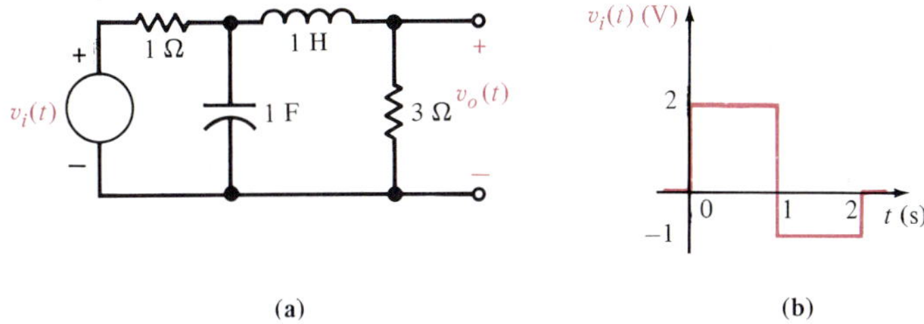

Figure D17.9

Ans: $v_o(t) = 2(\frac{3}{4} - \frac{3}{4}e^{-2t} - \frac{3}{2}te^{-2t})u(t) - 3[\frac{3}{4} - \frac{3}{4}e^{-2(t-1)} - \frac{3}{2}(t-1)e^{-2(t-1)}]u(t-1) + [\frac{3}{4} - \frac{3}{4}e^{-2(t-2)} - \frac{3}{2}(t-2)e^{-2(t-2)}]u(t-2)$ V.

17.3 Transfer Function

In Chapter 14 we introduced the concept of network or transfer function. It is essentially nothing more than the ratio of some output variable to some input variable. If both variables are voltages, the transfer function is a voltage gain. If both variables are currents, the transfer function is a current gain. If one variable is a voltage and the other is a current, the transfer function becomes a transfer admittance or impedance.

In deriving a transfer function, all initial conditions are set equal to zero. In addition, if the output is generated by more than one input source in a network, superposition can be employed in conjunction with the transfer function for each source.

To present this concept in a more formal manner, let us assume that the input/output relationship for a linear circuit is

$$b_n \frac{d^n y_o(t)}{dt^n} + b_{n-1} \frac{d^{n-1} y_o(t)}{dt^{n-1}} + \cdots + b_1 \frac{dy_o(t)}{dt} + b_0 y_o(t)$$

$$= a_m \frac{d^m x_i(t)}{dt^m} + a_{m-1} \frac{d^{m-1} x_i(t)}{dt^{m-1}} + \cdots + a_1 \frac{dx_i(t)}{dt} + a_0 x_i(t)$$

If all the initial conditions are zero, the transform of the equation is

$$(b_n s^n + \mathrm{b}_{n-1} s^{n-1} + \cdots + b_1 s + b_0)\mathbf{Y}_o(s)$$

$$= (a_m s^m + a_{m-1} s^{m-1} + \cdots + a_1 s + a_0)\mathbf{X}_i(s)$$

or

$$\frac{\mathbf{Y}_o(s)}{\mathbf{X}_i(s)} = \frac{a_m s^m + a_{m-1} s^{m-1} + \cdots + a_1 s + a_0}{b_n s^n + b_{n-1} s^{n-1} + \cdots + b_1 s + b_0}$$

This ratio of $\mathbf{Y}_o(s)$ to $\mathbf{X}_i(s)$ is called the *transfer* or *network function*, which we denote as $\mathbf{H}(s)$; that is,

$$\frac{\mathbf{Y}_o(s)}{\mathbf{X}_i(s)} = \mathbf{H}(s)$$

or

$$\mathbf{Y}_o(s) = \mathbf{H}(s)\mathbf{X}_i(s) \tag{17.15}$$

This equation states that the output response $\mathbf{Y}_o(s)$ is equal to the network function multiplied by the input $\mathbf{X}_i(s)$. Note that if $\mathbf{X}_i(t) = \delta(t)$ and therefore $\mathbf{X}_i(s) = 1$, the impulse response is equal to the inverse Laplace transform of the network function. This is an extremely important concept because it illustrates that if we know the impulse response of a network, we can find the response due to some other forcing function using Eq. (17.15).

At this point it is informative to review briefly the natural response of both first-order and second-order networks. We have demonstrated in Chapter 7 that if only a single storage element is present, the natural response of a network to an initial condition is always of the form

$$x(t) = X_0 e^{-t/T_c}$$

where $x(t)$ can be either $v(t)$ or $i(t)$, X_0 is the initial value of $x(t)$, and T_c is the time constant of the network.

As illustrated in Chapter 8, the natural response of a second-order network is controlled by the roots of the *characteristic equation,* which is of the form

$$s^2 + 2\zeta\omega_0 s + \omega_0^2 = 0$$

where ζ is the *damping ratio* and ω_0 is the *undamped natural frequency*. These two key factors, ζ and ω_0, control the response, and there are basically three cases of interest.

Case 1, $\zeta > 1$: Overdamped Network. The roots of the characteristic equation are $s_1, s_2 = -\zeta\omega_0 \pm \omega_0\sqrt{\zeta^2 - 1}$, and therefore, the network response is of the form

$$x(t) = K_1 e^{-(\zeta\omega_0 + \omega_0\sqrt{\zeta^2-1})t} + K_2 e^{-(\zeta\omega_0 - \omega_0\sqrt{\zeta^2-1})t}$$

Case 2, $\zeta < 1$: Underdamped Network. The roots of the characteristic equation are $s_1, s_2 = -\zeta\omega_0 \pm j\omega_0\sqrt{1 - \zeta^2}$, and therefore, the network response is of the form

$$x(t) = Ke^{-\zeta\omega_0 t} \cos(\omega_0\sqrt{1 - \zeta^2}\,t + \phi)$$

Case 3, $\zeta = 1$: Critically Damped Network. The roots of the characteristic equation are $s_1, s_2 = -\omega_0$, and hence the response is of the form

$$x(t) = K_1 t e^{-\omega_0 t} + K_2 e^{-\omega_0 t}$$

It is very important that the reader note that the characteristic equation is the denominator of the response function $\mathbf{X}(s)$, and the roots of this equation, which are the poles of the network, determine the form of the network's natural response.

A convenient method for displaying the network's poles and zeros in graphical form is the use of a pole–zero plot. A pole–zero plot of a function can be accomplished using what is commonly called the *complex* or *s-plane*. In the complex plane the abscissa is σ and the ordinate is $j\omega$. Zeros are represented by 0's and poles are represented by $\times$'s. Although we are concerned only with the finite poles and zeros specified by the network or response function, we should point out that a rational function must have the same number of poles and zeros. Therefore, if $n > m$, there are $n - m$ zeros at the point at infinity, and if $n < m$, there are $m - n$ poles at the point at infinity. A systems engineer can tell a lot about the operation of a network or system by simply examining its pole–zero plot.

In order to correlate the natural response of a network to an initial condition with the network's pole locations, we have illustrated in Fig. 17.12 the correspondence for all three cases: overdamped, underdamped, and critically damped. Note that if the network poles are real and unequal, the response is slow, and therefore, $x(t)$ takes a long time to reach zero. If the network poles are complex conjugates, the response is fast; however, it overshoots and is eventually damped out. The dividing line between the overdamped and underdamped cases is the critically damped case in which the roots are real and equal. In this case the transient response dies out as quickly as possible, with no overshoot.

EXAMPLE 17.11

If the impulse response of a network is $h(t) = e^{-t}$, let us determine the response $v_o(t)$ to an input $v_i(t) = 10e^{-2t}u(t)$ V.

The transformed variables are

$$\mathbf{H}(s) = \frac{1}{s + 1}$$

$$\mathbf{V}_i(s) = \frac{10}{s + 2}$$

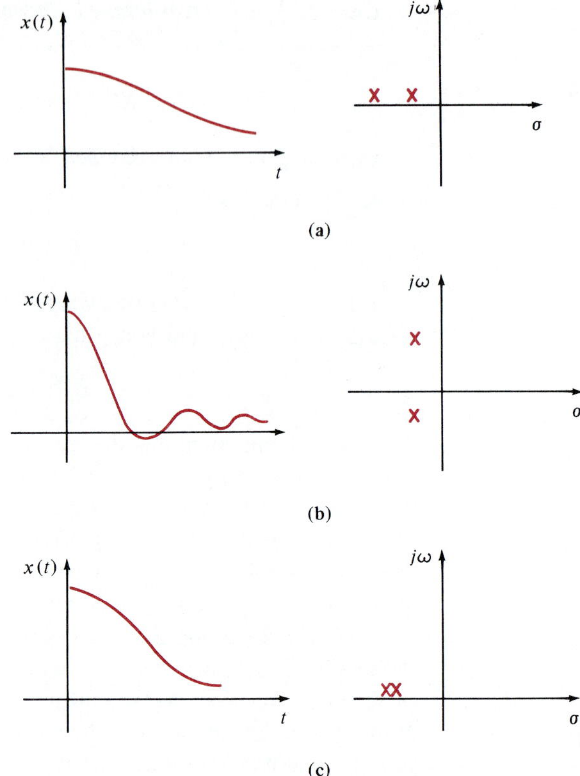

Figure 17.12 Natural response of a second-order network together with network pole locations for three cases: (a) overdamped; (b) underdamped; (c) critically damped.

Therefore,

$$\mathbf{V}_o(s) = \mathbf{H}(s)\mathbf{V}_i(s)$$

$$= \frac{10}{(s+1)(s+2)}$$

and hence

$$v_o(t) = 10(e^{-t} - e^{-2t})u(t) \text{ V}$$

The importance of the transfer function stems from the fact that it provides the systems engineer with a great deal of knowledge about the system's operation, since its dynamic properties are governed by the system poles.

EXAMPLE 17.12

Let us derive the transfer function $\mathbf{V}_o(s)/\mathbf{V}_i(s)$ for the network in Fig. 17.13a.

Our output variable is the voltage across a variable capacitor and the input voltage is a unit step. The transformed network is shown in Fig. 17.13b. The mesh equations for the

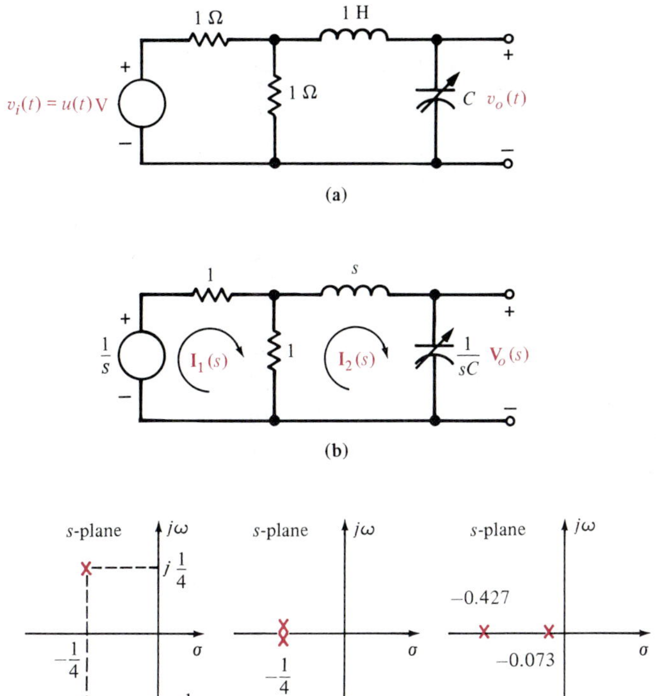

Figure 17.13 Networks and pole–zero plots used in Example 17.12.

network are

$$2\mathbf{I}_1(s) - \mathbf{I}_2(s) = \mathbf{V}_i(s)$$

$$-\mathbf{I}_1(s) + \left(s + \frac{1}{sC} + 1\right)\mathbf{I}_2(s) = 0$$

and the output equation is

$$\mathbf{V}_o(s) = \frac{1}{sC}\mathbf{I}_2(s)$$

From these equations we find that the transfer function is

$$\frac{\mathbf{V}_o(s)}{\mathbf{V}_i(s)} = \frac{1/2C}{s^2 + \frac{1}{2}s + 1/C}$$

Since the transfer function is dependent on the value of the capacitor, let us examine the transfer function and the output response for three values of the capacitor.

(a) $C = 8$ F:

$$\frac{\mathbf{V}_o(s)}{\mathbf{V}_i(s)} = \frac{\frac{1}{16}}{(s^2 + \frac{1}{2}s + \frac{1}{8})}$$

$$= \frac{\frac{1}{16}}{(s + \frac{1}{4} - j\frac{1}{4})(s + \frac{1}{4} + j\frac{1}{4})}$$

The output response is

$$\mathbf{V}_o(s) = \frac{\frac{1}{16}}{s(s + \frac{1}{4} - j\frac{1}{4})(s + \frac{1}{4} + j\frac{1}{4})}$$

As illustrated in Chapter 8, the poles of the transfer function, which are the roots of the characteristic equation, are complex conjugates, as shown in Fig. 17.13c, and therefore the output response will be *underdamped*. The output response as a function of time is

$$v_o(t) = \left[\frac{1}{2} + \frac{1}{\sqrt{2}} e^{-t/4} \cos\left(\frac{t}{4} + 135°\right)\right] u(t) \text{ V}$$

Note that for large values of time the transient oscillations, represented by the second term in the response, become negligible and the output settles out to a value of $\frac{1}{2}$ V. This can also be seen directly from the circuit since for large values of time the input looks like a dc source, the inductor acts like a short circuit, the capacitor acts like an open circuit, and the resistors form a voltage divider.

(b) $C = 16$ F:

$$\frac{\mathbf{V}_o(s)}{\mathbf{V}_i(s)} = \frac{\frac{1}{32}}{s^2 + \frac{1}{2}s + \frac{1}{16}}$$

$$= \frac{\frac{1}{32}}{(s + \frac{1}{4})^2}$$

The output response is

$$\mathbf{V}_o(s) = \frac{\frac{1}{32}}{s(s + \frac{1}{4})^2}$$

Since the poles of the transfer function are real and equal as shown in Fig. 17.13d, the output response will be *critically damped*. $v_o(t) = \mathcal{L}^{-1}[\mathbf{V}_o(s)]$ is

$$v_o(t) = \left[\frac{1}{2} - \left(\frac{t}{8} + \frac{1}{2}\right)e^{-t/4}\right] u(t) \text{ V}$$

(c) $C = 32$ F:

$$\frac{\mathbf{V}_o(s)}{\mathbf{V}_i(s)} = \frac{\frac{1}{64}}{s^2 + \frac{1}{2}s + \frac{1}{32}}$$

$$= \frac{\frac{1}{64}}{(s + 0.427)(s + 0.073)}$$

The output response is

$$\mathbf{V}_o(s) = \frac{\frac{1}{64}}{s(s + 0.427)(s + 0.073)}$$

The poles of the transfer function are real and unequal as shown in Fig. 17.13e, and therefore, the output response will be *overdamped*. The response as a function of time is

$$v_o(t) = (0.5 + 0.103e^{-0.427t} - 0.603e^{-0.073t})u(t) \text{ V}$$

Although the values selected for the network parameters are not very practical, the reader is reminded that both magnitude and frequency scaling, as outlined in Chapter 14, can be applied here also. ■

EXAMPLE 17.13

For the network in Fig. 17.14a let us compute (a) the transfer function, (b) the type of damping exhibited by the network, and (c) the unit step response.

Recall that the voltage across the op-amp input terminals is zero and therefore KCL at the node labeled $\mathbf{V}_1(s)$ in Fig. 17.14b yields the following equation:

$$\frac{\mathbf{V}_S(s) - \mathbf{V}_1(s)}{1} = s\mathbf{V}_1(s) + \frac{\mathbf{V}_1(s) - \mathbf{V}_o(s)}{1} + \frac{\mathbf{V}_1(s)}{1}$$

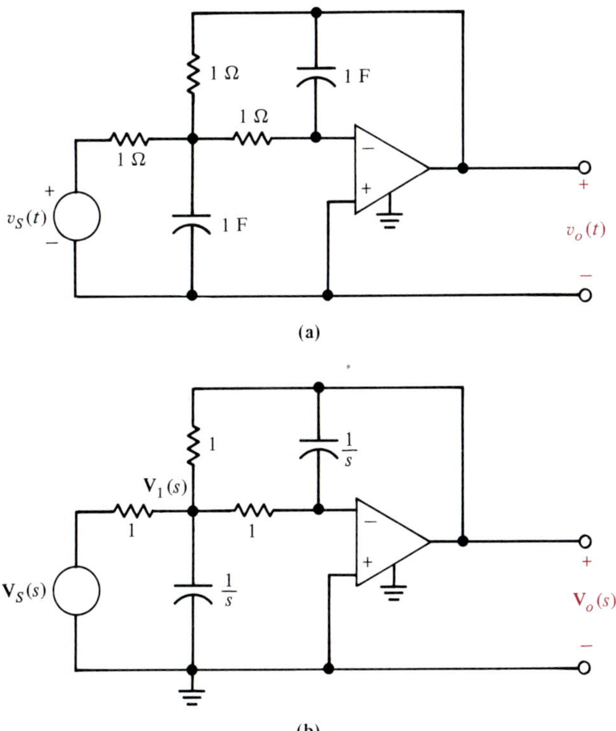

Figure 17.14 Circuits used in Example 17.13.

Since the current into the negative input terminal of the op-amp is zero, KCL requires that

$$s\mathbf{V}_o(s) = -\frac{\mathbf{V}_1(s)}{1}$$

Combining the two equations yields the transfer function

$$\frac{\mathbf{V}_o(s)}{\mathbf{V}_S(s)} = \frac{-1}{s^2 + 3s + 1}$$

which can be expressed in the form

$$\frac{\mathbf{V}_o(s)}{\mathbf{V}_S(s)} = \frac{-1}{(s + 2.62)(s + 0.38)}$$

Since the roots are real and unequal, the step response of the network will be overdamped. The step response is

$$\mathbf{V}_o(s) = \frac{-1}{s(s + 2.62)(s + 0.38)}$$

$$= \frac{-1}{s} + \frac{-0.17}{s + 2.62} + \frac{1.17}{s + 0.38}$$

Therefore,

$$v_o(t) = (-1 - 0.17e^{-2.62t} + 1.17e^{-0.38t})u(t) \text{ V}$$

■

DRILL EXERCISE

D17.10. If the unit impulse response of a network is known to be $\frac{10}{9}(e^{-t} - e^{-10t})$, determine the unit step response.

Ans: $x(t) = (1 - \frac{10}{9}e^{-t} + \frac{1}{9}e^{-10t})u(t)$.

D17.11. The transfer function for a network is

$$\mathbf{H}(s) = \frac{s + 10}{s^2 + 4s + 8}$$

Determine the pole–zero plot of $\mathbf{H}(s)$, the type of damping exhibited by the network, and the unit step response of the network.

Ans: The network is underdamped; $x(t) = [\frac{10}{8} + 1.46e^{-2t}\cos(2t - 210.96°)]u(t)$.

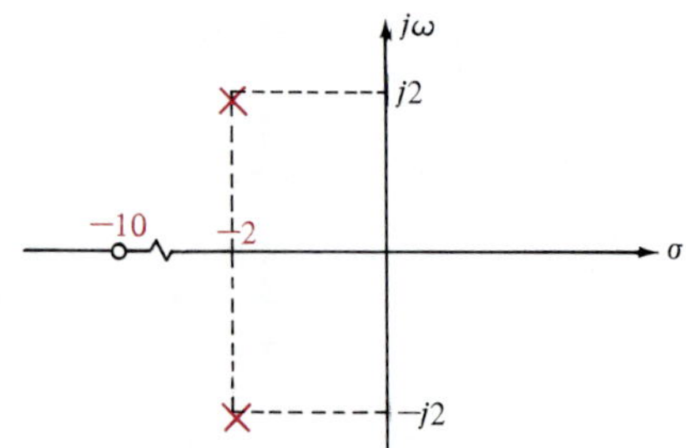

D17.12. Find the transfer function of the network shown in Fig. D17.12.

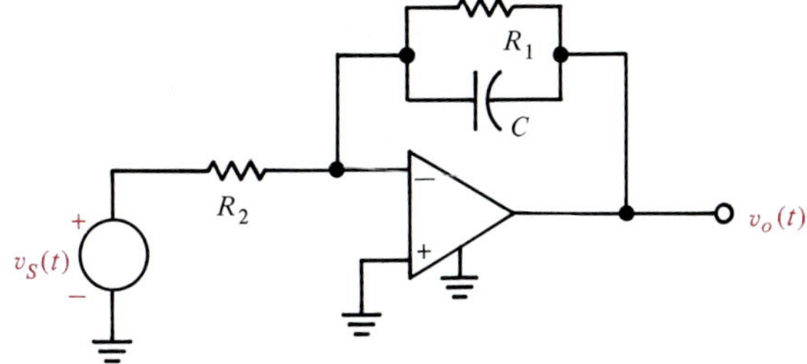

Figure D17.12

Ans: $\dfrac{\mathbf{V}_o(s)}{\mathbf{V}_S(s)} = \dfrac{-1/R_2C}{s + 1/R_1C}$.

Recall from our previous discussion that if a second-order network is underdamped, the characteristic equation of the network is of the form

$$s^2 + 2\zeta\omega_0 s + \omega_0^2 = 0$$

and the roots of this equation, which are the network poles, are of the form

$$s_1, s_2 = -\zeta\omega_0 \pm j\omega_0\sqrt{1-\zeta^2}$$

The roots s_1 and s_2, when plotted in the s-plane, generally appear as shown in Fig. 17.15, where

$$\zeta = \text{damping ratio}$$

$$\omega_0 = \text{undamped natural frequency}$$

and as shown in Fig. 17.15,

$$\zeta = \cos\theta$$

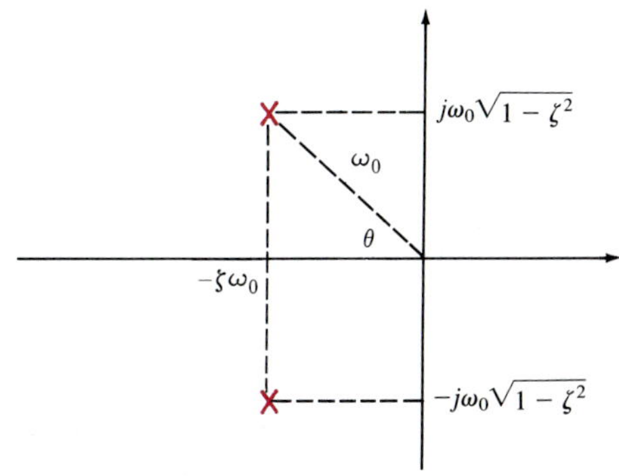

Figure 17.15 Pole locations for second-order underdamped network.

It is important for the reader to note that the damping ratio and the undamped natural frequency are exactly the same quantities as those employed in Chapter 14 when determining a network's frequency response. We find here that it is these same quantities that govern the network's transient response.

EXAMPLE 17.14

In the network in Fig. 17.16a we wish to select R to provide the desired step response of the network. Specifically, we wish to have a damping ratio of $\xi = 0.707$. We will find the value of R to provide this response, plot the pole–zero pattern for the network transfer function, and determine $v_o(t)$ for $t > 0$ if $v_i(t) = u(t)$ V.

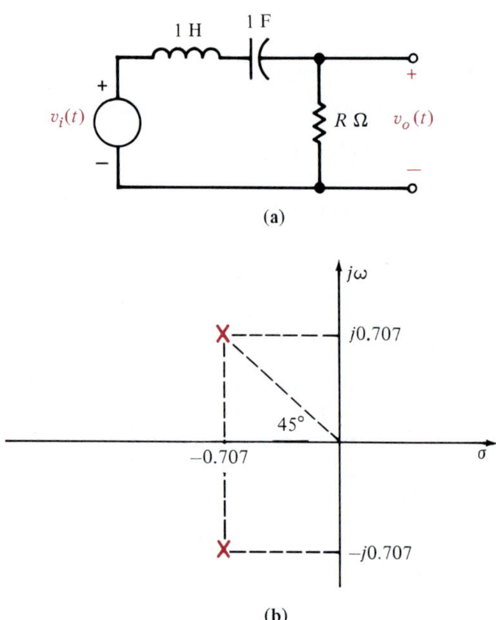

Figure 17.16 Network and pole locations for the problem in Example 17.14.

The network transfer function is

$$\frac{\mathbf{V}_o(s)}{\mathbf{V}_i(s)} = \frac{Rs}{s^2 + Rs + 1}$$

The characteristic equation is of the form $s^2 + 2\zeta\omega_0 s + \omega_0^2$, where $\omega_0 = 1$ and $2\zeta\omega_0 = R$. For $\zeta = 0.707$, $R = 1.414$. The pole–zero plot for the transfer function is shown in Fig. 17.16b.

The output $\mathbf{V}_o(s)$ for a unit step input is

$$\mathbf{V}_o(s) = \frac{1.414s}{s(s + 0.707 - j0.707)(s + 0.707 + j0.707)}$$

$$= \frac{K_1}{s + 0.707 - j0.707} + \frac{K_1^*}{s + 0.707 + j0.707}$$

$$= \frac{1 \angle -90^\circ}{s + 0.707 - j0.707} + \frac{1 \angle +90^\circ}{s + 0.707 + j0.707}$$

Hence

$$v_o(t) = [2e^{-0.707t} \cos (0.707t - 90^\circ)]u(t) \text{ V}$$

■

17.4 Steady-State Response

In Section 17.2 we have demonstrated, using a variety of examples, the power of the Laplace transform technique in determining the complete response of a network. This complete response is composed of transient terms which disappear as $t \to \infty$ and steady-state terms which are present at all times. Let us now examine a method by which to determine the steady-state response of a network directly. Recall from previous examples that the network response can be written as

$$\mathbf{Y}(s) = \mathbf{H}(s)\mathbf{X}(s) \tag{17.16}$$

where $\mathbf{Y}(s)$ is the output or response, $\mathbf{X}(s)$ is the input or forcing function, and $\mathbf{H}(s)$ is the network function or transfer function defined in Section 14.1. The transient portion of the response $\mathbf{Y}(s)$ results from the poles of $\mathbf{H}(s)$, and the steady-state portion of the response results from the poles of the input or forcing function.

As a direct parallel to the sinusoidal response of a network as outlined in Section 9.2, we assume that the forcing function is of the form

$$x(t) = X_M e^{j\omega_0 t} \tag{17.17}$$

which by Euler's identity can be written as

$$x(t) = X_M \cos \omega_0 t + jX_M \sin \omega_0 t \tag{17.18}$$

The Laplace transform of Eq. (17.17) is

$$\mathbf{X}(s) = \frac{X_M}{s - j\omega_0} \tag{17.19}$$

and therefore,

$$\mathbf{Y}(s) = \mathbf{H}(s)\left(\frac{X_M}{s - j\omega_0}\right) \tag{17.20}$$

At this point we tacitly assume that $\mathbf{H}(s)$ does not have any poles of the form $(s - j\omega_k)$. If, however, this is the case, we simply encounter difficulty in defining the steady-state response.

Performing a partial-fraction expansion of Eq. (17.20) yields

$$\mathbf{Y}(s) = \frac{X_M \mathbf{H}(j\omega_0)}{s - j\omega_0} + \text{terms that occur due to the poles of } \mathbf{H}(s) \tag{17.21}$$

The first term to the right of the equal sign can be expressed as

$$\mathbf{Y}(s) = \frac{X_M |\mathbf{H}(j\omega_0)| e^{j\phi(j\omega_0)}}{s - j\omega_0} + \cdots \tag{17.22}$$

since $\mathbf{H}(j\omega_0)$ is a complex quantity with a magnitude and phase that are a function of $j\omega_0$.

Performing the inverse transform of Eq. (17.22), we obtain

$$\begin{aligned} y(t) &= X_M |\mathbf{H}(j\omega_0)| e^{j\omega_0 t} e^{j\phi(j\omega_0)} + \cdots \\ &= X_M |\mathbf{H}(j\omega_0)| e^{j(\omega_0 t + \phi(j\omega_0))} + \cdots \end{aligned} \tag{17.23}$$

and hence the steady-state response is

$$y_{ss}(t) = X_M |\mathbf{H}(j\omega_0)| e^{j(\omega_0 t + \phi(j\omega_0))} \tag{17.24}$$

Since the actual forcing function is $X_M \cos \omega_0 t$, which is the real part of $X_M e^{j\omega_0 t}$, the steady-state response is the real part of Eq. (17.24).

$$y_{ss}(t) = X_M |\mathbf{H}(j\omega_0)| \cos [\omega_0 t + \phi(j\omega_0)] \tag{17.25}$$

In general, the forcing function may have a phase angle θ. In this case, θ is simply added to $\phi(j\omega_0)$ so that the resultant phase of the response is $\phi(j\omega_0) + \theta$.

EXAMPLE 17.15

For the circuit shown in Fig. 17.17a, we wish to determine the steady-state voltage $v_{oss}(t)$ for $t > 0$ if the initial conditions are zero.

As illustrated earlier, this problem could be solved using a variety of techniques, such as node equations, mesh equations, source transformation, and Thévenin's theorem. We will employ node equations to obtain the solution. The transformed network using the impedance values for the parameters is shown in Fig. 17.17b. The node equations for this network are

$$\left(\frac{1}{2} + \frac{1}{s} + \frac{s}{2}\right)\mathbf{V}_1(s) - \left(\frac{s}{2}\right)\mathbf{V}_o(s) = \frac{1}{2}\mathbf{V}_i(s)$$

$$-\left(\frac{s}{2}\right)\mathbf{V}_1(s) + \left(\frac{s}{2} + 1\right)\mathbf{V}_o(s) = 0$$

Solving these equations for $\mathbf{V}_o(s)$, we obtain

$$\mathbf{V}_o(s) = \frac{s^2}{3s^2 + 4s + 4}\mathbf{V}_i(s)$$

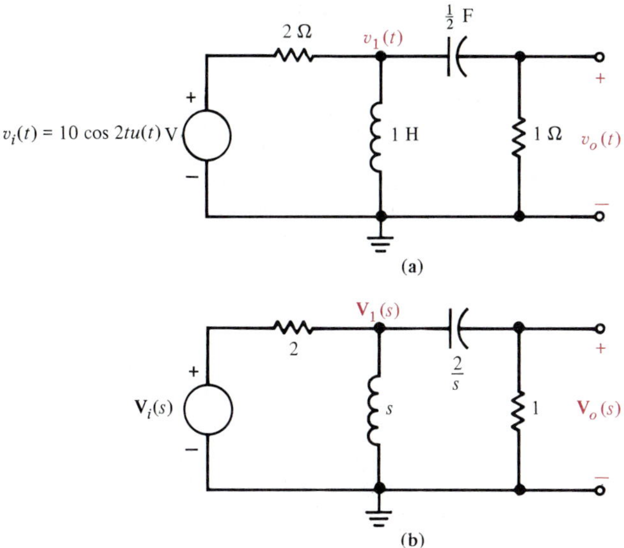

Figure 17.17 Circuit used in Example 17.15.

Note that this equation is in the form of Eq. (17.16), where $\mathbf{H}(s)$ is

$$\mathbf{H}(s) = \frac{s^2}{3s^2 + 4s + 4}$$

Since the forcing function is $10 \cos 2t\, u(t)$, then $V_M = 10$ and $\omega_0 = 2$. Hence

$$\begin{aligned}\mathbf{H}(j2) &= \frac{(j2)^2}{3(j2)^2 + 4(j2) + 4} \\ &= 0.354 \angle 45^\circ\end{aligned}$$

Therefore,

$$\begin{aligned}|\mathbf{H}(j2)| &= 0.354 \\ \phi(j2) &= 45^\circ\end{aligned}$$

and hence the steady-state response is

$$\begin{aligned}v_{oss}(t) &= V_M|\mathbf{H}(j2)| \cos [2t + \phi(j2)] \\ &= 3.54 \cos (2t + 45^\circ)\end{aligned}$$

The complete (transient plus steady-state) response can be obtained from the expression

$$\begin{aligned}\mathbf{V}_o(s) &= \frac{s^2}{3s^2 + 4s + 4}\mathbf{V}_i(s) \\ &= \frac{s^2}{3s^2 + 4s + 4}\left(\frac{10s}{s^2 + 4}\right) \\ &= \frac{10s^3}{(s^2 + 4)(3s^2 + 4s + 4)}\end{aligned}$$

Determining the inverse Laplace transform of this function using the techniques of Chapter 16, we obtain

$$v_o(t) = 3.54 \cos(2t + 45°) + 1.44e^{-(2/3)t} \cos\left(\frac{2\sqrt{2}}{3}t - 55°\right) \text{ V}$$

Note that as $t \to \infty$ the second term approaches zero, and thus the steady-state response is

$$v_{oss}(t) = 3.54 \cos(2t + 45°) \text{ V}$$

which can easily be checked using a phasor analysis. ■

DRILL EXERCISE

D17.13. Determine the steady-state voltage $v_{oss}(t)$ in the network in Fig. D17.13 for $t > 0$ if the initial conditions in the network are zero.

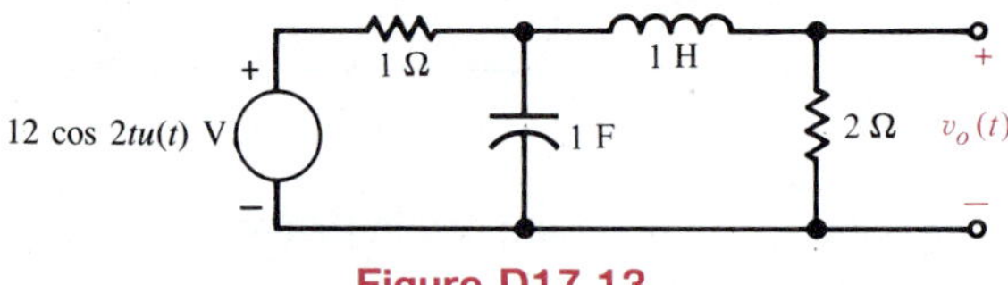

Figure D17.13

Ans: $v_{oss}(t) = 3.95 \cos(2t - 99.46°)$ V.

D17.14. Determine the steady-state voltage $v_{oss}(t)$ in the network in Fig. D17.14 for $t > 0$ if all initial conditions are zero.

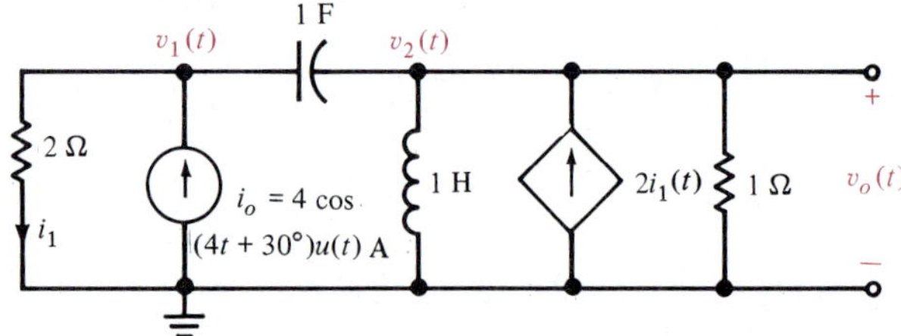

Figure D17.14

Ans: $v_{oss}(t) = 6.86 \cos(4t + 54.66°)$ V.

17.5 Summary

In this chapter we have applied the power of the Laplace transform to circuit analysis. We have demonstrated how to describe the circuit elements in the s-domain, and we have used all the network analysis techniques and network theorems derived in dc analysis to obtain a solution. Both the steady-state and transient analysis have been performed. The ease with which one can determine the response of a network to complex inputs using the Laplace transform was shown.

KEY POINTS

- Circuit elements and their initial conditions can be represented in the s-domain by an equivalent circuit.
- All the dc network analysis techniques can be applied to networks represented in the s-domain.
- The response of a network obtained using the Laplace transform is the complete response (i.e., transient plus steady state).
- The steady-state response of a network can be obtained via an evaluation of the network function at the forcing function frequency.
- Once the initial conditions in a network are found, the Laplace transform yields the transient response of the network by solving a set of algebraic equations in the s-domain and then determining the inverse Laplace transform of the desired variable.
- In a network, the roots of the characteristic equation determine the network response.
- By representing complex waveforms using the Laplace transform, the response of a network to complex inputs can be obtained in a straightforward manner.
- The inverse Laplace transform of the network transfer function is the impulse response of the network.

PROBLEMS

17.1. Find the input impedance $\mathbf{Z}(s)$ of the network in Fig. P17.1.

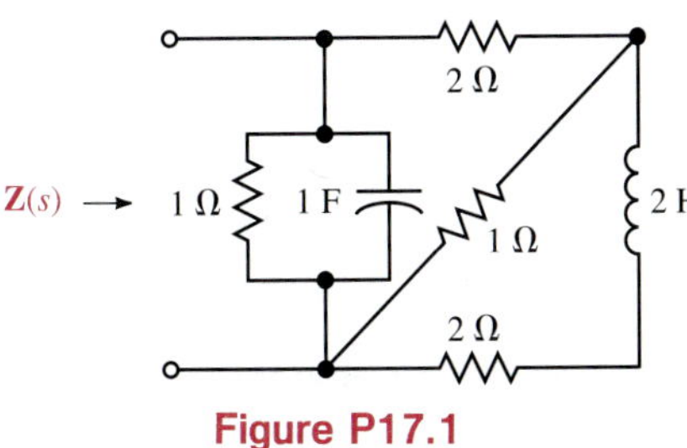

Figure P17.1

17.2. Find the input impedance $\mathbf{Z}(s)$ of the network in Fig. P17.2 (a) when the terminals B-B' are open circuited and (b) when the terminals B-B' are short circuited.

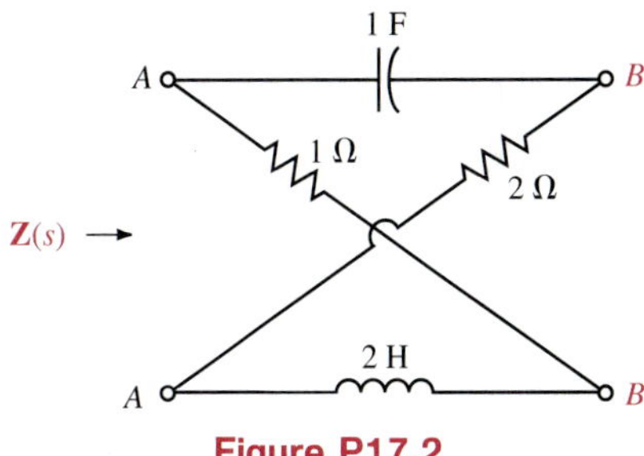

Figure P17.2

17.3. Given the network in Fig. P17.3, determine the value of the output voltage $v_o(t)$ as $t \rightarrow \infty$.

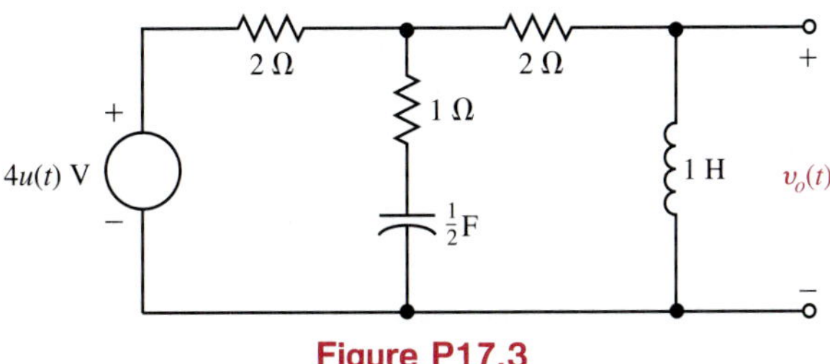

Figure P17.3

17.4. Given the network in Fig. P17.4, determine the value of the output voltage $v_o(t)$ as $t \to \infty$.

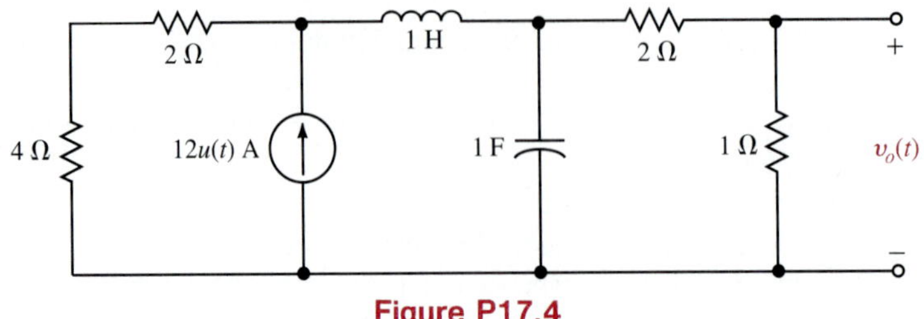

Figure P17.4

17.5. For the network shown in Fig. P17.5, find $v_o(t)$, $t > 0$, using node equations.

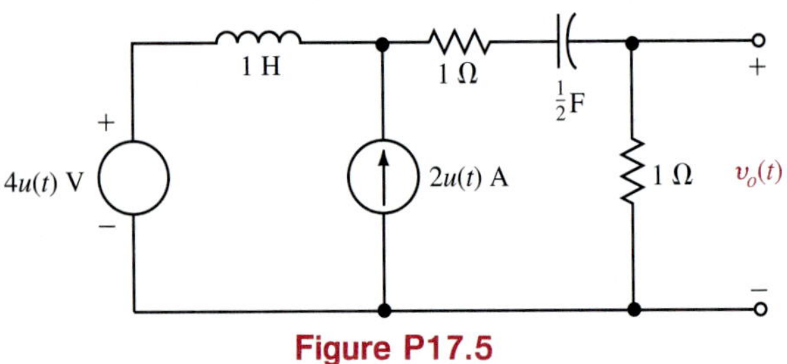

Figure P17.5

17.6. Use loop equations to solve Problem 17.5.

17.7. Use superposition to solve Problem 17.5.

17.8. Use source transformation to solve Problem 17.5.

17.9. Use Thévenin's theorem to solve Problem 17.5.

17.10. Use mesh equations to find $v_o(t)$, $t > 0$, in the network in Fig. P17.10.

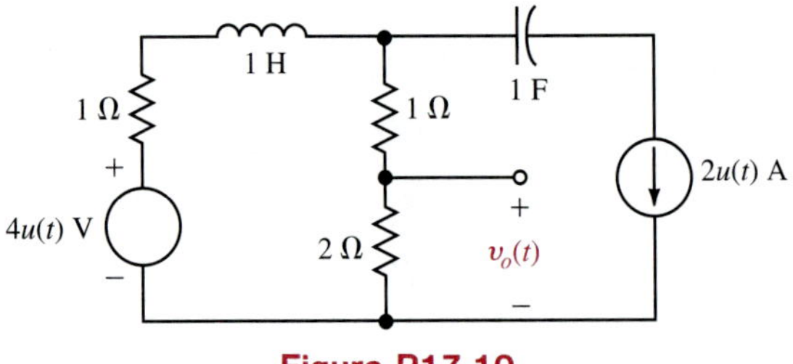

Figure P17.10

17.11. Use Thévenin's theorem to solve Problem 17.10.

17.12. For the network shown in Fig. P17.12, find $v_o(t)$, $t > 0$, using mesh equations.

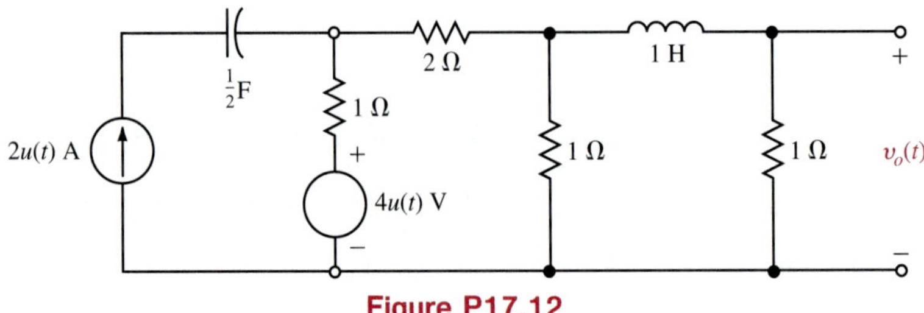

Figure P17.12

17.13. Use Thévenin's theorem to solve Problem 17.12.

17.14. Given the network in Fig. P17.14, find $i_o(t)$, $t > 0$, using mesh equations.

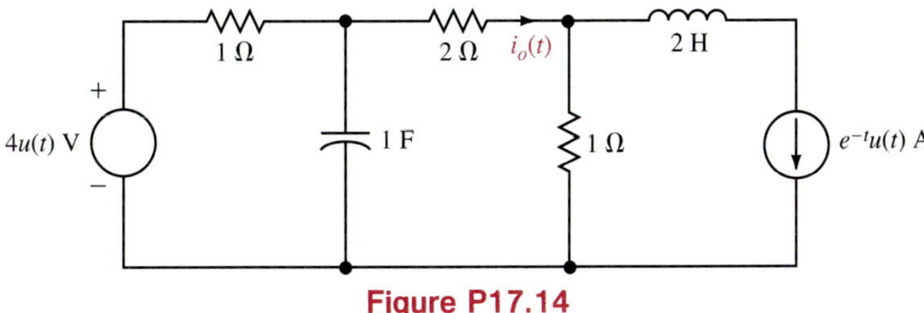

Figure P17.14

17.15. Use Thévenin's theorem to solve Problem 17.14.

17.16. For the network shown in Fig. P17.16, find $v_o(t)$, $t > 0$, using node equations.

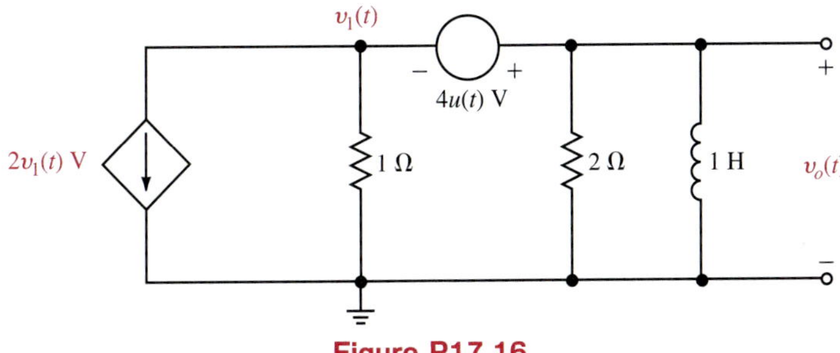

Figure P17.16

17.17. Use Thévenin's theorem to solve Problem 17.16.

17.18. For the network shown in Fig. P17.18, find $v_o(t)$, $t > 0$.

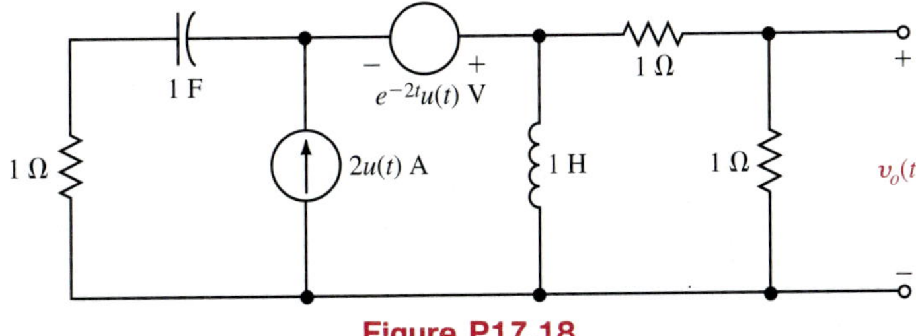

Figure P17.18

17.19. For the network shown in Fig. P17.19, find $i_o(t)$, $t > 0$.

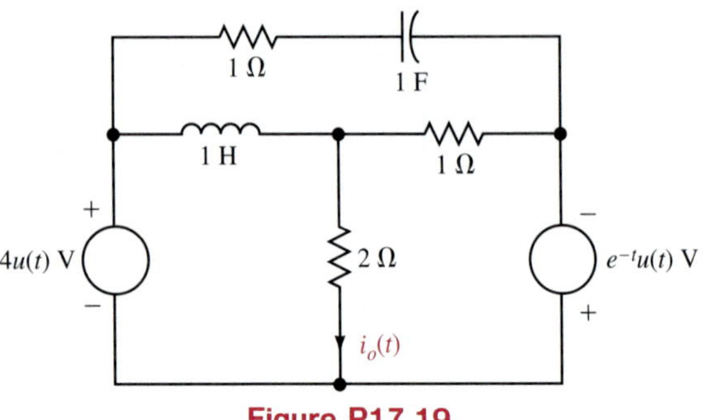

Figure P17.19

17.20. Use Thévenin's theorem to find $v_o(t)$, $t > 0$, in the network in Fig. P17.20.

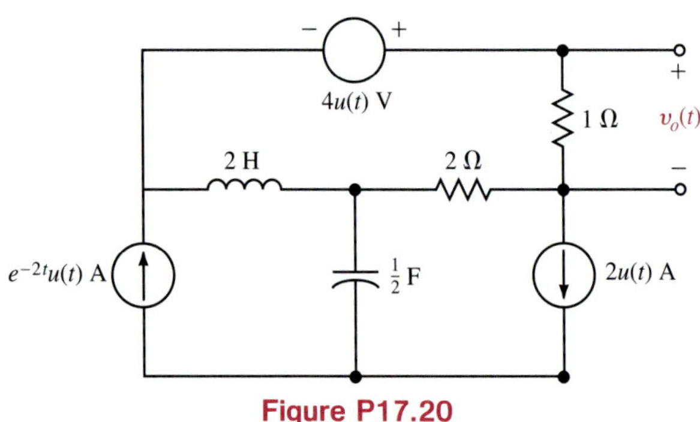

Figure P17.20

17.21. Use mesh analysis to find $v_o(t)$, $t > 0$, in the network in Fig. P17.21.

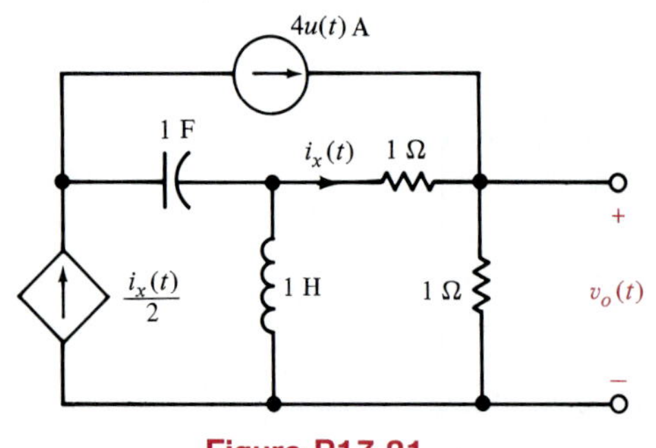

Figure P17.21

17.22. Use Thévenin's theorem to find $v_o(t)$, $t > 0$, in the network shown in Fig. P17.22.

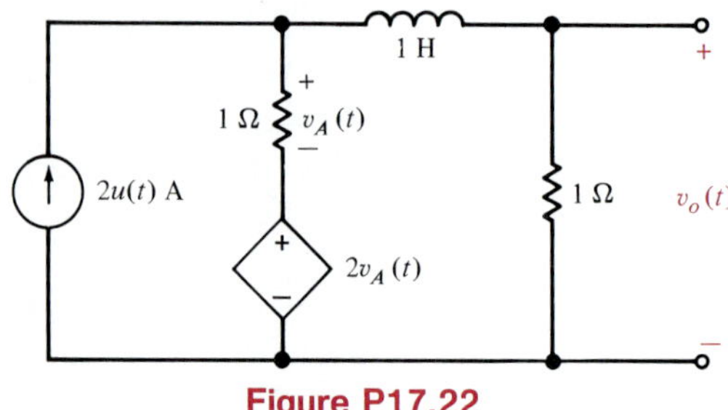

Figure P17.22

17.23. Find $v_o(t)$, $t > 0$, in the network in Fig. P17.23 using nodal analysis.

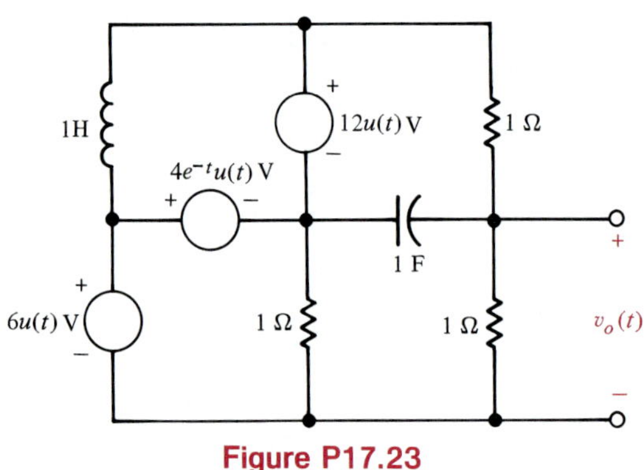

Figure P17.23

17.24. Use loop analysis to find $v_o(t)$ for $t > 0$ in the network in Fig. P17.24.

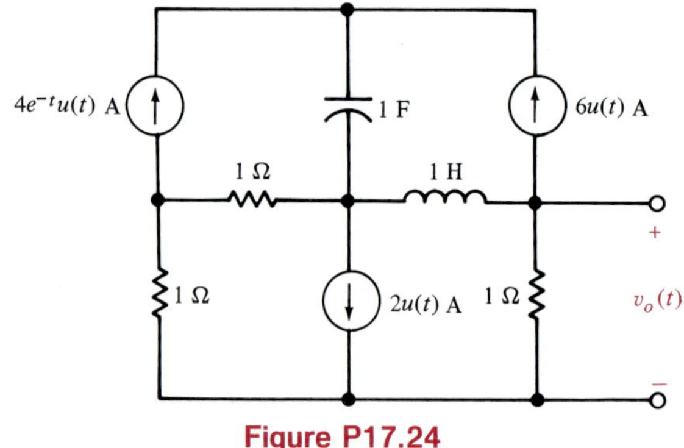

Figure P17.24

17.25. Find $v_o(t)$, $t > 0$, in the network in Fig. P17.25.

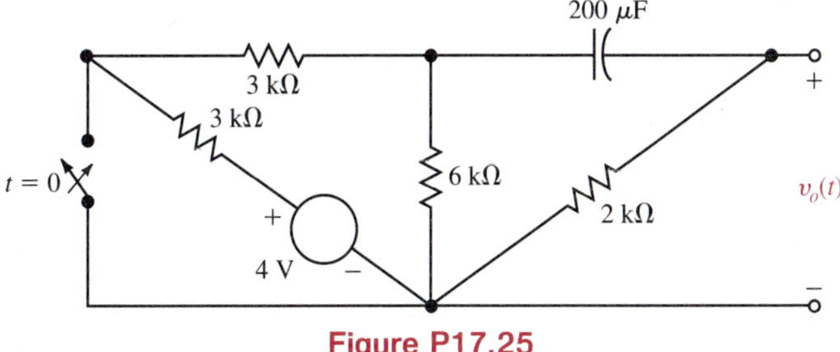

Figure P17.25

17.26. Find $i_o(t)$ for $t > 0$, in the network shown in Fig. P17.26.

17.27. Find $i_o(t)$, $t > 0$, in the network in Fig. P17.27.

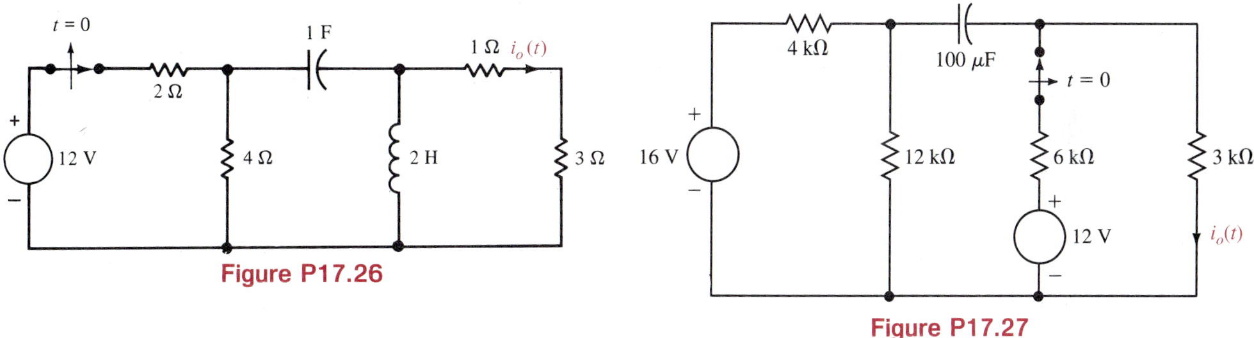

Figure P17.26

Figure P17.27

17.28. Find $i_o(t)$, $t > 0$, in the network in Fig. P17.28.

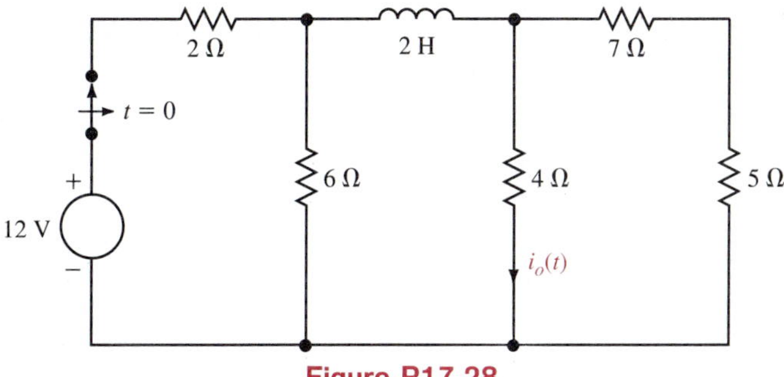

Figure P17.28

17.29. Find $v_o(t)$, $t > 0$, in the circuit in Fig. P17.29.

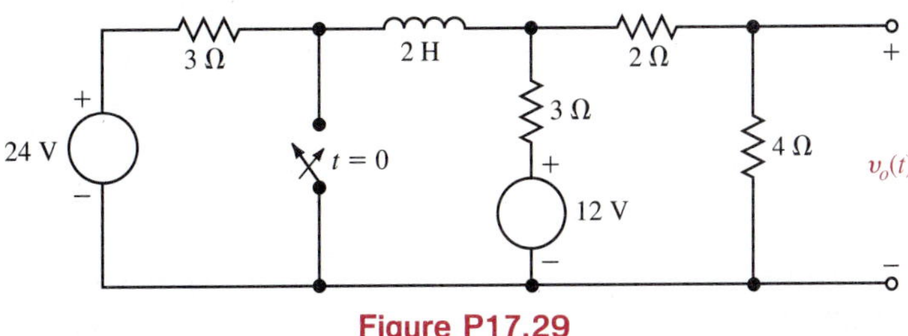

Figure P17.29

17.30. Find $v_o(t)$ for $t > 0$ in the network in Fig. P17.30.

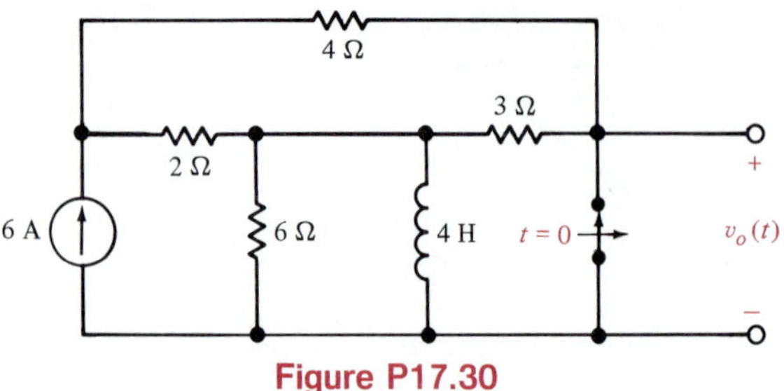

Figure P17.30

17.31. Find $v_o(t)$ for $t > 0$ in the network in Fig. P17.31.

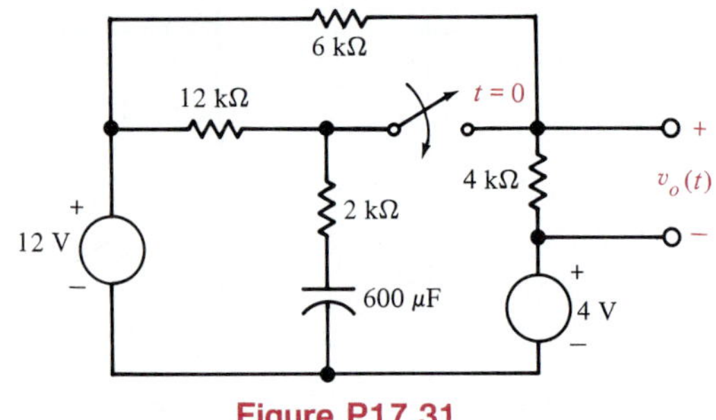

Figure P17.31

17.32. Find $v_o(t)$ for $t > 0$ in the network in Fig. P17.32.

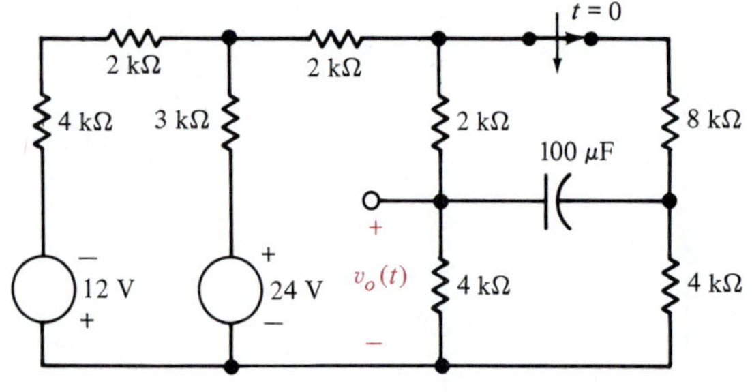

Figure P17.32

17.33. Find the transmission parameters for the network in Fig. P17.33.

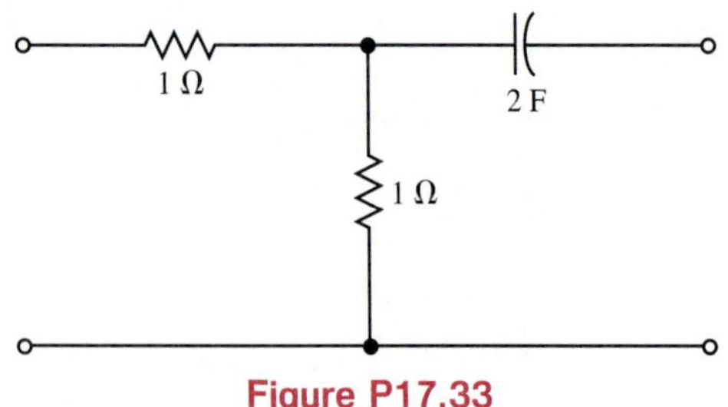

Figure P17.33

17.34. Find the Z parameters of the two-port network in Fig. P17.34a. Using these parameters, determine $i_2(t)$ in the network in Fig. P17.34b.

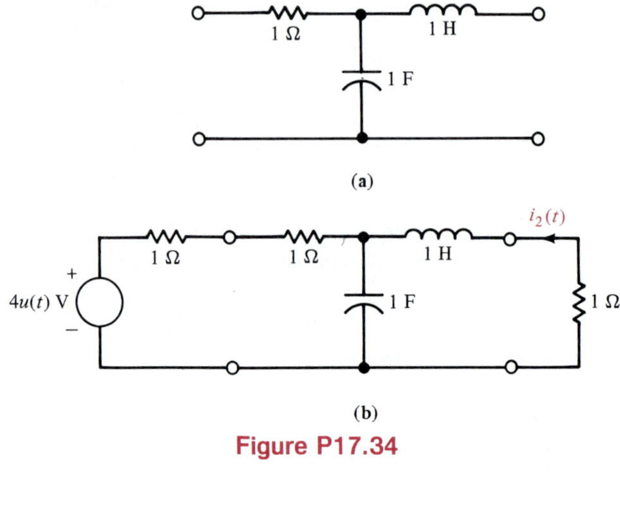

Figure P17.34

17.35. Find the transmission parameters of the network in Fig. P17.35.

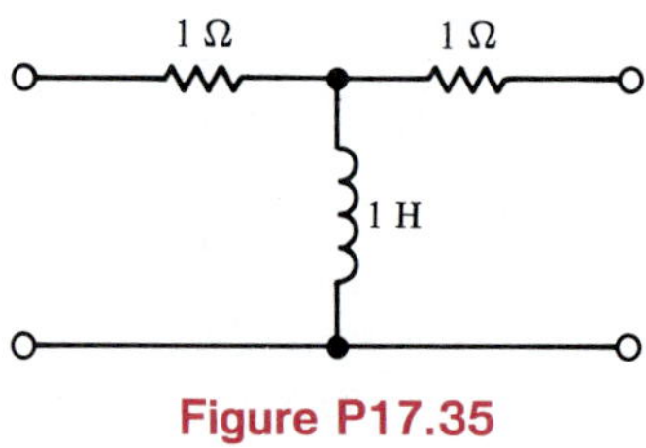

Figure P17.35

17.36. Using the results of Problem 17.35, find the transmission parameters of the network in Fig. P17.36.

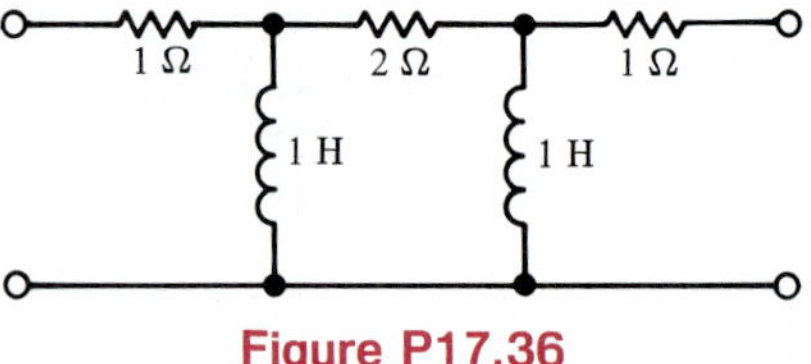

Figure P17.36

17.37. Find $v_o(t)$ for $t > 0$ in the network in Fig. P17.37.

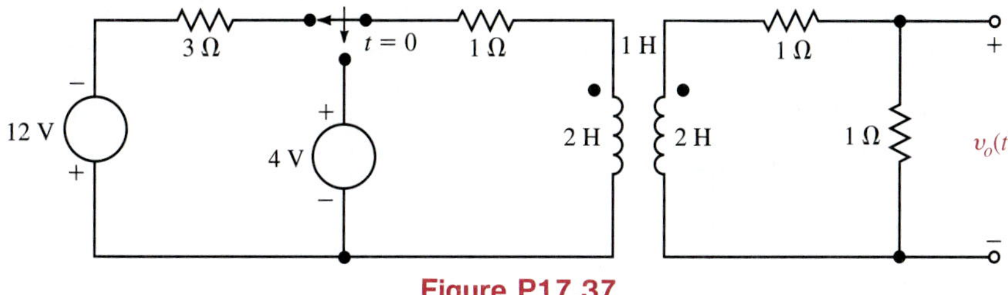

Figure P17.37

17.38. Find $v_o(t)$ for $t > 0$ in the network in Fig. P17.38.

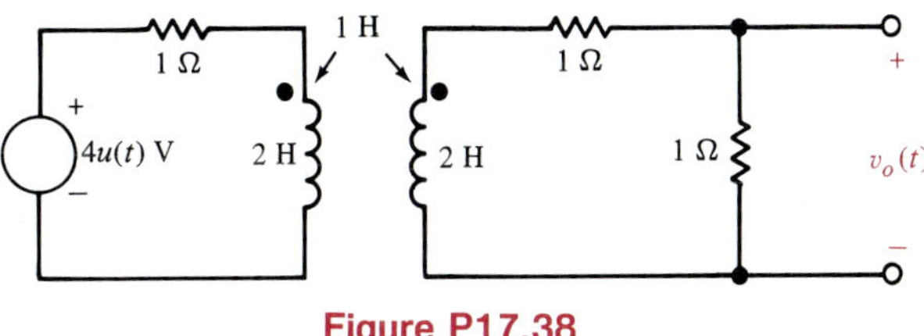

Figure P17.38

17.39. Find $v_o(t)$ for $t > 0$ in the network in Fig. P17.39.

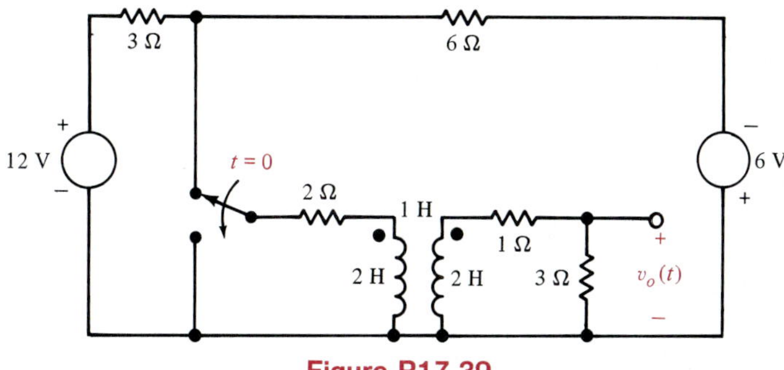

Figure P17.39

17.40. Find $i_o(t)$ for $t > 0$ in the network in Fig. P17.40.

17.41. Find $v_o(t)$ for $t > 0$ in the network in Fig. P17.41.

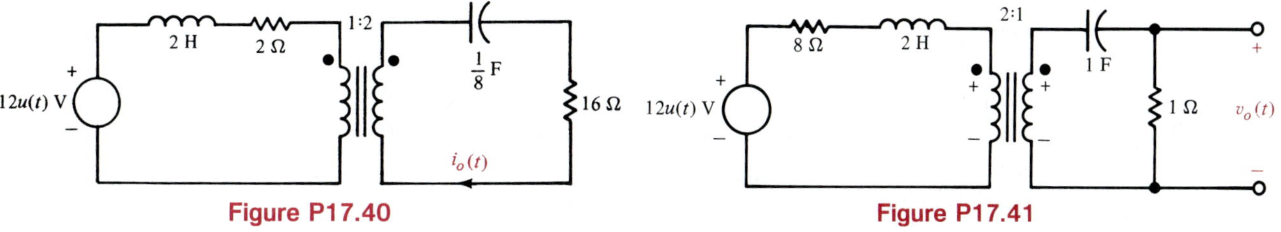

Figure P17.40

Figure P17.41

17.42. Determine the initial and final values of the voltage $v_o(t)$ in the network in Fig. P17.42.

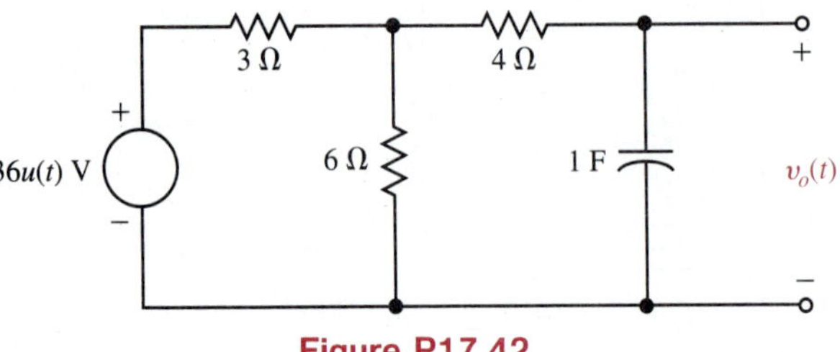

Figure P17.42

17.43. Determine the initial and final values of the voltage $v_o(t)$ in the network in Fig. P17.43.

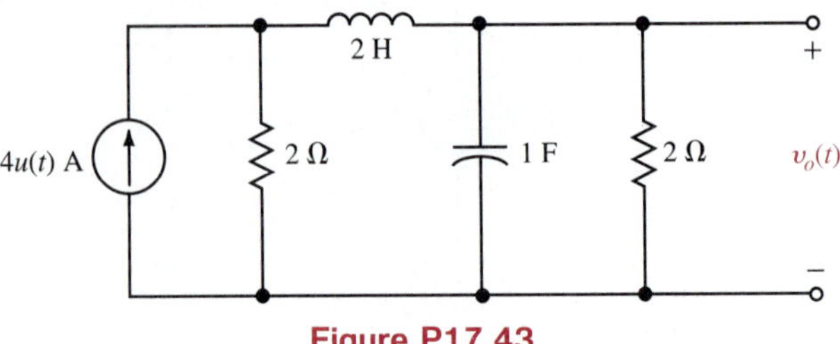

Figure P17.43

17.44. Find the initial and final values of the voltage $v_o(t)$ in the network in Fig. P17.44.

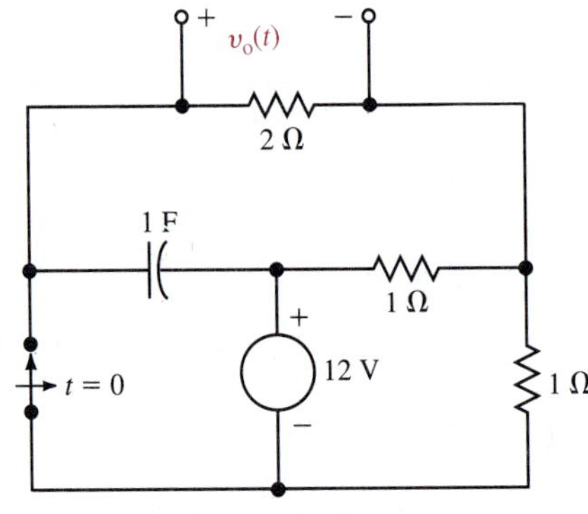

Figure P17.44

17.45. Find the output voltage $v_o(t)$, $t > 0$, in the network in Fig. 17.45a if the input is represented by the waveform shown in Fig. P17.45b.

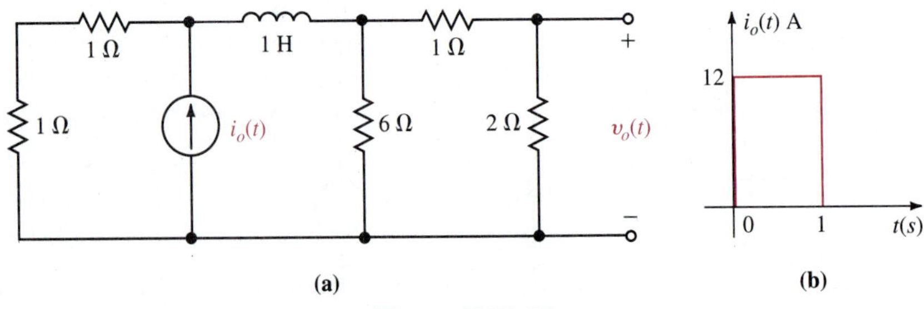

Figure P17.45

17.46. Solve Problem 17.45 if the input is as shown in Fig. P17.46.

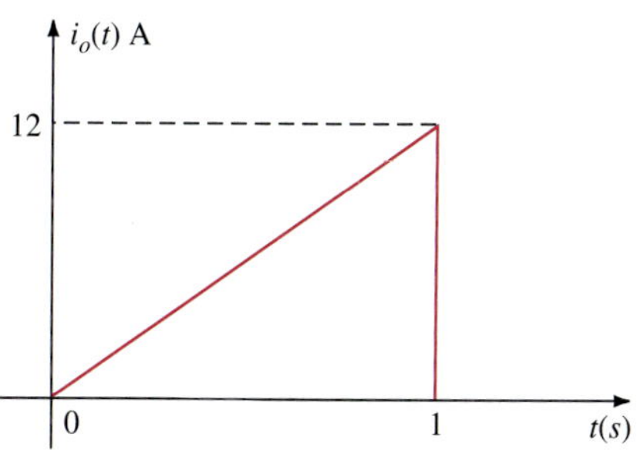

Figure P17.46

17.47. Solve Problem 17.45 if the input is as shown in Fig. P17.47.

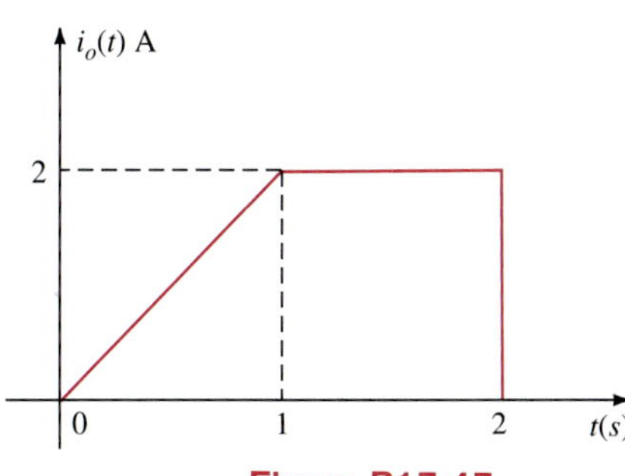

Figure P17.47

17.48. Find the transfer function for the network shown in Fig. P17.48.

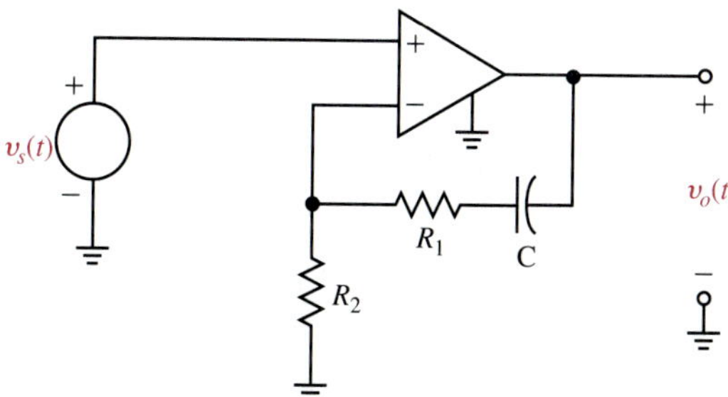

Figure P17.48

17.49. Find the transfer function for the network in Fig. P17.49.

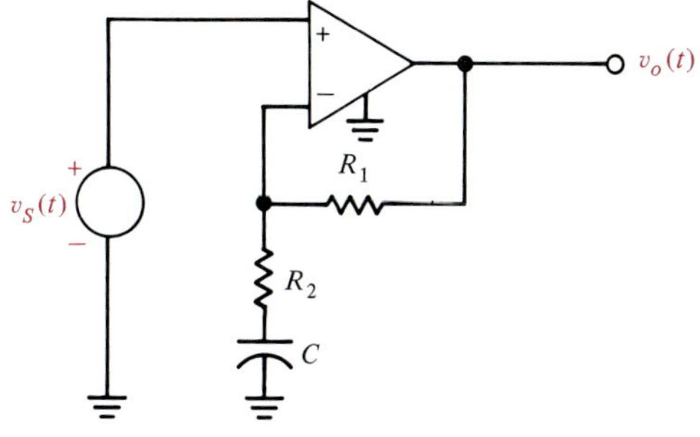

Figure P17.49

17.50. Determine the transfer function of the network in Fig. P17.50. If a step function is applied to the network, what type of damping will the network exhibit?

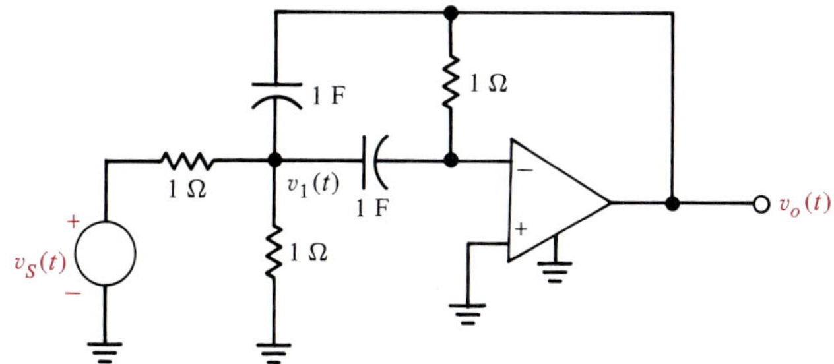

Figure P17.50

17.51. Find the transfer function for the network in Fig. P17.51. If a step function is applied to the network, will the response be overdamped, underdamped, or critically damped?

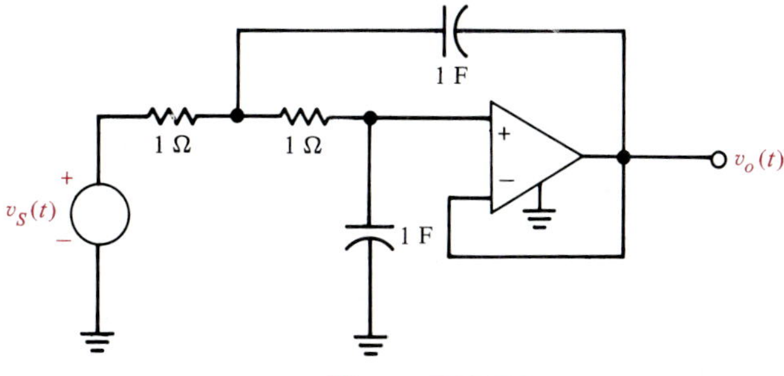

Figure P17.51

17.52. Find the transfer function for the network in Fig. P17.52.

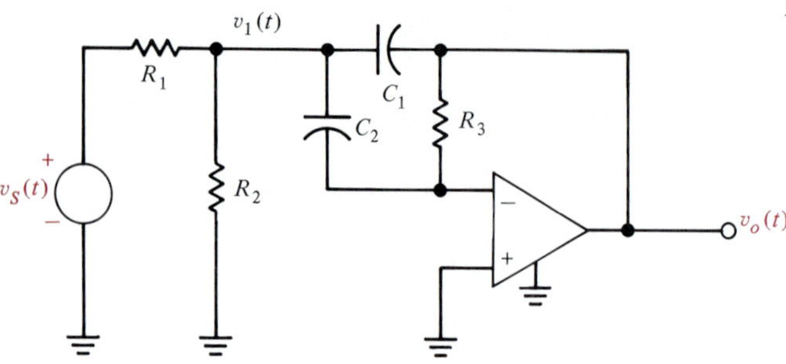

Figure P17.52

17.53. Find the steady-state response $v_o(t)$ for the network in Fig. P17.53.

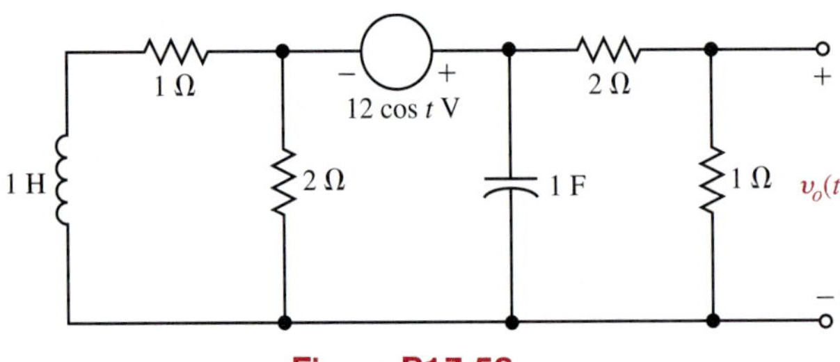

Figure P17.53

17.54. Determine the steady-state response $v_o(t)$ for the network in Fig. P17.54.

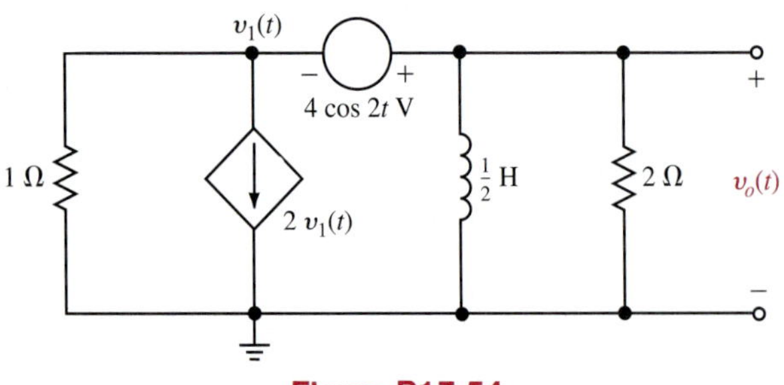

Figure P17.54

17.55. Determine the steady-state response $i_o(t)$ for the network in Fig. P17.55.

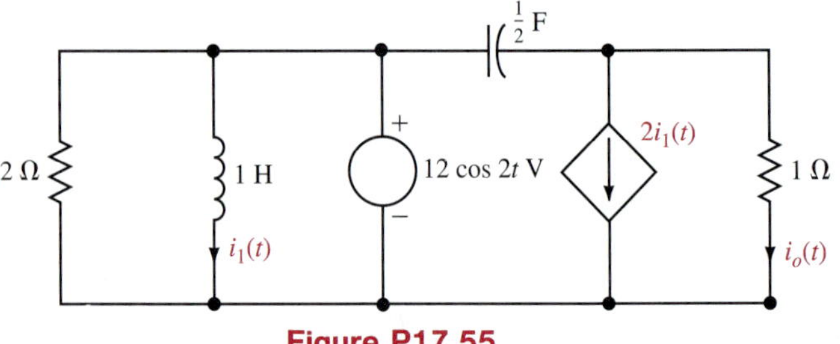

Figure P17.55

17.56. Find the steady-state response $v_o(t)$ for $t > 0$ in the network in Fig. P17.56.

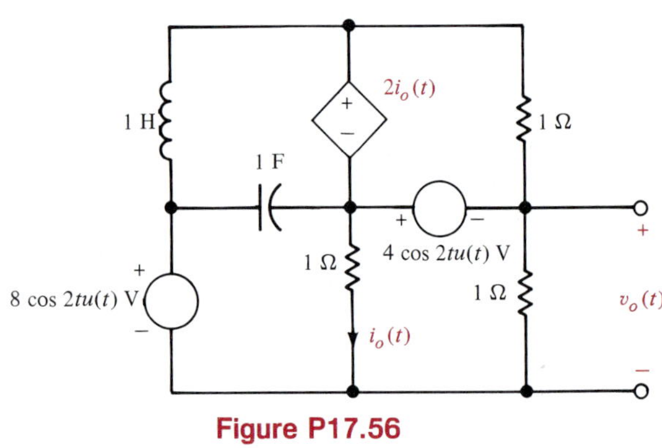

Figure P17.56

17.57. If the input to a network is $x(t) = u(t)$, and the network transfer function is $h(t) = e^{-t} - e^{-2t}$, find the network output $y(t)$.

17.58. The response of a network to the input $x(t) = [e^{-2t} - e^{-6t}]u(t)$ is $y(t) = 4[1 - te^{-t} - e^{-t}]u(t)$. Determine the impulse response of the network.

17.59. If the input to a network is $x_1(t) = e^{-t}u(t)$, the network response is $y_1(t) = [t - 1 + e^{-t}]u(t)$. Find the network response $y_2(t)$ if the input is $x_2(t) = [e^{-t} - e^{-2t}]u(t)$.

17.60. The voltage response of the network to a unit step input is

$$\mathbf{V}_o(s) = \frac{2(s+1)}{s(s^2 + 12s + 27)}$$

Is the response overdamped?

17.61. The current response of a network to a unit step input is

$$\mathbf{I}_o(s) = \frac{10(s+2)}{s(s^2+11s+30)}$$

Is the response underdamped?

17.62. The voltage response of a network to a unit step input is

$$\mathbf{V}_o(s) = \frac{10}{s(s^2+8s+16)}$$

Is the response critically damped?

CHAPTER

18

Fourier Analysis Techniques

In this chapter we examine two very important topics: the Fourier series and the Fourier transform. These two techniques vastly expand our circuit analysis capabilities because they provide a means of effectively dealing with nonsinusoidal periodic signals and aperiodic signals. Using the Fourier series, we show that we can determine the steady-state response of a network to a nonsinusoidal periodic input. The Fourier transform will allow us to analyze circuits with aperiodic inputs by transforming the problem to the frequency domain, solving it algebraically, and then transforming back to the time domain in a manner similar to that used with Laplace transforms.

18.1 Fourier Series

A periodic function is one that satisfies the relationship

$$f(t) = f(t + nT_0), \qquad n = \pm 1, \pm 2, \pm 3, \ldots$$

for every value of t where T_0 is the period. As we have shown in previous chapters, the sinusoidal function is a very important periodic function. However, there are many other periodic functions that have wide applications. For example, laboratory signal generators produce the pulse-train and square-wave signals shown in Fig. 18.1a and b, respectively,

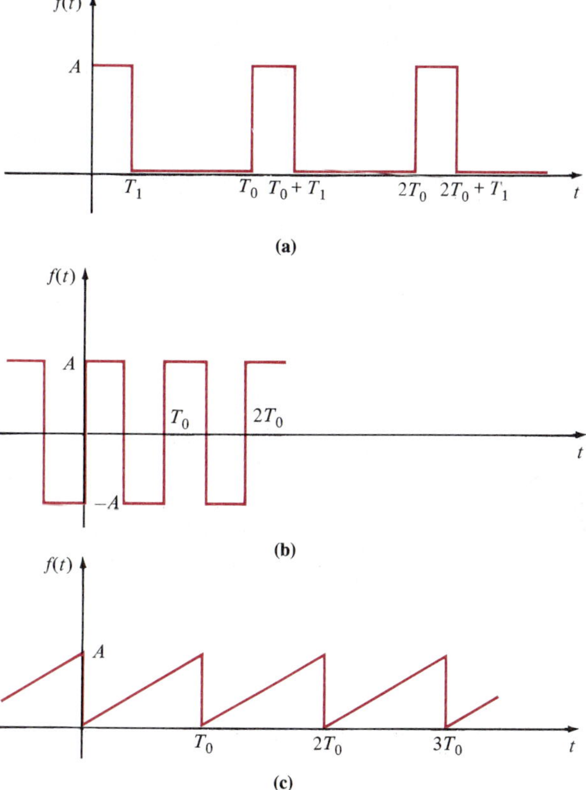

Figure 18.1 Some useful periodic signals.

which are used for testing circuits. The oscilloscope is another laboratory instrument and the sweep of its electron beam across the face of the cathode ray tube is controlled by a triangular signal of the form shown in Fig. 18.1c.

The techniques we will explore are based on the work of Jean Baptiste Joseph Fourier. Although our analyses will be confined to electric circuits, it is important to point out that the techniques are applicable to a wide range of engineering problems. In fact, it was Fourier's work in heat flow that led to the techniques which will be presented here.

In his work, Fourier demonstrated that a periodic function $f(t)$ could be expressed as a sum of sinusoidal functions. Therefore, given this fact and the fact that if a periodic function is expressed as a sum of linearly independent functions, each function in the sum must be periodic with the same period, and the function $f(t)$ can be expressed in the form

$$f(t) = a_0 + \sum_{n=1}^{\infty} D_n \cos(n\omega_0 t + \theta_n) \tag{18.1}$$

where $\omega_0 = 2\pi/T_0$ and a_0 is the average value of the waveform. An examination of this expression illustrates that all sinusoidal waveforms that are periodic with period T_0 have been included. For example, for $n = 1$, one cycle covers T_0 seconds and $D_1 \cos(\omega_0 t + \theta_1)$ is called the *fundamental*. For $n = 2$, two cycles fall within T_0 seconds and the term $D_2 \cos(2\omega_0 t + \theta_2)$ is called the *second harmonic*. In general, for $n = k$, k cycles fall within T_0 seconds and $D_k \cos(k\omega_0 t + \theta_k)$ is the *kth harmonic term*.

Since the function $\cos(n\omega_0 t + \theta_n)$ can be written in exponential form using Euler's identity or as a sum of cosine and sine terms of the form $\cos n\omega_0 t$ and $\sin n\omega_0 t$ as demonstrated in Chapter 9, the series in Eq. (18.1) can be written as

$$f(t) = a_0 + \sum_{\substack{n=-\infty \\ n\neq 0}}^{\infty} \mathbf{c}_n e^{jn\omega_0 t} = \sum_{n=-\infty}^{\infty} \mathbf{c}_n e^{jn\omega_0 t} \tag{18.2}$$

$$= a_0 + \sum_{n=1}^{\infty} a_n \cos n\omega_0 t + b_n \sin n\omega_0 t \tag{18.3}$$

Using the real-part relationship employed as a transformation between the time domain and the frequency domain, we can express $f(t)$ as

$$f(t) = a_0 + \sum_{n=1}^{\infty} \text{Re}\,[(D_n \underline{/\theta_n})e^{jn\omega_0 t}] \tag{18.4}$$

$$= a_0 + \sum_{n=1}^{\infty} \text{Re}\,(2\mathbf{c}_n e^{jn\omega_0 t}) \tag{18.5}$$

$$= a_0 + \sum_{n=1}^{\infty} \text{Re}\,[(a_n - jb_n)e^{jn\omega_0 t}] \tag{18.6}$$

These equations allow us to write the Fourier series in a number of equivalent forms. Note that the *phasor* for the nth harmonic is

$$D_n \underline{/\theta_n} = 2\mathbf{c}_n = a_n - jb_n \tag{18.7}$$

The approach we will take will be to represent a nonsinusoidal periodic input by a sum of complex exponential functions, which because of Euler's identity is equivalent to a sum of sines and cosines. We will then use (1) the superposition property of linear systems and (2) our knowledge that the steady-state response of a time-invariant linear system to a sinusoidal input of frequency ω_0 is a sinusoidal function of the same frequency to determine the response of such a system.

In order to illustrate the manner in which a nonsinusoidal periodic signal can be represented by a Fourier series, consider the periodic function shown in Fig. 18.2a. In Fig. 18.2b–d we can see the impact of using a specific number of terms in the series to represent the original function. Note that the series more closely represents the original function as we employ more and more terms.

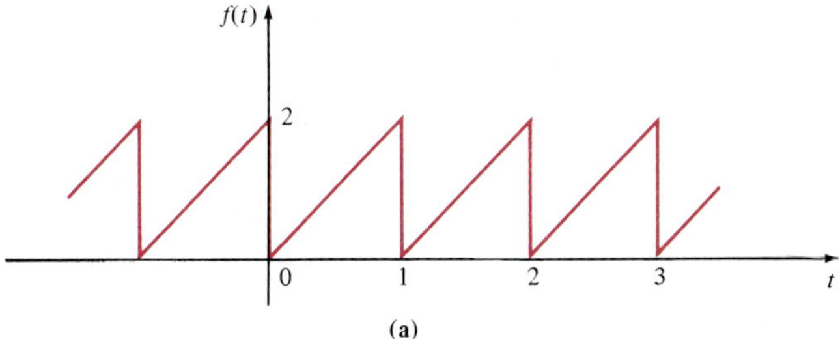

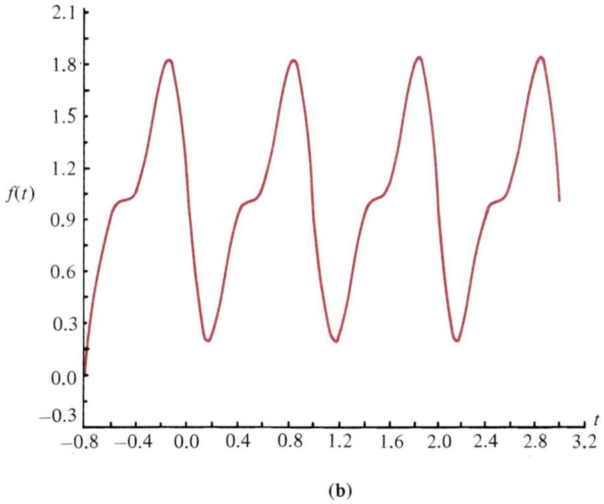

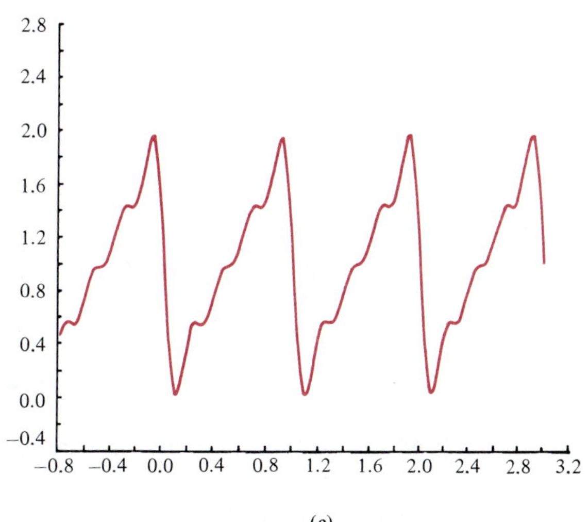

Figure 18.2 Periodic function (a) and its representation by a fixed number of Fourier series terms: (b) 2 terms; (c) 4 terms; (d) 100 terms.

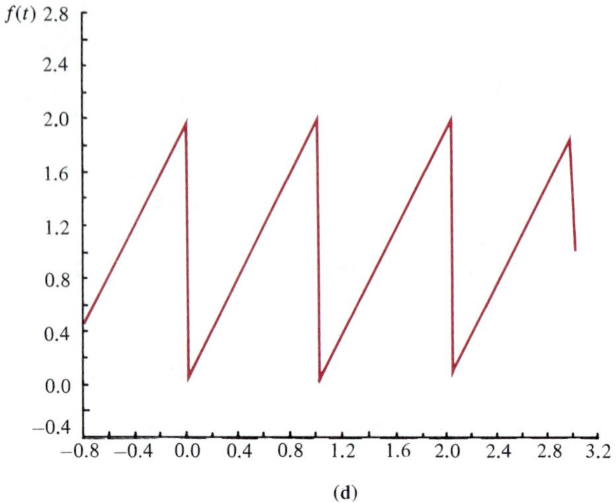

Figure 18.2 (continued)

Exponential Fourier Series

Any physically realizable periodic signal may be represented over the interval $t_1 < t < t_1 + T_0$ by the *exponential Fourier series*

$$f(t) = \sum_{n=-\infty}^{\infty} \mathbf{c}_n e^{jn\omega_0 t} \tag{18.8}$$

where the $\mathbf{c}_n$ are the complex (phasor) Fourier coefficients. These coefficients are derived as follows. Multiplying both sides of Eq. (18.8) by $e^{-jk\omega_0 t}$ and integrating over the interval t_1 to $t_1 + T_0$, we obtain

$$\int_{t_1}^{t_1+T_0} f(t)e^{-jk\omega_0 t}\,dt = \int_{t_1}^{t_1+T_0} \left(\sum_{n=-\infty}^{\infty} \mathbf{c}_n e^{jn\omega_0 t}\right) e^{-jk\omega_0 t}\,dt$$

$$= \mathbf{c}_k T_0$$

since

$$\int_{t_1}^{t_1+T_0} e^{j(n-k)\omega_0 t}\,dt = \begin{cases} 0 & \text{for } n \neq k \\ T_0 & \text{for } n = k \end{cases}$$

Therefore, the Fourier coefficients are defined by the equation

$$\mathbf{c}_n = \frac{1}{T_0}\int_{t_1}^{t_1+T_0} f(t)e^{-jn\omega_0 t}\,dt \tag{18.9}$$

The following example illustrates the manner in which we can represent a periodic signal by an exponential Fourier series.

EXAMPLE 18.1

We wish to determine the exponential Fourier series for the periodic voltage waveform shown in Fig. 18.3.

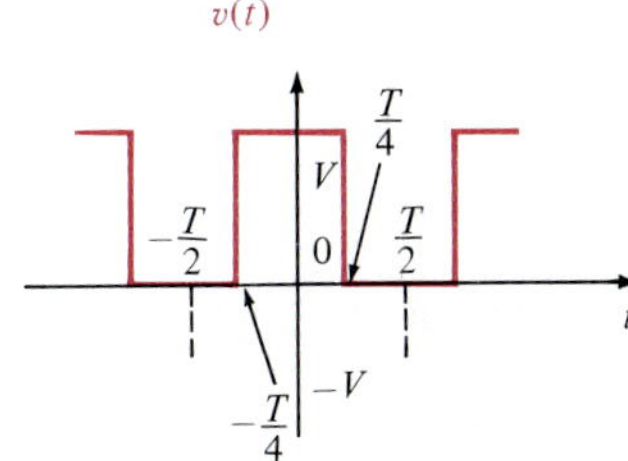

Figure 18.3 Periodic voltage waveform.

The Fourier coefficients are determined using Eq. (18.9) by integrating over one complete period of the waveform.

$$\mathbf{c}_n = \frac{1}{T}\int_{-T/2}^{T/2} f(t)e^{-jn\omega_0 t}\,dt$$

$$= \frac{1}{T}\int_{-T/2}^{-T/4} -Ve^{-jn\omega_0 t}\,dt + \int_{-T/4}^{T/4} Ve^{-jn\omega_0 t}\,dt + \int_{T/4}^{T/2} -Ve^{-jn\omega_0 t}\,dt$$

$$= \frac{V}{jn\omega_0 T}\left[+e^{-jn\omega_0 t}\Big|_{-T/2}^{-T/4} - e^{-jn\omega_0 t}\Big|_{-T/4}^{T/4} + e^{-jn\omega_0 t}\Big|_{T/4}^{T/2}\right]$$

$$= \frac{V}{jn\omega_0 T}(2e^{jn\pi/2} - 2e^{-jn\pi/2} + e^{-jn\pi} - e^{+jn\pi})$$

$$= \frac{V}{n\omega_0 T}\left[4\sin\frac{n\pi}{2} - 2\sin(n\pi)\right]$$

$$= 0 \qquad \text{for } n \text{ even}$$

$$= \frac{2V}{n\pi}\sin\frac{n\pi}{2} \qquad \text{for } n \text{ odd}$$

Since $\mathbf{c}_0$ corresponds to the average value of the waveform, $\mathbf{c}_0 = 0$. This term can also be evaluated using the original equation for $\mathbf{c}_n$. Therefore,

$$v(t) = \sum_{\substack{n=-\infty \\ n\neq 0 \\ n \text{ odd}}}^{\infty} \frac{2V}{n\pi}\sin\frac{n\pi}{2}\,e^{jn\omega_0 t}$$

This equation can be written as

$$v(t) = \sum_{\substack{n=1\\ n\text{ odd}}}^{\infty} \frac{2V}{n\pi} \sin\frac{n\pi}{2} e^{jn\omega_0 t} + \sum_{\substack{n=-1\\ n\text{ odd}}}^{-\infty} \frac{2V}{n\pi} \sin\frac{n\pi}{2} e^{jn\omega_0 t}$$

$$= \sum_{\substack{n=1\\ n\text{ odd}}}^{\infty} \left(\frac{2V}{n\pi} \sin\frac{n\pi}{2}\right) e^{jn\omega_0 t} + \left(\frac{2V}{n\pi} \sin\frac{n\pi}{2}\right)^* e^{-jn\omega_0 t}$$

Since a number plus its complex conjugate is equal to two times the real part of the number, $v(t)$ can be written as

$$v(t) = \sum_{\substack{n=1\\ n\text{ odd}}}^{\infty} 2\,\text{Re}\left(\frac{2V}{n\pi} \sin\frac{n\pi}{2} e^{jn\omega_0 t}\right)$$

or

$$v(t) = \sum_{\substack{n=1\\ n\text{ odd}}}^{\infty} \frac{4V}{n\pi} \sin\frac{n\pi}{2} \cos n\omega_0 t$$

Note that this same result could have been obtained by integrating over the interval $-T/4$ to $3T/4$. ■

DRILL EXERCISE

D18.1. Find the Fourier coefficients for the waveform in Fig. D18.1.

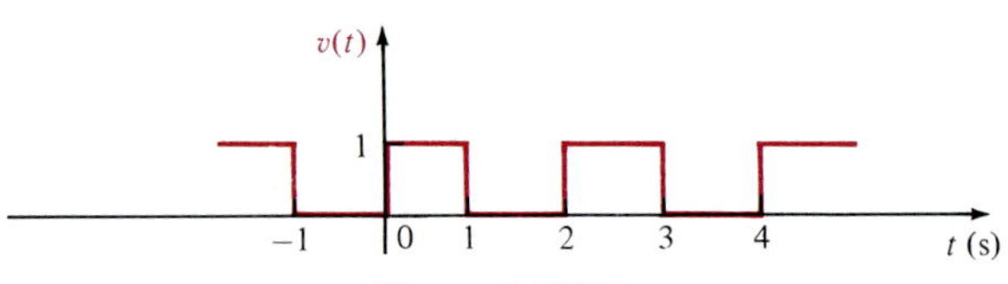

Figure D18.1

Ans: $\mathbf{c}_n = \dfrac{1 - e^{-jn\pi}}{j2\pi n}$.

D18.2. Find the Fourier coefficients for the waveform in Fig. D18.2.

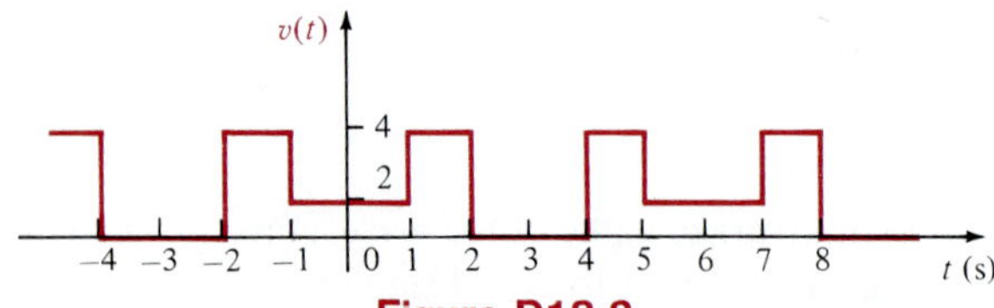

Figure D18.2

Ans: $\mathbf{c}_n = \dfrac{2}{n\pi}\left(2 \sin\dfrac{2\pi n}{3} - \sin\dfrac{n\pi}{3}\right)$, $\mathbf{c}_0 = 2$.

At this point it is both interesting and informative for us to compare the equation for the exponential Fourier series coefficients, that is,

$$\mathbf{c}_n = \frac{1}{T_0}\int_0^{T_0} f(t)e^{-jn\omega_0 t}\,dt \tag{18.10}$$

with those obtained in Chapter 16 for determining the Laplace transform of a periodic function. Recall that for a periodic function $f(t)$ with period T_0,

$$\mathscr{L}[f(t)] = \frac{\mathbf{F}_1(s)}{1 - e^{-sT_0}}$$

where $\mathbf{F}_1(s) = \mathscr{L}[f_1(t)]$, the Laplace transform of the first period of $f(t)$. In addition,

$$\mathbf{F}_1(s) = \int_0^{T_0} f(t)e^{-st}\,dt \tag{18.11}$$

Comparing Eqs. (18.10) and (18.11) indicates that

$$\mathbf{c}_n = \frac{\mathbf{F}_1(jn\omega_0)}{T_0} \tag{18.12}$$

That is, by means of the Laplace transform of the first period of the function, we can derive the Fourier coefficients $\mathbf{c}_n$.

EXAMPLE 18.2

Let us determine the Fourier series coefficients $\mathbf{c}_n$ for the waveform in Fig. 18.3 using the Laplace transform approach.

The first period of the function $v(t)$ is

$$v_1(t) = V\left[u(t) - 2u\left(t - \frac{T}{4}\right) + 2u\left(t - \frac{3T}{4}\right) - u(t - T)\right]$$

and hence

$$\mathbf{V}_1(s) = \frac{V}{s}(1 - 2e^{-(T/4)s} + 2e^{-(3T/4)s} - e^{-Ts})$$

Using Eq. (18.12) yields

$$\mathbf{c}_n = \frac{V}{Tjn\omega_0}(1 - 2e^{-(T/4)jn\omega_0} + 2e^{-(3T/4)jn\omega_0} - e^{-Tjn\omega_0})$$

$$= \frac{V}{j2\pi n}(1 - 2e^{-j(n\pi/2)} + 2e^{-j(3\pi/2)n} - e^{-j2\pi n})$$

This function is zero for n even, and for n odd it can be written as

$$\mathbf{c}_n = \frac{V}{j2\pi n}[2e^{+j(n\pi/2)} - 2e^{-j(n\pi/2)}]$$

$$= \frac{2V}{n\pi}\sin\frac{n\pi}{2} \qquad \text{for } n \text{ odd}$$

which is identical to that obtained in Example 18.1. ■

DRILL EXERCISE

D18.3. Repeat Drill Exercise D18.1 using the Laplace transform approach.

Ans: See answer to Drill Exercise D18.1.

Let us once again consider the impulse function. If $f(t)$ is continuous over the interval $-\infty < t < \infty$, then

$$\int_{-\infty}^{\infty} f(t)\delta(t - t_0)\,dt = f(t_0) \tag{18.13}$$

This result, which illustrates the sampling property of the impulse function, indicates that if t_0 were varied from $-\infty$ to $+\infty$, the entire function $f(t)$ would be reproduced as we *scan* from one limit to the other.

The derivative of the unit impulse $\delta'(t)$ consists of a pair of impulses as shown in Fig. 18.4. This doublet, as it is often called, has two important properties. First, the area under $\delta'(t)$ is equal to zero, that is,

$$\int_{-\infty}^{\infty} \delta'(t)\,dt = 0$$

and second,

$$\int_{-\infty}^{\infty} f(t)\delta'(t - t_0)\,dt = -f'(t_0) \tag{18.14}$$

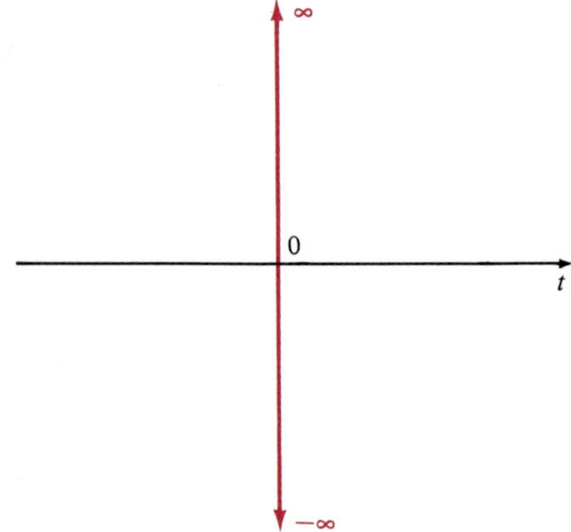

Figure 18.4 Derivative of the unit impulse, or doublet.

This last result can be derived by integrating the left side of the equation by parts.

$$\int_{-\infty}^{\infty} f(t)\delta'(t - t_0)\,dt = f(t)\delta(t - t_0)\Bigg|_{-\infty}^{\infty} - \int_{-\infty}^{\infty} f'(t)\delta(t - t_0)\,dt$$

$$= -f'(t_0)$$

In general, it can be shown that

$$\int_{-\infty}^{\infty} f(t)\delta^{(n)}(t - t_0)\,dt = (-1)^n f^{(n)}(t_0) \qquad (18.15)$$

where $\delta^{(n)}$ and $f^{(n)}$ denote nth-order derivatives.

We will now use these basic properties of the impulse function to compute the complex Fourier coefficients. When evaluating the Fourier coefficients, only a single period of the signal is considered in the analysis. In addition, the technique we illustrate does not compute the dc level of the signal, and therefore, this quantity must be evaluated by inspection, or through integration, whichever is easier. The following examples illustrate the approach.

EXAMPLE 18.3

Consider the waveform shown in Fig. 18.5a. The Fourier coefficients can be evaluated over the single period from $t = 0$ to $t = T_0$ as shown in Fig. 18.5b. Since the waveform contains no impulses, let us differentiate the signal over the single period $t = 0$ to $t = T_0$ to obtain $v_1'(t)$ as shown in Fig. 18.5c.

Since the signal $v_1'(t)$ consists of only impulses, we can easily evaluate the Fourier coefficients for this signal in the following manner. Since

$$v_1(t) = \sum_{n=-\infty}^{\infty} \mathbf{c}_n e^{jn\omega_0 t}$$

then

$$v_1'(t) = \sum_{n=-\infty}^{\infty} jn\omega_0 \mathbf{c}_n e^{jn\omega_0 t} \qquad (18.16)$$

We now define

$$\beta_n = jn\omega_0 \mathbf{c}_n \qquad (18.17)$$

so that

$$\mathbf{c}_n = \frac{\beta_n}{jn\omega_0} \qquad (18.18)$$

We will now evaluate the Fourier coefficients β_n for $v_1'(t)$ and then using Eq. (18.18) obtain the Fourier coefficients for the original waveform; $v_1'(t)$ can be written as

$$v_1'(t) = V\delta(t) - 2V\delta\left(t - \frac{T_0}{2}\right) + V\delta(t - T_0)$$

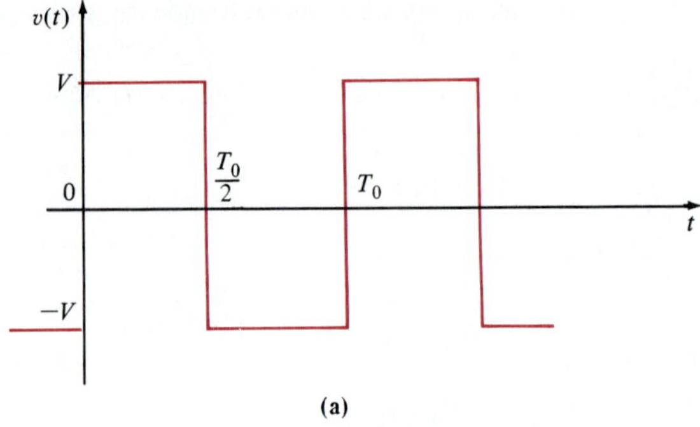

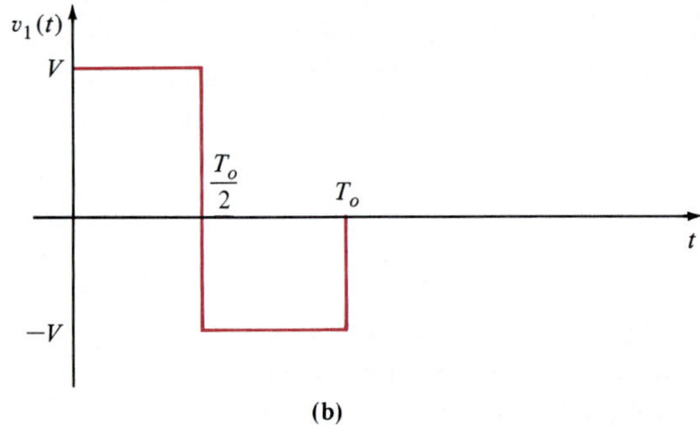

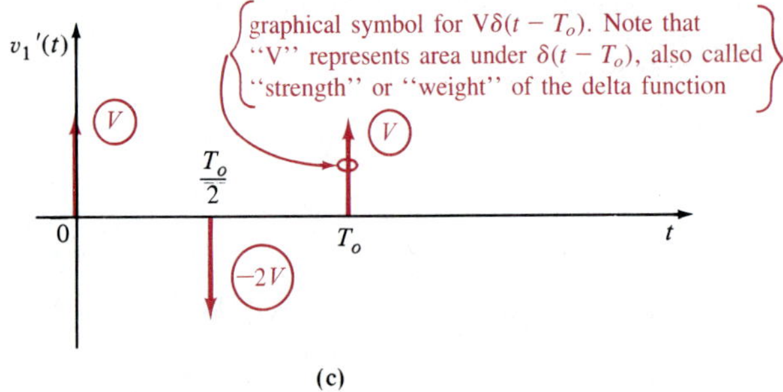

Figure 18.5 Figures used in Example 18.3.

The Fourier coefficients are then

$$\beta_n = \frac{1}{T_0} \int_0^{T_0} v_1'(t) e^{-jn\omega_0 t} \, dt$$

Using the sampling property of the impulse function as illustrated in Eq. (18.13) gives us

$$\beta_n = \frac{V}{T_0}(1 - 2e^{-jn\omega_0(T_0/2)} + e^{-jn\omega_0 T_0})$$

The Fourier coefficients for $v(t)$ are then

$$\mathbf{c}_n = \frac{\beta_n}{jn\omega_0}$$

$$= \frac{V}{jn\omega_0 T_0}(1 - 2e^{-jn\omega_0(T_0/2)} + e^{-jn\omega_0 T_0})$$

Using the fact that $\omega_0 = 2\pi/T_0$, we find that the equation reduces to

$$\mathbf{c}_n = \frac{V}{j2\pi n}(1 - 2e^{-jn\pi} + e^{-j2\pi n})$$

$$= 0 \qquad \text{for } n \text{ even}$$

$$= \frac{2V}{jn\pi} \qquad \text{for } n \text{ odd}$$

Since the dc level of the signal is zero, the Fourier coefficients above are complete. ■

EXAMPLE 18.4

Let us determine the Fourier coefficients for the waveform in Fig. 18.6a.

A single period of the waveform labeled as $v_1(t)$ is shown in Fig. 18.6b. $v_1'(t)$, which can be expressed as

$$v_1'(t) = u(t) - u(t - \pi) - 2\pi\delta(t - \pi) + \pi\delta(t - 2\pi)$$

is shown in Fig. 18.6c. $v_1''(t)$, which consists only of impulses and doublets as shown in Fig. 18.6d, is

$$v_1''(t) = \delta(t) - \delta(t - \pi) - 2\pi\delta'(t - \pi) + \pi\delta'(t - 2\pi)$$

If we let $\boldsymbol{\gamma}_n$ be the Fourier coefficients for $v_1''(t)$, then

$$\boldsymbol{\gamma}_n = \frac{1}{2\pi}\int_0^{2\pi} v_1''(t)e^{-jn\omega_0 t}\,dt$$

$$= \frac{1}{2\pi}\int_0^{2\pi} [\delta(t) - \delta(t - \pi) - 2\pi\delta'(t - \pi) + \pi\delta'(t - 2\pi)]e^{-jn\omega_0 t}\,dt$$

Using Eqs. (18.13) and (18.14) and the fact that $\omega_0 = 2\pi/2\pi = 1$, we obtain

$$\boldsymbol{\gamma}_n = \frac{1}{2\pi}(1 - e^{-jn\pi} - j2\pi n e^{-jn\pi} - jn\pi e^{-j2\pi n})$$

$$= \frac{1}{2\pi}[1 - (1 + j2\pi n)e^{-jn\pi} - jn\pi e^{-j2\pi n}]$$

However, from Eq. (18.16),

$$v(t) = \sum_{n=-\infty}^{\infty} (jn\omega_0)^2 \mathbf{c}_n e^{jn\omega_0 t}$$

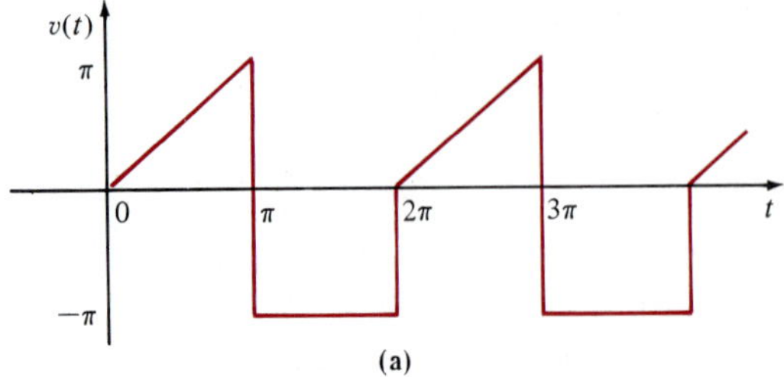

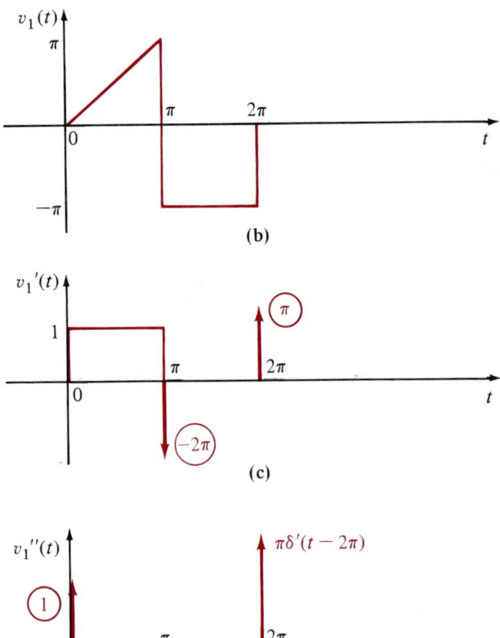

Figure 18.6 Figures used in Example 18.4.

Therefore,

$$\boldsymbol{\gamma}_n = (jn\omega_0)^2\mathbf{c}_n$$

and

$$\mathbf{c}_n = \frac{\boldsymbol{\gamma}_n}{(jn\omega_0)^2}$$

Hence the Fourier coefficients for the waveform in Fig. 18.6a are

$$\mathbf{c}_n = \frac{-1}{2\pi n^2}[1 - (1 + j2\pi n)e^{-jn\pi} - jn\pi e^{-j2\pi n}]$$

for all n except $n = 0$. $\mathbf{c}_0$, the dc level, is easily calculated as $\mathbf{c}_0 = -\pi/4$. ■

DRILL EXERCISE

D18.4. Use the derivative method to determine the Fourier coefficients for the waveform in Fig. D18.4.

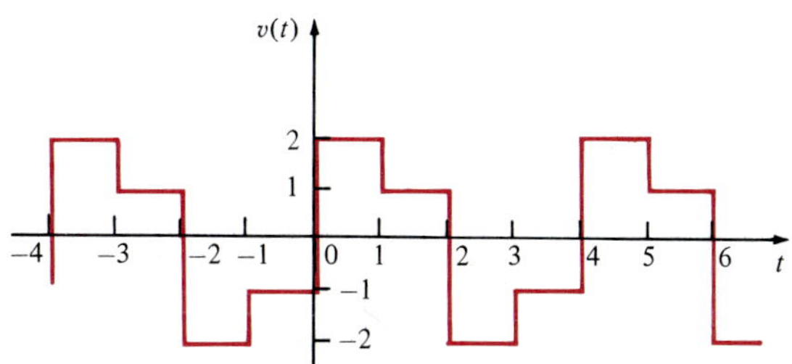

Figure D18.4

Ans: $\mathbf{c}_n = \dfrac{1}{j2\pi n}(2 - e^{-jn\pi/2} - 3e^{-jn\pi} + e^{-(3\pi n/2)} + e^{-j2\pi n})$, $n \neq 0$; $\mathbf{c}_0 = 0$.

Trigonometric Fourier Series

Let us now examine another form of the Fourier series. Since

$$2\mathbf{c}_n = a_n - jb_n \tag{18.19}$$

we will examine this quantity $2\mathbf{c}_n$ and separate it into its real and imaginary parts. Using Eq. (18.9), we find that

$$2\mathbf{c}_n = \frac{2}{T_0}\int_{t_1}^{t_1+T_0} f(t)e^{-jn\omega_0 t}\,dt \tag{18.20}$$

Using Euler's identity, we can write this equation in the form

$$\begin{aligned} 2\mathbf{c}_n &= \frac{2}{T_0}\int_{t_1}^{t_1+T_0} f(t)(\cos n\omega_0 t - j\sin n\omega_0 t)\,dt \\ &= \frac{2}{T_0}\int_{t_1}^{t_1+T_0} f(t)\cos n\omega_0 t\,dt - j\frac{2}{T_0}\int_{t_1}^{t_1+T_0} f(t)\sin n\omega_0 t\,dt \end{aligned}$$

From Eq. (18.19) we note then that

$$a_n = \frac{2}{T_0} \int_{t_1}^{t_1+T_0} f(t) \cos n\omega_0 t \, dt \tag{18.21}$$

$$b_n = \frac{2}{T_0} \int_{t_1}^{t_1+T_0} f(t) \sin n\omega_0 t \, dt \tag{18.22}$$

These are the coefficients of the Fourier series described by Eq. (18.3), which we call the *trigonometric Fourier series*. These equations are derived directly in most textbooks using the orthogonality properties of the cosine and sine functions. Note that we can now evaluate $\mathbf{c}_n$, a_n, b_n, and since

$$2\mathbf{c}_n = \mathrm{D}_n \angle \theta_n \tag{18.23}$$

we can derive the coefficients for the *cosine Fourier series* described by Eq. (18.1). This form of the Fourier series is particularly useful because it allows us to represent each harmonic of the function as a phasor.

From Eq. (18.9) we note that $\mathbf{c}_0$, which is written as a_0, is

$$a_0 = \frac{1}{T_0} \int_{t_1}^{t_1+T_0} f(t) \, dt \tag{18.24}$$

This is the average value of the signal $f(t)$ and can often be evaluated directly from the waveform.

Symmetry and the Trigonometric Fourier Series

If a signal exhibits certain symmetrical properties, we can take advantage of these properties to simplify the calculations of the Fourier coefficients. There are three types of symmetry: (1) even-function symmetry, (2) odd-function symmetry, and (3) half-wave symmetry.

Even-Function Symmetry. A function is said to be even if

$$f(t) = f(-t) \tag{18.25}$$

An even function is symmetrical about the vertical axis and a notable example is the function $\cos n\omega_0 t$. Note that the waveform in Fig. 18.3 also exhibits even-function symmetry. Let us now determine the expressions for the Fourier coefficients if the function satisfies Eq. (18.25).

If we let $t_1 = -T_0/2$ in Eq. (18.24), we obtain

$$a_0 = \frac{1}{T_0} \int_{-T_0/2}^{T_0/2} f(t) \, dt$$

which can be written as

$$a_0 = \frac{1}{T_0} \int_{-T_0/2}^{0} f(t) \, dt + \frac{1}{T_0} \int_{0}^{T_0/2} f(t) \, dt$$

If we now change the variable on the first integral (i.e., let $t = -x$), then $f(-x) = f(x)$, $dt = -dx$, and the range of integration is from $x = T_0/2$ to 0. Therefore, the equation above becomes

$$
\begin{aligned}
a_0 &= \frac{1}{T_0}\int_{T_0/2}^{0} f(x)(-dx) + \frac{1}{T_0}\int_{0}^{T_0/2} f(t)\,dt \\
&= \frac{1}{T_0}\int_{0}^{T_0/2} f(x)\,dx + \frac{1}{T_0}\int_{0}^{T_0/2} f(t)\,dt \\
&= \frac{2}{T_0}\int_{0}^{T_0/2} f(t)\,dt
\end{aligned} \tag{18.26}
$$

The other Fourier coefficients are derived in a similar manner. The a_n coefficient can be written

$$
a_n = \frac{2}{T_0}\int_{-T_0/2}^{0} f(t)\cos n\omega_0 t\,dt + \frac{2}{T_0}\int_{0}^{T_0/2} f(t)\cos n\omega_0 t\,dt
$$

Employing the change of variable that led to Eq. (18.26), we can express the equation above as

$$
\begin{aligned}
a_n &= \frac{2}{T_0}\int_{T_0/2}^{0} f(x)\cos(-n\omega_0 x)(-dx) + \frac{2}{T_0}\int_{0}^{T_0/2} f(t)\cos n\omega_0 t\,dt \\
&= \frac{2}{T_0}\int_{0}^{T_0/2} f(x)\cos n\omega_0 x\,dx + \frac{2}{T_0}\int_{0}^{T_0/2} f(t)\cos n\omega_0 t\,dt \\
&= \frac{4}{T_0}\int_{0}^{T_0/2} f(t)\cos n\omega_0 t\,dt
\end{aligned} \tag{18.27}
$$

Once again following the development above, we can write the equation for the b_n coefficient as

$$
b_n = \frac{2}{T_0}\int_{-T_0/2}^{0} f(t)\sin n\omega_0 t\,dt + \int_{0}^{T_0/2} f(t)\sin n\omega_0 t\,dt
$$

The variable change employed above yields

$$
\begin{aligned}
b_n &= \frac{2}{T_0}\int_{T_0/2}^{0} f(x)\sin(-n\omega_0 x)(-dx) + \frac{2}{T_0}\int_{0}^{T_0/2} f(t)\sin n\omega_0 t\,dt \\
&= \frac{-2}{T_0}\int_{0}^{T_0/2} f(x)\sin n\omega_0 x\,dx + \frac{2}{T_0}\int_{0}^{T_0/2} f(t)\sin n\omega_0 t\,dt \\
&= 0
\end{aligned} \tag{18.28}
$$

The preceding analysis indicates that the Fourier series for an even periodic function consists only of a constant term and cosine terms. Therefore, if $f(t)$ is even, $b_n = 0$ and from Eqs. (18.19) and (18.23), $\mathbf{c}_n$ are real and θ_n are multiples of 180°.

Odd-Function Symmetry. A function is said to be odd if

$$f(t) = -f(-t) \tag{18.29}$$

An example of an odd function is sin $n\omega_0 t$. Another example is the waveform in Fig. 18.7a. Following the mathematical development that led to Eqs. (18.26) to (18.28), we can show that for an odd function the Fourier coefficients are

$$a_0 = 0 \tag{18.30}$$

$$a_n = 0 \qquad \text{for all } n > 0 \tag{18.31}$$

$$b_n = \frac{4}{T_0}\int_0^{T_0/2} f(t) \sin n\omega_0 t \, dt \tag{18.32}$$

Therefore, if $f(t)$ is odd, $a_n = 0$ and, from Eqs. (18.19) and (18.23), $\mathbf{c}_n$ are pure imaginary and θ_n are odd multiples of 90°.

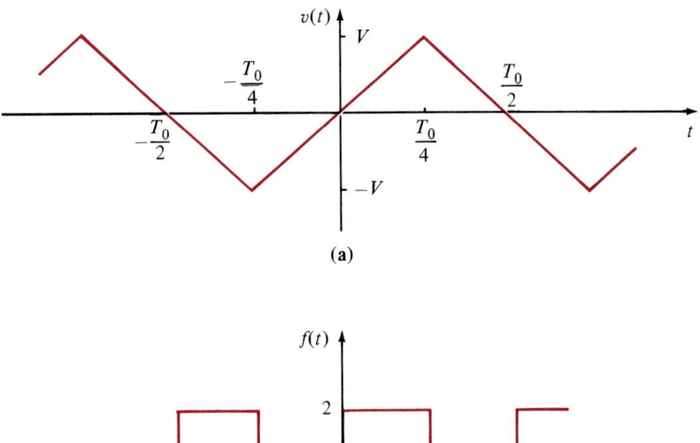

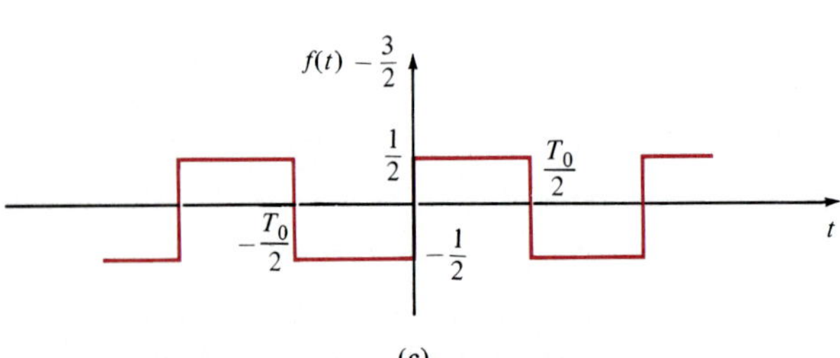

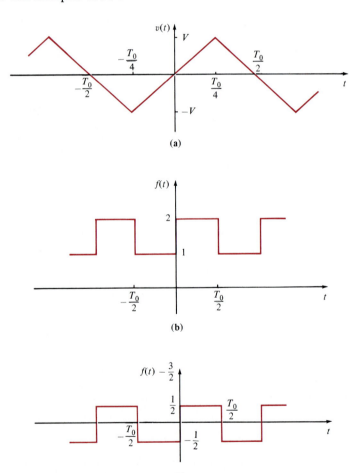

Figure 18.7 Three waveforms; (a) and (c) possess half-wave symmetry.

Half-Wave Symmetry. A function is said to possess *half-wave symmetry* if

$$f(t) = -f\left(t - \frac{T_0}{2}\right) \tag{18.33}$$

Basically, this equation states that each half-cycle is an inverted version of the adjacent half-cycle; that is, if the waveform from $-T_0/2$ to 0 is inverted, it is identical to the waveform from 0 to $T_0/2$. The waveform shown in Fig. 18.7a possesses half-wave symmetry.

Once again we can derive the expressions for the Fourier coefficients in this case by repeating the mathematical development that led to the equations for even-function symmetry using the change of variable $t = x + T_0/2$ and Eq. (18.33). The results of this development are the following equations:

$$a_0 = 0 \tag{18.34}$$

$$a_n = b_n = 0 \qquad \text{for } n \text{ even} \tag{18.35}$$

$$a_n = \frac{4}{T_0}\int_0^{T_0/2} f(t)\cos n\omega_0 t\, dt \qquad \text{for } n \text{ odd} \tag{18.36}$$

$$b_n = \frac{4}{T_0}\int_0^{T_0/2} f(t)\sin n\omega_0 t\, dt \qquad \text{for } n \text{ odd} \tag{18.37}$$

The following equations are often useful in the evaluation of the trigonometric Fourier series coefficients:

$$\begin{aligned}
\int \sin ax\, dx &= -\frac{1}{a}\cos ax \\
\int \cos ax\, dx &= \frac{1}{a}\sin ax \\
\int x\sin ax\, dx &= \frac{1}{a^2}\sin ax - \frac{1}{a}x\cos ax \\
\int x\cos ax\, dx &= \frac{1}{a^2}\cos ax + \frac{1}{a}x\sin ax
\end{aligned} \tag{18.38}$$

EXAMPLE 18.5

We wish to find the trigonometric Fourier series for the periodic signal in Fig. 18.3.

The waveform exhibits even-function symmetry and therefore

$$a_0 = 0$$

$$b_n = 0 \qquad \text{for all } n$$

and

$$a_n = \frac{4}{T}\int_0^{T/2} f(t)\cos n\omega_0 t\, dt \qquad n \neq 0$$

$$= \frac{4}{T}\left(\int_0^{T/4} V\cos n\omega_0 t\, dt - \int_{T/4}^{T/2} V\cos n\omega_0 t\, dt\right)$$

$$= \frac{4V}{n\omega_0 T}\left(\sin n\omega_0 t\Big|_0^{T/4} - \sin n\omega_0 t\Big|_{T/4}^{T/2}\right)$$

$$= \frac{4V}{n\omega_0 T}\left(\sin\frac{n\pi}{2} - \sin n\pi + \sin\frac{n\pi}{2}\right)$$

$$= \frac{8V}{n2\pi}\sin\frac{n\pi}{2} \qquad \text{for } n \text{ odd}$$

$$= \frac{4V}{n\pi}\sin\frac{n\pi}{2} \qquad \text{for } n \text{ odd}$$

The reader should compare this result with that obtained in Example 18.1. Note that this function possesses half-wave symmetry. ■

EXAMPLE 18.6

Let us determine the trigonometric Fourier series expansion for the waveform shown in Fig. 18.7a.

The function not only exhibits odd-function symmetry, but it possesses half-wave symmetry as well. Therefore, it is necessary to determine only the coefficients b_n for n odd. Note that

$$v(t) = \begin{cases} \dfrac{4Vt}{T_0} & 0 \leq t \leq T_0/4 \\[2ex] 2V - \dfrac{4Vt}{T_0} & T_0/4 < t \leq T_0/2 \end{cases}$$

The b_n coefficients are then

$$b_n = \frac{4}{T_0}\int^{T_0/4} 0\,\frac{4Vt}{T_0}\sin n\omega_0 t\, dt + \frac{4}{T_0}\int_{T_0/4}^{T_0/2}\left(2V - \frac{4Vt}{T_0}\right)\sin n\omega_0 t\, dt$$

The evaluation of these integrals is tedious but straightforward and yields

$$b_n = \frac{8V}{n^2\pi^2}\sin\frac{n\pi}{2} \qquad \text{for } n \text{ odd}$$

Hence the Fourier series expansion is

$$v(t) = \sum_{\substack{n=1 \\ n \text{ odd}}}^{\infty} \frac{8V}{n^2\pi^2}\sin\frac{n\pi}{2}\sin n\omega_0 t$$

■

EXAMPLE 18.7

We wish to find the trigonometric Fourier series expansion of the waveform in Fig. 18.7b. Note that this waveform has an average value of $\frac{3}{2}$. Therefore, instead of determining the Fourier series expansion of $f(t)$, we will determine the Fourier series for $f(t) - 3/2$, which is the waveform shown in Fig. 18.7c. The latter waveform possesses half-wave symmetry. The function is also odd and therefore

$$\begin{aligned} b_n &= \frac{4}{T_0} \int_0^{T_0/2} \tfrac{1}{2} \sin n\omega_0 t \, dt \\ &= \frac{2}{T_0}\left(\frac{-1}{n\omega_0} \cos n\omega_0 t \Big|_0^{T_0/2}\right) \\ &= \frac{-2}{n\omega_0 T_0}(\cos n\pi - 1) \\ &= \frac{2}{n\pi} \qquad n \text{ odd} \end{aligned}$$

Therefore, the Fourier series expansion for $f(t) - \frac{3}{2}$ is

$$f(t) - \tfrac{3}{2} = \sum_{\substack{n=1 \\ n \text{ odd}}}^{\infty} \frac{2}{n\pi} \sin n\omega_0 t$$

or

$$f(t) = \tfrac{3}{2} + \sum_{\substack{n=1 \\ n \text{ odd}}}^{\infty} \frac{2}{n\pi} \sin n\omega_0 t$$

DRILL EXERCISE

D18.5. Determine the type of symmetry exhibited by the waveforms in Figs. D18.2 and D18.4.

Ans: Figure D18.2, even symmetry; Fig. D18.4, half-wave symmetry.

D18.6. Find the trigonometric Fourier series for the voltage waveform in Fig. D18.2.

Ans: $v(t) = 2 + \sum_{n=1}^{\infty} \frac{4}{n\pi}\left(2 \sin \frac{2\pi n}{3} - \sin \frac{n\pi}{3}\right) \cos \frac{n\pi}{3} t$ V.

D18.7. Find the trigonometric Fourier series for the voltage waveform in Fig. D18.4.

Ans: $v(t) = \sum_{\substack{n=1 \\ n \text{ odd}}}^{\infty} \frac{2}{n\pi} \sin \frac{n\pi}{2} \cos \frac{n\pi}{2} t + \frac{2}{n\pi}(2 - \cos n\pi) \sin \frac{n\pi}{2} t$ V.

We have tried to indicate the advantage of recognizing symmetry in a waveform. However, it is interesting for us to note that any signal $f(t)$ can be resolved into two components: an even component $f_e(t)$ and an odd component $f_o(t)$. In other words,

$$f(t) = f_e(t) + f_o(t) \tag{18.39}$$

From this equation we can write

$$f(-t) = f_e(-t) + f_o(-t)$$

However, for an even function $f_e(-t) = f_e(t)$ and for an odd function $f_o(-t) = -f_o(t)$. Hence

$$f(-t) = f_e(t) - f_o(t) \tag{18.40}$$

Solving Eqs. (18.39) and (18.40) for the even and odd components of the signal yields

$$f_e(t) = \tfrac{1}{2}[f(t) + f(-t)] \tag{18.41}$$

$$f_o(t) = \tfrac{1}{2}[f(t) - f(-t)] \tag{18.42}$$

Let us now indicate how these equations can be used in determining the Fourier coefficients of a periodic waveform.

EXAMPLE 18.8

Let us determine the trigonometric Fourier series coefficients for the waveform in Fig. 18.8a via Eqs. (18.35) and (18.36).

One period of the function is shown in Fig. 18.8b. Figure 18.8b–g is used to compute the even and odd functions $f_e(t)$ and $f_o(t)$. Note that

$$f_e(t) = \frac{1}{2} - \frac{t}{\pi} \qquad 0 \le t \le \pi$$

and

$$f_o(t) = -\tfrac{1}{2} \qquad 0 \le t \le \pi$$

From the original waveform we note that $\omega_0 = 2\pi/2\pi = 1$ and the average value of $f(t) = 0$ and hence $a_0 = 0$. For $f_e(t)$,

$$\begin{aligned} a_n &= \frac{4}{2\pi}\int_0^{\pi}\left(-\frac{t}{\pi} + \frac{1}{2}\right)\cos nt\, dt \\ &= \frac{-2}{\pi^2}\int_0^{\pi} t\cos nt\, dt + \frac{1}{\pi}\int_0^{\pi}\cos nt\, dt \\ &= \frac{-2}{\pi^2}\left(\frac{1}{n^2}\cos nt + \frac{t}{n}\sin nt\right)\Bigg|_0^{\pi} + \frac{1}{\pi}\left(\frac{1}{n}\sin nt\right)\Bigg|_0^{\pi} \\ &= 0 \text{ for } n \text{ even} \\ &= \frac{4}{\pi^2 n^2} \qquad \text{for } n \text{ odd} \end{aligned}$$

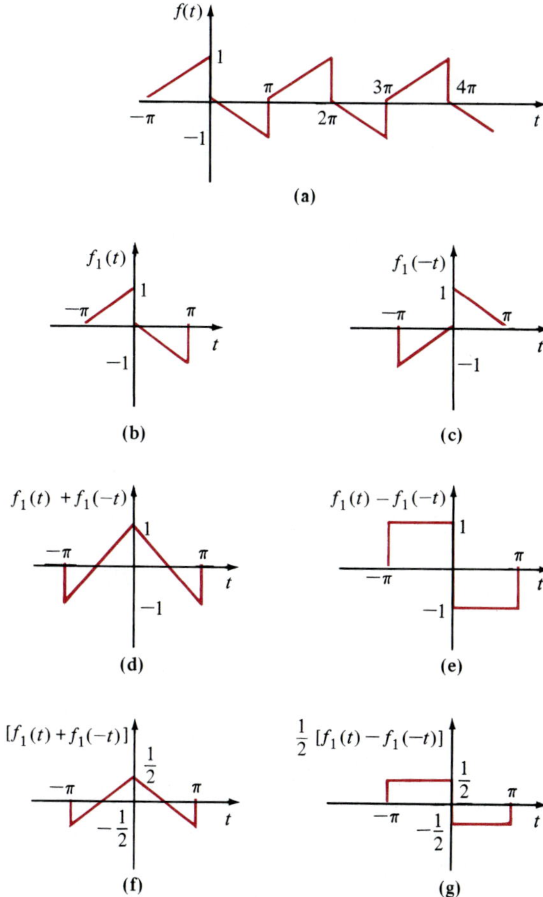

Figure 18.8 Figures used in Example 18.8.

For $f_o(t)$

$$b_n = \frac{4}{2\pi}\int_0^{\pi} \frac{-1}{2} \sin nt\, dt$$

$$= \frac{-1}{\pi}\int_0^{\pi} \sin nt\, dt$$

$$= \frac{1}{\pi n}(\cos nt)_0^{\pi}$$

$$= 0 \text{ for } n \text{ even}$$

$$= \frac{-2}{\pi n} \qquad \text{for } n \text{ odd}$$

■

Although the use of the technique employed in the example does not necessarily provide a feasible alternative for calculating the trigonometric Fourier coefficients, it does provide additional insight when examining waveforms for symmetry.

PSPICE Analysis

When a transient analysis is performed by PSPICE, a Fourier analysis can also be executed. The control statement for the Fourier analysis is

```
.FOUR FREQ OV1 <OV2 OV3 ...>
```

where .FOUR indicates that a Fourier analysis is requested, FREQ is the fundamental frequency of the waveform, and OV represents the variable(s) of interest. The control statement for the Fourier analysis follows the transient analysis control statement, which we recall from Chapter 7 is of the form

```
.TRAN TSTEP TSTOP TSTART TMAX
```

The Fourier analysis is performed over the interval from TSTOP − 1/FREQ to TSTOP. For maximum accuracy TMAX should be equal to the period of the waveform divided by 100 (i.e., 1/100 FREQ).

To facilitate the use of PSPICE, the network that is used in a Fourier series analysis is a voltage source in parallel with a 1-Ω resistor. The voltage source defines the particular waveform under investigation, and the output that results from the analysis is the dc level of the waveform and the first nine harmonics.

Because of the manner in which the PSPICE algorithm selects a reference angle, we must normalize the magnitude and phase of the components calculated analytically to obtain a direct comparison between these calculated components and those generated by PSPICE. The magnitude components are normalized by dividing all components by the magnitude of the fundamental, and the phase components are normalized by subtracting from the phase angle of each component the phase angle of the fundamental. An example of this normalization is shown below.

Harmonic Number	Fourier Component	Normalized Component	Phase Angle	Normalized Phase Angle
1	1.949	1	70.396	0
2	1.856	0.9524	50.792	− 19.6
3	1.708	0.8762	31.188	− 39.208
4	1.511	0.775	11.584	− 58.812

EXAMPLE 18.9

Consider the waveform in Fig. 18.9. Let us use PSPICE to find the dc level of the waveform and the first nine harmonics.

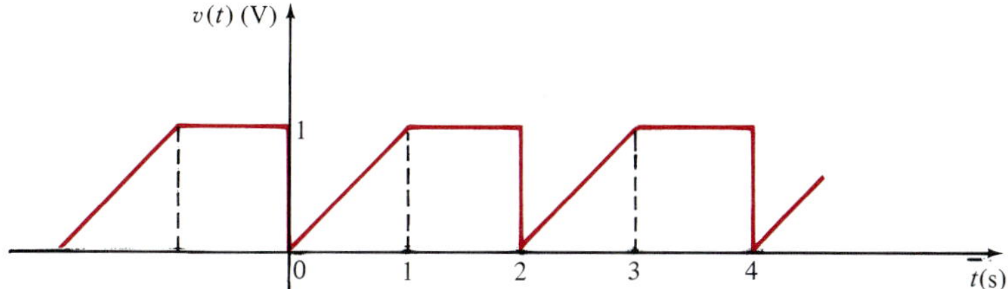

Figure 18.9 Waveform used in Example 18.9.

The PSPICE program is as follows.

```
EXAMPLE 18.9
V1 1 0 PWL(0 0 1 1 2 1)
R1 1 0 1
.OPTIONS LIMPTS=201 ITL5=1000
.TRAN   .01  2  0  .02
.FOUR   .5   V(1)
.END
```

The first statement indicates that 201 points are used to specify the waveform, and 1000 iterations are used to converge to a solution. Both of these numbers are minimums and larger numbers are normally required to obtain an accurate solution.

The program output is

```
FOURIER COMPONENTS OF TRANSIENT RESPONSE V(1)

DC COMPONENT=7.474811E-01
HARMONIC FREQUENCY  FOURIER   NORMALIZED    PHASE     NORMALIZED
   NO       (HZ)    COMPONENT  COMPONENT    (DEG)    PHASE (DEG)

   1     5.000E-01 3.800E-01  1.000E+00  -1.469E+02   0.000E+00
   2     1.000E+00 1.592E-01  4.189E-01  -1.782E+02  -3.131E+01
   3     1.500E+00 1.095E-01  2.882E-01  -1.654E+02  -1.856E+01
   4     2.000E+00 7.963E-02  2.095E-01  -1.764E+02  -2.950E+01
   5     2.500E+00 6.487E-02  1.707E-01  -1.683E+02  -2.144E+01
   6     3.000E+00 5.313E-02  1.398E-01  -1.746E+02  -2.768E+01
   7     3.500E+00 4.621E-02  1.216E-01  -1.686E+02  -2.168E+01
   8     4.000E+00 3.990E-02  1.050E-01  -1.727E+02  -2.587E+01
   9     4.500E+00 3.593E-02  9.454E-02  -1.679E+02  -2.102E+01
```

Note that although the dc level is clearly 0.75, the program yields 0.7475. Such inaccurracies can be corrected with more points and iterations. ■

EXAMPLE 18.10

Let us perform a Fourier analysis using PSPICE on the waveform shown in Fig. 18.6.

The PSPICE program is

```
EXAMPLE 18.10
.OPTIONS LIMP TS=701 ITL5=10000
V1   1         PWL(0 0 3.1415 3.1416 3.1416 0 3.1417 -3.1416
+6.2 831   -3.1416    6.2832 0)
R1   1   0    1
.TRAN .01     6.2832 0 .062832
.FOUR .1592V(1)
.END
```

The program output is

```
FOURIER COMPONENTS OF TRANSIENT RESPONSE V(1)
DC COMPONENT=-7.872032E-01

HARMONIC FREQUENCY  FOURIER   NORMALIZED    PHASE     NORMALIZED
   NO       (HZ)   COMPONENT  COMPONENT     (DEG)    PHASE (DEG)

    1     1.592E-01 3.065E+00 1.000E+00  -1.170E+01   0.000E+00
    2     3.184E-01 4.999E-01 1.631E-01  -1.794E+02  -1.677E+02
    3     4.776E-01 1.002E+00 3.269E-01  -3.188E+00   8.512E+00
    4     6.368E-01 2.500E-01 8.156E-02  -1.789E+02  -1.672E+02
    5     7.960E-01 6.001E-01 1.958E-01  -9.979E-01   1.070E+01
    6     9.552E-01 1.667E-01 5.438E-02  -1.783E+02  -1.666E+02
    7     1.114E+00 4.285E-01 1.398E-01   2.698E-01   1.197E+01
    8     1.274E+00 1.250E-01 4.079E-02  -1.777E+02  -1.660E+02
    9     1.433E+00 3.333E-01 1.088E-01   1.229E+00   1.293E+01
```

This analysis appears to be more accurate because of the larger number of points and iterations. ■

Time Shifting

Let us now examine the effect of time shifting a periodic waveform $f(t)$ defined by the equation

$$f(t) = \sum_{n=-\infty}^{\infty} \mathbf{c}_n e^{jn\omega_0 t}$$

Note that

$$f(t - t_0) = \sum_{n=-\infty}^{\infty} \mathbf{c}_n e^{jn\omega_0(t-t_0)}$$

$$= \sum_{n=-\infty}^{\infty} (\mathbf{c}_n e^{-jn\omega_0 t_0}) e^{jn\omega_0 t} \tag{18.43}$$

Since $e^{-jn\omega_0 t_0}$ corresponds to a phase shift, the Fourier coefficients of the time-shifted function are the Fourier coefficients of the original function with the angle shifted by an amount directly proportional to frequency. Therefore, time shift in the time domain corresponds to phase shift in the frequency domain.

EXAMPLE 18.11

Let us time delay the waveform in Fig. 18.3 by a quarter period and compute the Fourier series.

The waveform in Fig. 18.3 time delayed by $T_0/4$ is shown in Fig. 18.10. Since the time delay is $T_0/4$,

$$n\omega_0 t_d = n\frac{2\pi}{T_0}\frac{T_0}{4} = n\frac{\pi}{2} = n\,90°$$

Therefore, using Eq. (18.43) and the results of Example 18.1, the Fourier coefficients for the time-shifted waveform are

$$\mathbf{c}_n = \frac{2V}{n\pi}\sin\frac{n\pi}{2}\,\underline{/-n\,90°} \qquad n \text{ odd}$$

and therefore,

$$v(t) = \sum_{\substack{n=1\\ n \text{ odd}}}^{\infty} \frac{4V}{n\pi}\sin\frac{n\pi}{2}\cos\,(n\omega_0 t - n\,90°)$$

If we compute the Fourier coefficients for the time-shifted waveform in Fig. 18.10, we obtain

$$\begin{aligned}\mathbf{c}_n &= \frac{1}{T_0}\int_{-T_0/2}^{T_0/2} f(t)e^{-jn\omega_0 t}\,dt \\ &= \frac{1}{T_0}\int_{-T_0/2}^{0} -Ve^{-jn\omega_0 t}\,dt + \frac{1}{T_0}\int_{0}^{T_0/2} Ve^{-jn\omega_0 t}\,dt \\ &= \frac{2V}{jn\pi} \qquad \text{for } n \text{ odd}\end{aligned}$$

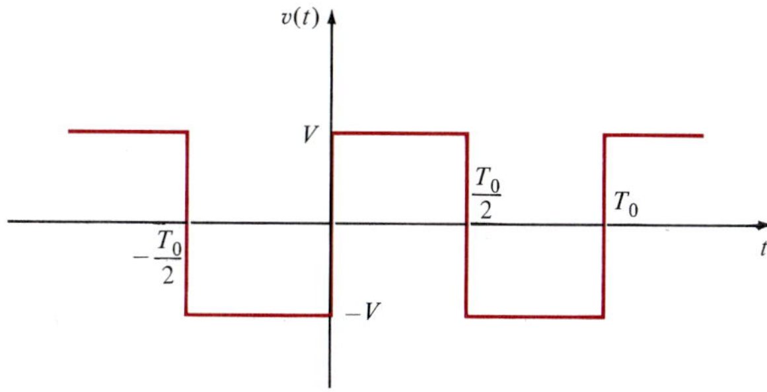

Figure 18.10 Waveform in Fig. 18.3 time shifted by $T_0/4$.

Therefore,

$$\mathbf{c}_n = \frac{2V}{n\pi} \angle -90^\circ \qquad n \text{ odd}$$

Since n is odd, we can show that this expression is equivalent to the one obtained above. ■

In general, we can compute the phase shift in degrees using the expression

$$\text{phase shift (deg)} = \omega_0 t_d = (360^\circ) \frac{t_d}{T_0} \tag{18.44}$$

so that a time shift of one-quarter period corresponds to a 90° phase shift.

As another interesting facet of the time shift, consider a function $f_1(t)$ which is nonzero in the interval $0 \leq t \leq T_0/2$ and is zero in the interval $T_0/2 < t \leq T_0$. For purposes of illustration, let us assume that $f_1(t)$ is the triangular waveform shown in Fig. 18.11a.

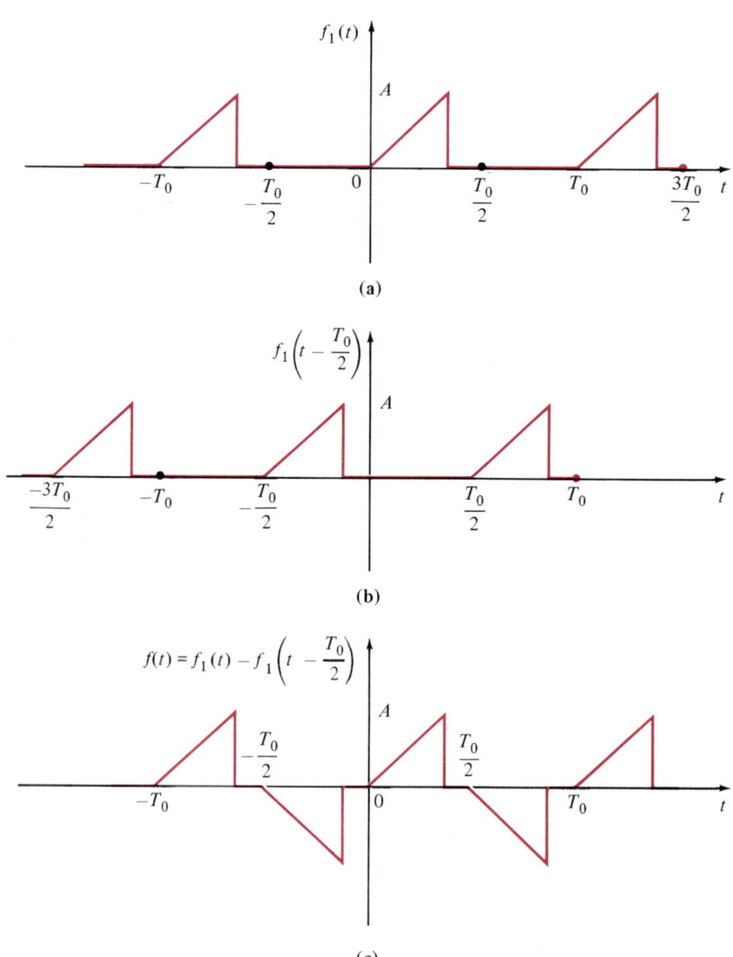

Figure 18.11 Waveforms that illustrate the generation of half-wave symmetry.

$f_1(t - T_0/2)$ is then shown in Fig. 18.11b. Then the function $f(t)$ defined as

$$f(t) = f_1(t) - f_1\left(t - \frac{T_0}{2}\right) \tag{18.45}$$

is shown in Fig. 18.11c. Note that $f(t)$ has half-wave symmetry. In addition, note that if

$$f_1(t) = \sum_{n=-\infty}^{\infty} \mathbf{c}_n e^{-jn\omega_0 t}$$

then

$$f(t) = f_1(t) - f_1\left(t - \frac{T_0}{2}\right) = \sum_{n=-\infty}^{\infty} \mathbf{c}_n(1 - e^{-jn\pi})e^{jn\omega_0 t}$$

$$= \begin{cases} \sum_{n=-\infty}^{\infty} 2\mathbf{c}_n e^{jn\omega_0 t} & n \text{ odd} \\ 0 & n \text{ even} \end{cases} \tag{18.46}$$

Therefore, we see that any function with half-wave symmetry can be expressed in the form of Eq. (18.45), where the Fourier series is defined by Eq. (18.46), and $\mathbf{c}_n$ is the Fourier coefficients for $f_1(t)$.

DRILL EXERCISE

D18.8. If the waveform in Fig. D18.1 is time-delayed 1 sec, we obtain the waveform in Fig. D18.8. Compute the exponential Fourier coefficients for the waveform in Fig. D18.8 and show that they differ from the coefficients for the waveform in Fig. D18.1 by an angle of $n(180°)$.

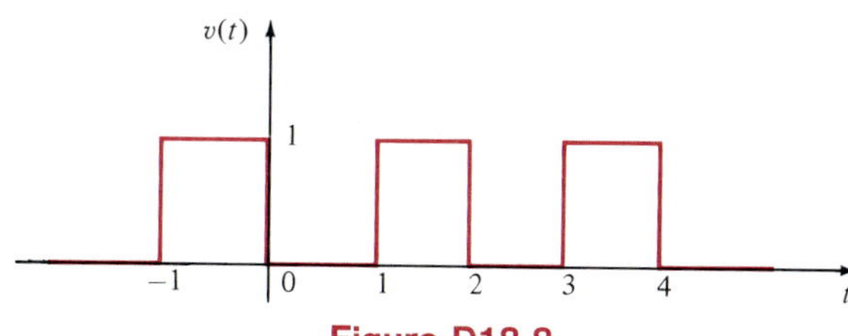

Figure D18.8

Ans: $\mathbf{c}_0 = \frac{1}{2}$, $\mathbf{c}_n = -\left(\frac{1 - e^{-jn\pi}}{j2\pi n}\right)$.

Waveform Generation

The magnitude of the harmonics in a Fourier series are independent of the time scale for a given wave shape. Therefore, the equations for a variety of waveforms can be given in tabular form without expressing a specific time scale. Table 18.1 is a set of commonly occurring periodic waves where the advantage of symmetry has been used to simplify the

Table 18.1 Fourier Series for Some Common Waveforms

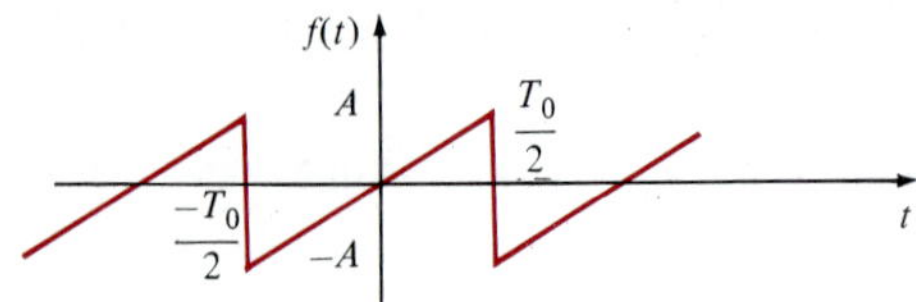

$$f(t) = \sum_{n=1}^{\infty} (-1)^{n+1} \frac{2A}{n\pi} \sin n\omega_0 t$$

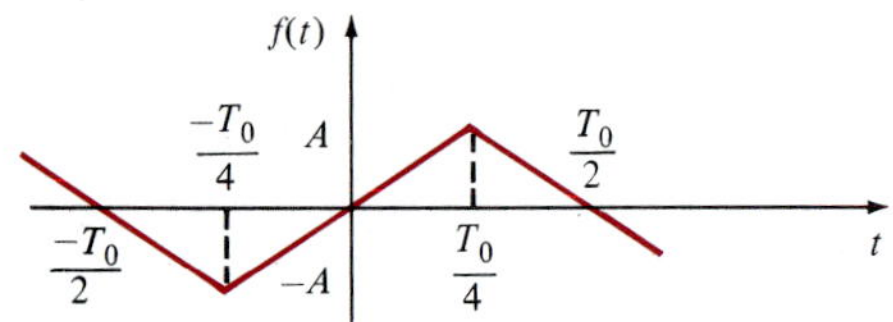

$$f(t) = \sum_{\substack{n=1 \\ n \text{ odd}}}^{\infty} \frac{8A}{n^2\pi^2} \sin \frac{n\pi}{2} \sin n\omega_0 t$$

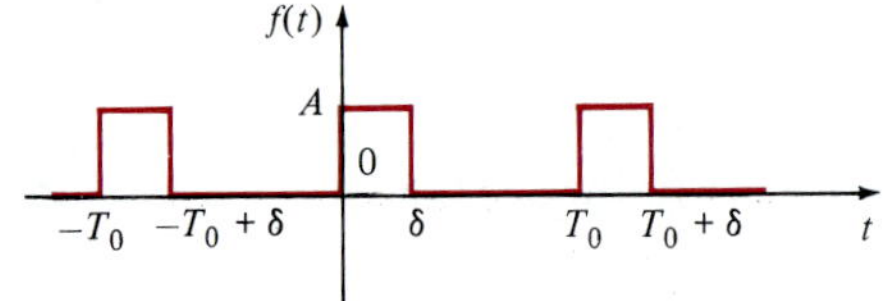

$$f(t) = \sum_{n=-\infty}^{\infty} \frac{A}{n\pi} \sin \frac{n\pi\delta}{T_0} e^{jn\omega_0[t-(\delta/2)]}$$

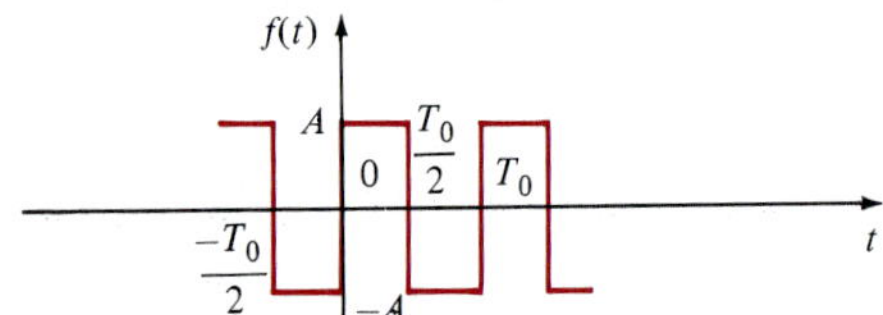

$$f(t) = \sum_{\substack{n=1 \\ n \text{ odd}}}^{\infty} \frac{4A}{n\pi} \sin n\omega_0 t$$

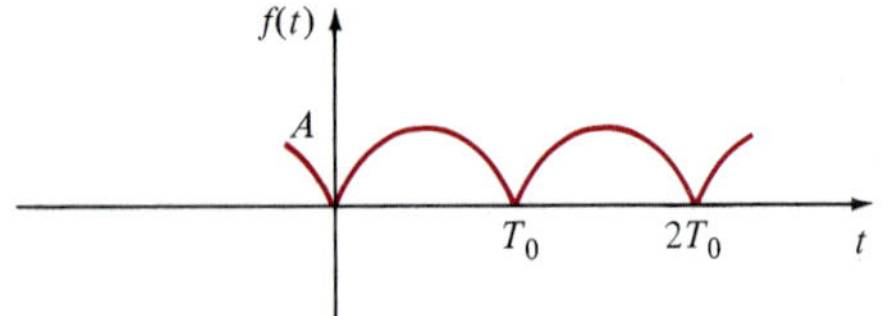

$$f(t) = \frac{2A}{\pi} + \sum_{n=1}^{\infty} \frac{4A}{\pi(1-4n^2)} \cos n\omega_0 t$$

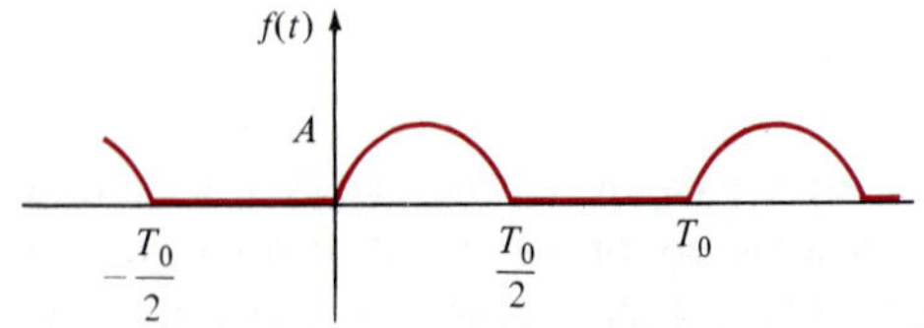

$$f(t) = \frac{A}{\pi} + \frac{A}{2} \sin \omega_0 t + \sum_{\substack{n=2 \\ n \text{ even}}}^{\infty} \frac{2A}{\pi(1-n^2)} \cos n\omega_0 t$$

Table 18.1 (continued)

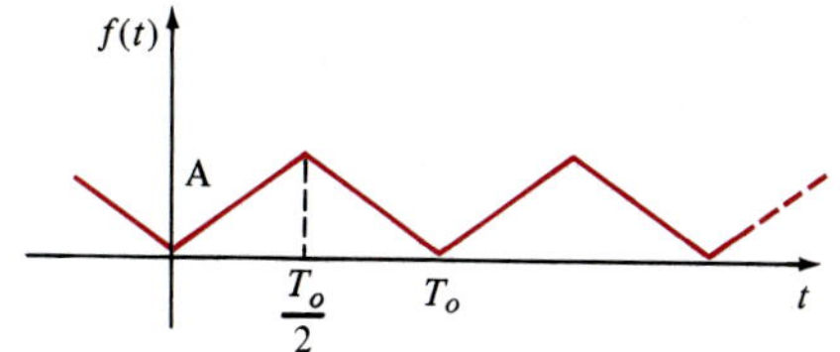

$$f(t) = \frac{A}{2} + \sum_{\substack{n=-\infty \\ n\neq 0 \\ n \text{ odd}}}^{\infty} \frac{-2A}{n^2\pi^2} e^{jn\omega_0 t}$$

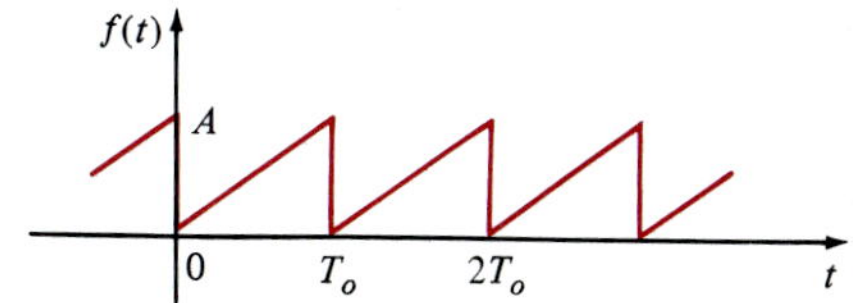

$$f(t) = \frac{A}{2} + \sum_{n=1}^{\infty} \frac{-A}{n\pi} \sin n\omega_0 t$$

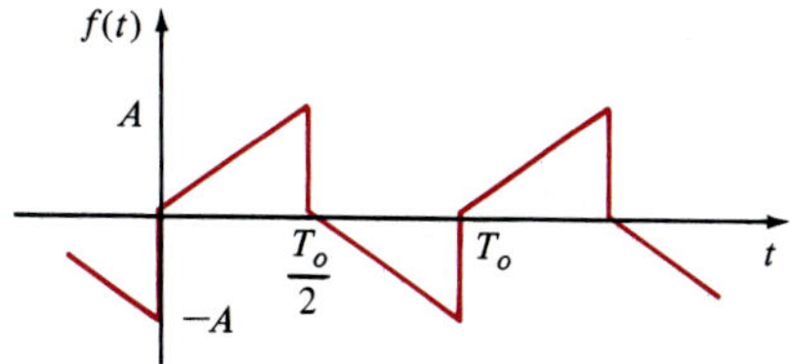

$$f(t) = \sum_{n=1}^{\infty} \frac{-4A}{\pi^2 n^2} \cos n\omega_0 t + \frac{2A}{\pi n} \sin n\omega_0 t$$

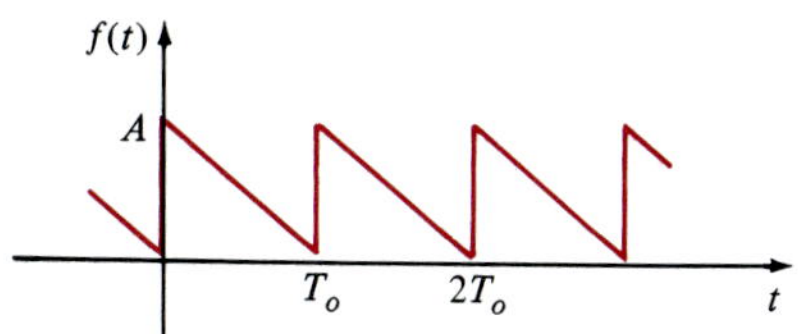

$$f(t) = \frac{A}{2} + \sum_{n=1}^{\infty} \frac{A}{\pi n} \sin n\omega_0 t$$

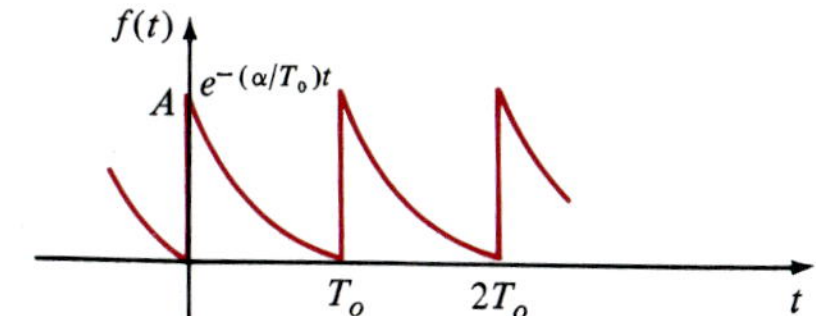

$$f(t) = \sum_{n=-\infty}^{\infty} \frac{A(1 - e^{-\alpha})}{\alpha + j2\pi n} e^{jn\omega_0 t}$$

coefficients. These waveforms can be used to generate other waveforms. The level of a wave can be adjusted by changing the average value component; the time can be shifted by adjusting the angle of the harmonics; and two waveforms can be added to produce a third waveform; for example, the waveforms in Fig. 18.12a and b can be added to produce the waveform in Fig. 18.12c.

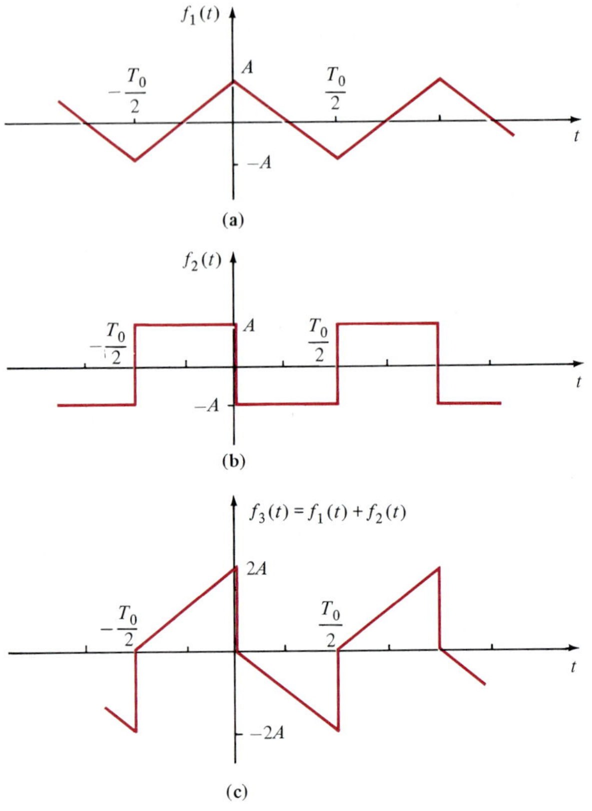

Figure 18.12 Example of waveform generation.

DRILL EXERCISE

D18.9. Two periodic waveforms are shown in Fig. D18.9. Compute the exponential Fourier series for each waveform, and then add the results to obtain the Fourier series for the waveform in Fig. D18.2.

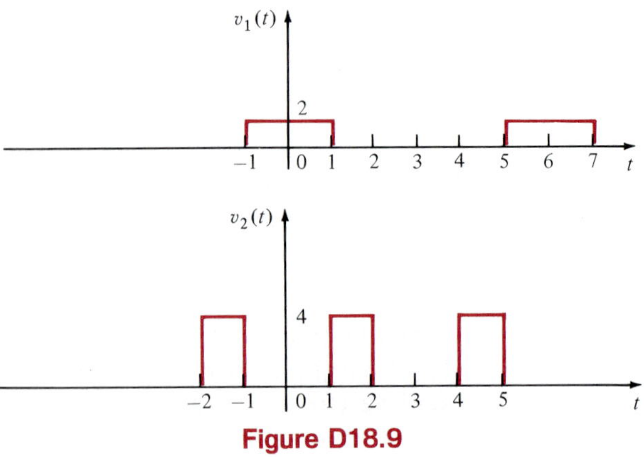

Figure D18.9

$$\textit{Ans:}\ v_1(t) = \frac{2}{3} + \sum_{\substack{n=-\infty \\ n\neq 0}}^{\infty} \frac{2}{n\pi} \sin \frac{n\pi}{3} e^{jn\omega_0 t},$$

$$v_2(t) = \frac{4}{3} + \sum_{n=-\infty}^{\infty} -\frac{4}{n\pi}\left(\sin \frac{n\pi}{3} - \sin \frac{2n\pi}{3}\right) e^{jn\omega_0 t}.$$

Frequency Spectrum

The *frequency spectrum* of the function $f(t)$ expressed as a Fourier series consists of a plot of the amplitude of the harmonics versus frequency, which we call the *amplitude spectrum,* and a plot of the phase of the harmonics versus frequency, which we call the *phase spectrum*. Since the frequency components are discrete, the spectra are called *line spectra*. Such spectra illustrate the frequency content of the signal. Plots of the amplitude and phase spectra are based on Eqs. (18.1), (18.3), and (18.7) and represent the amplitude and phase of the signal at specific frequencies.

EXAMPLE 18.12

The Fourier series for the triangular-type waveform shown in Fig. 18.12c with $A = 5$ is given by the equation

$$v(t) = \sum_{\substack{n=1 \\ n \text{ odd}}}^{\infty} \left(\frac{20}{n\pi} \sin n\omega_0 t - \frac{40}{n^2\pi^2} \cos n\omega_0 t\right)$$

We wish to plot the first four terms of the amplitude and phase spectra for this signal.

Since $D_n \underline{/\theta_n} = a_n - jb_n$, the first four terms for this signal are

$$D_1 \underline{/\theta_1} = -\frac{40}{\pi^2} - j\frac{20}{\pi} = 7.5 \underline{/-122^\circ}$$

$$D_3 \underline{/\theta_3} = -\frac{40}{9\pi^2} - j\frac{20}{3\pi} = 2.2 \underline{/-102^\circ}$$

$$D_5 \underline{/\theta_5} = -\frac{40}{25\pi^2} - j\frac{20}{5\pi} = 1.3 \underline{/-97^\circ}$$

$$D_7 \underline{/\theta_7} = -\frac{40}{49\pi^2} - j\frac{20}{7\pi} = 0.91 \underline{/-95^\circ}$$

Therefore, the plots of the amplitude and phase versus ω are as shown in Fig. 18.13. ■

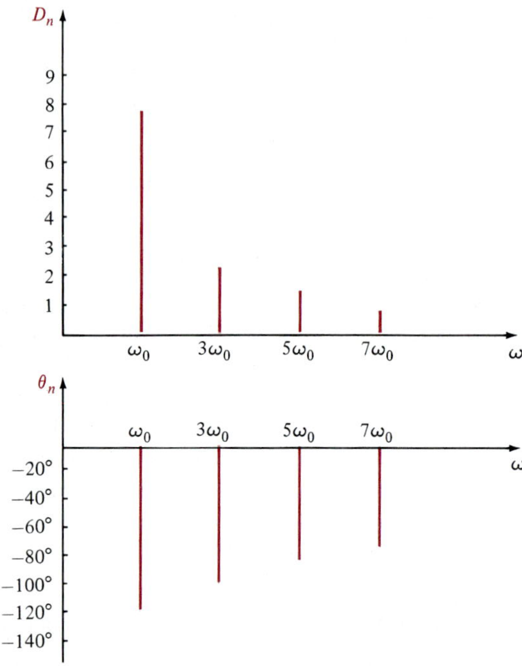

Figure 18.13 Amplitude and phase spectra.

DRILL EXERCISE

D18.10. Determine the trigonometric Fourier series for the voltage waveform in Fig. D18.10 and plot the first four terms of the amplitude and phase spectra for this signal.

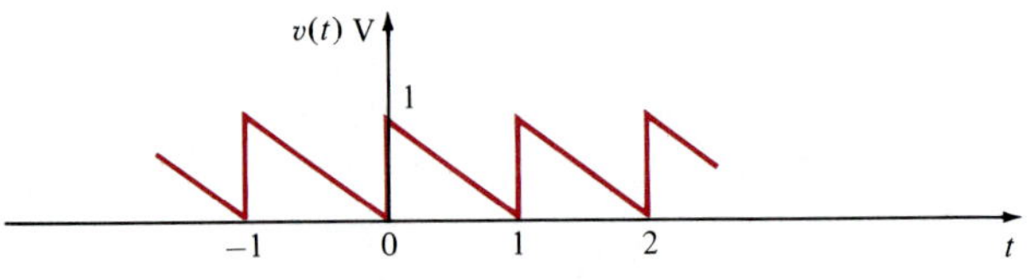

Figure D18.10

Ans: $a_0 = \frac{1}{2}$, $D_1 = -j(1/\pi)$, $D_2 = -j(1/2\pi)$, $D_3 = -j(1/3\pi)$, $D_4 = -j(1/4\pi)$.

Steady-State Network Response

If a periodic signal is applied to a network, the steady-state voltage or current response at some point in the circuit can be found in the following manner. First, we represent the periodic forcing function by a Fourier series. If the input forcing function for a network is a voltage, the input can be expressed in the form

$$v(t) = v_0 + v_1(t) + v_2(t) + \cdots$$

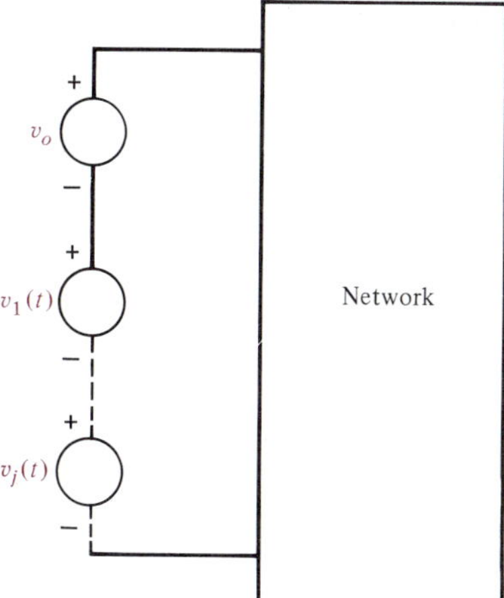

Figure 18.14 Network with a periodic voltage forcing function.

and therefore represented in the time domain as shown in Fig. 18.14. Each source has its own amplitude and frequency. Next we determine the response due to each component of the input Fourier series; that is, we use phasor analysis in the frequency domain to determine the network response due to each source. The network response due to each source in the frequency domain is then transformed to the time domain. Finally, we add the time-domain solutions due to each source using the principle of superposition to obtain the Fourier series for the total *steady-state* network response.

EXAMPLE 18.13

We wish to determine the steady-state voltage $v_o(t)$ in Fig. 18.15 if the input voltage $v(t)$ is given by the expression

$$v(t) = \sum_{\substack{n=1 \\ n \text{ odd}}}^{\infty} \left(\frac{20}{n\pi} \sin 2nt - \frac{40}{n^2\pi^2} \cos 2nt \right) \text{ V}$$

Note that this source has no constant term, and therefore its dc value is zero. The amplitude and phase for the first four terms of this signal are given in Example 18.12, and therefore, the signal $v(t)$ can be written as

$$v(t) = 7.5 \cos (2t - 122°) + 2.2 \cos (6t - 102°) + 1.3 \cos (10t - 97°) + 0.91 \cos (14t - 95°) + \cdots$$

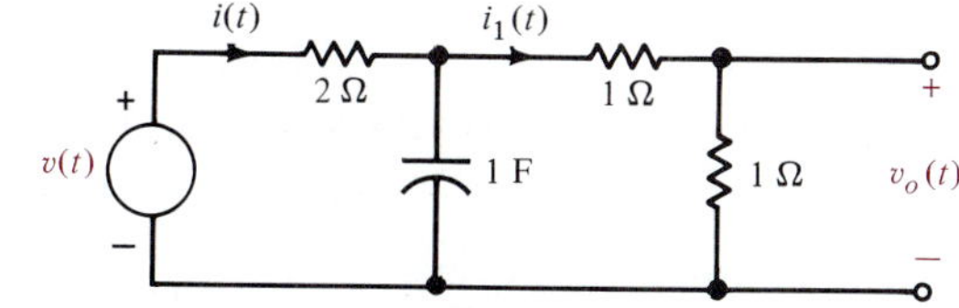

Figure 18.15 *RC* circuit employed in Example 18.13.

From the network we find that

$$\mathbf{I} = \frac{\mathbf{V}}{2 + \dfrac{2/j\omega}{2 + 1/j\omega}} = \frac{\mathbf{V}(1 + 2j\omega)}{4 + 4j\omega}$$

$$\mathbf{I}_1 = \frac{\mathbf{I}(1/j\omega)}{2 + 1/j\omega} = \frac{\mathbf{I}}{1 + 2j\omega}$$

$$\mathbf{V}_o = (1)\mathbf{I}_1 = 1 \cdot \frac{\mathbf{V}(1 + 2j\omega)}{4 + 4j\omega} \frac{1}{1 + 2j\omega} = \frac{\mathbf{V}}{4 + 4j\omega}$$

Therefore, since $\omega_0 = 2$,

$$\mathbf{V}_o(n) = \frac{\mathbf{V}(n)}{4 + j8n}$$

The individual components of the output due to the components of the input source are then

$$\mathbf{V}_o(\omega_0) = \frac{7.5 \underline{/-122^\circ}}{4 + j8} = 0.84 \underline{/-185.4^\circ} \text{ V}$$

$$\mathbf{V}_o(3\omega_0) = \frac{2.2 \underline{/-102^\circ}}{4 + j24} = 0.09 \underline{/-182.5^\circ} \text{ V}$$

$$\mathbf{V}_o(5\omega_0) = \frac{1.3 \underline{/-97^\circ}}{4 + j40} = 0.03 \underline{/-181.3^\circ} \text{ V}$$

$$\mathbf{V}_o(7\omega_0) = \frac{0.91 \underline{/-95^\circ}}{4 + j56} = 0.016 \underline{/-181^\circ} \text{ V}$$

Hence the steady-state output voltage $v_o(t)$ can be written as

$$v_o(t) = 0.84 \cos (2t - 185.4^\circ) + 0.09 \cos (6t - 182.5^\circ)$$
$$+ 0.03 \cos (10t - 181.3^\circ) + 0.016 \cos (14t - 181^\circ) + \cdots$$

DRILL EXERCISE

D18.11 Determine the expression for the steady-state current $i(t)$ in Fig. D18.11 if the input voltage $v_S(t)$ is given by the expression

$$v_S(t) = \frac{20}{\pi} + \sum_{n=1}^{\infty} \frac{-40}{\pi(4n^2 - 1)} \cos 2nt \text{ V}$$

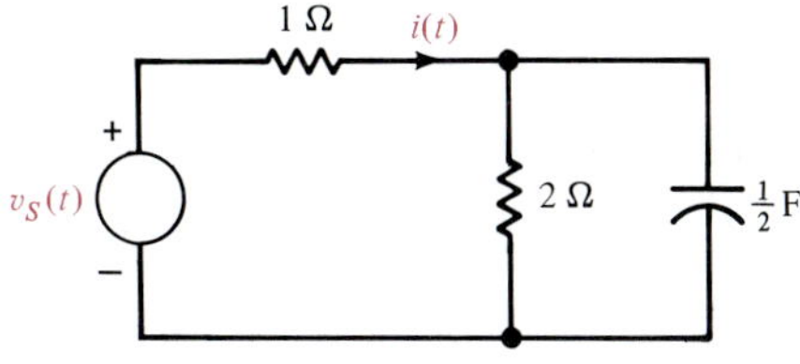

Figure D18.11

Ans: $i(t) = 2.12 + \sum_{n=1}^{\infty} \frac{-40}{\pi(4n^2 - 1)} \frac{1}{A_n} \cos(2nt - \theta_n)$ A.

Average Power

We have shown that when a linear network is forced with a nonsinusoidal periodic signal, voltages and currents throughout the network are of the form

$$v(t) = V_{\text{dc}} + \sum_{n=1}^{\infty} V_n \cos(n\omega_0 t - \theta_{v_n})$$

and

$$i(t) = I_{\text{dc}} + \sum_{n=1}^{\infty} I_n \cos(n\omega_0 t - \theta_{i_n})$$

If we employ the passive sign convention and assume that the voltage across an element and the current through it are given by the equations above, then from Eq. (11.6),

$$\begin{aligned} P &= \frac{1}{T}\int_{t_0}^{t_0+T} p(t)\, dt \\ &= \frac{1}{T}\int_{t_0}^{t_0+T} v(t)i(t)\, dt \end{aligned}$$

Note that the integrand involves the product of two infinite series. However, the determination of the average power is actually easier than it appears. First, note that the product $V_{\text{dc}}I_{\text{dc}}$ when integrated over a period and divided by the period is simply $V_{\text{dc}}I_{\text{dc}}$. Second, the product of V_{dc} and any harmonic of the current or I_{dc} and any harmonic of the voltage when integrated over a period yields zero. Third, the product of any two *different* harmonics of the voltage and the current when integrated over a period yields zero. Finally, only nonzero terms result from the products of voltage and current at the *same* frequency. Hence, using the mathematical development that follows Eq. (11.6), we find that

$$P = V_{\text{dc}}I_{\text{dc}} + \sum_{n=1}^{\infty} \frac{V_n I_n}{2} \cos(\theta_{v_n} - \theta_{i_n}) \tag{18.47}$$

EXAMPLE 18.14

In the network in Fig. 18.16, $v(t) = 42 + 16 \cos(377t + 30°) + 12 \cos(754t - 20°)$ V. We wish to compute the current $i(t)$ and determine the average power absorbed by the network.

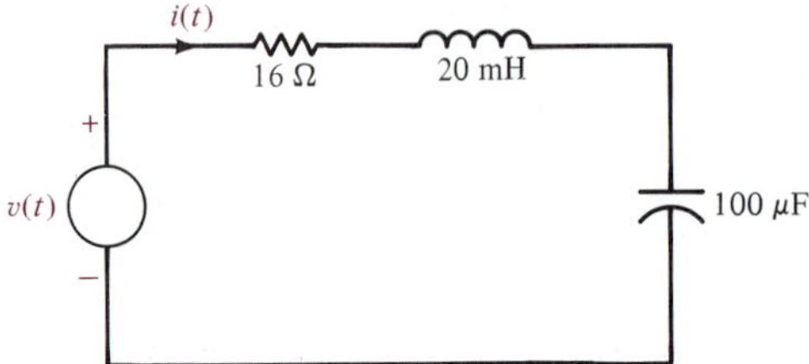

Figure 18.16 Network used in Example 18.14.

The capacitor acts as an open circuit to dc, and therefore, $I_{dc} = 0$. At $\omega = 377$ rad/sec,

$$\frac{1}{j\omega C} = \frac{1}{j(377)(100)10^{-6}} = -j26.53\ \Omega$$

$$j\omega L = j(377)(20)10^{-3} = j7.54\ \Omega$$

Hence

$$\mathbf{I}_{377} = \frac{16\ \underline{/30°}}{16 + j7.54 - j26.53} = 0.64\ \underline{/79.88°}\ \text{A}$$

At $\omega = 754$ rad/sec,

$$\frac{1}{j\omega C} = \frac{1}{j(754)(100)10^{-6}} = -j13.26\ \Omega$$

$$j\omega L = j(754)(20)10^{-3} = j15.08\ \Omega$$

Hence

$$\mathbf{I}_{754} = \frac{12\underline{/-20°}}{16 + j15.08 - j13.26} = 0.75\ \underline{/-26.49°}\ \text{A}$$

Therefore, the current $i(t)$ is

$$i(t) = 0.64 \cos(377t + 79.88°) + 0.75 \cos(754t - 26.49°)\ \text{A}$$

and the average power absorbed by the network is

$$P = (42)(0) + \frac{(16)(0.64)}{2} \cos(30° - 79.88°) + \frac{(12)(0.75)}{2} \cos(-20° + 26.49°)$$

$$= 7.77\ \text{W}$$

■

DRILL EXERCISE

D18.12. At the terminals of a two-port network, the voltage $v(t)$ and the current $i(t)$ are given by the following expressions:

$$v(t) = 64 + 36 \cos (377t + 60°) - 24 \cos (754t + 102°) \text{ V}$$

$$i(t) = 1.8 \cos (377t + 45°) + 1.2 \cos (754t + 100°) \text{ A}$$

Find the average power absorbed by the network.

Ans: $P_{\text{Ave}} = 16.91$ W.

18.2 Fourier Transform

The preceding sections of this chapter have illustrated that the exponential Fourier series can be used to represent a periodic signal for all time. We will now consider a technique for representing an aperiodic signal for all values of time.

Suppose that an aperiodic signal $f(t)$ is as shown in Fig. 18.17a. We now construct a new signal $f_p(t)$ which is identical to $f(t)$ in the interval $-T/2$ to $T/2$ but is *periodic* with period T as shown in Fig. 18.17b. Since $f_p(t)$ is periodic, it can be represented in the interval $-\infty$ to ∞ by an exponential Fourier series.

$$f_p(t) = \sum_{n=-\infty}^{\infty} \mathbf{c}_n e^{jn\omega_0 t} \tag{18.48}$$

where

$$\mathbf{c}_n = \frac{1}{T} \int_{-T/2}^{T/2} f_p(t) e^{-jn\omega_0 t} \, dt \tag{18.49}$$

and

$$\omega_0 = \frac{2\pi}{T} \tag{18.50}$$

At this point we note that if we take the limit of the function $f_p(t)$ as $T \to \infty$, the periodic signal in Fig. 18.17b approaches the aperiodic signal in Fig. 18.17a; that is, the repetitious signals centered at $-T$ and $+T$ in Fig. 18.17b are moved to infinity.

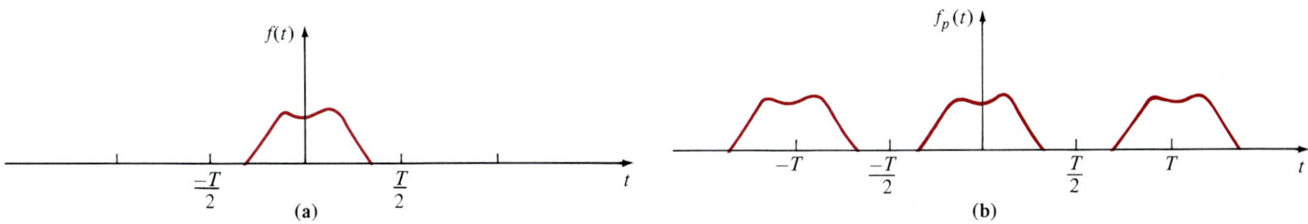

Figure 18.17 Aperiodic and periodic signals.

The line spectrum for the periodic signal exists at harmonic frequencies ($n\omega_0$) and the incremental spacing between the harmonics is

$$\Delta\omega = (n + 1)\omega_0 - n\omega_0 = \omega_0 = \frac{2\pi}{T} \tag{18.51}$$

As $T \rightarrow \infty$ the lines in the frequency spectrum for $f_p(t)$ come closer and closer together, $\Delta\omega$ approaches the differential $d\omega$, and $n\omega_0$ can take on any value of ω. Under these conditions the line spectrum becomes a continuous spectrum. Since as $T \rightarrow \infty$, $\mathbf{c}_n \rightarrow 0$ in Eq. (18.49), we will examine the product $\mathbf{c}_n T$ where

$$\mathbf{c}_n T = \int_{-T/2}^{T/2} f_p(t) e^{-jn\omega_0 t}\, dt$$

In the limit as $T \rightarrow \infty$,

$$\lim_{T\rightarrow\infty} (\mathbf{c}_n T) = \lim_{T\rightarrow\infty} \int_{-T/2}^{T/2} f_p(t) e^{-jn\omega_0 t}\, dt$$

which in view of the previous discussion can be written as

$$\lim_{T\rightarrow\infty} (\mathbf{c}_n T) = \int_{-\infty}^{\infty} f(t) e^{-j\omega t}\, dt$$

This integral is the Fourier transform of $f(t)$, which we will denote as $\mathbf{F}(\omega)$, and hence

$$\mathbf{F}(\omega) = \int_{-\infty}^{\infty} f(t) e^{-j\omega t}\, dt \tag{18.52}$$

Similarly, $f_p(t)$ can be expressed as

$$\begin{aligned} f_p(t) &= \sum_{n=-\infty}^{\infty} \mathbf{c}_n e^{jn\omega_0 t} \\ &= \sum_{n=-\infty}^{\infty} (\mathbf{c}_n T) e^{jn\omega_0 t} \frac{1}{T} \\ &= \sum_{n=-\infty}^{\infty} (\mathbf{c}_n T) e^{jn\omega_0 t} \frac{\Delta\omega}{2\pi} \end{aligned}$$

which in the limit as $T \rightarrow \infty$ becomes

$$f(t) = \frac{1}{2\pi} \int_{-\infty}^{\infty} \mathbf{F}(\omega) e^{j\omega t}\, d\omega \tag{18.53}$$

Equations (18.52) and (18.53) constitute what is called the *Fourier transform pair*. Since $\mathbf{F}(\omega)$ is the Fourier transform of $f(t)$ and $f(t)$ is the inverse Fourier transform of $\mathbf{F}(\omega)$, they are normally expressed in the form

$$\mathbf{F}(\omega) = \mathcal{F}[f(t)] = \int_{-\infty}^{\infty} f(t) e^{-j\omega t}\, dt \tag{18.54}$$

$$f(t) = \mathcal{F}^{-1}[\mathbf{F}(\omega)] = \frac{1}{2\pi} \int_{-\infty}^{\infty} \mathbf{F}(\omega) e^{j\omega t}\, d\omega \tag{18.55}$$

Some Important Transform Pairs

There are a number of important Fourier transform pairs and in the following material we will derive a number of them and then list some of the more common ones in tabular form.

EXAMPLE 18.15

We wish to derive the Fourier transform for the voltage pulse shown in Fig. 18.18a.

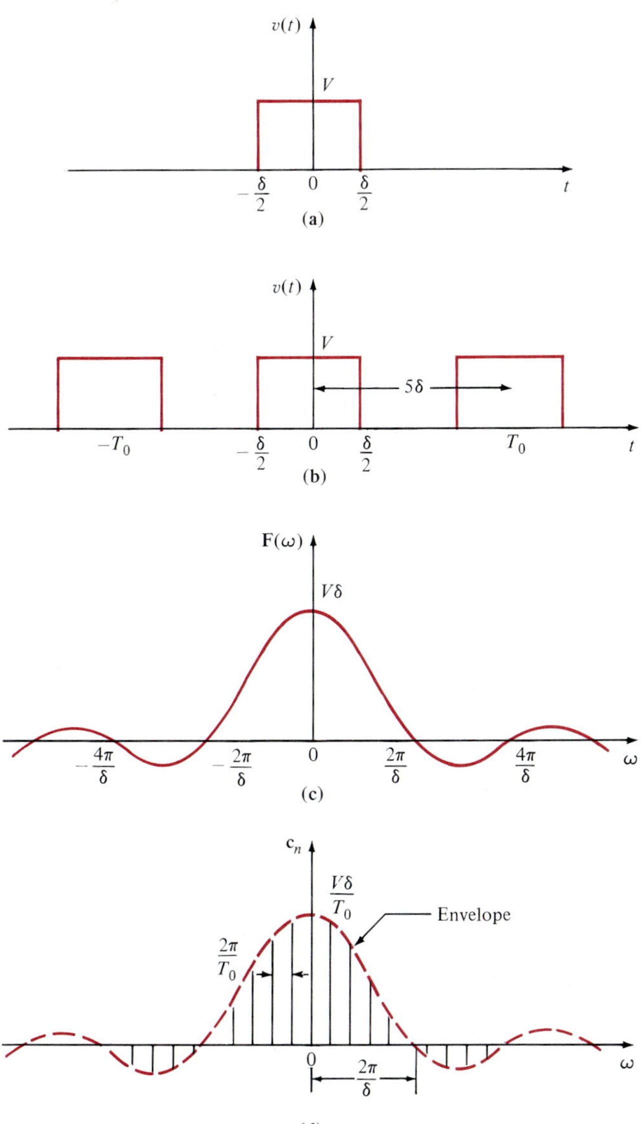

Figure 18.18 Pulses and their spectra.

Using Eq. (18.54), the Fourier transform is

$$\mathbf{F}(\omega) = \int_{-\delta/2}^{\delta/2} Ve^{-j\omega t}\,dt$$

$$= \frac{V}{j\omega}e^{-j\omega t}\bigg|_{-\delta/2}^{\delta/2}$$

$$= V\frac{e^{-j\omega\delta/2} - e^{+j\omega\delta/2}}{-j\omega}$$

$$= V\delta\frac{\sin(\omega\delta/2)}{\omega\delta/2}$$

Therefore, the Fourier transform for the function

$$f(t) = \begin{cases} 0 & -\infty < t \leq -\dfrac{\delta}{2} \\ V & -\dfrac{\delta}{2} < t \leq \dfrac{\delta}{2} \\ 0 & \dfrac{\delta}{2} < t < \infty \end{cases}$$

is

$$\mathbf{F}(\omega) = V\delta\frac{\sin(\omega\delta/2)}{\omega\delta/2}$$

A plot of this function is shown in Fig. 18.18c. Let us explore this example even further. Consider now the pulse train shown in Fig. 18.18b. Using the techniques that have been demonstrated earlier, we can show that the Fourier coefficients for this waveform are

$$\mathbf{c}_n = \frac{V\delta}{T_0}\frac{\sin(n\omega_0\delta/2)}{n\omega_0\delta/2}$$

The line spectrum for $T_0 = 5\delta$ is shown in Fig. 18.18d.

What these equations and figures in this example indicate are that as $T_0 \to \infty$ and the periodic function becomes aperiodic, the lines in the discrete spectrum become denser and the amplitude gets smaller, and the amplitude spectrum changes from a line spectrum to a continuous spectrum. Note that the envelope for the discrete spectrum has the same shape as the continuous spectrum. Since the Fourier series represents the amplitude and phase of the signal at specific frequencies, the Fourier transform also specifies the frequency content of a signal. ■

EXAMPLE 18.16

The Fourier transform of the unit impulse function $\delta(t - a)$ is

$$\mathbf{F}(\omega) = \int_{-\infty}^{\infty} \delta(t - a)e^{-j\omega t}\,dt$$

Using the sampling property of the unit impulse, we find that

$$\mathbf{F}(\omega) = e^{-j\omega a}$$

and if $a = 0$, then

$$\mathbf{F}(\omega) = 1$$

Note then that the $\mathbf{F}(\omega)$ for $f(t) = \delta(t)$ is *constant for all frequencies*. This is an important property, as we shall see later. ■

EXAMPLE 18.17

We wish to determine the Fourier transform of the function $f(t) = e^{j\omega_0 t}$.

In this case we note that if $\mathbf{F}(\omega) = 2\pi\delta(\omega - \omega_0)$, then

$$\begin{aligned} f(t) &= \frac{1}{2\pi}\int_{-\infty}^{\infty} 2\pi\delta(\omega - \omega_0)e^{j\omega t}\,d\omega \\ &= e^{j\omega_0 t} \end{aligned}$$

Therefore, $f(t) = e^{j\omega_0 t}$ and $\mathbf{F}(\omega) = 2\pi\delta(\omega - \omega_0)$ represent a Fourier transform pair. ■

EXAMPLE 18.18

The Fourier transform of the function $f(t) = \cos\omega_0 t$ is

$$\begin{aligned} \mathbf{F}(\omega) &= \int_{-\infty}^{\infty} \cos\omega_0 t e^{-j\omega t}\,dt \\ &= \frac{1}{2}\int_{-\infty}^{\infty} e^{j\omega_0 t}e^{-j\omega t}\,dt + \frac{1}{2}\int_{-\infty}^{\infty} e^{-j\omega_0 t}e^{-j\omega t}\,dt \end{aligned}$$

Using the results of Example 18.17, we obtain

$$\mathbf{F}(\omega) = \pi\delta(\omega - \omega_0) + \pi\delta(\omega + \omega_0)$$ ■

EXAMPLE 18.19

Let us determine the Fourier transform of the function $f(t) = e^{-at}u(t)$.

The Fourier transform of this function is

$$\begin{aligned} \mathbf{F}(\omega) &= \int_{-\infty}^{\infty} e^{-at}u(t)e^{-j\omega t}\,dt \\ &= \int_{0}^{\infty} e^{-at}e^{-j\omega t}\,dt \\ &= \frac{-1}{a + j\omega} e^{-at}e^{-j\omega t}\Big|_{0}^{\infty} \\ &= \frac{1}{a + j\omega} \end{aligned}$$

for $a > 0$ since $e^{-at} \to 0$ as $t \to \infty$. ■

A number of useful Fourier transform pairs are shown in Table 18.2.

Table 18.2 Fourier Transform Pairs

$f(t)$	$\mathbf{F}(\omega)$
$\delta(t-a)$	$e^{-j\omega a}$
A	$2\pi A\delta(\omega)$
$e^{j\omega_0 t}$	$2\pi\delta(\omega-\omega_0)$
$\cos\omega_0 t$	$\pi\delta(\omega-\omega_0)+\pi\delta(\omega+\omega_0)$
$\sin\omega_0 t$	$j\pi\delta(\omega+\omega_0)-j\pi\delta(\omega-\omega_0)$
$e^{-at}u(t),\quad a>0$	$\dfrac{1}{a+j\omega}$
$e^{-a\lvert t\rvert},\quad a>0$	$\dfrac{2a}{a^2+\omega^2}$
$e^{-at}\cos\omega_0 tu(t),\quad a>0$	$\dfrac{j\omega+a}{(j\omega+a)^2+\omega_0^2}$
$e^{-at}\sin\omega_0 tu(t),\quad a>0$	$\dfrac{\omega_0}{(j\omega+a)^2+\omega_0^2}$

DRILL EXERCISE

D18.13. If $f(t)=\sin\omega_0 t$, find $\mathbf{F}(\omega)$.

Ans: $\mathbf{F}(\omega)=\pi j[\delta(\omega+\omega_0)-\delta(\omega-\omega_0)]$.

Some Properties of the Fourier Transform

Let us examine now some of the properties of the Fourier transform defined by the equation

$$\mathbf{F}(\omega)=\int_{-\infty}^{\infty} f(t)e^{-j\omega t}\,dt$$

Using Euler's identity, we can write this function as

$$\mathbf{F}(\omega)=A(\omega)+jB(\omega) \tag{18.56}$$

where

$$A(\omega)=\int_{-\infty}^{\infty} f(t)\cos\omega t\,dt \tag{18.57}$$

$$B(\omega)=-\int_{-\infty}^{\infty} f(t)\sin\omega t\,dt \tag{18.58}$$

From Eq. (18.56) we note that

$$\mathbf{F}(\omega) = |\mathbf{F}(\omega)|e^{j\theta(\omega)} \tag{18.59}$$

where

$$|\mathbf{F}(\omega)| = \sqrt{A^2(\omega) + B^2(\omega)} \tag{18.60}$$

$$\theta(\omega) = \tan^{-1}\frac{B(\omega)}{A(\omega)} \tag{18.61}$$

Given the definitions above, we can show that

$$\begin{aligned} A(\omega) &= A(-\omega) \\ B(\omega) &= -B(-\omega) \\ \mathbf{F}(-\omega) &= \mathbf{F}^*(\omega) \end{aligned} \tag{18.62}$$

Equations (18.62) simply state that $A(\omega)$ is an even function of ω, $B(\omega)$ is an odd function of ω, and $\mathbf{F}(\omega)$ evaluated at $-\omega$ is the complex conjugate of $\mathbf{F}(\omega)$. Therefore, $|\mathbf{F}(\omega)|$ is an even function of ω and $\theta(\omega)$ is an odd function of ω. Some additional properties of the Fourier transform are now presented in rapid succession.

Linearity. If $\mathcal{F}[f_1(t)] = \mathbf{F}_1(\omega)$ and $\mathcal{F}[f_2(t)] = \mathbf{F}_2(\omega)$, then

$$\begin{aligned} \mathcal{F}[af_1(t) + bf_2(t)] &= \int_{-\infty}^{\infty} [af_1(t) + bf_2(t)]e^{-j\omega t}\,dt \\ &= \int_{-\infty}^{\infty} af_1(t)e^{-j\omega t}\,dt + \int_{-\infty}^{\infty} bf_2(t)e^{-j\omega t}\,dt \\ &= a\mathbf{F}_1(\omega) + b\mathbf{F}_2(\omega) \end{aligned} \tag{18.63}$$

Time Scaling. If $\mathcal{F}[f(t)] = \mathbf{F}(\omega)$, then if $a > 0$,

$$\mathcal{F}[f(at)] = \int_{-\infty}^{\infty} f(at)e^{-j\omega t}\,dt$$

If we now let $x = at$, then

$$\begin{aligned} \mathcal{F}[f(at)] &= \int_{-\infty}^{\infty} f(x)e^{-j\omega x/a}\,\frac{dx}{a} \\ &= \frac{1}{a}\mathbf{F}\left(\frac{\omega}{a}\right) \end{aligned} \tag{18.64}$$

and if $a < 0$, then following the previous development,

$$\begin{aligned} \mathcal{F}[f(at)] &= \int_{+\infty}^{-\infty} f(x)e^{-j\omega x/a}\,\frac{dx}{a} \\ &= -\frac{1}{a}\mathbf{F}\left(\frac{\omega}{a}\right) \end{aligned} \tag{18.65}$$

Time Shifting. If $\mathcal{F}[f(t)] = \mathbf{F}(\omega)$, then

$$\mathcal{F}[f(t - t_0)] = \int_{-\infty}^{\infty} f(t - t_0)e^{-j\omega t}\, dt$$

If we now let $x = t - t_0$, then

$$\mathcal{F}[f(t - t_0)] = \int_{-\infty}^{\infty} f(x)e^{-j\omega(x+t_0)}\, dx$$

$$= e^{-j\omega t_0}\mathbf{F}(\omega) \tag{18.66}$$

Modulation. If $\mathcal{F}[f(t)] = \mathbf{F}(\omega)$, then

$$\mathcal{F}[e^{j\omega_0 t}f(t)] = \int_{-\infty}^{\infty} f(t)e^{j\omega_0 t}e^{-j\omega t}\, dt$$

$$= \int_{-\infty}^{\infty} f(t)e^{-j(\omega-\omega_0)t}\, dt$$

$$= \mathbf{F}(\omega - \omega_0) \tag{18.67}$$

Differentiation. If $f(t)$ is a function that is continuous in any finite interval, is absolutely integrable, and its first derivative is piecewise continuous and absolutely integrable, then

$$\mathcal{F}[f'(t)] = \int_{-\infty}^{\infty} f'(t)e^{-j\omega t}\, dt$$

Using integration by parts, we obtain

$$\mathcal{F}[f'(t)] = f(t)e^{-j\omega t}\Big|_{-\infty}^{\infty} + \int_{-\infty}^{\infty} j\omega f(t)e^{-j\omega t}\, dt$$

Since $f(t)$ is absolutely integrable,

$$\lim_{t\to-\infty} |f(t)| = \lim_{t\to+\infty} |f(t)| = 0$$

and therefore,

$$\mathcal{F}[f'(t)] = j\omega\mathbf{F}(\omega) \tag{18.68}$$

Repeated application of this technique under the conditions listed above yields

$$\mathcal{F}[f^{(n)}(t)] = (j\omega)^n\mathbf{F}(\omega) \tag{18.69}$$

It is important to note that Eq. (18.68) does not guarantee the existence of $\mathcal{F}[f'(t)]$. It simply states that *if* the transform exists, it is given by Eq. (18.68).

Time Convolution. If $\mathcal{F}[f_1(t)] = \mathbf{F}_1(\omega)$ and $\mathcal{F}[f_2(t)] = \mathbf{F}_2(\omega)$, then

$$\mathcal{F}\left[\int_{-\infty}^{\infty} f_1(x)f_2(t - x)\, dx\right] = \int_{t=-\infty}^{\infty}\int_{x=-\infty}^{\infty} f_1(x)f_2(t - x)\, dx\, e^{-j\omega t}\, dt$$

$$= \int_{x=-\infty}^{\infty} f_1(x)\int_{t=-\infty}^{\infty} f_2(t - x)e^{-j\omega t}\, dt\, dx$$

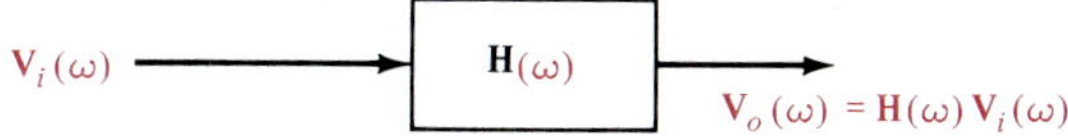

Figure 18.19 Representation of the time convolution property.

If we now let $u = t - x$, then

$$\mathcal{F}\left[\int_{-\infty}^{\infty} f_1(x)f_2(t-x)\,dx\right] = \int_{x=-\infty}^{\infty} f_1(x)\int_{u=-\infty}^{\infty} f_2(u)e^{-j\omega(u+x)}\,du\,dx$$

$$= \int_{x=-\infty}^{\infty} f_1(x)e^{-j\omega x}\int_{u=-\infty}^{\infty} f_2(u)e^{-j\omega u}\,du\,dx$$

$$= \mathbf{F}_1(\omega)\mathbf{F}_2(\omega) \tag{18.70}$$

We should note very carefully the time convolution property of the Fourier transform. With reference to Fig. 18.19, this property states that if $\mathbf{V}_i(\omega) = \mathcal{F}[v_i(t)]$, $\mathbf{H}(\omega) = \mathcal{F}[h(t)]$, and $\mathbf{V}_o(\omega) = \mathcal{F}[v_o(t)]$, then

$$\mathbf{V}_o(\omega) = \mathbf{H}(\omega)\mathbf{V}_i(\omega) \tag{18.71}$$

where $\mathbf{V}_i(\omega)$ represents the input signal, $\mathbf{H}(\omega)$ is the network transfer function, and $\mathbf{V}_o(\omega)$ represents the output signal. Equation (18.71) tacitly assumes that the initial conditions of the network are zero.

Table 18.3 provides a short list of some of the Fourier transform properties.

Table 18.3 Properties of the Fourier Transform

$f(t)$	$\mathbf{F}(\omega)$
$Af(t)$	$A\mathbf{F}(\omega)$
$f_1(t) \pm f_2(t)$	$\mathbf{F}_1(\omega) \pm \mathbf{F}_2(\omega)$
$f(at)$	$\frac{1}{a}\mathbf{F}\left(\frac{\omega}{a}\right),\ a > 0$
$f(t - t_0)$	$e^{-j\omega t_0}\mathbf{F}(\omega)$
$e^{j\omega t_0}f(t)$	$F(\omega - \omega_0)$
$\frac{d^n f(t)}{dt^n}$	$(j\omega)^n\mathbf{F}(\omega)$
$t^n f(t)$	$(j)^n \frac{d^n\mathbf{F}(\omega)}{d\omega^n}$
$f(t)\cos\omega_0 t$	$\frac{1}{2}[\mathbf{F}(\omega - \omega_0) + \mathbf{F}(\omega + \omega_0)]$
$\int_{-\infty}^{\infty} f_1(x)f_2(t-x)\,dx$	$\mathbf{F}_1(\omega)\mathbf{F}_2(\omega)$
$f_1(t)f_2(t)$	$\frac{1}{2\pi}\int_{-\infty}^{\infty}\mathbf{F}_1(x)\mathbf{F}_2(\omega - x)\,dx$

EXAMPLE 18.20

Let us determine the Fourier transform of the function $f(t - t_0) = e^{-(t-t_0)}u(t - t_0)$.

By definition,

$$\mathscr{F}[f(t - t_0)] = \int_{-\infty}^{\infty} f(t - t_0)e^{-j\omega t}\, dt$$

which can be written using the time-shifting property as

$$\mathscr{F}[f(t - t_0)] = e^{-j\omega t_0}\int_{-\infty}^{\infty} f(t)e^{-j\omega t}\, dt$$

$$= e^{-j\omega t_0}\int_{-\infty}^{\infty} e^{-t}u(t)e^{-j\omega t}\, dt$$

which from the results of Example 18.19 is

$$\mathscr{F}[f(t - t_0)] = \frac{e^{-j\omega t_0}}{1 + j\omega}$$

■

EXAMPLE 18.21

Let us determine the Fourier transform of the function

$$f(t) = \frac{d}{dt}[e^{-at}u(t)] = -ae^{-at}u(t) + e^{-at}\,\delta(t)$$

Then

$$\mathscr{F}[f(t)] = \int_{-\infty}^{\infty} [-ae^{-at}u(t) + e^{-at}\,\delta(t)]e^{-j\omega t}\, dt$$

$$= -\int_{-\infty}^{\infty} ae^{-at}u(t)e^{-j\omega t}\, dt + \int_{-\infty}^{\infty} e^{-at}\,\delta(t)e^{-j\omega t}\, dt$$

$$= \frac{-a}{a + j\omega} + 1$$

$$= \frac{j\omega}{a + j\omega}$$

The transform could also be evaluated using the differentiation property; that is, if

$$\mathscr{F}[e^{-at}u(t)] = \frac{1}{a + j\omega}$$

then

$$\mathscr{F}\left[\frac{d}{dt}\left(e^{-at}u(t)\right)\right] = j\omega\left(\frac{1}{a + j\omega}\right) = \frac{j\omega}{a + j\omega}$$

■

DRILL EXERCISE

D18.14. If $\mathscr{F}[e^{-t}u(t)] = 1/(1 + j\omega)$, use the time-scaling property of the Fourier transform to find $\mathscr{F}[e^{-at}u(t)]$, $a > 0$.

Ans: $\dfrac{1}{a + j\omega}$.

D18.15. Use the property $\mathscr{F}[t^n f(t)] = j^n[d^n\mathbf{F}(\omega)/d\omega^n]$ to determine the Fourier transform of $te^{-at}u(t)$, $a > 0$.

Ans: $\dfrac{1}{(a + j\omega)^2}$.

D18.16. Determine the output $v_o(t)$ in Fig. D18.16 if the input signal $v_i(t) = e^{-t}u(t)$, the network impulse response $h(t) = e^{-2t}u(t)$, and all initial conditions are zero.

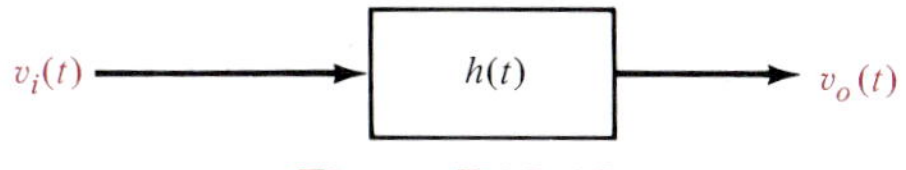

Figure D18.16

Ans: $v_o(t) = (e^{-t} - e^{-2t})u(t)$ V.

Parseval's Theorem

A mathematical statement of Parseval's theorem is

$$\int_{-\infty}^{\infty} f^2(t)\, dt = \frac{1}{2\pi}\int_{-\infty}^{\infty} |\mathbf{F}(\omega)|^2\, d\omega \tag{18.72}$$

This relationship can be easily derived as follows:

$$\begin{aligned}
\int_{-\infty}^{\infty} f^2(t)\, dt &= \int_{-\infty}^{\infty} f(t)\frac{1}{2\pi}\int_{-\infty}^{\infty} \mathbf{F}(\omega)e^{j\omega t}\, d\omega\, dt \\
&= \frac{1}{2\pi}\int_{-\infty}^{\infty} \mathbf{F}(\omega)\int_{-\infty}^{\infty} f(t)e^{-j(-\omega)t}\, dt\, d\omega \\
&= \frac{1}{2\pi}\int_{-\infty}^{\infty} \mathbf{F}(\omega)\mathbf{F}(-\omega)\, d\omega \\
&= \frac{1}{2\pi}\int_{-\infty}^{\infty} \mathbf{F}(\omega)\mathbf{F}^*(\omega)\, d\omega \\
&= \frac{1}{2\pi}\int_{-\infty}^{\infty} |\mathbf{F}(\omega)|^2\, d\omega
\end{aligned}$$

The importance of Parseval's theorem can be seen if we imagine that $f(t)$ represents the current in a 1-Ω resistor. Since $f^2(t)$ is power and the integral of power over time is energy, Eq. (18.72) shows that we can compute this 1-Ω energy or normalized energy in either the time domain or the frequency domain.

Applications

Let us now apply to circuit problems some of the things we have learned about the Fourier transform.

EXAMPLE 18.22

Using the transform technique, we wish to determine $v_0(t)$ in Fig. 18.20 if (a) $v_i(t) = 5e^{-2t}u(t)$ V and (b) $v_i(t) = 5 \cos 2t$ V.

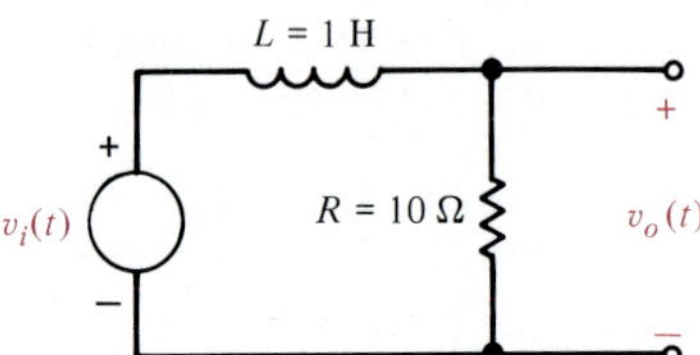

Figure 18.20 Simple *RL* circuit.

(a) In this case since $v_i(t) = 5e^{-2t}u(t)$ V, then

$$\mathbf{V}_i(\omega) = \frac{5}{2 + j\omega} \text{ V}$$

$\mathbf{H}(\omega)$ for the network is

$$\mathbf{H}(\omega) = \frac{R}{R + j\omega L}$$

$$= \frac{10}{10 + j\omega}$$

From Eq. (18.71),

$$\mathbf{V}_o(\omega) = \mathbf{H}(\omega)\mathbf{V}_i(\omega)$$

$$= \frac{50}{(2 + j\omega)(10 + j\omega)}$$

$$= \frac{50}{8}\left(\frac{1}{2 + j\omega} - \frac{1}{10 + j\omega}\right) \text{ V}$$

Hence, from Table 18.3, we see that

$$v_o(t) = 6.25[e^{-2t}u(t) - e^{-10t}u(t)] \text{ V}$$

(b) In this case, since $v_i(t) = 5 \cos 2t$,

$$\mathbf{V}_i(\omega) = 5\pi\delta(\omega - 2) + 5\pi\delta(\omega + 2) \text{ V}$$

The output voltage in the frequency domain is then

$$\mathbf{V}_o(\omega) = \frac{50\pi[\delta(\omega - 2) + \delta(\omega + 2)]}{(10 + j\omega)}$$

Using the inverse Fourier transform gives us

$$v_o(t) = \mathcal{F}^{-1}[\mathbf{V}_o(\omega)] = \frac{1}{2\pi}\int_{-\infty}^{\infty} 50\pi \frac{\delta(\omega - 2) + \delta(\omega + 2)}{10 + j\omega} e^{j\omega t}\, d\omega$$

Employing the sampling property of the unit impulse function, we obtain

$$\begin{aligned} v_o(t) &= 25\left(\frac{e^{j2t}}{10 + j2} + \frac{e^{-j2t}}{10 - j2}\right) \\ &= 25\left(\frac{e^{j2t}}{10.2e^{j11.31°}} + \frac{e^{-j2t}}{10.2e^{-j11.31°}}\right) \\ &= 4.90 \cos(2t - 11.31°) \text{ V} \end{aligned}$$

This result can be easily checked using phasor analysis. ■

EXAMPLE 18.23

Consider the network shown in Fig. 18.21a. This network represents a simple low-pass filter as shown in Chapter 14. We wish to illustrate the impact of this network on the input signal by examining the frequency characteristics of the output signal and the relationship between the 1-Ω or normalized energy at the input and output of the network.

The network transfer function is

$$\mathbf{H}(\omega) = \frac{1/RC}{1/RC + j\omega} = \frac{5}{5 + j\omega} = \frac{1}{1 + 0.2j\omega}$$

The Fourier transform of the input signal is

$$\mathbf{V}_i(\omega) = \frac{20}{20 + j\omega} = \frac{1}{1 + 0.05j\omega}$$

Then, using Eq. (18.71), the Fourier transform of the output is

$$\mathbf{V}_o(\omega) = \frac{1}{(1 + 0.2j\omega)(1 + 0.05j\omega)}$$

Using the techniques of Chapter 14, we note that the straight-line log-magnitude plot (frequency characteristic) for these functions is shown in Fig. 18.21b–d. Note that the low-pass filter passes the low frequencies of the input signal but attenuates the high frequencies.

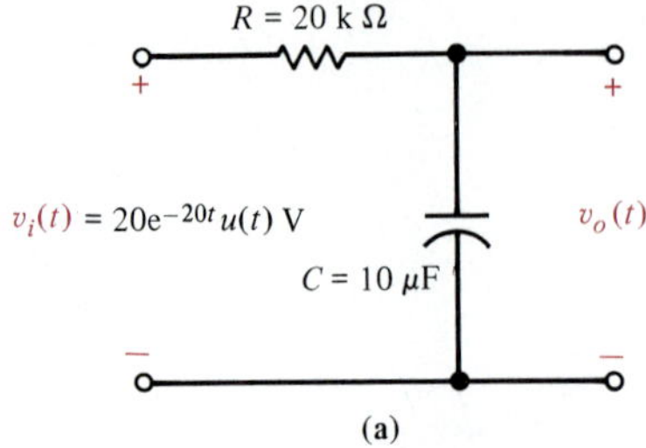

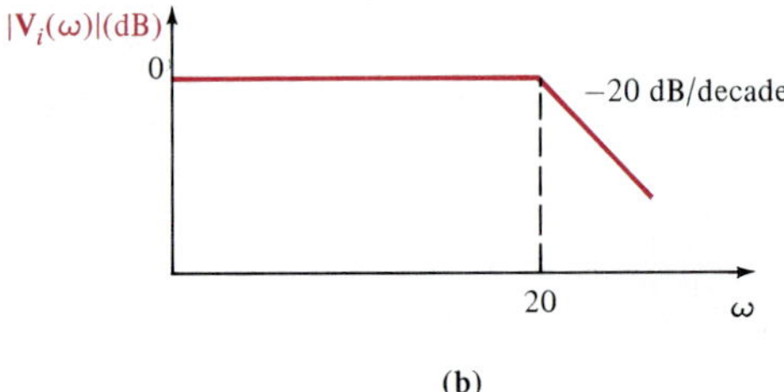

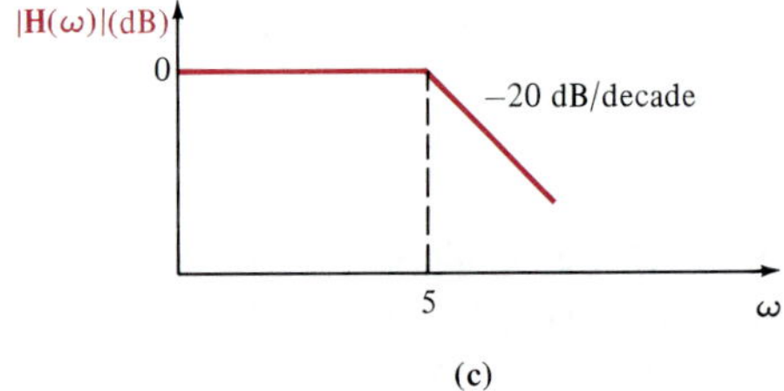

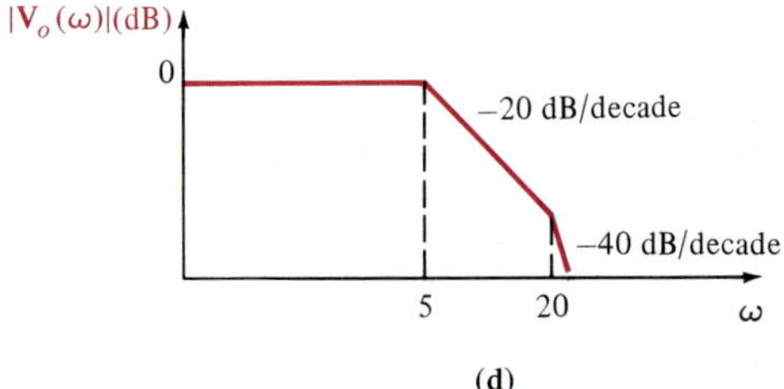

Figure 18.21 Low-pass filter, its frequency characteristic, and its input and output spectra.

The normalized energy at the filter input is

$$\begin{aligned} W_i &= \int_0^\infty (20e^{-20t})^2 \, dt \\ &= \frac{400}{-40} e^{-40t} \Big|_0^\infty \\ &= 10 \text{ J} \end{aligned}$$

The normalized energy at the filter output can be computed using Parseval's theorem. Since

$$\mathbf{V}_o(\omega) = \frac{100}{(5 + j\omega)(20 + j\omega)}$$

and

$$|\mathbf{V}_o(\omega)|^2 = \frac{10^4}{(\omega^2 + 25)(\omega^2 + 400)}$$

$|\mathbf{V}_0(\omega)|^2$ is an even function, and therefore,

$$W_o = 2\left(\frac{1}{2\pi}\right)\int_0^\infty \frac{10^4 \, d\omega}{(\omega^2 + 25)(\omega^2 + 400)}$$

However, we can use the fact that

$$\frac{10^4}{(\omega^2 + 25)(\omega^2 + 400)} = \frac{10^4/375}{\omega^2 + 25} - \frac{10^4/375}{\omega^2 + 400}$$

Then

$$\begin{aligned} W_o &= \frac{1}{\pi}\left(\int_0^\infty \frac{10^4/375}{\omega^2 + 25}\, d\omega - \int_0^\infty \frac{10^4/375}{\omega^2 + 400}\, d\omega\right) \\ &= \frac{10^4}{375}\left(\frac{1}{\pi}\right)\left[\frac{1}{5}\left(\frac{\pi}{2}\right) - \frac{1}{20}\left(\frac{\pi}{2}\right)\right] \\ &= 2.0 \text{ J} \end{aligned}$$

■

Example 18.23 illustrates the effect that $\mathbf{H}(\omega)$ has on the frequency spectrum of the input signal. In general, $\mathbf{H}(\omega)$ can be selected to shape that spectrum in some prescribed manner. As an illustration of this effect, consider the *ideal* frequency spectrums shown in Fig. 18.22. In Fig. 18.22a is shown an ideal input magnitude spectrum $|\mathbf{V}_i(\omega)|$. $|\mathbf{H}(\omega)|$ and the output magnitude spectrum $|\mathbf{V}_o(\omega)|$, which are related by Eq. (18.71), are shown in Fig. 18.22b–e for an *ideal* low-pass, high-pass, band-pass, and band-elimination filter, respectively.

DRILL EXERCISE

D18.17. Compute the total 1-Ω energy content of the signal $v_i(t) = e^{-2t}u(t)$ using both the time-domain and frequency-domain approaches.

Ans: $W_T = 0.25$ J.

D18.18. Compute the 1-Ω energy content of the signal $v_1(t) = e^{-2t}u(t)$ in the frequency range from 0 to 1 rad/sec.

Ans: $W = 0.07$ J.

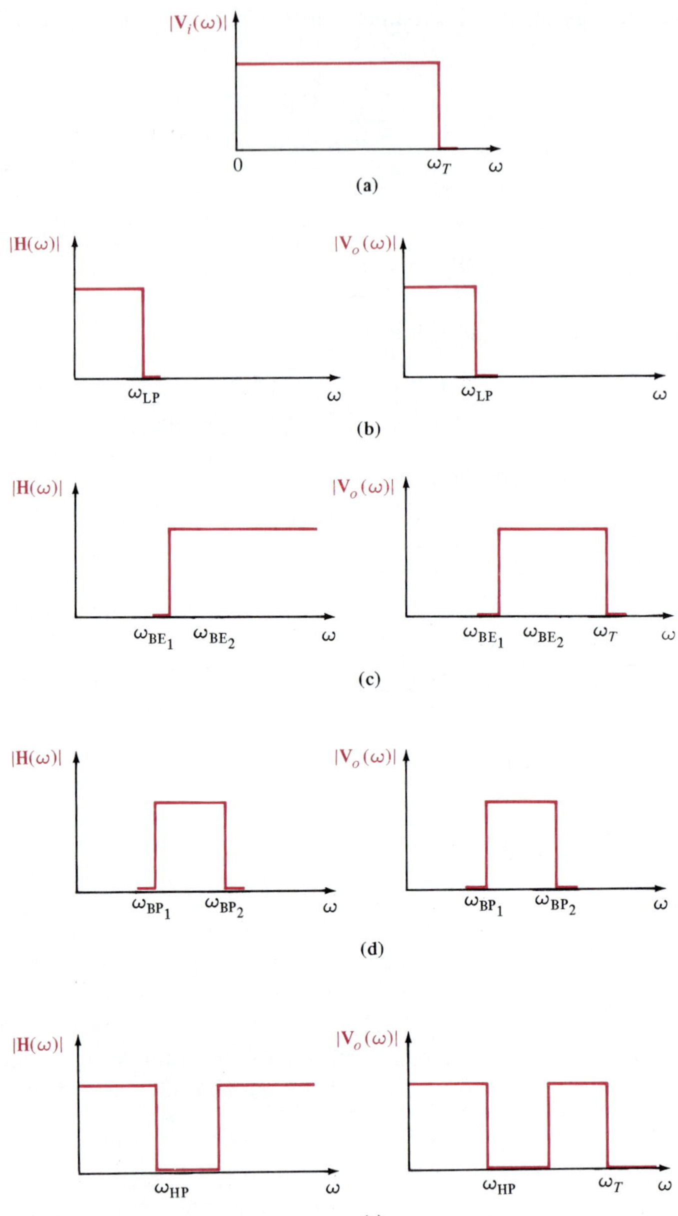

Figure 18.22 Frequency spectra for the input and output of ideal low-pass, high-pass, band-pass, and band-elimination filters.

We note that by using Parseval's theorem we can compute the total energy content of a signal using either a time-domain or frequency-domain approach. However, the frequency domain is more flexible in that it permits us to determine the energy content of a signal within some specified frequency band.

18.3 Summary

We have shown that the Fourier series technique provides us with an effective means to deal with electric circuits which are forced with inputs that are periodic. The Fourier series for a particular signal can be expressed in several different equivalent forms. If the periodic signal displays some form of symmetry, this symmetry can be used to simplify the calculations involved in determining the Fourier coefficients employed to describe the waveform. We found that the steady-state response of a network to a periodic input involved expressing the input as a Fourier series, determining the circuit response to each component of the input, and summing the response due to each input component to determine the total network response.

The Fourier transform provides us with a technique for analyzing circuits with aperiodic forcing functions. A number of Fourier transform pairs have been presented together with some important properties of the Fourier transform. Two important relationships from a circuit analysis point of view are Parseval's theorem and the equation $\mathbf{Y}(\omega) = \mathbf{H}(\omega)\mathbf{X}(\omega)$, where $\mathbf{Y}(\omega)$ and $\mathbf{X}(\omega)$ represent the Fourier transform of the circuit output and input, respectively, and $\mathbf{H}(\omega)$ is the circuit transfer function. We have shown that Parseval's theorem allows us to compute the energy in a signal in the frequency domain and even determine the amount of signal energy that is contained within a certain range of frequencies. The input–output relationship in the frequency domain was shown to be an effective means of computing the circuit response to aperiodic inputs.

KEY POINTS

- A periodic function satisfies the relationship $f(t) = f(t + nT_0)$, $n = \pm 1, \pm 2, \pm 3, \ldots$.
- A nonsinusoidal periodic function can be represented by a sum of sinusoidal functions.
- A Fourier series can be expressed in exponential or trigonometric form.
- A periodic function that exhibits even-function symmetry can be expressed by a Fourier series containing only a constant term and cosine terms.
- A periodic function that exhibits odd-function symmetry can be expressed by a Fourier series containing only sine terms.
- A time shift of a periodic signal in the time domain corresponds to a phase shift in the frequency domain.
- The frequency spectra of a function expressed as a Fourier series is called a line spectrum, since the frequency components are discrete.
- The network response of a nonsinusoidal periodic input can be obtained by expressing the input as a Fourier series, using phasor analysis to determine the response of each source in the series, then transforming the individual responses to the time domain and applying superposition.

- The Fourier transform pairs are expressed as

$$\mathbf{F}(\omega) = \mathcal{F}[f(t)] = \int_{-\infty}^{\infty} f(t)^{-j\omega t}\, dt$$

$$f(t) = \mathcal{F}^{-1}[\mathbf{F}(\omega)] = \frac{1}{2\pi}\int_{-\infty}^{\infty} \mathbf{F}(\omega)e^{j\omega t}\, d\omega$$

- $\mathbf{V}_o(j\omega) = \mathbf{H}(j\omega)\mathbf{V}_i(j\omega)$, where $\mathbf{V}_i(j\omega)$ represents the network input signal, $\mathbf{H}(j\omega)$ represents the network transfer function, and $\mathbf{V}_o(j\omega)$ represents the output signal of the network.
- Parseval's theorem can be used to compute the total energy content or the energy content over a specific frequency range of a signal using either a time-domain or frequency-domain approach.

PROBLEMS

18.1. Find the exponential Fourier series for the periodic pulse train shown in Fig. P18.1.

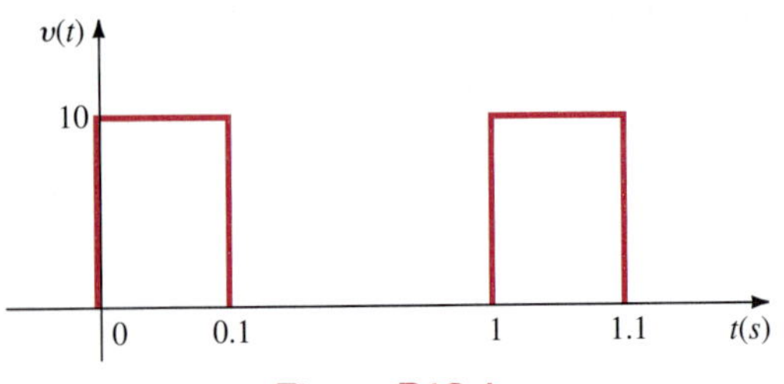

Figure P18.1

18.2. Find the exponential Fourier series for the signal shown in Fig. P18.2.

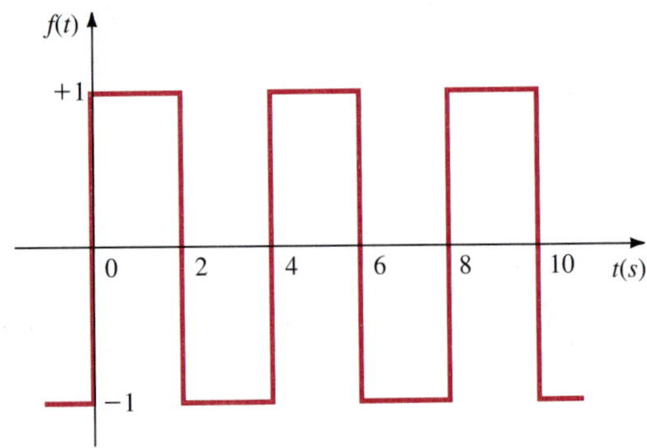

Figure P18.2

18.3. Find the exponential Fourier series for the periodic signal shown in Fig. P18.3.

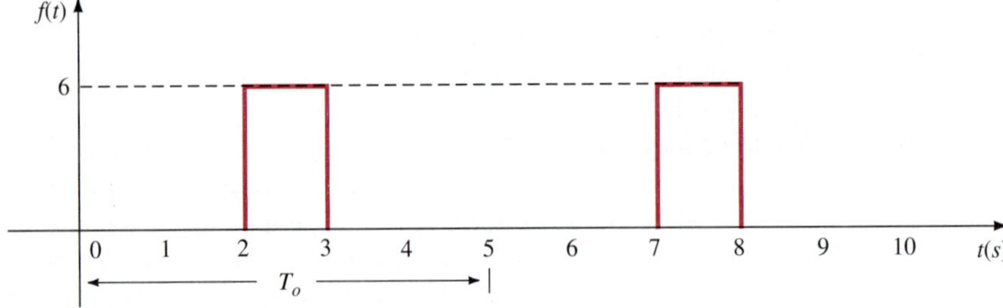

Figure P18.3

18.4. Compute the exponential Fourier series for the waveform that is the sum of the two waveforms in Fig. P18.4 by computing the exponential Fourier series of the two waveforms and adding them.

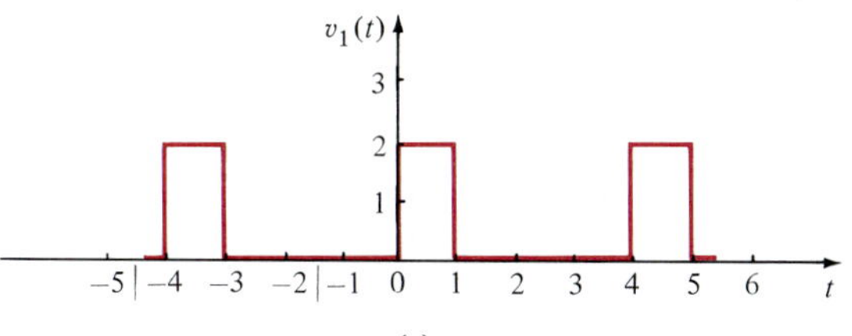

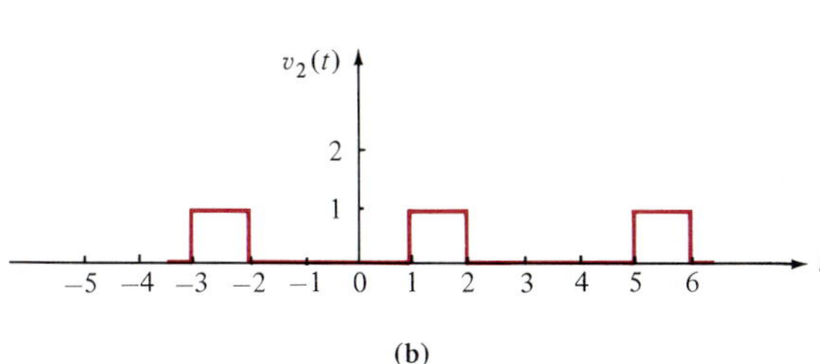

Figure P18.4

18.5. Use the Laplace transform technique to find the exponential Fourier series coefficients for the waveform in Fig. P18.2.

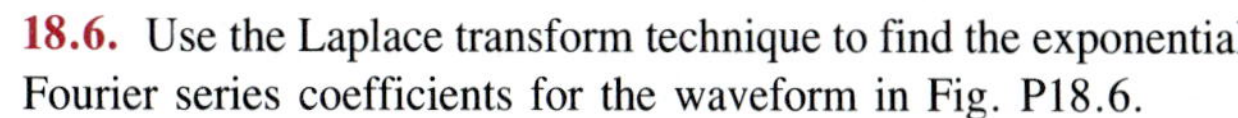
18.6. Use the Laplace transform technique to find the exponential Fourier series coefficients for the waveform in Fig. P18.6.

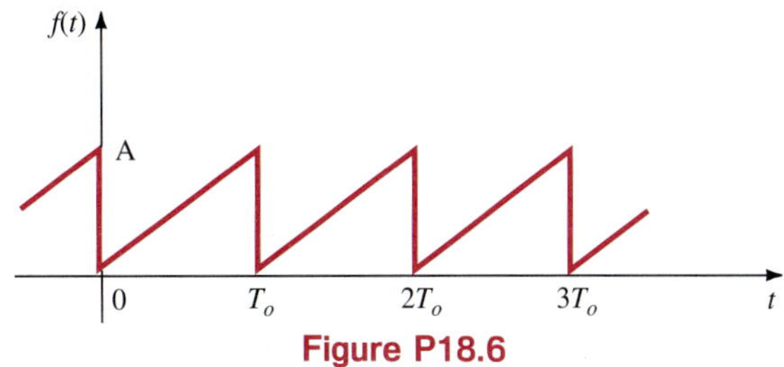

Figure P18.6

18.7. Use the derivative method to find the exponential Fourier series coefficients for the waveform in Fig. P18.3.

18.8. Use the derivative method to find the exponential Fourier series coefficients for the waveform in Fig. P18.8.

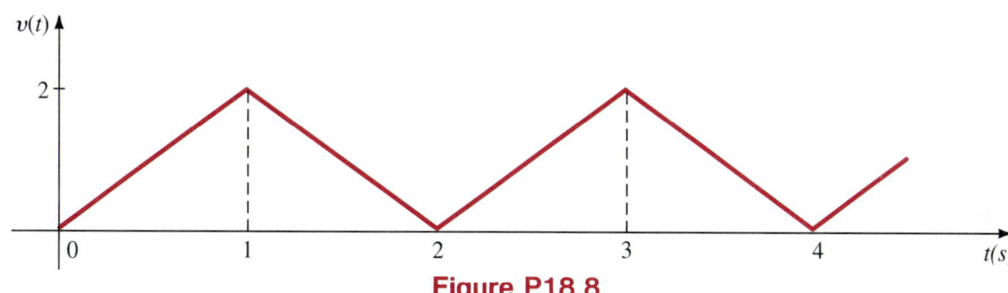

Figure P18.8

18.9. Use PSPICE to find the Fourier coefficients for the waveform in Fig. P18.16.

18.10. Use PSPICE to find the Fourier coefficients for the waveform in Fig. P18.25.

18.11. Use PSPICE to find the Fourier coefficients for the waveform in Fig. P18.24.

18.12. Given the waveform in Fig. P18.12, determine the type of symmetry that exists if the origin is selected at (a) l_1 and (b) l_2.

l_1 l_2

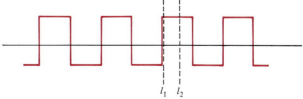

Figure P18.12

18.13. What type of symmetry is exhibited by the two waveforms in Fig. P18.13?

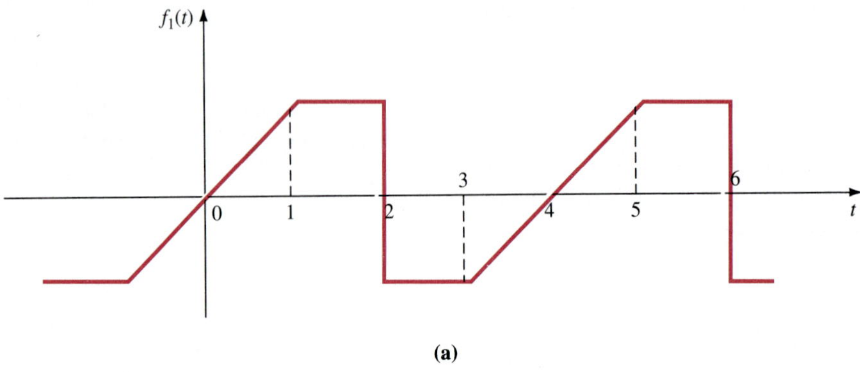

(a)

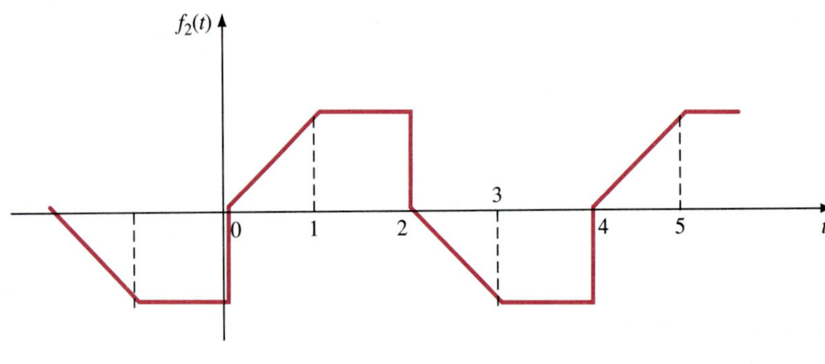

(b)

Figure P18.13

18.14. Derive Equation (18.36).

18.15. Given the waveform in Fig. P18.15, show that

$$f(t) = \frac{A}{2} + \sum_{n=1}^{\infty} \frac{-A}{n\pi} \sin \frac{2\pi n}{T_0} t$$

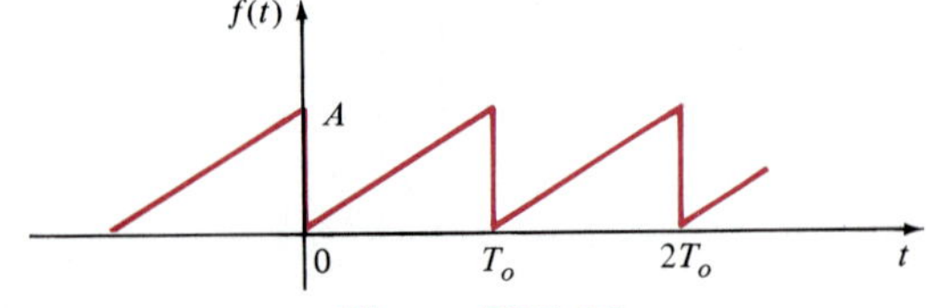

Figure P18.15

18.16. Derive the trigonometric Fourier series for the waveform shown in Fig. P18.16.

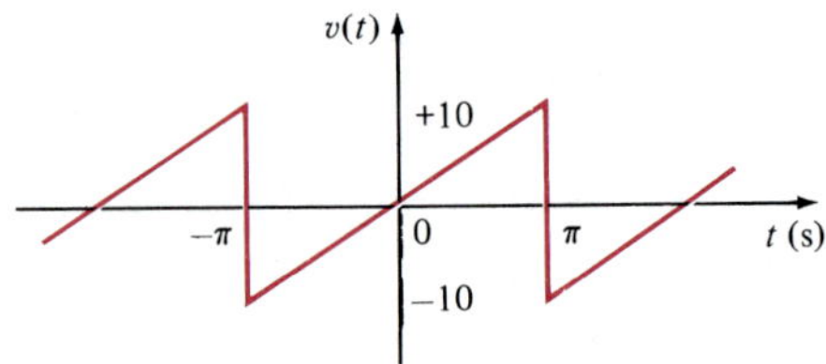

Figure P18.16

18.17. Find the trigonometric Fourier series coefficients for the waveform in Fig. P18.17.

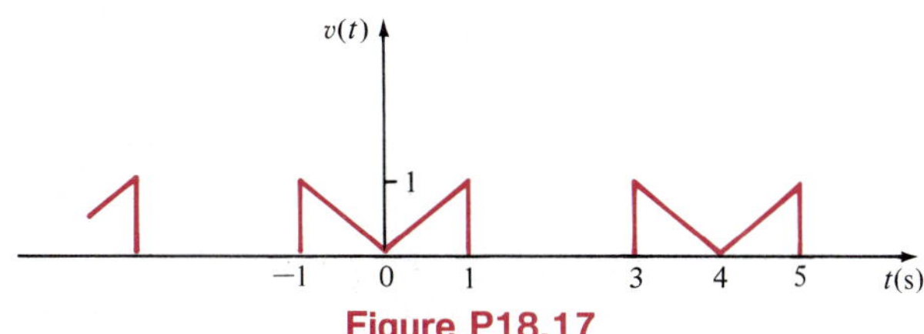

Figure P18.17

18.18. Find the trigonometric Fourier series coefficients for the waveform in Fig. P18.18.

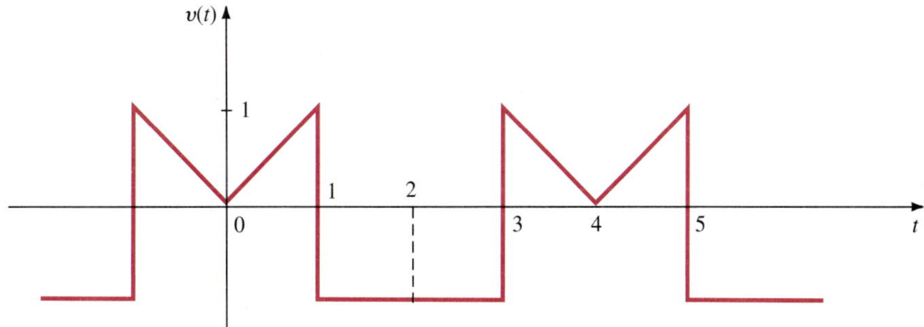

Figure P18.18

18.19. Find the trigonometric Fourier series coefficients for the waveform in Fig. P18.19.

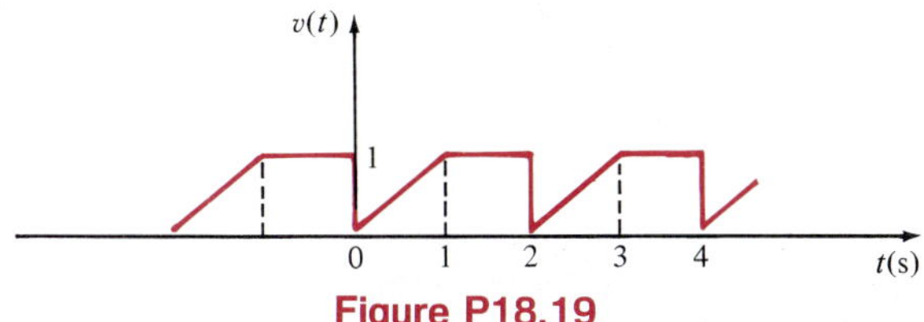

Figure P18.19

18.20. Find the trigonometric Fourier series coefficients for the waveform in Fig. P18.13a.

18.21. Derive the trigonometric Fourier series for the function $v(t) = A|\sin t|$ as shown in Fig. P18.21.

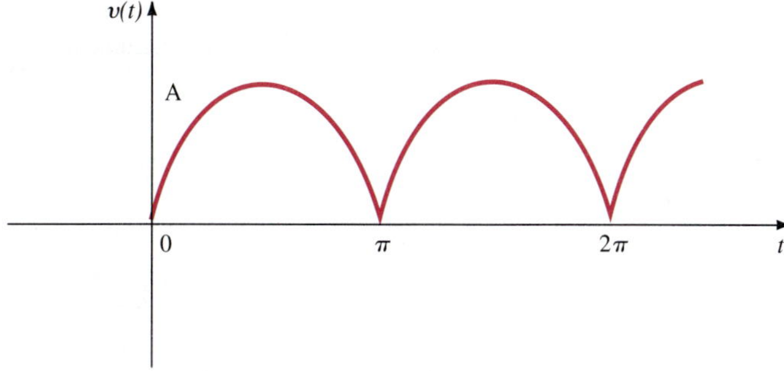

Figure P18.21

18.22. Derive the trigonometric Fourier series for the waveform shown in Fig. P18.22.

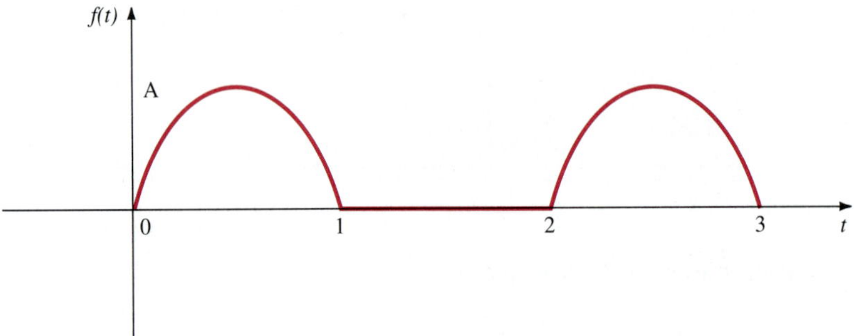

Figure P18.22

18.23. Find the trigonometric Fourier series for the waveform shown in Fig. P18.23.

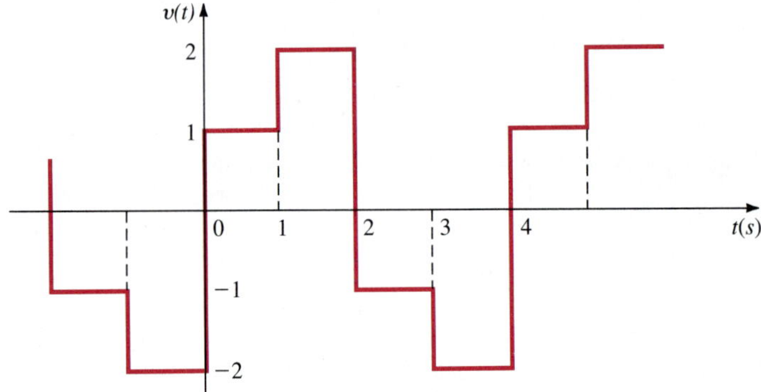

Figure P18.23

18.24. Derive the trigonometric Fourier series for the function shown in Fig. P18.24.

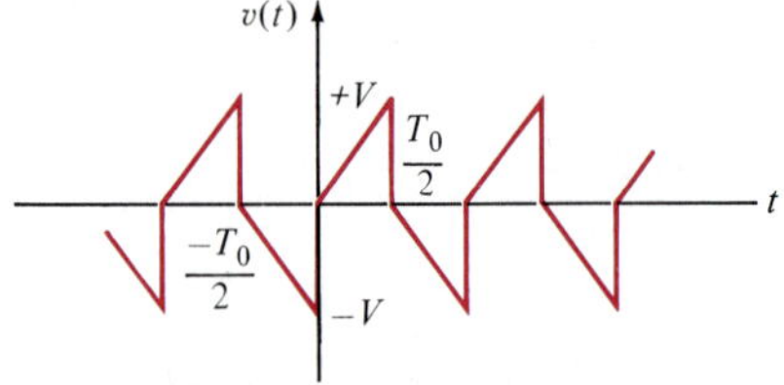

Figure P18.24

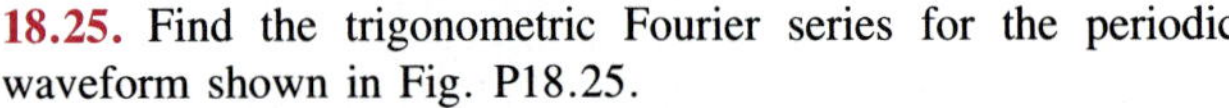

18.25. Find the trigonometric Fourier series for the periodic waveform shown in Fig. P18.25.

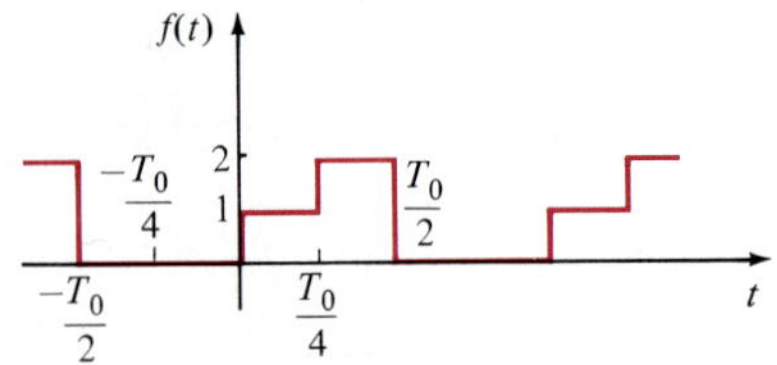

Figure P18.25

18.26. The discrete line spectrum for a periodic function $f(t)$ is shown in Fig. P18.26. Determine the expression for $f(t)$.

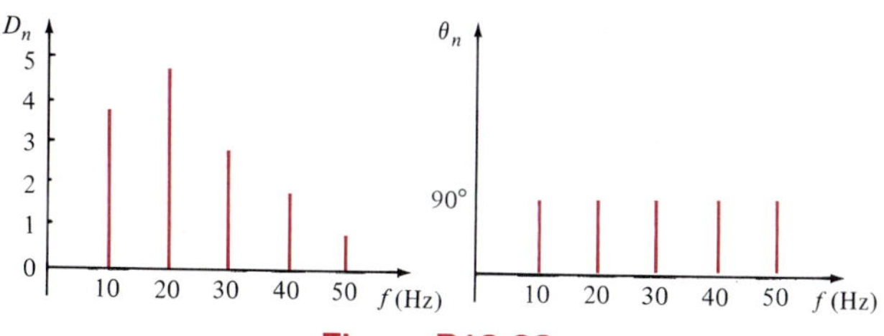

Figure P18.26

18.27. The amplitude and phase spectra for a periodic function $v(t)$ which has only a small number of terms is shown in Fig. P18.27. Determine the expression for $v(t)$ if $T_0 = 0.1$ sec.

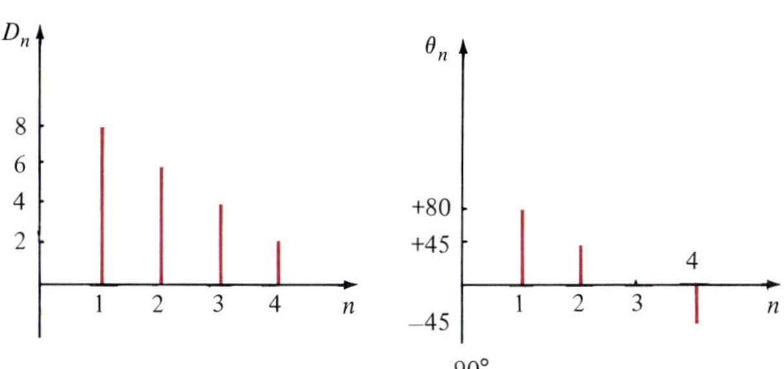

Figure P18.27

18.28. Plot the first four terms of the amplitude and phase spectra for the signal

$$f(t) = \sum_{\substack{n=1 \\ n \text{ odd}}}^{\infty} \frac{-2}{n\pi} \sin \frac{n\pi}{2} \cos n\omega_0 t + \frac{6}{n\pi} \sin n\omega_0 t$$

18.29. Determine the steady-state response of the current $i_o(t)$ in the circuit shown in Fig. P18.29 if the input voltage is described by the waveform shown in Problem 18.16.

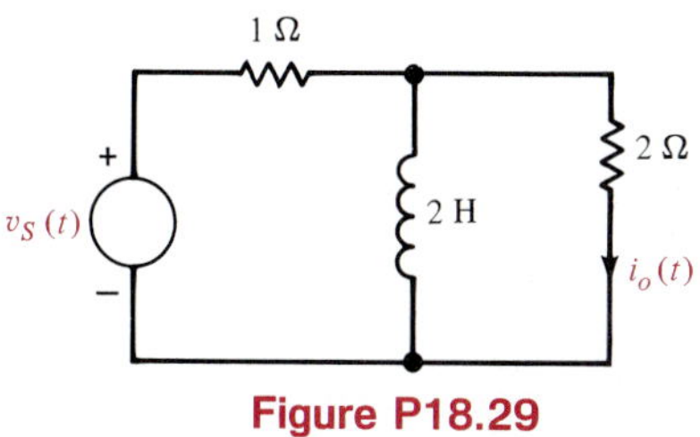

Figure P18.29

18.30. Find the steady-state current $i(t)$ in the network in Fig. P18.30a if the input signal is shown in Fig. P18.30b.

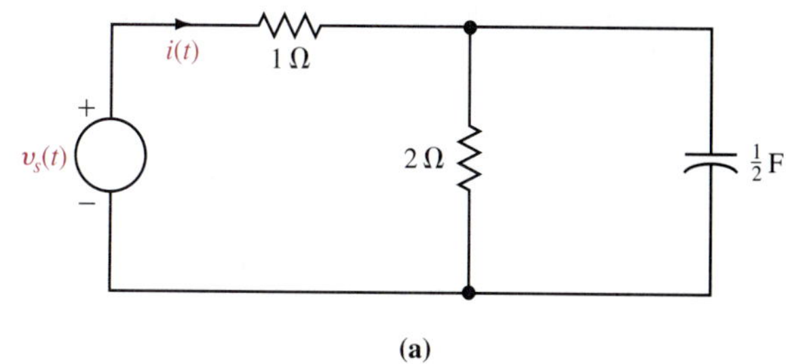

(a)

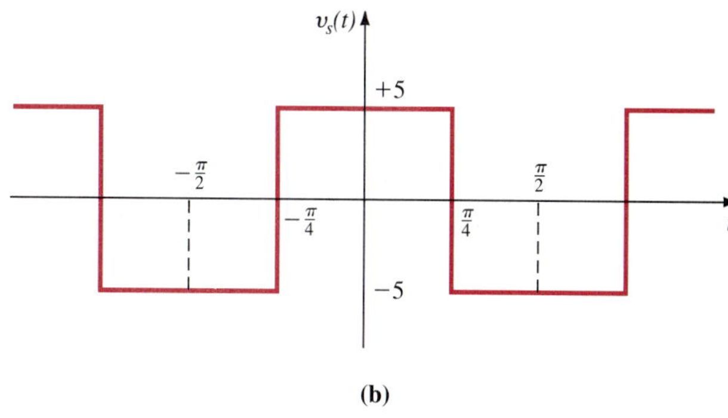

(b)

Figure P18.30

18.31. If the input voltage in Problem 18.29 is

$$v_S(t) = 1 - \frac{2}{\pi} \sum_{n=1}^{\infty} \frac{1}{n} \sin 0.2\, \pi n t \text{ V}$$

find the expression for the steady-state current $i_o(t)$.

18.32. Determine the first three terms of the steady-state voltage $v_o(t)$ in Fig. P18.32 if the input voltage is a periodic signal of the form

$$v(t) = \frac{1}{2} + \sum_{n=1}^{\infty} \frac{1}{n\pi} (\cos n\pi - 1) \sin nt \text{ V}$$

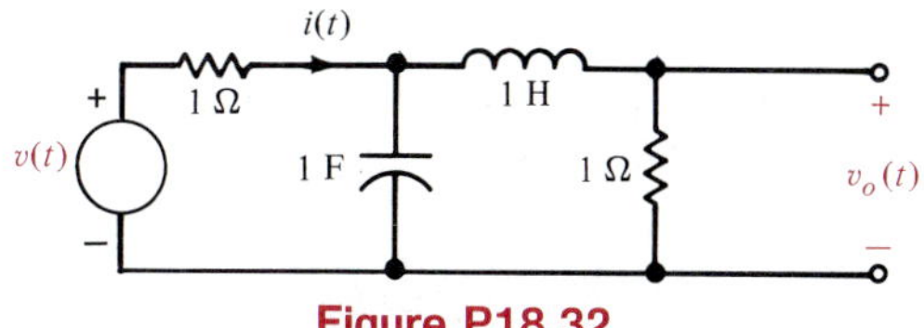

Figure P18.32

18.33. The current $i_S(t)$ shown in Fig. P18.33a is applied to the circuit shown in Fig. P18.33b. Determine the expression for the steady-state current $i_o(t)$ using the first four harmonics.

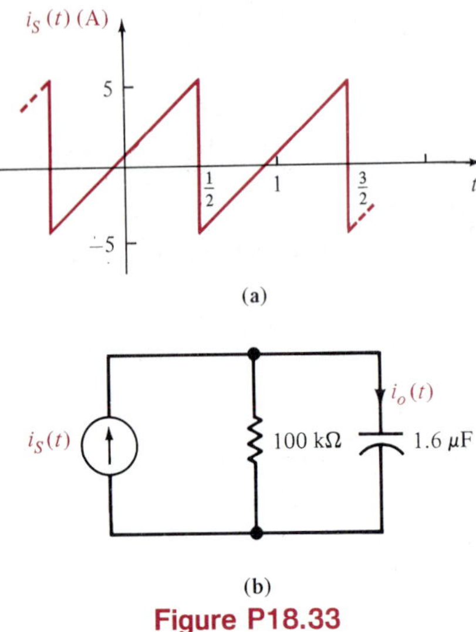

Figure P18.33

18.34. Determine the steady-state voltage $v_o(t)$ in the network in Fig. P18.34a if the input current is given in Fig. P18.34b.

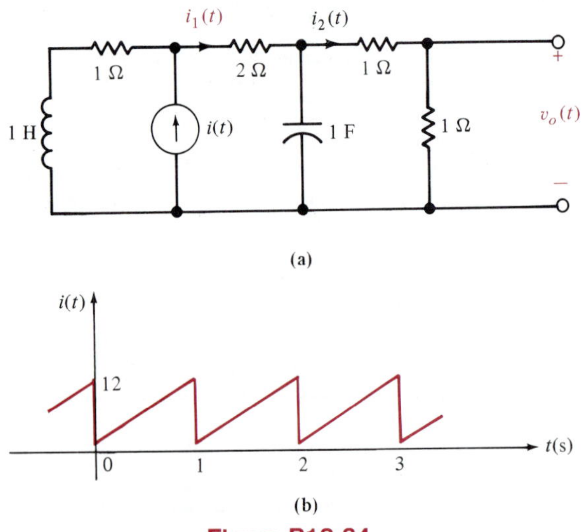

Figure P18.34

18.35. Find the average power absorbed by the network in Fig. P18.35 if

$$v(t) = 12 + 6 \cos (377t - 10°) + 4 \cos (754t - 60°) \text{ V}$$

$$i(t) = 0.2 + 0.4 \cos (377t - 150°) - 0.2 \cos (754t - 80°) + 0.1 \cos (1131t - 60°) \text{ A}$$

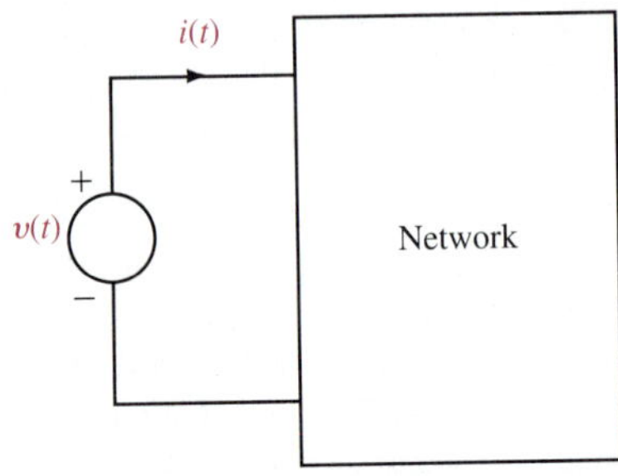

Figure P18.35

18.36. Find the average power absorbed by the network in Fig. P18.36 if $v(t) = 60 + 36 \cos (377t + 45°) + 24 \cos (754t - 60°)$ V.

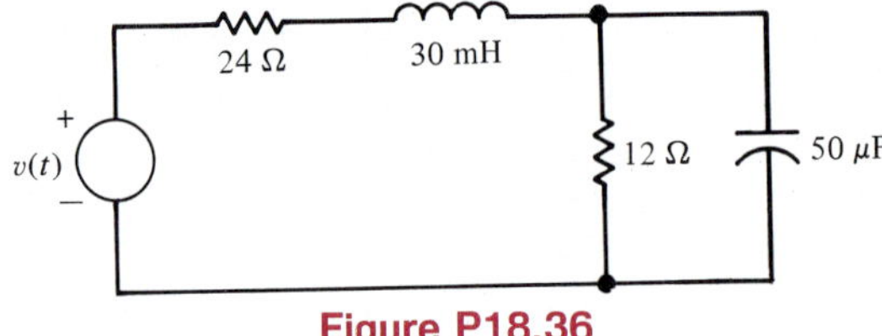

Figure P18.36

18.37. Find the average power absorbed by the 12-Ω resistor in the network in Fig. P18.36 if $v(t) = 50 + 25 \cos (377t - 45°) + 12.5 \cos (754t + 45°)$ V.

18.38. Determine the Fourier transform of the waveform shown in Fig. P18.38.

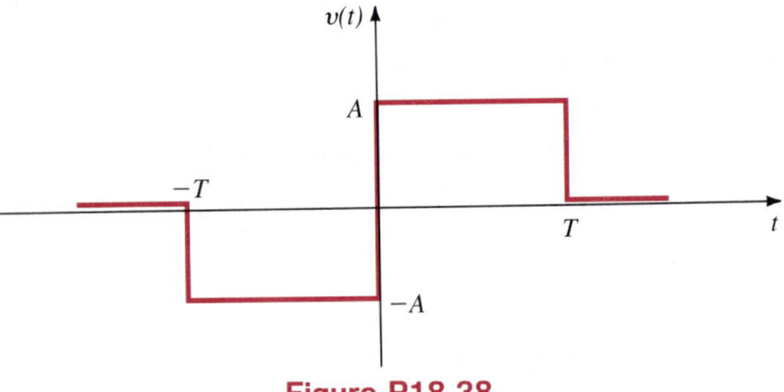

Figure P18.38

18.39. Derive the Fourier transform for the following functions:
(a) $f(t) = e^{-2t} \cos 4t\, u(t)$.
(b) $f(t) = e^{-2t} \sin 4t\, u(t)$.

18.40. Show that

$$\mathscr{F}[f_1(t)f_2(t)] = \frac{1}{2\pi}\int_{-\infty}^{\infty} \mathbf{F}_1(x)\mathbf{F}_2(\omega - x)\, dx$$

18.41. Find the Fourier transform of the function $f(t) = e^{-a|t|}$.

18.42. Find the Fourier transform of the function $f(t) = 12e^{-2|t|} \cos 4t$.

18.43. Determine the output signal $v_o(t)$ of a network with input signal $v_i(t) = 3e^{-t}u(t)$ and network impulse response $h(t) = e^{-2t}u(t)$. Assume that all initial conditions are zero.

18.44. The input signal to a network is $v_i(t) = e^{-3t}u(t)$. The transfer function of the network is $\mathbf{H}(j\omega) = 1/(j\omega + 4)$. Find the output of the network $v_o(t)$ if the initial conditions are zero.

18.45. Use the transform technique to find $v_o(t)$ in the network in Fig. P18.34a, if (a) $i(t) = 4(e^{-t} - e^{-2t})u(t)$ A and (b) $i(t) = 12 \cos 4t$ A.

18.46. The input signal for the network in Fig. P18.46 is $v_i(t) = 10e^{-5t}u(t)$ V. Determine the total 1-Ω energy content of the output $v_o(t)$.

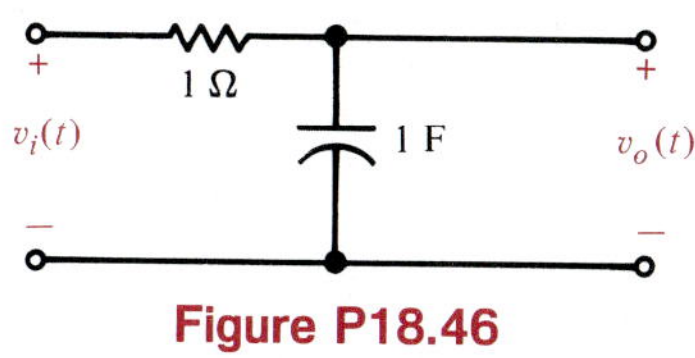

Figure P18.46

18.47. Use the Fourier transform to find $i(t)$ in the network in Fig. P18.47 if $v_i(t) = 2e^{-t}u(t)$.

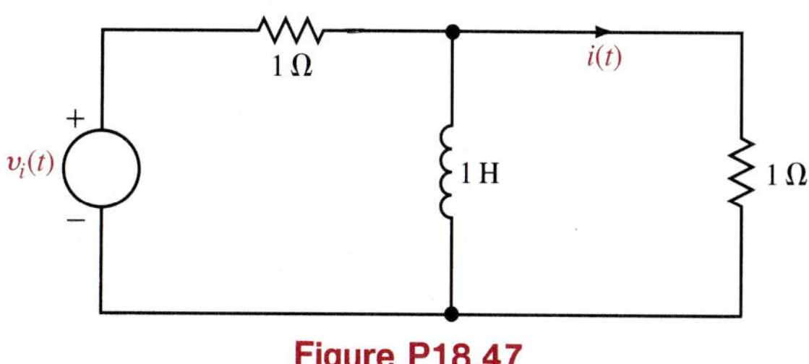

Figure P18.47

18.48. Use the transform technique to find $v_o(t)$ in the network in Fig. P18.48 if (a) $v_1(t) = 4e^{-t}u(t)$ V and (b) $v_1(t) = 4(e^{-2t} + 2e^{-4t})u(t)$ V.

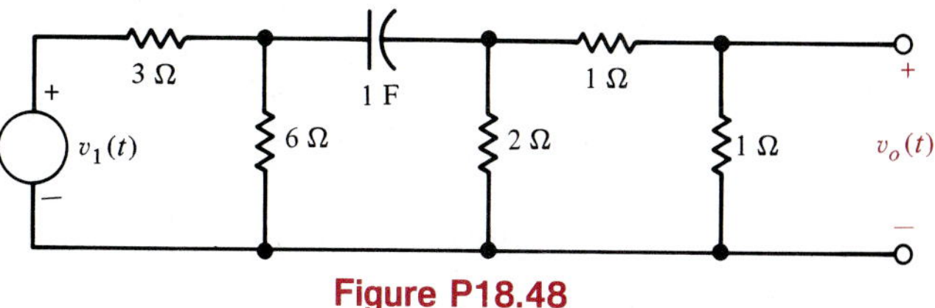

Figure P18.48

18.49. Compute the 1-Ω energy content of the signal $v_o(t)$ in Fig. P18.46 in the frequency range from $\omega = 2$ to $\omega = 4$ rad/sec.

18.50. Determine the 1-Ω energy content of the signal $v_o(t)$ in Fig. P18.46 in the frequency range from 0 to 1 rad/sec.

18.51. Determine the voltage $v_o(t)$ in the circuit shown in Fig. P18.51, using the Fourier transform if $i_i(t) = 2e^{-4t}u(t)$ A.

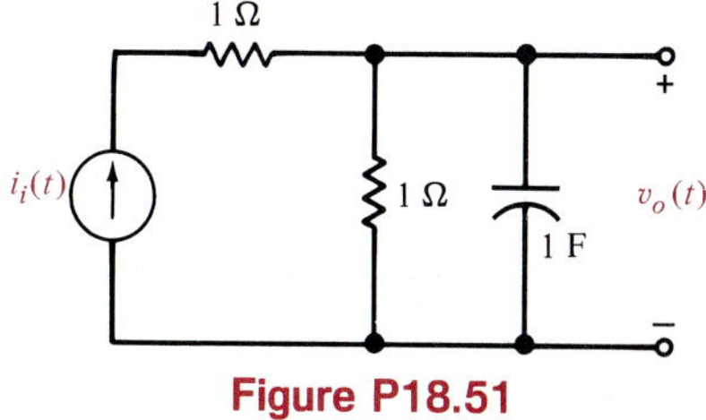

Figure P18.51

18.52. Compare the 1-Ω energy at both the input and output of the network in Problem 18.51 for the given input forcing function.

APPENDIX

Techniques for Solving Linearly Independent Simultaneous Equations

In the solution of various circuit problems we encounter a system of simultaneous equations of the form

$$\begin{aligned} a_{11}x_1 + a_{12}x_2 + \cdots + a_{1n}x_n &= b_1 \\ a_{21}x_1 + a_{22}x_2 + \cdots + a_{2n}x_n &= b_2 \\ \vdots \qquad \vdots \qquad\qquad \vdots \quad &\quad \vdots \\ a_{n1}x_1 + a_{n2}x_2 + \cdots + a_{nn}x_n &= b_n \end{aligned} \tag{A.1}$$

where the x's and b's are typically voltages and currents or currents and voltages, respectively.

As the title implies, we assume that the equations are linearly *independent*. As a brief reminder of the meaning of linear independence, consider the following KCL equations written for each node of a three-node network:

$$\tfrac{3}{2}V_1 - \tfrac{1}{2}V_2 - 4 = 0 \tag{A.2}$$

$$-\tfrac{1}{2}V_1 + \tfrac{5}{6}V_2 + 5 = 0 \tag{A.3}$$

$$-V_1 - \tfrac{1}{3}V_2 - 1 = 0 \tag{A.4}$$

where V_1 and V_2 are two node voltages that are measured with respect to the third (reference) node. Linear independence implies that we cannot find constants a_1, a_2, and a_3 such that

$$a_1(\tfrac{3}{2}V_1 - \tfrac{1}{2}V_2 - 4) + a_2(-\tfrac{1}{2}V_1 + \tfrac{5}{6}V_2 + 5) + a_3(-V_1 - \tfrac{1}{3}V_2 - 1) = 0 \tag{A.5}$$

However, in this case if we select $a_1 = a_2 = a_3 = 1$, we obtain

$$+\tfrac{3}{2}V_1 - \tfrac{1}{2}V_2 - 4 - \tfrac{1}{2}V_1 + \tfrac{5}{6}V_2 + 5 - V_1 - \tfrac{1}{3}V_2 - 1 = 0$$

$$0 = 0$$

Said another way, this means, for example, that Eqs. (A.2) and (A.3) can be used to obtain Eq. (A.4), and therefore, Eq. (A.4) is linearly *dependent* on Eqs. (A.2) and (A.3). Furthermore, any two of the equations could be used to obtain the third equation. Therefore, only two of the three equations are linearly *independent*.

We will now describe two techniques for solving linearly independent simultaneous equations—Gaussian elimination and matrices.

A.1 Gaussian Elimination

The following example will serve to demonstrate the steps involved in applying this technique.

EXAMPLE A.1

Let us find the solution to the following set of equations:

$$7X_1 - 4X_2 - X_3 = 4 \tag{A.6}$$

$$-4X_1 + 7X_2 - 2X_3 = 0 \tag{A.7}$$

$$-X_1 - 2X_2 + 3X_3 = -1 \tag{A.8}$$

The algorithm (i.e., step-by-step procedure) for applying the Gaussian elimination method proceeds in the following systematic way. First, we solve Eq. (A.6) for the variable X_1 in terms of the other variables in X_2 and X_3.

$$X_1 = \tfrac{4}{7} + \tfrac{4}{7}X_2 + \tfrac{1}{7}X_3 \tag{A.9}$$

We then substitute this result into Eqs. (A.7) and (A.8) to obtain

$$\tfrac{33}{7}X_2 - \tfrac{18}{7}X_3 = \tfrac{16}{7} \tag{A.10}$$

$$-\tfrac{18}{7}X_2 + \tfrac{20}{7}X_3 = -\tfrac{3}{7} \tag{A.11}$$

Continuing the reduction we now solve Eq. (A.10) for X_2 in terms of X_3:

$$X_2 = \tfrac{16}{33} + \tfrac{18}{33}X_3 \tag{A.12}$$

Substituting this expression for X_2 into Eq. (A.11) yields

$$\tfrac{336}{231}X_3 = \tfrac{189}{231}$$

or

$$X_3 = 0.563 \tag{A.13}$$

Now backtracking through the equations, we can determine X_2 from Eq. (A.12) as

$$X_2 = 0.792$$

and X_1 from Eq. (A.9) as

$$X_1 = 1.104$$

In this simple example we have not addressed such issues as zero coefficients or the impact of round-off errors. We have, however, illustrated the basic procedure. ■

Because of the very methodical manner in which the elimination takes place, the algorithm is easily adaptable to computer analysis, and efficient computer codes that implement the technique are available in standard software packages (e.g., see the linear system subroutines in the IMSL Subroutine Library). In addition, there are a large number of individual programs written to implement the Gaussian elimination technique (e.g., Stephen W. Director, *Circuit Theory: A Computational Approach*, John Wiley & Sons, Inc., New York, 1975, pp. 73–76).

A.2 Matrices

A *matrix* is defined to be a rectangular array of numbers arranged in rows and columns and written in the form

$$\begin{bmatrix} a_{11} & a_{12} & \cdots & a_{1n} \\ a_{21} & a_{22} & \cdots & a_{2n} \\ \vdots & \vdots & & \vdots \\ a_{m1} & a_{m2} & \cdots & a_{mn} \end{bmatrix}$$

This array is called an m by n ($m \times n$) matrix because it has m rows and n columns. A matrix is a convenient way of representing arrays of numbers; however, one must remember that the matrix itself has no numerical value. In the array above the numbers or functions a_{ij} are called the *elements* of the matrix. Any matrix that has the same number of rows as columns is called a *square matrix*. The sum of the diagonal elements of a square matrix is called the *trace* of the matrix.

EXAMPLE A.2

The following are matrices:

$$\begin{bmatrix} a \\ b \\ c \\ d \end{bmatrix}, \quad \begin{bmatrix} 1 & 3 \\ 2 & 4 \end{bmatrix}, \quad \begin{bmatrix} 4 & 3 & 2 & 1 \\ 5 & 6 & 7 & 8 \end{bmatrix}, \quad [1 \quad 2 \quad 3]$$

■

The *identity matrix* is a diagonal matrix in which all diagonal elements are equal to one.

EXAMPLE A.3

The following are identity matrices:

$$\begin{bmatrix} 1 & 0 \\ 0 & 1 \end{bmatrix}, \quad \begin{bmatrix} 1 & 0 & 0 \\ 0 & 1 & 0 \\ 0 & 0 & 1 \end{bmatrix}, \quad \ldots, \quad \begin{bmatrix} 1 & 0 & \cdot & \cdot & \cdots & 0 \\ 0 & 1 & 0 & \cdot & \cdots & 0 \\ 0 & 0 & 1 & 0 & \cdots & 0 \\ \vdots & & & & & \\ 0 & 0 & \cdot & \cdot & \cdots & 1 \end{bmatrix}$$

■

The matrices $\boldsymbol{A}$ and $\boldsymbol{B}$ are said to be *equal* if their corresponding elements are equal. In other words, $\boldsymbol{A} = \boldsymbol{B}$ if and only if $a_{ij} = b_{ij}$ for all i and j.

EXAMPLE A.4

If

$$\boldsymbol{A} = \begin{bmatrix} 3 & 2 & 1 \\ 4 & 5 & 6 \end{bmatrix} \quad \text{and} \quad \boldsymbol{B} = \begin{bmatrix} 3 & 2 & 1 \\ 4 & 5 & 6 \end{bmatrix}$$

then $\boldsymbol{A} = \boldsymbol{B}$ because $a_{ij} = b_{ij}$ for all i and j. ■

The addition and subtraction of two matrices of the same order (i.e., $m \times n$) is accomplished as follows:

$$\boldsymbol{C} = \boldsymbol{A} \pm \boldsymbol{B} \tag{A.14}$$

or

$$c_{ij} = a_{ij} \pm b_{ij} \qquad \text{for all } i \text{ and } j$$

That is, the elements of $\boldsymbol{C}$ are the sum or difference of the corresponding elements of $\boldsymbol{A}$ and $\boldsymbol{B}$.

$$\boldsymbol{C} = \begin{bmatrix} c_{11} \cdots & c_{1n} \\ \vdots & \vdots \\ c_{m1} \quad \cdots & c_{mn} \end{bmatrix} = \begin{bmatrix} a_{11} & \cdots & a_{1n} \\ \vdots & & \vdots \\ a_{m1} & \cdots & a_{mn} \end{bmatrix} \pm \begin{bmatrix} b_{11} & \cdots & b_{1n} \\ \vdots & & \vdots \\ b_{m1} & \cdots & b_{mn} \end{bmatrix}$$

$$= \begin{bmatrix} a_{11} \pm b_{11} & \cdots & a_{1n} \pm b_{1n} \\ \vdots & & \vdots \\ a_{m1} \pm b_{m1} & \cdots & a_{mn} \pm b_{mn} \end{bmatrix}$$

EXAMPLE A.5

If

$$\boldsymbol{A} = \begin{bmatrix} 1 & 3 \\ 2 & 4 \end{bmatrix}, \quad \boldsymbol{B} = \begin{bmatrix} -1 & -2 \\ 3 & 4 \end{bmatrix}, \quad \text{and} \quad \boldsymbol{C} = \begin{bmatrix} 1 & 3 \\ 4 & 2 \end{bmatrix}$$

then

$$\boldsymbol{A} + \boldsymbol{B} = \begin{bmatrix} 0 & 1 \\ 5 & 8 \end{bmatrix}, \quad \boldsymbol{A} - \boldsymbol{B} = \begin{bmatrix} 2 & 5 \\ -1 & 0 \end{bmatrix}, \quad \boldsymbol{A} - \boldsymbol{B} + \boldsymbol{C} = \begin{bmatrix} 3 & 8 \\ 3 & 2 \end{bmatrix}$$

■

Given a matrix $\boldsymbol{A}$ and a scalar λ, *the multiplication* of $\boldsymbol{A}$ by the scalar λ, written $\lambda\boldsymbol{A}$, is defined to be

$$\lambda\boldsymbol{A} = \begin{bmatrix} \lambda a_{11} & \cdots & \lambda a_{1n} \\ \vdots & & \\ \lambda a_{m1} & \cdots & \lambda a_{mn} \end{bmatrix} \tag{A.15}$$

EXAMPLE A.6

If $\lambda_1 = 2$, $\lambda_2 = -2$, and

$$\boldsymbol{A} = \begin{bmatrix} 2 & 4 & 6 \\ 1 & 3 & 5 \end{bmatrix}$$

then

$$\lambda_1\boldsymbol{A} = \begin{bmatrix} 4 & 8 & 12 \\ 2 & 6 & 10 \end{bmatrix}$$

$$\lambda_2\boldsymbol{A} = \begin{bmatrix} -4 & -8 & -12 \\ -2 & -6 & -10 \end{bmatrix}$$

■

Consider now the multiplication of two matrices. If we are given an $m \times n$ matrix $\boldsymbol{A}$ and an $n \times r$ matrix $\boldsymbol{B}$, the product $\boldsymbol{AB}$ is defined to be an $m \times r$ matrix $\boldsymbol{C}$ whose elements are given by the expression

$$c_{ij} = \sum_{k=1}^{n} a_{ik}b_{kj}, \qquad i = 1, \ldots, m, \quad j = 1, \ldots, r \tag{A.16}$$

Note that the product $\boldsymbol{AB}$ is defined only when the number of columns of $\boldsymbol{A}$ is equal to the number of rows of $\boldsymbol{B}$.

EXAMPLE A.7

Suppose that the matrices $\boldsymbol{A}$ and $\boldsymbol{B}$ are defined as follows:

$$\boldsymbol{A} = \begin{bmatrix} a_{11} & a_{12} \\ a_{21} & a_{22} \\ a_{31} & a_{32} \end{bmatrix} \qquad \boldsymbol{B} = \begin{bmatrix} b_{11} & b_{12} \\ b_{21} & b_{22} \end{bmatrix}$$

Note that in the formula above, $m = 3$, $n = 2$, $r = 2$. Using this formula, we can calculate

$$c_{11} = \sum_{k=1}^{2} a_{1k}b_{k1} = a_{11}b_{11} + a_{12}b_{21}$$

$$c_{12} = \sum_{k=1}^{2} a_{1k}b_{k2} = a_{11}b_{12} + a_{12}b_{22}$$

$$\vdots$$

These elements form the array

$$C = AB = \begin{bmatrix} a_{11}b_{11} + a_{12}b_{21} & a_{11}b_{12} + a_{12}b_{22} \\ a_{21}b_{11} + a_{22}b_{21} & a_{21}b_{12} + a_{22}b_{22} \\ a_{31}b_{11} + a_{32}b_{21} & a_{31}b_{12} + a_{32}b_{22} \end{bmatrix}$$ ■

A close inspection of the product above illustrates that multiplication is a "row-by-column" operation. In other words, each element in a row of the first matrix is multiplied by the corresponding element in a column of the second matrix and then the products are summed. This operation is diagrammed as follows:

$$\begin{bmatrix} c_{11} & \cdots & c_{1p} \\ \vdots & \boxed{c_{ij}} & \vdots \\ c_{m1} & \cdots & c_{mp} \end{bmatrix} = \begin{bmatrix} a_{11} & \cdots & a_{1n} \\ \boxed{a_{i1} \quad \cdots \quad a_{in}} \\ a_{m1} & \cdots & a_{mn} \end{bmatrix} \begin{bmatrix} b_{11} & \cdots & b_{1j} & \cdots & b_{1p} \\ \vdots & & \vdots & & \vdots \\ b_{n1} & \cdots & b_{nj} & \cdots & b_{np} \end{bmatrix} \tag{A.17}$$

The following examples will illustrate the computational technique.

EXAMPLE A.8

If

$$A = \begin{bmatrix} 1 & 3 \\ 2 & 4 \end{bmatrix} \quad \text{and} \quad B = \begin{bmatrix} 2 & 1 \\ 3 & 5 \end{bmatrix}$$

then

$$AB = \begin{bmatrix} (1)(2) + (3)(3) & (1)(1) + (3)(5) \\ (2)(2) + (4)(3) & (2)(1) + (4)(5) \end{bmatrix} = \begin{bmatrix} 11 & 16 \\ 16 & 22 \end{bmatrix}$$

If

$$C = \begin{bmatrix} 1 & 2 \\ 3 & 4 \end{bmatrix} \quad \text{and} \quad D = \begin{bmatrix} 1 \\ 2 \end{bmatrix}$$

then

$$CD = \begin{bmatrix} (1)(1) + (2)(2) \\ (3)(1) + (4)(2) \end{bmatrix} = \begin{bmatrix} 5 \\ 11 \end{bmatrix}$$ ■

The matrix of order $n \times m$ obtained by interchanging the rows and columns of an $m \times n$ matrix A is called the *transpose* of A and is denoted by A^T.

EXAMPLE A.9

If

$$A = \begin{bmatrix} 1 \\ 2 \\ 3 \end{bmatrix}$$

then

$$A^T = [1 \quad 2 \quad 3]$$

and if

$$A = \begin{bmatrix} 1 & 4 \\ 2 & 5 \\ 3 & 6 \end{bmatrix}$$

then

$$A^T = \begin{bmatrix} 1 & 2 & 3 \\ 4 & 5 & 6 \end{bmatrix}$$

■

If A is a square matrix of the form

$$A = \begin{bmatrix} a_{11} & a_{12} \\ a_{21} & a_{22} \end{bmatrix}$$

the determinant of A is defined to be

$$|A| = a_{11}a_{22} - a_{21}a_{12}$$

that is, it is equal to the product $a_{11}a_{22}$ of elements *down* the diagonal from left to right minus the product $a_{21}a_{12}$ of elements *up* the diagonal from left to right.

EXAMPLE A.10

If

$$A = \begin{bmatrix} 4 & -1 \\ -1 & 3 \end{bmatrix}$$

then

$$|A| = (4)(3) - (-1)(-1) = 11$$

The evaluation of larger determinants is accomplished by expanding the determinant in *cofactors*. The cofactor A_{ij} of the element a_{ij} of a determinant is equal to the product of $(-1)^{i+j}$ and the determinant remaining after row i and column j are deleted. ■

EXAMPLE A.11

Given the 3 × 3 determinant below, determine the cofactors A_{11}, A_{23}, and A_{31}.

$$A = \begin{vmatrix} a_{11} & a_{12} & a_{13} \\ a_{21} & a_{22} & a_{23} \\ a_{31} & a_{32} & a_{33} \end{vmatrix}$$

Since we could use any row or column to evaluate the determinant, we can use the second column, so that

$$\Delta = a_{12}A_{12} + a_{22}A_{22} + a_{32}A_{32}$$

$$= a_{12}(-1)^3 \begin{vmatrix} a_{21} & a_{23} \\ a_{31} & a_{33} \end{vmatrix} + a_{22}(-1)^4 \begin{vmatrix} a_{11} & a_{13} \\ a_{31} & a_{33} \end{vmatrix}$$

$$+ a_{32}(-1)^5 \begin{vmatrix} a_{11} & a_{13} \\ a_{21} & a_{23} \end{vmatrix}$$

or the third row so that

$$\Delta = a_{31}A_{31} + a_{32}A_{32} + a_{33}A_{33}$$

$$= a_{31}(-1)^4 \begin{vmatrix} a_{12} & a_{13} \\ a_{22} & a_{23} \end{vmatrix} + a_{32}(-1)^5 \begin{vmatrix} a_{11} & a_{13} \\ a_{21} & a_{23} \end{vmatrix}$$

$$+ a_{33}(-1)^6 \begin{vmatrix} a_{11} & a_{12} \\ a_{21} & a_{22} \end{vmatrix}$$

or some other row or column. ■

EXAMPLE A.12

Calculate the determinant of the following matrix.

$$\mathbf{A} = \begin{bmatrix} 7 & -4 & -1 \\ -4 & 7 & -2 \\ -1 & -2 & 3 \end{bmatrix}$$

Expanding the determinant using the first row, we obtain

$$|\mathbf{A}| = 7\begin{vmatrix} 7 & -2 \\ -2 & 3 \end{vmatrix} - (-4)\begin{vmatrix} -4 & -2 \\ -1 & 3 \end{vmatrix} - 1\begin{vmatrix} -4 & 7 \\ -1 & -2 \end{vmatrix}$$

$$= 7(17) + 4(-14) - (15)$$

$$= 48$$

■

As defined for determinants, the *cofactor* A_{ij} of the element a_{ij} of any square matrix $\mathbf{A}$ is equal to the product $(-1)^{i+j}$ and the determinant of the submatrix obtained from $\mathbf{A}$ by deleting row i and column j.

EXAMPLE A.13

Given the matrix

$$\mathbf{A} = \begin{bmatrix} a_{11} & a_{12} & a_{13} \\ a_{21} & a_{22} & a_{23} \\ a_{31} & a_{32} & a_{33} \end{bmatrix}$$

the cofactors A_{11}, A_{12}, and A_{22} are

$$A_{11} = (-1)^2 \begin{vmatrix} a_{22} & a_{23} \\ a_{32} & a_{33} \end{vmatrix} = a_{22}a_{33} - a_{32}a_{23}$$

$$A_{12} = (-1)^3 \begin{vmatrix} a_{21} & a_{23} \\ a_{31} & a_{33} \end{vmatrix} = -(a_{21}a_{33} - a_{31}a_{23})$$

$$A_{22} = (-1)^2 \begin{vmatrix} a_{11} & a_{13} \\ a_{31} & a_{33} \end{vmatrix} = a_{11}a_{33} - a_{31}a_{13}$$

■

The *adjoint* of the matrix $\boldsymbol{A}$ (adj $\boldsymbol{A}$) is the transpose of the matrix obtained from $\boldsymbol{A}$ by replacing each element a_{ij} by its cofactor A_{ij}. In other words, if

$$\boldsymbol{A} = \begin{bmatrix} a_{11} & a_{12} & \cdots & a_{1n} \\ a_{21} & a_{22} & \cdots & \cdot \\ \vdots & \vdots & & \vdots \\ a_{n1} & \cdots & \cdots & a_{nn} \end{bmatrix}$$

then

$$\text{adj } \boldsymbol{A} = \begin{bmatrix} A_{11} & A_{21} & \cdots & A_{n1} \\ A_{12} & A_{22} & \cdots & \cdot \\ \vdots & \vdots & & \vdots \\ A_{1n} & \cdots & \cdots & A_{nn} \end{bmatrix}$$

If $\boldsymbol{A}$ is a square matrix and if there exists a square matrix $\boldsymbol{A}^{-1}$ such that

$$\boldsymbol{A}^{-1}\boldsymbol{A} = \boldsymbol{A}\boldsymbol{A}^{-1} = \boldsymbol{I} \tag{A.18}$$

then $\boldsymbol{A}^{-1}$ is called the *inverse* of $\boldsymbol{A}$. It can be shown that the inverse of the matrix $\boldsymbol{A}$ is equal to the adjoint divided by the determinant (written here as $|\boldsymbol{A}|$); that is,

$$\boldsymbol{A}^{-1} = \frac{\text{adj } \boldsymbol{A}}{|\boldsymbol{A}|} \tag{A.19}$$

EXAMPLE A.14

Given

$$\boldsymbol{A} = \begin{bmatrix} 2 & 3 \\ 1 & 4 \end{bmatrix}$$

then

$$|\boldsymbol{A}| = (2)(4) - (1)(3) = 5$$

and

$$\text{adj } \boldsymbol{A} = \begin{bmatrix} 4 & -3 \\ -1 & 2 \end{bmatrix}$$

Therefore,

$$A^{-1} = \tfrac{1}{5}\begin{bmatrix} 4 & -3 \\ -1 & 2 \end{bmatrix}$$

Also, if

$$A = \begin{bmatrix} 2 & 3 & 1 \\ 1 & 2 & 3 \\ 3 & 1 & 2 \end{bmatrix}$$

then

$$\begin{aligned} |A| &= 2\begin{vmatrix} 2 & 3 \\ 1 & 2 \end{vmatrix} - 1\begin{vmatrix} 3 & 1 \\ 1 & 2 \end{vmatrix} + 3\begin{vmatrix} 3 & 1 \\ 2 & 3 \end{vmatrix} \\ &= 2 - 5 + 21 = 18 \end{aligned}$$

and

$$\text{adj}\, A = \begin{bmatrix} 1 & -5 & 7 \\ 7 & 1 & -5 \\ -5 & 7 & 1 \end{bmatrix}$$

Therefore,

$$A^{-1} = \tfrac{1}{18}\begin{bmatrix} 1 & -5 & 7 \\ 7 & 1 & -5 \\ -5 & 7 & 1 \end{bmatrix}$$

■

We now have the tools necessary to solve Eqs. (A.1) using matrices. The following example illustrates the approach.

EXAMPLE A.15

The node equations for a network are

$$\begin{aligned} 2V_1 + 3V_2 + V_3 &= 9 \\ V_1 + 2V_2 + 3V_3 &= 6 \\ 3V_1 + V_2 + 2V_3 &= 8 \end{aligned}$$

Note that this set of simultaneous equations can be written as a single matrix equation in the form

$$\begin{bmatrix} 2 & 3 & 1 \\ 1 & 2 & 3 \\ 3 & 1 & 2 \end{bmatrix}\begin{bmatrix} V_1 \\ V_2 \\ V_3 \end{bmatrix} = \begin{bmatrix} 9 \\ 6 \\ 8 \end{bmatrix}$$

or

$$AV = I$$

Multiplying both sides of the equation above through by $\boldsymbol{A}^{-1}$ yields

$$\boldsymbol{A}^{-1}\boldsymbol{A}\boldsymbol{V} = \boldsymbol{A}^{-1}\boldsymbol{I}$$

or

$$\boldsymbol{V} = \boldsymbol{A}^{-1}\boldsymbol{I}$$

$\boldsymbol{A}^{-1}$ was calculated in Example A.14. Employing that inverse here, we obtain

$$\boldsymbol{V} = \tfrac{1}{18}\begin{bmatrix} 1 & -5 & 7 \\ 7 & 1 & -5 \\ -5 & 7 & 1 \end{bmatrix}\begin{bmatrix} 9 \\ 6 \\ 8 \end{bmatrix}$$

or

$$\begin{bmatrix} V_1 \\ V_2 \\ V_3 \end{bmatrix} = \tfrac{1}{18}\begin{bmatrix} 35 \\ 29 \\ 5 \end{bmatrix}$$

and hence

$$V_1 = \tfrac{36}{18},\ V_2 = \tfrac{29}{18},\ \text{and } V_3 = \tfrac{5}{18}$$ ■

APPENDIX

PSPICE Installation Procedure

The installation procedure has two phases. The first is four steps that set up the PC to run PSPICE. The second is seven steps to set the default printer and display for your computer.

B.1 Set Up PC (or Clone) to Run PSPICE

1. Create a PSPICE directory:

```
C:\> MKDIR PSPICE
```

2. Copy each PSPICE disk to the new directory:

```
C:\> COPY A:*.* C:\PSPICE
```

3. Ensure that the following two lines are in the C:\CONFIG.SYS file:

```
FILES=20
BUFFERS=20
```

(Note: Values may be larger than 20, but not smaller.)

4. Edit the C:\AUTOEXEC.BAT file to include PSPICE in the path. For example,

```
PATH C:\;C:\DOS;C:\PSPICE
```

B.2

Set PSPICE Default Display and Printer

1. Execute PSPICE:

```
C:\> CD \PSPICE
C:\PSPICE> PS
```

2. Select 'Files' from the menu line.
3. Select 'Display/Prn Setup . . . ' from the submenu. This displays a window in which the user must select the display type, the port to which the printer is attached, and the type of printer.
4. Press F4 to get a list of the display choices. Select the appropriate choice for your type of monitor by pressing the arrow keys until the desired choice is highlighted. Press the Enter key.
5. Press F4 to get a list of port choices. Select the appropriate choice for your system (usually "prn" or "lptl") by pressing the arrow keys until the desired choice is highlighted. Press the Enter key.
6. Press F4 to get a list of printer types. The arrow keys must be used to highlight the desired choice. There are more choices than can be displayed in the window at one time. If you cannot find one which corresponds to your type of printer, refer to the printer manual to determine which choice is closest to your type of printer. (Note: Most dot matrix printers will emulate some type of Epson printer. Select the "ps" option for a Postscript printer.) Press Enter to select the highlighted choice.
7. Press the Esc key to return to the PSPICE menu line.

APPENDIX

Complex Numbers

The reader has normally already encountered complex numbers and their use in previous work, and therefore only a quick review of the elements employed in this book are presented here.

C.1 Complex Number Representation

Complex numbers are typically represented in three forms: *exponential, polar,* and *rectangular*. In the exponential form a complex number **A** is written as

$$\mathbf{A} = ze^{j\theta} \tag{C.1}$$

The nonnegative quantity z is known as the *amplitude* or *magnitude,* the real quantity θ is called the *angle,* and j is the *imaginary operator* $j = \sqrt{-1}$, where $j^2 = -1$, $j^3 = -\sqrt{-1} = -j$, and so on. As indicated in the main body of the text, θ is expressed in radians or degrees.

The *polar form* of a complex number **A**, which is symbolically equivalent to the exponential form, is written as

$$\mathbf{A} = z \angle\theta \tag{C.2}$$

Note that in this case the expression $e^{j\theta}$ is replaced by the angle symbol $\angle\theta$. The representation of a complex number **A** by a magnitude of z at a given angle θ suggests a representation using polar coordinates in a complex plane.

The rectangular representation of a complex number is written as

$$\mathbf{A} = x + jy \tag{C.3}$$

where x is the real part of $\mathbf{A}$ and y is the imaginary part of $\mathbf{A}$, which is usually expressed in the form

$$\begin{aligned} x &= \text{Re}\,(\mathbf{A}) \\ y &= \text{Im}\,(\mathbf{A}) \end{aligned} \tag{C.4}$$

The complex number $\mathbf{A} = x + jy$ can be graphically represented in the complex plane as shown in Fig. C.1. Note that the imaginary part of $\mathbf{A}$, y, is real. Note that $x + jy$ uniquely locates a point in the complex plane which could also be specified by a magnitude z, representing the straight-line distance from the origin to the point, and an angle θ, which represents the angle between the positive real axis and the straight line connecting the point with the origin.

The connection between the various representations of $\mathbf{A}$ can be seen via Euler's identity, which is

$$e^{j\theta} = \cos\theta + j\sin\theta \tag{C.5}$$

Using this identity the complex number $\mathbf{A}$ can be written as

$$\mathbf{A} = ze^{j\theta} = z\cos\theta + jz\sin\theta \tag{C.6}$$

which as shown in Fig. C.1 is equivalent to

$$\mathbf{A} = x + jy$$

Equating the real and imaginary parts of these two equations yields

$$\begin{aligned} x &= z\cos\theta \\ y &= z\sin\theta \end{aligned} \tag{C.7}$$

From these equations we obtain

$$x^2 + y^2 = z^2\cos^2\theta + z^2\sin^2\theta = z^2$$

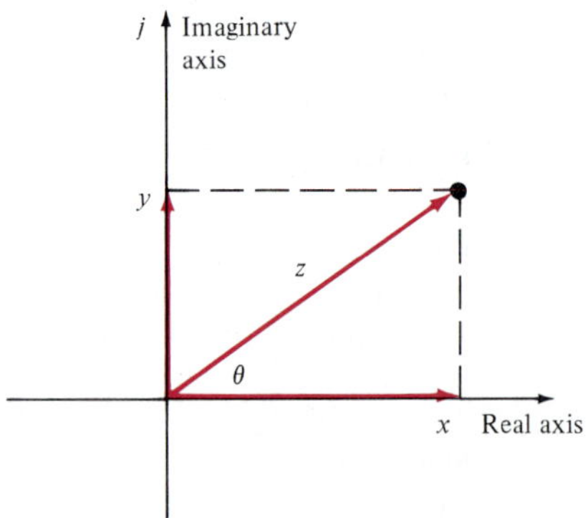

Figure C.1 Representation of a complex number in the complex plane.

Therefore,

$$z = \sqrt{x^2 + y^2} \geq 0 \tag{C.8}$$

Furthermore,

$$\frac{z \sin \theta}{z \cos \theta} = \tan \theta = \frac{y}{x}$$

and hence

$$\theta = \tan^{-1} \frac{y}{x} \tag{C.9}$$

The interrelationships among the three representations of a complex number are as follows.

Exponential	Polar	Rectangular
$ze^{j\theta}$	$z \underline{/\theta}$	$x + jy$
$\theta = \tan^{-1} \frac{y}{x}$	$\theta = \tan^{-1} \frac{y}{x}$	$x = z \cos \theta$
$z = \sqrt{x^2 + y^2}$	$z = \sqrt{x^2 + y^2}$	$y = z \sin \theta$

EXAMPLE C.1

If a complex number **A** in polar form is $\mathbf{A} = 10 \underline{/30^\circ}$, express **A** in both exponential and rectangular forms.

$$\mathbf{A} = 10 \underline{/30^\circ} = 10e^{j30^\circ} = 10[\cos 30^\circ + j \sin 30^\circ] = 8.66 + j5.0$$

■

EXAMPLE C.2

If $\mathbf{A} = 4 + j3$, express **A** in both exponential and polar forms. In addition, express $-\mathbf{A}$ in exponential and polar forms with a positive magnitude.

$$\mathbf{A} = 4 + j3 = \sqrt{4^2 + 3^2} \quad \underline{/\tan^{-1} \frac{3}{4}}$$

$$= 5 \underline{/36.9^\circ}$$

Also,

$$-\mathbf{A} = -5 \underline{/36.9^\circ} = 5 \underline{/36.9^\circ + 180^\circ} = 5 \underline{/216.9^\circ} = 5e^{j216.9^\circ}$$

or

$$-\mathbf{A} = -5 \underline{/36.9^\circ} = 5 \underline{/36.9^\circ - 180^\circ} = 5 \underline{/-143.1^\circ} = 5e^{-j143.1^\circ}$$

■

C.2 Mathematical Operations

We will now show that the operations of addition, subtraction, multiplication, and division apply to complex numbers in the same manner that they apply to real numbers. Before proceeding with this illustration, however, let us examine two important definitions.

Two complex numbers **A** and **B** defined as

$$\mathbf{A} = z_1 e^{j\theta_1} = z_1 \underline{/\theta_1} = x_1 + jy_1$$

$$\mathbf{B} = z_2 e^{j\theta_2} = z_2 \underline{/\theta_2} = x_2 + jy_2$$

are *equal* if and only if $x_1 = x_2$ and $y_1 = y_2$ or $z_1 = z_2$ and $\theta_1 = \theta_2 \pm n360°$, where $n = 0, 1, 2, 3, \ldots$.

EXAMPLE C.3

If $\mathbf{A} = 2 + j3$, $\mathbf{B} = 2 - j3$, $\mathbf{C} = 4 \underline{/30°}$, and $\mathbf{D} = 4 \underline{/750°}$, then $\mathbf{A} \neq \mathbf{B}$, but $\mathbf{C} = \mathbf{D}$, since $30° = 30° + 2(360°)$. ■

The *conjugate,* $\mathbf{A}^*$, of a complex number $\mathbf{A} = x + jy$ is defined to be

$$\mathbf{A}^* = x - jy \tag{C.10}$$

that is, j is replaced by $-j$ in the rectangular form (or polar form) to obtain the conjugate. Note that the magnitude of $\mathbf{A}^*$ is the same as that of **A**, since

$$z = \sqrt{x^2 + (-y)^2} = \sqrt{x^2 + y^2}$$

However, the angle is now

$$\tan^{-1}\frac{-y}{x} = -\theta$$

Therefore, the conjugate is written in exponential and polar form as

$$\mathbf{A}^* = ze^{-j\theta} = z \underline{/-\theta} \tag{C.11}$$

We also have the relationship

$$(\mathbf{A}^*)^* = \mathbf{A} \tag{C.12}$$

EXAMPLE C.4

If $\mathbf{A} = 10 \underline{/30°}$ and $\mathbf{B} = 4 + j3$, then $\mathbf{A}^* = 10 \underline{/-30°}$ and $\mathbf{B}^* = 4 - j3$. $(\mathbf{A}^*)^* = 10 \underline{/30°} = \mathbf{A}$ and $(\mathbf{B}^*)^* = 4 + j3 = \mathbf{B}$. ■

Addition

The sum of two complex numbers $\mathbf{A} = x_1 + jy_1$ and $\mathbf{B} = x_2 + jy_2$ is

$$\begin{aligned}\mathbf{A} + \mathbf{B} &= x_1 + jy_1 + x_2 + jy_2 \\ &= (x_1 + x_2) + j(y_1 + y_2)\end{aligned} \tag{C.13}$$

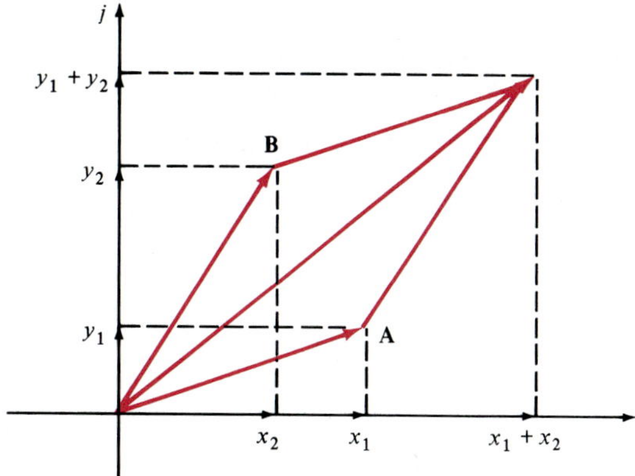

Figure C.2 Vector addition for complex numbers.

that is, we simply add the individual real parts, and we add the individual imaginary parts to obtain the components of the resultant complex number. This addition can be illustrated graphically by plotting each of the complex numbers as vectors and then performing the vector addition. This graphical approach is shown in Fig. C.2. Note that the vector addition is accomplished by plotting the vectors tail to head or simply completing the parallelogram.

EXAMPLE C.5

Given the complex numbers $\mathbf{A} = 4 + j1$, $\mathbf{B} = 3 - j2$, and $\mathbf{C} = -2 - j4$, we wish to calculate $\mathbf{A} + \mathbf{B}$ and $\mathbf{A} + \mathbf{C}$ (Fig. C.3).

$$\begin{aligned}\mathbf{A} + \mathbf{B} &= (4 + j1) + (3 - j2)\\ &= 7 - j1\\ \mathbf{A} + \mathbf{C} &= (4 + j1) + (-2 - j4)\\ &= 2 - j3\end{aligned}$$

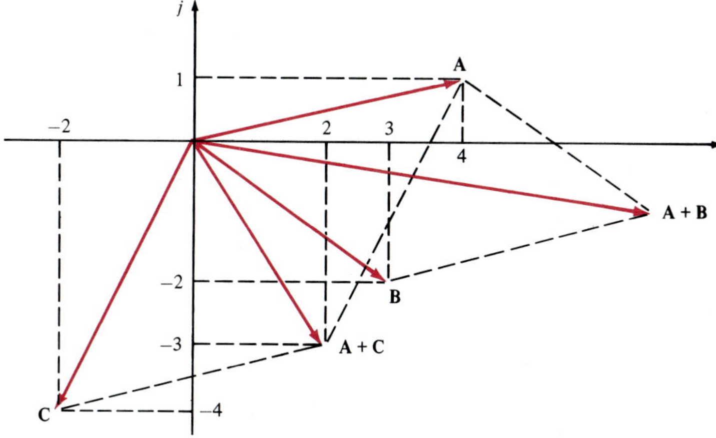

Figure C.3 Examples of complex number addition. ■

EXAMPLE C.6

We wish to calculate the sum **A** + **B** if $\mathbf{A} = 5\,\angle 36.9^\circ$ and $\mathbf{B} = 5\,\angle 53.1^\circ$.

We must first convert from polar to rectangular form.

$$\mathbf{A} = 5\,\angle 36.9^\circ = 4 + j3$$

$$\mathbf{B} = 5\,\angle 53.1^\circ = 3 + j4$$

Therefore,

$$\begin{aligned}\mathbf{A} + \mathbf{B} &= 4 + j3 + 3 + j4 = 7 + j7 \\ &= 9.9\,\angle 45^\circ\end{aligned}$$

Subtraction

The difference of two complex numbers $\mathbf{A} = x_1 + jy_1$ and $\mathbf{B} = x_2 + jy_2$ is

$$\begin{aligned}\mathbf{A} - \mathbf{B} &= (x_1 + jy_1) - (x_2 + jy_2) \\ &= (x_1 - x_2) + j(y_1 - y_2)\end{aligned} \tag{C.14}$$

that is, we simply subtract the individual real parts and we subtract the individual imaginary parts to obtain the components of the resultant complex number. Since a negative sign corresponds to a phase or angle change of 180°, the graphical technique for performing the subtraction (**A** − **B**) can be accomplished by drawing **A** and **B** as vectors, rotating the vector **B** 180° and then adding it to the vector **A**.

EXAMPLE C.7

Given $\mathbf{A} = 3 + j1$ and $\mathbf{B} = 2 - j2$, calculate the difference **A** − **B**.

$$\begin{aligned}\mathbf{A} - \mathbf{B} &= (3 + j1) - (2 - j2) \\ &= 1 + j3\end{aligned}$$

The graphical solution is shown in Fig. C.4.

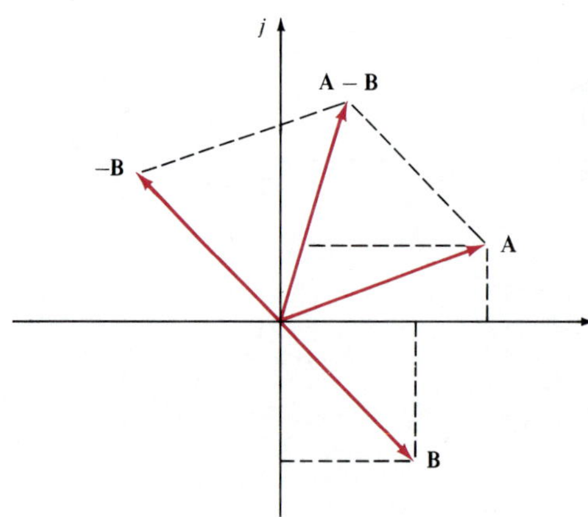

Figure C.4 Example of subtracting complex numbers.

EXAMPLE C.8

Let us calculate the difference $\mathbf{A} - \mathbf{B}$ if $\mathbf{A} = 5 \underline{/36.9^\circ}$ and $\mathbf{B} = 5 \underline{/53.1^\circ}$.

Converting both numbers from polar to rectangular form, we obtain

$$\mathbf{A} = 5 \underline{/36.9^\circ} = 4 + j3$$

$$\mathbf{B} = 5 \underline{/53.1^\circ} = 3 + j4$$

Then

$$\mathbf{A} - \mathbf{B} = (4 + j3) - (3 + j4) = 1 - j1 = \sqrt{2} \underline{/-45^\circ}$$

■

EXAMPLE C.9

Given the complex number $\mathbf{A} = 5 \underline{/36.9^\circ}$, calculate $\mathbf{A}^*$, $\mathbf{A} + \mathbf{A}^*$, and $\mathbf{A} - \mathbf{A}^*$.

If $\mathbf{A} = 5 \underline{/36.9^\circ} = 4 + j3$, then $\mathbf{A}^* = 5 \underline{/-36.9^\circ} = 4 - j3$. Hence $\mathbf{A} + \mathbf{A}^* = 8$ and $\mathbf{A} - \mathbf{A}^* = j6$. ■

Note that addition and subtraction of complex numbers is a straightforward operation if the numbers are expressed in rectangular form. Note also that the sum of a complex number and its conjugate is a real number, and the difference of a complex number and its conjugate is an imaginary number.

Multiplication

The product of two complex numbers $\mathbf{A} = z_1 e^{j\theta_1} = z_1 \underline{/\theta_1} = x_1 + jy_1$ and $\mathbf{B} = z_2 e^{j\theta_2} = z_2 \underline{/\theta_2} = x_2 + jy_2$ is

$$\mathbf{AB} = (z_1 e^{j\theta_1})(z_2 e^{j\theta_2}) = z_1 z_2 e^{j(\theta_1 + \theta_2)} = z_1 z_2 \underline{/\theta_1 + \theta_2} \tag{C.15}$$

or

$$\begin{aligned} \mathbf{AB} &= (x_1 + jy_1)(x_2 + jy_2) \\ &= x_1 x_2 + jx_1 y_2 + jx_2 y_1 + j^2 y_1 y_2 \\ &= (x_1 x_2 - y_1 y_2) + j(x_1 y_2 + x_2 y_1) \end{aligned} \tag{C.16}$$

If the two complex numbers are in exponential or polar form, multiplication is readily accomplished by multiplying their magnitudes and adding their angles. Multiplication is straightforward, although slightly more complicated, if the numbers are expressed in rectangular form.

The product of a complex number and its conjugate is a real number; that is

$$\mathbf{AA}^* = (ze^{j\theta})(ze^{-j\theta}) = z^2 e^{j0} = z^2 \underline{/0^\circ} = z^2 \tag{C.17}$$

Note that this real number is the square of the magnitude of the complex number.

EXAMPLE C.10

If $\mathbf{A} = 10 \underline{/30^\circ}$ and $\mathbf{B} = 5 \underline{/15^\circ}$, the products **AB** and **AA*** are

$$\mathbf{AB} = (10 \underline{/30^\circ})(5 \underline{/15^\circ}) = 50 \underline{/45^\circ}$$

and

$$\mathbf{AA}^* = (10 \underline{/30^\circ})(10 \underline{/-30^\circ}) = 100 \underline{/0^\circ} = 100$$

■

EXAMPLE C.11

Given $\mathbf{A} = 5 \underline{/36.9^\circ}$ and $\mathbf{B} = 5 \underline{/53.1^\circ}$, we wish to calculate the product in both polar and rectangular forms.

$$\begin{aligned}\mathbf{AB} &= (5 \underline{/36.9^\circ})(5 \underline{/53.1^\circ}) = 25 \underline{/90^\circ} \\ &= (4 + j3)(3 + j4) \\ &= 12 + j16 + j9 + j^2 12 \\ &= 25j \\ &= 25 \underline{/90^\circ}\end{aligned}$$

■

Division

The quotient of two complex numbers $\mathbf{A} = z_1 e^{j\theta_1} = z_1 \underline{/\theta_1} = x_1 + jy_1$ and $\mathbf{B} = z_2 e^{j\theta_2} = z_2 \underline{/\theta_2} = x_2 + jy_2$ is

$$\frac{\mathbf{A}}{\mathbf{B}} = \frac{z_1 e^{j\theta_1}}{z_2 e^{j\theta_2}} = \frac{z_1}{z_2} e^{j(\theta_1 - \theta_2)} = \frac{z_1}{z_2} \underline{/\theta_1 - \theta_2} \tag{C.18}$$

that is, if the numbers are in exponential or polar form, division is immediately accomplished by dividing their magnitudes and subtracting their angles as shown above. If the numbers are in rectangular form, or the answer is desired in rectangular form, then the following procedure can be used.

$$\frac{\mathbf{A}}{\mathbf{B}} = \frac{x_1 + jy_1}{x_2 + jy_2}$$

The denominator is rationalized by multiplying both numerator and denominator by $\mathbf{B}^*$:

$$\begin{aligned}\frac{\mathbf{AB}^*}{\mathbf{BB}^*} &= \frac{(x_1 + jy_1)(x_2 - jy_2)}{(x_2 + jy_2)(x_2 - jy_2)} \\ &= \frac{x_1 x_2 + y_1 y_2}{x_2^2 + y_2^2} + j\frac{x_2 y_1 - x_1 y_2}{x_2^2 + y_2^2}\end{aligned} \tag{C.19}$$

In this form the denominator is real and the quotient is given in rectangular form.

EXAMPLE C.12

If $\mathbf{A} = 10 \angle 30°$ and $\mathbf{B} = 5 \angle 53.1°$, determine the quotient $\mathbf{A}/\mathbf{B}$ in both polar and rectangular forms.

$$\frac{\mathbf{A}}{\mathbf{B}} = \frac{\mathbf{AB}^*}{\mathbf{BB}^*} = \frac{8.66 + j5}{3 + j4}\,\frac{3 - j4}{3 - j4} \qquad \text{or} \qquad \frac{\mathbf{A}}{\mathbf{B}} = \frac{10 \angle 30°}{5 \angle 53.1°}$$

$$= \frac{(8.66 + j5)(3 - j4)}{3^2 + 4^2} \qquad\qquad = 2 \angle -23.1°$$

$$= \frac{45.98 - j19.64}{25} \qquad\qquad = 1.84 - j0.79$$

$$= 1.84 - j0.79$$

■

As a final example, consider the following one, which requires the use of many of the techniques presented above.

EXAMPLE C.13

Given $\mathbf{A} = 10 \angle 30°$, $\mathbf{B} = 2 + j2$, $\mathbf{C} = 4 + j3$, and $\mathbf{D} = 4 \angle 10°$, calculate the expression for $\mathbf{AB}/(\mathbf{C} + \mathbf{D})$ in rectangular form.

$$\frac{\mathbf{AB}}{\mathbf{C} + \mathbf{D}} = \frac{(10 \angle 30°)(2 + j2)}{(4 + j3) + (4 \angle 10°)}$$

$$= \frac{(10 \angle 30°)(2\sqrt{2} \angle 45°)}{4 + j3 + 3.94 + j0.69}$$

$$= \frac{20\sqrt{2} \angle 75°}{7.94 + j3.69}$$

$$= \frac{20\sqrt{2} \angle 75°}{8.75 \angle 24.93°}$$

$$= 3.23 \angle 50.07°$$

$$= 2.07 + j2.48$$

■

References

[1] IRWIN, J. DAVID, *Basic Engineering Circuit Analysis,* 3rd ed., Macmillan Publishing Co., New York, 1990.

[2] TUINENGA, PAUL W., SPICE, *A Guide to Circuit Simulation and Analysis Using PSPICE,* Prentice Hall, Englewood Cliffs, N.J., 1988.

[3] VLADIMIRESCU, A., and LIU, SALLY, "The Simulation of MOS Integrated Circuits Using SPICE 2," Department of Electrical Engineering and Computer Science, University of California, Berkeley, Calif., UCB/ERL M80/7, February 1980.

[4] VLADIMIRESCU, A., NEWTON, A. R., and PEDERSON, D. O., "SPICE Version 2F.1 User's Guide," Department of Electrical Engineering and Computer Science, University of California, Berkeley, Calif., February 1980.

[5] NAGLE, L. W., "SPICE 2: A Program to Simulate Semiconductor Circuits," Department of Electrical Engineering and Computer Science, University of California, Berkeley, Calif., ERL-M520, May 1975.

[6] HAYT, W. H., JR., and KEMMERLY, JACK, E., *Engineering Circuit Analysis,* 4th ed., McGraw-Hill Book Company, New York, 1986.

[7] NILSSON, J. W., *Electric Circuits,* 2nd ed., Addison-Wesley Publishing Co., Inc., Reading, Mass., 1986.

[8] JOHNSON, D. E., HILBURN, J. L., and JOHNSON, J. R., *Basic Electric Circuit Analysis,* Prentice Hall, Englewood Cliffs, N.J., 1978.

[9] BOBROW, L. S., *Elementary Linear Circuit Analysis,* Holt, Rinehart and Winston, New York, 1981.

[10] Blackwell, W. A., and Grigsby, L. L., *Introductory Network Theory,* PWS Engineering, Boston, 1985.

[11] Balabanian, N., *Fundamentals of Circuit Theory,* Allyn & Bacon, Inc., Boston, 1961.

[12] Kirwin, G. J., and Grodzinsky, S. E., *Basic Circuit Analysis,* Houghton Mifflin Company, Boston, 1980.

[13] Aseltine, J. A., *Transform Methods in Linear System Analysis,* McGraw-Hill Book Company, New York, 1958.

[14] Chan, Shu-Park, Chan, Shu-Yan, and Chan, Shu-Gar, *Analysis of Linear Networks and Systems,* Addison-Wesley Publishing Co., Inc., Reading, Mass., 1972.

[15] Chen, Wai-Kai, *Linear Networks and Systems,* Brooks/Cole Engineering Division of Wadsworth, Inc., Monterey, Calif., 1983.

[16] Friedland, B., Wing, O., and Ash, R., *Principles of Linear Networks,* McGraw-Hill Book Company, New York, 1961.

[17] Kuo, F. F., *Network Analysis and Synthesis,* John Wiley & Sons, Inc., New York, 1962.

[18] Ley, B. J., Lutz, S. G., and Rehberg, C. F., *Linear Circuit Analysis,* McGraw-Hill Book Company, New York, 1959.

[19] Su, K. L., *Fundamentals of Circuit, Electronics and Signal Analysis,* Houghton Mifflin Company, Boston, 1978.

[20] Trick, T. N., *Introduction to Circuit Analysis,* John Wiley & Sons, Inc., New York, 1977.

[21] Van Valkenburg, M. E., and Kinariwala, B. K., *Linear Circuits,* Prentice Hall, Englewood Cliffs, N.J., 1982.

[22] Fitzgerald, A. E., Higginbotham, E. E., and Grabel, A., *Basic Electrical Engineering,* McGraw-Hill Book Company, New York, 1981.

[23] Carlson, A. B., and Gisser, D. G., *Electrical Engineering,* Addison-Wesley Publishing Co., Inc., Reading, Mass., 1981.

[24] Van Valkenburg, M. E., *Analog Filter Design,* Holt, Rinehart and Winston, New York, 1982.

[25] Cheng, D. K., *Analysis of Linear Systems,* Addison-Wesley Publishing Co., Inc., Reading, Mass., 1959.

[26] Weinberg, L., *Network Analysis and Synthesis,* McGraw-Hill Book Company, New York, 1962.

[27] Stremler, F. G., *Introduction to Communication Systems,* Addison-Wesley Publishing Co., Inc., Reading, Mass., 1977.

Answers to Selected Problems

Chapter 1

1.1 1080 C **1.3** 0.5 C **1.7** $6(1 - e^{-6})$ J **1.11** Absorbing 32 W **1.15** 8 V
1.18 24 V

Chapter 2

2.1 $I_o = 7$ A, $I_1 = 9$ A, $I_2 = 6$ A **2.4** 32 W **2.6** 4 A **2.10** −4 A
2.14 $I_1 = -6$ A, $V_o = -16/3$ V **2.18** −3/2 A **2.21** 32/3 V **2.25** −20 V
2.28 −2 A **2.32** 36 V **2.36** 324 W **2.40** 9 A **2.44** 9 A **2.48** 9 A
2.52 48 V **2.57** 4 A **2.59** 3 A **2.63** −16/3 A **2.67** 6 Ω **2.69** 72 V

Chapter 3

3.1 1 V **3.4** −12/7 A **3.8** 208/19 V **3.12** −4 V **3.15** 4 A
3.20 $V_2 = -4/3$ V, $V_3 = 68/3$ V **3.24** 24/5 V **3.27** 5/3 A **3.31** 2 V
3.35 8/3 V **3.39** 124/19 A **3.43** 7 A **3.47** 24 V **3.51** −4/5 A
3.56 11/2 V **3.59** 8/5 V **3.62** 0 V

3.66

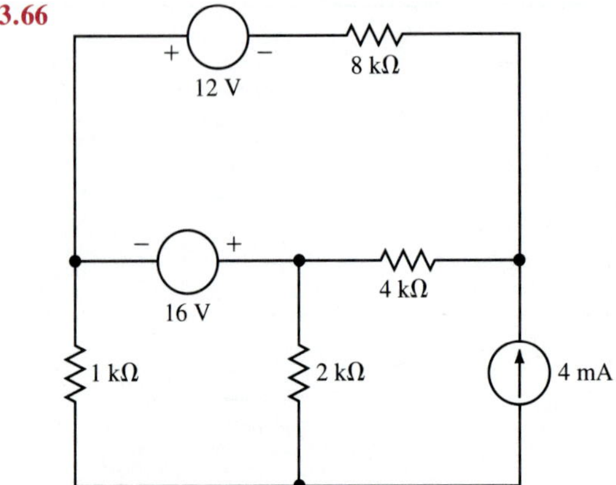

3.71 **(a)** Five tree branches, five links **(b)** five tree branches, four links

3.72 **(a)** $i_1 - i_2 = 0$
$i_4 + i_3 - i_2 = 0$
$i_4 + i_5 - i_7 = 0$
$i_6 - i_7 = 0$

(b) $i_1 + i_2 = 0$
$i_2 - i_3 - i_7 = 0$
$i_2 + i_4 + i_6 = 0$
$i_7 - i_5 + i_6 = 0$

3.75 **(a)** $v_3 - v_1 + v_5 + v_7 = 0$
$v_4 - v_1 + v_5 = 0$
$-v_5 + v_2 - v_6 = 0$
$-v_7 + v_8 + v_9 = 0$

(b) $-v_1 + v_3 + v_2 = 0$
$v_6 - v_3 + v_5 = 0$
$v_4 - v_2 - v_6 = 0$

3.79 $v_{out} = \left(1 + \frac{R_4}{R_3}\right)v_2 - \left(\frac{R_4}{R_3}\right)\left(1 + \frac{R_2}{R_1}\right)v_1$

3.82 $\frac{v_o}{v_1} = -\left(\frac{R_2}{R_1}\right)\left(1 + \frac{R_4}{R_2} + \frac{R_4}{R_3}\right)$

Chapter 4

4.1 4 A **4.3** 6 A **4.7** 2 mA **4.11** 0 A **4.15** 0.25 V **4.19** 8 V
4.23 1.4286 V **4.26** 6.5714 V **4.31** $V_2 = 5$ V, gain = 1 **4.35** 0 V **4.39** 10 V

4.42

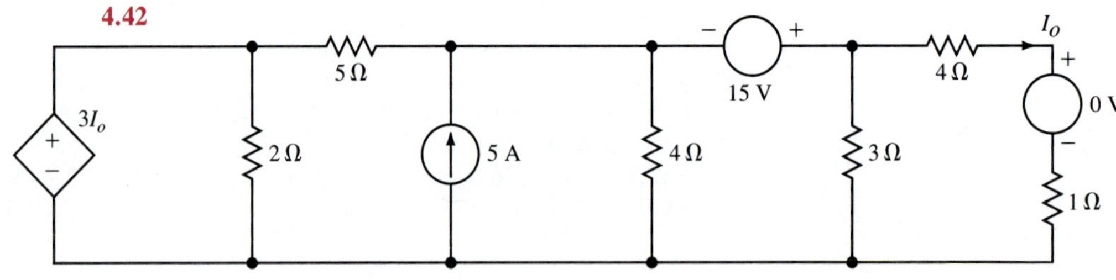

Chapter 5

5.1 4 A **5.4** 12/7 A **5.8** 9/5 A **5.12** 18 V **5.16** 0 A **5.19** −11/12 A
5.23 −1 A **5.27** 76/11 V **5.30** 9/5 A **5.33** −2 A **5.38** 9/2 V
5.42 12/13 V **5.45** 6 V **5.48** 9/2 A **5.51** 1/3 A **5.55** $R_{Th} = 20/9\ \Omega$
5.59 $R_L = 4\ \Omega$, $P_L = 9$ W **5.64** 5 Ω

Chapter 6

6.1 40 V **6.4** $\pm 2.92 \cos 377t$ A

6.8

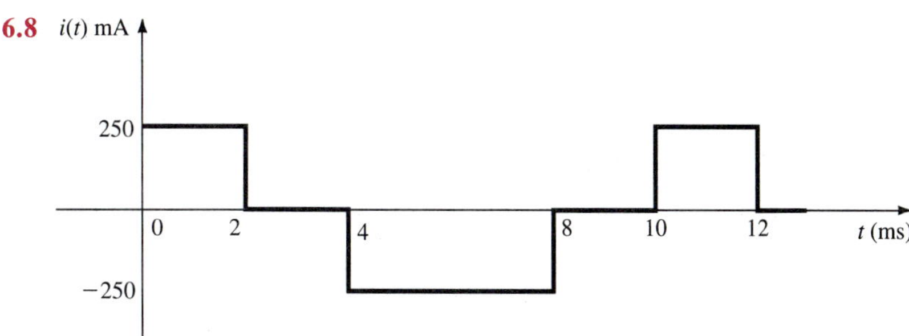

6.12 $\pm 130.63 \cos 377t$ V **6.15** $V(5) = -0.91$ V, $W(5) = 91.97$ J

6.19

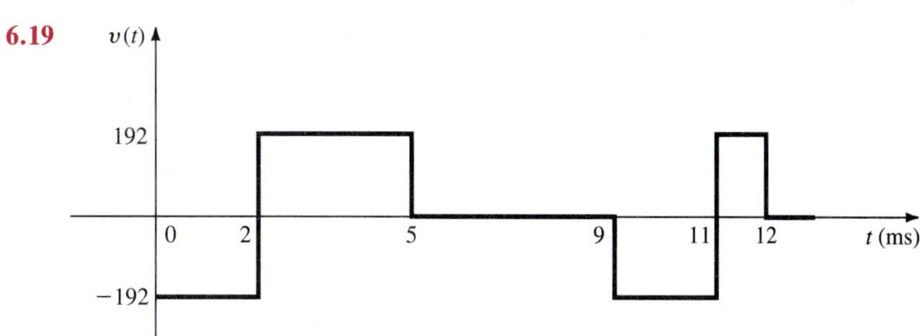

6.24 $C_{max} = 8\ \mu F$, $C_{min} = 1/2\ \mu F$ **6.28** 2 μF **6.31** 1.32 μF **6.35** 3.18 μF
6.40 3 mH **6.43** 0 H

Chapter 7

7.1 $4e^{-1.2t}$ V **7.4** $-3e^{-t/0.4}$ V **7.8** $8 + e^{-t/0.8}$ V **7.12** $3 + (8/3 - 3)e^{-t/0.3}$ mA
7.15 $3/4 + (9/19 - 3/4)e^{-t/0.6}$ mA **7.19** $-18/5e^{-t/0.125}$ V **7.24** $2e^{-t/1.2}$ mA
7.28 $14/5 + (-2 - 14/5)e^{-3t/0.4}$ mA **7.32** $54/13e^{-72/13t}$ V **7.35** $-12/5e^{-24/5t}$ A
7.39 $-6 + (3 + 6)e^{-6t}$ V **7.44** $6 + (5 - 6)e^{-t/3}$ A **7.49** $-24/5 + (4 + 24/5)e^{-20t}$ V
7.54 $1/3(10e^{-t} - 11e^{-5t})$ V
7.57 $8(1 - e^{-t/0.8})[u(t) - u(t - 1)] + 8(1 - e^{-1/0.8})e^{-(t-1)/0.8}u(t - 1)$ V

7.60

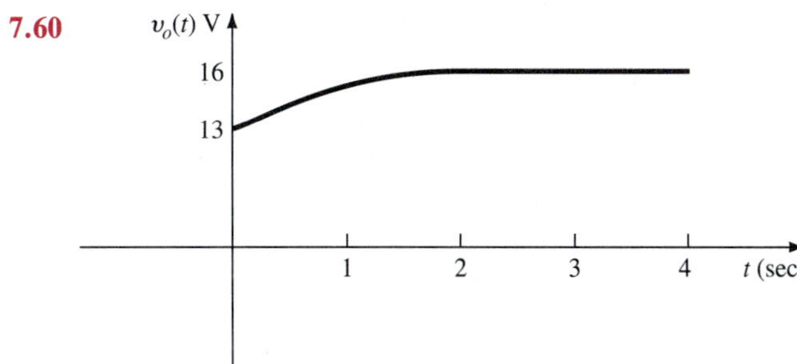

7.65 $5e^{-t/0.1}[u(t) - u(t - 0.05)] - 1.97e^{-(t-0.05)/0.1}u(t - 0.05)$ V

7.68

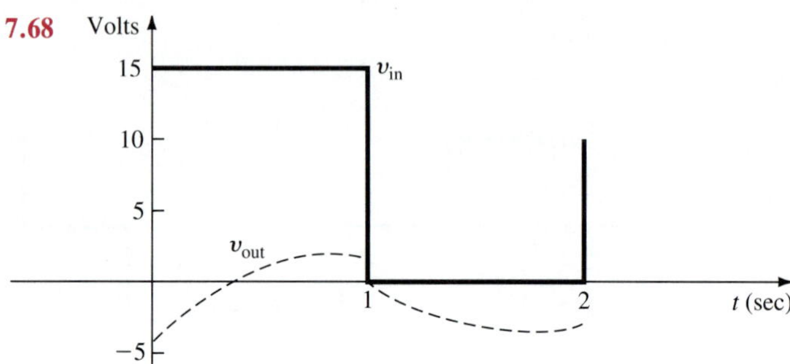

Chapter 8

8.1 **(a)** $s^2 + 2s + 5 = 0$
(b) $s = -1 \pm j2$
(c) $e^{-t}(k_1 \cos 2t + k_2 \sin 2t)$

8.4 **(a)** $s^2 + 6s + 9 = 0$
(b) $s_1 = -3,\ s_2 = -3$
(c) $k_1e^{-3t} + k_2te^{-3t}$

8.7 Underdamped **8.10** $e^{-2t} \cos (t/2)$ A **8.13** $5e^{-2t} - 4e^{-3t}$ V

8.17 $4 - 4e^{-t} - 3te^{-t}$ V **8.21** $10e^{-4t} \cos 2t - 40e^{-4t} \sin 2t$ V

8.25 $-4/7e^{-t} + 32/7e^{-8t}$ A

8.28

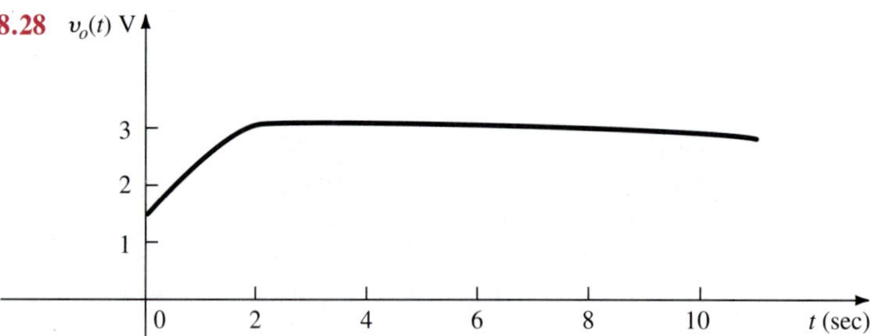

8.31

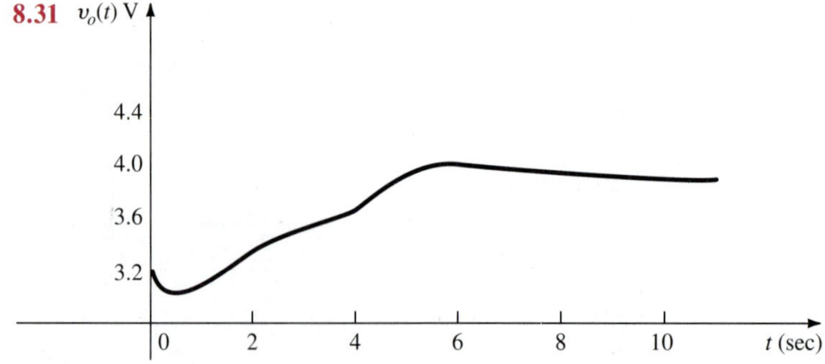

8.35

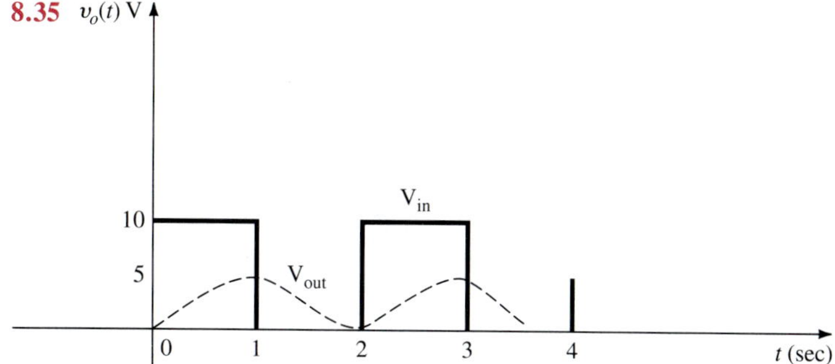

Chapter 9

9.1 The two voltages are in phase.

9.4 $v_1(t)$ leads $i_1(t)$ by -55°
$v_1(t)$ leads $i_2(t)$ by -230°

9.7 **(a)** $i_1(t) = 5 \cos (377t + 180^\circ)$ A or $\mathbf{I}_1 = 5\angle 180^\circ$ A
(b) $i_2(t) = 6 \sin (377t + 45^\circ)$ A or $\mathbf{I}_2 = 6\angle -45^\circ$ A

9.8 **(a)** $\mathbf{I}_1 = 8\angle 68^\circ$ A or $i_1(t) = 8 \cos (377t + 68^\circ)$ A
(b) $\mathbf{I}_2 = 4\angle 64^\circ$ A or $i_2(t) = 4 \cos (377t + 64^\circ)$ A

9.9 **(a)** $\mathbf{I}_1 = 6\angle -78^\circ$ A or $i_1(t) = 6 \cos (377t - 78^\circ)$ A
(b) $\mathbf{I}_2 = 4.5\angle -228^\circ$ A or $i_2(t) = 4.5 \cos (377t - 228^\circ)$ A

9.12 $5.09 + j4.96\ \Omega$ **9.16** $-j0.41$ S **9.21** $4 - j3\ \Omega$ **9.24** 5.31 mH

9.27

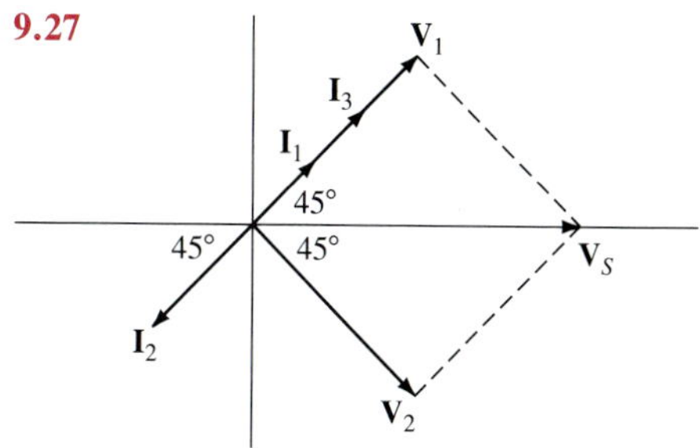

9.31 $\mathbf{I}_1 = 8.05\angle -3.95^\circ$ A, $\mathbf{I}_2 = 4.99\angle 93.17^\circ$ A, $\mathbf{I}_3 = 9.98\angle -33.69^\circ$ A
$\mathbf{I}_4 = 4.47\angle -60.26^\circ$ A, $\mathbf{I}_5 = 6.32\angle -15.26^\circ$ A

9.34 $9.41\angle 11.31^\circ$ V **9.39** $2.68\angle 63.43^\circ$ A **9.42** $13.66\angle -36.11^\circ$ V

9.45 $2.24\angle 161.57^\circ$ A **9.49** $-8 + j3$ V **9.53** $5.66\angle -45^\circ$ V **9.55** $8 + j4$ A

9.59 $1.98\angle 82.95^\circ\ \Omega$

9.63 $-\dfrac{R_F}{R_i}\left(\dfrac{1 + j\omega C_i R_i}{1 + j\omega C_F R_F}\right)$

Chapter 10

10.1 $8\angle -90^\circ$ V **10.4** $10.73\angle 206.57^\circ$ V **10.8** $10.6\angle -135^\circ$ A

10.12 $2.83\angle 45^\circ$ A **10.14** $3.09\angle 23.83^\circ$ V **10.18** $35.8\angle -71.58^\circ$ V

10.22 $26 - j2$ A **10.26** $9.05\angle 51.25^\circ$ V **10.30** $6\angle 0^\circ$ A **10.34** $3.69\angle 83.88^\circ$ V

10.38 $-1.47\angle -72.89^\circ$ V **10.41** $5.5\angle -104.04^\circ$ V **10.46** $3.09\angle 23.83^\circ$ V

10.50 $1.3\angle 12.53^\circ$ V **10.54** $35.8\angle -71.58^\circ$ V **10.59** $2.95\angle -38.03^\circ$ A

10.63 $\mathbf{Z}_{Th} = 1/3 + j2/3\ \Omega$ **10.67** $3.175\underline{/-56.05°}$ A **10.70** $0.2308\underline{/-133.5°}$ A
10.74 $2.526\underline{/-18.52°}$ V

Chapter 11

11.1 1.58 W **11.4** $P_s = 29.19$ W, $P_{4\Omega} = 9.77$ W **11.8** 119.99 W
11.11 32.49 W **11.15** $\mathbf{Z}_L = 5\ \Omega$, $P_L = 5.28$ W
11.19 $\mathbf{Z}_L = 0.6 - j0.8\ \Omega$, $P_L = 47.82$ W **11.22** $\mathbf{Z}_L = 3 - j2\ \Omega$, $P_L = 8/3$ W
11.26 $\mathbf{Z}_L = 0.4 - j0.2\ \Omega$, $P_L = 18.05$ W **11.30** 1.53 A **11.34** 0.94 A
11.37 2.67 V
11.41 **(a)** $50\underline{/20°}$ V
(b) $16\underline{/-30°}$ V
(c) $14.14\underline{/-80°}$ V
11.45 440 V **11.49** $\mathbf{Z}_{line} = 0.062 + j0.0826\ \Omega$, $\mathbf{V}_{in} = 236.95\underline{/1.19°}$ V
11.53 $303.73\underline{/12.43°}$ V **11.55** $309.53\underline{/11.42°}$ V **11.60** 305 μF **11.64** 60.97 W

Chapter 12

12.1 $\mathbf{I}_a = 5.56\underline{/6.34°}$ A rms, $\mathbf{I}_b = 5.56\underline{/-113.66°}$ A rms, $\mathbf{I}_c = 5.56\underline{/-233.66°}$ A rms,
$\mathbf{V}_{AN} = 111.2\underline{/59.47°}$ V rms, $\mathbf{V}_{BN} = 111.2\underline{/-60.53°}$ V rms, $\mathbf{V}_{CN} = 111.2\underline{/-180.53°}$ V rms
12.4 $120\sqrt{3}\underline{/40°}$ V rms **12.7** $\mathbf{V}_{AN} = 145.98\underline{/60.38°}$ V rms, $|\mathbf{V}_{an}| = 160$ V rms
12.11 $1.49 + j3.00\ \Omega$ **12.15** $19.95 + j21.93\ \Omega$ **12.19** $207.9\underline{/40°}$ V rms
12.23 $\mathbf{I}_{aA} = 11.10\underline{/-48.99°}$ A rms, $\mathbf{I}_{bB} = 11.10\underline{/-168.99°}$ A rms, $\mathbf{I}_{cC} = 11.10\underline{/71.01°}$ A rms,
$|\mathbf{V}_{AN}| = 111$ V rms
12.27 $32.18\underline{/25°}\ \Omega$
12.30 $\mathbf{I}_{AN} = 9.37\underline{/-4.4°}$ A rms, $\mathbf{I}_{BN} = 9.37\underline{/-124.4°}$ A rms, $\mathbf{I}_{CN} = 9.37\underline{/-244.4°}$ A rms
12.34 $\mathbf{I}_{ba} = 8.64\underline{/57.94°}$ A rms, $\mathbf{I}_{cb} = 8.64\underline{/-62.06°}$ A rms, $\mathbf{I}_{ac} = 8.64\underline{/-182.06°}$ A rms
12.38 $\mathbf{S}_{3\phi} = 522.8\underline{/67.9°}$ VA, $\mathbf{S}_{load} = P_{load} + jQ_{load} = 65.6 + j161.4$ VA
12.42 $16.65 + j11.07\ \Omega$ **12.46** 0.91 lagging **12.50** 18.5 kVA at 0.65 pf lagging
12.56 9409.96 W **12.58** 20,687 W **12.62** $94.1\underline{/30°}\ \Omega$ **12.64** 422.45 μF

Chapter 13

13.1 $5.33\underline{/-123.69°}$ V

13.4
$$-\mathbf{V}_1 + \mathrm{R}_1\mathbf{I}_1 + \frac{1}{j\omega C_1}(\mathbf{I}_1 - \mathbf{I}_2) + R_2(\mathbf{I}_1 - \mathbf{I}_2) + \mathbf{V}_2 = 0$$
$$\frac{1}{j\omega C_1}(\mathbf{I}_2 - \mathbf{I}_1) + j\omega L_1\mathbf{I}_2 + j\omega M\mathbf{I}_3 + R_3\mathbf{I}_2 + \frac{1}{j\omega C_2}(\mathbf{I}_2 - \mathbf{I}_3) = 0$$
$$-V_2 + R_2(\mathbf{I}_3 - \mathbf{I}_1) + \frac{1}{j\omega C_2}(\mathbf{I}_3 - \mathbf{I}_2) + R_4\mathbf{I}_3 + j\omega L_3\mathbf{I}_3 + j\omega M\mathbf{I}_2 = 0$$

13.8 $20.87\underline{/4.29°}$ V **13.11** $1.21\underline{/190.18°}$ V **13.15** 385.63 μJ **13.18** $6.5 + j8\ \Omega$
13.22 $0.005572\underline{/89.89°}$ V **13.25** $0.03014\underline{/-30.26°}$ V **13.29** $44.72\underline{/-153.43°}$ V
13.33 $15.94\underline{/175.2°}$ V **13.36** $2.61\underline{/-33.4°}$ A **13.40** $15.86\underline{/-43°}$ V
13.43 $-18.32\underline{/17.74°}$ V **13.47** 20 **13.52** 5737.72 W **13.56** $2.615\underline{/-33.40°}$ A
13.60 550 kVA

Chapter 14

14.1 $\dfrac{8s(s+1)}{2s^2 + 6s + 1}$ **14.4** $\dfrac{RCs + 3}{3LCs^2 + 2RCs + 3}$

14.8

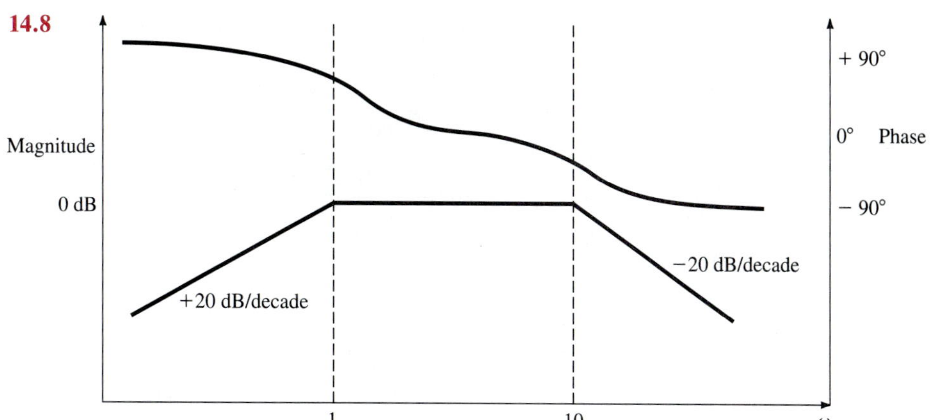

14.13

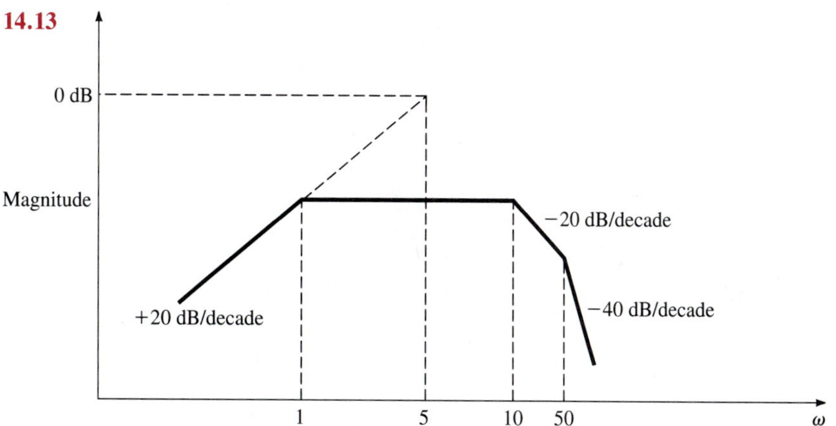

14.15

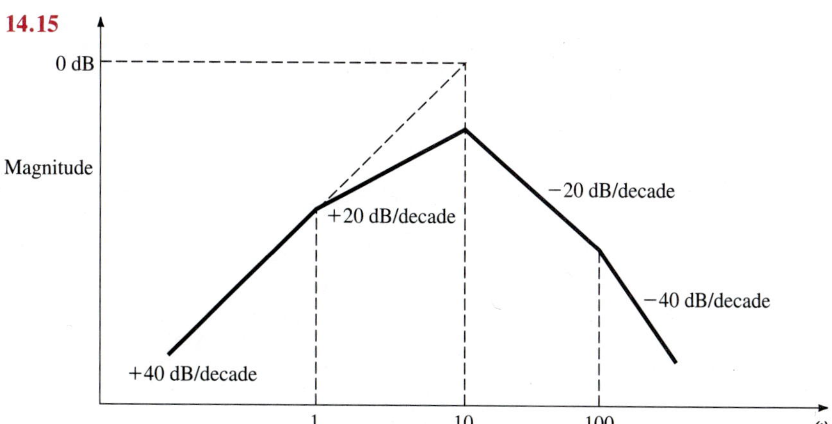

14.20 $\dfrac{100(j\omega + 1)\left(\dfrac{j\omega}{120} + 1\right)}{\left(\dfrac{j\omega}{10} + 1\right)\left(\dfrac{j\omega}{80} + 1\right)^2}$ **14.23** $\dfrac{40\left(\dfrac{j\omega}{50} + 1\right)\left(\dfrac{j\omega}{1000} + 1\right)}{(j\omega)\left(\dfrac{j\omega}{400} + 1\right)^2}$

14.26 $\dfrac{(20)^2\left(\dfrac{j\omega}{1.5} + 1\right)\left(\dfrac{j\omega}{10} + 1\right)}{-\omega^2\left(\dfrac{j\omega}{80} + 1\right)\left(\dfrac{j\omega}{200} + 1\right)}$

14.29 $L = 1.59$ mH, $C = 1.59$ nF

14.32 $\omega_o = 10^4$ rad/sec, $Q = 10$, $\omega_{max} = 9974.97$ rad/sec, $|\mathbf{V}_o|_{max} = 60.075$ V

14.36 $\omega_o = 500$ rad/sec, $Q = 12.5$ **14.40** $L = 19.53$ mH, $Q = 10.42$, BW $= 153.6$ rad/sec

14.43 $\omega_o = 2000$ rad/sec, $Q = 25$, BW $= 80$ rad/sec, $P_{res} = 18$ W

14.47 $\omega_o = 790.57$ rad/sec, $Q = 126.49$

14.51 $\omega_o = 1000$ rad/sec, BW $= 50$, $Q = 20$, $\mathbf{V}_o = 40\angle 0^\circ$ V

14.54

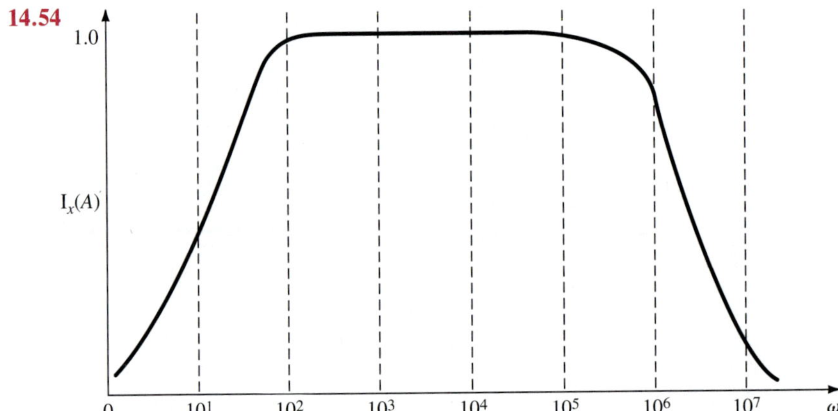

14.58 $R = 2000\ \Omega$, $L = 500$ H, $C = 1/8000$ F

14.62 $\dfrac{j\omega - (1/RC)}{j\omega + (1/RC)}$, all pass filter

14.66 $\dfrac{\mathbf{V}_o}{\mathbf{V}_i} = \dfrac{1}{(j\omega + 1)^2}$,

Magnitude

−40 dB/decade

0 1 ω

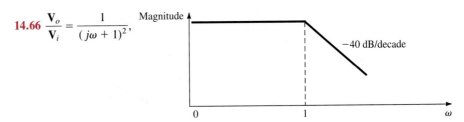

This is a low-pass filter.

14.69 $H_o = \dfrac{R_2R_4}{R_1R_3}$, $\zeta = \dfrac{1}{2}\left(\sqrt{\dfrac{R_4C_2}{R_2C_1}} + \sqrt{\dfrac{R_2C_1}{R_4C_2}}\right)$

$$\omega_c = \frac{1}{\sqrt{R_4R_2C_1C_2}}$$

Chapter 15

15.1 (a) $\begin{bmatrix} \dfrac{1}{\mathbf{Z}_L} & -\dfrac{1}{\mathbf{Z}_L} \\ -\dfrac{1}{\mathbf{Z}_L} & \dfrac{1}{\mathbf{Z}_L} \end{bmatrix}$ (b) $\begin{bmatrix} \mathbf{Z}_L & \mathbf{Z}_L \\ \mathbf{Z}_L & \mathbf{Z}_L \end{bmatrix}$

15.4 -4 A

15.8 $\begin{bmatrix} j\omega(C_1 + C_2) & -j\omega C_1 \\ \dfrac{A}{R} - j\omega C_1 & \dfrac{1}{j\omega L} + \dfrac{1}{R} + j\omega(C_1 + C_3) \end{bmatrix}$

15.12 1.384 MΩ **15.16** 884 Ω

15.20 $\begin{bmatrix} -\dfrac{1}{\gamma} & \dfrac{-\mathbf{Z}_2}{\gamma} \\ -\dfrac{1}{\gamma \mathbf{Z}_1} & \dfrac{-\mathbf{Z}_2}{\gamma \mathbf{Z}_1} \end{bmatrix}$

15.24 $\begin{bmatrix} \frac{1}{4} & 18.17 - j6.75 \\ 0 & 4 \end{bmatrix}$, $\mathbf{Z}_i = 4.84\underline{/-20.42^\circ}$ Ω

15.28 $\begin{bmatrix} \frac{5}{11} & -\frac{2}{11} \\ -\frac{2}{11} & \frac{3}{11} \end{bmatrix}$

15.32 $\begin{bmatrix} \frac{3}{2} & \frac{11}{2} \\ \frac{1}{2} & \frac{5}{2} \end{bmatrix}$

15.36 $\begin{bmatrix} 3 & j8 \\ 3 - j & 3 + j8 \end{bmatrix}$

15.40 $\begin{bmatrix} \frac{5}{3} + 2j\omega & -(\frac{4}{3} + j\omega) \\ -(\frac{4}{3} + j\omega) & \frac{5}{3} + 2j\omega \end{bmatrix}$

15.44 $\begin{bmatrix} \frac{1}{2} & \frac{49}{2} - j\frac{27}{2} \\ 0 & 2 \end{bmatrix}$, $\mathbf{I}_o = 0.30\underline{/75.62^\circ}$ A

15.48 $4.63\underline{/157.5^\circ}$ V

Chapter 16

16.2 $\mathbf{F}(s) = \dfrac{e^{-a}}{4x}\left(\dfrac{1}{s + 1}\right)$

16.5 $\mathbf{F}(s) = e^{-4s}\left[\dfrac{e^{-4a}}{(s + a)^2} + \dfrac{4e^{-4a}}{s + a}\right]$

16.8 $\mathbf{F}(s) = \dfrac{(s^2 + s + 1)e^{-s}}{s^2(s + 1)}$ **16.11** $\mathbf{F}(s) = \dfrac{A}{s}(1 - e^{-Ts})$

16.15 $\mathbf{F}(s) = \dfrac{2}{s^2(1 - e^{-2s})}(1 - e^{-s} - se^{-2s})$

16.19 $\mathbf{F}(s) = \dfrac{1}{1 - e^{-2s}}\left(\dfrac{2}{s^2}e^{-s} - \dfrac{2}{s^2}e^{-2s} - \dfrac{2}{s}e^{-2s}\right)$

16.23 $\mathbf{F}(s) = \dfrac{1}{1 - e^{-4s}}\left(\dfrac{1}{s^2} - \dfrac{1}{s^2}e^{-s} - \dfrac{1}{s}e^{-2s} - \dfrac{1}{s^2}e^{-2s} + \dfrac{1}{s^2}e^{-3s} + \dfrac{1}{s}e^{-4s}\right)$

16.26 $\mathbf{F}(s) = \dfrac{1}{1 - e^{-\pi s}}\left(\dfrac{10}{s^2 + 1} + \dfrac{10e^{-\pi s}}{s^2 + 1}\right)$

16.30 **(a)** $f(t) = (\frac{1}{4}e^{-2t} + \frac{3}{4}e^{-6t})\, u(t)$
(b) $f(t) = (\frac{3}{4} + \frac{1}{4}e^{-4t})\, u(t)$

16.34 **(a)** $f(t) = (2te^{-t} + e^{-t})\, u(t)$
(b) $f(t) = (6 - 5te^{-t} - 6e^{-t})\, u(t)$

16.39 **(a)** $f(t) = [-\frac{1}{4}e^{-2t} + 2(\frac{5}{8})e^{-2t} \cos 2t]\, u(t)$
(b) $f(t) = [\delta(t) - 4(2)(\frac{1}{4})e^{-2t} \cos (2t - 90°)]\, u(t)$

16.42 $f(t) = [-3 + 3t + \frac{12}{5}e^{-t} + \frac{2}{3}e^{-2t} \cos (2t - 26.56°)]\, u(t)$

16.47 $f(t) = (t - 1 + e^{-t})\, u(t) - [(t - 1) - 1 + e^{-(t-1)}]\, u(t - 1)$

16.51 **(a)** $f(0) = 2,\ f(\infty) = 4$
(b) $f(0) = 2,\ f(\infty) = 0$
(c) $f(0) = 2,\ f(\infty) = 0$

16.54 **(a)** $x(t) = (\frac{1}{2}e^{-t} + \frac{1}{2}e^{-3t})\, u(t)$
(b) $x(t) = (\frac{1}{2} + \frac{3}{2}e^{-4t})\, u(t)$

16.57 $y(t) = (2te^{-t} - 2e^{-t} + 2e^{-2t})\, u(t)$

Chapter 17

17.1 $\mathbf{Z}(s) = \dfrac{6s + 8}{6s^2 + 16s + 11}$ **17.4** 8 V **17.8** $[2\sqrt{2}e^{-t} \cos (t - 45°)]\, u(t)$ V

17.12 $\frac{6}{7}(1 - e^{-7/4t})\, u(t)$ V **17.16** $\frac{24}{7}(e^{-2/7t})\, u(t)$ V **17.21** $(2 - 6e^{-4t})\, u(t)$ V

17.24 $(-\frac{4}{3} + 2e^{-t} - \frac{20}{3}e^{-3t})\, u(t)$ V

17.30 $(4 + 12e^{-3/2t})$ V, $t > 0$

17.34 $\mathbf{Z}(s) = \begin{bmatrix} \dfrac{s+1}{s} & \dfrac{1}{s} \\ \dfrac{1}{s} & \dfrac{s^2+1}{s} \end{bmatrix}$

$$i_2(t) = \left[-\frac{4}{3} + 2(0.85)e^{-3/4t} \cos\left(\frac{\sqrt{15}}{4}t - 37.76°\right)\right] u(t) \text{ A}$$

17.38 $1.15(e^{-0.42t} - e^{-1.58t})\, u(t)$ V **17.40** $1.34(e^{-2.62t} - e^{-0.38t})\, u(t)$ A

17.43 $v_o(0) = 0$ V, $v_o(\infty) = 4$ V

17.48 $\dfrac{\mathbf{V}_o}{} = \dfrac{R_1 + R_2}{} \left[\dfrac{s + \dfrac{1}{(R_1 + R_2)C}}{}\right]$

17.52 $\dfrac{\mathbf{V}_o}{\mathbf{V}_S} = \dfrac{-\dfrac{s}{R_1C_1}}{s^2 + s\left(\dfrac{C_1}{C_1C_2R_3} + \dfrac{C_2}{C_1C_2R_3}\right) + \dfrac{R_1 + R_2}{C_1C_2R_1R_2R_3}}$

17.56 $5.23 \cos (2t + 97.76°)$ V, $t > 0$ **17.60** The circuit is overdamped.

Chapter 18

18.1 $f(x) = \dfrac{10}{\pi} \sum_{n=-\infty}^{\infty} \dfrac{1}{n} \sin (0.1n\pi)e^{-j0.2\pi} e^{jn\omega_0 t}$

18.5 $\mathbf{c}_n = 0$ for n even or 0

$= \dfrac{1}{j2\pi n}$ for n odd

18.8 $\mathbf{c}_n = -\dfrac{4}{(n\pi)^2}$ n odd

$= 0$ n even and $\mathbf{c}_o = 1$

18.12 **(a)** Odd function with half-wave symmetry **(b)** even function

18.16 $v(t) = \sum_{n=1}^{\infty} (-1)^{n+1} \dfrac{20}{n\pi} \sin nt$

18.19 $a_n = \dfrac{1}{n^2\pi^2} (\cos n\pi - 1)$

$b_n = -\dfrac{1}{n\pi}$

18.22 $f(t) = \dfrac{1}{\pi} + \dfrac{1}{2} \sin \pi t + \sum_{n=2}^{\infty} \dfrac{1 - \cos (1 + n)\pi}{\pi(1 - n^2)} \cos n\pi t$

18.26 $f(t) = 3 - 4 \sin 20\pi t - 5 \sin 40\pi t - 3 \sin 60\pi t - 2 \sin 80\pi t - \sin 100\pi t$

18.29 $A_n = \left| \dfrac{2 + j6n}{j2n} \right|$

$i_o(t) = \sum_{n=1}^{\infty} (-1)^{n+1} \dfrac{20}{n\pi} \left(\dfrac{1}{A_n}\right) \cos (nt - 90° - \theta_n)$ A

18.32 $\mathbf{V}_o(0) = 0.25\underline{/0°}$ V, $\mathbf{V}_o(1) = 0.28\underline{/26.6°}$ V, $\mathbf{V}_o(2) = 0$, and $\mathbf{V}_o(3) = 0.023\underline{/-49°}$ V

18.36 124.06 W

18.39 **(a)** $\dfrac{j\omega + 2}{(j\omega + 2)^2 + 16}$

(b) $\dfrac{4}{(j\omega + 2)^2 + 16}$

18.43 $3(e^{-t} - e^{-2t})\, u(t)$ V **18.47** $(2e^{-t} - e^{-1/2t})\, u(t)$ **A** **18.50** 0.99 J

Index

Commonly Used Prefixes

giga (G)	10^9
mega (M)	10^6
kilo (k)	10^3
milli (m)	10^{-3}
micro (μ)	10^{-6}
nano (n)	10^{-9}
pico (p)	10^{-12}